FISH NUTRITION

FISH NUTRITION

FOURTH EDITION

Edited by

Ronald W. Hardy
Sadasivam J. Kaushik

ELSEVIER

ACADEMIC PRESS
An imprint of Elsevier

Academic Press is an imprint of Elsevier
125 London Wall, London EC2Y 5AS, United Kingdom
525 B Street, Suite 1650, San Diego, CA 92101, United States
50 Hampshire Street, 5th Floor, Cambridge, MA 02139, United States
The Boulevard, Langford Lane, Kidlington, Oxford OX5 1GB, United Kingdom

Notices

Knowledge and best practice in this field are constantly changing. As new research and experience broaden our understanding, changes in research methods, professional practices, or medical treatment may become necessary.

Practitioners and researchers must always rely on their own experience and knowledge in evaluating and using any information, methods, compounds, or experiments described herein. In using such information or methods they should be mindful of their own safety and the safety of others, including parties for whom they have a professional responsibility.

To the fullest extent of the law, neither the Publisher nor the authors, contributors, or editors, assume any liability for any injury and/or damage to persons or property as a matter of products liability, negligence or otherwise, or from any use or operation of any methods, products, instructions, or ideas contained in the material herein.

Library of Congress Cataloging-in-Publication Data
A catalog record for this book is available from the Library of Congress

British Library Cataloguing-in-Publication Data
A catalogue record for this book is available from the British Library

ISBN: 978-0-12-819587-1

For information on all Academic Press publications visit our website at https://www.elsevier.com/books-and-journals

Publisher: Charlotte Cockle
Acquisitions Editor: Anna Valutkevich
Editorial Project Manager: Andrea Dulberger
Production Project Manager: Omer Mukthar
Cover Designer: Christian J. Bilbow

Typeset by TNQ Technologies

Contents

6. The minerals

Santosh P. Lall

7. Carbohydrates

Sadasivam J. Kaushik, Stéphane Panserat, and Johan W. Schrama

8. Nutritional physiology

Brian C. Small

9. Diet formulation and manufacture

Ronald W. Hardy and Andreas Brezas

10. Diet analysis and evaluation

S. Charles Bai, Ronald W. Hardy, and Ali Hamidoghli

11. Nutritional supplements and fish health

Delbert M. Gatlin, III and Fernando Y. Yamamoto

12. Antinutrients and adventitious toxins

Åshild Krogdahl, Trond M. Kortner, and Ronald W. Hardy

13. Nutritional pathology

Ronald J. Roberts

14. Feeding and fish husbandry

D. Allen Davis and Ronald W. Hardy

Contributors

Q.H. Ai Ocean University of China, Qingdao, Shandong, China

Andreas Brezas Hagerman Fish Culture Experiment Station, University of Idaho, Hagerman, ID, United States; CJ Europe GmbH, Main Airport Center, Frankfurt am Main, Germany

S. Charles Bai FAO World Fisheries University Pilot Program; Feeds and Foods Nutrition Research Center, Pukyong National University, Busan, South Gyeongsang Province, Republic of Korea

D. Allen Davis School of Fisheries, Aquaculture & Aquatic Sciences, Auburn University, Auburn, AL, United States

Zhen-Yu Du School of Life Sciences, East China Normal University, Shanghai, P.R. China

L. Feng Sichuan Agricultural University, Chengdu, Sichuan, China

David S. Francis School of Life and Environmental Sciences, Deakin University, Geelong, VIC, Australia

Delbert M. Gatlin, III Department of Ecology and Conservation Biology, Texas A&M University System, College Station, TX, United States

Ali Hamidoghli Feeds and Foods Nutrition Research Center, Pukyong National University, Busan, South Gyeongsang Province, Republic of Korea

Ronald W. Hardy Aquaculture Research Institute, Hagerman Fish Culture Experiment Station, University of Idaho, Hagerman, ID, United States

G. He Ocean University of China, Qingdao, Shandong, China

Sadasivam J. Kaushik INRAe, Univ Pau and Pays de l'Adour, Nutrition, Metabolism and Aquaculture Unit, Saint Pée sur Nivelle, France

Trond M. Kortner Faculty of Veterinary Medicine, Norwegian University of Life Sciences, Ås, Norway

Åshild Krogdahl Faculty of Veterinary Medicine, Norwegian University of Life Sciences, Ås, Norway

Santosh P. Lall Retired from National Research Council of Canada, Halifax, NS, Canada

Kangsen Mai Ocean University of China, Qingdao, Shandong, China

Rolf E. Olsen Institute of Biology, Norwegian University of Science and Technology, Trondheim, Norway

Stéphane Panserat INRAe, Univ Pau and Pays de l'Adour, Nutrition, Metabolism and Aquaculture Unit, Saint Pée sur Nivelle, France

Einar Ringø Norwegian College of Fishery Science, UiT The Arctic University of Norway, Tromsø, Norway

Ronald J. Roberts Center for Sustainable Aquaculture, Hagerman Fish Culture Experiment Station, University of Idaho, Hagerman, ID, United States; Institute of Aquaculture, University of Stirling, Scotland

Johan W. Schrama Aquaculture and Fisheries Group, Wageningen University, Wageningen, The Netherlands

Brian C. Small Aquaculture Research Institute, Hagerman Fish Culture Experiment Station, University of Idaho, Hagerman, ID, United States

Douglas R. Tocher Institute of Aquaculture, Faculty of Natural Sciences, University of Stirling, Stirling, Scotland, United Kingdom

Giovanni M. Turchini School of Life and Environmental Sciences, Deakin University, Geelong, VIC, Australia

R. Waagbø Institute of Marine Research/ University of Bergen, Bergen, Norway

S.Q. Xie Institute of Hydrobiology, Chinese Academy of Sciences, Wuhan, Hubei, China

M. Xue Institute of Feed Research, Chinese Academy of Agricultural Science, Beijing, China

Fernando Y. Yamamoto Department of Ecology and Conservation Biology, Texas A&M University System, College Station, TX, United States

X.Q. Zhou Sichuan Agricultural University, Chengdu, Sichuan, China

Preface

Over the past two decades since the previous edition of *Fish Nutrition* was published, production of fish and crustacea through aquaculture has increased from approximately 25 million metric tons (mmt) to over 65 mmt. Production of high-value, intensively reared species, such as shrimp and salmon, increased impressively, but more impressive has been the increase in production of lower-value, freshwater species, such as carp, catfish, and tilapia species, which now make up over three quarters of global fish production. The freshwater fish sector's growth is due to higher use of nutritionally complete feeds, as production has become more intensive to better utilize freshwater resources. Approximately 63% of global fish feed production is used to produce low trophic level freshwater species. Global fish feed production has also correspondingly increased fourfold, requiring feed producers to use sustainable and novel feed ingredients as alternatives to fishmeal and fish oil. The nutritional values of alternative feed ingredients are generally lower than fishmeal or fish oil due to contents of essential nutrients, antinutrients, or poorly digested components, creating challenges for fish feed formulators and producers. Given the expectation that aquaculture production will continue to increase, the need for advanced fish nutrition knowledge to enable precision formulation of feeds using sustainable and novel alternative feed ingredients has never been greater.

The pioneering research of John Halver laid the foundation of the science of fish nutrition, summarized in the first edition of *Fish Nutrition*, published in 1972. Over the next three decades, Halver produced two revised and updated editions of Fish Nutrition that documented progress in the evolving, expanding, and maturing fish nutrition ecosystem. The information in the new editions of *Fish Nutrition* built a science-based foundation for development of feeds for farmed fish and shrimp. Dr. Halver died in 2012, but his influence is evident throughout the fourth edition of *Fish Nutrition*, both in topics and on new knowledge needed for the aquaculture feed industry to meet the challenges of the coming decades. Since the previous edition of *Fish Nutrition* was released, the number of published papers in the field has more than doubled compared with the number published in the previous two decades. This reflects the growth of the aquaculture industry and the emphasis on economic, environmental, and social/societal concerns. Feeds are central to the production of safe, healthful, and sustainably produced fish and shrimp that contribute to food security and safety. However, basic information necessary to formulate precision feeds, such as essential nutrient requirements and sensitivity to antinutrients or adventitious toxins, are known for only a few farmed species. Therefore, a solid understanding of fish nutrition, physiology, feed manufacturing, and feed management is essential to improve production efficiencies across the hundreds of fish and crustacea species being raised.

As in previous editions of *Fish Nutrition*, the fourth edition's chapters are authored by internationally recognized experts. The new edition begins with a chapter describing the history of fish nutrition, starting with efforts to prevent disease conditions associated with nutritional deficiencies in early feeds, notably anemia, and continuing to identify essential nutrients in parallel with nutritional discoveries in domestic and laboratory animals. This chapter concludes with an assessment of the current status and future needs of fish nutrition. The second chapter, "Bioenergetics," has been updated with new information on bioenergetics and the application of the net energy system. The third chapter, "The Vitamins," has been completely revised and reorganized, adding new information on dietary requirements for optimum health and wellness, methods of vitamin analysis, and how novel feed formulations affect adequate supply of vitamins. The chapter, "Proteins and Amino Acids," presents information on dietary requirements and on how new feed formulations containing alternative and novel protein sources present challenges associated with different amino acid profiles and availability compared with feeds heavily relying on fishmeal. Of all of the challenges associated with increased fish feed production, the finite global supplies of fish oil, the primary source of omega-3 fatty acids in fish feeds, is the most pressing. The chapter, "The Lipids," presents a context for this challenge and reviews current knowledge of fatty acid requirements, fatty acid metabolism, and effects of fatty acids on fish health in several important farmed fish species. Alternate sources of essential omega-3 fatty acids for fish feeds, such as algae meals and GMO oilseeds, are reviewed and evaluated. In addition to providing an up-to-date review of the mineral requirements of fish, "The Minerals," addresses the pressing need to assess the effects of reducing fishmeal levels and increasing plant protein ingredients in fish feeds on meeting the requirements for minerals and trace elements. Finally, a new chapter, "The Carbohydrates," rounds out the first section of the fourth edition of *Fish Nutrition* by reviewing current knowledge of the implications of increasing feed levels of plant-derived feed ingredients that supply both simple and complex carbohydrates to fish and crustacea species. This chapter replaces the chapter "Intermediary Metabolism" in previous editions. A point worth mentioning is that over the past two decades, the use of integrative molecular tools has contributed significantly toward our understanding of the effects of specific nutrients and their interactions on metabolism and physiology affecting overall performance and health. The first seven chapters discuss the potential of such tools and criteria to determine nutrient requirements and their utilization.

The second section of this edition of *Fish Nutrition* focuses on practical aspects of the applications of nutrition science contributing toward the goal of developing precision feed formulations for an array of farmed species, beginning with "Nutritional Physiology." This chapter covers anatomy and digestive physiology of different categories of farmed fish and shrimp but goes further to discuss interactions of feeds with the endocrine system and how such interactions affect fish performance. The next two chapters, "Diet Formulation and Manufacture" and "Diet Analysis and Evaluation" explain how the information on nutritional requirements and physiology is put to practical use to grow fish in economical and environmentally suitable ways. Feed ingredient and diet analysis and evaluation are crucial elements of continuous improvement in feed quality. Feed ingredients must be selected and combined to produce feeds suited for the wide variety of goals of aquaculture, taking into consideration the feeding habits of various farmed species. Feed ingredient selection is based in vitro chemical and nutritional analysis, and in vivo ingredient and diet assessment. In addition to essential nutrients, feeds must contain other biologically active constituents to support fish growth and wellness. The chapter "Nutritional Supplements and Fish Health" presents current knowledge in this developing area of fish nutrition. Feeds can also have adverse effects on fish performance through introduction of undesirable materials that feed ingredients may contain. The chapter "Adventitious Toxins and Antinutrients" covers this important area of fish nutrition, focusing on common topics of concern in practical fish feeds. "Nutritional Pathology" presents a comprehensive survey of clinical signs of nutritional deficiencies and pathologies from a veterinary science perspective, serving as a guide to professionals charged with detection and diagnosis of nutritional problems in aquatic animals. The final chapter "Feeding and Fish Husbandry" covers the most important and often neglected area of fish nutrition, namely, how to manage the difficult task of feeding aquatic animals reared in large numbers. Appropriate feeding strategies and feed management are the keys for a viable, efficient, ecologically reliable, and socially beneficial aquaculture enterprise. Application of artificial intelligence and new modeling tools to enable precision feeding are emerging technological advancements that have the potential to greatly advance feed management for the aquaculture industry.

This edition of *Fish Nutrition* is intended to be a valuable resource for academics, professionals in the aquaculture industry, graduate students, and others wishing to explore the intricacies of applying principles of animal nutrition to aquatic animals. Fish feeds have undergone continuous improvement, especially in the past 30 years, but further improvement is needed to increase nutrient utilization efficiency, minimize losses to the environment, and develop feeds for the hundreds of farmed fish species for which little information on nutritional requirements exists. The authors of the chapters in this edition have reviewed huge numbers of scientific publications, but they have not simply written review articles. Rather they have summarized current knowledge with the intent of making it accessible to practitioners in the fish nutrition field. Fish nutrition is a practical discipline that integrates basic nutritional biochemistry and physiology with fish husbandry with the aim of providing high quality food for a growing global population with finite global resources. Chapter authors have deep understanding of their respective disciplines but also wide understanding of the aquaculture industry throughout the world, from low technology pond culture to highly intensive, technology-driven recirculating aquaculture systems. We greatly value their contributions to this edition of Fish Nutrition.

Ronald W. Hardy

Sadasivam J. Kaushik

Editor biographies

Ronald W. Hardy is a Distinguished Professor Emeritus in the Department of Animal and Veterinary Sciences and past Director of the Aquaculture Research Institute at the University of Idaho, Moscow. Dr Hardy received his PhD from the University of Washington in Fisheries, conducting his dissertation on the effects of dietary protein and pyridoxine levels on growth and disease resistance of Chinook salmon. He is a member of the American Institute of Nutrition and the World Aquaculture Society, serving as the Vice President of the latter from 2001 to 2003. In addition to coediting the previous edition of Fish Nutrition, he has authored over 50 books and book chapters, and over 300 scientific papers, proceedings, and popular articles on fish feed and fish nutrition. He served as chair of the committee that produced the 2011 National Research Council (NRC) bulletin Nutrient Requirements of Fish and Shrimp.

Sadasivam J. Kaushik earned his bachelor's and master's degrees at the prestigious Presidency College of Madras University in India. He received his Doctor of Science degree from the University of Bretagne Occidentale, France, for his work on the influence of salinity on nitrogen metabolism and arginine requirement of rainbow trout. As an active member of the committee on fish nutrition and production of the International Union of Nutritional Sciences, he was actively involved in fish nutrition research and training. He was the director of the INRA–IFREMER Joint research units on fish nutrition and also headed the hydrobiology research station at Saint Pée sur Nivelle, France. He has coauthored more than 300 scientific papers and books on fish nutrition and feeding. Has been active in coordinating European research projects in fish nutrition and metabolism. Further to his retirement from INRA, he held a European research area chair position in the University of Las Palmas, Spain.

CHAPTER

1

Fish nutrition—history and perspectives

Ronald W. Hardy[1], *Sadasivam J. Kaushik*[2], *Kangsen Mai*[3] *and S. Charles Bai*[4]

[1]Aquaculture Research Institute, Hagerman Fish Culture Experiment Station, University of Idaho, Hagerman, ID, United States; [2]INRAe, Univ Pau and Pays de l'Adour, Nutrition, Metabolism and Aquaculture Unit, Saint Pée sur Nivelle, France; [3]Ocean University of China, Qingdao, Shandong, China; [4]FAO World Fisheries University Pilot Program; Feeds and Foods Nutrition Research Center, Pukyong National University, Busan, South Gyeongsang Province, Republic of Korea

Aquaculture, the cultivation of aquatic organisms, began thousands of years ago when ancient people collected fish from rivers and the sea to hold in enclosures until they were harvested. Images from Egyptian, Assyrian, and Roman sources show these activities; however, recorded details of how they were carried out are lost. Records uncovered in oracle bone inscriptions from the Shang Dynasty (1395–1123 BCE) in Yin Ruins of Henan Province indicated that fish farming was being practiced. Clay depictions of carp ponds and the writings of Fan Li from around 500 BCE (Chinese Aquaculture Society (Li, 1986)) further document aquaculture activities in China. Fan Li's writings, available today as *"Fan Li on Pisciculture"* (also known as *The Art of Fish Breeding*), were included in a book by Jia Sixie entitled *"Key Technologies by People in the State of Qi,"* written during the Northern Wei dynasty (386–535 AD). While Fan Li's writings do not describe feeding of fish, the number of fish in ponds managed according to his instructions exceeded the biomass that such ponds could sustain without enrichment to increase production of natural food to support fish growth. Since that period, aquaculture systems in China, Southeast Asia, and India have been following this production system, classified today as extensive aquaculture. In extensive aquaculture, fish are not fed but obtain nutrients from natural food. Natural productivity can be enhanced by pond fertilization with manure, human waste, vegetation, and other materials. Extensive polyculture systems involve stocking ponds with complementary fish species that consumed a range of natural foods, e.g., fish that feed on organisms living in the sediment of ponds, on plankton, on aquatic insects, or on vegetation. Historical practices have been defined as "protoaquaculture" wherein some degree of management of the life

Fish Nutrition, Fourth Edition
https://doi.org/10.1016/B978-0-12-819587-1.00006-9

cycle of fish or shellfish is applied which increases food production (Beveridge and Little, 2002). Readers seeking further details on the history of aquaculture will find a number of sources in the scientific literature (Stickney, 1996; Nash, 2010; Stickney and Treece, 2012).

1.1 Early records of fish nutrition

he earliest known record describing fish feeds is found in Manasoltara, a Sanskrit encyclopedia by King Someswara III (1126-1138 AD) in India, where there is a chapter devoted to feeding fish (Hora, 1948; Bagchi and Jha, 2011). In Europe, writings about fish feeds emerged in the 19th century with development of farming of trout and carp. Feeds were made by farmers using locally available materials, including organ meats from animal slaughterhouses, milk curd, fresh chopped fish, and chicken eggs (Rumsey, 1994). Among other efforts to provide feed for trout, farmers in the United States placed the severed heads of cattle on poles over ponds to produce maggots that fell into the water. This practice fell out of favor due to concerns about sanitation and overall unpleasantness. Generally speaking, this early period of trout nutrition was characterized by widely divergent concepts, such as the assertion that trout were "*carnivorous*" and therefore could not be expected to flourish on feeds containing any plant material. Etymologically, the term "carnivore" means "meat eater" and refers to species that prey exclusively on other species in terrestrial food chains. The term is problematic when applied to aquatic organisms. Terrestrial and aquatic food chains differ significantly, and most farmed fish belong to higher trophic levels where food sources are aquatic animals rather than plants. Most fish species begin life as piscivores, and many freshwater species make a transition to omnivory as juveniles as their digestive systems develop. Eventually, fish culturists accepted the concept that fish foods should be assessed based upon their chemical composition rather than the source of ingredients (Page, 1985; cited by Rumsey, 1994). Early studies of the natural diet of fish provided some guidance for formulation of trout feeds. Researchers surveyed the natural diet of trout by catching wild fish and analyzing their stomach contents (Embody and Gordon, 1924). This enabled them to calculate the proximate composition of the natural diet of wild trout which they reported to be 49% crude protein, 15%–16% fat, 8% crude fiber, and 10% ash, expressed on a dry weight basis. As far as individual nutrient requirements were concerned, with the exception of a report showing that trout required iodine in the diet to prevent thyroid hyperplasia (goiter) (Gaylord et al., 1914), little was known.

Although it has been common knowledge since ancient times that certain foods prevent specific diseases, such as night blindness or scurvy, systematic studies of vitamins only began in the early 20th century with the discovery that nonpolar solvent or water extracts of foods contained substances that were essential in the diet to support life. They were named vitamins by the pioneering nutrition researcher, Dr. Casimir Funk, because they were thought to be derived from ammonia and were thus "vital amines" (Funk, 1912). The extracts were called fat-soluble A and water-soluble B. Further refinement of extracts revealed that each contained multiple substances essential for life. Fat-soluble A contained four vitamins, A, D, E, and K, now commonly called the fat-soluble vitamins. Water-soluble B was eventually found to contain a number of vitamins that were named in order of discovery, i.e., B_1 (thiamin), B_2 (riboflavin), B_3 (niacin), B_5 (pantothenic acid), B_6 (pyridoxine), B_7 (biotin), B_9 (folic acid), and B_{12} (cobalamin). Each vitamin was found to have one or more specific functions in metabolism that were essential for normal growth and health, and, for most, specific clinical deficiency signs associated with insufficient dietary intake.

1.2 Recent history of fish nutrition research

Systematic investigation into fish nutritional requirements originated in the United States with creation of a partnership with Cornell University, the New York State Conservation Department, and the US Bureau of Fisheries. The partnership established a trout nutrition research program at the Cortland Hatchery in New York in 1932. Research focused in two areas: (1) nutritional biochemistry and requirements and (2) development of practical diet formulations for an expanding trout hatchery system operated for fisheries conservation and enhancement by state and federal agencies. The Cortland Hatchery produced annual reports summarizing their work that provide the most complete documentation of the early years of fish nutrition and diet development research that exist.

Scientists at the Cortland hatchery were well aware of nutritional research being conducted with livestock, poultry, and laboratory animals and adopted their rigorous approaches and new technology in fish nutrition research. This represented a departure from previous fish nutrition research that consisted mainly of feeding trials to compare various feed formulations. However, the aquatic existence of fish complicated the direct application of experimental approaches and techniques used in animal and poultry nutritional studies. Fish are poikilothermic. Their metabolism varies with rearing water temperature, and they are sensitive to water quality parameters, such as oxygen and ammonia levels. Fish obtain some essential minerals directly from rearing water and, to complicate matters further, maintain internal homeostasis of electrolytes against either hypotonic or, for marine fish, hypertonic conditions. Digestibility studies involve collection of feces, a simple procedure with chickens or rats but a difficult one with fish. Furthermore, unlike rats or chickens that are fed ad libitum in nutrition studies, fish do not feed themselves. Fish were typically fed a percentage of body weight per day that varied with fish size and water temperature. However, in fish nutrition studies, fish must be fed to apparent satiation to ensure that restricted feeding does not mask differences in growth that might be due to diet. Feeding level, therefore, is a potentially confounding variable in nutrition research, especially if water temperature changes during a feeding trial. Concerning digestion and nutrient utilization, differences between the terrestrial and aquatic environments in food webs are a further complication. Terrestrial plants are the foundation of the food chain for land animals, supplying nutrition in seeds, leaves, fruits, tubers, and so on, with starch being the primary storage form of energy. These food sources are absent in the aquatic environment, resulting in a food chain in which lipids are the primary stored form of energy. This difference is reflected in their intestinal physiology, digestive capacity, nutritional requirements, and appropriate feeds.

An overlooked feature of fish with implications for fish nutrition research relates to reproductive strategies and fecundity. Unlike poultry, rats, or farm animals, fish species used in aquaculture or medical research have large clutch sizes, ranging from several hundred for zebrafish, 2000–10,000 for salmonids, to hundreds of thousands for many marine species. Producing many offspring results in sufficient genetic variation among offspring to cope with variable environmental conditions from year to year during crucial early life history stages, ensuring that some will survive. This accounts for the high variation in performance of fish within a single treatment group often observed in fish nutrition studies. More importantly, it complicates adoption of standard experimental designs with appropriate replication used in nutritional studies with other experimental animals that exhibit less phenotypic variation among cohorts in nutrition studies, such as chicks. Selecting the appropriate number replicate tanks per treatment and the number of fish per replicate tank for a fish nutrition experiment requires knowledge of expected variation among and within tanks, plus knowledge of social or behavioral traits that may confound feeding trials, such as dominance hierarchies or cannibalism. By today's standards, early fish nutrition research studies were often of short duration and lacked appropriate experimental design, sufficient replication, and statistical treatment of data. Nevertheless, early fish nutrition research findings are remarkable in that their conclusions have frequently been validated using modern experimental designs and performance assessment tools.

While the Cortland hatchery was focusing on trout nutrition, salmon nutrition became a focus of study in the western United States at the University of Washington, Oregon State University, and at regional state and federal fisheries agencies, notably the US Fish and Wildlife Service. Starting in the 1930s, many dams for irrigation and hydroelectric power generation were constructed on major rivers in the western United States, especially in the Columbia River system. The dams blocked a large proportion of Pacific salmon migratory corridors to upstream spawning and juvenile nursery areas. To mitigate loss of spawning and salmon nursery habitats, many hatcheries were constructed, creating a need to provide suitable feeds to rear juvenile Pacific salmon for release as smolts. Chinook (*Oncorhynchus tshawytscha*) and coho (*Oncorhynchus kisutch*) salmon were the main species raised in hatcheries, although sockeye salmon (*Oncorhynchus nerka*), chum salmon (*Oncorhynchus keta*), and pink salmon (*Oncorhynchus gorbuscha*) were also raised. Juveniles were raised to the smolt stage and released in rivers to migrate to the ocean. After spending several years at offshore feeding areas, salmon migrate to coastal areas where they are harvested in commercial, tribal, and recreational fisheries. A high smolt-to-adult return ratio was the primary goal of hatchery programs, and as a result, the focus of fish nutrition research was on developing feed formulations to produce healthy, robust juveniles. The cost of feed was secondary to the goal of producing smolts that survived hatchery release and contributed to fisheries. Feeds cannot be precisely formulated without information on nutritional requirement, so research focused on identifying dietary requirements of juvenile salmon for essential nutrients. The substantial investments in research capacity and information that was generated provided a foundation for subsequent research on essential nutrients and appropriate diets to rear fish to harvest size by the aquaculture industry decades later.

Prior to the middle of the 20th century, feeds for hatchery-reared trout and Pacific salmon were made on site at federal, state, and private hatcheries using locally available ingredients. Since the nutritional requirements of fish were not yet known, except in broad terms, feed formulations were developed by trial and error. Successful feed formulations were those that supported fish growth, maintained rearing water quality, and prevented clinical signs of nutritional deficiencies. Fresh by-products of animal slaughter were considered essential dietary constituents to prevent nutritional deficiencies. Feed formulations were developed empirically, rather than on the basis of meeting known nutritional requirements. For example, in the first feed evaluation trial conducted at the Cortland hatchery (Phillips et al., 1940), four diet formulations were tested:

- Diet 1: 50 parts cottonseed meal, 50 parts raw beef liver, and 50 parts dried skim milk
- Diet 2: 50 parts cottonseed meal, 50 parts raw beef liver, and 50 parts dried buttermilk
- Diet 3: 50 parts cottonseed meal, 50 parts raw beef heart, and 50 parts dried buttermilk
- Diet 4: 50 parts cottonseed meal, 50 parts raw beef spleen, and 50 parts dried buttermilk

The next round of trials was conducted using similar formulations but containing skim milk from different suppliers, substituting peanut meal for cottonseed meal, and replacing beef parts with sheep plucks (viscera) or with dried meat scraps. From this work, researchers concluded that skim milk and buttermilk were similar in value as ingredients and that trout health could not be maintained longer than 8 weeks unless feeds were supplemented with raw animal liver or heart (Cortland Hatchery Report Number 1, 1932). Over the next 10 years, many feed ingredients were tested in trout feeds at the Cortland Hatchery to develop formulations that hatchery personnel could use to prepare economical diets from commonly available ingredients. By 1942, 809 diet formulations had been tested. However, no matter what combination of dry ingredients was used, raw beef liver or other organ meat had to be included in the diet to prevent early mortality of the trout.

Concurrent with studies to develop feed formulations, research at the Cortland Hatchery was being conducted to explore the nutritional requirements of trout. For example, Tunison and McCay (1935) investigated the calcium requirement of trout, showing that the calcium content in the whole body of trout was higher than could be accounted for by the calcium content of feed they consumed. They concluded that trout could obtain calcium directly from rearing water. A major research topic was anemia, a widespread problem in hatchery-reared trout that was prevented by feeding beef liver. Liver was thought to contain an antianemia component that McCay named Factor H. Six B vitamins were hypothesized to comprise Factor H: thiamin, riboflavin, pyridoxine, pantothenic acid, niacin, and biotin. However, when these vitamins were supplemented into a dry trout diet, either alone or in various combinations, anemia was not prevented (Cortland Hatchery Report, No 13, 1944). Folic acid, an essential vitamin, was first extracted and isolated from spinach leaves in 1941 and given the name folic from the Latin word for leaf (folium). Folic acid was isolated from liver cells in 1943 by Dr. Lucy Wills working in India on leave from the Royal Free Hospital in London, and its structure identified a few years later. Beef liver is a rich source of folic acid and was therefore hypothesized to be Factor H, but supplementing purified folic acid to semipurified diets did not prevent anemia in trout. Later research identified vitamin B_{12} as the missing nutrient associated with Factor H, as described in the following.

The 1920 and 1930s were a period of rapid progress in discovery and characterization of vitamins, facilitated by using rats and mice as experimental animals (Carpenter, 2003). Researchers used semipurified diets to determine requirements for individual vitamins, and in the late 1940s, fish nutritionists adopted this approach, first using semipurified diet formulations that were routinely fed to rats. However, rat semipurified diets consisting of casein and sucrose with vitamin and mineral supplements proved unsuitable for trout unless the fish received dried beef liver (McLaren et al., 1946). After much trial and error, a diet consisted of casein, dextrin, salt mixture, crab meal, corn oil, cod liver oil, and a vitamin mixture that supported trout growth and health in short-term studies was developed (McLaren et al., 1947). Using this diet, B vitamins and ascorbic acid were omitted one-by-one from the vitamin mixture and added back in increments to experimental diet, enabling scientists to make broad estimates of dietary requirements of trout for thiamin (1−10 mg/kg), riboflavin (5−15 mg/kg), pyridoxine (1−10 mg/kg), pantothenic acid (10−100 mg/kg), nicotinic acid (1−5 mg/kg), choline (50−100 mg/kg), inositol (250−500 mg/kg), biotin (0.05−0.25 mg/kg), folic acid (1−5 mg/kg), and ascorbic acid (250−500 mg/kg). To estimate the vitamin requirements of trout more accurately, semipurified diets had to be vitamin free. The semipurified diets used in the McLaren et al. studies were not totally vitamin free; casein and crab meal both supplied vitamins to the diet. A modified semipurified diet was developed which contained casein, gelatin, lard, cooked potato starch, cellulose flour, and minerals (Wolf, 1951). This diet supported trout growth rates over a 25-week period equivalent to contemporary practical diets that contained beef liver. Commercial casein contained small amounts of B vitamins, but when vitamin-free casein, made by extracting casein with solvents, was substituted for commercial casein, trout refused to eat the diet. Regarding the need for vitamin supplementation, Wolf (1951) reported that deletion of pantothenic acid resulted in cessation of growth and gill disease after 8 weeks of feeding. Likewise, deletion of folic acid slowed fish growth and resulted in anemia, and deletion of vitamin B_{12} resulted in fish mortality and anemia.

Meanwhile, on the west coast of the United States, similar efforts were being made to develop diets for hatchery-reared fry and juvenile Pacific salmon. Normal hatchery practice was to freeze adult salmon that migrated back to freshwater to spawn at hatcheries and use them in fry and juvenile salmon feeds by mixing chopped salmon with dry ingredients. Although fish could be grown using such feeds, anemia was a continuous problem and could only be prevented by periodically feeding beef liver. In a pivotal event in the history of fish nutrition, Dr. Lauren Donaldson, a fisheries professor at the University of Washington's College of Fisheries, requested assistance from the Department of Biochemistry to determine if a dietary treatment for salmon anemia could be found. John E. Halver, a graduate student, was assigned to investigate this problem. The Department of Biochemistry had just obtained a few mg of purified folic acid. Halver injected salmon fingerlings afflicted with anemia with folic acid and vitamin B_{12}, curing anemia in 10 days. This experience led Halver to change his Ph.D. research from human to fish nutrition. For his Ph.D. dissertation research, he developed a semipurified diet formulation that was truly vitamin-free and supported normal fish growth (Halver and Coates, 1957) and then used this diet to demonstrate the essentiality of several B vitamins for juvenile chinook salmon (Halver, 1957). What followed was a series of studies that resulted in estimates of the quantitative vitamin and later amino acid requirements of chinook salmon, coho salmon, and rainbow trout (Halver, 1972; NRC, 1973). Work by Halver and associates established a model approach for fish nutrition studies that was used by researchers in the United States, Europe, and Japan in the 1960 and 1970s to further elucidate the dietary requirements of salmonids, channel catfish, carp, and other farmed fish. In the United States, the progression of catfish culture from extensive pond rearing to intensive production required the development of nutritionally complete feeds. Research conducted by university and government researchers led to development of economical catfish feeds, resulting in the tremendous expansion of the US catfish industry in the 1970 and 1980s to become the largest aquaculture industry in the United States. Robinson and Li (2019) have written an excellent historical review of the catfish nutrition research. By the early 1970s, estimates for most of essential nutrient requirements of juvenile salmon and catfish had been made (NRC, 1973).

Developments in fish nutrition in the United States were adopted in other countries and applied to other farmed fish species, particularly in Japan and Europe. Aquaculture has a long history in Japan and in the second half of the 20th century; researchers in Japan made substantial contributions to fish nutrition, although many of their studies were published in Japanese with an English abstract, thus limiting their recognition by fish nutritionists in western countries. Examples of early fish nutrition research by Japanese scientists included amino acid composition of hatchery diets (Ogino, 1957), digestibility (Nose and Toyama, 1966), the nutritional requirements of *Penaeus japonicus*, the Japanese prawn (Kanazawa et al., 1971), vitamin requirements of trout (Kitamura et al., 1965) (Kitamura et al., 1967), and dietary fatty acid requirements (Toyomizu et al., 1963; Kayama, 1964; Higashi et al., 1966). By the late 1970s, Japanese researchers had estimated the dietary requirements for amino acids, fatty acids, vitamins, and minerals for trout, carp, shrimp, and other farmed species. The Tokyo University of Fisheries and Kagoshima University under the leadership of T. Watanabe and S. Kanazawa established strong research programs that attracted and trained many international students who returned to their home countries to lead fish nutrition programs.

Fish nutrition research also took hold in European countries in the 1970s, particularly in the United Kingdom, France, Germany, Italy, Norway, Greece, Poland, and Hungary. These research programs were led by several research leaders, many of whom expanded their research focus from animal nutrition to fish. For example, in the United Kingdom, the Unit for Research into Fish Nutrition was established as a component of the National Institute for Research into Dairying at Reading by C. Cowey, who subsequently joined the Institute of Marine Biochemistry in Aberdeen, Scotland, and led a team of researchers who approached fish nutrition from a biochemical perspective. Cowey organized the first International symposium on Fish Nutrition and Feeding in Aberdeen, Scotland, in 1985. This initiative led to the establishment of biannual International symposia on fish nutrition and feeding which continues today. Fish nutrition continued at the Institute of Aquaculture at the University of Stirling, Scotland, with a focus on lipid nutrition and metabolism, and training many international students and scientists. In Germany, research on fish nutrition was undertaken in both the eastern and western parts. The first World Symposium on Finfish Nutrition and Fishfeed Technology was held in Hamburg, Germany, in 1979 under the umbrella of the European Inland Fisheries Advisory Committee (EIFAC) of the FAO (Halver and Tiews, 1979). The EIFAC also organized a dedicated workshop on Methodology for Determination of Nutrient Requirements in Fish in 1993 in Eichenhau, Germany. In Norway, the Institute of Nutrition in Bergen added fish nutrition to its programs on animal and human nutrition in the late 1970s. Basic and applied fish nutrition research continues to this day at the Institute of Marine Research in Bergen. In the 1970s, with the rapid development of Atlantic salmon farming in Norway, another major group dedicated to fish nutrition, Akvaforsk, now NOFIMA, was actively engaged in research and development in fish nutrition and feeding.

In France, Luquet established the first laboratory dedicated to fish nutrition as part of animal nutrition group in the National Institute for Agricultural Research (INRA) in Jouy en Josas. This laboratory then moved to the southwest of France in Saint Pee sur Nivelle in

1976, where the group became a leading laboratory conducting research on fish nutrition and metabolism. Currently, in almost all European countries, research teams as part of universities or public research institutes deal with dedicated work on various aspects of nutrition of different species of farmed fish at such locations as the Hellenic Centre for Marine Research, Athens, Greece; the University of Las Palmas, Spain; the University of Porto, Portugal; and the University of Udine in Italy, just to mention a few.

1.3 Fish nutrition developments in China

China now accounts for over 62% of global farmed fish and crustacea production, the result of rapid development of all aspects of aquaculture over the past 50 years. China's more than 3000-year history of aquaculture can be divided into three stages. From the Shang Dynasty to the 1950s is the first stage, characterized by capture of wild fingerings to stock ponds and reliance on natural productivity in ponds to provide nutrients for the fish. From the 1950–1980s is the second stage. In the early 1950s, China was the first to succeed in artificial propagation of four major fishes (black carp, grass carp, silver carp, and bighead carp). Hence, this stage is characterized by using artificially propagated fingerings, while still being largely dependent on natural pond productivity. Some mixtures of agricultural by-products were occasionally used as feeds, but this was not based on nutrient requirement data. The real research on quantitative nutrient requirements of fish did not start until around 1980. From the 1980s onward is the third stage, characterized by adoption of high-density intensive culture models, artificially propagated seedlings, and provision of formulated feed based on nutrient requirements. China is now the most important aquaculture country in the world. The total aquaculture production of China in 2018 was 49.91 million tons (fish 54%, crustaceans 10.3%, mollusks 29.3%, algae 4.7%, and others 1.6%), accounting for more than 60.7% of global production (FAO, 2020). Aquaculture in China is entering a new development stage characterized by integrating technologies from other fields, such as bioengineering, information engineering, and artificial intelligence.

The development of aquaculture nutrition research and aquafeed industry in China can also be divided into four periods: (1) the germination period from the early 1950s until 1979; (2) the starting period from 1980 to 1989; (3) the rapid development period from 1990 until 2000; and (4) the leapfrog period from 2001 until the present day.

At the end of the 1950s, after the success of artificial propagation of fishes in China, the government put forward a strategy of "prioritizing aquaculture, codeveloping farming, and capturing," which led to the rapid development of aquaculture. With the increase of farming density and production of cultured fish, farmers naturally wanted to feed them. However, until the late 1970s, there was no real aquaculture nutrition research in China to guide the preparation of aquafeed and to establish appropriate feeding practices. At that time, some silage feeds and agricultural by-products, such as grass, algae, crop vines, rice or wheat bran, soybean or peanut cakes, silkworm pupa, and earthworms, were simply fed separately or mixed in a certain proportion based on fish nutrition data published abroad, which could only be called mixed feed. Feed conversion ratios were 4–10 or even higher. Therefore, this stage belongs to the germination period of aquaculture nutrition in China.

In the late 1970s, a large number of fishes and crustaceans, represented by *Fenneropenaeus chinensis*, were successfully propagated artificially, and this promoted further rapid development of aquaculture in China. The method of directly feeding feed ingredients, randomly mixed feed or small trash fish, could not meet the demands of aquaculture development. There was an obvious and urgent need to define the nutrient requirements of aquaculture animals and to develop the aquafeed industry in China. Universities such as Sun Yat-sen University, Ocean University of China, and Shanghai Ocean University started to offer fish nutrition courses. Professors Ding Lin, Aijie Li, and Daozun Wang, the pioneers of aquaculture nutrition in China, took the lead in recruiting and supervising postgraduate students. At the same time, academia and industry also began to research and develop feed formulations, processing technologies and machinery to meet the market's demand. During the starting period of China's aquaculture nutrition, the research had the following characters. First, due to the large variety of farmed species, studies generally were not comprehensive, mainly focusing on the requirements of macronutrients, such as protein, lipid, and carbohydrate, and practical feed formulations. There were much less studies on micronutrient requirements and nutritional physiology, which were necessary for further optimization of feed formulation and improved feed utilization. Second, experimental designs used in research studies were not statistically robust, and feeding periods were generally too short to obtain reliable and comparable results. Third, a substantial portion of research effort focused on development of feed additives described as growth promoters due to the potential for short-term profits. Insufficient resources were devoted to determining functional nutritional or nonnutritive additives. Fourth, development of complete sets of processing equipment for aquafeed was insufficient. Finally, inadequate attention was paid to the exploitation and quality control of feed ingredients (Mai, 2020). Research during this period did make progress, however, as evidenced by overall feed conversion rations being reduced from 4–10 to 3–4.

With the rapid development of intensive aquaculture after 2000, the demand for the quantity and quality of aquatic feed increased rapidly. However, Chinese aquaculture is unique in many ways. There is a high diversity in farmed species, feeding habits, ecological distribution, and culture mode. Nutritional studies published in China have been conducted on 197 fish species (strains), 26 crustaceans, 16 mollusks, 5 echinoderms, 3 amphibians, and 2 reptiles. They have a wide geographic distribution, different feeding habits (herbivorous, carnivorous, omnivorous, and filter feeders), and different culture modes. With a late start of aquaculture nutrition research, insufficient nutrition research experts, and limited research funds, it is impossible to quantify the nutritional requirements of so many species even in a long period. Sporadic nutrition research cannot meet the needs of industry development either. To address this, an R&D strategy was developed in the early 1990s: "selecting representative species, focusing our efforts, standardizing research methods, studying systematically, and radiating research results." That is to say, the focus of research should be on species that represent larger classes of fish, based on classification status, geographic distribution, feeding habits, and culture modes, Chinese characteristics, meaning not studied in other countries and the potential to scale up production rapidly. The country's main researchers would gather to conduct continuous and systematic studies on the selected species, by using standardized research methodology, and even same feed ingredients and other experimental materials when necessary, to improve the comparability, accuracy, and generalizability of research data. A relatively complete database of nutrient requirements of the representative species would gradually be established. Then, results would be applied to other farmed species with similar taxonomic status, geographic distribution, feeding habits, and farming modes, respectively. Ten representative species were selected: grass carp (*Ctenopharyngodon idellus*), silver crucian carp (*Carassius auratus* gibelio), tilapia (*Oreochromis* spp.), large yellow croaker (*Pseudosciaena crocea*), sea bass (*Lateolabrax japonicus*), cobia (*Rachycentron canadum*), grouper (*Epinephelus* spp), prawn (*Penaeus* spp.), Chinese mitten crab (*Eriocheir sinensis*), and abalone (*Haliotis discus hannai*). This suggestion was accepted, and systematic nutrition studies of the representative species were conducted with the support of a series of national research projects. Subsequently, a public platform for the nutritional parameters of major aquaculture animals in China was gradually established. It provided a scientific basis for the design of feed formulations and laid the foundation for the rise and rapid development of China's aquafeed industry. The feed conversion ratio of aquafeed dropped to 2–3. This period belongs to the "rapid development period" of China's aquaculture nutrition research and aquafeed industry (Mai, 2020).

Entering the 21st century, aquaculture nutrition research in China began the "leapfrog period." With the continuous increase in aquaculture nutrition professionals and research funds, international cooperation and exchanges have been strengthened. Not only nutrient requirements of more and more farmed species but also new information on feeding behavior, regulation of nutritional metabolism, effects of nutrition on health, environment, and mechanisms that determine farmed species quality are being identified. A series of new findings have filled some gaps in aquaculture nutrition and laid a theoretical foundation for the design of precise feed formulations and feeding strategies. For example, studies have found that learning and memory factors can promote the effect of a taste receptor, T1R1, on the memory of food selection through the central nervous system in the process of mandarin fish (*Siniperca chuatsi*) domestication. This appears to cause the food addiction by stimulating the continuous expression of proenkephalin, and the domestication of mandarin fish that in nature only feeds on live prey to feeding on feed can be achieved by methylation of T1R1 gene (Duo et al., 2018). For the first time, it is proved that insufficient activation of the nutrient sensing pathway centering on mTOR is the main cause of the decline in fish growth after dietary fish meal is replaced by plant protein (Xu et al., 2016). Studies have demonstrated that lipophagy, an important lipolysis pathway, also exists in fish and that blocking the lipophagy flow reduces the lipid metabolism activity of fish overall, thus proving that lipophagy is a necessary cellular metabolic process for fish lipid metabolism (Wang et al., 2018). In addition, Peng et al. (2017) first reported that Cyp2r1 depletion results in 1,25(OH)2D3 deficiency, retarded growth, and excessive visceral adipose tissue in zebrafish. 1,25(OH)2D3 regulates lipid metabolism through the regulation of Pgc1a, controlling mitochondrial biogenesis and oxidative metabolism in zebrafish visceral adipose tissue. It is found that ω-6 polyunsaturated fatty acids (linoleic acid) can activate both autophagy and antioxidation in a synergistic feedback loop via TOR-dependent and TOR-independent signaling pathways in large yellow croaker (Yang et al., 2020). These are innovative research findings that illustrate new advances in fish nutrition made possible by application of high-throughput technologies to fish nutrition and nutritional physiology.

1.4 Evolution of fish feeds

As mentioned earlier, feeds for salmon raised in hatcheries for fisheries enhancement were made on-site, often using frozen, adult salmon carcasses as a feed constituent. Aside from detrimental effects on water quality, feeds containing salmon carcasses also were found to be a disease vector responsible for fish tuberculosis, caused by *Mycobacterium chelonei* (Wood and Ordahl, 1958). Although this disease is rare today, it was a significant problem for the Pacific salmon fisheries enhancement programs because its main clinical manifestation was undeveloped gonads in maturing fish that returned to hatcheries. This impacted the number of eggs needed to fulfill hatchery production goals (Wood and Ordahl, 1958). After years of research to establish the cause of undeveloped gonads in adult salmon and the discovery that the putative pathogen was not destroyed by freezing but was destroyed by heat, the use of frozen salmon carcasses or processing byproducts in feeds for juvenile salmon was discontinued and thermal processing of salmon carcasses or other fish used in feeds was mandated. This led to the development of an open-formula, semimoist salmon feed called the Oregon Moist Pellet (OMP), a cold-pelleted feed that required frozen storage. After extensive testing, the OMP became the standard feed used in hatchery programs (Hublou, 1963). The OMP was produced by commercial feed companies and distributed to hatcheries, eliminating the cost and complications associated with producing feed at each hatchery. The adoption of the OMP also improved hatchery water quality and provided consistent and predictable growth at each hatchery.

The OMP had one major drawback; it was a high-moisture feed (~28%) and therefore had to remain frozen during shipping and storage until it was fed. It was also relatively expensive. Research by the US Fish and Wildlife Service at the Cortland Hatchery and the Abernathy Fish Technology Center, located in Washington State, led to develop open-formula formulations for dry, pelleted trout and salmon feeds (Phillips et al., 1964; Fowler and Burrows, 1971). The open-formula feeds were produced by commercial feed companies supplying state and federal hatcheries, and commercial trout farmers. The formulations were later modified by feed companies as proprietary, closed-formula feeds that are the basis for feeds used today.

Intensive aquaculture, defined as systems in which fish obtain essential nutrients from nutritionally complete feeds, was limited to freshwater species until the development of marine farming of salmonids in the 1970s. This led to changes in fish nutrition research from developing feeds for juveniles to developing feeds to grow fish to postjuvenile, harvestable sizes. The economics of fish farming are simple—grow fish to harvest size economically and without significant losses to disease. Feed is the largest operating cost for fish and shrimp farmers, and losses to disease were often the difference between business success or failure. The development of economical feed formulations with efficient feed conversion ratios and prevention of diseases became the primary areas of fish nutrition research. Salmon farms used moist feeds similar to the OMP in early days but switched to pelleted feeds to eliminate logistical and economic issues associated with the use of moist feeds in marine pens. However, pelleted feeds had limitations, namely the need for relatively high starch levels to increase pellet strength, durability and water stability, buoyancy in seawater (pellets sank rapidly, and a significant portion was lost), and upper limits on fat content. Protein levels in feeds for juvenile salmonids were high, and protein was the most expensive component in fish feeds, but it was unknown if postjuvenile fish needed high protein diets. Protein in feeds was primarily supplied by fishmeal, but economics dictated fishmeal levels needed to be reduced by replacing a portion of fishmeal with alternative protein sources. Could this be done without reducing fish growth and health? Basic knowledge of fish digestion, metabolism and bioenergetics, amino acid requirements, and roles of micronutrients was needed to support the growing aquaculture industry, and the research focus in fish nutrition changed accordingly.

Today, aquaculture supplies more than 50% of the world's seafood, defined as food produced from freshwater and marine species. Today, over 95% of global fish feed production is used by the aquaculture industry, a reversal of the situation 30 years ago when most fish feed was used by agencies in fisheries enhancement efforts, at least for salmonids. Feeds for salmon are highly efficient, and salmon farming has become a net producer of seafood despite the use of marine-derived ingredients in feeds, the result of decades of nutrition research (Naylor, 2021). Feeds for most other farmed fish species lag behind salmonids in efficiency, mainly due to lack of knowledge of requirements for essential nutrients that prevents precision feed formulation. This is evident from the information presented in the NRC (2011) bulletin—tables listing nutritional requirements for salmonids are mostly complete, whereas those for marine fish and other many important freshwater species are either incomplete or mostly blank.

Feed ingredient supplies represent a future challenge for the aquaculture industry, as is best illustrated by the situation in China. In 1991, the year China began to record aquafeed production statistics, 750,000 tons were produced. Feed production reached 22.11 million tons in 2019, increasing about 30 times in 28 years, ranking first in the world and accounting for more than 45% of global production. China is not only the largest aquaculture producing country with a high degree of species diversity but also a country with a shortage of feed raw materials. The main feed ingredients, especially fishmeal and soybean meal, rely on imports for more than 85% of the amount currently used. China's aquafeed industry lacks high-quality raw materials and therefore uses a large number of nontraditional raw materials with low quality and low value. In response to these practical difficulties, high-efficient technologies of feed enzyme and microbial fermentation have been developed, which have greatly improved the bioavailability and feed efficiency of nontraditional raw materials (Cao et al., 2007; Zhou et al., 2013; Shao et al., 2019; Rahimnejada et al., 2019), expanding the selection of feed materials and contributing to the sustainable development of China's aquaculture.

1.5 Fish nutrition challenges

The fish and crustacean species currently produced by aquaculture, and probably all fish, likely require 40 or so essential nutrients, some of which they can obtain from water. Properly designed feeding trials to determine fish nutritional requirements involve experimental feeds containing at least five levels of the nutrient being evaluated with at least three replicate groups of each experimental feed. This makes the prospect of estimating the nutritional requirements of 220 or more farmed fish species daunting, especially considering that nutritional requirements may change with life history stage. However, there are some shortcuts and assumptions that lessen the task. First, as mentioned, fish obtain many essential minerals from rearing water. Therefore, research on the dietary requirements of minerals can focus on those that must be supplied in the diet and those that are likely to be affected by chemical form or interactions with other feed ingredients. Second, vitamin requirements of fish can be easily met by supplementing feeds with premixes. The cost of oversupplementing vitamins in feeds is not a crucial economic concern. Therefore, vitamin requirement research can focus on vitamins associated with specific metabolic requirements and those susceptible to loss during feed manufacturing and storage. Hence, it is important to evaluate the micronutrient metabolism in different fish species to estimate the precise requirement. Third, despite the incredible diversity in life history, habitat, and nutritional habits of farmed fish species, essential amino acid requirements do not vary much among species, especially among carnivorous (piscivorous) species. Amino acid content of muscle tissue is remarkably similar among species. This simplifies the task of estimating amino acid requirements of fish, although refining requirements will remain a need for many commercially important species, especially as the proportion of alternative protein ingredients from plant or microbial sources are used in feeds. Nevertheless, similarity in digestive physiology and natural diet among many carnivorous fish species suggests that amino acid requirements of salmonids or other species where information is known can be applied to species where there is little work on amino acid requirement. Also, new approaches have been proposed to assess essential amino acid requirements by the continuous measurement of plasma free amino acid concentrations. This method can be used after force feeding with the graded levels of targeted essential amino acids in surgically modified fish (Ok et al., 2001; Bai et al., 2003; Park et al., 2005; Bae et al., 2011, 2012; Yun et al., 2015, 2016). Fourth, fatty acid requirements are the most challenging area of fish nutritional requirement research. In addition to the well-known critical roles of the essential fatty acids ecosapentanoic acid (EPA) and docosahexaenoic acid (DHA), several other fatty acids, notably arachidonic acid, have been shown to be essential in some fish species (Shahkar et al., 2016). In addition, fish feeds are formulating using digestible energy (DE) values, whereas in livestock and poultry feeds, net energy (NE) and metabolizable energy (ME) are used as the true values. This difference is due to the difficulties in measuring nonfecal energy in fish (Groot et al., 2021). Development of fish models to calculate ME or NE for each species is an essential approach to support precision formulations for fish feeds. Fish larval nutrition is another challenging area of research due to the small size of organisms at this stage, sensitivity, and undeveloped organs. Hence, larviculture is the bottleneck of a successful aquaculture production. Often high mortalities (up to 90% in Japanese glass eel) and low growth rates are observed in hatcheries that are likely due to the lack of proper nutrition. It is important to understand the ontogeny of the digestive system and micronutrient requirements to produce diets with high ingestion rates that maximize growth and survival in larvae. Finally, the amount of protein in ingredients and feeds is measured based on the conventional conversion value of nitrogen content to crude protein of 6.25. However, the actual nitrogen value in protein ingredients used in fish feeds ranges from 5.556 to 6.667 depending on the amino acid composition of the protein ingredients. Feed formulations based upon protein contents of ingredients calculated using specific nitrogen values for major protein ingredients will improve feed efficiency, as will more precise values for amino acid requirements for many fish species. Sophisticated research models are required to study essential amino acid requirements in fish (Seong et al., 2018).

Convergence of complementary disciplines, e.g., biochemistry, genetics, physiology, and application of emerging technologies such as high-throughput "omics" to nutritional studies will accelerate progress in fish nutrition research. It is imperative that we use complementary approaches in fish nutrition research. Going beyond the yardsticks/criteria of conventional animal husbandry, fish nutrition research should benefit from applying analytical tools of physiology at the whole organism level, of biochemistry at the tissue, organ, or cellular levels and of expression levels of relevant genes. The application of high-throughput, integrative "omics" tools in fish/crustacean nutrition research should rely on pertinent, duly validated biomarkers of response to different nutrients and should be used to complement responses at the molecular level with responses at other levels.

1.6 Future needs

Aquaculture production is expected to continue to grow into the foreseeable future, creating a growing demand for fish feeds that support intensification of freshwater production, expansion of marine production, and development of intensive, land-based recirculating aquaculture systems (RASs). At the same time, the global supply of fishmeal and fish oil produced from marine resources cannot expand further (Naylor, 2021). Thus, the most important future need in fish nutrition will be to continue to reduce the percentages of fishmeal and fish oil in feeds with alternative ingredients without reducing fish growth, fish health, and nutritional value to the consumer. Continuously changing ingredient compositions of feeds, based on feed ingredient supplies and relative prices, societal issues such as food/feed safety, make optimal diet formulation a moving target. In fact, one of the key assumptions of feed formulation is that there is no "ideal" feed formulation for a particular species. Rather there are numerous feed formulations that meet specifications and support production goals. This points to the need for benchmarking feed formulations against a "REFERENCE" diet, which can be theoretically tested and validated by different groups and for different species. Reference diets should be an open-formula diets that closely resemble production diets for the species being investigated. Besides assessing the similarities or differences between species of finfish or shrimp, researchers should also consider possible feed × strain interactions, with the aim of identifying strains that flourish on feeds containing alternative ingredients. A final future need is to assess the potential to improve fish and shrimp growth, health, or value to consumers through other means, such as nutritional programming.

References

Bae, J.Y., Ok, I.H., Lee, S., Hung, S.S., Min, T.S., Bai, S.C., 2011. Reevaluation of dietary methionine requirement by plasma methionine and ammonia concentrations in surgically modified rainbow trout, *Oncorhynchus mykiss*. J. Appl. Ichthyol. 27 (3), 887–891.

Bae, J.Y., Park, G., Yun, H., Hung, S.S., Bai, S.C., 2012. The dietary valine requirement for rainbow trout, *Oncorhynchus mykiss*, can be estimated by plasma free valine and ammonia concentrations after dorsal aorta cannulation. J. Appl. Anim. Res. 40 (1), 73–79.

Bagchi, A., Jha, P., 2011. Fish and fisheries in Indian heritage and development of pisciculture in India. Rev. Fish. Sci. 2011, 85–118.

Bai, S.C., Ok, I.H., Park, G.J., Kim, K.W., Choi, S.M., 2003. Development of modeling system for assessing essential amino acid requirements using surgically modified rainbow trout. J. Aquacult. 16 (1), 1–7.
Beveridge, M., Little, D., 2002. The history of aquaculture in traditional societies. In: Costa-Pierce, B. (Ed.), Ecological Aquaculture, the Evolution of the Blue Revolution. Blackwell Science Ltd., Oxford, pp. 3–29.
Cao, L., et al., 2007. Applicaiton of microbial phytase in fish feed. Enzym. Microb. Technol. 40, 497–507.
Carpenter, K., 2003. A short history of nutritional science: Part 3 (1912-1944). J. Nutr. 133, 3023–3032.
Duo, Y., et al., 2018. Memory function in feeding habit transformation of Mandarin fish (*Siniperca chuatsi*). Ing. J. Mol. Sci. 19, 1254–1270.
Embody, G., Gordon, M., 1924. A comparative study of natural and artificial foods of brook trout. Trans. Am. Fish. Soc. 48, 185–200.
FAO, 2020. The State of World FIsheries and Aquaculture. Food and Aquaculture Organization, Rome.
Fowler, L., Burrows, R., 1971. The Abernathy salmon diet. Prog. Fish-Cult. 33, 67–75.
Funk, C., 1912. The etiology of the deficiency diseases. J. State Med. 20, 341–368.
Gaylord, H., Marsh, M., Busch, F., Simpson, M., 1914. Carcinoma of the thyroid in the salmonid fishes. Bull. U.S. Bur. Fish. 32, 363–524.
Groot, R., Lyons, P., Schrama, J.W., 2021. Digestible Energy versus Net Energy Approaches in Feed Evaluation for Rainbow Trout (*Oncorhynchus mykiss*). Animal Feed Science and Technology, p. 114893.
Halver, J., 1957. Nutrition of salmonid fishes. III. Water-soluble vitamin requirements of chinook salmon. J. Nutr. 62, 225–243.
Halver, J., 1972. The vitamins. In: Halver, J. (Ed.), Fish Nutrition. Academic Press, New York, pp. 29–103.
Halver, J., Coates, J., 1957. A vitamin test diet for long-term feeding studies. Prog. Fish-Cult. 19, 112–119.
Halver, J., Tiews, K. (Eds.), 1979. Proc. World Symp. On Finfish Nutrition and Feed Techology (Berlin, Heenemann Verlagsgesellschaft mbH).
Higashi, H., et al., 1966. Effect of ethyl linoleate, ethyl linolenate and ethyl esters of highly unsaturated fatty acids on essential fatty acid deficiency in rainbow trout. J. Vitaminol. 12, 74–79.
Hora, S., 1948. Knowledge of the ancient Hindus concerning fish and fisheries of India: I. References to fish in Arthashastra (ca. 300 B.C.). J. Royal Asiatic Soc. Bengal Sci. 14, 7–10.
Hublou, W., 1963. Oregon pellets. Prog. Fish-Cult. 25 (4), 175–180.
Kanazawa, A., Tanaka, N., Teshima, S., Kashiwada, K., 1971. Nutritional requirements of prawn. 2. Requirement for sterols. Bull. Jpn. Soc. Sci. Fish. 37, 211–215.
Kayama, M., 1964. Lipid metabolism in fish. J. Jpn. Oil Chem. Soc. 10, 511–519.
Kitamura, S., Ohara, S., Suwa, T., Nakagawa, K., 1965. Studies on vitamin requirements of rainbow trout, Salmo gairdneri. I. On the ascorbic acid. Bull. Jpn. Soc. Sci. Fish. 31, 818–825.
Kitamura, S., Ohara, S., Suwa, T., Nakagawa, K., 1967. Studies on vitamin requirements of rainbow trout. II. The deficiency symptoms of fourteen kinds of vitamin. Bull. Jpn. Soc. Sci. Fish. 33, 1120–1125.
Li, F., 1986. Fan Li on pisciculture. In: Chen, L. (Ed.), Fan Li on Pisciculture. Chinese Agriculture Publishers, Beijing.
Mai, K., 2020. History and prospect of aquaculture nutrition and aquafeed industry of China. Feed Indus. 41, 2–6.
McLaren, B., Herman, E., Elvehjem, C., 1946. Nutrition of rainbow trout; studies with purified rations. Arch. Biochem. 10, 443-441.
McLaren, B., Keller, E., O'Donnell, D., Ellehjem, C., 1947. The nutrition of trout. I. Studies of vitamin requirements. Arch. Biochem. 15, 169–178.
Nash, C., 2010. The History of Aquaculture. John Wiley & Sons, Inc., New York(NY).
Naylor, R., et al., 2021. A 20-year retrospective review of global aquaculture. Nature 591, 551–563. https://doi.org/https://doi.org/10.1038/s41586-021-03308-6.
Nose, T., Toyama, K., 1966. Protein digestibility of brown fish meal in rainbow trout. Bull. Freshwater Fish. Res. Lab. Tokyo 15 (2), 213–224.
NRC, 1973. Nutrient Requirements of Trout, Salmon and Catfish. National Academy Press, Washington, DC.
NRC, 2011. Nutrient Requirements of Fish and Shrimp. National Academy Press, Washington, DC.
Ogino, C., 1957. Amino acid composition of the protein in hatchery diet. Bull. Jpn. Soc. Sci. Fish. 23, 447–449.
Ok, I.H., Bai, S.C., Park, G.J., Choi, S.M., Kim, K.W., 2001. The patterns of plasma free amino acids after force-feeding in rainbow trout *Oncorhynchus mykiss* (Walbaum) with and without dorsal aorta cannulation. Aquacult. Res. 32 (s1), 70–75.
Park, G.J., Bai, S.C., Ok, I.H., Han, K.M., Hung, S.S., Rogers, Q.R., Min, T.S., 2005. Post prandial plasma free arginine concentrations increase in rainbow trout fed arginine deficient diets. Asian-Australas. J. Anim. Sci. 18 (3), 396–402.

Peng, X., et al., 2017. Fatty acid oxidation in zebrafish adipose tissue is promoted by 1,25(OH)2D3. Cell Rep. 19, 1444–1455.

Phillips, A.J., et al., 1964. Dry Concentrates as Complete Fish Foods. New York State Conservation Department, Albany, NY.

Phillips, A., et al., 1940. Feed Mixtures for the Production Hatchery. New York State Conservation Department, Albany.

Rahimnejada, S., et al., 2019. Replacemetn of fish meal with Bacillus pumillus SE5 and Pseudozyma aphidis ZR1 fermented soybean meal in diets for Japanese seabass (*Lateolabrax japonicus*). Fish Shellfish Immunol. 84, 978–997.

Robinson, E., Li, M., 2019. Channel catfish, *Ictalurus punctatus*, nutrition in the United States: a historical perspective. J. World Aquacult. Soc. 51, 93–119.

Rumsey, G., 1994. History of early diet development in fish culture, 1000 B.C to A.D. 1955. Prog. Fish-Cult. 56, 1–6.

Seong, M., Jung, J., Bai, S.C., 2018. A Preliminary Study to Reevaluate the Conventional Nitrogen-To-Protein (N:P) Conversion Factor (6.25) for the Major Aquafeed Ingredients. KOSFAS 2018 Conference. GWNU, Gangneung, Korea.

Shahkar, E., Yun, H., Lee, S., Kim, D.J., Kim, S.K., Lee, B.I., Bai, S.C., 2016. Evaluation of the optimum dietary arachidonic acid level and its essentiality based on growth and non-specific immune responses in Japanese eel, Anguilla japonica. Aquaculture 452, 209–216.

Shao, J., et al., 2019. Replacement of fishmeal by fermented soybean meal could enhance the growth performance but not significantly influence the intestinal microbiota of white shrimp Litopenaeus vannamei. Aquaculture 504, 354–360.

Stickney, R., 1996. Aquaculture in the United States - A Historical Survey. John Wiley and Sons, Inc., New York.

Stickney, R., Treece, G., 2012. History of aquaculture. In: Tidwell, J. (Ed.), Aquaculture Production Systems. Baton Rouge: World Aqua. Soc. Wiley-Blackwell.

Toyomizu, M., Kawasaki, K., Tomiyasu, Y., 1963. Effect of dietary oil and the fatty acid composition of rainbow trout. Bull. Jpn. Soc. Sci. Fish. 29, 957–961.

Wang, J., et al., 2018. Lipophagy is essential for lipid metabolism in fish. Sci. Bull. 63, 879–882.

Wolf, L., 1951. Diet experiments with trout. Prog. Fish-Cult. 13, 17–24.

Wood, J., Ordahl, E., 1958. Tuberculosis in Pacific Salmon and Steelhead Trout, Portland. Fish Commission of Oregon. Contribution No. 25.

Xu, D., et al., 2016. Postprandial nutrient sensing and metabolic responses after partial dietary fishmeal replacement by soybean meal in turbot (*Scophthalmus maximus* L.). Br. J. Nutr. 115, 379–388.

Yang, B., et al., 2020. ω-6 Polyunsaturated fatty acids (linoleic acid) activate both autophagy and antioxidation in a synergistic feedback loop via TOR-dependent and TOR-independent signaling pathways. Cell Death Dis. 11 (7), 1–19.

Yun, H., Park, G., Ok, I., Katya, K., Heung, S., Bai, S.C., 2015. Evaluation of optimum dietary threonine requirement by plasma free threonine and ammonia concentrations in surgically modified rainbow trout, *Oncorhynchus mykiss*. Asian-Australas. J. Anim. Sci. 28 (4), 551.

Yun, H., Park, G., Ok, I., Katya, K., Hung, S., Bai, S.C., 2016. Determination of the dietary lysine requirement by measuring plasma free lysine concentrations in rainbow trout *Oncorhynchus mykiss* after dorsal aorta cannulation. Fish. Aquat. Sci. 19 (1).

Zhou, Y., et al., 2013. Enhancement of growth and intestinal flora in grass carp: the effect of exogenous cellulase. Aquaculture 416–417, 1–7.

CHAPTER

2

Bioenergetics

Sadasivam J. Kaushik[1] *and Johan W. Schrama*[2]

[1]INRAe, Univ Pau and Pays de l'Adour, Nutrition, Metabolism and Aquaculture Unit, Saint Pée sur Nivelle, France; [2]Aquaculture and Fisheries Group, Wageningen University, Wageningen, The Netherlands

2.1 Introduction

Fish and shrimp, like all organisms, use ingested macronutrients for maintenance, growth, and reproduction. Part of these ingested macronutrients are used as fuel for ensuring different metabolic processes and physicochemical changes. The different aspects of metabolism, catabolism as fuels and anabolism as storage in tissues, can be represented in energy units. This approach is called bioenergetics or nutritional energetics. Following the first law of thermodynamics, the partition of energy-yielding components between anabolism as growth and storage in tissues and catabolism as fuels can be followed by the study of the balance between dietary energy intake and energy expenditure. All biological systems obey the laws of thermodynamics. Bioenergetics or biochemical thermodynamics is the study of the energy changes accompanying biochemical reactions in biological systems.

From a historical perspective, the pioneering works by Lavoisier and Laplace in the late 18th century can be considered as the foundation of studies on bioenergetics in animal nutrition. They recognized the importance of food consumption on oxygen uptake and heat production, which led to the classical contention that life is a process of combustion. By the end of the 19th century, Rubner in Germany demonstrated that the fundamental laws of thermodynamics are applicable to live animals and described the relationship between body surface area and fasting metabolic rates in homeothermic animals. Rubner was also the first to provide data on energy equivalents for the three macronutrients: proteins, fats and carbohydrates. The basis for studies on energy metabolism of animals was consolidated in two major books, entitled *Bioenergetics and Growth* by Brody in 1945 and entitled *The Fire of Life* by Kleiber in 1961.

Ege and Krogh (1914) were likely the first to apply the principles of bioenergetics to fish. Ivlev (1939) provided some bases on energy balance studies in carp. Over the past 50 years, many studies on energy expenditure and dietary energy utilization have been undertaken

Fish Nutrition, Fourth Edition
https://doi.org/10.1016/B978-0-12-819587-1.00011-2

with fish, but few have been undertaken with crustaceans. Several extensive reviews on bioenergetics of fish have also been made with particular relevance to aquaculture over the past five decades: Phillips (1972), Brett and Groves (1979), Cho and Kaushik (1985, 1990), Tytler and Calow (1985), Smith (1989), Cho and Bureau (1995), and Bureau et al. (2002). The focus here is to provide some bases and to update our knowledge in energy intake and utilization in fish and shrimp.

2.2 Energy production and transformation in biological systems/at the cellular level

Organisms, which are consuming food, derive their energy from oxidation of part of the digested macronutrients, carbohydrates, lipids, and proteins. During starvation, organisms obtain this from turnover of cells and tissues. During digestion of food, proteins and peptides are hydrolyzed to amino acids, triacylglycerols to fatty acids and glycerol, and carbohydrates to monosaccharides, mainly glucose. After absorption, these hydrolyzed end products are either incorporated into cellular structures after transformation through anabolic reactions leading to energy gain for the organism or catabolized through energy-generating reactions.

The first law of thermodynamics implies conservation of energy, meaning that the total energy of a system remains constant unless there is an input of energy (food intake) or energy output (excretion of feces, urine, gametes). This also implies that within a given system, energy is neither lost nor gained during any changes but that energy may be transferred from one part to another or may be transformed into another form of energy. Thermodynamic principles as they apply to biological systems are reviewed in several textbooks (Blaxter, 1989; Mayes and Botham, 2003). The terms "exergonic" and "endergonic" are used to describe the processes accompanied, respectively, by the loss or gain of free energy (Mayes and Botham, 2003). From a metabolic point of view, the endergonic/synthetic reactions correspond to anabolism, and catabolism is represented by exergonic reactions. In biological systems, adenosine triphosphate (ATP) is the main compound serving as a transducer of energy from a large number of reactions involving energy gain and to an equally large number of processes leading to energy loss, for instance, due to physical activity or osmoregulation.

ATP is a phosphorylated nucleotide containing adenine, ribose, and three phosphate groups and plays an important role in energy transfer, acting as a donor of high-energy phosphate to form compounds with lower free energies of hydrolysis such as adenosine diphosphate (ADP) or monophosphate (AMP). Similarly, ADP can be transformed into ATP by accepting high-energy phosphate, which is produced during metabolic processes. In fact, the ATP/ADP cycle connects those processes that liberate free energy to those processes that need energy. ATP is thus continuously consumed and regenerated, but at a very rapid rate, since the total ATP/ADP pool is sufficient to maintain an active tissue only for a few seconds (Mayes and Botham, 2003). The system that couples respiration to the generation of the high-energy intermediate, ATP, is termed oxidative phosphorylation, which allows aerobic organisms to capture the available free energy of respiratory substrates.

At the cellular level, energy production, generation of ATP, occurs in three stages: (1) formation of acetyl-coenzyme A (CoA) from fatty acids, amino acids, or pyruvate, an end product of glycolysis; (2) degradation of acetyl groups through the tricarboxylic acid (TCA) cycle (also called the citric acid or Krebs cycle); and (3) oxidative phosphorylation. Within the cells, this occurs mainly in the mitochondria, which contain all the necessary enzyme systems. The highest levels of oxidation are observed in tissues rich in mitochondria: liver, heart, and gills as well as the red muscle.

In fish, as in other vertebrates, the TCA cycle is the common pathway for the utilization of energy substrates under aerobic conditions. In contrast to birds and mammals, fish utilize amino acids rather than glucose as preferential energy substrates. Fatty acids also contribute toward production of ATP, with a practical consequence that when a high-fat diet is consumed, there is generally a decrease in amino acid oxidation. There are also differences between tissues: In the liver, keto acids resulting from amino acid catabolism constitute the main energy substrates along with acetyl groups arising from beta oxidation of fatty acids; in the white muscle, conversion of glucose into lactate appears to be the main source of energy during burst swimming. The oxidative and glycolytic capacities of muscle also appear to vary from one species to another, often linked to the capacity of the species for sustained swimming.

Biological processes are never 100% efficient. The production of ATP and the use of ATP for anabolic and vital live processes inevitably lead to heat production in all animals. From a nutritional bioenergetics point of view, it is recognized that the amount of heat produced by an animal depends on the chemical nature and consequent energy content of the compounds catabolized. Amino acid catabolism involves a number of reversible reactions of transamination and deamination with the end products of nitrogen (ammonia, urea) excreted into the environment by aquatic animals through gills or urine (Chapter 4). Once nitrogen has been removed, the carbon skeleton can be used either for gluconeogenesis via oxaloacetate or be oxidized in the TCA cycle. Although lipogenesis occurs mainly in the liver in fish, its degradation occurs in almost all tissues. Mechanisms for transport into the mitochondria and for fatty acid oxidation in fish have been dealt with in detail (Chapter 5). The fatty acids liberated from the hydrolysis of TAGs by lipases are transported into the mitochondria by acyl-carnitine transferase system. Successive cycles of beta-oxidation lead to the release of 2-carbon fragments in the form of acetyl-CoA, which enters the TCA cycle. Glycolysis is the main route for glucose catabolism in fish tissues, leading to the formation of pyruvate, also generating a small amount of energy. As in other animals, glycolytic reactions are all reversible and can thus function in the direction of gluconeogenesis (Chapter 7). Besides glycolysis, the pentose cycle also leads to lipogenesis.

As regards the TCA cycle, all the reactions that occur within the mitochondria function in a similar way in mammals and fish, although there are minor differences between species. Carbon flux through the TCA cycle is regulated by the NADH/NAD^+ and ATP/ADP ratios. The relative proportions of ATP and ADP in the cell are an indicator of the balance between energy-consuming and energy-generating reactions. The energetic status of a cell or an organ is generally described by a term called "energy charge," which is defined as EC = [ATP] + 1/2 [ADP]/[ATP] + [ADP] + [AMP]. There are differences between fish and land animals.

2.3 Partition of dietary energy and utilization (energy flow)

A schematic overview of the energy flow through an animal (i.e., energy budget), which results from the intake of energy (IE) by the consumption of food, is presented in Fig. 2.1.

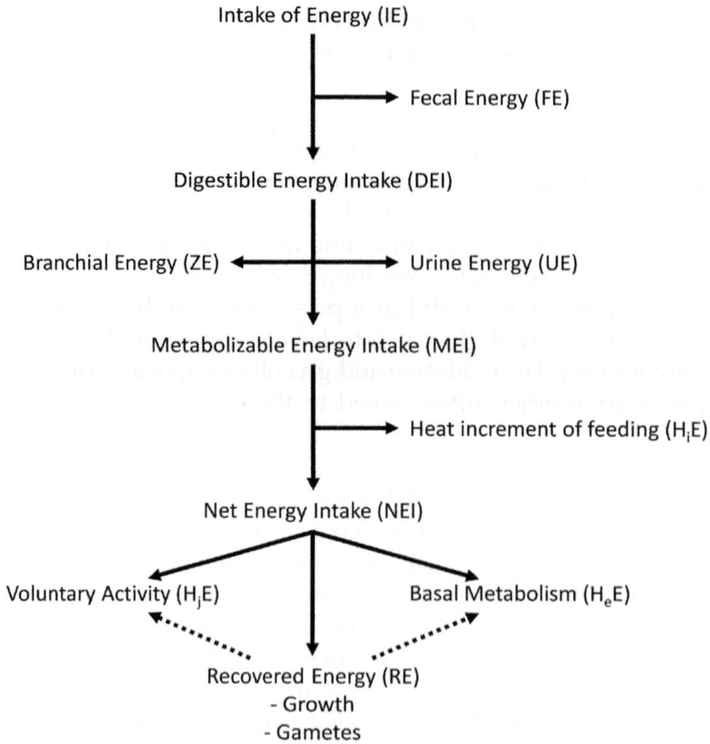

FIGURE 2.1 Energy budget: schematic overview of the flow of energy through fish and shrimp. *Adapted from National Research Council (NRC) 2011. Nutrient Requirements of Fish. National Academy Press, Washington D.C.*

Part of the IE is lost with the feces (FE), the urine (UE), and via the gills (ZE). The part of IE not lost via these routes is termed the metabolizable energy intake (MEI), being equal to IE – FE – UE – ZE. The level of anabolic and catabolic processes in an animal is related to the amount of food consumed and consequently to the amount of heat production/losses (HE). This part is also termed the "heat increment of feeding" (H_iE). The part of MEI that is not lost, as H_iE is the net energy intake (NEI). The NEI is used for basal metabolism (H_eE), voluntary activity (H_jE), growth, and reproduction, with the highest priority for basal metabolism (i.e., vital life processes). When the NEI is not sufficient to sustain the basal metabolism and voluntary activity, part of the body energy reserves are mobilized to cover these requirements, which is indicated by the dotted arrows in Fig. 2.1.

The units of expression of the energy budget are often kilojoules per fish per day (kJ/fish/d) or kJ per kilogram metabolic body weight per day (kJ/kg$^{0.8}$/d). In the feeds, the common units for the expression of energy content are kJ/g or MJ/kg on fresh or dry matter basis. One kJ is equivalent to 4.184 kilo calories. The energy content of a feed can be expressed as gross energy (GE); digestible energy (DE), being GE corrected for the fecal energy (FE) losses; metabolizable energy (ME), being DE corrected for branchial and urinary energy losses; and net energy (NE), being ME corrected for the heat increment of feeding. A summary of energy budget and dietary energy content nomenclature is given in Table 2.1.

TABLE 2.1 Nomenclature of the types of energy contents of feeds and the components of an energy budget.

	Abbreviation
Type of dietary energy:	
Gross energy	GE
Digestible energy	DE
Metabolizable energy	ME
Net energy	NE
Component of energy budget:	
Intake of energy	IE
Fecal energy losses	FE
Fecal energy losses of metabolic origin	F_mE
Digestible energy intake	DEI
Apparent digestibility of energy	ADC_E
Urinary energy losses	UE
Branchial energy losses	ZE
Metabolizable energy intake	MEI
Heat losses (heat production)	HE
Heat increment of feeding	H_iE
Energy losses induced by voluntary activity	H_jE
Basal metabolism	H_eE
Fasting heat losses	HE_f
Maintenance metabolizable energy requirement	ME_m
Heat loss at maintenance feeding (i.e., maintenance energy)	HE_m
Partial utilization efficiency of DE for growth	$k_{g:DE}$
Partial utilization efficiency of ME for growth	$k_{g:ME}$

Based on National Research Council (NRC) 2011. Nutrient Requirements of Fish. National Academy Press, Washington D.C.

2.3.1 Gross energy: dietary fuels

In animal nutrition, GE is the common term to describe the amount of energy released during combustion of nutrients. It is the same as enthalpy, a term used by physicists to describe the heat of combustion. The GE of feeds or feedstuffs are determined by bomb calorimetry, which measures the enthalpy of the complete oxidation of carbon, hydrogen and nitrogen to carbon dioxide, water, and nitrogen oxide. Except for nitrogen, this mimics the full oxidation of nutrients in vivo. The GE of a feed(stuff) depends on its macronutrient composition.

The mean GE values of crude ash, protein, lipids, and carbohydrates (including fiber) are 0, 23.6, 39.5, and 17.2 kJ/g, respectively (National Research Council (NRC), 2011). As an alternative to bomb calorimetry, the GE of a feed(stuff) can be calculated from its analyzed macronutrient content. Alternatively, GE content of a compound diet can also be calculated based on its ingredient composition and the measured values of GE of each of the ingredient of the diet. Intake energy (IE, Fig. 2.1) by an animal is calculated as the product of its feed intake and the dietary GE content.

2.3.2 Digestible energy

Once feed is consumed, it first needs to be digested followed by absorption from the digestive tract before nutrients can be used for anabolic and catabolic processes. The digestion and absorption of nutrients is sometimes termed "assimilation," but the use of this term should be discouraged. Absorption deals with the uptake of nutrients through the digestive tract, and assimilation involves processes of synthesizing new compounds from the absorbed nutrients within the cells. Part of the consumed feed resists digestion, passes through the digestive tract, and is voided as fecal material, which also contains energy. The total amount of energy egested as fecal matter is referred to as FE (Fig. 2.1). By definition, digestible energy intake (DEI) is IE − FE, and the apparent digestibility coefficient of energy (ADC_E) is DEI divided by IE. The DE content of a diet is its GE content corrected for FE losses and is calculated as GE times ADC_E. In general, the digestibility of energy in fish of a given nutrient is not substantially influenced by other dietary components, with the exception of carbohydrates (Chapter 7). The DE of a diet can be determined either by measuring ADC_E using an inert marker method and measuring the energy content in feed and feces, relative to an inert tracer; by measuring the digestibility of the proteins, lipids, and carbohydrates present in the diet followed by calculation using the energy content of these digested nutrients (respectively, 23.6, 39.5, and 17.2 kJ/g); or by calculating the weighted average of the DEs of the ingredients in the feed. Thus, if the interactions between ingredients regarding digestion are negligible, DE of a complete feed can be considered to be additive.

Variation in the digestibility of feed or feed ingredients is generally a major factor affecting the variation in their potential as energy sources to the animal, since FE is a major loss of ingested GE. Therefore, values for DE and values for the digestibility of individual nutrients should be used to estimate levels of available energy and nutrients (as opposed to GE or crude nutrient levels) in feed ingredients for diet formulation (Cho and Kaushik, 1990). Feeds should be formulated based on DE or digestible nutrient levels, rather than on the basis of crude nutrient levels (e.g., crude protein) or GE. Sufficient information is now available on DE values of common fish feed ingredients to allow feeds to be formulated on a DE or a digestible nutrient basis. It is, however, important to emphasize that DE level of a given diet is only an indication of the potential contribution of the energy from nutrients in the ingredient. These values do not serve as measures of the utilizable energy or of the productivity of the diet.

2.3.2.1 Measurement, methods, factors affecting digestible energy

To get data on DE of feeds, it is necessary to quantify with precision FE losses. This involves the measurement of digestibility of feeds and feedstuffs, which requires first the collection of representative samples of feces. In contrast to land animals where fecal collection is easy, separating fecal material from water and collecting fecal samples without contamination by uneaten feed and to avoid possible losses due to leaching of fecal matter necessitate specific methods and tools adapted to aquatic environment.

Given that quantitative collection of feces corresponding to a given meal is very difficult with aquatic organisms grown in groups, digestibility measurements using such a direct method involving quantification of intake and fecal losses are rarely used with fish or shrimp. Digestibility measurements in aquatic organisms rely on the collection of a representative fecal sample (free of uneaten feed particles) and the use of a digestion indicator, or tracer, to obviate the need to quantify dietary intake and fecal output (indirect method). The inclusion of an inert tracer in the diet facilitates the calculation of digestibility coefficients of nutrients or energy of a diet. The tracer should not affect the digestive processes, should not have any adverse effects on overall metabolism, and should be easy to detect and quantify.

To collect fecal matter, invasive or active techniques including collecting feces from the lower part of the intestine by abdominal stripping or dissecting the fish and collecting feces from the distal part are commonly employed. Such techniques cannot be used with all fish or shrimp and in any case lead to underestimation of digestibility. Passive techniques involve collection of feces naturally voided by the fish or shrimp based on collection of feces after decantation either using a metabolism chamber, using collection chambers attached to rearing tanks, or using continuous automatic fecal collectors. The two systems, such as the one designed and developed by Cho and Slinger (1979) in the University of Guelph, Canada, and the automatic fecal collector developed in INRA, Saint Pée sur Nivelle, France, by Choubert et al. (1979), yield meaningful estimates of digestibility of nutrients if used correctly. To obtain reliable data on digestibility, it is necessary that collected feces does not stay too long in the aquatic environment, does not break up or leach out, and is not contaminated with uneaten feed or digestive tissues. Estimates of apparent digestibility of macronutrients and energy measured with such techniques tend to be fairly stable, and when these techniques are used in a standardized fashion, reliable data on DE values can be obtained. The best or the most optimal fecal collection method differs though between fish species as, e.g., some fish species are anatomically difficult to strip (many cyprinids and tilapia), while some other species produce highly unstable feces (catfishes).

Fecal matter is composed of the undigested food components as well as unreabsorbed residues of digestive tract origin, such as mucosal cells, mucoproteins, digestive enzymes, and other secretions released into the digestive tract. The enthalpy of combustion of these materials represents a loss of energy, which is not derived from the food. This energy loss is designated FE of metabolic origin (F_mE) and can be influenced by the characteristics of the food and the level of feed/protein intake. Estimates of F_mE allow the description of "true" DE values, which are generally higher than the "apparent" digestible energy values (ADC_E). Measurement of F_mE of fish has received little attention. The fecal metabolic losses in fish are very much with endogenous protein/nitrogen losses (Ogino and Chen, 1973; Wilson et al., 1981; Anderson et al., 1992). As was shown by Ogino and Chen (1973), in fish fed a protein-adequate diet, the contribution of metabolic fecal nitrogen to total fecal nitrogen is very small. Under these conditions, the difference between the "true" and the apparent digestibility of protein is negligible, and as such, metabolic fecal losses will have little influence in terms of DE. Based on very little data available (Ogino and Chen, 1973), F_mE loss as nitrogen can be converted into energy terms to be about 0.4 kJ/100 g live body weight per day or 20 kJ/100 g dry matter intake, which is very small, representing less than 1% of the IE or about 10%–20% of the FE of fish. In summary, the apparent DEI will be equal to IE − FE and the true (or standardized) DEI will be as follows: IE − (FE − F_mE).

2.3.2.2 *Digestible energy of feeds and feedstuffs*

Reliable and precise data on DE values of diets is extremely important in studies on nutritional energetics. As described earlier, measurement of digestibility measurements in vivo with fish or shrimp requires specialized equipment. As a diet is made of a mixture of various feedstuffs, knowing the digestibility of all possible feedstuffs of potential interest may allow for the development of diets based on digestible protein (DP) and DE levels, for instance. This, however, requires that estimates of apparent digestibility of nutrients and energy of different ingredients are available and that the values of DE of the individual ingredients are additive, an assumption that generally holds true (Cho and Kaushik, 1990; Watanabe et al., 1996; Sklan et al., 2004). However, this will not be the case if some of the ingredients strongly interact with other ingredients such as those rich in nonstarch polysaccharides (Glencross et al., 2012; Maas et al., 2020).

To determine the DE contents of ingredients, a protocol was initially proposed by Cho and Slinger (1979). This protocol involves comparison of the digestibility of a reference diet with that of a test diet. The reference diet used should be palatable, water-stable, nutritionally adequate, and produced using ingredients generally used in practical fish feeds. Much care is also warranted to ensure that there are no possible interactions between ingredients used in the reference diet. Inclusion of the inert tracer in the reference diet allows the ADC of the energy (and nutrients) of the diet to be calculated from measurements of the ratios of nutrient to indicator in the diet and feces. The test diets are made by mixing a proportion of the reference diet and the test ingredient under assessment, generally at a 70:30 ratio. Details on how to calculate the apparent digestibility coefficients (ADCs) for the nutrients and energy of the test and reference diets have been described in detail by different authors (Cho and Slinger, 1979; Forster, 1999; Bureau et al. 2002). Some refinements to these calculations have been proposed to clarify and take careful account of the units of expression of data on feedstuffs, and finished feeds were also provided by Bureau and Hua (2006) and Glencross et al. (2007).

In general, the ingredient composition of a compounded diet is a major factor determining its DE content. Protein-rich feedstuffs, such as fish meals, processed animal proteins, and oilseed protein concentrates, have a high energy digestibility, and the values of DE are close to the GE values of such ingredients. Similarly, most fish and shrimp readily digest dietary lipids, and consequently, the amount of DE provided will be high. For ingredients having relatively high levels of starch, fibers, and other nonstarch polysaccharides, which are poorly digested by many teleosts, DE values will be lower. Given the possible effects of the type and level of carbohydrate sources in a diet on the digestibility of starch (Chapter 7), the provision of DE from such feedstuffs may not be additive as the ADCs of protein and lipid sources (Kim and Kaushik, 1992).

Many studies on factors affecting digestibility in fish (including ADC_E) often reported no effect of feeding level (Choubert et al., 1984; Storebakken and Austreng, 1987; Cho and Kaushik, 1990; Zhou et al., 2003; Bureau and Hua, 2006). However, when poor quality diets are used, a decline in ADC_E can occur with increasing ration size, as found in common carp (Phan et al., 2019), Nile tilapia (Haidar et al., 2016), and rainbow trout (Staessen et al., 2020).

Among factors affecting DE, the effect of water temperature has been studied to a certain extent. An increase in the level of inclusion of raw/native starch in pelleted diets can lead to a decrease in DE (Chapter 7). Very early studies on the effects of a sudden rise in temperature on digestibility (Choubert et al., 1982a,b) showed that the rate of transit of foodstuffs through the digestive tract as well as the digestibility of dry matter and energy was affected by water temperature. Many studies with salmonids under controlled conditions have shown a similar effect of an increase in water temperature on starch digestibility and hence on the provision of DE (Médale et al., 1991; Watanabe et al., 1996; Azevedo et al., 1998). Specific studies on the effects of other environmental factors such as salinity on digestibility of energy are scarce. Studies undertaken by Krogdahl et al. (2004) with euryhaline species such as the rainbow trout (*Oncorhynchus mykiss*) or Atlantic salmon (*Salmo salar*) did not show any effect of salinity on ADC of energy. In Nile tilapia, the digestibility of all macronutrients and thus energy was higher at 15 ppt compared to 0 ppt (Tran-Ngoc et al., 2017). Some recent work has looked into the effect of dissolved oxygen levels in the water on energy digestion. Tran-Ngoc et al. (2016) noted only a small effects of hypoxia on nutrient digestibility in Nile tilapia feed restricted ration levels. In contrast, Tran-Duy et al. (2008) found, in 200 g Nile tilapia, a 3%–5% difference in ADC_E at severe hypoxic condition compared with normoxic conditions but not in smaller (60g) fish. Despite the lack of data, it is likely that the impact of environmental conditions, especially suboptimal condition, will be higher and also affected by the quality of the diets.

2.3.3 Metabolizable energy

Digestion of a diet leads to the liberation of amino acids, fatty acids, and sugars, which are absorbed and serve as the principal metabolic fuels for the body. Catabolism of fats and carbohydrates results in the formation of carbon dioxide and water. Amino acids in excess of requirements for protein biosynthesis are deaminated with the end products of nitrogen (ammonia, urea) excreted into the environment by aquatic animals through gills or urine and the carbon residues oxidized via the TCA cycle or in some cases converted to fat or carbohydrates. Excretion of nitrogenous waste compounds, of which ammonia amounts to about 85% in most fish species (Kaushik and Cowey, 1991), results in non-FE losses. A small but significant proportion (5%–15%) of nonfecal nitrogen is also excreted as urea (Brett and Zala, 1975; Kaushik, 1980) in almost all teleosts studied so far. The end products of amino acid/purine catabolism are excreted mainly through the gills and some through the kidneys. In energy terms, the branchial energy loss (ZE) and through the kidneys as urinary energy loss (UE) are not accounted for by the DE value of a diet. Dietary metabolizable energy (ME) is dietary DE corrected for the energy lost by the fish via gills and urinary tract. By definition, the MEI (Fig. 2.1) corresponds to the physiologically available fuel value from the diet consumed by the fish calculated as follows: $ME = IE - (FE + UE + ZE)$.

Although nonfecal nitrogen losses contribute significantly to the environmental load in terms of ammonia nitrogen and also that high levels of ammonia in the water can be toxic to fish and shrimp, from the point of view of energy balance, their contribution is small. Generally in rainbow trout, $UE + ZE$ is less than 3% of MEI (Kaushik, 1998). In European sea bass, nonfecal nitrogen losses as percentage of digestible nitrogen intake ranged between 69.5% and 74.2%, while in terms of energy, UE + ZE as percentage of DEI ranged between 6.2% and 6.5% (Santos et al., 2010). Similarly, in Nile tilapia, nonfecal nitrogen losses as percentage of digested nitrogen varied between 57.9% and 68.3%, which equals to 5.3%–6.4% of DEI being excreted as UE + ZE (Schrama et al., 2012).

Endogenous branchial and urinary nitrogen excretion ($UN_e + ZN_e$) rates measured in rainbow trout after 3–4 days of fasting range between 80 and 130 mg N/kg BW/d, which equals to 2.0–3.2 kJ/kg BW/d (Watanabe and Ohta, 1995; Kaushik, 1980, 1998). These studies also showed that $UN_e + ZN_e$ were the most affected by water temperature and body weight. In marine fish such as the European sea bass (*Dicentrarchus labrax*), gilthead seabream (*Sparus aurata*), or turbot (*Psetta maximus*), values for $UN_e + ZN_e$ were found to be in the range of 100–160 mg N/kg/day (Ballestrazzi et al., 1994; Dosdat et al., 1996), which is comparable with the endogenous losses in rainbow trout (Kaushik, 1998).

2.3.3.1 *Measurement*

For the estimation of ME of the whole diet, we need to measure both branchial and urinary losses released into the aquatic environment in which the fish live. Smith (1971) used a specifically designed metabolism chamber to separate branchial and urinary excreta and analyzed the effects of a diet force-fed to rainbow trout. Using urinary cannula for measurement of the urine flow rate in conjunction with spot sampling of urine is another approach that has been used to estimate the urinary excretion of glucose and consequently urinary energy loss (Cho and Bureau et al., 1998; Deng et al., 2000). Data obtained on branchial and urinary energy losses under stressful confinement conditions of metabolism chambers most likely overestimate losses than would occur with unrestrained free-living fish feeding normally (Cho and Kaushik, 1990; Wood, 2001).

Monitoring nitrogenous wastes (ammonia and urea) released in the water in the rearing environment are a commonly used approach. Brett and Zala (1975) used a semistatic system to follow the diurnal pattern of nitrogen excretion of young sockeye salmon (*Oncorhynchus nerka*) by allowing ammonia to build up during alternate periods of closed–open-circuit cycles. Kaushik (1980) was the first to estimate the postprandial excretion rates in a flow-through system in a continuous manner using a peristaltic pump for sampling and an autoanalyzer to analyze the concentrations of ammonia nitrogen. This method allowed continuous monitoring of ammonia and urea nitrogen excretion under normal physiological conditions even in larval fish (Kaushik et al., 1982). It is necessary to ensure constant water flow rate and the precise measurement of low concentrations of ammonia in the outlet water. Application of such a technique has revealed postprandial patterns of ammonia nitrogen excretion to be similar among phylogenetically different species (Dosdat et al., 1996; Chakraborty and Chakraborty, 1998). Postprandial increase in ammonia excretion and oxygen consumption as affected by dietary proteins has also been shown in different crustaceans (Rosas et al., 1995).

As mentioned earlier, direct measurement of UE + ZE corresponding to a given meal requires dedicated equipment, specially designed tools and techniques. However, UE + ZE can also be indirectly estimated by combined measurements of energy and nitrogen budgets of fish by the comparative carcass analysis method as proposed by Cho and Kaushik (1985; 1990). Since UE + ZE occurs mainly as nitrogenous product losses, the total nonfecal nitrogen

TABLE 2.2 Comparative carcass analysis scheme for calculating the nitrogen (N) and energy budget.

Measurements:

1. Feed intake and growth (in g/d).
2. Macronutrient composition of the diet (N and energy content).
3. Apparent digestibility of N and energy (ADC_N and ADC_E respectively).
4. Start and end fish body content of N and energy.
5. Estimation of energy losses during fasting (HE_f):
   A. Measuring carcass energy losses during a fasting period of (ca. 7–10 days).
   B. Regressing of recovered energy (RE) on metabolizable energy intake (MEI) from an energy balance trail. HE_f is equal to the intercept of the estimated relationship (MEI = 0).

Calculation of nitrogen (N) budget (in g per time unit):

1. N intake (IN) from feed intake and dietary N content
2. Digestible N intake (DNI) as: $DNI = ADC_N \times IN$
3. Recovered nitrogen (RN; i.e., N gain) from start and end body weight and N content
4. Nonfecal N losses (UN + ZN; respectively urinary and branchial) as: $UN + ZN = RN - DN$

Calculation of energy budget (in kJ per time unit):

1. Energy intake (IE) from feed intake and dietary energy content
2. Digestible energy intake (DEI): $DEI = ADC_E \times IE$
3. Nonfecal energy losses (UE + ZE): $UE + ZE = 24.9 \times (UN + ZN)$
4. Metabolizable energy intake (MEI): $MEI = DEI - (UE + ZE)$
5. Recovered energy (RE; i.e., energy gain) from start and end body weight and energy content
6. Heat production (HE): $HE = MEI - RE$
7. Heat increment of feeding (H_iE): $H_iE = HE - HE_f$
8. Net energy intake (NEI): $NEI = MEI - H_iE$ or $NEI = RE + HE_f$

Based on Cho, C.Y., Kaushik, S.J., 1985. Effects of protein intake on metabolizable and net energy values of fish diets. in: C.B. Cowey, Mackie, A.M., (Eds.), A.J.G.B. (Eds.), In :Nutrition and Feeding of Fish,. Academic Press, London, pp. 95–117.

loss, branchial and urinary, is estimated by the difference between digested nitrogen and whole-body nitrogen gain as detailed in Table 2.2. In this approach, the feed intake, growth, nutrient digestibility (energy and nitrogen), and body composition at the start and end of the experiment are measured. From these data, first the nitrogen budget is calculated followed by calculation of the energy budget (see Table 2.2).

Although most fish and shrimp are mainly ammoniotelic, ammonia-N representing at least 85% of the nitrogenous wastes, there is also some urea-N excretion, besides other nitrogenous catabolites (Kaushik and Cowey, 1991). The energy of combustion value of ammonia and urea is 20.5 kJ/g (24.9 kJ/g N) and 10.5 kJ/g (22.5 kJ/g N), respectively (Brafield and Llewellyn, 1982). Because most nitrogen losses are as ammonia, and the difference in the amount of energy loss per gram of nitrogen between ammonia and urea is small, it has been proposed that the loss of 1 g of nitrogen by fish under normal conditions can be equated to an average energy loss of 24.9 kJ. So UE + ZE can be calculated as 24.9 times the calculated nonfecal nitrogen loss (in g).

2.3.3.2 Factors affecting metabolic waste output

In some studies with fish, dietary energy is expressed in terms of ME, by analogy with systems utilized for poultry. In the case of poultry, it is not possible to measure DE values, as it is difficult to separate feces and urine. Since in most cases, the amount of FE originating from the different ingredients is not influenced by the overall diet composition (i.e., additive), the DE of a diet can be calculated by adding the respective DE values of the dietary ingredients (weighted mean). In contrast, the metabolic losses via urine and gills (UE + ZE) are strongly affected by dietary factors as well as biotic factors. As a result, the ME content of a given ingredient is not independent of the diet composition, physiological status of fish, and rearing conditions. In other words, additivity of ingredient ME content of a complete feed does often not apply. As mentioned by Cho and Kaushik (1990), under aquaculture conditions, ME has significance only as long as it is measured with respect to an animal's response to a complete diet under a defined set of biological and environmental conditions. This together with the fact that UE + ZE are not easily measurable as the determination of digestibility is the major reason why the energy content of ingredients and complete diets have historically mostly been expressed on DE basis in aquaculture.

The main dietary factors affecting non-FE losses are the DP level, DE content, amino acid balance, and the DE/DP ratio as these influence the efficiency of protein retention. In an early study with rainbow trout, Kaushik and Oliva-Teles (1985) showed that an increase in dietary DE by an increase in dietary nonprotein energy sources led to a decrease in ammonia nitrogen excretion, UE + ZE, and hence an increase in ME. A decrease in the dietary DP/DE ratio (from 23 to 18 mg N/kJ) was shown to reduce nitrogen excretion (Kaushik, 1998). Available data indicate that as with the salmonids, N excretion is reduced with decreasing DP/DE ratios in species such as European sea bass and gilthead seabream (Kaushik, 1998; Lupatsch et al., 2003a). As a general rule, UE + ZE is linearly related to dietary protein levels at a given DP/DE ratio. Furthermore, UE + ZE are positively related to dietary DP/DE ratios, at least within a certain range of DP/DE. As an example, derived from a study of Haidar et al. (2018) on Nile tilapia, UE + ZE as percentage of DEI increased from 2.8% to 5.9% when the dietary DP/DE ratio increased from 16 to 27 mg/kJ (Fig. 2.2A), but in terms N losses, increase in the nonfecal losses (UN + ZN) as percentage of digestible N intake was much larger, from 46% to 64% (Fig. 2.2B). The amino acid composition of the diet is another factor that has a strong effect on UE + ZE and consequently on growth and efficiency of nitrogen utilization (Chapter 4). Rainbow trout fed two diets with equal protein and DE contents, but with different amino acid profiles, resulted in UN + ZN as percentage of digestible N of 55% and 66%, respectively, which equals to 4.5% and 5.3% in terms of energy (UE + ZE as percentage of DEI) (Saravanan et al., 2013b). The dietary protein level and form as well as supplementation with crystalline amino acids can affect total ammonia nitrogen excretion (Webb and Gatlin, 2003). Feeding amino acids in excess of the requirement will also lead to an increase in ammonia excretion with a consequent effect on ME, though small in terms of energy.

Under normal condition, it is assumed that the amount of nonnitrogenous energy losses via urine and branchia is minimal. As regards the dietary supply of nonprotein energy

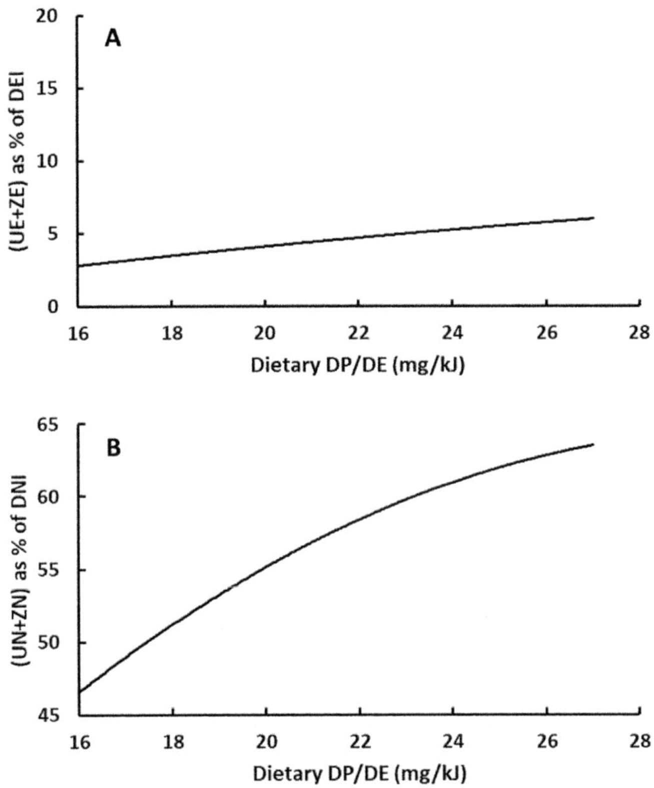

FIGURE 2.2 The effect of dietary DP/DE ratio on urinary and branchial energy losses (UE + ZU) as percentage of digestible energy (DE) intake (DEI) (panel A) and urinary and branchial nitrogen losses (UN and ZN) as percentage of digestible nitrogen intake (DNI) (panel B) in 5–44 g juvenile Nile tilapia feed equal amounts of digestible protein (DP). *Derived from Haidar et al. (2018).*

sources to modulate the DP/DE ratios, the beneficial effects of inclusion of gelatinized starch as compared with native starch on nitrogen excretion were shown long back by Kaushik and Oliva-Teles (1985). However, Bureau (1997) showed that rainbow trout fed high levels of digestible carbohydrates not only exhibited a postprandial hyperglycemia but also excreted significant amounts of glucose in their urine with significantly increased UE + ZE values. This might lead to an underestimation of the UE + ZE values of diets containing high levels of digestible carbohydrates when the ME content is calculated only on the basis of nitrogenous waste energy excretion (Table 2.2). Such a potential loss of glucose might explain the observed low or even almost zero energetic utilization efficiency of digested carbohydrates in some studies with Asian sea bass (Phan et al., 2019) or the snakehead (Phan et al., 2021). However, this requires further validation.

When fish are grown at their normal thermal range, water temperature does not appear to have an effect on the ME/DE ratio of diets (Kaushik, 1980; Azevedo et al., 1998; Rodehutscord and Pfeffer, 1999; Xie et al., 2011). However, Xie et al. (2011) found that tilapia grown at a high temperature had low metabolic losses. Comparisons between species with regard to ME are difficult to make because of confounding factors related to feed composition and rearing systems. Very few studies have attempted to analyze the possible differences in efficiency of N retention/N losses as affected by nondietary factors, such as between and within species differences. Dias et al. (1999) observed significant differences in efficiency of N retention in European sea bass and rainbow trout fed similar diets. Marine fish species appear to retain a much lower proportion of the DP fed to them than do salmonid fish species and therefore have significantly higher UE + ZE values (Kaushik, 1998). Grisdale-Helland et al. (2007) made a comparison of energy budget between Atlantic salmon, rainbow trout, and Atlantic cod (*Gadus morhua*) fed the same diets and found that non-FE losses differed between species. Available data do not appear to indicate any significant influence of genetic origin (strain, family, ploidy) on nitrogen excretion per unit N intake (Kaushik et al., 1984; Oliva-Teles and Kaushik, 1987a, b). From different studies with different species, it appears that the metabolic energy losses represent between 2% and 8% of the GE, depending on the protein source, amino acid profile, and DP/DE ratios. A point worth emphasizing is that ME does not relate to raw materials but only to a complete diet. Therefore, fish or shrimp feeds are commonly formulated on the basis of DE or on NE basis as described later.

2.3.4 Heat production

Heat is liberated by all animals as a consequence of: metabolic transformation of dietary substrates into tissue components (e.g., growth), tissue turnover, and immune defense, and in the case of aquatic organisms due to osmoregulation and physical activities such as swimming. Factors affecting these growth and maintenance processes determine the heat production (**HE**). Fluctuation in HE is dependent on the duration of the measurement/recording. Within-day variation in HE is determined by, e.g., moment of feeding or intensity of swimming etc. When HE is measured over period >1 day, the major factor influencing HE is feed intake. In Fig. 2.3, the relationship between HE and MEI is given (expressed per day). The rate

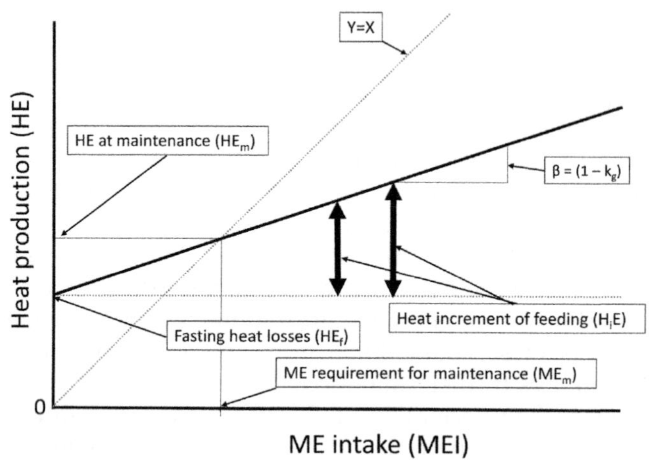

FIGURE 2.3 Schematic representation of the relation between MEI and HE. *HE*, heat production; *MEI*, metabolizable energy intake.

at which heat is liberated is an indication of the intensity of ongoing reactions and is designated as the metabolic rate (Kleiber, 1975). Basal metabolic rate, also termed "minimal metabolism," reflects the minimum rate of metabolic activity needed to sustain the structure and function of the body tissues. After feed intake, there is an increase in metabolic rate as a consequence of mechanical work (search for food, ingestion, digestive tract movements) as well as biochemical work required for digestion, absorption, and metabolic bioconversion processes of the nutrients supplied by the diet. These result in an increase in oxygen consumption after food intake, which was called the "specific dynamic effect" of food by Rubner at the beginning of the 20th century, also called "specific dynamic action" or SDA. Other terms such as calorigenic effect, thermogenic effect, and postprandial thermogenesis have been used to describe this energy expenditure related to feed intake, but the term "heat increment of feeding" (**H_iE**) is the most commonly accepted term. Normally SDA has been measured as the area under the curve of the HE after ingestion of a single meal, thus measuring HE within a day. In contrast, H_iE is based on measurements of over periods longer than 1 day, such as in energy budgets as indicated in Table 2.2. As depicted in Fig. 2.3, H_iE is defined as the difference between HE and the heat production at fasting (HE_f). The different components of metabolism lead to the release of heat from the ME derived from the diet. Since the laws of thermodynamics apply for animals, the part of MEI which is lost as HE will not contribute whole-body growth or nutrient/energy gain and thus the recovered energy as growth or invested in reproduction (RE) is equal to MEI − HE. In Fig. 2.4, the relationship between RE and MEI is given.

Animal calorimetry is the method applied by physiologists and nutritionists for measuring HE and can be undertaken over short time periods. HE can be measured by one of two methods: direct calorimetry that involves measurement of heat dissipated by fish in the aquatic environment, or indirect calorimetry based on measurement of oxygen consumption by fish. Alternatively, HE can also be calculated by determination of energy budgets (balances) by comparative carcass analysis by estimating the change in total body energy content

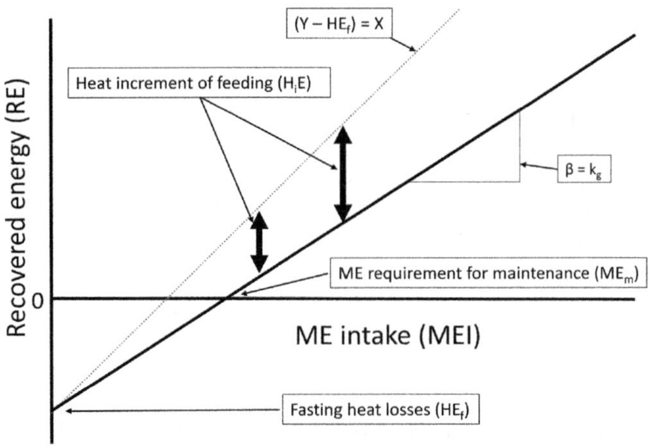

FIGURE 2.4 Schematic representation of the relation between MEI and RE. *MEI*, metabolizable energy intake; *RE*, recovered energy.

based on gain in whole-body mass and its chemical composition (see Table 2.2). This comparative carcass analysis technique requires data obtained over a sufficiently long period of growth (weeks). Some studies have evaluated the relative merits of both direct and indirect calorimetry to measure the HE of fish (Cho and Kaushik, 1985, 1990; Tytler and Calow, 1985). For bioenergetic studies with aquatic animals, the indirect calorimetric studies are reliable and more practical than direct calorimetry.

Direct determination of HE requires the measurement of very small changes in water temperature resulting from metabolic activities of fish. A direct calorimeter was designed by Smith et al. (1978a,b) to measure HE in fish, but the estimates of HE measured with this calorimeter, although appealing, were found to be unrealistically high compared with methods based on measurements of oxygen consumption (Brett and Groves, 1979; Cho et al., 1982). Besides, very precise instruments to record small changes in temperature were needed, not commonly available at that time. But with current progress in indwelling sensors available and information technology (Zion, 2012; Simbeye and Yang, 2014; Parra et al., 2018), one can envisage reactivation of direct calorimetric studies in situ under farming conditions.

The indirect method of estimating heat production relies on monitoring respiratory exchanges (oxygen consumption, carbon dioxide excretion) and nitrogenous waste excretion of animals. Specific respirometers for studying basal metabolism, swimming activity, and oxygen consumption were developed very early (Spoor, 1941; Fry, 1957; Blazka et al., 1960) to monitor oxygen uptake and estimate energy expenditure in the context of hydrobiology and ecology of fish. The amount of heat produced per each liter of oxygen consumed (oxycalorific coefficient) in metabolism varies little, depending on the substrate, which is oxidized (lipid, carbohydrate, or protein). Oxycalorific coefficients range between 19.2 kJ/L of oxygen for protein to 21.2 kJ/L of oxygen for carbohydrates. A value of 20.1 kJ/L of oxygen has been proposed to estimate the heat production in land animals (Blaxter, 1989). Because a larger proportion of HE in fish is derived from protein and lipid catabolism and less from carbohydrate catabolism than in terrestrial farm animals, the value of 19.4 kJ/L of oxygen (13.6 kJ/g oxygen) was adopted to calculate HE from oxygen consumption measurements in fish (Cho and Kaushik, 1990).

Quantification of oxygen consumption is done by combined measurement of dissolved oxygen content of the water at the inlet and outlet of tanks using electrodes and precise control of water flow over the tanks (Kaushik and Médale, 1994; Médale et al., 1995; Saravanan et al., 2012, 2013a). On the basis of these daily patterns, the prefeeding, basal levels, and the increase in metabolic rates due to feeding can be calculated for oxygen uptake. Additionally, the measurement of carbon dioxide (CO_2) in inlet and outlet water will allow determination of respiratory quotients RQ, volume, or mole of carbon dioxide produced per volume or mole of oxygen consumed (Gélineau et al., 1998). Knowledge of the RQ allows estimation of the relative proportions of fats and carbohydrates used for energy purposes. Combined with data on ammonia/nitrogen excretion rates converted into energy terms, values on nonprotein RQ can be calculated. This allows for estimation of the contribution of the relative proportions of three substrates utilized for meeting the energy demands (Bureau et al., 2002).

As it is not always feasible to undertake calorimetric studies with live fish reared under different systems, an alternative system based on the use of comparative carcass analyses was proposed by Cho and Kaushik (1985; 1990) and is reported in Table 2.2. This method is easily applicable for fish, because large numbers of fish can be sampled and easily homogenized for the analyses. The results have also been found to be in good agreement with those based on indirect calorimetry. Many studies have used such approaches to follow energy budgets in different species (Azevedo et al., 1998; Lupatsch et al., 1998, 2010; Rodehutscord and Pfeffer, 1999; Bailey and Alanara, 2001, 2006; Glencross et al., 2010, 2017; Xie et al., 2011; Schrama et al., 2012; Haidar et al., 2016).

Some other methods to determine HE, such as those based on the use of stable isotopes, are considered to hold potential to follow nutrient flux from feed to organisms and thus also draw energy budgets. Food or ingredients naturally or artificially enriched in certain stable isotopes have been used to study the nutrition of fish or shrimp (Parker et al., 1989; Gamboa Delgado, 2009; Beltran et al., 2009; Felip et al., 2012), but not fully exploited to follow energy flow through aquatic organisms under farming conditions.

2.3.4.1 *Minimal metabolism*

Regardless of whether feed is consumed, all animals have an energy requirement for maintaining life and the essential physiological functions. This is termed "minimal metabolism" or "standard metabolism" and is measured as fasting heat production (HE_f) in undisturbed fish or shrimp kept under short fasting conditions. In fish and shrimp, HE_f includes energy spent for "basal metabolism" (H_eE) and energy spent for voluntary or resting activity (H_jE), such as minor bodily movements and muscular activity (see Fig. 2.1). In the case of homeotherms, HE_f can also include energy spent for the regulation of body temperature (H_cE). Measurement of HE_f must be undertaken under standardized conditions of temperature, oxygen availability, and animal welfare in fish or shrimp held without feed over short periods. One way of measuring basal metabolic rate is to measure energy expenditure of fish swimming at different rates and extrapolating to zero activity (Smith, 1989). This requires specific flow-through respirometers or swimming tunnels (Spoor, 1941; Fry, 1957; Blazka et al., 1960), but this is not easily applicable to all species of fish. Another way is to measure oxygen consumption of free-swimming fish fasted for 3–7 days to eliminate the effect of the feed consumed, and its subsequent metabolism is recommended as the most suitable approach for determining HE_f (Cho and Kaushik, 1990; Cho and Bureau, 1995). Comparative carcass analysis of fasted fish over a sufficiently long duration is another method of estimating HE_f. Extrapolation of relationship between MEI and HE can also provide an estimate of HE_f (Fig. 2.3). The last two approaches appear to provide similar estimates of HE_f, at least for rainbow trout (Bureau, 1997).

When standardized for weight differences, HE_f values of fish are a 5- to 20-fold lower than HE_f of homeothermic terrestrial vertebrates. For rainbow trout, different studies suggest a HE_f of approximately 30–40 $kJ\ kg^{-0.824}/d$ at water temperatures between 15 and 18°C (Kaushik and Gomes, 1988; Cho, 1991; Bureau, 1997), which is much lower than HE_f values of 300–590 $kJ\ kg^{-0.75}/d$ reported for domestic homeothermic farm animals (Bureau et al., 2002). The lower HE_f values in fish can be attributed to (1) the ectothermic nature of these animals, which do not spend energy on maintaining a constant body temperature; (2) a lower sodium pump activity; (3) the aquatic mode of life eliminating the need to oppose gravity; and (4) excretion of nitrogenous wastes as ammonia directly into the external medium without having to convert this into less toxic forms such as the urea or uric acid, as is done by land animals and birds, respectively.

From the same comparative point of view, while the HE_f of fish is low, the participation of body protein degradation in meeting the energy needs, calculated as proportionate loss in terms of N per unit energy loss in fasted animals, is several fold higher in fish (2–5 mg N/kJ) compared with land animals (0.2 mg N/kJ). Variability in fish values is due to differences in the duration of fasting, water temperature, and fish size. Analysis of energy loss and the corresponding changes in energy-yielding substrates under fasting or protein-free feeding conditions shows that in all fish, amino acid catabolism contributes greatly to the provision of energy (Kaushik and Luquet, 1977; Cho and Kaushik, 1985).

Various factors can alter HE_f. In fish, both whole-body mass and water temperature are major factors influencing HE_f (Clarke and Johnston, 1999; Killen et al., 2010). The relationship between metabolic rate at fasting and body mass is generally described by equation $Y = aBM^b$, where Y is the metabolic rate; BW is body mass (in kg); a is the scaling coefficient, which differs between fish species and is strongly affected by water temperature; and b is the scaling exponent or slope of the allometric plot of log metabolic rate versus log body mass. The scaling coefficient "a" varies widely among taxa (White et al., 2006), whereas the exponent "b" is traditionally assumed to be around $^3/_4 = 0.75$ and to be relatively constant across different animal classes, often termed "universal scaling law." Some critical analyses of data from different taxa have been made (Clarke and Johnson, 1999; da Silva et al., 2006; White et al., 2006; White and Kearney, 2013) to address the value of this exponent ("b"). Killen et al. (2010) examined resting metabolic rates among 89 species of teleost fish from an ecological perspective, to assess the possible effects of lifestyle, swimming mode and temperature on the scaling exponent ("b"). This exponent, "b," decreased as the metabolic activity increased with temperature and was influenced by the type of natural habitats, which was possibly associated with different capacities for swimming and due to predator–prey interactions in nature. Even for fish under farming conditions, the exponent "b" varies considerably. Estimates of the exponent "b" between difference fish species range between 0.7 and 0.9 (Brett and Groves, 1979; Hepher, 1988; Cho, 1991; Lupatsch et al., 1998; Liu et al., 2000). It appears reasonable to assume that metabolic body weight (MBW) can, in practice, be calculated as $kg^{0.8}$ for most species. When expressing energy metabolism parameters per MBW ($kg^{0.8}$), one corrects for the impact of body weight on energy metabolism, and it makes comparison of fish with different weight classes possible.

As fish and shrimp are poikilotherms, their metabolic rates and energy expenditures are influenced by water temperature. Based on analysis of oxygen consumption data from several trials, Cho and Kaushik (1990) reported that the HE_f of rainbow trout as a function of water temperature could be described as follows: $HE_f = (-1.04 + 3.26 \times T - 0.05 \times T^2) \times BW^{0.824}$, where HE_f is the fasting heat production (kJ/d); T is the water temperature (°C); and BW is the body weight (kg). Based on an analysis of data from 69 species of teleosts, Clarke and Johnson (1999) reported that there was a curvilinear relationship between fasting metabolic rate and water temperature. This led them to suggest that the HE_f of tropical fish might be considerably higher than that of cold-water fish. A major caveat with this and similar studies based on literature data is that they mostly deal with individual data for one "representative" temperature per species. When one examines experimental evidence from different species of farmed fish, which show that the basal metabolisms of cold-, temperate-, and warm-water species appear to be fairly similar (Cho and Kaushik, 1990; Kaushik, 1998; Lupatsch et al., 1998; Liu et al., 2000; Glencross, 2008; Schrama et al., 2012). It is worth, however, mentioning that for a given eurythermic species, increasing water temperature results in a more or less linear increase in H_eE up to a certain temperature, after which further increases in temperature do not induce an equivalent increase in H_eE. Conversely, the metabolic rate of fish is reduced when the water temperature and consequently the body temperature of the fish are reduced.

The influence of water salinity on the basal metabolic rates is not well established, and experimental data on energy expenditure with euryhaline species maintained/reared at different salinities are not numerous. Tilapia reared at three different salinities exhibited lower metabolic rates at an isosmotic salinity of 12 parts per thousand (ppt) than when reared at 0 or 35 ppt (Ferby and Lutz, 1987). Euryhaline teleost fishes from different habitats can have different levels of isosmotic salinity ranging between 10.2 and 12.7 ppt (Sampaio and Bianchini, 2002). Measurements of oxygen consumption at the whole-body or tissue levels have shown some regulation of the energy supply during acclimation to different osmotic environments, which varies with species, level of acclimatization, and life habits (Tseng and Huang, 2008). At the tissue level, carbohydrate metabolism appears to play a major role in the energy supply for iono- and osmoregulation. The liver is the major source supplying carbohydrate metabolites to osmoregulatory organs, and the roles of lipids and proteins remain largely unclear. Most experimental data, however, suggest that the energy cost of osmoregulation is very low, equivalent to less than 5% of the basal metabolism.

As regards the effects of genetic origin on oxygen uptake and on basal metabolic rates, some early studies undertaken with rainbow trout did not show any differences between families (Kaushik et al., 1984) or due to triploidization (Oliva-Teles and Kaushik, 1987a, b). Since diploid and triploid fish are known to have different cell/nuclear sizes and that whole-body metabolic rate can be considered as the sum of metabolic rates of constituent cells, Maciak et al. (2011) analyzed the relation between cell size and metabolic rates in diploid and triploid model fish and found an inverse relationship between cell size and standard metabolic rate. Their data suggest that cell/genome/nucleus size is a possible determinant of variations in metabolic rates and possibly on allometric scaling. In salmonids, this does not appear to be the case at the cellular level; oxygen consumption rates of triploid blood cells were equal to those of diploid blood cells (Benfey, 1999). Atkins and Benfey (2008) found in two salmonids, Atlantic salmon and brook charr (*Salvelinus fontinalis*), that triploids had higher metabolic rates than diploids at lower temperatures, and lower metabolic rates than diploids at higher temperatures, suggesting that triploids have lower thermal optima than diploids. Studies undertaken with growth hormone transgenic Atlantic salmon (Cook et al., 2000a, b) showed that oxygen consumption rates of transgenic fish were higher than that of control salmon both under postabsorptive state (24 h after meal) and after weeks of starvation.

2.3.4.2 *Maintenance requirement*

The amount of metabolizable energy required for an animal to maintain zero energy balance (zero energy gain, RE = 0) is termed the maintenance metabolizable energy requirement (ME_m) (Fig. 2.4). The heat loss when an animal is fed at maintenance (HE_m) is equal to ME_m (Fig. 2.3). HE_m amounts to fasting heat production (HE_f) plus the heat increment of feeding (H_iE) associated with utilizing nutrients absorbed to meet energy losses associated with basal metabolism and to maintain energy balance. Sometimes, HE_m (i.e., ME_m) is erroneously equated to the energy losses observed in fasted fish (HE_f), which is different from

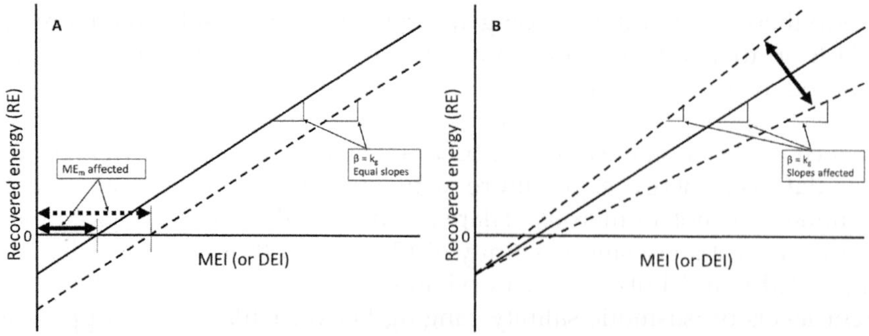

FIGURE 2.5 Conceptual scheme on how biotic and abiotic factors can alter the relationship between MEI (or DEI) and RE. Panel A shows the impact of factors only influencing the metabolizable energy requirements for maintenance (ME_m or DE_m). Panel B reflects the impact of factors that changing the slope of the relationship (i.e., the energy utilization efficiency for growth, k_g). *DEI*, digestible energy intake; *MEI*, metabolizable energy intake; *RE*, recovered energy.

maintaining energy homeostasis at the organismal level, which necessarily involves some energy intake and its utilization, reflecting heat increment of feeding (H_iE) for maintenance. The best way to measure HE_m (ME_m) consists of measuring the energy gain of fish fed graded levels of a nutritionally adequate feed of known energy content (Elliott, 1982; Brown et al., 1990; Cho and Kaushik, 1990). HE_m and thus ME_m are estimated by interpolating the response data to zero energy gain or using a regression analysis of retained energy (RE) as a function ME intake (MEI) to the level of intake that results in zero whole-body energy gain ($RE = 0$) as depicted in Figs. 2.4 and 2.5. Given the difficulties involved in quantifying ME in fish or shrimp, and as quantitatively the metabolic energy losses ($UE + ZE$) can be considered to be small, it is common to relate data on RE to DEI. Similar to ME_m, DEI at zero RE is termed the digestible energy requirements for maintenance (DE_m). Since DEI is not corrected for branchial and urinary energy losses, estimates of DE_m are larger than ME_m.

At the organism level, body weight is a major factor determining ME_m (DE_m). To make ME_m comparable between and within species, values are expressed per unit metabolic body weight, this being $BW^{0.8}$ (in $kg^{0.8}$) in fish. But even when expressed per unit metabolic body weight, energy requirements for maintenance can be influenced by biotic and abiotic factors. In Fig. 2.5, a conceptual scheme is provided on how these factors can affect the relationship between MEI and RE. Panel A reflects factors that only influence ME_m without altering the slope (e.g., water temperature). Panel B reflects the impact of factors that alter the relationship by altering the slope (i.e., dietary macronutrient composition; ratio between protein and fat retention). In general, suboptimal environmental conditions are often assumed to increase ME_m. Under suboptimal conditions, an animal has to spend more energy for maintaining its internal homeostasis. However, opposite response can also occur, like Glencross (2009) reported that reducing water-dissolved oxygen levels reduced DE_m of rainbow trout, from 48 $kJ/kg^{0.8}/d$ at normoxia to 38 $kJ/kg^{0.8}/d$ at hypoxia. Lupatsch et al. (2010) noted that DE_m of European sea bass was influenced by stocking density, 43 versus 51 $kJ/kg^{0.8}/d$ at high and low stocking densities, respectively. As suggested by Santos et al. (2010), energy requirements for maintenance might be reduced when swimming activity is reduced. Water temperature is, however, the most import environmental factor altering ME_m, which increases with increasing water temperatures. This effect of temperature is mainly on ME_m and not on the slope of the relationship between MEI and RE (Fig. 2.5A). This impact of temperature on the relationship according to Fig. 2.5A has been demonstrated in rainbow trout (e.g., Cho and Kaushik, 1990), mulloway (*Argyrosomus japonicus*) (Pirozzi et al., 2010), and white grouper (*Epinephelus aeneus*) (Lupatsch and Kissil, 2005).

Under optimal environmental conditions, the metabolizable energy requirements for maintenance (ME_m) in land animals and birds range from 460 to 670 kJ/kg$^{0.8}$/d, which is much higher than values reported for fish (Knap and Kause, 2018). A summary of ME_m (or DE_m) estimates from various fish species at optimal growth temperature is provided in Table 2.3. Generalizing for fish, ME_m ranges between approximately 40 to 80 kJ/kg$^{0.8}$/d. Between but also within fish species, there is still considerable variation in estimates of ME_m. Factors contributing to this variability are differences in diet composition; methodological aspects such as duration of trial number of ration levels; scaling factor used to calculate metabolic weight; or the statistical methods applied and environmental conditions. Efforts have been made to develop generalized equations for maintenance energy requirements (Zhou et al., 2005).

2.3.4.3 *Heat increment of feeding (H_iE) and energy utilization efficiencies*

Feed intake leads to an immediate increase in heat production (HE) in fish and shrimp (Fig. 2.3). As mentioned earlier, this increase in energy expenditure related to feed intake is referred to as the heat increment of feeding (H_iE), also termed SDA, calorigenic effect, or postprandial thermogenesis. The absolute value of H_iE is MEI times the slope of the relationship between HE and MEI in Fig. 2.3. The slope of the linear relation between HE to MEI (Fig. 2.3) is equal to 1 − slope of the linear relationship between RE and MEI (Fig. 2.4). For growth (excluding reproduction related processes), the slope of the linear relationship between RE and MEI is termed the partial utilization efficiency of ME for growth ($k_{g:ME}$). Similarly, the slope of the relationship between RE and DEI is termed the partial utilization efficiency of DE for growth ($k_{g:DE}$). Because the relationship of DEI and RE is not corrected for branchial and urinary energy losses (UE + ZE), $k_{g:DE}$ is smaller than $k_{g:ME}$.

In absolute terms, H_iE depends to a large extent upon the dietary nutrient composition, the type of energy retentions (fat vs. protein), and the plane of nutrition (Brody, 1945). The energetic costs associated with feeding are in fact the result of a number of factors linked with eating and ingestion of feed as well as absorptive and postabsorptive physiological processes. These costs vary with diet composition and are generally measured using indirect calorimetric measurements. The duration of this H_iE depends mainly upon the quantity and quality of the diet, which in itself can be governed by factors such as water temperature and growth rate. In an early study, Saunders (1963) found that oxygen consumption in Atlantic cod remained above the fasting level for 7 days following a meal at 10°C, whereas at 15°C, this effect lasted for only 4–5 days. Kaushik and Dabrowski (1983) similarly found that even in developing young larvae and juveniles, the increase in oxygen uptake corresponded more or less to the rate of transit of feedstuffs in the digestive tract. Special care and standardized procedures are needed to quantify H_iE to avoid artifacts due to nutritional imbalance and variable rearing conditions, which can affect overall growth and nutrient/energy gain.

TABLE 2.3 Data on maintenance requirements (HE_m, expressed as kJ ME/kg$^{0.8}$/d) in fish and shrimp.

Species	Body weight, g	Temp °C	HE_m	References
Atlantic salmon (*Salmo salar*)	95	10	32[a]	Helland et al. (2010)
Asian sea bass (*Lates calcarifer*)	15	30	35[a]	Glencross (2008)
	410	30	46[a]	idem
Channel catfish (*Ictalurus punctatus*)	8–10	27.5	25	Gatlin et al. (1986)
European sea bass (*Dicentrarchus labrax*)	15–140	24	45[a]	Lupatsch et al. (2001, 2003)
	74	22	43–51[a]	Lupatsch et al. (2010)
Gilthead sea bream (*Sparus aurata*)	30–160	24	48[a]	Lupatsch et al. (1998, 2003)
Nile tilapia (*Oreochromis niloticus*)	330	28	60	Lupatsch et al. (2010)
	10-300	28	26	Van Trung et al. (2011)
	75	28	56–62	Schrama et al. (2012)
	44	28	96–110[a]	Haidar et al. (2016)
Rainbow trout (*Oncorhynchus mykiss*)	150	18	58–75	Kaushik and Gomes (1988)
	158	9	19	Bureau and Hua (2006)
	300	15	33	Storebakken et al. (1991)
	96	14	42[a]	Glencross (2008)
	55	16	44[a]	Glencross (2009)
Red drum (*Sciaenops ocellatus*)	5.5	25	48	McGoogan and Gatlin (1998)
Snakehead (*Channa striata*)	29	28.5	7–48[a]	Phan et al. (2021)
Tra catfish (*Pangasianodon hypophthalmus*)	40	32	40[a]	Glencross et al. (2010)
Yellowtail (*Seriola quinqueradiata*)	750	15	77	Watanabe et al. (2000)
Yellowtail (*Seriola lalandi*)	50–2000	20–25	87	Booth et al. (2010)
Pacific white shrimp (*Litopenaeus vannamnei*)	1.5–7.5	28	110[a]	Lupatsch et al. (2008)

[a] *HE_m expressed on digestible energy (DE) basis.*

As regards the postprandial changes in heat loss, three factors are recognized as having an influence on the amplitude and duration of H_iE: digestion and absorption processes (H_dE), formation and excretion of metabolic wastes (H_wE), and transformation and interconversion of substrates and their retention in tissues (H_rE). Some studies have tried to separate the biochemical aspects of postabsorptive processes from the physical or mechanical aspects of feeding and digestion on H_iE in fish. Using nondigestible matter in feeds, differences in heat production resulting from a complete diet and that from a diluted diet were assessed to determine heat production referred to as "mechanical SDA" (Tandler and Beamish, 1979), which could not be confirmed subsequently by other studies with other species (Jobling and Davies, 1980). The absorption of amino acids, peptides, and glucose, by the intestinal mucosa, often occurs through an energy-dependent, active transport systems. Overall, the heat losses associated with diet ingestion and digestion (H_dE) are very small compared with that associated with metabolic work ($H_rE + H_wE$).

Both theoretical considerations and experimental evidence suggest that the heat losses associated with transformation of the substrates and their retention in tissues (H_rE) represent a large proportion of the H_iE in animals. The physiological bases of this increased heat production are clearly related to postabsorptive processes and the subsequent metabolic work required for the synthesis of proteins and lipids in target tissues, their metabolism, and elimination of nitrogenous wastes. The energy cost of synthesis of nitrogenous end products in land animals has been estimated to be 13 and 10 kJ/g N for urea and uric acid, respectively (Martin and Blaxter, 1965). These catabolites must be concentrated and excreted in terrestrial animals and birds, requiring the expenditure of further energy. In ammoniotelic fish and shrimp, ammonia efficiently excreted into the surrounding medium without a need for energy for detoxification or concentration. As a result, H_wE is considered to represent only a very small fraction of the H_iE of fish. Some teleosts under specific environmental stressful situations, such as tilapia living in alkaline lakes, are capable of shifting to ureagenesis and excretion (Wood et al., 1989), and in such exceptional cases, H_wE may be a significant contributor to H_iE. As mentioned earlier, there are no convincing data on within species differences between families or strains of a given species with regard to H_iE as affected by dietary factors. However, Cook et al. (2000b) found that postprandial oxygen uptake and the heat increment associated with feeding were 1.5- to 1.7-fold higher in transgenic Atlantic salmon compared with controls. As regards possible genetic determinism of metabolic rates, Kinghorn (1983) developed a system to simultaneously estimate food consumption of fish and energy expenditure (SDA) and found that variation in oxygen consumption was a highly heritable trait. Further work along these lines is definitely needed using nutritionally well-defined diets to assess feed × strain interactions, where determination especially of postprandial metabolic rates can serve as reliable biomarkers. Secondly, although useful in terms of energy utilization and expenditure, simple measurement of H_iE or SDA based on oxygen consumption rates do not really enable us to identify the underlying causes. There is also a definite need to develop estimates of the energetic contribution of specific physiological processes related to SDA (McCue, 2006).

Besides the impacts of feed intake, within- and between-day variation in heat losses (production) can be induced by heat losses associated with activity. Bioenergetic studies using respirometers or swimming tunnels provide precious information on the energetic cost of swimming for foraging purposes, for analyzing the mobilization of energy reserves during spawning migration over long distances (Webb, 1971; Brett and Groves, 1979; van den Thillart et al., 2004) and also for analyzing dietary energy flow through the organisms. Under farming conditions, energy losses associated with activity are difficult to separate from basal metabolism, and it is accepted that when constructing the energy budget of free-swimming fish under aquaculture conditions, the cost of activity is rather negligible and is already included in the estimate of HE_f (i.e., H_eE) (Cho et al., 1982).

The increase in H_iE per unit of DEI increase (or MEI), which was defined above as $1 - k_{gDE}$ ($1 - k_{gME}$), is not a fixed value (Fig. 2.5B). As reviewed by Schrama et al., 2012, k_{gDE} between different fish species ranged from 0.31 to 0.82, but also within fish species, there is variability in k_{gDE}. Factors that can alter k_{gDE} (and also k_{gME}) are the chemical composition of DEI (DP, digestible fat, and digestible carbohydrates); the composition of the RE (protein vs. fat gain); and type of production process (somatic vs. gonadal growth). It has been shown that dietary macronutrient composition alters k_{gDE} in Nile tilapia (Schrama et al., 2012), trout (Rodehutscord & Pfeffer, 1999), Asian sea bass (Glencross et al., 2017), and snakehead (Phan et al., 2021). Generalizing within fish species, replacing digestible carbohydrates in the diets by digestible fat increases k_{gDE}. In Nile tilapia, k_{gDE} altered when digestible starch was partially replaced by digestible nonstarch polysaccharides (Haidar et al., 2016).

The major form in which energy is retained with animals is either protein or fat. Only a minor portion is retained as carbohydrates (mainly glycogen). As first demonstrated by Kielanowski (1965), it is well accepted in farm animals that the partial efficiency of ME utilization for energy gain in the form of protein (k_p) is lower than that for energy gain in the form of fat (k_f). As first applied by Kielanowski (1965), k_p and k_f can be estimated by multiple regression explaining MEI as a function of energy gain as protein (RE_p) and energy gain as fat (RE_f) using the following equation: $MEI = HE_m + 1/k_p \times RE_p + 1/k_f \times RE_f$. In Table 2.4, estimates of k_p and k_f are given for different fish species. Quiet similar to terrestrial farm animals, k_p in fish ranges between 0.49 and 0.61 and is lower than k_f. Values of k_f in fish range between 0.72 and 0.91. Compared with k_p, the estimates of k_f have a wider range, which might be due to the source of nonprotein energy in the diet (fat vs. carbohydrates). Carbohydrate-rich diets will most likely result in lower k_f values. In general, the smaller value of k_p compared with k_f implies that protein gain induces more HE than fat deposition (both on energy basis).

TABLE 2.4 Estimation of the energy utilization efficiency of metabolizable energy for protein gain (k_p) and for lipid gain (k_f) in various fish species.

Species	Maintenance HE	k_p	k_f	References
Common carp	42 $kJ/kg^{0.75}/d$	0.56	0.72	Schwarz and Kirschgessner (1995)
European sea bass	40 $kJ/kg^{0.80}/d$	0.53	0.90	Lupatsch et al. (2003)
Gilthead sea bream	43 $kJ/kg^{0.80}/d$	0.53	0.76	Lupatsch et al. (2003)
Mulloway	0[a]	0.49	0.75	Pirozzi et al. (2010)
Rainbow trout	1.4 $kJ/kg^{0.39}/d$	0.54	0.90[a]	Rodehutscord and Pfeffer (1999)
White grouper	31 $kJ/kg^{0.80}/d$	0.56	0.91	Lupatsch et al. (2003)
Yellowtail kingfish	102 $kJ/kg^{0.80}/d$	0.61	0.83	Booth et al. (2010)

[a]*Value fixed in estimation.*

2.3.5 Net energy

The net energy intake (NEI) is defined as NEI = MEI − H_iE (Fig. 2.1). NEI is the amount of energy, which is available for maintenance, muscle activity, and somatic and gonadal growth. Since RE = MEI − HE = MEI − (HE_f + H_iE), another way to express NEI is NEI = RE + HE_f. So by definition, the NEI, which a fish has available, depends on all factors which affect MEI and H_iE. The amount of NEI a fish at a given moment (life time) has available from its MEI is dependent on (1) the chemical nature (composition) of MEI and (2) the ratio between protein and fat synthesis. Based on these theoretical considerations the NEI can be related (1) to the dietary nutrient composition of MEI (DEI) or (2) to the amount the energy used for protein or fat deposition. This later approach requires the estimated values of k_p and k_f as described in the previous section.

Similar to the values of DE and ME, the NE value can be applied to diets, which reflects the potential of a diet to cover the maintenance costs and somatic (or gonadal) growth. The efficiency of utilization of ME varies with the type of the absorbed nutrients (protein, fat, or carbohydrates) and whether these substrates are used for ATP production or for the deposition of protein, fat, or glycogen. The efficiency of ATP production is highest from glucose, followed by fat acids, and is lowest from amino acids (Blaxter, 1989). In land animals, the theoretical (maximal) efficiency of conversion and retention of the macronutrients (the stochiometric efficiencies) are well documented (Blaxter, 1989). For instance, with the conversion of glucose into glycogen, 5% of the energy of glucose is lost as H_iE, whereas if glucose is retained as lipids, 30% of its energy is lost as H_iE. Conversion of absorbed lipids into body lipids is theoretically 96% efficient, which means only 4% will be dissipated as H_iE. Conversion of absorbed amino acids into body proteins is, in theory, 86% efficient, entailing a H_iE of 14%, whereas conversion of amino acids into body lipids is, in theory, only 66% efficient, or 34% of energy is lost as H_iE. It should be realized that these stochiometric efficiencies given by Blaxter (1989) do not account for energetic costs (losses) related to food ingestion, digestion, and also aspects such as protein turnover connected to protein gain. This is also reflected by the fact that also in fish, the values of k_p and k_f given in Table 2.4 are all lower than the previously given stochiometric efficiencies.

For determining the cost of nutrient deposition and interconversion, specific studies to analyze the relations between MEI and RE into its different components are needed. The energy costs of protein and lipid deposition are simply defined as the ME required to promote unit increments in body protein or lipid. This can be done applying multiple regression analyses of the relation between MEI and the relative proportions of protein and fat energy deposition. The partial efficiency of ME utilization for whole-body growth (k_{gME}), protein deposition (k_p), and lipid deposition (k_f) is the ratio of the net energy retained to the corresponding ME intake components. NE systems are based on the utilization of ME for maintenance and for growth and are based on prediction equations taking into account either digestible nutrients or DE (or ME) and the chemical characteristics of feeds and growth. Some NE systems have been established from specific designed large-scale studies with energy balance measurements on animals (Noblet et al., 1994; Noblet, 2006), while other approaches involve analysis of literature data on energy balance studies, biochemical information, and/or theoretical assumptions (Emmans, 1994; Boisen and Verstegen, 1998). In monogastric land animals, NE is considered the most appropriate system to express energy values and requirements. Because of its importance, it typically serves as a reference for expressing the requirements of other nutrients (e.g., protein, amino acids, minerals, etc.). As an example, the dietary requirements for each essential amino acid are expressed as digestible amino acid per unit of NE in the Dutch pig evaluation system (CVB, 1993). Two approaches are applied in NE systems in monogastric land animals: (1) those relating ME to the type of gain (fat and protein) applying in different forms the relationship $ME = HEm + (RE_p/k_p) + (RE_f/k_f)$ and (2) those predicting NE from the absorbed (i.e., digested) macronutrients (protein, fat, starch, nonstarch polysaccharides, etc.). In this later approach, instead of linearly relating RE to MEI, RE is related by multiple regression to the different types of digestible nutrient intakes such as protein, fat, starch, and so on. In equation form: $RE = \mu + \beta_1 \times dCP + \beta_2 \times dFat + \beta_3 \times dStarch + \ldots$, where dCP is the DP intake; dFat is the digestible fat intake; dStarch is the digestible starch intake; μ is HE_f; and β_1 is the energetic utilization of digested protein for energy gain.

The system proposed for pigs by Noblet et al. (1994) and further developed and validated in subsequent studies (Noblett and van Milgen, 2004; Noblet, 2006) proposes NE prediction equations that are applicable to ingredients and compound feeds and different stages of pig production (Noblet, 2006). As mentioned previously, the efficiency of ME for NE differs widely between nutrients in pigs. A multivariate model was proposed by van Milgen and Noblet (1999), which is considered to provide more accurate and reliable estimates of k_p, k_f, and HE_f.

The NE system has not been explored well in aquatic animals, although its potential has been long recognized by some authors. For example, a few studies have used the factorial approach to estimate k_p and k_f in different species of fish (Table 2.4). Over the past decade, concerted efforts to apply the net energy (NE) approach in fish feed evaluation, which takes full account of the possible differences in the utilization of the different digestible macronutrients (DP, fat, carbohydrates), have been made in different species of fish (Schrama et al., 2012; 2018; Phan et al., 2019, 2021; Groot et al., 2021). In these studies, the NE is predicted from the digestible dietary nutrient content of protein, fat, and carbohydrate (respectively, dCP, dFat, and dCarb). Schrama et al. (2018) undertook a metaanalysis of available data from 17 studies with 68 diets on energy flow involving two species, the rainbow trout and Nile tilapia. In the studies of Phan et al. (2019, 2021) dealt with three species, the common carp Asian sea bass and snakehead, in which NE equations were estimated by energy balance studies using diets having a big contrasts in dCP, dFat, and dCarb content and feeding these diets at different levels. This resulted in NE equation described as follows:

$$NE = a \times dCP + b \times dFat + c \times dCarb$$

where a, b, and c are the amount of potential retained as energy (RE) when fed above maintenance per gram of DP, fat, and carbohydrate, respectively. If the values of a, b, and c are divided by the energy content of protein, fat, and carbohydrates (23.6, 39.5, and 17.2 kJ/g, respectively), the energetic efficiency of DP, digestible fat, and digestible carbohydrate for growth ($k_{NE;dCP}$, $k_{NE;dFat}$, and $k_{NE;dCarb}$, respectively) is obtained. In the fish NE formulas, no distinction is made between different types of carbohydrates (starch vs. nonstarch polysaccharides, NSPs) because in many fish species, NSPs are indigestible (see Chapter 7), which is in contrast to NE systems for pigs where carbohydrates are split out.

TABLE 2.5 Data on net energy coefficients in various fish species: estimation of the energy utilization efficiency of digested protein ($k_{NE;dCP}$), digested fat($k_{NE;dFat}$), and digested carbohydrates ($k_{NE;dCarb}$).

Species	$k_{NE;dCP}$	$k_{NE;dFat}$	$k_{NE;dCarb}$	References
Asian sea bass	0.64	0.94	0.18[a]	Phan et al. (2019)
Common carp	0.47	0.86	0.60	Phan et al. (2019)
Nile tilapia	0.49	0.91	0.66	Schrama et al. (2018)
Rainbow trout	0.64	0.89	0.70[a]	Schrama et al. (2018)
Snakehead	0.53	0.78	0.05	Phan et al. (2021)

[a]*For Asian sea bass and rainbow trout, the retained energy (RE) exhibited a curvilinear relationship with the amount of digested carbohydrates.*

Estimates of $k_{NE;dCP}$, $k_{NE;dFat}$, and $k_{NE;dCarb}$ for different fish species are given in Table 2.5, for applying only linear relationships with digested nutrients. Summary of some of the conclusions drawn from these initial studies applying an NE approach is as follows. The NE value of digested protein is comparable between tilapia, carp, and snakehead and lower than those in rainbow trout and Asian sea bass. The $k_{NE;dCP}$ estimates of these species ranged between 0.47 and 0.67. The NE value of digested fat is more or less constant across different species of fish. Estimates of $k_{NE;dFat}$ range between 0.78 and 0.94 and are much higher than the estimates of $k_{NE;dCP}$ and $k_{NE;dCarb}$. The NE value of digested carbohydrates exhibits the largest variability between fish species. Values of $k_{NE;dCP}$ varied between 0.05 and 0.70. The NE values of digested carbohydrates were comparable for tilapia, carp, and rainbow and very similar to values for pigs (Noblet et al., 1994; CVB, 1993). However, digested carbohydrates in both snakehead and Asian sea bass seemed to have no or minimal NE value. The larger variability in carbohydrate utilization was also reflected in the fact that RE was curvilinearly related to the digestible carbohydrate intake in rainbow trout and Asian sea bass. This curvilinear relationship was hypothesized to relate to the poor/limited capability of these species to metabolize/regulated blood glucose (See Chapter 7).

2.3.6 Retained energy as growth

The simplest way to estimate RE is to determine the energy content of truly representative samples of fish or shrimp at the beginning and at the end of a growth trial in which feed intake of the groups is measured (comparative carcass analysis) (Table 2.4). The difference between the enthalpy of combustion of the body at the beginning and that at the end of a period of time is referred to as "recovered energy" or energy retained (RE). Alternatively, RE can be estimated by the energy balance approach by calculating the difference between IE and FE, UE + ZE, and HE, which implies that all these parameters are really measured. The part of MEI, which is not dissipated as heat (HE), is theoretically retained in the body as new tissue and thus RE = MEI − HE. Most of the studies on fish or shrimp nutrition try to apply the comparative carcass analysis approach. RE represents the overall deposition of energy-yielding nutrients, mainly proteins and lipids, reflecting the outcomes of numerous processes involving synthesis, transformations, bioconversions, and degradations of nutrients at different levels, regulated by various factors. At the whole-body level, proteins and lipids are the main energy-yielding components, and glycogen generally represents only a small proportion (<1%).

Whole-body N content is a major parameter in mass balance models used to estimate energy gain or losses in terms of protein. Based on an analysis of a large number of published data on whole-body N content of more than 2800 samples from different species of fish, Ramseyer (2002) calculated that whole-body N content (Y in g) was related fish wet body weight (X in g) as $Y = 0.0271\ X - 0.0281$ ($n = 2811$; $R^2 = 0.989$). This equation implies that the N content of fish (Y/X) is only changing at low body weights. For fish larger than 100 g, this give an N content on wet basis of 2.7%, which is equal to a protein content of 17% on wet weight basis. In energy terms, this implies that about 4 kJ/g wet body weight is stored as protein. Also within species, several studies have shown that, in general, whole-body protein and ash contents (% wet weight) do not change much with body weight, but the relative proportions of water and fat contents vary in rainbow trout (Shearer, 1994; Bureau et al., 2003), Nile tilapia (Van Trung et al., 2011), European sea bass, gilthead seabream or grouper (Lupatsch et al., 2003b), and Pacific white shrimp (Lupatsch et al., 2008). Whole-body energy content increase is due mainly to increasing lipid content with increasing body weight. The whole-body fat content varies depending on a number of factors such as fish age (i.e., body weight), fish species, and dietary factors. Large excesses of DEI (MEI) and improper dietary DP/DE ratios will result in the deposition of a larger proportion of RE as lipid. Hartman and Brandt (1995) analyzed the whole-body GE content of a large number of fish and estimated that the whole-body energy content (Y, in J per g wet weight) was related to whole-body dry matter content (DW in %) as $Y = 45.29 \times DW^{1.507}$ ($n = 587$, $R^2 = 0.95$). Seasonal changes in body composition, in relation to specific physiological stages or endocrine status, are also known to occur. A general pattern of changes in whole-body energy content with body weight is given in Fig. 2.6. With increasing body weight, the energy stored as protein remains

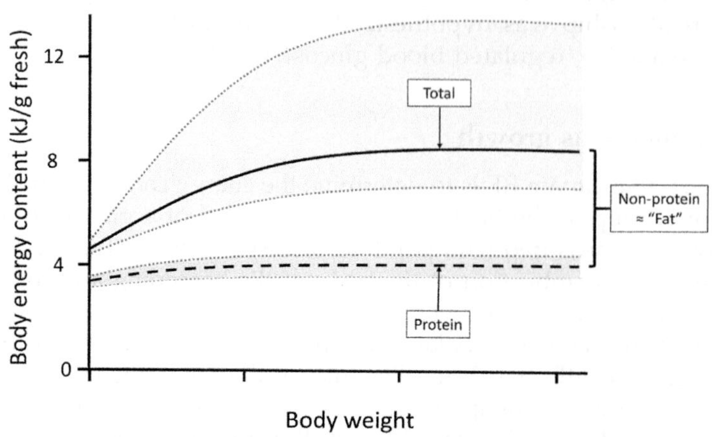

FIGURE 2.6 Generalization of relation between whole-body energy content and body weight (i.e., age). The solid *line* is the (total) body energy content, and the broken line is the energy content stored as protein. The shaded areas reflect the variability in these relations ships induced by biotic and abiotic factors.

relatively constant and is minimally affected by diet and environmental factors, whereas the total body energy content increases with age (body weight) and also the potential impact of abiotic factors.

2.3.6.1 *Energy demand and utilization for reproduction*

Reproduction is a critical part of the production cycle for farm fish and involves dramatic changes in energy acquisition and partitioning by aquatic organisms, be it under natural conditions or under farming. During the specific physiological changes occurring during maturation and spawning, the major portion of DEI/MEI is used for gonadal development and reproduction. The energy cost of reproduction in fish in nature exceeds that of simple gamete production. Indeed, development of external secondary sexual characters, production of mucus, nuptial behaviors and activities, construction of nest for certain species, and migration all involve expenditure of energy. For species undergoing anadromous or catadromous migration, an added challenge is that of osmoregulation. Some species cease feeding before undertaking upstream migration, development of ovaries, or secondary sexual characters and compete for spawning sites or mates. They undertake variable periods of migration over long distances across the oceans to riverine environments, which impose significant energy demands, which are met mainly from bodily reserves, mainly amino acids (Mommsen et al., 1980; Mommsen, 2004). As reproduction involves the synthesis and deposition of gonadal material that is formed regardless of the level of dietary energy intake, the necessary energy is withdrawn from other body tissues if the dietary supply is insufficient (Jonsson et al., 1991, 1997). Some very early studies had followed the mobilization of whole-body fat and its deposition into gonadal tissues from a quantitative point of view in rainbow trout (Nassour and Leger, 1989).

Among teleosts, egg size varies across a wide range, from 0.3 to 90 mm in diameter. Energy content of egg dry matter varies within a range of 20–30 kJ/g, with a mean value of about 23.5 kJ/kg dry matter, irrespective of egg diameter (Kamler, 2005). In some species, in females, gonad size can represent up to 30% of body mass. In contrast, male gonads at maturity represent less than 10% of body mass. In some multiple spawners such as the gilthead sea bream, which continue to feed during the spawning season, the total egg production over a single season can even reach 100% of the body mass (Tandler et al., 1995). In European sea bass, cumulative spawning over a reproductive cycle can lead up to 2.8 million eggs/kg female (Mylonas et al., 2003). Recent data have shown that in the gilthead seabream, the total number of eggs produced per kg of female through more than 75 spawnings spread over 3 months can reach up to 4 million eggs (Ferosekhan et al., 2020), which can amount to close to 2.8 kg of eggs/kg of female, equivalent to more than 8 MJ/kg of female.

Barneche et al. (2018) postulated that, in the wild, larger marine fish contribute disproportionately to stock recruitment and that there is a clear hyperallometric scaling of reproductive investment, meaning relative to whole-body mass, there was more energy deposited in the gonads during breeding season. Some recent work with farmed European sea bass (Sadoul et al., 2020) applied a "dynamic energy budget (DEB)" model (Kooijman, 2001), which theoretically provides a conceptual and quantitative framework to study the whole life cycle of an individual while making explicit use of energy and mass balances. Sadoul et al. (2020) confirmed such a hyperallometric relation between broodstock size and daily investment in reproduction in terms of energy with the exponents ranging above 1, although unfortunately, an energy budget could not be drawn, as the amount of energy intake was not taken into account.

There is a definite lack of precise quantitative data on energy acquisition and expenditure in farmed fish or shrimp during gametogenesis and reproduction. Early studies on the nutrition of rainbow trout broodstock (Takeuchi et al., 1981; Watanabe et al., 1984) showed that diets with low- but good-quality protein levels (35%–38% CP) with "high"-fat levels (18%) and a DE level of above 17 MJ/kg are capable of ensuring reproductive success and good-quality gametes. When dietary nutrients/energy supply is not sufficient during gametogenesis, most species favor gonadal development at the expense of somatic growth. For instance, during reproduction, marine flatfish, which do not consume feed during vitellogenic period, lose body mass with a concurrent increase in gonads, and the efficiency of transfer and deposition of energy from body to gonads is around 30% (Dygert, 1990), similar to salmonids. The amount of body energy stored in the eggs represents 8%–15% of the energy of the whole body, corresponding to the gonadosomatic index. While the composition and energy content of gonads is more or less characterized for several fish species, there are no studies on the actual cost of gonad formation (e.g., H_iE associated with nutrient deposition in gonads). Manipulations of DEI through changes in diet composition (fat level, DP/DE ratio) or through ration size (Knox et al., 1988) appear to yield consistent results, but further work is needed to partition energy flow. As there are species that stop feeding and those that continue to feed but have multiple spawnings over the season, quantitative analysis of the redistribution of DEI into RE for gametogenesis and for gonad development becomes complicated. Determination of this energy cost will necessarily involve long-term studies during which close monitoring of feed/energy intake and the quantity of energy deposited in the gametes is included.

2.4 Meeting energy requirements for growth

Overall energy requirements for growth can be defined as the quantity of DE/ME/NE necessary to produce 1 kilogram of fish, expressed as MJ/kg gain in body mass. A large number of researchers have tried to determine the protein/energy requirements of fish (Cho and Bureau, 1998; Bureau et al., 2000; Lupatsch et al., 1998, Lupatsch et al., 2003b, 2008). In terms of growth response, the composition of the gain in body mass (protein and lipid content) also has an influence on net energy requirements. As poikilotherms, voluntary feed intake, metabolic rates as well as growth, and energy flow through fish or shrimp are all affected by water temperature. Very early works of Elliott (1982) showed that both temperature and ration size have an impact on growth and energetics of salmonids kept in captivity. Bailey and Alanara (2006) analyzed the effect of body weight and temperature on DE requirements for unit weight gain in a number of fish species. They examined the effects of body weight and temperature on DE requirements using literature data for a large number of farmed fish species. They found that, in general, DE needs increased with increasing body weight in salmoniformes as well as in pleuronectiformes but found no effect of temperature in different species within these groups. Perciformes showed no effect of body weight or temperature on DE needs. They attributed such variable responses to possible differences in natural habitats and lifestyles of the species concerned. Given that the different species of fish are reared at different temperatures and the overall growth rates, and consequently, the energy increments per unit time, vary, it is indeed difficult to draw generalized conclusions.

Data on energy requirements expressed in absolute terms (example: kJ/fish/day) stand valid for the specific conditions of the studies, affected by genotype and environmental factors including diet composition and feeding practices/strategies. As fish grow at different rates and possibly deposit protein and nonprotein energy fractions differently, estimations of energy requirements should be undertaken with care to provide tools capable of predicting requirements on possible growth/nutrient gain trajectories.

Dietary DE content was and is still considered a major factor controlling feed intake in fish. There is evidence in the literature showing that when offered diets with varying DE levels, fish adjust their overall feed intake to maintain a given level of energy intake (see Houlihan et al., 2001; Yamamoto et al., 2000). In one of the early studies, an attempt was made to apply a relatively well-established technique with land animals, called "technique of separate feeding" involving feeding diets containing protein and nonprotein energy successively (Kaushik et al., 1981; Kaushik and Luquet, 1984) to rainbow trout. They found that there was an adjustment of the voluntary feed intake of the nonprotein energy complement diet in relation to the amount of protein fed and that the GE intake was relatively constant beyond a certain level of protein intake.

When fed to satiation by hand or by demand feeders, or using "eater-meter" devices, fish are capable of adjusting their feed intake depending on the dietary DE levels capable of ensuring the animal's growth potential. In some studies with fish, regulation of voluntary feed intake by dietary DE levels was not clearly demonstrated (see Houlihan et al., 2001). Geurden et al. (2006) could not detect any reduced feed intake with high-fat diets in rainbow trout. As in land animals, there is increasing evidence that feed intake of fish is also regulated by the lean growth potential of animals. The capacity of fish to adjust their intakes of diets of different DE is also believed to be determined by the physical capacity of the digestive tract (Lupatsch et al., 2001).

Cho (1991) proposed a bioenergetics-based approach to determine the feed requirements to reach a given growth target based on diet composition, water temperature, expected level of performance, and changes in body composition. This model has been updated by Cho and Bureau (1998) and extended to nonsalmonids (Kaushik, 1998). A rough scheme for the determination or estimation of DE, H_eE, H_iE, and UE + ZE and for calculating the total energy requirement and consequently the feed requirements is presented in Table 2.2. One main requirement is to have an accurate prediction of growth potential of a given species, strain, and genotype of fish or shrimp, under a given environmental/husbandry conditions. There have been a number of confirmatory studies to show that the use of the thermal unit growth coefficient (TGC) $= [FBW^{1/3} - IBW^{1/3}]/(\text{Sum degree days} \times 100)$ can facilitate the prediction of final body weight. This approach has been shown to represent faithfully the actual growth curves of a number of species of fish as well as shrimp and to be useful for predicting weight gain increment over a given period of time. The different calculations involved have been explained in detail by a number of authors with different species of fish and shrimp (Cho and Bureau, 1998; Bureau et al., 2000; Lupatsch and Kissil, 2005; Lupatsch et al., 2008; Van Trung et al., 2011).

Some recent work has explored the possibility of an "oxy-static control" of feed intake. The hypothesis that the voluntary feed intake in fish is possibly regulated by diet-induced differences in energy expenditure reflected by changes in oxygen consumption (Saravanan et al. 2012, 2013a). Groups of rainbow trout were fed with diets having a similar DP/DE ratio of 18 mg/kJ, but with different types of nonprotein energy sources (fat vs.) to create diet-induced differences in oxygen use. They found that energy expenditure (oxygen consumption) declined with increasing dietary nonprotein DE supplied as fat. The inverse relation between DE intake and the diet-induced oxygen use was considered to suggest a possible role of the latter in feed intake regulation. In another study, diets with different protein quality in terms of essential amino acid were fed to trout maintained under normoxic or hypoxic conditions and reared over a 6-week period (Saravanan et al., 2013b). Results suggested that feed intake in fish is possibly constrained by a set-point value of oxygen consumption.

2.5 Conclusions

The different aspects of partition of energy flow from feed to fish are similar to those known in land animals or birds. There are, however, differences between aquatic and land animals both at the cellular level and at the level of the whole organism. Ectothermy, ammoniotelism, and aquatic mode of life confer advantages in terms of energy demands to aquatic animals. Metabolizable energy provision by nutrients differs also among fish species from different trophic levels and habitats. The large number of species reared by aquaculture is a real challenge where any effort toward generation of basic knowledge, models, and tools transferable or adaptable from one species to another will benefit all, especially for improving nutrient resource utilization efficiency. There is also definitely a need for precisely quantifying the partition of dietary DE for somatic growth, reproduction, and other physiological changes due to environmental factors.

References

Anderson, J.S., Lall, S.P., Anderson, D.M., Chandrasoma, J., 1992. Apparent and true availability of amino acids from common feed ingredients for Atlantic salmon (*Salmo salar*) reared in sea water. Aquaculture 108, 111–124.

Atkins, M.E., Benfey, T.J., 2008. Effect of acclimation temperature on routine metabolic rate in triploid salmonids. Comp. Biochem. Physiol. a-Mole. Integrat. Physiol. 149, 157–161.

Azevedo, P.A., Cho, C.Y., Leeson, S., Bureau, D.P., 1998. Effects of feeding level and water temperature on growth, nutrient and energy utilization and waste outputs of rainbow trout (*Oncorhynchus mykiss*). Aquat. Living Resour. 11, 227–238.

Bailey, J., Alanara, A., 2001. A test of a feed budget model for rainbow trout, *Oncorhynchus mykiss* (Walbaum). Aquacult. Res. 32, 465–469.

Bailey, J., Alanara, A., 2006. Digestible energy need (DEN) of selected farmed fish species. Aquaculture 251, 438–455.

Ballestrazzi, R., Lanari, D., Dagaro, E., Mion, A., 1994. The effect of dietary protein level and source on growth, body composition, total ammonia and reactive phosphate excretion of growing sea bass (*Dicentrarchus labrax*). Aquaculture. 127, 197-206.

Barneche, D.R., Robertson, D.R., White, C.R., Marshall, D.J., 2018. Fish reproductive-energy output increases disproportionately with body size. Science 360, 642.

Beltran, M., Fernandez-Borras, J., Medale, F., Perez-Sanchez, J., Kaushik, S., Blasco, J., 2009. Natural abundance of (15) N and (13)C in fish tissues and the use of stable isotopes as dietary protein tracers in rainbow trout and gilthead sea bream. Aquacult. Nutr. 15, 9–18.

Benfey, T.J., 1999. The physiology and behavior of triploid fishes. Rev. Fish. Sci. 7, 36–67.

Blaxter, K., 1989. Energy Metabolism in Animals and Man. Cambridge University Press.

Blazka, P., Volf, M., Cepela, M., 1960. A new type of respirometer for the determination of the metabolism of fish in an active state. Physiol. Bohemoslov 9, 553–558.

Boisen, S., Verstegen, M.W.A., 1998. Evaluation of feedstuffs and pig diets. Energy or nutrient-based evaluation systems? II. Proposal for a new nutrient-based evaluation system. Acta Agric. Scandinavica, A, Animal Sci. 48, 95–102.

Booth, M.A., Allan, G.L., Pirozzi, I., 2010. Estimation of digestible protein and energy requirements of yellowtail kingfish *Seriola lalandi* using a factorial approach. Aquaculture 307, 247–259.

Brafield, A.E., Llewellyn, M.J., 1982. Animal Energetics. Blackie, Glasgow, UK, pp. 1–168.

Brett, J.R., Groves, T.D.D., 1979. Physiological energetics. In: Hoar, W.S., Rundall, D.J. (Eds.), Fish Physiology, vol. III. Academic Press, London, pp. 279–343.

Brett, J.R., Zala, C.A., 1975. Daily patterns of nitorgen excretion and oxygen consumption of sockeye salmon (*Oncorhynchus nerka*) under controlled conditions. J. Fish. Res. Board Can. 32, 2479–2486.

Brody, S., 1945. Bioenergetics and Growth. Reinhold, New York.

Brown, P.B., Neill, W.H., Robinson, E.H., 1990. Preliminary evaluation of whole body energy changes as a method of estimating maintenance energy needs of fish. J. Fish. Biol. 36, 107–108.

Bureau, D.P., 1997. The Partitioning of Energy from Digestible Carbohydrates by Rainbow Trout (*Oncorhynchus mykiss*). PhD Thesis. University of Guelph, ON, Canada, p. 170.

Bureau, D.P., Azevedo, P.A., Tapia-Salazar, M., Cuzon, G., 2000. Pattern and cost of growth and nutrient deposition in fish and shrimp: potential implications and applications. In: Cruz -Suárez, L.E., Ricque-Marie, D., Tapia-Salazar, M., Olvera-Novoa, M.A., Civera-Cerecedo, R. (Eds.), V Simposium Internacional de Nutrición Acuícola, Mérida, Yucatán, Mexico, pp. 111–140.

Bureau, D.P., Gunther, S.J., Cho, C.Y., 2003. Chemical composition and preliminary theoretical estimates of waste outputs of rainbow trout reared in commercial cage culture operations in Ontario. N. Am. J. Aquacult. 65, 33–38.

Bureau, D.P., Hua, K., 2006. Letter to the editor of aquaculture. Aquaculture 252, 103–105.

Bureau, D.P., Kaushik, S.J., Cho, C.Y., 2002. Bioenergetics. In: Halver, J.E.H., Hardy, R.W. (Eds.), Fish Nutrition. Academic Press.

Chakraborty, S.C., Chakraborty, S., 1998. Effect of dietary protein level on excretion of ammonia in Indian major carp, Labeo rohita, fingerlings. Aquacult. Nutr. 4, 47–51.

Cho, C.Y., Kaushik, S.J., 1990. Nutritional energetics in fish: energy and protein utilization in rainbow trout. World Rev. Nutr. Diet. 61, 132–172.

Cho, C.Y., 1991. The role of nutritional energetics research in formulating cost-effective diets for fish, with particular reference to salmonids. Proceedings of the Aquaculture Nutrition Workshop Salamander Bay. South Wales, New Australia, pp. 9–16.

Cho, C.Y., Bureau, D.P., 1995. Determination of the energy requirements of fish with particular reference to salmonids. J. Appl. Ichthyol. 11, 141–163.

Cho, C.Y., Bureau, D.P., 1998. Development of bioenergetic models and the Fish-PrFEQ software to estimate production, feeding ration and waste output in aquaculture. Aquat. Living Resour. 11, 199–210.

Cho, C.Y., Slinger, S.J., Bayley, H.S., 1982. Bioenergetics of salmonid fishes: energy intake, expenditure and productivity. Comp. Biochem. Physiol. 73B, 25-41.

Cho, C.Y., Kaushik, S.J., 1985. Effects of protein intake on metabolizable and net energy values of fish diets. In: Cowey, C.B., Mackie, A.M., G.B., A.J. (Eds.), In :Nutrition and Feeding of Fish. Academic Press, London, pp. 95–117.

Cho, C.Y., Slinger, S.J., 1979. Apparent digestibility measurement in feedstuffs for rainbow trout. In: Finfish Nutrition and Fishfeed Technology, Hamburg, pp. 239-247. Heenemann Verlagsgesellschaft; Berlin, Hamburg, pp. 239–247.

Choubert, G., Blanc, J.M., Luquet, P., 1984. Influence de la modification de la fréquence de distribution des repas sur la digestibilité du régime alimentaire chez la truite arc-en-ciel. Ann. Zootech. 33, 255–262.

Choubert, G., DeLaNoue, J., Luquet, P., 1979. Continuous quantitative automatic collector for fish feces. Progressive Fish-Culturist. 41, 64–67.

Choubert, G., DeLaNoue, J., Luquet, P., 1982a. Digestibility in fish: improved device for the automatic collection of feces. Aquaculture 29, 185–189.

Choubert, G., Fauconneau, B., Luquet, P., 1982b. Influence d'une élévation de la température de l'eau sur la digestibilité de la matiére sèche, de l'azote et de l'energie de l'aliment distribué à la truite arc-en-ciel (Salmo gairdneri Rich). Reprod. Nutr. Dev. 22, 941–949.

Clarke, A., Johnston, N.M., 1999. Scaling of metabolic rate with body mass and temperature in teleost fish. J. Anim. Ecol. 68, 893–905.

Cook, J.T., McNiven, M.A., Sutterlin, A.M., 2000a. Metabolic rate of pre-smelt growth-enhanced transgenic Atlantic salmon (*Salmo salar*). Aquaculture 188, 33–45.

Cook, J.T., Sutterlin, A.M., McNiven, M.A., 2000b. Effect of food deprivation on oxygen consumption and body composition of growth-enhanced transgenic Atlantic salmon (*Salmo salar*). Aquaculture 188, 47–63.

CVB (Centraal Veevoeder Bureau), 1993. Net Energy of Feedstuffs for Swine. CVB Report No. 7. CVB, Lelystad, The Netherlands.

da Silva, J.K.L., Garcia, G.J.M., Barbosa, L.A., 2006. Allometric scaling laws of metabolism. Phys. Life Rev. 3, 229–261.

Deng, D.F., Refstie, S., Hemre, G.I., Crocker, C.E., Chen, H.Y., Cech Jr., J.J., Hung, S.S.O., 2000. A new technique of feeding, repeated sampling of blood and continuous collection of urine in white sturgeon. Fish Physiol. Biochem. 22, 191–197.

Dias, J., Corraze, G., Arzel, J., Alvarez, M.J., Bautista, J.M., Lopez-Bote, C., Kaushik, S.J., 1999. Effects du rapport protéine/énergie des régimes alimentaires chez la truite et le bar en élevage. Perspectives de contrôle nutritionnel des dépôts lipidiques. Cybium 23 (suppl.), 127–137.

Dosdat, A., Servais, F., Métailler, R., Huelvan, C., Desbruyeres, E., 1996. Comparison of nitrogenous losses in five teleost fish species. Aquaculture 141, 107–127.

Dygert, P.H., 1990. Seasonal changes in energy content and proximate composition associated with somatic growth and reproduction in a representative age-class of female English sole. Trans. Am. Fish. Soc. 119, 791–801.

Ege, R., Krogh, A., 1914. On the relation between the temperature and the respiratory exchange in fishes. Int. Revue der gesamten Hydrobiol. Hydrograph. 7, 48–55.

Elliott, J.M., 1982. The effects of temperature and ration size on the growth and energetics of salmonids in captivity. Comp. Biochem. Physiol. 72B, 81–91.

Emmans, G., 1994. Effective energy: a concept of energy utilization applied across species. Br. J. Nutr. 71, 801–821.

Febry, R., Lutz, P., 1987. Energy partitioning in fish: the activityrelated cost of osmoregulation in a euryhaline cichlid. J. Exp. Biol. 128, 63–85.

Felip, O., Ibarz, A., Fernández-Borràs, J., Beltrán, M., Martín-Pérez, M., Planas, J.V., Blasco, J., 2012. Tracing metabolic routes of dietary carbohydrate and protein in rainbow trout (*Oncorhynchus mykiss*) using stable isotopes ([13 C] starch and [15 N] protein): effects of gelatinisation of starches and sustained swimming. Br. J. Nutr. 107, 834–844.

Ferosekhan, S., Xu, H., Turkmen, S., Gómez, A., Afonso, J.M., Fontanillas, R., Rosenlund, G., Kaushik, S., Izquierdo, M., 2020. Reproductive performance of gilthead seabream (*Sparus aurata*) broodstock showing different expression of fatty acyl desaturase 2 and fed two dietary fatty acid profiles. Sci. Rep. 10, 15547.

Forster, I., 1999. A note on the method of calculating digestibility coefficients of nutrients provided by single ingredients to feeds of aquatic animals. Aquacult. Nutr. 5, 143–145.

Fry, F.E.J., 1957. The aquatic respiration of fish. In: Brown, M.E. (Ed.), The Physiology of Fishes, vol. 1. Academic Press, New York, pp. 1–63.

Gamboa Delgado, J., 2009. Application of Natural Stable Isotopes in Aquaculture Nutrition. PhD Thesis. University of Wales-Bangor.

Gatlin, D.M., Poe, W.E., Wilson, R.P., 1986. Protein and energy requirements of fingerling channel catfish for maintenance and maximum growth. J. Nutr. 116, 2121–2131.

Gelineau, A., Medale, F., Boujard, T., 1998. Effect of feeding time on postprandial nitrogen excretion and energy expenditure in rainbow trout. J. Fish. Biol. 52, 655–664.

Geurden, I., Gondouin, E., Rimbach, M., Koppe, W., Kaushik, S., Boujard, T., 2006. The evaluation of energy intake adjustments and preferences in juvenile rainbow trout fed increasing amounts of lipid. Physiol. Behav. 88, 325–332.

Glencross, B.D., 2008. A factorial growth and feed utilisation model for barramundi, *Lates calcarifer* based on Australian production conditions. Aquacult. Nutr. 14, 360–373.

Glencross, B.D., 2009. Reduced water oxygen levels affect maximal feed intake, but not protein or energy utilization efficiency of rainbow trout (*Oncorhynchus mykiss*). Aquacult. Nutr. 15, 1–8.

Glencross, B., Rutherford, N., Bourne, N., 2012. The influence of various starch and non-starch polysaccharides on the digestibility of diets fed to rainbow trout (*Oncorhynchus mykiss*). Aquaculture 356–357, 141–146.

Glencross, B.D., Booth, M., Allan, G.L., 2007. A feed is only as good as its ingredients a review of ingredient evaluation strategies for aquaculture feeds. Aquacult. Nutr. 13, 17–34.

Glencross, B., Hien TTT, Phuong, N.T., et al., 2010. A factorial approach to defining the energy and protein requirements of Tra catfish, *Pangasianodon hupothalalus*. Aquacult. Nutr. 17, e396–e405.

Glencross, B.D., Blyth, D., Bourne, N., Cheers, S., Irvin, S., Wade, N.M., 2017. An analysis of partial efficiencies of energy utilisation of different macronutrients by barramundi (*Lates calcarifer*) shows that starch restricts protein utilisation in carnivorous fish. Br. J. Nutr. 117, 500–510.

Grisdale-Helland, B., Shearer, K.D., Helland, J., 2007. Energy and nutrient utilization of Atlantic cod, Atlantic salmon and rainbow trout fed diets differing in energy content. Aquacult. Nutr. 13, 321–334.

Groot, R., Lyons, P., Schrama, J., 2021. Digestible energy versus net energy approaches in feed evaluation for rainbow trout (*Oncorhynchus mykiss*). Anim. Feed Sci. Technol. 274, 114893.

Haidar, M.N., Bleeker, S., Heinsbroek, L.T.N., Schrama, J.W., 2018. Effect of constant digestible protein intake and varying digestible energy levels on energy and protein utilization in Nile tilapia. Aquaculture 489, 28–35.

Haidar, M.N., Petie, M., Heinsbroek, L.T.N., Verreth, J.A.J., Schrama, J.W., 2016. The effect of type of carbohydrate (starch vs. nonstarch polysaccharides) on nutrients digestibility, energy retention and maintenance requirements in Nile tilapia. Aquaculture 463, 241–247.

Hartman, K.J., Brandt, S.B., 1995. Estimating energy density of fish. Trans. Am. Fish. Soc. 124, 347–355.

Helland, S.J., Hatlen, B., Grisdale-Helland, B., 2010. Energy, protein and amino acid requirements for maintenance and efficiency of utilization for growth of Atlantic salmon post-smolts determined using increasing ration levels. Aquaculture 305, 150–158.

Hepher, B., 1988. Nutrition of Pond Fishes. Cambridge University Press, Cambridge, UK.

Houlihan, D.F., Boujard, D., Jobling, M., 2001. Food Intake in Fish. Blackwell Science, Oxford, UK.

Ivlev, V., 1939. Energy balance in the carp. Zool. zh. Mosk. 18, 303–318 (In Russian).

Jobling, M., Davies, P.S., 1980. Effects of feeding on metabolic rate, and the specifie dynamic action in plaice, *Pleuronectes platessa* L. J. Fish. Biol. 16, 629–638.

Jonsson, N., Jonsson, B., Hansen, L.P., 1997. Changes in proximate composition and estimates of energetic costs during upstream migration and spawning in Atlantic salmon *Salmo salar*. J. Anim. Ecol. 66 (3), 425–436.

Jonsson, N.J., Hansen, B., L. P., 1991. Energetic cost of spawning in male and female Atlantic salmon (*Salmo salar* L.). J. Fish. Biol. 39, 739–744.

Kamler, E., 2005. Parent–egg–progeny relationships in teleost fishes: an energetics perspective. Rev. Fish Biol. Fish. 15, 399–421.

Kaushik, S.J., Médale, F., 1994. Energy requirements, utilization and dietary supply to salmonids. Aquaculture 124, 81–97.

Kaushik, S.J., 1980. Influence of nutritional status on the daily patterns of nitrogen excretion in the carp (*Cyprinus carpio* L.) and the rainbow trout (*Salmo gairdneri* R.). Reprod. Nutr. Dev. 20, 1751–1765.

Kaushik, S.J., 1998. Nutritional bioenergetics and estimation of waste production in non-salmonids. Aquat. Living Resour. 11, 211–217.

Kaushik, S.J., Cowey, C.B., 1991. Dietary factors affecting nitrogen excretion by fish. In: Cowey, C.B., Cho, C.Y. (Eds.), Nutritional Strategies and Aquaculture Waste. Univ. Guelph, Canada, pp. 3–19.

Kaushik, S.J., Dabrowski, K., 1983. Postprandial metabolic changes in larval and juvenile carp (*Cyprinus carpio*). Reprod. Nutr. Dev. 23, 223–234.

Kaushik, S.J., Dabrowski, K., Luquet, P., 1982. Patterns of nitrogen excretion and oxygen consumption during ontogenesis of common carp (*Cyprinus carpio*). Can. J. Fish. Aquat. Sci. 39, 1095–1105.

Kaushik, S.J., Fauconneau, B., Blanc, J.M., 1984. A study of nitrogen excretion and oxygen consumption in five half-sib families of rainbow trout. Reprod. Nutr. Dev. 24, 431–438.

Kaushik, S.J., Gomes, E.F., 1988. Effect of frequency of feeding on nitrogen and energy balance in rainbow trout under maintenance conditions. Aquaculture 73, 3–4.

Kaushik, S.J., Luquet, P., 1977. Endogenous nitrogen loss from the muscle of rainbow trout starved in fresh water and sea water. Ann. Hydrobiol. 8, 129–134.

Kaushik, S.J., Luquet, P., 1984. Relationship between protein intake and voluntary energy intake as affected by body weight with an estimation of maintenance needs in rainbow trout. Z. für Tierphysiol. Tierernaehrung Futtermittelkd. 51, 57–69.

Kaushik, S.J., Luquet, P., Blanc, D., 1981. Usefulness of feeding protein and non-protein calories apart in studies on energy-protein interrelationships in rainbow trout. Ann. Zootech. 30, 3–11.

Kaushik, S.J., Oliva-Teles, A., 1985. Effect of digestible energy on nitrogen and energy balance in rainbow trout. Aquaculture 50, 89–101.

Kielanowski, J., 1965. Estimates of the energy cost of protein deposition in growing animals. In: Blaxter, K.L. (Ed.), Proceedings of the 3rd Symposium on Energy Metabolism. Academic Press, London, pp. 13–20.

Killen, S.S., Atkinson, D., Glazier, D.S., 2010. The intraspecific scaling of metabolic rate with body mass in fishes depends on lifestyle and temperature. Ecol. Lett. 13, 184–193.

Kim, J.D., Kaushik, S.J., 1992. Contribution of digestible energy from carbohydrates and estimation of protein/energy requirements for growth of rainbow trout (*Oncorhynchus mykiss*). Aqauculture 106, 161–169.

Kinghorn, B., 1983. Genetic variation in food conversion efficiency and growth in rainbow trout. Aquaculture 32, 141–155.

Kleiber, M., 1975. In: Robert E, K. (Ed.), The Fire of Life (New York).

Knap, P.W., Kause, A., 2018. Phenotyping for genetic improvement of feed efficiency in fish: lessons from pig breeding. Front. Genet. 9 (184), 1.

Knox, D., Bromage, N.R., Cowey, C.B., Springate, J.R.C., 1988. The effect of broodstock ration size on the composition of rainbow trout eggs (Salmo gairdneri). Aquaculture 69 (1–2), 93–104, 1988.

Kooijman, S.A.L.M., 2001. Quantitative aspects of metabolic organization: a discussion of concepts. Phil. Trans. R. Soc. Lond. B 356, 331–349.

Krogdahl, A., Sundby, A., Olli, J.J., 2004. Atlantic salmon (*Salmo salar*) and rainbow trout (*Oncorhynchus mykiss*) digest and metabolize nutrients differently. Effects of water salinity and dietary starch level. Aquaculture 229, 335–360.

Liu, J.S., Cui, Y.B., Liu, J.K., 2000. Resting metabolism and heat increment of feeding in mandarin fish (*Siniperca chuatsi*) and Chinese snakehead (*Channa argus*). Comp. Biochem. Physiol. Mol. Integr. Physiol. 127, 131–138.

Lupatsch, I., Cuthbertson, L., Davies, S., Shields, R.J., 2008. Studies on energy and protein requirements to improve feed management of the Pacific white shrimp, *Litopenaeus vannamei*. In: Elizabeth Cruz Suárez, L., D.R.M., Tapia Salazar, M., Nieto López, M.G., Villarreal Cavazos, D.A., Juan Pablo Lazò, y M., Teresa, V. (Eds.), Avances en Nutrición Acuicola IX, pp. 281–295.

Lupatsch, I., Kissil, G.W., 2005. Feed formulations based on energy and protein demands in white grouper (Epinephelus aeneus). Aquaculture 248, 83–95.

Lupatsch, I., Kissil, G.W., Sklan, D., 2001. Optimization of feeding regimes for European sea bass Dicentrarchus labrax: a factorial approach. Aquaculture 202, 289–302.

Lupatsch, I., Kissil, G.W., Sklan, D., 2003. Comparison of energy and protein efficiency among three fish species gilthead sea bream (*Sparus aurata*), European sea bass (*Dicentrarchus labrax*) and white grouper (*Epinephelus aeneus*): energy expenditure for protein and lipid deposition. Aquaculture 225, 175–189.

Lupatsch, I., Kissil, G.W., Sklan, D., 2003. Defining energy and protein requirements of gilthead seabream (*Sparus aurata*) to optimize feeds and feeding regimes. Isr. J. Aquacult. 55, 243–257.

Lupatsch, I., Kissil, G.W., Sklan, D., Pfeffer, E., 1998. Energy and protein requirements for maintenance and growth in gilthead seabream (*Sparus aurata* L.). Aquacult. Nutr. 4, 165–173.

Lupatsch, I., Santos, G.A., Schrama, J.W., Verreth, J.A.J., 2010. Effect of stocking density and feeding level on energy expenditure and stress responsiveness in European sea bass *Dicentrarchus labrax*. Aquaculture 298, 245–250.

Maas, R.M., Verdegem, M.C.J., Wiegertjes, G.F., Schrama, J.W., 2020. Carbohydrate utilisation by tilapia: a meta-analytical approach. Rev. Aquacult. 12, 1851–1866.

Maciak, S., Janko, K., Kotusz, J., Choleva, L., Boroń, A., Juchno, D., Kujawa, R., Kozłowski, J., Konarzewski, M., 2011. Standard metabolic rate (SMR) is inversely related to erythrocyte and genome size in allopolyploid fish of the *Cobitis taenia* hybrid complex. Funct. Ecol. 25, 1072–1078.

Martin, A.K., Blaxter, K.L., 1965. The energy cost of urea synthesis in sheep. In: Blaxter, K.L. (Ed.), Energy Metabolism. Academic Press, New York, pp. 83–91.

Mayes, P.A., Botham, K.M., 2003. Bioenergetics: the role of ATP. In: Murray, R.K., Graner, D.K., Mayes, P.A., Rodwell, V.W. (Eds.), Harper's Illustrated Biochemistry, 26th Edition. Lange Medical Books/McGraw-Hill, New York, pp. 80–87.

McCue, M.D., 2006. Specific dynamic action: a century of investigation. Comp. Biochem. Physiol. a-Mol. Integr. Physiol. 144, 381–394.

McGoogan, B.B., Gatlin, D.M., 1998. Metabolic requirements of red drum, *Sciaenops ocellatus*, for protein and energy based on weight gain and body composition. J. Nutr. 128, 123–129.

Médale, F., Aguirre, P., Kaushik, S.J., 1991. Utilization of dietary carbohydrates by rainbow trout at two water temperatures. In: Wenk, C., Boessinger, M. (Eds.), Energy Metabolism of Farm Animals, 58. European Association for Animal Production, pp. 392–395.

Médale, F., Brauge, C., Vallée, F., Kaushik, S.J., 1995. Effects of dietary protein/energy ratio, ration size, dietary energy source and water temperature on nitrogen excretion in rainbow trout. Water Sci. Technol. 31, 185–194.

Mommsen, T.P., 2004. Salmon spawning migration and muscle protein metabolism: the August Krogh principle at work. Comp. Biochem. Physiol. B Biochem. Mol. Biol. 139, 383–400.

Mommsen, T.P., French, C.J., Hochachka, P.W., 1980. Sites and atterns of protein and amino acid utilization during th spawning migraton of salmon. Can. J. Zool. 58, 1785–1799.

Mylonas, C.C., Sigelaki, I., Divanach, P., Mananõs, E., Carrillo, M., Afonso-Polyviou, A., 2003. Multiple spawning and egg quality of individual European sea bass (*Dicentrarchus labrax*) females after repeated injections of GnRHa. Aquaculture 221, 605–620.

Nassour, I., Léger, C.L., 1989. Deposition and mobilisation of body fat during sexual maturation in female trout (Salmo gairdneri Richardson). Aquat. Living Resour. 2, 153–159.

National Research Council (NRC), 2011. Nutrient requirements of fish and shrimp. National Academy of Sciences. Washington, DC, USA.

Noblet, J., 2006. Recent advances in energy evaluation of feeds for pigs. In: Garnsworthy, J.W. (Ed.), Recent Advances in Animal Nutrition 2005. Nottingham University Press, Nottingham, UK, pp. 1–26.

Noblet, J., Fortune, H., Shi, X., Dubois, S., 1994. Prediction of net energy value of feeds for growing pigs. J. Anim. Sci. 72, 344–354.

Noblet, J., van Milgen, J., 2004. Energy value of pig feeds: effect of pig body weight and energy evaluation system. J. Anim. Sci. 82 (13, E. Suppl.), E229–E238.

Ogino, C., Chen, M.S., 1973. Protein nutrition in fish .3. Apparent and true digestibility of dietary proteins in carp. Bull. Jpn. Soc. Sci. Fish. 39, 649–651.

Oliva-Teles, A., Kaushik, S.J., 1987a. Nitrogen and energy metabolism during the early ontogeny of diploid and triploid rainbow trout (Salmo gairdneri R.). Comp. Biochem. Physiol. A 87A, 157–160.

Oliva-Teles, A., Kaushik, S.J., 1987b. Metabolic utilization of diets by polyploid rainbow trout (Salmo gairdneri). Comp. Biochem. Physiol. A 88, 45–47.

Parker, P., Anderson, R., Lawrence, A., 1989. A δ 13 C and δ 15 N Tracer Study of Nutrition in Aquaculture: *Penaeus Vannamei* in a Pond Growout System, Stable Isotopes in Ecological Research. Springer, pp. 288–303.

Parra, L., Lloret, G., Lloret, J., Rodilla, M., 2018. Physical sensors for precision aquaculture: a review. IEEE Sensor. J. 18 (10), 3915–3923. https://doi.org/10.1109/JSEN.2018.2817158.

Pirozzi, I., Booth, M.A., Allan, G.L., 2010. Protein and energy utilization and the requirements for maintenance in juvenile mulloway (*Argyrosomus japonicus*). Fish Physiol. Biochem. 36, 109–121.

Phan, L., Groot, R., Konnert, G.D., Masagounder, K., Figueiredo-Silva, A.C., Glencross, B.D., Schrama, J.W., 2019. Differences in energy utilisation efficiencies of digestible macronutrients in common carp (*Cyprinus carpio*) and barramundi (*Lates calcarifer*). Aquaculture 511, 734738.

Phan, L., Masagounder, K., Mas-Muñoz, J., Schrama, J., 2021. Differences in energy utilization efficiency of digested protein, fat and carbohydrates in snakehead (*Channa striata*). Aquaculture. 532, 736066.

Phillips, A.J., 1972. Calorie and energy requirements. In: Halver, J.E. (Ed.), Fish Nutrition. Academic Press, New York, pp. 2–28.

Ramseyer, L.J., 2002. Predicting whole-fish nitrogen content from fish wet weight using regression analysis. N. Am. J. Aquacult. 64, 195–204.

Rodehutscord, M., Pfeffer, E., 1999. Maintenance requirement for digestible energy and efficiency of utilisation of digestible energy for retention in rainbow trout, *Oncorhynchus mykiss*. Aquaculture 179, 95–107.

Rosas, C., Sanchez, A., Díaz, E., Soto, L.A., Gaxiola, G., Brito, R., Baes, M., Pedroza, R., 1995. Oxygen consumption and ammonia excretion of *Penaeus setiferus, P. schmitti, P. duorarum* and *P. notialis* postlarvae fed purified test diets: effect of protein level on substrate metabolism. Aquat. Living Resour. 8, 161–169.

Sadoul, B., Geffroy, B., Lallement, S., Kearney, M., 2020. Multiple working hypotheses for hyperallometric reproduction in fishes under metabolic theory. Ecol. Model. 433, 109228.

Santos, G.A., Schrama, J.W., Mamauag, R.E.P., Rombout, J.H.W.M., Verreth, J.A.J., 2010. Chronic stress impairs performance, energy metabolism and welfare indicators in European seabass (*Dicentrarchus labrax*): the combined effects of fish crowding and water quality deterioration. Aquaculture 299, 73–80.

Sampaio, L.A., Bianchini, A., 2002. Salinity effects on osmoregulation and growth of the euryhaline flounder Paralichthys orbignyanus. J. Exp. Mar. Biol. Ecol. 269, 187–196.

Saravanan, S., Geurden, I., Figueiredo-Silva, A.C., Nusantoro, S., Kaushik, S., Verreth, J., Schrama, J.W., 2013b. Oxygen consumption constrains food intake in fish fed diets varying in essential amino acid composition. PloS One 8 (8), e72757.

Saravanan, S., Schrama, J., Figueiredo-Silva, A.C., Kaushik, S.J., Verreth, J.A.J., 2012. Constraints on energy intake in fish: the link between diet composition, energy metabolism, and energy intake in rainbow trout. PloS One 7 (4), e34743.

Saravanan, S., Geurden, I., Figueiredo-Silva, A.C., Kaushik, S., Verreth, J., Schrama, J.W., 2013a. Voluntary feed intake in rainbow trout is regulated by diet-induced differences in oxygen use. J. Nutr. 143, 781–787.

Saunders, R.L., 1963. Respiration of the Atlantic cod. J. Fish. Res. Board Can. 20, 373–386.

Schrama, J.W., Haidar, M.N., Geurden, I., Heinsbroek, L.T., Kaushik, S.J., 2018. Energy efficiency of digestible protein, fat and carbohydrate utilisation for growth in rainbow trout and Nile tilapia. Br. J. Nutr. 119, 782–791.

Schrama, J.W., Saravanan, S., Geurden, I., Heinsbroek, L.T., Kaushik, S.J., Verreth, J.A., 2012. Dietary nutrient composition affects digestible energy utilisation for growth: a study on Nile tilapia (*Oreochromis niloticus*) and a literature comparison across fish species. Br. J. Nutr. 108, 277–289.

Schwarz, F.J., Kirchgessner, M., 1995. Effects of different diets and levels of feeding on retention and efficiency of utilization of energy and protein by carp (*Cyprinus carpio* L.). J. Appl. Ichthyol. 11, 363–366.

Shearer, K.D., 1994. Factors affecting the proximate composition of cultured fishes with emphasis on salmonids. Aquaculture 119, 63–88.

Simbeye, D.S., Yang, S.F., 2014. Water quality monitoring and control for aquaculture based on wireless sensor networks. J. Network. 9, 840.

Sklan, D., Prag, T., Lupatsch, I., 2004. Apparent digestibility coefficients of feed ingredients and their prediction in diets for tilapia *0* x *Oreochromis aureus* (Teleostei, Cichlidae). Aquacult. Res. 35, 358–364.

Smith, R.R., 1971. A method for measuring digestibility and metabolizable energy of fish feeds. Progressive Fish-Culturist. 33, 132–134.

Smith, R.R., 1989. Nutritional energetics. In: Halver, J.E.H. (Ed.), Fish Nutrition. Academic Press, San Diego, pp. 1–29.

Smith, R.R., Rumsey, G.L., Scott, M.L., 1978a. Heat increment associated with dietary protein, fat, carbohydrate and complete diets in salmonids: comparative energetic efficiency. J. Nutr. 108, 1025–1032.

Smith, R.R., Rumsey, G.L., Scott, M.L., 1978b. Net energy maintenance requirements of salmonids as measured by direct calorimetry: effect of body size and environmental temperature. J. Nutr. 108, 1017–1024.

Spoor, W.A., 1941. A method for measuring the activity of fishes. Ecology 22 (3), 329–331. Brett, J.R., 1973. Energy expenditure of Sockeye salmon, *Oncorhynchus nerka*, during sustained performances. Journal of Fisheries Research Board of Canada 30, 1799-1809.

Staessen, T.W.O., Verdegem, M.C.J., Weththasinghe, P., Schrama, J.W., 2020. The effect of dietary non-starch polysaccharide level and bile acid supplementation on fat digestibility and the bile acid balance in rainbow trout (*Oncorhynchus mykiss*). Aquaculture 523, 735174.

Storebakken, T., Austreng, E., 1987. Ration level for salmonids. II. Growth, feed intake, protein digestibility, body composition, and feed conversion in rainbow trout weighing 0.5 - 1.0 kg. Aquaculture 60, 207–221.

Storebakken, T., Hung, S.S.O., Calvert, C.C., Plisetskaya, E.M., 1991. Nutrient partitioning in rainbow trout at different feeding rates. Aquaculture 96, 191–203.

Takeuchi, T., Watanabe, T., Ogino, C., Saito, M., Nishimura, K., Nose, T., 1981. A long-term feeding with rainbow trout by a low protein diet with a high energy value. Bull. Jpn. Soc. Sci. Fish. 47, 637–643.

Tandler, A., Beamish, F.W.H., 1979. Mechanical and biochemical components of apparent specifie dynamic action in largemouth bass, *Micropterm salmoides* Laceepede. J. Fish. Biol. 14, 343–350.

Tandler, A., Harel, M., Koven, W.M., Kolkovski, S., 1995. Broodstock and larvae nutrition in gilthead seabream *Sparus aurata* - new findings on its mode of involvement in improving growth, survival and swimbladder inflation. Isr. J. Aquacult.Bamidgeh 47, 95–111.

Tran Duy, A., Schrama, J.W., Dam, A.A. van, Verreth, J.A.J., 2008. Effects of oxygen concentration and body weight on maximum feed intake, growth and hematological parameters of Nile tilapia. *Oreochromis niloticus*. Aquaculture 275, 152–162.

Tran-Ngoc, T.K., Dinh, N.T., Tin, N.H., Roem, A.J., Schrama, J.W., Verreth, J.A.J., 2016. Interaction between dissolved oxygen concentration and diet composition on growth, digestibility and intestinal health of Nile tilapia (*Oreochromis niloticus*). Aquaculture 462, 101–108.

Tran-Ngoc, K., Schrama, J.W., Le, M.T.T., Nguyen, T.H., Roem, A.J., Verreth, J.A.J., 2017. Salinity and diet composition affect digestibility and intestinal morphology in Nile tilapia (*Oreochromis niloticus*). Aquaculture 469, 36–43.

Tseng, Y.C., Hwang, P.P., 2008. Some insights into energy metabolism for osmoregulation in fish. Comp. Biochem. Physiol. 148C, 419–429.

Tytler, P., Calow, P., 1985. Fish Energetics. New Perspectives. Croom Helm, London.

Van Den Thillart, G., Van Ginneken, V., Körner, F., Heijmans, R., Van der Linden, R., Gluvers, A., 2004. Endurance swimming of European eel. J. Fish. Biol. 65, 312–318.

van Milgen, J., Noblet, J., 1999. Energy partitioning in growing pigs: the use of a multivariate model as an alternative for the factorial analysis. J. Anim. Sci. 77, 2154–2162.

Van Trung, D., Diu, N.T., Hao, N.T., Glencross, B., 2011. Development of a nutritional model to define the energy and protein requirements of tilapia. *Oreochromis niloticus*. Aquaculture 320, 69–75.

Watanabe, K., Aoki, H., Yamagata, Y., Kiron, V., Satoh, S., Watanabe, T., 2000. Energy and protein requirements of yellowtail during winter season. Fish. Sci. 66, 521–527.

Watanabe, T., Ohta, M., 1995. Endogenous nitrogen excretion and non-fecal energy losses in carp and rainbow trout. Fish. Sci. 61, 53–60.

Watanabe, T., Takeuchi, T., Saito, M., Nishimura, K., 1984. Effect of low protein-high calory or essential fatty acid deficiency diet on reproduction of rainbow trout. Bull. Jpn. Soc. Sci. Fish. 50, 1207–1215.

Watanabe, T., Takeuchi, T., Satoh, S., Kiron, V., 1996. Digestible energy methodological influences and a mode of calculation. Fish. Sci. 62, 288–292.

Webb, K.A., Gatlin, D.M., 2003. Effects of dietary protein level and form on production characteristics and ammonia excretion of red drum Sciaenops ocellatus. Aquaculture 225, 17–26.

Webb, P., 1971. The swimming energetics of trout: II. Oxygen consumption and swimming efficiency. J. Exp. Biol. 55, 521–540.

White, C.R., Kearney, M.R., 2013. Determinants of inter-specific variation in basal metabolic rate. J. Comp. Physiol. B 183, 1–26.

White, C.R., Phillips, N.F., Seymour, R.S., 2006. The scaling and temperature dependence of vertebrate metabolism. Biol. Lett. 2, 125–127.

Wilson, R.P., Robinson, E.H., Poe, W.E., 1981. Apparent and true availability of amino acids from common feed ingredients for channel catfish. J. Nutr. 111, 923–929.

Wood, C., Perry, S., Wright, P., Bergman, H., Randall, D., 1989. Ammonia and urea dynamics in the Lake Magadi tilapia, a ureotelic teleost fish adapted to an extremely alkaline environment. Respir. Physiol. 77, 1–20.

Wood, C.M., 2001. Influence of feeding, exercise, and temperature on nitrogen metabolism and excretion. Fish Physiol. 20, 201–238.

Xie, S., Zheng, K., Chen, J., Zhang, Z., Zhu, X., Yang, Y., 2011. Effect of water temperature on energy budget of Nile tilapia, *Oreochromis niloticus*. Aquacult. Nutr. 17, e683–e690.

Yamamoto, T., Shima, T., Unuma, T., Shiraishi, M., Akiyama, T., Tabata, M., 2000. Voluntary intake of diets with varying digestible energy contents and energy sources, by juvenile rainbow trout Oncorhynchus mykiss, using self-feeders. Fish. Sci. 66 (3), 528–534.

Zhou, Z., Cui, Y., Xie, S., Zhu, X., Lei, W., Xue, M., Yang, Y., 2003. Effect of feeding frequency on growth, feed utilization, and size variation of juvenile gibel carp (*Carassius auratus gibelio*). J. Appl. Ichthyol. 19, 244–249.

Zhou, Z.G., Xie, S.Q., Lei, W., Zhu, X.M., Yang, Y.X., 2005. A bioenergetic model to estimate feed requirement of gibel carp, *Carassius auratus* gibelio. Aquaculture 248, 287–297.

Zion, B., 2012. The use of computer vision technologies in aquaculture – a review. Comput. Electron. Agric. 88, 125–132.

CHAPTER

3

Vitamins

Kangsen Mai[1], R. Waagbø[2], X.Q. Zhou[3], Q.H. Ai[1] and L. Feng[3]

[1]Ocean University of China, Qingdao, Shandong, China; [2]Institute of Marine Research/ University of Bergen, Bergen, Norway; [3]Sichuan Agricultural University, Chengdu, Sichuan, China

3.1 Introduction

3.1.1 Vitamins and their sources

Knowledge of the existence and essentiality of vitamins had its start from the study of human nutrient deficiencies when certain foods or food fractions were omitted from the diet. A classic example was the discovery of a link between consumption of polished (dehulled) white rice and beriberi, a progressive disorder characterized by numbness and weakness in limbs that plagued Japan, China, and other Southeast Asian countries. Eijkman, working in Jakarta, created an animal model of beriberi by feeding chickens a diet of polished rice, allowing the disease condition to be studied in depth. Funk showed that beriberi in pigeons could be cured by feeding a concentrate made from rice polishings in combination with polished rice and proposed that the essential factor in rice polishing extracts was an amine. This result was published in a paper on the etiology of deficiency diseases in 1912, and he introduced the term "vitamin" (from vital amine) to describe this category of essential micronutrients. The active component of the water-soluble factor from rice polishings was later identified as thiamin, or vitamin B_1. Eijkman received the Nobel Prize for Physiology and Medicine in 1938 for his work on beriberi. Other vitamins were subsequently discovered in a variety of foods. For example, Waugh and King identified vitamin C (ascorbic acid) as the compound in limes and lemons that cured scurvy in 1932. In general, the focus of vitamin research in both human medicine and animal science was on curing diseases associated with insufficient vitamin intake. This approach influenced the way in which dietary requirements were established, i.e., the dietary requirement was the level at which deficiency signs were not detected and animal growth was normal. It also led to the concept of one nutrient–one disease, meaning that each vitamin deficiency was associated with one specific deficiency syndrome. Continuous advances in understanding of cell and organism metabolism have

Fish Nutrition, Fourth Edition
https://doi.org/10.1016/B978-0-12-819587-1.00014-8

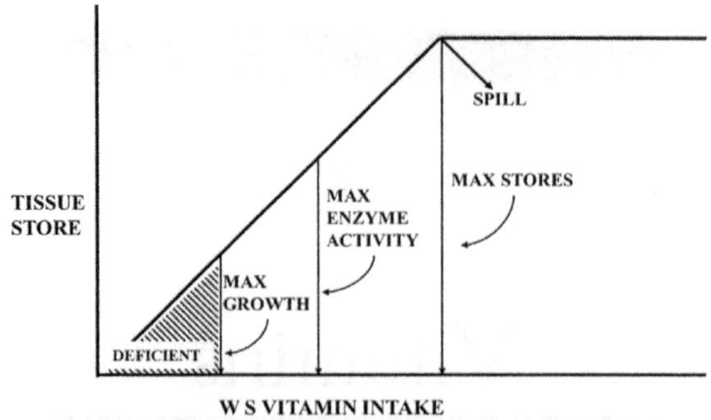

FIGURE 3.1 The assessment of water-soluble vitamin requirements.

shown that the definition of a dietary requirement depends on the response variable being measured (Fig. 3.1). The intake of individual B-vitamins associated with absence of deficiency signs and normal growth is generally less than the intake level required for tissue saturation. If activity of enzymes for which specific vitamins are essential cofactors is measured, higher dietary intake is often necessary to reach a level where further increases do not increase enzyme activity. The maximum expression of genes related to the functions of a certain vitamin may be a more sensitive indicator for determining the vitamin requirement. However, so far, the research in this area is still very limited. Studies on Jian carp (Wu et al., 2011) and grass carp (Li et al., 2017), respectively, found that the requirements of choline and inositol based on gene expression levels were higher than those based on growth. Identifying essential vitamins and estimating optimum dietary intake of essential vitamins and vitamin-like substances was a productive and extremely important area of research in livestock and human medicine for the first half of the 20th century. In the case of fish nutrition, much remains to be discovered.

Vitamins are a collective term for a group of organic compounds which are essential micronutrients required by organisms. They do not yield energy like sugars, proteins, and fats, but they are essential cofactors for many cellular processes and pathways, and essential for cell membrane integrity, endocrine system operation, and immune system functions, to name a few. There are many natural sources of various vitamins, including plant, animal, and microbial sources (Table 3.1). Vitamins are also supplemented to the feed in different chemical forms. The vitamins in feed ingredients and additives appear in different chemical forms or vitamers that may differ in vitamin activity. However, the crystalline and coated forms of water-soluble vitamins loss at different degrees during feed production (pelleting or extrusion) and storage (Table 3.2).

3.1.2 Classification and functions

Vitamins were first categorized as water-soluble or fat-soluble based on the original extraction of food sources using water or nonpolar solvents that yielded two crude extracts having vitamin-like activity. Subsequent research revealed that both fractions contained several compounds having distinct vitamin-like properties. Water-soluble vitamins are readily soluble in water but not soluble in nonpolar organic solvents. After absorption, they are not stored in significant amounts in the body and excreted in the urine. The water-soluble vitamins include eight well-recognized members of the vitamin B complex, plus choline,

TABLE 3.1 Overview over B-vitamins in selected ingredients used in fish and crustaceans' feeds and historical ranges of requirement estimates in fish and shrimp.

		Vitamins								
Source	**Specification**	**Thiamin**	**Riboflavin**	**Niacin**	**Pantothenate**	**Pyridoxine**	**Biotin**	**Folic acid**	**Cobalamin**	**References**
Complete diets										
Test diets	Test diet H-440	5	20	75	50	5	0.5	1.5	0.01	Halver and Hardy (2002)
Commercial diet	Typical complete feed	15	25	180	50	15	0.6	8	0.03	Gabaudan and Hardy (2000)
Live feeds										
Rotifers		2–57	22–44	191–249	86	20–25	1.6–1.8	4.0–5.7	23–43	Rønnestad and Hamre (2001)
Artemia		3–12	27–60	160–250	60–180	2–33	2–5	6–11	2–5	Hamre et al. (2008, 2013)
Copepods		13–23	14–27	100–150	20–40	2–6	0.6–0.9	3–5	1–2	
Feed ingredients										
Fish meal	Herring	0.4	9.7	85	17.3	4.8	0.5	0.3	0.43	NRC (2011)
Blue mussel	Meat meal								0.55	Arnason et al. (2015)
Kelp meal	*Ascophyllum nodosum*	1	5	23	7	–	0.1	0.1	0	NRC (2011)
Spirulina meal	Meal	0.5–5.8	4.5–4.7	15	0.1	0.9–1.0	0.01	9.9	0.16–0.18	Ragaza et al. (2020)
Microalgae	Four species	29–109	25–50			3.6–17	1.1–1.9	17–24	1.7–1.95	Brown et al. (1999)
Poultry	Byproduct meal	5.7	9.1	178	46.6	6.4	0.3	0.7	0	NRC (2011)
Blood meal	Spray-dried	0.3	2.9	22	3.2	4.5	0.3	0	0.013	NRC (2011)
Corn gluten	60 CP	0.3	2	60	3.5	6.9	0.2	0.3	0	NRC (2011)
Soy	SPC	6	2.9	28	16.3	6	0.3	0.6	0	NRC (2011)
Wheat	Grain	4.5	1.3	53	10.1	3	0.1	0.4	0	NRC (2011)
Insect meal	Four species	<0.01–11	16–77	34–91	27–45	1.7–6.1	0.35–0.68	0.8–2.7	0.005–0.237	Finke (2013)

Continued

TABLE 3.1 Overview over B-vitamins in selected ingredients used in fish and crustaceans' feeds and historical ranges of requirement estimates in fish and shrimp.—cont'd

		Vitamins								
Source	**Specification**	**Thiamin**	**Riboflavin**	**Niacin**	**Pantothenate**	**Pyridoxine**	**Biotin**	**Folic acid**	**Cobalamin**	**References**
Vitamin requirements										
Fish species										
	NRC (2011)	0.5–15	2.7–25	1–200	10–50	1–20	0.05–2.49	0.8–10	0.015–0.094	NRC (2011)
	Freshwater	0.5–10	4–9	10–150	10–30	3–15	0.06–1	1–1.5	0.02	NRC (2011)
	Seawater	11	11	12	36	12	0.67	1.2	0.05	NRC (2011)
Crustacean species										
Penaeid shrimps		120	40	150	100	120	1	5	<0.1	Conklin (1989)
Penaeid shrimp spp		13–120	23–80	7–400	nd	80–120	nd	2–8	0.2	Shiau (1998)
	NRC (2011)	14–120	23–80	7.2–400	100–750	72–120	0.4–2.4	1.9–5	0.01–0.2	NRC (2011)
Penaeus indicus	Commercial diets	20	16	26	30	20	0.4	4	0.04	Evans (2009)
P. indicus	Commercial diets	20	40	100	40	20	0.5	8	0.04	Rotosquare (2009)[a]
P. indicus	Commercial diets	90	75	210	150	75	0.3	24	0.045	GreenLabel (2004)[b]
Penaeus monodon	Commercial diets	26	50	140	52	32	0.6	9	0.056	Rotosquare (2009)[a]
P. monodon	Commercial diets	14	22.5	7.2	139	40.5	2.4	2.1	0.2	FAO.org
Shrimp	CIBA (2002)	15–30	15–60	60–120	60–120	60–120	0.05–0.5	6–10	0.01–0.02	CIBA (2002) [c]
Giant river prawn—*Macrobrachium rosenbergii*					Probably similar to other crustaceans					FAO.org

[a]*Rotosquare Enterprises Ltd., Taiwan (2009) (www.FAO.org).*
[b]*Green Label (see www.FAO.org).*
[c]*Central Institute of Brackishwater Aquaculture, Chennai, India (see www.FAO.org).*
CP: Crude Protein; SPC: Soy protein concentrate.

TABLE 3.2 Stability (% of added) of B-vitamins in pelleted and extruded diets after 3 months of storage at room temperature (Gabaudan and Hardy, 2000).

Vitamin	Supplemented vitamin form	Steam pellets	Extruded pellets
Thiamin	Mononitrate or hydrochloride salts	85–100	60–80
Riboflavin	Spay-dried powder of crystalline or fermented products	90–100	90–100
Niacin	Nicotinamide or nicotinic acid products	90–100	90–100
Pantothenic acid	Calcium D-pantothenate salt	90–100	80–100
Pyridoxine	Crystalline hydrochloride salt	90–100	80–90
Biotin	D-Biotin product	90–100	70–90
Folate	Crystalline or spray-dried folic acid products	70–90	50–65
Cobalamin	Fermented product	60–90	40–80

inositol, and ascorbic acid for a total of 11. Historical literature shows that all fish and shrimp species require water-soluble vitamins. The dietary requirement for the eight members of vitamin B complex is small (Table 3.1), but the roles of these vitamins in the growth, physiology, and metabolism are quite important. Fat-soluble vitamins are easily soluble in nonpolar organic solvents but are not easily soluble in water. In contrast to water-soluble vitamins, they are stored in the body and have a low excretion rate. The fat-soluble vitamins A, D, E, and K occur in different chemical forms to exert different physiological functions. There are also several vitamin-like compounds with less-defined activities in fish, such as lipoic acid (LA), astaxanthin, p-aminobenzoic acid, and citrine.

In general, vitamins not only act as coenzymes and participate in biochemical reactions of the body but also serve as the raw materials of metabolic substrates to regulate the body's metabolism or are components of structural compounds in cells and tissues. For example, most of the vitamin B complex are coenzymes and essential for enzymes in the glycolysis, tricarboxylic acid cycle, and the final oxidative phosphorylation, reducing the nutrient substrates to energy, carbon dioxide, and water (Table 3.3; Fig. 3.2). Vitamins A and K are involved in maintaining the functions of epithelial cells and the synthesis of messenger RNA in the process of blood-clotting protein synthesis, respectively. Besides, feeding studies have examined the potential of supplementing vitamins beyond requirement levels to positively influence fish and shrimp physiology, stress resistance, and immunity (Nakagawa et al., 2007; Waagbø and Remø, 2020). Furthermore, novel functions of the water-soluble vitamins are included in the expression *nutrigenomics* which is concerned with the impact of dietary vitamins on the system biology, or the genome, the transcriptome, the proteome, and the metabolome (Martin and Król, 2017). In human nutrition, for example, the vitamins folate (substrate) and riboflavin (coenzyme) can modify the reaction that regulates methylation of cytosine in DNA and control of gene expression through modulating methylenetetrahydrofolate reductase. In the same line, deficiency of vitamins taking part in the one-carbon (1C) metabolism in parental zebrafish affected fecundity and the embryonic gene expressions

TABLE 3.3 B-vitamins overview—active coenzyme forms, biochemical roles, status biomarkers, and clinical deficiency signs in fish and crustaceans.

Vitamin	Trivial name	Active coenzyme	Main biochemical roles	Main importance for nutrient	Vitamin status biomarker	Exposed organs at deficiency	Specific signs of deficiency in fish	Specific signs of deficiency in shrimp
Thiamin	Vitamin B1	Thiamin pyrophosphate (TPP)	Oxidative decarboxylation of α-keto acids, interconvention of phosphorylated carbohydrates	Carbohydrates	Tissue/blood transketolase, pyruvate dehydrogenase, α-ketoglutarate dehydrogenase activity; liver/muscle thiamin status	Neural and muscle tissues	Hyperirritability, convulsions, skin and fin hemorrhage	Reduction in growth and feed efficiency, mortality
Riboflavin	Vitamin B2	Flavin adenine dinucleotide (FAD) and flavin mononucleotide (FMN)	Electron and hydrogen transfer reactions in numerous redox reactions in nutrient metabolism and energy production, antioxidation	Lipids, proteins, carbohydrates, energy	Tissue/plasma erythrocyte glutathione reductase, D-amino acid oxidase activity; liver/muscle riboflavin status	Skin, mucosal tissue, neural tissue, eye	Cataract, skin, and fin erosions/hemorrhage	Light pigmentation, exoskeleton malformation, irritability
Niacin	Vitamin B3 (Nicotinic acid, nicotinamide)	Nicotinamide adenine dinucleotide (NAD) and nicotinamide adenine dinucleotide phosphate (NADP)	Electron and hydrogen transfer reactions in numerous redox reactions in nutrient metabolism and energy production; ADP ribose transfer	Lipids, proteins, carbohydrates, energy	Tissue niacin status	Skin, mucousal tissue, neural tissue, muscle	Skin and fin erosions/hemorrhage, fatigue, exophthalmia	
Pantothenic acid	Vitamin B5	Coenzyme A (CoA)	Acyl transfer in fatty acid metabolism; acetylation/acylation of substrates; gene expression	Carbohydrates, proteins, and fatty acids	Liver/gill pantothenic acid status	Gills, skin, neural tissue	Clubbed gills, skin, and fin hemorrhage, exophthalmia	Reduction in growth, mortality, thin and soft shell, light coloration, irritability

Pyridoxine	Vitamin B6 (pyridoxine, pyridoxal, pyridoxamine)	Pyridoxal phosphate (PLP) and pyridoxamine phosphate (PMP)	Metabolism of amino acids, glycogen, and sphingoid bases for energy production; gene expression	Protein, amino acids, lipids	Tissue aspartate aminotransferase/alanine aminotransferase (ASAT/ALAT) activity; liver/muscle vitamin B6 status	Muscle, skin, neural tissue, blood cells	Convulsions, erratic swimming
Biotin	Vitamin B7, Vitamin H	Carboxybiotin	Carbon dioxide transfer/carboxylations in nutrient metabolism and energy production; gene expression	Lipids, carbohydrates	Liver pyruvate carboxylase activity; Liver/muscle biotin status	Skin, gills, liver, neural tissue	Clubbed gills, muscle atrophy, nervous dysfunction
Folate	Vitamin B9	Tetrahydrofolate glutamate (THF) derivatives	Transfer of one-carbon compounds for nutrient metabolism and energy production; nucleotide synthesis for DNA and RNA; hematopoiesis; gene expression	Amino acids, nucleic acids	Blood hemoglobin; blood smear; liver folate status	Blood cells, skin, neural tissue, cell growth	Macrocytic/megaloblastic anemia, exophthalmia
Cobalamin	Vitamin B12	Methylcobalamin, adenosylcobalamin	Methylations in nutrient metabolism, nucleic acids, and energy production; hematopoiesis	Amino acids, nucleic acids, fatty acids	Blood hemoglobin; blood smear; liver cobalamin status	Blood cells, neural tissue, cell growth	Pernicious/megaloblastic anemia

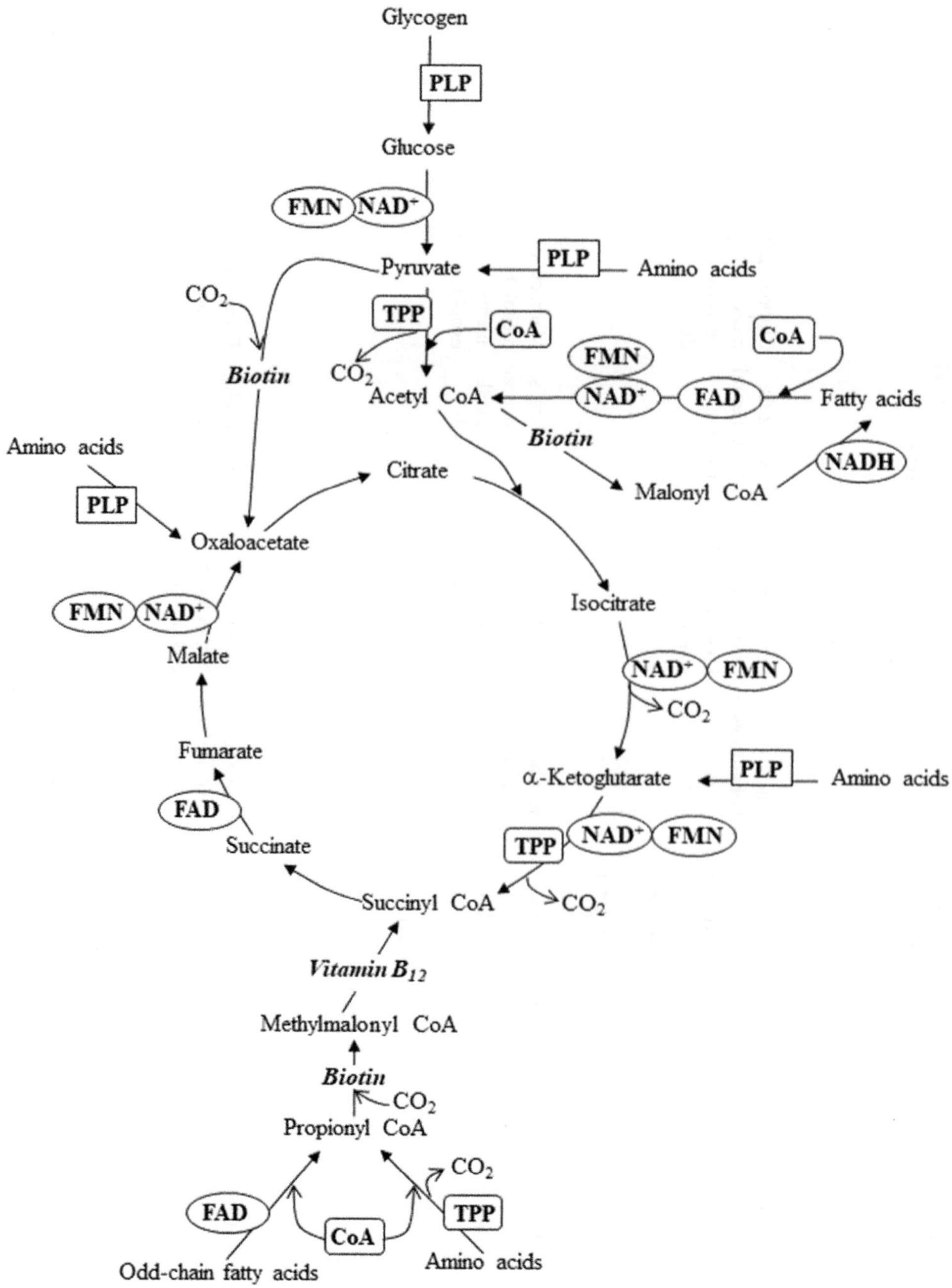

FIGURE 3.2 The B-vitamins participate and catalyze many biochemical reactions in cellular aerobic and anaerobic metabolism of the energy-giving substrates from amino acids, fatty acids, and glucose (see Table 3.3 for vitamin coenzyme abbreviations). Reprinted from Mæland (2000) Water-soluble vitamins in the nutrition of Atlantic halibut *Hippoglossus hippoglosssus* L. larvae and juveniles. PhD thesis, University of Bergen, Norway; with personal permission.

in biological processes related to overall health and regulation developmental processes and was also visualized as inherited deviations in the 1C cycle and lipid metabolism (Skjærven et al., 2018).

3.1.3 Avitaminosis and hypervitaminosis

Because vitamins play an indispensable role in life activities, a deficiency in one or more can lead to clinical signs of disease. Typical avitaminosis signs can be general, such as reduced feed intake, reduced growth and anemia, or specific to an individual vitamin, such as Chastek-type paralysis, cataracts, convulsions, scoliosis, slime patch disease, and clubbed gills. Although severe B-vitamin deficiencies are rarely observed in aquaculture today, recent studies have shown that subclinical deficiencies or suboptimal feed levels occur more frequently (Hemre et al., 2016).

Although vitamins are essential, an excess dietary supply, called hypervitaminosis, can cause adverse effects. Hypervitaminosis of vitamins A and D was commonly induced when seal and whale liver, both very rich sources of these vitamins, were used as one of the fresh meat components in salmon diets. For shrimp species, overfortification of some B-vitamins, like riboflavin, niacin, and pyridoxine, can result in reduced growth.

3.1.4 History of vitamin requirement studies in fish

During the early stages of development of purified and practical diets for fish, anemia was a common problem unless fresh animal organ meat, such as liver, was included in the diet. In 1927, McCay and Dilley at Cornell University worked to identify the compound or compounds in fresh land animal meat that prevented anemia, which they termed antianemic factor H. Twenty years elapsed, however, before this water-soluble factor was identified as a combination of vitamin B_{12} and folic acid. In 1934, McCay and Tunison observed scoliosis and lordosis in brook trout fed formalin-preserved meat, but the symptoms took nearly 1 year to develop and did not correlate with dietary intake of the recently identified vitamin C. Thereafter, Tunison et al. measured the level of thiamin, riboflavin, and nicotinic acid in the liver, pyloric caecums, kidney, and muscle to establish baselines for the requirements of trout for these vitamins in 1942.

McLaren et al. (1947a) developed a vitamin test diet containing crystalline vitamins, casein, dextrin, and oils, with crab meal or dried liver as the source of the antianemic factor. These pioneer fish nutritionists were able first to report tentative, qualitative, and later estimates of quantitative requirements of rainbow trout for thiamin, riboflavin, pyridoxine, pantothenic acid, inositol, biotin, folic acid, choline, and niacin, using fish weight gain and feed conversion ratios as response variables in feeding trials (McLaren et al., 1947b). Their test diet did furnish some control over many vitamins but was not entirely vitamin-free. In 1951, Wolf developed a test diet for rainbow trout which contained commercial casein, gelatin, potato starch, hydrogenated cottonseed oil, α-cellulose flour, minerals, cod liver oil, and crystalline vitamins. The diet was used successfully to induce several vitamin deficiency syndromes by deleting one vitamin at a time from the vitamin supplement and feeding the resultant diets to trout. The growth performance of Chinook salmon fingerlings feeding this diet was similar to those feeding Wolf's test diet (Halver, 1957). Later this original, complete vitamin test diet, improved by lowering the protein content, was used for short-term feeding studies with coho salmon and sockeye salmon, and for long-term feeding studies for at least three reproductive cycles with rainbow trout. The tool was forged to accurately determine the quantitative vitamin requirements of salmonids. Further, specific deficiency syndromes for many vitamins were described that provided diagnostic tools to identify vitamin deficiencies in hatcheries and farms. Eleven water-soluble vitamins were identified as being required by salmon (Halver, 1957). Qualitative water-soluble vitamin requirements were also identified for carp (Ogino, 1967), Japanese sea bass (Ai et al., 2004), large yellow croaker (Ai et al., 2006), and cobia (Mai et al., 2009). Modifications of these diets have been used for many other fish species to determine their qualitative or quantitative vitamin requirements. A summary of vitamin requirements and functions in fish metabolism can be found in successive US National Academy of Sciences (NAS)/National Research Council (NRC) bulletins, and standard techniques for conducting vitamin requirement studies were included in the European Inland Fisheries and Aquaculture Advisory Commission (EIFAAC) workshop on methodology for determination of nutrient requirement of fish (NRC, 2011).

The assessment of vitamin requirements includes somatic performance (maximum growth) and more or less specific pathological deficiency signs, and more specific biomarkers, such as specific enzyme activities and gene markers. The requirements for vitamins are normally given as mg vitamin/kg dry feed. Since many of the vitamins participate in intermediary metabolism, it has been suggested that the requirement should be given on energy basis, relative to growth, or as a daily requirement intake (mg vitamin/kg fish/day).

Despite the historical information on the individual vitamin requirements in farmed fish and shrimp species (NRC, 2011), there is a gap in the knowledge, and the estimates vary considerably with species, developmental stage, overall nutrition, and rearing conditions. It is generally assumed that fish larvae and juveniles have a higher vitamin requirement than later stages. Part of the uncertainty around estimation of requirement can also be ascribed to the possible supply of some B-vitamins from intestinal microorganisms, such as for biotin and cobalamin. Meanwhile, the micronutrient requirements are different between ploidies. Besides, the feed ingredients and water temperatures also contribute to the variation in the vitamin requirements.

Many farmed fish and shrimp species require live prey as first feed at their larvae stage, mostly due to their immature digestive tract and the difficulties with nutrient bioavailability, and leakage of nutrients from small, formulated feed particles. Using live feed is, however, a delicate balance between prey size, nutrient composition, and feed density. The nutritional profiles of different types of commercially available live feed organisms, like rotifers and *Artemia*, may be imbalanced, and, therefore, live prey may need to be modified or enriched with nutrients before use. Of the water-soluble vitamins, for example, low levels of vitamin C and thiamin are seen in rotifers enriched with yeast medium, while the other B-vitamins are in high concentrations. Harvested copepods are often regarded as the golden standard for nutrient composition (Table 3.1). Analysis of vitamins in the yolk sac, larvae body and the whole yolk sac larvae under way to start feeding can give important information on rate of transfer, retention, and use of nutrients during ontogeny (Mæland et al., 2003; Waagbø, 2010). Finally, since pond-cultured copepods seem to be natural prey and complete feed source for farmed larvae, vitamin analysis of the copepods may indicate semiquantitatively the requirement for the fish larvae (Table 3.1).

3.2 The water-soluble vitamins

The water-soluble vitamins include eight vitamin B complex, ascorbic acid, inositol, and choline, which are normally slightly to readily soluble in water. However, these nutrients have little in common from the chemical point of view. The water-soluble vitamins participate in metabolism regulation as coenzymes or serve as the raw materials of metabolic substrates or components of structural compounds in cells and tissues. Although some of water-soluble vitamins can be synthesized in animal cells or by intestinal microflora, water-soluble vitamins are not stored extensively in the tissues and in most cases excreted in urine. This means that the fish, like other animals, need a continuous supply of these vitamins.

3.2.1 Thiamin (vitamin B_1)

The history of thiamin started with an experimentally induced neurological disease called beriberi by Eijkman in the late 1800s, and an isolation and purification of an antiberiberi factor from rice bran by Funk in 1912. The first use of thiamin in fish was reported by Schneberger in 1941, who injected crystalline thiamin to cure a "diet-related disease" in rainbow trout (Halver and Hardy, 2002).

3.2.1.1 Chemical structure and characteristics

The chemical structure of the thiamin molecule consists of two ring structures, pyrimidine and thiazole moieties bound by a methylene bridge, and is relatively stable to heat. The thiamin hydrochloride salts are colorless and are more soluble in weak alkaline solutions. Both the hydrochloride and mononitrate thiamin salts have been used as the active vitamin in test diets for fish nutrition studies and for fish diet formulations. Thiamin acts as a coenzyme in the form of thiamin pyrophosphate (TPP) that includes an energy-rich diphosphate group (Fig. 3.3).

Thiamin is absorbed after a dephosphorylation of TPP in the proximate intestine by transporters (in human thiamin transporters ThTr1 and ThTr2), transported in blood, mainly in the erythrocytes, and taken up in the target tissues by the transporters (Gropper et al., 2018).

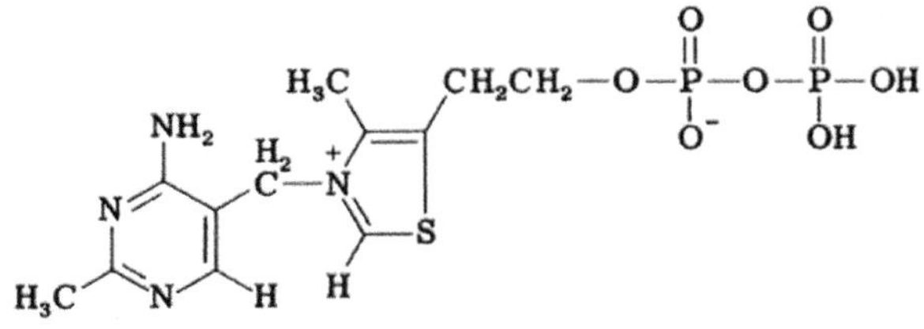

FIGURE 3.3 The structure of the active coenzyme form, thiamin pyrophosphate (TPP). The enzyme *thiaminase* splits the methylene bridge between the two ring structures and destroys the vitamin activity.

3.2.1.2 Positive functions

Thiamin participates in enzymatic decarboxylation reactions in carbohydrate metabolism. The active coenzyme compound, TPP, participates in oxidative decarboxylation reactions (cocarboxylase) of α-keto acids, like pyruvate in the glycolysis (*pyruvate dehydrogenase*) entering the Krebs' cycle for further energy extraction, and *α-ketoglutarate dehydrogenase* in the Krebs' cycle (Fig. 3.2). Another classical biochemical reaction that needs thiamin as coenzyme is the *transketolase* reaction, for cytoplasmic glucose oxidation in the hexose monophosphate shunt (pentose phosphate pathway).

It is not yet known if thiamin takes part in non-coenzyme roles in the nervous system in fish as in humans, like affecting neural cell sodium permeability and chloride transporter activation (Gropper et al., 2018).

3.2.1.3 Deficiency syndromes

Deficiency of thiamin affects carbohydrate metabolism, and organs like brain, eye lens, nerves, and muscle tissues based on carbohydrate energy are particularly sensitive for thiamin deficiency (Halver and Hardy, 2002; Harder et al., 2018). Since TPP is retained longer in muscle than in the brain tissue, deficiency symptoms include severe neurological disorders, like hyperirritability, shock syndrome (light or physical tank disturbances), and the so-called "truck-winding symptom" seen in eel. More general symptoms include poor growth, anorexia, dark pigmentation, poor appetite, muscle atrophy, fatigue, convulsions, instability and loss of equilibrium, abnormal behavior, edema, and increased mortality (Halver and Hardy, 2002; NRC, 2011; Harder et al., 2018, Table 3.4).

Thiamin deficiency occurs because of low dietary intakes, unbalanced diets (high in carbohydrates), or due to dietary antithiamin factors. The latter include chemical cleavage of thiamin by sulfiting agents (antioxidants additives), polyphenolic compounds with tannins and catechins (in coffee and tea), cyanogenic glycosides (like linamarin/lotaustralin in cassava), and the antimetabolites, oxythiamine and butyl thiamine (Halver and Hardy, 2002). However, historically occurring episodes of deficiency were related to enzymatic cleavage of the thiamin molecule at the methylene bridge, by the enzyme *thiaminase*, occurring among others in raw fish ingredients and certain algae (Harder et al., 2018).

A special case of thiamin deficiency has been related to the so-called early mortality syndrome (EMS) in wild Atlantic salmon (Harder et al., 2018). It is speculated if there is a relationship between thiaminase activity of algal origin in the prey fish and the development of thiamin deficiency in Atlantic salmon broodfish, resulting in critical low thiamin status and mortalities in the eggs and offspring. This implies that thiaminase activity may be a part of the etiology behind the thiamin deficiency–related reproduction disorders in salmonids, known as the Cayuga Syndrome in Great Lakes in the United States (Finger Lakes of New York), and the M74 syndrome in rivers entering the Baltic Sea (Wistbacka et al., 2009). Among supporting findings, critical low thiamin concentrations and thiamin deficiency–like symptoms have been observed in salmon broodfish and in eggs and alevins suffering from the M74 syndrome by Fisher et al. in 1998. Brood stock injected with thiamin were cured. Further, alevins suffering from the M74 syndrome were effectively cured by exposure to dilute thiamin solutions in the water (Wooster et al., 2000; Amcoff et al., 2002). Finally, physiologically inactive thiamin analogs, like oxythiamine (Amcoff et al., 1999) and pyrithiamine (Amcoff et al., 2002) partly reproduced the deficiency symptoms seen in the M74 syndrome. Tiger shrimp fed a thiamin-free diet showed reduced growth and feed conversion and low survival, but not specific neurological symptoms (Chen et al., 1991).

TABLE 3.4 Thiamin deficiency symptoms observed in fish and crustaceans beyond mortalities and reduced growth.

Deficiency symptom	Description	Species	References
Anorexia	Reduced feed intake	Rainbow trout	Kitamura et al. (1967a); Masumoto et al. (1987)
		Channel catfish	Murai and Andrews (1978b)
		Chinook salmon	Halver (1957)
		Japanese eel	Arai et al. (1972)
		Japanese parrotfish	Ikeda et al. (1988)
		Common carp	Aoe et al. (1969)
		Yellowtail	Ishihara et al. (1978)
		Tiger shrimp	Chen et al. (1991)
Changed skin coloration	Darker or lighter color than normal	Channel catfish	Murai and Andrews (1978b)
		Common carp	Aoe et al. (1969)
		Japanese eel	Arai et al. (1972)
		Rainbow trout	Masumoto et al. (1987)
		Yellowtail	Ishihara et al. (1978)
Changed behavior	Abnormal swimming	Japanese eel	Arai et al. (1972)
		Rainbow trout	Kitamura et al. (1967a)
	Tetany, ataxia	Chinook salmon	Halver (1957)
		Japanese eel	Arai et al. (1972)
		Rainbow trout	Masumoto et al. (1987)
	Hyperirritability	Japanese parrotfish	Ikeda et al. (1988)
	Lethargy	Channel catfish	Dupree (1966)
	Loss of equilibrium	Channel catfish	Dupree (1966)
		Chinook salmon	Halver (1957)
		Rainbow trout	Kitamura et al. (1967a)
		Yellowtail	Ishihara et al. (1978)
Damage to organs	Muscle (atrophy)	Chinook salmon	Halver (1957)
	Skin and fins (hemorrhage, congestion, necrosis)	Common carp	Aoe et al. (1969)
		Japanese eel	Arai et al. (1972)
	Gill (necrosis)	Rainbow trout	McLaren et al. (1947b)

Modified from Hansen, A., Waagbø, R., Hemre, G., 2015. New B vitamin recommendations in fish when fed plant-based diets. Aquacult. Nutr. 21, 507–527.

TABLE 3.5 Requirement of thiamin in fish and crustaceans.

Species	Requirement (mg/kg)	Response parameter	References
Salmonids			
Rainbow trout[b]	1	Growth	NRC (2011)
	2.5–3.1	Liver storage	Phillips et al. (1946)
Pacific salmon	10–15[a]	Liver storage	Halver and Hardy (2002)
Atlantic salmon	<3.4	Growth, muscle thiamin, mRNA apoenzyme expression	Hemre et al. (2016)
Freshwater species			
Channel catfish	1[a]	Growth	Murai and Andrews (1978b)
Common carp	0.5[a]	Growth, liver storage	Aoe et al. (1969)
Jian carp	1.0	Growth, immune parameters	Huang et al. (2011); Feng et al. (2011)
Grass carp	1.3	Growth, feed/protein efficiency, liver thiamin, clinical chemistry	Jiang et al. (2014)
Nile tilapia	3.5	Growth, feed intake, neurological symptoms, hematology, body composition	Lim et al. (2011)
Indian major carp (mrigal)	1.8–3.3	Growth, protein retention, RNA/DNA ratio, hematology, liver thiamin	Zehra and Khan (2017)
Snakehead	2.3–2.6	Growth, feed conversion, liver thiamin, liver oxidative status, blood transketolase activity	Zehra and Khan (2018b)
Grouper	>0.5	Liver translocase activity	Huang et al. (2007)
	<0.5	Increased TBARs	
Marine species			
Turbot	0.6	Growth	Cowey et al. (1975)
	2.6	Erythrocyte transketolase activity	
Yellowtail	1.2	Growth	Masumoto (2002)
	11.2	Liver storage	
Gilthead sea bream	5–10[a]	Growth	Morris and Davies (1995)

TABLE 3.5 Requirement of thiamin in fish and crustaceans.—cont'd

Species	Requirement (mg/kg)	Response parameter	References
Crustaceans			
Juvenile marine shrimp	14	Haemolymph TPP, transketolase	Chen et al. (1994)
Penaeus japonicus	60	Growth, whole body thiamin	Deshimaru and Kuroki (1979)
Pacific white shrimp	54.2	Growth, feed utilization, hepatopancreas thiamin, nonspecific immunity	Huang et al. (2015)

[a]*Added amount; dietary level of the vitamin was not analyzed in these studies.*
[b]Latin names: *Gilthead sea bream* (Sparus aurata), *yellowtail* (Seriola quinqueradiata), *Turbot* (Scophtalmus maximus), *Nile tilapia* (Oreochromis niloticus), *Grouper* (Carassius auratus), *Common carp* (Cyprinus carpio), *Jian carp* (Cyprinus carpio *var. Jian*), *Grass carp* (Ctenopharyngodon idella), *Japanese eel* (Anguilla japonica), *Channel catfish* (Ictalurus punctatus), *Pacific salmon* (Onchorhyncus *spp.*), *Rainbow trout* (Oncorhynchus mykiss), *Snakehead* (Channa punctata), *Juvenile marine shrimp* (Penaeus monodon), *Pacific white shrimp* (Litopenaeus vannamei), *Indian major carp* (Cirrhinus mrigala).
Modified from Hansen, A., Waagbø, R., Hemre, G., 2015. New B vitamin recommendations in fish when fed plant-based diets. Aquacult. Nutr. 21, 507–527.

3.2.1.4 *Requirements*

The minimum requirement of thiamin is determined by the level supporting normal appetite, digestion, growth, fertility, and lack of neurological deficiency symptoms. Historical dose-response feeding studies in trout, salmon, carp, channel catfish, and sea bream defining their respective requirements are given by Halver and Hardy (2002). The requirements differ widely among the fish species, in the range from 0.5 to 15 mg/kg feed, while from 14 to 60 mg/kg feed for shrimps (Table 3.5). On energy basis, the requirement is relatively less in carnivorous fishes and is higher in omnivorous and herbivorous fish per energy unit. Since the protein requirement for growth differs with fish size and water temperature, it is important to include and consider the relative macrocomposition in the evaluation of the vitamin requirements. Groups of Atlantic salmon smolt reared at increasing temperatures from 10 to 20°C showed a 5% decrease in liver thiamin while muscle thiamin was increased by 27%, indicating that thiamin status in these organs reflected the change toward anaerobic metabolism differently (Waagbø et al., 2020). The thiamin requirement seems to be higher for juveniles than for later stages, both due to higher growth rates and less capacity for vitamin storage. Increasing plant ingredients with starch may increase the thiamin requirement (Morris and Davies, 1995).

Aquatic crustaceans are slow bottom feeders, where leaching of water-soluble nutrients may be a general challenge. In general, thiamin requirement estimates in crustaceans are an order of magnitude higher than for fish species (Tables 3.1 and 3.5), which may relate to leaching loss. For *Penaeus japonicus* fed a semisynthetic diet, the thiamin requirement was estimated to 60 mg/kg based on growth and whole animal thiamin status (Deshimaru and Kuroki, 1979).

3.2.1.5 *Sources and protection*

Both marine and plant feed ingredients commonly used in aquaculture feeds are sources for thiamin, often sufficiently to cover the requirement for growth and welfare (Table 3.1). In moist feed, thiamin may be exposed to chemical or biological destruction. Thiamin is supplemented feed through the vitamin mixture in form of crystalline mononitrate or hydrochloride salts. Thiamin is relatively stable toward dry heat during feed production, where 60%–80% retain after extrusion and storage at room temperature for 3 months (Table 3.2). Thiamin in yeast (torula) and corn can be made more easily available after pretreatments like steaming and chemical treatment (Halver and Hardy, 2002). As a salt additive, leaching of thiamin can be a problem in microparticulate and liposome diets.

3.2.1.6 *Antimetabolites and inactivation*

Of the antithiamin factors, *thiaminase I* (EC 2.5.1.2) has been the most focused nutritional cause of deficiency, both in wild and farmed fish fed raw fish. The chemical reaction includes nucleophiles that bind covalently to the pyrimidine moiety of the broken thiamin molecule, whereas the thiazole moiety is released. Among such nucleophiles are common compounds such as aromatic amines, heterocyclic compounds, and sulfhydryl compounds (like nicotinic acid, amino acids, AA derivatives, and organic bases). While novel plant and insect feed ingredients in addition to raw fish may contain thiaminase I, heat treatment of the ingredients and complete feeds denature the enzyme activity. An old study suggests that acetylcholine is an antagonist to thiamin, and a relationship between thiamin and lipoic acid has been reported (Halver and Hardy, 2002).

3.2.1.7 *Clinical assessment*

Tissue levels of thiamin often reflect dietary levels below saturation levels (Masumoto, 2002). Thiamin status (TPP) in farmed species can be measured in liver and muscle tissues, or as thiamin-dependent *transketolase* activity in erythrocytes, liver, and kidney (Cowey et al., 1975; Masumoto et al., 1987; Huang et al., 2007). Huang et al. (2007) measured increased oxidative status (liver thiobarbituric acid reactive substances [TBARS]) at low thiamin status. Blood hematology and clinical analyses are often used as subclinical health markers to support the evaluation of early thiamin deficiency and sufficiency. Muscle tissue saturation (0.7 mg/kg) was achieved in Atlantic salmon fed 3.4 mg thiamin/kg in a plant-based diet (Hemre et al., 2016). Reference values for liver thiamin status are between 8 and 20 μg/g wet weight in salmonid species (Halver and Hardy, 2002), while carp species show considerably lower liver saturation levels, like 0.7 mg/kg in grass carp (Jiang et al., 2014) and 2.5 mg/kg in Indian major carp (Zehra and Khan, 2017). Example of whole-body tissue saturation in shrimps showed 1.4 mg thiamin/kg (fed 120 mg/kg), as compared to 1.0 mg thiamin/kg when fed 60 mg/kg (Conklin, 1989).

3.2.2 Riboflavin (vitamin B_2)

The early history of riboflavin includes isolation and crystallization of a growth factor that was identified as riboflavin by Kuhn et al., in 1933. Flavin compounds from other sources (lactoflavin, hepatoflavin, and ovoflavin) were shown to be equal to pure riboflavin. During 2 decades from 1947, riboflavin deficiency was described in several fish species, trout, salmon, carp, and catfish, as summarized by Halver and Hardy (2002).

3.2.2.1 Chemical structure and characteristics

Riboflavin has its name from the ribose-like side chain and its strong yellow (Lat. *flavus*) color in crystalline form. It has three ring structures (Fig. 3.4) in its structural formula ($C_{17}H_{20}N_4O_6$). It is slightly soluble in water and emits a yellow-green fluorescence. Solubility can be increased in alkali solutions and alcohols. The compound is stable to oxidation and dry heat, while decomposing to lumiflavin in visible and UV light.

Riboflavin acts as coenzyme in enzymes taking part in the cellular redox respiration system, where it appears as the nucleotide derivatives, flavin adenine dinucleotide (FAD) and flavin mononucleotide (FMN).

Riboflavin attached or bound to proteins and in active coenzyme forms needs to be hydrolyzed to free riboflavin before intestinal absorption. In humans, free riboflavin is taken up through an energy-dependent riboflavin vitamin transporter (RFVT3). The further transport to the circulation depends on a transporter RFVT1 and riboflavin is transported to the target tissues mainly by plasma albumin and immunoglobulins (Gropper et al., 2018).

3.2.2.2 Positive functions

The biochemical reactions involving riboflavin coenzymes are related to many flavoenzymes (oxidases and reductases) in cell respiration, catalyzing oxidation, and transportation of hydrogen and electrons to the reduced energy carrying purine nucleotides $FADH_2$ and $FMNH_2$. The coenzymes are involved in the metabolism of keto acids, fatty acids, and amino acids, and include, among others, enzymes in the electron transport chain (*cytochrome c reductase)*, dehydrogenation of amino acids (*D- and L-amino acid oxidases, D-AAO)*, purine catabolism (*xanthine oxidase)*, vitamer B_6 metabolism (*pyridoxine phosphate oxidase*), pyruvate oxidative decarboxylation (*pyruvate dehydrogenase* complex), forming fumarate from succinate (*succinate dehydrogenase*), fatty acids β-oxidation (*acetyl-CoA dehydrogenase)*, forming acids (like pyridoxic acid and retinoic acid) from aldehydes (*aldehyde oxidase*), synthesis of folate vitamer 5-methyl tetrahydrofolic acid (THF) (*methylene-THF reductase*), taking part as an antioxidant by reducing glutathione GSSG to GSH (*glutathione reductase*), and more

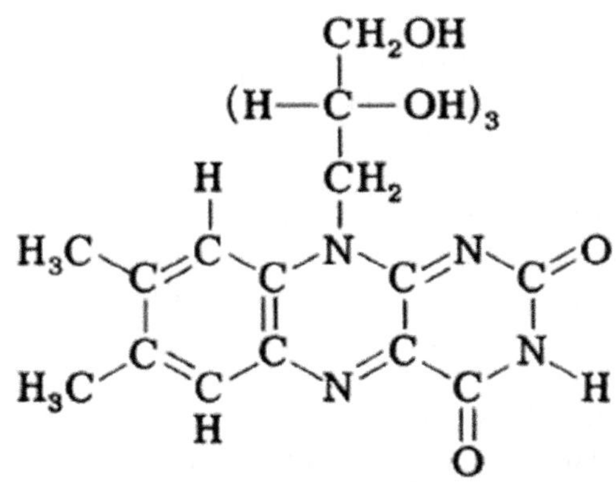

FIGURE 3.4 Structural formula of riboflavin. In the active coenzyme forms, flavin mononucleotide (FMN) and flavin adenine dinucleotide (FAD), the hydroxy group is esterified with a phosphate group and adenine diphosphate group, respectively.

(Saedisomeolia and Ashoori, 2018). Riboflavin is involved in metabolism of pyridoxine, niacin, and folate, including the conversion of tryptophan to nicotinic acid. For the latter, most fish species lack this ability (NRC, 2011).

Novel functions for riboflavin in health risks like prevention of cancers and neurological degenerative disorders in humans have been reviewed (Saedisomeolia and Ashoori, 2018).

3.2.2.3 *Deficiency syndromes*

Besides general symptoms as reduced survival, appetite, feed efficiency, and growth, more specific riboflavin deficiency symptoms include corneal damages, photophobia, cataract, nervous swimming behavior, anemia, fin and snout erosions, and abnormal skin pigmentation (Table 3.6). The symptoms are mostly reversible when fed riboflavin supplemented diets, except for mature cataracts since the lens tissue grows and develops inward (Bjerkås et al., 2006). Due to the avascular tissues of the eye, disorders here are among the first developed at deficiency and are more specific symptoms of riboflavin deficiency. The vital role of riboflavin in antioxidation means that some of the observed deficiency symptoms could be the origin for increased lipid peroxidation. Recently, Chen et al. (2015a) showed increased malondialdehyde, a marker on lipid peroxidation, in gills and intestine of grass carp fed a riboflavin-deficient diet.

Deletion of riboflavin from a semisynthetic diet did not affect the growth of *P. monodon* in 8 weeks feeding study (Reddy et al., 1999). On the contrary, deletion of riboflavin led to increased growth in *P. monodon* (Catacutan and Cruz, 1989). However, symptoms like light coloration, irritability, and exoskeletal malformations like protuberant cuticle at intersomites and short-head dwarfism were seen in shrimps fed a purified diet without supplemented riboflavin (Chen and Hwang, 1992).

3.2.2.4 *Requirements*

Riboflavin is easily taken up in the intestine, the liver and muscle tissues are rapidly saturated, and excess is excreted in the urine (NRC, 2011). Highest concentrations are found in active tissues with aerobic oxidation, like the liver, heart, kidney, and red muscle.

Requirement estimates are most often based upon growth performance, organ riboflavin status, and organ or plasma flavoenzyme activity (D-AAO, ERG) (Table 3.7). The vitamin is added in increasing feed concentrations and response parameters recorded after approximately 10–12 weeks. The requirement is fairly stable for fish species and ranges between 3 and 11 mg/kg, while somewhat higher estimates (15–80 mg riboflavin/kg) for shrimps (Tables 3.1 and 3.7). The requirement depends on the overall feed composition, like lipid energy concentration (Brønstad et al., 2002). For most species, the requirement has been established in juveniles since they are assumed to have higher requirement than at later stages, due to rapid growth and development, gut immaturity, and less vitamin storage capacity. Based on muscle riboflavin saturation, Hemre et al. (2016) suggested that a plant-based practical diet for Atlantic salmon should contain 5 mg riboflavin/kg, especially in the juvenile freshwater period. Like for several essential nutrients, riboflavin supplementation up to the requirement increased intestinal enzymatic activities in Jian carp (Li et al., 2010a). The riboflavin requirement seems not to be dependent on temperature or genetic differences in growth rate (NRC, 2011). In Atlantic salmon fed a commercial diet and reared at temperatures between 10 and 20°C, liver riboflavin status did not change with increasing temperature

TABLE 3.6 Riboflavin deficiency symptoms observed in fish and crustaceans beyond mortalities and reduced growth.

Deficiency symptom	Description	Species	References
Anorexia	Reduced feed intake	Chinook salmon	Halver (1957)
		Common carp	Ogino (1967); Aoe (1967); Takeuchi et al. (1980)
		Japanese eel	Arai et al. (1972)
		Rainbow trout	Takeuchi et al. (1980)
		Channel catfish	Serrini et al. (1996)
		Japanese parrotfish	Ikeda et al. (1988)
		Red hybrid tilapia	Lim et al. (1993)
		Sunshine bass	Deng and Wilson (2003)
		Tilapia	Soliman and Wilson (1992b)
Anemia	Reduced hematocrit	Tilapia	Soliman and Wilson (1992b)
Changed skin coloration	Darker color than normal	Chinook salmon	Halver (1957)
		Sunshine bass	Deng and Wilson (2003)
		Tilapia	Soliman and Wilson (1992b)
Changed behavior	Abnormal swimming	Japanese eel	Arai et al. (1972)
	In coordination	Chinook salmon	Halver (1957)
	Lethargy	Japanese parrotfish	Ikeda et al. (1988)
		Tilapia	Soliman and Wilson (1992b)
	Nervous disorders	Common carp	Ogino (1967)
		Red hybrid tilapia	Lim et al. (1993)
Damage to organs	Skin and fins (hemorrhage, lesions, dermatitis, erosion)	Common carp	Ogino (1967); Aoe et al. (1967); Takeuchi et al. (1980)
		Japanese eel	Arai et al. (1972)
		Tilapia	Soliman and Wilson (1992b)
	Eyes (hemorrhage)	Chinook salmon	Halver (1957)
	Urinary tubule (necrosis)	Common carp	Aoe et al. (1967)
Other	Photophobia	Chinook salmon	Halver (1957)
		Common carp	Ogino (1967)
		Japanese eel	Arai et al. (1972)
	Body dwarfism	Channel catfish	Murai and Andrews (1978a); Serrini et al. (1996)

Continued

TABLE 3.6 Riboflavin deficiency symptoms observed in fish and crustaceans beyond mortalities and reduced growth.—cont'd

Deficiency symptom	Description	Species	References
		Red hybrid tilapia	Lim et al. (1993)
	Cataract	Channel catfish	Dupree (1966)
		Chinook salmon	Halver (1957)
		Rainbow trout	Takeuchi et al. (1980)
		Red hybrid tilapia	Lim et al. (1993)
		Silver salmon	Coates and Halver (1958)
		Sunshine bass	Deng and Wilson (2003)
		Tilapia	Soliman and Wilson (1992b)
	Protruding cuticle at intersomites, short-head dwarfism	Juvenile marine shrimp	Chen and Hwang (1992)

Modified from Hansen, A., Waagbø, R., Hemre, G., 2015. New B vitamin recommendations in fish when fed plant-based diets. Aquacult. Nutr. 21, 507–527.

TABLE 3.7 Requirement for riboflavin in different fish and crustacean species.

Species[d]	Requirement (mg/kg)	Response parameter	References
Salmonids			
Atlantic salmon	<6.5	Growth	Brønstad et al. (2002)
	6.5	Liver riboflavin	
	5	Muscle riboflavin	Hemre et al. (2016)
Rainbow trout	6[a]	Liver riboflavin	Takeuchi et al. (1980)
	3.6	Growth	Woodward (1985)
	4.6	Liver riboflavin	
	6.6	Spleen riboflavin	
	5.6	Liver D-AAO activity[b]	
	2.7	Liver and heart riboflavin Liver D-AAO activity	Amezaga and Knox (1990)
	1.7	Erythrocyte EGR[c] activity	

TABLE 3.7 Requirement for riboflavin in different fish and crustacean species.—cont'd

Species[d]	Requirement (mg/kg)	Response parameter	References
Freshwater species			
Channel catfish	9[a]	Growth	Murai and Andrews (1978a)
	4.3[a]	Growth	Serrini et al. (1996)
	6[a]	Liver D-AAO activity	
Common carp	4[a]	Growth	Aoe et al. (1967)
	6.2[a]	Liver riboflavin	
	1[a]	Liver riboflavin	Ogino (1967)
	6[a]	Liver riboflavin	Takeuchi et al. (1980)
Tilapia	6[a]	Liver D-AAO activity	Soliman and Wilson (1992b)
Red hybrid tilapia	5[a]	Growth	Lim et al. (1993)
Sunshine bass	4.1 5	Growth Liver D-AAO activity	Deng and Wilson (2003)
Jian carp	5	Growth, feed, and protein efficiency, intestinal enzymes	Li et al. (2010a)
Snakehead	6	Growth, feed, and protein efficiency, liver riboflavin, antioxidation, body composition	Zehra and Kahn (2018a)
Marine species			
Yellowtail	2.9 11	Growth Liver riboflavin	Hosokawa (2000)
Crustaceans			
Juvenile marine shrimp	22.3	Body riboflavin	Chen and Hwang (1992)
Chinese mitten crab *Eriocheir sinensis*			Wang et al. (2010)

[a]*Added amount; dietary level of the vitamin was not analyzed in these studies.*
[b]*D-Amino acid oxidase.*
[c]Glutathione reductase.
[d]*Latin names: Atlantic salmon (*Salmo salar*), Rainbow trout (*Oncorhynchus mykiss*), Channel catfish (*Ictalurus punctatus*), Common carp (*Cyprinus carpio*), Jian carp (*Cyprinus carpio *var. Jian), Tilapia (*Oreochromis aureus*), Red hybrid tilapia (*Oreochromis mossambicus x Oreochromis niloticus*), Snakehead (*Channa punctata*), Yellowtail (*Seriola quinqueradiata*), Sunshine bass (*Morone chrysops x Morone saxatilis*), Juvenile marine shrimp (*Penaeus monodon*), Chinese mitten crab (*Eriocheir sinensis*).*
Modified from Hansen, A., Waagbø, R., Hemre, G., 2015. New B vitamin recommendations in fish when fed plant-based diets. Aquacult. Nutr. 21, 507–527.

while muscle status decreased with 36%. This decline was probably related to change in muscle metabolism at elevated temperatures (Waagbø et al., 2020). Water-soluble vitamins are rarely toxic for fish in high concentrations, and as much as 600 riboflavin mg/kg was shown not to be toxic to rainbow trout (NRC, 2011).

3.2.2.5 *Sources and protection*

Traditional marine and animal feed ingredients normally contain adequate amounts of riboflavin, while refined plant products contain less (Table 3.1). Besides the natural content in feed ingredients, riboflavin is added as a crystalline salt. Since the vitamin is sensitive to light, moisture, and heat, ingredients, vitamin mixes, and complete diets should be stored in a cool and dark place. The laboratories conducting analysis of B-vitamins often have artificial light to protect the vitamins during analysis. Riboflavin is, however, relatively unaffected by extrusion and 3 months storage with losses up to 10% (Table 3.2).

3.2.2.6 *Antimetabolites and inactivation*

Analogs of riboflavin, like galactoflavin and the hydroxyethyl analog, seem to be antagonists to riboflavin in experimental animals and microorganisms (Halver and Hardy, 2002), probably by competitively inhibiting the flavoenzymes. This has, however, not been evaluated in fish and crustaceans.

3.2.2.7 *Clinical assessment*

Riboflavin levels can easily be measured in liver and muscle tissue or in erythrocytes, and results are used to assess riboflavin status. Since excess riboflavin is excreted in the urine, urinary riboflavin relative to creatinine has been used as a measure of sufficiency. The activation coefficient of enzymes like *glutathione reductase* in erythrocytes has been used to indicate riboflavin status in fish (Hughes et al., 1981). Similarly, hepatic *D-amino acid oxidase* seems to be a sensitive marker for riboflavin status in rainbow trout (Woodward, 1984) and hybrid striped bass (Deng and Wilson, 2003). In juvenile marine shrimp, hemolymph *glutathione reductase* activity failed as a riboflavin status indicator, so body riboflavin status was used for the estimation (Chen and Hwang, 1992). With the novel role of riboflavin as a key component in the integrated antioxidant defense, oxidative stress markers should be included in the evaluation of riboflavin deficiency (Chen et al., 2015a).

3.2.3 Niacin (vitamin B_3)

Niacin deficiency symptoms in fish were first described by McLaren et al. (1947b). Niacin was identified a decade earlier as essential to prevent black tongue disease in dogs, used as model animal for humans. Pellagra is a skin disease in humans suffering from niacin deficiency, today known as the 4D disease—diarrhea, dermatitis, dementia, and death. Historical pellagra outbreaks from the 18th to 20th century, mostly among the poorest in the European and American populations, were related to the introduction of maize as their main stable food. Interestingly, Latin American populations are also using maize as a stable food but did not experience pellagra, and this was later found to be related to their pretreatment of the maize prior to cooking, presoaking it in alkaline limewater which increased the bioavailability of niacin by releasing its binding to glycoproteins (Kirkland and Meyer-Ficca, 2018).

3.2.3.1 *Chemical structure and characteristics*

The niacin molecule (chemical formula $C_6H_5O_2N$) has a simple heterocyclic hexagonal ring structure with a nitrogen in first position and the carbon in third position linked to a carboxyl acid group (pyridine-3-carboxylic acid or nicotinic acid) or an amide group (nicotinamide) (Fig. 3.5). Of several more vitamers, these two have the highest biological niacin activity. Like riboflavin, niacin appears in active coenzyme forms, with the nitrogen esterified to a pyridine nucleotide (nicotinamide adenine dinucleotide, NAD, and nicotinamide adenine dinucleotide phosphate, NADP) with high-energy diphosphate bindings. Niacin is a white crystalline compound, soluble in water and alcohol, and more soluble in alkali. The simple rings structure is stable to dry heat in acid and alkaline solutions (Kirkland and Meyer-Ficca, 2018).

3.2.3.2 *Positive functions*

Niacin appears as coenzyme in many enzymatic reactions participating in cellular redox reactions of energy-giving substrates, carbohydrates, lipids, and proteins, both through anaerobic and aerobic metabolism (Fig. 3.2). The major function is the transfer of hydrogen and electrons to the next step in the redox transport chain. It is linked to the mitochondrial oxidative phosphorylation (the respiratory chain) of the substrates to the endpoint energy in the form of adenosine triphosphate (ATP) and end products, carbon dioxide and water.

Novel-examined functions of niacin are related to NAD serving as a cosubstrate for many ADP-ribosylation enzymes with varied functions. NAD-consuming enzymes are suggested as important genetic and epigenetic regulators, like poly-(ADP ribose) polymerases and sirtuins (Kirkland and Meyer-Ficca, 2018). The latter is a family of histone proteins that seems to regulate organismal lifespan, oxidative stress, and DNA damage.

3.2.3.3 *Deficiency syndromes*

Participating in intermediary metabolism and energy extraction of the macronutrients, deficiency symptoms of niacin in fish are diffused (Table 3.8). Beyond reduced growth, feed utilization, poor appetite, and increased mortalities, symptoms like skin and fin diseases, light sensitivity, intestinal lesions, tissue edema and muscular weakness, spasms, anemia, jaw deformity, as well as abnormal swimming behavior have been observed (Halver and Hardy, 2002; NRC, 2011; Davies and Rawling, 2013; Hansen et al., 2015).

Shrimp fed with niacin-deficient diets show lower growth performance and feed efficiency than those fed with the niacin-sufficient diets, while no other specific signs of niacin deficiency were observed (Shiau and Suen, 1994; Xia et al., 2015).

FIGURE 3.5 Structural formulas of niacin and nicotinamide, where the latter forms the active coenzymes nicotinamide adenine dinucleotide (NAD) and nicotinamide adenine dinucleotide phosphate (NADP).

TABLE 3.8 Niacin deficiency signs in fish and crustacean species beyond mortalities and reduced growth.

Deficiency symptom	Description	Species	References
Anorexia	Reduced feed intake	African catfish	Morris et al. (1998)
		Chinook salmon	Halver (1957)
		Channel catfish	Ng et al. (1997)
		Japanese eel	Arai et al. (1972)
		Japanese parrotfish	Ikeda et al. (1988)
Anemia	Reduced hematocrit	African catfish	Morris et al. (1998)
		Channel catfish	Ng et al. (1997); Andrews and Murai (1978)
		Japanese eel	Arai et al. (1972)
Changed behavior	Abnormal swimming	African catfish	Morris et al. (1998)
		Chinook salmon	Halver (1957)
		Japanese eel	Arai et al. (1972)
	Tetany	Channel catfish	Ng et al. (1997)
		Chinook salmon	Halver (1957)
	Lethargy	Channel catfish	Dupree (1966)
	In coordination	Channel catfish	Dupree (1966)
Damage to organs	Stomach and colon (lesions, edema)	Chinook salmon	Halver (1957)
	Swollen gills	Rainbow trout	McLaren et al. (1947b)
	Skin and fins (haemorrhages, lesions, erosion, dermopathy)	African catfish	Morris et al. (1998)
		Channel catfish	Andrews and Murai (1978)
		Hybrid tilapia	Shiau and Suen (1992)
		Japanese eel	Arai et al. (1972)
		Rainbow trout	Poston and Wolfe (1985)
Other	Exophthalmia	Channel catfish	Andrews and Murai (1978)

Modified from Hansen, A., Waagbø, R., Hemre, G., 2015. New B vitamin recommendations in fish when fed plant-based diets. Aquacult. Nutr. 21, 507–527.

3.2.3.4 *Requirements*

According to historical information, the requirement of niacin in fish varies considerably between 14 and 200 mg/kg among species (Table 3.9). This may relate to diet composition, with uncertainties in niacin bioavailability in plant ingredients and interactions with macronutrient composition.

TABLE 3.9 Requirement for niacin in different fish and crustacean species.

Species[b]	Requirement (mg/kg)	Response parameter	References
Salmonids			
Rainbow trout	10[a]	Growth	Poston and Wolfe (1985)
Atlantic salmon	65	Whole body niacin	Hemre et al. (2016)
Pacific salmon	150[a]	Growth	Halver and Hardy (2002)
	200[a]	Liver niacin	
Freshwater species			
African catfish	33.1	Growth	Morris et al. (1998)
Channel catfish	14.0[a]	Growth	Andrews and Murai (1978)
	7.4[a]	Growth	Ng et al. (1997)
	>24[a]	Liver niacin	
Hybrid tilapia	26[a] (380 g/kg glucose)	Growth	Shiau and Suen (1992)
	121[a] (380 g/kg dextrin)		
Indian catfish	25.4	Growth Liver niacin	Mohamed and Ibrahim (2001)
Rohu	33	Growth	Ahmed (2011)
Mrigal	30	Growth	Ahmed (2011)
Genetically Improved Farmed Tilapia (GIFT)	20.4 (85)[c]	Growth, liver niacin, clinical chemistry	Jiang et al. (2014)
Marine species			
Yellowtail	12	Growth	Masumoto (2002)
Golden pompano	30	Growth, feed efficiency, clinical chemistry, antioxidants	Xun et al. (2019)
Crustaceans			
Pacific white shrimp	110	Growth, feed, and protein efficiency	Xia et al. (2015)
Tiger shrimp (grass shrimp)	7.2	Growth, feed conversion ratio	Shiau and Suen (1994)
Indian white prawn	250	Growth	Boonyaratpalin (1998)

[a]*Added amount, not confirmed by analysis.*
[b]*Latin names of fish species; Rainbow trout* (Oncorhynchus mykiss)*, Pacific salmon* (Oncorhynchus *spp.*)*, African catfish* (Clarias gariepinus)*, Channel catfish* (Ictalurus punctatus)*, Hybrid tilapia* (Oreochromis niloticus X O. Aureus)*, Indian catfish* (Heteropneustes fossilis)*, Rohu* (Labeo rohita)*, Mrigal* (Cirrhinus mrigala)*, Yellowtail* (Seriola quinqueradiata)*, Golden pompano* (Trachinotus ovatus)*, Pacific white shrimp* (Litopenaeus vannamei)*, Kuruma shrimp* (Marsupenaeus japonicus)*, Tiger shrimp* (Penaeus monodon)*, Indian white prawn* (Fenneropenaeus indicus)*.*
[c]*Liver niacin saturation.*
Modified from Hansen, A., Waagbø, R., Hemre, G., 2015. New B vitamin recommendations in fish when fed plant-based diets. Aquacult. Nutr. 21, 507–527.

Source of carbohydrates in the feed may alter niacin requirements, like in hybrid tilapia where the requirement was 26 or 121 mg/kg depending in groups fed 38% glucose or dextrin (Shiau and Suen, 1992). There was, however, no correlation between dietary carbohydrate level and niacin requirement for maximum growth in fish species, which may relate to variation in carbohydrate utilization capacity among the species (Hansen et al., 2015). Based on whole body status of niacin, revised requirement of niacin in Atlantic salmon fed plant-based diets was suggested to be 65 mg/kg (Hemre et al., 2016). Both liver (50%) and white muscle tissue (30%) increased in niacin status in Atlantic salmon smolt fed a commercial diet at elevated water temperature (20°C), compared to 10°C (Waagbø et al., 2020). This illustrates that niacin status varies with metabolism, which may confound requirement studies.

The wide and mostly high niacin requirement estimates for crustaceans (7–400 mg/kg; Table 3.9) probably reflect feed composition, technical quality, and feeding, leaving uncertainty in the estimates to niacin bioavailability and leaching loss.

Extreme excess of niacin in the diet (10,000 mg/kg) affected lipid metabolism in juvenile brook trout, with increased lipid liver but decreased body fat (Poston, 1969b).

3.2.3.5 *Sources and protection*

Niacin requirements are influenced by dietary niacin concentration and bioavailability (NRC, 2011). Traditional animal and plant feed ingredients used in aquafeeds normally contain sufficient niacin to cover the requirement for fish growth and welfare (Table 3.1). Niacin in fish meal reaches 85 mg/kg, while soybean meal contains 28 mg/kg. Niacin from plant ingredients may, however, be less bioavailable for fish (Ng et al., 1997), especially from untreated corn (30%) as compared to 100% for fish meal. Niacin is supplemented as nicotinic acid or niacinamide in the vitamin premix and is stable, with less than 10% losses under feed production and prolonged storage (Table 3.2).

While tryptophan can be converted to niacin in land animals and humans, this does not occur in fish species (NRC, 2011). For example, excess tryptophan did not restore growth and deficiency markers in channel catfish fed a niacin-deficient diet (Ng et al., 1997). Similarly, Xia et al. (2015) suggested that the tryptophan—niacin conversion—is minimal or absent in Pacific white shrimp.

3.2.3.6 *Antimetabolites and inactivation*

In cereal grains, most of niacin is bound to glycoproteins, leaving only 30% available for absorption. Pretreatment with heat and alkali may release the vitamin and increase the bioavailability.

3.2.3.7 *Clinical assessment*

Dietary niacin is directly reflected by organ status of niacin. Whole body niacin (increasing from 24 to 55 mg/kg) did, however, not reach saturation level in Atlantic salmon fed increasing niacin concentrations in a plant ingredient–based diet in the range of 36–110 mg/kg (Hemre et al., 2016). Niacin can alternatively be analyzed in erythrocytes, as a nonlethal assessment. No studies in fish have used niacin-specific enzymatic activities in the status assessment, probably due to the ubiquitous dependence of niacin in coenzymes of cellular metabolism (Hansen et al., 2015).

3.2.4 Pantothenic acid (vitamin B_5)

During the decade from 1935 to 1945, a factor curing dermatitis was identified, isolated from food sources, and later pantothenic acid (from the Greek, meaning "from everywhere") was synthetized in crystalline form. During the decade from 1945 to 1955, Phillips et al. and Rucker et al. observed clubbed gills fed a pantothenic acid–deficient diet in trout and salmon, respectively.

3.2.4.1 *Chemical structure and characteristics*

D-Pantothenic acid (chemical formula is $C_9H_{12}O_5N$; Fig. 3.6) is a part of acetyl coenzyme A which is needed in cellular acylation and acetylation enzymatic reactions. While the free acid is a yellow viscous oil, the calcium salt is normally used as vitamin additive in feed. The white salt is soluble in water, stable to oxidation and reducing agents, but labile to dry heat and hot acid or alkali solutions. Pantothenol has nearly similar vitamin B_5 activity as pantothenic acid.

When using microbiological assay to analyze for pantothenic acid, one should keep in mind that esters of vitamers B_5 (acetate, benzoate, and diphosphate esters) may be bioactive for animals but not for lactic acid bacteria, and the bioassay may, therefore, underestimate the feed levels. The optical isomer L-pantothenic acid appears to be physiologically inert.

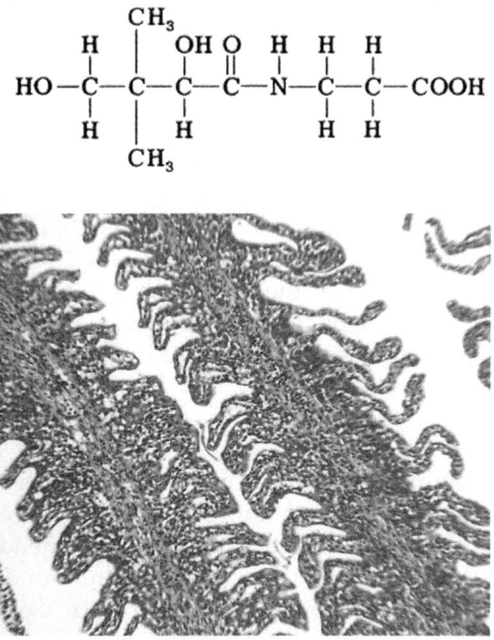

FIGURE 3.6 Structural formula of pantothenic acid, a component in the active coenzyme form coenzyme A (CoA). The section shows gill histopathology with epithelial hyperplasia in pantothenic acid–deficient rainbow trout. *From Olsvik, P., Hemre, G., Waagbø, R., 2013. Exploring early micronutrient deficiencies in rainbow trout* (Oncorhynchus mykiss)*, by nex-generation sequencing technology-from black box to functional genomics. PloS One 8 (7), e69461; Photos: Dr. Anne-Berit Olsen, Norwegian Veterinary Institute, Bergen, Norway. https://doi.org/10.1371/journal.pone.0069461.g001.*

3.2.4.2 *Positive functions*

Pantothenic acid, as a part of acetyl coenzyme A, participates in numerous enzymatic acetylation and acylation reactions in the intermediary energy metabolism of glucose, fatty acids, and amino acids through the Krebs cycle (Fig. 3.2, Table 3.3). The reactions include the oxidation of pyruvate and acetaldehyde for entering Krebs' cycle, condensation reactions for synthesis of acetate, fatty acids, and citrate in the fatty acid metabolism, and acetylation of aromatic amines and choline to neurotransmitters. The active acetyl CoA is the essential intermediate in metabolism and participates in most acyl transfer reactions, including succinate, benzoate, propionate, and butyrate (Fig. 3.2). *Branched-chain amino acid dehydrogenase* degrading branched-chained amino acids (isoleucine, leucine, and valine) in the mitochondria requires CoA as a cofactor. Pantothenic acid is involved in biosynthesis of phospholipids, isoprenoids, steroids, as well as fatty acid elongation (Halver and Hardy, 2002).

The role of pantothenic acid in immunity is unclear, but examples show that the vitamin affects the complement system in red sea bream, *Pagrus major* (Yano et al., 1988), and enhanced disease resistance and nonspecific and specific immune responses in Jian carp (Wen et al., 2010).

3.2.4.3 *Deficiency syndromes*

Due to the major involvement in intermediate metabolism of carbohydrates, lipids, and protein, pantothenic acid is a key nutrient for normal physiology and metabolism of growing fish (NRC, 2011). This makes energy-demanding mitochondria rich tissues especially susceptible to deficiency. Besides common deficiency symptoms like reduced feed intake and growth and sluggishness, more specific symptoms appear as a gill disorder within few weeks on a vitamin-deficient diet. The gill filaments show proliferation, swelling, and fusion of filaments and lamellae ("clubbed gills" Fig. 3.6), and the surface may be covered with an exudate. The gill disorder correlates with low pantothenic acid status and has been observed in most farmed fish species (salmon, rainbow trout, eel, carp, yellowtail, red sea bream, catfish, and grouper) (Halver and Hardy, 2002). Later symptoms are change in behavior, anemia, and mortality. Table 3.10 shows the deficiency symptoms observed in different fish species. According to the concept of vitamin deficiency, recovery or alleviation from the symptoms is seen when pantothenic acid is reintroduced in the diet, given that the fish are eating.

3.2.4.4 *Requirements*

The pantothenic acid requirement for normal growth and development ranges from 6 to 40 mg/kg, with higher requirements in more active fish species than in sedate species (Table 3.11). In crustaceans, far higher estimates are recommended, which probably relates to feed, feeding, and leaching losses of water-soluble compounds (Table 3.1).

While tissue pantothenic acid is clearly correlated with dietary concentrations, requirement estimates using maximum CoA concentration in liver are not accurate because the concentration seems to be strictly regulated and may not always correlate with pantothenic acid deficiency (Hansen et al., 2015). More nonspecific measures of pantothenic acid deficiency and sufficiency are intestinal organ relative weights and functions (Wen et al., 2009).

TABLE 3.10 Pantothenic acid deficiency signs in fish and crustacean species beyond mortalities and reduced growth.

Deficiency symptom	Description	Species	References
Anorexia	Reduced feed intake	Channel catfish	Murai and Andrews (1979)
		Japanese parrotfish	Ikeda et al. (1988)
		Common carp	Ogino (1967)
		Grouper	Lin et al. (2012)
		Rainbow trout	Kitamura et al. (1967a)
Anemia	Reduced red blood cell (RBC) count, hematocrit (Hct) and hemoglobin (Hb) concentrations	Channel catfish	Murai and Andrews (1979)
		Common carp	Ogino (1967)
		Grouper	Lin et al. (2012)
		Tilapia	Soliman and Wilson (1992a)
Changed behavior	Abnormal swimming	Grouper	Lin et al. (2012)
		Japanese eel	Arai et al. (1972)
		Rainbow trout	Kitamura et al. (1967a)
		Common carp	Ogino (1967)
		Tilapia	Soliman and Wilson (1992a)
	Tetany, ataxia	Japanese eel	Arai et al. (1972)
		Japanese parrotfish	Ikeda et al. (1988)
	Lethargy	Channel catfish	Dupree (1966)
		Japanese parrotfish	Ikeda et al. (1988)
Damage to organs	Skin and fins (erosion, haemorrhages, dermatitis)	Blue tilapia	Roem et al. (1991)
		Channel catfish	Murai and Andrews (1979)
		Common carp	Ogino (1967)
		Japanese eel	Arai et al. (1972)
		Tilapia	Soliman and Wilson (1992a)
	Gills (lesions, erosion)	Channel catfish	Dupree (1966); Wilson (1983)
		Rainbow trout	Kitamura et al. (1967a)
	Clubbed gills	Blue tilapia	Roem et al. (1991)
		Channel catfish	Dupree (1966); Murai and Andrews (1979)

Continued

TABLE 3.10 Pantothenic acid deficiency signs in fish and crustacean species beyond mortalities and reduced growth.—cont'd

Deficiency symptom	Description	Species	References
		Grouper	Lin et al. (2012)
		Rainbow trout	Karges and Woodward (1984)
		Silver salmon	Coates and Halver (1958)
		Tilapia	Soliman and Wilson (1992a)
Other	Exophthalmia	Common carp	
	Oxidative stress Intestinal dysfunction	Blunt snout bream	Qian et al. (2015)

Modified from Hansen, A., Waagbø, R., Hemre, G., 2015. New B vitamin recommendations in fish when fed plant-based diets. Aquacult. Nutr. 21, 507–527.

When feeding Atlantic salmon increasing pantothenic acid as a part of the micronutrient mixture in plant-based diets, improved growth was obtained when increasing the level of B-vitamins up to 1.5 times the NRC (2011) recommendations, with 25 mg/kg. Maximum gill tissue saturation of pantothenic acid was 5.5 mg/kg (Hemre et al., 2016). They confirmed a starting pantothenic acid deficiency from the low muscle pantothenic acid status (4 mg/kg) versus 10 mg/kg in fish fed micronutrient supplementation following the NRC (2011).

3.2.4.5 *Sources and protection*

Pantothenic acid is found in cereals and animal feed ingredients to a different degree (Table 3.1). Fish flesh and fish meal contain relatively high levels (17 mg/kg). As a supplement in the vitamin mixture, pantothenic acid is added in form of the relatively stable calcium or sodium D-pantothenate salts. Since the L-isomer is not biologically active, a racemic mixture of calcium DL-pantothenate salt has 50% activity. Approximately 20% loss of pantothenate can be expected during feed production and storage (Table 3.2).

3.2.4.6 *Antimetabolites and inactivation*

Some cereal brans contain chemically bound pantothenic acid in a form having low bioavailability to fish that may be released during ingredient or feed production. This may, however, add to uncertainty as to whether the requirement will be met. The key role of pantothenic acid in metabolism may be challenged by chemical compounds with antivitamin effects, like tumor inhibiting agents, or antagonistic effects by pantoyltaurine (where taurine replaces β-alanine in the molecule) and methyl-pantothenic acid. While the antagonists may be used to accelerate deficiency symptoms in the animal, excess pantothenic acid can overcome the inhibitory effects.

3.2.4.7 *Clinical assessment*

Analysis of pantothenic acid in the feed should include proper hydrolysis in the preparation step, preferably with enzymes in weak acid (Mæland et al., 2000). Similarly, the

TABLE 3.11 Requirement for pantothenic acid in different fish and crustacean species.

Species	Requirement (mg/kg)	Response parameter	References
Salmonids			
Rainbow trout[c]	20	Growth	NRC (2011)
Pacific salmon	20	Growth	NRC (2011)
Atlantic salmon	25	Growth, Gill pantothenic acid	Hemre et al. (2016)
Freshwater species			
Channel catfish	15[a]	Growth	Murai and Andrews (1979)
	10[a]	Growth	Wilson et al. (1983)
Tilapia	10[a]	Growth	Soliman and Wilson (1992a)
Blue tilapia	6[a]	Growth	Roem et al. (1991)
	10[a]	Survival	
Common carp	30[a]	Growth	Ogino (1967)
	40[a]	Liver pantothenic acid	
Jian carp	23.0	Growth	Wen et al. (2009)
	25.6	Enzyme activity (intestinal trypsin and lipase)	
Blunt snout bream	24.1	Growth, protein efficiency, liver CoA and pantothenic acid, intestinal enzymes, liver fatty acids, and oxidative status	Qian et al. (2015)
Marine species			
Yellowtail	13.5	Growth	Masumoto (2002)
	35.9	Liver pantothenic acid	
Grouper	11.0	Growth Liver CoA concentration[b]	Lin et al. (2012)
Asian sea bass	15 90	Growth Tissue saturation	Boonyaratpalin and Williams (2002)
Crustaceans			
Grass shrimp	101–139	Growth, feed efficiency, hepatopancreas lipid, pantothenic acid, and CoA	Shiau and Hsu (1999)

[a]*Added amount; not analyzed dietary level.*
[b]*Coenzyme A.*
[c]*Latin names of the fish species: Rainbow trout* (Oncorhynchus mykiss)*, Pacific salmon* (Oncorhynchus *spp.*)*, Channel catfish* (Ictalurus punctatus)*, Tilapia* (Oreochromis aureus)*, Blue tilapia* (Tilapia aurea)*, Common carp* (Cyprinus carpio)*, Jian carp* (Cyprinus carpio *var. Jian*)*, Blunt snout bream* (Megalobrama amblycephala)*, Yellowtail* (Seriola quinqueradiata)*, Grouper* (Epinephelus malabaricus)*, Asian sea bass* (Lates calcarifer)*, Grass shrimp* (Penaeus monodon)*.*
Modified from Hansen, A., Waagbø, R., Hemre, G., 2015. New B vitamin recommendations in fish when fed plant-based diets. Aquacult. Nutr. 21, 507–527.

preparation from fish tissue stores should be liberated to pantothenic acid, pantetheine, phosphopantothenic acid, phosphopantotheine, and bound coenzyme A. Since *branched-chain amino acid dehydrogenase* degrading branched-chained amino acids (isoleucine, leucine, and valine) in the mitochondria requires CoA as a cofactor, both reduced tissue activity of the enzyme and elevated plasma levels of branched amino acids can be used as suitable markers for pantothenic acid deficiency (Masumoto et al., 1999).

3.2.5 Pyridoxine (vitamin B_6)

In the 1930s, a factor reported to cure dermatitis in rats was isolated, named pyridoxine by Paul György, and synthesized chemically. The first quantitative requirement study of pyridoxine in fish was conducted as early as 1944. In the coming decades from 1947 to 1981, pyridoxine deficiency was reported in trout, salmon, catfish, carp, sea bream, yellowtail, and eel (Halver and Hardy, 2002).

3.2.5.1 *Chemical structure and characteristics*

Vitamin B_6 or pyridoxine appears in seven different vitamer forms (Fig. 3.7), of which are 2-methylpyridine ring structure with an alcohol (pyridoxine, PN), aldehyde (pyridoxal, PL), or amine group (pyridoxamine, PM) in the fourth position, their respective phosphorylated (in fifth position) forms [pyridoxine phosphate (PNP), pyridoxal phosphate (PLP), and pyridoxamine phosphate (PMP)], and the carboxyl acid form (pyridoxic acid). Excess pyridoxine is excreted as inactive pyridoxic acid in the urine.

Pyridoxine hydrochloride ($C_8H_{11}O_3N_7$ HCl) is a stable form added to feed which is readily water soluble and heat stable in either acid or alkaline solutions, while it is sensitive to UV light. The vitamers PL and PM are more chemical labile in solutions by exposure to air, heat, and light.

PLP and PMP are the active coenzyme forms, and in animal tissues, the active vitamer forms are readily converted to the coenzyme forms.

3.2.5.2 *Positive functions*

Vitamin B_6 participate in the metabolism of proteins, lipids, and carbohydrates serving as coenzyme in more than 150 enzymatic reactions within many different enzyme classes (oxidoreductases, transferases, hydrolases, lyases, and isomerases), but the mostly known reactions relate to protein metabolism with transamination and deamination reactions of

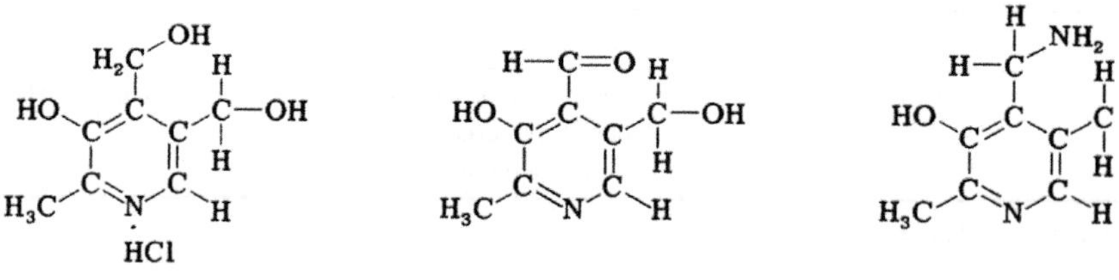

FIGURE 3.7 Structural formulas of the vitamers B_6, pyridoxine hydrochloride, pyridoxal, and pyridoxamine. The three vitamers appear in active phosphorylated coenzyme forms, while the pyridoxic acid form is inactive. Glycosidic pyridoxine appears in plant products.

amino acids, for example, *aspartate aminotransferase* (ASAT) among 22 transaminases (Bird, 2018). It is involved in the synthesis of γ-aminolevulinic acid and porphyrins and neurotransmitters like serotonin through the decarboxylation of 5-hydroxytryptophan to 5-hydroxytryptamine or serotonin. Similarly, PNP has coenzyme function in the metabolism of glutamic acid, lysine, methionine, histidine, cysteine, and alanine. PN has roles in lipid metabolism, including the catabolism of glycogen, as well as in mRNA synthesis and polypeptide synthesis.

In Jian carp, dietary vitamin B_6 had a protective effect on infectious diseases, while PN up to the requirement 5 mg/kg improved resistance to pathogen challenge and leukocyte phagocytic activity (Feng et al., 2011). Excess PN above the requirement did not affect immunity and resistance to bacterial challenge in Atlantic salmon (Albrektsen et al., 1995). PL has important roles in lipid metabolism, illustrated by increased 22:6 n-3 in muscle of rainbow trout fed practical diets supplemented excess vitamin B_6 (Maranesi et al., 2005).

Novel roles of vitamin B_6 relate to positive effects on inflammation and overall immunity in animals (Bird, 2018). In line with this, diet supplemented with excess pyridoxine (100 mg/kg) reduced the stress while triggering the immune response in rohu fingerlings exposed to the pesticide endosulfan (Akhtar et al., 2010). It is suggested that vitamin B_6 also interferes with the effects of other steroid hormones, like estrogen during vitellogenesis in Atlantic salmon (Albrektsen, 1994).

Recent focus on water-soluble vitamins in fish has been on the role of brood stock nutrition and feeding on next generations, focusing on nutritional programming and epigenetic mechanisms (Skjærven et al., 2018). This includes especially nutrients related to the 1C metabolism, like vitamin B_6, vitamin B_{12}, folate, and the amino acid, methionine. These studies have among others demonstrated that a parental low 1C nutrient diet influenced zebrafish offspring through nutritional programming and changed epigenetic regulation through DNA methylation, gene transcription, and liver lipid accumulation to a fatty liver–like phenotype (Skjærven et al., 2018).

3.2.5.3 *Deficiency syndromes*

Since PN has a major role in protein and amino acid metabolism, the requirement is high in carnivorous fish, and the muscle and liver storages may be easily and rapidly exhausted (Hardy et al., 1979). Acute vitamin B_6 deficiency symptoms occur after 2–3 weeks in salmon fed a PN-deficient, high protein (50%) diet at 12–15°C, and after 4 weeks, total mortalities occur (Halver and Hardy, 2002).

Reduced growth, feed intakes (anorexia), and feed utilization are nonspecific vitamin B_6 deficiency symptoms. Among the more specific deficiency symptoms in fish species are nervous disorders (epileptic type, hyperirritability, and dark/green pigmentation in the skin), gasping breathing (hyperventilation), peritoneal edema, and premortem rigor (Tables 3.3 and 3.12). Depending on severity of symptoms, injection or feeding the active vitamer form PLP to fish results in rapid recovery in fish within days. Analysis of transaminases in plasma, erythrocytes, liver, and muscle tissues reflects the state of deficiency and correlates with vitamers B_6 status in the fish (Smith et al., 1974; Adron et al., 1978; Kissil et al., 1981; Albrektsen et al., 1993). Despite growth depression, mortalities, and low vitamin status, vitamin B_6–deficient shrimps do not display nervous disorders (Deshimaru and Kuroki, 1979).

TABLE 3.12 Pyridoxine deficiency signs in fish and crustacean species beyond mortalities and reduced growth.

Deficiency symptom	Description	Species	References
Anorexia	Reduced feed intake	Channel catfish	Andrews and Murai (1979)
		Chinook salmon	Halver (1957)
		Hybrid tilapia	Shiau and Hsieh (1997)
		Indian catfish	Mohamed (2001b)
		Japanese parrotfish	Ikeda et al. (1988)
		Grouper	Huang et al. (2005)
Anemia	Normocytic, Normochromic anemia	Rainbow trout	Smith et al. (1974)
		Grouper	Huang et al. (2005)
Changed skin coloration	Darker or lighter color than normal	Chinook salmon	Halver (1957)
		Indian catfish	Mohamed (2001b)
		Rainbow trout	Smith et al. (1974)
Changed behavior	Abnormal swimming	Channel catfish	Dupree (1966)
		Japanese parrotfish	Ikeda et al. (1988)
		Rainbow trout	Kitamura et al. (1967a); Smith et al. (1974)
	Hyperirritability	Chinook salmon	Halver (1957)
		Gilthead sea bream	Kissil et al. (1981)
		Rainbow trout	Smith et al. (1974)
	Tetany, ataxia	Channel catfish	Dupree (1966); Andrews and Murai (1979)
		Chinook salmon	Halver (1957)
		Hybrid tilapia	Lim et al. (1995); Shiau and Hsieh (1997)
		Indian catfish	Mohamed (2001b)
		Japanese eel	Arai et al. (1972)
		Rainbow trout	Smith et al. (1974)
	Lethargy	Hybrid tilapia	Lim et al. (1995)
		Indian catfish	Mohamed (2001b)
	Loss of equilibrium	Japanese parrotfish	Ikeda et al. (1988)
	Nervous disorders	Channel catfish	Andrews and Murai (1979)

TABLE 3.12 Pyridoxine deficiency signs in fish and crustacean species beyond mortalities and reduced growth.—cont'd

Deficiency symptom	Description	Species	References
		Chinook salmon	Halver (1957)
		Rainbow trout	McLaren et al. (1947b); Smith et al. (1974)
		Common carp	Ogino (1965)
		Japanese eel	Arai et al. (1972)
		Grouper	Huang et al. (2005)
	Neurological disorders	Gilthead sea bream	Kissil et al. (1981)
		Hybrid tilapia	Lim et al. (1995)
	Rapid and gasping breathing	Chinook salmon	Halver (1957)
Damage to organs	Intestine (edema, lesions, haemorrhages)	Chinook salmon	Halver (1957)
		Indian catfish	Mohamed (2001b)
		Rainbow trout	Smith et al. (1974)
	Kidney, liver, and pancreas (lesions, haemorrhages)	Indian catfish	Mohamed (2001b)
		Rainbow trout	McLaren et al. (1947b); Smith et al. (1974)
	Skin and fins (edema, haemorrhages, erosion)	Common carp	Ogino (1965)
		Hybrid tilapia	Lim et al. (1995)
Other	Exophthalmia	Common carp	Ogino (1965)

Modified from Hansen, A., Waagbø, R., Hemre, G., 2015. New B vitamin recommendations in fish when fed plant-based diets. Aquacult. Nutr. 21, 507–527.

3.2.5.4 *Requirements*

Normally the requirement is in the range of 5–20 mg/kg, in the lower range for freshwater and planktivorous species, and in the higher range for juveniles and carnivorous species (Shiau and Hsieh, 1997). Status measures are visual deficiency symptoms, maximum growth performance, and maximum organ pyridoxine storage and enzyme activity.

The requirement depends on growth and protein metabolism, of which both are higher in juveniles, at elevated temperatures and at higher dietary protein levels (NRC, 2011). Woodward (1994) suggested that the requirement of vitamin B_6 is directly related to growth, by a factor of 3.1 μg/g weight gain on dry weight (15 nmol/g weight gain). Interestingly, a value of 3.3 μg/g weight gain was estimated in developing halibut yolk sac larvae until first feeding (Rønnestad et al., 1997). High tryptophan diets have also been shown to increase the vitamin B_6 requirement (Halver and Hardy, 2002). No changes in the liver and muscle levels of pyridoxine were, however, observed in Atlantic salmon smolt fed a commercial diet at temperatures between 10 and 20°C, as compared to other B-vitamins (Waagbø et al., 2020).

Feng et al. (2010) showed that pyridoxine up to the requirement of 5 mg/kg also improved survival in Jian carp after pathogen challenge and immune cell activities.

For crustaceans, the pyridoxine requirement estimates are generally higher than in fish species, ranging from 60 to 200 mg/kg (Table 3.13). For *Penaeus japonicus* fed a semisynthetic

TABLE 3.13 Requirement for pyridoxine in different fish and crustacean species.

Species	Requirement (mg/kg)	Response parameter	References
Salmonids			
Atlantic salmon[d]	2–3	Growth	Albrektsen et al. (1993)
	6–8	Muscle ASAT[b] activity	
Rainbow trout	25[a]	Muscle storage	Maranesi et al. (2005)
Silver salmon	6[a]	Growth	Coates and Halver (1958)
Freshwater fish			
Channel catfish	3[a]	Growth	Andrews and Murai (1979)
Indian catfish	2.4	Growth	Mohamed (2001b)
	3.4	Liver storage Liver ASAT activity	
Common carp	5.4	Growth	Ogino (1965)
	>40	Liver storage Liver activity of ASAT and ALAT[c]	
Indian major carp	6	Growth, feed efficiency, RNA/DNA ratio, hematology	Zehra and Khan (2017)
	8	Liver pyridoxine	
Jian carp	6	Growth	He et al. (2009)
	5	Plasma ammonia, Intestinal amylase activity	
Hybrid tilapia	1.7/16.5 (280/360 g/kg protein)	Growth	Shiau and Hsieh (1997)
	9.5/15.0 (280/360 g/kg protein)	Liver ASAT activity	
Nile tilapia	11 (380 g/kg protein)	Growth	Teixeira et al. (2012)
Marine fish			
Gilthead sea bream	2	Absence of deficiency symptoms	Kissil et al. (1981)
Turbot	0.5–1.0	Growth	
	2.5	Liver and muscle activity of ASAT and ALAT	Adron et al. (1978)

TABLE 3.13 Requirement for pyridoxine in different fish and crustacean species.—cont'd

Species	Requirement (mg/kg)	Response parameter	References
Golden pompano	8.8–9.3	Growth, feed efficiency, liver pyridoxine, max ALAT, ASAT enzyme activities, oxidative status, intestinal flora	Huang et al. (2019c)
Grouper	2	Survival, Hematology, Liver activity of transaminases	Huang et al. (2005)
Crustaceans			
Penaeus japonicus	120	Growth, body pyridoxine	Deshimaru and Kuroki (1979)
Penaeus monodon	72–89	Survival, transaminase activity	Shiau and Wu (2003)
Pacific white shrimp	120 107–152	Growth, survival, transaminase activity	Li et al. (2010b)
	110	Growth	Cui et al. (2016)
	168	Hepatopancreas pyridoxine	

[a]*Added amount (not analyzed).*
[b]*Aspartate aminotransferase.*
[c]*Alanine aminotransferase.*
[d]*Latin names of fish species: Atlantic salmon* (Salmo salar), *Rainbow trout* (Oncorhynchus mykiss), *Silver salmon* (Oncorhynchus kisutch), *Channel catfish* (Ictalurus punctatus), *Indian catfish* (Heteropneustes fossilis), *Common carp* (Cyprinus carpio), *Indian major carp* (Cirrhinus mrigala, *Hamilton*), *Jian carp* (Cyprinus carpio *var. Jian*), *Hybrid tilapia* (Oreochromis niloticus X O. aureus), *Nile tilapia* (Oreochromis niloticus), *Gilthead sea bream* (Sparus aurata), *Turbot* (Scophthalmus maximus), *Golden pompano* (Trachinotus ovatus), *Pacific white shrimp* (Litopenaeus vannamei).
Modified from Hansen, A., Waagbø, R., Hemre, G., 2015. New B vitamin recommendations in fish when fed plant-based diets. Aquacult. Nutr. 21, 507–527.

diet, pyridoxine requirement was estimated to 60 mg/kg based on growth and whole animal vitamin status (Deshimaru and Kuroki, 1979). Both maximum tissue PN and enzyme activities indicate, however, considerable higher pyridoxine requirements (Table 3.13). Recently, Cui et al. (2016) showed that vitamin B_6 not only improved growth performance and feed utilization but also had a significant impact on the innate immunity of juvenile Pacific white shrimp.

3.2.5.5 Sources and protection

Traditional feed resources like marine meals, plant ingredients, and animal byproducts are good sources of vitamin B_6 vitamers (Table 3.1). Plants contain mainly the pyridoxine, including the glycosylated form, while animal tissues contain the pyridoxal and pyridoxamine vitamers (NRC, 2011). In the body, they can all be easily converted to their active phosphorylated coenzyme forms PLP and PMP. The vitamers are generally unstable to UV lights, and feed ingredients should be stored and protected from sunlight, while PL and PLP are also labile to oxidation in air and heat during feed production.

Even though feed ingredients may contain enough vitamin B_6, one must ensure that the requirement is covered by supplementation of crystalline PN HCl (82.3% activity). The vitamin is relatively unstable in micronutrient mixtures kept under unfavorable conditions, while feed ingredients and complete diets may lose up to 10%–20% during feed production and storage (Table 3.2).

3.2.5.6 *Antimetabolites and inactivation*

Vitamin B_6 antagonists can either competitively replace or inactivate the coenzyme vitamer, where deoxypyridoxine is an example of the former (Halver and Hardy, 2002). Adding this antagonist to the diet may accelerate a vitamin B_6 deficiency state. Similarly, methoxypyridoxine and the pharmaceutical isonicotinic acid hydrazide (used in treatment of tuberculosis in man) act as vitamin B_6 antagonists. The irreversible conversion of excess vitamers B_6 to pyridoxic acid inactivates the vitamin.

3.2.5.7 *Clinical assessment*

Among easy measurable vitamin B_6 biomarkers are plasma enzyme activity, while tissue vitamer B_6 concentrations and examination for classical deficiency symptoms support a potential suboptimal vitamin status or deficiency. Both analysis of transaminases in plasma and tissues, and their activation coefficient (α = Transaminase activity $_{\text{PNP added}}$/Transaminase activity $_{\text{Native}}$) reflect the state of vitamin B_6 deficiency and correlate with vitamers B_6 status in the fish.

Whole-body pyridoxine levels in shrimp (*Penaeus japonicus*) were 0.08, 0.43, and 0.54 mg/kg when fed a deficient diet, fed the reported requirement level and at levels twice the requirement, respectively (Deshimaru and Kuroki, 1979). In addition to growth parameters, changes in enzymatic activity have been used as an indicator of pyridoxine status in white shrimp (Li et al., 2010b).

3.2.6 Biotin (vitamin B_7)

The nutritional history of biotin was initiated with injuries in rats caused by feeding egg white in the early 19th century, where the active identified component was in the 1930s and was called coenzyme R or vitamin H. Biotin was isolated in 1937, its biological functions defined in 1940, and the vitamin synthesized 2 years later (Halver and Hardy, 2002). In 1945, the classical blue slime patch disease in trout was found to be related to biotin deficiency (McLaren et al., 1947b).

3.2.6.1 *Chemical structure and characteristics*

δ-Biotin, hexahydro-2-oxo-l-thieno-3,4-imidazole-4-valeric acid, has the chemical formula $C_{10}H_{16}O_3N_2S$, and its structural formula was given in Fig. 3.8. While D-biotin is the active form, the L-isomer has no biological activity. This carboxylic acid is slightly soluble in water, while the salts are more soluble. The compound is stable to heat and light but destroyed in acid and alkaline solutions and by strong oxidants. Biotin appears active bound to lysine moieties in proteins (the ε-amino group of lysine and the carboxyl of biotin is combined in a peptide bond), to form the coenzyme biocytin. The vitamin is liberated by enzymatic hydrolysis (biotinidase) as well as other peptidases before intestinal uptake or prior to biotin analyses.

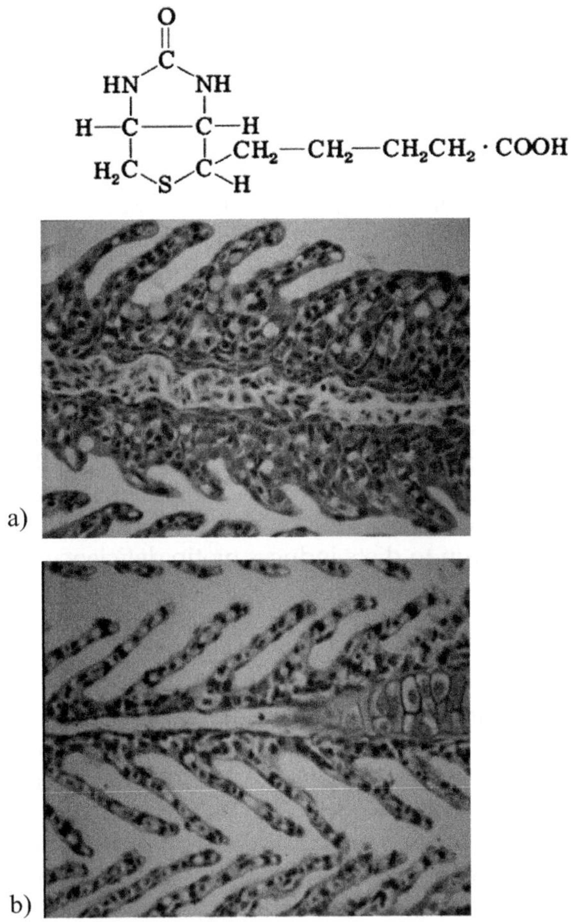

FIGURE 3.8 Structural formula of D-biotin. Biotin binds to a protein-bound lysin residue to form the active coenzyme biocytin. The sections are: (A) gills from juvenile Atlantic salmon fed a fishmeal-based diet with 10% spray-dried hen's egg white for 18 weeks show hypertrophy and hyperplasia with fusion of secondary lamellae and (B) normal gill tissue. *From Mæland, A., Waagbø, R., Sandnes, K., Hjeltnes, B., 1998. Biotin in practical fish-meal based diet for Atlantic salmon Salmo salar L. fry. Aquacult. Nutr. 4, 241–247, with permission by John Wiley and sons.*

3.2.6.2 *Positive functions*

Biocytin as coenzyme is required in carboxylation and decarboxylation reactions, fixing CO_2 in compounds in the Krebs cycle (tricarboxylic acids), purine synthesis, elongation of fatty acids, general in lipid synthesis, citrulline synthesis, and cis isomerization of unsaturated fatty acids to biologically active fatty acids. Enzymes include *acetyl-CoA carboxylase, pyruvate carboxylase (PC),* and *propionyl-CoA carboxylase* (Fig. 3.2). Supplementation of biotin above the requirement for growth may have beneficial stress ameliorating effects (Waagbø et al., 1994). Biotin also functions in the regulation of gene expression, cell development, and reproduction through mechanisms other than the role as coenzyme (Yossa et al., 2015). On the other hand, Zhao et al. (2012) examined how dietary biotin affected intestinal enzyme activities and microbiota in Jian carp, showing improved digestive capacity with increased dietary biotin.

3.2.6.3 *Deficiency syndromes*

Besides reduced growth, anorexia, and mortality, biotin deficiency symptoms in fish are seen as gill, skin, and intestinal lesions, fragmentation of erythrocytes, and spasms (Halver and Hardy, 2002). In brook trout, blue slime patch disease is a classical biotin deficiency symptom. Deficiency is experienced after 8–12 weeks on a biotin-deficient diet or a diet added a minor amount of hen egg white containing avidin, a biotin antagonist, as part of the protein (Table 3.14).

One of the biotin-dependent enzymes, pyruvate carboxylase (PC), converts pyruvic acid to oxaloacetic acid in glycolysis, and the activity of this enzyme has been showed to be a sensitive indicator of biotin status in fish (Woodward and Frigg, 1989).

Avidin, a glycoprotein in egg white, binds strongly to biotin and makes the vitamin inaccessible for absorption in the intestine (Mock, 2007). Biotin deficiency due to intake of raw egg white is a part of the history on the discovery of biotin as a vitamin (Mock, 2007). Supplementation of raw egg white or avidin to diets induces biotin deficiency symptoms in several fish species (Poston, 1976; Mæland et al., 1998, Fig. 3.8).

3.2.6.4 *Requirements*

Among the B-vitamins, the biotin requirement is quantitatively low and relatively equal among fish species, mostly ranging from 0.2 to 1.0 mg/kg (Tables 3.1 and 3.15). The minimum dietary requirement for biotin for optimal growth in salmonids, such as the rainbow trout and the lake trout, has been estimated to be 0.1–0.3 mg/kg, respectively (Poston, 1976; Woodward and Frigg, 1989). Koppe (1993) demonstrated that biotin requirement was met in young rainbow trout fed a practical diet without any supplementation of biotin (0.11 mg/kg diet), while somewhat higher levels were needed for optimal lysozyme levels in serum and mucus (0.15 mg/kg diet). Considerably higher dietary concentrations (0.7 mg/kg) were found useful during seawater adaptation (smoltification) of Atlantic salmon (Waagbø et al., 1994).

Biotin requirement estimates in shrimp species range from 0.3 to 2.4 mg/kg (Table 3.15). For tissue biotin saturation and maximum enzymatic activities, considerably higher estimates are needed.

Dietary lipid level affects the biotin requirement, as demonstrated by *PC* activity in fish fed with a lipid-free and biotin-deficient diet (NRC, 2011). Feeding a commercial diet to Atlantic salmon at temperatures from 10 to 20°C, liver biotin concentration rose 150% while muscle status remained the same (Waagbø et al., 2020). This probably reflects the relatively higher importance for biotin-dependent reactions in liver than in muscle with temperature.

In Atlantic salmon fed plant-based diets with biotin ranging from 0.25 to 1.14 mg/kg, whole body biotin status did not reflect the dietary concentration (Hemre et al., 2016). It was concluded that it is not necessary to add biotin to salmon feeds based on the given plant ingredients used.

TABLE 3.14 Biotin deficiency signs in fish and crustacean species beyond mortalities and reduced growth.

Deficiency symptom	Description	Species	References
Anorexia	Reduced feed intake	Asian catfish	Mohamed et al. (2000)
		Japanese eel	Arai et al. (1972)
		Japanese parrotfish	Ikeda et al. (1988)
		Japanese seabass	Li et al. (2010c)
		Lake trout	Poston and Page (1982)
		Rainbow trout	Castledine et al. (1978)
		Catla catla	Khan and Khan (2019)
Anemia	Reduced hematocrit (Hct)	Common carp	Ogino et al. (1970a)
Changed skin coloration	Darker or lighter color than normal	Asian catfish	Mohamed et al. (2000)
		Indian catfish	Mohamed and Ibrahim (2001)
		Japanese eel	Arai et al. (1972)
		Japanese seabass	Li et al. (2010c)
		Rainbow trout	Woodward and Frigg (1989)
		Catla catla	Khan and Khan (2019)
Changed behavior	Abnormal swimming	Indian catfish	Mohamed (2001a)
		Japanese eel	Arai et al. (1972)
	Lethargy	Rainbow trout	Woodward and Frigg (1989)
	Tetany	Asian catfish	Mohamed et al. (2000)
		Indian catfish	Mohamed (2001a)
	Hyperirritability	Japanese parrotfish	Ikeda et al. (1988)
Damage to organs	Gills (histopathological changes, lesions)	Common carp	Ogino et al. (1970a)
		Lake trout	Poston and Page (1982)
		Rainbow trout	Woodward and Frigg (1989)
		Atlantic salmon	Mæland et al. (1998)
		Zebrafish	Yossa et al. (2011)
	Liver (lesions)	Lake trout	Poston and Page (1982)
	Pancreas and kidney (histopathological changes)	Common carp	Ogino et al. (1970a)
	Skin and fins (atrophy)	Common carp	Ogino et al. (1970a)
		Japanese seabass	Li et al. (2010c)
Other	Deposition of glycogen in kidney tubules	Lake trout	Poston and Page (1982)
	Rapid ocular pulsation	Rainbow trout	Woodward and Frigg (1989)

Modified from Hansen, A., Waagbø, R., Hemre, G., 2015. New B vitamin recommendations in fish when fed plant-based diets. Aquacult. Nutr. 21, 507–527.

TABLE 3.15 Requirement for biotin in different fish and crustacean species.

Species	Requirement (mg/kg)	Response parameter	References
Salmonids			
Atlantic salmon[d]	<0.3	Growth Liver storage Enzyme activity (liver PC[b])	Mæland et al. (1998)
	0.7	Seawater tolerance	Waagbø et al. (1994)
	<0.2	Growth Liver storage	
Lake trout	0.1[a]	Growth	Poston (1976); Poston and Page (1982)
	0.5–1.0[a]	Optimal swimming stamina	
Rainbow trout	0.25	Growth Enzyme activity (liver PC)	Castledine et al. (1978)
	0.08	Growth Liver storage	Woodward and Frigg (1989)
	0.14	Enzyme activity (muscle PC)	
Freshwater fish			
Channel catfish	0.33[a]	Growth Enzyme activity (liver PC)	Lovell and Buston (1984)
Common carp	1[a]	Growth	Ogino et al. (1970a)
	10[a]	Liver storage	
Hybrid tilapia	0.05	Growth Enzyme activity (liver PC and AC[c])	Shiau and Chin (1999)
	>5	Whole body storage	
Indian catfish	0.25	Growth Whole body storage Enzyme activity (liver PC and AC)	Mohamed (2001a)
Jian carp	0.15	Growth, intestinal digestive enzymes, proximal intestine fold height	Zhao et al. (2012)
Asian catfish	2.5	Growth	Mohamed et al. (2000)
	>5.15	Liver storage	
	>0.1	Enzyme activity (liver PC and AC)	
Zebrafish	0.5	Growth	Yossa et al. (2011)

TABLE 3.15 Requirement for biotin in different fish and crustacean species.—cont'd

Species	Requirement (mg/kg)	Response parameter	References
Channa punctata	0.44	Growth, protein retention, feed efficiency	Zehra and Khan (2019b)
	1.02	Liver biotin	
Catla catla	0.41–0.87	Growth, protein retention, feed efficiency, hematology Liver PC activity and biotin	Khan and Khan (2019)
Marine fish			
Japanese seabass	0.05	Growth	Li et al. (2010c)
	>6.22	Liver storage	
Yellowtail	0.22	Growth	Masumoto (2002)
	0.67	Liver storage	
Crustaceans			
Grass shrimp	2.0–2.4	Growth, feed, and protein efficiency	Shiau and Chin (1998)
	10–20	Hepatopancreas biotin, PC, and AC	

[a]*Added amount (not analyzed).*
[b]*Pyruvate carboxylase.*
[c]*Acetyl-CoA carboxylase.*
[d]*Latin names for the fish species: Atlantic salmon* (Salmo salar), *Lake trout* (Salvelinus namaycush), *Rainbow trout* (Oncorhynchus mykiss), *Channel catfish* (Ictalurus punctatus), *Hybrid tilapia* (Oreochromis niloticus X O. Aureus), *Indian catfish* (Heteropneustes fossilis), *Common carp* (Cyprinus carpio), *Jian carp* (Cyprinus carpio *var.* Jian), *Asian catfish* (Clarias batrachus), *Zebrafish* (Danio rerio), *Japanese sea bass* (Lateolabrax japonicus), *Yellowtail* (Seriola quinqueradiata), *Grass shrimp* (Penaeus monodon).
Modified from Hansen, A., Waagbø, R., Hemre, G., 2015. New B vitamin recommendations in fish when fed plant-based diets. Aquacult. Nutr. 21, 507–527.

3.2.6.5 Sources and protection

Biotin is widely found in low concentrations in animal and plant ingredients (Table 3.1). Traditionally used feed ingredients normally cover considerable parts of the requirement for biotin (Table 3.15), ranging from 0.4 to 0.8 mg/kg in fish meal (Mæland and Sandnes, 1999), as observed in fish species fed practical feeds (Mæland et al., 1998; Hemre et al., 2016).

Biotin added to the vitamin mixture as the active D-biotin form is one of the most expensive vitamin supplements. The relative stability of biotin during feed production and storage at room temperature for 3 months ranges from 70% to 90% (Table 3.2).

3.2.6.6 Antimetabolites and inactivation

The chemical form oxybiotin has partial vitamin activity, while, for example, the oxybiotin sulfonic acid analog acts as an antivitamin and can generate deficiency symptoms by inhibiting metabolism and bacterial growth (Halver and Hardy, 2002). Excess biotin may competitively overcome this inhibition.

The protein avidin, found in raw egg white, binds irreversibly to biotin and makes it unavailable for fish. Bound biotin is, however, released after heating and denaturing of the avidin. Diets including 10% raw hen egg white as part of the protein caused biotin deficiency and mortality in Atlantic salmon fry after 10 weeks (Mæland et al., 1998).

Feed ingredients and diets should be protected against oxidation agents. Raw hen egg white powder should not be used due to the antivitamin avidin. Heat treatment of the egg white before use will inactivate the avidin binding.

3.2.6.7 Clinical assessment

Liver biotin is a good marker of status, ranging from 6 to 12 μg/g tissue in healthy salmon. Liver PC activity reflects the biotin status in fish and has for a long time been used as marker of biotin status (Table 3.15). Liver PC activity has also proved useful to evaluate the biotin status in tiger shrimp (Shiau and Chin, 1998).

3.2.7 Folate (vitamin B_9)

Folic acid was synthesized in 1946 after an extract was isolated and used to cure megaloblastic anemia. The active component was called vitamin M and vitamin B_C. Folic acid was used by several fisheries research groups in the 1940s to cure anemia in fish fed purified diets, and folic acid was incorporated in fish test diets for trout in 1950 (Naderi and House, 2018).

3.2.7.1 Chemical structure and characteristics

Folic acid or folacin has the chemical formula $C_{19}H_{19}N_7O_6$ (molecular weight 441.4 g/mol) and structural formula given in Fig. 3.9, consisting of a pteridine ring linked to *p*-aminobenzoic acid by a methylene group (pteroic acid) and linked to glutamic acid as an amide. Folic acid appears as a yellow flaky crystal soluble in water and alcohol. It is easily destroyed by heat and sunlight, as well as prolonged storage.

The term folates (or vitamers folacin) cover many analogs that have biological activity, including pteroic acid, rhizopterin, folinic acid, xanthopterin, and several formyltetrahydropteroylglutamic acid derivatives. The xanthoptherin was early identified as an antianaemic factor H for fish which is an expression of the combination of folate and cobalamin (Halver and Hardy, 2002). Folic acid is reduced to the active form THF in the tissues.

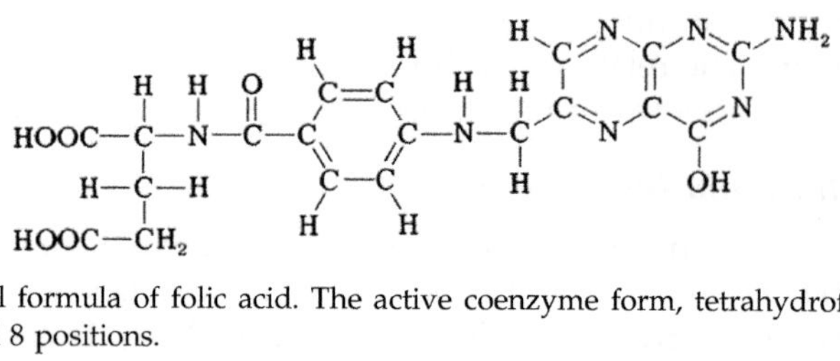

FIGURE 3.9 Structural formula of folic acid. The active coenzyme form, tetrahydrofolic acid (THF), is hydrogenated in the 5, 6, 7, and 8 positions.

3.2.7.2 *Positive functions*

Folate is activated (reduced) by ascorbic acid to its active coenzyme form, tetrahydrofolate [5-formyl-5,6,7,8-tetrahydropteroylglutameric acid (5-formyl-THF)]. The active form of folate functions as a carrier and donor of 1C units in the intermediary metabolism of lipids, amino acids, and DNA, involving several other essential nutrients, like cobalamin, biotin, niacin, methionine, choline, betaine, and homocysteine (NRC, 2011). THF is involved in the conversion of serine to glycine in the mitochondria and synthesis of histidine. Folate-dependent reactions are found in the methionine-homocysteine metabolism and the biosynthesis of purines and pyrimidines for nucleotide bases of DNA and RNA (Naderi and House, 2018). These biochemical roles of folate are essential functions in cell division, cell membrane function, regulation of gene expression, and the intermediary metabolism (NRC, 2011). These functions of folate can easily be seen in lack of red blood cell formation and disturbed cell maturation in blood smears from deficient fish (Halver and Hardy, 2002).

Several studies have demonstrated the role of folate in the integrated antioxidant system, immunity, and disease resistance (NRC, 2011). For example, Lin et al. (2011) showed that dietary level of folate affected nonspecific immune responses and measures of tissue oxidation and activity of antioxidant enzymes in grouper. The dietary level of folate to achieve maximum immune response was, however, similar to the level recommended for growth.

For channel catfish, neither liver saturation nor any sign of toxicity was detected with dietary levels as high as 10 mg/kg (Duncan et al., 1993), whereas a similar dietary level caused a moderate reduction in hematological parameters in rainbow trout while several clinical immune markers were stimulated (Firouz et al., 2013).

Folate has been in focus for roles in epigenetic regulation through intermediate metabolites, together with other B-vitamins and amino acids that take part in 1C metabolism through the folate and methionine cycles (Ducker and Rabinowitz, 2017). Feeding studies in fish indicate that nutrients can adjust the epigenetic chromatin and alter gene expression and phenotype even after feeding regimes end, both within one generation, but more interestingly, through generations (Skjærven et al., 2018; Adam et al., 2019).

3.2.7.3 *Deficiency syndromes*

General symptoms in fish include growth retardation, reduced feed intake and efficiency, and later more classical macrocytic anemia (Tables 3.3 and 3.16). The anemia related to folate deficiency is classified by macrocytic normochromic red blood cells, with anisocytosis and poikilocytosis. Both blood smears (Fig. 3.11) and anterior kidney imprints show macrocytic cells and proerythrocytes arrested in cell division (Halver and Hardy, 2002). Other signs that have been observed are poor growth, anorexia, general anemia, lethargy, fragile fins, dark skin pigmentation, and infarction of the spleen in various species of fish (Table 3.16).

As folic acid and vitamin B_{12} have complementary roles in metabolism, the deficiency of folate has similar symptoms to vitamin B_{12} deficiency (NRC, 2011). Combined deficiency of folate and vitamin B_{12} accelerated the onset of anemia (Waagbø et al., 2001).

Thus, supplementation of folate in excess may mask vitamin B_{12} deficiency. On the other hand, a deficiency of vitamin B_{12} can cause a conditional folate deficiency, by blocking folate metabolism and causing accumulation of 5-methyl folate ("methyl trap") (Naderi and House, 2018).

TABLE 3.16 Folate deficiency signs in fish and crustacean species beyond mortalities and reduced growth.

Deficiency symptom	Description	Species	References
Anorexia	Reduced feed intake.	Channel catfish	Dupree (1966)
		Japanese eel	Arai et al. (1972)
		Japanese parrotfish	Ikeda et al. (1988)
Anemia	Megaloblastic anemia with large, fragmented cells, nucleus in erythrocytes	Atlantic salmon	Sandnes (1994); Waagbø et al. (2001)
		Channel catfish	Duncan et al. (1993)
		Coho salmon	Smith (1968); Smith and Halver (1969)
		Rainbow trout	Kitamura et al. (1967a)
		Rohu	John & Mahajan (1979)
Changed skin coloration	Darker skin pigmentation	Japanese eel	Arai et al. (1972)
Changed behavior	Abnormal swimming	Rainbow trout	Kitamura et al. (1967a)
	Lethargy	Channel catfish	Dupree (1966)
	Hyperirritability	Japanese parrotfish	Ikeda et al. (1988)
Damage to organs	Fins (atrophy)	Rainbow trout	Kitamura et al. (1967a)
	Gill barrier function Gill oxidative status	Grass carp	Shi et al. (2015)
Immunity	Gill immune status	Grass carp	Shi et al. (2015)
Other	Increased relative liver weight	Rainbow trout	Kitamura et al. (1967a)
	Vulnerable to bacterial infections	Chinese mitten crab	Wei et al. (2016)

Modified from Hansen, A., Waagbø, R., Hemre, G., 2015. New B vitamin recommendations in fish when fed plant-based diets. Aquacult. Nutr. 21, 507–527.

Recently, folate deficiency was studied in gill tissue of young grass carp, causing oxidative damage, cellular membrane disruptions, and expression of genes markers indicative to decreased immune status (Shi et al., 2015). This supports historical studies reporting reduced growth and increased sensitivity to oxidative stress and bacterial infection in folate-deficient fish (Lin et al., 2011; NRC, 2011) and Chinese mitten crab (Wei et al., 2016).

Several warm water fish and crustacean species fed a folate free diet did not show deficiency symptoms, probably due to synthesis of folate from intestinal microbial production (NRC, 2011).

3.2.7.4 *Requirements*

The requirement seems to be similar among fish species, mostly ranging between 1 and 2 mg/kg (Table 3.17). A recent study on fingerling *Catla catla* fed a semisynthetic diet indicated a requirement of 0.4 mg/kg, based on growth and feed performance, protein retention, and antioxidant and immune parameters, while 0.6 mg/kg was needed for liver folate saturation (Khan and Khan, 2020a).

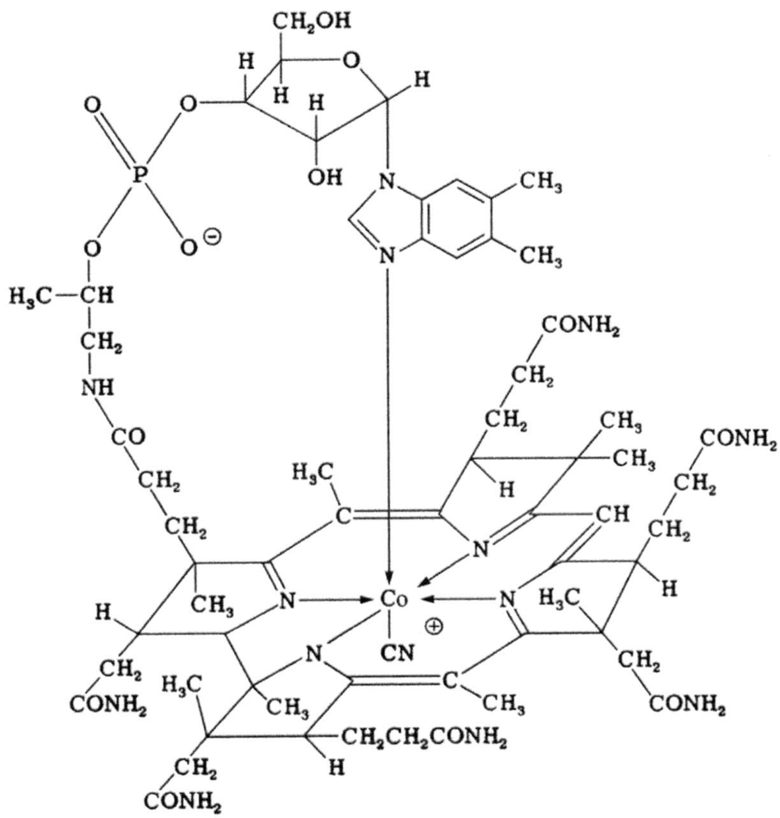

FIGURE 3.10 Structural formula of vitamin B_{12} (cyanocobalamin), a planar corrin ring structure with a monovalent cobalt at the center. The cyanogroup (-CN) can be replaced by other groups to form hydroxy, nitrite, 5′-deoxyadenosyl or methylcobalamin.

In Atlantic salmon smolts exposed to temperatures between 10 and 20°C for 60 days, liver folate increased by 22% with high temperature, while muscle folate decreased with 67% (Waagbø et al., 2020). It can be questioned if the folate requirement increases with temperature, since the liver status was at the low end of a requirement study, while a folate requirement estimates at 3.3 mg/kg for salmon fed a plant-based diet (Hemre et al., 2016).

The requirement recommendation for folate seems to be considerably higher for crustaceans than for fish (Tables 3.1 and 3.17). This may relate to leaching problems due to the physical properties of the crustacean feed and the slow feeding. However, with assumed favorable roles of folate in the integrated antioxidant defense, nonspecific host defense, and resistance to bacterial infections, a higher requirement for crustacean may be justified (Wei et al., 2016).

3.2.7.5 *Sources and protection*

Folates are susceptible to oxidation during processing, enhanced by oxygen, light, and heat. Heat-treated dried feed ingredients have low levels of folate and need to be

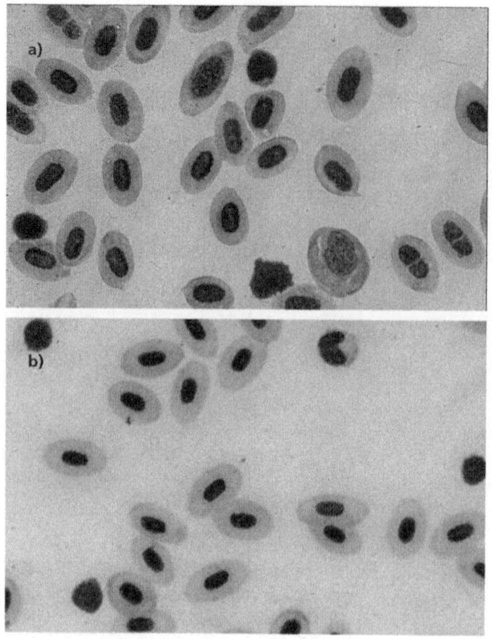

FIGURE 3.11 (A) Blood smears showing megaloblastic anemia in Atlantic salmon suffering from a combined folate and cobalamin deficiency, with large and fragmented nucleus in erythrocytes and white blood cells, as compared to (B) normal blood cells. *From Waagbø, R., Graff, I., Hamre, K., 2001. Vitaminer. In: Waagbø, R., et al. (Eds.), Fiskeernæring, 94–123, with permission from Norsk Fiskeoppdrett AS.*

supplemented through stable folic acid products. Insects contain xanthopterin, which has folic acid activity and the same nuclear structure as folic acid. Some folic acid may be synthesized by gut bacteria, especially in some warm-water fish species (NRC, 2011). In pond culture, where the diet is supplementary to nature and fish eat aquatic and terrestrial insects, algae, and other natural foods, folic acid in the supplement is not as critical as in intensive aquaculture.

Folate is added to the vitamin premix in a dry crystalline or spray-dried form. It is among the most unstable vitamins, and 35%–50% activity is lost during feed extrusion and extended storage, and when material is exposed to sunlight (Table 3.2). Feed should be kept in a cool, dark place, and protected from moist.

3.2.7.6 *Antimetabolites and inactivation*

One antagonist for folacin, 4-aminopteroylglutamic acid or aminopterin, is used in cancer treatment. The compound amethopterin (4-amino-N10-methylpteroylglutamic acid) can be used to induce folate deficiency by suppressing 4-amino coenzymes, resulting in reduced biosynthesis of purine, pyrimidine, and nucleic acids. When nucleic acid synthesis is inhibited, macrocytic anemia eventually occurs (Halver and Hardy, 2002).

3.2.7.7 *Clinical assessment*

Analysis of folate in feed and tissues can be performed by microbiological assay, which covers many folate vitamers (Mæland et al., 2000). Assessment of folate status in muscle,

TABLE 3.17 Requirement for folate in different fish and crustacean species.

Species[b]	Requirement (mg/kg)	Response parameter	References
Salmonids			
Atlantic salmon	2.1 3.3	Liver folate	Sandnes (1994); Waagbø et al. (2001) Hemre et al. (2016)
Rainbow trout	0.6	Growth	Cowey and Woodward (1993)
	1.1	Liver storage	
Freshwater fish			
Channel catfish	1.0[a]	Growth	Duncan et al. (1993)
	1.2[a]	Hematocrit	
	≥10.0[a]	Liver storage	
Hybrid tilapia	0.8[a]	Growth	Shiau and Huang (2001a)
	6.0[a]	Liver storage Hepatosomatic index	
Grass carp	3.6–4.3	Growth	Zhao et al. (2008)
Green catfish	2	Growth, survival, feed and protein efficiency	Hien and Doolgindachbaporn (2011)
Nile tilapia	1.0	Growth, hematology	Barros et al. (2009)
Grass carp	1.6 2.1	Growth, gill lysozyme activity, gill oxidative status	Shi et al. (2015)
Catla catla	0.4 0.6	Growth, feed and protein efficiency, antioxidant and immune parameters, liver folate	Khan and Khan (2020a)
Blunt snout bream	0.68–0.82	Growth, digestive enzymes, immunity, antioxidation	Sesay et al. (2016)
Siberian sturgeon	1.8–4.4	Growth, composition, hematology, antioxidation, immunity	Falah et al. (2020)
Marine fish			
Yellowtail	0.8	Growth	Masumoto (2002)
	1.2	Liver storage	
Grouper	0.7	Growth Liver storage Nonspecific immune response Hepatic oxidative status	Lin et al. (2011)

Continued

TABLE 3.17 Requirement for folate in different fish and crustacean species.—cont'd

Species[b]	Requirement (mg/kg)	Response parameter	References
Crustaceans			
Penaeus chinensis	5–10	Growth, survival	Liu et al. (1995)
Grass shrimp	1.9–2.1	Growth, Hepatopancreas folate, hepatosomatic index	Shiau and Huang (2001b)
Chinese mitten crab	2.3 2.9	Growth, antioxidant enzymes, hepatopancreas folate, hemocyte count	Wei et al. (2016)

[a]*Added amount (not confirmed by analysis).*
[b]*Latin names of fish species: Atlantic salmon* (Salmo salar)*, Rainbow trout* (Oncorhynchus mykiss)*, Channel catfish* (Ictalurus punctatus)*, Hybrid tilapia* (Oreochromis niloticus x O-aureus)*, Blunt snout bream* (Megalobrama amblycephala)*, Grass carp* (Ctenopharyngodon idella)*, Green catfish* (Mystus nemurus*, Valenciennes, 1840*)*, Siberian sturgeon* (Acipenser baerii)*, Yellowtail* (Seriola quinqueradiata)*, Grouper* (Epinephelus malabaricus)*, Freshwater prawn* (Macrobrachium rosenbergii)*, Grass shrimp* (Penaeus monodon)*, Chinese mitten crab* (Eriocheir sinensis)*.*
Modified from Hansen, A., Waagbø, R., Hemre, G., 2015. New B vitamin recommendations in fish when fed plant-based diets. Aquacult. Nutr. 21, 507–527.

liver, and plasma in Atlantic salmon postsmolts fed diets differing in folate (0.4 or 9.2 mg/kg) showed that plasma folate was the most sensitive status indicator, with 2, 1.6, and 7.7 times increase between the diets, respectively (Waagbø et al., 2001). From here, representative folate status in muscle, liver, and plasma are 0.2 mg/kg, 10.6 mg/kg, and 37 ng/mL, respectively. Classical hematological analyses, including blood and head kidney imprints, are very useful in diagnosing folate deficiency. Although relative liver weight (hepatosomatic index, HSI) is not a specific marker, it has been a parameter for assessing folate requirement in both fish and shrimp species (Shiau and Huang, 2001a,b; NRC, 2011).

3.2.8 Cobalamin (vitamin B_{12})

An antipernicious anemia factor was isolated and crystallized in 1948, and shortly after in 1949, Halver and Norris successfully injected salmon with vitamin B_{12} alone or in combination with folate and xanthopterin to cure anemia. One year later, Wolf included both vitamin B_{12} and folic acid in the trout test diet (NRC, 2011).

3.2.8.1 Chemical structure and characteristics

Vitamin B_{12} or cyanocobalamin (molecular weight of 1355 and structural formula in Fig. 3.10) is the vitamin with the most complex structure, including a planar corrin ring structure and a perpendicular nucleotide group. With the ring structures and a cobalt atom in the center, the vitamin appears as a red crystal compound. The corrin ring structure is similar to porphyrin and chlorophyll, with their respective central metals in form of iron and magnesium atoms. The cobalt molecule can link to a cyanogroup (cyanocobalamin) or other small chemical groups forming hydroxocobalamin, hydrocobalamin, 5′-deoxyadenosylcobalamin, and methylcobalamin. The two latter are the active coenzymes. However, several vitamers B_{12} with varying degrees of vitamin activity exist. Mild in vitro treatment with cyanide converts these analogs into cyanocobalamin, which can be useful in the analysis of total vitamin

B_{12} (Mæland et al., 2000). The compound is unstable during storage and in heated acid or alkaline solutions. Consequently, feed should be stored cold and for shorter periods before fed to the fish. Moist diets should be maintained at a neutral pH.

3.2.8.2 *Positive functions*

Cobalamin is needed for cell division, like erythropoiesis, the nervous system, and in the energy metabolism of fatty acids and amino acids (NRC, 2011, Table 3.3; Fig. 3.1). Vitamin B_{12} is involved as coenzyme in two main biochemical pathways through the enzyme *methylmalonyl-CoA mutase* in fatty acid metabolism, converting methylmalonyl coenzyme A to succinyl coenzyme A, and the enzyme *methionine synthase* in transmethylation of homocysteine to form methionine, and recycling of folate. These are processes in cell metabolism, growth, and differentiation (Smith et al., 2018). Thus, vitamin B_{12} functions together with folate, which is classically seen in hemopoiesis in the development and maturation of erythrocytes. A deficiency leads to large cells arrested during cell division, visualized as a megalocytic anemia (NRC, 2011, Fig. 3.11).

Other biochemical reactions where vitamin B_{12} participates as coenzyme are isomerization of methyl aspartate to glutamate, fatty acid metabolism, and several 1C metabolism reactions and in the synthesis of labile methyl compounds, methylation of the purine ring during thymine synthesis, cholesterol metabolism, purine and pyrimidine biosynthesis, and the metabolism of glycols (Hansen et al., 2015).

There is an increased focus on 1C nutrients (methionine, vitamin B_6, vitamin B_{12}, and folate) in fish energy metabolism. A recent cell model study showed that lipid metabolism in isolated liver cells from Atlantic salmon was affected by the 1C diets fed to the fish prior to cell isolation. Cells isolated from fish fed low dietary 1C nutrients seem to have a higher assembly of triacylglycerol (TAG), as well as synthesis of fatty acids. Expression of selected genes indicated higher lipid synthesis and oxidation as compared to cells isolated from fish fed the surplus 1C diets (Espe et al., 2020). Previously, this group showed that parental feeding with a 1C nutrient-deficient diet epigenetically changed DNA methylation and expression of lipid genes associated with disturbed lipid metabolism and fatty liver in the offspring, as compared to parents fed a diet sufficient in 1C nutrients (Skjærven et al., 2018).

3.2.8.3 *Deficiency syndromes*

Fish and crustacean species with vitamin B_{12} deficiency symptoms show poor growth, and reduced feed intake, feed efficiency, and high mortalities (Li et al., 2016b; Zehra and Khan, 2019a; Khan and Khan, 2020b). Blood smears from fish show classical macrocytic anemia (pernicious anemia) with fragmentation of erythrocytes ("shoe print"). Anemia appears as fragmented erythrocytes with many aberrant forms present. Since vitamin B_{12} functions closely to folate (THF), they share many deficiency symptoms (Tables 3.3, 3.16 and 3.18).

It is assumed that fish, like humans, need a small molecular weight glycoprotein (intrinsic factor, IF) for intestinal absorption and other special proteins for cobalamin transportation (transcobalamin) in the body (Smith et al., 2018). A state of deficiency may, therefore, appear both because of missing IF and lack of dietary cobalamin. However, despite that transcripts of genes for several cobalamin-related proteins were found (haptocorrin, transcobalamin, cubilin, amnionless, and magalin), no IF or transcobalamin receptors were detected in the

TABLE 3.18 Cobalamin deficiency signs in fish and crustacean species beyond mortalities and reduced growth.

Deficiency symptom	Description	Species	References
Anorexia	Reduced feed intake	Chinook salmon	Halver (1957)
		Japanese eel	Arai et al. (1972)
		Japanese parrotfish	Ikeda et al. (1988)
Anemia	Megaloblastic anemia; large, fragmented nucleus in circulating blood cell populations	Atlantic salmon	Sandnes (1994); Waagbø et al. (2001)
		Salmonids	Woodward (1994)
		Brown trout	Phillips et al. (1994)
		Channel catfish	Limsuwan and Lovell (1981)
		Chinook salmon	Halver (1957)
		Rohu	John and Mahajan (1979)

Modified from Hansen, A., Waagbø, R., Hemre, G., 2015. New B vitamin recommendations in fish when fed plant-based diets. Aquacult. Nutr. 21, 507–527.

zebrafish (Hansen et al., 2013). Due to reabsorption, the cobalamin stores are slowly (12–16 weeks) exhausted in fish fed a deficient diet (Halver, 1957).

In humans, neurological signs of cobalamin deficiency can often appear in the absence of anemia, and anemia should, therefore, not be the sole marker for vitamin B_{12} deficiency or insufficiency (Smith et al., 2018). The deficiency may be related to appearance of inactive vitamin B_{12} analogs that associate with neurological abnormalities (Carmel et al., 1988). This may be the case for fish and crustaceans as well.

3.2.8.4 Requirements

Synthesis of vitamin B_{12} by the intestinal microflora covered the requirement for several warm water species (NRC, 2011). The data until now demonstrate qualitative requirement for most species, while quantitatively estimates are between 0.015 and 0.02 mg/kg dry diet for salmonids (Table 3.19).

Tiger shrimp seem to require 0.2 mg/kg feed to prevent anemia (Shiau and Lung, 1993a). Similarly, the optimum dietary vitamin B_{12} requirement recommended for juvenile Chinese mitten crab is 0.20–0.22 mg/kg diet based on growth rate and total hemocyte count (Wei et al., 2014). Thus, the cobalamin requirement seems to be higher in crustaceans than in fish species. As for fish species, dietary folic acid and cobalamin interact and may mask deficiency signs in crustacean (Shiau and Lung, 1993b). While no adverse effects were identified in Atlantic salmon fed a dietary level of 0.7 mg/kg feed (Hemre et al., 2016), cobalamin supplementations far above the requirement may lead to growth depression in crabs (Wei et al., 2014).

TABLE 3.19 Requirement for cobalamin in different fish and crustacean species.

Species[a]	Requirement (mg/kg)	Response parameter	References
Salmonids			
Pacific salmon	0.015–0.02	Liver cobalamin storage	Halver and Hardy (2002)
Atlantic salmon	<0.2 0.17	Liver cobalamin storage	Sandnes (1994) Hemre et al. (2016)
Rainbow trout	0.007	Growth	Woodward (1994)
Freshwater fish			
Zebrafish	>0.05	Whole body cobalamin storage	Hansen et al. (2013)
Major carp	0.04	Growth, protein retention, hematology, antioxidant, and immune parameters	Khan and Khan (2020b)
Blunt snout bream	0.06	Growth, digestive and absorptive capacity, endocrine response	Li et al. (2016b)
Snakehead	0.067	Growth, liver cobalamin, antioxidant capacity, intestinal enzymes, and immune parameters	Zehra and Khan (2019a)
Channel catfish	0.02	Growth	Limsuwan and Lovell (1981)
	>0.32	Liver cobalamin storage	
Marine fish			
Yellowtail	0.05	Growth, liver cobalamin	Shimeno (1991); Masumoto (2002)
Crustaceans			
Grass shrimp	0.2	Growth, hematology	Shiau and Lung (1993b)
Penaeus chinensis	0.01–0.02	Growth, survival	Liu and Li (1995)
Chinese mitten crab	0.20–0.22	Growth, survival, hepatopancreas cobalamin, whole-body protein, hemocyte count, nonspecific immunity	Wei et al. (2014)

[a]*Latin names of fish species: Pacific salmon* (Oncorhynchus *spp.*), *Atlantic salmon* (Salmo salar), *Rainbow trout* (Oncorhynchus mykiss), *Zebrafish* (Danio rerio), *Channel catfish* (Ictalurus punctatus), *Yellowtail* (Seriola quinqueradiata), *Major carp* (Catla catla, *Hamilton*), *Grass carp* (Ctenopharyngodon idella), *Blunt snout bream* (Megalobrama amblycephala), *Snakehead* (Channa punctata), *Grass shrimp* (Penaeus monodon), *Chinese mitten crab* (Eriocheir sinensis).

Modified from Hansen, A., Waagbø, R., Hemre, G., 2015. New B vitamin recommendations in fish when fed plant-based diets. Aquacult. Nutr. 21, 507–527.

3.2.8.5 Sources and protection

Neither plants nor animals can synthesize vitamin B_{12}, so the vitamin is mainly originated in animals and animal products from microbial synthesis (Smith et al., 2018). Cobalamin does not generally occur in plant ingredients, while animal feed ingredients, like fish meal, can be rich in the vitamin (0.2–0.4 mg/kg). Thus, fish meal–based diets do not need to be supplemented as vitamin B_{12} (Table 3.1). Despite a dispute on the origin of cobalamin, algae like spirulina have been shown to be rich in cobalamin (Ragaza et al., 2020). Vitamin B_{12} is a fermented product and supplemented in dry form (1% activity). The stability of vitamin B_{12} during feed extrusion and storage ranges from 40% to 88% (Table 3.2).

3.2.8.6 Antimetabolites and inactivation

While most cobalamin forms can be converted to the active coenzymes, adenosylcobalamin and methylcobalamin, some corrinoids and cobalamins are considered inactive vitamin B_{12} analogs that by analysis can mask a cobalamin deficiency (Carmel et al., 1988). The cobalamin molecule is very unstable to light and in acid or alkali solutions (Halver and Hardy, 2002). This is important to consider, both in moist feed preparation and during cobalamin analysis (Mæland et al., 2000).

3.2.8.7 Clinical assessment

Tissue cobalamin concentration has been used for quantifying vitamin requirements of aquatic animals and recording vitamin status (Table 3.19). Typical liver and muscle cobalamin concentrations are 0.6 and 0.07 mg/kg in Atlantic salmon fed sufficient folate and cobalamin (Waagbø et al., 2001). Interestingly, increasing dietary cobalamin above the requirement, from 0.2 to 0.4 mg/kg, did not increase liver and muscle levels in Atlantic salmon, while plasma cobalamin rose from 39 to 62 ng/mL. Plasma concentration may, therefore, be a nonlethal way to assess cobalamin status in fish, as in human clinical medicine. Hematological assessments, like blood hemoglobin, blood cell count, and blood smears, may give different indications on the progression of macrocytic anemia, depending on the cause and severity of the cobalamin deficiency. Since the biochemical functions of vitamin B_{12} and folate are tightly connected in metabolism, differential diagnosis of cobalamin or folate deficiency can be difficult. Hematology in fish with vitamin B_{12} deficiency is characterized by varied fragmented erythrocytes and varying red blood cell count, whereas folic acid deficiency shows only senile cells, some with pycnotic nuclei, characteristic of macrocytic normochromic anemia (Halver and Hardy, 2002). In humans, increased levels of homocysteine, due to reduced remethylation, is used as a biomarker for suboptimal vitamin B_{12} and folate status. Homocysteine is mostly used as a biochemical marker in the assessment of cardiovascular and renal diseases, as well as other organ dysfunctions (Kozat and Okman, 2017).

3.2.9 Ascorbic acid

The beneficial effect of curing scurvy with fruit juice was first identified in 1753. The factor behind was named vitamin C in 1920, identified in 1932, and renamed as ascorbic acid in 1933. Then, its interest for animal nutrition was recognized to prevent guinea pigs from the scurvy in 1933 and to be essential for trout in 1935. Thereafter, a very large number of studies have focused on the deficiency syndromes and requirements of ascorbic acid in aquatic animals during 1940–2000s (Dabrowski, 2000). In the last 2 decades, deeper insight has been obtained on the antioxidant, antistress, and immunomodulatory roles of ascorbic acid and its interaction with other nutrients in aquatic animals.

3.2.9.1 *Chemical structure and characteristics*

The chemical formula of ascorbic acid (2,3-didehydro-l-threohexano-1,4-lactone; also L-ascorbic acid) is $C_6H_8O_6$ (Fig. 3.12). The term vitamin C describes all compounds exhibiting the biological activity of ascorbic acid. There are two major forms, reduced (L-ascorbic acid) and oxidized (L-dehydroascorbic acid) forms present in any diet. The reduced form of vitamin C is white, crystalline, odorless, and soluble in water but insoluble in fat solvents. Ascorbic acid is very unstable, sensitive to heat, light and heavy-metal ions, and rapidly oxidized (Halver and Hardy, 2002).

3.2.9.2 *Positive functions*

Vitamin C is absorbed by the small intestine and reabsorbed by the renal tubules and then circulated in the blood to enter the target organs/tissues of the body, where ascorbic acid and dehydroascorbic acid can transform each other (Wilson, 2005). In humans, vitamin C is absorbed through passive transport and active transport at high and low doses, respectively. Reduced and oxidized forms are uptaken by the Na^+-dependent vitamin C transporters (SVCTs) and glucose transporters (GLUTs), respectively (Corpe et al., 2013). Two SVCTs (SVCT1 and SVCT2) have been cloned in human and Senegalese sole (Jiménez-Fernández et al., 2012). Vitamin C is taken up by the intestine, liver, and kidney via SVCT1, and by the other organs and tissues via SVCT2 in human (MacDonald et al., 2002). In fish, SVCTs-depended Na^+-coupled ascorbic acid transport mechanism at the brush border membrane of enterocytes is similar to those found in mammals (Maffia et al., 2000).

Vitamin C has important physiological functions as follows.

3.2.9.2.1 Antioxidant functions

Ascorbic acid acts primarily as an electron donor, which can react with oxidants, like free radicals (Bae et al., 2012a). Such quenching of oxidants can reduce the toxicities of heavy metals or toxins, regenerate the metabolically active form of vitamin E, affect reproduction, protect glutathione in its reduced form, and prevent cataracts caused by cumulative photo-oxidants of the lens in fish (Sealey and Gatlin, 2002; Gao et al., 2014). Ascorbic acid enhances antioxidant protection to maintain the structural integrity of cells, which is related to the

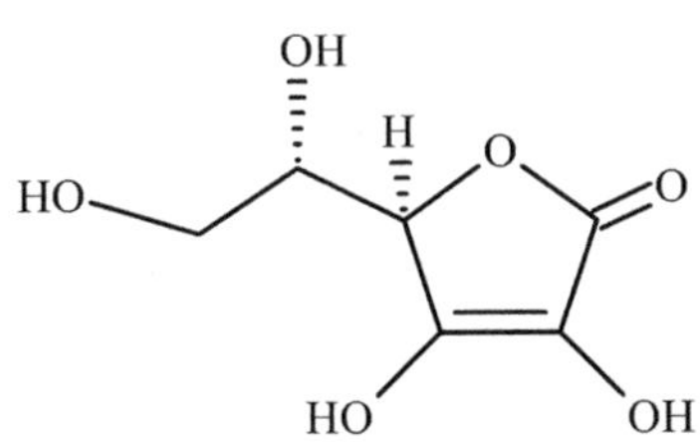

FIGURE 3.12 The structural formula of ascorbic acid.

increase of nonenzymic and enzymic antioxidant abilities modulated by the Nrf2/Keap1 signaling in various organs of grass carp (Xu et al., 2016). This antioxidant protection action also occurs in various organs of some aquatic animals like rainbow trout (Dabrowski et al., 2004), prawn (Asaikkutti et al., 2016), and Pacific abalone (Wu et al., 2014).

3.2.9.2.2 Promotion of collagen synthesis

Collagens, the basis of connective tissues, are found in the skin, ligaments, cartilages, vertebral discs, joint linings, bones, teethes, and capillary walls in humans. Vitamin C acts a cofactor in the hydroxylation of proline and lysine to hydroxyproline and hydroxylysine in procollagen and, thus, is necessary for the formation of connective tissue, normal cartilage, bone formation, wound repair, and healthy blood vessels in animals. Vitamin C can increase bone collagen content, facilitate bone health, prevent hemorrhages, and regenerate the skin and muscle at the wound sites in channel catfish (Lim and Lovell, 1978), as well as promote wound healing in rainbow trout (Wahli et al., 2003).

3.2.9.2.3 Improving iron absorption and utilization

Ascorbic acid can reduce insoluble ferric iron (Fe^{3+}) to the soluble ferrous form (Fe^{2+}) to facilitate enteric absorption of iron and later participates in releasing iron from transferrin and ferritin, and later incorporated into hemoglobin in human. Vitamin C improves iron absorption in grass shrimp (Hsu and Shiau, 1999).

3.2.9.2.4 Enzyme cosubstrate functions and strengthening stress resistance

Vitamin C functions as a cosubstrate for at least 10 enzymes that function in electron transport reactions involved in the synthesis of norepinephrine, peptide hormones, and carnitine, and the metabolism of tyrosine, xenobiotics, and steroids. These hormones or metabolites are involved in stress responses. Vitamin C can enhance the resistance against various stresses in aquatic animals, such as high temperature stress in Wuchang bream (Ming et al., 2012) and salinity stress in Nile tilapia (Caxico et al., 2018).

3.2.9.2.5 Immunomodulation

Vitamin C enhances innate immune defense including promotion of leukocyte proliferation in pirarucu (de Andrade et al., 2007), macrophage chemotaxis in channel catfish (Lim et al., 2000), and respiratory burst of phagocytes in the head kidney of sea bream (Ortuño et al., 2001). Meanwhile, vitamin C increases at least one of antibacterial compound activities or contents, like lysozyme, acid phosphatase (ACP), complements, hemagglutination, antimicrobial peptide in the serum/plasma, and (or) immune organs in fish like Atlantic salmon (Waagbø et al., 1993a). Vitamin C enhances acquired immune response by promotion of antibody production in Indian major carp (Nayak et al., 2007) and increase in serum total immunoglobulin concentration in juvenile cobia (Zhou et al., 2012). Furthermore, vitamin C can mitigate the inflammatory response, which might be related to the inhibition of NF-κB signaling and enhancement of target of rapamycin (TOR) signaling pathways in immune organs of grass carp (Xu et al., 2016).

3.2.9.3 *Deficiency syndromes*

Most terrestrial animals (except human, monkeys, guinea pigs, some of birds, insects, and invertebrates) can de novo synthesize ascorbic acid from glucose by the key synthase, L-gulonolactone oxidase (Dabrowski, 2000) (Fig. 3.13). In fish, only nonteleost taxa including lungfishes, sharks and rays, sturgeons and gars can de novo self-synthesize vitamin C in the kidney (Moreau and Dabrowski, 2000; Cho et al., 2007). However, most fish especially teleost fish and crustaceans cannot synthesize vitamin C due to lack of L-gulonolactone oxidase (Zempleni et al., 2013). Thus, an adequate external supply is required to meet their requirements.

As shown in Table 3.20, dietary inadequate vitamin C causes growth retardation, increase of mortality, deformities, and clinical and pathological changes in aquatic animals.

Excess vitamin C causes scoliosis, lordosis and kyphosis, prehaemal lordosis and deformed proximal pterygiophores of the dorsal and anal fins, pugheadness, prominent lower jaw, deformed caudal fin complex in ossified specimen, and deformed caudal fin complex in nonossified specimen indicating that deformities in this region developed before ossification in the larvae of European sea bass (Darias et al., 2011).

3.2.9.4 *Requirements*

The requirements of vitamin C for aquatic animals are listed in Table 3.21, which can be affected by the following factors: **(1) Species:** Requirements of vitamin C are various among some aquatic animal species, which range from 8.3 to 750 mg/kg diet; **(2) Vitamin C sources:** Generally, the use of unstable, heat-labile, easily water-soluble forms like "ascorbic acid" in the feeds is often overestimation of vitamin C requirements, while the use of stable forms like ascorbate phosphate forms of vitamin C is the most reliable estimation of the vitamin C requirements (Woodward, 1994; Gouillou-Coustans and Kaushik, 2000); **(3) Response criteria:** The vitamin C requirements based on vitamin C nutritional status in serum or tissue are higher than these based on growth.

3.2.9.5 *Sources and protection*

Vitamin C is widely distributed in plants and animals. The natural rich sources of vitamin C are citrus, green leafy plants, bell peppers, broccoli, spinach, tomatoes, potatoes, and animal liver and kidney. The synthetic vitamin C sources include L-ascorbyl-2-sulphate (C2S), L-ascorbyl-2-monophosphate-Mg (C2MP-Mg), L-ascorbyl-2-monophosphate-Ca (C2MP–Ca), L-ascorbyl-2-monophosphate-Na/Mg (C2MP-Na/Mg), L-ascorbyl-2-polyphosphate

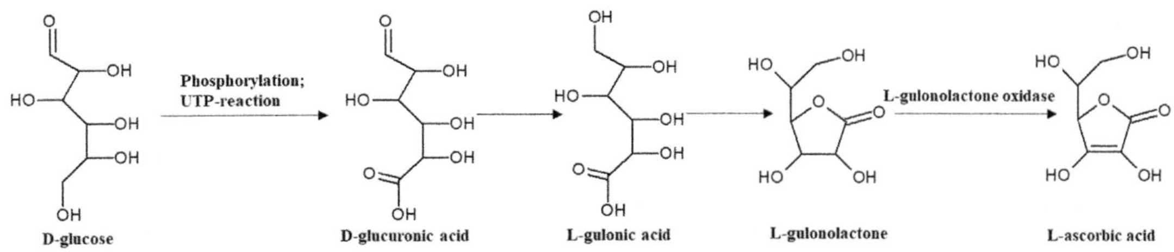

FIGURE 3.13 Ascorbic acid biosynthesis in animals.

TABLE 3.20 The vitamin C deficiency syndromes.

Deficiency syndromes	Species	References
Growth retardation	Grass carp	Xu et al. (2016)
	Tilapia	Wu et al. (2015)
	Catfish	Adham et al. (2000)
	Bass	Darias et al. (2011)
	Red sea bream	El Basuini et al. (2017)
	Japanese eel	Shahkar et al. (2015)
	Grouper	Lin and Shiau (2005a)
	Cobia	Zhou et al. (2012)
	Prawn	Asaikkutti et al. (2016)
	Grass shrimp	Hsu and Shiau (1998)
Increase of mortality and/or decrease of disease resistance ability	Grass carp	Xu et al. (2016)
	India major carp	Mahajan and Agrawal (1980)
	Nile tilapia	Ibrahem et al. (2010)
	Bass	Kop (2019)
	Japanese eel	Bae et al. (2012a)
	Large yellow croaker	Ai et al. (2006)
	Grouper	Lin and Shiau (2005a)
	Cobia	Zhou et al. (2012)
	Prawn	Asaikkutti et al. (2016)
Bone deformities (spinal, jaws, or snout)	Carp	Mahajan and Agrawal (1980)
	Catfish	Yildirim-Aksoy et al. (2008)
	Bass	Kop (2019)
	Red drum	Aguirre and Gatlin (1999)
	Olive flounder	Wang et al. (2002)
Disturbances in moulting	Shrimp	Niu et al. (2009)
Dystropy (anorexia, lethargy, and abnormal swimming)	India major carp	Mahajan and Agrawal (1980)
	Largemouth bass	Chen et al. (2015b)
	European sea bass	Kop (2019)
	Japanese eel	Ren et al. (2005)
Loss of pigmentation	India major carp	Mahajan and Agrawal (1980)

TABLE 3.20 The vitamin C deficiency syndromes.—cont'd

Deficiency syndromes	Species	References
Dark skin coloration	Channel catfish	Lim and Lovell (1978)
	Grouper	Lin and Shiau (2005a,b)
	Largemouth bass	Chen et al. (2015b)
	European sea bass	Kop (2019)
Hemorrhages	India major carp	Mahajan and Agrawal (1980)
	Channel catfish	Lim and Lovell (1978)
	Japanese eel	Ren et al. (2005)
Lesions of the skin with loss of scales and necrosis/erosion of the fins	India major carp	Mahajan and Agrawal (1980)
	Catfish	Lim and Lovell (1978)
	Japanese sea bass	Ai et al. (2004)
Exophthalmia	Largemouth bass	Chen et al. (2015b)
Histopathological changes (gills, liver, and kidney)	Catfish	Adham et al. (2000)
Anemia	Catfish	Adham et al. (2000)
	Great sturgeon	Falahatkar et al. (2006)
	Cobia	Zhou et al. (2012)
	Rainbow trout	Waagbø et al. (1989)
	Striped bass	Fazio et al. (2020)

TABLE 3.21 The vitamin C requirements reported in aquatic animals.

Species	Initial weight (g)	Forms	Requirements (mg/kg diet or mg AA equivalent/kg diet)	Response criteria	References
Grass carp (*Ctenopharyngodon idella* Val.)	264.37	Ascorbic acid	92.8	PWG	Xu et al. (2016)
Common carp (*Cyprinus carpio* L.)	First-feeding larvae	C2PP	45 354	Growth BAAC	Gouillou-Coustans et al. (1998)
India major carp (*Cirrhinus mrigala*)	Hatchlings	Ascorbic acid	650–750	WG	Mahajan and Agrawal (1980)
Jian carp (*Cyprinus carpio* var. Jian)	12.63	LAPP	40.9	SGR	Liu et al. (2011)

Continued

TABLE 3.21 The vitamin C requirements reported in aquatic animals.—cont'd

Species	Initial weight (g)	Forms	Requirements (mg/kg diet or mg AA equivalent/kg diet)	Response criteria	References
Gibel carp (*Carassius auratus gibelio* var. CAS III)	77.2	LAPP	193.2	PAAC	Shao et al. (2018)
Tilapia (*Oreochromis niloticus*)	220.06	LAPP	53.05 180.53 185.86	WG LAAC MAAC	Wu et al. (2015)
African catfish (*Clarias gariepinus*)	19.9	Ascorbic acid	46	ADS, C/H	Eya (1996)
Channel catfish (*Ictalurus punctatus*)	2.3	Ascorbic acid	50	PWG, FE	Andrews and Murai (1975)
Channel catfish (*Ictalurus punctatus*)	2.3	Ascorbic acid	60	WG, FE, C/H	Lim and Lovell (1978)
Yellow catfish (*Pelteobagrus fulvidraco*, Richardson)	2.00	LAPP	114.5 102.5	SGR LA	Liang et al. (2017)
Atlantic salmon (*Salmo salar*)	0.16	C2MP–Ca	10–20	Growth	Sandnes et al. (1992)
Angelfish (*Pterophyllum eimekei*)	1.12	C2MP-Mg	360	PWG	Blom et al. (2000)
Japanese sea bass (*Lateolabrax japonicus*)	6.26	LAPP	53.5 93.4 207.2	SGR LAAC MAAC	Ai et al. (2004)
Grouper (*Epinephelus malabaricus*)	7.58	Ascorbic acid	45.3	PWG	Lin and Shiau (2005a)
	3.20	2 MP-Mg C2MP-Na	17.9 (2 MP-Mg) 8.3 (C2MP-Na)	PWG	Lin and Shiau (2004)
	6.69	C2S C2PP	46.2 (C2S) 17.8 (C2PP)	PWG	Lin and Shiau (2005b)
Large yellow croaker (*Pseudosciaena crocea*)	17.82	LAPP	28.2 87.0	Survival LAAC	Ai et al. (2006)
Juvenile Japanese eel (*Anguilla japonica*)	11.0		27.0	Growth	Ren et al. (2005)
	15	AMP	41.1 41.2 43.9 43.1	PWG SGR FE PER	Bae et al. (2012a)
Brood stock Japanese eel (*Anguilla japonica*)	360	AMP	410.7 911.7	LAAC TAAC	Shahkar et al. (2015)

TABLE 3.21 The vitamin C requirements reported in aquatic animals.—cont'd

Species	Initial weight (g)	Forms	Requirements (mg/kg diet or mg AA equivalent/kg diet)	Response criteria	References
Largemouth bass (*Micropterus salmoides*)	6.67	AMP	109.5 102.6 147.8	WG SGR LAAC	Chen et al. (2015b)
Golden pompano (*Trachinotus ovatus*)	13.57	LAPP	49.73	SGR	Zhang et al. (2019)
Freshwater prawn (*Macrobrachium malcolmsonii*)	0.25	AMP	94.11	WG	Asaikkutti et al. (2016)
White shrimp (*Litopenaeus vannamei*)	0.001	LAPP	191	WG	Niu et al. (2009)
Grass shrimp (*Palaemonetes pugio*)	0.55	C2MP-Na C2MP-Mg	26.7 (C2MP-Na) 22.5 (C2MP-Mg)	PWG	Hsu and Shiau (1998)

AMP, L-ascorbyl-2-monophosphate; *BAAC*, body ascorbic acid content; *C2D*, ascorbate-2-glucose; *C2MP—Ca*, L-ascorbyl-2-monophosphate-Ca; *C2MP-Mg*, L-ascorbyl-2-monophosphate-Mg; *C2MP-Na/Mg*, L-ascorbyl-2-monophosphate-Na/Mg; *C2PP*, L-ascorbyl-2-polyphosphate; *C2S*, L-ascorbyl-2-sulphate; *FE*, feed efficiency; *HK*, head kidney; *LA*, lysozyme activity; *LAAC*, liver ascorbic acid content; *LAPP*, L-ascorbyl-2-polyphosphate; *MAAC*, muscle ascorbic acid content; *PAAC*, plasma ascorbic acid content; *PER*, protein efficiency ratio; *PWG*, percent weight gain; *SGR*, specific growth rate; *TAAC*, testes ascorbic acid content; *WG*, weight gain.

(C2PP), and ascorbate-2-glucose (C2D) (Dawood and Koshio, 2018). Among these, the C2S is nearly not available in some fishes like Atlantic salmon (Sandnes et al., 1990). The biopotencies of the various vitamin C forms are different as following: C2Mp-Na (100%) > C2PP (84%) > C2Mp-Mg (46%) > C2S (32%) in grouper (NRC, 2011). Ascorbic acid is unstable, and most of its activity is lost during processing and storage as a result of exposure to high temperature, oxygen, and light. Meanwhile, copper, iron, or other metals can catalyze ascorbic acid atmospheric oxidation into the biologically inactive form. Thus, when ascorbic acid is used, the feed must be protected from aerobic oxidation, and any moist feed must be carefully protected from oxidizing agents, air, and copper, iron, and other metals. In addition, stable form like phosphate derivatives of ascorbic acid are generally used in feed manufacture.

3.2.9.6 *Antimetabolites and inactivation*

D-ascorbic acid, 6-deoxy-L-ascorbic acid, and L-glucoascorbic acid (without or with low activities) can compete with vitamin C for sites of several enzyme reactions. Meanwhile, hexoses (e.g., glucose and fructose) and flavonoids (e.g., phloretin and quercetin) can competitively inhibit the uptake of dehydroascorbic acid due to co-use of the transporters of GLUTs (Corpe et al., 2013).

3.2.9.7 *Clinical assessment and analysis*

The vitamin C nutritional status in aquatic animals can be determined by lack of deficiency signs coupled with growth performances, and analysis of ascorbic acid concentration in the serum/plasma/blood, various organs, and carcass (Mai, 1998; Ai et al., 2004; Zhou et al., 2012; Shahkar et al., 2015). In addition, bone collagen levels were used to measure the vitamin C status in large yellow croaker (Ai et al., 2006). Ascorbic acid and its derivatives can be quantified by high-performance liquid chromatography (HPLC) techniques (Dabrowski, 2000).

3.2.10 Inositol

Inositol was discovered by Scherer in 1850, characterized in 1887, and recognized as an essential nutrient for mouse in 1940. Right from 1947, the importance of dietary inositol supply in fish has been recognized. Thereafter, a few studies focused on the deficiency syndromes and requirements of this vitamin in some aquatic animals during 1940–2000s. In the last 2 decades, researches shed light on the nutritional roles of inositol in antioxidant, osmoregulation, immunomodulation, lipid metabolism, and requirements for numerous aquatic animals (Shirmohammad et al., 2016).

3.2.10.1 *Chemical structure and characteristics*

Inositol is a cyclohexitol with the chemical formula of $C_6H_{12}O_6$ (Fig. 3.14). It occurs in nine stereoisomers, including seven natural forms (*myo-*, *scyllo-*, *muco-*, D-*chiro-*, L-*chiro-*, *neo-*, and *cis*-inositol) and two unnatural forms (*allo-* and *epi*-inositol) (Schneider, 2015). Among these, *myo*-inositol (cis-1,2,3,5 trans-4,6-cyclohexane-hexanol) is the most prevalent biologically active and the only nutritionally important form (Su and Shiau, 2004). *Myo*-inositol is a white, crystalline powder, soluble in water, and insoluble in alcohol and ether (Combs and McClung, 2017).

3.2.10.2 *Positive functions*

The predominantly absorptive and transport form is free *myo*-inositol, which depends on active transport by transporters in mammals. To date, three *myo*-inositol transporters, including Na^+-coupled *myo*-inositol transporter-1 (SMIT1), SMIT2, and H^+-dependent, HMIT, have been identified in mammals (Schneider, 2015). *Myo*-inositol is absorbed from the chyme into the enterocytes by SMIT2 and then is taken up by the liver via carrier-mediated diffusion, by the brain via SMIT1 and SMIT2, by the renal medulla via SMIT1, and by proximal tubule cells in the kidney cortex via SMIT2 in human (Aouameur et al., 2007). Homologs of the SMITs and HMIT are also present in teleost genomes

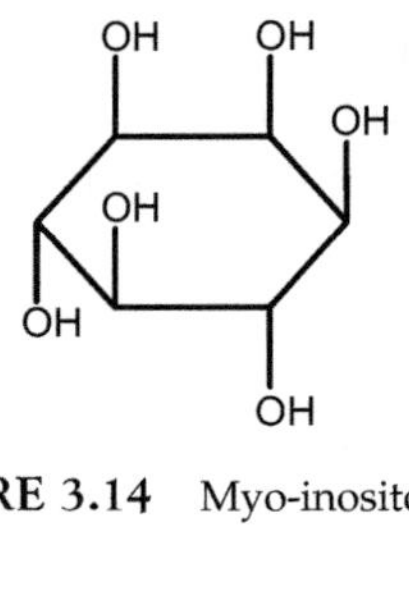

FIGURE 3.14 Myo-inositol form.

(Kalujnaia et al., 2016). Within eukaryote cells, *myo*-inositol can be converted to phosphatidylinositol (PI) by PI synthetase.

Myo-inositol and its metabolites have important physiological functions as follows.

3.2.10.2.1 Antioxidant and metal chelate actions

The free alcoholic hydroxyl group in *myo*-inositol can donate hydrogen or chelate metal ions (Santoro et al., 2007). *Myo*-inositol increases reactive oxygen species (ROS) scavenging ability and decreases oxidative damage by increasing antioxidant ability, which might be related to upregulation of PKC (protein kinase C)/Nrf2 signaling in the intestine of grass carp (Li et al., 2017) and in the muscle of Jian carp (Jiang et al., 2015a). The antioxidant action of *myo*-inositol also occurs in the liver/hepatopancreas of hybrid sturgeon (Li et al., 2019), white shrimp (Chen et al., 2019), and Chinese mitten crab (Bu et al., 2020).

3.2.10.2.2 Osmoregulation

Myo-inositol is a major osmolyte, which can directly regulate osmotic balance in many mammalian cells. *Myo*-inositol modulates the osmotic response to salinity changes in tilapia (Sacchi et al., 2014), turbot (Cui et al., 2020), and European eel (Kalujnaia et al., 2013).

3.2.10.2.3 Maintains membrane structure and function

Myo-inositol is incorporated into PI which is an important component of all cell membranes. The polar head group and highly nonpolar fatty acyl chains of PI facilitate the recruitment of proteins and, ultimately, control the cytoskeleton and membrane dynamics during mitosis (Combs and McClung, 2017). *Myo*-inositol promotes enterocyte proliferation in Jian carp (Jiang et al., 2013a); regulates left–right development in the zebrafish (Sarmah et al., 2005); maintains the structural integrity of the intestine/enterocytes; and, thus, promotes the digestive and absorptive ability of Jian carp (Jiang et al., 2009, 2013a).

3.2.10.2.4 Signal transduction

The inositol-containing phospholipids act as vital elements in transmembrane signaling. Inositol-1,4,5-triphosphate (IP_3) is an intracellular second messenger that stimulates the release of Ca^{2+} and, thus, activates a variety of Ca^{2+}-dependent protein kinases to phosphorylate a large number of proteins important to cells (Wu, 2017). *Myo*-inositol can enhance tight junction, which might be related to the increase of TOR and inhibition of myosin light-chain kinase (MLCK) signaling in the intestine of grass carp (Li et al., 2018).

3.2.10.2.5 Modulates lipid transport

Myo-inositol plays an important role by promoting the export of fat from liver (Halver and Hardy, 2002). Lipotropic action to prevent lipid deposition in the liver by *myo*-inositol occurs in fishes like tilapia (Zaki et al., 2010) and grouper (Su and Shiau, 2004).

3.2.10.2.6 Immunomodulation

The character of PI in the cell membrane allows it as a regulator of endosome–lysosome trafficking and induction of pathogen killing and antigen-processing pathways in lymphoid cells of mammals (Combs and McClung, 2017). *Myo*-inositol enhances innate immune and/or acquired immune, as well as attenuates inflammatory response in various immune organs like the intestine of grass carp (Li et al., 2018), Jian carp (Jiang et al., 2010), and Chinese mitten crab (Bu et al., 2020).

3.2.10.3 *Deficiency symptoms*

Myo-inositol can be synthesized de novo from glucose in most terrestrial animals (except gerbil) and some aquatic animals (Fig. 3.15). The de novo synthesis of *myo*-inositol occurs in the intestine, liver, kidney, brain, and gill of fish like common carp (Aoe and Masuda, 1967), channel catfish (Burtle and Lovell, 1989), and Mozambique tilapia (Gardell et al., 2013), as well as the visceral tissue of abalone (Mai et al., 2001). However, for some aquatic species (like common carp), de novo synthesis is not possible or synthesis is insufficient to meet the metabolic requirements, and exogenous supplementation of *myo*-inositol is necessary.

As shown in Table 3.22, inadequate dietary inositol could cause growth retardation, increase of mortality, clinical, and pathological changes, as well as lipodystrophy in aquatic animals.

3.2.10.4 *Requirements*

The requirements of *myo*-inositol for aquatic animals are listed in Table 3.23, which can be affected by the following factors: **(1) Species:** Generally, shrimp appear to require more than teleosts. Requirement of *myo*-inositol for grass shrimp is 3397 mg/kg diet, which is much higher than that of 94.3−617.7 mg/kg in fish; **(2) De novo synthetic ability:** Juvenile sunshine bass and abalone can synthesize de novo inositol to meet the requirements, whereas juvenile common carp cannot produce enough inositol de novo, thus requiring 440 mg/kg diet.

3.2.10.5 *Sources and protection*

Seeds of plant (wheat germ, dried peas, and beans) and organs of animal (brain, liver, heart, and glandular) are the natural rich sources of *myo*-inositol. The predominant form of *myo*-inositol occurring in plant sources is phytic acid. However, except for ruminants, most animal (including fish and shrimp) have little or no intestinal phytase activity, and thus phytic acid is poorly utilized as a source of *myo*-inositol. In animal products, inositol is present both in its free form and as inositol-containing phospholipid (primarily PI). *Myo*-inositol can be synthesized or easily isolated from biological material in free or combined forms. Inositol is stable during feed processing.

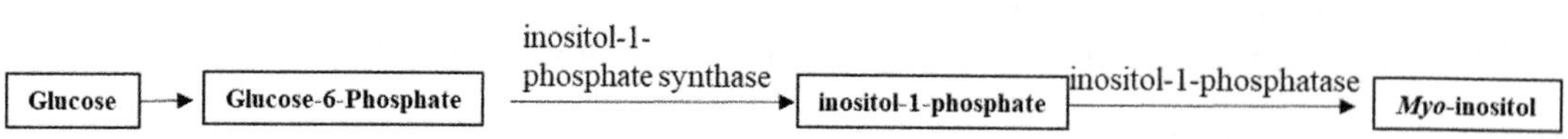

FIGURE 3.15 The de novo synthesis of inositol.

TABLE 3.22 The inositol deficiency syndromes.

Deficiency syndromes	Species	References
Growth retardation	Grass carp	Li et al. (2017)
	Common carp	Aoe and Masuda (1967)
	Gibel carp	Gong et al. (2014)
	Tilapia	Shiau and Su (2005)
	Parrot fish	Khosravi et al. (2015)
	Chinook salmon	Halver (1957)
	Channa punctata	Zehra and Khan (2019c)
	Grouper	Su and Shiau (2004)
	Olive flounder	Lee et al. (2009)
	Amur sturgeon	Wang et al. (2018)
	Prawn	Kanazawa et al. (1976)
	Grass shrimp	Shiau and Su (2004)
	Chinese mitten crab	Bu et al. (2020)
Increased mortality	Gibel carp	Gong et al. (2014)
	Olive flounder	Lee et al. (2009)
	Grass shrimp	Shiau and Su (2004)
Hemorrhage in skin, fin, or muscle	Common carp	Aoe and Masuda (1967)
	Channa punctata	Zehra and Khan (2019c)
Distended stomach and slow gastric emptying	Chinook salmon	Halver (1957)
White–gray intestine	Japanese eel	Arai et al. (1972)
Anemia	Rainbow trout	Phillips and Brockway (1957)
Decrease of moulting frequency	Chinese mitten crab	Bu et al. (2020)
Liver abnormal lipid accumulation	Tilapia	Zaki et al. (2010)
	Grouper	Su and Shiau (2004)
	Parrot fish	Khosravi et al. (2015)
	Amur sturgeon	Wang et al. (2018)

3.2.10.6 *Antimetabolites and inactivation*

There are eight optically or biologically inactive stereoisomers of inositol. Biologically inactive forms (*scyllo*-inositol) do not compete for critical sites in metabolism. Only chemically synthesized inactive stereoisomers (*epi*-inositol and *allo*-inositol) might interfere with the

TABLE 3.23 The inositol recommendation levels reported in aquatic animals.

Species	Initial weight (g)	Recommendation levels (mg/kg diet)	Response criteria	References
Grass carp (*Ctenopharyngodon idella* Val.)	221.8	276.7	PWG	Li et al. (2017)
Common carp (*Cyprinus carpio* L.)	6.3	440	PWG	Aoe and Masuda (1967)
Jian carp (*Cyprinus carpio* var. Jian)	22.28	518.0	PWG	Jiang et al. (2009)
Gibel carp (*Carassius auratus gibelio*)	3.38	165.3	PWG	Gong et al. (2014)
Tilapia (*Oreochromis niloticus* × *Oreochromis aureus*)	0.51	408 404	PWG LIC	Shiau and Su (2005)
Nile tilapia (*Oreochromis niloticus*)	5.8	NN	Growth	Peres et al. (2004)
Parrot fish (*Oplegnathus fasciatus*)	11.5	94.3 121.3	PWG LIC	Khosravi et al. (2015)
Atlantic salmon (*Salmo salar*)	0.19	531	Growth	Waagbø et al. (1998)
Olive flounder (*Paralichthys olivaceus*, Temminch et Schlegel)	10	617.7	PWG	Lee et al. (2009)
Grouper (*Epinephelus malabaricus*)	6.02	365 335	PWG LIC	Su and Shiau (2004)
Amur sturgeon (*Acipenser schrenckii*)	11.9	336.1	PWG	Wang et al. (2018)
Channa punctata (Bloch)	3.78	504.9	LIC	Zehra and Khan (2019c)
Hybrid sturgeon (*Acipenser baerii* ♀ × *A. schrenckii* ♂)	28.28	335.8 590.7	PWG LIC	Li et al. (2019)
Grass shrimp (*Penaeus monodon*)	0.44	3397 3275	PWG MGIC	Shiau and Su (2004)
Chinese mitten crab (*Eriocheir sinensis*)	0.40	1707 1672	PWG LIC	Bu et al. (2020)
Abalone (*Haliotis discus hannai* Ino)	0.1446	NN	NN	Mai et al. (2001)

LIC, liver inositol concentration; *MGIC*, midgut gland inositol concentration; *NN*, not need; *PWG*, percent weight gain.

metabolism of inositol (Schneider, 2015). However, the active cis–trans isomer does compete and introduces errors in the structural configuration of essential components in cell membranes. Salts of the hexaphosphate or phytin forms of inositol are unavailable to some animals. High levels of glucose can inhibit the absorption and transportation of *myo*-inositol.

3.2.10.7 Clinical assessment and analysis

The *myo*-inositol status in aquatic animals can be determined by analysis of the liver *myo*-inositol concentration and the absence of deficiency signs coupled with growth performance. The inositol concentrations in tissues can be determined by an enzymatic assay (Ashizawa et al., 2000), a microbiological assay method (Waagbø et al., 1998), or HPLC (Li et al., 2017).

3.2.11 Choline

Choline was first isolated from pig and cattle bile by Adolph Strecker in 1862, chemically synthesized in 1865, and recognized as a nutrient for terrestrial animals in 1920s. After then, fish choline deficiency syndromes were first reported in trout and salmon during 1940–1960s. From 1970 to 2000s, the nutritional importance of choline in aquatic animals mainly focused on lipid metabolism and requirements for some fishes and shrimps (Kanazawa et al., 1976; Zhang and Wilson, 1999). In the last 2 decades, the nutritional roles of choline in aquatic animals have been focused on antioxidant, antistress, immunoregulation, interaction with other dietary factors, and requirements for more fishes and shrimps (Wu and Davis, 2005).

3.2.11.1 Chemical structure and characteristics

Choline (2-hydroxy-N,N,N-trimethylethanaminium, $C_5H_{15}NO_2$) is a quaternary ammonium compound (Fig. 3.16). It is a colorless, viscous, and very hydroscopic liquid, soluble in water and ethanol but insoluble in organic solvents. As a strong base, choline reacts with many chemical compounds and decomposes to trimethylamine in alkaline solution (Combs and McClung, 2017).

3.2.11.2 Positive functions

Choline can be absorbed and transported as choline-containing phospholipid (mainly lysophosphatidylcholine [lysoPC] and sphingomyelin) by micelle-dependent diffusion and free choline by both saturable carrier-mediated process and passive diffusion (less efficient) in terrestrial animals (Zeisel, 1981). Within the enterocytes, lysoPC and free choline are reesterified to form PC, some of which is packaged into chylomicrons and very low-density lipoproteins (VLDLs) for export into the portal vein in fish. The unesterified free choline is transported to the tissues by high-affinity choline transporters CHT1s (in brain), intermediate-affinity choline transporters–like proteins (in many tissues), and polyspecific organic cation transporters (in liver, intestine, kidney, and brain) in mammals (Michel et al., 2006).The main fate of choline is the synthesis of PC, which is catalyzed by choline kinase (Fig. 3.17). CDP (cytidine diphosphate)-choline pathway is operative in the formation of liver lecithin in rainbow trout (Holub et al., 1975). Besides, choline is oxidized to form betaine catalyzed by choline oxidase in mammalian kidney and liver (Zeisel, 1981), and in some fish (Treberg and Driedzic, 2007). Meanwhile, a small fraction of choline can be acetylated to form acetylcholine catalyzed by choline acetyltransferase in the nervous system and immune cells (Tyagi et al., 2008). Choline acetyltransferase is found in the spleen of Nile tilapia (Toledo-Ibarra et al., 2014).

Choline and its derivatives have many physiological functions as follows.

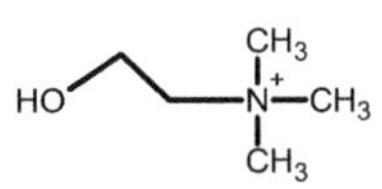

FIGURE 3.16 The chemical structure of choline.

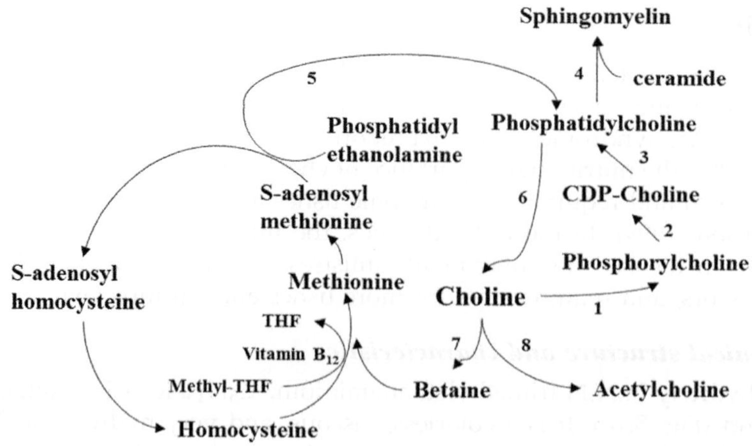

FIGURE 3.17 The metabolism of choline. The enzymes that catalyze the reactions are: (1) choline kinase; (2) CTP: phosphocholine cytidylyltransferase; (3) CDP (cytidine diphosphate)-choline: DAG (diacylglycerol) choline-phosphotransferase; (4) phosphatidylcholine: ceramide cholinephosphotransferase; (5) phosphatidylethanolamine N-methyltransferase (PEMT); (6) a series of enzymes for phosphatidylcholine (PC) lipolysis; (7) choline oxidase; (8) choline acetyltransferase. *THF*, tetrahydrofolate.

3.2.11.2.1 Maintain membrane structure

Choline is the precursor for the phospholipids including PC and sphingomyelin, which are the essential components of all biomembranes. In aquatic animals, choline could maintain the tissue structural integrity in Jian carp (Wu et al., 2016) and shrimp (Reddy et al., 1999).

3.2.11.2.2 Modulate lipid transport

The choline derivative, phosphatidylcholine, is a major component of VLDL which is necessary for exporting fat from liver. Choline is helpful in improving VLDL production in blunt snout bream (Li et al., 2014a), and decreased liver lipid content in many fish such as cobia (Mai et al., 2009), as well as gut mucosa lipid accumulation in Atlantic salmon (Hansen et al., 2020).

3.2.11.2.3 Signal transduction and neurotransmission

Choline-containing phospholipids act as vital elements in transmembrane signaling. Hydrolysis of PC produces diacylglycerol, an activator of PKC which phosphorylates many target proteins in the cell. Choline could modulate the gene and protein expression, and phosphorylated level of TOR signaling in grass carp (Yuan et al., 2020). Choline is also the precursor of acetylcholine which is a key neurotransmitter and affects the functions of several organs through cholinergic neurons.

3.2.11.2.4 Methyl donors

The triplet of methyl groups of choline and its oxidized metabolite betaine enables them to serve as methyl donors, which affect methyl metabolism. Choline could maintain the normal global DNA methylation in mammals (Niculescu et al., 2006).

3.2.11.2.5 Antioxidation

The intramolecular hydroxy amine group of choline participates in decomposing hydro-peroxides and decreasing ROS content in sardine oil in vitro (Saito and Ishihara, 1997).

Moreover, choline could increase ROS scavenging ability and decrease oxidative damage by increasing antioxidant ability probably via Nrf2 signaling pathway in tissues of grass carp (Zhao et al., 2016).

3.2.11.2.6 Immunoregulation

Choline enhances immunity including alpha-2-macroglobulin and complement C3 contents in orange-spotted grouper (Shiu et al., 2017), phagocytosis, lysozyme activity, C3 content, antimicrobial peptide gene expressions, and tissues immunoglobulin contents in carp (Wu et al., 2013; Zhao et al., 2016); and attenuates intestinal inflammation probably through Janus kinase/signal transducer and activator of transcription (JAK/STAT) and NF-κB/IκB signaling in Jian carp (Wu et al., 2016).

3.2.11.3 *Deficiency syndromes*

Most animals synthesize de novo choline via the sequential, S-adenosylmethionine—dependent methylation of phosphatidylethanolamine by phosphatidylethanolamine N-methyltransferase (PEMT) to produce PC (Zeisel, 1981), which occurs mostly in the liver. Choline biosynthesis is found in channel catfish (Zhang and Wilson, 1999) and trout (Rumsey, 1991). However, the rate or capacity for choline biosynthesis may be various among fishes, and the synthesis of choline cannot meet the requirements for most fish.

Inadequate dietary choline induces deficiency symptoms like growth retardation, abnormal liver lipid deposition, and pathological changes in aquatic animal (Table 3.24). However, excess dietary choline also reduced growth performance of grouper (Qin et al., 2017) and induced higher mortality in hybrid striped bass (Griffin et al., 1994) and leukocyte infiltration in the intestine of Jian carp (Wu et al., 2016).

3.2.11.4 *Requirements*

The choline requirements for aquatic animals are listed in Table 3.25, which could be affected by the following factors: **(1) Species:** As shown in Table 3.25, the choline requirements vary among species, ranging from 429 to 4000 mg/kg diet in fish and shrimp; **(2) Growth stage:** Generally, the younger fish needs more choline than older one does. The choline requirement for juvenile grass carp (1283.4—1330.7 mg/kg diet) is slightly higher than that for ongrowing ones (1136.5—1210.7 mg/kg diet); **(3) Response criterion:** Estimates of requirements for fish based on choline content in liver and body are higher than those based on weight gain, as showed in blunt snout bream, hybrid tilapia, and cobia; **(4) Dietary factors:** Betaine spares choline requirement in channel catfish fingerlings (Wu and Davis, 2005). Meanwhile, methionine compensated for the choline deficiency in kuruma shrimp (Michael et al., 2006).

3.2.11.5 *Sources and protection*

Choline is widely present in many feedstuffs, mostly in the form of PC. The natural rich sources of choline include wheat germ, soybean, and other bean meal, liver tissue, and egg yolk. The synthetic choline sources include choline chloride (the most commonly used supplementary form) and choline bitartrate. Choline is stable during feed processing. However, choline should not be mixed in vitamin premixes because of its strong hydroscopicity and alkalinity.

TABLE 3.24 Choline deficiency syndromes.

Deficiency syndromes	Species	References
Growth retardation	Grass carp	Yuan et al. (2020)
	Blunt snout bream	Jiang et al. (2013b)
	Hybrid tilapia	Shiau and Lo (2000)
	Channel catfish	Zhang and Wilson (1999)
	Rainbow trout	Rumsey (1991)
	Chinook salmon	Halver (1957)
	Yellow perch	Twibell and Brown (2000)
	Red drum	Craig and Gatlin (1996)
	Hybrid striped bass	Griffin et al. (1994)
	Cobia	Mai et al. (2009)
	Grouper	Qin et al. (2017)
	Grass shrimp	Shiau and Lo (2001)
Higher mortalities	Hybrid tilapia	Shiau and Lo (2000)
	Rainbow trout	Poston (1991)
	Cobia	Mai et al. (2009)
	Grass shrimp	Shiau and Lo (2001)
	Penaeus japonicus	Kanazawa et al. (1976)
Increased liver fat deposition	Channel catfish	Wilson and Poe (1988)
	Giant grouper	Yeh et al. (2015)
Decreased liver fat contents	Hybrid tilapia	Shiau and Lo (2000)
	Red drum	Craig and Gatlin (1996)
Anemia, projected eyes, light-colored body	Rainbow trout	Kitamura et al. (1967a)
Extended abdomen	Blunt snout bream	Jiang et al. (2013b)
	Rainbow trout	Kitamura et al. (1967a)
Light yellow–colored liver	Blunt snout bream	Jiang et al. (2013b)
	Rainbow trout	Kitamura et al. (1967a)
Hemorrhagic abdomens	Blunt snout bream	Jiang et al. (2013b)
Hemorrhagic kidney and intestine	Chinook salmon	Halver (1957)

TABLE 3.24 Choline deficiency syndromes.—cont'd

Deficiency syndromes	Species	References
Vacuolization of hepatic cells	Carp	Ogino et al. (1970b)
	White sturgeon	Hung (1989)
Sloughing of mucosal epithelium, focal degeneration of exocrine pancreas, inflammatory infiltration in liver periadventitial veins	White sturgeon	Hung (1989)
Inflammatory infiltration in intestine	Jian carp	Wu et al. (2016)

TABLE 3.25 The recommended levels of choline reported in aquatic animal.

Species	Initial weight (g)	Recommendation levels (mg/kg diet)	Response criteria	References
Grass carp (*Ctenopharyngodon idella*)	9.3	1331 1283	WG FE	Yuan et al. (2020)
	266.6	1137	PWG	Zhao et al. (2015)
Blunt snout bream (*Megalobrama amblycephala*)	1.80	1198 1525 1365	WG LCC MCC	Jiang et al. (2013b)
Gibel carp (*Carassius auratus gibelio*)	5.45	2500	SGR	Duan et al. (2012)
Jian carp (*Cyprinus carpio* var. Jian)	7.94	566	SGR	Wu et al. (2011)
Hybrid tilapia (*Oreochromis niloticus* × *O. aureus*)	0.62	880 920 923	WG BCC LLC	Shiau and Lo (2000)
Yellow perch (*Perca flavescens*)	15.7	598 634	WG FI	Twibell and Brown (2000)
Rainbow trout (*Oncorhynchus mykiss*)	0.12	4000	Growth	Poston (1991)
	1.4 3.2	813 774	Growth	Rumsey (1991)
Hybrid striped bass (*Morone saxatilis* × *Morone chrysops*)	10.4	500	WG LLC	Griffin et al. (1994)
Red drum (*Sciaenops ocellatus*)	54.7	588 ± 35	WG	Craig and Gatlin (1996)
Cobia (*Rachycentron canadum*)	4.2	696 877 950	WG LCC MCC	Mai et al. (2009)

Continued

TABLE 3.25 The recommended levels of choline reported in aquatic animal.—cont'd

Species	Initial weight (g)	Recommendation levels (mg/kg diet)	Response criteria	References
Orange-spotted grouper (*Epinephelus coioides*)	87.87	1094 1580	SGR LCC	Qin et al. (2017)
Grass shrimp (*Penaeus monodon*)	1.17	6000	WG BCC	Shiau and Lo (2001)
Litopenaeus vannamei	0.72	871	IGR	Gong et al. (2003)
Prawn (*Penaeus japonicus*)	0.87	600	Survival and WG	Kanazawa et al. (1976)

BCC, body choline concentration; *FE*, feed efficiency; *FI*, feed intake; *IGR*, instantaneous growth rate; *LCC*, liver choline concentration; *LLC*, liver lipid content; *MCC*, muscle choline concentration; *SGR*, specific growth rate; *WG/PWG*, weight gain %.

3.2.11.6 Clinical assessment and analysis

The choline nutritional status in fish and shrimp can be indicated by the liver lipid and choline content, and the absence of deficiency signs (Shiau and Lo, 2000; Halver and Hardy, 2002). There are several methods to measure liver choline content, including HPLC method (Rumsey, 1991) and enzymatic assay (Zhang and Wilson, 1999). Weight gain is a more sensitive indicator than liver fat in rainbow trout (Rumsey, 1991). In mammals, choline nutritional status can be also estimated by assaying the liver phosphocholine content and serum creatine phosphokinase activity.

3.3 The fat-soluble vitamins

Fat-soluble vitamins are all isoprene derivatives, which are insoluble in water and soluble in fat and fat-soluble solvents such as ether and chloroform. They often coexist with lipids in feed, generally exist in feed materials rich in fat, and their absorption rely on the presence of fat. Differing from the water-soluble vitamins, fat-soluble vitamins can be stored in large quantities in the liver (hepatopancreas) of animals. Thus, long-term oversupply may cause hypervitaminosis symptoms. Sometimes these hypervitaminosis symptoms mimic hypovitaminosis signs, as in the case of the vitamins A and D. The toxicity symptoms observed when excess vitamin E or K is ingested are more discrete. Fish rations often include large quantities of fish meal or fish viscera and are often enriched with fish oils to increase the caloric density or content of polyunsaturated fatty acids (PUFAs) of the ration. In these cases, excessive intake of the fat-soluble vitamins is often encountered (Halver and Hardy, 2002).

3.3.1 Vitamin A

Vitamin A was first discovered as "fat-soluble A" in rat feeding studies by McCollum's group and Osborne's group in 1913, and renamed as vitamin A in 1920. Soon after, the vitamin A activity of carotene from plant materials was discovered in 1919.The nutritional importance of vitamin A in fish was recognized since vitamin A deficiency induced xerophthalmia and cataracts in fish (Halver and Hardy, 2002). From 1960 to 2000s, studies on vitamin A nutrition in aquatic animals mainly focused on deficiency syndrome and requirements for fish, shrimp, and abalone. In the last 2 decades, research on the nutritional role of vitamin A in aquatic animals has been extended to reproduction, skeletal development, and immunity, reviewed recently by Hernandez and Hardy (2020).

3.3.1.1 *Chemical structure and characteristics*

Vitamin A is an unsaturated monohydric alcohol containing β-ionone ring (Fig. 3.18) and has three basic forms, i.e., retinol, retinal (also retinaldehyde), and retinoic acid. Retinol and retinal are interconverted by NAD(P)H-dependent retinaldehyde reductase. Retinoic acid can be formed from retinal but cannot be converted back to retinal or retinol.

Carotenoids are plant-derived vitamin A sources. There are approximately 500 different carotenoids known to exist, of which only about 60 of these carotenoids have provitamin A activity (Combs and McClung, 2017). β-carotene is known as the most efficient provitamin A carotenoids, which is a tail-conjoined retinoid dimer with the chemical formula $C_{40}H_{56}$ (Fig. 3.19).

Vitamin A is a pale yellow crystalline solid with very low melting points (62–65°C) and soluble in ethanol and organic solvents including fats and oils but insoluble in water. Carotenoids have the color of yellow, orange, or red. Both vitamin A and carotenoids are very sensitive to oxygen and heat; however, the esterified retinoids are quite stable.

International unit (IU) and retinol equivalent (RE) are used to describe in a standardized manner the biological activity of vitamin A and provitamin A carotenoids in various forms (Combs and McClung, 2017). One IU of vitamin A is defined to the activity of 0.30 μg crystalline all-trans-retinol, while 1 μg RE = 1 μg retinol = 3.33 IU.

3.3.1.2 *Positive functions*

Retinoid metabolism is fundamentally similar among all vertebrate classes. Most studies focusing on vitamin A metabolism have been carried out in higher vertebrates, while only a few works in fish. In the intestine, retinol can be released from retinyl esters by retinyl ester hydrolases and from some of the provitamin A carotenoids by β-carotene 15,15′-dioxygenase. After that, retinol along with carotenoids mix with lipid and bile salts to form micelles which are absorbed into epithelial cells via passive diffusion. Once inside the enterocytes, retinols are esterified by lecithin retinol acyltransferase (LRAT) (Harrison, 2005), and then along with carotenoids that were not cleaved incorporated into chylomicra which travel through

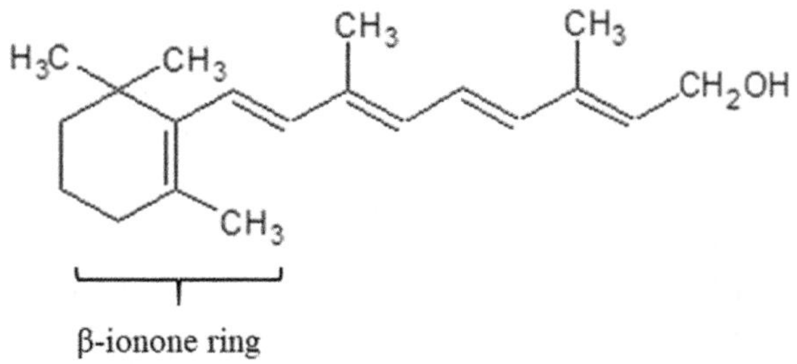

FIGURE 3.18 The chemical structure of vitamin A.

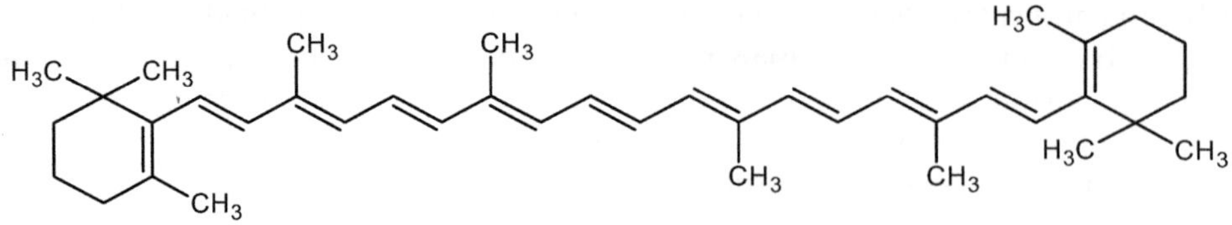

FIGURE 3.19 The chemical structures of provitamin β-carotene.

the intestinal lymphatics, release into the circulatory system, and are taken up by tissues. Most of the retinyl esters and carotenoids are taken up by the liver, while the remaining is taken up by extrahepatic tissues. In the liver, retinol can be released from retinyl esters by retinyl ester hydrolase and can be either complexed with retinol-binding protein (RBP) and released into the plasma or reesterified by LRAT for long-term storage, whereas the carotenoids can be cleaved to retinol, stored, or packaged into VLDLs, and released into the circulation (Tanumihardjo et al., 2016). In the eye, retinol is taken up by the retinal pigment epithelium and stored in the pigment epithelial and Müller cells as esterified form in lipid droplets. The stored retinol is transported to rod cells by interphotoreceptor retinol-binding protein (IRBP) for discharge of the visual function (Combs and McClung, 2017).

In fish intestinal mucosa or liver (hepatopancreas), β-carotene can be converted to vitamin A by β-carotene dioxygenase and retinaldehyde reductase (Fig. 3.20). One molecule of β-carotene can theoretically produce two molecules of retinol. However, β-carotene only has one-sixth vitamin A value of retinol in human (Combs and McClung, 2017), while one-nineteenth in hybrid tilapia (Hu et al., 2006). Although the conversion ratio of β-carotene to vitamin A is low, hybrid tilapia is able to utilize β-carotene to fulfill the vitamin A requirements (Hu et al., 2006).

Vitamin A plays an important role in most of the key biological processes. The positive functions are as follows.

3.3.1.2.1 Vision formation

Retinal is a component of rhodopsin which is an ingredient of the rod cells of the retina and important for the vision formation. For proper vision, retinol is conversed into retinal. When rhodopsin is exposed to light, it breaks down to form all-trans-retinal and opsin, coupled with the rapid flux of calcium in the membrane of the rod cell, which triggers a nerve impulse and allows light to be perceived by the vision centers of the brain (Fig. 3.21).

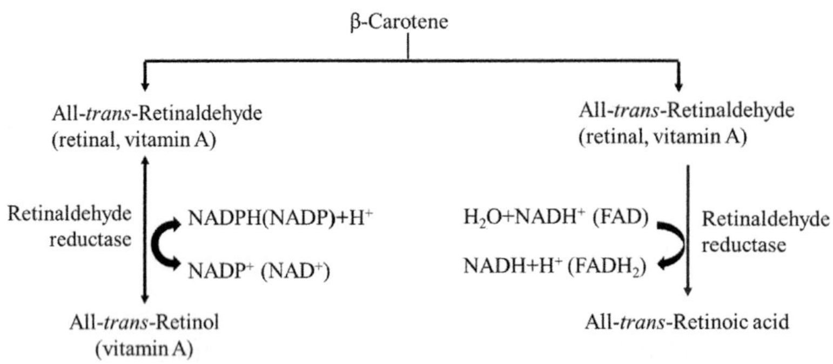

FIGURE 3.20 Conversion of β-carotene into vitamin A.

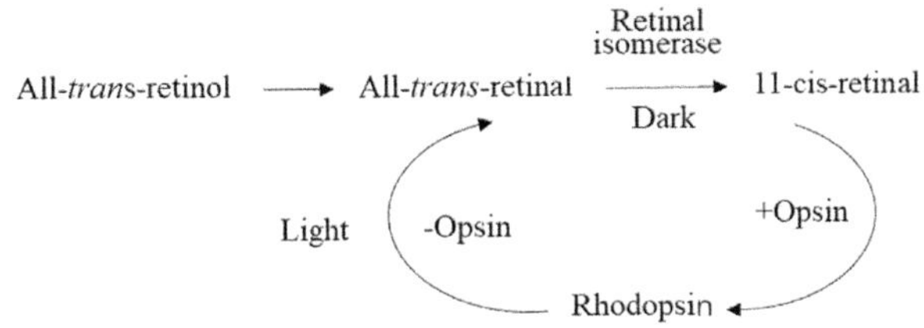

FIGURE 3.21 The role of vitamin A in the visual cycle.

3.3.1.2.2 Cell proliferation and differentiation

Vitamin A is required for the cell proliferation and differentiation. In fish, vitamin A improves intestinal cell proliferation and differentiation in juvenile sunshine bass (Hemre et al., 2004) and Atlantic halibut (Moren et al., 2004), and increases intestinal fold height in Jian carp (Yang et al., 2008). In vitro, retinoids also stimulate the differentiation of pituitary cells (Combs and McClung, 2017).

3.3.1.2.3 Immunoregulation

Vitamin A has an important role in immunoregulation that has been well documented in humans. The immunoregulation role of vitamin A appears to be discharged by ligating retinoic acid receptor (RAR)-retinoid X receptor (RXR) to alter the expression of immune-related genes. Both RAR and RXR receptors are constitutively expressed in B cells and T cells. In mammals, retinoic acid is identified as the key factor to induce intestinal Th17 cells (Cassani et al., 2012). Vitamin A improves the innate immune function as shown by the enhanced nonspecific immunity in gilthead sea bream (Cuesta et al., 2002) and Nile tilapia (Guimarães et al., 2014), as well as the acquired immune function in Jian carp (Yang et al., 2008). Further, vitamin A also mitigates intestinal inflammation which was partly connected to the NF-κB canonical signaling pathway and the p38 mitogen-activated protein kinase signaling pathway in grass carp (Zhang et al., 2017a).

3.3.1.2.4 Reproduction

Studies on the relationship between vitamin A and reproductive performance have mainly focused on mammals, with little work in aquatic animals. Retinoic acid is required for spermatogenesis in chicken, rat, and cow (Combs and McClung, 2017), embryonic survival, fetal growth, and development in mammals (Wu, 2017), and also plays a fundamental role in the reproduction of female zebrafish (Alsop et al., 2008). Further, vitamin A improves the fecundity of female brood stock in rainbow trout (Fontagné-Dicharry et al., 2010), promotes gonad development and hatchability in bighead carp (Santiago and Gonzal, 2000), as well as increases the percentage of buoyant eggs and hatching rate (Furuita et al., 2001) and normal larvae (Furuita et al., 2003) in Japanese flounder.

3.3.1.2.5 Antioxidation

Vitamin A (retinol) and its derivatives, such as retinal and retinoic acid, are known for their ability to inhibit lipid peroxidation due to chain breaking by scavenging of peroxyl radicals, formation of radical adducts, and also scavenge peroxyl radicals by electron transfer

(TaRóżanowska et al., 2005). In addition, β-carotene can stabilize organic peroxide free radicals within its conjugated alkyl structure (Wu, 2017). In fish, vitamin A was found to be beneficial to counteract oxidative damage in the intestine of grass carp (Jiang et al., 2019) and in Siberian sturgeon larvae (Fontagné et al., 2006), which was partly due to the increased activities and mRNA levels of antioxidant enzymes, probably via the Nrf2/Keap1a signaling pathway in grass carp (Jiang et al., 2019).

3.3.1.2.6 Bone metabolism and skeletogenesis

Vitamin A has an essential role in chondrogenic and osteogenic development during early life stages of gilthead sea bream (Fernández et al., 2008) and had a remarkable impact in the skeleton morphogenesis in Senegalese sole larvae (Fernández et al., 2009). Optimal level of retinol for harmonious ontogenesis fluctuated along sea bass larvae development, and inadequate dietary retinol levels altered morphogenesis of European sea bass (Mazurais et al., 2009).

3.3.1.3 Syndromes of deficiency or excess

Fish are not able to synthesize vitamin A and have to obtain it from their diet. Hypovitaminosis A usually includes growth retardation, increased mortality, and abnormal morphology and histopathological changes (Table 3.26).

Excess dietary vitamin A is toxic and induces several undesired syndromes, such as growth retardation, increased mortality, skeletal deformities, and histopathological changes (Table 3.27). Marine fish are very sensitive to hypervitaminosis A during early larval stages but less sensitive at later life history stages (Fernández et al., 2008).

3.3.1.4 Requirements

The vitamin A requirements for aquatic animals reported in the last 2 decades are listed in Table 3.28 and are affected by the following factors: **(1) Species**: Vitamin A requirement for some fish range from 0.66 to 30 mg/kg diet; **(2) Growth stage**: Generally, higher vitamin A requirement is needed in brood stock and early developmental stage as shown in rainbow trout (Fontagné-Dicharry et al., 2010); **(3) Dietary factors:** Requirement of vitamin A is high when the diet contained oxidized lipid (Fontagné et al., 2006); **(4) Response criteria:** Vitamin A requirements based on disease resistance and maximum liver storage are slightly higher than that based on growth performance in grass carp and hybrid tilapia, respectively.

3.3.1.5 Sources and protection

Vitamin A exists in natural products in many different forms. In animals, it exists in the form of retinyl esters and mainly in adipose tissue, liver, and milk. Provitamin A carotenoids are rich in plant that are dark green, orange, and yellow. Besides, the synthetic vitamin A sources such as retinyl palmitate and retinyl acetate are also available for inclusion in the feeds. Both vitamin A and carotenoids are very sensitive to air, light, and heat, so they must be kept away from heat, oxygen, metal ions, peroxide lipids, and ultraviolet light. Moreover, antioxidants, esterified and coated forms of vitamin A, are widely used in aquatic feed to keep the activity of vitamin A.

TABLE 3.26 Vitamin A deficiency syndromes.

Deficiency syndromes	Species	References
Growth retardation	Grass carp	Zhang et al. (2017a)
	Jian carp	Yang et al. (2008)
	Tilapia	Hu et al. (2006); Guimarães et al. (2014)
	Rainbow trout	Fontagné-Dicharry et al. (2010)
	Greasy grouper	Mohamed et al. (2003)
	Sunshine bass	Hemre et al. (2004)
	Japanese flounder	Hernandez et al. (2005)
Increased mortality	Sunshine bass	Hemre et al. (2004)
	Greasy grouper	Mohamed et al. (2003)
	Grass carp	Zhang et al. (2017a)
Small liver	Japanese flounder	Hernandez et al. (2005)
Amorphous and granulomatous spleen and liver	Nile tilapia	Campeche et al. (2009)
Bone and operculum deformities	Jian carp	Yang et al. (2008)
	Nile tilapia	Guimarães et al. (2014)
Hemorrhages	Tilapia	Campeche et al. (2009); Guimarães et al. (2014)
	Sunshine bass	Hemre et al. (2004)
	Greasy grouper	Mohamed et al. (2003)
	Atlantic halibut	Moren et al. (2004)
	Grass carp	Zhang et al. (2017a)
Anemia	Rainbow trout	Kitamura et al. (1967b)
Caudal peduncle erosion	Greasy grouper	Mohamed et al. (2003)

3.3.1.6 *Clinical assessment and analysis*

The vitamin A can be accurately quantified in biological specimens using high-performance liquid–liquid partition chromatography. In fish, vitamin A status can be assessed by the absence of deficiency signs, liver, serum, and whole fish vitamin A content (Mohamed et al., 2003).

3.3.2 Vitamin D

Vitamin D was first found by using cod liver oil to cure rickets in dogs in 1919 and was identified in 1931. The structure and synthesis of VD_2 and VD_3 were achieved during

TABLE 3.27 Vitamin A excessive syndromes.

Excessive syndromes	Species	References
Growth retardation	Rainbow trout	Hilton (1983)
	Japanese flounder	Hernandez et al. (2005)
Increased mortality	Rainbow trout	Fontagné-Dicharry et al. (2010)
Reduced the number of vertebral segments and disturbed bone formation	European sea bass	Villeneuve et al. (2006)
Abnormal vertebral growth	Atlantic salmon fry	Ørnsrud et al. (2002)
Scoliosis and lordosis	Rainbow trout	Hilton (1983)
Skeletal deformities	Gilthead sea bream	Fernández et al. (2008)
Reduced enterocyte cell proliferation	Atlantic halibut	Hemre et al. (2004)
Pale fragile livers	Japanese flounder	Hernandez et al. (2005)
Necrotic fins	Rainbow trout	Hilton (1983)

TABLE 3.28 The vitamin A requirements reported in aquatic animal.

Species	Initial weight (g)	Forms	Requirements (mg/kg diet)	Response criteria	References
Grass carp (*Ctenopharyngodon idella*)	262	Retinyl acetate	0.66 0.71	PWG EM	Zhang et al. (2017a)
Gibel carp (*Carassius auratus gibelio var.*CAS III)	69.39	Retinyl acetate	0.81	PC	Shao et al. (2016)
Common carp (*Cyprinus carpio*)	–	–	1.2–6	WG,MLS	Aoe et al. (1968)
Hybrid tilapia (*Oreochromis niloticus* × *O. aureus*)	1.6	Retinyl acetate	1.76 2.09	WG MLS	Hu et al. (2006)
Nile tilapia (*Oreochromis niloticus*)	5.26	Retinyl acetate	1.06	PGDS	Guimarães et al. (2014)
Dourado (*Salminus brasiliensis*)	18	Retinyl acetate	2.64	WG	Koch et al. (2018)
Atlantic halibut (*Hippoglossus hippoglossus*)	0.4	Retinyl acetate	2.5	MLS, ADS	Moren et al. (2004)
Sunshine bass (*Morone chrysops* × *M. saxatilis*)	7.64	Retinol	0.51–40.52	WG	Hemre et al. (2004)

TABLE 3.28 The vitamin A requirements reported in aquatic animal.—cont'd

Species	Initial weight (g)	Forms	Requirements (mg/kg diet)	Response criteria	References
European sea bass (*Dicentrarchus labrax*)	0.0008	Retinyl acetate	31	WG	Villeneuve et al. (2005)
Yellowtail (*Seriola lalandi*)	–	–	5.68	WG,MLS	Shimeno (1991)
Pacific salmon (*Oncorhynchus* spp.)	–	–	R	WG	Halver (1972)
Greasy grouper (*Epinephelus tauvina*)	5.84	Retinyl acetate	0.93	WG	Mohamed et al. (2003)
Japanese flounder (*Paralichthys olivaceus*)	1.59	Retinyl palmitate	2.7	WG	Hernandez et al. (2005)
Tiger shrimp (*Penaeus monodon*)	0.97	Retinyl acetate	2.51	WG	Shiau and Chen (2000)
Penaeid shrimp (*Penaeus vannamei*)	0.04	Vitamin A acetate	1.44	WG	He et al. (1992)

ADS, absence of deficiency signs; *Dph*, day post hatching; *EM*, enteritis morbidity; *MLS*, maximum liver storage; *PC*, plasma vitamin A content; *PGDS*, prevent the gross deficiency signs; *R*, required but no value determined.

1932–36. Meanwhile, the first interest of its nutritional role in aquatic animals was conducted in catfish and goldfish in 1933. Thereafter, a few studies focused on deficiency syndromes, requirements, and effectiveness of vitamin D in some aquatic animals during 1930–2000s. In the last 2 decades, researches shed light on vitamin D in immune and requirements for some other aquatic animals.

3.3.2.1 *Chemical structure and characteristics*

There are several biological active forms of vitamin D, which are derivatives from substitution in the rings or of functional groups in the side chain. Two of the most important forms are vitamin D_2 (ergocalciferol, $C_{28}H_{44}O$, in plants) and vitamin D_3 (cholecalciferol, $C_{27}H_{44}O$, in animals) (Fig. 3.22). Both vitamin D_2 and D_3 are white-yellow powders, moderately soluble in fats, oils, and ethanol, and freely soluble in acetone, ether, and petroleum ether. One IU of vitamin D has been defined to equal to 0.025 μg of vitamin D_2 or vitamin D_3.

3.3.2.2 *Positive functions*

The absorption of vitamin D relies on micelle-dependent passive diffusion in animals. The transport of vitamin D or its metabolites relies on the vitamin D binding protein (DBP) in most animals, except some birds and fish with cartilaginous skeletons such as sharks and rays. In the liver, vitamin D is converted by hydroxylation of side chain carbon C-25 to yield 25-hydroxyvitamin D_3 [25(OH)D_3, calcidiol] which is in turn transported to the kidney for a second hydroxylation, and then travels in blood plasma of human (Fig. 3.23). However, different from the human, these two hydroxylation processes occur only in the liver of common carp and bastard halibut (Takeuchi et al., 1991). The hydroxylation products of 25(OH)

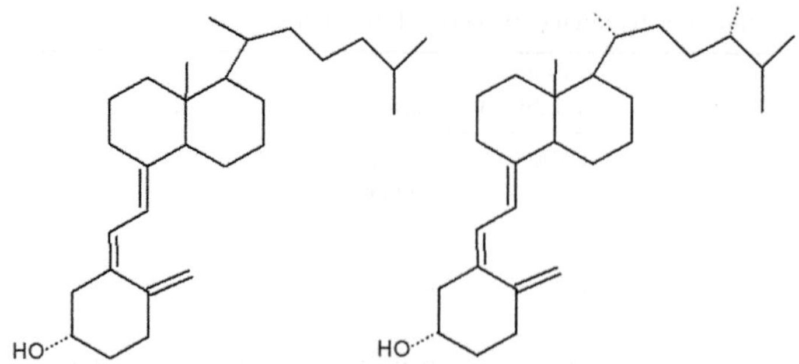

Vitamin D_2 (ergocalciferol) Vitamin D_3 (cholecalciferol)

FIGURE 3.22 The structural formula of vitamin D.

Cholecalciferol (Vitamin D_3)

Vitamin D_3-25-hydroxylase ↓

Calcidiol (25(OHD$_3$))

25-(OH)D_3-1α-hydroxylase ↓

Calcitriol (1α, 25(OH)$_2$D$_3$)

Vitamin D_3 metabolic pathway

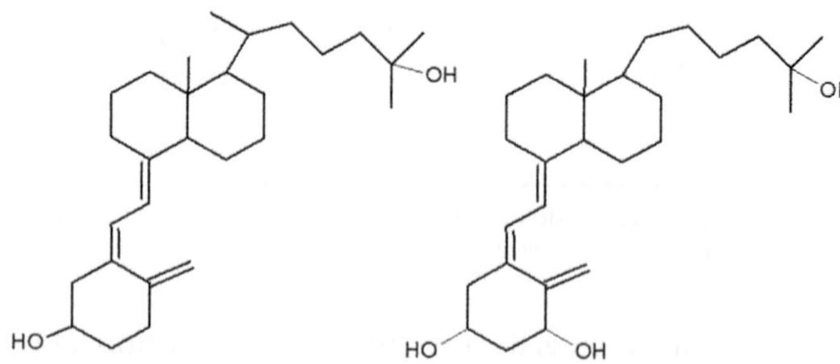

25-OH-vitamin D_3 1,25-(OH)$_2$-vitamin D_3

FIGURE 3.23 The metabolic pathway and metabolites involved in VD_3.

D_3 include 1,25-dihydroxyvitamin D_3 (1,25(OH)$_2$D$_3$, calcitriol) and 24,25-dihydroxyvitamin D_3 (24,25(OH)$_2$D$_3$), which are all found in mammals and fish.

The main physiological functions of vitamin D are as follows.

3.3.2.2.1 Maintaining homeostasis of calcium and phosphate

The best known action of vitamin D is that of maintaining homeostasis of calcium and phosphate and, thus, protecting skeletal integrity in fish (Lovell and Li, 1978; Takeuchi et al., 1991). The vitamin D_3 hormone/metabolite, 1,25(OH)$_2$D$_3$, performs its physiological functions, mostly via the specific receptor, vitamin D receptor (VDR). Briefly, 1,25(OH)$_2$D$_3$ binds the VDR (VD/VDR signaling) in the nucleus to upregulate the transcription of genes coded for calcium- and phosphate-binding proteins which act as activation of the vitamin D–dependent calcium and phosphate transport systems of the enterocytes, stimulation of the osteoclasts for the release of calcium and phosphate from bone, and enhancement of calcium and phosphate resorption by the kidney (Christakos et al., 2016). VDR has been found in various fish like Atlantic salmon (Lock et al., 2007). VD_3 or 1,25(OH)$_2$D$_3$ enhanced the intestinal calcium absorption in European sea bass (Darias et al., 2010), increased serum phosphate concentration in European eel (MacIntyre et al., 1976), promoted bone mineralization, maintained normal skeletal development in European sea bass (Darias et al., 2010), and controlled mineral homeostasis during seaward migration of Atlantic salmon (Eide-Graff et al., 2004).

3.3.2.2.2 Lipid metabolism regulation

Besides bone and mineral homeostasis function, the VD/VDR signaling play important roles in lipid metabolism, chronic metabolism, obesity, etc., in terrestrial animals. $1,25(OH)_2D_3$ promoted the fatty acid oxidation in the adipose tissue by VD/VDR signaling–induced expression of Pgc1a which is a key regulator of oxidative metabolism in zebrafish (Peng et al., 2017).

3.3.2.2.3 Immunomodulation

Vitamin D can enhance the innate immunity by promotion of phagocytic ability of leukocytes in sea bream (Cerezuela et al., 2009). Meanwhile, it attenuates inflammatory response associated with decreasing the expression of proinflammatory cytokines, which partly related to downregulating TLR4, Myd88, and NF-κBp65 mRNA abundance in the intestine/enterocytes of Jian carp (Jiang et al., 2015b).

3.3.2.3 *Syndromes of deficiency or excess*

Mammals are able to photosynthesize vitamin D in the skin to meet partial requirement, whereas aquatic animals do not synthesize vitamin D following UV exposure and exogenous supplementation of vitamin D is necessary (Lock et al., 2010). As shown in Table 3.29, inadequate dietary supply of vitamin D could cause growth retardation, deformities, and clinical and pathological changes in aquatic animals.

The European Food Safety Authority (EFSA) Panel on Additives and Products or Substances used in Animal Feed (EFSA FEEDAP Panel) has estimated a safe upper dietary level of 1.5 mg vitamin D_3/kg feed for salmonids (EFSA, 2017). As shown in Table 3.30,

TABLE 3.29 The vitamin D deficiency syndromes.

Deficiency syndromes	Species	References
Growth retardation	Blunt head bream	Miao et al. (2015)
	Hybrid tilapia	Shiau and Hwang (1993)
	Channel catfish	Lovell and Li (1978)
	Rainbow trout	Barnett et al. (1982)
	Amago salmon	Taveekijakarn et al. (1996a)
	European sea bass	Darias et al. (2010)
	Grass shrimp	Shiau and Hwang (1994)
Vertebral, branchiostegal, pugheadness, and caudal-fin deformities	European sea bass	Darias et al. (2010)
Lordosis-like droop tail	Rainbow trout	Barnett et al. (1982)
Epidermal pathology (skin and underlying musculature destruction, erosion of the tips of the dorsal, pectoral, and caudal fins)	Amago salmon	Taveekijakarn et al. (1996a)

TABLE 3.30 The vitamin D excessive syndromes.

Excessive syndromes	Species	References
Growth retardation	Brook trout	Poston (1969a)
Increased mortality	Japanese flounder	Haga et al. (2004)
Vertebral and (or) branchiostegal deformities	European sea bass	Darias et al. (2010)
	Japanese flounder	Haga et al. (2004)
Lethargy and dark coloration	Brook trout	Poston (1969a)
Hypermelanosis on the blind side	Japanese flounder	Yasuyuki et al. (1998)
Hypercalcemia	Freshwater catfish	Najmi et al. (1979)
Hypercalcemia and hyperphosphatemia	Catfish	Swarup et al. (1984)
	Common carp	Takeuchi et al. (1991)

high levels of vitamin D_3 or its metabolite $1\alpha,25\text{-}(OH)_2D_3$ result in adverse effects, like growth retardation, increase of mortality, deformities, and clinical and pathological changes in aquatic animals.

3.3.2.4 *Requirements*

The requirements for vitamin D of aquatic animals are listed in Table 3.31, as affected by the following factors: **(1) Species:** Requirements of vitamin D are various among some

TABLE 3.31 The vitamin D requirements reported in aquatic animals.

Species	Initial weight (g)	Requirements (IU/kg diet)	Response criteria	References
Wuchang bream (*Megalobrama amblycephala*)	17.7	5430 4970	SGR FCR	Miao et al. (2015)
Hybrid tilapia (*Oreochromis niloticus* × *O. aureus*)	0.78	374.8	PWG	Shiau and Hwang (1993)
Channel catfish (*Ictalurus punctatus*)	0.5	500	WG	Lovell and Li (1978)
Siberian sturgeon (*Acipenser baerii*)	3.47	1683.30 1403.27	PWG Osteocalcin	Wang et al. (2017)
Rainbow trout (*Salmo gairdneri*)	3.0	800−1600	WG, inhibit tetany	Barnett et al. (1982)
European sea bass (*Dicentrarchus labrax*)	Nine-day-old larvae	27,600	Growth and development	Darias et al. (2010)
Grass shrimp (*Penaeus monodon*)	0.19	4000	WG	Shiau and Hwang (1994)

PWG, percent weight gain; *SGR*, specific growth rate; *WG*, weight gain.

aquatic animal species, which range from 374.8 to 5430 IU/kg diet in juveniles (Table 3.31). **(2) Growth stages:** The requirement of vitamin D for nine-day-old larvae European sea bass (Darias et al., 2010) is 11.5 times the amount recommended for juveniles (NRC, 2011).

3.3.2.5 Sources and protection

Vitamin D is rarely present in plants, except in sun-dried roughages. The natural rich sources of vitamin D are oily fishes and their products, dairy products, mushrooms, and fortified foods oils. Among these, fish liver and oils are the richest natural sources of vitamin D (Wu, 2017). Salmon is an excellent source for vitamin D. However, vitamin D contents between wild and farmed species, and wild salmon in different waters really have large variation in vitamin D contents. The synthetic vitamin D sources include VD_2 and VD_3. VD_3 is more effective than VD_2 in rainbow trout (Barnett et al., 1982) and grass shrimp (Shiau and Hwang, 1994). Meanwhile, both vitamin D_3 and vitamin D_2 are unable upon overirradiation and oxidized during storage. Antioxidant addition plus encapsulation within a beadlet provide protection of vitamin D against oxidation.

3.3.2.6 Clinical assessment and analysis

Vitamin D nutritional status in aquatic animals can be determined by lack of deficiency signs coupled with growth performances and analysis of the concentrations of cholecalciferol (vitamin D_3), 25-OH-D_3, and 1,25-$(OH)_2$-D_3 in the serum and liver (Barnett et al., 1982; Takeuchi et al., 1991). Vitamin D concentrations can be detected by chromatographing on a straight phase HPLC, while 25-OH-D_3 and 1,25-$(OH)_2$-D_3 can be measured simultaneously by using a Nichols Advantage chemiluminescent assay developed by Nichols Institute Diagnostics (San Clemente, CA, US) (Wang et al., 2017).

3.3.3 Vitamin E

In 1922, Evans and Bishop firstly discovered an antisterility vitamin in rats and termed it as "vitamin E." Studies of vitamin E were conducted in Chinook salmon since 1964. Before 1988, researches on vitamin E nutrition in aquatic animals mainly focused on deficiency syndromes, requirements, and possible effects on reproduction. In the last 3 decades, the nutritional roles of vitamin E have also focused on immune regulation, antioxidant capacity, and requirements for new species.

3.3.3.1 Chemical structure and characteristics

Natural forms of vitamin E are all d-stereoisomers which include a substituted aromatic ring and a long isoprenoid side chain. Four natural forms of tocopherol [d-α-tocopherol (5,7,8-trimethyltocol), D-β-tocopherol (5,8-dimethyltocol), D-γ-tocopherol (5,7-dimethyltocol), d-δ-tocopherol (7,8-dimethyltocol)] differ in the number and position of methyl groups at its aromatic ring. Compared with tocopherols, tocotrienols have the same methyl groups but three different unsaturated double carbon bonds in the side chain. Hence, vitamin E has eight possible stereoisomers (Fig. 3.24). However, tocotrienols are very rare in nature.

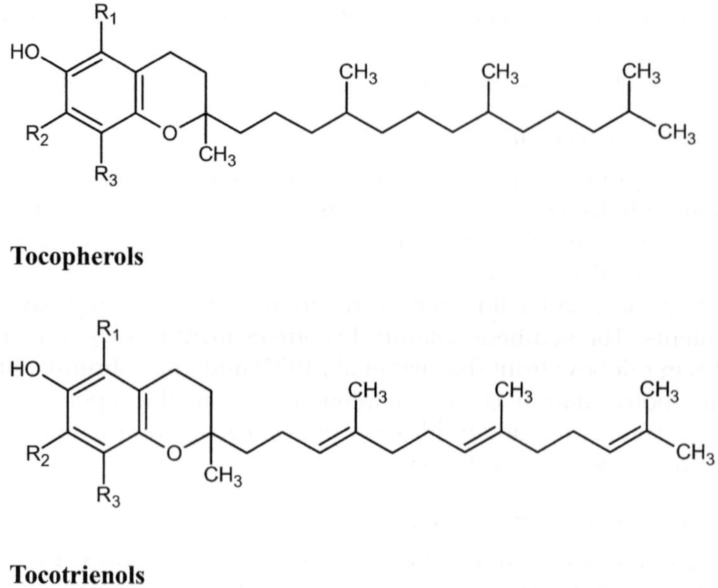

FIGURE 3.24 Chemical structure of tocopherols and tocotrienols.

The term "dl-α-tocopheryl acetate" was adopted as the international standard for assessing the biopotencies of vitamin E for more than 30 years. One IU of vitamin E is defined as the biological activity of 1 mg of dl-α-tocopheryl acetate. Currently, R,R,R-α-tocopherol is used as the new international standard. 0.67 IU of R,R,R-α-tocopherol equal to 1 equivalent (mg).

3.3.3.2 *Positive functions*

In fish, the α-tocopheryl acetate is hydrolyzed in the proximal intestine and then absorbed into the intestinal lymphatics and then the blood as chylomicrons and VLDLs via passive diffusion. The lipoproteins containing vitamin E are transported into the liver via a receptor-mediated process (Kayden and Traber, 1993). In hepatocytes, tocopherol transfer protein (TTP) carries vitamin E to the nascent VLDLs and then the vitamin E-containing VLDLs are released into the blood (Ulatowski and Manor, 2013). In the circulation, about half of the VLDLs are delipidated and returned to the liver, and the other half is converted to low-density lipoprotein (LDL), which transports tocopherols to peripheral tissues by receptor-mediated endocytosis or by exchange (Cohn et al., 1992).

The main physiological functions of vitamin E are listed as follows.

3.3.3.2.1 Antioxidant capacity

Because of the reactivity of the phenolic hydrogen on its C-6 hydroxyl group and the ability of the chromanol ring system to stabilize an unpaired electron, vitamin E can terminate chain reactions among PUFA in membranes, which can protect fish from oxidant damage. Vitamin E could increase the antioxidant enzyme activities in serum and/or various organs of grass carp (Pan et al., 2017) and yellow catfish (Lu et al., 2016). In grass carp, the increased activity of antioxidant enzymes might be related to the upregulated mRNA levels of antioxidant enzymes via Nrf2/Keap1 signaling in head kidney (Pan et al., 2017).

3.3.3.2.2 Reproductive functions

The capacity of vitamin E in scavenging ROS can provide an important protective role for the sperm cells (Izquierdo et al., 2001). Vitamin E improves reproductive performance as shown by improving the sperm quality of male rainbow trout (Canyurt and Akhan, 2008), increasing gonadosomatic index of yellow perch males (Lee and Dabrowski, 2004), and enhancing spawning performance, fertilization, and egg hatchability in *Litopenaeus vannamei* (Du et al., 2006).

3.3.3.2.3 Detoxification

Vitamin E attenuates the genotoxicity and cytotoxicity of heavy metals such as Cd, Cu, Pb, and Zn in Nile tilapia (Harabawy and Mosleh, 2014), mitigates the deleterious effects of Cd in common carp (Mohiseni et al., 2017), and reduces MeHg toxicity effects in muscle, liver, and kidney of olive flounder (Moniruzzaman et al., 2017).

3.3.3.2.4 Immunomodulation

In fish, vitamin E enhances innate immune responses such as phagocytic index, activities of lysozyme and ACP, expression of antimicrobial peptide, complement contents, and acquired immune response (immunoglobulin contents) in the head kidney, spleen, skin, and serum or plasma of grass carp (Pan et al., 2017), grouper (Lin and Shiau, 2005c), and cobia (Zhou et al., 2013), in the haemolymph or the coelomic fluid of grass shrimp (Lee and Shiau, 2004) and sea cucumber (Li et al., 2020). Meanwhile, vitamin E could attenuate inflammatory response (indicated by upregulated antiinflammatory cytokines like IL-10 and downregulated proinflammatory cytokines like TNF-α) via NF-κB/IκB and TOR signaling in tissues of grass carp (Pan et al., 2017).

3.3.3.3 *Syndromes of deficiency or excess*

Vitamin E cannot be synthesized de novo in aquatic animals. Hence, lack of exogenous supplement of vitamin E usually causes growth retardation, decreased disease resistance capacity, and some other histopathological changes (Table 3.32). Meanwhile, excess vitamin E can also cause poor growth performance of grass carp (Pan et al., 2017) and toxic liver reaction in common carp (Watanabe et al., 1970).

3.3.3.4 *Requirements*

Vitamin E requirements of selected aquatic animals are presented in Table 3.33, which can be affected by several factors as follows. **(1) Species:** The vitamin E requirements range from 21.2 to 480 mg/kg diet among some aquatic species; **(2) Growth stage:** The requirements for juvenile grass carp (200 mg/kg diet) are higher than that for subadult (116.2 mg/kg diet); **(3) Response criteria:** The vitamin E requirements based on disease resistance and hepatopancreatic vitamin E concentration are higher than that based on weight gain in grass carp and grass shrimp; **(4) Dietary factor:** A high level of dietary lipid or PUFA could increase the requirements of vitamin E in several fish.

TABLE 3.32 Vitamin E deficiency syndromes.

Deficiency syndromes	Species	References
Growth retardation	Grass carp	Pan et al. (2017); Li et al. (2014b)
	Blunt snout bream	Zhang et al. (2017b)
	Grouper	Lin and Shiau (2005c)
	Eel	Bae et al. (2012b)
	Atlantic salmon	Lall et al. (1988)
	Yellow catfish	Lu et al. (2016)
	Turbot	Niu et al. (2014)
	Cobia	Zhou et al. (2013)
	Grass shrimp	Lee and Shiau (2004)
	Sea cucumber	Li et al. (2020)
Decreased disease resistance capacity	Grass carp	Pan et al. (2017)
	Yellow catfish	Lu et al. (2016)
	Atlantic salmon	Waagbø et al. (1993b)
Anemia	Atlantic salmon	Poston et al. (1976)
	Rainbow trout	Cowey et al. (1983)
Yellowish liver	Atlantic salmon	Hamre et al. (1994)
Hepatocyte necrosis	Blue tilapia	Roem et al. (1990)
	Channel catfish	Lovell et al. (1984); Murai and Andrews (1974)
	Atlantic salmon	Hamre et al. (1994); Poston et al. (1976)
	Rainbow trout	Moccia et al. (1984)
Muscle dystrophy	Grass carp	Takeuchi et al. (1992)
	Channel catfish	Lovell et al. (1984)
	Rainbow trout	Cowey et al. (1984); Hilton et al. (1980)
	Atlantic salmon	Poston et al. (1976)
Lordosis	Grass carp	Takeuchi et al. (1992)

3.3.3.5 *Sources and protection*

Most animal-derived products (such as butter, lard, fish, eggs, and beef) and plant-derived oils, such as wheat germ oil, soybean oil and corn oil, seeds, cereal grains (except corn and wheat grains), and green fodder, are rich in vitamin E. Chemically synthesized α-tocopherol is a racemic DL-α-tocopherol mixture. α-Tocopheryl acetate is widely used as vitamin E source because of its stability in the air oxidation or the presence of labile compounds (for example, polyunsaturated fish oil).

TABLE 3.33 The vitamin E requirements reported in aquatic animals.

Species	Initial weight (g)	Requirements (mg/kg diet)	Response criteria	References
Grass carp (*Ctenopharyngodon idella*)	266.39	116.2	PWG	Pan et al. (2017)
		130.9	Skin lesion morbidity	
	11.2	100.36	PWG	Li et al. (2014b)
	4.5	200	Muscular dystrophy and lordosis	Takeuchi et al. (1992)
Blunt snout bream (*Megalobrama amblycephala*)	0.59	55.53	PWG	Zhang et al. (2017b)
Yellow catfish (*Pelteobagrus fulvidraco*)	2.01	33	SGR	Lu et al. (2016)
Channel catfish (*Ictalurus punctatus*)	143	50	Ascorbic acid stimulated lipid peroxidation	Wilson et al. (1984)
Tilapia (*Oreochromis niloticus* × *O. aureus*)	0.67	44 (50g lipid per kg diets) 66 (120g lipid per kg diets)	PWG	Shiau and Shiau (2001)
Turbot (*Scophthalmus maximus*)	113.08	480	FBW	Niu et al. (2014)
Atlantic salmon (*Salmo salar*)	0.25	60	FBW, blood hemoglobin, survival	Hamre & Lie (1995)
Grouper (*Epinephelus malabaricus*)	7.80	61 (4% lipid) 104 (9% lipid)	PWG	Lin and Shiau (2005c)
Hybrid striped bass (*Morone chrysops* female × *M. saxatilis* male)	1.8	28	PWG	Kocabas and Gatlin (1999)
Cobia (*Rachycentron canadum*)	6.1	78	PWG	Zhou et al. (2013)
		111	Lysozyme	
Red drum (*Sciaenops ocellatus*)	12.2	31	Liver TBARS	Li and Gatlin (2009)
Eel (*Anguilla japonica*)	15	21.2	PWG	Bae et al. (2012b)
		21.6	SGR	
		21.2	FE	
		21.5	Protein efficiency ratio	
Grass shrimp (*Penaeus monodon*)	0.29	84.5	PWG	Lee and Shiau (2004)
		89.0	Total hemocyte count	
		178.7	Hepatopancreatic vitamin E concentration	

Continued

TABLE 3.33 The vitamin E requirements reported in aquatic animals.—cont'd

Species	Initial weight (g)	Requirements (mg/kg diet)	Response criteria	References
Sea cucumber (*Apostichopus japonicus*)	7.96	91.68	SGR	Wang et al. (2015)
		88.33	Total hemocyte count	
		114.68	Body wall TBARS values	
	15.36	187.72	PWG	Li et al. (2020)

AWG, absolute weight gain; *FBW*, final body weight; *FE*, feed efficiency; *PWG*, percent weight gain; *SGR*, specific growth rate.

3.3.3.6 Clinical assessment and analysis

The standard colorimetric and fluorometric methods can be easily carried out for analysis of vitamin E in animal tissues for biological research and clinical testing. However, for more sophisticated and sensitive detection of α-tocopherol and other forms of tocopherol and tocotrienol in animal tissues, high-performance liquid–liquid partition chromatography is more suitable. The erythrocyte fragility test has been used as an indicator of the vitamin E physiological state in fish since 1964. Further, erythrocyte peroxide hemolysis can also be used as a criterion for determining vitamin E status (Sau et al., 2004).

3.3.4 Vitamin K

Vitamin K was discovered by Henrik Dam in 1929, and then named in 1935. Soon after, vitamin K_1 and K_2 were isolated from alfalfa and putrefied fish, respectively, in 1939 by Edward Doisy's group, who later confirmed the structure of vitamin K_1 and K_2. Since then, the role of vitamin K in hemorrhage in humans and rats was found. During 1960–2000s, researchers began to explore the nutritional effects of vitamin K on fish and shellfish, focusing on deficiency syndromes and requirements. In the last 2 decades, studies on the role of vitamin K in bone mineralization and skeletal development in aquatic animals have been pursued (Krossøy et al., 2011).

3.3.4.1 Chemical structure and characteristics

The term vitamin K represents a family of compounds that have a common 2-methyl-1,4-naphthoquinone ring but differ in the structure at the 3-position (Fig. 3.25). It contains phylloquinone (vitamin K_1, 2-methyl-3-phytyl-1,4-naphthoquinone, mainly in plants), menaquinones (vitamin K_2, 2-methyl-1,4-naphthoquinones, MKs, mainly in microbial origin), and menadione (vitamin K_3, 2-methyl-1,4-naphthoquinone, synthetic form). Vitamin K_1 is a yellow oil, whereas other K vitamers are yellow crystals. Meanwhile, vitamin K_1 and K_2 are fat soluble, but vitamin K_3 is water soluble. All K vitamers are sensitive to light and alkali, but relatively stable to heat.

3.3.4.2 Positive functions

The intestinal absorption of naturally occurring vitamin K (vitamin K_1 and K_2) is mainly mediated by a well-established process, which requires the formation of mixed micelles and noncarrier-mediated passive diffusion. Once inside the enterocyte, vitamin K_1 and K_2

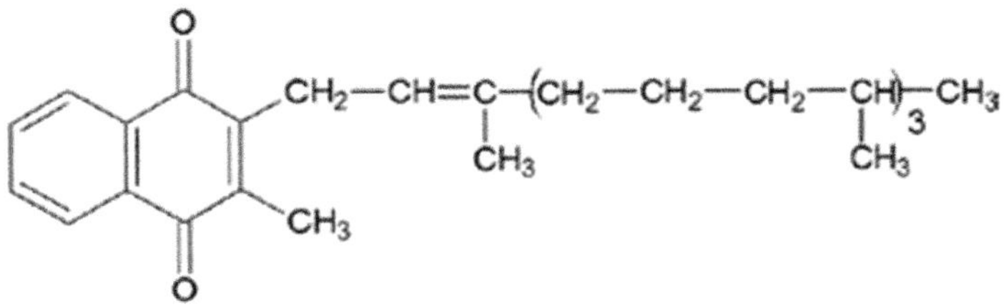

Phylloquinone

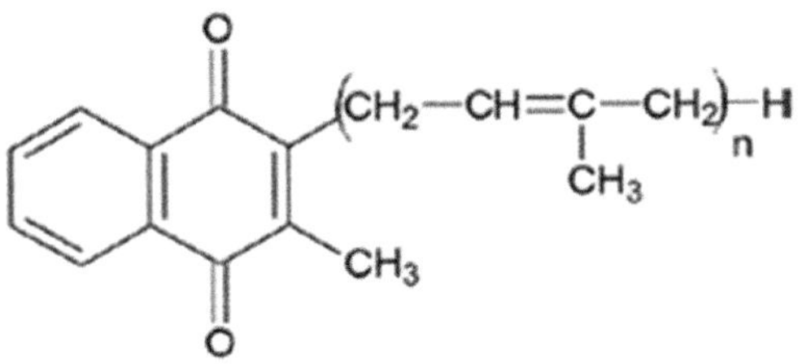

Menaquinones (n may equal 6, 7, 9)

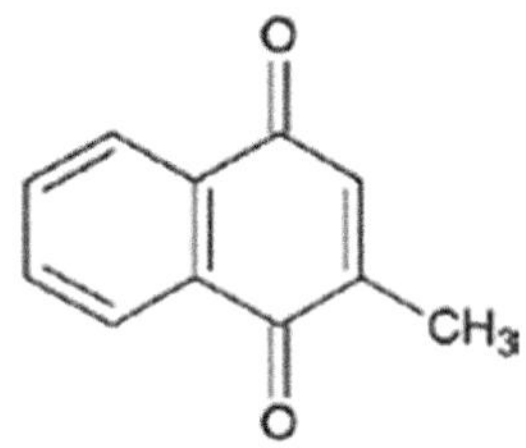

Menadione

FIGURE 3.25 The chemical structure of phylloquinone, menaquinones (n may equal 6, 7, 9), and menadione.

are assembled into chylomicrons, which then enter into the lymphatics and then the bloodstream in fish (Wu, 2017). In animal cells, vitamin K undergoes cyclic oxidation and reduction, which involves in the production and oxidation of vitamin K hydroquinone catalyzed by vitamin K reductase and vitamin K γ-glutamyl carboxylase, respectively, and reduction of vitamin K-2, 3-epoxide catalyzed by vitamin K epoxide reductase (Fig. 3.26). In this cycle, various functional γ-carboxylated proteins are produced. Vitamin K γ-glutamyl carboxylase showed a widespread tissue distribution, and vitamin K epoxide reductase is differentially expressed in adult tissues of zebrafish (Fernández et al., 2015).

Vitamin K is a cofactor of vitamin K–dependent γ-glutamyl carboxylase that catalyzes the posttranslational modification of glutamate residues to form γ-carboxyglutamate (Gla), which is essential for the biological activity of vitamin K–dependent proteins. Accordingly, vitamin K directly and as vitamin K–dependent Gla proteins has the following physiological functions.

3.3.4.2.1 Blood clotting

Blood clotting is produced by a complex system of proteins including prothrombin, proconvertin, blood clotting factors IX and X, the activation of which requires a cascade of vitamin K–dependent reactions. Vitamin K could reduce coagulation time in salmon (Taveekijakarn et al., 1996b).

3.3.4.2.2 Gla proteins–related other functions

Osteocalcin and matrix Gla protein, two important proteins in regulating calcification, are also vitamin K–dependent Gla proteins. The osteocalcin, matrix Gla protein, and a novel Gla-rich protein have been found in turbot (Roberto et al., 2009), Atlantic salmon (Krossøy et al., 2009a), and Adriatic sturgeon (Viegas et al., 2008), respectively. Vitamin K decreases

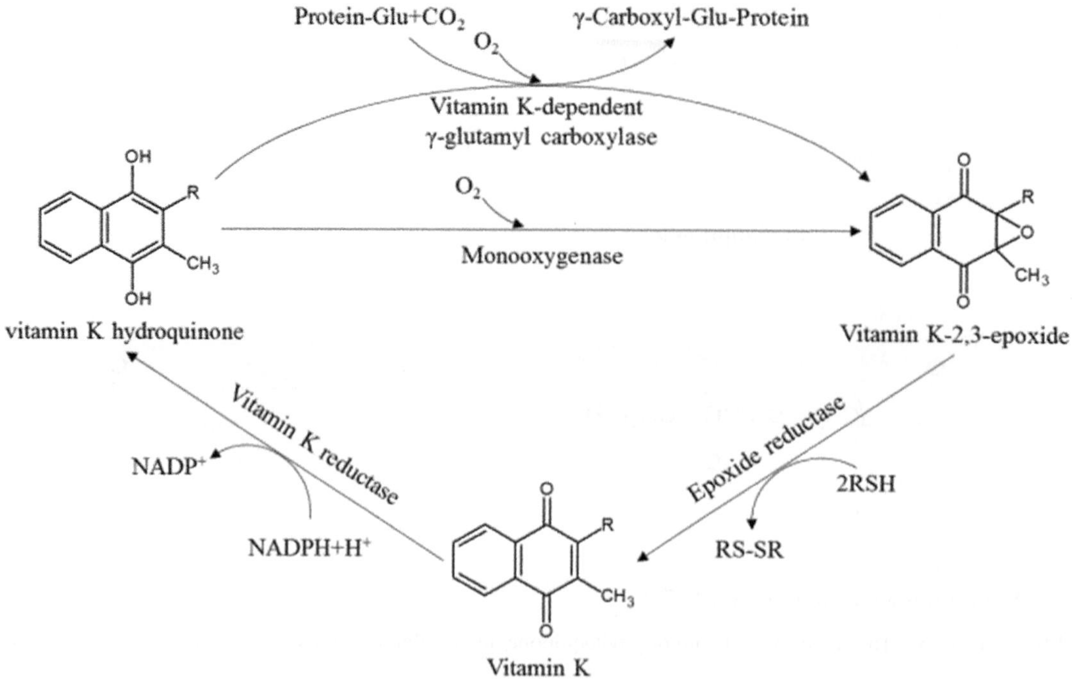

FIGURE 3.26 The vitamin K cycle.

skeletal deformity incidences in mummichog (Udagawa, 2001) and Senegalese sole (Richard et al., 2014). Besides, some other Gla proteins have been identified in several tissues including nervous tissue and sperm. In Senegalese sole, vitamin K could improve gonad maturation and sperm quality (Fernández et al., 2019).

3.3.4.2.3 Antioxidation

Vitamin K hydroquinone, a product from the vitamin K cycle, could act as a potent antioxidant in rat liver microsomes (Vervoort et al., 1997). Vitamin K could increase activities of several antioxidant enzymes such as superoxide dismutase (SOD), glutathione S-transferase, and glutathione reductase in tissues of Jian carp (Yuan et al., 2016).

3.3.4.3 Syndromes of deficiency or excess

Although some animals do not require supplementation of vitamin K because of intestinal bacterial synthesis, vitamin K synthesis microflora has not been described in aquatic animals. The vitamin K deficiency syndromes in aquatic animals mainly include growth retardation, hematological index impairment, and bone structure abnormality (Table 3.34). However, a massive dose of menadione sodium bisulfite (MSB) increased incidence of bone deformity in mummichog (Udagawa, 2001).

3.3.4.4 Requirements

The vitamin K requirements reported in aquatic animal are listed in Table 3.35. Although the vitamin K is considered to be essential for Senegalese sole (Richard et al., 2014), trout, and salmon (Salte and Norberg, 1991), the exact requirements for these fishes were not determined.

TABLE 3.34 Vitamin K deficiency syndromes.

Deficiency syndromes	Species	References
Growth retardation	Jian carp	Yuan et al. (2016)
	Amago salmon	Taveekijakarn et al. (1996b)
	Penaeus chinensis	Shiau and Liu (1994a)
	P. monodon	Shiau and Liu (1994b)
Increased mortality	Mummichog	Udagawa and Hirose (1998)
	Amago salmon	Taveekijakarn et al. (1996b)
Prolonged coagulation time	Amago salmon	Taveekijakarn et al. (1996b)
Anemia	Rainbow trout	Kitamura et al. (1967a)
	Amago salmon	Taveekijakarn et al. (1996b)
Hemorrhage	Jian carp	Yuan et al. (2016)
	Amago salmon	Taveekijakarn et al. (1996b)
Higher incidence of bone deformity	Mummichog	Udagawa (2001, 2004)
Thin and weak bone	Mummichog	Udagawa (2004)
Decreased calcium deposition	*P. monodon*	Shiau and Liu (1994b)
Histopathological changes (necrotic hepatocytes and renal tubules, slight telangiectasis in the gill lamellae, empty sinusoids in the splenic pulp)	Amago salmon	Taveekijakarn et al. (1996b)

TABLE 3.35 The vitamin K requirements reported in fish and shrimp.

Species	Initial weight (g)	Forms	Requirements (mg/kg diet)	Response criteria	References
Jian carp (*Cyprinus carpio* var. Jian)	8.9	MNB	3.13	PWG	Yuan et al. (2016)
Penaeus chinensis	0.37	Menadione	185	WG	Shiau and Liu (1994a)
Penaeus monodon	1.33	Menadione	31.8 40.2	PER VK-dependent protein precursor content	Shiau and Liu (1994b)

MNB, menadione nicotinamide bisulfite; *PER*, protein efficiency ratio; *WG/PWG*, weight gain %.

3.3.4.5 Sources and protection

Vitamin K_1 is most abundant in green leafy vegetables and herbs, such as kale, parsley, spinach, and green cabbage. The synthetic vitamin K includes four forms of menadione salt: menadione sodium bisulfite (MSB), menadione nicotinamide bisulfite (MNB), menadione sodium bisulfite complex (MSBC), and menadione dimethlypyrimidinol bisulfite (MPB). It has been found that phylloquinone and/or MNB are more suitable than MSB as a vitamin K source for fish (Grisdale-Helland et al., 1991; Udagawa, 2001). Both MSB and MNB can be converted to MK-4 in some fish (like Atlantic salmon) (Krossøy et al., 2009b) and abalone (Tan and Mai, 2001). Since vitamin K can be affected by moisture and the presence of trace minerals, it should not be mixed with minerals, and the vitamin K–contained diets should be kept dry and fed as soon as practicable after manufacture.

3.3.4.6 *Antimetabolites and inactivation*

Dicumarol [3,3′-methylbis-(4-hydroxycoumarin)], isolated from spoiled sweet clover, was found to induce hemorrhages and prolonged clotting times via blocking the normal function of vitamin K. Several substituted 4-hydroxycoumarin have been used as anticoagulant drug in humans and animals (Combs and McClung, 2017). Warfarin [3-(a-acetonylbenzyl)-4-hydroxycoumarin], an analog of dicumarol, is a vitamin K antagonist that works by directly inhibiting the recycling of vitamin K. It depleted plasma factor X and prolonged prothrombin time in trout and salmon (Salte and Norberg, 1991), and caused brain hemorrhages, skeletal deformities, and triggered ectopic calcifications in zebrafish, which may be due to an altered γ-carboxylation of vitamin K–dependent proteins and/or pregnance X receptor signaling (Fernández et al., 2014). Furthermore, the use of certain antibiotics containing an N-methylthiotetrazole side chain (e.g., cefamandole, cefoperazone) can result in a vitamin K antagonism in mammalian (Booth and Al Rajabi, 2008). Supplementation with nifurazolidon and suphaguanidine also caused lower growth rate in cod (Grahl-Madsen and Lie, 1997) and abalone (Tan and Mai, 2001), respectively.

3.3.4.7 *Clinical assessment and analysis*

In aquatic animals, the vitamin K nutritional status can be indicated by blood clotting time, γ-glutamyl carboxylase activity, and vitamin K epoxide reductase genes expression (Fernández et al., 2015). In mammals, the useful biomarkers of vitamin K status include plasma phylloquinone, prothrombin time, plasma undercarboxylated vitamin K–dependent proteins (e.g., undercarboxylated prothrombin and osteocalcin) which is considered to be a more sensitive indicator of vitamin K deficiency, and urinary Gla excretion (Booth and Al Rajabi, 2008). Plasma phylloquinone can be measured by HPLC method.

3.4 Vitamin-like compound

3.4.1 Lipoic acid

LA was first isolated from bovine liver by Lester J. Reed in 1951 and found to be synthesized in the mitochondria of rats by Jean-Paul Carreau in 1979. There were few studies about the nutritional effects of LA in aquatic animals until early 2000s (Terjesen et al., 2004; Park et al., 2006). During the last 2 decades, researches in aquatic animals mainly focused on the effects of LA on lipid metabolism, antioxidant, and immunity (Kütter et al., 2012; Liu et al., 2018).

3.4.1.1 *Chemical structure and characteristics*

LA (1,2-dithiolane-3-pentanoic acid) is a disulfide-containing compound with the formula of $C_8H_{12}O_2S_2$ (Fig. 3.27). It is a white crystalline, and both water soluble and fat soluble. The

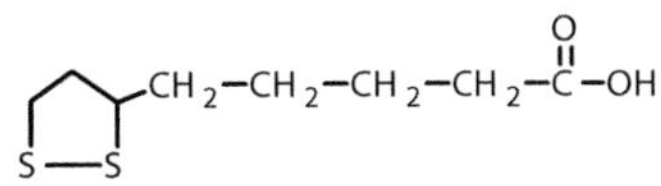

FIGURE 3.27 The chemical structure of lipoic acid.

C6 atom is the chiral center; thus, the molecule has two enantiomers, R-lipoic acid and S-lipoic acid.

3.4.1.2 *Positive functions*

The protein-bound LA is released by protease degradation of lipoyl-lysine in the intestinal lumen. The enteric absorption of LA is facilitated by the monocarboxylate transporter and/or Na^+-dependent multivitamin transporter, which also may be involved in the tissue uptake of LA from the blood plasma (Shay et al., 2009). However, the absorption and transport of LA in aquatic animal are not clearly defined. The major metabolic fate of LA is oxidized by extensive β-oxidation in the mitochondria in mammals. In addition, LA is rapidly reduced to dihydro LA in a nicotinamide adenine dinucleotide (NADH)−dependent manner by lipoamide dehydrogenase and in a nicotinamide adenine dinucleotide phosphate (NADPH)−dependent manner by glutathione reductase and thioredoxin reductase in mammal cells (Jones et al., 2002).

LA has three major physiological functions as follows.

3.4.1.2.1 Coenzyme

The amide form of LA, lipoamide, is a coenzyme for several mitochondrial enzymes, including pyruvate oxidation decarboxylation, α-ketoglutarate dehydrogenase, and branched-chain ketoacid dehydrogenase, which play important roles in energy and amino acid metabolism, respectively (Wu, 2017). Thus, LA could modulate protein metabolism as shown by altered plasma-free amino acid profile in pacu (Terjesen et al., 2004) and increased muscle protein content in grass carp probably through activating the TOR pathway (Shi et al., 2018); regulate lipid metabolism as shown by increasing tissue contents of cholesterol and/or triglyceride in common carp (Santos et al., 2016) and Chinese mitten crab (Xu et al., 2019b), respectively; and promote tissue lipolysis and fatty acid β-oxidation which were, respectively, attributed to the activated AMPKα-ATGL and AMPKα-CPT-1α pathways in grass carp (Shi et al., 2018).

3.4.1.2.2 Antioxidation and metal chelation

LA and its reduced form dihydrolipoic acid are known to create a potent redox couple that enables them to increase regeneration of glutathione, vitamin E, and vitamin C, and scavenge hydroxyl radicals and singlet oxygen, and chelate metal ions (Shay et al., 2009). LA could improve tissue antioxidant ability consequently decreasing oxidative damage in common carp (Santos et al., 2016), tilapia (Xu et al., 2019a), Pacific white shrimp (Da Silva Martins et al., 2018), abalone (Zhang et al., 2010), and Chinese mitten crab (Xu et al., 2019b), and elevate mRNA expression of antioxidant enzymes probably via Nrf2/Keap1 signaling pathways in grass carp (Shi et al., 2017), enhance the detoxified metabolization of As and decrease accumulation of As and/or Cd in tissues of Pacific white shrimp (Lobato et al., 2018) and abalone (Huang et al., 2019a).

3.4.1.2.3 Immunoregulation

LA could improve immunity, i.e., activities of lysozyme and ACP, contents of C3, C4, and IgM, and expression of antimicrobial peptide, and attenuate inflammatory response (indicated by upregulated antiinflammatory cytokines like IL-10 and downregulated proinflammatory cytokines like TNF-α) probably via NF-κB/IκB and TOR signaling in tissues of grass carp (Liu et al., 2018).

3.4.1.3 Supplementation

LA can be synthesized in the mitochondria from unsaturated fatty acids and cysteine in rats. However, this process is not clearly defined in aquatic animals. Dietary supplementations of LA for aquatic animal are also reported (Table 3.36). Generally, the recommended supplementations of LA based on disease resistance and antioxidant are slightly higher than that based on weight gain, as shown in grass carp and Nile tilapia.

However, excessive LA presented several negative effects in aquatic animal. Firstly, high levels of LA decreased growth performance of grass carp (Liu et al., 2018), tilapia (Xu et al., 2019a), and plata pompano (Kütter et al., 2012). Secondly, excessive LA decreased total antioxidant capacity in hepatopancreas of abalone (Zhang et al., 2010) and increased malondialdehyde content in hepatopancreas of Chinese mitten crab (Xu et al., 2019b). Finally, excessive LA impaired immune function including decrease of lysozyme and ACP activities, C3 and C4 contents in immune organs of grass carp (Liu et al., 2018).

3.4.1.4 Sources and protection

The natural sources of LA are tissues rich in mitochondria including glandular tissues, liver, or chloroplasts including garden peas, rice bran, and spinach. Most LA in these sources exists in protein-bound form. The synthetic LA is a racemic mixture of R-lipoic acid and S-lipoic acid. Generally, LA is stable at ambient temperature conditions.

TABLE 3.36 The recommended supplementation of lipoic acid for reported aquatic animal.

Species	Initial weight (g)	Supplementation (mg/kg diet)	Response criteria	References
Grass carp (*Ctenopharyngodon idella*)	226	315.37 382.33	WG Morbidity	Liu et al. (2018)
Nile tilapia (*Oreochromis niloticus*)	0.47	439 460	WG Liver SOD activity	Xu et al. (2019a)
Plata pompano (*Trachinotus marginatus*)	1.54	316–524	Growth, antioxidant status	Kütter et al. (2012)
Abalone (*Haliotis discus hannai* Ino)	0.42	709	WG	Zhang et al. (2010)
Chinese mitten crab (*Eriocheir sinensis*)	1.14	1339.13 1574.09	WG SGR	Xu et al. (2019b)

SGR, specific growth rate; *SOD*, superoxide dismutase; *WG*, weight gain.

3.4.2 Astaxanthin

Astaxanthin was originally isolated from a lobster and was employed initially only for the pigmentation of salmonids or shrimp in aquaculture. Later in 1991, astaxanthin gained approval as a feed supplement when its biological activities, antioxidative features, and physiological performance as a precursor of vitamin A in rats and fish were reported. Given the similarity in action of astaxanthin and canthaxanthin compared to vitamin E and vitamin A, it has been suggested that these two carotenoids should be listed among the fat-soluble vitamins (Torrissen and Christiansen, 1995). Studies assessing the effect of different sources of carotenoids on pigmentation in crustaceans, and recommended levels for optimal pigmentation, were summarized by Wade et al. (2017a).

3.4.2.1 *Chemical structure and characteristics*

Astaxanthin is a highly unsaturated terpenoid compound, with the chemical formula $C_{40}H_{52}O_4$ (Fig. 3.28). It has a molecular structure similar to β-carotene and other carotenoids. It has an extended structure, with polar region at either end of the molecule ionone rings, which enables it to possess a suitable capacity of neutralizing free radicals. It also has a nonpolar zone in the middle, which is made up of a series of carbon—carbon double bond termed "conjugated." Astaxanthin exists in stereoisomers, geometric isomers, and free and esterified forms (Higuera-Ciapara et al., 2006). There are several astaxanthin stereoisomers in nature, including (3S, 3′S), (3R, 3′R), and (3R, 3′S). The stereoisomers are different in the configuration of the two hydroxyl groups on the molecule. (3S, 3′S) and (3R, 3′R) are the most abundant forms in nature.

3.4.2.2 *Positive functions*

Astaxanthin has an antioxidant activity as high as 10 times more than other carotenoids such as zeaxanthin, lutein, canthaxanthin, and β-carotene, and 100 times more than α-tocopherol and so has even been given the title of "super vitamin E." The presence of the hydroxyl and keto moieties on each ionone ring is responsible for this high antioxidant activity. The oxo function is capable to resonance-stabilize carbon-centered radicals, which may explain the powerful antioxidative properties of astaxanthin without prooxidative contributions. The polyene chain in astaxanthin traps radicals in the cell membrane, while the terminal ring of astaxanthin could scavenge radicals at the outer and inner parts of cell membrane. It has also been reported that astaxanthin enhances antioxidant activity in fish (Song et al., 2016). Apart from the antioxidant capability, astaxanthin can also promote immunity by increasing the production of IgA, IgG, and IgM in response to T-dependent stimuli

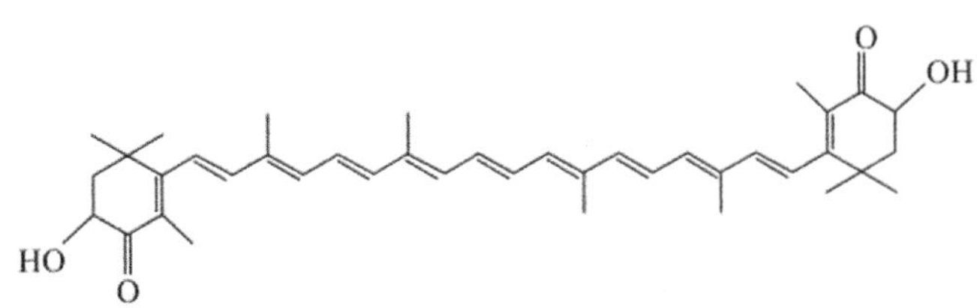

FIGURE 3.28 Structural formula of astaxanthin.

(Liu et al., 2016). In addition, as astaxanthin has better absorption compared to other carotenoids such as lutein, zeaxanthin, isozeaxanthin, and beta-carotene, it is often incorporated into ornamental fish diets with the aim to enhance coloration (Nurhadi et al., 2019). Other than these main functions, astaxanthin also has several biological functions, such as in improving reproductive capacity and stress tolerance (Yuan et al., 2011).

3.4.2.3 *Supplementation*

Considerable researches have suggested the importance of dietary astaxanthin as a factor in stimulating growth and survival of aquatic animals. For example, large yellow croaker juveniles receiving astaxanthin concentrations of 30, 50, and 100 mg/kg, respectively, gain considerable body weight and higher survival after 66-day feeding (Li et al., 2014c). Apart from those fish species, the supplementation of astaxanthin in the feed of giant tiger prawn could also enhance growth and improve survival (Wade et al., 2017b). However, the suitable supplementation level of astaxanthin varies among different species. Based on recent studies, appropriate levels of astaxanthin for different aquatic animals are listed in the Table 3.37.

TABLE 3.37 The recommended levels of astaxanthin in aquatic animals.

Astaxanthin	Initial weight	Inclusion level/range (mg/kg diet)	Response criteria	References
Atlantic salmon (*Salmo salar*)	1.7 g	36–190	Growth and survival	Christiansen and Torrissen (1996)
Red porgy (*Pagrus pagrus*)	44.5 g	40	Growth	Kalinowski et al. (2005)
Rainbow trout (*Oncorhynchus mykiss*)	113–148 mg	12.5–92.9	Growth	Lakeh et al. (2010)
Asian sea bass (*Lates calcarifer*)	28.07 g	50–150	Growth	Lim et al. (2019b)
Large yellow croaker (*Larimichthys crocea*)	5.57 g	30–100	Growth	Li et al. (2014c)
Atlantic cod (*Gadus morhua*)	4.54 kg	50–100	Growth and survival	Hansen et al. (2016)
Kuruma prawn (*Penaeus japonicus*)	0.4 g	50 and 100	Survival	Chien and Shiau (2005)
Pacific white shrimp (*Penaeus vannamei*)	1.01 g	125–150	Growth	Zhang et al. (2013)
Giant tiger prawn (*Penaeus monodon*)	7.25 g	25–100	Weight gain	Wade et al. (2017b)
Red king crab (*Paralithodes camtschaticus*)	5.1 mg	0–380	Growth and survival	Daly et al. (2013)

There are many factors that influence the effect of astaxanthin, including source, form, and feeding method. Studies have shown that natural astaxanthin can induce better change in body color and increase the total astaxanthin content in the body compared with synthetic astaxanthin (Kurnia et al., 2007). Apart from the source, the form of astaxanthin affects the absorption and utilization of astaxanthin. When feeding rainbow trout esterified astaxanthin, the absorption efficiency of astaxanthin is lower than the groups feeding free astaxanthin (White et al., 2002). Besides, feeding mode and feeding period are also important factors affecting astaxanthin utilization.

3.4.2.4 Sources and protections

The natural sources of astaxanthin are algae, yeast, salmon, trout, krill, shrimp, and crayfish (Ambati et al., 2014). Currently, commercial astaxanthin is generally derived from chemical synthesis or from natural resources rich in astaxanthin, such as *Haematococcus pluvialis* and *Phaffia rhodozyma*. The freshwater microalga, *Haematococcus pluvialis*, is one of the most promising sources of natural astaxanthin (Han et al., 2013). In addition, muscle of both wild and farmed salmonid is rich in astaxanthin, which is an important source of astaxanthin (EFSA, 2006).

Astaxanthin is widely known to be extremely sensitive to heat, intense light, and oxidative conditions due to its highly unsaturated molecular structure (Armenta and Guerrero-Legarreta, 2009). In the process of feed processing, mixing may incorporate air into the blender and cause undesirable oxidation of carotenoids. Therefore, the use of a vacuum mixer and the addition of secondary antioxidants (BHT and BHA) have been applied to improve the oxidation stability of astaxanthin (Orapint and Bundit, 2012). In addition, overdrying should also be avoided, and optimal processing temperature (60–80°C) should be applied to reduce the loss of pigment (Anderson and Sunderland, 2002). If necessary, postliquid application of fats or oils via coating will avoid the risk of damaging heat-sensitive carotenoids (Dethlefsen et al., 2016).

A good storage condition is essential for the final feed products. The best storage condition for microalgal dry biomass of carotenoids is to keep them under vacuum and nitrogen atmosphere in the dark place with high retentions (Gouveia and Empis, 2003). Furthermore, astaxanthin can be added to formulas that contain more sources of fat, such as feed oil, for better stability. The reason is that some vegetable oils contain flavonoids, polyphenols, and tocopherols, which are considered to have stable activity (Rao et al., 2007). Peng et al. (2010) also observed improved stability and bioavailability when astaxanthin is encapsulated in liposomes.

3.4.2.5 Application assessment

Analysis and quantitative determination of astaxanthin in aquatic animals is accomplished by spectrophotometry, co-thin layer chromatography (TLC) and co-HPLC with standards.

Astaxanthin mainly exists in free state or esterified state in nature. Zhou et al. (2017) indicated that the absorption, transformation, and accumulation of astaxanthin in animals have selectivity. Studies have shown that astaxanthin is absorbed in a similar way to esters, that is, separated from its binding by digestive enzymes in the gastrointestinal tract of animals, emulsified with other lipids to form chylomicrons, which are simple diffusion to the surface of the intestinal wall, and then absorbed by intestinal mucosal cells (Tyssandier et al., 2001).

The absorption of astaxanthin is also influenced by the astaxanthin origin level, feed composition, and genetic factors of animals themselves (Baker et al., 2002). White et al. (2002) also find that different intestinal sites have different absorption rate, and the anterior intestine might play the most important role. In addition, adding lipids to feeds is considered beneficial to the absorption of astaxanthin in salmonids (Chimsung et al., 2013). As for transformation of astaxanthin, study has shown that astaxanthin is transported by all lipoprotein fractions of various densities in different amounts (Choubert et al., 1994). However, the absorption and transformation of astaxanthin in animals are quite complex, and many processes need to be further explored.

Apart from the absorption and transformation, the sources and contents of astaxanthin in feed will also have different effects on astaxanthin concentration in different tissues of experimental animals. The serum and muscle astaxanthin concentration of rainbow trout fed with synthetic astaxanthin were higher than that with astaxanthin from the green microalgae (Choubert et al., 2006). However, astaxanthin concentration in the skin of red sea bream fed with natural astaxanthin was higher than fish fed with synthetic astaxanthin (Kurnia et al., 2007). Astaxanthin deposition in tissues is not an exact value; it is linearly correlated with the content of astaxanthin in the feed (Kouakou and Georges, 2006). Even within the same organism, astaxanthin levels can vary widely among tissues. Assessment method of biologically active astaxanthin in diet is the same as the tissues. However, astaxanthin is easily lost during processing and storage of feed (Lim et al., 2019a). Hence, special consideration should be given during analysis of these pigments.

3.5 Interactions between vitamins and other nutrients

Although each vitamin has its specific physiological functions, vitamins also interact with other nutrients in terms of function and metabolism. Thus, it is important to understand the interactions between vitamins and other nutrients.

3.5.1 Interactions between vitamins

3.5.1.1 Synergistic effect

3.5.1.1.1 Vitamin C and vitamin E, lipoic acid

Vitamin C and E can spare each other as shown in two aspects. Vitamin C reduces the incidence and severity of vitamin E deficiency signs in yellow perch (Lee and Dabrowski, 2004) and hybrid striped bass (Sealey and Gatlin, 2002). Vitamin E also protects fish from vitamin C deficiency symptoms as shown in channel catfish (Yildirim-Aksoy et al., 2008). Vitamin C or E increases α-tocopherol levels in plasma or tissue and ascorbic acid contents in liver or viscera of several fishes (like channel catfish) (Yildirim-Aksoy et al., 2008) and kuruma shrimp (Nguyen et al., 2012), respectively. Besides, vitamin C and E exert a synergistic effect in improving intestinal protease activity and antioxidant defense in discus fish (Liu et al., 2019), innate immunity in gilthead sea bream (Ortuño et al., 2001), and reproductive performance in kuruma shrimp (Nguyen et al., 2012). Furthermore, LA partially reverses tissue ascorbic acid depletion induced by vitamin C deficiency in pacu (Park et al., 2006), which is due to that LA and dihydrolipoic acid could enable the regeneration of vitamin C (Shay et al., 2009).

3.5.1.1.2 Other vitamins

In *Labeo rohita*, a combined deficiency of vitamin B_{12} and folic acid results in an accelerated anemia (John and Mahajan, 1979), which can be ascribed to the fact that vitamin B_{12} plays an important role in converting 5-methyl-tetrahydrofolic acid (5-methyl-FH_4) to THF, and its deficiency induces a functional folate deficiency by increasing proportion of 5-methyl-FH_4 (Combs and McClung, 2017). Choline and inositol could compensate deficiency of each other in juvenile kuruma shrimp (Michael and Koshio, 2008). Besides, vitamin A increases viscera 25-hydroxyvitamin D_3 and 1α,25-dihydroxyvitamin D_3 contents in abalone (Zhang et al., 2007). Astaxanthin intake increases liver vitamin E levels; meanwhile, both vitamin E and astaxanthin enhances skin carotenoid content in large yellow croaker (Yi et al., 2018). Furthermore, vitamin E acts in synergy with vitamin A or astaxanthin to improve the antioxidant capacity in rainbow trout (Kelestemur and Özdemir, 2013) and large yellow croaker (Yi et al., 2018), respectively.

3.5.1.2 Antagonistic effects

3.5.1.2.1 Vitamin A and vitamin K

High levels of vitamin A might interfere with the synthesis of vitamin K_2 by intestinal flora, the absorption, or metabolism of vitamin K in rat (Walker et al., 1947). In diets with normal levels of vitamin K diet, supplementation with high levels of vitamin A results in high mortality in Atlantic salmon (Grisdale-Helland et al., 1991).

3.5.2 Interactions with protein

3.5.2.1 Affecting protein digestibility

Many vitamins including vitamin K (Yuan et al., 2016), thiamin (Huang et al., 2011), riboflavin (Li et al., 2010a), pantothenic acid (Wen et al., 2009), niacin (Li et al., 2016a), biotin (Zhao et al., 2012), ascorbic acid (Liu et al., 2011), and choline (Wu et al., 2011) promote protein digestive enzyme activities (trypsin and/or chymotrypsin) in the intestine of Jian carp and/or grass carp. These actions can be associated with acetylcholine which as a neurotransmitter stimulates digestive enzyme secretion in animals. The mentioned vitamins (except vitamin K) can increase the acetylcholine contents by various ways: (1) choline can be metabolized to acetylcholine in animals (Combs and McClung, 2017); (2) riboflavin, pantothenic acid, niacin, and ascorbic acid promote the synthesis of acetylcholine (Torda and Wolff, 1945); (3) thiamin can inhibit acetylcholinesterase activity which can hydrolyze acetylcholine (Alspach and Ingraham, 1977). Besides, *myo*-inositol also increases the pancreatic chymotrypsin activities in Jian carp (Jiang et al., 2009), and this action is associated with the facilitation of inositol phosphatide on transmembrane movement of chymotrypsinogen (Rothman, 1978).

3.5.2.2 *Affecting protein metabolism*

Most B-vitamins serve as cofactors or components of coenzymes of protein or amino acid metabolism. In gilthead sea bream fry, high level of vitamin B_6 can promote the protein utilization in both low-protein and high-protein diets (Baker and Davies, 1995). However, low-protein diet supplemented with high level of vitamin B_6 causes a normocytic normochromic anemia in juvenile tilapia (Shiau and Hsieh, 1997). High-protein diet with pantothenic acid supplementations can improve the growth and feed intake in Nile tilapia (Ayyat et al., 2011). Choline can interact with methionine in three aspects: (1) choline and methionine can compensate deficiency of each other in juvenile kuruma shrimp (Michael et al., 2006); (2) choline can spare a portion of methionine in channel catfish (Wu and Davis, 2005) and juvenile shrimp *Penaeus monodon* (Richard et al., 2011); (3) choline and methionine synergistically promote the growth, while high levels of choline and methionine synergistically aggravate growth retardation in tilapia (Kasper et al., 2000). LA has a protein-sparing effect from energy production by promoting lipolysis and fatty acid β-oxidation in grass carp (Shi et al., 2018).

3.5.3 Interactions with lipid

3.5.3.1 *Vitamins and lipid peroxidation*

When fishes were fed with oxidized fish oil, antioxidant vitamins such as vitamin C or E could decrease tissue lipid peroxidation in Atlantic halibut (Lewis-McCrea and Lall, 2007) and sea bream (Gao et al., 2013), but high levels of vitamin C and E have also been shown to exacerbate liver lipid peroxidation in Japanese flounder (Gao et al., 2014). A high level of dietary oxidized fish oil reduces liver ascorbic acid and/or α-tocopherol levels in rainbow trout (Hung et al., 1981) and Japanese flounder (Gao et al., 2014).

When fish are fed with high fat or highly unsaturated fatty acid (HUFA), LA or vitamin E, respectively, decreases tissue lipid peroxidation in grass carp (Shi et al., 2017), tilapia (Xu et al., 2019a), and grouper (Lin and Shiau, 2005c). Besides, high levels of dietary lipid result in poor growth in grouper fed with low levels of vitamin E (Lin and Shiau, 2005c). This also reduces tissue α-tocopherol retention in Japanese flounder (Wang et al., 2006) and grouper (Lin and Shiau, 2005c). However, high levels of dietary lipid and vitamin E synergistically improve growth in blue tilapia (Roem et al., 1990).

3.5.3.2 *Vitamins and lipid metabolisms*

3.5.3.2.1 Vitamin C, lipoic acid, vitamin E, and lipid metabolism

When fish are fed with oxidized oil or high levels of PUFA, vitamin E slows the decrease of liver fatty acids in turbot (Stéphan et al., 1995); vitamin E or C or LA, respectively, increases liver and/or muscle total n-3 PUFA contents in red sea bream (Gao et al., 2013), Japanese flounder (Wang et al., 2006), and grass carp (Shi et al., 2017); while vitamin E or LA, respectively, decreases liver lipid contents in channel catfish (Murai and Andrews, 1974) and grass carp (Huang et al., 2019b). Furthermore, vitamin C or LA improves lipid digestion in Jian carp (Liu et al., 2011) and Chinese mitten crab (Xu et al., 2018), respectively.

3.5.3.2.2 Other vitamins and lipid metabolism

Thiamin (Huang et al., 2011), riboflavin (Li et al., 2010a), pantothenic acid (Wen et al., 2009), niacin (Li et al., 2016a), biotin (Zhao et al., 2012), and vitamin K (Yuan et al., 2016) increase body lipid content or lipid utilization by improving lipid digestion in Jian carp and grass carp, respectively. *Myo*-inositol also increases lipid utilization in Jian carp (Jiang et al., 2009) and decreases liver lipid deposition in some fishes (Wang et al., 2018) and muscle monounsaturated fatty acid content in Pacific white shrimp fed high oxidized fish oil diet (Chen et al., 2019). Meanwhile, choline decreases liver lipid content via enhancing lipid transport from the liver by increasing VLDL production in blunt snout bream fed with high-lipid diet (Li et al., 2014a).

3.5.4 Interactions with carbohydrates

Most B-vitamins help the body to convert carbohydrates into glucose in animals including fish. High levels of thiamin in high-carbohydrate diets elevate liver glycogen accumulation in gilthead sea bream (Morris and Davies, 1995). Dextrin diet supplemented with high level of niacin further promotes the growth of tilapia compared with glucose diet (Shiau and Suen, 1992). Dietary starch aggravates the biotin deficiency in rainbow trout (Walton et al., 1984). LA enhances dietary starch utilization in common carp (Santos et al., 2016).

3.5.5 Interactions with minerals

3.5.5.1 Vitamin D, calcium, and phosphorus

Vitamin D can promote the absorption and metabolism of calcium (Ca) and phosphorus (P). Vitamin D and Ca synergistically increase the serum Ca levels in freshwater mud eel (Srivastav, 1983). Either lack of vitamin D or P can aggravate the serum Ca/P ratio abnormity and decrease mantle P concentrations in abalone (Zhang et al., 2003). High levels of vitamin D_3 cause hypercalcemia in male walking catfish (Swarup and Srivastav, 1982).

3.5.5.2 Vitamin E and selenium

Vitamin E and selenium (Se) can compensate for the lack of each other to prevent muscle myopathy in yellowtail kingfish (Le et al., 2014). Vitamin E inhibits the effects of Se deficiency syndrome in rainbow trout (Bell et al., 1985) and Atlantic salmon (Poston et al., 1976). Simultaneous deficiencies of vitamin E and Se cause a variety of deficiency signs (death, growth retardation, anemia, severe myopathy, and exudative diathesis) in fingerling channel catfish (Gatlin et al., 1986) and decrease the liver microsome packed cell volume in rainbow trout (Bell et al., 1985). Furthermore, vitamin E and Se can synergistically protect rainbow trout liver microsome (Bell et al., 1985) and largemouth bass muscle (Chen et al., 2013) from lipid peroxidation, and increase plasma Se content in rainbow trout, liver vitamin E content, and plasma-specific glutathione peroxidase activities in Atlantic salmon (Poston et al., 1976), as well as serum lysozyme and bactericidal activities in yellowtail kingfish (Le et al., 2014).

3.5.5.3 *Ascorbic acid, iron, and other minerals*

Ascorbic acid and iron synergistically promote the metamorphosis, growth, body vitamin C contents, and total antioxidant capacity in the larvae of Senegalese sole (Jiménez-Fernández et al., 2018). However, high level of vitamin C aggravates iron deficiency (increase of mortality and anemia) in channel catfish (Lim et al., 2000). Ascorbic acid and Se synergistically enhance growth performance and physiological response (serum growth hormone and hematological indexes) in mahseer fish (Khan et al., 2017), as well as antioxidative ability and immune response in Nile tilapia (Dawood et al., 2020). In addition, high levels of vitamin C attenuate high copper (Cu)–caused poor growth performances and Cu accumulations in the hepatopancreas, muscle, and heamolymph of juvenile grass shrimp (Lee and Shiau, 2003).

3.5.5.4 *Thiamin and magnesium*

Thiamin participates in energy metabolism as a coenzyme with Mg as a cofactor in mammals (Fujiwara et al., 1976). Mg attenuates thiamin deficiency–caused high mortality and prolongs the survival time of juvenile lake trout (Lee et al., 2012).

References

Adam, A., et al., 2019. Profiling DNA methylation patterns of zebrafish liver associated with parental high dietary arachidonic acid. PloS One 14, e0220934.

Adham, K., Hashem, H., Abu-Shabana, M., Kamel, A., 2000. Vitamin C deficiency in the catfish. *Clarias gariepinus*. Aquacult. Nutr. 6, 129–139.

Adron, J., Knox, D., Cowey, C., 1978. Studies on nutrition of marine flatfish- Pyridoxine requirement of turbot *Scophthalmus maximus*. Br. J. Nutr. 40, 261–268.

Aguirre, P., Gatlin, D., 1999. Dietary vitamin C requirement of red drum *Sciaenops ocellatus*. Aquacult. Nutr. 5, 247–249.

Ahmed, I., 2011. Effect of dietary niacin on growth and body composition of two Indian major carps rohu, *Labeo rohita*, and mrigal, *Cirrhinus mrigala* (Hamilton), fingerlings based on dose-response study. Aquacult. Int. 19, 567–584.

Ai, Q., et al., 2004. Effects of dietary vitamin C on growth and immune response of Japanese seabass. Lateolabrax Japonicus. Aquacult. 242, 489–500.

Ai, Q., et al., 2006. Effects of dietary vitamin C on survival, growth, and immunity of large yellow croaker, *Pseudosciaena crocea*. Aquaculture 261, 327–336.

Akhtar, M., et al., 2010. Stress mitigating and immunomodulatory effect of dietary pyridoxine in *Labeo rohita* (Hamilton), fingerlings. Aquacult. Res. 41, 991–1002.

Albrektsen, S., 1994. Studies on Vitamin B6 Nutrition in Atlantic Salmon (*Salmo salar*). PhD thesis. at the Department of Fisheries and Marine Biology. University of Bergen, Bergen, Norway.

Albrektsen, S., Waagbø, R., Sandnes, K., 1993. Tissue vitamin B6 concentrations and aspartate aminotransferase (AspT), activity in Atlantic salmon (*Salmo salar*), fed graded dietary levels of vitamin B6. Fiskeridirektoratets skrifter, Ernæring 6, 21–34.

Albrektsen, S., Sandnes, K., Glette, J., Waagbø, R., 1995. Influence of dietary vitamin B6 on tissue vitamin B6 contents and immunity in Atlantic salmon, *Salmo salar* L. Aquacult. Res. 26, 331–339.

Alsop, D., Matsumoto, J., Brown, S., Van Der Kraak, G., 2008. Retinoid requirements in the reproduction of zebrafish. Gen. Comp. Endocrinol. 156, 51–62.

Alspach, J., Ingraham, L., 1977. Inhibition of acetylcholinesterase by thiamine. A structure-function study. J. Med. Chem. 20, 161–164.

Ambati, R., Phang, S., Ravi, S., Ravishankar Gokare, A., 2014. Astaxanthin: sources, extraction, stability, biological activities and its commercial applications—a review. Mar. Drugs 12, 128–152.

Amcoff, P., et al., 1999. Physiological and morphological effects of microinjection of oxythiamine and PCBs in embryos of Baltic salmon (*Salmo salar*): a comparison with the M74 syndrome. Ambio 28, 55–66.

Amcoff, P., et al., 2002. Physiological, biochemical and morphological studies of Baltic salmon yolk-sac fry with experimental thiamine deficiency: relations to the M74 syndrome. Aquat. Toxicol. 61, 15–33.

Amezaga, M., Knox, D., 1990. Riboflavin requirements on on-growing rainbow trout, *Oncorhynchus mykiss*. Aquaculture 88, 87–98.

Anderson, J., Sunderland, R., 2002. Effect of extruder moisture and dryer processing temperature on vitamin C and E and astaxanthin stability. Aquaculture 207, 137–149.

Andrews, J., Murai, T., 1975. Studies on the vitamin C requirements of channel catfish (*Ictalurus punctatus*). J. Nutr. 105, 557–561.

Andrews, J., Murai, T., 1978. Dietary niacin requirement for channel catfish. J. Nutr. 108, 1508–1511.

Andrews, J., Murai, T., 1979. Pyridoxine requirements of channel catfish. J. Nutr. 109, 533–537.

Aoe, H., Masuda, I., 1967. Water soluble vitamin requirements of carp-II. Requirements for para-aminobenzoic acid and inositol. Bull. Jpn. Soc. Sci. Fish. 33, 674–680.

Aoe, H., Masuda, I., Saito, T., Komo, A., 1967. Water-soluble vitamin requirements of carp-I. Requirement for vitamin B2. Nippon Suisan Gakk 33, 355–360.

Aoe, H., et al., 1968. Requirement of young carp for vitamin A. Bull. Jpn. Soc. Sci. Fish. 35, 459–465.

Aoe, H., et al., 1969. Water-Soluble vitamin requirements of Carp-VI requirement for thiamine and effects of antithiamines. Nippon Suisan Gakkaishi 35, 459–465.

Aouameur, R., et al., 2007. SMIT2 mediates all myo-inositol uptake in apical membranes of rat small intestine. Am. J. Physiol.-Gastr. L. 293, G1300–G1307.

Arai, S., Nose, T., Hasimoto, Y., 1972. Qualitative requirements of young eels, *Anguilla japonika*, for water-soluble vitamins and their dificiency symptoms. Bull. Freshwater Res. Lab. Tokyo. 22, 69–83.

Armenta, R., Guerrero-Legarreta, I., 2009. Stability studies on astaxanthin extracted from fermented shrimp byproducts. J. Agric. Food Chem. 57, 6095–6100.

Arnason, J., et al., 2015. Local Fish Feed Ingredients for Competitive and Sustainable Production of High-Quality Aquaculture Feed, vol. 72. Nordic Innovation Publication.

Asaikkutti, A., et al., 2016. Effect of different levels dietary vitamin C on growth performance, muscle composition, antioxidant and enzyme activity of freshwater prawn, *Macrobrachium malcolmsonii*. Aquacult. Rep. 3, 229–236.

Ashizawa, N., Yoshida, M., Aotsuka, T., 2000. An enzymatic assay for *myo*-inositol in tissue samples. J. Biochem. Biophys. Methods 44, 89–94.

Ayyat, M., El-Marakby, H., Sharaf, S., 2011. Effect of dietary protein level, stocking density, and dietary pantothenic acid supplementation rate on performance and blood components of Nile tilapia *Oreochromis niloticus*. J. Applied Aquacult. 23, 122–135.

Bae, J., et al., 2012a. Re-evaluation of the optimum dietary vitamin C requirement in juvenile eel, *Anguilla japonica* by using L-ascorbyl-2-monophosphate. *Asian-Australas*. J. Anim. Sci. 25, 98–103.

Bae, J., et al., 2012b. Evaluation of optimum dietary vitamin E requirements using DL-α-tocopheryl acetate in the juvenile eel, *Anguilla japonica*. J. Appl. Ichthyol. 29, 213–217.

Baker, R., Davies, S., 1995. The effect of pyridoxine supplementation on dietary protein utilization in gilthead seabream fry. Anim. Sci. 60, 157–162.

Baker, R., Pfeiffer, A., Schoner, F., Smith-Lemmon, L., 2002. Pigmenting efficacy of astaxanthin and canthaxanthin in fresh-water reared Atlantic salmon, *Salmo salar*. Anim. Feed Sci. Technol. 99, 97–106.

Barnett, B., Cho, C., Slinger, S., 1982. Relative biopotency of dietary ergocalciferol and cholecalciferol and the role of and requirement for vitamin D in rainbow trout (*Salmo gairdneri*). J. Nutr. 112, 2011–2019.

Barros, M., et al., 2009. Haematological response and growth performance of Nile tilapia (*Oreochromis niloticus* L.) fed diets containing folic acid. Aquacult. Res. 40, 895–903.

Bell, J., Cowey, C., Adron, J., Shanks, A., 1985. Some effects of vitamin E and selenium deprivation on tissue enzyme levels and indices of tissue peroxidation in rainbow trout (*Salmo gairdneri*). Br. J. Nutr. 53, 149–157.

Bird, R., 2018. Chap. 4 the emerging role of vitamin B6 in inflammation and carcinogenesis. In: Advances in Food and Nutrition Research, vol. 83. Elsevier Inc., pp. 151–194

Bjerkås, E., Breck, O., Waagbø, R., 2006. The Role of Nutrition in Cataract Formation in Farmed Fish. CAB Reviews: Perspectives in Agriculture, Veterinary Science, Nutrition and Natural Resources 1 No. 033, pp. 1–16.

Blom, J., Dabrowski, K., Ebeling, J., 2000. Vitamin C requirements of the angelfish *Pterophylum scalare*. J. World Aquacult. Soc. 31, 115–118.

Boonyaratpalin, M., 1998. Nutrition of *Penaeus merguiensis* and *Penaeus indicus*. Rev. Fish. Sci. 6 (1 & 2), 69–78.

Boonyaratpalin, M., Williams, K., 2002. Asian sea bass, *Lates calcarifer*. In: Webster, C.D., Lim, C. (Eds.), Nutrient Requirements and Feeding of Finfish for Aquaculture. CAB International, pp. 40–50.

Booth, S., Al Rajabi, A., 2008. Determinants of vitamin K status in humans. Vitam. Horm. 78, 1–22.

Brønstad, I., Bjerkås, I., Waagbø, R., 2002. The need for riboflavin supplementation in high and low energy diets for Atantic salmon *Salma salar* L. Parr. Aquacult. Nutr. 8, 209–220.

Brown, M., et al., 1999. The vitamin content of microalgae used in aquaculture. J. Appl. Phycol. 11, 247–255.

Bu, X., et al., 2020. Effects of *myo*-inositol on growth performance, body composition, antioxidant status, non-specific immunity and lipid metabolism of juvenile Chinese mitten crab (*Eriocheir sinensis*). Aquacult. Nutr. 26, 1623–1635.

Burtle, G., Lovell, R., 1989. Lack of response of channel catfish (*Ictalurus punctatus*) to dietary *myo*-inositol. Can. J. Fish. Aquat. Sci. 46, 218–222.

Campeche, D., Catharino, R., Godoy, H., Cyrino, J., 2009. Vitamin A in diets for nile tilapia. Sci. Agric. 66, 751–756.

Canyurt, M., Akhan, S., 2008. Effect of dietary vitamin E on the sperm quality of rainbow trout (*Onchorynchus mykiss*). Aquacult. Res. 39, 1014–1018.

Carmel, R., Karnaze, D., Weiner, J., 1988. Neurologic abnormalities in cobalamin deficiency are associated with higher cobalamin "analogue" values than are hematologic abnormalities. J. Lab. Clin. Med. 111, 57–62.

Cassani, B., et al., 2012. Vitamin A and immune regulation: role of retinoic acid in gut-associated dendritic cell education, immune protection and tolerance. Mol. Aspect. Med. 33, 63–76.

Castledine, A., et al., 1978. Influence of dietary biotin level on growth, metabolism and pathology of rainbow trout. J. Nutr. 108, 698–711.

Catacutan, M., Cruz, D., 1989. Growth and mid-gut cells profile of *Penaeus monodon* fed water-soluble vitamin deficient diets. Aquaculture 81, 37–144.

Caxico, V., et al., 2018. Expression of genes related to antioxidant activity in Nile tilapia kept under salinity stress and fed diets containing different levels of vitamin C. J. Toxicol. Environ. Health 81, 20–30.

Cerezuela, R., Cuesta, A., Meseguer, J., Ángeles Esteban, M., 2009. Effects of dietary vitamin D_3 administration on innate immune parameters of seabream (*Sparus aurata* L.). Fish Shellfish Immunol. 26, 243–248.

Chen, H., Hwang, G., 1992. Estimation of the dietary riboflavin required to maximize tissue riboflavin concentration in juvenile shrimp (*Penaeus monodon*). J. Nutr. 122, 2474–2478.

Chen, H., Wu, F., Tang, S., 1991. Thiamin requirement of juvenile shrimp (*Penaeus monodon*). J. Nutr. 12, 1984–1989.

Chen, H., Wu, F., Tang, S., 1994. Sensitivity of transketolase to the thiamin status of juvenile marine shrimp (*Penaeus monodon*). Comp. Biochem. Physiol. A 109, 655–659.

Chen, Y., et al., 2013. Effect of dietary vitamin E and selenium supplementation on growth, body composition, and antioxidant defense mechanism in juvenile largemouth bass (*Micropterus salmoide*) fed oxidized fish oil. Fish Physiol. Biochem. 39, 593–604.

Chen, L., et al., 2015a. Dietary riboflavin deficiency decreases immunity and antioxidant capacity, and changes tight junction proteins and related signaling molecules mRNA expression in the gills of young grass carp. Fish Shellfish Immunol. 45, 307–320.

Chen, Y., et al., 2015b. Dietary vitamin C requirement and its effects on tissue antioxidant capacity of juvenile largemouth bass. *Micropterus salmoides*. Aquaculture 435, 431–436.

Chen, S., et al., 2019. Changes in growth performance, haematological parameters, hepatopancreas histopathology and antioxidant status of pacific white shrimp (*Litopenaeus vannamei*) fed oxidized fish oil: regulation by dietary *myo*-inositol. Fish Shellfish Immunol. 88, 53–64.

Chien, Y., Shiau, W., 2005. The effects of dietary supplementation of algae and synthetic astaxanthin on body astaxanthin, survival, growth, and low dissolved oxygen stress resistance of kuruma prawn, *Marsupenaeus japonicus* Bate. J. Exp. Mar. Biol. Ecol. 318, 201–211.

Chimsung, N., et al., 2013. Effects of dietary cholesterol on astaxanthin transport in plasma of Atlantic salmon (*Salmo salar*). Comp. Biochem. Physiol. B 165, 73–81.

Cho, Y., et al., 2007. Isolation and characterization of cDNA sequences of l-gulono-gamma-lactone oxidase, a key enzyme for biosynthesis of ascorbic acid, from extant primitive fish groups. Comp. Biochem. Physiol. B 147, 178–190.

Choubert, G., Milicua, J., Gomez, R., 1994. The transport of astaxanthin in immature rainbow trout *Oncorhynchus mykiss* serum. Comp. Biochem. Physiol. A 108, 245–248.

Choubert, G., Mendespinto, M., Morais, R., 2006. Pigmenting efficacy of astaxanthin fed to rainbow trout *Oncorhynchus mykiss*: effect of dietary astaxanthin and lipid sources. Aquaculture 257, 429–436.

Christakos, S., et al., 2016. Vitamin D: metabolism, molecular mechanism of action, and pleiotropic effects. Physiol. Rev. 96, 365–408.

Christiansen, R., Torrissen, O., 1996. Growth and survival of Atlantic salmon, *Salmo salar* L. fed different dietary levels of astaxanthin. Juveniles. Aquacult. Nutr. 2, 55–62.

CIBA, 2002. Training Manual on Shrimp and Fish Nutrition and Feed Management. CIBA Special Bulletin No. 15, vol. 98. Central Institute of Brackish water Aquaculture, Chennai, India.

Coates, J., Halver, J., 1958. Water-soluble Vitamin Requirements of Silver Salmon (No. 281). US Department of the Interior, Fish and Wildlife Service.

Cohn, W., Goss-Sampson, M., Grun, H., Muller, D., 1992. Plasma clearance and net uptake of α-tocopherol and low-density lipoprotein by tissues in WHHL and control rabbits. Biochem. J. 287, 247–254.

Combs Jr., G., McClung, J., 2017. The Vitamins, Fifth Edition Fundamental Aspects in Nutrition and Health. Elsevier Academic Press.

Conklin, D., 1989. Vitamin requirements of juvenuile penaeid shrimp. Advances in tropical aquaculture, aquacop IFREMER. Actes Colloq. 9, 287–308.

Corpe, C., et al., 2013. Intestinal dehydroascorbic acid (DHA) transport mediated by the facilitative sugar transporters, GLUT2 and GLUT8. J. Biol. Chem. 288, 9092–9101.

Cowey, C., Woodward, B., 1993. The dietary requirement of young rainbow trout *(Oncorhynchus mykiss)*, for folic acid. J. Nutr. 123, 1594–1600.

Cowey, C., Adron, J., Knox, D., Ball, G., 1975. Studies on nutrition of marine flatfish-thiamin requirement of turbot (*Scophthalmus maximus*). Br. J. Nutr. 34, 383–390.

Cowey, C., Adron, J., Youngson, A., 1983. The vitamin E requirement of rainbow trout (*Salmo gairdneri*) given diets containing polyunsaturated fatty acids derived from fish oil. Aquaculture 30, 85–93.

Cowey, C., et al., 1984. The effect of vitamin E and oxidized fish oil on the nutrition of rainbow trout (*Salmo gairdneri*) grown at natural, varying water temperatures. Br. J. Nutr. 51, 443–451.

Craig, S., Gatlin, D., 1996. Dietary choline requirement of juvenile red drum (*Sciaenops ocellatus*). J. Nutr. 126, 1696–1700.

Cuesta, A., et al., 2002. Changes in some innate defence parameters of seabream (*Sparus aurata* L.) induced by retinol acetate. Fish Shellfish Immunol. 13, 279–291.

Cui, P., Zhou, Q., Huang, X., Xia, M., 2016. Effect of dietary vitamin B6 on growth, feed utilization, health and non-specific immune of juvenile Pacific white shrimp, *Litopenaeus vannamei*. Aquacul. Nutr. 22, 941–1159.

Cui, W., Ma, A., Wang, X., Huang, Z., 2020. *Myo*-inositol enhances the low-salinity tolerance of turbot (*Scophthalmus maximus*) by modulating cortisol synthesis. Biochem. Bioph. Res. Co. 526, 913–919.

Dabrowski, K., 2000. Ascorbic Acid in Aquatic Organisms: Status and Perspectives. CRC press.

Dabrowski, K., et al., 2004. Effects of dietary ascorbic acid on oxygen stress (hypoxia or hyperoxia), growth and tissue vitamin concentrations in juvenile rainbow trout (*Oncorhynchus mykiss*). Aquaculture 233, 383–392.

Daly, B., Swingle, J., Eckert, G., 2013. Dietary astaxanthin supplementation for hatchery-cultured red king crab, *Paralithodes camtschaticus*, juveniles. Aquacult. Nutr. 19, 312–320.

Darias, M., et al., 2010. Dietary vitamin D_3 affects digestive system ontogenesis and ossification in European sea bass (*Dicentrachus labrax*, Linnaeus, 1758). Aquaculture 298, 300–307.

Darias, M., et al., 2011. Imbalanced dietary ascorbic acid alters molecular pathways involved in skeletogenesis of developing European sea bass (*Dicentrarchus labrax*). Comp. Biochem. Physiol. A 159, 46–55.

Davies, S., Rawling, M., 2013. Niacin: One of.the Key B Vitamins for Sustaining Healthy Fish Growth and Production. International AquaFeed. Perendale Publishers Ltd, UK, pp. 21–23.

Dawood, M., Koshio, S., 2018. Vitamin C supplementation to optimize growth, health and stress resistance in aquatic animals. Rev. Aquacult. 10, 334–350.

Dawood, M., et al., 2020. The potential role of nano-selenium and vitamin C on the performances of Nile tilapia (*Oreochromis niloticus*). Environ. Sci. Pollut. Res. 27, 9843–9852.

de Andrade, J., et al., 2007. Influence of diets supplemented with vitamins C and E on pirarucu (*Arapaima gigas*) blood parameters. Comp. Biochem. Physiol. A 146, 576–580.

Deng, D., Wilson, R., 2003. Dietary riboflavin requirement of juvenile sunshine bass *(Morone chrysops♀×Morone saxatilis♂)*. Aquaculture 218, 695–701.

Deshimaru, O., Kuroki, K., 1979. Requirement of prawn for dietary thiamine, pyridoxine, and choline chloride. Bull. Jpn. Soc. Sci. Fish. 45, 363–367.

Dethlefsen, M., et al., 2016. Effect of storage on oxidative quality and stability of extruded astaxanthin-coated fish feed pellets. Anim. Feed Sci. Technol. 221, 157–166.

Du, S., et al., 2006. Effect of different levels of dietary α-tocopherol on ovarian maturation and reproductive performance of broodstock *Litopenaeus vannamei* (Boone). J. Shellfish Res. 25, 589–593.

Duan, Y., et al., 2012. Dietary choline requirement in slight methionine-deficient diet for juvenile gibel carp (*Carassius auratus gibelio*). Aquacult. Nutr. 18, 620–627.

Ducker, G., Rabinowitz, J., 2017. One-carbon metabolism in health and disease. Cell Metabol. 25, 27–42.

Duncan, P., et al., 1993. Dietary folate requirement determined for channel catfish, *Ictalurus punctatus*. J. Nutr. 123, 1888–1897.

Dupree, H., 1966. Vitamins Essential for Growth of Channel Catfish, vol. 7. US Department of the Interior, Fish and Wildlife Service, Bureau of Sport Fisheries and Wildlife.

EFSA, 2006. Opinion of the panel on additives and products or substances used in animal feed (FEEDAP) on the safety of use of colouring agents in animal nutrition PART II. Capsanthin, Citranaxanthin, and Cryptoxanthin. EFSA J. 4, 386.

EFSA, 2017. Safety of vitamin D_3 addition to feedingstuffs for fish. EFSA J. 15, e4713.

Eide-Graff, I., Stefansson, S., Aksnes, L., Lie, Ø., 2004. Plasma levels of vitamin D_3 metabolites during parr-smolt transformation of Atlantic salmon *Salmo salar* L. Aquaculture 240, 617–622.

El Basuini, M., et al., 2017. Effects of dietary copper nanoparticles and vitamin C supplementations on growth performance, immune response and stress resistance of red sea bream, *Pagrus major*. Aquacult. Nutr. 23, 1329–1340.

Espe, M., et al., 2020. The level of 1C diets fed prior to cell isolation affects lipid metabolism in primary liver cells isolated from Atlantic salmon *(Salmo salar)*. Aquacult. Nutr. https://doi.org/10.1111/anu.13058.

Evans, L., 2009. The rise and demise of the commercial shrimp farm, amatikulu prawns (pty), Ltd., 1989 to 2004 (Kwazulu natal, South Africa). In: Aquaculture Compendium. CABI, Wallingford, United Kingdom.

Eya, J., 1996. 'Broken-skull disease'in african catfish *Clarias gariepinus* is related to a dietary deficiency of ascorbic acid. J. World Aquacult. Soc. 27, 493–498.

Falah, F.J., Islami, H.R., Mehrgan, M.S., 2020. Dietary folic acid improved growth performance, immuno-physiological response and antioxidant status of fingerling Siberian sturgeon, *Acipenser baerii* (Brandt 1896). Aquacult. Rep. 17, 100391.

Falahatkar, B., et al., 2006. Effects of dietary vitamin C supplementation on performance, tissue chemical composition and alkaline phosphatase activity in great sturgeon (*Huso huso*). J. Appl. Ichthyol. 22, 283–286.

Fazio, F., et al., 2020. Effect of dietary vitamin C supplementation on the blood parameters of striped bass morone saxatilis (Walbaum, 1752). Turk. J. Fish. Aquat. Sci. 20, 491–497.

Feng, L., et al., 2010. Effects of dietary pyridoxine on disease resistance, immune responses and intestinal microflora in juvenile Jian carp (*Cyprinus carpio* var. Jian). Aquacult. Nutr. 16, 254–261.

Feng, L., et al., 2011. Effect of dietary thiamin supplement on immune responses and intestinal microflora in juvenile Jian carp (*Cyprinus carpio* var. Jian). Aquacult. Nutr. 17, 557–569.

Fernández, I., et al., 2008. Larval performance and skeletal deformities in farmed gilthead sea bream (*Sparus aurata*) fed with graded levels of vitamin A enriched rotifers (*Brachionus plicatilis*). Aquaculture 283, 102–115.

Fernández, I., et al., 2009. Effect of dietary vitamin A on Senegalese sole (*Solea senegalensis*) skeletogenesis and larval quality. Aquaculture 295, 250–265.

Fernández, I., et al., 2014. Warfarin, a potential pollutant in aquatic environment acting through Pxr signaling pathway and g-glutamyl carboxylation of vitamin K-dependent proteins. Environ. Pollut. 194, 86–95.

Fernández, I., et al., 2015. Zebrafish vitamin K epoxide reductases: expression in vivo, along extracellular matrix mineralization and under phylloquinone and warfarin in vitro exposure. Fish Physiol. Biochem. 41, 745–759.

Fernández, I., et al., 2019. Circulating small non-coding RNAs provide new insights into vitamin K nutrition and reproductive physiology in teleost fish. BBA. Gen. Subjects 1863, 39–51.

Finke, M., 2013. Complete nutrient content of four species of feeder insects. Zoo Biol. 32, 27–36.

Firouz, A., et al., 2013. The effects of folic acid treatment on biometric and blood parameters of fingerling rainbow trout fishes (*Oncorhynchus mykiss*). J. Aquacult. Res. Dev. 4, 1000175.

Fontagné, S., et al., 2006. Effects of dietary oxidized lipid and vitamin A on the early development and antioxidant status of Siberian sturgeon (*Acipenser baeri*) larvae. Aquaculture 257, 400–411.

Fontagné-Dicharry, S., et al., 2010. Effects of dietary vitamin A on broodstock performance, egg quality, early growth and retinoid nuclear receptor expression in rainbow trout (*Oncorhynchus mykiss*). Aquaculture 303, 40–49.

Fujiwara, M., Itokawa, Y., Kimura, M., 1976. Experimental Studies on the Relationships between Thiamine and Divalent Cations, Calcium and Magnesium. Thiamine. John Wiley and Sons, NY, pp. 63–80.

Furuita, H., et al., 2001. Effects of high dose of vitamin A on reproduction and egg quality of Japanese flounder. *Paralichthys olivaceus*. Fisheries Sci. 67, 606–613.

Furuita, H., et al., 2003. Supplemental effect of vitamin A in diet on the reproductive performance and egg quality of the Japanese flounder *Paralichthys olivaceus* (T & S). Aquacult. Res. 34, 461–468.

Gabaudan, J., Hardy, R., 2000. Vitamin sources for fish feeds. In: Stickney, R. (Ed.), Encyclopedia of Aquaculture. John Wiley and Sons, Inc., New York, NY, pp. 961–965.

Gao, J., et al., 2013. Effect of dietary oxidized fish oil and vitamin C supplementation on growth performance and reduction of oxidative stress in red sea bream *Pagrus major*. Aquacult. Nutr. 19, 35–44.

Gao, J., et al., 2014. Interactive effects of vitamin C and E supplementation on growth performance, fatty acid composition and reduction of oxidative stress in juvenile Japanese flounder *Paralichthys olivaceus* fed dietary oxidized fish oil. Aquaculture 422–423, 84–90.

Gardell, A., et al., 2013. Tilapia (*Oreochromis mossambicus*) brain cells respond to hyperosmotic challenge by inducing *myo*-inositol biosynthesis. J. Exp. Biol. 216, 4615–4625.

Gatlin III, D., Poe, W., Wilson, R., 1986. Effects of singular and combined dietary deficiencies of selenium and vitamin E on fingerling channel catfish (*Ictalurus punctatus*). J. Nutr. 116, 1061–1067.

Gong, H., Lawrence, A., Jiang, D., Gatlin, D., 2003. Effect of dietary phospholipids on the choline requirement of *Litopenaeus vannamei* juveniles. J. World Aquacult. Soc. 34, 289–299.

Gong, W., et al., 2014. Dietary *myo*-inositol requirement for juvenile gibel carp (*Carassius auratus gibelio*). Aquacult. Nutr. 20, 514–519.

Gouillou-Coustans, M., Kaushik, S., 2000. Ascorbic acid requirement in freshwater and marine fish: is there a difference. In: Dabrowski, K. (Ed.), Ascorbic Acid in Aquatic Organisms: Status and Perspectives. CRC press, pp. 49–68.

Gouillou-Coustans, M., Bergot, P., Kaushik, S., 1998. Dietary ascorbic acid needs of common carp (*Cyprinus carpio*) larvae. Aquaculture 161, 453–461.

Gouveia, L., Empis, J., 2003. Relative stabilities of microalgal carotenoids in microalgal extracts, biomass and fish feed: effect of storage conditions. Innovat. Food Sci. Emerg. Technol. 4, 227–233.

Grahl-Madsen, E., Lie, Ø., 1997. Effects of different levels of vitamin K in diets for cod (*Gadus morhua*). Aquaculture 151, 269–274.

Griffin, M., Wilson, K., White, M.R., Brown, P.B., 1994. Dietary choline requirement of juvenile hybrid striped bass. J. Nutr. 124, 1685–1689.

Grisdale-Helland, B., Helland, S., Åsgård, T., 1991. Problems associated with the present use of menadione sodium bisulfite and vitamin A in diets for Atlantic salmon. Aquaculture 92, 351–358.

Gropper, S., Smith, J., Carr, T., 2018. Chapt. 9 water-soluble vitamins. In: Advanced Nutrition and Human Metabolism, seventh ed. Cengage Learning, Boston, MA, US, pp. 299–368.

Guimarães, I., et al., 2014. Effects of dietary levels of vitamin A on growth, hematology, immune response and resistance of Nile tilapia (*Oreochromis niloticus*) to Streptococcus iniae. Anim. Feed Sci. Tech. 188, 126–136.

Haga, Y., et al., 2004. Vitamin D_3 compounds induce hypermelanosis on the blind side and vertebral deformity in juvenile Japanese flounder *Paralichthys olivaceus*. Fish. Sci. 70, 59–67.

Halver, J., 1957. Nutrition of salmonid fishes. III. Water-soluble vitamin requirements of Chinook salmon. J. Nutr. 62, 225–243.

Halver, J., 1972. The vitamins. In: Halver, J. (Ed.), Fish Nutrition. Academic Press, New York, pp. 29–103.

Halver, J., Hardy, R., 2002. Fish Nutrition. Academic Press, San Diego, USA.

Hamre, K., et al., 1994. Decreased concentration of hemoglobin, accumulation of lipid oxidation products and unchanged skeletal muscle in Atlantic salmon (*Salmo salar*) fed low dietary vitamin E. Fish Physiol. Biochem. 12, 421–429.

Hamre, K., et al., 2008. Several micronutrients in the rotiferBrachionus sp. may not fulfil the nutritional requirements of marine fish larvae. Aquacult. Nutr. 14, 51–60.

Hamre, K., et al., 2013. Fish larval nutrition and feed formulation: knowledge gaps and bottlenecks for advances in larval rearing. Rev. Aquacult. 5, 26–58.

Hamre, K., Lie, Ø., 1995. Minimum requirement of vitamin E for Atlantic salmon, *Salmo salar*, at startfeeding. Aquacult. Res. 26, 175–184.

Han, D., Li, Y., Hu, Q., 2013. Astaxanthin in microalgae: pathways, functions and biotechnological implications. Algae 28, 131–147.

Hansen, A., Olsvik, P., Hemre, G., 2013. Effect of different dietary B12 levels on their retention in the body of zebrafish *Danio rerio* and on the gene expression of vitamin B12 binding proteins. Aquacult. Nutr. 19, 413–420.

Hansen, A., Waagbø, R., Hemre, G., 2015. New B vitamin recommendations in fish when fed plant-based diets. Aquacult. Nutr. 21, 507–527.

Hansen, Y., Puvanendran, V., Bangera, R., 2016. Broodstock diet with water and astaxanthin improve condition and egg output of brood fish and larval survival in Atlantic cod, *Gadus morhua* L. Aquacult. Res. 47, 819–829.

Hansen, A., et al., 2020. Choline supplementation prevents diet induced gut mucosa lipid accumulation in post-smolt Atlantic salmon (*Salmo salar* L.). BMC Vet. Res. 16, 32.

Harabawy, A., Mosleh, Y., 2014. The role of vitamins A, C, E and selenium as antioxidants against genotoxicity and cytotoxicity of cadmium, copper, lead and zinc on erythrocytes of Nile tilapia, *Oreochromis niloticus*. Ecotoxicol. Environ. Saf. 104, 28–35.

Harder, A., et al., 2018. Thiamine deficiency in fishes: causes, consequences, and potential solutions. Rev. Fish Biol. Fish. 28, 865–886.

Hardy, R., Halver, J., Brannon, E., 1979. Effect of dietary protein level on the pyridoxine requirement and disease resistance of chinook salmon. In: Halver, J., Tiews, K. (Eds.), Proceedings of the World Symposium on Finfish Nutrition and Fishfeed Technology. Heenemann Verlagsgesellschaft, Berlin, pp. 253–260.

Harrison, E., 2005. Mechanisms of digestion and absorption of dietary vitamin A. Annu. Rev. Nutr. 25, 87–103.

He, H., Lawrence, A., Liu, R., 1992. Evaluation of dietary essentiality of fat-soluble vitamins, A, D, E and K for penaeid shrimp (*Penaeus vannamei*). Aquaculture 103, 177–185.

He, W., et al., 2009. Dietary pyridoxine requirement of juvenile Jian carp (*Cyprinus carpio* var. Jian). Aquacult. Nutr. 15, 402–408.

Hemre, G., Deng, D., Wilson, R., Berntssen, M., 2004. Vitamin A metabolism and early biological responses in juvenile sunshine bass (*Morone chrysops* × *M. saxatilis*) fed graded levels of vitamin A. Aquaculture 235, 645–658.

Hemre, G., et al., 2016. Atlantic salmon *(Salmo salar)* require increased dietary levels of B-vitamins when fed diets with high inclusion of plant based ingredients. PeerJ 4, e2493.

Hernandez, L., Hardy, R., 2020. Vitamin A functions and requirements in fish. Aquacult. Res. 51, 3061–3071.

Hernandez, L., et al., 2005. Dietary vitamin A requirements of juvenile Japanese flounder *Paralichthys olivaceus*. Aquacult. Nutr. 11, 3–9.

Hien, D., Doolgindachbaporn, S., 2011. Effect of niacin and folic acid in feed rations on growth and live weights of green catfish (*Mystus nemurus* Valenciennes 1840). Pakistan J. Biol. Sci. 14, 64–68.

Higuera-Ciapara, I., Felix-Valenzuela, L., Goycoolea, F., 2006. Astaxanthin: a review of its chemistry and applications. Crit. Rev. Food Sci. Nutr. 46, 185–196.

Hilton, J., 1983. Hypervitaminosis A in rainbow trout (*Salmo gairdneri*): toxicity signs and maximum tolerable level1. J. Nutr. 113, 1737–1745.

Hilton, J., Hodson, P., Slinger, S., 1980. The requirement and toxicity of selenium in rainbow trout (*Salmo gairdneri*). J. Nutr. 110, 2527–2535.

Holub, B., Nilsson, K., Piekarski, J., Slinger, S., 1975. Biosynthesis of lecithin by the CDP-choline pathway in liver microsomes of rainbow trout, *Salmo gairdneri*. J. Fish. Res. Board Can. 32, 1633–1637.

Hosokawa, H., 2000. Studies on the vitamin nutrition in yellowtail (*Seriola quinqueradiata*). Mem. Facul. Agric. Kochi Univers. 71, 1–156.

Hsu, T., Shiau, S., 1998. Comparison of vitamin C requirement for maximum growth of grass shrimp, *Penaeus monodon*, with L-ascorbyl-2-monophosphate-Na and L-ascorbyl-2-monophosphate-Mg. Aquaculture 163, 203–213.

Hsu, T., Shiau, S., 1999. Influence of dietary ascorbate derivatives on tissue copper, iron and zinc concentrations in grass shrimp, *Penaeus monodon*. Aquaculture 179, 457–464.

Hu, C., Chen, S., Pan, C., Huang, C., 2006. Effects of dietary vitamin A or β-carotene concentrations on growth of juvenile hybrid tilapia, *Oreochromis niloticus* × *O. aureus*. Aquaculture 253, 602–607.

Huang, J., et al., 2005. Pyridoxine deficiency of grouper, *Epinephelus coioides*: physiological and biochemical alteration. Fish Physiol. Biochem. 31, 331–337.

Huang, J., et al., 2007. Effects of dietary thiamine on the physiological status of the grouper *Epinephelus coioides*. Fish Physiol. Biochem. 33, 167–172.

Huang, H., et al., 2011. Effects of dietary thiamin supplement on growth, body composition and intestinal enzyme activities of juvenile Jian carp (*Cyprinus carpio* var. Jian). Aquacult. Nutr. 17, e233–e240.

Huang, X., et al., 2015. Dietary thiamin could improve growth performance, feed utilization and non-specific immune response for juvenile Pacific white shrimp (*Litopenaeus vannamei*). Aquacult. Nutr. 21, 364–372.

Huang, D., et al., 2019a. Protective effects of dietary α-lipoic acid on abalone *Haliotis discus hannai* against the oxidative damage under waterborne cadmium stress. Aquacult. Nutr. 25, 263–270.

Huang, C., et al., 2019b. Influence of dietary alpha-lipoic acid and lipid level on the growth performance, food intake and gene expression of peripheral appetite regulating factors in juvenile grass carp (*Ctenopharyngodon idellus*). Aquaculture 505, 412–422.

Huang, Q., et al., 2019c. Effect of dietary vitamin B6 supplementation on growth and intestinal microflora of juvenile golden pompano (*Trachinotus ovatus*). Aquacult. Res. 50, 2359–2370.

Hughes, S., Rumsey, G., Nichum, J., 1981. Riboflavin requirement of fingerling rainbow trout. Prog. Fish-Cult. 43, 167.

Hung, S., 1989. Choline requirement of hatchery-produced juvenile white sturgeon (*Acipenser transmontunus*). Aquaculture 78, 183–194.

Hung, S., Cho, C., Slinger, S., 1981. Effect of oxidized fish oil, DL-alpha-tocopheryl acetate and ethoxyquin supplementation on the vitamin E nutrition of rainbow trout (*Salmo gairdneri*) fed practical diets. J. Nutr. 111, 648–657.

Ibrahem, M., Fathi, M., Mesalhy, S., Abd El-Aty, A., 2010. Effect of dietary supplementation of inulin and vitamin C on the growth, hematology, innate immunity, and resistance of Nile tilapia (*Oreochromis niloticus*). Fish Shellfish Immunol. 29, 241–246.

Ikeda, S., et al., 1988. Qualitative requirements of the Japanese parrot fish for water-soluble vitamins. Nippon Suisan Gakk 54, 2029–2035.

Ishihara, T., Hara, K., Yagi, M., Yasuda, M., 1978. Studies on thiaminase-1 in marine fish. 8. Thiamine requirement of Yellowtail fed anchovy. Nippon Suisan Gakk 44, 659–664.

Izquierdo, M., Fernandez-Palacios, H., Tacon, A., 2001. Effect of broodstock nutrition on reproductive performance of fish. Aquaculture 197, 25–42.

Jiang, W., et al., 2009. Growth, digestive capacity and intestinal microflora of juvenile Jian carp (*Cyprinus carpio* var. Jian) fed graded levels of dietary inositol. Aquacult. Res. 40, 955–962.

Jiang, W., et al., 2010. Effects of graded levels of dietary *myo*-inositol on non-specific immune and specific immune parameters in juvenile Jian carp (*Cyprinus carpio* var. Jian). Aquacult. Res. 41, 1413–1420.

Jiang, W., et al., 2013a. Effects of *myo*-inositol on proliferation, differentiation, oxidative status and antioxidant capacity of carp enterocytes in primary culture. Aquacult. Nutr. 19, 45–53.

Jiang, G., et al., 2013b. Dietary choline requirement for juvenile blunt snout bream, *Megalobrama amblycephala*. Aquacult. Nutr. 19, 499–505.

Jiang, M., et al., 2014. Dietary thiamin requirement of juvenile grass carp, *Ctenopharyngodon idella*. J. World Aquacult. Soc. 45, 461–468.

Jiang, W., et al., 2015a. Copper exposure induces toxicity to the antioxidant system via the destruction of Nrf2/ARE signaling and caspase-3-regulated DNA damage in fish muscle: amelioration by myo-inositol. Aquat. Toxicol. 159, 245–255.

Jiang, J., et al., 2015b. Vitamin D inhibits lipopolysaccharide-induced inflammatory response potentially through the Toll-like receptor 4 signalling pathway in the intestine and enterocytes of juvenile Jian carp (*Cyprinus carpio* var. Jian). Br. J. Nutr. 114, 1560–1568.

Jiang, W., et al., 2019. Vitamin A deficiency impairs intestinal physical barrier function of fish. Fish Shellfish Immunol. 87, 546–558.

Jiménez-Fernández, E., et al., 2012. Molecular characterization and transcriptional regulation of the sodium-dependent vitamin C transporter genes (slc23a1 and slc23a2) in a teleost fish, the Senegalese sole (*Solea senegalensis*). Comp. Biochem. Physiol. B 161, 208–218.

Jiménez-Fernández, E., et al., 2018. Assessing the role of vitamin C and iron in early larvae stages of *Solea senegalensis* fed enriched artemia. Aquaculture 488, 145–154.

John, M., Mahajan, C., 1979. The physiological response of fishes to a deficiency of cyanocobalamin and folic acid. J. Fish. Biol. 14, 127–133.

Jones, W., et al., 2002. Uptake, recycling, and antioxidant actions of α-lipoic acid in endothelial cells. Free Radic. Biol. Med. 33, 83–93.

Kalinowski, C., et al., 2005. Effect of different carotenoid sources and their dietary levels on red porgy (*Pagrus pagrus*) growth and skin colour. Aquaculture 244 (1–4), 223–231.

Kalujnaia, S., et al., 2013. Seawater acclimation and inositol monophosphatase isoform expression in the European eel (*Anguilla anguilla*) and Nile tilapia (*Orechromis niloticus*). Am. J. Physiol.-Reg. I. 305, R369–R384.

Kalujnaia, S., Hazon, N., Cramb, G., 2016. *myo*-Inositol phosphate synthase expression in the European eel (*Anguilla anguilla*) and Nile tilapia (*Oreochromis niloticus*): effect of seawater acclimation. Am. J. Physiol.-Reg. I. 311, R287–R298.

Kanazawa, A., Teshima, S., Tanaka, N., 1976. Nutritional requirements of prawn V: requirements for choline and inositol. Mem. Fac. Fish. Kagoshima Univ. 25, 47–51.

Karges, R., Woodward, B., 1984. Development of lamellar epithelial hyperplasia in gills of pantothenic acid-deficient rainbow trout, *Salom gairdneri* R. J. Fish. Biol. 25, 57–62.

Kasper, C., White, M., Brown, P., 2000. Choline is required by tilapia when methionine is not in excess. J. Nutr. 130, 238–242.

Kayden, H., Traber, M., 1993. Absorption, lipoprotein transport, and regulation of plasma concentrations of vitamin E in humans. J. Lipid Res. 34, 343–358.

Kelestemur, G., Özdemir, Y., 2013. Effects of dietary vitamin A and E on growth performance and antioxidant status in blood of juvenile rainbow trout (*Oncorhynchus mykiss*, W. 1792) exposed to flow rate stress. J. Anim. Plant Sci. 23, 821–827.

Khan, Y., Khan, M., 2019. Dietary biotin requirement of fingerling *Catla catla* (Hamilton), based on growth, feed conversion efficiency, and liver biotin concentration. J. World Aquacult. Soc. 50, 674–683.

Khan, Y., Khan, M., 2020a. Dietary folic acid requirement of fingerling Catla, Catla catla (Hamilton). Aquacult. Nutr. 26, 1035–1045.

Khan, Y., Khan, M., 2020b. Effects of dietary cyanocobalamin on growth performance, non-specific immune response, antioxidant capacity, haematological parameters, body composition and liver cyanocobalamin concentration of fingerling major carp, *Catla catla* (Hamilton). Aquacult. Nutr. https://doi.org/10.1111/anu.13210.

Khan, K., et al., 2017. Synergistic effects of dietary nano selenium and vitamin C on growth, feeding, and physiological parameters of mahseer fish (*Tor putitora*). Aquacult. Rep. 5, 70–75.

Khosravi, S., et al., 2015. Dietary myo-inositol requirement of parrot fish, *Oplegnathus fasciatus*. Aquaculture 436, 1–7.

Kirkland, J., Meyer-Ficca, M., 2018. Niacin. Adv. Food Nutr. Res. 83, 83–149.

Kissil, G., Cowey, C., Adron, J., Richards, R., 1981. Pyridoxine requrements of gilthead bream, *Spartus aurata*. Aquaculture 23, 243–255.

Kitamura, S., Suwa, T., Ohara, S., Nakagawa, K., 1967a. Studies on vitamin requirements of rainbow trout. 2. The deficiency symptoms of fourteen kinds of vitamin. Bull. Jpn. Soc. Sci. Fish. 33, 1120–1125.

Kitamura, S., Suwa, T., Ohara, S., Nakagawa, K., 1967b. Studies on vitamin requirements of rainbow trout - III. Requirement for vitamin A and deficiency symptoms. Bull. Jpn. Soc. Sci. Fish. 33, 1126–1131.

Kocabas, A., Gatlin, D., 1999. Dietary vitamin E requirement of hybrid striped bass (*Morone chrysops femaleM. saxatilis male*). Aquacult. Nutr. 5, 3–7.

Koch, J., et al., 2018. Vitamin A requirements of dourado (*Salminus brasiliensis*): growth performance and immunological parameters. Aquaculture 491, 86–93.

Kop, A., 2019. The effects of dietary vitamin C levels on tissue ascorbic acid concentration and growth in European sea bass (*Dicentrarchus labrax*, L.1758). Indian J. Anim. Res. 53, 1635–1638.

Koppe, W., 1993. Auswirkungen einer unterschiedlichen Biotinversorgung auf Wachstum, Biotinstatus sowie Parameter der Immun-und stressrespons bei Regenbogenforellen (*Oncorhynchus mykiss*). Institut für Tierphysiologie und Tierernärung, Georg-August-Universität zu Göttingen, Göttingen, Germany.

Kouakou, N., Georges, C., 2006. Effect of varying the concentration of dietary astaxanthin on rainbow trout, *Oncorhynchus mykiss* muscle pigmentation. J. Appl. Aquacult. 18, 61–73.

Kozat, S., Okman, E., 2017. Homocystein: a new biochemical marker in livestock sector. J. Adv. Vet. Anim. Res. 4 (4), 319–332.

Krossøy, C., Ørnsrud, R., Wargelius, A., 2009a. Differential gene expression of bgp and mgp in trabecular and compact bone of Atlantic salmon (*Salmo salar* L.) vertebrae. J. Anat. 215, 663–672.

Krossøy, C., et al., 2009b. Dietary menadione nicotinamide bisulphite (vitamin K_3) does not affect growth or bone health in first-feeding fry of Atlantic salmon (*Salmo salar* L.). Aquacult. Nutr. 15, 638–649.

Krossøy, C., Waagbø, R., Ørnsrud, R., 2011. Vitamin K in fish nutrition. Aquacult. Nutr. 17, 585–594.

Kuhn, R., Gyorgy, P., Wagstaff, L., 1933. Über Ovoflavin, den Farbstoff des Eiklars. Ber. Deutsch. Chem. Ges. 66, 576–580.

Kurnia, A., Satoh, S., Kuramoto, D., Hanzawa, S., 2007. Effect of different astaxanthin sources on skin pigmentation of red sea bream (*Pagrus major*). Aquac. Sci. 55, 441–447.

Kütter, M., et al., 2012. Effects of dietary α-lipoic acid on growth, body composition and antioxidant status in the Plata pompano *Trachinotus marginatus* (Pisces, Carangidae). Aquaculture 368–369, 29–35.

Lakeh, A., et al., 2010. Growth performance, mortality and carotenoid pigmentation of fry offspring as affected by dietary supplementation of astaxanthin to female rainbow trout (*Oncorhynchus mykiss*) broodstock. J. Appl. Ichthyol. 26, 35–39.

Lall, S., Olivier, G., Hines, J., Ferguson, H., 1988. The role of vitamin E in nutrition and immune response of Atlantic salmon (*Salmo salar*). Bull. Aquacult. Ass.Can. 88, 76–78.

Le, K., Fotedar, R., Partridge, G., 2014. Selenium and vitamin E interaction in the nutrition of yellowtail kingfish (*Seriola lalandi*): physiological and immune responses. Aquacult. Nutr. 20, 303–313.

Lee, K., Dabrowski, K., 2004. Long-term effects and interactions of dietary vitamins C and E on growth and reproduction of yellow perch, *Perca flavescens*. Aquaculture 230, 377–389.

Lee, M., Shiau, S., 2003. Increase of dietary vitamin C improves haemocyte respiratory burst response and growth of juvenile grass shrimp, *Penaeus monodon*, fed with high dietary copper. Fish Shellfish Immunol. 14, 305–315.

Lee, M., Shiau, S., 2004. Vitamin E requirements of juvenile grass shrimp, *Penaeus monodon*, and effects on non-specific immune responses. Fish Shellfish Immunol. 16, 475–485.

Lee, B., Lee, K., Lim, S., Lee, S., 2009. Dietary *myo*-inositol requirement for olive flounder, *Paralichthys olivaceus* (Temminch et Schlegel). Aquac. Res. 40, 83–90.

Lee, B., et al., 2012. Effects of dietary vitamin B1 (thiamine) and magnesium on the survival, growth and histological indicators in lake trout (*Salvelinus namaycush*) juveniles. Comp. Biochem. Physiol. A 162, 219–226.

Lewis-McCrea, L., Lall, S., 2007. Effects of moderately oxidized dietary lipid and the role of vitamin E on the development of skeletal abnormalities in juvenile Atlantic halibut (*Hippoglossus hippoglossus*). Aquaculture 262, 142–155.

Li, P., Gatlin, D., 2009. Dietary vitamin E requirement of the red drum *Sciaenops ocellatus*. Aquacult. Nutr. 15, 313–319.

Li, W., et al., 2010a. Effect of dietary riboflavin on growth, feed utilization, body composition and intestinal enzyme activities of juvenile Jian carp (*Cyprinus carpio* var. Jian). Aquacult. Nutr. 16, 137–143.

Li, E., et al., 2010b. Dietary vitamin B6 requirement of the Pacific white shrimp, *Litopenaeus vannamei*, at low salinity. J. World Aquacult. Soc. 41, 756–763.

Li, J., et al., 2010c. Estimation of dietary biotin requirement of Japanese seabass, *Lateolabrax japonicus* C. Aquacult. Nutr. 16, 231–236.

Li, J., et al., 2014a. Effects of dietary choline supplementation on growth performance and hepatic lipid transport in blunt snout bream (*Megalobrama amblycephala*) fed high-fat diets. Aquaculture 434, 340–347.

Li, J., et al., 2014b. Effects of vitamin E on growth performance and antioxidant status in juvenile grass carp *Ctenopharyngodon idellus*. Aquaculture 430, 21–27.

Li, M., et al., 2014c. Comparison effect of dietary astaxanthin and *Haematococcus pluvialis* on growth performance, antioxidant status and immune response of large yellow croaker Pseudosciaena crocea. Aquaculture 434, 227–232.

Li, S., et al., 2016a. Deficiency of dietary niacin decreases digestion and absorption capacities via declining the digestive and brush border enzyme activities and downregulating those enzyme gene transcription related to TOR pathway of the hepatopancreas and intestine in young grass carp (*Ctenopharyngodon idellus*). Aquacult. Nutr. 22, 1267–1282.

Li, F., et al., 2016b. Dietary cyanocobalamin requirement of fingerling blunt snout bream *Megalobrama amblycephala* determined by growth performance, digestive and absorptive capability and status of the GH-IGF-I axis. Aquaculture 464, 647–653.

Li, S., et al., 2017. Dietary *myo*-inositol deficiency decreased the growth performances and impaired intestinal physical barrier function partly relating to nrf2, jnk, e2f4 and mlck signaling in young grass carp (*Ctenopharyngodon idella*). Fish Shellfish Immunol. 67, 475–492.

Li, S., et al., 2018. Dietary myo-inositol deficiency decreased intestinal immune function related to NF-κB and TOR signaling in the intestine of young grass carp (*Ctenopharyngodon idella*). Fish Shellfish Immunol. 76, 333–346.

Li, Y., et al., 2019. Dietary *myo*-inositol requirements of juvenile hybrid sturgeon (*Acipenser baerii* ♀ × *A. schrenckii* ♂). Aquacult. Res. 51, 1143–1151.

Li, B., et al., 2020. Requirement of vitamin E of growing sea cucumber Apostichopus japonicus Selenka. Aquacult. Res. 51, 1284–1292.

Liang, X., et al., 2017. Effect of dietary vitamin C on the growth performance, antioxidant ability and innate immunity of juvenile yellow catfish (*Pelteobagrus fulvidraco* Richardson). Aquacult. Res. 48, 149–160.

Lim, C., Lovell, R., 1978. Pathology of the vitamin C deficiency syndrome in channel catfish (*Ictalurus punctatus*). J. Nutr. 108, 1137–1146.

Lim, C., LeaMaster, B., Brock, J., 1993. Riboflavin requirement of fingerling red hybrid tilapia grown in Seawater. J. World Aquacult. Soc. 24, 451–458.

Lim, C., LeaMaster, B., Brock, J., 1995. Pyridoxine requirement of fingerling red hybrid tilapia grown in seawater. J. Appl. Aquacult. 5, 49–60.

Lim, C., Klesius, P., Li, M., Robinson, E., 2000. Interaction between dietary levels of iron and vitamin C on growth, hematology, immune response and resistance of channel catfish (*Ictalurus punctatus*) to *Edwardsiella ictaluri* challenge. Aquaculture 185, 313–327.

Lim, C., Yildirim-Aksoy, M., Barros, M., Klesius, P., 2011. Thiamin requirement of nile tilapia, *Oreochromis niloticus*. J. World Aquac. Soc. 42, 824–833.

Lim, K., Fatimah, M., Mohamed, S., Mohd-Salleh, K., 2019a. Astaxanthin as feed supplement in aquatic animals. Rev. Aquc. 10, 738–773.

Lim, K., Fatimah, M., Mohamed, S., Mohd-Salleh, K., 2019b. Dietary administration of astaxanthin improves feed utilization, growth performance and survival of Asian seabass, *Lates calcarifer* (Bloch, 1790). Aquacult. Nutr. 25, 1410–1421.

Limsuwan, T., Lovell, R., 1981. Intestinal synthesis and absorption of vitamin B12 in Channel catfish. J. Nutr. 111, 133–140.

Lin, M., Shiau, S., 2004. Requirements of vitamin C (L-ascorbyl-2-monophosphate-Mg and L-ascorbyl-2-monophosphate-Na) and its effects on immune responses of grouper, Epinephelus malabaricus. Aquacult. Nutr. 10, 327–333.

Lin, M., Shiau, S., 2005a. Dietary L-ascorbic acid affects growth, nonspecific immune responses and disease resistance in juvenile grouper, *Epinephelus malabaricus*. Aquaculture 244, 215–221.

Lin, M., Shiau, S., 2005b. Requirements of vitamin C (l-ascorbyl-2-sulphate andl-ascorbyl-2-polyphosphate) and its effects on non-specific immune responses of grouper, *Epinephelus malabaricus*. Aquacult. Nutr. 11, 183–189.

Lin, Y., Shiau, S., 2005c. Dietary vitamin E requirement of grouper, *Epinephelus malabaricus*, at two lipid levels, and their effects on immune responses. Aquaculture 248, 235–244.

Lin, Y., Lin, H., Shiau, S., 2011. Dietary folic acid requirement of grouper, *Epinephelus malabaricus*, and its effects on non-specific immune responses. Aquaculture 317, 133–137.

Lin, Y., Lin, H., Shiau, S., 2012. Estimation of dietary pantothenic acid requirement of grouper, *Epinephelus malabaricus* according to physiological and biochemical parameters. Aquaculture 324, 92–96.

Liu, T., Li, A.J., 1995. Studies on the optimal requirements of pantothenic acid, biotin, folic acid and vitamin B12 in the shrimp *Penacus chinensis*. J. Fish. Sci. China 2, 48–55.

Liu, T., Jiameng, Z., Li, A., 1995. Studies on the optimal requirements of pantothenic acid, biotin, folic acid and vitamin B_{12} in the shrimp *Penaeus chinensis*. J. Fish. Sci. China 2, 48–55.

Liu, Y., et al., 2011. Effects of graded levels of dietary vitamin C on the growth, digestive capacity and intestinal microflora of juvenile Jian carp (*Cyprinus carpio* var. Jian). Aquacult. Res. 42, 534–548.

Liu, F., et al., 2016. Effects of astaxanthin and emodin on the growth, stress resistance and disease resistance of yellow catfish (*Pelteobagrus fulvidraco*). Fish Shellfish Immunol. 51, 125–135.

Liu, H., et al., 2018. Optimal α-lipoic acid strengthen immunity of young grass carp (*Ctenopharyngodon idella*) by enhancing immune function of head kidney, spleen and skin. Fish Shellfish Immunol. 80, 600–617.

Liu, H., et al., 2019. Effects of dietary vitamin C and vitamin E on the growth, antioxidant defence and digestive enzyme activities of juvenile discus fish (*Symphysodon haraldi*). Aquacult. Nutr. 25, 176–183.

Lobato, R., et al., 2018. The effect of diet enriched with lipoic acid in the accumulation and metabolization of metals in different organs of Litopenaeus vannamei. Aquacult. Res. 49, 3702–3710.

Lock, E., et al., 2007. The vitamin D receptor and its ligand 1 ,25-dihydroxyvitamin D_3 in Atlantic salmon (*Salmo salar*). J. Endocrinol. 193, 459–471.

Lock, E., Waagbo, R., Wendelaar Bonga, S., Flik, G., 2010. The significance of vitamin D for fish: a review. Aquacult. Nutr. 16, 100–116.

Lovell, R., Buston, J., 1984. Biotin supplementation of practical diets for channel catfish. J. Nutr. 114, 1092–1096.

Lovell, R., Li, Y., 1978. Essentiality of vitamin D in diets of channel catfish (*Ictalurus punctatus*). Trans. Am. Fish. Soc. 107, 809–811.

Lovell, R., Miyazaki, T., Rabegnator, S., 1984. Requirement for α-tocopherol by channel catfish fed diets low in polyunsaturated triglycerides. J. Nutr. 114, 894–901.

Lu, Y., et al., 2016. Effects of dietary vitamin E on the growth performance, antioxidant status and innate immune response in juvenile yellow catfish (*Pelteobagrus fulvidraco*). Aquaculture 464, 609–617.

MacDonald, L., Thumser, A., Sharp, P., 2002. Decreased expression of the vitamin C transporter SVCT1 by ascorbic acid in a human intestinal epithelial cell line. Br. J. Nutr. 87, 97–100.

MacIntyre, I., et al., 1976. Regulation of vitamin D: an evolutionary view. Clin. Endocrinol. 5, s85–s95.

Maffia, M., Verri, T., Storelli, C., 2000. In vitro methods and results of ascorbic acid absorption in epithelial tissues of fish. In: Dabrowski, K. (Ed.), Ascorbic Acid in Aquatic Organisms: Status and Perspectives, vol. 211. CRC press.

Mahajan, C., Agrawal, N., 1980. Nutritional requirement of ascorbic acid by Indian major carp, *Cirrhina mrigala*, during early growth. Aquaculture 19, 37–48.

Mai, K., 1998. Comparative studies on the nutrition of two species of abalone, *Haliotis tuberculata* L. and *Haliotis discus* hannai Ino.: VII. Effects of dietary vitamin C on survival, growth and tissue concentration of ascorbic acid. Aquaculture 161, 383–392.

Mai, K., Wu, G., Zhu, W., 2001. Abalone, *Haliotis discus hannai* Ino, can synthesize *myo*-inositol *de novo* to meet physiological needs. J. Nutr. 131, 2898–2903.

Mai, K., et al., 2009. Dietary choline requirement for juvenile cobia, *Rachycentron canadum*. Aquaculture 289, 124–128.

Mæland, A., Sandnes, K., 1999. Determination of biotin in low-temperature fishmeal processed from different species by means of a microbiological method using *Lactobacillus plantarum* as test organism. J. Sci. Food Agric. 79, 1298–1300.

Mæland, A., Waagbø, R., Sandnes, K., Hjeltnes, B., 1998. Biotin in practical fish-meal based diet for Atlantic salmon *Salmo salar* L. fry. Aquacult. Nutr. 4, 241–247.

Mæland, A., et al., 2000. Water-soluble vitamins in natural plankton (copeods), during two consecutive spring blooms compared to vitamins in *Artemia franciscana* nauplii and metanauplii. Mar. Biol. 136, 765–772.

Mæland, A., Rønnestad, I., Waagbø, R., 2003. Short communication-Folate in eggs and developing larvae of Atlantic halibut, *Hippoglossus hippoglossus* L. Aquacult. Nutr. 9, 185–188.

Maranesi, M., Marchetti, M., Bochicchio, D., Cabrini, L., 2005. Vitamin B6 supplementation increases the docosahexaenoic acid concentration of muscle lipids of rainbow trout (*Oncorhynchus mykiss*). Aquacult. Res. 36, 431–438.

Martin, S., Krol, E., 2017. Nutrigenomics and immune function in fish: new insights from omics technologies. Dev. Comp. Immunol. 75, 86–98.

Masumoto, T., 2002. Yellowtail, serola quinqueradiata. In: Webster, C., Lim, C. (Eds.), Nutrient Requirements and Feeding of Finfish for Aquaculture. CAB International, Oxon, pp. 131–146.

Masumoto, T., Hardy, R., Casillas, E., 1987. Comparison of transketolase activity and thiamine pyrophosphate levels in erytrocytes and liver of rainbow trout (*Salmo gairdneri*), as indicators of thiamine status. J. Nutr. 117, 1422–1426.

Masumoto, T., et al., 1999. Effect of pantothenic acid deficiency of plasma free amino acid profile in rainbow trout *Oncorhynchus mykiss*. Fish. Sci. 65, 794–795.

Mazurais, D., et al., 2009. Optimal levels of dietary vitamin A for reduced deformity incidence during development of European sea bass larvae (*Dicentrarchus labrax*) depend on malformation type. Aquaculture 294, 262–270.

McLaren, B., Keller, E., O'Donnell, D., Elvehjem, C., 1947a. The nutrition of rainbow trout .2. Further studies with purified rations. Arch. Biochem. Biophys. 15, 179.

McLaren, B., Keller, E., O'Donnell, D., Elvehjem, C., 1947b. The nutrition of rainbow trout. 1. Studies of vitamin requirements. Arch. Biochem. Biophys. 15, 169–178.

Miao, L., et al., 2015. Dietary vitamin D3 requirement of Wuchang bream (*Megalobrama amblycephala*). Aquaculture 436, 104–109.

Michael, F., Koshio, S., 2008. Biochemical studies on the interactive effects of dietary choline and inositol in juvenile kuruma shrimp, *Marsupenaeus japonicus* Bate. Aquaculture 285, 179–183.

Michael, F., et al., 2006. Effect of choline and methionine as methyl group donors on juvenile kuruma shrimp, *Marsupenaeus japonicus* Bate. Aquaculture 258, 521–528.

Michel, V., Yuan, Z., Ramsubir, S., Bakovic, M., 2006. Choline transport for phospholipid synthesis. Exp. Biol. Med. 231, 490–504.

Ming, J., et al., 2012. Effects of emodin and vitamin C on growth performance, biochemical parameters and two HSP70s mRNA expression of Wuchang bream (*Megalobrama amblycephala* Yih) under high temperature stress. Fish Shellfish Immunol. 32, 651–661.

Moccia, R., Hung, S., Slinger, S., Ferguson, H., 1984. Effect of oxidized fish oil, vitamin E and ethoxyquin on the histopathology and haematology of rainbow trout, *Salmo gairdneri* Richardson. J. Fish. Dis. 7, 269–282.

Mock, D.M., 2007. In: Zempleni, J., Rucker, R.B., McCormick, D.B., Suttie, J. (Eds.), Handbook of Vitamins, fourth ed. CRC Press.

Mohamed, J., 2001a. Dietary biotin requirement determined for Indian catfish, *Heteropneustes fossilis* (Bloch), fingerlings. Aquacult. Res. 32, 709–716.

Mohamed, J., 2001b. Dietary pyridoxine requirement of the Indian catfish, Heteropneustes fossilis. Aquaculture 194, 327–335.

Mohamed, J., Ibrahim, A., 2001. Quantifying the dietary niacin requirement of the Indian catfish, *Heteropneustes fossilis* (Bloch), fingerlings. Aquacult. Res. 32, 157–162.

Mohamed, J., Ravisankar, B., Ibrahim, A., 2000. Quantifying the dietary biotin requirement of the catfish, *Clarias batrachus*. Aquacult. Int. 8, 9–18.

Mohamed, J., et al., 2003. Dietary vitamin A requirement of juvenile greasy grouper (*Epinephelus tauvina*). Aquaculture 219, 693–701.

Mohiseni, M., et al., 2017. Comparative effects of Shirazi thyme and vitamin E on some growth and plasma biochemical changes in common carp (*Cyprinus carpio*) during cadmium exposure. Aquacult. Res. 48, 4811–4821.

Moniruzzaman, M., et al., 2017. Synergistic effects of dietary vitamin E and selenomethionine on growth performance and tissue methylmercury accumulation on mercury-induced toxicity in juvenile olive flounder, *Paralichthys olivaceus* (Temminck et Schlegel). Aquac. Res. 48, 570–580.

Moreau, R., Dabrowski, K., 2000. Gluconolactone oxidase presence in fishes: activity and significance. In: Dabrowski, K. (Ed.), Ascorbic Acid in Aquatic Organisms: Status and Perspectives. CRC press, pp. 28–47.

Moren, M., et al., 2004. An optimum level of vitamin A supplements for Atlantic halibut (*Hippoglossus hippoglossus* L.) juveniles. Aquaculture 235, 587–599.

Morris, P., Davies, S., 1995. Thiamin supplementation of diets containing varied lipid: carbohydrate ratio given to gilthead seabream (*Sparus aurata* L.). Anim. Sci. 61, 597–603.

Morris, P., Baker, R., Davies, S., 1998. Nicotinic acid supplementation of diets for the African catfish, *Clarias gariepinus* (Burchell). Aquacult. Res. 29, 791–799.

Murai, T., Andrews, J., 1974. Interactions of dietary alpha-tocopherol, oxidized menhaden oil and ethoxyquin on channel catfish (*Ictalurus punctatus*). J. Nutr. 104, 1416–1431.

Murai, T., Andrews, J., 1978a. Riboflavin requirement of channel catfish fingerlings. J. Nutr. 108, 1512–1517.

Murai, T., Andrews, J., 1978b. Thiamine requirement of channel catfish fingerlings. J. Nutr. 108, 176–180.

Murai, T., Andrews, J., 1979. Pantothenic acid requirements of channel catfish fingerlings. J. Nutr. 109, 1140–1142.

Naderi, N., House, J., 2018. Recent developments in folate nutrition. In: Advances in Food and Nutrition Research, vol. 83. Elsevier, Amsterdam, pp. 195–213.

Najmi, A., Ahmad, N., Swarup, K., 1979. Corpuscles of Stannius of Mystus vittatus in relation to calcium and sodium rich environments. Arch. Biol. 90, 1–22.

Nakagawa, H., Sato, M., Gatlin III, D., 2007. Dietary supplements for the health and quality of cultured fish. Usp. Khim.: Cab Int. 133–167.

Nayak, S., Swain, P., Mukherjee, S., 2007. Effect of dietary supplementation of probiotic and vitamin C on the immune response of Indian major carp, *Labeo rohita* (Ham.). Fish Shellfish Immunol. 23, 892–896.

Ng, W., Serrini, G., Zhang, Z., Wilson, R., 1997. Niacin requirement and inability of tryptophan to act as a precursor of NAD(+), in channel catfish, *Ictalurus punctatus*. Aquaculture 152, 273–285.

Nguyen, B., et al., 2012. Effects of dietary vitamins C and E and their interactions on reproductive performance, larval quality and tissue vitamin contents in kuruma shrimp, *Marsupenaeus japonicus* Bate. Aquaculture 334–337, 73–81.

Niculescu, M., Craciunescu, C., Zeisel, S., 2006. Dietary choline deficiency alters global and genespecific DNA methylation in the developing hippocampus of mouse fetal brains. Faseb. J. 20, 43–49.

Niu, J., et al., 2009. Nutrient values of dietary ascorbic acid (l-ascorbyl-2-polyphosphate) on growth, survival and stress tolerance of larval shrimp, Litopenaeus vannamei. Aquacult. Nutr. 15, 194–201.

Niu, H., et al., 2014. Effect of dietary vitamin E on the growth performance and nonspecific immunity in sub-adult turbot (*Scophthalmus maximus*). Fish Shellfish Immunol. 41, 501–506.

NRC (National Research Council), 2011. Nutrient Requirement of Fish and Shellfish. The National Academies Press, Washington.

Nurhadi, T., et al., 2019. Effects of astaxanthin and canthaxanthin addition to ranchu goldfish (*Carassius auratus*) diet related to rate of color quality enhancement. World News Nat. Sci. 24, 178–183.

Ogino, C., 1965. B vitamin requirements of carp *Cyprinus carpio*. 1. Deficiency symptoms and requirements of vitamin B6. Bull. Jpn. Soc. Sci. Fish. 31, 546–551.

Ogino, C., 1967. B Vitamin requirements of carp. 2. Requirements of riboflavin and pantothenic acid. Bull. Jpn. Soc. Sci. Fish. 33, 351–354.

Ogino, C., et al., 1970a. B vitamin requirements of carp. 3. Requirement for biotin. Nippon Suisan Gakk 36, 734–740.

Ogino, C., et al., 1970b. B vitamin requirements of carp-IV. requirement for choline. Bull. Jpn. Soc. Sci. Fish. 36, 1140–1146.

Orapint, J., Bundit, Y., 2012. Stability of carotenoid diets during feed processing and under different storage conditions. Molecules 17, 5651–5660.

Ørnsrud, R., et al., 2002. Hypervitaminosis A in first-feeding fry of the Atlantic salmon (*Salmo salar* L.). Aquacult. Nutr. 8, 7–13.

Ortuño, J., Cuesta, A., Angeles Esteban, M., Meseguer, J., 2001. Effect of oral administration of high vitamin C and E dosages on the gilthead seabream (*Sparus aurata* L.) innate immune system. Vet. Immunol. Immunopathol. 79, 167–180.

Pan, J., et al., 2017. Vitamin E deficiency depressed fish growth, disease resistance, and the immunity and structural integrity of immune organs in grass carp (*Ctenopharyngodon idella*): referring to NF-kappaB, TOR and Nrf2 signaling. Fish Shellfish Immunol. 60, 219–236.

Park, K., et al., 2006. α-Lipoic acid-enrichment partially reverses tissue ascorbic acid depletion in pacu (*Piaractus mesopotamicus*) fed vitamin C-devoid diets. Fish Physiol. Biochem. 32, 329–338.

Peng, C., Chang, C., Peng, R., Chyau, C., 2010. Improved membrane transport of astaxanthine by liposomal encapsulation. Eur. J. Pharm. Biopharm. 75, 154–161.

Peng, X., et al., 2017. Fatty acid oxidation in zebrafish adipose tissue is promoted by 1α,25(OH) $_2$ D $_3$. Cell Rep. 19, 1444–1455.

Peres, H., Lim, C., Klesius, P., 2004. Growth, chemical composition and resistance to *Streptococcus iniae* challenge of juvenile Nile tilapia (*Oreochromis niloticus*) fed graded levels of dietary inositol. Aquaculture 235, 423–432.

Phillips Jr., A., Brockway, D., 1957. The nutrition of trout: IV. Vitamin requirements. Progress. Fish Cult. 19, 119–123.

Phillips, A., et al., 1946. The nutrition of trout. Fish. Res. Bull. 9, 3–21.

Phillips, A., et al., 1994. The production of anemia in brown trout. Fish. Res. Bull. 27, 66–70.

Poston, H., 1969a. Effects of massive doses of vitamin D_3 on fingerling brook trout. Fish. Res. Bull. 32, 48–50.

Poston, H., 1969b. The effect of excess levels of niacin on the lipid metabolism of fingerling brook trout. Fish. Res. Bull. 32, 9–12.

Poston, H., 1976. Optimum level of dietary biotin for growth, feed-utilisation, and swimming stamina of fingerling lake trout (*Salvelinus namaycush*). J. Fish. Res. Board Can. 33, 1803–1806.

Poston, H., 1991. Choline requirement of swim-up rainbow trout fry. Progress. Fish Cult. 53, 220–223.

Poston, H., Page, J., 1982. Gross and histological signs of dietary deficiencies of biotin and pantothenic acid in lake trout, *Salvelinus namaycush*. Cornell Vet. 72, 242–261.

Poston, H., Wolfe, M., 1985. Niacin requirement for optimal growth, feed conversion and protection of rainbow trout, *Salmo gairdneri* R., from ultraviolet B irradation. J. Fish. Dis. 8, 451–460.

Poston, H., Combs Jr., G., Leibovitz, L., 1976. Vitamin E and selenium interrelations in the diet of Atlantic salmon (*Salmo salar*): gross, histological and biochemical deficiency signs. J. Nutr. 106, 892–904.

Qian, Y., et al., 2015. Effects of dietary pantothenic acid on growth, intestinal function, anti-oxidative status and fatty acids synthesis of juvenile blunt snout bream *Megalobrama amblycephala*. PloS One 10, e0119518.

Qin, D., et al., 2017. Effects of dietary choline on growth performance, lipid deposition and hepatic lipid transport of grouper (*Epinephelus coioides*). Aquacult. Nutr. 23, 453–459.

Ragaza, J., et al., 2020. A review on Spirulina: alternative media for cultivation and nutritive value as an aquafeed. Rev. Aquacult. 12, 2371–2395.

Rao, A., Sarada, R., Ravishankar, G., 2007. Stabilization of astaxanthin in edible oils and its use as an antioxidant. J. Sci. Food Agric. 87, 957–965.

Reddy, H., Naik, M., Annappaswamy, T., 1999. Evaluation of the dietary essentiality of vitamins for *Penaeus monodon*. Aquacult. Nutr. 5, 267–275.

Ren, T., et al., 2005. Optimum dietary level of L-ascorbic acid for Japanese eel, *Anguilla japonica*. J. World Aquacult. Soc. 36, 437–443.

Richard, L., et al., 2011. The effect of choline and cystine on the utilisation of methionine for protein accretion, remethylation and trans-sulfuration in juvenile shrimp *Penaeus monodon*. Br. J. Nutr. 106, 825–835.

Richard, N., et al., 2014. Dietary supplementation with vitamin k affects transcriptome and proteome of Senegalese sole, improving larval performance and quality. Mar. Biotechnol. 16, 522–537.

Roberto, V., et al., 2009. Matrix Gla protein in turbot (*Scophthalmus maximus*): gene expression analysis and identification of sites of protein accumulation. Aquaculture 294, 202–211.

Roem, A., Kohler, C., Stickney, R., 1990. Vitamin E requirements of the blue tilapia, *Oreochromis aureus* (Steindachner), in relation to dietary lipid level. Aquaculture 87, 155–164.

Roem, A., Stickney, R., Kohler, C., 1991. Dietary pantothenic acid requirement of the blue tilapia. Progress. Fish-cult. 53, 216–219.

Rønnestad, I., Lie, Ø., Waagbø, R., 1997. Vitamin B6 in Atlantic halibut, *Hippoglossus hippoglossus* - endogenous utilization and retention in larvae fed natural zooplankton. Aquaculture 157, 337–345.

Rønnestad, I., Hamre, K., 2001. Chap 18. Marine Larver - spesielle ernæringskrav. In: Waagbø, R., et al. (Eds.), Kystnæringen Forlag and Bokklubb AS. Norway (In Norwegian), Bergen, pp. 298–312.

Rothman, S., 1978. Chymotrypsinogen inositol phosphatide complexes and the transport of digestive enzyme across membranes. BBA.-Biomembranes 509, 374–383.

Rumsey, G., 1991. Choline-betaine requirements of rainbow trout (*Oncorhynchus mykiss*). Aquaculture 95, 107–116.

Sacchi, R., Gardell, A., Chang, N., Kültz, D., 2014. Osmotic regulation and tissue localization of the myo-inositol biosynthesis pathway in tilapia (*Oreochromis mossambicus*) larvae. J. Exp. Zool. Part A 321, 457–466.

Saedisomeolia, A., Ashoori, M., 2018. Riboflavin in human health: a review of current evidences. Adv. Food Nutr. Res. 83, 57–81.

Saito, H., Ishihara, K., 1997. Antioxidant activity and active sites of phospholipids as antioxidants. J. Am. Oil Chem. Soc. 74, 1531–1536.

Salte, R., Norberg, K., 1991. Effects of warfarin on vitamin K-dependent coagulation factors in Atlantic salmon and rainbow trout with special reference to factor X. Thromb. Res. 63, 39–45.

Sandnes, K., 1994. Vannløselige vitaminer (Watersoluble vitamines). Project report. Grant 104914/110 Ernæringsinstitutt, F. In: Research Council of Norway, Bergen, Norway (In Norwegian).

Sandnes, K., Hansen, T., Killie, J., Waagbø, R., 1990. Ascorbate-2-sulfate as a dietary vitamin C source for Atlantic salmon (*Salmo salar*): 1. Growth, bioactivity, haematology and humoral immune response. Fish Physiol. Biochem. 8, 419–427.

Sandnes, K., Torrissen, O., Waagbø, R., 1992. The minimum dietary requirement of vitamin C in Atlantic salmon (*Salmo salar*) fry using Ca ascorbate-2-monophosphate as dietary source. Fish Physiol. Biochem. 10, 315–319.

Santiago, C., Gonzal, A., 2000. Effect of prepared diet and vitamins A, E and C supplementation on the reproductive performance of cage-reared bighead carp *Aristichthys nobilis* (Richardson). J. Appl. Ichthyol. 16, 8–13.

Santoro, M., et al., 2007. Kinetics and mechanism of the chromic oxidation of myo-inositol. Polyhedron 26, 169–177.

Santos, R., et al., 2016. Effects of lipoic acid on growth and biochemical responses of common carp fed with carbohydrate diets. Fish Physiol. Biochem. 42, 1699–1707.

Sarmah, B., Latimer, A., Appel, B., Wente, S., 2005. Inositol polyphosphates regulate zebrafish left-right asymmetry. Dev. Cell 9, 133–145.

Sau, S., Paul, B., Mohanta, K., Mohanty, S., 2004. Dietary vitamin E requirement, fish performance and carcass composition of rohu (*Labeo rohita*) fry. Aquaculture 240, 359–368.

Schneberger, E., 1941. Fisheries research in Wisconsin. Prog. Fish Cult. 56, 14.

Schneider, S., 2015. Inositol transport proteins. FEBS Lett. 589, 1049–1058.

Sealey, W., Gatlin, D., 2002. In vitro manipulations of vitamin C and vitamin E concentrations alter intracellular O^{-2} production of hybrid striped bass (*Morone chrysops*×*Morone saxatilis*) head-kidney cells. Fish Shellfish Immunol. 12, 131–140.

Serrini, G., Zhang, Z., Wilson, R., 1996. Dietary riboflavin requirement of fingerling channel catfish (*Ictalurus punctatus*). Aquaculture 139, 285–290.

Sesay, D., et al., 2016. Effects of dietary folic acid on the growth, digestive enzyme activity, immune response and antioxidant enzyme activity of blunt snout bream (*Megalobrama amblycephala*), fingerling. Aquaculture 452, 142–150.

Shahkar, E., et al., 2015. Effects of dietary vitamin C levels on tissue ascorbic acid concentration, hematology, non-specific immune response and gonad histology in broodstock Japanese eel, *Anguilla japonica*. Aquaculture 438, 115–121.

Shao, L., et al., 2016. Effects of dietary vitamin A on growth, hematology, digestion and lipometabolism of on-growing gibel carp (*Carassius auratus* gibelio var. CAS III). Aquaculture 460, 83–89.

Shao, L., et al., 2018. Effects of dietary vitamin C on growth, gonad development and antioxidant ability of on-growing gibel carp (*Carassius auratus* gibelio var. CAS III). Aquacult. Res. 49, 1242–1249.

Shay, K., et al., 2009. Alpha-lipoic acid as a dietary supplement: molecular mechanisms and therapeutic potential. BBA. Gen. Subjects 1790, 1149–1160.

Shi, L., et al., 2015. Folic acid deficiency impairs the gill health status associated with the NF-kB, MLCK and Nrf2 signaling pathways in the gills of young grass carp (*Ctenopharyngodon idella*). Fish Shellfish Immunol. 47, 289–301.

Shi, X., et al., 2017. α-lipoic acid ameliorates n-3 highly-unsaturated fatty acids induced lipid peroxidation via regulating antioxidant defenses in grass carp (*Ctenopharyngodon idellus*). Fish Shellfish Immunol. 67, 359–367.

Shi, X., et al., 2018. The protein-sparing effect of α-lipoic acid in juvenile grass carp, *Ctenopharyngodon idellus*: effects on lipolysis, fatty acid β-oxidation and protein synthesis. Br. J. Nutr. 120, 977–987.

Shiau, S., 1998. Nutrient requirements of penaeid shrimps. Aquaculture 164, 77–93.

Shiau, S., Chen, Y., 2000. Estimation of the dietary vitamin A requirement of juvenile grass shrimp, *Penaeus monodon*. J. Nutr. 130, 90–94.

Shiau, S., Chin, Y., 1998. Dietary biotin requirement for maximum growth of juvenile grass shrimp, *Penaeus monodon*. J. Nutr. 128, 2494–2497.

Shiau, S., Chin, Y., 1999. Estimation of the dietary biotin requirement of juvenile hybrid tilapia, *Oreochromis niloticus* x O-aureus. Aquaculture 170, 71–78.

Shiau, S., Hsieh, H., 1997. Vitamin B6 requirements of tilapia *Oreochromis niloticus x O. aureus* fed two dietary protein concentrations. Fish. Sci. 63, 1002–1007.

Shiau, S., Hsu, C., 1999. Dietary pantothenic acid requirement of juvenile grass shrimp, *Penaeus monodon*. J. Nutr. 129, 718–721.

Shiau, S., Huang, S., 2001a. Dietary folic acid requirement for maximum growth of juvenile tilapia *Oreochromis niloticus* x *O-aureus*. Fish. Sci. 67, 655–659.

Shiau, S., Huang, S., 2001b. Dietary folic acid requirement determined for grass shrimp, *Penaeus monodon*. Aquaculture 200, 339–347.

Shiau, S., Hwang, J., 1993. Vitamin D requirements of juvenile hybrid tilapia *Oreochromis niloticus* x *O. aureus*. Nippon Suisan Gakk 59, 553–558.

Shiau, S., Hwang, J., 1994. The dietary requirement of juvenile grass shrimp (*Penaeus monodon*) for vitamin D. J. Nutr. 124, 2445–2450.

Shiau, S., Liu, J., 1994a. Estimation of the dietary vitamin K requirement of juvenile *Penaeus chinensis* using menadione. Aquaculture 126, 129–135.

Shiau, S., Liu, J., 1994b. Quantifying the vitamin K requirement of juvenile marine shrimp (*Penaeus monodon*) with menadione. J. Nutr. 124, 277–282.

Shiau, S., Lo, P., 2000. Dietary choline requirements of juvenile hybrid tilapia, *Oreochromis niloticus* × *O. aureus*. J. Nutr. 130, 100–103.

Shiau, S., Lo, P., 2001. Dietary choline requirement of juvenile grass shrimp (*Penaeus monodon*). Anim. Sci. 72, 477–482.

Shiau, S., Lung, C., 1993a. Estimation of the vitamin B12 requirements of grass shrimp, *Penaeus monodon*. Aquaculture 117, 157–163.

Shiau, S., Lung, C., 1993b. No dietary vitamin B12 required for juvenile tilapia *Oreochromis niloticus* × *O. aureus*. Comp. Biochem. Physiol. 105A, 147150.

Shiau, S., Shiau, L., 2001. Re-evaluation of the vitamin E requirements of juvenile tilapia (*Oreochromis niloticus* × *O. aureus*). Anim. Sci. 72, 529–534.

Shiau, S., Su, S., 2004. Dietary inositol requirement for juvenile grass shrimp, *Penaeus monodon*. Aquaculture 241, 1–8.

Shiau, S., Su, S., 2005. Juvenile tilapia *(Oreochromis niloticus* × *Oreochromis aureus)* requires dietary *myo*-inositol for maximal growth. Aquaculture 243, 273–277.

Shiau, S., Suen, G., 1992. Estimation of the niacin requirements for tilapia fed diets containing glucose or dextrin. J. Nutr. 122, 2030–2036.

Shiau, S., Suen, G., 1994. The dietary requirement of juvenile grass shrimp (*Penaeus monodon*), for niacin. Aquaculture 125, 139–145.

Shiau, S., Wu, M., 2003. Dietary vitamin B6 requirement of grass shrimp *Penaeus monodon*. Aquaculture 225, 397–404.

Shimeno, S., 1991. Yellowtail, *Seriola quinqueradiata*. In: Wilson, R. (Ed.), Handboook of Nutrient Requirements of Finfish, pp. 181–191.

Shirmohammad, F., Mehri, M., Joezy-Shekalgorabi, S., 2016. A review on the role of inositol in aquaculture. Iran. J. Fish. Sci. 15, 1388–1409.

Shiu, Y., et al., 2017. Plasma immune protein analysis in the orange-spotted grouper *Epinephelus coioides*: evidence for altered expressions of immune factors associated with a choline-supplemented diet. Fish Shellfish Immunol. 65, 235–243.

Silva Martins, D., Á., et al., 2018. Antioxidant effects of nanoencapsulated lipoic acid in tissues and on the immune condition in haemolymph of Pacific white shrimp *Litopenaeus vannamei* (Boone, 1931). Aquacult. Nutr. 24, 1255–1262.

Skjærven, K., et al., 2018. Parental micronutrient deficiency distorts liver DNA methylation and expression of lipid genes associated with a fatty-liver-like phenotype in offspring. Sci. Rep. 8, 3055.

Smith, C., 1968. Hematological changes in coho salmon fed a folic acid deficient diet. J. Fish. Res. Board Can. 25, 151–156.

Smith, C., Halver, J., 1969. Folic acid anemia in coho salmon. J. Fish. Res. Board Can. 26, 111–114.

Smith, C., Brin, M., Halver, J., 1974. Biochemical, physiological, and pathological changes in pyridoxine-deficient Rainbow trout (*Salmo gairdneri*). J. Fish. Res. Board Can. 31, 1893–1898.

Smith, D., Warren, M., Refsum, H., 2018. Chap. 6 vitamin B12. In: Advances in Food and Nutrition Research, vol. 83. Elsevier Inc., pp. 215–279

Soliman, A., Wilson, R., 1992a. Water-soluble vitamin requirements of tilapia. 1. Pantothenic acid requirement of blue tilapia, *Oreochromis aureus*. Aquaculture 104, 121–126.

Soliman, A., Wilson, R., 1992b. Water-soluble vitamin requirements of tilapia. 1. Riboflavin requirement of blue tilapia, *Oreochromis aureus*. Aquaculture 104, 309–314.

Song, X., et al., 2016. Dietary astaxanthin improved the body pigmentation and antioxidant function, but not the growth of discus fish (*Symphysodon spp.*). Aquacult. Res. 48, 1359–1367.

Srivastav, A., 1983. Calcaemic responses in the freshwater mud eel, *Amphipnous cuchia*, to vitamin D3 administration. J. Fish. Biol. 23, 301–303.

Stéphan, G., Guillaume, J., Lamour, F., 1995. Lipid peroxidation in turbot (*Scophthahus maximus*) tissue: effect of dietary vitamin E and dietary n-6 or n-3 polyunsaturated fatty acids. Aquaculture 130, 251–268.

Su, S., Shiau, S., 2004. Requirements of dietary *myo*-inositol of juvenile grouper, Epinephelus malabaricus. J. Fish. Soc. Taiwan 31, 311–317.

Swarup, K., Srivastav, S., 1982. Vitamin D3-induced hypercalcaemia in male catfish, *Clarias batrachus*. Gen. Comp. Endocrinol. 46, 271–274.

Swarup, K., Norman, A., Srivastav, A., Srivastav, S., 1984. Dose-dependent vitamin D_3 and 1, 25-dihydroxyvitamin D_3-induced hypercalcemia and hyperphosphatemia in male catfish *Clarias batrachus*. Comp. Biochem. Physiol. B 78, 553–555.

Takeuchi, L., Takeuchi, T., Ogino, C., 1980. Riboflavin requirements in carp and rainbow trout. Nippon Suisan Gakk 46, 733–737.

Takeuchi, A., Okano, T., Kobayashi, T., 1991. The existence of 25-hydroxyvitamin D3-1α-hydroxylase in the liver of carp and bastard halibut. Life Sci. 48, 275–282.

Takeuchi, T., Watanabe, K., Satoh, S., Watanabe, T., 1992. Requirement of grass carp fingerlings for α-Tocopherol. Nippon Suisan Gakkai Shi. 58, 1743–1749.

Tan, B., Mai, K., 2001. Effects of dietary vitamin K on survival, growth, and tissue concentrations of phylloquinone (PK) and menaquinone-4 (MK-4) for juvenile abalone, *Haliotis discus hannai* Ino. J. Exp. Mar. Biol. Ecol. 256, 229–239.

Tanumihardjo, S., et al., 2016. Biomarkers of nutrition for development (BOND)-Vitamin A review. J. Nutr. 146, 1816S–1848S.

TaRóżanowska, M., et al., 2005. Pulse radiolysis study of the interaction of retinoids with peroxyl radicals. Free Radic. Biol. Med. 39, 1399–1405.

Taveekijakarn, P., Miyazaki, T., Matsumoto, M., Aral, S., 1996a. Histopathological and haematological changes in amago salmon, *Oncorhynchus rhodurus* (Jordan & McGregor), fed a vitamin-D-free diet. J. Fish. Dis. 19, 289–294.

Taveekijakarn, P., Miyazaki, T., Matsumoto, M., Arai, S., 1996b. Studies on vitamin K deficiency in amago salmon, *Oncorhynchus rhodurus* (Jordan & McGregor). J. Fish. Dis. 19, 209–214.

Teixeira, C., et al., 2012. Growth performance of Nile tilapia, *Oreochromis niloticus*, fed diets containing levels of pyridoxine and haematological response under heat stress. Aquacult. Res. 43, 1081–1088.

Terjesen, B., et al., 2004. Lipoic acid and ascorbic acid affect plasma free amino acids selectively in the teleost fish pacu *(Piaractus mesopotamicus)*. J. Nutr. 134, 2930–2934.

Toledo-Ibarra, G., Díaz-Resendiz, K., Pavónb, L., Girón-Pérez, M., 2014. Cholinergic activity in mononuclear cells of Nile tilapia (*Oreochromis niloticus*) fish. Adv. Neuroimmune Biol. 5, 229–234.

Torda, C., Wolff, H., 1945. Effect of vitamins on acetylcholine synthesis. The apparently specific action of vitamin E. P. Soc. Exp. Biol. Med. 58, 163–165.

Torrissen, O., Christiansen, R., 1995. Requirements for carotenoids in fish diets. J. Appl. Ichthyol. 11, 225–230.

Treberg, J., Driedzic, W., 2007. The accumulation and synthesis of betaine in winter skate (*Leucoraja ocellata*). Comp. Biochem. Physiol. A 147, 475–483.

Twibell, R., Brown, P., 2000. Dietary choline requirement of juvenile yellow perch (*Perca flavescens*). J. Nutr. 130, 95–99.

Tyagi, E., Agrawal, R., Nath, C., Shukla, R., 2008. Influence of LPS-induced neuroinflammation on acetylcholinesterase activity in rat brain. J. Neuroimmunol. 205, 51–56.

Tyssandier, V., et al., 2001. Main factors governing the transfer of carotenoids from emulsion lipid droplets to micelles. Biochim. Biophys. Acta 1533, 285–292.

Udagawa, M., 2001. The effect of dietary vitamin K (phylloquinone and menadione) levels on the vertebral formation in mummichog *Fundulus heteroclitus*. Fish. Sci. 67, 104–109.

Udagawa, M., 2004. The effect of parental vitamin K deficiency on bone structure in mummichog *Fundulus heteroclitus*. J. World Aquacult. Soc. 35, 366–371.

Udagawa, M., Hirose, K., 1998. Physiological abnormality and the tissue vitamin k level of mummichog *Fundulus heteroclitus* fed with a vitamin K free diet. Fish. Sci. 64, 612–616.

Ulatowski, L., Manor, D., 2013. Vitamin E trafficking in neurologic health and disease. Annu. Rev. Nutr. 33, 87–103.

Vervoort, L., Ronden, J., Thijssen, H., 1997. The potent antioxidant activity of the vitamin K cycle in microsomal lipid peroxidation. Biochem. Pharmacol. 54, 871–876.

Viegas, C., et al., 2008. Gla-rich protein (GRP), a new vitamin K-dependent protein identified from sturgeon cartilage and highly conserved in vertebrates. J. Biol. Chem. 283, 36655–36664.

Villeneuve, L., et al., 2005. Dietary levels of all-trans retinol affect retinoid nuclear receptor expression and skeletal development in European sea bass larvae. Br. J. Nutr. 93, 791–801.

Villeneuve, L., et al., 2006. Intake of high levels of vitamin A and polyunsaturated fatty acids during different developmental periods modifies the expression of morphogenesis genes in European sea bass (*Dicentrarchus labrax*). Br. J. Nutr. 95, 677–687.

Waagbø, R., 2010. Water-soluble vitamins in fish ontogeny. Aquacult. Res. 41, 433–744.

Waagbø, R., Remø, S., 2020. Chap 7. Functional diets in fish health management. In: Kibenge, F., Powell, M. (Eds.), Aquaculture Health Management, pp. 187–234.

Waagbø, R., Thorsen, T., Sandnes, K., 1989. Role of dietary ascorbic acid in vitellogenesis in rainbow trout (*Salmo gairdneri*). Aquaculture 80, 301–314.

Waagbø, R., Glette, J., Raa-Nilsen, E., Sandnes, K., 1993a. Dietary vitamin C, immunity and disease resistance in Atlantic salmon (*Salmo salar*). Fish Physiol. Biochem. 12, 61–73.

Waagbø, R., et al., 1993b. Health aspects of dietary lipid sources and vitamin E in Atlantic salmon (*Salmo salar*). II. Spleen and erythrocyte phospholipid fatty acid composition, nonspecific immunity and disease resistance. Fiskidir. Skr. Ser. Ernaer. 6, 63–80.

Waagbø, R., Sandnes, K., Mæland, A., 1994. Biotinbehov hos smolt (Biotin requirements for smolts). In: Norsk Fiskeoppdrett, 2A, pp. 24–25.

Waagbø, R., Sandnes, K., Lie, O., Roem, A., 1998. Effects of inositol supplementation on growth, chemical composition and blood chemistry in Atlantic salmon, *Salmo salar* L., fry. Aquacult. Nutr. 4, 53–59.

Waagbø, R., Graff, I., Hamre, K., 2001. Vitaminer. In: Waagbø, R., et al. (Eds.), Fiskeernæring, pp. 94–123.

Waagbø, R., Olsvik, P., Remø, S., 2020. Chapter 6. Nutritional and metabolic disorders. In: Woo, P., Iwama, G. (Eds.), Climate Change and Non-infectious Fish Disorders (CCNFD), pp. 102–124.

Wade, N., Gabaudan, J., Glencross, B., 2017a. A review of carotenoid utilisation and function in crustacean aquaculture. Rev. Aquacult. 9, 141–156.

Wade, et al., 2017b. Dietary astaxanthin levels affect colour, growth, carotenoid digestibility and the accumulation of specific carotenoid esters in the giant tiger shrimp, *Penaeus monodon*. Aquacult. Res. 48, 395–406.

Wahli, T., et al., 2003. Influence of dietary vitamin C on the wound healing process in rainbow trout (*Oncorhynchus mykiss*). Aquaculture 225, 371–386.

Walker, S., Eylenburg, E., Moore, T., 1947. The action of vitamin K in hypervitaminosis A. Biochem. J. 41, 575–580.

Walton, M., Cowey, C., Adron, J., 1984. Effects of biotin deficiency in rainbow trout (*Salmo gairdneri*) fed diets of different lipid and carbohydrate content. Aquaculture 37, 21–38.

Wang, X., Kim, K., Bai, S., 2002. Effects of different dietary levels of L-ascorbyl-2-polyphosphate on growth and tissue vitamin C concentrations in juvenile olive flounder, *Paralichthys olivaceus* (*Temminck et Schlegel*). Aquacult. Res. 33, 261–267.

Wang, Z., et al., 2006. Effect of high dietary intakes of vitamin E and n-3 HUFA on immune responses and resistance to Edwardsiella tarda challenge in Japanese flounder (*Paralichthys olivaceus*, Temminck and Schlegel). Aquacult. Res. 37, 681–692.

Wang, Y., Li, E., Sun, X., Chen, L., 2010. Dietary vitamin B2 requirement of juvenile Chinese mitten crab, Eriocheir sinensis. J. Feed Ind. 31, 13–16.

Wang, J., et al., 2015. Vitamin E requirement of sea cucumber (*Apostichopus japonicus*) and its' effects on nonspecific immune responses. Aquacult. Res. 46, 1628–1637.

Wang, L., et al., 2017. Effects of the supplementation of vitamin D_3 on the growth and vitamin D metabolites in juvenile Siberian sturgeon (*Acipenser baerii*). Fish Physiol. Biochem. 43, 901–909.

Wang, C., et al., 2018. Effects of dietary *myo*-inositol on growth, chemical composition and plasma chemistry of Amur sturgeon Acipenser schrenckii. Aquacult. Int. 26, 1481–1492.

Watanabe, T., Takashima, F., Ogino, C., Hibiya, T., 1970. Requirement of young carp for α-tocopherol. Nippon Suisan Gakkai Shi. 36, 972–976.

Wei, J., et al., 2014. Dietary cyanocobalamin requirement and its effect on non-specific immunity and disease resistance in juvenile Chinese mitten crab Eriocheir sinensis. Aquaculture 434, 179–183.

Wei, J., et al., 2016. Effects of dietary folic acid on growth, antioxidant capacity, non-specific immune response and disease resistance of juvenile Chinese mitten crab *Eriocheir sinensis* (Milne-Edwards, 1853). Aquacult. Nutr. 22, 567–574.

Wen, Z., et al., 2009. Effect of dietary pantothenic acid supplement on growth, body composition and intestinal enzyme activities of juvenile Jian carp (*Cyprinus carpio* var. Jian). Aquacult. Nutr. 15, 470–476.

Wen, Z., et al., 2010. Immune response, disease resistance and intestinal microflora of Jian carp (*Cyprinus carpio* var. Jian), fed graded levels of pantothenic acid. Aquacult. Nutr. 16, 430–436.

White, D., et al., 2002. Effect of esterification on the absorption of astaxanthin in rainbow trout, *Oncorhynchus mykiss* (Walbaum). Aquacult. Res. 33, 343–350.

Wilson, J., 2005. Regulation of vitamin C transport. Annu. Rev. Nutr. 25, 105–125.

Wilson, R., Poe, W., 1988. Choline nutrition of fingerling channel catfish. Aquaculture 68, 65–71.

Wilson, R., Bowser, P., Poe, W., 1983. Dietary pantothenic acid requirement of fingerling channel catfish. J. Nutr. 113, 2124–2128.

Wilson, R., Bowser, P., Poe, W., 1984. Dietary vitamin E requirement of fingerling channel catfish. J. Nutr. 114, 2053–2058.

Wistbacka, S., Lönnström, L., Bonsdorff, E., Bylund, G., 2009. Thiaminase activity of crucian carp *Carassius carassius* injected with a bacterial fish pathogen, *Aeromonas salmonicida subsp.* salmonicida. J. Aquat. Anim. Health 21, 217–228.

Woodward, B., 1984. Symptoms of severe riboflavin deficiency without ocular opacity in rainbow trout (*Salmo gairdneri*). Aquaculture 39, 275–281.

Woodward, B., 1985. Riboflavin requirement for growth, tissue saturation and maximal flavin-dependent ezyme activiety in young rainbow trout (*Salmo gardneri*), at two temperatures. J. Nutr. 115, 78–84.

Woodward, B., 1994. Dietary vitamin requirements of cultured young fish, with emphasis on quantitative estimates for salmonids. Aquaculture 124, 133–168.

Woodward, B., Frigg, M., 1989. Dietary biotin requirements of young rainbow trout (*Salmo gairdneri*), determined by weight-gain, hepatic biotin concentration and maximal biotin dependent enzyme activities in liver and white muscle. J. Nutr. 119, 54–60.

Wooster, G., Bowser, P., Brown, S., Fisher, J., 2000. Remediation of Cayuga syndrome in landlocked Atlantic salmon *Salmo salar* using egg and sac-fry bath treatments of thiamine-hydrochloride. J. World Aquacult. Soc. 31, 149–157.

Wu, G., 2017. Principles of Animal Nutrition. CRC Press, Boca Raton, FL, USA.

Wu, G., Davis, D., 2005. Interrelationship among methionine, choline, and betaine in channel catfish, *Ictalurus punctatus*. J. World Aquacult. Soc. 36, 337–345.

Wu, P., et al., 2011. Effect of dietary choline on growth, intestinal enzyme activities and relative expressions of target of rapamycin and eIF4E-binding protein2 gene in muscle, hepatopancreas and intestine of juvenile Jian carp (*Cyprinus carpio* var. Jian). Aquaculture 317, 107–116.

Wu, P., et al., 2013. Dietary choline modulates immune responses, and gene expressions of TOR and eIF4E-binding protein2 in immune organs of juvenile Jian carp (*Cyprinus carpio* var. Jian). Fish Shellfish Immunol. 35, 697–706.

Wu, C., et al., 2014. Dietary ascorbic acid modulates the expression profile of stress protein genes in hepatopancreas of adult Pacific abalone *Haliotis discus hannai* Ino. Fish Shellfish Immunol. 41, 120–125.

Wu, F., et al., 2015. Vitamin C requirement of adult genetically improved farmed tilapia, *Oreochromis niloticus*. Aquacult. Int. 23, 1203–1215.

Wu, P., et al., 2016. Dietary choline deficiency and excess induced intestinal inflammation and alteration of intestinal tight junction protein transcription potentially by modulating NF-κB, STAT and p38 MAPK signaling molecules in juvenile Jian carp. Fish Shellfish Immunol. 58, 462–473.

Xia, M., et al., 2015. Dietary niacin levels in practical diets for *Litopenaeus vannamei* to support maximum growth. Aquacult. Nutr. 21, 853–860.

Xu, H., et al., 2016. Dietary vitamin C deficiency depresses the growth, head kidney and spleen immunity and structural integrity by regulating NF-κB, TOR, Nrf2, apoptosis and MLCK signaling in young grass carp (*Ctenopharyngodon idella*). Fish Shellfish Immunol. 52, 111–138.

Xu, C., et al., 2018. Effects of α-lipoic acid on growth performance, body composition, antioxidant status and lipid catabolism of juvenile Chinese mitten crab *Eriocheir sinensis* fed different lipid percentage. Aquaculture 484, 286–292.

Xu, F., et al., 2019a. Effects of α-lipoic acid on growth performance, body composition, antioxidant profile and lipid metabolism of the GIFT tilapia (*Oreochromis niloticus*) fed high-fat diets. Aquacult. Nutr. 25, 585–596.

Xu, C., et al., 2019b. α-lipoic acid regulate growth, antioxidant status and lipid metabolism of Chinese mitten crab *Eriocheir sinensis*: optimum supplement level and metabonomics response. Aquaculture 506, 94–103.

Xun, P., et al., 2019. Effects of dietary niacin on growth performance, serum biochemistry, antioxidant ability of liver, intestinal digestion, and absorption in juvenile golden pompano. Isr. J. Aquacult. 71, 1603.

Yang, Q., Zhou, X., Jiang, J., Liu, Y., 2008. Effect of dietary vitamin A deficiency on growth performance, feed utilization and immune responses of juvenile Jian carp (*Cyprinus carpio* var. Jian). Aquacult. Res. 39, 902–906.

Yano, T., Nakao, M., Furuichi, M., Yone, Y., 1988. Effects of dietary choline, pantothenic acid and vitamin C on the serum complement activity of red sea bream. Nippon Suisan Gakk 54, 141–144.

Yasuyuki, H., Takeuchi, T., Itagaki, E., Fukunaga, T., 1998. Relationship between fat soluble vitamins in diet and the occurrence of colour abnormality on the blind side of juvenile Japanese flounder. Aquacult. Sci. 46, 279–286.

Yeh, S., et al., 2015. Improvement in lipid metabolism and stress tolerance of juvenile giant grouper, *Epinephelus lanceolatus* (Bloch), fed supplemental choline. Aquacult. Res. 46, 1810–1821.

Yi, X., et al., 2018. Effects of dietary vitamin E and astaxanthin on growth, skin colour and antioxidative capacity of large yellow croaker *Larimichthys crocea*. Aquacult. Nutr. 24, 472–480.

Yildirim-Aksoy, M., Lim, C., Li, M., Klesius, P., 2008. Interaction between dietary levels of vitamins C and E on growth and immune responses in channel catfish, *Ictalurus punctatus* (Rafinesque). Aquacult. Res. 39, 1198–1209.

Yossa, R., et al., 2011. Effects of dietary biotin and avidin on growth, survival, feed conversion, biotin status and gene expression of zebrafish *Danio rerio*. Comp. Biochem. Physiol. B 160, 150–158.

Yossa, R., et al., 2015. Current knowledge on biotin nutrition in fish and research perspectives. Rev. Aquacult. 7, 59–73.

Yuan, J., Peng, J., Yin, K., Wang, J., 2011. Potential health-promoting effects of astaxanthin: a high-value carotenoid mostly from microalgae. Mol. Nutr. Food Res. 55, 150–165.

Yuan, J., et al., 2016. Effects of dietary vitamin K levels on growth performance, enzyme activities and antioxidant status in the hepatopancreas and intestine of juvenile Jian carp (*Cyprinus carpio* var. Jian). Aquacult. Nutr. 22, 352–366.

Yuan, Z., et al., 2020. Choline deficiency decreased the growth performances and damaged the amino acid absorption capacity in juvenile grass carp (*Ctenopharyngodon idella*). Aquaculture 518, 734829.

Zaki, M., et al., 2010. Impact of different levels of dietary *myo*-inositol on the growth performance, histological structure of gonads and liver of red tilapia reared in brackish water. Afr. J. Biotechnol. 9, 4808–4817.

Zehra, S., Khan, M., 2017. Dietary thiamin and pyridoxine requirements of fingerling Indian major carp, Cirrhinus mrigala (Hamilton). Aquacult. Res. 48, 4945–4957.

Zehra, S., Khan, M., 2018a. Dietary riboflavin requirement of fingerling *Channa punctatus* (Bloch) based on growth, conversion efficiencies, protein retention, liver riboflavin storage, RNA/DNA ratio and carcass composition. Aquacult. Nutr. 24, 269–276.

Zehra, S., Khan, M., 2018b. Dietary thiamin requirement of fingerling *Channa punctatus* (Bloch), based on growth, protein gain, liver thiamin storage, RNA/DNA ratio and biochemical composition. Aquacult. Nutr. 24, 1015–1023.

Zehra, S., Khan, M., 2019a. Effects of different levels of dietary cyanocobalamin on growth, liver cyanocobalamin concentration, antioxidant capacity, intestinal enzymes and non-specific immune response for optimum inclusion in the commercial feeds of fingerling Channa punctatus (Bloch). Aquaculture 511, 734272.

Zehra, S., Khan, M., 2019b. Dietary biotin requirement of fingerling Channa punctatus (Bloch). J. Appl. Aquacult. 31, 236–253.

Zehra, S., Khan, M., 2019c. Quantification of dietary inositol requirement for fingerling *Channa punctatus* (Bloch) based on growth, antioxidant status, hematological tools and liver inositol concentration. Aquaculture 512, 734280.

Zeisel, S., 1981. Dietary choline: biochemistry, physiology, and pharmacology. Ann. Rev. Nutr. 1, 95–121.

Zempleni, J., Suttie, J., Gregory III, J., Stover, P., 2013. Handbook of Vitamins. CRC Press.

Zhang, Z., Wilson, R., 1999. Reevaluation of the choline requirement of fingerling channel catfish (*Ictalurus punctatus*) and determination of the availability of choline in common feed ingredients. Aquaculture 180, 89–98.

Zhang, W., Mai, K., Xu, W., Ma, H., 2003. Metabolic responses to dietary cholecalciferol and phosphorus in abalone *Haliotis discus hannai* Ino. J. Exp. Zool. Part A. 299, 110–117.

Zhang, W., et al., 2007. Interaction between vitamins A and D on growth and metabolic responses of abalone *Haliotis discus hannai*, Ino. J. Shellfish Res. 26, 51–58.

Zhang, W., et al., 2010. Effects of dietary α-lipoic acid on the growth and antioxidative responses of juvenile abalone *Haliotis discus hannai* Ino. Aquacult. Res. 41, e781–e787.

Zhang, J., et al., 2013. Effects of dietary astaxanthin on growth, antioxidant capacity and gene expression in Pacific white shrimp Litopenaeus vannamei. Aquacult. Nutr. 19, 917–927.

Zhang, L., et al., 2017a. Vitamin A deficiency suppresses fish immune function with differences in different intestinal segments: the role of transcriptional factor NF-κB andp38 mitogen-activated protein kinase signalling pathways. Br. J. Nutr. 117, 67–82.

Zhang, Y., Li, Y., Liang, X., Gao, J., 2017b. Effects of dietary vitamin E supplementation on growth performance, fatty acid composition, lipid peroxidation and peroxisome proliferator-activated receptors (PPAR) expressions in juvenile blunt snout bream *Megalobrama amblycephala*. Fish Physiol. Biochem. 43, 913–922.

Zhang, G., et al., 2019. Effects of dietary vitamin C on growth, flesh quality and antioxidant capacity of juvenile golden pompano Trachinotus ovatus. Aquacult. Res. 50, 2856–2866.

Zhao, Z., Wen, H., Wu, F., Liu, A., 2008. Dietary folic acid requirement for grass carp fingerling, Ctenopharyngodon idella. J. Shanghai Fish. Univ. 17, 187–192.

Zhao, S., et al., 2012. Effects of dietary biotin supplement on growth, body composition, intestinal enzyme activities and microbiota of juvenile Jian carp (*Cyprinus carpio* var. Jian). Aquacult. Nutr. 18, 400–410.

Zhao, H., et al., 2015. Flesh shear force, cooking loss, muscle antioxidant status and relative expression of signaling molecules (Nrf2, Keap1, TOR, and CK2) and their target genes in young grass carp (*Ctenopharyngodon idella*) muscle fed with graded levels of choline. PloS One 10, e142915.

Zhao, H., et al., 2016. Dietary choline regulates antibacterial activity, inflammatory response and barrier function in the gills of grass carp (*Ctenopharyngodon idella*). Fish Shellfish Immunol. 52, 139–150.

Zhou, Q., et al., 2012. Effect of dietary vitamin C on the growth performance and innate immunity of juvenile cobia (*Rachycentron canadum*). Fish Shellfish Immunol. 32, 969–975.

Zhou, Q., et al., 2013. Dietary vitamin E could improve growth performance, lipid peroxidation and non-specific immune responses for juvenile cobia (*Rachycentron canadum*). Aquacult. Nutr. 19, 421–429.

Zhou, Q., Liu, T., Lu, Y., 2017. Recent progress in sources, biological function, absorption and metabolism characteristics of astaxanthin. Food Res. Dev. 38, 214–219.

CHAPTER

4

Protein and amino acids

Kangsen Mai[1], M. Xue[2], G. He[1], S.Q. Xie[3] and Sadasivam J. Kaushik[4]

[1]Ocean University of China, Qingdao, Shandong, China; [2]Institute of Feed Research, Chinese Academy of Agricultural Science, Beijing, China; [3]Institute of Hydrobiology, Chinese Academy of Sciences, Wuhan, Hubei, China; [4]INRAe, Univ Pau and Pays de l'Adour, Nutrition, Metabolism and Aquaculture Unit, Saint Pée sur Nivelle, France

4.1 Nutritional biochemistry of proteins and amino acids

Proteins and amino acids (AAs) are the most important components for the animal to maintain the normal physiological/biochemical processes and growth of tissues. The interest of chemists for substances which we now call "proteins" started more than 200 years ago. It was Antoine Fourcroy (1755–1809), a French chemist, who recognized three distinct varieties of proteins from animal sources in 1789: albumin, fibrin, and gelatin and at least two more from vegetable sources. They were often collectively called "albumins." In 1838, Jacob Berzelius, a Swedish chemist, coined the name "protein" derived from the Greek word "πρoτειoσ" (Proteios), meaning "primary," to this class of molecules.

AAs contain a central carbon atom (C_α), which is attached to a hydrogen atom, an amino group (NH_2), and a carboxyl group (COOH). Each AA is distinguished by the presence of a side chain, which differs from one AA to another (Fig. 4.1). Proteins, also often called polypeptides, are made up of AA residues linked by peptide bonds. A simple peptide bond is formed by the condensation of two AAs with the elimination of water. Addition of further AAs to the chain leads to tripeptides, tetrapeptides, and so on. Chains of up to 20 AAs are called oligopeptides, and longer ones are termed polypeptides (Fig. 4.2). Polypeptides range in size from a few AAs to thousands. Proteins consist either of a single polypeptide chain, or they are formed from separate polypeptide chains called subunits. Each protein can thus be defined as a polypeptide with specific biological function(s).

The number and order of all of the AA residues in a polypeptide constitute its primary structure. Proteins can be classified based on their composition, their structures, or their functions. Examples of proteins classified based on their composition are either simple proteins such as albumins, globulins, and glutenins or conjugated proteins such as glycoproteins,

Fish Nutrition, Fourth Edition
https://doi.org/10.1016/B978-0-12-819587-1.00012-4

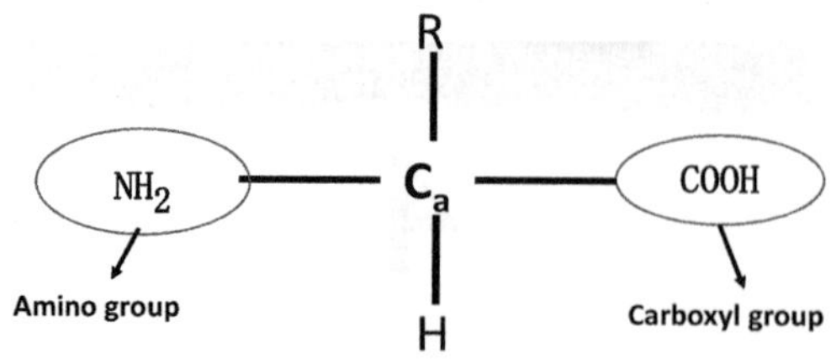

FIGURE 4.1 Schematic representation of an amino acid. Amino acids contain a central carbon atom (C_α), which is attached to a hydrogen atom, an amino group (NH_2) and a carboxyl group (COOH). The letter R indicates the presence of a side chain, which distinguishes each amino acid.

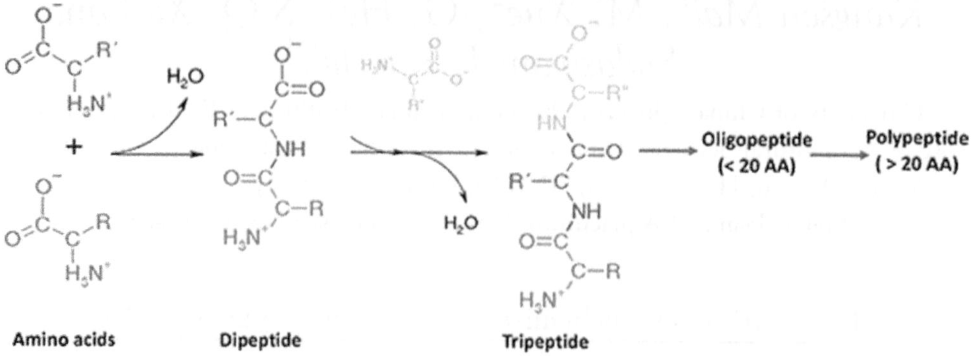

FIGURE 4.2 Formation of di-, tri-, oligo-, and polypeptides through condensation of amino acids. *Based on Buxbaum (2015).*

lipoproteins, and chromoproteins. They can also be classified based on structure such as globular proteins with polypeptide chains folded into a spherical or globular shape (e.g., casein, egg albumin) or fibrous proteins having polypeptide chains arranged in long strands (e.g., collagen, keratin). A third way of classification of proteins is based on specific functions (e.g., enzymes, hormones, transporters). Some proteins contain other covalently bound components and prosthetic groups and undergo posttranslational modifications.

Proteins and AAs constitute the most critical molecules because of the roles they play in the maintenance and growth of bodily structure of all living organisms. There are tens of thousands of different proteins produced by biological organisms with each one of the different proteins having a specific structure, function, and a unique AA sequence. Proteins are involved in essentially every process that takes place within the cells of all organisms and have a remarkable diversity of functions. Some proteins, such as actin and tubulin, confer stiffness and rigidity to otherwise fluid biological components. Other proteins, such as myosin, have a mechanical function and are capable of generating mechanical forces, such as those exerted by contracting muscles. Some others such as collagen and elastin are components of connective tissue, such as the cartilage. Proteins are thus essential components for every type of cell in the body, including muscles, bones, organs, tendons, and ligaments. Proteins have also a large array of physiological functions: as enzymes, they catalyze biochemical reactions, and as transporters, they allow the entry and exit of molecules through cells. Some proteins are important in immune responses, cell adhesion, cell signaling, and functioning of the cell cycle. From the nutritional physiological point of view, digestive proteases and AA-catabolizing enzymes play a major role.

Of the over 300 naturally occurring AAs, a fixed set of 20 AAs constitute the monomer units, which serve as building blocks of proteins. To describe the AAs that are constituents of proteins, both three-letter and one-letter abbreviations have been assigned to AAs and are often used for writing out primary sequences. Although all AAs contain nitrogen, the relative proportion of nitrogen in each of the 20 AAs varies and so does the molecular mass of each AA (Table 4.1).

After feed intake, proteins are digested and hydrolyzed into AAs or peptides of different sizes. The AAs are absorbed and transported to different target tissues and used for maintenance of bodily functions and for somatic growth. AAs are incorporated primarily into

TABLE 4.1 The names, codes, molecular masses, and nitrogen content of the 20 amino acids that make up the proteins.

Amino acid	Three-letter code	One-letter code	Molecular mass (g/mol)	Nitrogen content, %
Alanine	Ala	A	89	15.72
Arginine[a]	Arg	R	155	32.16
Asparagine	Asn	N	132	21.2
Aspartate	Asp	D	133	10.52
Cysteine	Cys	C	121	11.56
Glutamate	Glu	E	146	9.52
Glutamine	Gln	Q	146	19.17
Glycine	Gly	G	75	18.66
Histidine	His	H	174	27.08
Isoleucine	Ile	I	131	10.68
Leucine	Leu	L	131	10.68
Lysine	Lys	K	146	19.16
Methionine	Met	M	149	9.39
Phenylalanine	Phe	F	165	8.48
Proline	Pro	P	115	12.16
Serine	Ser	S	105	13.33
Threonine	Thr	T	119	11.76
Tryptophan	Trp	W	204	13.72
Tyrosine	Tyr	Y	181	7.73
Valine	Val	V	117	11.96

[a]*The amino acids with bold format indicates that they are essential/indispensable amino acids.*

protein or degraded to provide energy. Degradation of AAs involves removal of the nitrogen and catabolism of the carbon skeleton, mainly through the tricarboxylic acid (TCA) cycle for gluconeogenesis or lipogenesis (Fig. 4.3).

In all animals including fish and crustaceans, body tissues are continuously formed and are broken down (Fig. 4.4). In growing animals, protein synthesis exceeds degradation, and the balance between these two processes results in protein deposition or accretion. Tissue-free AA pools are always in a dynamic status with the relative importance of AAs supplied through dietary protein intake, and those coming through endogenous protein breakdown can vary very much between species, stage of growth, and metabolic status.

AAs enter the tissue-free AA pool via the dietary protein intake or from the degradation of body proteins. AAs leave the free AA pool through protein synthesis and oxidation.

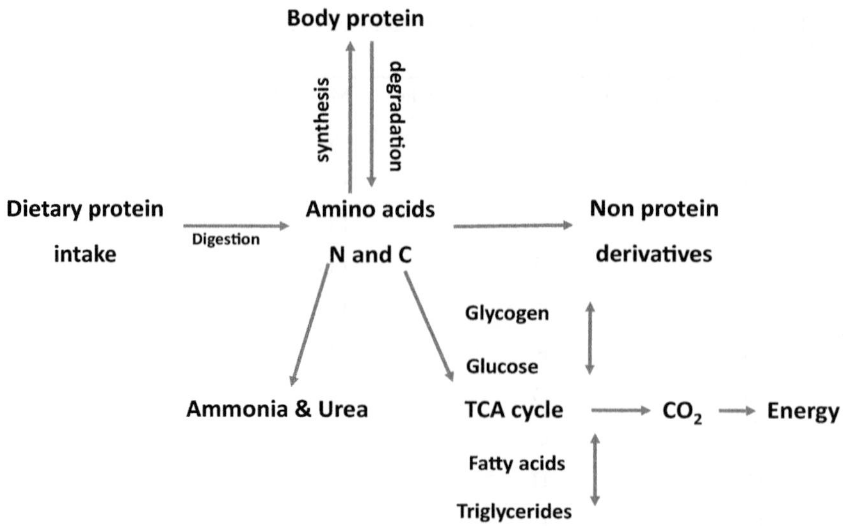

FIGURE 4.3 Overview of protein and amino acid metabolism.

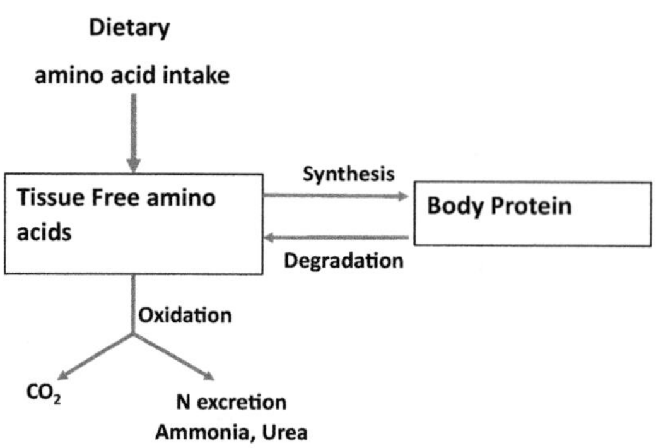

FIGURE 4.4 A simplified two-compartment model of protein turnover. Amino acids enter the tissue free amino acid pool via the dietary protein intake or from the degradation of body proteins. Amino acids leave the free amino acid pool through protein synthesis and oxidation.

In aquaculture as in all food-producing systems, the major objective is the provision of food protein for humans. Protein is the major nitrogenous macronutrient in foods and the fundamental component of animal tissues. Proteins are the major organic material in fish or shrimp, making up about 65%–75% of the total on a dry-weight basis. The major proteins in fish consist of myofibrillar, sarcoplasmic, and connective tissue proteins.

Fish and shrimp, very much like the humans or land animals, cannot synthesize all the 20 proteogenic AAs in adequate amounts to support growth or maintain their physiological well-being. Consequently, all aquatic animals, like their terrestrial counterparts, must acquire several of these in their diet, through the consumption of different feeds containing proteins. The pioneering studies on protein and AA nutrition of fish were undertaken by Halver and coworkers in the late 1950s and early 1960s with Pacific salmons. Fish, like other animals, do not have a true protein requirement per se but have a requirement for a well-balanced mixture of AAs. The basic function of proteins in the nutrition of any organism is to supply adequate amounts of the needed AAs. The protein quality, which is the main yardstick for measuring the biological, nutritional, or nutritive value of a food, depends on its AA composition and on the physiological utilization of the AAs after digestion, absorption, and minimal obligatory rates of oxidation. Dietary protein sources can include both animal and plant sources, each containing varying proportions of different functional proteins with specific AA profiles.

Molecular tools and techniques have paved the way for rapid determination of DNA sequences and, by inference, the AA sequences of proteins from structural genes. The number of such sequences of proteins is growing continuously, and they are deposited in protein sequence databases. Progress has also been made to set up a single worldwide database such as the UniProt, to combine data from many of the publicly available databases of protein sequences and functions, most useful for undertaking comparative studies on the functional roles of proteins.

4.2 Protein and amino acids

AAs provide the basic building blocks for protein synthesis and precursors for hormones, neurotransmitters, and metabolites such as polyamines, creatine, and phosphatidylserine. AAs also act as the major source of metabolic energy for fish. Understanding the routes and mechanisms of protein and AA metabolism is critical for optimal growth and homeostasis of animals.

4.2.1 Digestion and absorption of proteins

4.2.1.1 Protein digestion

The digestion of feed protein is a dynamic process of simultaneous release and absorption of small peptides and AAs. Protein is digested and broken into small peptides and free AAs by a variety of proteases and absorbed by intestinal epithelial cells through transporters across the brush border membrane. The processes of protein digestion and absorption in fish are influenced by many factors, including fish species, developmental stage, dietary protein level and composition, and feed processing technologies.

Protein digestion efficiency mainly relies on the type and function of digestive enzymes, which include pepsin, trypsin, elastase, chymotrypsin, aminopeptidase, carboxypeptidase, dipeptidase, tripeptidase, and so on. The digested peptides are then further digested into AAs by peptidases such as leucine aminopeptidase (LAP) and dipeptidyl aminopeptidase (DAP). These peptidases are secreted by the small intestine and have high activity in the brush border of intestinal villus (Tengjaroenkul et al., 2000).

Most of the digestive enzymes are secreted into the extracellular matrix in the form of zymogen. The conversion of zymogen to its active form requires a series of conformational changes and peptide bond cleavages, which remove the propeptide region and expose the active sites of the enzyme (Richter et al., 1998). The postactivated proteases are modulated by environmental factors and specific inhibitors. The change of environmental temperature directly affects protease activities and thus protein digestibility. The optimal pH of most intestinal proteases in fish is 6.5–9.5, which is the pH value of intestinal fluid after satiety (Hidalgo et al., 1999). There was no significant difference in the optimal pH of intestinal proteases among fish with different feeding habits and fish with or without stomach (El Kertaoui et al., 2019). However, the optimal pH of protease activities varies with the growth stage and the development of the digestive system. High-quality feed provides better stimulation toward protease activities in fish than low-quality feed (Krogdahl et al., 2015).

In terrestrial animals, intestinal microbiota plays an important role in host digestive tract function and metabolism. For fish, there is still much debate as regards the potential contribution of gut microbiota to the nutrition of the host organism (Ray et al., 2012). Some in vitro studies have shown the presence of proteolytic bacteria in the digestive tract of fish species such as tilapia, grass carp, and common carp (Bairagi et al., 2002). *Cetobacterium* and protease-producing bacteria are dominant in the intestinal microbiota of channel catfish (*Ictalurus punctatus*), largemouth bass (*Micropterus salmoides*), and bluegill (*Lepomis macrochirus*) (Larsen et al., 2014). Whether the functions of gut microbiota in vitro reflect what happens in vivo needs further investigations (Ray et al., 2012).

4.2.1.2 Amino acid and peptide transporters in digestive tract

The end products of digestive proteases are peptides and AAs, which have to be transported across the intestinal epithelia to the target tissues and organs. This process requires the assistance of AA and peptide transporters (Ferraris and Ahearn, 1984; Sire and Vernier, 1992). Amino acid transporter (AAT) is a class of proteins involved in the transmembrane transport of AAs. AAT not only participates in the transport of AAs but also regulates signal pathways and various cellular physiological and metabolic activities (Miguel-Aliaga, 2012). AAT can be divided into Na^+-dependent AAT system and Na^+-independent AAT system according to whether it depends on the electrochemical potential energy produced by Na^+ on the cell membrane. According to the properties of transported AAs, it can also be divided into neutral AAT system, acidic AAT system, basic AAT system, imino transfer system, and β-AAT system (Xu et al., 2016b).

AAT determines the absorption rate of AAs and affects the utilization of dietary proteins (Hundal and Taylor, 2009). Most of the essential neutral AAs are transported through B^0AT1 and ASCT2 in fish, while cationic Lys is mainly transported through the amino acid transporter B^{0+} ($B^{0+}AT$) (Poncet and Taylor, 2013). Neutral and cationic AATs, such as B0AT1, LAT2, TAT1, PAT1, and CAT2, are abundantly expressed in the whole intestine of turbot (*Scophthalmus maximus* L.) to facilitate maximum absorption of AAs in a short digestive tract

(Xu et al., 2016b). The expression of AAT is also regulated by dietary protein sources (Liu et al., 2014a; Xu et al., 2017; Zhang et al., 2019a).

Digested proteins are transported not only in the form of free AAs but also in the form of dipeptides or tripeptides through peptide transporters (Verri et al., 2010, 2011, 2017; Wang et al., 2017). In vertebrates, five peptide transporters have been identified, including PepT1, PepT2, PTR3, PHT1, and HPT1. Among them, PepT1 is the main and most studied peptide transporter in teleost fish (Ostaszewska et al., 2010a). It can transport more than 400 different dipeptides and 8000 tripeptides (Samsudin et al., 2016). The expression of PepT1 in the intestine of fish is reported to be influenced by diets (Ostaszewska et al., 2010b; Xu et al., 2017).

4.2.2 Free amino acid pools

There are two nitrogen pools in the organism, free AA pools and protein pools, which are intercommuted with protein synthesis and protein degradation (Sugden and Fuller, 1991). Compared with the AAs in protein pool, free AAs represent only 2%–3% of the total amount of AAs in the whole fish (Houlihan et al., 1995b). The input of the free AA pool includes digestion and absorption of ingested proteins, protein turnover, and synthesis of nonessential amino acids (NEAAs). The output includes protein synthesis, nonprotein molecule synthesis, and ammonia oxidation (Fraser et al., 2007). The changes of free AAs pools can be sensed by correlated signal transduction systems, which regulate the expression of downstream molecules, as well as protein synthesis and degradation, ultimately growth and development of animals (Gallinetti et al., 2013).

Postprandial AA kinetics is influenced by the digestion and absorption rates after ingestion. Generally, free AAs are absorbed and transported to various organs through circulation after digestion, with free AAs elevated (Nose, 1973; Plakas et al., 1980; Houlihan et al., 1995b), and dropped to the basal level in a short time (Kaushik and Luquet, 1977, 1979; Walton and Wilson, 1986; Lyndon et al., 1993).

In addition to protein synthesis, the major pathway of AA utilization in fish is deamination and oxidation to provide energy. It is recognized that most of the energy requirement of fish is met from protein and lipid (Cho and Kaushik, 1990). Brown et al. (1991) found that approximately 21% of the nitrogen in an essential amino acid (EAA) mixture was excreted as ammonia 24 h following infusion in catfish (Brown and Cameron, 1991).

Protein synthesis and degradation are major factors to regulate the homeostasis of free AAs. Different tissues have different levels of protein synthesis rates (Carter and Houlihan, 2001). The liver has the highest protein synthesis rate among all tissues, but the protein deposition rate is relatively low. Data showed that 57% of the newly synthesized proteins were degraded and returned to free AA pool, which are used for oxidation to provide energy or recycling for protein synthesis (Houlihan et al., 1995a). Muscle is the main site of protein deposition and the largest AA pool of the body. Protein degradation is the main means to maintain free AA homeostasis under starvation conditions.

4.2.3 Amino acid metabolism

After digestion and absorption, AAs enter different organs and tissues to supply for anabolic and catabolic metabolism. Although most of the AA metabolism processes of fish are similar to that of terrestrial animals, fish possess some unique AA metabolic characters.

As mentioned earlier, the limited utilization of carbohydrates requires more AAs for energy supply. Free AAs undergo transdeamination in liver. Kidney is the second most important organ for gluconeogenesis using AAs as substrate. Skeletal muscle is the largest reservoir of free AAs in fish. In addition, digestive tract is also found to be one of the most active organs for AA metabolism in fish (Brosnan, 2003; Breuillard et al., 2015).

The main anabolic pathway of AAs is to synthesize various tissue proteins, peptides, and other nitrogen-containing substances, such as Gln. Many substances with important biological functions such as enzymes, peptide hormones, and antibodies are all products of AAs. The catabolism of AAs is generally divided into three steps. The first step is deamination or transdeamination. Removed amino group is either converted to ammonia or converted to the amino group of Asp or Glu. The second step is the excretion of ammonia directly (in fish) or the formation of urea by combining with the nitrogen atom of aspartic acid (in mammals). The third step is to convert the carbon skeleton of the AA into a general metabolic intermediate into the TCA cycle or form other biological macromolecules. For example, Ala produces α-ketoglutarate through the action of Ala aminotransferase and aspartate through the action of aspartate aminotransferase. Ser and Gly can enter the citric acid cycle to produce acetyl CoA and ultimately glutamic acid. Important metabolites of AAs include ammonia, carbon dioxide, long-chain and short-chain fatty acids, glucose, hydrogen sulfide, ketone bodies, nitric oxide (NO), urea, uric acid, polyamines, and other nitrogenous substances with biological functions (Blachier et al., 2007; Morris, 2007; Rider et al., 2007; Sugita et al., 2007; Montanez et al., 2008).

AAs are important energy sources for fish. AAs regulate energy homeostasis in fish through AA catabolism or regulation of other energy metabolism pathways. Ala regulates gluconeogenesis and glycolysis by inhibiting alanine kinase (Meijer, 2003), and glutamate and aspartate mediate the transfer of reducing equivalents across the mitochondrial membrane, thereby regulating glycolysis and cells oxidation status (Brosnan, 2001). AMP-activated protein kinase (AMPK) is an evolutionarily conserved serine/threonine-protein kinase, which plays a crucial role in maintaining cellular energy homeostasis (Foretz and Viollet, 2011) and is activated when ATP is insufficiently supplied (Appuhamy et al., 2014). Activated AMPK inhibits energy-consuming processes such as protein synthesis, while enhancing ATP production processes (such as fatty acid oxidation and glucose uptake) (Kudchodkar et al., 2007). A deficiency in EAAs such as Met activates the AMPK activity of turbot (*Scophthalmus maximus* L.) muscle cells while promoting the TCA and oxidative phosphorylation (Jiang et al., 2017a).

AA metabolism is influenced by both endogenous and exogenous factors (Fig. 4.5). To meet normal growth and cope with various stresses, the dynamic balance of AA metabolism is precisely controlled. When fish are in a rapid growth stage, for example, the anabolism of AAs exceeds catabolism to ensure effective protein accretion (Ronnestad et al., 1999). For example, the requirement for Met and Lys in small-sized tilapia (*Oreochromis mossambicus*) was also significantly higher than that in large scale (Liu et al., 2013). In contrast, when fish are under stress conditions such as starvation, the catabolism of AAs should be kept at higher level than anabolism level to ensure an adequate supply of energy to cope with stress. In addition, hormones also play an important role in regulating AA metabolism. Hormones such as growth hormone/insulin/IGF can activate mechanistic target of rapamycin (mTOR) signaling to promote AA anabolism (Liu et al., 2019). Cortisol level is upregulated after exercise in rainbow trout (*Salmo gairdnerii* R.),

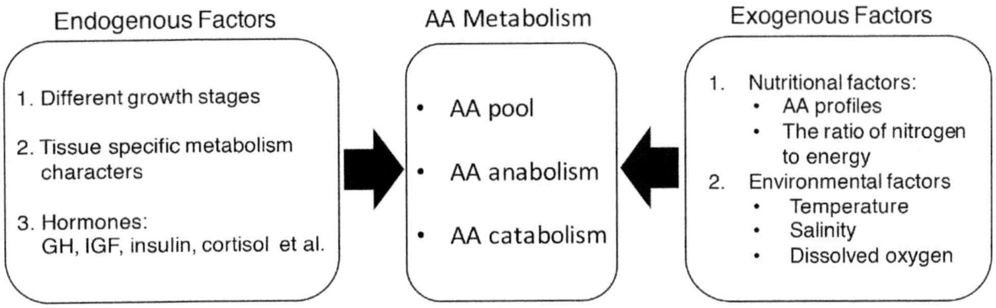

FIGURE 4.5 Endogenous and exogenous regulatory factors of amino acids metabolism.

and drugs inhibiting the increase of cortisol significantly change the AA content (Milligan, 1997). Among dietary factors, amino acid profile and the protein to energy ratio are known to affect amino acid utilization. At the target tissue level, all EAAs must be present in the correct proportions for cellular protein synthesis to proceed. An imbalance of amino acids in feed can significantly promote the oxidation of amino acids and reduce protein deposition (Aragao et al., 2004; Kurpad, 2018).

Environmental factors, including temperature, salinity, dissolved oxygen, and so on, play a significant role in the regulation of fish AA metabolism. Temperature stress induces protein degradation of the Asian sea bass (*Lates calcarifer*) and inhibits the feeding and protein synthesis and ultimately hinders growth of cobia (*Rachycentron canadum*) (Glencross and Bermudes, 2011; Sun and Chen, 2014). High salinity stress promoted AA catabolism in *Eleginops maclovinus* (Vargas-Chacoff et al., 2016). Low salinity stress inhibited the specific growth rate (SGR) of juvenile American shad (*Alosa sapidissima*) and European sea bass (*Dicentrarchus labrax*) (Rubio et al., 2005; Liu et al., 2017b). Hypoxia significantly inhibited the feeding behavior of turbot, channel catfish, and rainbow trout (Person-Le Ruyet et al., 2003; Abdel-Tawwab et al., 2019).

4.2.4 Protein turnover

4.2.4.1 Protein synthesis

Protein synthesis is regulated by intracellular signaling pathways that sense extracellular signals, such as growth factors, and AAs. At the cellular level, two major signaling pathways could sense the AA levels. One is the mTOR signaling system that can sense AA abundance. mTOR is a serine/threonine protein kinase that is highly conserved among eukaryotes. Associated with different binding proteins, it forms two complexes, mTORC1 and mTORC2. mTORC1 is sensitive to rapamycin and regulates protein synthesis through the activities of 4EBP1 and p70S6K. mTORC2 is insensitive to rapamycin and mainly involved in the regulation of cytoskeleton and cell morphology (Reiling and Sabatini, 2006). Dietary proteins, AA, and insulin can activate mTORC1 in rainbow trout (Seiliez et al., 2008a, 2011). The other is the AA response (AAR) signaling pathway with general regulation of repressed protein kinase 2 (GCN2) protein as the core, which can sense AA balance. GCN2 is an evolutionarily conserved serine/threonine kinase that senses AA deficiency (Hao et al., 2005; Maurin et al., 2005).

Under AA starved condition, uncharged tRNA activates GCN2, which phosphorylates eIF2α protein and significantly reduces the rate of initiation of general translation for protein biosynthesis by regulating translational factors such as ATF4, a transcription factor that promotes the expression of a range of genes (Kilberg et al., 2009).

Insulin/insulin-like growth factor (IGF) is the major hormone axis that regulates intracellular protein synthesis and metabolism in fish (Reinecke et al., 2005). Insulin promotes the uptake of AAs and the synthesis of proteins in cells and inhibits the breakdown of proteins and facilitating growth (Nelson and Sheridan, 2006). Insulin treatment increased the protein synthesis by 13% and reduced the protein degradation in the primary rainbow trout myocytes (Cleveland and Weber, 2010). In Atlantic salmon, IGF-I inhibited protein catabolism through MAFbx/atrogin-1 in muscle (Valente et al., 2012). In rainbow trout, refeeding leads to coordinated actions of growth hormone and IGF-I, which activates PI3K/Akt pathway and increases protein translation (Gabillard et al., 2006) (Fig. 4.6).

4.2.4.2 *Protein degradation*

The growth of animal depends on the balance between protein synthesis and protein degradation (Borsook and Dubnoff, 1951). Protein degradation plays crucial roles in processing damaged or misfolded proteins, unwanted components, posttranslationally modified proteins, and the degradation of proteins into AAs for intracellular use and maintenance of cells self-balance (Houlihan, 1991). The degradation of protein in fish mainly includes three pathways, ubiquitin–proteasome, calpain, and lysosome-related degradation pathways (Ciechanover et al., 1984; Goll et al., 2003; Mommsen, 2004; Cassidy et al., 2016). The contribution of each pathway to the protein breakdown depends on endogenous and exogenous stimuli and would work together to response to nutritional disturbances (Pasiakos and Carbone, 2014).

Ubiquitin–proteasome is a class of highly conserved pathway related to protein degradation, and their existence can be found from bacteria to mammals (Johnston et al., 2011). In the ubiquitin–proteasome system, damaged proteins or proteins with a short half-life will

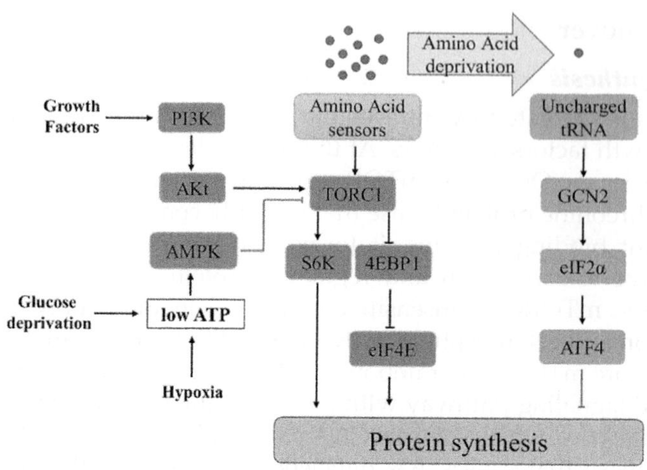

FIGURE 4.6 The availability of amino acids control protein synthesis through mTOR and GCN2 signaling pathways.

be modified by ubiquitination and degraded by the proteasome. Proteasome is a protein complex composed of 10–20 subunits, with a core of 20S proteasome and two regulatory 19S proteasomes. The 20S proteasome has a hollow structure and is a large catalytically active protease, consisting of two outer rings (7 α subunits) and two inner rings (7 β subunits) (Luo et al., 2010). Ubiquitin is a highly conserved polypeptide chain composed of 76 AAs. The target protein is covalently linked to several ubiquitin molecules under the action of ubiquitin-activating enzyme E1, ubiquitin-binding enzyme E2, and protein ubiquitin ligase E3 and then recognized and rapidly degraded by the 26S proteasome (Joazeiro and Weissman, 2000). Ubiquitin ligases MuRF1 and MAFbx (or Atrogin-1) and the activity of 20S proteasome are regarded as important markers for fish muscle atrophy and degradation (Cassidy et al., 2016; Song et al., 2016). In most vertebrates, the ubiquitin–proteasome system plays a major role in protein degradation (Lecker et al., 2004). However, the ubiquitin–proteasome system accounts for only a relatively small amount of total protein degradation in fish, especially only about 4% in muscle (Nemova et al., 2016).

Calpain is an intracellular Ca^{2+}-dependent cysteine protease and presents in all vertebrates. Typical calpain is a heterodimer consisting of a large catalytic subunit of 80 kD and a small regulatory subunit of 30 kD (Goll et al., 2003). Calpain has a wide range of substrates, including membrane proteins (such as receptors and transporters), cytoskeletal proteins, transcription factors (such as p53), and enzymes (such as phosphatase, caspases, PKC, and phospholipase C) (Laszlo and Yunchao, 2014). Multiple calpain genes are identified in fish species including zebrafish (Lepage and Bruce, 2008), catfish (Preziosa et al., 2013), rainbow trout (Salem et al., 2005b), and Atlantic halibut (Macqueen et al., 2010). The levels of calpain expression were influenced by nutritional conditions in Atlantic halibut (Macqueen et al., 2010) and rainbow trout (Salem et al., 2005b). Studies on the muscles of carp and rainbow trout found that the calcium concentration required for calpain activation is much higher than that of mammals (Toyohara and Makinodan, 1989; Salem et al., 2005a).

Lysosomes are vesicular organelles surrounded by monolayer and containing a variety of acidic hydrolases that break down various cellular and noncellular components. These enzymes include lipase, protease, nuclease, and sulfate esterase. The process in which excess or abnormal organelles in cells are transported to lysosomes for degradation is known as autophagy. Autophagy refers to the formation of autophagosomes by the inclusion of excessive or abnormal organelles, proteins, and part of cytoplasm in the cell, which will be fused with lysosomes and degraded (Wang and Klionsky, 2003). There are two mechanisms that induce autophagosome formation in animals in response to dietary factors and nutritional status. The first is a slower one that requires gene expression and is thought to produce autophagosomes after prolonged starvation (Mizushima et al., 2004; Sandri, 2010). The second is rapid and transient induction mediated by mTOR and AMPK complexes, which are key nutrients and energy sensors in eukaryotic cells, respectively (Kim et al., 2011). In recent years, more and more attention has been paid to the study of autophagy system in fish. Similar to mammals, the autophagy–lysosome degradation system accounts for about 40% of the body protein degradation in fish (Seiliez et al., 2014). Studies have shown that long-term starvation can increase the expression of autophagy-related genes such as gabarapl1, LC3B, atg4b, and atg12l in rainbow trout muscle (Seiliez et al., 2012). Apart from the autophagy pathways, a category of cysteine proteases called cathepsin are also involved in protein degradation in lysosomes. At least 10 cathepsins have been detected in fish (Nielsen and Nielsen, 2001).

4.2.4.3 Protein deposition

The relative balance between protein synthesis and protein degradation determines the rate of protein deposition. In fish, muscle is the main site for protein deposition (Houlihan, 1991). In mammals, the muscle growth mainly occurs through hypertrophy, i.e., the increase of size of muscle cells, after birth. In fish, skeletal muscle growth can be divided into two developmental stages. The first is stratified hyperplasia, forming a specific type of muscular layer (fast muscle and slow muscle). Stratified hyperplasia mainly occurs in the early stage of juveniles (Johnston, 2001). The second is mosaic hyperplasia: the number of muscle fibers in each muscular layer, especially in the white muscle layer, increases (Silva et al., 2009). Mosaic hyperplasia is the main form of increasing the number of rapid fibers in the juvenile and adult stages of most species (Weatherley et al., 1988).

Protein deposition rates vary with age and body mass. In rainbow trout, the protein turnover rate and protein retention efficiency (PRE) were found to be the highest in the early stages, and as fish grow, the protein accumulation and protein synthesis rates decrease (Peragón, 2001). Larval fish grow rapidly with protein accretion rates more than 30%/day (Conceição et al., 1997). Before sexual maturity, the nutrients absorbed by the fish are mainly used for protein deposition and promoting growth. After sexual maturity, most of the nutrients absorbed by the fish body are converted into gonadal development and overwintering material accumulation (Fernandes and McMeans, 2019).

Protein turnover rate varies among different tissues. In general, hepato-pancreatin protein turnover rate is the fastest, and muscle protein turnover rate is lower (Siekevitz, 1972). Temperature has a great influence on the protein metabolism of fish. Both low and high temperature will affect the protein synthesis rate. At low temperature, protein synthesis rate is low and proteasome activity is high, which ultimately resulted in lower PRE and lower growth efficiency (Lamarre et al., 2009). The short-headed fleecy blenny in South Pole decreased protein synthesis (Ks) and RNA/protein ratio in the process of thermal adaptation (Storch et al., 2005). The same results were obtained in studies of Atlantic cod (Treberg et al., 2005). The highest protein synthesis rate in white muscle of Asia sea bass is achieved only at optimum temperature (Katersky and Carter, 2010). In European crucian carp, the amounts of protein synthesis in heart, muscle, liver, and brain were significantly reduced under low oxygen and reduced overall metabolism (Smith et al., 1996). A study on the Amazonian cichlid (*Astronotus ocellatus*) also found that under the condition of severe hypoxia, Ks of this fish was significantly reduced, while Kd was not detected to have significant changes (Cassidy et al., 2018b). Protein deposition decreases under high or low salinity conditions in larva of giant freshwater prawn (*Macrobrachium rosenbergii*) (Intanai et al., 2009). High stocking density has negative effects on the growth of fish. The ubiquitin–proteasome system and autophagy system were activated, protein degradation was enhanced, and growth was inhibited under long-term crowding stress in flatfish (Valenzuela et al., 2018).

The nutrients in food directly or indirectly regulate muscle protein accumulation by changing protein synthesis and degradation (Song et al., 2016). The protein deposition rate increases as the food intake increases in greenback flounder (*Rhombosolea tapirine*) (Carter and Bransden, 2001) and carp (Meyer-Burgdorff and Rosenow, 1995). With the increase of feeding rate and frequency, the concentration of free AAs in the plasma substrate is used for protein synthesis increases (Cleveland and Burr, 2011). With the increase of AA concentration, the phosphorylation levels of S6K1 and 4EBP1 are gradually increased, and protein synthesis is enhanced (Fox et al., 1998; Hara et al., 1998; Kimball et al., 1998; Wang et al., 1998; Shigemitsu et al., 1999;

Krause et al., 2002). Under natural conditions, many fish species might go through a period of long-term lack of food or low food intake, during which time the overall metabolic rate of the fish decreases. On the other hand, lack of nutrients rapidly inhibits mTOR signaling and stimulates the overall protein breakdown by regulating the ubiquitin–proteasome and autophagy (Rousseau and Bertolotti, 2016; Zhao and Goldberg, 2016). In the stage of long-term starvation, the activity and content of related enzymes in the calpain and ubiquitination pathways increased in rainbow trout (Salem et al., 2006, 2007; Seiliez et al., 2008b) and Arctic charr (Cassidy et al., 2018a). After refeeding, the rate of protein synthesis increases rapidly, and the rate of protein degradation decreases (Lyndon et al., 1992; Carter et al., 1993; Robertson et al., 2001).

The AA composition in feed affects the composition of the tissue-free AA pool and thus affects the protein deposition rate in fish. Lack of Met in the feed will reduce the activation of the translation initiation factor ribosomal protein S6 and increase the phosphorylation of eIF2α and the transcription factor 4 (ATF4) and its target gene asparagine synthase (ASNS) to activate the GCN2 pathway (Belghit et al., 2014; Wang et al., 2016b). On the other hand, the addition of AAs such as Leu and Arg in the feed can significantly increase the protein deposition rate in gibel carp (*Carassius auratus gibelio*) (Tu et al., 2015b) and rainbow trout (Lansard et al., 2011). The relationship between protein and energy intake is critical for the efficiency with which dietary protein can be distributed into growth. In rainbow trout, increasing carbohydrate ratio in feed resulted in decreased phosphorylation of S6 in liver and decreased activation of AKT, S6K, 4EBP1, and S6 in muscle after feeding, resulting in reduced protein synthesis (Skiba-Cassy et al., 2013).

4.2.5 Methodologies for protein metabolism analyses

During the past decade, our understanding of protein metabolism in fish has been vastly advanced because of combination of methodologies, ranging from physiological characterization of phenotypes using traditional nutritional research methods such as indirect calorimetry, nutrient balance, and body composition assessment (Pinto et al., 2007; Kou et al., 2015) to isotopic tracer methods (Lokman and Symonds, 2014; Lamarre et al., 2015; Rise et al., 2019), and -omics approaches at multiple spatial, temporal, and dimensionality scales (Mohanty et al., 2019), which make it possible to research the important regulatory factors of protein metabolism in aquatic organisms and promote the solution of a variety of problems in aquaculture, including nutrition, health, quality, safety, and animal welfare (Petro-Sakuma et al., 2020).

4.2.5.1 *Stable isotope labeling techniques*

Stable isotopes are atoms with the same number of protons and different numbers of neutrons, which enable them to be used as tracers to distinguish between endogenous and exogenous nutrient and to track the metabolic kinetics of exogenous nutrients. In recent years, stable isotope-labeled AAs also has a lot of applications in protein metabolism measurements of fish, crustaceans, mollusks, and other aquatic organisms (Hawkins, 1985; Carter and Mente, 2014). Commonly used stable isotopes include ^{15}N, ^{13}C, ^{2}H, and ^{18}O (Gasier et al., 2010). In general, there are three methods for determining the rate of protein synthesis, including (1) constant infusion, (2) flooding dose, (3) stochastic endpoint.

The method of constant infusion is to slowly and continuously inject stable isotope-labeled AAs into the plasma of fish under anesthesia through veins, usually at a dosage of 5%–10%

of the tracer concentration in plasma and for about 4–16 h (Fraser and Rogers, 2007). Constant infusion method requires long-term puncture and catheterization of animals and keeps them motionless (Haschemeyer and Smith, 1979).

The flooding dose method is to inject organisms with a very large dose of AA labeled with stable isotopes and without labeling. The dosage of labeled AA should be several times of the content of the AA in the plasma-free AA pool and ensure that the concentration of the AA in the free AA pool tends to be consistent in a short time. Fish can be injected intraperitoneally or intravenously (Fraser and Rogers, 2007). The flooding dose method can only measure the protein anabolism in a short time, and it may lead to the change of protein synthesis rate in a short time because a large number of AAs enter the blood and affect the accuracy of results. It is widely used in the measurement of protein metabolic rate in fish (Gasier et al., 2009; Robinson et al., 2011; Felip et al., 2012, 2013) and crustaceans (Gamboa-Delgado and Le Vay, 2009; Gamboa-Delgado et al., 2011, 2013).

The stochastic endpoint method usually measures labeled nitrogen flux rates and protein synthesis rates. Animals are fed AAs or proteins with stable isotopes and then measure the content of oxidized labeled AA in the associated excretory pool. When AAs enter the body, they are either oxidized out of the body or incorporated into proteins. Thus, when ^{15}N-labeled AAs are used as tracers, the amount of AAs used for incorporation into protein synthesis can be calculated by the following formula (Waterlow et al., 1978):

$$e^{*}d^{*-1} = E_xQ^{-1} = E_x(Z + E_{TN})^{-1}$$

where e^* is the cumulative excretion of ^{15}N-ammonia after time t, d^* is the total amount of ^{15}N-AAs or proteins fed to animals, E_x is the nitrogen in total ammonia excreta after time t, Q is the nitrogen flux, Z is the nitrogen used for protein synthesis, and E_{TN} is the total nitrogen excretion after time t, including ammonia and urea.

At present, the stochastic endpoint method to determine the protein synthesis rate has been widely used in aquatic animals (Hawkins, 1985; Wang et al., 2019a). Without sacrificing the experimental animals, the protein synthesis rate of the same individual can be determined repeatedly. However, the random endpoint method can only determine the protein synthesis rate of the whole animal, unlike the two methods mentioned earlier, which can determine the protein synthesis rate of different tissues.

4.2.5.2 Omics approaches

Omics technologies have been widely used recently to study the systematic changes of protein metabolism (Prieto-Álamo et al., 2012; Mohanty et al., 2019; Rise et al., 2019). Transcriptome analysis has become a powerful tool for research and has been widely used in the study of fish biology (Hook, 2010). Through the microarray analysis on rainbow trout, it found that many functional genes related to protein turnover increased in the process of regaining muscle growth after refeeding, and the expression of genes related to protein synthesis in fish liver was downregulated during starvation, but the expression of genes related to protein degradation was not significantly changed (Panserat and Kaushik, 2010). Proteomics provides a new approach to study the mechanism of protein metabolism and regulation (Aebersold and Mann, 2016). Previous studies have used shotgun proteomic approach to identify the proteome profile of head kidney and spleen in rainbow trout, and the parts related to protein metabolism can provide a basis for subsequent studies (Kumar et al., 2018). Studies have compared the skeletal muscles of growth hormone transgenesis and growth-selected strains of salmon with their wild-type counterparts, which had striking proteomic differences, and researchers found that many unique molecular mechanisms are the basis of growth-enhanced phenotypes (Causey et al., 2019). Traditional proteomics methods provide a "snapshot" of the proteome but may cover up differences in the rates of synthesis and/or degradation of specific proteins that implicated physiological significance (Doherty et al., 2009). In recent years, a new method for calculating the turnover rate of a single protein using stable isotope labeling and mass spectrometry has been developed (Doherty et al., 2005; Price et al., 2010). The feasibility of this approach has been demonstrated through proteomics-based strategies to study protein dynamics in carp tissues and through the synthesis and degradation of isoforms of proteins (Doherty et al., 2012).

Metabolomics, newly developed after genomics and proteomics, refers to the multiparameter quantitative analysis of the dynamic metabolic response of biological systems to pathological or physiological stimuli or genetic changes (Nicholson et al., 1999). Compared with nutrigenomics, nutritional metabolomics has many advantages (Dunn and Ellis, 2005; Rezzi et al., 2007). The number of small-molecule metabolites in a cell is much fewer than the number of genes and proteins; therefore the research complexity is relatively reduced, and the experiment requires less time and cost. In addition, small changes in enzyme concentration or activity in an organism can significantly change the final metabolites of biochemical reactions, so the study of metabolomics is easier and more direct. At present, AA metabolism can be studied through the research of nutritional metabolites, and then the effects of nutrients on human and animal physiological functions can be studied at the molecular level (Elliott and Ong, 2002). Nuclear magnetic resonance (NMR) metabolomics was used to analyze the effect of taurine on the intestinal metabolism of groupers. The possible mechanism to accelerate the growth of fish is through improving energy utilization and AA intake, and promoting the synthesis of proteins, lipids, and purines (Shen et al., 2019). In addition, studies on the red drum (*Sciaenops ocellatus*) using NMR metabolomics have found that when fish are fed a soybean-based diet, the concentrations of Gly, Ser, and Thr metabolites in the liver and muscle are increased, which is related to the catabolism pathway of proteins (Casu et al., 2017).

4.3 Protein requirement

Fish and shrimp, like other land animals, do not have a requirement for proteins per se but rather have a requirement for a well-balanced mixture of essential (or indispensable) and nonessential (or dispensable) amino acids obtained from the diet by digestion of proteins (Wilson and Halver, 1986). Compared to land animal feeds, fish and shrimp feeds contain relatively higher dietary protein levels (NRC, 2011). When dietary protein level is too low, fish or shrimp cannot sustain their normal physiological functions, with reduced growth and health. Excess dietary protein intake does not necessarily lead to more growth but will lead to increased nitrogen losses and even decreased growth (Fig. 4.7). It is important to supply adequate levels of dietary proteins with the correct amino acid profiles in order to maintain normal growth and good health and reduce nitrogenous losses.

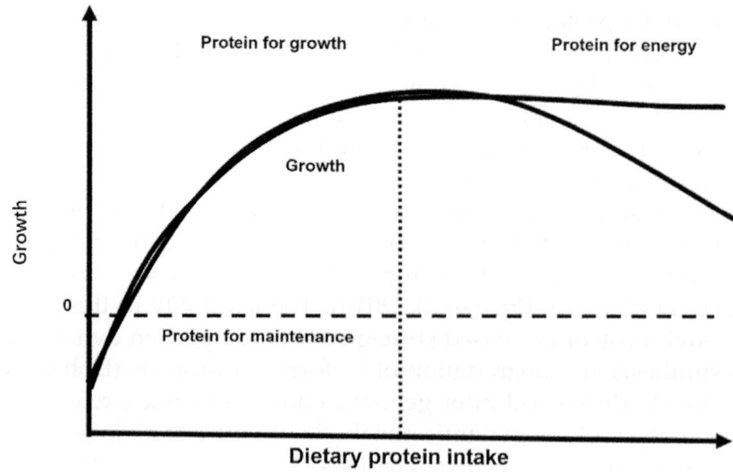

FIGURE 4.7 The relationship between fish growth rate and dietary protein levels (Shearer, 2000).

4.3.1 Methodological considerations

The earliest studies on protein requirements of fish were undertaken with Pacific salmon in the late 1950s. A commonly used method for the determination of protein requirements of fish or shrimp is to provide diets with graded levels of dietary proteins and to measure the growth response of the animals. The dose–response relationship obtained enables to estimate the protein requirement. In such studies based on a single factor, gradient concentrations, at least 6 dietary levels must be used to obtain a good response curve (Cowey, 1992). Essential nutrient levels in experimental diets must meet or exceed the nutritional requirements of the test animals. The methods used for the estimation of protein requirements of fish vary widely including different protein sources, feeding levels and regime, water temperature, trial duration (Oliva-Teles et al., 2019). The dietary protein range and levels used for the trial could also affect the estimation of the requirement (Rodehutscord and Pack, 1999; Shearer, 2000). The most widely used response criterion for the determination of protein requirement is growth, i.e., weight gain (WG) or specific growth rate (SGR). Feed utilization including feed efficiency (FE) or feed conversion ratio (FCR) and protein utilization (protein retention efficiency (PRE), or protein productive value (PPV)) are also used as response criteria. Protein or nitrogen gain is a more robust and rational criterion rather than overall gain in body mass (NRC, 2011).

Data on protein requirements are commonly expressed as dietary protein concentration (% of diet) though it is recommended to express them on a per unit body weight per day or as unit protein per unit energy contents (Cowey, 1995; Kaushik and Seiliez, 2010; Oliva-Teles et al., 2019).

Different statistical methods have been used for the estimation of protein requirements based on data from dose–response studies in fish (Cowey, 1995; Shearer, 2000; Hernandez-Llamas, 2009; Oliva Teles et al., 2019).

(1) Multiple comparisons tests following an Analysis of Variance (ANOVA) consider all data as noncontinuous values while the dietary protein levels should be treated as a continuous variable (Oliva-Teles et al., 2019). This method is sometimes not really valid for dose–response analyses.
(2) Broken line analysis. The "broken line" model implies that there is a linear response to dietary protein levels up to a certain level, after which there is a change in response which can be of three kinds: a) the response does not change with increasing doses, b) there is only a slight positive response, and c) the response shows a decline with increasing dietary levels. "Broken line" analysis is considered to be more suitable for the evaluation of responses of individuals rather than population (Shearer, 2000; Pesti et al., 2009).
(3) Polynomial regression analysis is used when the real dose–response is not linear but bears little biological significance.
(4) Multi-parameter saturation kinetics model (Mercer et al., 1989; Mercer, 1992), which is widely used in livestock and poultry has also been used in fish and shrimp (Richard et al., 2010).

Given the differences in species and the methodologies applied, it is difficult to compare results of protein requirement from different studies (Shearer, 2000; Ye et al., 2015, 2017).

4.3.2 Data on protein requirements of fish and crustaceans

Available data on dietary protein requirements relating to the levels which can achieve the maximum growth are reported in Tables 4.2–4.4. Most data are available for juveniles and on-growing fishes while the reports are very limited for larvae or brood stock. The reported requirements vary from 24% to 62% and even as high as 70%. Protein requirements range for larval fishes ranged from 25% to 60% while 20% to 60% for on-growing ones.

When the data on protein requirements are expressed in terms of unit protein intake per growth, there are no big differences between species (Tacon and Cowey, 1985). Ye et al. (2015, 2017) compared the protein requirements of juveniles and on-growing gibel carp reared under similar experimental conditions. When calculated as g protein required per kg fish per day, the requirements for juveniles and on-growing were respectively 12.2 and 6.5 g/kg BW/d, while the specific growth rates (SGR) were 3.44% and 1.72%/d. The protein requirements expressed per unit WG are 3.5 and 3.8 g, respectively.

Analyzing literature data on protein requirements of fish, Oliva-Teles et al. (2019) calculated the protein intake using as a function of thermal growth coefficient (TGC), which is considered to be a better descriptor of growth rate of poikilothermic animals (Cho, 1992), and found the good correlation between estimated protein intake and observed protein intake.

Fish are considered to require higher protein in their diets than other land warm-blooded animals. On the other hand, fish show higher protein utilization than other animals, as they spend less energy for the maintenance of body temperature, keeping position, and excretion of nitrogen. Protein requirements per unit weight, protein retention, and protein growth efficiency are considered to be similar to other monogastric animals (Table 4.5) (Bowen, 1987; NRC, 2011; Oliva Teles et al., 2019).

TABLE 4.2 Protein requirements of seawater fishes (modified from Oliva-Teles et al., 2019).

Species	Latin name	Trophic level	Dietary protein source	Initial body weight (g)	Protein requirement (%)	Protein requirement (g/kgBW/d)	Adjustment model	Reference
Atlantic halibut	*Hippoglossus hippoglossus*	4	FM-based	559	41	2.14	ANOVA	Cited from Oliva-Teles et al. (2019)
		4	FM-based	980	35	1.65	ANOVA	Cited from Oliva-Teles et al. (2019)
Black sea bass	*Centropristis striata*	3.9	FM-based	6.7	45.3	15.13	Broken line	Cited from Oliva-Teles et al. (2019)
Black sea bream	*Acanthopagrus schlegelii*	3.2	FM-based	13.1	41.4	10.88	Quadratic	Cited from Oliva-Teles et al. (2019)
Brazilian sardine	*Sardinella brasiliensis*	3.1	FM-casein-gelatin	0.93	36.7	25.36	Quadratic	Sterzelecki (2018)
Striped catfish	*Pangasianodon hypophthalmus*	3.1	FM-based	0.2	25	9.99	ANOVA	Cited from Oliva-Teles et al. (2019)
Chu's croaker	*Nibea coibor*	3.5	FM−casein	75.9	45.6	11.30	Quadratic	Cited from Oliva-Teles et al. (2019)
Cobia	*Rachycentron canadum*	4	FM−casein	32.5	44.5	15.09	Quadratic	Cited from Oliva-Teles et al. (2019)
Coho salmon	*Oncorhynchus kisutch*	4.2		14.7	40	2.11	ANOVA	Cited from Oliva-Teles et al. (2019)
European seabass	*Dicentrarchus labrax*	3.5	FM-based	99	43.9	4.89	ANOVA	Cited from Oliva-Teles et al. (2019)
		3.5	FM-based	160	43.9	2.53	ANOVA	Cited from Oliva-Teles et al. (2019)
Giant croaker	*Argyrosomus japonicus*	4.5	FM-based	1.9	45	15.06	ANOVA	Cited from Oliva-Teles et al. (2019)
			FM−SBM−yeast	44.4	48	10.03	ANOVA	Cited from Oliva-Teles et al. (2019)
Giant croaker	*Nibea japonica*	4.5	FM-SBM-brewer's yeast-casein	44.4	48	11.45	ANOVA	Cited from Oliva-Teles et al. (2019)
Giant grouper	*Epinephelus lanceolatus*	4	FM-casein	5.37	56	14.13	Quadratic	Gao et al. (2019)
Gilthead sea bream	*Sparus aurata*	3.7	FM-based	0.8	55	19.38	ANOVA	Cited from Oliva-Teles et al. (2019)
			Casein	2.6	40	12.82	No model or ANOVA	Cited from Oliva-Teles et al. (2019)

Golden pompano	*Trachinotus ovatus*	3.7	FM–SBM–RSM–yeast	4.7	45	15.32	ANOVA	Cited from Oliva-Teles et al. (2019)
Grey mullet	*Chelon ramada*	2.3	FM–casein	2.5	24	7.28	ANOVA	Cited from Oliva-Teles et al. (2019)
Grouper	*Epinephelus coioides*	4	FM-based	10.7	48	13.28	Broken line	Cited from Oliva-Teles et al. (2019)
			FM-Casein	274.17	43.84	7.43	Broken line	Yan et al. (2019)
Grouper	*Epinephelus malabaricus*	4.1	FM-based	9.2	50.2	10.61	Broken line	Cited from Oliva-Teles et al. (2019)
			FM-based	63.5	50	7.61	Quadratic	Cited from Oliva-Teles et al. (2019)
			Casein	3.8	47.8	11.10	ANOVA	Cited from Oliva-Teles et al. (2019)
Haddock	*Melanogrammus aeglefinus*	4	FM–casein–krill	6.9	45	8.77	ANOVA	Cited from Oliva-Teles et al. (2019)
Hybrid grouper	*Epinephelus fuscoguttatus* ♀ × *E. lanceolatus* ♂	4.1	FM-casein	3.58	53.5	15.95	Quadratic	Jiang et al. (2016e)
Korean rockfish	*Sebastes schlegeli*	3.8	FM–casein	7.3	48.6	14.04	Broken line	Cited from Oliva-Teles et al. (2019)
Leopard coral grouper	*Platichthys stellatus*	4.4	FM-wheat gluten-soybean protein concentrate	37.4	50	7.19	ANOVA	Xia et al. (2019)
Mangrove red snapper	*Lutjanus argentimaculatus*	3.6	FM-based	8	42.8	2.22	Broken line	Cited from Oliva-Teles et al. (2019)
			FM–SBM–shrimp meal	12.3	43	5.57	Broken line	Cited from Oliva-Teles et al. (2019)
Marbled spinefoot rabbitfish	*Siganus rivulatus*	2	FM-based	1.5	40	36.62	ANOVA	Cited from Oliva-Teles et al. (2019)
Milkfish	*Chanos chanos*	2.4	FM–casein–groundnut oilcake	0.3	40	6.51	ANOVA	Cited from Oliva-Teles et al. (2019)
			FM–casein–groundnut oilcake	4.4	40	10.65	ANOVA	Cited from Oliva-Teles et al. (2019)
Mullet	*Mugil liza*	2	FM–casein–gelatin	1.2	35	29.26	ANOVA	Cited from Oliva-Teles et al. (2019)
Olive flounder	*Paralichthys olivaceus*	3.7	FM-based	4.1	46	9.91	Broken line	Cited from Oliva-Teles et al. (2019)

Continued

TABLE 4.2 Protein requirements of seawater fishes (modified from Oliva-Teles et al., 2019).—cont'd

Species	Latin name	Trophic level	Dietary protein source	Initial body weight (g)	Protein requirement (%)	Protein requirement (g/kgBW/d)	Adjustment model	Reference
			FM-based	13.3	44	10.83	Broken line	Cited from Oliva-Teles et al. (2019)
			FM-based	5.9	50	16.83	ANOVA	Cited from Oliva-Teles et al. (2019)
			FM-based	8.1	35	8.66	ANOVA	Cited from Oliva-Teles et al. (2019)
			FM-based	8.1	45	10.49	ANOVA	Cited from Oliva-Teles et al. (2019)
			FM-based	13.3	44	10.83	Broken line	Cited from Oliva-Teles et al. (2019)
			FM-based	22.7	45	10.15	ANOVA	Cited from Oliva-Teles et al. (2019)
			FM-based	4.1	51.2	10.98	Broken line	Cited from Oliva-Teles et al. (2019)
			FM−casein	4.1	57.7	13.50	Quadratic	Cited from Oliva-Teles et al. (2019)
Pacific threadfin	*Polydactylus sexfilis*	3.4	FM−casein−gelatin	3.4	41	18.22	Quadratic	Cited from Oliva-Teles et al. (2019)
Plaice	*Pleuronectes platessa*	3.2	FM-based	14.2	50		Quadratic	Cited from Oliva-Teles et al. (2019)
			Casein	2.2	70	9.36	No model or ANOVA	Cited from Oliva-Teles et al. (2019)
Rainbow trout	*Oncorhynchus mykiss*	4.1	Casein−gelatin	6.3	45	7.51	Broken line	Cited from Oliva-Teles et al. (2019)
Red porgy	*Pagrus pagrus*	3.9	FM-based	2.8	50.8	12.40	Broken line	Cited from Oliva-Teles et al. (2019)
Red spotted grouper	*Epinephelus akaara*	4	FM-based	7.9	50.8	14.19	Quadratic	Cited from Oliva-Teles et al. (2019)
Redlip mullet	*Planiliza haematocheila*	2.5	Casein	1.2	40	16.17	ANOVA	Cited from Oliva-Teles et al. (2019)
Senegalese sole	*Solea senegalensis*	3.3	FM−wheat gluten	11.9	53	5.99	ANOVA	Cited from Oliva-Teles et al. (2019)
Sharpsnout sea bream	*Diplodus puntazzo*	3.2	FM-based	49.3	42.9	6.39	Quadratic	Cited from Oliva-Teles et al. (2019)

		3.2	FM-based	49.3	42.9	6.24	Quadratic	Cited from Oliva-Teles et al. (2019)
Shi drum	*Umbrina cirrosa*	3.4	FM-based	86.3	47	6.48	ANOVA	Cited from Oliva-Teles et al. (2019)
Silver pomfret	*Pampus argenteus*	3.3	FM-based	12.5	49	15.77	Quadratic	Cited from Oliva-Teles et al. (2019)
Southern flounder	*Paralichthys lethostigma*	3.5	FM–casein–gelatin	32.9	51.52	16.64	Broken line	Cited from Oliva-Teles et al. (2019)
Spotted sand bass	*Paralabrax maculatofasciatus*	4.2	Casein	2.4	55	18.10	ANOVA	Cited from Oliva-Teles et al. (2019)
Starry flounder	*Planiliza haematocheila*	3.6	FM-based	90	50	6.83	ANOVA	Cited from Oliva-Teles et al. (2019)
Sub-antarctic notothenioid	*Eleginops maclovinus*	3.5	FM-based	40	34.7	3.99	Exponential	Cited from Oliva-Teles et al. (2019)
Summer flounder	*Paralichthys dentatus*	4.5	FM-based	11.7	56	10.06	ANOVA	Cited from Oliva-Teles et al. (2019)
Tiger puffer	*Takifugu rubripes*	3.6	Casein	17	41	12.57	ANOVA	Cited from Oliva-Teles et al. (2019)
Turbot	*Scophthalmus maximus*	4.4	FM-based	38.2	57	5.18	Broken line	Cited from Oliva-Teles et al. (2019)
			FM-based	89	49.4	5.65	Broken line	Cited from Oliva-Teles et al. (2019)
			FM-based	34.5	50	7.28	ANOVA	Cited from Oliva-Teles et al. (2019)
Two-banded seabream	*Diplodus vulgaris*	3.5	FM-based	6.1	35.7	7.38	Quadratic	Cited from Oliva-Teles et al. (2019)
White sea bass	*Atractoscion nobilis*	4.3	FM–SBM–soy protein concentrate–corn gluten	3.6	43	12.24	ANOVA	Cited from Oliva-Teles et al. (2019)
White sea bream	*Diplodus sargus*	3.4	FM-based	22	27	5.36	Exponential	Cited from Oliva-Teles et al. (2019)
		3.4	FM-based	1.5	40	13.12	ANOVA	Cited from Oliva-Teles et al. (2019)
Yellow puffer	*Takifugu obscurus*	3.4	FM-based	18.5	50	11.46	Broken line	Cited from Oliva-Teles et al. (2019)
Zebra sea bream	*Diplodus cervinus*	3	FM-based	7.7	43.8	6.15	Curvilinear-plateau model	Cited from Oliva-Teles et al. (2019)

**FM*, fishmeal; *RSM*, rapeseed meal; *SBM*, soybean meal. Trophic level are from: https://www.fishbase.se/search.php.

TABLE 4.3 Protein requirements of freshwater fishes (modified from Oliva-Teles et al., 2019).

Species	Latin name	Trophic level	Dietary protein source	Initial body weight (g)	Protein requirement (%)	Protein requirement (g/kgBW/d)	Adjustment model	Reference
African catfish	*Clarias gariepinus*	3.8	FM-based	39.7	30	8.71	ANOVA	Cited from Oliva-Teles et al. (2019)
			Casein–gelatin	8.2	39.2	15.81	Quadratic	Cited from Oliva-Teles et al. (2019)
African cichlid	*Pseudotropheus socolofi*	2.7	FM-based	0.1	40	27.37	Broken line	Cited from Oliva-Teles et al. (2019)
Air-breathing catfish	*Heteropneustes fossilis*	3.6	Casein	0.8	35.4	10.82	ANOVA	Cited from Oliva-Teles et al. (2019)
			Casein–gelatin	10	40	9.29	Quadratic	Cited from Oliva-Teles et al. (2019)
American eel	*Anguilla rostrata*	3.8	FM-based	8.1	47	6.12	ANOVA	Cited from Oliva-Teles et al. (2019)
Arctic charr	*Salvelinus alpinus*	4.4	FM-based	2.6	33	6.44	Broken line	Cited from Oliva-Teles et al. (2019)
Asian red-tailed catfish	*Hemibagrus wyckioides*	3.7	FM-based	1.9	44.1	10.05	Broken line	Cited from Oliva-Teles et al. (2019)
Bagrid catfish	*Horabagrus brachysoma*	3.1	SBM–groundnut	2.3	39.1	12.93	Quadratic	Cited from Oliva-Teles et al. (2019)
Bagrid catfish	*Mystus nemurus*	3.6	FM-casein-gelatin	7.6	44	10.44	Broken line	Cited from Oliva-Teles et al. (2019)
Barbless carp	*Cyprinus pellegrini*	2.7	FM-based	12.4	39.3	6.60	Broken line	Cited from Oliva-Teles et al. (2019)
Barred knifejaw	*Oplegnathus fasciatus*	3.6	FM-based	7.1	45.2	9.37	Broken line	Cited from Oliva-Teles et al. (2019)
Beluga	*Huso huso*	4.4	FM-based	1.3	38.9	11.82	Broken line	Cited from Oliva-Teles et al. (2019)

Bighead carp	*Aristichthys nobilis*	4.3	FM-based	0	30.1	34.40	Quadratic	Cited from Oliva-Teles et al. (2019)
Black bufalo	*Ictiobus niger*	3.4	FM-corn gluten meal-shrimp meal	151	41	4.83	ANOVA	Guy et al. (2018)
Blackbelt cichlid	*Vieja melanura*	2.6	FM-based	0.3	40.8	15.27	Broken line	Cited from Oliva-Teles et al. (2019)
Bluegill sunfish	*Lepomis macrochirus*	3.2	FM-based	1.2	42.3	10.79	Broken line	Cited from Oliva-Teles et al. (2019)
Blunthead bream	*Megalobrama amblycephala*	3.4	Casein	8.1	32.6	8.17	Broken line	Cited from Oliva-Teles et al. (2019)
			FM–casein	16.1	34	7.45	ANOVA	Cited from Oliva-Teles et al. (2019)
Brook trout	*Salvelinus fontinalis*	3.3	FM-based	28.4	50	6.16	Broken line	Cited from Oliva-Teles et al. (2019)
			FM-based	29.1	48	6.01	Broken line	Cited from Oliva-Teles et al. (2019)
Brown trout	*Salmo trutta*	3.4	Casein/SFPC	1.2	53	0.00	ANOVA	Cited from Oliva-Teles et al. (2019)
Cachara catfish	*Pseudoplatystoma reticulatum*	4.2	FM-based	16.1	49.2	12.88	Quadratic	Cited from Oliva-Teles et al. (2019)
Channel catfish	*Ictalurus punctatus*	4.2	Egg powder	200	40	9.80	ANOVA	Cited from Oliva-Teles et al. (2019)
Chinese sucker	*Myxocyprinus asiaticus*	3.1	FM-based	13.5	46.5	8.89	Broken line	Cited from Oliva-Teles et al. (2019)
Common carp	*Cyprinus carpio* var. Jian	3.1	FM-based	16.7	34.1	8.74	Broken line	Cited from Oliva-Teles et al. (2019)
Crucian carp	*Carassius auratus*	2	FM-based	0.2	29.9	10.01	ANOVA	Cited from Oliva-Teles et al. (2019)
Discus fish	*Symphysodona equifasciata*	3.3	FM-based	4.5	50.1	17.44	Quadratic	Cited from Oliva-Teles et al. (2019)

Continued

TABLE 4.3 Protein requirements of freshwater fishes (modified from Oliva-Teles et al., 2019).—cont'd

Species	Latin name	Trophic level	Dietary protein source	Initial body weight (g)	Protein requirement (%)	Protein requirement (g/kgBW/d)	Adjustment model	Reference
Fresh-water catfish	*Hemibagrus nemurus*	3.6	FM-based	7.6	44	10.44	Broken line	Cited from Oliva-Teles et al. (2019)
			FM-based	25.4	42	5.95	ANOVA	Cited from Oliva-Teles et al. (2019)
Freshwater murrel	*Channa punctatus*	3.8	Casein–gelatin	4.6	46.2	11.43	Quadratic	Cited from Oliva-Teles et al. (2019)
Fringe-lipped carp	*Labeo fimbriatus*	2	FM–SBM–groundnut cake	4.8	30.1	10.34	Quadratic	Cited from Oliva-Teles et al. (2019)
Fuxian minnow	*Anabarilius grahami*	3.2	FM-based	4.1	38.6	4.74	Broken line	Cited from Oliva-Teles et al. (2019)
Gibel carp	*Carassius auratus gibelio* var. CAS III	2.5	SPC–casein	3.2	40.2	13.55	Broken line /quadratic	Cited from Oliva-Teles et al. (2019)
			SPC–casein	86.7	33.7	4.63	Broken line /quadratic	Cited from Oliva-Teles et al. (2019)
			FM–casein	3.7	41.4	11.41	Quadratic	Cited from Oliva-Teles et al. (2019)
			FM–casein	84.8	36.5	7.01	Quadratic	Cited from Oliva-Teles et al. (2019)
Golden shiners	*Notemigonus crysoleucas*	2.7	FM-based	0.2	28.9	11.72	ANOVA	Cited from Oliva-Teles et al. (2019)
Golden-line barbell	*Sinocyclocheilus grahami*	2.9	FM-based	7.6	38.6	6.49	Broken line	Cited from Oliva-Teles et al. (2019)
Grass carp	*Ctenopharyngodon idellus*	2	Casein–gelatin	4.3	40	8.99	ANOVA	Cited from Oliva-Teles et al. (2019)
			FM–casein–gelatin	264	28.6	6.86	ANOVA	Cited from Oliva-Teles et al. (2019)
Holland's spini barbell	*Spinibarbus hollandi*	3.2	FM-based	8.5	32.7	6.13	Broken line	Cited from Oliva-Teles et al. (2019)
Hybrid snakehead	*Channa maculata* ♀ × *Channa argus* ♂	3.7	FM-SBM	27.8	47.9	10.69	Quadratic	Zhang et al. (2017)

Hybrid sturgeon	*Acipenser baerii* ♀×*A. gueldenstaedtii* ♂	3.3	FM-SBM	25.1	34	25.83	Quadratic	Guo et al. (2012)
Ide	*Leuciscus idus*	3.8	FM–casein–gelatin	33	36.9	10.35	ANOVA	Cited from Oliva-Teles et al. (2019)
Jundia	*Rhamdia quelen*	3.9	Casein–gelatin	1.5	37.3	7.93	Broken line	Cited from Oliva-Teles et al. (2019)
Kutum fish	*Rutilus kutum*	3.2	FM-based	0.5	41.6	16.03	Quadratic	Cited from Oliva-Teles et al. (2019)
			FM-based	0.5	46.4	17.88	Quadratic	Cited from Oliva-Teles et al. (2019)
Largemouth bass	*Micropterus salmoides*	3.8	FM–SBM	14.5	43.5	4.39	Broken line	Cited from Oliva-Teles et al. (2019)
Mahseer	*Tor putitora*	2.9	FM-based	12.2	45.3	7.03	Broken line	Cited from Oliva-Teles et al. (2019)
Malaysian mahseer	*Tor tambroides*	2	FM-based	20.9	48	16.22	Quadratic	Cited from Oliva-Teles et al. (2019)
Manchurian trout	*Brachymystax lenok*	3.8	FM-based	3.4	43.6	5.22	Broken line	Cited from Oliva-Teles et al. (2019)
Mexican cichlid	*Cichlasoma urophthalmus*	3.9	FM-based	0.3	38.3	11.35	Broken line	Cited from Oliva-Teles et al. (2019)
			FM-based	0.2	38	11.80	ANOVA	Cited from Oliva-Teles et al. (2019)
Mossambic tilapia	*Oreochromis mossambicus*	2.2	FM-based	1.8	40	18.99	ANOVA	Cited from Oliva-Teles et al. (2019)
Murray cod	*Maccullochella peelii peelii*	4.2	FM–SBM	21.5	50	9.62	ANOVA	Cited from Oliva-Teles et al. (2019)
Nile tilapia	*Oreochromis niloticus*	2	Casein, gelatin, SBM, CSM, RSM	3.8	35	13.65	Quadratic	Cited from Oliva-Teles et al. (2019)
			SBM–casein	216.7	29.3	6.49	Quadratic	Cited from Oliva-Teles et al. (2019)

Continued

TABLE 4.3 Protein requirements of freshwater fishes (modified from Oliva-Teles et al., 2019).—cont'd

Species	Latin name	Trophic level	Dietary protein source	Initial body weight (g)	Protein requirement (%)	Protein requirement (g/kgBW/d)	Adjustment model	Reference
Pacu	*Piaractus mesopotamicus*	2	FM–SBM	15.5	27	4.91	ANOVA	Cited from Oliva-Teles et al. (2019)
Pike perch	*Sander lucioperca*	4	FM-based	1.05	52.9	24.95	Broken line	Schulz et al. (2007)
Piracanjuba	*Brycon orbignyanus*	2.5	Casein–gelatin	8.4	29	5.81	Exponential	Cited from Oliva-Teles et al. (2019)
Rainbow trout		4.1	FM-based	3.5	40	13.13	No model or ANOVA	Cited from Oliva-Teles et al. (2019)
			FM-SBM	63.17	45	13.48	ANOVA	Sevgili et al. (2012)
Rockfish		3.8	FM-based	7.3	50.9	12.09	Broken line	Cited from Oliva-Teles et al. (2019)
			FM-based	7.3	49.3	10.53	Broken line	Cited from Oliva-Teles et al. (2019)
			FM-based	7.3	46.2	9.66	Broken line	Cited from Oliva-Teles et al. (2019)
			FM-based	7.3	45.1	9.40	Broken line	Cited from Oliva-Teles et al. (2019)
Siberian sturgeon	*Acipenser baeri*	3.3	FM-based	22	40	4.60	Quadratic	Cited from Oliva-Teles et al. (2019)
Silver barb	*Barbonymus gonionotus*	2.4	Casein–gelatin	0.9	31.8	10.03	Broken line	Cited from Oliva-Teles et al. (2019)
			FM–based	1.1	35	14.34	ANOVA	Cited from Oliva-Teles et al. (2019)
Silver barb	*Puntius gonionotus*	2.4	Casein-gelatin	0.88	31.77	10.02	Broken line	Cited from Oliva-Teles et al. (2019)
Silver perch	*Bidyanus bidyanus*	3	FM-based	2.7	42.1	16.12	Broken line	Cited from Oliva-Teles et al. (2019)
Skin carp	*Hemibarbus labeo*	3.4	FM–SBM	7.6	44	5.95	Broken line	Cited from Oliva-Teles et al. (2019)
Snakehead	*Channa micropeltes*	3.9	FM-based	139.6	52		Broken line	Cited from Oliva-Teles et al. (2019)

Snakehead	*Channa striata*	3.4	FM-groundnut oil cake	0.6	55	20.63	ANOVA	Cited from Oliva-Teles et al. (2019)
			SBM-wheat gluten-FM-poultry meal-feather meal	31	51.6	10.19	Quadratic	Hua et al. (2019)
Snakehead	*Parachanna obscura*	3.4	FM−casein	0.1	55	28.94	ANOVA	Cited from Oliva-Teles et al. (2019)
			FM−casein	4.1	50	8.06	ANOVA	Cited from Oliva-Teles et al. (2019)
Spotted barbel	*Hemibarbus maculates* Bleeker	3.5	FM-Casein	1.6	37.4	10.24	Quadratic	Cited from Oliva-Teles et al. (2019)
			FM-Casein	9.49	35.5	8.41	Quadratic	Cited from Oliva-Teles et al. (2019)
			FM−casein	1.6	35	9.59	ANOVA	Cited from Oliva-Teles et al. (2019)
			FM−casein	9.4	30	6.15	ANOVA	Cited from Oliva-Teles et al. (2019)
Striped catfish		3.1	Casein−gelatin	4.2	37.1	7.70	Quadratic	Cited from Oliva-Teles et al. (2019)
Swordtail fish	*Xiphophorus helleri*	3.2	FM-Casein-Gelatin	1.13	30	7.17	ANOVA	Chong et al. (2004)
Tambaqui	*Colossoma macropomum*	2	FM−SBM	1.5	50	32.54	ANOVA	Cited from Oliva-Teles et al. (2019)
			FM−SBM	30.1	40	20.55	ANOVA	Cited from Oliva-Teles et al. (2019)
			FM−SBM	95.9	40	13.26	ANOVA	Cited from Oliva-Teles et al. (2019)
Tench	*Tinca tinca*	3.7	FM-based	0.4	52.7	12.82	Quadratic	Cited from Oliva-Teles et al. (2019)
Tiger puffer	*Fugu rubripes*	3.6	Casein	2	50	20.18	No model or ANOVA	Cited from Oliva-Teles et al. (2019)

Continued

TABLE 4.3 Protein requirements of freshwater fishes (modified from Oliva-Teles et al., 2019).—cont'd

Species	Latin name	Trophic level	Dietary protein source	Initial body weight (g)	Protein requirement (%)	Protein requirement (g/kgBW/d)	Adjustment model	Reference
Tin foil barb	*Barbodes altus*	2.4	Casein−gelatin	0.8	41.7	13.87	Quadratic	Cited from Oliva-Teles et al. (2019)
Triangular bream	*Megalobrama terminalis*	3.3	FM-based	1.3	44	10.57	ANOVA	Cited from Oliva-Teles et al. (2019)
White sturgeon	*Acipenser transmontanus*	3.3	Casein	145.2	40.5	4.82	Broken line	Cited from Oliva-Teles et al. (2019)
Yellow tail cichlid	*Pseudotropheus acei*		FM-based	0.9	35	10.66	ANOVA	Cited from Oliva-Teles et al. (2019)
Zillii tilapia	*Tilapia zillii*	2.5	Casein	1.8	35	12.31	No model or ANOVA	Cited from Oliva-Teles et al. (2019)

*FM: fishmeal; SBM: soybean meal; RSM: rapeseed meal; CSM: cotton seed meal; SFPC: Soluble fish protein concentrate; Trophic level are from: https://www.fishbase.se/search.php

TABLE 4.4 Protein requirements of brackish water fishes.

Species	Latin name	Dietary protein source	Initial body weight (g)	Protein requirement (%)	Protein requirement (g/kgBW/d)	Adjustment model	References
Coho salmon	*Oncorhynchus kisutch*	Casein–gelatin	14.3	40	2.97	ANOVA	Zeitoun et al. (1974)
Gray mullet	*Mugil cephalus* L.	FM–GNC–SBM	2.7	32.7	8.07	Quadratic	Talukdar et al. (2020)
Mexican Silverside	*Menidia estor*	FM-based	69.2	40.9	9.49	Broken line	Martinez-Palacios et al. (2007)
Nile tilapia	*Oreochromis niloticus*	Casein–gelatin	2.68	37.86	10.20	Broken line/ quadratic	Singha et al. (2020)
Obscure puffer	*Takifugu obscurus*	FM-based	12.4	37	9.64	Quadratic	Ye et al. (2017a)
Rainbow trout	*Oncorhynchus mykiss*	Casein–gelatin	6.2	40	6.91	Broken line	Zeitoun et al. (1973)
Southern flounder	*Paralichthys lethostigma*	FM-based	24.5	50	12.43	Broken line	Gonzalez et al. (2005)

ANOVA, analysis of variance; *FM*, fish meal; *GNC*, groundnut cake; *SBM*, soybean meal.

TABLE 4.5 Comparison of protein requirement between fishes and other land animals.

Species	Fish[a]	Other vertebrates
Specific growth rate (%/d)	2.29	2.44
Protein requirement (%)	41.9	20.0
Protein intake at maximum growth (mg protein ingested/g body weight/d)	11.1	12.0
Protein retention (100*g protein retained/g protein ingested)	32.4	29.0
Protein growth efficiency (g growth/g protein ingested)	1.87	1.96
Feed conversion efficiency (g growth/g diet ingested)	0.69	0.26
References		Bowen (1987)

[a] *The mean values for fish are calculated based on the data in Tables 4.2–4.4.*

The data on requirements show differences between fish having different feeding habits/trophic levels in nature (carnivorous, omnivorous, and herbivorous), living habitats (freshwater, marine water, and diadromous) and water temperatures (cold, warm, and tropic), and different growing periods (larvae, on-growing, and adult).

Dietary protein requirements for crustaceans have been studied in a few species including marine and freshwater shrimps and crabs. The optimal dietary protein for maximum growth ranges between 24% and 51.5% (Table 4.6). As regards protein requirements of some other aquatic animals including bivalves, echinoderms, amphibians, and reptiles, only limited data are available as summarized in Table 4.7.

4.3.3 Factors affecting protein requirement

4.3.3.1 *Genetic characteristics*

4.3.3.1.1 Feeding habits

Generally, carnivorous fishes are thought to require higher dietary protein (Tacon and Cowey, 1985; Tacon et al., 2010; NRC, 2011; Bowyer et al., 2013). Oliva-Teles et al. (2019) compared different published data and found that fishes of higher trophic levels showed higher dietary protein requirement. Pei et al. (2004) found higher nitrogen retention in carnivorous longsnout catfish (*Leiocassis longirostris*) compared to omnivorous gibel carp. One possible explanation for the high apparent protein requirement of fish is due to a limited supply of non-protein energy sources. Table 4.8 shows that there is no direct relationship among protein requirements in cyprinids having different feeding habits. However, it is still not clear whether there is a clear link between difference in protein requirements and different feeding habits in nature.

4.3.3.1.2 Fish strains

Genetic improvement strategies in aquaculture have focused mainly on improved growth rates. Whether this has an implication in protein requirements and utilization is not well established. For instance, the use of genetically improved farmed Nile tilapia (*Oreochromis*

TABLE 4.6 Protein requirements of crustaceans.

Species	Latin name	Dietary protein source	Initial body weight (g)	Protein requirement (%)	Protein requirement (g/kgBW/d)	Adjustment model	Reference
Chinese mitten crab	*Eriocheir sinensis*	Shrimp meal	1.07	39	12.23	ANOVA	Mu et al. (1998)
Chinese shrimp	*Penaeus chinensis*	FM, SBM	0.9	40		Broken-line	Kim and Hur (1993)
Freshwater prawn	*Macrobrachium rosenbergii*	Casein	0.15	35.5	7.1	ANOVA	Teshima et al. (2006)
Giant tiger shrimp	*Penaeus monodon*	Casein	1.32	40	70.03	ANOVA	Alava and Lim (1983)
Indian white shrimp	*Fenneropenaeus indicus*	FM-shrimp meal-clam meal	<1	45		ANOVA	Noverian and Gopal (2005)
		FM-shrimp meal-clam meal	1-5	40		ANOVA	Noverian and Gopal (2005)
		FM-shrimp meal-clam meal	5-10	35		ANOVA	Noverian and Gopal (2005)
Indian white shrimp	*Penaeus indicus*	Prawn meal, FM	0.4-1.1	43		ANOVA	Colvin (1976)
		Casein	20	35-37.5		ANOVA	Gopal and Raj (1990)
Mud crab	*Scylla paramamosain*	FM, Casein	0.04	47.06		Quadratic	Zheng et al. (2020)
	Scylla serrata	FM-squid meal-shrimp meal-clam meal	0.25	46.9	19.53	Quadratic	Unnikrishnan and Paulraj (2010)
		Soy protein concentrate-crab meal	20.43	44.15		ANOVA	Nguyen et al. (2014)
Pink shrimp	*Farfantepenaeus paulensis*	FM-soybean bran-gelatin	0.072	35	28.93	ANOVA	Ballester et al. (2010)
Red swamp crayfish	*Procambarus clarkii*	FM, SBM	12	39.6	13.1	Broken-line	Lu et al. (2020)

Continued

TABLE 4.6 Protein requirements of crustaceans.—cont'd

Species	Latin name	Dietary protein source	Initial body weight (g)	Protein requirement (%)	Protein requirement (g/kgBW/d)	Adjustment model	Reference
Swimming crab	*Portunus trituberculatus*	FM-SBM	2.5	51.5	16.29	Broken line /quadratic	Jin et al. (2013)
		FM-SBM-krill meal	3.75	51	18.06	ANOVA	Huo et al. (2014)
White shrimp	*Litopenaeus vannamei*	FM-SBM-peanut meal-squid meal-shrimp crumb-beer yeast	0.42	44	17.82	ANOVA	Zhu et al. (2010)
		FM-SBM-peanut meal-squid meal-shrimp crumb-beer yeast	0.56	38	15.62	ANOVA	Zhu et al. (2010)
		FM-peanut meal-corn gluten meal-cottonseed gluten meal	6.2	43	12.89	Quadratic	Xia et al. (2010)
		FM-SBM-rice grits	5.3	32.9		ANOVA	Jatoba et al. (2014)
		FM-based	1.3	33.4	19.34	Broken-line	Shahkar et al. (2014)
		FM, SBM	1.48	24	6.40	ANOVA	Panigrahi et al. (2018)

**FM*, fishmeal; *SBM*, soybean meal.

TABLE 4.7 Protein requirements of other aquaculture animals.

Species	Latin name	Dietary protein source	Initial body weight (g)	Protein requirement (%)	Protein requirement (g/kgBW/d)	Adjustment model	Reference
American Bullfrog	*Rana catesbeiana*	FM-based	8.84	40	14.96	Quadratic	Olvera-Novoa et al. (2007)
Ivory shell	*Babylonia areolata*	FM-SBM-shrimp meal	0.29	35	16.39	ANOVA	Chaitanawisuti et al. (2010)
		FM-casein-gelatin	0.54	45	10.04	Quadratic	Zhou et al. (2007a)
Sea cucumber	*Apostichopus japonicus*	FM-SBM	5	11	2.43	ANOVA	Bai et al. (2016)
Sea urchin	*Lytechinus variegatus*	Algae kelp-krill	22.8	20	7.37	ANOVA	Hammer et al. (2006)

*FM, fishmeal; RSM, rapeseed meal; SBM, soybean meal.

niloticus, GIFT) is widely practiced, but published data on requirements show no significant differences between original Nile tilapia and its GIFT strain (Table 4.9). There have been several improved strains of gibel carp, and the protein requirements for the juveniles show some differences in the requirement for gibel carp CAS V (Table 4.10).

4.3.3.2 *Diet composition*

4.3.3.2.1 Protein sources

Different dietary protein sources have been used to evaluate the protein requirement. In the first study on protein requirement for chinook salmon, De Long et al. (1958) used purified ingredients including casein, gelatin and other amino acids as protein sources and whole egg protein was used as referenced amino acid profile. A recent analysis by Oliva-Teles et al. (2019) showed that close to 50% of studies used fishmeal as the only dietary protein and more than one-third of studies included fishmeal as one of the protein sources, while only around 15% of studies were based on diets containing semi-purified ingredients such as casein or casein-gelatin. Dietary amino acid (including essential and non-essential amino acids) balance is important for the highest utilization of dietary protein. An ideal protein is one that provides the exact balance of amino acids needed for maximum growth and protein deposition. A feed formulated based on the ideal protein concept is an effective way of using less dietary protein to meet the amino acid requirements and to obtain better protein efficiency. For most fishes, content of essential amino acids (EAA) and non-essential amino acids (NEAA) should be around 50:50 and it could make the best proportion for efficiency body protein synthesis (Green et al., 2002; Oliva Teles et al., 2019).

Within a given species, Ye et al. (2015, 2017) compared the protein requirements of juvenile and post-juvenile gibel carp fed diets based on fishmeal protein and plant protein sources in gibel carp under the same experimental conditions. They found that the estimates of protein

TABLE 4.8 Protein requirements of several carp strains with different food habits.

Species	Trophic level	Diet type	IBW (g)	Protein requirement (%)	Protein requirement (g/kgBW/d)	Adjustment model	References
Grass carp	2	Casein	0.2	43	/	Broken line	Dabrowski (1977)
		FM–SBM	913	30	/	ANOVA	Khan et al. (2004)
		Casein–gelatin	4.3	40	8.99	ANOVA	Jin et al. (2015a)
		FM–casein –gelatin	264	28.6	6.86	ANOVA	Xu et al. (2016b)
Gibel carp CAS III	2.5	SPC–casein	3.2	40.2	13.55	Broken line /quadratic	Ye et al. (2015)
		SPC–casein	86.7	33.7	4.63	Broken line /quadratic	Ye et al. (2015)
		FM–casein	3.7	41.4	11.41	Quadratic	Ye et al. (2017)
		FM–casein	84.8	36.5	7.01	Quadratic	Ye et al. (2017)
Common carp	3.1	Casein	4.3	31		No model or ANOVA	Takeuchi et al. (1979)
		Casein	6	38		Broken line	Ogino and Saito (1970)
		Casein–gelatin	2.6	34		ANOVA	Murai et al. (1985)

**CSM*, cottonseed meal; *FM*, fishmeal; *RSM*, rapeseed meal; *SBM*, soybean meal.

TABLE 4.9 Protein requirements of Nile tilapia.

Species	Diet type	IBW (g)	Protein requirement (%)	Protein requirement (g/kgBW/d)	Adjustment model	References
Nile tilapia	FM-based	0.012	45	33.82	ANOVA	El-Sayed and Teshima (1992)
	FM-based	0.024	30.4	11.91	ANOVA	De Silva and Perera (1985)
	FM-based	0.2	50	11.32	ANOVA	Larumbe-Moran et al. (2010)
	FM–SBM	0.5	45	17.38	ANOVA	Abdel-Tawwab et al. (2010)
	FM–SBM–yeast	6.3	29	7.61	ANOVA	Jover Cerda et al. (1998)
	FM–SBM–yeast	12.2	39	9.33	ANOVA	Jover Cerda et al. (1998)
	FM–SBM	13.85	35.3	9.24	ANOVA	Mohammadi et al. (2014)
	FM–SBM	19.5	35	7.63	ANOVA	Abdel-Tawwab et al. (2010)
	FM–SBM	22.9	30	6.59	ANOVA	Sweilum et al. (2005)
	FM–SBM–BM–PBM	34.5	36	10.87	ANOVA	Hooley et al. (2014)
	FM–SBM	38.8	25	2.61	ANOVA	Sweilum et al. (2005)
	FM–based	40	30	6.39	ANOVA	Siddiqui et al. (1988)
	FM–SBM	40	35	5.40	ANOVA	Abdel-Tawwab et al. (2010)
Nile tilapia, GIFT	Casein–gelatin	2.68	37.37	10.26	Broken line	Singha et al. (2020)
	Casein–gelatin–PPB	3.8	35	13.65	Quadratic	Kpundeh et al. (2015)
	SBM–casein	216.7	29.3	6.49	Quadratic	Liu et al. (2017a)

ANOVA, analysis of variance; *BM*, blood meal; *FM*, fish meal; *PPB*, plant protein blend; *SBM*, soybean meal.

TABLE 4.10 Protein requirements for different strains of gibel carp.

Strain	IBW (g)	SGR (%/d)	Protein requirement (%)	References
Gibelio carp	2.53	1.35	39.3	He et al. (1988)
Gibelio carp CAS III	3.7	3.41	41.4	Ye et al. (2017)
Gibelio carp CAS V	6.14	2.85	30.7	Xu, unpublished

IBW, initial body weight; *SGR*, specific growth rate.

requirements of juvenile gibel carp based on fishmeal protein and plant protein were 36.0% and 40.2%, respectively, while the values for on-growing fish were 33.0% and 34.0%, demonstrating that growth performance of gibel carp fed plant protein diets improves with age.

4.3.3.2.2 Protein/energy ratio

As dietary protein is also used by fish to supply energy, dietary digestible protein to digestible energy ratios have a strong influence on protein utilization and dietary protein levels (Cho and Kaushik, 1985, 1990). An increase in dietary nonprotein energy within a suitable range can spare the protein and reduce the dietary protein levels required for growth (Wee and Tacon, 1982; Wang et al., 1985; Wee, 1986). Based on an extensive review of published data, Oliva-Teles et al. (2019) found that protein intake (g) per unit of energy intake (MJ) required per unit of weight gain was about 24.3 g protein per MJ dietary energy and was not related to fish body weight. It would be worth having such data analyzed and reported in terms of digestible protein and digestible energy values.

4.3.3.2.3 Other nutrients

Other dietary nutrients may also affect protein and energy utilization and therefore affect dietary protein requirement. Shiau and Hsieh (1997) found dietary vitamin B_6 deficiency could affect fish growth and protein utilization. Zehra and Khan (2012) found that increased dietary vitamin C increased protein retention in Indian common carp, mrigal (*Cirrhinus mrigala*). There were also different results showing no effects of dietary vitamin on protein requirement in juvenile yellow catfish (*Pelteobagrus fulvidraco*) (Wang et al., 2005). Such results likely depend on the nutrient levels and dietary formulation.

4.3.3.3 *Size and age*

It is generally considered that dietary protein requirement decreases with increasing fish size (Wilson and Halver, 1986; NRC, 2011) though there is no significant relationship between dietary protein requirements and initial body weight when put all species together (Oliva-Teles et al., 2019). Small fish are reported to require higher dietary protein levels as in Nile tilapia (Siddiqui et al., 1988; Jover Cerda et al., 1998; Abdel-Tawwab et al., 2010) and pacu (*Colossoma macropomum*) (Van der Meer et al., 1995). Tu et al. (2015a) fed graded levels of dietary protein to broodstock fish and reported a dietary protein requirement of about 37% of diet which was higher than that for on-growing gibel carp (33%) and similar with the juvenile (36%). Chong et al. (2004) found that the reproductive performance of female swordtail broodstock was improved with increased dietary protein level.

4.3.3.4 *Environmental factors*

Most environmental factors including water temperature, salinity, aquaculture system, feeding regime that affect fish growth and feed utilization, can have an impact on dietary protein requirement. However, systematic studies with the same species at different environmental conditions are very scarce. De Long et al. (1958) found dietary protein requirements for *Oncorhynchus tshawytscha* juveniles increased from 40% at 8.3°C to 55% at 14.4°C, and Amin et al. (2014) found the requirements increased from 40% at 9°C to 44% at 15°C in *Salvelinus fontinalis*. Increased dietary protein requirements with water temperature were also found in striped bass (*Morne saxatilis*) and red drum (Millikin, 1983; Daniels and Robinson, 1986). Reviewing literature data available, Oliva-Teles et al. (2019) found no clear indication of a change in dietary protein requirement as affected by water temperature.

Very few studies have reported the effect of water salinity on protein requirements. One of the earliest studies on the effect of salinity in a euryhaline teleost was made by Zeitoun et al. (1973) who observed that optimal growth of rainbow trout at 10 ppt salinity was achieved with a dietary protein level of 40% whereas a level of 45% was required at a salinity of 20 ppt. In the mullet, gray mullet (*Liza ramada*) fry, reared at 0–24‰ water salinity, best growth rate was observed at 40% dietary protein while similar high growth was achieved at 30%–40% dietary protein at 36‰ (El-Sayed, 1991). Li et al. (2004) and Larumbe-Moran et al. (2010) found that water salinity did not affect protein requirements of Nile tilapia fry and juveniles. Protein requirement for juvenile sunshine bass (*Morone chrysops* ♀ × *Morone saxatilis*♂) was reported around 41% of dry diet in both fresh and brackish water (Brown et al., 1992). De Silva and Perera (1985) compared the effects of dietary protein on young Nile tilapia at four water salinities and found that, the highest SGR in freshwater was achieved at 30.4% dietary protein, while the highest SGR was obtained at 28.0% dietary protein at 5‰, 10‰, and 15‰ water salinity. Analysis of literature data did not show any significant effect of water salinity on protein requirement (Oliva-Teles et al., 2019).

4.3.4 Maintenance requirements

The dietary protein requirement for maintenance is the amount of protein required by the animal to maintain normal health and physiological functions without gaining or losing weight. Maintenance requirement for proteins is the amount of protein/nitrogen required to maintain nitrogen balance. The additional protein is used for growth and reproduction. Maintenance requirements for nitrogen/protein can be determined using specific methods.

4.3.4.1 *Methodology*

One direct method used to estimate the maintenance protein requirement is to feed fish with a protein-free diet (or one with very low dietary nitrogen levels) or to fast the fish and the protein loss can be approximately considered to be the nitrogen for maintenance (Lupatsch et al., 1998).

When feeding fish with a low protein or a protein-free diet or when fish are fasted, the animal was not under normal situation, the calculated maintenance requirement could be different from the real one. It was reported that endogenous body nitrogen loss was higher (Kaushik et al., 1995).

The relationship between whole-body weight or nitrogen gain and dietary nitrogen/protein intake is used to calculate the maintenance protein requirement when fish body nitrogen gain or growth is zero.

Another method is to feed the fish with different levels of dietary protein (Fournier et al., 2002; Ozorio et al., 2009; Richard et al., 2010) or to feed the fish the same feed at different feeding levels (Peres and Oliva-Teles, 2005; Helland et al., 2010; Lupatsch et al., 2010; Yuan et al., 2010; Zimmermann et al., 2019) and backcalculates the dietary protein intake at "zero growth." Gatlin III et al. (1986) concluded that the back-calculation maintenance protein requirements based on growth and protein retention were similar.

4.3.4.2 Estimated maintenance requirements

Data available on maintenance protein requirements of different fishes are reported in Table 4.11. Although fish body size affects metabolism, growth, and feed utilization, data show similar maintenance protein requirements in different fishes. As regards body size, there is limited data and they do not indicate any clear effect of body size on maintenance requirements in different species of fish (Lupatsch et al., 1998; Liebert et al., 2006).

The estimation of maintenance protein requirement can be affected by the methods used for the estimation (Liebert et al., 2006). Pirozzi et al. (2010) found that maintenance requirement of digestible protein for juvenile mulloway (*Argyrosomus japonicus*) was not affected by water temperature while maintenance requirements for digestible energy increased with increasing temperature. Pirozzi et al. (2019) found that the maintenance protein for juvenile yellowtail kingfish (*Seriola lalandi*) increased with increasing water temperature. As regards possible effect of genotype, Mambrini et al. (2004) found that there was no difference in maintenance protein requirement between selected and the control strain of brown trout (*Salmo trutta*). Table 4.12 shows that percentage of protein requirements for maintenance to maximum growth for seawater fishes are around 30 while 14 for freshwater fishes.

4.4 Qualitative and quantitative amino acid requirement

Amino acids (AA) are classified as nutritionally EAA (or IAA), conditionally essential AA (CEAA), and NEAA (or DAA). The term "EAA" was coined in 1912, and these AAs must be provided to animals to maintain their growth and nitrogen balance. In all species of fish and shrimp studied so far, the EAAs are His, Ile, Leu, Lys, Met, Phe, Thr, Trp, Val, and Arg. CEAAs are AAs that can be normally synthesized in adequate amounts by animals but must be provided in the diet to meet optimal needs under certain conditions, like in the diets of broodstock and larvae, for improving the health status of pathologically challenges animals and/or when fed protein sources with large AA imbalances (Li and Wu, 2018). The CEAAs are Gln, Glu, Gly, Pro, and a free sulfur containing AA, Tau for fish. As in most animals, dietary NEAAs for fish and shrimp are Ala, Asn, Asp, Cys, Glu, Gln, Gly, Pro, Ser, and Tyr. Although the concepts of EAAs and NEAAs have been used for over a century, growing evidence from studies in near two decades has shown that an adequate dietary supply of NEAAs is also needed to ensure growth (Cowey, 1994; Green and Hardy, 2002; Wu, 2018).

4.4.1 Methodology

The initial studies on fish for qualitative AA requirements were conducted by Halver's team in the late 1950s with salmonids (Halver, 1957). This approach involves feeding a complete basal diet or a tested AA-deficient diet composed by either all crystalline AAs or a

TABLE 4.11 Maintenance protein requirements for some aquaculture species.

Species	Latin name	Initial body weight (g)	Water temperature (°C)	Maintenance protein requirement	References
Air-breathing catfish	*Heteropneustes fossilis*	7.9	20.5–23.5	24 g CP/kg/d	Ahmed (2010)
Atlantic salmon	***Salmo salar***	96	10	117.5 mg DP/kg$^{0.7}$/d	Helland et al. (2010)
		1.5	14.7	54 mg CP/kg$^{0.75}$	Abboudi et al. (2006)
Black porgy	*Acanthopagrus schlegeli*	50–200	18	0.48 mgN/g/d	Zhu et al. (2009)
Black sea turbot	*Psetta maxima*	5	17	127.3 mg/kg metabolic BW/d	Fournier et al. (2002)
Brown trout	*Salmo trutta* (selected)	3.7	11–12	8.1 g CP/kg$^{0.75}$/d	Mambrini et al. (2004)
	(unselected)	3.8	11–12	8.5 g CP/kg$^{0.75}$/d	Mambrini et al. (2004)
Channel catfish	*Ictalurus punctatus*	8.5	26.7	1.32 g CP/kg/d	Gatlin III et al. (1986)
Cobia	*Rachycentron canadum*	10	21–33	1.9 g CP/kg/d	Sun and Chen (2014)
European sea bass	*Dicentrarchus labrax*	22	25	1.15 g DP/kg/d	Peres and Oliva-Teles (2005)
		5	24	45.1 mg/kg metabolic BW/d	Fournier et al. (2002)
Giant croaker	*Argyrosomus japonicus*	40, 129	20, 26	0.47g DP/kg$^{0.7}$/d	Pirozzi et al. (2010)
Gilthead seabream	*Sparus aurata*	7.4	24	84.7 mg/kg metabolic BW/d	Fournier et al. (2002)
		30.1–92.5	23–24	0.86 g CP/kg$^{0.70}$/d	Lupatsch et al. (1998)
Hybrid tilapia	*Oreochromis mossambicus* × *Oreochromis hornorum*	12.4	28	75 mg N/kg$^{0.67}$/d	Liebert et al. (2006)
		12.3	28	74 mg N/kg$^{0.67}$/d	Liebert et al. (2006)
Japanese yellowtail	*Seriola quinqueradiata*	31	29.8	3.4g CP/kg/d	Watanabe (2000)
		94	27.1	3.1 g CP/kg/d	Watanabe (2000)
		506	18.8	2.7 g CP/kg/d	Watanabe (2000)
Large yellowtail kingfish	*Seriola lalandi*	187	15	0.28 g CP/kg$^{0.7}$/d	Pirozzi et al. (2019)
			24	0.66 g CP/kg/d	Pirozzi et al. (2019)

Continued

TABLE 4.11 Maintenance protein requirements for some aquaculture species.—cont'd

Species	Latin name	Initial body weight (g)	Water temperature (°C)	Maintenance protein requirement	References
Nile tilapia	*Oreochromis niloticus*	12.3	28	72 mg N/kg$^{0.67}$/d	Liebert et al. (2006)
		330	28	0.98 g DP/kg$^{0.70}$	Lupatsch et al. (2010)
		8	28	2 g CP/kg/d	Kaushik et al. (1995)
Pikeperch	*Sander lucioperca*	252	23	0.58 g/kg$^{0.70}$/d	Zimmermann et al. (2019)
Rainbow trout	*Oncorhynchus mykiss*	6.7	17	37.8 mg/kg metabolic BW/d	Fournier et al. (2002)
Rohu	*Labeo rohita*	0.55	26.5–28	32 g CP/kg/d	Ahmed (2007)
Two-banded seabream	*Diplodus vulgaris*	6–20	24.1	2.3 g CP/kg/d	Ozorio et al. (2009)

TABLE 4.12 Comparison of maintenance protein requirement and optimal protein for maximum growth in some fishes.

Species	Latin name	Maintenance protein requirement (g CP/kg/d)	References	Protein requirement (g CP/kg/d)	References	Percentage of protein requirements for maintenance to maximum growth
Black porgy	*Acanthopagrus schlegeli*	3.02	Zhu et al. (2009)	10.88	Zhang et al. (2010)	27.76
Channel catfish	*Ictalurus punctatus*	1.32	Gatlin III et al. (1986)	9.80	Garling and Wilson (1976)	13.47
Common two-banded sea bream	*Diplodus vulgaris*	2.3	Ozorio et al. (2009)	7.38	Ozorio et al. (2009)	31.17
Japanese yellowtail	*Seriola quinqueradiata*	3.4	Watanabe (2000)	24.89	Takeuchi et al. (1992)	13.66
Nile tilapia	*Oreochromis niloticus*	2	Kaushik et al. (1995)	13.65	Kpundeh et al. (2015)	14.65

mixture of casein, gelatin, and AAs formulated and bound with some chemical binder. The most AA profile of the complete basal diet is identical to that of yolk-sac fry or casein–gelatin protein component (Halver, 1957; Halver and Shanks, 1960). For the definition of AAs essential for the growth of fish or shrimp, the method involved the use of radioactively labeled carbon source, such as uniformly labeled [U–^{14}C] glucose or acetate. A significant amount of the radioactive carbon will be incorporated only into those compounds that the animal can itself synthesize. By analyzing the AAs from the tissue proteins, the EAAs are identified as those that are not radioactive. Such methods were applied with fish and shrimp (Cowey et al., 1970; Cowey and Forster, 1971).

Subsequently, determination of the quantitative requirements for the 10 EAAs of salmonids was undertaken by Halver and his collaborators. The development of an AA test diet was useful to determine AA requirements in different species with some modifications of the test diets (Nose, 1963; Nose et al., 1974; Wilson et al., 1977). Thereafter, more and more studies turn to use practical or semipurified basal diet with fishmeal or plant proteins as main protein sources (Michelato et al., 2013; Xie et al., 2014b; Niu et al., 2018) to make sure fish is under a comparable growth rate with those in practice.

The classic method to quantify the AA requirements of animals involves growth performance by dose–response studies or nitrogen balance tests. Several other methods have also been applied, which are based on analyses of physiological and metabolic profiles, such as plasma or tissue AAs, ammonia and other metabolites concentrations, and rates of AA oxidation using labeled AAs. Besides, minimal requirements of AA requirements can also be estimated by factorial models (Rodehutscord et al., 1997; Richard et al., 2010; Hua and Bureau, 2019). Each method has its own strengths and weaknesses, and the selection of a method should vary with species, body size, and the availability of facilities. A number of biotic and environmental factors can have an influence on the robustness of data obtained.

4.4.1.1 *Growth response assays*

Most data for AAs requirements have been obtained using the conventional growth trial during a given period of time. In these studies, a basal diet (semipurified or practical formulation) is designed with a deficiency in the target AA but assumed meeting all other known nutrient requirements of the animal, and graded levels of the tested AA are supplemented in several other test diets. Replicate groups of fish are fed the diets under a controllable feeding system until measurable differences appear in the WG. Fry and juveniles are very sensitive to the dietary provision of AAs, and they are selected as target animals in most studies, although the method can also be applied to the larger-size life stages (Berge et al., 1997; Dairiki et al., 2013; Gan et al., 2014; Wang et al., 2016a).

Several reviews have identified a large variety of EAA requirements among and within species, depending on the differences in feed formulations, variability in ingredient sources, experimental methods, and models used to analyze the data (Bowen, 1987; Cowey, 1992, 1993, 1994, 1995; Mambrini and Kaushik, 1995a; NRC, 2011; Peres and Oliva-Teles., 2017; Salze et al., 2017).

4.4.1.2 *Nitrogen balance*

A nitrogen (N) balance trial by deletion method is an efficient way to estimate protein or various EAA requirements in one growth trial, which is also named as deletion method. In tissues or the whole body, N almost only presents in proteins and AA, and a little in nucleotides. The N atom is neither degraded nor synthesized in animals. Thus, the N balance theory is conceptualized that dietary N intake (input) should be equal to N excretion (output) by the animal (Bender, 1965). Besides, the method is also based on the concept that the reduction or removal of a nonlimiting AA has no effect on N gain. Conversely, when a single AA is limiting in the diet, the rate of body protein accretion is causally related to that one AA. The changes in N gain measured on removal of a proportion of each IAA in turn are used to calculate a dietary AA pattern in which all EAAs are equally limiting. This was first tested with rats (Bender, 1965) and adapted to derive the optimum AA balance in pigs (Wang and Fuller, 1989). In fish, this approach was first used for deriving the optimum EAA pattern for juvenile Atlantic salmon (Rollin et al., 2003) and later with European seabass (Peres and Oliva-Teles, 2009) and Nile tilapia (Diógenes et al., 2016; Do Nascimento et al., 2020). For the deletion of EAA, the experimental diets are formulated to be isonitrogenous through the supplementation of one or several NEAA mixture. The formulation of experimental diets should ensure that all the nutrients are in the same concentration as the positive control diet, with the exception of the test EAA. Thus, the application of deletion technique results in a reduction of the specific EAA in the experimental diets at the same proportion while maintaining the optimum relationship among all others.

4.4.1.3 *Physiological and metabolic profiles*

Except for the endpoints related with growth performance (e.g., gain in body mass, survival, feed or protein utilization, health and eventually reproductive performance), complementary physiological criteria specific to the AA under test or overall protein metabolism are useful and have been used in studies on EAA requirements. The parameters, such as the concentrations of plasma or tissue free AAs, ammonia and other metabolites (Keembiyehetty et al., 1992; Wen et al., 2014; Fujimoto et al., 2019) hemoglobin (Khan and Abidi, 2014), antioxidant response (Zhao et al., 2013; Wang et al., 2015b), and cataract score (Remø et al., 2014), are applied in aquatic animal nutrition. Measuring postprandial-free AA levels in plasma/serum or muscle is an efficient approach on the AA requirements within a short-term or combining with a long-term growth trial (Kaushik, 1979). Under an optimal intake level, the tested EAA should remain low content in blood or muscle until the requirement for the EAA is met and then increase to high levels, and accompany with high blood ammonia when overdose (Yun et al., 2016). However, such an approach has not always been found to be reliable for assessing EAA requirements, and a number of factors can affect the validity, such as the time of sampling after the meal, nature itself of the AAs tested, and its possible agonistic/antagonistic interactions with other AAs (Wu, 2018).

4.4.1.4 *Amino acid oxidation (tracer studies)*

AA oxidation depends on its concentration in the free pools, the nutritional status, the physiological requirement under a specific condition, and, most importantly, the dietary level of the first limiting AA (Kim et al., 1983). The general considerations of the interrelationships between AA oxidation and dietary AA intake in animals are as follows: (1) regardless of AA profile in the diet, the oxidation of AA will be increased when the AA intake is increased, because the excessive AAs (possibly except for glutamine) must be oxidized to CO_2, water, ammonia, and other nitrogenous metabolites; (2) when the dietary intake of AA is below the requirement of an animal, the oxidation of AA is reduced to spare AA for protein synthesis; (3) when the dietary intake of AA is at the optimal level for protein synthesis, the

AA oxidation is at a minimum level; and (4) with a protein-adequate diet, an excess of an EAA would increase its oxidation, but not necessarily the oxidation of other AAs. On the contrary, the deficiency of one EAA can limit protein synthesis while increasing oxidation of other AAs progressively with increasing dietary intake of AA or protein (Kim et al., 1983; Wu, 2018).

Thus, the direct AA oxidation (DAAO) method is based on the principle that, when dietary intake of a test EAA is below the physiological requirement of an animal, the rate of its oxidation should remain relatively low until reaching the physiological requirement, it would increase sharply. The optimal requirement is obtained by the rate of test AA oxidation rate maintained at an optimal level and varies little within a narrow range of AA contact in the diet (the *dotted line* in Fig. 4.8). The word "direct" of DAAO is used because the oxidation of a test AA (e.g., lysine) is determined to estimate its dietary requirement (dietary lysine requirement). Walton et al. (1984a) used the DAAO method to confirm the lysine requirement of rainbow trout based on the WG rate. Following the growth study, fish from each treatment were injected intraperitoneally with a tracer dose of [U$-^{14}$C]-lysine over a 20 h. The expired $^{14}CO_2$ was collected and used as a direct measurement of the oxidation rate of lysine in the fish. The lysine requirement of rainbow trout obtained with the DAAO method (20 g/kg diet) was similar to the value of 19 g/kg diet obtained from growth data. Lall et al. (1994) obtained a consistent value of arginine requirement (16 g/kg diet or 41 g/kg protein) for Atlantic salmon by a 14 w growth trial and DAAO method. However, studies on tryptophan and phenylalanine failed to successfully obtain a credible requirement value (Walton et al., 1984b; Kim et al., 1992b). The mode of administration of labeled AA (intraperitoneal injection vs. dietary inclusion) differed between the two studies, which makes it difficult to make a direct comparison of results.

The indirect (or indicator) AA oxidation (IAAO) method is based on the concept that when an EAA is deficient, then all other AAs, including the indicator AAs, will be oxidized. With an increasing intake of the limiting AA, IAAO will decrease, indicating increasing protein synthesis. Once the requirement for the limiting AA is met, there will be no further change

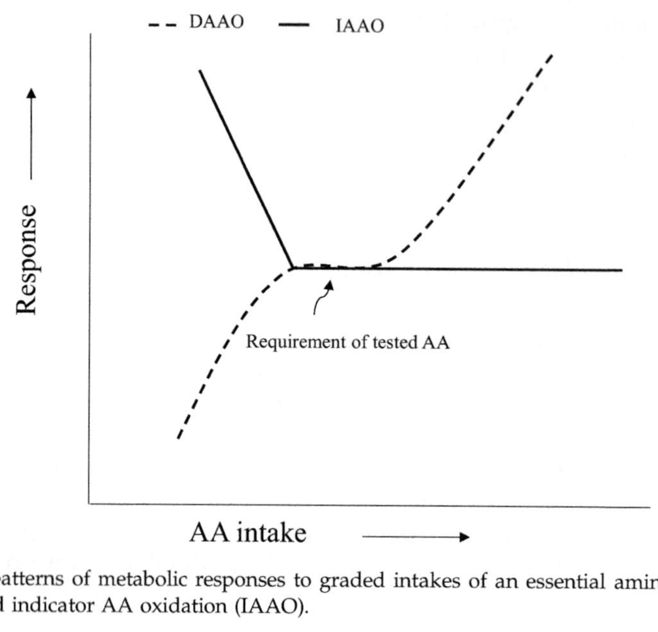

FIGURE 4.8 The patterns of metabolic responses to graded intakes of an essential amino acid with direct AA oxidation (DAAO) and indicator AA oxidation (IAAO).

in the IAAO (the *solid line* in Fig. 4.8). The word "indicator" is used before "AA oxidation" because the oxidation of a different AA (e.g., phenylalanine is mostly used as indicator AA) than the test AA (e.g., histidine) is determined to assess the dietary requirement of the test AA (Kim et al., 1983).

AA oxidation methods are used to rapidly assess the dietary protein and AA requirements for mammals across the life stage (Elango et al., 2008; Moehn et al., 2008; Mansilla et al., 2018; Templeman et al., 2019) but have met with little success in fish. Fauconneau et al. (1992) used the oxidation of phenylalanine and threonine as indicators for estimating the arginine requirement of rainbow trout, but the results were not conclusive. The common disadvantages of the AA oxidation methods include that (1) the rates of oxidation of the different EAAs are variable; (2) dietary intakes of both protein and energy greatly affect AA oxidation; (3) a breakpoint in the oxidation of the labeled AA may not be easily identified with precision in a short period of time; (4) a short-term measurement may not reflect the dynamic changes in a long-term growth trial; and (5) the production of labeled CO_2 from a ^{14}C- or ^{13}C-labeled AA may not well reflect the true oxidation AA in specific cells and tissues. The discrepancies between the AA oxidation method and growth data could be also associated with label dilution, isotope exchange, determination of intracellular specific activities of immediate precursors, and isotope steady states (Mambrini and Kaushik, 1995a; Wu, 2018).

4.4.1.5 Factorial models

Factorial models are useful for describing by quantitatively compartmentalizing nutrient or energy flows into the growth and maintenance components at the whole-body level and has been applied in some fish (Shearer, 1995; Lupatsch et al., 1998, 2001). The factorial models have also been used to estimate dietary EAA requirements of fish (Abboudi et al., 2007; Hua and Bureau, 2019) or shrimp (Richard et al., 2010). A primary requisite is that we have precise data on maintenance requirements. The processes involved in maintenance include renewal of protein lost via the skin and gastrointestinal tract, together with losses due to oxidation of AAs, synthesis of other N-compounds from AAs, and protein turnover (Cowey, 1994). Secondly, we need precise data on protein growth and the composition of growth, besides fecal and metabolic losses. Novel variants of such factorial mode (Zhang et al., 2013) ls have been developed, which take into account the possible roles and participation also of CEAA (Wu, 2018).

Such a factorial model has been tested with swine, for example, for determination of the physiological requirement for glutamine at the whole-body or specific tissue levels. Use of such factorial models allows requirements to be estimated for animals differing in their productive state (Fuller, 1994) and can be, in theory, applied to a wide range of conditions (Shearer, 1995). One major limitation in its application to fish is that we have only limited information on the maintenance AA requirements of fish (Mambrini and Kaushik, 1995a; Rodehutscord et al., 1997; Fournier et al., 2002; Abboudi et al., 2007) or shrimp (Richard et al., 2010). One of the drawbacks of the conventional factorial modes for estimating EAA requirements is that they assume that there are no interactions among EAA and other nutrients and assume constant efficiency of utilization of AAs regardless of physiological state (NRC, 2011). Hua (2013) provided a detailed outline on a model for estimating maintenance AA requirements of fish. Hua and Bureau (2019) developed this further to estimate the EAA requirements of rainbow trout and Atlantic salmon at different life stages based on expected body WG, body protein deposition, AA profile in body deposition, maintenance requirement, and inevitable AA catabolism or nitrogenous losses. In this study, besides data on the absolute amount of EAA requirement (g/d) estimated based on the aforementioned steps, the EAA requirements data could be expressed in different units (% diet, % protein, g/MJ DE). As regards AA availability, compared with terrestrial animals, there is little data available on the true digestibility of AA (true AA availability) in fish (Wilson et al., 1981; Yamamoto et al., 1998). This lack of data on AA availability can lead to variations in the estimation of EAA requirements. It is, however, worth stating that such values of dietary AA requirements calculated using factorial models should be verified in vivo by growth trials.

4.4.1.6 Whole-body amino acid profile and A/E ratio

Given the large number of farmed fish and crustacean species, it is difficult to establish the quantitative requirements for all the 10 EAA for each of the species concerned by long-term feeding and growth trial. The EAA profile of fish eggs, muscle, or whole body has been commonly used as an indicator of fish AA requirements (Mambrini and Kaushik, 1995a), especially for those species, whose EAA requirements have not been determined using conventional dose–response analyses (NRC, 2011). This method is useful to develop early guidelines or rough estimates of the EAA requirements of fish or shrimp. However, because of the low retention of the digested AAs (generally less than 50%), the profile of retained EAA in the whole body may not fully correspond to the EAA requirements of growing animals. Certain AA requirements could be overestimated, which are preferentially retained in tissues, such as leucine and lysine, whereas others (e.g., Met, Thr, His, and Arg) could be underestimated because they may play more active roles in different metabolic pathways and be less proportionally retained.

The A/E ratio (ratio of individual EAA to the sum of EAA) provides a simple means of estimating the requirements of all EAA when only one is known by relating the level of each EAA based on whole-body AA profile to each other (Arai, 1981). The A/E ratio method is less cost and can quickly obtain results of AA requirement.

During the past decades, researchers used whole-body patterns or A/E ratios to estimate the EAA requirements on several species, including rainbow trout and Atlantic salmon (*Salmo salar*) (Wilson and Cowey, 1985), common carp (*Cyprinus carpio*) (Nose, 1979), Siberian sturgeon (*Acipenser baeri*) and white sturgeon (*Acipenser transmontanus*) (Kaushik et al., 1991; Ng and Hung, 1995), plata pompano (*Trachinotus marginatus*) (Tesser et al., 2014), gilthead bream (*Sparus aurata*) (Gaber et al., 2016), and several potential species for aquaculture, such as meagre (*Argyrosomus regius*) (Saavedra et al., 2015), silver pomfret (*Pampus argenteus*) (Hossain et al., 2011), and pebbly fish (*Alestes baremoze*) (Kasozi et al., 2019). The average requirements of 10 EAA of fish and crustaceans were calculated based on their whole-body AA composition and shown in Table 4.13.

It appears that the ideal protein would be the one that reflects the whole-body EAA profile for many species. Fish potentially retains more His, Lys, Phe, Thr, and Val than those in whole shrimp, and other AAs in body are similar among them during growth.

4.4.2 Requirements of fish for essential amino acids

AA requirements of fish and nutritional regulation of AA metabolism have been subject to some reviews (Cowey and Wlaton, 1989; Tibaldi and Kaushik, 2005; Kaushik and Seiliez, 2010; Peres and Oliva-Teles, 2017; Salze et al., 2017).

Quantitative data for all 10 EAAs requirements are available for several species and are reported in Tables 4.14–4.23. For some species, such as rainbow trout and Nile tilapia, the

TABLE 4.13 Whole-body amino acid composition of different fish and crustaceans (expressed as g 16 g/N).

	Fish[a]	Shrimp[b]
EAA		
Arg	6.4 ± 0.6	6.6 ± 1.2
His	2.3 ± 0.4[a]	1.9 ± 0.2[b]
Ile	4.1 ± 0.6	3.6 ± 0.3
Leu	7.1 ± 0.6[a]	6.1 ± 0.5[b]
Lys	8.4 ± 0.9[a]	6.4 ± 0.5[b]
Met	2.8 ± 0.5	2.2 ± 0.2
Phe	4.4 ± 0.4[a]	3.4 ± 0.3[b]
Thr	4.5 ± 0.6[a]	3.2 ± 0.1[b]
Val	5.0 ± 0.8[a]	4.0 ± 0.6[b]
Trp	0.8 ± 0.2	0.9 ± 0.2
NEAA		
Ala	5.7 ± 0.7	4.9 ± 0.6
Asp	9.0 ± 1.1	8.4 ± 0.3
Cys	1.0 ± 0.2	0.8 ± 0.1
Glu	13.3 ± 1.3	12.6 ± 1.2
Gly	6.0 ± 1.4	5.0 ± 1.3
Pro	4.3 ± 1.1	3.1 ± 0.4
Ser	4.0 ± 0.5	3.3 ± 0.3
Tyr	3.4 ± 0.6	3.3 ± 0.4

[a]*Data sources include: Wilson and Cowey (1985); Mambrini and Kaushik (1995); Ng and Hung (1995); Kim and Lall (2000); Monentcham et al. (2010); Kaushik et al., 2011. Hossain et al., (2011); Tesser et al. (2014); Saavedra et al. (2015); Gaber et al., 2016; Kasozi et al. (2019).*
[b]*Data sources include Deshimaru and Shigeno (1972), Peñaflorida (1989), Akiyama et al. (1997).*
EAA, essential amino acid; *NEAA*, nonessential amino acid.

EAA requirements have been determined by different research groups for more than once using different basal diets and AA sources with different life stages of fish with different regression models. Measured AA requirements of different species, expressed as a proportion of the diet, show a large degree of variation. The major reason for this variability in EAA requirements data can be attributed to differences in methodological approaches as well as to the modes of expression of data.

For the species, which requirements for EAAs are known, it is possible to express the EAA requirements directly as a percentage of the diet or in g/16 gN for practical the formulation. AAs play an important role in meeting the energy requirements of fish and shrimp. The EAA/DE ratio is also used for AA requirement estimation for in fish nutrition (Kaushik, 1995; Encarnação et al., 2004).

4.4.2.1 *Arginine requirements*

Arginine (Arg) is an α-AA with a side chain consisting of a 3-carbon aliphatic straight chain, and the distal end of which is capped by a complex guanidinium group. Arg is a

TABLE 4.14 Arginine requirements of fish and crustaceans.

Species	IBW (g)	Requirement (% diet)	Requirement (g/16 g N)	Endpoints	Methods/Models	References
African bonytongue (*Heterotis niloticus*)	0.2		3.9	A/E ratios	Egg/whole-body AAs	Monentcham et al. (2010)
	2.0–60.0		4.0			
	400		4.1			
	4000–5000		3.9			
African catfish	16.6	1.8	4.5	WG, SGR, FCR	Growth/broken line	Fagbenro et al. (1999)
Channel catfish	0.6	1.8–2.2	4.5–5.0	WG, FCR	Growth/polynomial	Singh and Khan (2007)
	138	0.8–0.9	3.3–3.8	WG, FR	Growth/least squares	Buentello and Gatlin (2000)
	200	1.0	4.3	WG	Growth/broken line	Robinson et al. (1981)
Arctic charr	20.0–35.0	2.4	–	A/E ratios	Whole-body AAs	Gurure et al. (2007)
Asian sea bass	2.5	1.8	3.8	WG	Growth/broken line	Murillo-Gurrea et al. (2001)
Atlantic salmon	0.2–20.0	1.5–1.9	4.3–4.4		Factorial models	Hua and Bureau (2019)
	20.0–100	1.5–1.9	4.2–4.4			
	100–500	1.5–1.9	4.1–4.2			
	500–1500	1.4–1.7	3.8–3.9			
	1500–2500	1.2–1.5	3.3–3.4			
	2.6	1.8	4.0	PPV	N balance	Rollin et al. (2003)
	114	1.6	4.1	WG, expired $^{14}CO_2$	Growth/broken line	Lall et al. (1994)
	383	2.1–2.2	5.0–5.1	WG, PPV	Growth/exponential	Berge et al. (1997)
Black sea bream	10.5	2.8–3.1	6.8–8.1	SGR, PPV	Growth/polynomial	Zhou et al. (2010a)
Blunt snout bream	2.6	2.3–2.5	6.7–7.2	SGR, FR, PV	Growth/polynomial	Ren et al. (2013)
	20.0	1.8–1.9	5.6	FCR, SGR	Growth/polynomial	Liang et al. (2016)

Bottom-feeder curimbata (*Prochilodus lineatus*)	–	–	4.0	A/E ratios	Whole-body AAs	De Almeida Bicudo and Cyrino (2009)
Chinook salmon (*Oncorhynchus tshawytscha*)	3.5	2.3	5.9	WG	Growth/broken line	Klein and Halver (1970)
Cobia	3.4	2.8–2.9	6.1–6.2	SGR, PPV	Growth/polynomial	Ren et al. (2012)
Coho salmon	0.9	2.7–3.0	4.9–5.5	WG	Growth/broken line	Luzzana et al. (1998)
	2.9	2.4	6.0	WG	Growth/broken line	Klein and Halver (1970)
Colliroja (*Astyanax fasciatus*)	16.2	2.0	4.6	A/E ratios	Whole-body AAs	Furuya et al. (2015)
Common carp	6.3	1.8	5.5	SGR	Growth/polynomial	Chen et al. (2012)
	6.3	1.9–2.2	5.6–6.4	serum LC, HT	Growth/polynomial	Chen et al. (2015)
	76.4	1.3–1.5	3.8–4.4	A/E ratios	Whole-body AAs	Ogino (1980)
Dourado (*Salminus brasiliensis*)	324	1.4–1.5	3.3–3.4	FBW, WG, SGR, FCR	Growth/broken line	Dairiki et al. (2013)
European sea bass	–	–	4.6	A/E ratios	Whole-body AAs	Kaushik (1998)
	0.8	1.9	4.6	WG	Growth/broken line	Tibaldi et al., (1993)
	2.1	1.8	3.9	WG	Growth/broken line	Tibaldi et al., (1994)
Gibel carp	51.6 148	1.6 1.3	5.3 4.2	SGR SGR	Growth/broken line	Tu et al. (2015b)
Gilthead seabream	–	–	5.4	A/E ratios	Whole-body AAs	Kaushik (1998)
	2.2	1.2	3.1	A/E ratios	Whole-body AAs	Gaber et al. (2016)
	4.6	2.3	5.6	A/E ratios	Whole-body AAs	Peres and Oliva-Teles (2009)
Golden pompano	18.8	2.7	6.3	WG	Growth/polynomial	Lin et al. (2015)
Plata pompano	25.0	–	4.0	A/E ratios	Whole-body muscle AAs	Tesser et al. (2014)
Grass carp	6.2	1.5	5.5	SGR	Growth/polynomial	Chen et al. (2019)
	279	1.7	5.7	MDA	Growth/polynomial	Wang et al. (2015a)

Continued

TABLE 4.14 Arginine requirements of fish and crustaceans.—cont'd

Species	IBW (g)	Requirement (% diet)	Requirement (g/16 g N)	Endpoints	Methods/Models	References
Grouper	16.1	2.7	5.5	WG	Growth/broken line	Luo et al. (2007)
Grouper hybrid (*Epinephelus fuscoguttatus* ♀ × *Epinephelus lanceolatus* ♂)	11.7	3.6–3.7	6.6–6.8	WG, PPV	Growth/polynomial	Wu et al. (2018a)
Yellow grouper (*Epinephelus awoara*)	4.3	2.8	6.5	SGR, WG	Growth/polynomial	Zhou et al. (2012a)
Indian major carp, Catla	0.6	1.7	5.1	AWG	Growth/polynomial	Zehra and Khan (2013a)
	1.2	1.5	4.8	WG	Growth/broken line	Ravi and Devaraj (1991)
Indian major carp, mrigal	0.6	1.8–1.8	4.6	FCR, PPV	Growth/polynomial	Ahmed and Khan (2004a)
Indian major carp, roho	0.6	1.2–1.4	3.1–3.5	SGR, FCR, PPV, ERE	Growth/broken line Polynomial	Abidi and Khan (2009)
	1.7–1.9	2.3	5.8	WG	Growth/broken line	Murthy and Varghese (1995)
Japanese flounder	1.9	2.0–2.1	4.1–4.2	WG, FCE	Growth/broken line	Alam et al. (2002)
	3.0	1.6	3.4	A/E ratios	Whole-body AAs	Forster and Ogata (1998)
Matrinxã (*Brycon amazonicus*)	–	–	2.9	A/E ratios	Whole-body AAs	De Almeida Bicudo and Cyrino (2009)
Milkfish	0.7	2.1	5.3	WG	Growth/broken line	Borlongan (1991)
	0.7	2.1	5.3	WG	Growth/broken line	Borlongan and Coloso (1993)
Nile tilapia	0.02	1.2	4.2	WG	Growth/broken line	Santiago and Lovell (1988)
	3.0	1.4	4.9	WG	Growth/polynomial	Neu et al. (2016)
	6.0	1.8	6.2	WG	Growth/polynomial	Yue et al. (2015)
	63.0	2.3–3.5	6.6–10.0	ALT, LC	Growth/ANOVA	Fujimoto et al. (2019)

Nile tilapia, GIFT	159	–	4.9	PPV	N balance	Do Nascimento et al. (2020)
Mozambique tilapia	1.7	1.6	4.0	SGR	Growth/ANOVA	Jackson and Capper (1982)
Pacu	54.3 823 1563	1.4 1.4 1.3	– – –	A/E ratios	Whole-body AAs	Khan et al. (2019)
Pebbly fish (*Alestes baremoze*)	–	–	3.3–3.6	A/E ratios	Whole-body AAs	Kasozi et al. (2019)
Rainbow trout (*Oncorhynchus mykiss*)	0.2–20.0	1.3–1.7	3.7–4.7		Factorial models	Hua and Bureau (2019)
	20.0–100	1.3–1.7	3.7–3.8			
	100–500	1.2–1.5	3.5–3.9			
	500–1500	0.9–1.2	2.6–2.7			
	1.0	2.5–2.8	5.4–5.9	WG	Growth/ANOVA	Ketola (1983)
	1.3 2.5	1.2	3.0	DGC	Growth/exponential	Cho et al. (1992)
	12.6	1.4	4.0	WG	Growth/broken line	Kim et al. (1992a)
	50.0–200	1.4	4.1	PPV	Growth/polynomial	Pack et al. (1995)
Rainbow trout (*Salmo gairdneri*)	1.5	1.4	3.5	WG, FE	Growth/polynomial	Chiu et al. (1988)
	4.7	1.7	4.2	WG, FE		
	7.2	1.6–1.8	3.6–4.0	WG	Growth/broken line	Walton et al. (1986)
	100	1.2	2.6	FCR	Growth/polynomial	Kaushik et al. (1988)
Red drum	56.6	1.4–1.8	4.1–5.0	FE, PPV, WG	Growth/least squares	Barziza et al. (2000)
	–	1.3	3.7	A/E ratios	Muscle AAs	Moon and Gatlin (1991)
Red sea bream (*Pagrus major*)	–	2.5	4.2	LG	Growth/broken line	López-Alvarado and Kanazawa (1994)
	13.3	2.4	4.7	WG	Growth/broken line	Rahimnejad and Lee (2014)

Continued

TABLE 4.14 Arginine requirements of fish and crustaceans.—cont'd

Species	IBW (g)	Requirement (% diet)	Requirement (g/16 g N)	Endpoints	Methods/Models	References
Silver perch	0.4	2.7	6.8	WG	Growth/broken line	Ngamsnae (1999)
Silver pomfret	–	–	3.5	A/E ratios	Whole-body AAs	Hossain et al. (2011)
Striped bass	111 790	1.9 1.9	3.9 3.9	A/E ratios	Whole-body AAs	Small and Soares (1998)
Striped bass hybrid (*Morone saxatilis* × *Morone chrysops*)	3.0 7.0	1.5 1.5–1.6	4.4 4.1–4.4	FE WG, FE	Growth/broken line	Griffin et al. (1994a)
Turbot	7.4	1.6–3.0	3.0–5.4	WG	Growth/ANOVA	Fournier et al., 2002
White sturgeon	66.7	2.1	4.8	A/E ratios	Whole-body AAs	Ng and Hung (1995)
Yellow perch (*Perca flavescens*)	10.6–11.0	1.4	4.3	WG	Growth/polynomial	Twibell and Brown (1997)
Yellow tail	32.0	1.4–1.6	3.4–3.9	WG	Growth/ANOVA	Ruchimat et al. (1998)
Crustaceans						
Kuruma shrimp (*Marsupenaeus japonicas*)	0.3	2.7	5.3	WG	Growth/broken line	Alam et al. (2004)
Pacific white shrimp	0.5	2.0	4.8	WG	Growth/broken line	Zhou et al. (2012b)
Tiger shrimp	0.02	1.9	5.3	WG	Growth/broken line	Millamena et al. (1998)
	0.3	2.5	5.5	WG	Growth/broken line	Chen et al. (1992)

AAs, amino acids; *ALT*, alanine aminotransferase; *ANOVA*, analysis of variance; *AWG*, absolute weight gain; *DGC*, daily growth coefficient; *ERE*, energy retention efficiency; *FBW*, final body weight; *FCE*, feed conversion efficiency; *FCR*, feed conversion ratio; *FE*, feed efficiency; *HT*, hemagglutination titer; *LC*, leukocyte count; *LG*, length gain; *MDA*, malondialdehyde; *PPV*, protein productive value; *SGR*, specific growth rate; *WG*, weight gain.

precursor for creatine and nitric oxide (NO) synthesis in the ornithine cycle. Arg is also a precursor of glutamine, glutamate, and proline, required for the de novo synthesis of proteins and of putrescine, essential to the synthesis of polyamines involved in cell growth and proliferation. Arg is also degraded by nitric oxide synthases (NOS) into citruline and nitric oxide (NO), which regulates different physiological functions such as blood flow, respiration, and glucose homeostasis. Arg can be synthesized from citrulline, glutamine, glutamate, and proline via the intestinal–renal axis in humans and most other mammals. Arg is an EAA in fish and crustaceans, like in some carnivorous mammals, such as cat and ferrets (Morris and Rogers, 1978; Deshmukh and Shope, 1983), which are unable to produce Arg endogenously due to the lack of the key synthase enzymes, and this is also considered as the reason for a lack of Arg synthesis in fish (Andersen et al., 2016).

An analog of *N*-acetylglutamate (NAG), *N*-carbamylglutamate (NCG), functions by activating intestinal pyrroline-5-carboxylic acid synthase, and carbamyl phosphate synthase (the rate-limiting enzymes responsible for both the ornithine cycle and the Arg synthetic pathway) promotes the synthesis of endogenous Arg, which had been confirmed in mammals and fish (Frank et al., 2007; Wu et al., 2010; Wang et al., 2019b; Huang et al., 2019). Arg is stable and not toxic to cells, in general. However, overdose Arg-HCl (>9 g/day) occasionally induces nausea, gut discomfort, and diarrhea in humans (Grimble, 2007), which may be induced by a rapid and excess NO production and impaired intestinal absorption of other dietary basic AA (lysine and histidine). The antagonism between Arg and Lys is explained in the later part. Arg is also reported to be a potent secretagogue of several pancreatic hormones in fish (Plisetskaya et al., 1991; Mommsen et al., 2001). However, a close analysis shows that changes in plasma insulin level occur only when fish are either given pharmacological doses of arginine or fed diets with arginine levels far exceeding the requirement levels (Lall et al., 1994).

Arg deficiency has been shown to cause growth reduction and poor protein retention in most fish. Available data on Arg requirement for fish show great variability ranging from 0.8% to 3.7% of the diet or 3.3–10.0 g/16 gN by growth studies (Table 4.14). In a review article by Peres and Oliva-Teles (2017), authors illustrated a recommendation for arginine requirement of marine fish as 4.5–4.7 g/16 g N by metaanalysis based on 16 studies. A few studies have reported the Arg requirement of crustaceans, with a range of 1.9%–2.7% of diet, or 4.8–5.5 g/16 gN (Millamena et al., 1998; Chen et al., 1992; Zhou et al., 2012b).

4.4.2.2 *Histidine requirements*

Histidine (His) is classified as a basic AA. It has a positively charged imidazole functional group that can act as both an acid and a base, i.e., it can both donate and accept protons under some conditions (NRC, 2011). His is an essential AA for hemoglobin synthesis in all animals, important for growth, tissue formation and repair, and the maintenance of osmoregulation and myelin sheaths that act as protectors for nerve cells, reducing the incidence of cataracts (Bjerkas and Sveier, 2004; Remø et al., 2014), and participate in one-carbon metabolism, consequently affecting DNA and protein synthesis (Wu, 2018). The study of Waagbø et al. (2010) shows the risk for cataract development for adult Atlantic salmon, 1 year after the transfer of salmon smolts from freshwater to full-strength seawater, mostly can be prevented by histidine supplementation just before and during the early phase of cataract development.

Data on His requirements of fish are presented in Table 4.15. Most of the growth studies on His requirement have mainly focused on small size fish (initial body weight < 50 g). Considerable variation appears to exist among the species studied, with a range of 0.4%–1.6% of diet

TABLE 4.15 Histidine requirements of fish and crustaceans.

Species	IBW (g)	Requirement (% diet)	Requirement (g/16 g N)	Biomarkers	Methods/Models	References
African bonytongue	0.2	–	2.0	A/E ratios	Egg/whole-body AA	Monentcham et al. (2010)
	2.0–60.0	–	2.1			
	400	–	2.3			
	4000–5000	–	2.5			
African catfish	0.2	0.4	1.0–1.1	WG, FCR, PPV, ERE	Growth/polynomial	Khan and Abidi (2009)
Channel catfish	200	0.6	1.5	WG	Growth/broken line	Wilson et al. (1980)
Indian/stinging catfish (*Heteropneustes fossilis*)	4.7	0.5–0.6	1.3–1.5	WG, FCR, PER, PPV	Growth/polynomial	Ahmed (2013)
	5.3	1.5–1.6	3.5–3.6	AWG, PPV, Hb, Ht	Growth/polynomial	Khan and Abidi (2014)
	6.6	0.9–1.0	2.4–2.6	AWG, FCR, PG, histidine retention	Growth/polynomial	Farhat and Khan (2013a)
Arctic charr	20.4	0.9	–	A/E ratios	Whole-body AAs	Gurure et al. (2007)
Atlantic salmon	0.2–20.0	1.0–1.2	2.6–2.7		Factorial models	Hua and Bureau (2019)
	20.0–100	0.9–1.2	2.6–2.7			
	100–500	0.9–1.2	2.6			
	500–1500	0.8–1.1	2.3–2.4			
	1500–2500	0.7–0.9	2.0–2.1			
	1.4	0.7	1.5	PPV	N balance	Rollin et al. (2003)
	71.4	1.4	3.2	Cataract score	Growth/ANOVA	Remø et al. (2014)
Blunt snout bream	23.3	1.1	3.6	WG, SGR	Growth/polynomial	Wilson-Arop et al. (2018)
Bottom-feeder curimbata	–	–	1.5	A/E ratios	Whole-body AAs	De Almeida Bicudo and Cyrino (2009)
Chinook salmon	2.9	0.7	1.8	WG	Growth/least squares	Klein and Halver (1970)
Coho salmon	1.6	0.7	1.6	WG	Growth/broken line	Akiyama et al. (1985)
	3.2	0.7	1.7	WG	Growth/least squares	Klein and Halver (1970)

Colliroja	16.2	0.5	1.2	A/E ratios	Whole-body AAs	Furuya et al. (2015)
Common carp	8.8	0.8	2.4	SGR	Growth/polynomial	Zhao et al. (2012a)
European sea bass	–	–	1.6	A/E ratios	Whole-body AAs	Kaushik (1998)
Gilthead seabream	–	–	1.7	A/E ratios	Whole-body AAs	Kaushik (1998)
	2.2	1.6	3.5	A/E ratios	Whole-body/muscle AAs	Gaber et al. (2016)
	4.6	0.8	1.9	A/E ratios	Whole-body AAs	Peres and Oliva-Teles (2009)
Grass carp	3.7	1.2	3.2	FBW, FCR	Growth/polynomial	Gao et al. (2015)
	280	0.8–0.9	2.5–3.0	GPx, PWG, ROS, MDA, PC	Growth/polynomial	Jiang et al., (2016b,c)
Indian major carp, catla	0.7	0.6–0.7	1.9–2.1	AWG, FCR, PG, Hb, Ht, RBCs	Growth/broken line	Zehra and Khan (2016a)
	1.2	1.0	2.5	WG	Growth/broken line	Ravi and Devaraj (1991)
Indian major carp, mrigal	0.6	0.9	2.1–2.2	FCR, PPV	Growth/polynomial	Ahmed and Khan (2005a)
Indian major carp, roho	0.4	0.8–0.9	2.1–2.3	WG, FCR	Growth/polynomial	Khan and Abidi (2007a)
	1.7–1.9	0.9	2.3	WG	Growth/broken line	Murthy and Varghese (1995)
Japanese flounder	3.0	0.6	1.3	A/E ratios	Whole-body AAs	Forster and Ogata (1998)
Matrinxã	–	–	1.1	A/E ratios	Whole-body AAs	De Almeida Bicudo and Cyrino (2009)
Milkfish	0.7	0.8	2.0	WG	Growth/broken line	Borlongan and Coloso (1993)
Nile tilapia	0.1	0.5	1.7	WG	Growth/broken line	Santiago and Lovell (1988)
	4.3	0.7	2.1	A/E ratios	Whole-body AAs	Ogunji et al. (2005)
	4.8	0.8	3.1	WG	Growth/polynomial	Michelato et al. (2016a)

Continued

TABLE 4.15 Histidine requirements of fish and crustaceans.—cont'd

Species	IBW (g)	Requirement (% diet)	Requirement (g/16 g N)	Biomarkers	Methods/Models	References
Nile tilapia, GIFT	20.6	0.5	1.7	A/E ratios	Whole-body AAs	Diógenes et al. (2016)
	159	–	1.7	PPV	N balance	Do Nascimento et al. (2020)
Pacu	54.3 823 1563	0.4 0.4 0.4	–	A/E ratios	Whole-body AAs	Khan et al. (2019)
Pebbly fish	–	–	1.4–1.5	A/E ratios	Whole-body AAs	Kasozi et al. (2019)
Plata pompano	25.0	–	1.2–1.3	A/E ratios	Whole-body/muscle AAs	Tesser et al. (2014)
Rainbow trout	0.2–20.0	0.8–1.0	2.3–2.4		Factorial models	Hua and Bureau (2019)
	20.0–100	0.8–1.0	2.3			
	100–500	0.8–1.0	2.1–2.2			
	500–1500	0.6–0.7	1.6–1.7			
	47.0	0.6	1.7	WG	Growth/polynomial	Rodehutscord et al. (1997)
Red drum	–	0.6	1.7	A/E ratios	Muscle AAs	Moon and Gatlin (1991)
Red sea bream	1.7	0.6	1.4	A/E ratios	Whole-body AAs	Forster and Ogata (1998)
Silver pomfret	–	–	1.4	A/E ratios	Whole-body AAs	Hossain et al. (2011)
Striped bass	111 790	0.6	1.7	A/E ratios	Whole-body AAs	Small and Soares (1998)
Turbot	–	–	1.5	A/E ratios	Whole-body AAs	Kaushik (1998)
White sturgeon	66.7	1.0	2.3	A/E ratios	Whole-body AAs	Ng and Hung (1995)
Crustaceans						
Tiger shrimp	0.2	0.8	2.2	WG	Growth/broken line	Millamena et al. (1999)

AAs, amino acids; *ANOVA*, analysis of variance; *AWG*, absolute weight gain; *ERE*, energy retention efficiency; *FBW*, final body weight; *FCR*, feed conversion ratio; *GPx*, glutathione peroxidase; *Hb*, hemoglobin; *Ht*, hematocrit; *MDA*, malondialdehyde; *PC*, protein carbonyls; *PER*, protein efficiency ratio; *PG*, protein gain; *PPV*, protein productive value; *PWG*, percent weight gain; *RBCs*, red blood cell counts; *ROS*, reactive oxygen species; *SGR*, specific growth rate; *WG*, weight gain.

or 1.0–3.6 g/16 gN for the requirement values estimated by dose–response growth studies. Low dietary His levels have also been linked to cataract development in Atlantic salmon (Waagbø et al., 2010; Remø et al., 2014). His requirement of Tiger shrimp, *Penaeus monodon*, was estimated to be 0.8% of the diet (2.2 g/16 gN) by Millamena et al. (1999).

4.4.2.3 Branched-chain amino acid requirements

Branched-chain amino acids (BCAAs) refer to leucine (Leu), isoleucine (Ile), and valine (Val) having aliphatic side chains that are nonlinear. Leu is an α-AA with an isobutyl R group. Ile has a secondary butyl side chain that is a large aliphatic hydrophobic chiral side chain. Four stereoisomers of Ile are possible although, in nature, only one enantiomeric form, (2S, 3S)-2-amino-3-methylpentanoic acid, exists. Val has an isopropyl side chain (NRC, 2011).

Leu is considered to be the key to the normal growth and reproductive potential of the fish because it stimulates protein synthesis in muscle tissues (NRC, 2011). It is a ketogenic AA with specificity in BCAAs, which can stimulate the release of insulin from islet cells in vitro (Panten et al., 1974). Leu can activate protein synthesis and decrease proteolysis, thus controlling the maintenance of nitrogen balance. This ability was first documented in the 1970s (Buse and Reid, 1975), showing the unique role of Leu in the modulation of skeletal muscle metabolism. Leu works with Val and Ile to protect and replenish the muscles. It plays an important role in producing hemoglobin, maintaining plasma glucose levels, and increasing growth hormone production, anti-stress response, and energy metabolism (Zhang et al., 2020). Leu deficiency in the diet can cause severe biochemical malfunction including growth retardation (Wilson and Halver, 1986). Ile plays a major role in the formation of many metabolites, generally concentrated in the muscle tissues of the body. Ile, together with the other two BACCs, plays an important role in protein synthesis and degradation and participates in insulin biosynthesis and secretion (Kimball and Jefferson, 2006). In addition, it also contributes to energy production in the body and has been found to reduce twitches and tremors in animals (Braverman et al., 2003). Dietary Ile deficiency or excess can cause retardation of growth and feed conversion in fish (Huang et al., 2015; Liu et al., 2014a; Zhao et al., 2012b; Sharf and Khan, 2020). Val plays an important role in many metabolic pathways, such as tissue repair and growth and maintaining nitrogen balance in the body (Nose, 1979; Shimomura et al., 2006; Brosnan and Brosnan, 2006a). Dietary Val deficiency can cause growth retardation and poor FE (Nose, 1979; Ravi and Devaraj, 1991; Borlongan and Coloso, 1993; Ahmed and Khan, 2006). It has been observed in different animals and humans that Ile, Leu, and Val produce antagonistic effects when the proportion of these three AAs in diets are ill-balanced. The antagonistic effects among BCAAs are discussed later.

4.4.2.3.1 Isoleucine requirement

The requirement for Ile appears to range between 0.5% and 1.8% of diet (1.5–4.3 g/16 gN) in different species of fish estimated by the dose–response growth studies (Table 4.16). The lowest value was observed in lake trout (*Salvelinus namaycush*) of 2.7 g body weight (0.5% of diet) and Channel catfish with 200g body weight (0.6% of diet), and the highest value was 1.7%–1.8% of diet (4.0–4.1 g/16 gN) for juvenile golden pompano (*Trachinotus ovatus*) (Huang et al., 2015), and grass carp (*Cenopharyngodon idellus*) requires 1.1%–1.3% of diet (3.5–4.3 g/16 gN) based on intestinal claudin 3 gene expression (Feng et al., 2017a). Applying the factorial method, Hua and Bureau (2019) found the Ile requirement of Atlantic salmon with various body sizes to range between 2.7 and 3.5 g/16 gN. For the Pacific white shrimp

TABLE 4.16 Isoleucine requirement of fish and crustaceans.

Species	IBW (g)	Requirement (% diet)	Requirement (g/16 g N)	Biomarkers	Methods/Models	References
African bonytongue	0.20		2.9	A/E ratios	Egg/whole-body AAs	Monentcham et al. (2010)
	2.0–60.0		2.9			
	400		2.9			
	4000–5000		2.9			
Arctic charr	20.0–35.0	1.2	–	A/E ratios	Whole-body AAs	Gurure et al. (2007)
Atlantic salmon	0.2–20.0	1.2–1.6	3.4–3.6		Factorial models	Hua and Bureau (2019)
	20.0–100	1.2–1.6	3.4–3.5			
	100–500	1.2–1.5	3.3–3.4			
	500–1500	1.1–1.4	3.1–3.2			
	1500–2500	1.0–1.2	2.7–2.8			
Bottom-feeder curimbata	–	–	2.9	A/E ratios	Whole-body AAs	De Almeida Bicudo and Cyrino (2009)
Channel catfish	200	0.6	2.6	WG	Growth/broken line	Wilson et al. (1980)
Chinook salmon	1.0	1.0	2.6	WG	Growth/polynomial	Chance et al. (1964)
Colliroja	16.2	1.4	3.3	A/E ratios	Whole-body AAs	Furuya et al. (2015)
Common carp	6.9	1.3	3.8	SGR	Growth/polynomial	Zhao et al. (2012b)
	6.9	1.3	3.7–4.0	LA, MDA	Growth/polynomial	Zhao et al. (2013)
	6.9	1.0	2.9–3.0	Intestinal health	Growth/polynomial	Zhao et al. (2014)
	76.4	0.8	2.3	A/E ratios	Whole-body AAs	Ogino (1980)
European sea bass	–	–	2.6	A/E ratios	Whole-body AAs	Kaushik (1998)
Gilthead sea bream	–	–	2.6	A/E ratios	Whole-body AAs	Kaushik (1998)
	2.2	0.5	1.1	A/E ratios	Whole-body AAs	Gaber et al. (2016)
	4.6	1.0	2.3	A/E ratios	Whole-body AAs	Peres and Oliva-Teles (2009)

Golden pompano	6.4	1.7–1.8	4.0–4.1	WG, FCR, PPV	Growth/polynomial	Huang et al. (2015)
Plata pompano	25.0	–	1.6–1.8	A/E ratios	Whole-body /muscle AAs	Tesser et al. (2014)
Grass carp	257	1.2	3.9	PWG	Growth/polynomial	Gan et al. (2014)
	257	1.1–1.3	3.5–4.3	gill PC, claudin-3 mRNA level	Growth/polynomial	Feng et al. (2017a)
Indian major carp, catla	0.6	1.1–1.2	3.4–3.6	AWG, PPV, IRE, RNA/DNA ratio, carcass protein	Growth/polynomial	Zehra and Khan (2013b)
	1.2	0.9	2.4	WG	Growth/broken line	Ravi and Devaraj (1991)
Indian major carp, mrigal	0.9–1.0	1.3	3.1	WG	Growth/broken line	Benakappa and Varghese (2003)
	0.6	1.2–1.3	3.0–3.3	WG, FCR, PPV	Growth/polynomial	Ahmed and Khan (2006)
Indian major carp, roho	0.4	1.5–1.6	3.8–4.0	WGR, FCR, PPV, SGR	Growth/polynomial	Khan and Abidi (2007b)
Japanese flounder	3.0	1.0	2.0	A/E ratios	Whole-body AAs	Forster and Ogata (1998)
Lake trout (*Salvelinus namaycush*)	2.7	0.5–0.9	1.5–2.6	FCR	Growth/ANOVA	Hughes et al. (1983)
Matrinxã	–	–	3.1	A/E ratios	Whole-body AAs	De Almeida Bicudo and Cyrino (2009)
Milkfish	0.3	1.6	4.0	WG	Growth/broken line	Borlongan and Coloso (1993)
Nile tilapia	0.1	0.9	3.1	WG	Growth/broken line	Santiago and Lovell (1988)
	4.3	1.2	3.5	A/E ratios	Whole-body AAs	Ogunji et al. (2005)
	18.1	0.7	2.5	WG, FE, PPV	Growth/ANOVA	Neu et al. (2017)
Nile tilapia, GIFT	20.4	0.9	2.8	A/E ratios	Whole-body AAs	Diógenes et al. (2016)
	159	–	3.3	PPV	N balance	Do Nascimento et al. (2020)
Pacu	54.3 823 1563	0.8 0.8 0.9	–	A/E ratios	Whole-body AAs	Khan et al. (2019)

Continued

TABLE 4.16 Isoleucine requirement of fish and crustaceans.—cont'd

Species	IBW (g)	Requirement (% diet)	Requirement (g/16 g N)	Biomarkers	Methods/Models	References
Pebbly fish	–	–	2.4	A/E ratios	Whole-body AAs	Kasozi et al. (2019)
Rainbow trout	0.2–20.0	1.1–1.4	3.0–3.1		Factorial models	Hua and Bureau (2019)
	20.0–100	1.1–1.4	3.0–3.1			
	100–500	1.0–1.3	2.8			
	500–1500	0.8–1.0	2.1–2.2			
	47.0	1.4	4.0	WG	Growth/polynomial	Rodehutscord et al. (1997)
	126	0.9	2.4	A/E ratios	Whole-body AAs	Ogino (1980)
Red drum	–	1.0	2.9	A/E ratios	muscle AAs	Moon and Gatlin (1991)
	3.7	1.1	3.0	WG	Growth/polynomial	Castillo and Gatlin (2018)
Red sea bream	1.7	1.1	2.2	A/E ratios	Whole-body AAs	Forster and Ogata (1998)
Silver pomfret	–	–	2.0	A/E ratios	Whole-body AAs	Hossain et al. (2011)
Snakehead (*Channa punctatus*)	6.7	1.8	4.0–4.1	AWG, FCR, PPV, IRE	Growth/polynomial	Sharf and Khan (2020)
Striped bass	111 790	0.9	2.5	A/E ratios	Whole-body AAs	Small and Soares (1998)
Turbot	–	–	2.6	A/E ratios	Whole-body AAs	Kaushik (1998)
White sturgeon	66.7	1.3	3.0	A/E ratios	Whole-body AAs	Ng and Hung (1995)
Crustaceans						
Pacific white shrimp	0.4	1.6	3.9	WG, FE, PPV	Growth/polynomial	Liu et al. (2014b)
Tiger shrimp	0.02	1.0	2.7	WG	Growth/broken line	Millamena et al. (1999)

ANOVA, analysis of variance; *AWG*, absolute weight gain; *FCR*, feed conversion ratio; *FE*, feed efficiency; *IRE*, isoleucine retention efficiency; *LA*, lysozyme activity; *MDA*, malondialdehyde; *PC*, protein carbonyls; *PPV*, protein productive value; *PWG*, percent weight gain; *SGR*, specific growth rate; *WG*, weight gain.

reared in low salinity water, Ile requirement was estimated to be 1.6% of the diet or 3.9 g/16 gN (Liu et al., 2014a). In postlarval tiger shrimp, the requirement of Ile was found to be 1.0% of the diet or 2.7 g/16 gN (Millamena et al., 1999).

4.4.2.3.2 Leucine requirements

Leu requirement of fish is in the range of 0.9%–3.4% of diet or 3.5–8.0 g/16 N based on the growth studies (Table 4.17). The most studied species is Nile tilapia (including GIFT strain) with various body sizes and by different study methods. The Leu requirement of Nile tilapia was at a range of 1.0%–2.0% of diet (3.4–5.1 g/16 gN). Do Nascimento et al. (2020) estimated the Leu requirement of GIFT strain of Nile tilapia as 4.8 g/16 gN by a nitrogen balance test. The Leu requirement of crustaceans is 1.7%–2.4% of diet or 4.3–5.9 g/16 gN for Pacific white shrimp and tiger shrimp (Liu et al., 2014b; Millamena et al., 1999).

4.4.2.3.3 Valine requirements

Most studies on Val requirement of aquatic animals are undertaken in combination with those on the other two BCAAs. The Val requirement of fish is distributed in the range of 0.6%–2.0% of diet or 1.8–4.8 g/16 gN based on dose–response growth trials (Table 4.18). For the most studied species, rainbow trout and Nile tilapia (including GIFT strain), the optimum levels of dietary Leu are 1.0%–1.8% of diet (2.6–4.6 g/16 gN) and 0.8%–1.3% of diet (2.8–4.5 g/16 gN), respectively, estimated by growth studies and factorial models. Lake trout was found to have the lowest BCAAs among the species in the report of Hughes et al. (1983). The Val requirement of postlarval tiger shrimp is in 1.5% of diet or 3.7 g/16 gN (Millamena et al., 1999).

4.4.2.4 *Lysine requirements*

Lys is an α-AA and can exist in the *L*- or *D*-stereoisomeric forms. But only the *L*-form is present in biological proteins and is also commercially produced. Lys plays an important role in maintaining the normal growth and development of aquatic animals and the function of the nervous system, so it is also called "growth AA." Lys is one of the most limiting AAs in fish feed, especially when dietary fishmeal was replaced by cereal protein sources (Gatlin et al., 2007). Lack of Lys can lead to reduced growth and FE. One of the major roles of Lys is to regulate the carnitine synthesis in skeletal muscle cells and liver, which further helps convert fatty acids into energy and maintain cholesterol balance in the blood (Walton et al., 1984a). Besides, Lys deficiency causes the dorsal and caudal fin erosion and higher mortality rates in rainbow trout (Ketola, 1983) and affects myocytes development and IGF signaling in gilthead sea bream (Azizi et al., 2016).

The Lys requirement values for fish are summarized in Table 4.19. The requirements show a great variation among aquatic animals, ranging from 1.2% to 3.4% of diet or 3.3–8.6 g/16 gN for fish species, and1.6%–2.6% of diet or 4.0–5.8 g/16 gN for crustaceans based on the growth studies. However, for most fish, the requirement appears to range from 4 to 6 g/16 gN estimated in the review by Kaushik and Seiliez (2010).

A metaanalysis of literature data on Lys requirement of different species of fish and crustaceans was estimated to be 1.6% of diet or 4.4% of protein by a broken line regression model (Fig. 4.9) and lower than the 4.9% of protein by Mercer's nutrient kinetic model in marine fish (Peres and Oliva-Teles, 2017).

TABLE 4.17 Leucine requirement of fish and crustaceans.

Species	IBW (g)	Requirement (% diet)	Requirement (g/16 g N)	Biomarkers	Methods/Models	References
African bonytongue	0.2		4.5	A/E ratios	Egg/whole-body AAs	Monentcham et al. (2010)
	2.0–60.0		4.4			
	400		4.4			
	4000–5000		4.4			
Arctic charr	20.0–35.0	2.6	–	A/E ratios	Whole-body AAs	Gurure et al. (2007)
Atlantic salmon	0.2–20.0	2.2–2.8	6.2–6.4		Factorial models	Hua and Bureau (2019)
	20.0–100	2.2–2.8	6.2–6.4			
	100–500	2.2–2.7	6.0–6.2			
	500–1500	2.0–2.5	5.6–5.7			
	1500–2500	1.7–2.2	4.8–5.0			
Black carp (*Mylopharyngodon piceus*)	2.9	2.4	5.9–6.0	WG, FCR	Growth/polynomial	Wu et al. (2017a)
Bottom-feeder curimbata	–	–	4.9	A/E ratios	Whole-body AAs	De Almeida Bicudo and Cyrino (2009)
Catfish hybrid (*Pelteobagrus vachelli*♀ × *Leiocassis longirostris*♂)	23.2	2.8	7.3	PWG	Growth/broken line	Zhao et al. (2020)
Channel catfish	200	0.9	3.5	WG	Growth/broken line	Wilson et al. (1980)
Chinook salmon	2.5	1.6	3.9	WG	Growth/polynomial	Chance et al. (1964)
Colliroja	16.2	2.4	5.6	A/E ratios	Whole-body AAs	Furuya et al. (2015)
Common carp	76.4	1.4	4.1	A/E ratios	Whole-body AAs	Ogino (1980)
European sea bass	–	–	4.3	A/E ratios	Whole-body AAs	Kaushik (1998)
Gibel carp	5.2	1.9	5.4	Muscle AA, TOR pathway genes	Growth/ANOVA	Zou et al. (2018)
Gilthead sea bream	–	–	4.5	A/E ratios	Whole-body AAs	Kaushik (1998)
	2.2	2.4	5.3	A/E ratios	Whole-body/muscle AAs	Gaber et al. (2016)
	4.6	2.0	4.8	A/E ratios	Whole-body AAs	Peres and Oliva-Teles (2009)

Golden pompano	5.8	2.9–3.3	7.2–8.0	WG, SGR, PPV	Growth/polynomial	Tan et al. (2016)
Plata pompano	25.0	–	4.0–4.3	A/E ratios	Whole-body /muscle AAs	Tesser et al. (2014)
Grass carp	296	1.3	4.2–4.3	WGR, PAC, intestinal MDA	Growth/polynomial	Deng et al. (2014)
	296	1.3	3.9	Muscle MDA	Growth/polynomial	Deng et al. (2016)
	296	1.3	3.9–4.3	Gill MDA, gill PC	Growth/polynomial	Jiang et al. (2017c)
Grouper hybrid	6.9	3.3–3.4	6.1–6.4	WG, PPV	Growth/polynomial	Zhou et al. (2019)
Indian catfish	6.8	1.6–1.7	4.3–4.4	AWG, FCR, PUE, LRE	Growth/polynomial	Farhat and Khan (2014a)
Indian major carp, catla	0.7	1.6	4.8	AWG, FCR	Growth/polynomial	Zehra and Khan (2015a)
	1.2	1.5	3.7	WG	Growth/broken line	Ravi and Devaraj (1991)
Indian major carp (*Cirrhinus cirrhosus*)	1.1–1.2	1.7	4.3	WG	Growth/broken line	Benakappa and Varghese (2003)
Indian major carp, mrigal	0.6	1.5–1.6	3.8–3.9	WG, FCR, PPV	Growth/polynomial	Ahmed and Khan (2006)
Indian major carp, roho	0.4	1.5–1.6	3.8–3.9	WGR, FCR, PPV	Growth/polynomial	Abidi and Khan (2007)
Japanese flounder	3.0	1.9	3.9	A/E ratios	Whole-body AAs	Forster and Ogata (1998)
Japanese seabass (*Lateolabrax japonicus*)	8.0	2.4	5.7	WG	Growth/polynomial	Li et al. (2015b)
Lake trout	3.2	1.0–1.3	2.7–3.7	FCR	Growth/ANOVA	Hughes et al. (1983)
Large yellow croaker (*Pseudosciaena crocea* R.)	6.0	2.9	6.8	FW	Growth/polynomial	Li et al. (2010)
Matrinxã	–	–	5.4	A/E ratios	Whole-body AAs	De Almeida Bicudo and Cyrino (2009)
Milkfish	0.4	2.0	5.1	WG	Growth/broken line	Borlongan and Coloso (1993)
Nile tilapia	0.1	1.0	3.4	WG	Growth/broken line	Santiago and Lovell (1988)
	1.9	1.3	4.3	WG	Growth/polynomial	Gan et al. (2015)
Nile tilapia, GIFT	159	–	4.8	PPV	N balance	Do Nascimento et al. (2020)
Pacu	54.3 823 1563	1.3 1.3 1.3	– – –	A/E ratios	Whole-body AAs	Khan et al. (2019)

Continued

TABLE 4.17 Leucine requirement of fish and crustaceans.—cont'd

Species	IBW (g)	Requirement (% diet)	Requirement (g/16 g N)	Biomarkers	Methods/Models	References
Pebbly fish	–	–	4.3–4.5	A/E ratios	Whole-body AAs	Kasozi et al. (2019)
Rainbow Trout (*Oncorhynchus mykiss*)	0.2–20.0	2.0–2.5	5.4–5.6		Factorial models	Hua and Bureau (2019)
	20.0–100	1.9–2.4	5.4–5.5			
	100–500	1.8–2.3	5.0–5.1			
	500–1500	1.4–1.7	3.8–4.8			
	2	3.4	9.1	WG	Growth/polynomial	Choo et al. (1991)
	49.0	1.4	4.0	WG, PPV	Growth/polynomial	Rodehutscord et al. (1997)
Red drum	–	1.7	4.7	A/E ratios	Muscle AAs	Moon and Gatlin (1991)
	1.4	1.6	4.2	WG	Growth/polynomial	Castillo and Gatlin (2018)
Red sea bream	–	–	4.2	A/E ratios	Whole-body AAs	Forster and Ogata (1998)
White sturgeon	66.7	1.9	4.3	A/E ratios	Whole-body AAs	Ng and Hung (1995)
Silver pomfret	–	–	4.0	A/E ratios	Whole-body AAs	Hossain et al. (2011)
Striped bass	111 790	1.9	5.3	A/E ratios	Whole-body AAs	Small and Soares (1998)
Turbot	–	–	4.6	A/E ratios	Whole-body AAs	Kaushik (1998)
Crustaceans						
Pacific white shrimp	0.4	2.4	5.8–5.9	WG, FE, PPV	Growth/polynomial	Liu et al. (2014c)
Tiger shrimp	0.02	1.7	4.3	WG	Growth/broken line	Millamena et al. (1999)

ANOVA, analysis of variance; *AWG*, absolute weight gain; *FCR*, feed conversion ratio; *FE*, feed efficiency; *LRE*, leucine retention efficiency; *MDA*, malondialdehyde; *PAC*, plasma ammonia concentration; *PC*, protein carbonyls; *PPV*, protein productive value; *PUE*, protein utilization efficiency; *PWG*, percent weight gain; *SGR*, specific growth rate; *WG*, weight gain.

TABLE 4.18 Valine requirement.

Species	IBW (g)	Requirement (% diet)	Requirement (g/16 g N)	Biomarkers	Methods/Models	References
African bonytongue	0.2	–	3.5	A/E ratios	Egg/whole-body AAs	Monentcham et al. (2010)
	2.0–60.0	–	3.4			
	400	–	3.2			
	4000–5000	–	3.2			
Arctic charr	20.0–35.0	1.6	–	A/E ratios	Whole-body AAs	Gurure et al. (2007)
Atlantic salmon	0.2–20.0	1.6–2.0	4.4–4.5		Factorial models	Hua and Bureau (2019)
	20.0–100	1.6–2.0	4.4–4.5			
	100–500	1.5–1.9	4.3–4.4			
	500–1500	1.4–1.8	3.9–4.1			
	1500–2500	1.2–1.6	3.4–3.5			
	2.6	1.4	3.1	PPV	N balance	Rollin et al. (2003)
Blunt snout bream	10.2	1.3	3.7–3.9	SGR, FER	Growth/broken line	Ren et al. (2015a)
Bottom-feeder curimbata	–	–	3.5	A/E ratios	Whole-body AAs	De Almeida Bicudo and Cyrino (2009)
Channel catfish	200	0.7	3.0	WG	Growth/broken line	Wilson et al. (1980)
Chinook salmon	2.5	1.3	3.2	WG	Growth/polynomial	Chance et al. (1964)
Colliroja	16.2	1.5	3.5	A/E ratios	Whole-body AAs	Furuya et al. (2015)
Common carp	9.7	1.4	4.0	SGR	Growth/polynomial	Dong et al. (2012)
European sea bass	–	–	2.9	A/E ratios	Whole-body AAs	Kaushik (1998)
Gilthead sea bream	–	–	3.0	A/E ratios	Whole-body AAs	Kaushik (1998)
	2.2	1.2	2.7	A/E ratios	Whole-body /muscle AAs	Gaber et al. (2016)
	4.6	1.4	3.2	A/E ratios	Whole-body AAs	Peres and Oliva-Teles (2009)

Continued

TABLE 4.18 Valine requirement.—cont'd

Species	IBW (g)	Requirement (% diet)	Requirement (g/16 g N)	Biomarkers	Methods/Models	References
Golden pompano	5.3	2.0	4.6–4.7	WG, PPV	Growth/polynomial	Huang et al. (2017)
Plata pompano	25.0	–	1.8–2.1	A/E ratios	Whole-body /muscle AAs	Tesser et al. (2014)
Grass carp	268	1.4–1.5	4.8	Intestinal lysozyme activity, PWG	Growth/polynomial	Luo et al. (2014)
Indian major carp, catla	1.2	1.4	3.6	WG	Growth/broken line	Ravi and Devaraj (1991)
Indian major carp, mrigal	0.6	1.5–1.6	3.7–3.8	WGR, FCR, PPV	Growth/polynomial	Ahmed and Khan (2006)
	1.0–1.0	1.7	3.9	WG	Growth/broken line	Benakappa and Varghese (2003)
Indian major carp, roho	0.2	1.5	3.8	FCR	Growth/polynomial	Abidi and Khan (2004)
	–	1.5	3.8	FR, SGR	Growth/broken line	Murthy and Varghese (1997a)
Japanese flounder	3.0	1.2	2.5	A/E ratios	Whole-body AAs	Forster and Ogata (1998)
Lake trout	2.7	0.6–0.8	1.8–2.2	FCR	Growth/ANOVA	Hughes et al. (1983)
Matrinxã	–	–	3.3	A/E ratios	Whole-body AAs	De Almeida Bicudo and Cyrino (2009)
Milkfish	0.4	1.4	3.6	WG	Growth/broken line	Borlongan and Coloso (1993)
Nile tilapia	0.1	0.8	2.8	WG	Growth/broken line	Santiago and Lovell (1988)
	6.5	1.2–1.3	4.1–4.5	WG, PPV	Growth/broken line	Xiao et al. (2017)
	21.4	0.9	3.5	WG	Growth/polynomial	Rodrigues et al. (2019)
Nile tilapia, GIFT	159	–	3.5	PPV	N balance	Do Nascimento et al. (2020)
Pacu	54.3 823 1563	0.9 0.9 0.9	– – –	A/E ratios	Whole-body AAs	Khan et al. (2019)
Pebbly fish	–	–	2.6–2.7	A/E ratios	Whole-body AAs	Kasozi et al. (2019)
Rainbow trout (*Oncorhynchus mykiss*)	0.2–20.0	1.4–1.8	3.9–4.0		Factorial models	Hua and Bureau (2019)
	20.0–100	1.3–1.7	3.7–3.9			
	100–500	1.3–1.6	3.5–3.6			

	500–1500	1.0–1.2	2.6–2.7			
	49.0	1.6	4.6	WG	Growth/polynomial	Rodehutscord et al. (1997)
	493	1.4–1.5	3.9–4.1	PPval, PAval, PPA	Growth/broken line	Bae et al. (2012)
Red drum	2.8	1.2	3.4	WG	Growth/polynomial	Castillo and Gatlin (2018)
Red sea bream	32.0	0.9–1.0	2.0–2.2	WG, PPV	Growth/broken line	Rahimnejad and Lee (2013)
Silver pomfret	–	–	2.8	A/E ratios	Whole-body AAs	Hossain et al. (2011)
Striped bass	111 790	1.0	2.8	A/E ratios	Whole-body AAs	Small and Soares (1998)
Turbot	–	–	2.9	A/E ratios	Whole-body AAs	Kaushik (1998)
White sturgeon	–	–	3.3	A/E ratios	Whole-body AAs	Ng and Hung (1995)
Crustaceans						
Tiger shrimp	0.01	1.5	3.7	WG	Growth/broken line	Millamena et al. (1999)

ANOVA, analysis of variance; *FCR*, feed conversion ratio; *FER*, feed efficiency ratio; *PAval*, postabsorptive free valine concentrations; *PPA*, postprandial plasma ammonia concentrations; *PPV*, protein productive value; *PPval*, postprandial plasma-free valine concentrations; *PWG*, percent weight gain; *SGR*, specific growth rate; *WG*, weight gain.

TABLE 4.19 Lysine requirement of fish and crustaceans.

Species	IBW(g)	Requirement (% diet)	Requirement (g/16 g N)	Biomarkers	Methods/Models	References
African catfish	15.4	2.3	5.7	WG	Growth/broken line	Fagbenro et al. (1998)
Asian sea bass	13.1	2.1	4.5	WG	Growth/broken line	Murillo-Gurrea et al. (2001)
Atlantic cod (*Gadus morhua*)	120	2.4–2.6	4.7–5.1	PPV, lysine gain	Growth/exponential	Grisdale- et al. (2011)
Atlantic salmon	0.2–20.0	2.3–2.9			Factorial models	Hua and Bureau (2019)
	20.0–100	2.3–2.9				
	100–500	2.2–2.8				
	500–1500	2.0–2.6				
	1500–2500	1.8–2.2				
	642	2.2	5.0	PPV	Growth/exponential	Abboudi et al. (2006)
	4.7	2.0	4.0	WG	Growth/broken line	Anderson et al. (1993)
	642	2.2	5.0	WG	Growth/exponential	Espe et al. (2007)
	376	1.6–1.8	3.3–3.7	PER, free lysine in plasma and muscle	Growth/ANOVA	Berge et al. (1998)
	42.1	2.3	5.3	WG	Growth/ANOVA	Hauler et al. (2007)
Black sea bream	9.1	3.3	8.6	SGR	Growth/polynomial	Zhou et al. (2010b)
Channel catfish	200	1.2	5.1	WG	Growth/ANOVA	Wilson et al. (1977)
	200	1.5	5.0	WG	Growth/ANOVA	Robinson et al. (1980a)
Clum salmon (*Oncorhynchus keta*)	1.5	1.9	4.8	WG	Growth/ANOVA	Akiyama et al. (1985)
Cobia	1.3	2.3	5.3	SGR	Growth/broken line	Zhou et al. (2007b)
Common carp (Jian carp)	1.5	2.2	4.6	WG	Growth/broken line	Nose (1979)
	7.9	1.9	5.9	WG	Growth/broken line	Zhou et al. (2008)

Dourado	11.4	2.2	5.0	WG, SGR	Growth/broken line	Dairiki et al. (2013)
European sea bass	0.9	2.2	4.8	WG	Growth/broken line	Tibaldi and Lanari (1991)
Freshwater catfish (*Mystus nemurus* Cuv. & Val.)	4.0	1.2	3.5	SGR	Growth/polynomial	Tantikitti and Chimsung (2001)
Gilthead sea bream	3.5	2.2	5.0	PPV	Growth/polynomial	Marcouli et al. (2006)
Grass carp	3.2	2.1	5.4	WG, FE	Growth/polynomial	Wang et al. (2005)
Grouper	15.8	2.8	5.6	WG	Growth/broken line	Luo et al. (2006)
Indian major carp, catla	0.6	1.7–1.9	5.2–5.7	WG, FCR PPV, LRE, RNA/DNA ratio	Growth/polynomial	Zehra and Khan (2013c)
	275	1.5	5.8	FY	Growth/polynomial	Michelato et al. (2016b)
Indian major carp, mrigal	0.6	2.3	5.8	WG, FE	Growth/polynomial	Ahmed and Khan (2004b)
Japanese flounder	3.0	1.6–2.2	3.3–4.6	SGR, FE, PPV	Growth/broken line	Forster and Ogata (1998)
Japanese sea bass	5.5	2.5–2.6	5.8–6.1	SGR, FER, PER	Growth/broken line	Mai et al. (2006a)
Jundiá	1.5	1.5–1.7	4.5–5.1	WG	Growth/broken line/ Polynomial	Montes-Girao and Fracalossi (2006)
Largemouth bass	1.3	1.7–2.1	3.9–4.9	WG, FCR	Growth/broken line	Dairiki et al. (2007)
Large yellow croaker	1.2	2.4–2.5	5.7–5.8	SGR, FER, PER	Growth/broken line	Zhang et al. (2008)
Nile tilapia	1.4	1.5–1.6	5.4–5.7	WG, PPV	Growth/polynomial	Furuya et al. (2012)
	0.04	1.4	5.1	WG	Growth/broken line	Santiago and Lovell (1988)
	118	1.3–1.4	5.4–5.7	WG	Growth/polynomial	Furuya et al. (2004)
	5.7	1.4	5.2	WG	Growth/broken line	Furuya et al. (2006)
	1.9	2.1	6.2	WG	Growth/ANOVA	El-Saidy and Gaber (2002)

Continued

TABLE 4.19 Lysine requirement of fish and crustaceans.—cont'd

Species	IBW(g)	Requirement (% diet)	Requirement (g/16 g N)	Biomarkers	Methods/Models	References
Nile tilapia, GIFT	5.2	1.8–2.0	5.6–6.1	WG, PPV	Growth/broken line/ Polynomial	Prabu et al. (2019)
Tilapia	1.7	1.6	4.1	WG	Growth/ANOVA	Jackson and Capper (1982)
Pacific threadfin	3.0	1.8	5.1	SGR	Growth/broken line/ Polynomial	Deng et al. (2010)
Pacu	8.7	1.6	7.1	WG	Growth/broken line	Abimorad et al. (2010)
Rainbow trout	0.2–20.0	2.0–2.5			Factorial models	Hua and Bureau (2019)
	20.0–100	2.0–2.5				
	100–500	1.8–2.3				
	500–1500	1.4–1.8				
	1.0	2.9	6.1	WG	Growth/ANOVA	Ketola (1983)
	5.00	1.9	4.2	WG	Growth/broken line	Walton et al. (1984a)
		2.0	4.4	IAAO	Amino acid oxidation	
	7.0	1.9	4.3	WG	Growth/broken line	Walton et al. (1986)
	512	2.2–2.3	6.0–6.4	PA	Physiological and metabolism Broken line	Yun et al. (2016)
	24.0	1.7–2.3	4.4–5.8	WG	Growth/broken line/ Exponential	Encarnação et al. (2004)
	13.7	1.3	3.7	WG	Growth/broken line	Kim et al. (1992a)
	0.8	1.7–2.5	5.0–6.3	lysine gain	Growth/broken line/ SKM	Bodin et al. (2009)
	14.9	1.8	4.3	WG	Growth/ANOVA	Cheng et al. (2003)
	51.0	1.4–2.5	3.0–5.3	WG, PPV	Growth/exponential	Rodehutscord et al. (1997)
	7.0	2.0	4.2	PPV	Growth/exponential	Pfeffer et al. (1992)

Red sea bream	1.7	1.7–2.1	3.6–4.4	SGR, FE, PPV	Growth/broken line	Forster and Ogata (1998)
Silver perch	4.6	2.3	6.0	WG	Growth/broken line	Yang et al. (2011)
Stinging catfish	7.0	2.0–2.3	5.3–6.1	AWG; FCR; PPV; lysine deposition	Growth/broken line/ Polynomial	Farhat and Khan (2013b)
Yellow catfish	1.5	3.3	8.3	SGR	Growth/broken line/ Polynomial	Cao et al. (2012)
Striped bass	1.5/6.0	2.1	4.9	WG, SGR	Growth/least squares	Small and Soares JR (2000)
Striped bass hybrid	8.0	1.4	4.0	WG, plasma free lysine	Growth/broken line	Keembiyehetty and Gatlin (1992)
	5.0–12.0	1.4	4.0	WG	Growth/broken line	Griffin et al. (1992)
Turbot	18.1	2.5	5.0	WG	Growth/SKM	Peres and Oliva-Teles, (2008)
Crustaceans						
Atlantic ditch shrimp (*Palaemonetes varians*)	0.02	2.4–2.6	5.4–5.8	WG	Growth/broken line	Palma et al. (2015)
Pacific white shrimp	0.5	1.6	4.0	SGR	Growth/broken line	Xie et al. (2012)
Swimming crab	7.9	2.2	4.4	SGR	Growth/broken line	Jin et al. (2015b)
Tiger shrimp	0.02	2.1	5.2	WG	Growth/broken line	Millamena et al. (1998)

ANOVA, analysis of variance; *AWG*, absolute weight gain; *FCR*, feed conversion ratio; *FE*, feed efficiency; *FER*, feed efficiency ratio; *FY*, fillet yield; *IAAO*, indirect amino acid oxidation; *LRE*, leucine retention efficiency; *PA*, plasma amino acid; *PER*, protein efficiency ratio; *PPV*, protein productive value; *SGR*, specific growth rate; *SKM*, saturation kinetic model; *WG*, weight gain.

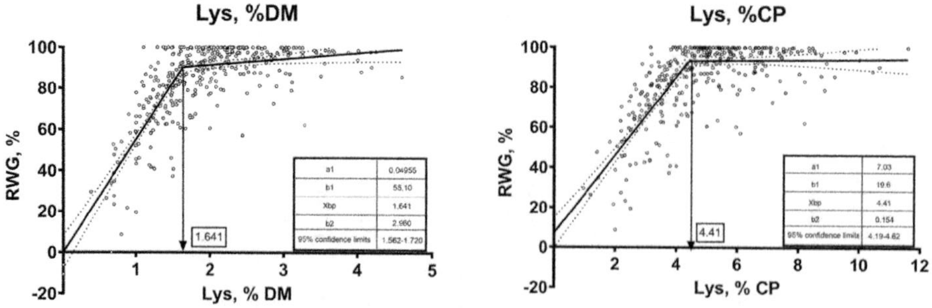

FIGURE 4.9 Metaanalysis of literature data on lysine requirement of different species of fish and crustaceans.

4.4.2.5 *Sulfur amino acid requirements*

Methionine (Met) and cystine (Cys) are classified as sulfur-containing protein-bound amino acids (SAAs). Except for the need for proper protein synthesis, Met and Cys have a number of physiological and biochemical functions in animals. Met is a nonpolar AA, which plays an important role in binding/recognition of hydrophobic ligands, such as lipids (Brosnan and Brosnan 2006b). Met not only acts as a protein synthesis precursor, but also participates in a wide range of other metabolic reactions including the production of S-adenosylmethionine (SAM), L-cysteine, glutathione, taurine, sulfate, phospholipids, and may participate in the formation of D-glucose and glycogen (Baker, 2006; Brosnan and Brosnan 2006b).

With a thiol side chain, Cys is a hydrophilic AA and acts as an important structural and functional component of many proteins and enzymes, forming disulfide bridges (dimers) in some proteins, which are important in folding the protein chain (Buxbaum, 2007). Cys is considered dispensable because it can be synthesized by fish from Met, but not vice versa. When fed a diet without or with low cysteine level, a portion of Met is directly incorporated into protein, and a portion is converted to cysteine. Conversely, if enough cysteine is included in the food, it reduces the dietary Met requirement. Thus, to determine the total SAA (Met plus Cys) or Met requirement, the dietary Cys content should be considered.

A deficiency in dietary Met is shown to lead to poor growth as well as specific disorders such as cataracts (Walton et al., 1982; Simmons et al., 1999; Keembiyehetty and Gatlin, 1993). The Met requirements of fish and other aquatic animals are shown in Table 4.20. The Met requirement of fish is distributed in the range of 0.5%–1.7% of diet or 1.2–4.5 g/16 gN with various levels of cysteine (0.03%–1.0% of diet) and depends on the Met sources used in the trials as well. Various commercial sources of methionine, such as *DL*-Met, 2-hydroxy-4-methylthiobutanamide salt (or named as methionine hydroxy analog [MHA]), or as a dipeptide Met-Met, have been tested. The efficiency of protein-bound Met could produce higher efficiency than crystallized Met (discussed in later part). Although there has been considerable debate over the biological efficacity of MHA in comparison to DL-Met with various species, it is assumed that the biological efficacy of MHA for fish about 75%–80% that of DL-Met on an equimolar basis (NRC, 2011). Kaushik and Seiliez (2010) had made a metaanalysis of Met requirement of fish, and it appears that most fish have a requirement of 2–3g/16 gN. Peres and Oliva-Teles (2017) observed that a dietary Met level of 2.4–2.7 g/16 gN was recommended for marine fish. Besides growth, a number of other biomarkers reflecting diverse physiological functions of Met have been used as criteria for

TABLE 4.20 Methionine requirements of fish and crustaceans.

Species	IBW(g)	Requirement (% diet)	Requirement (g/16 g N)	Biomarkers	Methods/Models	References
Asian seabass	2.6	1.0 0.3% Cys	2.2	WG	Growth/broken line	Coloso et al. (1999)
Atlantic salmon	0.2–20.0	0.9–1.1	–	–	Factorial models	Hua and Bureau (2019)
	20.0–100	0.9–1.1	–			
	100–500	0.9–1.1	–			
	500–1500	0.8–1.0	–			
	1500–2500	0.7–0.9	–			
	493	0.7 0.6% Cys	1.6	WG	Growth/ANOVA	Espe et al. (2008)
Black sea bream	14.2	1.7 0.3% Cys	4.5	SGR, PPV	Growth/polynomial	Zhou et al. (2011a)
Blunt snout bream	3.3	0.8–0.9 0.2% Cys	2.5	PPV, SGR	Growth/broken line	Liao et al. (2014)
Cobia	11.6	1.2 0.7% Cys	2.6	SGR	Growth/polynomial	Zhou et al. (2006)
Common carp (Jian carp)	8.2	0.8 0.8% Cys	2.6	PPV	Growth/polynomial	Xiao et al. (2012)
	9.9	1.2 0.3% Cys	3.3	WG	Growth/polynomial	Tang et al. (2009)
European sea bass	13.4	0.8–0.9 0.4% Cys	1.8–2.0	WG, PPV	Growth/broken line	Tulli et al. (2010)
Gibel carp	51.0	0.7–1.0 0.4% Cys	1.7–2.3	WG	Growth/polynomial/ broken line	Wang et al. (2016a)
Golden pompano	12.4	1.1–1.3 0.2% Cys	2.5–3.0	WG, PPV	Growth/polynomial	Niu et al. (2013)

Continued

TABLE 4.20 Methionine requirements of fish and crustaceans.—cont'd

Species	IBW(g)	Requirement (% diet)	Requirement (g/16 g N)	Biomarkers	Methods/Models	References
Grass carp	451	0.5–0.7 0.2%Cys	1.9–2.5	WG, intestinal trypsin, antioxidant	Growth/polynomial	Wu et al. (2017b)
Grouper	13.3	1.3 0.3% Cys	2.7	WG	Growth/broken line	Luo et al. (2005)
Large yellow croaker	1.2	1.4 0.1% Cys	3.2–3.3	SGR, FCR	Growth/polynomial	Mai et al. (2006b)
Indian major carp, mrigal	0.5	1.2 1.0% Cys	3.0	WG , FCR	Growth/polynomial	Ahmed et al. (2003)
	0.2	1.6–1.7 0.9% Cys	4.1–4.2	WG, FCR, PPV	Growth/broken line/ Polynomial	Khan and Abidi (2013)
Japanese flounder	2.8	1.4–1.5 0.1% Cys	2.9–3.0	WG, FCR	Growth/broken line	Alam et al. (2000)
Nile tilapia	5.6	0.5 0.4% Cys	1.8	WG	Growth/broken line	Nguyen and Davis (2009)
	0.1	0.8 0.2% Cys	2.7	WG	Growth/broken line	Santiago and Lovell (1988)
	563	0.9–1.0 digestible Met + Cys	3.4–3.8 digestible Met + Cys	WG, FY	Growth/ANOVA	Michelato et al. (2013)
Pangasius catfish (*Pangasius bocourti*)	162	0.6 0.4% Cys	1.9	SGR	Growth/polynomial	Yuangsoi et al. (2016)
Rainbow trout	0.2–20.0 20.0–100 100–500 500–1500	0.8–1.0 0.8–1.0 0.7–0.8 0.6–0.7	– – – –	–	Factorial models	Hua and Bureau (2019)
	15.4	0.5 0.5% Cys	1.2	WG	Growth/broken line	Kim et al. (1992b)
	2.7	0.6% 0.2% Cys	1.5	WG	Growth/ANOVA	Cowey et al. (1992)

Rockfish	43.6	1.4 0.1% Cys	2.8	SGR	Growth/polynomial	Yan et al. (2007)
Striped bass hybrid	4.8/8.6	0.9 0.1% Cys	2.5	WG	Growth/broken line	Keembiyehetty and Gatlin (1993)
	6.6	0.7 0.03%Cys	1.9–2.1	WG, FE, serum Met concentration	Growth/ANOVA	Griffin et al. (1994b)
White sea bass	5.6	0.9 0.6% Cys	2. 2	TGC	Growth/SKM	Salze et al. (2017a)
Yellow perch	4.5	1.0–1.1 0.03% Cys	3.0–3.3	WG, FCR	Growth/broken line	Twibell et al. (2000)
Yellowtail	23.3	1.1 0.3%	2.6	WG	Growth/broken line	Ruchimat et al. (1997)
Crustaceans						
Pacific white shrimp	3.0	0.8 0.4% Cys	2.0	WG	Growth/ANOVA	Niu et al. (2018)
Tiger shrimp	0.02	0.9 0.5% Cys	2.5	WG	Growth/broken line	Millamena et al. (1996)

ANOVA, analysis of variance; *FCR*, feed conversion ratio; *FE*, feed efficiency; *FY*, fillet yield; *PPV*, protein productive value; *SGR*, specific growth rate; *SKM*, saturation kinetic model; *TGC*, thermal-unit growth coefficient; *WG*, weight gain.

determining Met requirements: immune, antioxidant responses in serum and tissues, intestinal enzyme activity, microflora, hepatic glucose, lipid, and AA metabolism have been used to relate the effects of graded levels of dietary Met. Based on the use of such stringent criteria, Met requirements are found to be higher than those data based on whole-body growth (Liao et al., 2014; Wu et al., 2017b).

As mentioned earlier, dietary Cys can save the amount of dietary Met needed for growth performance. The cysteine replacement value for Met has been determined to be about 60% for channel catfish (Harding et al., 1977), 42% for rainbow trout (Kim et al., 1992b), and 40% for red drum (Moon and Gatlin, 1991) and hybrid striped bass (*Morone saxatilis* × *Morone chrysops*) (Griffin et al., 1994b).

4.4.2.6 Aromatic amino acids requirements

Aromatic AAs include phenylalanine (Phe) and tyrosine (Tyr), which are constituents of proteins. Phe is a nonpolar α-AA containing a phenyl group, and tyrosine contains a 4-hydroxy phenyl group. Tyr (4-hydroxyphenylalanine) is considered as a conditionally EAA (NRC, 2011). Phe is readily converted to Tyr in fish liver and kidney, whereas Tyr cannot be converted back to Phe. The enzyme Phe hydroxylase, which hydroxylates Phe to form Tyr, has been found in the livers of several species of fish (Voss and Waisman, 1966). Thus, Tyr becomes essential when dietary Phe level is below the minimum requirement level (Kim, 1993), and an adequate amount of dietary Tyr spares Phe from Tyr synthesis (NRC, 2011; Wu, 2013). Reduced growth performance and feed utilization have been observed in several fish species fed with phenylalanine deficient diets (Ahmed, 2009; Borlongan, 1992; Kim, 1993; Zehra and Khan, 2014).

In addition to protein synthesis, aromatic AAs have some other vital roles in metabolism: e.g., the precursors of important hormones and neurotransmitters, such as epinephrine and norepinephrine, thyroid hormone, and dopamine (Borlongan, 1992; Owen et al., 1999). Furthermore, Tyr has a special role by virtue of the phenol functionality; it serves as a receiver of phosphate groups and then changes the activity of the target protein (Cowey and Walton, 1989).

The dietary Phe requirements vary with different fish species and the dietary Tyr level, and the Phe or total aromatic AA requirement values for fish and crustacean are presented in Table 4.21. The dietary Phe requirement of fish is distributed in the range of 0.5%–2.3% of diet or 2.0–5.5 g/16 gN, respectively, and the optimum level of dietary Phe + Tyr for fish is 1.2%–2.5% of diet (4.3%–6.6% g/16 gN) estimated from the results of growth studies. For crustaceans, Phe requirement was estimated to be 1.4%–1.6% of diet or 3.7–3.9 g/16 gN based on the growth studies. The Tyr replacement value for Phe is between 39.5% and 60% in the fish species studied to date. Furthermore, it is obvious that the optimal inclusion of Phe for fish decreased with the increase of body weight (Hua and Bureau, 2019).

4.4.2.7 Threonine requirements

Apart from protein synthesis, threonine has a major role in intermediary metabolism (Walton and Cowey, 1982; Wu, 2018). The optimal inclusion of dietary Thr improves growth performance and immune response in fish (Abidi and Khan, 2008; Bodin et al., 2008; Gao et al., 2014; Li et al., 2007; Tibaldi and Tulli, 1999). Thr deficiency resulted in reduction of feed intake and growth in some species such as Japanese flounder (*Paralichthys olivaceus*) (Alam et al., 2003), rainbow trout (Rodehutscord et al., 1995b), Indian catfish (*Heteropneustes fossilis* (Bloch) (Ahmed, 2007)), and Indian major carp, roho (*Labeo rohita*) (Abidi and Khan, 2008). Reduced digestive and brush border enzyme activities were also found in fish intestines (Feng et al., 2013; Hong et al., 2015).

TABLE 4.21 Phenylalanine or total aromatic amino acid requirements of fish and crustaceans.

Species	IBW (g)	Requirement (% diet)	Requirement (g/16 g N)	Biomarkers	Method/Models	References
Atlantic salmon	0.2–20.0	1.4–1.8	3.9–4.0		Factorial models	Hua and Bureau (2019)
	20.0–100	1.4–1.8	3.9–4.0			
	100–500	1.4–1.7	3.8–3.9			
	500–1500	1.3–1.6	3.5–3.6			
	1500–2500	1.1–1.4	3.1–3.2			
	1.4	P + T: 2.5	P + T: 5.6	PPV	N balance	Rollin (2003)
Blunt snout bream	10.2	P + T: 1.9–2.0	P + T: 5.5–6.0	SGR, FER	Growth/polynomial	Ren et al. (2015b)
Indian major carp, catla	0.7	1.0 P + T: 1.7 TRV: 40.2%	3.1 P + T: 5.1	SGR	Growth/polynomial	Zehra and Khan (2014)
	1.2	1.5 P + T: 2.5 TRV: 40%.	3.7 P + T: 6.2	WG	Growth/broken line	Ravi and Devaraj (1991)
Indian major carp, mrigal	0.6	P + T:2.2 TRV*: 39.5%	P + T:5.3	WG	Growth/polynomial	Ahmed (2009)
	0.9	1.8 P + T: 2.3	4.4 P + T: 5.7	WG	Growth/broken line	Benakappa and Varghese (2004)
Indian major carp, roho	0.2	1.2 P + T: 2.2	3.0 P + T: 5.4–5.6	SGR, FCR, PER, PPV	Growth/polynomial	Khan and Abidi (2007c)
Channel catfish	6.9	0.5 P + T:1.2 TRV: 50%	2.1 P + T:5.0	WG, FE	Growth/broken line	Robinson et al. (1980b)
Chinook salmon	fingerling	1.7 P + T: 2.1	4.1 P + T: 5.1	WG	Growth/ –	Chance et al. (1964)
European sea bass	3.3	–	P + T: 2.6	A/E ratios	Whole-body AAs	Kaushik (1998)
	233					
Gilthead sea bream	4.6	P + T: 2.5	P + T: 5.8	A/E ratios	Whole-body AAs	Peres and Oliva-Teles (2009)
	40.0	–	P + T: 2.9	A/E ratios	Whole-body AAs	Kaushik (1998)
	173					

Continued

TABLE 4.21 Phenylalanine or total aromatic amino acid requirements of fish and crustaceans.—cont'd

Species	IBW (g)	Requirement (% diet)	Requirement (g/16 g N)	Biomarkers	Method/Models	References
Grass carp	3.6	P + T: 2.4	P + T: 6.6	WG, PER	Growth/polynomial	Gao et al. (2016)
	256	1.0	3.4	PWG	Growth/polynomial	Li et al. (2015a)
	256	0.8–1.2	2.8–3.9	LA, MDA	Growth/polynomial	Feng et al. (2015), (2017b)
Milkfish	0.6	P + T:2.4 TRV: 46%	P + T: 5.2	WG	Growth/broken line	Borlongan (1992)
Nile tilapia	0.01	1.1 P + T:1.6	3.8 P + T:5.5	WG	Growth/broken line	Santiago and Lovell (1988)
Rainbow trout	0.2–20.0 20.0–100 100–500 500–1500	1.2–1.6 1.2–1.5 1.1–1.4 0.9–1.0	3.4–3.6 3.4–3.5 3.1–3.3 2.4–2.5		Factorial models	Hua and Bureau (2019)
	12.7	0.7 P + T:1.5 TRV: 53%	2.0 P + T: 4.3	WG	Growth/broken line	Kim (1993)
Red drum	1.4	1.7 P + T: 2.1 TRV: 60%	4.8 P + T: 6.0	WG	Growth/broken line	Castillo et al. (2015)
Silver perch	10.0	2.3	5.5	WG	Growth/broken line	Ngamsnae et al. (1999)
Turbot	55.0 530	–	P + T: 5.3	A/E ratios	Whole-body AAs	Kaushik (1998)
Crustaceans						
Pacific white shrimp	0.4	1.6 P + T: 2.2	3.9 P + T: 5.5	WG, FCR, PPV	Growth/polynomial	Yan et al. (2019)
Prawn (*Marsupenaeus japonicus*)	0.8	1.3	2.6	EAA profile of the whole body	Whole-body AAs	Teshima et al. (2002)
Tiger shrimp	0.02	1.4	3.7 P + T: 4.9	WG	Growth/polynomial	Millamena et al. (1999)

EAA, essential amino acid; *FCR*, feed conversion ratio; *FE*, feed efficiency; *FER*, feed efficiency ratio; *LA*, lysozyme activity; *MDA*, malondialdehyde; *PER*, protein efficiency ratio; *PPV*, protein productive value; *PWG*, percent weight gain; *SGR*, specific growth rate; *TRV*, Tyr replacement value for Phe; *WG*, weight gain.

TABLE 4.22 Threonine requirements of fish and crustaceans.

Species	IBW (g)	Requirement (% diet)	Requirement (g/16 g N)	Biomarkers	Method/Models	References
Atlantic salmon	0.2–20.0 20.0–100 100–500 500–1500 1500–2500	1.3–1.6 1.2–1.6 1.2–1.5 1.1–1.4 1.0–1.2	3.5 3.4–3.5 3.4–3.5 3.1–3.2 2.7–2.8		Factorial models	Hua and Bureau (2019)
	0.8	1.1	2.7	Thr accretion	Growth/broken line	Bodin et al. (2008)
	1.4	1.2	2.7	PPV	N balance	Rollin et al. (2003)
	79.0	1.1–1.3	2.1–2.5	TGC, FER, PPV, TAA gain, threonine gain	Growth/exponential	Helland and Grisdale-Helland (2011)
Blunt snout bream	3.0	1.6	4.6	SGR	Growth/polynomial	Habte-Tsion et al. (2015)
Indian major carp, catla	0.6	1.4–1.5	4.1–4.5	AWG, FCR, PPV, threonine deposition, carcass protein, RNA/DNA ratio, Hb, Ht, RBCs	Growth/polynomial	Zehra and Khan (2016b)
	1.2	2.0	5.0	WG	Growth/broken line	Ravi and Devaraj (1991)
Indian major carp, mrigal	0.5	1.8	4.5–4.6	SGR, FCR, PPV	Growth/broken line	Ahmed et al. (2004)
	1.1	1.7	4.2	SGR	Growth/broken line	Benakappa and Varghese (2002)
Indian major carp, roho	0.6	1.5–1.7	3.8–4.2	WG, FCR, PER, PPV	Growth/broken line	Abidi and Khan (2008)
Channel catfish	12.5	0.5	2.2	WG	Growth/broken line	Wilson et al. (1978)
Common carp (Jian carp)	13.6	1.6	5.1	PWG	Growth/polynomial	Feng et al. (2013)
European sea bass	3.3 233	–	2.7	A/E ratios	Whole-body AAs	Kaushik (1998)
	7.5	1.2–1.5	2.5–3.0	WG, PPV, plasma-free threonine concentrations	Growth/polynomial	Tibaldi et al. (1999)

Continued

TABLE 4.22 Threonine requirements of fish and crustaceans.—cont'd

Species	IBW (g)	Requirement (% diet)	Requirement (g/16 g N)	Biomarkers	Method/Models	References
Gilthead sea bream	4.6	1.3	3.0	A/E ratios	Whole-body AAs	Peres and Oliva-Teles (2009)
	40.0 173	–	2.8	A/E ratios	Whole-body AAs	Kaushik (1998)
Grass carp	4.0	1.4	3.6	SGR	Growth/polynomial	Gao et al. (2014)
	9.5	1.5	4.5–4.7	SGR, enteritis morbidity, IgM content	Growth/polynomial	Dong et al. (2017)
	9.5	1.5–1.6	4.7–4.8	Gill rot morbidity, LA, MDA	Growth/polynomial	Dong et al. (2018)
	442	1.2	4.2	WG	Growth/polynomial	Hong et al. (2015)
Striped bass hybrid	3.0 9.8	0.9–1.0 0.8	2.4–2.9 2.4	WG, FE, PPV, WG	Growth/polynomial	Keembiyehetty et al. (1997)
Indian catfish	3.6	1.2–1.4	3.1–3.4	LWG, FCR, PER, PPV	Growth/polynomial	Ahmed (2007)
Japanese flounder	2.0	1.6	3.1–3.2	FCE, WG, PPV	Growth/broken line	Alam et al. (2003)
Milkfish	1.3	1.8	4.5	WG	Growth/broken line	Borlongan (1991)
Nile tilapia	0.1	1.1	3.8	WG	Growth/broken line	Santiago and Lovell (1988)
	3.0	1.3	4.7	WG	Growth/polynomial	Yue et al. (2014)
	563	1.2	4.0	FY	Growth/polynomial	Michelato et al. (2016c)
Rainbow trout	0.2–20.0	1.1–1.4	3.0–3.1		Factorial models	Hua and Bureau (2019)
	20.0–100	1.1–1.4	3.0–3.1			
	100–500	1.0–1.3	2.8			
	500–1500	0.8–1.0	2.1–2.2			
	1.8	1.1	2.6	Thr accretion	Growth/broken line	Bodin et al. (2008)
	506	1.0	2.7–3.1	PPthr, PAthr, PPA	Physiological and metabolism/ broken line	Yun et al. (2015)

Red drum	2.8	0.8	2.3	WG, PER	Growth/broken line	Boren and Gatlin (1995)
Striped bass	2.7	0.9–1.1	2.2–2.7	WG, SGR, FCR, ANU	Growth/polynomial	Small and Soares (1999)
Turbot	55.0 530	–	2.9	A/E ratios	Whole-body AAs	Kaushik (1998)
Crustaceans						
Pacific white shrimp	0.5	1.4	3.8	SGR	Growth/broken line	Huai et al. (2009)
	0.5	1.2	2.8	SGR	Growth/broken line	Zhou et al. (2013)
Prawn	0.8	1.2	2.3	EAA, CP retained	Whole-body AAs	Teshima et al. (2002)
Tiger shrimp	0.1	1.4	3.5	WG	Growth/polynomial	Millamena et al. (1997)

ANU, apparent nitrogen utilization; *AWG*, absolute weight gain; *EAA*, essential amino acid; *FCE*, feed conversion efficiency; *FCR*, feed conversion ratio; *FE*, feed efficiency; *FER*, feed efficiency ratio; *FY*, fillet yield; *Hb*, hemoglobin; *Ht*, hematocrit; *LA*, lysozyme activity; *LWG*, live weight gain; *MDA*, malondialdehyde; *PAthr*, postabsorptive free threonine concentrations; *PER*, protein efficiency ratio; *PPA*, postprandial plasma ammonia concentrations; *PPthr*, postprandial plasma free threonine concentrations; *PPV*, protein productive value; *PWG*, percent weight gain; *RBCs*, red blood cell counts; *SGR*, specific growth rate; *TAA*, total amino acid; *TGC*, thermal-unit growth coefficient; *WG*, weight gain.

The dietary Thr requirement values in the diet for fish species are summarized in Table 4.22. The values range from 0.5% to 2.0% of diet or 2.2–5.1 g/16 gN, based on growth trials. The most studied species are grass carp and Atlantic salmon by various methodologies. For the grass carp, the optimum level of Thr in diet is between 1.2% and 1.6% of diet (3.6–4.8 g/16 gN). For Atlantic salmon, the data are within the range of 1.0%–1.6% of diet or 2.2–3.5 g/16 gN. For crustaceans, dietary Thr requirement is at range of 1.2%–1.4% of diet or 2.3–3.8 g/16 gN.

4.4.2.8 *Tryptophan requirements*

Tryptophan is an α-AA containing an indole functional group, and only the L-stereoisomer of Trp is used in structural or enzyme proteins (NRC, 2011). In addition to protein synthesis, Trp acts as the precursor of several compounds, including serotonin (5-HT) and neurotransmitter (Hseu et al., 2003), which are involved in the regulation of gastrointestinal secretions, feeding behavior, and sensation (Jacobs and Azmitia, 1992). Several early studies had reported that diets with Trp deficiency or excess depressed growth performance and induced spinal deformities, such as scoliosis or lordosis, eye cataracts, and disruption of mineral metabolism in salmonids (Akiyama et al., 1986; Halver and Shanks, 1960; Walton et al., 1984b). A reduction in growth has also been observed with an excess of dietary Trp in rainbow trout (Poston and Rumsey, 1983), Nile tilapia (Santiago and Lovell, 1988), mrigal (Ahmed and Khan 2005b), and Indian catfish (Ahmed, 2012). Furthermore, supplementation of Trp in diet was shown to reduce aggressive behavior and cannibalism in Atlantic cod (*Gadus morhua*), grouper (*Epinephelus coioides*), rainbow trout, and freshwater crayfish (*Astacus leptodactylus*) (Hoglund et al., 2007; Hseu et al., 2003; Lepage et al., 2002, 2005; Winberg et al., 2001; Harlioglu et al., 2014) and decrease stress-induced cortisol production (Lepage et al., 2002, 2003) and immune suppression (Hoseini et al., 2016) in various fish species.

Data on the optimal dietary Trp levels in fish diets are presented in Table 4.23. The recommend inclusion of Trp appears to be within the range of 0.1%–0.5% of diet or 0.3–1.8 g/16 gN for fish estimated by growth studies. It is interesting to note that the optimal inclusion of Trp for Indian major carp catla (*Catla, catla*) increases with the increase of body weight, which is quite different from other species possibly due to methodological issues (Ravi and Devaraj, 1991; Zehra and Khan, 2015b). For crustaceans, data on Trp requirement are within the range of 0.2%–0.4% of diet or 0.5–1.7 g/16 gN.

4.4.3 Conditional essential amino acids

4.4.3.1 *Glycine*

Glycine (Gly) is the simplest AA in nature and is a major AA in animals. In some proteins such as collagen, Gly can represent 30% of total AAs together with proline (Pro) and hydroxyproline (HyPro) (Kimura et al., 1988). The body collagen content is about 0.3%–0.6% fresh weight in Atlantic salmon (Eckhoff et al., 1998; Aidos et al., 1999) or 0.5%–0.6% in gilthead sea bream (Suarez et al., 2005). Beyond being a proteinaceous AA, glycine is a major inhibitory neurotransmitter in the animal spinal cord and brain stem and also participates in the biosynthesis of purines, porphyrin, glutathione, creatine, and uric acid (Li and Wu, 2018). Besides, glycine is considered to have a sweet taste and can have an important role in feed intake regulation (Hara, 1994; Xue and Cui, 2001). An early study with rainbow trout (Adron and Mackie, 1978), however, did not find glycine to be a feeding stimulant. Gly is abundant in conventional high fishmeal diet, but it could be deficient when fishmeal is replaced by plant

TABLE 4.23 Tryptophan requirements of fish and crustaceans.

Species	IBW (g)	Requirement (% diet)	Requirement (g/16 g N)	Biomarkers	Method/Models	References
Atlantic salmon	0.2–20.0	0.3	0.7–0.8		Factorial models	Hua and Bureau (2019)
	20.0–100	0.3	0.7–0.8			
	100–500	0.3	0.7			
	500–1500	0.2–0.3	0.6–0.7			
	1500–2500	0.2–0.3	0.6			
	1.4	0.3	0.7	PPV	N balance	Rollin et al. (2003)
African catfish	11.5	0.4	1.1	SGR	Growth/broken line	Fagbenro and Nwanna (1999)
Asian seabass	5.3	0.2	0.4	WG	Growth/broken line	Coloso et al. (2004)
Indian major carp, catla	0.6	0.2–0.3	0.6–0.8	AWG, FCR, PG, RNA/DNA ratio	Growth/exponential	Zehra and Khan (2015b)
	1.2	0.4	1.0	WG	Growth/broken line	Ravi and Devaraj (1991)
Indian major carp, mrigal	0.6	0.4	1.0	WG	Growth/polynomial	Ahmed and Khan (2005b)
Indian major carp, roho	0.4	0.4	0.9–1.0	LWG, PER, PPV, FCR	Growth/polynomial	Abidi and Khan (2010)
	1.5	0.5	1.1	SGR	Growth/broken line	Murthy and Varghese (1997b)
Channel catfish	12.5	0.1	0.5	WG	Growth/broken line	Wilson et al. (1978)
Common carp	0.2	0.1	0.3	WG, PG	Growth/polynomial	Dabrowski (1981)
Common carp (Jian carp)	7.7	0.4	1.1	WG	Growth/polynomial	Tang et al. (2013)
European sea bass	3.3 233	–	0.6	A/E ratios	Whole-body AAs	Kaushik (1998)

Continued

TABLE 4.23 Tryptophan requirements of fish and crustaceans.—cont'd

Species	IBW (g)	Requirement (% diet)	Requirement (g/16 g N)	Biomarkers	Method/Models	References
Gilthead sea bream	40.0 173	–	0.6	A/E ratios	Whole-body AAs	Kaushik (1998)
Grass carp	287	0.4	1.3	PWG, PAC	Growth/polynomial	Wen et al. (2014)
		0.4	1.3–1.4	ROS, MDA, PC	Growth/polynomial	Jiang et al. (2015b)
		0.3–0.4	1.1–1.2	Muscle cooking loss, shear force, and MDA	Growth/polynomial	Jiang et al. (2016d)
Hybrid striped bass	42.4	0.2–0.3	0.6–0.8	WG	Growth/4SKM and 5SKM	Gaylord et al. (2005)
Indian catfish	4.4	0.3	0.8	FCR, PER, PPV	Growth/polynomial	Ahmed (2012)
	6.7	0.2–0.3	0.6–0.7	AWG, PPV, Hb, Ht	Growth/broken line	Farhat and Khan (2014b)
Japanese flounder	0.5	0.5	1.8	WG	Growth/ANOVA	Han et al. (2013)
Milkfish	7.7	0.3	0.6	WG	Growth/broken line	Coloso et al. (1992)
Nile tilapia	0.1	0.3	1.0	WG	Growth/broken line	Santiago and Lovell (1988)
	1.4	0.3	1.0	WG	Growth/polynomial	Wu et al. (2018b)
	3.4	0.3	1.1	WG	Growth/polynomial	Zaminhan et al. (2018)
	7.9	0.3	0.8–1.1	WG, Trp deposition, FE, PPV	Growth/broken line	Nguyen et al. (2019)
	38.2	0.3	1.0	WG	Growth/polynomial	Zaminhan et al. (2017)
Rainbow trout (*Oncorhynchus mykiss*)	0.2–20.0	0.2–0.3	0.6–0.7		Factorial models	Hua and Bureau (2019)
	20.0–100	0.2–0.3	0.6–0.7			
	100–500	0.2–0.3	0.6			
	500–1500	0.2	0.4–0.5			

Rainbow trout (*Salmo gairdneri*)	7.0	0.3	0.4	WG	Growth/broken line	Walton et al. (1986)
	13.0	0.2–0.3	0.6–0.7	WG	Growth/broken line	Kim et al. (1987)
	14.0	0.2	0.5	WG	Growth/broken line	Walton et al. (1984b)
Red drum	2.3	0.3	0.8	WG	Growth/polynomial	Pewitt et al. (2017)
Silver catfish (*Rhamdia quelen*)	4.7	0.3–0.4	0.9	WG, PPV, AP	Growth/polynomial	Pianesso et al. (2015)
		0.3	0.7–0.8		Growth/broken line	
Turbot	55.0	–	0.6	A/E ratios	Whole-body AAs	Kaushik (1998)
	530					
Crustaceans						
Pacific white shrimp	0.4	0.4	0.9–1.0	WG, FCR; PPV	Growth/polynomial	Jin et al. (2017)
Prawn	0.8	0.3	0.6	EAA and CP retained	Whole-body AAs	Teshima et al. (2002)
Tiger shrimp	0.02	0.2	0.5	WG	Growth/polynomial	Millamena et al. (1999)

ANOVA, analysis of variance; *AP*, acidic protease; *AWG*, absolute weight gain; *EAA*, essential amino acid; *FCR*, feed conversion ratio; *FE*, feed efficiency; *Hb*, hemoglobin; *Ht*, hematocrit; *LWG*, live weight gain; *MDA*, malondialdehyde; *PAC*, plasma ammonia concentration; *PC*, protein carbonyls; *PER*, protein efficiency ratio; *PPV*, protein productive value; *PWG*, percent weight gain; *ROS*, reactive oxygen species; *SGR*, specific growth rate; *SKM*, saturation kinetic model; *WG*, weight gain.

ingredients, because of the low content of serine, choline, threonine, and hydroxyproline for glycine synthesis (Li et al., 2011). A beneficial effect of glycine supplementation was also shown in juvenile white shrimp fed diets containing about 15% fish meal (Xie et al., 2014a).

4.4.3.2 *Glutamate and glutamine*

Glutamate (Glu) or glutamic acid is widely distributed in animal, dairy, and plant proteins (Beyreuther et al., 2007). Although an NEAA, Glu has critical roles in metabolism and regulates several physiological processes (Brosnan and Brosnan, 2013). Glutamate is a precursor of several other AAs, including alanine, aspartate, ornithine, and proline in the enterocytes (Reeds et al., 2000). Thus, L-glutamate can play a role both locally inside enterocytes and through the production of other AAs from the perspective of interorgan metabolism (Blachier et al., 2009). A number of in vitro studies have shown the beneficial effects of an addition of glutamine in improving the proliferation, function, and antioxidant capacity of primary cultures of fish enterocytes (Hu et al., 2014; Jiang et al., 2009, 2015a, 2016a, 2017b). A dietary supplementation with Glu was shown to improve muscle flesh quality by reducing oxidative damage and improving nonenzymatic and enzymatic antioxidant capacity in fish muscle (Zhao et al., 2019). A dietary supplementation of Gln was also found to enhance the activity of the antioxidant defense system and the serum nonspecific immunity in juvenile hybrid sturgeon (Zhu et al., 2011).

4.4.3.3 *Proline and hydroxyproline*

Proline and hydroxyproline are unique AAs both chemically and biochemically. They are key regulators of multiple biochemical and physiological processes in cells (Wu et al., 2011). Hydroxyproline is formed from proline through hydroxylation of proline in proteins, primarily collagen (Wu et al., 2017c). Hydroxyproline is crucial for collagen synthesis. Addition of 0.6%–1.2% of hydroxyproline to high plant protein diets was found to enhance collagen content in the muscle of turbot and large yellow croaker (*Larimichthys crocea*) and increase muscle texture and quality (Zhang et al., 2013, 2015; Liu et al., 2014c; Wei et al., 2016). As proline is a substrate for the synthesis of glycine, pyruvate, arginine, and glucose (Wu et al., 2011; Wu and Meininger, 2009). It can contribute toward the regulation of the redox status of cells (Phang et al., 2008, 2010). Compared with fishmeal, plant protein sources are low in proline content (Aksnes et al., 2006). Dietary hydroxyproline supplementation was found to improve muscle texture and flesh quality in turbot (Liu et al., 2014d) and in yellow croaker (Wei et al., 2016).

4.4.3.4 *Taurine*

Taurine is an intracellular AA only present in the free AA fraction in most animal tissues. Being a β-AA, it is not incorporated into any protein but is widely distributed in all animal tissues and truly little, if at all, present in the plant kingdom. Taurine is recognized to have a large range of biological roles being involved in several metabolic pathways (Huxtable, 1992). Taurine is involved in bile salt formation, antioxidation, cell membrane stability, cell volume regulation, osmoregulation, neurotransmitter function, glycolysis and glycogenesis, hypolipidemia, growth modulation, vision, and maintenance of calcium homeostasis (Ripps and Shen, 2012). Some of the early studies on the importance of inclusion of taurine in the diets of fish were to address issues linked with a so-called green liver syndrome in yellowtail (Goto et al., 2001; Takagi et al., 2006, 2008). In some marine finfish species, endogenous production of taurine seems to be limited, especially during the early developmental stages (Takeuchi et al., 2001). In these species, dietary supplementation with taurine has been found to have beneficial effects in terms of growth and health, leading to suggestions of conditional essentiality of dietary taurine in these species and for larval stages of fish. Only a couple of studies have attempted to determine the dietary essentiality of taurine. As regards the level of free taurine in fish meals, data available indicate that this ranges from 0.1% to 3.1% of diet or 0.3–7.5 g/16 gN. Besides, Brill et al. (2019) found that European seabass requires 5% of taurine in diet to meet the requirement of visual functions (Table 4.24).

4.4.4 Nutritional interactions of amino acids

Nutritional interactions of AA (e.g., imbalance, antagonism, and toxicity) influence the efficiency with which EAA are used for protein synthesis and other productive purposes. AA imbalance depresses feed intake, thereby causing reduced performance, which could be due to metabolic changes leading to altered free AA pool in blood, and also influences the appetite and feeding behaviors regulated by the peripheral and central nervous system (Fuller, 1994; Wu, 2018). Although the mechanism of these interactions is not fully elucidated, at least the competitive antagonistic receptors are widely studied.

4.4.4.1 *The agonistic and antagonistic effects among amino acids*

AA antagonisms can be defined as specific interactions in which the excess consumption of a structurally related AA increases the requirement for another EAA.

The two well-known antagonistic interactions are the BCAA antagonism involving leucine, isoleucine, and valine in which an excess of one (or two) of these AAs increases the requirement for the other BCAAs and the lysine–arginine antagonism in which excess dietary lysine increases the requirement for arginine.

4.4.4.1.1 Antagonisms among branched-chain amino acids

The metabolism of BCAAs has three important characteristics: first, the catabolic enzymes for BCAAs are distributed widely in body tissues, including the kidney, muscle, and the central nervous system. Second, these three AAs share the same common transporter for intestinal absorption. Finally, the first steps in the oxidation of each of three BCAAs are catalyzed by two common enzymes, in addition to the metabolism of the three AAs using the same enzymatic system. It has been observed in most land animals and humans that BCAAs could produce antagonistic effects when the relative proportion of these three AAs in the diets are ill-balanced. The reductions in plasma levels of isoleucine and valine, which are induced by leucine, have mainly been due to competitive inhibition during intestinal absorption and increased oxidation through BCKA dehydrogenase activation.

In fish, antagonistic effects relative BCAAs on growth and nutrient utilization have been assessed only in a limited number of species, and the results show some inconsistencies. Some early work (Chance, 1964; Nose, 1979; Hughes et al., 1983) indicated possible antagonisms between BCAA in salmonids. Later studies with channel catfish (Robinson et al., 1984) and rainbow trout (Rodehutscord et al., 1997) did not show any clear antagonistic reactions. Even a very high excess of dietary Leu did not induce any antagonistic effect in rainbow trout (Choo et al., 1991), contrary to observations of Yamamoto et al. (2004) who observed a clear anatagonism between BCAA in rainbow trout (*Oncorhynchus mykiss*) fed a diet rich in Leu.

4.4.4.1.2 Antagonisms between arginine and lysine

Both Lys and Arg are basic AAs and share a common dibasic AA carrier, and competitive inhibition between these two AA can affect their absorption, transport, and metabolism. In Atlantic salmon, Berge et al. (1998) observed that a dietary Lys excess led to a decrease in

TABLE 4.24 Taurine requirements of fish and crustaceans.

Species	IBW (g)	Requirement (% diet)	Requirement (g/16 g N)	Biomarkers	Method/Models	References
Asian seabass	26.8	0.5	1.0	WG	Growth/5SKM	Poppi et al. (2018)
Black carp	5.9	0.1	0.3	WG, enzyme activity, AOC	Growth/ANOVA	Zhang et al. (2018a)
California yellowtail (*Seriola lalandi*)	4.9	0.3	0.6	TGC, PPV	Growth/4SKM	Salze et al. (2018a)
Channel catfish	5.6	0.2	0.7	Growth, FE	Growth/polynomial	Peterson and Li (2018)
Common carp	1.0	1.5	4.2	Growth, enzyme activity, AOC	Growth/polynomial	Abdel-Tawwa b and Monier (2018)
Common dentex (*Dentex dentex*)	39.1	0.2	0.4	SGR	Growth/ANOVA	Chatzifotis et al. (2008)
European sea bass	0.8	0.2	0.4	SGR	Growth/ANOVA	Martinez et al. (2004)
Florida pompano (*Trachinotus carolinus*)	7.7	0.5–3.1 0.3–1.0	1.2–7.5 0.6–2.3	TGC FE, whole-body taurine	Growth/4SKM and 5SKM and broken quadratic model	Salze et al. (2018b)
Japanese flounder	0.2	1.5	2.4	WG, FE	Growth/ANOVA	Kim et al. (2005)
	0.9	1.4	2.5	WG, FE, PER	Growth/ANOVA	Park et al. (2002)
	19.5	0.9–1.3	1.8–2.7	FCR, PER, hematological parameters, immune response	Growth/ANOVA	Kim et al. (2017)
Nile tilapia	0.02	1.0	2.4	WG	Growth/polynomial	AL-Feky (2015)
Rock bream (*Oplegnathus fasciatus*)	2.7	0.5–0.6	1–1.2	WG, FE	Growth/ANOVA and Broken line	Ferreira et al. (2014)
Parrot fish (*Oplegnathus fasciatus*)	13.5	0.9	1.9	WG	Growth/broken line	Lim et al. (2013)

Red sea bream	2.5	0.5	0.8	SGR, WG, FE	Growth/ANOVA	Matsunari et al. (2008a)
	4.7	0.5	0.8	WG, FE	Growth/polynomial	Matsunari et al. (2008b)
	39.0	1.0	1.8	Epidermal thickness, scale loss	Growth/ANOVA	Kato et al. (2013)
Rice field eel (*Monopterus albus*)	25.1	0.2	0.4	AOC, immune enzyme activities	Growth/ANOVA	Hu et al. (2018)
Sablefish (*Anoplopoma fimbria*)	52.1	1.5	2.8	WG	Growth/5SKM	Johnson et al. (2015)
Tongue sole (*Cynoglossus semilaevis*)	3.3	1.0	1.7	SGR, enzyme activity	Growth/ANOVA	Zheng et al. (2015)
Totoaba (*Totoaba macdonaldi*)	10.0	0.5	0.9	Blood biochemistry, GBSI , ADC	Growth/4 SKM	Satriyo et al. (2017)
Turbot	3.7	1.4	2.8	SGR, FER, blood glucose level	Growth/ANOVA	Zhang et al. (2018b)
	3.7	1.4	2.8	Hyperplasia fiber growth and textural properties of muscle	Growth/ANOVA	Sampath et al. (2020)
White sea bass	4.0	1.0	2.5	TGC	Growth/4SKM	Jirsa et al. (2014)
Yellow catfish	5.2	1.1	2.7	WG, immunity, hyperammonemia	Growth/broken line	Li et al. (2016)
Crustaceans						
Pacific white shrimp	0.5	0.2	0.4	WG	Growth/polynomial	Yue et al. (2012)
Tiger shrimp	0.7	0.1	0.2	WG	Growth/ANOVA	Shiau and Chou (1994)

ADC, apparent digestibility coefficient; *ANOVA*, analysis of variance; *AOC*, antioxidant capacity; *FCR*, feed conversion ratio; *FE*, feed efficiency; *FER*, feed efficiency ratio; *GBSI*, gallbladder-somatic index; *PER*, protein efficiency ratio; *PPV*, protein productive value; *SGR*, specific growth rate; *SKM*, saturation kinetic model; *TGC*, thermal-unit growth coefficient; *WG*, weight gain.

plasma Arg levels and a reduction in hepatic arginase activity in vitro. A dietary excess Arg also was shown to decrease plasma Lys levels (Berge et al., 1997). Kaushik et al. (1988) discovered that the levels of plasma Arg significantly decreased because of increasing the dietary lysine level from 1.8% to 3.0% of the diet in rainbow trout. But this effect was perhaps due to a possible marginal Lys deficiency in the control diet used by the foresaid authors (Encarnacăo et al., 2004).

In the studies of channel catfish, European sea bass, and Japanese flounder, there was no negative effect of feeding excess lysine on plasma arginine levels and/or growth. However, an antagonist effect had been reported by Dai et al. (2015) in turbot. The Arg/Lys level significantly affected the growth of juvenile dark barbel catfish, black sea bream (Zhou et al., 2011b), and Pacific white shrimp (Dong et al., 2018b). In the work of Shen et al. (2017) reported when concentrations of Arg and Lys were low, there was no significant effect of different Arg/Lys ratios, but when the levels of dietary Lys and Arg were increased by 20%, any increase in Lys level had a negative effect on the growth of all-male yellow catfish, showing an interaction between Lys and Arg depending on their dietary levels. Higher dietary Lys level leads to stronger antagonism to Arg with a number of effects on enzyme activities linked to the ornithine urea cycle (Zhou et al., 2011b).

4.4.4.2 *Ideal protein concept*

The concept of "ideal protein" was defined for diets of chickens in the primary study of Fisher and Scott (1954), as the AA profile that meets exactly the requirement of the animal with no excess or deficiency based on the famous barrel analogy as so-called Liebig's law (Whitson & Walster, 1912). It was thought that the composition of EAAs in diets should be the same as that in hen's eggs or casein. Under the ideal protein concept, EAA requirements are considered as proportions relative to total EAAs, rather than as proportions to the whole diet (Wang and Fuller, 1989). One of the ideal AA patterns is stated as the ratio of each of the nine other EAAs to lysine, which is supposed the arbitrary value of 100. A refinement of the ideal AA profile with the inclusion of some of the CEAA (Cys, Gly, Pro, and Gln) led to a significant improvement (Baker and Han, 1994). The ideal protein concept has been well and widely used as a guideline to establish an AAs profile to formulate practical diets for poultry and swine, but only a few studies conducted on several fish species (Mambrini and Kaushik, 1995b; Green and Hardy, 2002; Rollin et al., 2003; Xue et al., 2012; Furuya et al., 2015; Turchini et al., 2019).

Table 4.25 proposes an ideal AA profile in dietary protein for teleost fish and penaeid shrimp. Although these ideal protein profiles are not perfect, they provide a reasonable starting point for formulating diets on an ideal protein basis (NRC, 2011).

Cowey and Cho (1993) tried to see if a basal diet with a dietary protein level that would just meet the EAA requirements of rainbow trout and see the effects of supplementing such a diet with different mixtures of EAA or NEAA on growth or enzyme activities. The addition of NEAA to a diet containing 251 g protein/kg diet did not improve growth and was without effect on the activities of liver enzymes involved in AA metabolism. A few studies have tried to determine an appropriate ratio between dietary EAA and NEAA (Hughes, 1985; Mambrini and Kaushik, 1995b). Green et al. (2002) recommended an optimal EAA/NEAA ratio as 1.33 for rainbow trout. Similarly, in gilthead sea bream, a dietary EAA/NEAA ratio of 1.1 was better than a ratio of 0.8 (Gomez-Requeni et al., 2003). In a study with European seabass, it was suggested that a dietary EAA/NEAA ratio of 50/50 is necessary to promote maximum

TABLE 4.25 Ideal amino acid profile for fish and penaeid shrimp derived from a review of the literatures.

	Fish[a]	Shrimp[b]
EAA		
Lys	100	100
Arg	84.6 ± 2.50	95
His	26.8 ± 7.24	38
Ile	53.4 ± 6.19	54
Leu	92.1 ± 11.2	70
Met	39.4 ± 7.72	38
Cys	12.8 ± 3.62	16
Phe	53.1 ± 1.68	55
Tyr	38.6 ± 5.86	35
Thr	60.6 ± 7.33	56
Val	63.5 ± 2.11	61
Trp	11.6 ± 1.46	14

[a]*Data sources include NRC, 2011; Furuya et al. (2015).*
[b]*Data from NRC, 2011.*
EAA, essential amino acid.

growth performance, and this ratio should be increased to 60/40 if maximization of feed, protein, and energy utilization is to be considered (Peres and Oliva-Teles, 2006).

One possible drawback in relying on the whole-body AA profile is that it does not take into account extensive catabolism of AAs by the digestive tract itself (Wu, 2014); a second point is that the CEAAs and NEAAs are not fully taken into account (Kim et al., 2001). There is a need to consider all proteinogenic AAs in proper proportions and adequate amounts (Wu, 2018).

4.4.4.3 *Efficacy of supplementation with crystalline amino acids*

To meet the dietary requirements for all EAAs, the main strategy behind feed formulation is to mix different complementary protein sources with or without supplementing with crystalline amino acids (CAAs). CAAs have been used commercially to meet the EAA requirement of land animals for more than 60 years. All the 10 EAAs, in various chemical forms produced by chemical synthesis or microbial fermentation, are available for use in farm animal feeds. In aquaculture, the same principles are applicable and have been demonstrated since long (Ogino, 1980; Nose et al., 1974). There have been concerns that special care is needed to protect CAAs for use in aquafeeds and that the postprandial uptake patterns of CAAs and protein bound AAs will differ (Cowey and Walton, 1988). Several studies have shown that CAAs are utilized as efficiently as those of intact protein origin in meeting

EAA requirements of fish (Espe and Lied, 1994; Rodehutscord et al., 1995a,b; Xue et al., 2012). Some other studies have found that the use of CAAs caused lower growth and FE of fish, which could be related to palatability, pH change, leaching, and rapid uptake (Refstie et al., 2001; Zhou et al., 2007b; Nguyen and Davis, 2016; Zhang et al., 2019b). The most convincing cognition is that CAA may be absorbed more rapidly and/or earlier in the gastrointestinal tract than protein/peptide-bound AAs (Cowey and Walton, 1988; Niu et al., 2018). The faster and/or earlier absorption may result in a greater proportion of the CAAs being catabolized, but not for protein biosynthesis (Cowey and Walton, 1988; Zarate et al., 1999).

Several studies had convinced the improved metabolic utilization of CAAs when fed more frequently (Yamada et al., 1981; Zarate et al., 1999; Zhao et al., 2010). Using coating, encapsulation, and polymerization techniques to reduce the solubility and absorption rate of CAA can improve their efficiency in fish feed. The leaching of dietary CAAs may explain the poor availability of CAA in crustaceans (Fox et al., 2006). In consideration of the available evidence, a suggestion was advised to give a slightly higher "safety margin" when meeting an EAA requirement using CAAs compared with using a highly digestible protein-bound or peptide AAs sources (NRC, 2011).

4.4.4.4 *Maillard reaction*

Maillard or Schiff reaction, which can occur during heat processing of fish feed or ingredients, is known to increase the susceptibility of proteins or free AAs to heat damage in the presence of carbohydrates or free sugars and reduce the bioavailability of AAs.

Maillard reactions can lead to a reduction in the biological value of proteins, due to the nonavailability of EAA and eventually production of antinutritional and toxic components (Friedman et al., 1989). Plakas et al. (1985, 1988) investigated the effect of the Maillard reaction in the diet of rainbow trout on growth and AA availability; the results showed that arginine and lysine exhibited the greatest losses in the diet after Maillard reaction, the apparent digestibility and absorption of individual AAs, particularly lysine, was lower, and plasma lysine levels were significantly depressed. The same result was found in the diet, growth, and AA availability of juvenile white sturgeon (Deng et al., 2005).

References

Abboudi, et al., 2006. Protein and lysine requirements for maintenance and for tissue accretion in Atlantic salmon (*Salmo salar*) fry. Aquaculture 261, 369–383.

Abboudi, T., Ooghe, W., Larondelle, Y., Rollin, X., 2007. Determination of the threonine requirement for maintenance in Atlantic salmon (*Salmo salar* L.) fry with the diet dilution procedure. Aquac. Nutr. 13, 281–290.

Abdel-Tawwab, M., Ahmad, M., Khattab, Y., Shalaby, A., 2010. Effect of dietary protein level, initial body weight, and their interaction on the growth, feed utilization, and physiological alterations of Nile tilapia, *Oreochromis niloticus* (L.). Aquaculture 298, 267–274.

Abdel-Tawwab, M., Monier, M., 2018. Stimulatory effect of dietary taurine on growth performance, digestive enzymes activity, antioxidant capacity, and tolerance of common carp, *Cyprinus carpio* L., fry to salinity stress. Fish Physiol. Biochem. 44, 639–649.

Abdel-Tawwab, M., Monier, M., Hoseinifar, S., Faggio, C., 2019. Fish response to hypoxia stress: growth, physiological, and immunological biomarkers. Fish Physiol. Biochem. 45, 997–1013.

Abidi, S., Khan, M., 2004. Dietary valine requirement of Indian major carp, *Labeo rohita* (Hamilton) fry. J. Appl. Ichthyol. 20, 118–122.

Abidi, S., Khan, M., 2007. Dietary leucine requirement of fingerling Indian major carp, *Labeo rohita* (Hamilton). Aquac. Res. 38, 478–486.

Abidi, S., Khan, M., 2008. Dietary threonine requirement of fingerling Indian major carp, *Labeo rohita* (Hamilton). Aquac. Res. 39, 1498–1505.

Abidi, S., Khan, M., 2009. Dietary arginine requirement of fingerling Indian major carp, *Labeo rohita* (Hamilton) based on growth, nutrient retention efficiencies, RNA/DNA ratio and body composition. J. Appl. Ichthyol. 25, 707–714.

Abidi, S., Khan, M., 2010. Dietary tryptophan requirement of fingerling rohu, *Labeo rohita* (Hamilton), based on growth and body composition. J. World Aquac. Soc. 41, 700–709.

Abimorad, E., Favero, G., Squassoni, G., Carneiro, D., 2010. Dietary digestible lysine requirement and essential amino acid to lysine ratio for pacu *Piaractus mesopotamicus*. Aquac. Nutr. 16, 370–377.

Adron, J., Mackie, A., 1978. Studies on the chemical nature of feeding stimulants for rainbow trout, *Salmo gairdneri* Richardson. J. Fish. Biol. 12, 303–310.

Aebersold, R., Mann, M., 2016. Mass-spectrometric exploration of proteome structure and function. Nature 537, 347–355.

Ahmed, I., Khan, M., 2004a. Dietary arginine requirement of fingerling Indian major carp, *Cirrhinus mrigala* (Hamilton). Aquac. Nutr. 10, 217–225.

Ahmed, I., Khan, M., 2004b. Dietary lysine requirement of fingerling Indian major carp, *Cirrhinus mrigala* (Hamilton). Aquaculture 235, 499–511.

Ahmed, I., Khan, M., 2005a. Dietary histidine requirement of fingerling Indian major carp, *Cirrhinus mrigala* (Hamilton). Aquac. Nutr. 11, 359–366.

Ahmed, I., Khan, M., 2005b. Dietary tryptophan requirement of fingerling Indian major carp, *Cirrhinus mrigala* (Hamilton). Aquac. Res. 36, 687–695.

Ahmed, I., Khan, M., 2006. Dietary branched-chain amino acid valine, isoleucine and leucine requirements of fingerling Indian major carp, *Cirrhinus mrigala* (Hamilton). Br. J. Nutr. 96, 450–460.

Ahmed, I., 2007. Dietary amino acid L-threonine requirement of fingerling Indian catfish, *Heteropneustes fossilis* (Bloch) estimated by growth and biochemical parameters. Aquac. Int. 15, 337–350.

Ahmed, I., 2009. Dietary total aromatic amino acid requirement and tyrosine replacement value for phenylalanine in Indian major carp: *Cirrhinus mrigala* (Hamilton) fingerlings. J. Appl. Ichthyol. 25, 719–727.

Ahmed, I., 2010. Response to the ration levels on growth, body composition, energy, and protein maintenance requirement of the Indian catfish (*Heteropneustes fossilis*—Bloch 1974). Fish Physiol. Biochem. 36, 1133–1143.

Ahmed, I., 2012. Dietary amino acid L-tryptophan requirement of fingerling Indian catfish, *Heteropneustes fossilis* (Bloch), estimated by growth and haemato-biochemical parameters. Fish Physiol. Biochem. 38, 1195–1209.

Ahmed, I., 2013. Dietary amino acid L-histidine requirement of fingerling Indian catfish, *Heteropneustes fossilis* (Bloch), estimated by growth and whole body protein and fat composition. J. Appl. Ichthyol. 29, 602–609.

Ahmed, I., Khan, M., Jafri, A., 2003. Dietary methionine requirement of fingerling Indian major carp, *Cirrhinus mrigala* (Hamilton). Aquac. Int. 11, 449–462.

Ahmed, I., Khan, M., Jafri, A., 2004. Dietary threonine requirement of fingerling Indian major carp, *Cirrhinus mrigala* (Hamilton). Aquac. Res. 35, 162–170.

Aidos, I., Lie, Ø., Espe, M., 1999. Collagen content in farmed Atlantic salmon (*Salmo salar* L.). J. Agric. Food Chem. 47, 1440–1444.

Akiyama, T., Arai, S., Murai, T., Nose, T., 1985. Threonine, histidine and lysine requirements of chum salmon fry. Nippon Suisan Gakk. 51, 635–639.

Akiyama, T., Murai, T., Mori, K., 1986. Role of tryptophan metabolism in inhibition of spinal deformity of chum salmon fry caused by tryptophan deficiency. Bull. Jpn. Soc. Sci. Fish. 52, 1255–1259.

Akiyama, T., Ooharal, I., Yamamoto, T., 1997. Comparison of essential amino acid requirements with A/E ratio among fish species. Fish. Sci. 63, 963–970.

Aksnes, A., et al., 2006. Size-fractionated fish hydrolysate as feed ingredient for rainbow trout (*Oncorhynchus mykiss*) fed high plant protein diets. I: growth, growth regulation and feed utilization. Aquaculture 261, 305–317.

Alam, M., et al., 2004. Dietary arginine requirement of juvenile kuruma shrimp *Marsupenaeus japonicus* (Bate). Aquac. Res. 35, 842–849.

Alam, M., Teshima, S., Ishikawa, M., Koshio, S., 2000. Methionine requirement of juvenile Japanese flounder *Paralichthys olivaceus*. J. World Aquac. Soc. 31, 618–626.

Alam, S., Teshima, S., Koshio, S., Ishikawa, M., 2003. Optimum dietary threonine level for juvenile Japanese flounder *Paralichthys olivaceus*. Asian Fish Sci. 16, 175–184.

Alam, S., Teshima, S., Koshio, S., Ishikawa, M., 2002. Arginine requirement of juvenile Japanese flounder *Paralichthys olivaceus* estimated by growth and biochemical parameters. Aquaculture 205, 127–140.

Alava, V., Lim, C., 1983. The quantitative dietary protein requirements of *penaeus monodon* juveniles in a controlled environment. Aquaculture 30, 53–61.

Al-Feky, S., El-Sayed, A., Ezzat, A., 2015. Dietary taurine enhances growth and feed utilization in larval Nile tilapia (*Oreochromis niloticus*) fed soybean meal-based diets. Aquac. Nutr. 22, 457–464.

Amin, M., Barnes, R., Adams, L., 2014. Effects of different protein and carbohydrate levels on growth performance and feed utilisation of brook trout, *Salvelinus fontinalis* (Mitchill, 1814), at two temperatures. J. Appl. Ichthyol. 30, 340–349.

Andersen, S., Waagbø, R., Espe, M., 2016. Functional amino acids in fish health and welfare. Front. Biosci. 8, 143–169.

Anderson, J., Lall, S., Anderson, D., McNiven, M., 1993. Quantitative dietary lysine requirement of Atlantic salmon (*Salmo salar*) Fingerlings. Can. J. Fish. Aquat. Sci. 50, 316–322.

Appuhamy, J., et al., 2014. Effects of AMP-activated protein kinase (AMPK) signaling and essential amino acids on mammalian target of rapamycin (mTOR) signaling and protein synthesis rates in mammary cells. J. Dairy Sci. 97, 419–429.

Aragao, C., et al., 2004. A balanced dietary amino acid profile improves amino acid retention in post-larval Senegalese sole (*Solea senegalensis*). Aquaculture 233, 293–304.

Arai, S., 1981. A purified test diet for Coho salmon. Bull. Jpn. Soc. Sci. Fish. 47, 547–550.

Azizi, S., et al., 2016. Lysine and leucine deficiencies affect myocytes development and IGF signaling in gilthead sea bream (*Sparus aurata*). PLoS One 11, e0147618.

Bae, J., et al., 2012. The dietary valine requirement for rainbow trout, *Oncorhynchus mykiss*, can be estimated by plasma free valine and ammonia concentrations after dorsal aorta cannulation. J. Appl. Anim. Res. 40, 73–79.

Bai, Y., et al., 2016. Effects of dietary protein levels on the growth, energy budget, and physiological and immunological performance of green, white and purple color morphs of sea cucumber, *Apostichopus japonicas*. Aquaculture 450, 375–382.

Bairagi, A., et al., 2002. Enzyme producing bacterial flora isolated from fish digestive tracts. Aquac. Int. 10, 109–121.

Baker, D., Han, Y., 1994. Ideal amino acid profile for broiler chicks during the first three weeks post-hatching. Poultry Sci. 73, 1441–1447.

Baker, D., 2006. Comparative species utilization and toxicity of sulfur amino acids. J. Nutr. 136, 1670–1675.

Ballester, E., 2010. Effect of practical diets with different protein levels on the performance of *Farfantepenaeus paulensis* juveniles nursed in a zero exchange suspended microbial flocs intensive system. Aqua. Nutri. 16, 163–172.

Barziza, D., Buentello, J., Gatlin, D., 2000. Dietary arginine requirement of juvenile red drum (*Sciaenops ocellatus*) based on weight gain and feed efficiency. J. Nutr. 130, 1796–1799.

Belghit, I., et al., 2014. Dietary methionine availability affects the main factors involved in muscle protein turnover in rainbow trout (*Oncorhynchus mykiss*). Br. J. Nutr. 112, 493–503.

Benakappa, S., Varghese, T., 2002. Dietary threonine requirement of Indian major carp, *Cirrhinus mrigala* (Hamilton), juveniles. Isr. J. Aquac. Bamidgeh 54, 183–188.

Benakappa, S., Varghese, T., 2003. Isoleucine, leucine, and valine requirement of juvenile Indian major carp, *Cirrhinus cirrhosus* (Bloch, 1795). Acta Ichthyol. Piscatoria 33, 161–172.

Benakappa, S., Varghese, T., 2004. Total aromatic amino acid requirement of the Indian major carp, *Cirrhinus mrigala* (Hamilton-Buchanan). Isr. J. Aquac. Bamidgeh 56, 129–135.

Bender, A., 1965. The balancing of amino acid mixtures and proteins. Proc. Nutr. Soc. 24, 190–197.

Berge, G., Lied, E., Sveier, H., 1997. Nutrition of Atlantic salmon (*Salmo salar*): the requirement and metabolism of arginine. Comp. Biochem. Physiol. Physiol. 117, 501–509.

Berge, G., Sveier, H., Lied, E., 1998. Nutrition of Atlantic salmon (*Salmo salar*); the requirement and metabolic effect of lysine. Comp. Biochem. Physiol. A 120, 477–485.

Beyreuther, K., et al., 2007. Consensus meeting: monosodium glutamate-an update. Eur. J. Clin. Nutr. 61, 304–313.

Bjerkas, E., Sveier, H., 2004. The influence of nutritional and environmental factors on osmoregulation and cataracts in Atlantic salmon (*Salmo salar* L.). Aquaculture 235, 101–122.

Blachier, F., Mariotti, F., Huneau, J., Tome, D., 2007. Effects of amino acid-derived luminal metabolites on the colonic epithelium and physiopathological consequences. Amino Acids 33, 547–562.

Blachier, F., Boutry, C., Bos, C., Tomé, D., 2009. Metabolism and functions of L-glutamate in the epithelial cells of the small and large intestines. Am. J. Clin. Nutr. 90, 814–821.

Bodin, N., et al., 2008. Threonine requirements for rainbow trout (*Oncorhynchus mykiss*) and Atlantic salmon (*Salmo salar*) at the fry stage are similar. Aquaculture 274, 353–365.

Bodin, N., et al., 2009. Protein level affects the relative lysine requirement of growing rainbow trout (*Oncorhynchus mykiss*) fry. Br. J. Nutr. 102, 37–53.

Boren, R., Gatlin, D., 1995. Dietary threonine requirement of juvenile red drum *Sciaenops ocellatus*. J. World Aquac. Soc. 26, 279–283.

Borlongan, I., Coloso, R., 1993. Requirements of juvenile milkfish (*Chanos chanos forsskal*) for essential amino acids. J. Nutr. 123, 125–132.

Borlongan, I., 1991. Arginine and threonine requirements of milkfish (*Chanos chanos* Forsskal) juveniles. Aquaculture 93, 313–322.

Borlongan, I., 1992. Dietary requirement of milkfish (*Chanos chanos* Forsskal) juveniles for total aromatic amino acids. Aquaculture 102, 309–317.

Borsook, H., Dubnoff, J., 1951. The metabolism of proteins and amino acids. Annu. Rev. Biochem. 20, 209.

Bowen, S., 1987. Dietary protein requirements of fishes-a reassessment. Can. J. Fish. Aquat. Sci. 44, 1995–2001.

Bowyer, J., Qin, J., Stone, D., 2013. Skeletal anomalies in reared European fish larvae and juveniles. Part 1: normal and anomalous skeletogenic processes. Rev. Aquac. 5, S99–S120.

Braverman, E., Pfeiffer, C., Blum, K., 2003. The healing nutrients within: facts, findings, and new research on amino acids. In: Hirsch, C. (Ed.), Branched Chain Amino Acids, third ed. Basic Health Publications, Inc. Laguna, Beach, CA, p. 434.

Breuillard, C., Cynober, L., Moinard, C., 2015. Citrulline and nitrogen homeostasis: an overview. Amino Acids 47, 685–691.

Brill, R., et al., 2019. Effects of dietary taurine level on visual function in European sea bass (*Dicentrarchus labrax*). PLoS One 14, e214347.

Brosnan, J., 2001. Amino acids, then and now – a reflection on Sir Hans Krebs' contribution to nitrogen metabolism. IUBMB Life 52, 265–270.

Brosnan, J., 2003. Interorgan amino acid transport and its regulation. J. Nutr. 133, 2068S–2072S.

Brosnan, J., Brosnan, M., 2006a. Branched-chain amino acids: enzyme and substrate regulation. J. Nutr. 136, 207–211.

Brosnan, J., Brosnan, M., 2006b. The sulfur-containing amino acids: an overview. J. Nutr. 136, 1636–1640.

Brosnan, J., Brosnan, M., 2013. Glutamate: a truly functional amino acid. Amino Acids 45, 413–418.

Brown, C., Cameron, J., 1991. The induction of specific dynamic action in channel catfish by infusion of essential amino acids. Physiol. Zool. 64, 276–297.

Brown, M., Nematipour, G., Gatlin, D., 1992. Dietary protein requirement of juvenile sunshine bass at different salinities. Prog. Fish-Cult. 54, 148–156.

Buentello, J., Gatlin, D., 2000. The dietary arginine requirement of channel catfish (*Ictalurus punctatus*) is influenced by endogenous synthesis of arginine from glutamic acid. Aquaculture 188, 311–321.

Buse, M., Reid, S., 1975. Leucine. A possible regulator of protein turnover in muscle. J. Clin. Invest. 56, 1250–1261.

Buxbaum, E., 2007. Fundamentals of Protein Structure and Function. Springer, New York, pp. 3–13.

Buxbaum, E., 2015. Fundamentals of Protein Structure and Function, second ed. Springer International Publishing, Switzerland, p. 521.

Cao, J., et al., 2012. A study on dietary L-lysine requirement of juvenile yellow catfish *Pelteobagrus fulvidraco*. Aquac. Nutr. 18, 35–45.

Carter, C., Bransden, M., 2001. Relationships between protein–nitrogen flux and feeding regime in greenback flounder, *Rhombosolea tapirina* (Günther). Comp. Biochem. Physiol. A. Mol. Integr. Physiol. 130, 799–807.

Carter, C., Houlihan, D., 2001. Protein synthesis. Fish Physiol. 20, 31–75.

Carter, C., Mente, E., 2014. Protein synthesis in crustaceans: a review focused on feeding and nutrition. Open Life Sci. 9, 1–10.

Carter, C., Houlihan, D., Brechin, J., McCarthy, I., 1993. The relationships between protein intake and protein accretion, synthesis, and retention efficiency for individual grass carp, *Ctenopharyngodon idella* (*Valenciennes*). Can. J. Zool. 71, 392–400.

Cassidy, A., et al., 2018a. Effects of fasting and refeeding on protein and glucose metabolism in Arctic charr. Comp. Biochem. Physiol. A. Mol. Integr. Physiol. 226, 66–74.

Cassidy, A., et al., 2018b. Protein synthesis is lowered by 4EBP1 and eIF2-α signaling while protein degradation may be maintained in fasting, hypoxic *Amazonian cichlid, Astronotus ocellatus*. J. Exp. Biol. 221, 221–231.

Cassidy, A., Saulnier, R., Lamarre, S., Planas, J., 2016. Adjustments of protein metabolism in fasting Arctic charr, *Salvelinus alpinus*. PLoS One 11, e0153364.

Castillo, S., Gatlin, D., 2018. Dietary requirements for leucine, isoleucine and valine (branched-chain amino acids) by juvenile red drum *Sciaenops ocellatus*. Aquac. Nutr. 24, 1056–1065.

Castillo, S., Halligan, S., Gatlin, D., 2015. Growth responses of juvenile red drum *Sciaenops ocellatus* to dietary phenylalanine and tyrosine can be used to calculate the total aromatic amino acid requirement. J. Nutr. 145, 2341–2346.

Casu, F., et al., 2017. Metabolomics analysis of effects of commercial soy-based protein products in red drum (*Sciaenops ocellatus*). J. Proteome Res. 16, 2481–2494.

Causey, D., et al., 2019. Proteomic comparison of selective breeding and growth hormone transgenesis in fish: unique pathways to enhanced growth. J. Proteom 192, 114–124.

Chaitanawisuti, N., et al., 2010. Effects of dietary calcium and phosphorus supplementation on the growth performance of juvenile spotted babylon *Babylonia areolata* culture in a recirculating culture system. Int. J. Fish. Aquac. 18, 303–313.

Chance, R., Mertz, E., Halver, J., 1964. Nutrition of salmonoid fishes: XII. isoleucine, leucine, valine and phenylalanine requirements of chinook salmon and interrelations between isoleucine and leucine for growth. J. Nutr. 83, 177–185.

Chatzifotis, S., Polemitou, I., Divanach, P., Antonopoulou, E., 2008. Effect of dietary taurine supplementation on growth performance and bile salt activated lipase activity of common dentex, *Dentex dentex*, fed a fish meal/soy protein concentrate-based diet. Aquaculture 275, 201–208.

Chen, G., et al., 2012. Effect of dietary arginine on growth, intestinal enzyme activities and gene expression in muscle, hepatopancreas and intestine of juvenile Jian carp (*Cyprinus carpio* var. Jian). Br. J. Nutr. 108, 195–207.

Chen, G., et al., 2015. Effect of dietary arginine on the immune response and gene expression in head kidney and spleen following infection of Jian carp with *Aeromonas hydrophila*. Fish Shellfish Immunol. 44, 195–202.

Chen, H., Leu, Y., Roelants, I., 1992. Quantification of arginine requirements of juvenile marine shrimp, *Penaeus monodon*, using microencapsulated arginine. Mar. Biol. 114, 229–233.

Chen, J., et al., 2019. Growth performance, intestinal morphology, hepatopancreatic antioxidant capacity and growth-related gene mRNA expressions of juvenile grass carp (*Ctenopharyngodon idellus*) as affected by graded levels of dietary arginine. Aquac. Nutr. 25, 1124–1134.

Cheng, Z., Hardy, R., Usry, J., 2003. Effects of lysine supplementation in plant protein-based diets on the performance of rainbow trout (*Oncorhynchus mykiss*) and apparent digestibility coefficients of nutrients. Aquaculture 215, 255–265.

Chiu, Y., Austic, R., Rumsey, G., 1988. Effect of feeding level and dietary electrolytes on the arginine requirement of rainbow trout (*Salmo gairdneri*). Aquaculture 69, 79–91.

Cho, C., 1992. Feeding systems for rainbow trout and other salmonids with reference to current estimates of energy and protein requirements. Aquaculture 100, 107–123.

Cho, C., Kaushik, C., 1985. Effect of protein intake on metabolizable and net energy values of fish diets. In: Cowey, C.B., Mackie, A., Bell, J. (Eds.), Nutrition and Feeding of Fish, pp. 95–117.

Cho, C., Kaushik, C., 1990. Nutritional energetics in fish: energy and protein utilization in rainbow trout (*Salmo gairdneri*). World Rev. Nutr. Diet. 61, 132–172.

Cho, C., Kaushik, S., Woodward, B., 1992. Dietary arginine requirement of young rainbow trout (*Oncorhynchus mykiss*). Comp. Biochem. Physiol. A: Phys. 102, 211–216.

Chong, A., Ishak, S., Osman, Z., Hashim, R., 2004. Effect of dietary protein level on the reproductive performance of female swordtails *Xiphophorus helleri* (Poeciliidae). Aquaculture 234, 381–392.

Choo, P., Smith, T., Cho, C., Ferguson, H., 1991. Dietary excesses of leucine influence growth and body composition of rainbow trout. J. Nutr. 121, 1932–1939.

Ciechanover, A., Finley, D., Varshavsky, A., 1984. The ubiquitin-mediated proteolytic pathway and mechanisms of energy-dependent intracellular protein degradation. J. Cell. Biochem. 24, 27–53.

Cleveland, B., Burr, G., 2011. Proteolytic response to feeding level in rainbow trout (*Oncorhynchus mykiss*). Aquaculture 319, 194–204.

Cleveland, B., Weber, G., 2010. Effects of insulin-like growth factor-I, insulin, and leucine on protein turnover and ubiquitin ligase expression in rainbow trout primary myocytes. Am. J. Physiol. Regul. Integr. Comp. Physiol. 298, R341–R350.

Coloso, R., Murillo-Gurrea, D., Borlongan, I., 1999. Sulphur amino acid requirement of juvenile Asian sea bass *Lates calcarifer*. J. Appl. Ichthyol. 15, 54–58.

Coloso, R., Tiro, L., Benitez, L., 1992. Requirement for tryptophan by milkfish (*Chanos chanos* Forsskal) juveniles. Fish Physiol. Biochem. 10, 35–41.

Coloso, R., Murillo-Gurrea, D., Borlongan, I., Catacutan, M., 2004. Tryptophan requirement of juvenile Asian sea bass *Lates calcarifer*. J. Appl. Ichthyol. 20, 43–47.

Colvin, P., 1976. Nutritional studies on penaeid prawns: protein requirements in compounded diets for juvenile *Pentzeus indicus* (Milne Edwards). Aquaculture 7, 315–326.

Conceição, L., et al., 1997. Amino acid metabolism and protein turnover in larval turbot (*Scophthalmus maximus*) fed natural zooplankton or Artemia. Mar. Biol. 129, 255–265.

Cowey, C., Cho, C., 1993. Nutritional requirements of fish. Proc. Nutr. Soc. 52, 417–426.

Cowey, C., Forster, J., 1971. The essential amino-acid requirements of the prawn *Palaemon serratus*. The growth of prawns on diets containing proteins of different amino-acid compositions. Mar. Biol. 10, 77–81.

Cowey, C., Walton, M., 1988. Studies on the uptake of (^{14}C) amino acids derived from both dietary (^{14}C) protein and dietary (^{14}C) amino acids by rainbow trout, *Salmo gairdneri* Richardson. J. Fish. Biol. 33, 295–305.

Cowey, C., 1994. Amino acid requirements of fish: a critical appraisal of present values. Aquaculture 124, 1–11.

Cowey, C., et al., 1992. Methionine intake in rainbow Trout (*Oncorhynchus mykiss*), relationship to cataract formation and the metabolism of methionine. J. Nutr. 122, 1154–1163.

Cowey, C., Adron, J., Blair, A., 1970. Studies on the nutrition of marine flatfish. The essential amino acid requirements of plaice and sole. J. Mar. Biol. Assoc. U. K. 50, 87–95.

Cowey, C., Walton, M., 1989. Intermediary metabolism. In: Halver, J.E. (Ed.), Fish Nutrition, second ed. Academic Press, New York, pp. 259–329.

Cowey, C., 1992. Nutrition: estimating requirements of rainbow trout. Aquaculture 100, 177–189.

Cowey, C., 1995. Protein and amino acid requirements: a critique of methods. J. Appl. Ichthyol. 11, 199–204.

Dabrowski, K., et al., 1977. Protein requirements of grass carp fry (*Ctenopharyngodon idella* Val.). Aquaculture 12, 63–73.

Dabrowski, K., 1981. Tryptophan requirement of common carp (*Cyprinus carpio* L.) fry. Zeitschrift für Tierphysiologie Tierernährung und Futtermittelkunde 46, 64–71.

Dai, W., et al., 2015. Effects of lysine-arginine interaction on growth performance, body composition, and muscle amino acid levels of juvenile turbot (*Scophthalmus maximus* L.). J. Fish. China 39, 876–887.

Dairiki, J., Borghesi, R., Dias, C., Cyrino, J., 2013. Lysine and arginine requirements of *Salminus brasiliensis*. Pesqui. Agropecu. Bras. 48, 1012–1020.

Dairiki, J., Dias, C., Cyrino, J., 2007. Lysine requirements of largemouth bass, *Micropterus salmoides:* a comparison of methods of analysis of dose-response trials data. J. Appl. Aquac. 19, 1–27.

Daniels, W., Robinson, E., 1986. Protein and energy requirements of juvenile red drum (*Sciaenops ocellatus*). Aquaculture 53, 243–252.

De Almeida Bicudo, A., Cyrino, J., 2009. Estimating amino acid requirement of Brazilian freshwater fish from muscle amino acid profile. J. Word Aquac. Soc. 40, 818–823.

De Long, D., Halver, J., Mertz, E., 1958. Nutrition of salmonoid fishes. VI Protein requirements of chinook salmon at two water temperatures. J. Nutr. 65, 589–599.

De Silva, S., Perera, M., 1985. Effects of dietary protein level on growth, food conversion, and protein use in young tilapia nilotica at four salinities. Trans. Am. Fish. Soc. 114, 584–589.

Deng, D., et al., 2010. Dietary lysine requirement of juvenile Pacific threadfin (*Polydactylus sexfilis*). Aquaculture 308, 44–48.

Deng, D., et al., 2005. Utilization of diets with hydrolyzed potato starch, or glucose by juvenile white sturgeon (*Acipenser transmontanus*), as affected by Maillard reaction during feed processing. Aquaculture 248, 103–109.

Deng, Y., et al., 2014. Differential growth performance, *intestinal* antioxidant status and relative expression of Nrf2 and its target genes in young grass carp (*Ctenopharyngodon idella*) fed with graded levels of leucine. Aquaculture 434, 66–73.

Deng, Y., et al., 2016. Dietary leucine improves flesh quality and alters mRNA expressions of Nrf2-mediated antioxidant enzymes in the muscle of grass carp (*Ctenopharyngodon idella*). Aquaculture 452, 380–387.

Deshimaru, O., Shigeno, K., 1972. Introduction to the artificial diet for prawn *Penaeus japonicus*. Aquaculture 1, 115–133.

Deshmukh, D., Shope, T., 1983. Arginine requirement and ammonia toxicity in ferrets. J. Nutr. 113, 1664–1667.

Diógenes, A., et al., 2016. Establishing the optimal essential amino acid ratios in juveniles of Nile tilapia (*Oreochromis niloticus*) by the deletion method. Aquac. Nutr. 22, 435–443.

Do Nascimento, T., et al., 2020. Determination of the optimum dietary essential amino acid profile for growing phase of Nile tilapia by deletion method. Aquaculture 523, 735204.

Doherty, M., et al., 2005. Proteome dynamics in complex organisms: using stable isotopes to monitor individual protein turnover rates. Proteomics 5, 522–533.

Doherty, M., et al., 2009. Turnover of the human proteome: determination of protein intracellular stability by dynamic SILAC. J. Proteome Res. 8, 104–112.

Doherty, M., et al., 2012. A proteomics strategy for determining the synthesis and degradation rates of individual proteins in fish. J. Proteom 75, 4471–4477.

Dong, M., et al., 2012. Growth, body composition, intestinal enzyme activities and microflora of juvenile Jian carp (*Cyprinus carpio* var. Jian) fed graded levels of dietary valine. Aquac. Nutr. 19, 1–14.

Dong, X., et al., 2018b. Effects of different arginine/lysine level on growth performance, body composition and digestive enzyme activity of *Macrobrachium rosenbergii*. Aquac. Nutr. 24, 1101–1111.

Dong, Y., et al., 2017. Threonine deficiency decreased intestinal immunity and aggravated inflammation associated withNF-κB and target of rapamycin signalling pathways in juvenile grass carp (*Ctenopharyngodon idella*) after infection with *Aeromonas hydrophila*. Br. J. Nutr. 118, 92–108.

Dong, Y., et al., 2018. Dietary threonine deficiency depressed the disease resistance, immune and physical barriers in the gills of juvenile grass carp (*Ctenopharyngodon idella*) under infection of *Flavobacterium columnare*. Fish Shellfish Immunol. 72, 161–173.

Dunn, W., Ellis, D., 2005. Metabolomics: current analytical platforms and methodologies. Trac. Trends Anal. Chem. 24, 285–294.

Eckhoff, K., Aidos, I., Hemre, G., Lie, O., 1998. Collagen content in farmed Atlantic salmon (*Salmo salar*, L.) and subsequent changes in solubility during storage on ice. Food Chem. 62, 197–200.

El Kertaoui, N., et al., 2019. Key nutritional factors and interactions during larval development of pikeperch (*Sander lucioperca*). Sci. Rep. 9, 2–13.

Elango, R., Ball, R., Pencharz, P., 2008. Indicator amino acid oxidation: concept and application. J. Nutr. 138, 243–246.

Elliott, R., Ong, T., 2002. Science, medicine, and the future - nutritional genomics. Br. Med. J. 324, 1438–1442.

El-Saidy, D., Gaber, M., 2002. Complete replacement of fish meal by soybean meal with dietary L-lysine supplementation for Nile tilapia *Oreochromis niloticus* (L.) Fingerlings. J. World Aquac. Soc. 33, 297–306.

El-Sayed, A., 1991. Protein requirements for optimum growth of Liza ramada fry (Mugilidae) at different water salinities. Aquat. Living Resour. 4, 117–123.

El-Sayed, A., Teshima, S., 1992. Protein and energy requirements of Nile tilapia, *Oreochromis niloticus* fry. Aquaculture 103, 55–63.

Encarnação, P., et al., 2004. Diet digestible energy content affects lysine utilization, but not dietary lysine requirements of rainbow trout (*Oncorhynchus mykiss*) for maximum growth. Aquaculture 235, 569–586.

Espe, M., Lied, E., 1994. Do Atlantic salmon (*Salmo salar*) utilize mixtures of free amino acids to the same extent as intact protein sources for muscle protein synthesis? Comp. Biochem. Physiol. 177A, 249–254.

Espe, M., Lemme, A., Petri, A., El-Mowafi, A., 2007. Assessment of lysine requirement for maximal protein accretion in Atlantic salmon using plant protein diets. Aquaculture 263, 168–178.

Espe, M., et al., 2008. Methionine intake affect hepatic sulphur metabolism in Atlantic salmon, *Salmo salar*. Aquaculture 274, 132–141.

Fagbenro, O., Nwanna, L., 1999. Dietary tryptophan requirement of the African catfish, *Clarias gariepinus*. J. Appl. Aquac. 9, 65–72.

Fagbenro, O., Balogun, A., Bello-Olusoji, O., Fasakin, E., 1998. Dietary lysine requirement of the African catfish, *Clarias gariepinus*. J. Appl. Aquac. 8, 71–77.

Fagbenro, O., Nwanna, L., Adebayo, O., 1999. Dietary arginine requirement of the African catfish, *Clarias gariepinus*. J. Appl. Aquac. 9, 59–64.

Farhat, Khan, M., 2013a. Effects of varying levels of dietary L-histidine on growth, feed conversion, protein gain, histidine retention, hematological and body composition in fingerling stinging catfish *Heteropneustes fossilis* (Bloch). Aquaculture 404–405, 130–138.

Farhat, Khan, M., 2013b. Dietary l-lysine requirement of fingerling stinging catfish, *Heteropneustes fossilis* (Bloch) for optimizing growth, feed conversion, protein and lysine deposition. Aquac. Res. 44, 523–533.

Farhat, Khan, M., 2014a. Response of fingerling stinging catfish, *Heteropneustes fossilis* (Bloch) to varying levels of dietary L-leucine in relation to growth, feed conversion, protein utilization, leucine retention and blood parameters. Aquac. Nutr. 20, 291–302.

Farhat, Khan, M., 2014b. Dietary L-tryptophan requirement of fingerling stinging catfish, *Heteropneustes fossilis* (Bloch). Aquac. Res. 45, 1224–1235.

Fauconneau, B., Basseres, A., Kaushik, S., 1992. Oxidation of phenylalanine and threonine in response to dietary arginine supply in rainbow trout (*Salmo gairdneri* R.). Comp. Biochem. Physiol. 101, 395–401.

Felip, O., et al., 2012. Tracing metabolic routes of dietary carbohydrate and protein in rainbow trout (*Oncorhynchus mykiss*) using stable isotopes ([^{13}C]starch and [^{15}N]protein): effects of gelatinisation of starches and sustained swimming. Br. J. Nutr. 107, 834–844.

Felip, O., et al., 2013. Beneficial effects of sustained activity on the use of dietary protein and carbohydrate traced with stable isotopes 15N and 13C in gilthead sea bream (*Sparus aurata*). J. Comp. Physiol. B 183, 223–234.

Feng, L., et al., 2013. Threonine affects intestinal function, protein synthesis and gene expression of TOR in Jian carp (*Cyprinus carpio* var. Jian). PLoS One 8, e69974.

Feng, L., et al., 2015. Dietary phenylalanine-improved intestinal barrier health in young grass carp (*Ctenopharyngodon idella*) is associated with increased immune status and regulated gene expression of cytokines, tight junction proteins, antioxidant enzymes and related signalling molecules. Fish Shellfish Immunol. 45, 495–509.

Feng, L., et al., 2017a. Gill structural integrity changes in fish deficient or excessive in dietary isoleucine: towards the modulation of tight junction protein, inflammation, apoptosis and antioxidant defense via NF-kB, TOR and Nrf2 signaling pathways. Fish Shellfish Immunol. 63, 127–138.

Feng, L., et al., 2017b. Protective role of phenylalanine on the ROS-induced oxidative damage, apoptosis and tight junction damage via Nrf2, TOR and NF-κB signalling molecules in the gill of fish. Fish Shellfish Immunol. 60, 185–196.

Fernandes, T., McMeans, B., 2019. Coping with the cold: energy storage strategies for surviving winter in freshwater fish. Ecography 42, 2037–2052.

Ferraris, R., Ahearn, G., 1984. Sugar and amino acid transport in fish intestine. Comp. Biochem. Physiol. A Physiol. 77, 397–413.

Ferreira, F., et al., 2014. Effects of taurine supplementation on the growth performance of juvenile rock bream *Oplegnathus fasciatus*. Fish Aquat. Sci. 17, 255–261.

Fisher, H., Scott, H., 1954. The essential amino acid requirements of chicks as related to their proportional occurrence in the fat-fee carcass. Arch. Biochem. Biophys. 51, 517–519.

Foretz, M., Viollet, B., 2011. Regulation of hepatic metabolism by AMPK. J. Hepatol. 54, 827–829.

Forster, I., Ogata, H., 1998. Lysine requirement of juvenile Japanese flounder *Paralichthys olivaceus* and juvenile red sea bream *Pagrus major*. Aquaculture 161, 131–142.

Fournier, V., et al., 2002. Protein and arginine requirements for maintenance and nitrogen gain in four teleosts. Br. J. Nutr. 87, 459–469.

Fox, H., Kimball, S., Jefferson, L., Lynch, C., 1998. Amino acids stimulate phosphorylation of p70S6k and organization of rat adipocytes into multicellular clusters. Am. J. Physiol. Cell Physiol. 274, 206–213.

Fox, J., Davis, D., Wilson, M., Lawrence, A., 2006. Current status of amino acid requirement research with marine penaeid shrimp. In: Cruz-Suarez, L.E., Ricque, D., Puello Cruz, A.C., Garcia-Ortega, A. (Eds.), Avances en nutricion acuicola, VIII simposium internacional de nutricion acuicola. Mazatlan, Sinaloa, Mexico, pp. 182–196.

Frank, J., et al., 2007. Oral N-carbamylglutamate supplementation increases protein synthesis in skeletal muscle of piglets. J. Nutr. 137, 315–319.

Fraser, K., Rogers, A., 2007. Protein metabolism in marine animals: the underlying mechanism of growth. Adv. Mar. Biol. 52, 267–362.

Fraser, K., Clarke, A., Peck, L., 2007. Growth in the slow lane: protein metabolism in the Antarctic limpet Nacella concinna (Strebel 1908). J. Exp. Biol. 210, 2691–2699.

Friedman, M., Brandon, D., Bates, A., Hymowitz, T., 1989. Effect of heat on the nutritional quality and safety of soybean cultivars. In: Friedman, M. (Ed.), Nutritional and Toxicological Consequences of Browning during Simulated Crust Baking. Protein Quality and Effects of Processing. Springer, Boston, MA. USA, pp. 189–217.

Fujimoto, R., et al., 2019. Supplementation with arginine in the diet of Nile tilapia reared in net cages. Pesqu. Agropecu. Bras. 54, e01099.

Fuller, M., 1994. Amino acid requirements of maintenance, body protein accretion and reproduction in pigs. In: D'Mello, J.P.F. (Ed.), Amino Acids in Farm Animal Nutrition. CAB International, Wallingford, UK, pp. 155–184.

Furuya, W., et al., 2004. Lysine requirement of Nile tilapia (*Oreochromis nilotucus*) for grow-out phase. Cienc. Rural. 34, 1571–1577.

Furuya, W., et al., 2006. Digestible lysine requirement of Nile tilapia juveniles. Rev. Bras. Zootec. 3, 937–942.

Furuya, W., et al., 2012. Digestible lysine requirement of Nile tilapia fingerlings fed arginine-to-lysine-balanced diets. Rev. Bras. Zootec. 3, 485–490.

Furuya, W., et al., 2015. Estimation of the dietary essential amino acid requirements of colliroja *Astyanax fasciatus* by using the ideal protein concept. Lat. Am. J. Aquac. Res. 43, 888–894.

Gaber, M., Salem, M., Zaki, M., Nour, A., 2016. Amino acid requirements of gilthead bream (*Sparus aurata*) juveniles. World J. Eng. Technol. 4, 18–24.

Gabillard, J., Kamangar, B., Montserrat, N., 2006. Coordinated regulation of the GH/IGF system genes during refeeding in rainbow trout (*Oncorhynchus mykiss*). J. Endocrinol. 191, 15–24.

Gallinetti, J., Harputlugil, E., Mitchell, J., 2013. Amino acid sensing in dietary-restriction-mediated longevity: roles of signal-transducing kinases GCN2 and TOR. Biochem. J. 449, 1–10.

Gamboa-Delgado, J., Le Vay, L., 2009. Natural stable isotopes as indicators of the relative contribution of soy protein and fish meal to tissue growth in Pacific white shrimp (*Litopenaeus vannamei*) fed compound diets. Aquaculture 291, 115–123.

Gamboa-Delgado, J., Peña-Rodríguez, A., Ricque-Marie, D., Cruz-Suárez, L.E., 2011. Assessment of nutrient allocation and metabolic turnover rate in Pacific white shrimp *Litopenaeus vannamei* co-fed live macroalgae *Ulva clathrata* and inert feed: dual stable isotope analysis. J. Shellfish Res. 30, 969–978.

Gamboa-Delgado, J., Rojas-Casas, M., Nieto-López, M., Cruz-Suárez, L., 2013. Simultaneous estimation of the nutritional contribution of fish meal, soy protein isolate and corn gluten to the growth of Pacific white shrimp (*Litopenaeus vannamei*) using dual stable isotope analysis. Aquaculture 380–383, 33–40.

Gan, L., et al., 2014. Flesh quality loss in response to dietary isoleucine deficiency and excess in fish: a link to impaired Nrf2-dependent antioxidant defense in muscle. PLoS One 9, e115129.

Gan, L., Zhou, L., Li, X., Yue, Y., 2015. Dietary leucine requirement of juvenile Nile tilapia, *Oreochromis niloticus*. Aquac. Nutr. 22, 1040–1046.

Gao, Y., et al., 2014. Effects of graded levels of threonine on growth performance, biochemical parameters and intestine morphology of juvenile grass carp *Ctenopharyngodon idella*. Aquaculture 424–425, 113–119.

Gao, Y., et al., 2015. Effects of graded levels of histidine on growth performance, digested enzymes activities, erythrocyte osmotic fragility and hypoxia-tolerance of juvenile grass carp *Ctenopharyngodon idella*. Aquaculture 452, 388–394.

Gao, Y., et al., 2016. Total aromatic amino acid requirement of juvenile grass carp (*Ctenopharyngodon idella*). Aquac. Nutr. 22, 865–872.

Gao, Y., et al., 2019. Effects of dietary protein levels on growth, feed utilization and expression of growth related genes of juvenile giant grouper (*Epinephelus lanceolatus*). Aquaculture 504, 369–374.

Garling, D., Wilson, R., 1976. Optimum dietary protein to energy ratio for channel catfish fingerlings, *Ictalurus punctatus*. J. Nutr. 106, 1368–1375.

Gasier, H., et al., 2009. A novel approach for assessing protein synthesis in channel catfish, *Ictalurus punctatus*. Comp. Biochem. Physiol. B Biochem. Mol. Biol. 154, 235–238.

Gasier, H., Fluckey, J., Previs, S., 2010. The application of ^2H_2O to measure skeletal muscle protein synthesis. Nutr. Metab. 7, 31.

Gatlin III, D., Poe, W., Wilson, R., 1986. Protein and energy requirements of fingerling channel catfish for maintenance and maximum growth. J. Nutr. 116, 2121–2131.

Gatlin, D., et al., 2007. Expanding the utilization of sustainable plant products in aquafeeds: a review. Aquac. Res. 38, 551–579.

Gaylord, T., Rawles, S., Davis, K., 2005. Dietary tryptophan requirement of hybrid striped bass (*Morone chrysops* × *M. saxatilis*). Aquac. Nutr. 11, 367–374.

Glencross, B., Bermudes, M., 2011. The effect of high water temperatures on the allometric scaling effects of energy and protein starvation losses in juvenile barramundi, *Lates calcarifer*. Comp. Biochem. Physiol. A. Mol. Integr. Physiol. 159, 167–174.

Goll, D., et al., 2003. The calpain system. Physiol. Rev. 83, 731.

Gomez-Requeni, P., et al., 2003. Effects of dietary amino acid prole on growth performance, key metabolic enzymes and somatotropic axis responsiveness of gilthead sea bream (*Sparus aurata*). Aquaculture 220, 749–767.

Gonzalez, S., et al., 2005. Dietary protein requirement of southern flounder, *Paralichthys lethostigma*. J. Appl. Aquacult. 17, 37–50.

Gopal, C., Raj, R., 1990. Protein requirement of juvenile*Penaeus indicus*. 1. Food consumption and growth. Proc. Ani. Sci. 99, 401–409.

Goto, T., et al., 2001. Studies on the green liver in cultured red sea bream fed low level and non-fish meal diets: relationship between hepatic taurine and biliverdin levels. Fish. Sci. 67, 58–63.

Green, J., Hardy, R., 2002. The optimum dietary essential amino acid pattern for rainbow trout (*Oncorhynchus mykiss*), to maximize nitrogen retention and minimize nitrogen excretion. Fish Physiol. Biochem. 27, 97–108.

Green, J., Hardy, R., Brannon, E., 2002. The optimum dietary essential: nonessential amino acid ratio for rainbow trout (*Oncorhynchus mykiss*), which maximizes nitrogen retention and minimizes nitrogen excretion. Fish Physiol. Biochem. 27, 109–115.

Griffin, M., Brown, P., Grant, A., 1992. The dietary lysine requirement of juvenile hybrid striped bass. J. Nutr. 122, 1332–1337.

Griffin, M., Wilson, K., Brown, P., 1994a. Dietary arginine requirement of juvenile hybrid striped bass. J. Nutr. 124, 888–893.

Griffin, M., White, M., Brown, P., 1994b. Total sulfur amino acid requirement and cysteine replacement value for juvenile hybrid striped bass (*Morone saxatilis* × *M. chrysops*). Comp. Biochem. Physiol. 108A, 423–429.

Grimble, G., 2007. Adverse gastrointestinal effects of arginine and related amino acids. J. Nutr. 137, 1693S–1701S.

Grisdale-Helland, B., Hatlen, B., Mundheim, H., Helland, S., 2011. Dietary lysine requirement and efficiency of lysine utilization for growth of Atlantic cod. Aquaculture 315, 260–268.

Guo, Z., et al., 2012. Effects of dietary protein level on growth performance, nitrogen and energy budget of juvenile hybrid sturgeon, *Acipenser baerii female* × A. *gueldenstaedtii* male. Aquaculture 338, 89–95.

Gurure, R., Atkinson, J., Moccia, R., 2007. Amino acid composition of Arctic charr, *Salvelinus alpinus* (L.) and the prediction of dietary requirements for essential amino acids. Aquac. Nutr. 13, 266–272.

Guy, E., Li, M., Allen, P., 2018. Effects of dietary protein levels on growth and body composition of juvenile (age-1) black buffalo *Ictiobus niger*. Aquaculture 492, 67–72.

Habte-Tsion, H., et al., 2015. Dietary threonine requirement of juvenile blunt snout bream (*Megalobrama amblycephala*). Aquaculture 437, 304–311.

Halver, J., Shanks, W., 1960. Nutrition of salmonid fishes. VIII. Indispensable amino acids for sockeye salmon. J. Nutr. 72, 340–346.

Halver, J., 1957. Nutrition of salmonoid fishes. IV. An amino acid test diet for Chinook salmon. J. Nutr. 10, 245–254.

Hammer, H., et al., 2006. The effect of dietary protein on consumption, survival, growth and production of the sea urchin *Lytechinus variegatus*. Aquaculture 254, 483–495.

Han, Y., et al., 2013. Optimum supplementations and interactive effects of methionine and tryptophan on growth and health status of juvenile Japanese flounder *Paralichthys olivaceus*. Aquac. Sci. 61, 239–251.

Hao, S., et al., 2005. Uncharged tRNA and sensing of amino acid deficiency in mammalian piriform cortex. Science 307, 1776–1778.

Hara, K., et al., 1998. Amino acid sufficiency and mTOR regulate p70 S6 kinase and eIF-4E BP1 through a common effector mechanism. J. Biol. Chem. 273, 14484–14494.

Hara, T., 1994. Olfaction and gustation in fish: an overview. Acta Physiol. Scand. 152, 207–217.

Harding, D., Allen, O., Wilson, R., 1977. Sulfur amino acid requirement of channel catfish: L-methionine and L-cystine. J. Nutr. 107, 2031–2035.

Harlioglu, M., Harlioglu, A., Yonar, S., Duran, T., 2014. Effects of dietary L-tryptophan on the agonistic behavior, growth, and survival of freshwater crayfish *Astacus leptodactylus* Eschscholtz. Aquac. Int. 22, 733–748.

Haschemeyer, A., Smith, M., 1979. Protein synthesis in liver, muscle and gill of mullet (*Mugil cephalus* L.) in vivo. Biol. Bull. 156, 93–102.

Hauler, R., Carter, C., Edwards, S., 2007. Feeding regime does not influence lysine utilisation by Atlantic salmon, *Salmo salar* L. Aquaculture 273, 545–555.

Hawkins, A., 1985. Relationships between the synthesis and breakdown of protein, dietary absorption and turnovers of nitrogen and carbon in the blue mussel, Mytilus edulis L. Oecologia 66, 42–49.

He, X, Jia, L., Li, Z., Yang, Y., 1988. Nutrient requirements of juvenile allogenetic crucian carp, *Carassius auratus gibelio*. In: Ryans, R. (Ed.), Proceeding of an International Symposium, Guangzhou, P.R.C., Fish Physiology, Fish Toxicology and Fisheries Management. Environmental Research Laboratory, Office of Research and Development, U.S. Environmental Protection Agency, Athens, Georgia, 30613., Athens, Georgia, pp. 73–87.

Helland, S., Grisdale-Helland, B., 2011. Dietary threonine requirement of Atlantic salmon smolts. Aquaculture 321, 230–236.

Helland, S., Hatlen, B., Grisdale-Helland, B., 2010. Energy, protein and amino acid requirements for maintenance and efficiency of utilization for growth of Atlantic salmon post-smolts determined using increasing ration levels. Aquaculture 305, 150–158.

Hernandez-Llamas, A., 2009. Conventional and alternative dose–response models to estimate nutrient requirements of aquaculture species. Aquaculture 292, 207–213.

Hidalgo, M., Urea, E., Sanz, A., 1999. Comparative study of digestive enzymes in fish with different nutritional habits. Proteolytic and amylase activities. Aquaculture 170, 267–283.

Hoglund, E., et al., 2007. Attenuation of stress-induced anorexia in brown trout (*Salmo trutta*) by pre-treatment with dietary L-tryptophan. Br. J. Nutr. 97, 786–789.

Hong, Y., et al., 2015. Growth, digestive and absorptive capacity and antioxidant status in intestine and hepatopancreas of sub-adult grass carp *Ctenopharyngodonidella* fed graded levels of dietary threonine. J. Anim. Sci. Biotechnol. 6, 1–11.

Hook, S., 2010. Promise and progress in environmental genomics: a status report on the applications of gene expression-based microarray studies in ecologically relevant fish species. J. Fish. Biol. 77, 1999–2022.

Hooley, C., Barrows, F., Paterson, J., Sealey, W., 2014. Examination of the effects of dietary protein and lipid levels on growth and stress tolerance of juvenile tilapia, *Oreochromis niloticus*. J. World Aquacult. Soc. 45, 115–126.

Hoseini, S., Mirghaed, A., Mazandarani, M., Zoheiri, F., 2016. Serum cortisol, glucose, thyroid hormones and non-specific immune responses of Persian sturgeon, *Acipenser persicus* to exogenous tryptophan and acute stress. Aquaculture 462, 17–23.

Hossain, M., Almatar, S., James, C., 2011. Whole body and egg amino acid composition of silver pomfret, *Pampus argenteus* (Euphrasen, 1788) and prediction of dietary requirements for essential amino acids. J. Appl. Ichthyol. 27, 1067–1071.

Houlihan, D., 1991. Protein turnover in ectotherms and its relationships to energetics. In: Houlihan, D., Livingstone, D., Lee, R. (Eds.), Adv. Comp. Environ. Physiol. Springer Berlin Heidelberg, Berlin, Heidelberg, pp. 1–43.

Houlihan, D., Carter, C., Mccarthy, I., 1995a. Chapter 8 Protein synthesis in fish. Biochem. Mol. Biol. Fishes 4, 191–220.

Houlihan, D., Mccarthy, I., Carter, C., Marttin, F., 1995b. Protein turnover and amino acid flux in fish larvae. Int. Counc. Explor. Sea (ICES) Mar. Sci. Symp. 201, 87–99.

Hseu, J., et al., 2003. Effect of exogenous tryptophan on cannibalism, survival and growth in juvenile grouper *Epinephelus coioides*. Aquaculture 218, 251–263.

Hu, K., et al., 2014. Oxidative damage repair by glutamine in fish enterocytes. Fish Physiol. Biochem. 40, 1437–1445.

Hu, Y., et al., 2018. Effect of dietary taurine supplementation on growth, digestive enzyme, immunity and resistant to dry stress of rice field eel (*Monopterus albus*) fed low fish meal diets. Aquac. Res. 49, 2108–2118.

Hua, K., et al., 2019. Effects of dietary protein and lipid levels on growth, body composition and nutrient utilization of *Channa striata*. Aquaculture 501, 368–373.

Hua, K., Bureau, D., 2019. Estimating changes in essential amino acid requirements of rainbow trout and Atlantic salmon as a function of body weight or diet composition using a novel factorial requirement model. Aquaculture 513, 734440.

Hua, K., 2013. Estimating maintenance essential amino acids requirements of fish through a nonlinear mixed modeling approach. Aquac. Res. 44, 542–553.

Huai, M., et al., 2009. Quantitative dietary threonine requirement of juvenile Pacific white shrimp, *Litopenaeus vannamei* (Boone) reared in low-salinity water. Aquac. Res. 40, 904–914.

Huang, H., et al., 2019. Dietary N-Carbamylglutamate (NCG) alleviates liver metabolic disease and hepatocyte apoptosis by suppressing ERK1/2-mTOR-S6K1 signal pathway via prompting endogenous arginine synthesis in Japanese seabass (*Lateolabrax japonicus*). Fish Shellfish Immunol. 90, 338–348.

Huang, Z., et al., 2015. Effects of dietary isoleucine levels on the growth performance, feed utilization, and serum biochemical indices of juvenile golden pompano, *Trachinotus ovatus*. Isr. J. Aquac.Bamid. 1226, 1–9.

Huang, Z., et al., 2017. Effect of dietary valine levels on the growth performance, feed utilization and immune function of juvenile golden pompano. Trachinotus ovatus. Aquac. Nutr. 24, 74–82.

Hughes, S., 1985. Evaluation of glutamic acid and glycine as sources of nonessential amino acids for lake trout (*Salvelinus namaycush*) and rainbow trout (*Salmo gairdneri*). Comp. Biochem. Physiol. 81, 669–671.

Hughes, S.G., Rumsey, G.L., Nesheim, M.C., 1983. Dietary requirements for essential branched-chain amino acids by lake trout. Trans. Am. Fish. Soc. 112, 812–817.

Hundal, H., Taylor, P., 2009. Amino acid transceptors: gate keepers of nutrient exchange and regulators of nutrient signaling. Am. J. Physiol. Endocrinol. Metab. 296, 603–613.

Huo, Y., et al., 2014. Effects of dietary protein and lipid levels on growth, feed utilization and body composition of juvenile swimming crab, *Portunus trituberculatus*. Aquaculture 434, 151–158.

Huxtable, R., 1992. Physiological actions of taurine. Physiol. Rev. 72, 101–163.

Intanai, I., Taylor, E., Whiteley, N., 2009. Effects of salinity on rates of protein synthesis and oxygen uptake in the post-larvae and juveniles of the tropical prawn *Macrobrachium rosenbergii* (de Man). Comp. Biochem. Physiol. A. Mol. Integr. Physiol. 152, 372–378.

Jackson, A., Capper, B., 1982. Investigations into the requirements of the tilapia *Sarotherodon mossambicus* for dietary methionine, lysine and arginine in semi-synthetic diets. Aquaculture 29, 289–297.

Jacobs, B., Azmitia, E., 1992. Structure and function of the brain serotonin system. Physiol. Rev. 72, 165–228.

Jatoba, A., et al., 2014. Protein levels for *Litopenaeus vannamei* in semi-intensive and biofloc systems. Aquaculture 43, 356–371.

Jiang, J., et al., 2009. Influence of glutamine and vitamin E on growth and antioxidant capacity of fish enterocytes. Aquac. Nutr. 15, 409–414.

Jiang, J., et al., 2015a. Effects of glutamate on growth, antioxidant capacity, and antioxidant-related signaling molecule expression in primary cultures of fish enterocytes. Fish Physiol. Biochem. 41, 1143–1153.

Jiang, W., et al., 2015b. The tight junction protein transcript abundance changes and oxidative damage by tryptophan deficiency or excess are related to the modulation of the signalling molecules, NF-κB p65, TOR, caspase-(3,8,9) and Nrf2 mRNA levels, in the gill of young grass carp (*Ctenopharyngodon idellus*). Fish Shellfish Immunol. 46, 168–180.

Jiang, J., et al., 2016a. Glutamate ameliorates copper-induced oxidative injury by regulating antioxidant defences in fish intestine. Br. J. Nutr. 116, 70–79.

Jiang, W., et al., 2016b. Changes in integrity of the gill during histidine deficiency or excess due to depression of cellular anti-oxidative ability, induction of apoptosis, inflammation and impair of cell-cell tight junctions related to Nrf2, TOR and NF-kB signaling in fish. Fish Shellfish Immunol. 56, 111–122.

Jiang, W., et al., 2016c. Histidine prevents cu-induced oxidative stress and the associated decreases in mRNA from encoding tight junction proteins in the intestine of grass carp (*Ctenopharyngodon idella*). PLoS One 11, e0157001.

Jiang, W., et al., 2016d. Enhanced muscle nutrient content and flesh quality, resulting from tryptophan, is associated with anti-oxidative damage referred to the Nrf2 and TOR signalling factors in young grass carp (*Ctenopharyngodon idella*): avoid tryptophan deficiency or excess. Food Chem. 199, 210–219.

Jiang, S., et al., 2016e. Optimal dietary protein level and protein to energy ratio for hybrid grouper (*Epinephelus fuscoguttatus* female x *Epinephelus lanceolatus* male) juveniles. Aquaculture 465, 28–36.

Jiang, H., et al., 2017a. Nutrient sensing and metabolic changes after methionine deprivation in primary muscle cells of turbot (*Scophthalmus maximus* L.). J. Nutr. Biochem. 50, 74–82.

Jiang, J., et al., 2017b. Glutamate attenuates lipopolysaccharide-induced oxidative damage and mRNA expression changes of tight junction and defensin proteins, inflammatory and apoptosis response signaling molecules in the intestine of fish. Fish Shellfish Immunol. 70, 473–484.

Jiang, W., et al., 2017c. Towards the modulation of oxidative damage, apoptosis and tight junction protein in response to dietary leucine deficiency: a likely cause of ROS-induced gill structural integrity impairment. Fish Shellfish Immunol. 70, 609–620.

Jin, M., et al., 2013. Dietary protein requirements of the juvenile swimming crab, *Portunus trituberculatus*. Aquaculture 414, 303–308.

Jin, Y., et al., 2015a. Interactions between dietary protein levels, growth performance, feed utilization, gene expression and metabolic products in juvenile grass carp (*Ctenopharyngodon idella*). Aquaculture 437, 75–83.

Jin, M., et al., 2015b. Dietary lysine requirement of juvenile swimming crab, *Portunus trituberculatus*. Aquaculture 448, 1–7.

Jin, Y., et al., 2017. Dietary tryptophan requirements of juvenile Pacific white shrimp, *Litopenaeus vannamei* (Boone) reared in low-salinity water. Aquac. Int. 25, 955–968.

Jirsa, D., et al., 2014. Taurine requirement for juvenile white seabass (*Atractoscion nobilis*) fed soy-based diets. Aquaculture 422–423, 36–41.

Joazeiro, C., Weissman, A., 2000. RING finger proteins: mediators of ubiquitin ligase activity. Cell 102, 549–552.

Johnston, I., 2001. Muscle Development and Growth. Academic Press, America.

Johnston, I., Bower, N., Macqueen, D., 2011. Growth and the regulation of myotomal muscle mass in teleost fish. J. Exp. Biol. 214, 1617–1628.

Johnson, R., et al., 2015. Effects of dietary taurine supplementation on growth, feed efficiency, and nutrient composition of juvenile sablefish (*Anoplopoma fimbria*) fed plant based feeds. Aquaculture 445, 79–85.

Jover Cerdá, C., et al., 1998. Crecimiento de Tilapias (*Oreochromis niloticus*, L.) con piensos extrusionados de diferente nivel proteico. Arch. Zootec. 47, 11–20.

Kasozi, N., et al., 2019. Dietary amino acid requirements of pebbly fish, *Alestes baremoze* (Joannis, 1835) based on whole body amino acid composition. Aquac. Rep. 14, 100197.

Katersky, R., Carter, C., 2010. The effect of temperature on post-prandial protein synthesis in juvenile barramundi, *Lates calcarifer*. Comp. Biochem. Physiol. A. Mol. Integr. Physiol. 156, 529–536.

Kato, K., et al., 2013. Effects of dietary taurine levels on epidermal thickness and scale loss in red sea bream, *Pagrus major*. Aquac. Res. 45, 1818–1824.

Kaushik, S., 1979. Application of a biochemical method for the estimation of amino acid needs in fish: quantitative arginine requirements of rainbow trout in different salinities. In: Halver, J.E., Tiew, K. (Eds.), Finfish Nutrition and Fishfeed technology., Heenemann Verlagsgesellschaft; Berlin (GFR). International Council for the Exploration of the Sea, Charlottenlund (Denmark), pp. 197–207. Schr. Bundesforschungsanst. Fisch. Hamb., v. 14/15.

Kaushik, S., 1995. Protein and amino acid nutrition of fish in comparison to terrestrial vertebrates. In: Nunes, A.F., Portugal, A.V., Costa, J.P., Ribeiro, J.R. (Eds.), Protein Metabolism and Nutrition. INIA, Portugal, pp. 47–56.

Kaushik, S., 1998. Whole body amino acid composition of European seabass (*Dicentrarchus labrax*), gilthead seabream (*Sparus aurata*) and turbot (*Psetta maxima*) with an estimation of their IAA requirement profiles. Aquat. Living Resour. 11, 355–358.

Kaushik, S., Luquet, P., 1977. Study of free amino acids in rainbow trout in relation to salinity changes. II. Muscle free amino acids during starvation. Ann. hydrobiol. 8, 375–387.

Kaushik, S., Luquet, P., 1979. Influence of dietary amino acid patterns on the free amino acid contents of blood and muscle of rainbow trout (*Salmo gairdnerii* R.). Comp. Biochem. Physiol. B 64, 175–180.

Kaushik, S., Georga, I., Koumoundouros, G., 2011. Growth and body composition of zebrafish (*Danio rerio*) larvae fed a compound feed from first feeding onward: toward implications on nutrient requirements. Zebrafish 8, 87–95.

Kaushik, S., Breque, J., Blanc, D., 1991. Requirement for protein and essential amino acids and their utilization by Siberian sturgeon (*Acipenser baeri*). In: Williot, P. (Ed.), Proceedings of the First International Symposium on the Sturgeon. Bordeaux-France. CEMAGREF, pp. 25–39.

Kaushik, S.J., Fauconneau, B., Terrier, L., Gras, J., 1988. Arginine requirement and status assessed by different biochemical indices in rainbow trout (*Salmo gairdneri* R.). Aquaculture 70, 75–95.

Kaushik, S., et al., 1995. Protein and energy needs for maintenance and growth of Nile tilapia (*Oreochromis niloticus*). J. Appl. Ichthyol. 11, 290–296.

Kaushik, S.J., Seiliez, I., 2010. Protein and amino acid nutrition and metabolism in fish: current knowledge and future needs. Aquac. Res. 41, 322–332.

Keembiyehetty, C., Gatlin III, D., 1992. Dietary lysine requirement of juvenile hybrid striped bass (*Morone chrysops* × *M. saxatilis*). Aquaculture 104, 271–277.

Keembiyehetty, C., Gatlin III, D., 1993. Total sulfur amino acid requirement of juvenile hybrid striped bass. Aquaculture 110, 331–339.

Keembiyehetty, C.N., Gatlin III, D., 1997. Dietary threonine requirement of juvenile hybrid striped bass (*Morone chrysops*♀×*M. saxatilis*♂). Aquac. Nutr. 3, 217–221.

Ketola, H., 1983. Requirement for dietary lysine and arginine by fry of rainbow trout. J. Anim. Sci. 56, 101–107.

Khan, M., Jafri, A., Chadha, N., 2004. Growth, reproductive performance, muscle and egg composition in grass carp, *Ctenopharyngodon idella* (Valenciennes), fed hydrilla or formulated diets with varying protein levels. Aquacul. Res. 35, 1277–1285.

Khan, K., et al., 2019. Whole-body amino acid pattern of juvenile, preadult, and adult pacu, *Piaractus mesopotamicus*, with an estimation of its dietary essential amino acid requirements. J. World Aquac. Soc. 51, 224–234.

Khan, M., Abidi, S., 2007a. Dietary histidine requirement of fingerling Indian major carp, *Labeo rohita* (Hamilton). Aquac. Nutr. 13, 424–430.

Khan, M., Abidi, S., 2007b. Dietary isoleucine requirement of fingerling Indian major carp, *Labeo rohita* (Hamilton). Aquac. Nutr. 13, 424–430.

Khan, M., Abidi, S., 2007c. Total aromatic amino acid requirement of Indian major carp *Labeo rohita* (Hamilton) fry. Aquaculture 267, 111–118.

Khan, M., Abidi, S., 2009. Optimum histidine requirement of fry African catfish, *Clarias gariepinus* (Burchell). Aquac. Res. 40, 1000–1010.

Khan, M., Abidi, S., 2013. Dietary methionine requirement of Indian major carp fry, *Cirrhinus mrigala* (Hamilton) based on growth, feed conversion and nitrogen retention efficiency. Aquac. Res. 44, 268–281.

Khan, M., Abidi, S., 2014. Dietary histidine requirement of Singhi, *Heteropneustes fossilis* fry (Bloch). Aquac. Res. 45, 1341–1354.

Kilberg, M., Shan, J., Su, N., 2009. ATF4-dependent transcription mediates signaling of amino acid limitation. Trends Endocrinol. Metabol. 20, 436–443.

Kim, K., 1993. Requirement for phenylalanine and replacement value of tyrosine for phenylalanine in rainbow trout (*Oncorhynchus mykiss*). Aquaculture 113, 243–250.

Kim, H., Hur, S., 1993. Optimum level of protein and the possibility of replacement of fish meal by soybean meal in feeds for flashy shrimp, *Penaeus chinensis*. J. Aquacult. 6, 173–186.

Kim, K., Kayes, T., Amundson, C., 1987. Effects of dietary tryptophan levels on growth, feed/gain, carcass composition and liver glutamate dehydrogenase activity in rainbow trout (*Salmo gairdneri*). Comp. Biochem. Physiol. 88, 737–741.

Kim, K., Kayes, T., Amundson, C., 1992a. Requirements for lysine and arginine by rainbow trout (*Oncorhynchus mykiss*). Aquaculture 106, 333–344.

Kim, K., Kayes, T., Amundson, C., 1992b. Requirements for sulfur amino acids and utilization of D-methionine by rainbow trout (*Oncorhynchus mykiss*). Aquaculture 101, 95–103.

Kim, J., Lall, S., 2000. Amino acid composition of whole body tissue of Atlantic halibut (*Hippoglossus hippoglossus*), yellowtail founder (*Pleuronectes ferruginea*) and Japanese founder (*Paralichthys olivaceus*). Aquaculture 187, 367–373.

Kim, K., McMillan, I., Bayley, H., 1983. Determination of amino acid requirements of young pigs using an indicator amino acid. Br. J. Nutr. 50, 369–382.

Kim, S., Baker, D., Easter, R., 2001. Dynamic ideal protein and limiting amino acids for lactating sows: the impact of amino acid mobilization. J. Anim. Sci. 79, 2356–2366.

Kim, K., et al., 2004. Effect of dietary cadmium on growth and haematological parameters of juvenile rockfish, *Sebastes schlegeli* (Hilgendorf). Aquacult. Res. 35, 80–86.

Kim, S., et al., 2005. Effect of taurine supplemented practical diet on growth performance and taurine contents in whole body and tissues of juvenile Japanese flounder *Paralichthys olivaceus*. Fish. Sci. 71, 627–632.

Kim, J., Kundu, M., Viollet, B., Guan, K., 2011. AMPK and mTOR regulate autophagy through direct phosphorylation of Ulk1. Nat. Cell Biol. 13, 132–141.

Kim, J., et al., 2017. Taurine supplementation in diet for olive flounder at low water temperature. Fish Aquat. Sci. 20, 1–8.

Kimball, S., Horetsky, R., Jefferson, L., 1998. Signal transduction pathways involved in the regulation of protein synthesis by insulin in L6 myoblasts. Am. J. Physiol. Cell Physiol. 274, 221–228.

Kimball, S., Jefferson, L., 2006. New functions for amino acids: effects on gene transcription and translation. Am. J. Clin. Nutr. 83, 501–507.

Kimura, S., et al., 1988. Characterization of fish muscle type I collagen. J. Food Sci. 53, 1315–1318.

Klein, R., Halver, J., 1970. Nutrition of salmonoid fishes: arginine and histidine requirements of chinook and coho salmon. J. Nutr. 100, 1105–1109.

Kou, H., Xu, S., Wang, A., 2015. Effect of replacing canola meal for fish meal on the growth, digestive enzyme activity, and amino acids, of ovate pompano, *Trachinotus ovatus*. Isr. J. Aquac. Bamidgeh 67, 10.

Kpundeh, M., et al., 2015. Effects of dietary protein levels on growth performance and haemato-immunological parameters of juvenile genetically improved farmed tilapia (GIFT), *Oreochromis niloticus*. Aquacult. Int. 23, 1189–1201.

Krause, U., et al., 2002. Signalling pathways and combinatory effects of insulin and amino acids in isolated rat hepatocytes. FEBS J. 269, 3742–3750.

Krogdahl, A., Sundby, A., Holm, H., 2015. Characteristics of digestive processes in Atlantic salmon (*Salmo salar*). Enzyme pH optima, chyme pH, and enzyme activities. Aquaculture 449, 27–36.

Kudchodkar, S., Del Prete, G., Maguire, T., Alwine, J., 2007. AMPK-mediated inhibition of mTOR kinase is circumvented during immediate-early times of human cytomegalovirus infection. J. Virol. 81, 3649–3651.

Kumar, G., Hummel, K., Razzazi-Fazeli, E., El-Matbouli, M., 2018. Proteome profiles of head kidney and spleen of rainbow trout (*Oncorhynchus mykiss*). Proteomics 18, 4.

Kurpad, A., 2018. 90th anniversary commentary: amino acid imbalances: still in the balance. J. Nutr. 148, 1647–1649.

Lall, S., et al., 1994. Quantitative arginine requirement of Atlantic salmon (*Salmo salar*) reared in sea water. Aquaculture 124, 13–25.

Lamarre, S., Le François, N., Driedzic, W., Blier, P., 2009. Protein synthesis is lowered while 20S proteasome activity is maintained following acclimation to low temperature in juvenile spotted wolffish (*Anarhichas minor* Olafsen). J. Exp. Biol. 212, 1294–1301.

Lamarre, S., Saulnier, R., Blier, P., Driedzic, W., 2015. A rapid and convenient method for measuring the fractional rate of protein synthesis in ectothermic animal tissues using a stable isotope tracer. Comp. Biochem. Physiol. B Biochem. Mol. Biol. 182, 1–5.

Lansard, M., et al., 2011. l-Leucine, l-methionine, and l-lysine are involved in the regulation of intermediary metabolism-related gene expression in rainbow trout hepatocytes. J. Nutr. 141, 75–80.

Larsen, A., Mohammed, H., Arias, C., 2014. Characterization of the gut microbiota of three commercially valuable warmwater fish species. J. Appl. Microbiol. 116, 1396–1404.

Larumbe-Moran, E., et al., 2010. Protein requirements of Nile tilapia (*Oreochromis niloticus*) fry cultured at different salinities. Aquac. Res. 41, 1150–1157.

Laszlo, K., Yunchao, S., 2014. The critical role of calpain in cell proliferation. J. Biomol. Res. Therapeut. 03, 4431–4437.

Lecker, S., et al., 2004. Multiple types of skeletal muscle atrophy involve a common program of changes in gene expression. FASEB J 18, 39–51.

Lepage, O., et al., 2003. Timecourse of the effect of dietary L-tryptophan on plasma cortisol levels in rainbow trout *Oncorhynchus mykiss*. J. Exp. Biol. 206, 3589–3599.

Lepage, O., Larson, E., Mayer, I., Winberg, S., 2005. Serotonin, but not melatonin, plays a role in shaping dominant-subordinate relationships and aggression in rainbow trout. Horm. Behav. 48, 233–242.

Lepage, O., Tottmar, O., Winberg, S., 2002. Elevated dietary intake of L-tryptophan counteracts the stress-induced elevation of plasma cortisol in rainbow trout (*Oncorhynchus mykiss*). J. Exp. Biol. 205, 3679–3687.

Lepage, S., Bruce, A., 2008. Characterization and comparative expression of zebrafish calpain system genes during early development. Dev. Dynam. 237, 819–829.

Li, M., et al., 2016. Effects of dietary taurine on growth, immunity and hyperammonemia in juvenile yellow catfish *Pelteobagrus fulvidraco* fed all-plant protein diets. Aquaculture 450, 349–355.

Li, P., Wu, G., 2018. Roles of dietary glycine, proline, and hydroxyproline in collagen synthesis and animal growth. Amino Acids 50, 29–38.

Li, P., et al., 2007. Amino acids and immune function. Br. J. Nutr. 98, 237–252.

Li, Y., et al., 2010. Dietary leucine requirement for juvenile large yellow croaker *Pseudosciaena crocea* (Richardson, 1846). J. Ocean Univ. China 9, 371–375.

Li, X., Rezaei, R., Li, P., Wu, G., 2011. Composition of amino acids in feed ingredients for animal diets. Amino Acids 40, 1159–1168.

Li, Y., et al., 2004. Growth and optimal protein requirement of tilapia fed casein as sole protein sources in both fresh water and sea. Fish. Sci. 23, 1–4.

Li, W., et al., 2015a. Effects of dietary phenylalanine on growth, digestive and brush border enzyme activities and antioxidant capacity in the hepatopancreas and intestine of young grass carp (*Ctenopharyngodon idella*). Aquac. Nutr. 21, 913–925.

Li, Y., Cheng, Z., Mai, K., Ai, Q., 2015b. Dietary leucine requirement of juvenile Japanese seabass (*Lateolabrax japonicus*). J. Ocean Univ. China 14, 121–126.

Liang, H., et al., 2016. Dietary arginine affects growth performance, plasma amino acid contents and gene expressions of the TOR signaling pathway in juvenile blunt snout bream, *Megalobrama amblycephala*. Aquaculture 461, 1–8.

Liao, Y., et al., 2014. Dietary methionine requirement of juvenile blunt snout bream (*Megalobrama amblycephala*) at a constant dietary cystine level. Aquac. Nutr. 20, 741–752.

Liebert, F., Sünder, A., Mohamed, K., 2006. Assessment of nitrogen maintenance requirement and potential for protein deposition in juvenile tilapia genotypes by application of an exponential nitrogen utilization model. Aquaculture 261, 1346–1355.

Lim, S., et al., 2013. Taurine is an essential nutrient for juvenile parrot fish *Oplegnathus fasciatus*. Aquaculture 414–415, 274–279.

Lin, H., et al., 2015. Effect of dietary arginine levels on the growth performance, feed utilization, non-specific immune response and disease resistance of juvenile golden pompano *Trachinotus ovatus*. Aquaculture 437, 382–389.

Liu, Z., et al., 2013. Characterization of oligopeptide transporter (PepT1) in grass carp (*Ctenopharyngodon idella*). Comp. Biochem. Physiol. B Biochem. Mol. Biol. 164, 194–200.

Liu, Z., et al., 2014a. Characterization of L-type Amino acid transporter 2 from crucian carp, *Carassius auratus*. J. World Aquac. Soc. 45, 709–718.

Liu, F., et al., 2014b. Quantitative dietary isoleucine requirement of juvenile Pacific white shrimp, *Litopenaeus vannamei* (Boone) reared in low-salinity water. Aquac. Int. 22, 1481–1497.

Liu, F., et al., 2014c. Quantitative dietary leucine requirement of juvenile Pacific white shrimp, *Litopenaeus vannamei* (Boone) reared in low-salinity water. Aquac. Nutr. 20, 332–340.

Liu, Y., et al., 2014d. Hydroxyproline supplementation on the performances of high plant protein source based diets in turbot (*Scophthalmus maximus* L.). Aquaculture 433, 476–480.

Liu, W., et al., 2017a. Dietary protein level affects the growth performance of large male genetically improved farmed tilapia, *Oreochromis niloticus*, reared in fertilized freshwater cages. J. World Aquacult. Soc. 48, 718–728.

Liu, Z., et al., 2017b. Effects of different salinities on growth performance, survival, digestive enzyme activity, immune response, and muscle fatty acid composition in juvenile American shad (*Alosa sapidissima*). Fish Physiol. Biochem. 43, 761–773.

Liu, C., et al., 2019. Recent advances in amino acid sensing and new challenges for protein nutrition in aquaculture. Mar. Life. Sci. Technol. 1, 1–10.

Lokman, P., Symonds, J., 2014. Molecular and biochemical tricks of the research trade: -omics approaches in finfish aquaculture. N. Z. J. Mar. Freshw. Res. 48, 492–505.

López-Alvarado, J., Kanazawa, A., 1994. Effect of dietary arginine levels on growth of red sea bream larvae fed diets supplemented with crystalline amino acids. Fish. Sci. 60, 435–439.

Luo, H., Wong, J., Wong, B., 2010. Protein degradation systems in viral myocarditis leading to dilated cardiomyopathy. Cardiovasc. Res. 85, 347–356.

Lu, X., et al., 2020. Effects of dietary protein levels on growth, muscle composition, digestive enzymes activities, hemolymph biochemical indices and ovary development of pre-adult red swamp crayfish (*Procambarus clarkii*). Aquacult. Rep. 18, 100542.

Luo, Z., et al., 2005. Dietary L-methionine requirement of juvenile grouper *Epinephelus coioides* at a constant dietary cystine level. Aquaculture 249, 409–418.

Luo, Z., et al., 2006. Quantitative L-lysine requirement of juvenile grouper *Epinephelus coioides*. Aquac. Nutr. 12, 165–172.

Luo, Z., et al., 2007. Effects of dietary arginine levels on growth performance and body composition of juvenile grouper *Epinephelus coioides*. J. Appl. Ichthyol. 23, 252–257.

Luo, J., et al., 2014. The impaired intestinal mucosal immune system by valine deficiency for young grass carp (*Ctenopharyngodon idella*) is associated with decreasing immune status and regulating tight junction proteins transcript abundance in the intestine. Fish Shellfish Immunol. 40, 197–207.

Lupatsch, I., Kissil, G., Sklan, D., Pfeffer, E., 1998. Energy and protein requirements for maintenance and growth in gilthead seabream (*Sparus aurata* L.). Aquac. Nutr. 4, 165–173.

Lupatsch, I., Kissil, G., Sklan, D., 2001. Optimization of feeding regines for European seabass *Dicentrarchus labrax*: a factorial approach. Aquaculture 202, 289–302.

Lupatsch, I., Deshev, R., Magen, I., 2010. Energy and protein demands for optimal egg production including maintenance requirements of female tilapia *Oreochromis niloticus*. Aquac. Res. 41, 763–769.

Luzzana, U., Hardy, R., Halver, J., 1998. Dietary arginine requirement of fingerling coho salmon (*Oncorhynchus kisutch*). Aquaculture 163, 137–150.

Lyndon, A., Houlihan, D., Hall, S., 1992. The effect of short-term fasting and a single meal on protein synthesis and oxygen consumption in cod, *Gadus morhua*. J. Comp. Physiol. B Biochem. Syst. Environ. Physiol. 162, 209–215.

Lyndon, A., Davidson, I., Houlihan, D., 1993. Changes in tissue and plasma free amino acid concentrations after feeding in Atlantic cod. Fish Physiol. Biochem. 65–75.

Macqueen, D., et al., 2010. Characterisation of capn1, capn2-like, capn3 and capn11 genes in Atlantic halibut (*Hippoglossus hippoglossus* L.): transcriptional regulation across tissues and in skeletal muscle at distinct nutritional states. Gene 453, 45–58.

Mai, K., et al., 2006a. Dietary lysine requirement of juvenile Japanese seabass, *Lateolabrax japonicus*. Aquaculture 258, 535–542.

Mai, K., et al., 2006b. Dietary methionine requirement of large yellow croaker, *Pseudosciaena crocea* R. Aquaculture 253, 564–572.

Mambrini, M., Kaushik, S., 1995a. Effect of temperature on sulfur amino acid requirements for maintenance and growth of juvenile rainbow trout. In: Nunes, A., Portugal, J., Costa, J., Ribeiro, J. (Eds.), Proc. 7 Int. Symp. on Protein Metabolism and Nutrition. EAAP Publ, 81, pp. 117–122.

Mambrini, M., Kaushik, S., 1995b. Indispensable amino acid requirements of fish: correspndence between quantitative data and amino acid profiles of tissue proteins. J. Appl. Ichthyol. 11, 240–247.

Mambrini, M., et al., 2004. Selection for growth in brown trout increases feed intake capacity without affecting maintenance and growth requirements. J. Anim. Sci. 82, 2865–2875.

Mansilla, W., Gorman, A., Fortener, L., Shoveller, A., 2018. Dietary phenylalanine requirements are similar in small, medium, and large breed adult dogs using the direct amino acid oxidation technique. J. Anim. Sci. 96, 3112–3120.

Marcouli, P., Alexis, M., Andriopoulou, A., Iliopoulou-georgudaki, J., 2006. Dietary lysine requirement of juvenile gilthead seabream *Sparus* aurata L. Aquac. Nutr. 12, 25–33.

Martinez, J., Chatzifotis, S., Divanach, P., Takeuchi, T., 2004. Effect of dietary taurine supplementation on growth performance and feed selection of sea bass *Dicentrarchus labrax* fry fed with demand-feeders. Fish. Sci. 70, 74–79.

Martinez-Palacios, C., et al., 2007. Dietary protein requirement of juvenile Mexican Silverside (*Menidia estor* Jordan 1879), a stomachless zooplanktophagous fish. Aquacult. Nutr. 13, 304–310.

Matsunari, H., et al., 2008a. Effect of dietary taurine and cystine on growth performance of juvenile red sea bream *Pagrus major*. Aquaculture 274, 142–147.

Matsunari, H., et al., 2008b. Optimum dietary taurine level in casein-based diet for juvenile red sea bream *Pagrus major*. Fish. Sci. 74, 347–353.

Maurin, A., et al., 2005. The GCN2 kinase biases feeding behavior to maintain amino acid homeostasis in omnivores. Cell Metabol. 1, 273–277.

Meijer, A., 2003. Amino acids as regulators and components of nonproteinogenic pathways. J. Nutr. 133, 2057S–2062S.

Mercer, L., 1992. The determination of nutritional requirements: mathematical modeling of nutrient-response curves. J. Nutr. 122, 706–708.

Mercer, L., May, H., Dodds, S., 1989. The determination of nutritional-requirements in rats-mathematical-modeling of sigmoidal, inhibited nutrient-response curves. J. Nutr. 119, 1465–1471.

Meyer-Burgdorff, K., Rosenow, H., 1995. Protein turnover and energy metabolism in growing carp. J. Anim. Physiol. Anim. Nutr. 73, 113–122.

Michelato, M., et al., 2013. Digestible methionine+cystine requirement for Nile tilapia from 550 to 700 g. Rev. Bras. Zootec. 42, 7–12.

Michelato, M., et al., 2016a. Dietary histidine requirement of Nile tilapia juveniles based on growth performance, expression of muscle-growth-related genes and haematological responses. Aquaculture 467, 63–70.

Michelato, M., et al., 2016b. Dietary lysine requirement to enhance muscle development and fillet yield of finishing Nile tilapia. Aquaculture 457, 124–130.

Michelato, M., et al., 2016c. Dietary threonine requirement to optimize protein retention and fillet production of fast-growing Nile tilapia. Aquac. Nutr. 22, 759–766.

Miguel-Aliaga, I., 2012. Nerveless and gutsy: intestinal nutrient sensing from invertebrates to humans. Semin. Cell Dev. Biol. 23, 614–620.

Millamena, O., Bautista-Teruel, M., Kanazawa, A., 1996. Methionine requirement of juvenile tiger shrimp *Penaeus monodon* Fabricius. Aquaculture 143, 403–410.

Millamena, O., Bautista, M., Reyes, O., Kanazawa, A., 1997. Threonine requirement of juvenile marine shrimp *Penaeus monodon*. Aquaculture 151, 9–14.

Millamena, O., Bautistateruel, M., Reyes, O., Kanazawa, A., 1998. Requirements of juvenile marine shrimp, *Penaeus monodon* (Fabricius) for lysine and arginine. Aquaculture 164, 95–104.

Millamena, O., Teruel, M., Kanazawa, A., Teshima, S., 1999. Quantitative dietary requirements of postlarval tiger shrimp, *Penaeus monodon*, for histidine, isoleucine, leucine, phenylalanine and tryptophan. Aquaculture 179, 169–179.

Milligan, C., 1997. The role of cortisol in amino acid mobilization and metabolism following exhaustive exercise in rainbow trout (*Oncorhynchus mykiss* Walbaum). Fish Physiol. Biochem. 16, 119–128.

Millikin, M., 1983. Interactive effects of dietary protein and lipid on growth and protein utilization of age-0 striped bass. Trans. Am. Fish. Soc. 112, 185–193.

Mizushima, N., et al., 2004. In vivo analysis of autophagy in response to nutrient starvation using transgenic mice expressing a fluorescent autophagosome marker. Mol. Biol. Cell 15, 1101–1111.

Moehn, S., Shoveller, A., Rademacher, M., Ball, R., 2008. An estimate of the methionine requirement and its variability in growing pigs using the indicator amino acid oxidation technique. J. Anim. Sci. 86, 364–369.

Mohammadi, M., et al., 2014. Optimization of dietary protein in all male nile tilapia (*Oreochromis niloticus*) reared in inland saline water. Anim. Nutr. Feed Techn. 91–99.

Mohanty, B., et al., 2019. Omics technology in fisheries and aquaculture. In: Mohanty, B. (Ed.), Advances in Fish Research. Narendra Publishing House, Delhi, pp. 1–30.

Mommsen, T., Moon, T., Plisetskaya, E., 2001. Effects of arginine on pancreatic hormones and hepatic metabolism in rainbow trout. Physiol. Biochem. Zool. 74, 668–678.

Mommsen, T., 2004. Salmon spawning migration and muscle protein metabolism: the August Krogh principle at work. Comp. Biochem. Physiol. B Biochem. Mol. Biol. 139, 383–400.

Monentcham, S., Whatelet, B., Pouomogne, V., Kestemont, P., 2010. Egg and whole-body amino acid profile of African bonytongue (*Heterotis niloticus*) with an estimation of their dietary indispensable amino acids requirements. Fish Physiol. Biochem. 36, 531–538.

Montanez, R., Rodriguesz-Caso, C., Sanchez-Jimenez, F., Medina, M., 2008. In silico analysis of arginine catabolism as a source of nitric oxide or polyamines in endothelial cells. Amino Acids 34, 223–229.

Montes-Girao, P., Fracalossi, D., 2006. Dietary lysine requirement as basis to estimate the essential dietary amino acid profile for Jundiá, rhamdia quelen. J. World Aquac. Soc. 37, 388–396.

Moon, H., Gatlin III, D., 1991. Total sulfur amino acid requirement of juvenile red drum, *Sciaenops ocellatus*. Aquaculture 95, 97–106.

Morris, J., Rogers, Q., 1978. Arginine: an essential amino acid for the cat. J. Nutr. 108, 1944–1953.

Morris, S., 2007. Arginine metabolism: boundaries of our knowledge. J. Nutr. 137, 1602S–1609S.

Mu, Y., et al., 1998. Effects of protein level in isocaloric diets on growth performance of the juvenile Chinese hairy crab, *Eriocheir sinensis*. Aquaculture 165, 139–148.

Murai, T., et al., 1985. Effects of dietary protein and lipid levels on performance and carcass composition of fingerling carp. Bull. Jpn. Soc. Sci. Fish. 51, 605–608.

Murillo-Gurrea, D., Coloso, R., Borlongan, I., Serrano, A., 2001. Lysine and arginine requirements of juvenile Asian sea bass (*Lates calcarifer*). J. Appl. Ichthyol. 17, 49–53.

Murthy, H., Varghese, T., 1995. Arginine and histidine requirements of the Indian major carp, *Labeo rohita* (Hamilton). Aquac. Nutr. 1, 235–239.

Murthy, H., Varghese, T., 1997a. Quantitative dietary requirements of the Indian major carp *Labeo rohita* for the essential amino acid valine. Indian J. Anim. Sci. 67, 1028–1030.

Murthy, H., Varghese, T., 1997b. Dietary tryptophan requirement for the growth of rohu, *Labeo rohita*. J. Appl. Aquac. 7, 71–79.

Nelson, L., Sheridan, M., 2006. Gastroenteropancreatic hormones and metabolism in fish. Gen. Comp. Endocrinol. 148, 116–124.

Nemova, N., Lysenko, L., Kantserova, N., 2016. Degradation of skeletal muscle protein during growth and development of salmonid fish. Russ. J. Dev. Biol. 47, 161–172.

Neu, D., et al., 2016. Growth performance, biochemical responses, and skeletal muscle development of juvenile Nile tilapia, *Oreochromis niloticus*, fed with increasing levels of arginine. J. World Aquac. Soc. 47, 248–259.

Neu, D., et al., 2017. Growth performance, hematology, and muscle growth in isoleucine fed Nile tilapia. Bol. Inst. Pesca. 43, 231–242.

Ng, W., Hung, S., 1995. Estimating the ideal dietary indispensable amino acid pattern for growth of white sturgeon, *Acipenser transmontanus* (Richardson). Aquac. Nutr. 1, 85–94.

Ngamsnae, P., De Silva, S., Gunasekera, R., 1999. Arginine and phenylalanine requirement of juvenile silver perch *Bidyanus bidyanus* and validation of the use of body amino acid composition for estimating individual amino acid requirements. Aquac. Nutr. 5, 173–180.

Nguyen, L., Davis, D., 2016. Comparison of crystalline lysine and intact lysine used as a supplement in practical diets of channel catfish (*Ictalurus punctatus*) and Nile tilapia (*Oreochromis niloticus*). Aquaculture 464, 331–339.

Nguyen, N., et al., 2014. Feed intake, molt frequency, tissue growth, feed efficiency and energy budget during a molt cycle of mud crab juveniles, *Scylla serrata* (Forskal, 1775), fed on different practical diets with graded levels of soy protein concentrate as main source of protein. Aquaculture 434, 499–509.

Nguyen, L., et al., 2019. Tryptophan requirement in semi-purified diets of juvenile Nile tilapia *Oreochromis niloticus*. Aquaculture 502, 258–267.

Nguyen, T., Davis, D., 2009. Methionine requirement in practical diets of juvenile Nile tilapia, *Oreochromis niloticus*. J. World Aquac. Soc. 40, 410–416.

Nicholson, J., Lindon, J., Holmes, E., 1999. 'Metabonomics': understanding the metabolic responses of living systems to pathophysiological stimuli via multivariate statistical analysis of biological NMR spectroscopic data. Xenobiotica 29, 1181–1189.

Nielsen, L., Nielsen, H., 2001. Purification and characterization of cathepsin D from herring muscle (*Clupea harengus*). Comp. Biochem. Physiol. B Biochem. Mol. Biol. 128, 351–363.

Niu, J., et al., 2013. Quantitative dietary methionine requirement of juvenile golden pompano *Trachinotus ovatus* at a constant dietary cystine level. Aquac. Nutr. 19, 677–686.

Niu, J., et al., 2018. Assessing the bioavailability of the Novel Met-Met product (AQUAVI® Met-Met) compared to DL-methionine (DL-Met) in white shrimp (*Litopenaeus vannamei*). Aquaculture 484, 322–332.

Nose, T., 1973. Changes in pattern of free plasma amino acid in rainbow trout after feeding. Bull. Freshw. Fish. Res. Lab. 22, 137–144.

Nose, T., 1979. Summary report on the requirements of essential amino acids for carp. In: Halver, J.E., Tiews, K. (Eds.), Finfish Nutrition and Fishfeed Technology. Germany Heenemann GmbH, Berlin, pp. 145–156.

Nose, T., 1963. Determination of nutritive value of food protein on fish. II. Effect of amino acid composition of high protein diets on growth and protein utilization of the rainbow trout. Freshwater Fish. Res. Lab. 13, 41–50.

Nose, T., Arai, D., Lee, L., Hashimoto, Y., 1974. A not on amino acids essential for growth of young carp. Bull. Jpn. Soc. Sci. Fish. 40, 903–908.

Noverian, H., Gopal, P., 2005. Effects of different levels of protein, lipid and their interaction on growth factors of Indian white shrimp (*Fenneropenaeus indicus*) post larvae. Casp. J. Env. Sci. 3, 43–47.

NRC (Naitonal Research Council), 2011. Nutrient Requirements of Fish and Shrimp. The National Academies Press, Washington.

Ogino, C., Saito, K., 1970. Protein nutrition in fish. 1. The utilization of dietary protein by young carp. Bull. Jpn. Soc. Sci. Fish. 36, 250–254.

Ogino, C., 1980. Requirements of carp and rainbow trout for essential amino acids. Bull. Jpn. Soc. Sci. Fish. 46, 171–174.

Ogunji, J., Wirth, M., Rennert, B., 2005. Assessing the dietary amino acid requirements of tilapia, *Oreochromis niloticus* fingerlings. J. Aquac. Trop. 20, 241–250.

Oliva-Teles, A., et al., 2019. Dietary protein requirements of fish-a meta-analysis. Rev. Aquac. r12391.

Olvera-Novoa, M.A., et al., 2007. Optimum protein level for growth in juvenile bullfrog (*Rana catesbeiana* Shaw, 1802). Aquaculture 266, 191–199.

Ostaszewska, T., et al., 2010a. The effect of plant protein-based diet supplemented with dipeptide or free amino acids on digestive tract morphology and PepT1 and PepT2 expressions in common carp (*Cyprinus carpio* L). Comp. Biochem. Physiol. A. Mol. Integr. Physiol. 157, 158–169.

Ostaszewska, T., et al., 2010b. The effect of peptide absorption on PepT1 gene expression and digestive system hormones in rainbow trout (*Oncorhynchus mykiss*). Comp. Biochem. Physiol. A. Mol. Integr. Physiol. 155, 107–114.

Owen, S., et al., 1999. In vivo rates of protein synthesis in Atlantic salmon (*Salmo salar* L.) smolts determined using a stable isotope flooding dose technique. Fish Physiol. Biochem. 20, 87–94.

Ozorio, R., et al., 2009. Protein requirement for maintenance and maximum growth of two-banded seabream (*Diplodus vulgaris*) juveniles. Aquac. Nutr. 15, 85–93.

Pack, M., Rodehutscord, M., Jacobs, S., Pfeffer, E., 1995. Amino acid requirements of rainbow trout (*Oncorhynchus mykiss*) II. Protein deposition as function of dietary methionine, threonine and arginine. J. Appl. Ichthyol. 11, 390–393.

Palma, J., Andrade, J., Lemme, A., Bureau, D., 2015. Quantitative dietary requirement of juvenile Atlantic ditch shrimp *Palaemonetes* varians for lysine, methionine and arginine. Aquac. Res. 46, 1822–1830.

Panigrahi, A., et al., 2018. Carbon:Nitrogen (C:N) ratio level variation influences microbial community of the system and growth as well as immunity of shrimp (*Litopenaeus vannamei*) in biofloc based culture system. Fish Shellfish Immun. 81, 329–337.

Panserat, S., Kaushik, S., 2010. Regulation of gene expression by nutritional factors in fish. Aquac. Res. 41, 751–762.

Panten, U., et al., 1974. Studies on the mechanism of L-leucine and alpha-ketoisocaproic acid- induced insulin release from perifused isolated pancreatic islets. Diabetologia 10, 149–154.

Park, G., Takeuchi, T., Yokoyama, M., Seikai, T., 2002. Optimal dietary taurine level for growth of juvenile Japanese flounder *Paralichthys olivaceus*. Fish. Sci. 68, 824–829.

Pasiakos, S., Carbone, J., 2014. Assessment of skeletal muscle proteolysis and the regulatory response to nutrition and exercise. IUBMB Life 66, 478–484.

Pei, Z., et al., 2004. Comparative study on the effect of dietary lipid level on growth and feed utilization for gibel carp (*Carassius auratus gibelio*) and Chinese longsnout catfish (*Leiocassis longirostris* Günther). Aquac. Nutr. 10, 209–216.

Peñaflorida, V., 1989. An evaluation of indigenous protein sources as potential component in the diet formulation for tiger prawn, *Penaeus monodon*, using essential amino acid index (EAAI). Aquaculture 83, 319–330.

Peragón, J., 2001. Growth, protein-turnover rates and nucleic-acid concentrations in the white muscle of rainbow trout during development. Int. J. Biochem. Cell Biol. 33, 1227–1238.

Peres, H., Oliva-Teles, A., 2005. Protein and energy metabolism of European seabass (*Dicentrarchus labrax*) juveniles and estimation of maintenance requirements. Fish Physiol. Biochem. 31, 23–31.

Peres, H., Oliva-Teles, A., 2006. Effect of the dietary essential to non-essential amino acid ratio on growth, feed utilization and nitrogen metabolism of European sea bass (*Dicentrarchus labrax*). Aquacuture 256, 395–402.

Peres, H., Oliva-Teles, A., 2008. Lysine requirement and efficiency of lysine utilization in turbot (*Scophthalmus maximus*) juveniles. Aquaculture 275, 283–290.

Peres, H., Oliva-Teles, A., 2009. The optimum dietary essential amino acid profile for gilthead seabream (*Sparus aurata*) juveniles. Aquaculture 296, 81–86.

Peres, H., Oliva-Teles, A., 2017. Protein and amino acid nutrition of marine fish. In: Cruz-Suárez, L., Ricque-Marie, D., Tapia-Salazar, M., Nieto-López, M., Villareal-Cavazos, D., Gamboa-Delgado, J., López Acuña, L., yGalaviz-Espinoza, M. (Eds.), Investigación y Desarrollo en Nurición Acuícola Universidad Autónoma de Nuevo León, San Nicolás de los Garza, Nuevo León, México, pp. 438–492.

Person-Le Ruyet, J., et al., 2003. Effects of repeated hypoxic shocks on growth and metabolism of turbot juveniles. Aquat. Living Resour. 16, 25–34.

Pesti, G., Vedenov, D., Cason, J., Billard, L., 2009. A comparison of methods to estimate nutritional requirements from experimental data. Br. Poultry Sci. 50, 16–32.

Peterson, B., Li, M., 2018. Effect of supplemental taurine on juvenile channel catfish, *Ictalurus punctatus* growth performance. Aquac. Nutr. 24, 310–314.

Petro-Sakuma, C., Celino-Brady, F., Breves, J., Seale, A., 2020. Growth hormone regulates intestinal gene expression of nutrient transporters in tilapia (*Oreochromis mossambicus*). Gen. Comp. Endocrinol. 292, 113464.

Pewitt, E., Castillo, S., Velásquez, A., Gatlin III, D., 2017. The dietary tryptophan requirement of juvenile red drum, *Sciaenops ocellatus*. Aquaculture 469, 112–116.

Pfeffer, E., Al Sabty, H., Haverkamp, R., 1992. Studies on lysine requirements of rainbow trout (*Oncorhynchus mykiss*) fed wheat gluten as only source of dietary protein. J. Anim. Physiol. Anim. Nutr. 67, 74–82.

Phang, J., Donald, S., Pandhare, J., Liu, Y., 2008. The metabolism of proline, a stress substrate, modulates carcinogenic pathways. Amino Acids 35, 681–690.

Phang, J., Liu, W., Zabirnyk, O., 2010. Proline metabolism and microenvironmental stress. Annu. Rev. Nutr. 30, 441–463.

Pianesso, D., et al., 2015. Determination of tryptophan requirements for juvenile silver catfish (*Rhamdia quelen*) and its effects on growth performance, plasma and hepatic metabolites and digestive enzymes activity. Anim. Feed Sci. Technol. 210, 172–183.

Pinto, W., et al., 2007. Growth, stress response and free amino acid levels in Senegalese sole (*Solea senegalensis* Kaup 1858) chronically exposed to exogenous ammonia. Aquac. Res. 38, 1198–1204.

Pirozzi, I., Benito, M., Booth, M., 2019. Protein, amino acid and energy utilisation of juvenile yellowtail kingfish (*Seriola lalandi*): quantifying abiotic influences. Aquaculture 513, 734439.

Pirozzi, I., Booth, M., Allan, G., 2010. Protein and energy utilization and the requirements for maintenance in juvenile mulloway (*Argyrosomus japonicus*). Fish Physiol. Biochem. 36, 109–121.

Plakas, S., Katayama, T., Tanaka, Y., Deshimaru, O., 1980. Changes in the levels of circulating plasma free amino acids of carp (*Cyprinus carpio*) after feeding a protein and an amino acid diet of similar composition. Aquaculture 21, 307–322.

Plakas, S., Lee, T., Wolke, R., 1988. Bioavailability of lysine in Maillard browned protein as determined by plasma lysine response in rainbow trout (*Salmo gairdneri*). J. Nutr. 118, 19–22.

Plakas, S., Lee, T., Wolke, R., Meade, T., 1985. Effect of Maillard browning reaction on protein utilization and plasma amino acid response by rainbow trout (*Salmo gairdneri*). J. Nutr. 115, 1589–1599.

Plisetskaya, E., Buchelli Narvaez, L., Hardy, R., Dickhoff, W., 1991. Effects of injected and dietary arginine on plasma insulin levels and growth of Pacific salmon and rainbow trout. Comp. Biochem. Physiol. 98, 165–170.

Poncet, N., Taylor, P., 2013. The role of amino acid transporters in nutrition. Curr. Opin. Clin. Nutr. Metab. Care 16, 57–65.

Poppi, D., Moore, S., Glencross, B., 2018. The effect of taurine supplementation to a plant-based diet for barramundi (*Lates calcarifer*) with varying methionine content. Aquac. Nutr. 24, 1340–1350.

Poston, H., Rumsey, G., 1983. Factors affecting dietary requirement and deficiency signs of L-tryptophan in rainbow trout. J. Nutr. 113, 2568–2577.

Prabu, E., Felix, N., Uma, A., Praveenraj, J., 2019. Effects of dietary L-lysine supplementation on growth, body composition and muscle-growth-related gene expression with an estimation of lysine requirement of GIFT tilapia. Aquac. Nutr. 26, 568–578.

Preziosa, E., et al., 2013. Effect of nutrient restriction and re-feeding on calpain family genes in skeletal muscle of channel catfish (*Ictalurus punctatus*). PLoS One 8, e59404.

Price, J., et al., 2010. Analysis of proteome dynamics in the mouse brain. Proc. Natl. Acad. Sci. U.S.A. 107, 14508–14513.

Prieto-Álamo, M., et al., 2012. Omics methodologies: new tools in aquaculture studies. In: Muchlisin, Z. (Ed.), Aquaculture. IntechOpen, Rijeka, pp. 361–390.

Rahimnejad, S., Lee, K., 2013. Dietary valine requirement of juvenile red sea bream *Pagrus major*. Aquaculture 416–417, 212–218.

Rahimnejad, S., Lee, K., 2014. Dietary arginine requirement of juvenile red sea bream *Pagrus major*. Aquaculture 434, 418–424.

Ravi, J., Devaraj, K., 1991. Quantitative essential amino acid requirements for growth of catla, *Catla catla* (Hamilton). Aquaculture 96, 281–291.

Ray, A., Ghosh, K., Ringo, E., 2012. Enzyme-producing bacteria isolated from fish gut: a review. Aquac. Nutr. 18, 465–492.

Reeds, P., Burrin, D., Stoll, B., Jahoor, F., 2000. Intestinal glutamate metabolism. J. Nutr. 130, 978–982.

Refstie, S., Storebakken, T., Baeverfjord, G., Roem, A., 2001. Long-term protein and lipid growth of Atlantic salmon (*Salmo salar*) fed diets with partial replacement of fish meal by soy protein products at medium or high lipid level. Aquaculture 193, 91–106.

Reiling, J., Sabatini, D., 2006. Stress and mTORture signaling. Oncogene 25, 6373–6383.

Reinecke, M., et al., 2005. Growth hormone and insulin-like growth factors in fish: where we are and where to go. Gen. Comp. Endocrinol. 142, 20–24.

Remø, S.C., et al., 2014. Dietary histidine requirement to reduce the risk and severity of cataracts is higher than the requirement for growth in Atlantic salmon smolts, independently of the dietary lipid source. Br. J. Nutr. 111, 1759–1772.

Ren, M.C., Ai, Q.H., Mai, K.S., 2012. Dietary arginine requirement of juvenile cobia (*Rachycentron canadum*). Aquac. Res. 45, 1–9.

Ren, M., et al., 2013. Dietary arginine requirement of juvenile blunt snout bream, *Megalobrama amblycephala*. Aquaculture 414, 229–234.

Ren, M., et al., 2015a. Dietary valine requirement of juvenile blunt snout bream (*Megalobrama amblycephala* Yih, 1955). J. Appl. Ichthyol. 31, 1086–1092.

Ren, M., et al., 2015b. Dietary phenylalanine requirement and tyrosine replacement value for phenylalanine of juvenile blunt snout bream, *Megalobrama amblycephala*. Aquaculture 442, 51–57.

Rezzi, S., Ramadan, Z., Fay, L., Kochhar, S., 2007. Nutritional metabonomics: applications and perspectives. J. Proteome Res. 6, 513–525.

Richard, L., et al., 2010. Maintenance and growth requirements for nitrogen, lysine and methionine and their utilisation efficiencies in juvenile black tiger shrimp, *Penaeus monodon*, using a factorial approach. Br. J. Nutr. 103, 984–995.

Richter, C., Tanaka, T., Yada, R., 1998. Mechanism of activation of the gastric aspartic proteinases: pepsinogen, progastricsin and prochymosin. Biochem. J. 335, 481–490.

Rider, J., et al., 2007. Spermine and spermidine mediate protection against oxidative damage caused by hydrogen peroxide. Amino Acids 33, 231–240.

Ripps, H., Shen, W., 2012. Review: taurine: A "very essential" amino acid. Mol. Vis. 18, 2673–2686.

Rise, M., Martyniuk, C., Chen, M., 2019. Comparative physiology and aquaculture: toward omics-enabled improvement of aquatic animal health and sustainable production. Comp. Biochem. Physiol. Genom. Proteonomics 31, 100603.

Robinson, E., Wilson, R., Poe, W., 1981. Arginine requirement and apparent absence of a lysine-arginine antagonist in fingerling channel catfish. J. Nutr. 111, 46–52.

Robinson, E., Poe, W., Wilson, R., 1984. Effects of feeding diets containing an imbalance of branched chain amino acids on fingerling channel catfish. Aquaculture 37, 51–62.

Robinson, E., Wilson, R., Poe, W., 1980a. Re-evaluation of the lysine requirement and lysine utilization by fingerling channel catfish. J. Nutr. 110, 2313–2316.

Robinson, E., Wilson, R., Poe, W., 1980b. Total aromatic amino acid requirement, phenylalanine requirement and tyrosine replacement value for fingerling channel catfish. J. Nutr. 110, 1805–1812.

Robertson, R., El-Haj, A., Clarke, A., Taylor, E., 2001. Effects of temperature on specific dynamic action and protein synthesis rates in the Baltic isopod crustacean, *Saduria entomon*. J. Exp. Mar. Biol. Ecol. 262, 113–129.

Robinson, J., Yanke, D., Mirza, J., Ballantyne, J., 2011. Plasma free amino acid kinetics in rainbow trout (*Oncorhynchus mykiss*) using a bolus injection of 15N-labeled amino acids. Amino Acids 40, 689–696.

Rodehutscord, M., Pack, M., 1999. Estimates of essential amino acid requirements from dose-response studies with rainbow trout and broiler chicken: effect of mathematical model. Arch. Anim. Nutr. 52, 223–244.

Rodehutscord, M., Becker, A., Pack, M., Pfeffer, E., 1997. Response of rainbow trout (*Oncorhynchus mykiss*) to supplements of individual essential amino acids in a semipurified diet, including an estimate of the maintenance requirement for essential amino acids. J. Nutr. 127, 1166–1175.

Rodehutscord, M., Jacobs, S., Pack, M., Pfeffer, E., 1995a. Response of rainbow trout (*Oncorhynchus mykiss*) growing from 50 to 150 g to supplements of DL-Methionine in a semipurified diet containing low or high levels of cystine. J. Nutr. 4, 964–969.

Rodehutscord, M., Jacobs, S., Pack, M., Pfeffer, E., 1995b. Response of rainbow trout (*Oncorhynchus mykiss*) growing from 50 to 170 g to supplements of either L-arginine or L-threonine in a semipurified diet. J. Nutr. 125, 970–975.

Rodrigues, R., et al., 2019. Valine in diets for juvenile Nile tilapia (*Oreochromis niloticus*): growth performance, chemical composition, blood parameters and skeletal muscle development. Spanish J. Agric. Res. 17, 602–610.

Rollin, X., et al., 2003. The optimum dietary indispensable amino acid pattern for growing Atlantic salmon (*Salmo salar* L.) fry. Br. J. Nutr. 90, 865–876.

Ronnestad, I., Thorsen, A., Finn, R., 1999. Fish larval nutrition: a review of recent advances in the roles of amino acids. Aquaculture 177, 201–216.

Rousseau, A., Bertolotti, A., 2016. An evolutionarily conserved pathway controls proteasome homeostasis. Nature 536, 184–189.

Rubio, V., Sanchez-Vazquez, F., Madrid, J., 2005. Effects of salinity on food intake and macronutrient selection in European sea bass. Physiol. Behav. 85, 333–339.

Ruchimat, T., Macumto, T., Hosokawa, H., Shimeno, S., 1997. Quantitative methionine requirement of yellowtail (*Seriola quinqueradiata*). Aquaculture 150, 113–122.

Ruchimat, T., Masumoto, T., Itoh, Y., Shimeno, S., 1998. Quantitative arginine requirement of juvenile yellowtail *Seriola quinqueradiata*. Fish. Sci. 64, 348–349.

Saavedra, M., et al., 2015. Amino acid profiles of meagre (*Argyrosomus regius*) larvae: towards the formulation of an amino acid balanced diet. Aquaculture 448, 315–320.

Salem, M., Kenney, P., Killefer, J., Nath, J., 2005a. Isolation and characterization of calpains from rainbow trout muscle and their role in texture development. J. Muscle Foods 15, 245–255.

Salem, M., et al., 2005b. Identification and molecular characterization of the rainbow trout calpains (Capn1 and Capn2): their expression in muscle wasting during starvation. Comp. Biochem. Physiol. B Biochem. Mol. Biol. 140, 0–71.

Salem, M., Kenney, P., Rexroad, C., Yao, J., 2006. Molecular characterization of muscle atrophy and proteolysis associated with spawning in rainbow trout. Comp. Biochem. Physiol. Genom. Proteonomics 1, 227–237.

Salem, M., Silverstein, J., Rexroad, C., Yao, J., 2007. Effect of starvation on global gene expression and proteolysis in rainbow trout (*Oncorhynchus mykiss*). BMC Genom. 8, 328.

Salze, G., Davis, D., 2017. Methionine requirement for juvenile white seabass, *Atractoscion nobilis*, using nonlinear models. J. World Aquac. Soc. 48, 729–740.

Salze, G., Quinton, M., Bureau, D., 2017. A meta-analysis of essential amino acid requirement of fish. In: Cruz-Suárez, L.E., Ricque-Marie, D., Tapia-Salazar, M., Nieto-López, M.G., Villareal-Cavazos, D.A., Gamboa-Delgado, J., López Acuña, L.M., yGalaviz-Espinoza, M. (Eds.), Investigación y Desarrollo en Nurición Acuícola Universidad Autónoma de Nuevo León, San Nicolás de los Garza, Nuevo León, México, pp. 303–337.

Salze, G., et al., 2018a. Quantitative dietary taurine requirement for California yellowtail, *Seriola lalandi*. J. World Aquac. Soc. 49, 113–126.

Salze, G., Rhodes, M., Davis, D., 2018b. Quantitative taurine requirement for juvenile Florida pompano (*Trachinotus carolinus*) fed soy-based practical diets. Aquac. Nutr. 25, 388–395.

Sampath, W., et al., 2020. Dietary taurine improves muscle growth and texture characteristics in juvenile turbot (*Scophthalmus maximus*). Aquac. Rep. 17, 100305.

Samsudin, F., et al., 2016. Accurate prediction of ligand affinities for a proton-dependent oligopeptide transporter. Cell Chem. Biol. 23, 299–309.

Sandri, M., 2010. Autophagy in skeletal muscle. FEBS Lett. 584, 1411–1416.

Santiago, C., Lovell, R., 1988. Amino acid requirements for growth of Nile tilapia. J. Nutr. 118, 1540–1546.

Satriyo, T., Galaviz, M., Salze, G., López, L., 2017. Assessment of dietary taurine essentiality on the physiological state of juvenile *Totoaba macdonaldi*. Aquac. Res. 48, 5677–5689.

Schulz, C., Bohm, M., Wirth, M., Rennert, B., 2007. Effect of dietary protein on growth, feed conversion, body composition and survival of pike perch fingerlings (*Sander lucioperca*). Aquacult. Nutr. 13, 373–380.

Seiliez, I., et al., 2008a. An in vivo and in vitro assessment of TOR signaling cascade in rainbow trout (*Oncorhynchus mykiss*). Am. J. Physiol. Regul. Integr. Comp. Physiol. 295, 329–335.

Seiliez, I., et al., 2008b. Feeding status regulates the polyubiquitination step of the ubiquitin-proteasome-dependent proteolysis in rainbow trout (*Oncorhynchus mykiss*) muscle. J. Nutr. 138, 487–491.

Seiliez, I., et al., 2011. Dietary carbohydrate-to-protein ratio affects TOR signaling and metabolism-related gene expression in the liver and muscle of rainbow trout after a single meal. Am. J. Physiol. Regul. Integr. Comp. Physiol. 300, 733–743.

Seiliez, I., et al., 2012. Amino acids downregulate the expression of several autophagy-related genes in rainbow trout myoblasts. Autophagy 8, 364–375.

Seiliez, I., Dias, K., Cleveland, B., 2014. Contribution of the autophagy-lysosomal and ubiquitin-proteasomal proteolytic systems to total proteolysis in rainbow trout (*Oncorhynchus mykiss*) myotubes. Am. J. Physiol. Regul. Integr. Comp. Physiol. 307, 1330–1337.

Sevgili, H., Hossu, B., Emre, Y., Kanyılmaz, M., 2012. Compensatory growth after various levels of dietary protein restriction in rainbow trout, *Oncorhynchus mykiss*. Aquaculture 344-349, 126–134.

Shahkar, E., et al., 2014. Evaluation of optimum dietary protein level for juvenile whiteleg shrimp (*Litopenaeus vannamei*). J. Crust. Biol. 34, 552–558.

Sharf, Y., Khan, M., 2020. Effect of dietary isoleucine level on growth, protein retention efficiency, haematological parameter, lysozyme activity and serum antioxidant status of fingerling *Channa punctatus* (Bloch). Aquac. Nutr. 26, 908–920.

Shearer, K., 1995. The use of factorial modeling to determine the dietary requirements for essential elements in fishes. Aquaculture 133, 57–72.

Shearer, K., 2000. Experimental design, statistical analysis and modelling of dietary nutrient requirement studies for fish: a critical review. Aquac. Nutr. 6, 91–102.

Shen, Y., et al., 2017. Effects of dietary arginine/lysine on growth performance, body composition, serum biochemical Indices and amino acid deposition rate of all-male yellow catfish. Chinese J. Anim. Nutr. 29, 2575–2586.

Shen, G., et al., 2019. Metabolic effect of dietary taurine supplementation on grouper (*Epinephelus coioides*): a ^{1}H-NMR-Based metabolomics study. Molecules 24, 2253.

Shiau, S., Hsieh, H., 1997. Vitamin B-6 requirements of tilapia, *Oreochromis niloticus* × *O. aureus*, varies with two dietary protein concentrations. Fish. Sci. 63, 1002–1007.

Shiau, S., Chou, B., 1994. Grass shrimp, *Penaeus monodon*, growth as influenced by dietary taurine supplementation. Comp. Biochem. Physiol. Physiol. 108, 137–142.

Shigemitsu, K., et al., 1999. Structural requirement of leucine for activation of p70 S6 kinase. FEBS Lett. 447, 303–306.

Shimomura, Y., et al., 2006. Nutraceutical effects of branched-chain amino acids on skeletal muscle. J. Nutr. 136, 529–532.

Siddiqui, A., Howlader, M., Adam, A., 1988. Effects of dietary protein levels on growth, feed conversion and protein utilization in fry and young Nile tilapia, *Oreochromis niloticus*. Aquaculture 70, 63–73.

Siekevitz, P., 1972. The turnover of proteins and the usage of information. J. Theor. Biol. 37, 321–334.

Silva, P., et al., 2009. Hyperplastic and hypertrophic growth of lateral muscle in blackspot seabream *Pagellus bogaraveo* from hatching to juvenile. J. Fish. Biol. 74, 37–53.

Silva, P., et al., 2010. Individual differences in cortisol levels and behaviour of Senegalese sole (*Solea senegalensis*) juveniles: evidence for coping styles. Appl. Anim. Behav. Sci. 124, 75–81.

Simmons, L., Moccia, R., Bureau, D., 1999. Dietary methionine requirement of juvenile Arctic charr *Salvelinus alpinus* (L.). Aquac. Nutr. 5, 93–100.

Singh, S., Khan, M., 2007. Dietary arginine requirement of fingerling hybrid Clarias (*Clarias gariepinus* × *Clarias macrocephalus*). Aquac. Res. 38, 17–25.

Singha, K., et al., 2020. Feeding graded levels of protein to genetically improved farmed tilapia (GIFT) juveniles reared in inland saline water: Effects on growth and gene expression of IGF-I, IGF-IR and IGF-BPI. Aquaculture 525, 735306.

Sire, M., Vernier, J., 1992. Intestinal absorption of protein in teleost fish. Comp. Biochem. Physiol. A Physiol. 103, 771–781.

Skiba-Cassy, S., et al., 2013. Apparent low ability of liver and muscle to adapt to variation of dietary carbohydrate:-protein ratio in rainbow trout (*Oncorhynchus mykiss*). Br. J. Nutr. 109, 1359–1372.

Small, B., Soares, J., 1998. Estimating the quantitative essential amino acid requirements of striped bass *Morone saxatilis*, using fillet A/E ratios. Aquac. Nutr. 4, 225–232.

Small, B., Soares, J., 1999. Quantitative dietary threonine requirement of juvenile striped bass *Morone saxatilis*. J. World Aquac. Soc. 30, 319–323.

Small, B., Soares JR., J., 2000. Quantitative dietary lysine requirement of juvenile striped bass *Morone saxatilis*. Aquac. Nutr. 6, 207–212.

Smith, R., Houlihan, D., Nilsson, G., Brechin, J., 1996. Tissue-specific changes in protein synthesis rates in vivo during anoxia in crucian carp. Am. J. Physiol. Regul. Integr. Comp. Physiol. 271, 897–904.

Song, F., et al., 2016. Comparative study on the cellular and systemic nutrient sensing and intermediary metabolism after partial replacement of fishmeal by meat and bone meal in the diet of turbot (*Scophthalmus maximus* L.). PLoS One 11, e0165708.

Sterzelecki, F., 2018. Effects of increasing protein level on the performance, enzyme activity and body composition of the Brazilian sardine, *Sardinella brasiliensis* (Steindachner, 1879). Aquacult. Nutr. 24, 366–374.

Storch, D., Lannig, G., Pörtner, H., 2005. Temperature-dependent protein synthesis capacities in Antarctic and temperate (North Sea) fish (Zoarcidae). J. Exp. Biol. 208, 2409–2420.

Suarez, M., et al., 2005. Changes in muscle collagen content during post mortem storage of farmed sea bream (*Sparus aurata*): influence on textural properties. Aquac. Int. 13, 315–325.

Sugden, P., Fuller, S., 1991. Regulation of protein turnover in skeletal and cardiac muscle. Biochem. J. 273, 21–37.

Sugita, Y., Takao, K., Toyama, Y., Shirahata, A., 2007. Enhancement of intestinal absorption of macromolecules by spermine in rats. Amino Acids 33, 253–260.

Sun, L., Chen, H., 2014. Effects of water temperature and fish size on growth and bioenergetics of cobia (*Rachycentron canadum*). Aquaculture 426, 172–180.

Sweilum, M., Abdella, M., El-Din, S., 2005. Effect of dietary protein-energy levels and fish initial sizes on growth rate, development and production of Nile tilapia, *Oreochromis niloticus* L. Aquacult. Res. 36, 1414–1421.

Tacon, A., Cowey, C., 1985. Protein and amino acid requirements. In: Tytler, P., Calow, P. (Eds.), Fish Energetics: New Perspectives. London & Sidney: Croom Helm, pp. 155–183.

Tacon, A., Metian, M., Turchini, G., De Silva, S., 2010. Responsible aquaculture and trophic level implications to global fish supply. Rev. Fish. Sci. 18, 94–105.

Takagi, S., et al., 2006. Efficacy of taurine supplementation for preventing green liver syndrome and improving growth performance in yearling red sea bream pagrus major fed low-fishmeal diet. Fish. Sci. 72, 1191–1199.

Takagi, S., et al., 2008. Taurine is an essential nutrient for yellowtail *Seriola quinqueradiata* fed non-fish meal diets based on soy protein concentrate. Aquaculture 280, 198–205.

Takeuchi, T, Watanabe, T., Ogino, C., 1979. Optimum ratio of dietary energy to protein for carp. Bull. Jpn. Soc. Sci. Fish. 44, 729–732.

Takeuchi, T., et al., 1992. Suitable protein and lipid-levels in diet for fingerlings of yellowtail. Bull. Jpn. Soc. Sci. Fish. 58, 1333–1339.

Takeuchi, T., Park, G., Seikai, T., Yokoyama, M., 2001. Taurine content in Japanese flounder *Paralichthys olivaceus* T. and S. and red sea bream Pagrus major T. and S. during the period of seed production. Aquac. Res. 32, 244–248.

Talukdar, A., et al., 2020. Effects of dietary protein on growth performance, nutrient utilization, digestive enzymes and physiological status of grey mullet, *Mugil cephalus* L. fingerlings reared in inland saline water. Aquacult. Nutr. 26, 921–935.

Tan, X., et al., 2016. Effects of dietary leucine on growth performance, feed utilization, non-specific immune responses and gut morphology of juvenile golden pompano *Trachinotus ovatus*. Aquaculture 465, 100–107.

Tang, L., et al., 2013. Effect of tryptophan on growth, intestinal enzyme activities and TOR gene expression in juvenile Jian carp (*Cyprinus carpio* var. Jian): studies in vivo and in vitro. Aquaculture 412, 23–33.

Tang, L., et al., 2009. Effect of methionine on intestinal enzymes activities, microflora and humoral immune of juvenile Jian carp (*Cyprinus carpio* var. Jian). Aquac. Nutr. 15, 477–483.

Tantikitti, C., Chimsung, N., 2001. Dietary lysine requirement of freshwater catfish (*Mystus nemurus* Cuv. & Val.). Aquac. Nutr. 15, 477–483.

Templeman, J., Mansilla, W., Fortener, L., Shoveller, A., 2019. Tryptophan requirements in small, medium, and large breed adult dogs using the indicator amino acid oxidation technique. J. Anim. Sci. 97, 3274–3285.

Tengjaroenkul, B., Smith, B., Caceci, T., Smith, S., 2000. Distribution of intestinal enzyme activities along the intestinal tract of cultured Nile tilapia, *Oreochromis niloticus* L. Aquaculture 182, 317–327.

Teshima, S., et al., 2002. Assessment of requirement values for essential amino acids in the prawn, *Marsupenaeus japonicus* (Bate). Aquac. Res. 33, 395–402.

Teshima, S., et al., 2006. Protein requirements of the fresh water prawn *Macrobrachium rosenbergii* evaluated by the factorial method. J. World Aquacult. Soc. 37, 145–153.

Tesser, M., Silva, E., Sampaio, L., 2014. Whole-body and muscle amino acid composition of plata pompano (*Trachinotus marginatus*) and prediction of dietary essential amino acid requirements. Rev. Colombiana Ciencias Pecuarias 27, 299–305.

Tibaldi, E., Kaushik, S., 2005. Amino acid requirement of Mediterranean fish species. Cah. Options Mediterr. 63, 59–65.

Tibaldi, E., Lanari, 1991. Optimal dietary lysine levels for growth and protein utilisation of fingerling sea bass (*Dicentrarchus labrax* L.) fed semipurified diets. Aquaculture 95, 297–304.

Tibaldi, E., Tulli, F., 1999. Dietary threonine requirement of juvenile European sea bass (*Dicentrarchus labrax*). Aquaculture 175, 155–166.

Tibaldi, E., et al., 1993. Preliminary evaluation of the arginine requirement of fingerling sea bass (*Dicentrarchus labrax*). Riv. Tal. Acquacol 28, 105–115.

Tibaldi, E., Tulli, F., Lanari, D., 1994. Arginine requirement and effect of different dietary arginine and lysine levels for fingerling sea bass (*Dicentrarchus labrax*). Aquaculture 127, 207–218.

Toyohara, H., Makinodan, Y., 1989. Comparison of calpain I and calpain II from carp muscle. Comp. Biochem. Physiol. B Comp. Biochem. 92, 577–581.

Treberg, J., Hall, J., Driedzic, W., 2005. Enhanced protein synthetic capacity in Atlantic cod (*Gadus morhua*) is associated with temperature-induced compensatory growth. Am. J. Physiol. Regul. Integr. Comp. Physiol. 288, 205–211.

Tu, Y., et al., 2015a. Growth performance, digestive enzyme, transaminase and GH-IGF-I axis gene responsiveness to different dietary protein levels in broodstock allogenogynetic gibel carp (*Carassius auratus gibelio*) CAS III. Aquaculture 446, 290–297.

Tu, Y., et al., 2015b. Dietary arginine requirement for gibel carp (*Carassius auratus gibelio* var. CAS III) reduces with fish size from 50 g to 150 g associated with modulation of genes involved in TOR signaling pathway. Aquaculture 449, 37–47.

Tulli, F., Messina, M., Calligaris, M., Tibaldi, E., 2010. Response of European sea bass (*Dicentrarchus labrax*) to graded levels of methionine (total sulfur amino acids) in soya protein-based semi-purified diets. Br. J. Nutr. 104, 664–673.

Turchini, G., Trushenski, J., Glencross, B., 2019. Thoughts for the future of Aquaculture Nutrition: realigning perspectives to reflect contemporary issues related to judicious use of marine resources in aquafeeds. N. Am. J. Aquac. 81, 13–39.

Twibell, R., Brown, P., 1997. Dietary arginine requirement of juvenile yellow perch. J. Nutr. 127, 1838–1841.

Twibell, R., Wilson, K., Brown, P., 2000. Dietary sulfur amino acid requirement of juvenile yellow perch fed the maximum cystine replacement value for methionine. J. Nutr. 130, 612–616.

Unnikrishnan, U., Paulraj, R., 2010. Dietary protein requirement of giant mud crab *Scylla serrata* juveniles fed iso-energetic formulated diets having graded protein levels. Aquac. Res. 41, 278–294.

Valente, L., Bower, N., Johnston, I., 2012. Postprandial expression of growth-related genes in Atlantic salmon (*Salmo salar* L.) juveniles fasted for 1 week and fed a single meal to satiation. Br. J. Nutr. 108, 2148–2157.

Valenzuela, C., et al., 2018. Chronic stress inhibits growth and induces proteolytic mechanisms through two different nonoverlapping pathways in the skeletal muscle of a teleost fish. Am. J. Physiol. Regul. Integr. Comp. Physiol. 314, 102–113.

Van der Meer, M., Machiels, M., Verdegem, M., 1995. The effect of dietary protein level on growth, protein utilization and body composition of *Colossoma macropomum* (Cuvier). Aquac. Res. 26, 901–909.

Vargas-Chacoff, L., et al., 2016. Metabolic responses to salinity changes in the subantarctic notothenioid teleost *Eleginops maclovinus*. Polar Biol. 39, 1297–1308.

Verri, T., et al., 2010. Transport of di-and tripeptides in teleost fish intestine. Aquac. Res. 41, 641–653.

Verri, T., et al., 2017. Di-and tripeptide transport in vertebrates: the contribution of teleost fish models. J. Comp. Physiol. B 187, 395–462.

Verri, T., Terova, G., Dabrowski, K., Saroglia, M., 2011. Peptide transport and animal growth: the fish paradigm. Biol. Lett. 7, 597–600.

Voss, J.C., Waisman, H.A., 1966. The phenylalanine hydroxylase content of livers of various vertebrates. Camp. Biochem. Physiol. 17, 49–58.

Waagbø, R, Tröße, C., Koppe, W., Fontanillas, R., Breck, O., 2010. Dietary histidine supplementation prevents cataract development in adult Atlantic salmon, Salmo salar L., in seawater. Br. J. Nutr. 104, 1460–1470.

Walton, M., Cowey, C., 1982. Aspects of intermediary metabolism in salmonid fish. Comp. Biochem. Physiol. B Comp. Biochem. 73, 59–79.

Walton, M., Wilson, R., 1986. Postprandial changes in plasma and liver free amino acids of rainbow trout fed complete diets containing casein. Aquaculture 51, 105–115.

Walton, M., Cowey, C., Adron, J., 1984a. The effect of dietary lysine levels on growth and metabolism of rainbow trout *(Salmo gairdneri*). Br. J. Nutr. 52, 115–122.

Walton, M., et al., 1984b. The effects of dietary tryptophan levels on growth and metabolism of rainbow trout (*Salmo gairdneri*). Br. J. Nutr. 51, 279–287.

Walton, M., Cowey, C., Coloso, R., Adron, J., 1986. Dietary requirements of rainbow trout for tryptophan, lysine and arginine determined by growth and biochemical measurements. Fish Physiol. Biochem. 2, 161–169.

Wang, C., Klionsky, D., 2003. The molecular mechanism of autophagy. Mol. Med. 9, 65–76.

Wang, K., Takeuchi, T., Watanabe, T., 1985. Optimum protein and digestible energy-levels in diets for tilapia-nilotica. Bull. Jpn. Soc. Sci. Fish. 51, 141–146.

Wang, T., Fuller, M., 1989. The optimum dietary amino acid pattern for growing pigs. 1. experiments by amino acid deletion. Br. J. Nutr. 62, 77–89.

Wang, G., et al., 2008. Effect of dietary protein on growth and hormone of *Culture alburnus* juveniles. Acta Hydrobiol. Sin. 32, 544–550.

Wang, X., Campbell, L., Miller, C., Proud, C., 1998. Amino acid availability regulates p70 S6 kinase and multiple translation factors. Biochem. J. 334, 261–267.

Wang, S., et al., 2005. Quantitative dietary lysine requirement of juvenile grass carp *Ctenopharyngodon idella*. Aquaculture 249, 419–429.

Wang, B., et al., 2015a. Copper-induced tight junction mRNA expression changes, apoptosis and antioxidant responses via NB-κB, TOR and Nrf2 signaling molecules in the gills of fish: preventive role of arginine. Aquat. Toxicol. 158, 125–137.

Wang, Q., et al., 2015b. Effect of Taurine on growth, muscle quality and antioxidant capacity of juvenile grass carp (*Ctenopharyngodon idella*) fed with high-fat diets. J. Northwest A & F Univ. 43, 49–56.

Wang, X., et al., 2016a. Dietary methionine requirement of the pre-adult gibel carp (*Carassius auratus gibelio*) at a constant dietary cystine level. Aquac. Nutr. 22, 509–516.

Wang, Z., et al., 2016b. Dietary methionine level influences growth and lipid metabolism via GCN2 pathway in cobia (*Rachycentron canadum*). Aquaculture 454, 148–156.

Wang, J., et al., 2017. Peptide transporter 1 (PepT1) in fish: a review. Aquac. Fish. 2, 193–206.

Wang, S., Fitzgibbon, Q., Carter, C., Smith, G., 2019a. Effect of protein synthesis inhibitor cycloheximide on starvation, fasting and feeding oxygen consumption in juvenile spiny lobster *Sagmariasus verreauxi*. J. Comp. Physiol. B 189, 351–365.

Wang, L., et al., 2019b. Effect of N-carbamoylglutamate supplementation on the growth performance, antioxidant status and immune response of mirror carp (*Cyprinus carpio*) fed an arginine-deficient diet. Fish Shellfish Immunol. 84, 280–289.

Wang, Y., et al., 2006. Effects of dietary protein and energy levels on growth, feed utilization and body composition of cuneate drum (*Nibea miichthioides*). Aquaculture 252, 421–428.

Watanabe, K., 2000. Energy and protein requirements of yellowtail for maximum growth and maintenance of body weight. Fish. Sci. 66, 1053–1061.

Waterlow, J., Garlick, P., Mill, W., 1978. Protein Turnover in Mammalian Tissues and in the Whole Body. Elsevier/North-Holland Biomedical Press, Amsterdam.

Weatherley, A., Gill, H., Lobo, A., 1988. Recruitment and maximal diameter of axial muscle fibres in teleosts and their relationship to somatic growth and ultimate size. J. Fish. Biol. 33, 851–859.

Wee, K., Tacon, A., 1982. A preliminary study on the dietary protein requirement of juvenile snakehead. Bull. Jpn. Soc. Sci. Fish. 48, 1463–1468.

Wee, K., 1986. A preliminary study on the dietary protein requirements of juvenile snakehead. In: Proceedings of International Conference Dev Managment Trop Living Aqua Resources, Serdang, Malaysia. 2–5 Aug 1983, pp. 131–136.

Wei, Z., et al., 2016. Dietary hydroxyproline improves the growth and muscle quality of large yellow croaker *Larimichthys crocea*. Aquaculture 464, 497–504.

Wen, H., et al., 2014. Dietary tryptophan modulates intestinal immune response, barrier function, antioxidant status and gene expression of TOR and Nrf2 in young grass carp (*Ctenopharyngodon idella*). Fish Shellfish Immunol. 40, 275–287.

Whitson, Walster, H., 1912. Soil and soil fertility. St. Paul, MN. 73.

Wilson, R., Cowey, C., 1985. Amino acid composition of whole body tissue of rainbow trout and Atlantic salmon. Aquaculture 48, 373–376.

Wilson, R., Halver, J., 1986. Protein and amino acid requirements of fishes. Annu. Rev. Nutr. 6, 225–244.

Wilson, R., Allen, O., Robinson, E., Poe, W., 1978. Tryptophan and threonine requirements of fingerling channel catfish. J. Nutr. 108, 1595–1599.

Wilson, R., Harding, D., Garling, D., 1977. Effect of dietary pH on amino acid utilization and the lysine requirement of fingerling channel catfish. J. Nutr. 107, 166–170.

Wilson, R., Poe, W., Robinson, E., 1980. Leucine, isoleucine, valine and histidine requirements of fingerling channel catfish. J. Nutr. 110, 627–633.

Wilson, R., Robinson, E., Poe, W., 1981. Apparent and true availability of amino acids from common feed ingredients for channel catfish. J. Nutr. 111, 923–929.

Wilson, Z., Carter, C., Purser, G., 2006. Nitrogen budgets for juvenile big-bellied seahorse *Hippocampus abdominalis* fed Artemia, mysids or pelleted feeds. Aquaculture 255, 233–241.

Wilson-Arop, O., et al., 2018. Dietary histidine requirement of juvenile blunt snout bream (*Megalobrama amblycephala*). Aquac. Nutr. 24, 1122–1132.

Winberg, S., Overli, O., Lepage, O., 2001. Suppression of aggression in rainbow trout (*Oncorhynchus mykiss*) by dietary L-tryptophan. J. Exp. Biol. 204, 3867–3876.

Wu, C., et al., 2017a. The effects of dietary leucine on the growth performances, body composition, metabolic abilities and innate immune responses in black carp *Mylopharyngodon piceus*. Fish Shellfish Immunol. 67, 419–428.

Wu, G., Meininger, C., 2009. Nitric oxide and vascular insulin resistance. Biofactors 35, 21–27.

Wu, G., 2013. Amino Acids Biochemistry and Nutrition. CRC. Press, Boca Raton (FL), pp. 402–404.

Wu, G., 2014. Dietary requirements of synthesizable amino acids by animals: a paradigm shift in protein nutrition. J. Anim. Sci. Biotechnol. 5, 34.

Wu, G., 2018. Principles of Animal Nutrition. CRC Press. Taylor & Francis Group, US, p. pp149.

Wu, G., et al., 2011. Proline and hydroxyproline metabolism: implications for animal and human nutrition. Amino Acids 40, 1053–1063.

Wu, M., et al., 2018a. Dietary arginine affects growth, gut morphology, oxidation resistance and immunity of hybrid grouper (*Epinephelus fuscoguttatus*♀ × *Epinephelus lanceolatus*♂) juveniles. Br. J. Nutr. 120, 269–282.

Wu, P., et al., 2017b. The relationship between dietary methionine and growth, digestion, absorption, and antioxidant status in intestinal and hepatopancreatic tissues of sub-adult grass carp (*Ctenopharyngodon idella*). J. Anim. Sci. Biotechnol. 63, 1–14.

Wu, X., et al., 2010. Dietary supplementation with l-arginine or N-carbamylglutamate enhances intestinal growth and heat shock protein-70 expression in weanling pigs fed a corn- and soybean meal-based diet. Amino Acids 39, 831–839.

Wu, Y., et al., 2018b. Effect of dietary tryptophan levels on growth and hemato-immunological parameters in juvenile Nile tilapia *Oreochromis niloticus*. Isr. J. Aquac. Bamidgeh 1519, 1–10.

Wu, Z., et al., 2017c. Metabolism, nutrition, and redox signaling of hydroxyproline. Antioxidants Redox Signal. 30, 1–30.

Xiao, W., et al., 2017. Dietary valine requirement of juvenile Nile tilapia, *Oreochromis niloticus*. Aquac. Nutr. 24, 315–323.

Xia, S., et al., 2010. Influence of dietary protein levels on growth, digestibility, digestive enzyme activity and stress tolerance in white-leg shrimp, *Litopenaeus vannamei* (Boone, 1931), reared in high-density tank trials. Aquacult. Res. 12, 1845–1854.

Xia, S., et al., 2019. Influence of dietary protein level on growth performance, digestibility and activity of immunity-related enzymes of leopard coral grouper, *Plectropomus leopardus* (Lacepede, 1802). Aquac. Nutr. 26, 242–247.

Xiao, W., et al., 2012. Lipid peroxidation, protein oxidant and antioxidant status of muscle and serum for juvenile Jian carp (*Cyprinus carpio var.* Jian) fed grade levels of methionine hydroxy analogue. Aquac. Nutr. 18, 90–97.

Xie, F., et al., 2012. Dietary lysine requirement of juvenile Pacific white shrimp, *Litopenaeus vannamei*. Aquaculture 358–359, 116–121.

Xie, S.W., et al., 2014a. Effect of glycine supplementation on growth performance, body composition and salinity stress of juvenile Pacific white shrimp, *Litopenaeus vannamei* fed low fishmeal diet. Aquaculture 418–419, 159–164.

Xie, Z., et al., 2014b. Effect of dietary taurine levels on growth performance and taurine content of *Nibea albiflora* larvae. Aquac. Int. 22, 1851–1862.

Xu, J, et al., 2016a. Effects of dietary protein levels on the disease resistance, immune function and physical barrier function in the gill of grass carp (*Ctenopharyngodon idella*) after challenged with *Flavobacterium columnare*. Fish Shellfish Immunol. 57, 1–6.

Xu, D., et al., 2016b. Expression pattern of peptide and amino acid genes in digestive tract of transporter juvenile turbot (*Scophthalmus maximus* L.). J. Ocean Univ. China 15, 334–340.

Xu, D., et al., 2017. Effect of fish meal replacement by plant protein blend on amino acid concentration, transportation and metabolism in juvenile turbot (*Scophthalmus maximus* L.). Aquac. Nutr. 23, 1169–1178.

Xue, M., Cui, Y., 2001. Effect of several feeding stimulants on diet preference by juvenile gibel carp (*Carassius auratus, gibelio*), fed diets with or without partial replacement of fish meal by meat and bone meal. Aquaculture 198, 281–292.

Xue, M., et al., 2012. Performance, body compositions, input and output of nitrogen and phosphorus in Siberian sturgeon, *Acipenser baerii* Brandt, as affected by dietary animal protein blend replacing fishmeal and protein levels. Aquac. Nutr. 18, 493–501.

Yamada, S., Tanaka, Y., Katayama, T., 1981. Feeding experiments with carp fry. Bull. Jpn. Soc. Sci. Fish. 47, 1247.

Yamamoto, T., et al., 1998. Apparent and true availabilities of amino acids from several protein sources for fingerling rainbow trout, common carp, and red sea bream. Fish. Sci. 64, 448–458.

Yamamoto, T., Shima, T., Furuita, H., 2004. Antagonistic effects of branched-chain amino acids induced by excess protein-bound leucine in diets for rainbow trout (*Oncorhynchus mykiss*). Aquaculture 232, 539–550.

Yan, J., Fu, J., Yong, J., Li, X., 2019. Dietary phenylalanine requirement of the juvenile pacific white shrimp *Litopenaeus vannamei* (Boone) reared in low-salinity water. J. Shellfish Res. 38, 35–41.

Yan, Q., et al., 2007. Dietary methionine requirement for juvenile rockfish, *Sebastes schlegeli*. Aquac. Nutr. 13, 163–169.

Yang, S., Liu, F., Liou, C., 2011. Assessment of dietary lysine requirement for silver perch (*Bidyanus bidyanus*) juveniles. Aquaculture 312, 102–108.

Yang, L., et al., 2018. Effects of dietary protein levels on the growth performance, activity of non-specific immunity and protein synthesis capacity of *Rhynchocypris lagowskii* Dybowski. Acta Hydrobiol. Sin. 42, 212–218.

Ye, W., et al., 2015. Comparative studies on dietary protein requirements of juvenile and on-growing gibel carp (*Carassius auratus gibelio*) based on fishmeal-free diets. Aquac. Nutr. 21, 286–299.

Ye, W., et al., 2017. Comparative study on dietary protein requirements for juvenile and pre-adult gibel carp (*Carassius auratus gibelio* var. CAS III). Aquac. Nutr. 23, 755–765.

Yuan, Y., et al., 2010. Effects of feeding levels on growth performance, feed utilization, body composition and apparent digestibility coefficients of nutrients for juvenile Chinese sucker, *Myxocyprinus asiaticus*. Aquac. Res. 41, 1030–1042.

Yuangsoi, B., Wongmaneeprateep, S., Sangsue, D., 2016. The optimal dietary DL-methionine on growth performance, body composition and amino acids profile of pangasius catfish (*Pangasius bocourti*). AACL Bioflux 9, 369–378.

Yue, Y., et al., 2014. Dietary threonine requirement of juvenile Nile tilapia, *Oreochromis niloticus*. Aquac. Int. 22, 1457–1467.

Yue, Y., et al., 2012. The effect of dietary taurine supplementation on growth performance, feed utilization and taurine contents in tissues of juvenile white shrimp (*Litopenaeus vannamei*, Boone, 1931) fed with low-fishmeal diets. Aquac. Res. 44, 1317–1325.

Yue, Y., et al., 2015. Effects of dietary arginine on growth performance, feed utilization, haematological parameters and non-specific immune responses of juvenile Nile tilapia (*Oreochromis niloticus* L.). Aquac. Res. 46, 1801–1809.

Yun, H., et al., 2015. Evaluation of optimum dietary threonine requirement by plasma free threonine and ammonia concentrations in surgically modified rainbow trout, *Oncorhynchus mykiss*. Asian-Australas. J. Anim. Sci. 28, 551–558.

Yun, H., et al., 2016. Determination of the dietary lysine requirement by measuring plasma free lysine concentrations in rainbow trout *Oncorhynchus mykiss* after dorsal aorta cannulation. Fish Aquat. Sci. 19, 1–7.

Zaminhan, M., et al., 2017. Dietary tryptophan requirements of juvenile Nile tilapia fed corn-soybean meal-based diets. Anim. Feed Sci. Technol. 227, 62–67.

Zaminhan, M., et al., 2018. Total and available tryptophan requirement of Nile tilapia, *Oreochromis niloticus*, fingerlings. Aquac. Nutr. 24, 1553–1562.

Zarate, D., Lovell, R., Payne, M., 1999. Effects of feeding frequency and rate of stomach evacuation on utilization of dietary free and protein-bound lysine for growth by channel catfish *Ictalurus punctatus*. Aquac. Nutr. 5, 17–22.

Zehra, S., Khan, M., 2012. Dietary protein requirement for fingerling *Channa punctatus* (Bloch), based on growth, feed conversion, protein retention and biochemical composition. Aquac. Int. 20, 383–395.

Zehra, S., Khan, M., 2013a. Dietary arginine requirement of fingerling Indian major Carp, *Catla catla* (Hamilton). J. World Aquac. Soc. 44, 363–373.

Zehra, S., Khan, M., 2013b. Dietary isoleucine requirement of fingerling catla, *Catla catla* (Hamilton), based on growth, protein productive value, isoleucine retention efficiency and carcass composition. Aquac. Int. 21, 1243–1259.

Zehra, S., Khan, M., 2013c. Dietary lysine requirement of fingerling *Catla catla* (Hamilton) based on growth, protein deposition, lysine retention efficiency, RNA/DNA ratio and carcass composition. Fish Physiol. Biochem. 39, 503–512.

Zehra, S., Khan, M., 2014. Dietary phenylalanine requirement and tyrosine replacement value for phenylalanine for fingerling *Catla catla* (Hamilton). Aquaculture 433, 256–265.

Zehra, S., Khan, M., 2015a. Dietary leucine requirement of fingerling *Catla catla* (Hamilton) based on growth, feed conversion ratio, RNA/DNA ratio, leucine gain, blood indices and carcass composition. Aquac. Int. 23, 577–595.

Zehra, S., Khan, M., 2015b. Dietary tryptophan requirement of fingerling Catla *catla* (Hamilton) based on growth, protein gain, RNA/DNA ratio, haematological parameters and carcass composition. Aquac. Nutr. 21, 690–701.

Zehra, S., Khan, M., 2016a. Dietary histidine requirement of fingerling *Catla catla* (Hamilton) based on growth, protein gain, histidine gain, RNA/DNA ratio, haematological indices and carcass composition. Aquac. Res. 47, 1028–1039.

Zehra, S., Khan, M., 2016b. Dietary threonine requirement of fingerling Indian major carp, *Catla catla* (Hamilton) estimated by growth, protein retention efficiency, threonine deposition, haematological parameters and carcass composition. Aquac. Res. 47, 253–265.

Zeitoun, I., et al., 1974. Influence of salinity on protein requirements of coho salmon *Oncorhynchus kisutch* smolts. J. Fish. Res. Bd. Can. 31, 1145–1148.

Zeitoun, I., Halver, J., Ullrey, D., Tack, P., 1973. Influence of salinity on protein requirements of rainbow trout (*Salmo gairdneri*) fingerlings. J. Fish. Res. Board Can. 30, 1867–1873.

Zhang, C., et al., 2008. Dietary lysine requirement of large yellow croaker, *Pseudosciaena crocea* R. Aquaculture 283, 123–127.

Zhang, K., et al., 2013. Effects of dietary hydroxyproline on growth performance, body composition, hydroxyproline and collagen concentrations in tissues in relation to prolyl 4-hydroxylase α (I) gene expression of juvenile turbot, *Scophthalmus maximus* L. fed high plant protein diets. Aquaculture 404–405, 77–84.

Zhang, K., et al., 2015. Proline with or without hydroxyproline influences collagen concentration and regulates prolyl 4-hydroxylase α (I) gene expression in juvenile turbo (*Scophthalmus maximus* L.). J. Ocean U. China 14, 541–548.

Zhang, Y., et al., 2017. Effects of dietary protein and lipid levels on growth, body and plasma biochemical composition and selective gene expression in liver of hybrid snakehead (*Channa maculata* ♀ × *Channa argus* ♂) fingerlings. Aquaculture 468, 1–9.

Zhang, J., et al., 2018a. Effect of dietary taurine supplementation on growth performance, digestive enzyme activities and antioxidant status of juvenile black carp (*Mylopharyngodon piceus*) fed with low fish meal diet. Aquac. Res. 49, 3187–3195.

Zhang, Y., et al., 2018b. Synergistic effects of dietary carbohydrate and taurine on growth performance, digestive enzyme activities and glucose metabolism in juvenile turbot *Scophthalmus maximus* L. Aquaculture 499, 32–41.

Zhang, J., et al., 2010. Dietary protein requirement of juvenile black sea bream, *Sparus microcephalus*. J. World Aquac. Soc. 41, 151–164.

Zhang, Y., et al., 2019a. Soybean glycinin decreased growth performance, impaired intestinal health, and amino acid absorption capacity of juvenile grass carp (*Ctenopharyngodon idella*). Fish Physiol. Biochem. 45, 1589–1602.

Zhang, Y., et al., 2019b. Metabolic disorder induces fatty liver in Japanese seabass, *Lateolabrax japonicas* fed a full plant protein diet and regulated by cAMP-JNK/NFκB-caspase signal pathway. Fish Shellfish Immunol. 90, 223–234.

Zhang, L., et al., 2020. Leucine supplementation: a novel strategy for modulating lipid metabolism and energy homeostasis. Nutrients 12, 1299.

Zhao, H., et al., 2010. Fishmeal can be completely replaced by soy protein concentrate by increasing feeding frequency in Nile tilapia (*Oreochromis niloticus* GIFT strain) less than 2g. Aquac. Nutr. 16, 648–653.

Zhao, B., et al., 2012a. Effects of dietary histidine levels on growth performance, body composition and intestinal enzymes activities of juvenile Jian carp (*Cyprinus carpio* var. Jian). Aquac. Nutr. 18, 220–232.

Zhao, J., et al., 2012b. Effects of dietary isoleucine on growth, the digestion and absorption capacity and gene expression in hepatopancreas and intestine of juvenile Jian carp (*Cyprinus carpio* var. jian). Aquaculture 368–369, 117–128.

Zhao, J., et al., 2013. Effects of dietary isoleucine on the immune response, antioxidant status and gene expression in the head kidney of juvenile Jian carp (*Cyprinus carpio* var. jian). Fish Shellfish Immunol. 35, 572–580.

Zhao, J., et al., 2014. Effect of dietary isoleucine on the immunity, antioxidant status, tight junctions and microflora in the intestine of juvenile Jian carp (*Cyprinus carpio* var. Jian). Fish Shellfish Immunol. 41, 663–673.

Zhao, J., Goldberg, A., 2016. Coordinate regulation of autophagy and the ubiquitin proteasome system by MTOR. Autophagy 12, 1967–1970.

Zhao, Y., et al., 2019. Effects of dietary glutamate supplementation on flesh quality, antioxidant defense and gene expression related to lipid metabolism and myogenic regulation in Jian carp (*Cyprinus carpio* var. Jian). Aquaculture 502, 212–222.

Zhao, Y., et al., 2020. Leucine improved growth performance, muscle growth, and muscle protein deposition through AKT/TOR and AKT/FOXO3a signaling pathways in hybrid catfish *Pelteobagrus vachelli* × *Leiocassis longirostris*. Cells 9, 327–348.

Zheng, P., et al., 2020. Dietary protein requirement of juvenile mud crab *Scylla paramamosain*. Aquaculture 518, 734852.

Zheng, K., Qin, B., Chang, Q., 2015. Effect of graded levels of taurine on growth performance and Ptry expression in the tongue sole (*Cynoglossus semilaevis*) postlarvae. Aquac. Nutr. 22, 1361–1368.

Zhou, F., et al., 2010a. Optimum arginine requirement of juvenile black sea bream, *Sparus macrocephalus*. Aquac. Res. 41, 418–430.

Zhou, F., et al., 2010b. Quantitative L-lysine requirement of juvenile black sea bream (*Sparus macrocephalus*). Aquac. Nutr. 16, 194–204.

Zhou, F., et al., 2011a. Dietary L-methionine requirement of juvenile black sea bream (*Sparus macrocephalus*) at a constant dietary cystine level. Aquac. Nutr. 17, 469–481.

Zhou, F., et al., 2011b. Effects of dietary arginine and lysine levels on growth performance, nutrient utilization and tissue biochemical profile of black sea bream, Acanthopagrus schlegelii, fingerlings. Aquaculture 319, 72–80.

Zhou, J., et al., 2007a. Optimal dietary protein requirement for juvenile ivory shell, *Babylonia areolate*. Aquaculture 270, 186–192.

Zhou, Q., et al., 2006. Optimal dietary methionine requirement for juvenile cobia (*Rachycentron canadum*). Aquaculture 258, 551–557.

Zhou, Q., et al., 2007b. Dietary lysine requirement of juvenile cobia (*Rachycentron canadum*). Aquaculture 273, 634–640.

Zhou, Q., et al., 2012a. Dietary arginine requirement of juvenile yellow grouper *Epinephelus awoara*. Aquaculture 350–353, 175–182.

Zhou, Q., et al., 2012b. Dietary arginine requirement of juvenile Pacific white shrimp, *Litopenaeus vannamei*. Aquaculture 364–365, 252–258.

Zhou, Q., Wang, Y., Wang, H., Tan, B., 2013. Dietary threonine requirements of juvenile Pacific white shrimp, *Litopenaeus vannamei*. Aquaculture 392–395, 142–147.

Zhou, X., et al., 2008. Dietary lysine requirement of juvenile Jian carp (*Cyprinus carpio* var. Jian). Aquac. Nutr. 14, 381–386.

Zhou, Z., et al., 2019. Effects of dietary leucine levels on growth, feed utilization, neuro-endocrine growth axis and TOR-related signaling genes expression of juvenile hybrid grouper (*Epinephelus fuscoguttatus*♀ × *Epinephelus lanceolatus*♂). Aquaculture 504, 172–181.

Zhu, Q., Qian, G., Gao, Y., Wang, C., 2009. Assessment of nitrogen maintenance requirement in juvenile black porgy *acanthopagrus schlegeli*. Chin. J. Oceanol. Limnol. 27, 63–68.

Zhu, X., et al., 2010. Effects of dietary protein and lipid levels on growth and energy productive value of pacific white shrimp, *Litopenaeus vannamei*, at different salinities. Aqua. Nutri. 16, 392–399.

Zhu, Q., et al., 2011. Dietary glutamine supplementation improves tissue antioxidant status and serum non-specific immunity of juvenile Hybrid sturgeon (*Acipenser schrenckii*♀ × *Huso dauricus*♂). J. Appl. Ichthyol. 27, 715–720.

Zimmermann, J., Steinberg, K., Meyer, S., Schulz, C., 2019. Maintenance requirement and utilisation efficiency of energy and protein for on growing pikeperch (*Sander lucioperca* (L.)). Aquac. Int. 27, 1109–1123.

Zou, T., et al., 2018. Effects of dietary leucine levels on growth, tissue protein content and relative expression of genes related to protein synthesis in juvenile gibel carp (*Carassius auratus gibelio* var. CAS III). Aquac. Res. 49, 2240–2248.

CHAPTER

5

The lipids

Giovanni M. Turchini[1], David S. Francis[1], Zhen-Yu Du[2], Rolf E. Olsen[3], Einar Ringø[4] and Douglas R. Tocher[5]

[1]School of Life and Environmental Sciences, Deakin University, Geelong, VIC, Australia; [2]School of Life Sciences, East China Normal University, Shanghai, P.R. China; [3]Institute of Biology, Norwegian University of Science and Technology, Trondheim, Norway; [4]Norwegian College of Fishery Science, UiT The Arctic University of Norway, Tromsø, Norway; [5]Institute of Aquaculture, Faculty of Natural Sciences, University of Stirling, Stirling, Scotland, United Kingdom

5.1 Introduction

Water is the molecule of life, but life as we know it would not have been possible without lipids. The lipid bilayer, a double-sheeted membrane composed of amphiphilic phospholipids (PL), is a spontaneous structure that permitted the formation of a barrier to separate an inner and an outer world, enveloping water and the first self-replicating genetic molecules (Ricardo and Szostak, 2009). It can therefore be said that lipids define life, which started with the self-assembling of lipid membranes most likely in freshwater, aquatic environments (Deamer, 2017). This is why the study of lipids in aquatic ecosystems is so multifaceted and fascinating, and the likely reason as to why lipids and lipid nutrition in aquatic animals has attracted such a high level of research interest (Sargent et al., 2002). Another reason for this intensity is linked to the fact that aquatic lipids differ substantially from terrestrial lipids. Photosynthetic organisms are the primary producers of lipids and fatty acids, and while animals are able to biosynthesize some fatty acids, they are unable to de novo synthesize polyunsaturated fatty acids (PUFA), which are therefore essential nutrients that need to originate from primary producers (Sargent et al., 1995). Therefore, animals need PUFA, particularly the long-chain polyunsaturated fatty acids (LC-PUFA), as these play a myriad of structural and functional roles, and are well-established critical components of the brain and nervous system (Sinclair, 2019), and key metabolic and physiological regulators involved in reproduction, development, growth, and health (Chow, 2008).

PUFA and LC-PUFA are predominantly classified into two distinct series, the omega-3 (n-3) and the omega-6 (n-6) fatty acids. The relative distribution and abundance of these fatty

Fish Nutrition, Fourth Edition
https://doi.org/10.1016/B978-0-12-819587-1.00003-3

acids in different environments and food webs is highly variable and has affected, via adaptation and evolution, the overall metabolism of the animal inhabitants (Colombo et al., 2017). The first essential PUFA formed in the fatty acid biosynthetic pathway of primary producing organisms such as plants is linoleic acid (LNA, 18:2n-6), which can be converted to α-linolenic acid (ALA, 18:3n-3) in chloroplasts. Interestingly, the intrinsic metabolic characteristics of terrestrial and aquatic plants have determined the differences found in their respective lipids, which in turn, have modeled and modified the nutritional physiology and requirements of both terrestrial and aquatic animals. Terrestrial plants possess two distinct fatty acid biosynthetic pathways, the prokaryotic and the eukaryotic pathways. The eukaryotic pathway occurs in nonphotosynthetic tissues (such as seeds), and results primarily in the production of LNA, whereas the prokaryotic pathway takes place in photosynthetic tissue (green leaves) and results in the production of both ALA and LNA (Tocher et al., 1998). As a result of this, ALA is likely one of the most common fatty acids on earth, comprising the predominant fatty acid in leaves, but commonly found in highly diluted forms in the terrestrial food web. On the contrary, LNA is typically found in highly concentrated forms, in energy-dense food items, such as seeds, and thus dominates terrestrial food webs (Turchini et al., 2012).

In contrast to the terrestrial realm, most plants/macroalgae in aquatic environments possess different reproductive strategies and do not produce seeds (Haynes, 1988). In addition, aquatic environments contain an abundance of unicellular photosynthetic microalgae (phytoplankton). Notably, both micro- and macroalgae can synthesize LC-PUFA, whereas terrestrial plants cannot (Harwood, 1996). However, marine phytoplankton can make considerably more LC-PUFA, particularly docosahexaenoic acid (DHA, 22:6n-3), in comparison to their freshwater counterparts (Brett and Muller-Navarra, 1997; Hixson et al., 2015). Consequently, aquatic food webs are predominantly characterized by an abundance of n-3 PUFA, with n-3 LC-PUFA being particularly abundant in marine ecosystems. As such, the world's oceans are considered as the primary source of the latter fatty acids for the continually expanding human population (Tocher et al., 2019).

Accordingly, it is possible to sketch the following, albeit simplistic, summary of different PUFA in different ecosystems and food webs. In marine environments, the presence of n-6 PUFA and ALA are relatively limited, whereas there is an abundance of eicosapentaenoic acid (EPA, 20:5n-3) and DHA originating from primary producers, which are then stored and concentrated in higher trophic level species. In freshwater environments, both n-6 and n-3 PUFA are present, with a relatively higher abundance of ALA and lesser amounts of EPA and DHA emanating from primary producers. Subsequently, freshwater animals are relatively efficient at converting ALA to n-3 LC-PUFA, and then storing and concentrating EPA and DHA in their tissues. In terrestrial environments, PUFA with 18 carbon atoms are abundant in primary producers, while LC-PUFA are not present. Therefore, terrestrial animals are either capable of bioconverting ALA to n-3 LC-PUFA or require access to aquatic environments to obtain these fatty acids given their essential nature in animal physiology and health.

Within this context, it should now be clear to readers why fish and seafood, and their high n-3 LC-PUFA content, are key items for the human diet. Global landings from wild capture fisheries are stagnating, if not declining, whereas the volume of aquaculture products is constantly rising, currently accounting for more than half of available fish and seafood for human consumption (FAO, 2020). Consequently, one of the key objectives of fish lipid nutrition is that of providing sufficient n-3 LC-PUFA in the edible tissues of cultured aquatic animals to fulfill the dietary needs of a growing global population (Tocher, 2015).

However, lipids include a great variety of molecules and nutrients, extending well beyond n-3 LC-PUFA and other essential fatty acids. Lipids play several structural and functional roles in living organisms, spanning from being a concentrated source of metabolic energy to acting as intracellular signaling molecules. Aquatic animals, in particular, evolved in environments characterized by a relatively limited availability of simple carbohydrates. Consequently, in comparison to terrestrial animals, aquatic animal metabolism has adapted to rely on lipids, and not carbohydrates, as a primary energy source (Hemre et al., 2002; Sargent et al., 2002; Polakof et al., 2012). Thus, the provision of adequate metabolic energy is therefore another key objective of fish lipid nutrition.

After a general overview of lipid biochemistry, this chapter is structured into three main sections. The first one follows the journey of lipids from feed intake, to digestion and absorption, the metabolic fates of lipids, including catabolic and anabolic processes, and interactions with the microbiome. The second section expands on some specific lipid classes characterized by unique nutritional roles, such as sterols, PL, and pigments. The final section covers a series of broader aspects of fish lipid nutrition in aquaculture, analyzing resource availability, challenges and opportunities, consumers' nutritional expectations, and the interactions between lipids in aquaculture feeds and the environment.

Following the previous three editions of this chapter, "The Lipids" in 1972, 1989, and 2002 in the book *Fish Nutrition*, an incredible amount of research on fish lipid nutrition has taken place. Research on this topic appears to be growing exponentially in terms of quantity and quality year by year, and this chapter cannot realistically aim at providing a comprehensive and systematic review and analysis of all available knowledge generated by the sector. Instead, the objective of this chapter is that of providing a general overview and an update of fish lipid nutrition, focusing on what is known in terms of general concepts, what remains unknown, and what the key challenges and likely future research directions and needs are for the sector. For a greater, in-depth study of the subject, readers are suggested to refer to the many review papers, book chapters, and books published on the topic of lipid nutrition of aquatic animals in recent years including the following, likely incomplete, suggestions (Sargent et al., 1989, 1995, 1999, 2002; Tocher et al., 1998; Izquierdo et al., 2000; Leaver et al., 2008a; Glencross, 2009; Turchini et al., 2009, 2011c; Tocher, 2010, 2015; Nasopoulou and Zabetakis, 2012).

5.2 Lipid biochemistry

Although the term "lipid" has no widely accepted definition, it is commonly used to describe a diverse and heterogeneous group of hydrophobic molecules. Thus, from a chemical viewpoint, a lipid encompasses any molecule soluble in nonpolar solvents, which results in a very large range of chemical compounds. Additionally, such a definition could be considered as being partially misleading as it excludes other amphiphilic molecules, possibly more soluble in water, that are functionally considered as being lipids. Therefore, the following more "functionally-restricted" definition has been proposed: *"Lipids are fatty acids and their derivatives, and substances related biosynthetically or functionally to these compounds*" (Christie and Han, 2010), and it is hereto recommended. From a nutritional viewpoint, lipids are very distinct from the other macronutrients, namely proteins, carbohydrates, and nucleic acids (Bell and Koppe, 2011). Lipids are ubiquitous in living organisms, playing a variety of essential functions, ranging from being a store of metabolic energy, to being structural components of membranes, precursors of oxylipin signaling molecules, hormones, pigments, and vitamins, as well as acting as enzyme cofactors and carriers of fat soluble nutrients (Sargent et al., 2002). To make sense out of such great chemical and functional variety, lipids are typically categorized into a series of lipid classes, which, with the exception of free sterols such as cholesterol, all contain one or more fatty acid molecules.

5.2.1 Fatty acids

Fatty acids are carboxylic acids (−COOH) with an aliphatic chain of variable length ($-CH_2-$) and a methyl end ($-CH_3$). The fatty acids most commonly found in living organisms typically have a chain composed of an even number of carbon atoms, varying from 12 to 22. Their carbon chain can be fully saturated by hydrogen atoms or unsaturated, with the resulting formation of a double bond, most commonly in the *cis* configuration, between two adjacent carbon atoms (Fig. 5.1). Common fatty acids can contain zero to six double bonds and, generally, when more than one double bond is present in the same fatty acid molecule, these are separated by a single methylene (CH_2) group, to generate the so-called

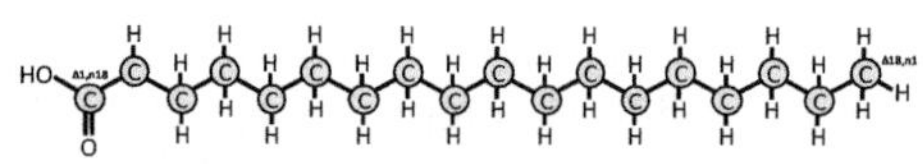

Stearic acid (18:0)

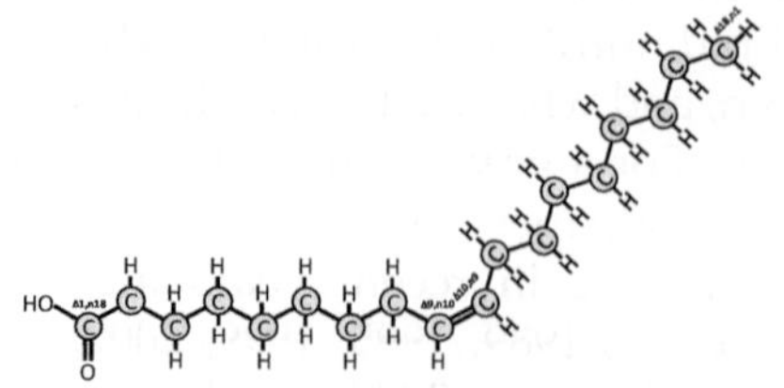

Oleic acid ($18{:}1^{\Delta 9}$; 18:1n-9)

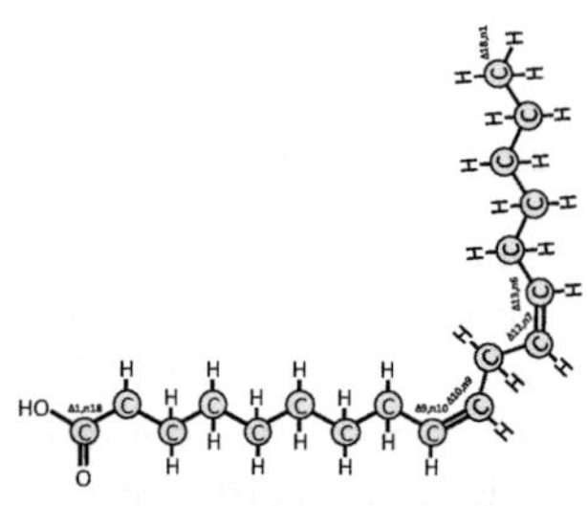

Linoleic acid ($18{:}1^{\Delta 9,12}$, 18:2n-6)

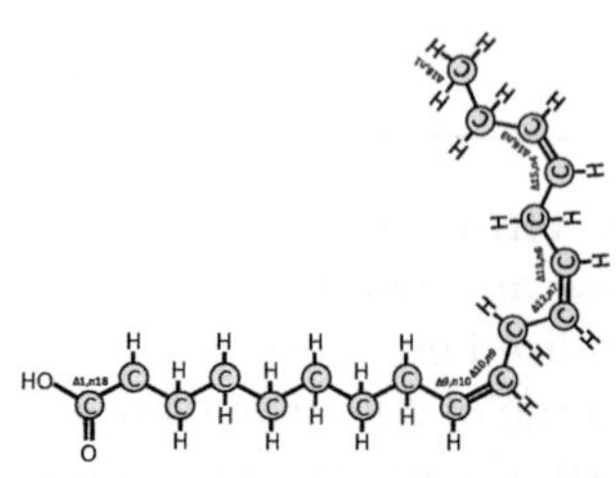

α-Linolenic acid ($18{:}1^{\Delta 9,12,15}$, 18:3n-3)

FIGURE 5.1 A visual representation of four fatty acids, their names, and formulae (in the "Δ" and "n" annotations), and how their shape is modified by the presence of double bonds.

"methylene-interrupted" PUFA. The above description encompasses the largest number of fatty acids commonly found in plants and animals, and these chemical characteristics are determined by the most common biosynthetic pathways found in living organisms. However, in nature and in artifacts resulting from processing, fatty acids with shorter, longer, branched, cyclic or oxygenated chains, chains composed of an odd number of carbon atoms, with double bonds in the *trans* configuration or conjugated (not methylene-interrupted), or with combinations of these characteristics, also exist. While the biological and nutritional roles of these "unusual" fatty acids is certainly important and worthy of investigation, the following sections focus on the most common fatty acids, which quantitatively represent the largest part of lipids.

5.2.1.1 Physical characteristics of fatty acids

The chemical characteristics of a given fatty acid determines its physical properties, and consequently its physiological roles. For example, and following the melting point principle (as the molecular weight increases, the melting point increases), the length (number of carbon atoms) of a fatty acid affects its melting point. However, the melting point of a fatty acid of a given chain length is also modified by its level of unsaturation. SFA have low reactivity and can be thought of as a "stick"; the carbon chain is straight, stiff, occupies minimal space, and can "pack" tightly. When unsaturation is present, the double bond provides a site for a variety of chemical reactions (such as oxidation) and will also modify the three-dimensional shape of the chain by bending it and forming an angle of 133 degrees (Fig. 5.1). When multiple double bonds are present, the reactivity of the fatty acid is further increased with simultaneous increased three-dimensional complexity, and the fatty acid occupies more space, reducing the possibility to pack tightly, and thus increasing its fluidity. Accordingly, the specific chemical structure of a fatty acid affects its overall physical properties, such as melting point, crystallization, fluidity, thermal properties, and optical properties in addition to its chemical properties and reactiveness, such as oxidation and hydrolysis, therefore impacting its overall physiological and nutritional roles (Chow, 2008).

5.2.1.2 Fatty acid formulae

Due to the great variety of fatty acids, the use of chemical formulae to summarize and accurately describe their characteristics is advantageous. Two simplified and specific formulae are routinely utilized: the so-called "n" or "ω" (omega) annotated formula and the "Δ" (delta) annotated formula. Both include information about chain length (the number of carbons), degree of unsaturation (number of double bonds), and position of the double bonds, with the difference between them being which end of the chain the carbons are counted from. Taking LNA as an example [octadecadienoic acid, $C_{18}H_{32}O_2$, $CH_3(CH_2)_4CH{=}CHCH_2CH{=}CH(CH_2)_7CO_2H$], the two formulae are 18:2n-6 and 18:2$\Delta^{9,12}$, respectively. In both formulae, the first number indicates the number of carbons in the chain (18 in this case), the second number specifies the number of double bonds (2 in this case), and the third number (n-6) or set of numbers ($\Delta^{9,12}$) is where the difference between the two systems lies. In the "n" annotated formula, the third number (n-6) denotes the position of the first double bond counted from the methyl end of the molecule, and no further information is provided. In the "Δ" annotated formula, the third set of numbers ($\Delta^{9,12}$) specifies the positions of all double bonds, counted from the carboxyl end of the chain (Fig. 5.1).

Clearly, the "Δ" annotated formula appears more precise, as it provides information on the position of all double bonds along the carbon chain. However, the "n" annotated formula is by far the most commonly used as, if not specified differently, the double bonds along the chain are separated by a methylene group, and thus once the first one is known, the others can be deduced. Secondly, the position of the first double bond from the methyl end of the chain is the one dictating the physiological and nutritional properties of the fatty acids. As such, this annotation greatly simplifies reasoning and discussions of metabolic and physiological aspects. Formulae of some of the most common fatty acids are presented in Table 5.1.

TABLE 5.1 Some of the most common fatty acids, their names, abbreviations, formulae, and fatty acid classes.

Fatty acid names	Abbreviation	Formula: "n-" (or "ω-") annotated	Formula: "Δ" annotated	Fatty acid classes
Palmitic acid, *or* hexadecanoic acid	PAM	16:0	16:0	• SFA (saturated fatty acid)
Stearic acid, *or* octadecanoic acid	STA	18:0	18:0	• SFA
Oleic acid, *or* octadecenoic acid	OLA	18:1n-9	$18{:}1^{\Delta 9}$	• MUFA (monounsaturated fatty acid)
Linoleic acid, *or* octadecadienoic acid	LNA	18:2n-6	$18{:}2^{\Delta 9,12}$	• PUFA (polyunsaturated fatty acid) • C_{18} PUFA (PUFA with 18 carbon atoms) • n-6 PUFA (omega-6 PUFA) • C_{18} n-6 PUFA
α-Linolenic acid, *or* octadecatrienoic acid	ALA	18:3n-3	$18{:}3^{\Delta 9,12,15}$	• PUFA • C_{18} PUFA • n-3 PUFA (omega-3 PUFA) • C_{18} n-3 PUFA
Arachidonic acid, *or* eicosatetraenoic acid	ARA	20:4n-6	$20{:}4^{\Delta 5,8,11,14}$	• PUFA • LC-PUFA (PUFA with 20 or more carbon atoms) • n-6 PUFA • n-6 LC-PUFA
Eicosapentaenoic acid	EPA	20:5n-3	$20{:}5^{\Delta 5,8,11,14,17}$	• PUFA • LC-PUFA • n-3 PUFA • n-3 LC-PUFA
Docosahexaenoic acid	DHA	22:6n-3	$22{:}6^{\Delta 4,7,10,13,16,19}$	• PUFA • LC-PUFA • n-3 PUFA • n-3 LC-PUFA

5.2.1.3 *Fatty acid names*

In addition to chemical formulae, specific fatty acid names, and related abbreviations, are often utilized. Interestingly, and similarly to animal species, fatty acids have two names: a common name and a systematic ("scientific") name. The latter describes, in Latin-derived nomenclature, the chemical formula of given fatty acids. For example, the SFA with 16 carbons (16:0) is called hexadecanoic acid, and the monounsaturated fatty acid with 18 carbons (18:1n-9) is called octadecenoic acid. On the other hand, the common name is often derived from their "traditional" sources and, as such, 16:0 is called palmitic acid and 18:1n-9 is called oleic acid, as they are abundant in the oils derived from palm fruits and olives, respectively. Abbreviations of the common or systematic names are often utilized and so a list of suggested names and abbreviations for the most common fatty acids is provided in Table 5.1, based on the recommendations of the American Oil Chemists' Society (AOCS).

5.2.1.4 *Fatty acid classes*

Different fatty acids with similar chemical characteristics often possess similar nutritional aspects, and accordingly it is often convenient to group them into classes. The classes are typically defined by the level of unsaturation, the position of the first double bond, and the overall chain length (number of carbon atoms). Accordingly, the three major fatty acid classes that encompass all fatty acids are: (i) saturated fatty acids (SFA: all the fatty acids with no double-bonds), (ii) monounsaturated fatty acids (MUFA: all the fatty acids with one double-bond), and (iii) polyunsaturated fatty acids (PUFA: all the fatty acids with two or more double-bonds). PUFA are then commonly subdivided further according to two different characteristics: (i) the position of the first double bond from the methyl end: n-3 PUFA and n-6 PUFA (often referred to as "omega-3" and "omega-6," respectively) and (ii) their chain lengths: C_{18} PUFA (PUFA with 18 carbon atoms) and long-chain PUFA (LC-PUFA: with 20 or more carbon atoms). A list of suggested names and abbreviations for the most common fatty acid classes is reported in Table 5.1, based on the recommendations from the book "Fish oil replacement and alternative lipid sources in aquaculture feeds" (Turchini et al., 2010).

5.2.2 Lipid classes

Fatty acids are seldom found as individual, free molecules (free fatty acids) and are typically found esterified (bound) to other molecules, such as glycerol, fatty alcohol, or phosphatidic acid, forming many different compounds, all generally referred to as lipids. These diverse chemical compounds are classified using two different systems: one based on their polarity, differentiating neutral (nonpolar) lipids from polar lipids, and the other based on their complexity, differentiating simple lipids from complex lipids. Using the former approach, and focusing on polarity, neutral lipids include triacylglycerols (TAG), diacylglycerols (DAG), monoacylglycerols (MAG), free fatty acids (FFA), steryl esters (SE), and wax esters (WE), whereas polar lipids include glycerophospholipids (aka phospholipids, PLs) and sphingolipids. Using the latter approach, and thus focusing on the number of compounds making up the final molecule, simple lipids include FFA, TAG, DAG, MAG, WE, and SE (all composed of one or two different compounds), whereas complex lipids are composed of three or more different compounds, such as PL, glycolipids, and so forth.

5.2.2.1 Monoacylglycerols and diacylglycerols

Glycerol is a simple polyol containing three carbon atoms that acts as the backbone in most lipids. When a single fatty acid is esterified to a molecule of glycerol, a MAG is obtained. Though the glycerol is a symmetric molecule, it is prochiral with respect to the reactions (bonds) that can take place on one of its primary alcohols. Accordingly, in its derivatives the *sn*-prefix is utilized to specify the stereospecificity of the final molecule. Thus, two types of MAG exist, those where the fatty acid is esterified to a primary alcohol (*sn*-1-MAG), or to the secondary alcohol (*sn*-2-MAG). When a further fatty acid is esterified to another alcohol of the glycerol backbone of MAG, a DAG is formed. Consequently, two possible forms of DAG exist, *sn*-1, *sn*-2-DAG and *sn*-1, *sn*-3-DAG. Visual representations of a glycerol, a MAG and a DAG are presented in Fig. 5.2. MAG and DAG are generally found at trace levels in animals and plant tissues as they are intermediates in the biosynthesis of TAG and other lipids, and intermediates in catabolic processes including the digestion of TAG, although they also can act as intracellular messengers.

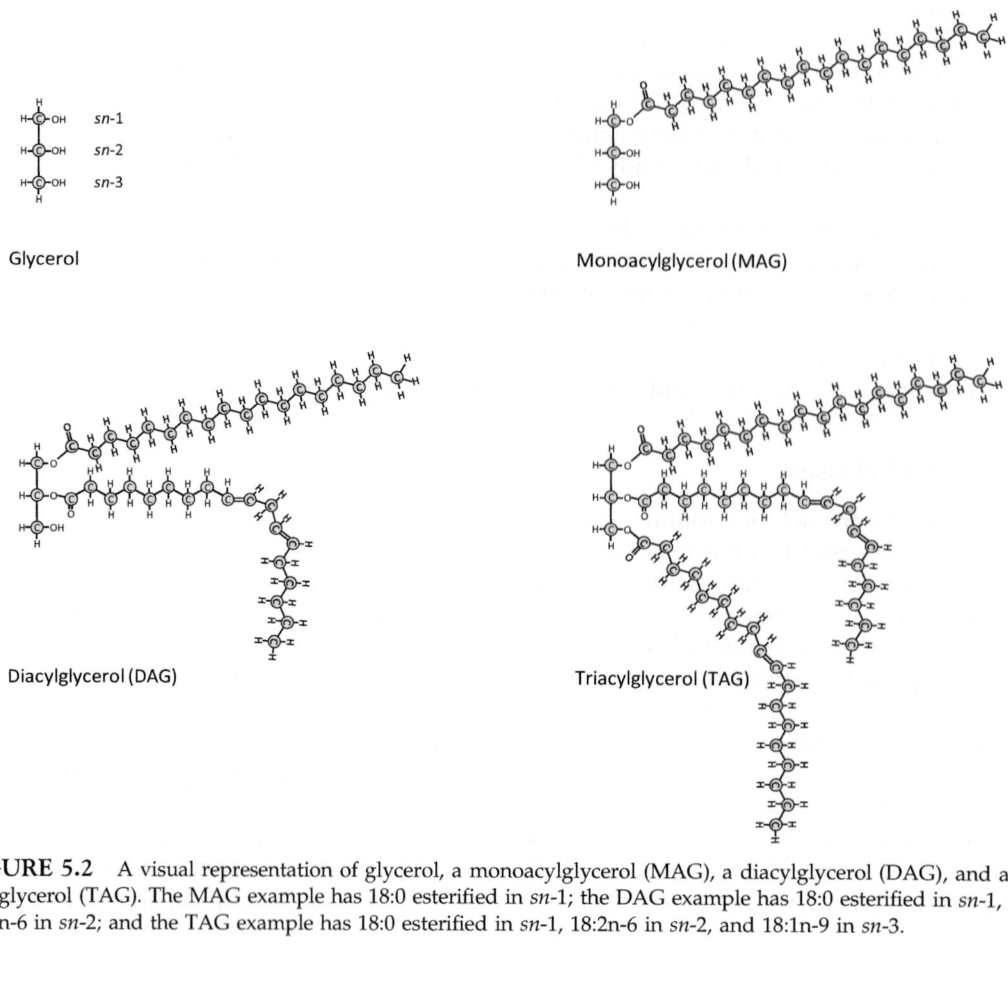

FIGURE 5.2 A visual representation of glycerol, a monoacylglycerol (MAG), a diacylglycerol (DAG), and a triacylglycerol (TAG). The MAG example has 18:0 esterified in *sn*-1; the DAG example has 18:0 esterified in *sn*-1, and 18:2n-6 in *sn*-2; and the TAG example has 18:0 esterified in *sn*-1, 18:2n-6 in *sn*-2, and 18:1n-9 in *sn*-3.

5.2.2.2 *Triacylglycerols*

Triacylglycerols (TAG), also termed triglycerides, are comprised of three fatty acids esterified to the *sn*-1, *sn*-2, and *sn*-3 hydroxyl groups of glycerol and are often the most abundant lipid class found in biological matrices Fig. 5.2. TAG are the most efficient way of storing energy (storage lipids) and are typically the predominant lipid class in fish tissues, dietary oils, and fats. The same or different fatty acids can be esterified to any of the three *sn* positions, creating a great variety of different enantiomeric forms. In nature, TAG are synthesized by enzymatic systems, differing across taxa and species, which are responsible for determining the center of asymmetry in the *sn*-2 position of the glycerol moiety (Christie and Han, 2010). To describe these forms, specific stereospecific numbering and naming systems exist but, for simplicity, these details are not reported here, and interested readers are suggested to refer to specialized literature available from the IUPAC-IUBMB Joint Commission on Biochemical Nomenclature (JCBN) (https://iubmb.org/biochemical-nomenclature/). The same fatty acid may be esterified to all three positions of glycerol. For example, triolein (1,2,3-trioleoyl-*sn*-glycerol), is a TAG comprised of three molecules of oleic acid (18:1n-9, OLA) and found naturally in some vegetable oils (VO), or 1,2,3- tridocosahexaenoyl-*sn*-glycerol, a TAG with three molecules of DHA, found as a major lipid component in the retina of some fish (Sargent et al., 2002). Alternatively, two different fatty acids may be esterified to the glycerol, as in 1,2-dioleoyl-3-palmitoyl-*sn*-glycerol (two molecules of 18:1n-9, one in the *sn*-1 position, and the other in the *sn*-2, and one 16:0, in the *sn*-3), which is an enantiomer of a 1-palmitoyl-2,3-dioleoyl-*sn*-glycerol. The third option is that three different fatty acids are esterified at the three *sn* positions, as in the example 1-stearoyl-2-linoleoyl-3-oleoyl-*sn*-glycerol graphically represented in Fig. 5.2. Considering the many different fatty acids and that they can be esterified to any *sn* position of glycerol, a great variety of TAG combinations (or molecular species) are possible. However, SFA and MUFA are most commonly found in the *sn*-1 and *sn*-2 positions, whereas PUFA and LC-PUFA are generally preferentially located in the *sn*-2 position of glycerol (Bell and Koppe, 2011). Accordingly, most of the EPA (50%) and DHA (80%) in Atlantic salmon (*Salmo salar*) was reported to be located in the *sn*-2 position of TAG, and this was apparently unaffected by the dietary lipid composition (Ruiz-Lopez et al., 2015).

5.2.2.3 *Wax esters*

Wax esters (WE) are naturally occurring molecules composed of a single fatty acid molecule esterified to a long-chain (fatty) alcohol (Fig. 5.3). However, some naturally occurring WE can have a more complicated composition and structure (Christie and Han, 2010). WE are particularly widespread in marine environments where they are present in animal, plant, and microbial tissues, playing a variety of functions ranging from energy storage to acting as waterproofing agents. Some marine zooplankton, such as calanoid copepods, contain high quantities of WE. Through the food chain, these become important nutrients for marine fish where, in some fish species, they accumulate in high concentrations in tissues and eggs (Sargent et al., 2002). In marine WE, fatty acids present can range from SFA to LC-PUFA, whereas the fatty alcohols are generally saturated or monounsaturated. In copepods found at high latitudes in the northern hemisphere, the abundant long-chain monounsaturated fatty alcohols, 20:1n-9 and 22:1n-11, in WE are converted to fatty acids in the fish

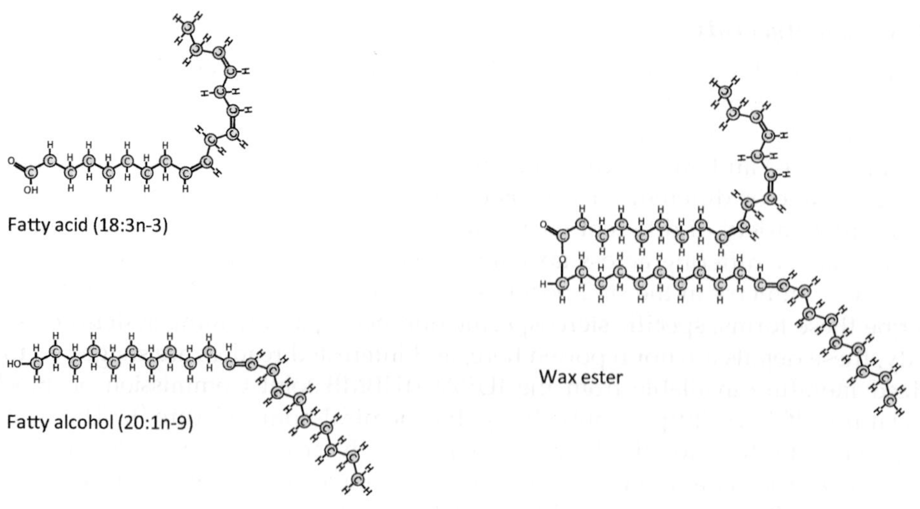

FIGURE 5.3 A visual representation of a wax ester (WE) composed of one fatty acid (18:3n-3) esterified to one fatty alcohol (20:1n-9).

species that feed on them, resulting in northern hemisphere fish oils (FO) being richer in these long-chain MUFA compared to southern hemisphere FO that tend to be richer in n-3 LC-PUFA.

5.2.2.4 Sterols and steryl esters

Sterols, of which cholesterol is the most commonly found in animals, are tetracyclic hydrocarbon ring system compounds. Cholesterol has a double bond in the second ring and one free hydroxyl group (Fig. 5.4) and is found only in trace levels in plants where other sterols, such as ergosterol, campesterol, stigmasterol, sitosterol, and brassicasterol are present and commonly referred to as phytosterols. Both cholesterol and phytosterols can be free or esterified to a fatty acid to form SE. In fish tissues, as with all animals, free cholesterol is found in association with PL in plasma lipoproteins and in cell membranes, where it plays an important role in the regulation of membrane fluidity.

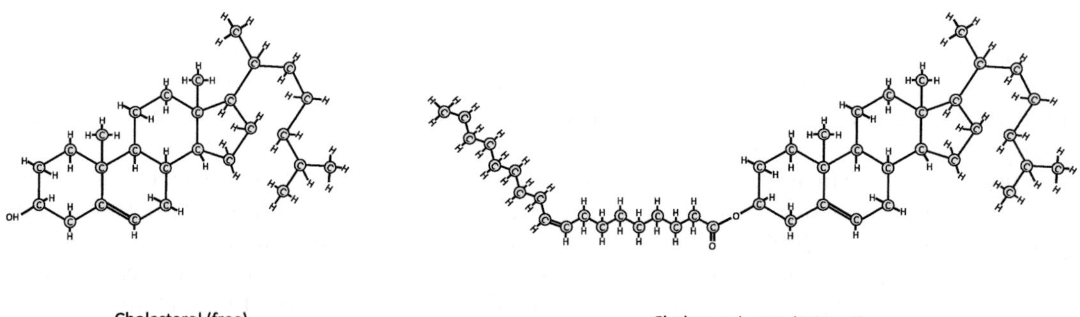

FIGURE 5.4 A visual representation of a molecule of cholesterol (free) and a cholesterol ester (esterified to an oleic acid molecule).

5.2.2.5 Phospholipids

Phospholipids (PL), the most commonly used term to refer to molecules more accurately termed phosphoglycerides or glycerophospholipids, represent the major group of polar lipids, as well as complex lipids. They contain phosphorus and are amphiphilic molecules, being hydrophilic and lipophilic in different parts of their structure. The basic PL backbone is phosphatidic acid, which is composed of a glycerol, two fatty acids in the *sn*-1 and *sn*-2 positions, and a phosphate group in the *sn*-3 (1,2-diacyl-*sn*-glycero-3-phosphate; Fig. 5.5). Phosphatidic acid is found in only trace amounts in tissues, but is metabolically important as the precursor of PL and as a signaling molecule (Christie and Han, 2010). The most common PL, and a major constituent of cellular membranes, is phosphatidylcholine, also referred to as lecithin, which is composed of a phosphatidic acid incorporating choline as a head group (Fig. 5.5). Other common bases can be incorporated such as ethanolamine, serine, and inositol to form phosphatidylethanolamine, phosphatidylserine, and phosphatidylinositol, respectively. The occurrence and abundance of these PL is variable in different tissues, as they play a series of different and specific metabolic functions (Bell and Koppe, 2011). Most commonly, SFA and MUFA are esterified at position *sn*-1 of the glycerol, whereas PUFA are preferentially esterified at position *sn*-2; however, as is the case with TAG, there are many exceptions to this generalization (Sargent et al., 2002).

5.2.2.6 Sphingolipids and other complex lipids

Sphingolipids are lipids comprising of a sphingoid base backbone and a set of long-chain aliphatic amino alcohols. Among them, sphingosines, and specifically sphingomyelin, are the most abundant, and are integral components of cell membranes. Other sphingolipids include ceramides and glycosphingolipids, such as cerebrosides and gangliosides. Sphingomyelin and cerebrosides are particularly abundant in neural tissues, but they have not yet been

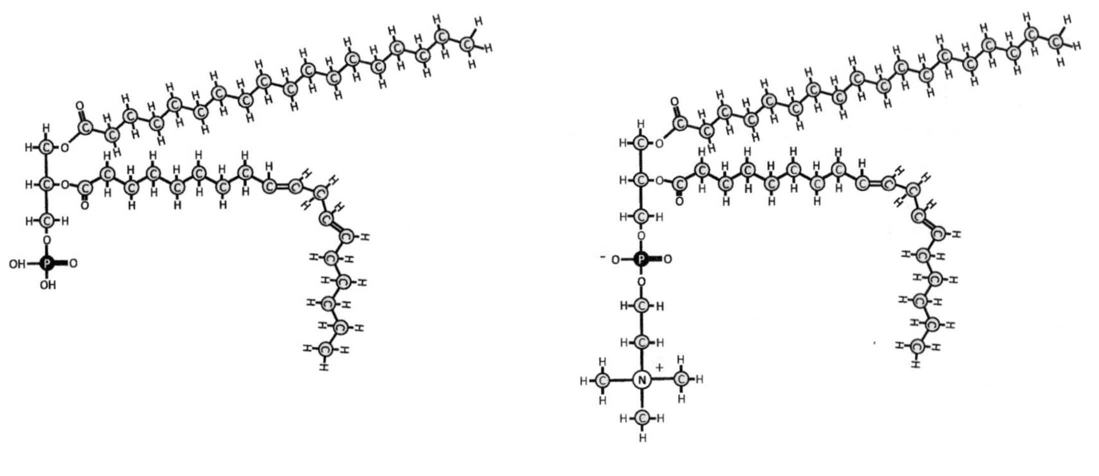

Phosphatidic acid Phosphatidylcholine

FIGURE 5.5 A visual representation of a molecule of phosphatidic acid and a phosphatidylcholine. In both cases, 18:0 is esterified in *sn*-1, and 18:2n-6 is esterified in *sn*-2.

extensively studied in fish (Bell and Koppe, 2011). Glyceroglycolipids are another group of complex lipids commonly found in small concentrations in plant tissues, comprised of a 1,2-diacyl-*sn*-glycerol joined by a glycosidic linkage at position *sn*-3 to a carbohydrate moiety (Christie and Han, 2010). Many other complex lipids exist but they have thus far received little research attention in fish nutrition.

5.3 Lipid intake

Lipids are present in fish diets either as intrinsic components of natural food items or as specifically added nutrients into artificial feeds. In feeds, lipids originate from the lipid residues of raw materials, such as protein sources, and from specifically added oils and fats, primarily aimed at satisfying the energy and essential fatty acid requirements of the target species. The total lipid intake is determined by the concentration of lipid in the feed and the amount of feed consumed by the animal. Feed intake is regulated by a multitude of factors, ranging from feed availability through to a series of other determinants, such as species-specific chemosensory responses, palatability, feed size, shape and appearance, biotic and abiotic environmental factors, feeding methodologies, times and rhythms, farming systems and conditions, nutritional factors, physiological conditions, anorexigenic and orexigenic neuropeptides, and hormonal regulation (Houlihan et al., 2001; Meyerhof and Korsching, 2009; Volkoff et al., 2010; Conde-Sieira and Soengas, 2017; Morais, 2017; Soengas et al., 2018).

5.3.1 Dietary lipids, energy, and food intake

Within such a complex system, dietary lipids are major factors affecting total energy intake and, as such, the modulation of total food intake. Though there is conflicting evidence concerning the exact role and mechanism by which dietary energy modulates feed intake in fishes, it is accepted that the total lipid content of the diet influences feed intake. A decrease in food intake usually occurs in fish fed diets containing higher lipid levels, likely as a result of higher levels of fatty acids in plasma and/or larger lipid stores (Soengas et al., 2018). Specifically, the density of digestible energy, independent of the source, or a combination of energy and nutrient utilization, and thus dependent on the source of energy, or even a function of heat production, which is modulated by oxidative metabolism and physiological limits, have all been shown to contribute to the regulation of voluntary feed intake in cultured fish (Saravanan et al., 2012). Thus, when it comes to lipid, the dietary quantity and digestibility are both important factors affecting food intake, as the overall energetic status of an animal is determined by the total available energy, and not simply the gross total energy content of the feed (Turchini et al., 2009).

In addition, the different oils utilized in feed formulations can affect overall palatability of the feed itself, directly affecting feeding behavior and total food intake. Furthermore, once lipids have been incorporated into the organism, specific nutrient sensing and nutritional status signaling can further modify feeding behavior. However, accurately measuring feed intake in aquatic animals, and thus studying feed palatability and fish appetite regulation, is a challenging exercise and therefore many questions remain unanswered.

5.3.2 Dietary lipid and feed palatability

In fish, feed palatability is determined by the interaction of feed components and specific chemosensory perception, involving both olfactory and taste receptors. Lipids, being by definition insoluble in water, are generally considered to play minor roles in defining the overall palatability of a diet, as aquatic animals are fundamentally attracted by water-soluble, nitrogen-containing, small molecules (Kasumyan and Doving, 2003; Morais, 2017). Nevertheless, impurities in the raw oils utilized in feed formulations may have an effect on feed palatability (Turchini et al., 2009) where, for example, highly oxidized oils reportedly act as a deterrent, negatively affecting palatability and feed intake (Houlihan et al., 2001). This has been demonstrated in gilthead seabream (*Sparus aurata*), where the extent of the response can be reinforced by increased oxidative stress status of the animal (Montoya et al., 2011).

Once in the mouth, lipids can be detected as fish taste receptors respond to various compounds, including fatty acids (Meyerhof and Korsching, 2009). In most studies where different dietary oils were tested in a variety of species, feed intake was seldom affected to a large extent, although the accurate assessment of voluntary feed intake was often overlooked and not a focus of the experimental design. However, this observation suggests that the type of dietary oil used in formulations has a relatively limited impact on feed palatability (Turchini et al., 2011c). Nevertheless, when studies specifically focusing on palatability and feed intake have been implemented, some intriguing evidence of direct effects of dietary oils and feed intake have been noted. The use of on-demand self-feeders has been shown to be a useful tool toward answering questions about feed palatability and nutritional regulation of voluntary feed intake. For example, rapid screening of feed palatability has been successfully demonstrated in Arctic charr (*Salvelinus alpinus*) (Carlberg et al., 2015), and gilthead seabream have been shown capable of selecting between nutritionally incomplete feeds and combining them to consume a balanced diet (Montoya et al., 2012). Focusing on the role of lipids, the feeding behavior of rainbow trout (*Oncorhynchus mykiss*) has been shown to be mediated by dietary oils, with fish being able to discriminate between different oils, clearly showing preference for some oils over others (Geurden et al., 2005, 2007). Consistently, rainbow trout have been shown to prefer a diet containing only FO as the added dietary lipid source, compared with diets containing some rapeseed oil (Pettersson et al., 2009). Likewise, European seabass (*Dicentrarchus labrax*) have been shown to be capable of discriminating between diets containing different oils (Luz et al., 2018). Therefore, it appears evident that different oils utilized in feed formulations can have an impact on the palatability of the resulting pellet. However, compared to other nutrients and raw materials used in formulated diets, such as protein sources, the potential effects of different oils on feed palatability are limited. Additionally, minor variations in feed palatability are likely detectable only when a choice between different feed is available, and are often overcome by prolonged exposure to that feed. Importantly, palatability is only one of the factors regulating feed intake, and specific nutrient sensing and nutritional regulation are also envisaged to be important contributors.

5.3.3 Dietary lipid, nutritional signaling, and fatty acid sensing

The roles of sensory perception (feed palatability) and postingestive signals (nutritional regulation) in the modulation of voluntary feed intake in fish likely overlap and interact and, in general, are poorly understood. An interesting study in sharpsnout seabream (*Diplodus puntazzo*), where encapsulated pellets were used to mask the organoleptic properties of experimental diets, showed that fish possess the ability to select a diet containing FO compared to a fat-free diet based on postingestive signals, but were not able to discriminate between diets containing different dietary oils (Almaida-Pagan et al., 2011). However, rainbow trout have been shown to discriminate between feeds containing different amounts of EPA and DHA, clearly preferring (consuming more) the feeds richer in n-3 LC-PUFA (Roy et al., 2020). Similarly, the ability to detect and modulate feed intake based on specific fatty acids was shown in Senegalese sole (*Solea senegalensis*) broodstock, which were able to detect dietary arachidonic acid (ARA, 20:4n-6) content and modify the relative intake of two feeds containing different amounts of ARA in response to seasonal/temperature changes and fluctuating nutritional requirements (Norambuena et al., 2012).

Changes in nutritional status can be detected by a series of mechanisms in the brain so that food intake can be modified accordingly (Conde-Sieira and Soengas, 2017). The existence of a hypothalamus fatty acid sensing system and its ability to detect 18:0 and OLA, and accordingly modulate voluntary feed intake, has been shown in rainbow trout (Libran-Perez et al., 2014). The hypothalamus can also detect LC-PUFA via a series of fatty acid sensing mechanisms, with interesting differences in fish compared to mammals (Soengas et al., 2018). Specifically, fish have an apparent capacity to respond to changes in the levels of specific fatty acids with different chain lengths and degree of unsaturation. In Senegalese sole, the hypothalamus has been show to detect OLA and ALA, and subsequently trigger an enhanced anorexigenic capacity, whereas 18:0 and EPA did not activate these fatty acid sensing systems (Conde-Sieira et al., 2015). Unfortunately, information remains limited despite evidence of the potential importance of these sensing systems. For example, increased abundance of dietary LNA in diets for Atlantic salmon was shown to not have an immediate effect on palatability, but a longer-term negative effect on appetite and feed intake was observed (Bell et al., 1993; Liland et al., 2013b). In addition to the existing knowledge gap concerning specific fatty acid sensing abilities in fish, relatively few attempts have been made to identify possible relationships between dietary lipids and orexigenic and anorexigenic factors, such as ghrelin and leptin (Ganga et al., 2005; Sissener et al., 2013; Francis et al., 2014; Velasco et al., 2016a; Bonacic et al., 2017). In general, these studies suggest the existence of possible effects of different dietary oils, but a robust understanding of these complex phenomena has not yet been achieved.

5.3.4 Lipid intake

The amount of lipid consumed by cultured aquatic animals is therefore the result of the lipid content of the diet and all of the above factors that modulate voluntary feed intake. Dietary lipid enters the body after passing through the digestive system and is processed prior to becoming biologically available to the organism, as presented in the following sections.

The study of food intake in aquatic animals is technically and logistically challenging, whereas sensory perception, and postingestive and nutritional signaling systems and regulation are complex and scientifically challenging. Nevertheless, it is evident that increased knowledge of how dietary lipids affect palatability, nutritional sensing and signaling, and modify food intake, as well as longer-term effects and possible adaptation to a specific diet, is important. This knowledge will help to understand the effects, at times contradictory and other times relatively limited, observed in the many empirical studies testing dietary oils in different species (Turchini et al., 2011c), and will also help refine aquafeed formulations toward achieving further improvements in their nutritional efficiency.

5.4 Lipid digestion

5.4.1 Digestive lipases

5.4.1.1 Lingual/gastric lipases

In many terrestrial animals, lipid digestion starts with lingual and/or gastric lipases. The main function of the gastric lipase being to partially degrade dietary TAG to increase levels of DAG and FFA, partitioning lipid droplets into smaller vesicles that are easier to digest once entering the intestinal system. As TAG is the key energy source for many fish species, some lipolytic activity would be expected in preparation for bulk hydrolysis when the digesta reaches the intestine. Although not extensively studied in fish, lingual, esophageal, and gastric lipases have been reported in many fish species, but it is likely that the activities are species specific. In primitive Pacific hagfish (*Eptatretus stoutii*), high neutral lipase activity was detected (with bile salt added) in the buccal cavity and pharyngocutaneous duct (Weinrauch et al., 2019). Some neutral lipase activity (no bile salt) was also reported in the esophagus and (nonacid) stomach of freshwater milkfish (*Chanos chanos*) (Borlongan, 1990), and stomachs of Asian bony tongue (*Scleropages formosus*) (Natalia et al., 2004) and Tibetan catfish (*Glyptosternum maculatum*) (Xiong et al., 2011). In the latter, the highest activity was obtained at pH6 (tested pH5–9) although stomach pH was not reported. There are few reports of gastric lipases in marine fish, but limited gastric lipolytic activity has been indicated in juvenile turbot (*Scophthalmus maximus*) (Koven et al., 1997). An acid lipase with an amino acid sequence similar to lysosomal, gastric, and lingual lipase has also been indicated in stomach tissue of winter flounder (*Pseudopleuronectes americanus*) by RT-PCR. However, inability to locate this by *in situ* hybridization suggests it may be a lysosomal lipase (Murray et al., 2003).

5.4.1.2 Pancreatic lipases

After leaving the stomach, or the anterior part of the intestine in stomach-less fish, lipid is usually solubilized by bile salts and PL to further reduce droplet size and increase lipolytic efficiency. In mammals, droplet size is reduced to around 1 μm with a core of TAG, DAG, WE, and a surface composed of bile salts, PL, and amphipathic peptides. In general, this process is assumed to be rather similar in most fishes although some species like Californian anchovy (*Engraulis mordax*), jack mackerel (*Trachurus symmetricus*), and leopard shark (*Triaizus sernifusciata*) do not appear to excrete a PL-rich bile (Patton et al., 1975, 1977). This will most certainly affect the physical properties of the droplets, including reduction in micellar size, subsequently increasing the surface area (Nichols and Ozarowski, 1990). There is also a great variability in bile salt composition that also will affect micellar solubility and lipase activities. For example, while taurocholate is the predominant bile salt in most fish species, cyprinol sulfate is the main bile salt in carps, and taurochenodeoxycholate in large-scale mullet (*Liza macrolepis*) (Yeh and Hwang, 2001).

The micellar droplets are then acted upon by at least three pancreatic lipases in fish, pancreatic triacylglycerol lipase (PTL), bile stimulated triacylglycerol lipase (BAL), and phospholipase A_2. PTL is activated when it encounters the water/oil interface and acts on both TAG and DAG substrates. However, binding is inhibited in the presence of bile salt concentrations above the critical micellar concentration and, to overcome this inhibition, a colipase is required to enable PTL to absorb on to bile salt covered interfaces and regain activity (Kurtovic et al., 2009). PTL is probably the enzyme causing the most controversy over the years. Originally it was understood to be the main TAG lipase in fish, as in mammals, due to its high specific activity and *sn*-1,3-specificity producing 2-MAG and FFA as the main products. The 2-MAG would then be a template for very efficient TAG reconstitution once absorbed into the enterocyte. The challenge of proving the existence of this enzyme has been discussed in detail elsewhere (Olsen and Ringø, 1997). As more fish species were studied, PTL or its colipase were mainly found or indicated in primitive fish such as hagfish (*Myxine glutinosa*) and cartilaginous fishes like ratfish (*Chimera monstrosa*), rayfish (*Raja radiata*), skates (*Raja erinacea*), and Greenland shark (*Somniosus microcephalus*) (Brockerhoff and Hoyle, 1965; Patton and Quinn, 1973; Sternby et al., 1983). A PTL with 58.3% identity to human PTL has also been cloned in a more primitive Actinopterygii, Japanese eel (*Anguilla japonica*) (Kurokawa et al., 2002). In more advanced bony fish, there is little clear evidence of PTL activity. One of the few examples may be Pacific bluefin tuna (*Thunnus orientalis*), where a partially purified lipase from caeca was inhibited by bile salts (Parra et al., 2007). Protein alignment studies of other fish species like Atlantic cod (*Gadus morhua*) and red seabream (*Pagrus major*) have suggested amino acid replacements in central active sites of the enzyme rendering it inactive as a lipase (Sæle et al., 2010). NCBI blasts have identified around 18 colipases in Actinopterygii, again mostly in primitive species or lungfish (Lepidosireniformes), with 50% or more identity to the human colipase.

It is therefore a distinct possibility that PTL has become redundant in higher fishes; however, the reason for this remains elusive. One possible clue may come from the fatty acid specificity of the enzyme. The typical PTL has low activity towards TAG containing longer acyl chains ($>C_{18}$) or with double bonds close to the carboxyl group (Arishima et al., 2009). As most aquatic forms of TAG would be of this type, it is possible that fish PTL would produce incomplete digested TAG with significant levels of DAG rather than 2-MAG that would require further degradation before absorption. This would make it less useful as a digestive lipase. Results showing that a PTL (1,3-specific) in blue shark (*Prionace glauca*) poorly hydrolyzes marine LC-PUFA (Patton and Nevenzel, 1974; Patton et al., 1977) may confirm this possibility.

The main TAG lipase in most higher fish species explored to date appears to be of the BAL type, which requires primary bile salts for activity against bulkier substrates. Most enzymes characterized, including those from Atlantic cod (Gjellesvik et al., 1992) chinook salmon (*Oncorhynchus tshawytscha*) and New Zealand hoki (*Macruronus novaezelandiae*) (Kurtovic et al., 2010), have amino acid sequences similar to that of mammalian counterparts. The enzyme has broad specificity and hydrolyzes TAG, DAG and MAG, SE, WE, and other esters. Furthermore, the enzyme will act on many fatty acids, including LC-PUFA (Kurtovic et al., 2009). There is some fatty acid specificity that may vary between species, but the general selectivity has been suggested to be PUFA > MUFA > SFA, and medium-chain > long-chain with low activity toward very short-chain fatty acids (Olsen and Ringø, 1997). Interestingly, DHA appears to have higher resistance to hydrolysis in structured lipids than other LC-PUFA (Bogevik et al., 2008a). Efficiency of BAL is usually several times higher toward TAG than toward WE and SE in most species examined, including Atlantic salmon (Bogevik et al., 2008b). Similar to mammalian BAL, most fish BAL do not appear to have positional specificities which means that the final products would be glycerol and FFA, producing a variety of 1-MAG, 2-MAG, and 3-MAG in the process, as observed in Atlantic salmon (Bogevik et al., 2008a). There are, however, exceptions to this. In Atlantic cod, a partially purified BAL showed distinct 1,3-specificity (Gjellesvik et al., 1989), while a tendency for 1,3-specificity in BAL assays (67% 2-MAG and the remaining 1/3-MAG) was observed in intestinal preparations from rainbow trout (Gøttsche et al., 2005). It is therefore possible that some fish have evolved partial positional specificity in TAG digestion but, in most cases, results appear to be nondiscriminatory.

5.4.1.3 *Other lipases*

The possibility that many fish completely hydrolyze TAG to glycerol and FFA rather than to 2-MAG and FFA was noted in the early 1970s and it was suggested that this complete degradation could be aided by other specific lipases (Babin and Vernier, 1989). This would be an alternative to avoid import of 1(3)-MAG that could pose problems for TAG synthesis by MGAT (acyl-CoA:monoacylglycerol acyltransferase) and DGAT (acyl-CoA:diacylglycerol acyltransferase). The issue was not followed up, but results from trials in human colorectal adenocarcinoma cells ($Caco_2$) demonstrated the capacity of these cells to excrete extracellular enzymes to hydrolyze MAG and lysophosphatidylcholine (lyso-PC) (Inaba et al., 2014). Similar capacity to hydrolyze both MAG and lyso-PC by extracellular and membrane-bound enzymes was recently shown in primary cultures of brown trout (*Salmo trutta*) enterocytes (Li et al., 2018b). The significance of these observations remains to be clarified.

5.4.2 Lipid absorption

Following primary digestion, the partially digested lipids, acylglycerols (DAG, 1,3- and 2-MAG), lyso-PL, FFA, and cholesterol will then reconstitute into smaller disc-shaped mixed micelles that will diffuse through the unstirred water layer for absorption (Fig. 5.6). Once delivered to the surface of enterocytes, FFA and acylglycerols are absorbed across the enterocyte membrane. Previously, this absorption was assumed to occur mainly by simple diffusion. One challenge with this theory was that FFA might accumulate in the cell wall and the cell itself and, as FFA are cytotoxic in a concentration-dependent manner, this would be detrimental to cells and membranes. For example, FFA are one of the major cytotoxic factors produced in the intestines following shock, causing apoptosis and necrosis (Penn and Schmid-Schönbein, 2008). Furthermore, such processes are highly dependent on concentration gradients meaning that uptake of lipid would diminish when the luminal concentration was reduced following digestion, leading to possible backflow of FFA. Consequently, it has become evident that a significant part of FFA uptake into enterocytes is indeed controlled at various levels by membrane-associated proteins.

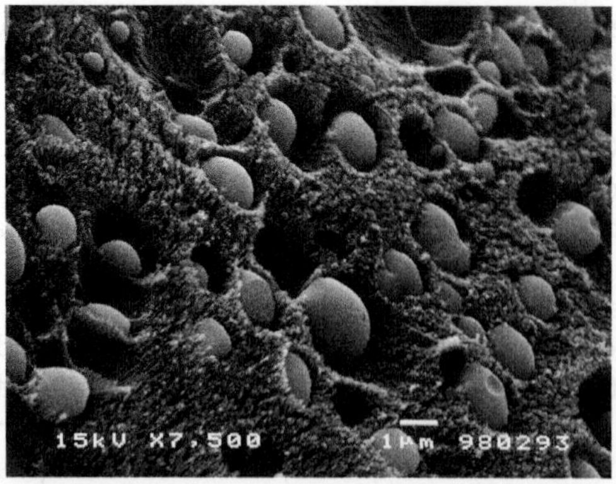

FIGURE 5.6 A scanning electron microscope image of mixed micelles on the enterocyte surface in Arctic charr (*Salvelinus alpinus*). *Photo by Reidar Myklebust, Einar Ringø, and Rolf Erik Olsen, with permission.*

The proteins responsible for this transport have only begun to be revealed, but we should expect much more information to become available in future. For the most part, the transporters that have been identified seem to be conserved through evolution and between species. The main intestinal fatty acid transport proteins appear to be scavenger receptors class B, SR-B1 and SR-B2 (also known as cluster of differentiation CD36), and one or several of six fatty acid transport proteins (FATP1-6). SR-B2 is often referred to as multifunctional as it binds a wide range of ligands, recognizing similar molecular patterns rather than specific epitopes in the ligands, and is involved in a range of processes including immune function. Typical ligands include fatty acids, lipoproteins, and collagen (Cifarelli and Abumrad, 2018). The protein promotes the partitioning of fatty acids into the outer leaflet of the lipid bilayer. After making their way to the inner leaflet (flip-flop) the fatty acids are bound to a cytoplasmic fatty acid binding protein ($FABP_c$). SR-B2 appears to facilitate the transport by providing a docking site for $FABP_c$ or for enzymes that act on fatty acids such as acyl-CoA synthetase. SR-B2 also shows protein–protein interactions with plasma membrane $FABP_{pm}$, which suggests that clusters of membrane proteins may participate in fatty acid uptake (Glatz and Luiken, 2017). SR-B1 shares many of the properties of SR-B2, but appears to have a wider specificity and also has a role in absorption of cholesterol (Cifarelli and Abumrad, 2018). In line with their conservation with evolution, SR-B1 and SR-B2 have been cloned and/or characterized in several fish species (Sundvold et al., 2011; Zhang et al., 2013b; Fink et al., 2015; Tian et al., 2017b; Zhou et al., 2018b; Huang and Cai, 2019; Ou et al., 2019). Sr-b1 and its paralog Sr-b1-2 are highly expressed in the intestine of Atlantic salmon (Kleveland et al., 2006; Sundvold et al., 2011) and moderately in rare minnow (*Gobiocypris rarus*) when compared to other tissues (Ou et al., 2019). Sr-b1 does not appear to be regulated by fatty acids in Atlantic salmon (Kleveland et al., 2006), but there are few data available from other species. The expression of Sr-b2 in fish appears to vary somewhat with studies. In grass carp (*Ctenopharyngodon idella*), intestinal expression was reported to be very high when compared to other tissues (Tian et al., 2017b), while being similar in all tested tissues in another study (Zhou et al., 2018b). Intestinal expression of Sr-b2 was reported to be very low in carp (*Cyprinus carpio*) and zebrafish (*Danio rerio*) (Fink et al., 2015) or, conversely, high in zebrafish

(Wang et al., 2019c). In amphioxus (*Branchiostoma japonicum*), Sr-b2 was high in hepatic caecum and intermediate in hindgut (Zhang et al., 2013b). The dietary regulation of Sr-b2 is also unclear. Intestinal mRNA expression was not affected in Atlantic salmon juveniles when dietary fishmeal (FM) was replaced by soybean meal from first feeding (Sahlmann et al., 2013), while transcriptomic data suggested upregulation of Sr-b2 when juveniles were fed VO compared to FO (Jin et al., 2018d). In line with these data, transcriptomic data revealed that Sr-b2 in pyloric caeca was upregulated in larger fish fed plant meals, antinutrients, or insect meals (Gu et al., 2014; Li et al., 2020d). In rainbow trout, expression was modest in all intestinal segments and much lower than in liver (Sánchez-Gurmaches et al., 2012). In Senegalese sole juveniles, intestinal Sr-b2 was more sensitive to elevated lipid levels than oil source (Bonacic et al., 2016). In zebrafish, intestinal Sr-b2 was reduced with increased lipid level (6%–18%), and was upregulated in fish fed olive oil compared to FO, soybean oil, or palmitic acid (Wang et al., 2019c). The reason for these discrepancies awaits further research.

After entering the enterocytes, FFA and acylglycerols including MAG are transported to subcellular compartments (ER) or metabolic sites by cytosolic fatty acid binding proteins (FABP). Mammalian enterocytes contain two proteins termed liver FABP or FABP1 and intestinal FABP or FABP2 that are generally assumed to be the main fatty acid transporters. The names were originally given from the tissue they were first identified. The different FABP appear to have specific, independent functions as knockout of FABP1 does not lead to a compensatory increase in FABP2 activity (Rodriguez Sawicki et al., 2017) suggesting different transport pathways. FABP1 binds up to two fatty acids and has a similar affinity for saturated and unsaturated fatty acids (Richieri et al., 1994). FABP1 also has a broad range of ligands including bile salts, lyso-PL, cholesterol, acyl-CoA, MAG, and endocannabinoids (Cifarelli and Abumrad, 2018). FABP2, on the other hand, has only one binding site for long-chain and, possibly, SFAs (Richieri et al., 1994; Hodsdon et al., 1996; Storch and Corsico, 2008). It has been suggested that FABP2 channels fatty acids for TAG synthesis while FABP1 transports MAG for TAG synthesis and is involved in β-oxidation (Yen et al., 2015).

FABP has proven rather different and more complicated in fish compared to other animals. The mammalian distribution of *fabp* is only found in primitive fish like the spotted gar, *Lepisosteus oculatus* (Venkatachalam et al., 2018). Teleosts went through a third whole genome duplication (WGD) 230–400 mill years ago, and salmonids went through a fourth WGD 25–120 mill years ago. This led to a massive increase in potential copies of *fabp*, some apparently performing different functions in cells, while others may have been lost through nonfunctionalization (Venkatachalam et al., 2017). For example, spotted gar (*Lepisosteus oculatus*) intestine (second WGD) appears to have only one copy of *fabp1* and *fabp2* (Venkatachalam et al., 2018), zebrafish (third WGD) has *fabp1a/fabp1b.1/fabp1b.2* and *fabp2*, while salmonids (fourth WGD) have at least *fabp1a/fabp1b, fabp2a.1/fabp2a.2/fabp2b* (Bayır et al., 2015; Venkatachalam et al., 2017). Another significant feature in fish is the tissue-wide expression of *fabp* genes. In spotted gar intestine, transcripts for seven *fabp* genes have been found, *fabp1* (moderate), *fabp2* (high), *fabp3* (low), *fabp6* (high), *fabp7* (low), *fabp10* (low), and *fabp11* (low) (Venkatachalam et al., 2018). In rainbow trout, there are 14 *fabp* genes. In intestinal segments, relatively high transcript abundances were found for *fabp2a.1, fabp2a.2, fabp2b, fabp3, fabp6*, and *fabp11b*, while very low abundances were found for *fabp1b, fabp7b.1, fabp7b.2, fabp10a.1*, and *fabp11a* (Bayır et al., 2015). Transcripts of *fabp2a.1* were mainly localized to pyloric caeca and stomach while *fabp2a.2* transcripts were mainly localized to pyloric caeca.

Fabp6 was mainly found in stomach, while *fabp11b* was mainly found in caeca. The pattern of numerous *fabp* gene transcripts in fish intestines appears common between species although the pattern of transcript expression may change. For example, in intestines, transcript levels of *fabp3* were very low in rainbow trout but high in spotted gar (Bayır et al., 2015; Venkatachalam et al., 2017, 2018). In Japanese seabass (*Lateolabrax japonicus*), *fabp1*, *fabp2*, and *fabp3* have the highest level of mRNA in the intestine, followed by *fabp4* and *fabp7* (Xu et al., 2017b).

There is relatively little information on the function of the different Fabp and their nutritional regulation. Hake (*Merluccius hubbsi*) Fabp10, which is only expressed in liver, has a high affinity for bilirubin and lyso-PL, but a low affinity for fatty acids (Crovetto and Córdoba, 2016). Knockout of *fabp2* in zebrafish led to an accumulation of intestinal TAG and reduced LC-PUFA level, suggesting impaired fatty acid uptake and transport (Zhao et al., 2020b). There are several levels of regulation of Fabp. One major pathway is through peroxisome proliferator-activated receptor (Ppar), mainly Pparα and Pparγ, where differential and tissue specific regulation has been observed for Fabp1a and Fabp1b.1, and various forms of Fabp7, Fabp10, and Fabp11 (Laprairie et al., 2017; Venkatachalam et al., 2017). In zebrafish, diets high in ALA increased intestinal Fabp1b.1 and Fabp7b, but not Fabp1a, Fabp7a, or Fabp11*a* (Karanth et al., 2009). In another trial, feeding an ALA-rich diet increased intestinal Fabp2 mRNA, particularly when compared to a low lipid diet, but Fabp3 and Fabp6 were not affected (Venkatachalam et al., 2013). Long-term feeding of oxidized FO to loach (*Misgurnus anguillicaudatus*) reduced intestinal Fabp2 transcripts (Zhang et al., 2019c) and, in gilthead sea bream, Fabp2 was reduced by increased carbohydrate while exchanges of VO and FO had no effect (Castro et al., 2016a). In general, dietary lipid does not appear to be a major regulator of Fabp in fish.

Fatty acid transport proteins (FATP), also known as solute carrier family 27, contains six members (FATP1-6). The exact mechanism of their action is still unresolved, but the products are CoA esters that trap fatty acids inside cells. Some mammalian data support the possibility that they are acyl-CoA synthetases directly esterifying incoming fatty acids to acyl CoA esters. Other data suggest that they are transmembrane proteins that work in concert with acyl-CoA synthetases. Recent work suggests that mammalian FATP4, the only member to be expressed in the intestinal brush border membrane, is localized to the ER where esterification drives increased fatty acid uptake (Xu et al., 2012; Cifarelli and Abumrad, 2018).

Fish seem to express a variety of Fatp proteins in the intestine. Furthermore, due to the gene duplications that have taken place in fish, more variants of Fatp are found in fish. Fatp1-α, Fatp1-β, Fatp4, and Fatp6 have been cloned and characterized in Japanese seabass (Xu et al., 2017c). The Fatp6 appeared not to have a transmembrane domain and may not be involved in transmembrane lipid transport. In this study, Fatp4 was reported to be the predominant intestinal Fatp, with lower mRNA levels of Fatp1-α. The expression of Fatp4 mRNA increased with increased doses of ARA, while the effect on Fatp1-α was less pronounced (Xu et al., 2017c). In zebrafish, *fatp1-β* and *fatp4* were the major intestinal *fatp* genes followed by relative high expression of *fatp2*. There is also lower but appreciable expression levels of *fatp1-α* and *fatp6* giving a total of five *fatp* genes expressed in zebrafish intestine (Wang et al., 2019c). The expression of *fatp1-β*, *fatp2*, and *fatp4* genes was reduced with increased dietary lipid level, while that of *fatp1-α* tended to increase, which could indicate reduced fatty acid uptake. In a study by Wang et al. (2019c), *fatp* expression was compared in zebrafish fed diets supplemented with either FO, palmitic acid, olive oil, soybean oil, or perilla oil. Fish fed FO generally had low expression of all *fatp* mRNAs, and replacing FO with palmitic acid, olive oil, and perilla oil, but not soybean oil (rich in LNA) increased expression of *fatp1-α*. Expression of *fatp2* and *fatp6* increased in fish fed soybean oil only, while *fatp4* expression was increased in fish fed palmitic acid. *fatp4* is also well expressed in the intestines of yellow catfish, *Pelteobagrus fulvidraco* (Chen et al., 2018). In Atlantic salmon expression pattern of *fatp2* and *fatp4* did not appear to be affected by replacement of FO with VO (Liland et al., 2018).

5.4.2.1 Lipid digestibility and fatty acid uptake

Lipid digestibility is the sum of processes occurring until the lipid is absorbed into enterocytes and on its way to target tissues. In fish it is usually studied as apparent digestibility where fecal matter is analyzed for remaining nutrients and the digestibility calculated via the accumulation of an indigestible marker in the feces. The availability of lipid and individual fatty acids will depend on several environmental factors, the food matrix used, selectivity in lipolytic activity, the type of lipid used, the fatty acid composition of the lipid, and their micellar solubility and selectivity in membrane transport. Notably, it is difficult to make direct comparisons between studies given the lack of standardization in dietary markers, reference diets, and fecal sampling methodology (Glencross et al., 2007). However, some intriguing linear models have been proposed to predict the influence of each of these factors on lipid digestibility (Sales, 2009).

Regardless of approach, it is well established that the apparent digestibility of LC-PUFA is generally highest of all practical fatty acids, increasing with increased unsaturation, and possibly aided by selective uptake of some EFA. Apparent digestibility of MUFA is also generally very high, with SFA having the overall lowest bioavailability. Furthermore, shorter chain fatty acids generally have better availability than longer chain fatty acids. Typically, the digestibility of PUFA is around 95%, with MUFA up to C_{18} being fairly similar, and then decreasing with increasing chain length. Digestibility of SFA are usually around 85%–95%, but there is more variability depending on several factors (Olsen and Ringø, 1997; Hansen et al., 2001; Ng et al., 2004; Trullàs et al., 2017; Betancor et al., 2018). An interesting discussion on this can be found in a study on Murray cod (*Maccullochella peelii peelii*) fed graded levels of VO (Francis et al., 2007). Various modeling approaches have shown that SFAs are one major cause for deceased apparent lipid digestibility in salmonids (Hua and Bureau, 2009). For example, apparent fatty acid digestibility in Atlantic salmon was only 45% when fed a palm oil supplemented diet (45% SFA) compared to 91.4% and 90.2% when fed capelin (20.8% SFA) and sunflower oil (12.5% SFA) diets, respectively (Torstensen et al., 2000). SFA availability in feeds can, however, be increased by synergistic interaction with unsaturated fatty acids (Hua and Bureau, 2009). In Atlantic salmon, apparent digestibility of 16:0 increased from 69.1% to 80.7% when the inclusion level of palm oil was reduced from 25% to 10% of the dietary lipid source (Ng et al., 2004). Furthermore, when Atlantic salmon were fed a diet with 84% of the lipid as palm oil, apparent digestibility of 16:0 was only 15.7%. When the palm oil was replaced by capelin oil containing 20.8% 16:0, its digestibility increased to 84.8% (Torstensen et al., 2000).

In mammals, having SFA in the *sn*-2 position of TAG increased SFA digestibility given they are subsequently absorbed as 2-MAG following pancreatic lipase digestion. However, no effect on fatty acid digestibility was observed in rainbow trout and gilthead seabream fed reesterified palm and rapeseed oils with higher SAT in the *sn*-2 position (Trullàs et al., 2015, 2016). This further indicated that uptake of 2-MAG may not be a major route of lipid digestion in these fish species.

In recent years, there has been much research into replacement of dietary FO with other TAG-rich oil sources in aquaculture. Provided that the SFA content is controlled, the lipid digestibility of most of these oils is good, or better, than FO controls. In Atlantic halibut (*Hippoglossus hippoglossus*), digestibility of flax, canola, and sunflower oils were reported to be around 93% and much better than poultry fat and herring oil (89%) (Martins et al., 2009). Part of this could be ascribed to a relatively high SFA content in herring and poultry oils (27%–29%) compared to the VO (10%–12%). In rainbow trout, digestibility of rapeseed oil with 11.1% SFA was better than that of the FO control containing 24.3% SFA (Trullàs et al., 2015). In Atlantic salmon, Camelina oil had similar digestibility as the FO control (around 94%) (Betancor et al., 2018). There was, however, a reduction in digestibility to 90.7% when regular Camelina oil was exchanged with transgenic *Camelina sativa* oil containing over 18% EPA+DHA. The issue could also be related to elevations in SFA in the latter diet. In European seabass, well-formulated VO mixes were reported to give over 99% fatty acid digestibility (Castro et al., 2015a). Furthermore, in the same species, up to 75% of dietary FO could be replaced with land animal fats high in MUFA without affecting lipid digestibility (Monteiro et al., 2018). In silvery-black porgy (*Sparidentex hasta*), replacing 50% of dietary FO with tallow (high in SFA), apparent lipid digestibility was similar (96%–98%) to that when FO was replaced with canola oil or sunflower oil. Replacing 100% of dietary FO with tallow led to only a marginal and nonsignificant decrease in lipid digestibility to 94.6% (Mozanzadeh et al., 2016). However, it should be noted that the temperature in this trial ranged from 25 to 31°C. Tallow has received increased attention from the salmon industry, particularly in Tasmania (Australia), with positive results being reported (Emery et al., 2014, 2016). In Tasmania, salmon are farmed at relatively higher water temperatures compared to other countries, and the marginally reduced digestibility resulting from tallow inclusion did not compromise performance and, in addition, resulted in improved utilization and retention of n-3 LC-PUFA. More recently, the positive effects of the possible inclusion of tallow in different seasons and, therefore, at different water temperatures, was further investigated (Mock, 2018). In this work, an important comment was raised with respect to the use of dietary lipids with a lower digestibility: when diets with different oils are compared, they should be formulated not to be iso-lipidic, but to provide the same amount of digestible lipid, as performed routinely for protein sources. This approach would provide more useful information about the actual potentiality of different oils. Similarly, the industry could use this approach when using oils rich in SFA and, therefore, with lower lipid digestibility. Of the more recent alternative lipid sources, 20% substitution of FO with oil fractions from black soldier fly larvae (*Hermetia illucens*) had no effect on lipid digestibility in rainbow trout (Dumas et al., 2018). The same was the case in Atlantic salmon juveniles when up to 100% FO was replaced with lipid extracts from heterotrophic microalgae *Schizochytrium* sp. (Tibbetts et al., 2020).

Delivery matrices are also important for lipid digestibility. New lipid sources including WE from lower trophic level animals have been suggested as alternatives to FO as n-3 LC-PUFA sources in fish feeds (Olsen et al., 2011). However, as noted above, the lipolytic activity toward WE is low in salmonids (Bogevik et al., 2008b), and its intrinsic hydrophobicity would pose a problem in digestive micellar structures. In line with this, oils rich in WE from the calanoid copepod *Calanus finmarchicus* have been shown to be digested more poorly by Atlantic salmon than FO-derived TAG (Oxley et al., 2009; Bogevik et al., 2010). This reduction of apparent digestibility was particularly noticeable at high inclusion levels where WE contributed more than 40% of the dietary lipid. It is also noticeable that Atlantic salmon, to some extent, is capable of compensating for reduced digestibility by increasing biliary volume and intestinal lipolytic activity. This increase is particularly evident at intermediate levels (30%–37%) of WE (Olsen et al., 2004; Oxley et al., 2009; Bogevik et al., 2010). Furthermore, long-chain fatty alcohols like 22:1n-11 tend to have lower digestibility than the corresponding fatty acids, while shorter chain saturated alcohols often have higher digestibility (Olsen et al., 2004; Oxley et al., 2009; Bogevik et al., 2010; Colombo-Hixson et al., 2011). The enterocyte lipid metabolism is however not affected by these diets (Oxley et al., 2005a).

The digestibility of dietary lipids is also under the influence of numerous nutritional, environmental, and species-dependent factors. Increasing water temperatures are generally assumed to improve lipid digestibility (Hua and Bureau, 2009), although data in some cases may not be as clear as expected (Bureau et al., 2008). The improvement is particularly prominent for SFA and is believed to be related to melting point limitations and incorporation into mixed micelles. In Atlantic salmon fed 25% palm oil, fatty acid digestibility increased from 84.6% to 87.3% and SFA digestibility increased from 64.4% to 72.0% when the temperature increased from 6 to 11°C (Ng et al., 2004). Similar tendencies were found in salmon maintained at 3 and 12°C where total fatty acid digestibility increased marginally from 95.9% to 96.5%, and SFA digestibility increased from 90.8% to 93.1% when fed a FO diet. In juvenile Atlantic salmon in freshwater, lipid (and energy) digestibility increased from 94.6% at 10°C to 96.2% at 20°C when fed a combined EPA/ARA enriched diet (Huguet et al., 2015). The digestibility of 16:0, the predominant SFA, increased from 85.8% to 92.0% while EPA digestibility (99.4%–99.5%) was unaffected. The effect was even more striking in Atlantic salmon fed WE-rich copepod oil from *Calanus finmarchicus* at 3 and 12°C. In this case, fatty alcohol digestibility increased with temperature from 73% to 81% with a significant effect also on saturated acids and alcohols (Bogevik et al., 2010). It is therefore likely that many high temperature species, such as silvery-black porgy mentioned above (Mozanzadeh et al., 2016), can be fed higher amounts of cost-effective saturated lipids such as lard from mammalian by-products without impacting performance.

Over the past decades, there has also been a drive toward replacement of dietary FM with other protein sources including plant meals, insect larvae, krill, copepods, microalgae, macroalgae, and animal by-products. Each of these have several challenges that may have major effects on nutrient digestibility and fish health. Many will affect the digestive process as a whole and will therefore reduce nutrient digestibility, including lipid, while others may target lipid utilization more specifically (Bureau, 2006; Drew et al., 2007; Henry et al., 2015; Krogdahl and Bakke, 2015; Sprague et al., 2017; Nogales-Mérida et al., 2019).

Early replacements for FM explored in the 1990s consisted mostly of plant proteins and, in particular, soymeal. Many plant proteins contain carbohydrates that are not digested in carnivorous animals, thus lowering their nutritive value (Gatlin et al., 2007). It also soon became evident that fish were much more sensitive to the plethora of antinutritional factors found in soybeans and many other plant proteins than mammals (Francis et al., 2001; Krogdahl and Bakke, 2015). Such diets could lead to inflammatory, enteritis-like, conditions and a general reduction in the digestibility of nutrients. In Atlantic salmon, lipid digestibility was reduced from 91.2% to 78.7% along with reduced digestibility of protein and energy when 35% of FM protein was replaced with standard soybean meal (Krogdahl et al., 2003). To reduce the level of antinutrients, some research has focused on the more expensive soybean concentrates, including alcohol-extracted defatted soybean meal. Some antinutrients like saponin are more or less eliminated while many others are reduced, except for phytin which prevents uptake of phosphorous (Zhou et al., 2018c). Soybean concentrates do however retain significant amounts of antinutrients, impacting feed efficiency if included in higher amounts in dietary formulations (Drew et al., 2007). Other plant protein sources may be less complicated than soybeans but many, such as canola, pea, and lupin, contain various antinutrients and carbohydrates that requires some form of processing to improve utilization in fish (Drew et al., 2007). A common factor appearing to reduce lipid digestibility specifically across various plant protein sources is cellulose. In Atlantic salmon, lipid digestibility was reduced in a linear manner with increased dietary cellulose content (Aslaksen et al., 2007). However, in a contemporary study, replacement of cellulose with soybean nonstarch polysaccharides at levels of 10% of the diet had no major effect on lipid digestibility, which was around 97%–98% (Kraugerud et al., 2007).

There has also been a major research effort into exploring the suitability and usability of various protein treatments with less complicated antinutrients, including isolates, which are water extracts of meals (Glencross et al., 2004). Recently, the breeding of special cultivars devoid of antigens like Kunitz trypsin inhibitor, lectin, and P34/Gly m Bd 30 k antigen have been shown to improve lipid digestibility (but lower protein digestibility and enteritis) in Atlantic salmon when compared to soybean concentrate (Krogdahl et al., 2020). However, most alternatives still pose challenges when used at high inclusion levels compared to the "gold standard" of FM, often resulting in impaired nutrient and lipid digestibility. Fermentation of plant protein is an efficient method of removing antinutrients (Dawood and Koshio, 2020). However, various forms of fermentation and bioprocessing still lower digestibility of many nutrients in several fish species when the alternative protein level approaches 40%–50% of the total dietary protein (Refstie et al., 2005). The sensitivity to plant proteins also varies with fish species and there are some major differences even within the salmonid family with respect to the tolerance of many of these proteins (Glencross et al., 2004).

A further area of interest has been in protein and lipid sources that do not compete with human nutrition and may also contain lower antinutrients. These include insects and single cell organisms such as yeast, bacteria, macroalgae, and microalgae (Roy and Pal, 2015; Hua et al., 2019; Jones et al., 2020). Their main advantage is that they can be grown on nonedible biomass. Microalgae is also cultured as an alternative source of n-3 LC-PUFA (Ferreira et al., 2019). Most algae are good sources for EPA only, but some species, *Emilia huxleyi* and *Pavlova lutheri,* will also produce DHA (Harwood, 2019). Microalgae are good protein and lipid sources but cell walls in some species such as *Nannochloropsis* will contain higher levels of indigestible carbohydrates that will reduce total digestibility. Many microalgae also contain thick cell walls that require crushing to increase nutrient availability (Shah et al., 2018). The best effect is often seen at moderate inclusion levels, typically 10%–20% of total protein. In Atlantic salmon, lipid digestibility fell progressively from 95.8% to 86.4% when whole cell *Chlorella vulgaris* replaced FM from 0% to 30%, while corresponding values for crushed *C. vulgaris* ranged from 95.8% to 90.7% (Tibbetts et al., 2017). There was also a reduction in apparent protein and energy digestibility. Lipid digestibility was also reduced in salmon fed a diet supplemented with 10% of preextruded *Nannochloropsis oceanica* (Gong et al., 2020). Curiously, using up to 6% of the biomass with heterotrophic microalgae *Schizochytrium limacinum* (Kousoulaki et al., 2020) led to a significant reduction in SFA digestibility from 80.5% to 62.7%, but a significant increase in MUFA and LC-PUFA digestibility. Reduction of lipid and nutrient digestibility in fish fed high levels of insect meals is often attributed to chitin that has been shown to reduce digestibility in several fish species like rainbow trout and tilapia (*Oreochromis* sp.) (Henry et al., 2015). Chitin was also suggested as the causative factor for the reduced lipid digestion in Atlantic salmon fed high levels of chitin-rich krill meal from *Euphausia superba* (Olsen et al., 2006). However, a later study showed that feeding Atlantic cod and Atlantic salmon up to 5% purified chitin derived from shrimp did not affect lipid digestibility, with Atlantic cod appearing capable of chitin digestion (Karlsen et al., 2010). It is therefore likely that there are species differences or different chitin matrices that affect lipid digestibility rather than the chitin itself.

5.4.2.2 TAG and PL resynthesis and intracellular fates

After transport to the ER, TAG resynthesis may proceed by one of two pathways, the MAG acyltransferase (MGAT) and glycerol-3-phosphate acyltransferase (GPAT) pathways. The mammalian MGAT pathway is assumed to use 2-MAG as template and starts with MGAT using acyl-CoA thioesters as substrate to produce 1,2-DAG. This can either be converted to TAG by DAG acyltransferase (DGAT), or to a PL, most commonly phosphatidylcholine (PC) via the cytidine diphosphate (CDP)-choline pathway through the action of choline kinase and choline/ethanolamine phosphotransferase, with the former as the rate limiting enzyme (Tocher et al., 2008) (Fig. 5.7).

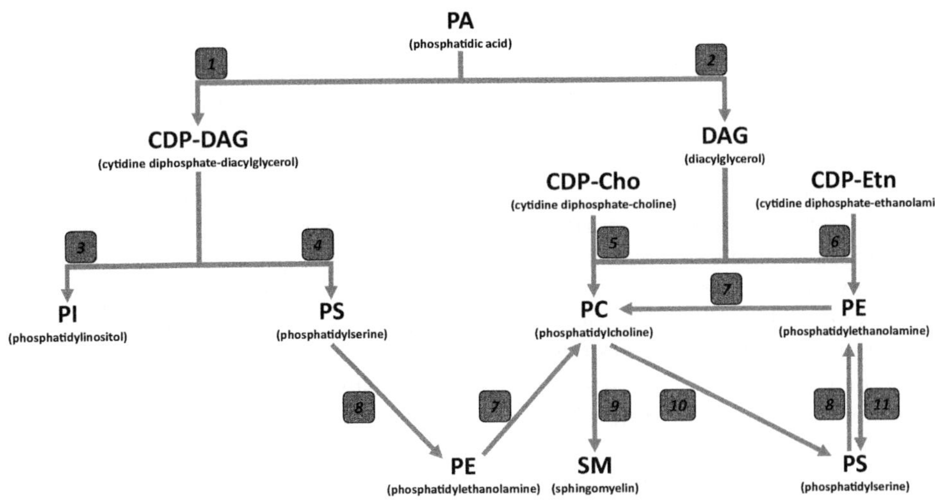

FIGURE 5.7 Biosynthetic pathways for the major phospholipid classes. The enzymes are: (1) CDP–DAG synthases, (2) PA-phosphatases, (3) PI synthase, (4) PS synthase, (5) CDP-choline: DAG phosphotransferase, (6) CDP-ethanolamine: DAG phosphotransferase, (7) PE methyltransferase, (8) PS decarboxylase, (9) SM synthase, (10) and (11) PS synthases via base exchange.

If the MAG template is not available, as with WE feeding, or if there is complete degradation of dietary TAG, then TAG must be resynthesized through the GPAT pathway. This pathway uses glycerol-3-phosphate (G-3-P) as template, which is acylated to *sn*-1-lysophosphatidic acid by G-3-P acyltransferase, followed by another acylation to phosphatidic acid by the *sn*-1-acylglycerol-3-phosphate O-acyltransferase enzyme (AGPAT). Phosphatidic acid is then dephosphorylated by phosphatidic acid phosphohydrolase to produce DAG, which can then be processed to TAG or PC by the DGAT or CDP choline pathways, respectively.

Most fish species appear to possess all the enzymes for the MGAT, GPAT, and CDP-choline pathways, although the contribution of each pathway to TAG resynthesis and possible species differences are unclear. As noted above, luminal lipolytic digestion probably produces FFA, and a mixture of MAGs rather than 2-MAG (Bogevik et al., 2008a). Complete hydrolysis to glycerol and FFA would mean that the GPAT pathway should predominate in fish. This does not seem to be the case in Atlantic salmon as the rate of TAG and PC resynthesis proceeds much faster through the MGAT pathway using 2-MAG as a substrate compared to the GPAT pathways using FFA and glycerol (Oxley et al., 2007). The alternative is that lipid reassembly proceeds using a variety of 1(3) and 2-MAG. As MGAT is normally positional specific, the DAG products would consist mainly of 1,3-DAG and some 1,2-DAG and, although DGAT can utilize 1,2-DAG for TAG synthesis, DGAT is not known to utilize 1,3-DAG. Accordingly, 1,3-DAG must be processed through an alternative route to TAG, but how this occurs in fish is not clear. Studies in brown trout enterocytes showed that 1(3)-MAG is converted to TAG without significant accumulation of 1,3-DAG, albeit at a much slower rate than with 2-MAG (Li and Olsen, 2017). Incubation of microsomes with 1-MAG only produced 1,3-DAG in Atlantic salmon (Oxley et al., 2005b). It is therefore possible that TAG synthesis proceeds through some rearrangement or reacylation, or that one or more MGAT/DGAT has a wider specificity than previously thought. In some mammals for example, MGAT3 has a unique and significant DGAT activity and can produce TAG from 1(3)-MAG (Cao et al., 2007). Genes for Mgat3 have so far been reported in zebrafish (Holmes, 2010) and Atlantic salmon (Król et al., 2016), but their specificities are not known.

Once TAG has been resynthesized in the ER, it is thought to be released into the PL bilayer creating a monolayer PL membrane with a core of TAG and, as this grows, it will be partitioned into the cytosol and will have one of two fates. It can bud off into the cytosol as lipid droplets, or it can be processed further and become incorporated into lipoproteins and exocytozed. Lipid droplets serve as intracellular lipid stores to be incorporated into lipoproteins in a fasting state. They will expand in size requiring more PL to prevent their coalescence. The single PL membrane of lipid droplets also contains many TAG synthesis enzymes including acyl-CoA synthetases, MGAT and GPAT, as well as perilipin proteins that protect the droplets from lipases. Microsomal transfer protein (MTP) is also implicated in lipid droplet formation in mammals (Yen et al., 2015; D'Aquila et al., 2016; Carreiro and Buhman, 2019) and probably also in fish.

Lipid droplets are regularly seen in enterocytes of many fish species after feeding. In fish like Arctic charr the droplets can fuse to form very large vesicles that can fill more than 50% of the epithelial volume (Olsen et al., 2000). There appears to be significant dietary influences involved as dietary PL reduce or eliminate such large aggregates in salmonids (Olsen et al., 1999, 2000, 2003; Taylor et al., 2015), carp (Fontagne et al., 1998), zebrafish (Sæle et al., 2018), and gilthead seabream (Salhi et al., 1999), and similar effects have been seen in Atlantic salmon fed choline to increase PC synthesis (Hansen et al., 2020). This suggests that exogenous PL, probably under high lipid loads, is required for proper lipoprotein assembly and synthesis, and dietary PL deficiency leads to surplus lipid being accumulated in cytosol. It has been suggested that the accumulation occurs at the ER level where low levels of PC prevents the formation of the apoB particle and halts further lipoprotein synthesis (Sæle et al., 2018). The tendency of lipid droplet accumulation appears to be highest in larval or juvenile fish (Salhi et al., 1999; Taylor et al., 2015), partly explained by immature and insufficient enterocyte PL synthesis in juvenile fish (Carmona-Antoñanzas et al., 2015; Jin et al., 2018d), indirectly leading to temporal PL deficiency. Lipid droplet accumulation is also affected by dietary fatty acids, with replacement of FO (and meals containing PL) with VO stimulating enterocyte lipid droplet accumulation, which can be reversed by supplementation of 16:0, which is often low in plant oils (Olsen et al., 2000). It has therefore been suggested that a lack of 16:0 could prevent endogenous PC synthesis as the pathway may require SFA in the *sn*-1 position of PC. Other studies have shown that OLA and LNA, characteristic of VO, stimulate TAG synthesis over PC synthesis in Atlantic salmon intestine while 16:0 to some extent stimulates PC synthesis (Oxley et al., 2007).

5.4.2.3 *Lipoprotein synthesis and export*

TAG destined for export are embedded into a lipoprotein with chylomicrons being the main intestinal lipoprotein followed by HDL (high-density lipoprotein) in mammals. Chylomicrons are formed in a two-step pathway. First, apolipoprotein B48 (apoB48) is translated and translocated into the ER lumen. MTP packs cholesterol, PL, and some TAG onto the apoB48 protein forming prechylomicron particles (Cifarelli and Abumrad, 2018; Carreiro and Buhman, 2019). Then, large non-apoB48 neutral lipid assemblages accumulating in ER lumen are transferred to the growing prechylomicron particle in a process involving MTP. The prechylomicron is then transported by budding from ER to Golgi in prechylomicron transport vesicles (PCTV) with the budding initiated by a protein complex including VAMP7, CD36, ApoB48, and FABP1. After fusion with the Golgi, prechylomicrons undergo further processing including glycosylation of apoB48, and apolipoprotein A1 (apoA1), originally synthesized in the ER, is transported to Golgi in COPII vesicles where it fuses with prechylomicrons. The mature chylomicrons also require apoA-IV before exiting into the cytosol and eventually being exocytozed from the cell and transported through the body (Cifarelli and Abumrad, 2018). Mammalian intestine is also a major producer of HDL, which is the main transporter of cholesterol. Initially ApoA-I is synthesized in enterocytes and then interacts with ATP-binding cassette subfamily A member 1 (ABCA1) that transfers PL and cholesterol in an MTP-independent manner to ApoA-I (Brunham et al., 2006). Free cholesterol must be esterified to enable it to move from the surface to the core of the maturing HDL, and the enzyme lecithin-cholesterol acyltransferase (LCAT) facilitates the transfer of fatty acids from PL to cholesterol. The serum lipoprotein pattern is affected both by dietary lipid load and the type of fatty acid. High lipid diets increase chylomicron and very-low density lipoprotein (VLDL) content and reduce HDL, and tend to increase low-density lipoprotein (LDL) size (Roche and Gibney, 2000; Guay et al., 2012). SFA, MUFA, n-6 PUFA, and n-3 PUFA have significant effects on lipoprotein profiles. For example, n-3 LC-PUFA reduce LDL and VLDL and increase HDL (DiNicolantonio and O'Keefe, 2018) and FO in particular will increase HDL size/numbers and reduce VLVL and LDL (Erkkilä and Lankinen, 2016).

The synthesis of lipoproteins in fish appears to have many similarities to that found in mammals and, although details are lacking, it is expected that more information will emerge with the increasing availability of molecular tools (e.g., CRISPR/CAS8, proteomics, RNAseq). Emerging studies in zebrafish confirm that at least parts of the lipoprotein assembly process are similar to that of mammals (Sæle et al., 2018; Morishita et al., 2019). While some earlier reports indicated that lipoproteins did not need processing in Golgi apparatus before exocytosis, it was shown that mutant zebrafish with impaired Golgi complex had severely impaired vesicular trafficking and, ultimately, lipid transport (Ho et al., 2006). One significant difference from mammals is the inability of fish intestine to synthesize apoB48 that is essential for chylomicron formation. ApoB48 is synthesized following posttranscriptional editing of full-length apoB100 mRNA by introducing a stop codon that terminates translation. The mRNA editing complexes or mRNA codons involved have not been found in nonmammalian vertebrates (Damsteegt et al., 2018). As apoB48 lacks LDL receptor binding sites (Hinsdale et al., 2002), efficient uptake of chylomicrons depends on the highly expressed hepatic apolipoprotein E receptor (Linton et al., 1993). It has therefore been suggested that the development of apoB48 ensures more rapid clearance of TAG from the circulation (Hodges and Scott, 1992). In humans, residence time for chylomicron TAG is 5–10 min (Roche and Gibney, 2000) while it can be hours in southern black seabream, *Acanthopagrus butcheri* (Arnold Reed et al., 1997). Mutated mice producing only apoB100 have elevated blood TAG, LDL, and VLDL levels that appears to support this (Farese et al., 1996). The lack of efficient blood clearance in fish also agrees with the general observation that fish are classified as hyperlipidemic and hypercholesterolemic compared to mammals. Fish also have smaller size chylomicrons more resembling VLDL (Babin and Vernier, 1989). This means that fish intestines probably produce VLDL and HDL that is transported to the liver and other tissues as opposed to mammals where VLDL mainly delivers lipids from the liver to tissues.

In primitive fishes like chondrichthyan, class Chondrichthyes, the major blood lipoprotein appears VLDL-like (Babin and Vernier, 1989) and relatively high levels of VLDL/LDL are also seen in primitive ray-finned fish like eels (Ando et al., 2005). Higher ray finned fish, and especially teleosts, have HDL as the prime lipoprotein followed by variable contents of VLDL and LDL (Babin and Vernier, 1989; Kondo et al., 2005; Kjær et al., 2009). ApoA-I and apoA-IV were the earliest lipoproteins to appear in primitive species with later forms occurring following gene duplications. ApoE arose in Coelacanthiformes, ApoA-II in Holostei, and ApoC-I in teleosts (Liu et al., 2019b). In addition, there are several paralogs of each apolipoprotein with, for example, four paralogs of ApoA-IV have been found in zebrafish (Quinlivan and Farber, 2017). It is therefore likely that the detailed mechanisms and pathways for intestinal lipoprotein synthesis and transport will vary between species. Most mRNAs required for generating VLDL and HDL in higher fishes such as apoA-I, apoA-II, apoA-IV, apoC-I, apoC-II, apoE, and apoB have been found in species like zebrafish and salmonids (Morais et al., 2011; De Santis et al., 2015; Quinlivan and Farber, 2017; Jin et al., 2018d). It is also worth noting that many Chondrichthyes and teleosts lack albumin or similar proteins (De Smet et al., 1998; Metcalf and Gemmell, 2005; Andreeva, 2019). This means that lipoproteins must have taken over their transport role.

Blood lipoprotein composition in fish is affected by many factors, not least diet and time of postprandial sampling. In Atlantic salmon, both HDL and LDL increased when fed high-fat diets (Torstensen et al., 2001), but the effect of lipid source on lipoprotein proportions was not so clear. Some studies suggested that replacement of FO with VO reduced HDL and total protein (Torstensen et al., 2004), while others suggested a reduction of VLDL and LDL only (Jordal et al., 2007), or have not found any effect of dietary VO on lipoprotein profiles (Torstensen et al., 2000). This may indicate that lipoprotein regulation in fish is different to that in mammals. Analysis of transcription levels using QPCR, microarray or RNAseq are also inconsistent. For example, replacing FO with VO in European seabass increased ApoA-I, ApoA-IV, ApoB, and mtb (Castro et al., 2015a) while in Atlantic salmon larvae and juveniles, ApoA-I_1, ApoA-IIa_1, and mtp_1 were reduced by dietary VO compared to FO (Jin et al., 2018c). ApoA-IV also increased in zebrafish fed low PL levels (Sæle et al., 2018), which may indicate increased lipoprotein turnover, or changes in lipoprotein size. Varying dietary inclusions of EPA/DHA/ARA did not affect mRNA transcripts of lipoprotein synthesis (Liland et al., 2018; Magalhães et al., 2020). Interestingly, rainbow trout lines with fatty muscle had higher capacity for chylomicron (or VLDL) synthesis, with upregulation of ApoA-IV and mtp, than lean lines (Kamalam et al., 2013).

5.5 Lipid synthesis and deposition

5.5.1 Biosynthesis of fatty acids

While most fatty acids present in fish lipids are obtained from the diet, SFA and MUFA can also be synthesized de novo from nonlipid precursors such as carbohydrate or protein (Bou et al., 2016; Liu et al., 2019a). Generally, the basic pathways involved in the synthesis of fatty acids from nonlipid precursors in fish are similar to those occurring in mammals (Henderson, 1996), where a series of coordinated enzymatic reactions are involved in the flow of carbons to fatty acids (Ameer et al., 2014; Sanders and Griffin, 2016; Song et al., 2018b). Initially, carbohydrates or proteins undergo catabolism to generate citrate in mitochondria, which is then transported to the cytosol where ATP-citrate lyase (ACLY) releases acetyl-CoA that is converted to malonyl-CoA by acetyl-CoA carboxylase (ACC). Finally, malonyl-CoA is utilized by the multienzyme complex fatty acid synthase (FAS) to produce palmitic acid (16:0) and stearic acid (18:0) that can be subsequently desaturated to palmitoleic acid (16:1n-7) and OLA (18:1n-9).

The *acc1* gene, as the first rate-limiting enzyme of fatty acid synthesis, has been characterized in fish and is widely distributed in tissues in different species (Cheng et al., 2011; Tang et al., 2013; Zheng et al., 2013c). Similarly, the nucleotide sequence of Fas has also been cloned in many species (Wang et al., 2010; Leng et al., 2012; Tang et al., 2013; Zheng et al., 2013c). The mRNA expression of *fas* is an important index of lipogenesis, with the liver demonstrated as the organ with highest expression (Wang et al., 2010; Leng et al., 2012). Purified Fas from liver of plaice (*Pleuronectes platessa*) had high efficiency to synthesize mainly palmitic acid and stearic acid (Wilson and Williamson, 1970).

Acyl-CoA desaturases are known for their roles in synthesizing unsaturated fatty acids (Ardiyanti et al., 2012). Among them, stearoyl-CoA desaturase (SCD) is responsible for inducing the first double bond between carbons 9 and 10 of palmitoyl (16:0)- and stearoyl (18:0)-CoA to form the MUFA, 16:1n-7 and 18:1n-9, respectively (Tocher et al., 1998; Castro et al., 2011). In mammals, five isoforms of the SCD gene have been identified, *SCD1, SCD2, SCD3, SCD4,* and *SCD5*, displaying different tissue distributions (Evans et al., 2008; Castro et al., 2011). However, only *scd1* and *scd2* isoforms have been identified in fish, with both having high expression in liver and low expression in brain in grass carp (Chang et al., 2001; Polley et al., 2003; Evans et al., 2008) and large yellow croaker (*Larimichthys crocea*) (Xu et al., 2015a). In Mozambique tilapia (*Oreochromis mossambicus*), *scd1* was only expressed in the liver (Hsieh et al., 2004), whereas in zebrafish (Hsieh et al., 2003) and milkfish (Hsieh et al., 2001), *scd1* was widely expressed in various tissues. The differences in *scd/SCD* expression profiles between fish and mammals may suggest different regulatory mechanisms.

5.5.2 Biochemical synthesis of glycerolipids

5.5.2.1 Triacylglycerol synthesis and fatty acid esterification

The basic principles of TAG synthesis have been introduced above (Section 5.4.2.2), and the two alternative (MGAT and GPAT) pathways have been presented. Overall, the synthetic routes for the two pathways use different substrates. In general, the MGAT pathway mainly utilizes 2-MAG as substrate and, as this is mostly derived from the digestion and absorption of lipids in the intestinal tract, this pathway shows significant activity in the intestine both in mammals and fish (Yasushi et al., 2010; Li et al., 2016b). In contrast, the GPAT pathway is primarily responsible for de novo synthesis of TAG utilizing FFA and glycerol as substrates. Accordingly, this pathway is an energy-consuming process, which occurs mainly in liver rather than intestine (Lehner and Kuksis, 1996; Li and Olsen, 2017).

Many studies have investigated the efficiency of fatty acid esterification into TAG in different fish tissues. Atlantic salmon hepatocytes incubated with 20:5n-3 accumulated significantly more cellular lipid than cells treated with 18:1n-9 or 22:6n-3 (Vegusdal et al., 2005). Moreover, 18:1n-9 was mainly esterified into TAG while 20:3n-6 and 20:4n-3 were primarily esterified into PL (Moya-Falcón et al., 2005). Incubation with 18:3n-3 induced higher TAG esterification than 18:2n-6 in grass carp hepatocytes whereas in Nile tilapia (*Oreochromis niloticus*) hepatocytes, TAG esterification was comparable with 18:3n-3 and 18:2n-6 (Jiao et al., 2020), highlighting the specificity of fatty acid esterification in affecting TAG biosynthesis in different species. However, in aquaculture, fish are fed oils that are mixtures of TAG containing many different fatty acids. After intestinal degradation and absorption of dietary oils/TAG, how the exogenous fatty acids are reesterified into TAG is of considerable interest. Using lipidomics, a pioneer study identified the molecular species of fillet lipids in Nile tilapia fed different oils showing that high dietary n-3 LC-PUFA intake increased their deposition at *sn*-1/3 of TAG and 16:0 was distributed preferentially at the outer positions of the glycerol backbone irrespective of dietary oil used (Liu et al., 2019c). More 18:2n-6 accumulated at the *sn*-2 position of TAG while 18:3n-3 was deposited at *sn*-1/3 positions in tilapia fed n-3 PUFA-enriched oils. This study brought insight to how dietary oils change the composition and positional distribution of fatty acids on the glycerol backbone, and potentially change the nutritional value of farmed fish.

5.5.2.2 Phospholipid biosynthesis

PL have biological functions in both mammals and fish including the structural scaffold of cell membranes, cell membrane transport, and the provision of metabolic energy (Tocher et al., 2008), and further details are provided in Section 5.10.

In general, PL biosynthesis pathways in fish are similar to those in mammals (Tocher, 1995), and all the enzymes and proteins involved in PL biosynthesis have been found in fish genomes, such as fugu/tiger puffer (*Takifugu rubripes*) and zebrafish (Lykidis, 2007). As mentioned briefly above, the biosynthesis of PL can occur via two main pathways, the cytidine diphosphate (CDP)-DAG and DAG pathways (Tocher, 1995; Lykidis, 2007) (Fig. 5.7). PC and PE (phosphatidylethanolamine) are synthesized from DAG and CDP-choline or CDP-ethanolamine, respectively, through the action of choline phosphotransferase (CPT) and ethanolamine phosphotransferase (EPT) (Holub et al., 1975; Leslie and Buckley, 1976; Hazel, 1990). In contrast, PS (phosphatidylserine) and PI (phosphatidylinositol) are synthesized from CDP-DAG and serine and inositol, respectively. However, PC can also be synthesized by methylation of PE (Hazel and Williams, 1990; Oxley et al., 2005b; Oxley et al., 2005a), and PE, PS, and PC can also be interconverted (Tocher et al., 2008) (Fig. 5.7). Oxley et al. (2007) and Li et al. (2016b) found that PC can also be synthesized from 2-MAG in fish enterocytes. In Atlantic cod larvae, the 2-MAG (MGAT) pathway dominated over the G-3-P (GPAT) pathway in TAG and PC synthesis (Li et al., 2016b). Accordingly, cod larvae synthesized PC to similar levels as TAG (Li et al., 2016b), agreeing with the relatively fast reacylation process of MAG to TAG in salmon enterocytes (Oxley et al., 2005b, 2007), and enterocytes and hepatocytes in brown trout (Li and Olsen, 2017).

De novo synthesis of PL is largely dependent on the amount and type of dietary lipid, with PL and TAG having different contributions to endogenous PL synthesis in fish larvae (Tocher et al., 2008). Comparing the lipid profiles of cod larvae and their live feed, most DHA in larval PL originated from feed PL rather than feed TAG and there was no correlation between n-3 LC-PUFA content of dietary TAG and larval PL (Olsen et al., 2014; Li et al., 2015). This suggested that the principal precursor of PL synthesis in larval enterocyte is dietary PL, with lyso-PL and FFA taken up by enterocytes and reacylated to reform PL (Li et al., 2015, 2016b). Dietary choline can also affect de novo synthesis of PL. In Atlantic cod larvae fed 2-oleoyl-[1,2,3-^{3}H]glycerol and [^{14}C(U)] glycerol by tube, higher percentages of ^{3}H- or ^{14}C-labeled PC, PE, PS, and PI, and a higher PC:TAG ratio were found in fish also fed choline chloride compared to fish without choline (Li et al., 2016b).

Several studies indicated that, in the development of fish larvae, the dietary fatty acid composition and position in dietary PL, and the specificity of the reacylation enzymes, all influence the fatty acid profile of larval PL, and it has also been suggested that DHA could be selectively incorporated and retained in PL, especially PE (Lehner and Kuksis, 1992; Olsen et al., 2000, 2014; Oxley et al., 2007; Li et al., 2015, 2016b). In addition to these dietary considerations modulating de novo PL biosynthesis, it has been shown that environmental temperature also affects the proportions of PE and PC (Hazel, 1984; Dey et al., 1993; Cordier et al., 2002) and, in fish, PC biosynthesis is especially sensitive with the PC/PE ratio being positively correlated with temperature (Dey et al., 1993; Cordier et al., 2002). Compared to TAG synthesis, more biochemical pathways, intermediates, and enzymes are involved in PL synthesis. However, research into fish PL synthesis is still limited and several topics require more attention in future including (1) the effects of dietary components including fatty acid composition, TAG/PL ratio, and choline on de novo synthesis of PL; (2) studies to evaluate the precise functions of enzymes in TAG/PL synthesis and the fates of different intermediate metabolites such as DAG, MAG, and G-3-P; (3) TAG/PL synthesis in different organs, especially intestine and liver, in different fish species; (4) effects of fish developmental stage and environmental factors on TAG/PL synthesis.

5.5.2.3 *Synthesis of sterols, sphingolipids, and wax esters*

Steroids include sterols like cholesterol, bile acids, and steroid hormones (Goedeke and Fernández-Hernando, 2012). Cholesterol, necessary for eukaryotic cell growth and development, is also the precursor for the synthesis of steroid hormones, bile acids and other physiologically active compounds. Almost all cells can synthesize cholesterol but liver is the main synthetic site (Repa and Mangelsdorf, 2000). The biosynthesis pathway consists of a series of complex reactions involving more than 20 enzymes and is energetically expensive requiring significant inputs of acetyl-CoA, ATP, oxygen, and reducing factors such as NADPH and NADH. In short, acetyl CoA is converted to lanosterol via the mevalonate pathway with mevalonic acid produced by 3-hydroxy-3-methylglutaryl-CoA reductase (HMGCR), which is generally regarded as the major irreversible and rate-limiting step in cholesterol biosynthesis. Lanosterol is then converted to cholesterol via a series of sequential demethylation and double bond reductions, through either the Bloch or Kandutsch—Russell pathways (Ikonen, 2008; Lee et al., 2013). The roles and functions of cholesterol and the impacts of dietary supply in fish are further discussed in Section 5.9.

Sphingolipids are divided into two main groups, sphingophospholipids and sphingoglycolipids, and are essential components of cell membranes and play important roles in membrane trafficking, cell proliferation, differentiation, apoptosis, and cell migration (Simons and Ikonen, 1997; Hannun et al., 2001). In mammals, sphingolipids are synthesized via de novo synthetic and salvage pathways. De novo synthesis of sphingolipids begins with the condensation of palmitoyl-CoA and L-serine by serine palmitoyltransferase to produce 3-ketodihydrosphingosine, which is then reduced by 3-ketodihydrosphingosine reductase (KDSR) to generate dihydrosphingosine that is converted to ceramides by ceramide synthases (Merrill Jr, 2011; Don et al., 2014). The salvage pathway starts with the degradation of sphingomyelin (SM) or glycosylated ceramides to ceramide and then to sphingosine (Sph), which is secreted into the cytosol and used to synthesize ceramide or sphingosine-1-phosphate (S1P) (Kitatani et al., 2008). In the Golgi and ER, ceramide is converted to SM or glucosylceramide (GlcCer) by SM synthase or GlcCer synthase, respectively, and GlcCer is further converted to more complex glycosphingolipids (Ardail et al., 2003; Kohyama-Koganeya et al., 2004).

Sphingolipid biosynthesis pathways in fish are similar to those in mammals with the Kdsr enzyme being crucial in de novo sphingolipid synthesis. Mutation of the *kdsr* gene in zebrafish caused compensatory activation of the sphingolipid salvage pathway, resulting in significant accumulation of sphingolipids including ceramides, Sph and S1P (Park et al., 2019). In addition, sphingosine kinase 2 mutant zebrafish experienced toxic effects of high levels of Sph because they are unable to convert it to S1P, although mutant embryos showed increased production of ceramide synthase 2b, which converts Sph to ceramide (Mendelson et al., 2017), indicating that ceramide synthase 2b can function as a sphingolipid-responsive factor mediating a feedback regulatory mechanism. In another study in zebrafish cells, a novel mitochondrial sphingomyelinase was characterized that catalyzed the hydrolysis of SM to generate ceramide (Yabu et al., 2009). Higher concentrations of glucosylceramide and SM, and ratios of glucosylceramide/ceramide and SM/ceramide were detected in skin of n-3 LC-PUFA-deficient Atlantic salmon, showing that dietary EPA and DHA had an effect on sphingolipid synthesis, although the specific mechanism was unclear (Cheng and Russell, 2004).

WE synthesis in fish has received only limited attention. In mammals, it is known that WE synthesis occurs in two main steps with fatty acyl-CoA reductase reducing fatty acyl-CoA to fatty alcohol and CoA using NADPH followed by WE synthetase catalyzing the transesterification of fatty acid and fatty alcohol to form WE (Cheng and Russell, 2004). Carp hepatopancreas homogenates contained enzymes for the hydrolysis and synthesis of WE after activation by the addition of ATP, CoA, and NADH (Kayama et al., 1979). In myctophid fishes, considerable amounts of WE with odd-numbered fatty acids and/or alcohols were produced in liver homogenate after addition of n-heptadecanol (17:0-alcohol) and/or 10-cis-heptadecenoic acid (17:1-acid) with cofactors such as NADPH and ATP, verifying the existence of WE synthase activity (Seo et al., 2001).

5.5.3 Lipid deposition and distribution

5.5.3.1 Lipid deposition in liver

Liver is also an important site for lipid deposition in fish, although the contribution of liver to whole body lipid storage capacity varies greatly with species. For example, in Atlantic cod, Atlantic halibut and tiger puffer where liver is the main energy storage organ, liver lipid content can reach more than 50% under natural conditions (Holdway and Beamish, 1984; Kikuchi et al., 2009; Zeng et al., 2010). In contrast, liver lipid contents in Atlantic salmon and rainbow trout are normally less than 5%, much lower than the lipid contents of muscle (Bou et al., 2017; Shahpar and Johari, 2019), and most other farmed fish have intermediate lipid contents (Table 5.2). Liver lipid deposition is related to age, sex, seasonal variation, and especially the abundance of food. In Atlantic halibut, liver lipid content of 800 g adult fish was significantly higher than that of 50 g juvenile fish, and hepatic lipid accumulation in the older fish was not sensitive to energy intake (Hazra et al., 1998). In zebrafish, liver lipid metabolism was different between male and female fish, with females having higher lipogenic activities and fat deposition than males (Sun et al., 2020). Interestingly, in male zebrafish exposed to estrogens, liver lipid metabolism converted to the female pattern and they accumulated more lipid (Sun et al., 2020). Liver lipid content also varies with gonad development and spawning period. For example, in female Arctic charr, liver lipid content increased from 3.5% to 14.7% during ovarian development and then decreased to 3.1% in postspawning females (Jøsrgensen et al., 2005). The SE content in fish liver was also reported to be lower during spawning than in the postspawning season (Shirai et al., 2001). Liver lipid contents in large yellow croaker and silvery pomfret (*Pampus argenteus*) in winter were 23.4% and 13.1%, respectively, but were only 12.4% and 4.9% in summer, respectively (Geng et al., 2015). Similarly, seasonal variations have been reported in other species (Hazra et al., 1998; Shirai et al., 2001) and feeding regimes also directly affected liver lipid content (Bransden et al., 2005, Du et al., 2006b; Zheng et al., 2016; Zhou et al., 2019). In summary, these studies suggest that lipid deposition in fish liver is a dynamic process under natural conditions.

TABLE 5.2 Lipid contents (% of wet weight) of whole body, liver, and muscle of farmed fish species.

Fish species	Whole body	Muscle	Liver	References
Herbivore				
Blunt snout bream *Megalobrama amblycephala*	3.11–11.07	1.43–1.60	7.46–10.09	Li et al. (2010b) Lu et al. (2014b) Liu et al. (2016) Qi et al. (2016)
Grass carp *Ctenopharyngodon idella*	2.70–8.20	0.40–3.48	4.65–26.05	Bransden et al. (2005) Gong et al. (2017) Zhou et al. (2017) Jin et al. (2018b)
Omnivore				
African catfish *Clarias gariepinus*	2.50–9.71	2.23–6.20	7.03–14.10	Lim et al. (2001) Ali and Jauncey (2004) Matter et al. (2004) Turan and Akyurt (2005)
Black carp *Mylopharyngodon piceus*	2.96–5.23	0.56–0.60	5.86–6.68	Sun et al. (2010) Hong et al. (2014) Wu et al. (2016) Zhang et al. (2018a)
Common carp *Cyprinus carpio*	3.60–12.05	4.00–8.20	3.30–8.90	Shimeno et al. (1997) Siddhuraju and Becker (2002) Desai and Singh (2009) Ljubojević et al. (2015)
Crucian carp *Carassius auratus*	6.18–10.32	1.89–4.41	1.20–5.17	Chen et al. (2016) Shi et al. (2016) Shi et al. (2017a) Sun et al. (2018b)
Tiger puffer *Takifugu rubripes*	5.70–6.80	1.10–1.20	34.40–48.80	Kikuchi et al. (2009) Xu et al. (2019b)
Tilapia *Oreochromis niloticus*	2.21–5.57	1.20–3.58	5.90–9.40	Ali et al. (2007) Deng et al. (2010) Siddhuraju and Becker (2003) Takeuchi et al. (2008)
Carnivore				
Atlantic cod *Gadus morhua*	3.47–7.30	0.80–1.81	35.69–58.99	Holdway and Beamish (1984) Zeng et al. (2010a)
Atlantic salmon *Salmo salar*	10.20–13.10	2.37–18.55	3.70–5.70	Aursand et al. (1994) Katikou et al. (2001) Nordgarden et al. (2002) Bou et al. (2017)

TABLE 5.2 Lipid contents (% of wet weight) of whole body, liver, and muscle of farmed fish species.—cont'd

Fish species	Whole body	Muscle	Liver	References
Gilthead seabream *Sparus aurata*	11.26–20.77	4.23–9.85	4.67–13.95	Mazzeo et al. (2013) Pérez-Jiménez et al. (2013) Castro et al. (2016b) Karapanagiotidis et al. (2019)
Japanese seabass *Lateolabrax japonicus*	1.58–11.60	0.96–1.27	4.52–28.07	Xu et al. (2011) Song et al. (2016) Yuan et al. (2016) Liang et al. (2017b) Wang et al. (2017a) Xu et al. (2017a)
Large yellow croaker *Larimichthys crocea*	5.45–16.06	6.46–13.08	10.81–37.41	Cai et al. (2017) Du et al. (2017) Ren et al. (2017) Wang et al. (2017c)
Orange-spotted grouper *Epinephelus coioides*	3.23–9.80	0.61–3.22	2.9–16.79	Zhou et al. (2007) Ren et al. (2015) Wang et al. (2016) Wang et al. (2017b)
Rainbow trout *Oncorhynchus mykiss*	4.26–9.40	0.75–12.31	3.61–4.61	Leatherland and Nuti (2006) Kennedy et al. (2007) Tasbozan et al. (2016) Acar et al. (2018) Song et al. (2018a) Shahpar and Johari (2019)
Turbot *Scophthalmus maximus*	3.34–6.51	2.97–3.45	6.91–19.47	Ren et al. (2017) Wang et al. (2019a)

Liver lipid content in farmed fish is often higher than those observed in wild fish (Du, 2014) and, when lipid deposition in liver exceeds its normal physiological level, it can cause fatty liver (Musso et al., 2009). Severe fatty liver, or fatty liver disease, in cultured fish is associated with poor growth and can cause immune suppression making fish vulnerable to pathogen infection, which can lead to disease (Bransden et al., 2005; Lu et al., 2014c). Currently, the frequent occurrence of fatty liver has become an important factor restricting the development of aquaculture for various species. One main reason for severe fat deposition is the wide use of high-fat diets (HFD) in many farmed fish species, with long-term HFD feeding increasing liver lipid accumulation (Santos et al., 1993; Rueda-Jasso et al., 2004). In a series of studies on grass carp, long-term HFD induced cellular lipotoxicity and impaired mitochondrial and peroxisomal fatty acid β-oxidation efficiency, leading to depressed lipid catabolism and severe lipid accumulation in liver (Du et al., 2005, 2006a, 2008). These adverse effects may be exacerbated by feeding increased LC-PUFA, for example high FO. Higher dietary LC-PUFA can increase the risk of peroxidation, which impairs the integrity of cell membranes and DNA (Du et al., 2008). Recently, studies suggested that long-term HFD caused leptin resistance, selectively lowering lipid catabolism (Liu et al., 2018b), and impairing lipophagic activities in Nile tilapia, causing reduced efficiency of lipid droplet elimination in liver (Han et al., 2020). Moreover, HFD also altered the expression profile of microRNA (miR) and genes related to lipid metabolism in blunt snout bream (*Megalobrama amblycephala*), inducing lipid deposition in liver (Tang et al., 2019). Based on the above studies, a model for HFD-induced fish fatty liver was suggested by Du et al. (2008) (Fig. 5.8) and, subsequently, the nutritional, physiological, and toxicological causes of fatty liver in fish, including excess feed/energy intake, essential nutrient deficiency, oxidized dietary ingredients, and toxicants in feeds and/or water were reviewed (Du, 2014). Combining all this knowledge, a comprehensive model (Fig. 5.9) to describe the complex causes and development of fatty liver that can explain most fatty liver types caused by single or multiple factors in many farmed fish was proposed (Du, 2014).

5.5.3.2 Lipid deposition in muscle

Lipid stored in muscle is an important energy source for specific physiological conditions such as prolonged muscular exercise (Jeukendrup, 2010) or fasting (Jaromowska and Górski, 1985). Muscle lipid also contributes to meat quality, influencing sensory properties such as flesh texture, flavor, and juiciness (Johansson et al., 2000; Katikou et al., 2001), as well as nutritional value (Robb et al., 2002; Wood et al., 2004). The distribution and content of muscle lipid varies widely and depends on fish species, type, and region of muscle (Katikou et al., 2001). Lipid content of fish muscle ranges from about 0.4% to around 20%, reflecting large differences in physiological and metabolic characteristics of fish species (Table 5.2). Species including Salmonids (e.g., Atlantic salmon), Scombrids (e.g., Atlantic mackerel, *Scomber scombrus*), Clupeids (e.g., European sardine/pilchard, *Sardina pilchardus* and Atlantic herring, *Clupea harengus*), and Engraulids (e.g., Peruvian anchoveta, *Engraulis ringens*) that store a high amount of lipid in fillet are generally termed "oily fish," whereas species with low

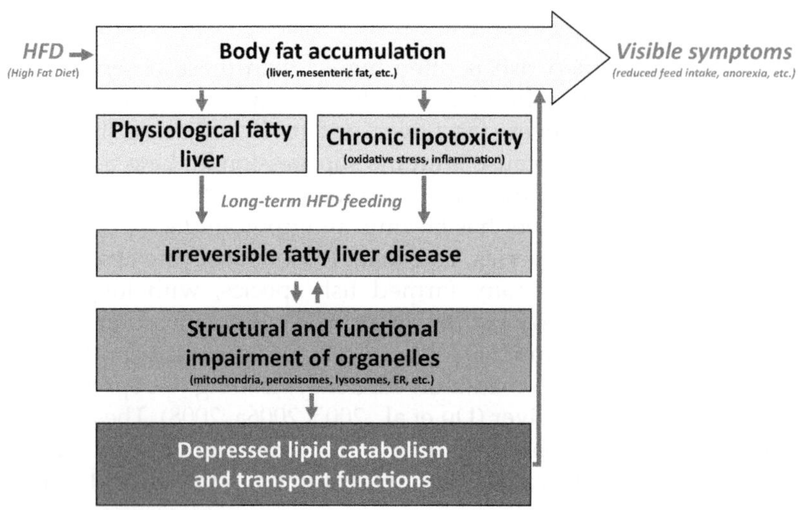

FIGURE 5.8 Development of fatty liver in fish with long-term feeding of a high-fat diet (HFD).

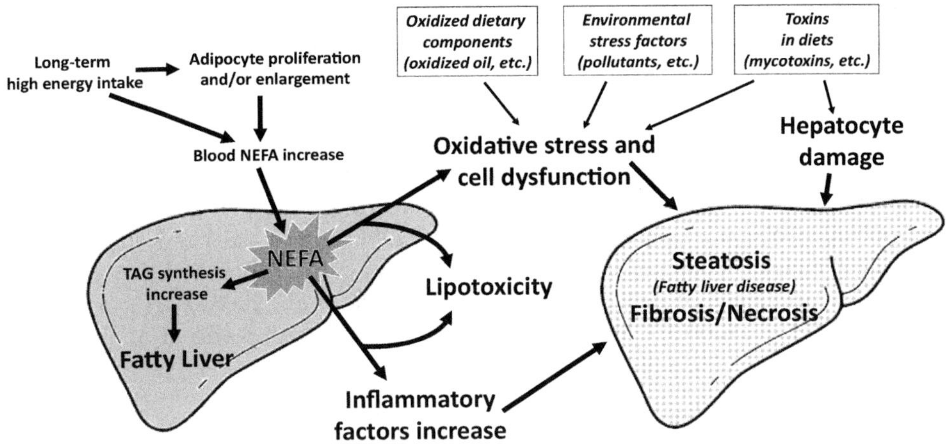

FIGURE 5.9 Schematic representation of the mechanisms involved in the development of fatty liver in farmed fish. *NEFA*, nonesterified fatty acids; *TAG*, triacylglycerol.

muscle lipid are generally termed "lean fish." In oily fish, different parts of the muscle have different lipid distribution patterns (Ackman, 1967). For example, highest lipid contents in Atlantic salmon fillets were obtained in ventral (belly flap) parts with lowest lipid in the middle and tail areas, while the dorsal part of the muscle had intermediate lipid contents (Katikou et al., 2001). In addition, lipid level was more than fourfold higher in red muscle than white muscle in Atlantic salmon (Nanton et al., 2007). Therefore, it is necessary to either specify the exact area of sampling when reporting muscle lipid contents of fish species or to sample the entire fillet.

5.5.3.3 Adipose tissue distribution and development

As the main fat deposition organ for most fish, adipose tissues have relatively fixed anatomic location although distribution varies with species. In general, adipose tissues of most species are largely distributed in viscera and are termed visceral adipose tissue (VAT) (Yuan et al., 2016; Wang et al., 2017f). The distribution of adipose tissue in several fish species was determined using a novel 3D-MRI technique for live fish (Wu et al., 2015), which showed that VAT was the main adipose tissue in species such as grass carp, Nile tilapia, large yellow croaker, and golden pompano (*Trachinotus ovatus*), but a large mass of subcutaneous adipose tissue (SCAT) was found in turbot. Similarly, other studies also found that the adipose tissues of flatfish species were mainly distributed at the roots of the dorsal and anal fins (Aubourg et al., 2010; Khieokhajonkhet et al., 2014). The VAT and SCAT have different structures and metabolic functions in Nile tilapia with VAT the preferred site for lipid deposition, whereas SCAT exhibited higher lipid catabolic activity (Wang et al., 2017f). Although adipose tissue in zebrafish was regulated by both developmental stage and fish size (Imrie and Sadler, 2010), knowledge concerning the mechanisms of adipose tissue development in different locations in fish are still lacking. Nutrient supply impacts the accumulation and mobilization of lipid in adipocyte development in fish, as in mammals

(Faust et al., 1980). In the absence of feed, zebrafish failed to form visceral preadipocytes, indicating that exogenous nutrition was required for adipocyte development (Flynn et al., 2009), with dietary lipid necessary to fully induce differentiation of primary preadipocytes (Salmerón et al., 2013). In addition, differentiation of adipocytes can be affected by unsaturated fatty acids, especially LC-PUFA, which have antilipogenic effects in Atlantic salmon, with EPA and DHA lowering TAG accumulation in mature adipocytes compared to OLA (Todorčević et al., 2010). This might be attributed to differences in fatty acid metabolism during adipocyte differentiation as most EPA and DHA were oxidized in preadipocytes, while they were mainly stored in TAG in mature adipocytes, whereas OLA was primarily stored in TAG at all stages of differentiation (Todorčević et al., 2008a). In grass carp, DHA decreased lipid accumulation by inducing apoptosis of adipocytes (Jin et al., 2018a, 2018c).

5.5.4 Regulation of lipid synthesis and deposition in fish

Lipid accumulation in vertebrates occurs due to imbalance between absorption from the diet and de novo lipogenesis, and catabolism via lipolysis and β-oxidation (Schutz, 2004). In fish, lipogenesis and lipid deposition has been confirmed to be tightly correlated to dietary components and endocrinological factors, and some regulatory mechanisms have been illustrated.

5.5.4.1 Nutritional regulation

Lipid deposition in fish depends largely on the amount and type of dietary lipid (Jordal et al., 2007; Kjær et al., 2008b; Betancor et al., 2015b; Castro et al., 2015b). Long-term HFD feeding in fish significantly increased lipid deposition and caused a series of physiological and biochemical changes (Du et al., 2005, 2006a) with the effects of HFD on liver lipid deposition described above. Magnetic resonance imaging showed that the volume of VAT and SCAT in Nile tilapia were both significantly increased by feeding HFD (Wang et al., 2017f). Thus, fish have adaptive mechanisms to regulate lipid metabolism in response to low- and high-fat diets. For example, low dietary lipid in Nile tilapia enhanced sterol response element binding protein-1 (Srebp1)-mediated expression of lipogenic (*acca*, *fas*, and ATP-citrate lyase [*acyl*]) and intracellular *fabp* genes to upregulate fatty acid biosynthesis accompanied with higher glycolysis to increase production of acetyl-CoA for fatty acid synthesis as adaptations to the low lipid intake (He et al., 2015b; Li et al., 2016a) (Fig. 5.10). Liver has been verified as the main lipogenic organ, with adipose tissue playing important roles in deposition of excess lipid intake after feeding Nile tilapia with HFD (He et al., 2015b). This study indicated that HFD feeding activated Pparγ in adipocytes through high concentration of circulating FFA, which are natural ligands of Pparγ (Spiegelman, 1998). Activated Pparγ triggered glycerol synthesis and lipogenesis activities, accelerating esterification of FFA into TAG inducing adipocyte proliferation (Fig. 5.10) (He et al., 2015b).

In aquaculture, it is well known that fish fatty acid compositions reflect dietary compositions (Kjær et al., 2008b; Leaver et al., 2008c; Morais et al., 2012b; Castro et al., 2015b). Moreover, dietary lipid sources can also affect fat deposition by regulating gene expression and activities of key enzymes involved in lipid metabolism, including anabolic and catabolic processes (Turchini et al., 2003). For example, replacing dietary FO by VO affected gene

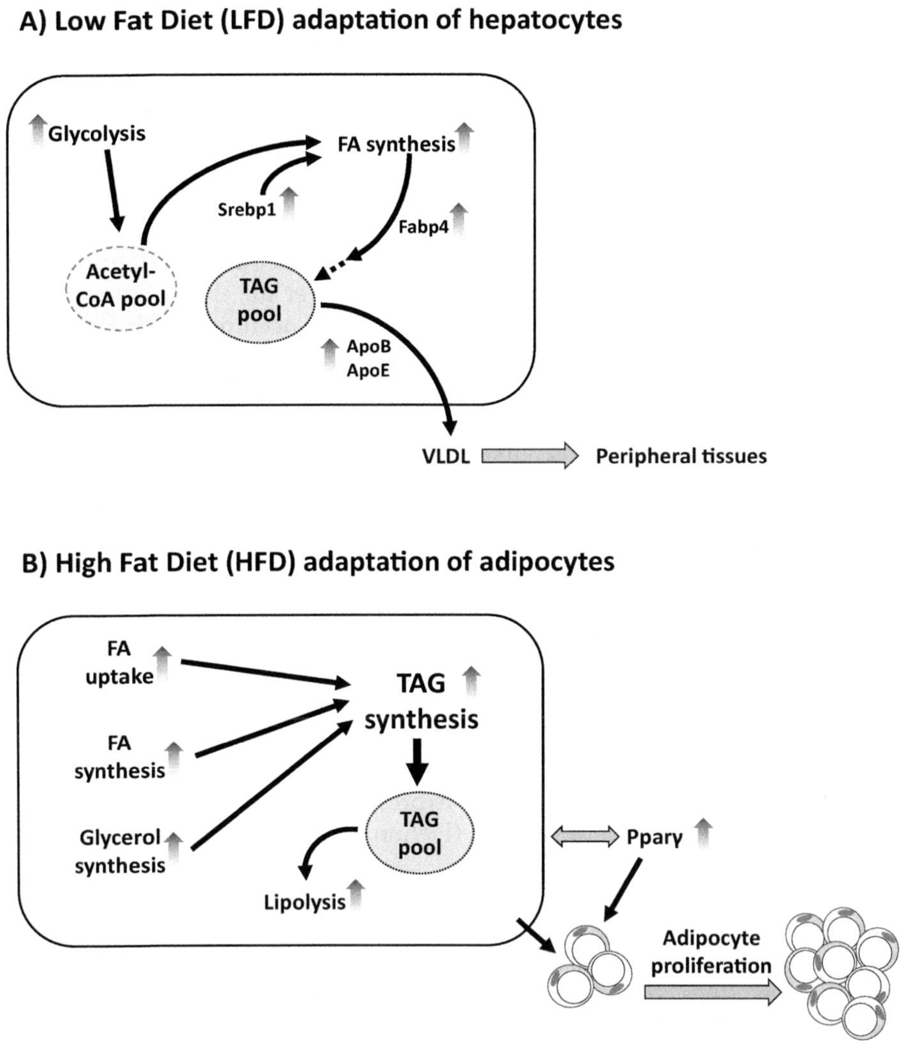

FIGURE 5.10 Proposed model of the metabolic adaptations to low-or high-fat diets in the liver and adipose tissue of Nile tilapia (He et al., 2015b). (A) Feeding a low-fat diet (LFD), glycolysis is accelerated in hepatocytes to provide acetyl-CoA for fatty acid (FA) and triacylglycerol (TAG) synthesis to maintain lipid homeostasis. This process is upregulated by Srebp1. The newly synthesized TAG can be exported to peripheral tissues mediated by increased very-low density lipoprotein (VLDL) containing ApoB and ApoE. (B) Feeding a high-fat diet (HFD), the uptake of diet-sourced FA, de novo synthesis of FA and glycerol are upregulated in adipocytes, increasing TAG synthesis. The excess cellular TAG is hydrolyzed by accelerated lipolysis and the released free fatty acid (FFA), together with the diet-sourced circulating FFA, upregulate Pparγ to trigger the differentiation and proliferation of visceral adipocytes.

expression and activity of lipogenic enzymes, and correspondingly altered lipid deposition in tissues (Menoyo et al., 2004; Panserat et al., 2008). In Atlantic salmon, mature adipocytes treated with OLA accumulated more lipid than those incubated with the same dose of EPA or DHA, suggesting that OLA leads to higher lipid deposition (mainly in the form of TAG) and adipogenesis than EPA or DHA (Todorčević et al., 2008a). The antiadipogenic effects of n-3 LC-PUFA, particularly EPA and DHA, have been verified in many other studies (Alvarez et al., 2000; Wang et al., 2012a) with similar results reported in carp and Atlantic salmon fed n-3 PUFA-enriched diets (Shikata and Shimeno, 1994; Menoyo et al., 2003). In a number of fish studies, n-3 PUFA have been reported to play beneficial roles by preventing excess fat accumulation (Ji et al., 2011; Chen et al., 2015b), whereas excess dietary SFA, such as 16:0, caused lipotoxicity and high lipid accumulation in tissues (Xue et al., 2006; Han et al., 2013; Ding et al., 2018). However, hepatic lipogenic enzyme activities were not modified by replacement of FO by VO in some studies in turbot (Regost et al., 2003), gilthead seabream (Menoyo et al., 2004), and Atlantic salmon (Torstensen et al., 2004), suggesting that the effects of dietary lipid sources on lipogenesis and lipid accumulation may vary among fish species and/or with experimental conditions.

In rainbow trout (Seiliez et al., 2011) and blackspot seabream (*Pagellus bogaraveo*) (Figueiredo-Silva et al., 2009), diets containing high protein and low carbohydrate enhanced lipogenic gene expression and/or enzymatic activities, suggesting that high-protein diets stimulate lipogenic pathways in these carnivorous species. However, in most fish nutrition studies, changing one nutrient in the diet causes corresponding changes in other nutrient levels and, hence, it is often difficult to distinguish the contribution of each nutrient in inducing metabolic changes (Dias et al., 1998; Seiliez et al., 2011; Kamalam et al., 2012). However, hepatic fatty acid biosynthetic gene expression in rainbow trout was more responsive to dietary protein and amino acids than to dietary carbohydrate and glucose (Dai et al., 2016). Furthermore, high levels of amino acids promoted hepatic fatty acid biosynthesis in trout hepatocytes through upregulating the expression of genes related to lipogenesis, NADPH production, and fatty acid bioconversion, accompanied with increased amino acid catabolic pathways (Dai et al., 2015). Further studies indicated that dietary protein and amino acid intake stimulated fatty acid biosynthetic gene expression in an mTORC1 (mechanistic target of rapamycin complex 1)-dependent manner (Pichon et al., 2006; Uebanso et al., 2009; Abu-Elheiga et al., 2012; Rietman et al., 2014).

Dietary protein source is also known to affect lipid deposition and fatty acid bioconversion potential, and alter serum and liver lipid levels in higher vertebrates including fish (Terasawa et al., 1994; Potter, 1995; Aoyama et al., 2000). In some studies, plant meals have had hypolipidemic effects, lowering liver and plasma TAG and cholesterol levels (Kaushik et al., 2004; Sitjà-Bobadilla et al., 2005; Maita et al., 2006; Ye et al., 2019). In contrast, other studies have shown hyperlipidemia and/or hepatic fat deposition in fish fed diets containing plant protein (Yaghoubi et al., 2016; Yin et al., 2018). Moreover, higher lipogenic enzyme activities were obtained in European seabass fed FM compared to other protein sources, suggesting dietary amino acid composition affected lipogenic potential (Dias et al., 2005). These contradictions may stem from differences in fish species, growth status, rearing conditions, feeding strategy, and dietary composition, specifically the plant protein ingredients and antinutritional factors.

Limited studies have investigated how dietary amino acid balance impacts the regulation of lipid metabolism. A recent study in juvenile blunt snout bream showed that 1.62% dietary arginine supplementation significantly upregulated mRNA levels of *acc*, *fas*, and glucose-6-phosphate dehydrogenase (*g6pd*) involved in lipid synthesis and increased plasma TAG content (Liang et al., 2017a). Likewise, many studies showed that supplemental L-leucine up-regulated lipogenic genes and lipid content in fish serum and tissues (Zhang et al., 2007;

Lansard et al., 2010a, 2010b; Freudenberg et al., 2013; Liang et al., 2019b). In contrast, deficiency of some amino acids, especially the essential amino acids, also caused fat accumulation. For example, feeding salmon a low-methionine diet increased Fas activity and lipid accumulation in liver (Espe et al., 2010), probably related to lowered methionine-sourced taurine synthesis that disturbs bile acid functions and, subsequently, impairs lipid metabolism.

It is commonly accepted that most fish species have a lower ability to utilize carbohydrate than mammals (NRC, 2011). However, dietary carbohydrate has been used widely in aquaculture because of its relatively low cost and wide availability. As a result, high-carbohydrate diet-induced fat accumulation has been frequently reported in fish (Brauge et al., 1994; Mohanta et al., 2009; Ren et al., 2011; Bou et al., 2016; Liu et al., 2018b). Studies examining the recovery of label from radioactive/stable isotope-labeled glucose or starch have indicated that the amount of lipid derived from glucose is limited in carnivorous fish (Hemre and Storebakken, 2000; Rawles et al., 2008; Ekmann et al., 2013). Thus, increased fat accumulation in fish fed a high-carbohydrate diet may be due to preferential oxidization of carbohydrate for energy instead of dietary lipid, reducing lipid catabolism (Hellerstein et al., 1996). It has also been suggested that carbohydrate plays a more important role in production of cytosolic reducing equivalents (NADPH) than delivery of carbon backbones (acetyl-CoA) for lipogenesis (Hemre et al., 2002). Furthermore, a recent study in yellow catfish revealed that high dietary glucose activated lipophagy and induced changes in lipid metabolism via enhanced carbohydrate response element-binding protein (Chrebp) DNA binding capacity at the Pparγ promoter region that, in turn, induced transcriptional activation of key genes related to lipogenesis (Zhao et al, 2020a). Other than dietary level, lipogenesis and lipid deposition in fish are also affected by the complexity of carbohydrate, which ranges from digestible mono-, di-, and polysaccharides to indigestible hemicelluloses and cellulose (Krogdahl et al., 2005).

Several studies have shown that mineral elements, including magnesium (Mg), zinc (Zn), and copper (Cu), influence lipid metabolism and deposition in fish (Liu et al., 2010; Chen et al., 2013a; Zheng et al., 2013a; Huang et al., 2014). While Mg plays major roles in numerous metabolic pathways (Lall, 2002), the effects of dietary Mg on lipid deposition are still debated (Dabrowska et al., 1991; Wang et al., 2011; Liang et al., 2012). Adequate dietary Mg increased lipid deposition in whole body of rainbow trout (Ogino et al., 1978) and grass carp (Liang et al., 2012), and muscle of grass carp (Wang et al., 2011) and Japanese seabass (Zhang et al., 2016a). However, dietary Mg reduced hepatic lipid accumulation in yellow catfish by suppressing transcriptional expression and activity of enzymes related to lipid metabolism, including hepatic 6-phosphogluconate dehydrogenase (6Pgd) and malic enzyme (Me), through Pparα, JAK-STAT, and AMPK signaling pathways (Wei et al., 2017b). Increasing dietary Cu has also been shown to regulate lipid metabolism, decreasing whole body lipid in great sturgeon/beluga (*Huso huso*) (Wang et al., 2009b; Mohseni et al., 2014), and hepatic lipid content in yellow catfish (Chen et al., 2015b) but, in Nile tilapia, a high Cu diet increased fat deposition in liver (Shaw and Handy, 2006). Excess dietary Zn reduced lipid content in liver of yellow catfish, accompanied by downregulation of *fas, acca,* and *pparγ* mRNA levels, whereas dietary Zn deficiency upregulated *fas* mRNA levels (Zheng et al., 2015a). Similar results were observed in Siberian sturgeon, *Acipenser baerii* (Moazenzadeh et al., 2018), where increased dietary Zn reduced muscle lipid content.

Dietary vitamin A promoted lipase activity and decreased lipid deposition in liver of Senegalese sole (Darias et al., 2012) and gilthead seabream larvae (Fernández and Gisbert, 2011), whereas high vitamin A increased liver lipid content in gibel carp (*Carassius gibelio*) (Shao et al., 2016), and increased Fas mRNA expression levels in juvenile orange-spotted grouper (*Epinephelus coioides*) (Yang et al., 2017b). In juvenile olive flounder (*Paralichthys olivaceus*), whole body lipid content increased in a dose-dependent manner as dietary vitamin C level increased (Lee et al., 2016), but vitamin C had no effect on whole body lipid content in other species (Falahatkar et al., 2006; Yang et al., 2017b). Dietary vitamin E supplementation reduced whole body lipid content of black sea bream (*Acanthopagrus schlegeli*) by reducing lipid peroxidation and promoting lipid assimilation and transportation (Peng et al., 2009), whereas vitamin E had no major effect on lipid content of meager (*Argyrosomus regius*) (Lozano et al., 2017). Dietary inositol (*myo*-inositol) at 100 mg/kg diet or higher reduced liver lipid content in Nile tilapia (Peres et al., 2004), but no lipid accumulation occurred in liver or kidney of channel catfish (*Ictalurus punctatus*) when myo-inositol was removed from the diet (Burtle and Lovell, 1989). Overall, studies on the regulation of lipid metabolism by vitamins in aquatic animals are limited and further research is required.

5.5.4.2 Endocrinological regulation

Insulin is well known as the most potent anabolic hormone with fundamental roles in the regulation of somatic growth and metabolism in vertebrates (Hernández-Sánchez et al., 2006). The interaction between insulin and its receptors (Insr) amplifies and diversifies insulin action (Saltiel and Kahn, 2001; Kitamura et al., 2003). However, the regulation of lipid metabolism by insulin in fish does not seem to follow the mammalian pattern (Mommsen and Plisetskaya, 1991). In rainbow trout, despite upregulation of liver X receptor (Lxr) and Srebp1c transcripts, insulin injection did not increase mRNA levels of the key enzymes involved in lipogenesis, i.e., G6pd, Acyl, Acc, and Fas (Jin et al., 2014). However, *fas* gene expression in trout primary hepatocytes was induced by insulin when accompanied by glucose (Plagnes-Juan et al., 2008). The effect of insulin on lipid metabolism in fish was dose-dependent as a physiological dose (1.4 IU/kg) upregulated G6pd and Fas transcript levels in rainbow trout (Polakof et al., 2010a), whereas a pharmacological dose (10 IU/kg) had no effect on mRNA expression of enzymes involved in lipogenesis (Plagnes-Juan et al., 2008). The regulation of lipid metabolism by insulin is also influenced by feeding status. Chronic insulin treatment inhibited hepatic lipogenic potential by reducing expression of Fas and Srebp1c in fasted rainbow trout, whereas hepatic lipogenesis was enhanced by insulin in trout fed a high-carbohydrate diet (Polakof et al., 2011).

Leptin is a pleiotropic hormone identified in all vertebrate classes that plays key roles in body weight regulation, energy homeostasis, and lipid utilization in mammals (Denver et al., 2011; Friedman-Einat et al., 2014; Seroussi et al., 2015). The amino acid sequence of leptin is not well conserved among or within classes (Londraville et al., 2014) with identity variation between human and teleost leptins ranging from 13% to 25% (Gorissen and Flik, 2014) although in silico modeling showed secondary and tertiary structures were evolutionarily conserved (Rønnestad et al., 2010; Denver et al., 2011). In teleosts, the liver was believed to be the primary site for leptin production (Douros et al., 2014; Londraville et al., 2014), but a recent study showed SCAT was another important tissue for leptin production in Nile tilapia (Liu et al., 2018b). Although rainbow trout adipocytes were reported to secrete leptin in vitro

(Salmerón et al., 2015a), whether mesenteric adipose tissue can secrete leptin remains questionable in fish. Intraperitoneal injection of recombinant leptin reduced hepatic lipid content, activities of lipogenic enzymes (Fas, 6pgd, G6pd, Me, and isocitrate dehydrogenase [Icdh]), and expression of *6pgd, g6pd, fas, ppar*γ, and *srebp-1* genes in yellow catfish (Song et al., 2015), and reduced lipogenesis and lipid contents in blood and other tissues in Nile tilapia (Liu et al., 2018b). The JAK-STAT and IRS-PI3K signaling pathways mediate leptin actions in regulating lipid metabolism, as specific inhibitors (AG490 and wortmannin) of JAK-STAT and IRS-PI3K signaling pathways suppressed the regulatory functions of leptin in reducing lipid deposition and expression of lipogenic genes in rainbow trout (Aguilar et al., 2011), grass carp (Lu et al., 2012), and yellow catfish (Song et al., 2015).

Growth hormone (GH) is a pluripotent hormone produced by the pituitary gland in vertebrates including teleosts, which binds to the transmembrane GH receptor regulating numerous processes including growth promotion and lipid mobilization (Björnsson et al., 2004; Nørrelund, 2005). Activities of enzymes related to fatty acid metabolism and NADPH production, including hydroxyacyl CoA dehydrogenase (Had) and Me, respectively, in muscle were lower, while the activity of 6Pgd in liver was increased, indicating an increased requirement for NADPH for increased lipid synthesis in GH transgenic coho salmon (*Oncorhynchus kisutch*) compared to wild type (Leggatt et al., 2009). However, an earlier study found no effects of GH transgenesis on muscle Had activity in coho salmon (Hill et al., 2000), suggesting the effects of GH on lipid synthesis requires further study.

Ghrelin, an endogenous ligand for GH secretagogue receptor, has been found to promote fat deposition in rodents by reducing utilization (Tschöp et al., 2000). The primary structure of ghrelin has been determined in teleosts including Japanese eel, zebrafish, black seabream, rainbow trout, Mozambique tilapia, channel catfish, and goldfish (*Carassius auratus*) (Kaiya et al., 2003a, 2003b; Yeung et al., 2006a, 2008). In Mozambique tilapia, two forms of ghrelin have been identified: n-octanoic (ghrelin-C8) and n-decanoic (ghrelin-C10) ghrelins (Riley et al., 2005). Intraperitoneal administration of ghrelin-C10 stimulated appetite and increased weight gain in tilapia, possibly due to increased lipid content in liver and muscle (Riley et al., 2005). In contrast, in rainbow trout, intraperitoneal injection of ghrelin either did not affect food intake (Jönsson et al., 2007) or decreased food intake compared with uninjected fish (Jönsson et al., 2010). However, ghrelin significantly increased expression of lipoprotein lipase (*lpl*) and *fas* genes, and thus stimulated TAG synthesis in rainbow trout adipocytes (Salmerón et al., 2015b). Similarly, intracerebroventricular injection of ghrelin in rainbow trout increased abundance of Srebp1c mRNA, suggesting enhanced lipogenesis, and was reported to act on the hypothalamus regulating fatty acid sensing mechanisms (Velasco et al., 2016a). This suggests ghrelin acts centrally by modulating lipid metabolism, resulting in increased potential for lipogenesis.

5.5.4.3 Molecular regulation

Mentioned above, SREBP-1 is a membrane-bound transcription factor that regulates the gene expression of enzymes involved in fatty acid biosynthesis (Minghetti et al., 2011). The role of SREBP-1c in mediating the hormonal (insulin and glucagon) and nutritional (dietary lipid and carbohydrate) regulation of genes of lipogenic and NADPH yielding enzymes has been demonstrated in mammals (Horton et al., 2002), and Srebp-1 has been cloned in Atlantic salmon and rainbow trout (Polakof et al., 2010a; Minghetti et al., 2011). Expression of Srebp-1 was upregulated by insulin in trout (Polakof et al., 2010a, 2010b), and dietary SFA and MUFA increased expression in liver of Japanese seabass (Dong et al., 2015). Indeed, *srebp-1* mRNA expression correlated positively with both the enzyme activities and mRNA levels of lipogenic genes including *fas*, *g6pd*, and *6pgd* (Chen et al., 2013b; Zheng et al., 2013a; Zhuo et al., 2014). However, in some fish studies, *srebp1* gene expression was reduced with increased dietary lipid, possibly due to the elevated intake of lipid reducing endogenous lipid synthesis to maintain a dynamic balance between absorption and endogenous lipid synthesis (Leng et al., 2012). However, dietary lipid level did not affect *srebp1* and *fas* expression levels in grass carp adipose tissue (Yuan et al., 2016), indicating that Srebp-1-regulated lipogenesis may be dependent on fish species and dietary lipid level.

PPARγ is found in all vertebrates and is a central factor in sensing fatty acid levels and regulating fatty acid metabolism (Rosen and Spiegelman, 2001). It is a ligand-activated transcription factor that heterodimerizes with retinoid X receptor (RXR) to bind to peroxisome proliferator response elements (PPRE) in DNA (Kota et al., 2005; Cho et al., 2009). In mammals, deficiency in PPARγ can cause severe metabolic syndromes including obesity, type II diabetes, and fatty liver (Tsukahara, 2013). In fish, *ppar*γ has been cloned from various fish species including zebrafish (Ibabe et al., 2002), Atlantic salmon (Leaver et al., 2005a), brown trout (Batista-Pinto et al., 2005), yellow catfish (Zheng et al., 2015b), and Nile tilapia (He et al., 2015a). In tilapia, Pparγ collaborates with Rxr to transcriptionally regulate target genes such as *fabp4* (He et al., 2015a) and plays a critical role in regulating lipid metabolism and adipocyte differentiation in response to fatty acid levels (Dutchak et al., 2012). However, some compounds that activate mammalian PPARγ do not activate piscine Pparγ (Maglich et al., 2003; Leaver et al., 2005a) and, thus, piscine Pparγ may have structural differences to the mammalian proteins. Despite this, the overall structural similarity of piscine Pparγ to all other PPAR strongly indicates that it is likely to be activated by compounds such as lipids and fatty acids that activate mammalian PPARγ. For example, high-dose LNA or ALA suppressed *ppar*γ expression in loach fin cells (Yun et al., 2018), although lipid intake increased hepatic *ppar*γ expression and a number of Pparγ-target genes involved in adipocyte differentiation and lipid storage (Chen et al., 2015b; Yuan et al., 2016b).

Recently, the roles of mTOR (mechanistic target of rapamycin) in lipogenic regulation have been reported in fish, with postprandial regulation of hepatic lipogenesis requiring the activation of the mechanistic target of rapamycin complex 1 (mTORC1) signaling in rainbow trout (Dai et al., 2013). This study showed that inhibition of hepatic mTORC1 decreased *srebp1c* gene expression and suppressed *fas* at gene expression and activity levels, indicating that Fas was controlled at a transcriptional level in an mTORC1-dependent manner. Similarly, in Nile tilapia, supplementation of the mTOR inhibitor, rapamycin, in a high-fat, high-carbohydrate diet was shown to alleviate dietary-induced lipid accumulation in liver. This was accompanied by significant depression of mTOR expression and phosphorylation (Han et al., 2020), consistent with mTOR being a regulatory factor maintaining lipid homeostasis in fish.

The ubiquitous transcription factor Yin Yang 1 (YY1) has diverse and complex cellular functions, with the activation of YY1 downregulating C/EBP (CCAAT/enhancer binding protein) homologous protein 10 (CHOP-10) by binding to YY1 sites in the CHOP-10 promoter, enabling transactivation of transcription factors required for lipogenesis in the early phase of adipocyte differentiation in mammals (Tang and Lane, 2000; Li et al., 2006;

Huang et al., 2008). The association of Yy1 with hepatic lipid accumulation was reported in zebrafish through downregulating Chop-10, which repressed expression of Pparγ and CCAAT-enhancer-binding protein α (C/ebp-α) in the lipogenic pathway (Her et al., 2013).

MicroRNA (miR) have been shown to participate in almost every metabolic homeostatic process, including lipogenesis (Song et al., 2014; Shen et al., 2015; Lie et al., 2016), and glucose-stimulated insulin secretion (Baffy, 2015; Massart et al., 2016). The expression of miR-27b decreased during adipogenesis of human adipose-derived stem cells, and overexpression of miR-27b repressed PPARγ and C/EBP-α expression during early onset of adipogenesis and TAG accumulation at later stages (Karbiener et al., 2009; Zou et al., 2015). Accordingly, PPAR-γ has a highly conserved binding site for miR-27b in the 3′ UTR (Karbiener et al., 2009). Recently, miR-27b depletion increased weight gain and expression of Pparγ, C/ebp-α, and Srebp-1c, promoting lipogenesis and lipid accumulation via adipogenesis and adipocyte hyperplasia in zebrafish (Hsu et al., 2018).

5.6 Long-chain polyunsaturated fatty acids

Polyunsaturated fatty acids (PUFA) cannot be synthesized de novo by any vertebrate, as they lack the $\omega6$ (or $\Delta12$) desaturase responsible for the production of 18:2n-6 (LNA) from 18:1n-9 (OLA), and the $\omega3$ (or $\Delta15$) desaturase that further converts LNA to 18:3n-3 (ALA) (Tocher, 2003). However, while PUFAs are thus essential dietary nutrients for all vertebrates, the biologically active PUFAs are actually LC-PUFA (Sargent et al., 2002). The C_{18} PUFA, ALA and LNA, can be converted to LC-PUFA in fish although this varies with species dependent upon the presence and expression of genes of fatty acid desaturation and elongation (Tocher, 2003) (Fig. 5.11). In consequence, which PUFA can satisfy essential fatty acid (EFA) requirements varies similarly with species (Tocher, 2010).

5.6.1 Roles and functions of LC-PUFA

5.6.1.1 *Structural and energy storage*

Some functions of LC-PUFA are common to all fatty acids. As constituents of PL that play an important structural role in cellular biomembranes, LC-PUFA influence physicochemical properties such as fluidity, and modulate activities of membrane proteins including receptors, carriers, and enzymes, thus affecting the functional properties of membranes (Tocher, 1995). Similarly, LC-PUFA can be β-oxidized in mitochondria to provide metabolic energy and, therefore, can also serve as an energy reserve as components of storage lipids such as TAG and WE (Tocher, 2003).

5.6.1.2 *Metabolic and physiological regulation*

In contrast to the above generic fatty acid roles, LC-PUFA and their derivatives have more specific functions as key metabolic regulators, which underpin their essentiality. Within the cell, LC-PUFA are ligands for various transcription factors such as PPAR and thereby have a role in regulating metabolic pathways through influencing gene expression (Leaver et al., 2008a). The n-6 and n-3 series LC-PUFA have divergent, often opposite, impacts on

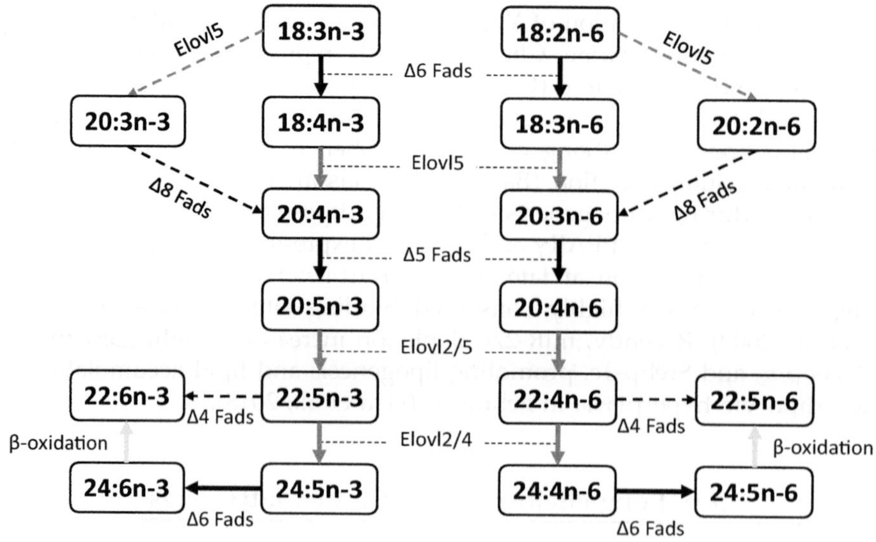

FIGURE 5.11 Pathways for the biosynthesis of long-chain polyunsaturated fatty acids in fish. Desaturations (*black arrows*) are catalyzed by fatty acyl desaturases (Fads) with Δ6, Δ5, and Δ4 indicating the position in the carbon chain at which the double bond is introduced counted from the carboxylic group. Elongations (*gray arrows*) are catalyzed by elongation of very long-chain fatty acid proteins (Elovl). *Solid arrows* represent the predominant pathway in most fish species (dependent upon the presence or otherwise of the genes), whereas *dashed arrows* represent pathways possible in some species. All reactions occur in endoplasmic reticulum other than β-oxidation, which occurs in peroxisomes.

metabolism and physiology that underpin the concept of balance between the two series, which cannot be interconverted in vertebrates (Fig. 5.11), and the importance of dietary LC-PUFA in both health and disease (Lands, 2014).

The LC-PUFA have further regulatory functions at an extracellular level, with derivatives of LC-PUFA having important physiological roles in homeostasis and stress response. These functions are mediated through eicosanoids, oxygenated products of ARA, EPA, and DHA, produced by cyclooxygenase (COX-1, COX-2) and lipoxygenase (5-LOX, 12-LOX, 15-LOX) enzymes. Products of ARA (n-6 series) include prostaglandin E_2 (PGE_2) and leukotriene B_4 (LTB_4), while EPA (n-3 series) counterparts are PGE_3 and LTB_5 (Schmitz and Ecker, 2008). Normally, n-6 eicosanoids are very potent with PGE_2 being proinflammatory and increasing vascular permeability, while LTB_4 stimulates synthesis of proinflammatory cytokines like tumor necrosis factor TNF-α, interleukins IL-1β and IL-6 while, in contrast, n-3 homologs like PGE_3 and LTB_5 have low potency (Dennis and Norris, 2015). As the LC-PUFA compete for the same COX and LOX enzymes in these pathways, EPA also blocks the ARA effects, further highlighting the need for balance between the n-3 and n-6 series LC-PUFA (Lands, 2014). The roles of LC-PUFA in eicosanoid pathways leading to prostaglandin, leukotriene, and other hydroxy-derivatives in fish have been reviewed previously (Knight et al., 1995; Tocher, 1995) including the role of prostaglandins in fish immunity (Gomez-Abellan and Sepulcre, 2016). In general, while both mammals and fish appear to have many of the same elements of eicosanoid cascades (Gomez-Abellan and Sepulcre, 2016), their actual effects in fish are still rather elusive.

More recently, the range of LC-PUFA derivatives has expanded due to the discovery of the specialized proresolving mediators (SPM) including lipoxins, resolvins, protectins, and maresins that are anti-inflammatory compounds that terminate (resolve) inflammatory responses (Serhan, 2014). While lipoxins are derivatives of ARA, the other SPM are derivatives of n-3 LC-PUFA, with EPA converted to resolvins E1/E2, while DHA is converted to resolvins D1-4, protectin D1 and maresin 1 (Dennis and Norris, 2015). There may also be resolvins produced from docosapentaenoic acid (DPA; 22:5n-3) (Serhan and Levy, 2018). Research into SPM and their role in resolving inflammation has produced some significant advances in the mammalian field in recent years (Serhan et al., 2015). In summary, normal inflammatory responses in health are acute and self-limited as they resolve on their own, and thus can be divided into initiation and resolution phases (Serhan and Levy, 2018). In the initiation phase, n-6 (ARA-derived) prostaglandins and leukotrienes are the critical eicosanoids for stimulating edema and neutrophil (polymorphonuclear leukocyte; PMN) infiltration. While PMN are key to the neutralization of potential pathogens via phagocytosis, excess PMN can result in release of antimicrobial compounds that can lead to tissue damage and chronic inflammation (Serhan and Levy, 2018). Resolution is an active host-driven process based on a temporal lipid mediator class switch, with PMN exposed to PGE_2 switching from production of LTB_4 to lipoxins (e.g., LXA_4) that are potent, active stop signals for PMN, limiting further infiltration (Levy et al., 2001). At the same time there is increased recruitment of monocytes/macrophages for repair and renewed function, and the production of the superfamily of n-3 LC-PUFA-derived SPM, each member of which possesses potent proresolving and anti-inflammatory actions (Serhan and Levy, 2018). The range of SPM, the enzyme pathways involved, and their cell type- and organ-specific properties are outside the scope of this review but have been reviewed extensively previously (Serhan, 2014; Serhan et al., 2015; Serhan and Levy, 2018).

In fish, the inflammatory process in general, and specifically, the role of SPM are very poorly studied. So far, there have been very few studies that have been limited to little more than identifying the presence of SPM in fish tissues. One early study reported that rainbow trout brain cells challenged in vitro with calcium ionophore A231287 biosynthesized a range of di- and trihydroxy products from endogenous DHA including resolvins D1, D2 and D5, and neuroprotectin D1 (Hong et al., 2005). Similarly, resolvin D1, protectin D1, and mono-hydroxy-DHA products were identified in trout head-kidney (Hong et al., 2007). The presence of resolvin D5 and protectin D1 as well as other endogenous DHA derivatives have been reported in the marine oily fish, Peruvian anchovy (*E. ringens*) (Oh et al., 2011), and resolvins D1 and D2 have been extracted from raw Atlantic salmon (Raatz et al., 2011).

5.6.2 Essential fatty acid requirements of fish

The EFA requirements of any animal can be described in both quantitative and qualitative terms, neither of which are fully established in detail for many species of fish, although quantitative requirements can probably be summarized more easily than qualitative requirements (Glencross, 2009). However, it is important to be clear on exactly what quantitative requirement levels actually mean. Historically, EFA requirement was reported as the dietary level required to prevent classical nutritional pathology (i.e., EFA deficiency signs) and this is low, often around 1% or lower of the diet, depending upon the fatty acids (PUFA v. LC-PUFA) being tested (Tocher, 2010). The reported EFA requirement data for fish species are probably reasonable estimates of this physiological requirement level (Glencross, 2009; Tocher, 2010). However, using humans as an example, while it is unlikely that many humans are actually EFA deficient, dietary supplementation with EFA, especially n-3 LC-PUFA ("Omega-3"), still has significant nutritional benefits and, therefore, a higher level of EFA is required to support optimum development and health (Tocher, 2015). Farmed fish are arguably in an analogous situation as early EFA studies, performed when growth expectations were lower, were largely done in small fish in short-term trials using diets with relatively low lipid level (Tocher, 2015). In contrast, modern sustainable feeds not only promote much higher growth rates due to their high-energy (lipid) formulations, but also, like human diets, have much higher n-6:n-3 PUFA ratios due to the replacement of marine ingredients with terrestrial ingredients lacking LC-PUFA (NRC, 2011; Turchini et al., 2011c). Consequently, it has been suggested that EFA requirements should be reassessed to establish the EFA requirement to support optimal growth and health of fish (NRC, 2011). While this higher level of requirement is not precisely defined for any fish species and will undoubtedly vary dependent upon other dietary factors, environmental conditions, and fish metabolism, it is not likely to be very high (Tocher, 2015). There is a third level of EFA that is not actually a requirement of the fish itself, but is one based on human requirements (Simopoulos, 2008), and that is the level required to ensure nutritional quality of the farmed fish products based on n-3 LC-PUFA content of flesh (Tocher, 2015; Tocher et al., 2019). To satisfy this, EFA, more specifically the LC-PUFA, EPA, and DHA, need to be supplied in feeds at levels that promote their deposition in fish tissues, and these are levels well in excess of those required for optimal growth and health of the fish.

Twenty years ago, at the time of the third edition of the book "Fish Nutrition," qualitative EFA requirements were relatively easily summarized as being satisfied generally by C_{18} PUFA, LNA, and ALA, in freshwater and salmonid species, while marine species require LC-PUFA, EPA, DHA, and ARA (Sargent et al., 2002). The differences between species being based on their ability, or not, for the endogenous biosynthesis of LC-PUFA from C_{18} PUFA (Tocher, 2010). However, in recent years, there have been considerable advances in our knowledge of LC-PUFA biosynthesis in fish species that has led us to completely reevaluate this generalization (Castro et al., 2016c; Monroig et al., 2016, 2018). The last edition of this book focused on the impact of developmental life stage (larvae, juvenile, adult, and broodstock) on EFA requirements (Sargent et al., 2002), and detailed qualitative and quantitative EFA requirements of fish have been reviewed more recently (Glencross, 2009; Tocher, 2010; NRC, 2011). The following sections focus on the molecular and genetic components of the LC-PUFA biosynthesis pathway, phylogenetic relationships, evolutionary history, and how they have provided further insight into our understanding of EFA requirements of fish.

5.6.3 LC-PUFA biosynthesis

In vertebrates, the biosynthesis of LC-PUFA from dietary essential C_{18} PUFA can occur through a number of potential pathway options, which consist of a series of alternating elongation and desaturation steps in microsomes that introduce 2C units and double bonds, respectively, into existing fatty acyl chains (Fig. 5.11) (Monroig et al., 2011b). However, production of DHA from EPA can require an extra elongation and a subsequent peroxisomal chain shortening step in a pathway termed the "Sprecher shunt" (Sprecher, 2000). Differences between species in their ability to produce LC-PUFA are dependent upon the presence or absence of the genes for the fatty acyl desaturase and elongation of very long chain fatty acid enzymes (Monroig et al., 2018). In mammals, the pathway consists of two desaturase genes, *fads1* and *fads2*, coding for Δ5 FADS (fatty acyl desaturase) and Δ6 FADS desaturases (Castro et al., 2012), and two elongase genes, *elovl2* and *elovl5*, coding for ELOVL2 (Elongation of very long chain fatty acids 2) and ELOVL5 elongases, respectively (Jakobsson et al., 2006), and production of DHA from EPA is via the Sprecher shunt (Sprecher, 2000). The desaturase and elongase enzymes operate on both n-6 and n-3 PUFA, and so there is competition between the two PUFA series in these pathways (Monroig et al., 2016). The Δ5 and Δ6 fatty acyl desaturases (FADS1 and FADS2) are termed "front-end" desaturases as they introduce double bonds into fatty acyl chains between the α carbon in the carboxyl group and the middle of the molecule (Napier et al., 1999). FADS2 catalyzes the desaturation of ALA to 18:4n-3 (and LNA to 18:3n-6), and FADS1 catalyzes the desaturation of 20:4n-3 to EPA (and 20:3n-6 to ARA) (see Fig. 5.11). While there are at least seven families of fatty acid elongases, ELOVL1-7, only two have major roles in the biosynthesis of LC-PUFA in mammals, ELOVL2 and ELOVL5. While ELOVL5 elongates mainly C_{18} and to a lesser extent C_{20} PUFA, ELOVL2 has specificity toward C_{20} and C_{22} PUFA (Fig. 5.11) (Jakobsson et al., 2006). However, relatively few mammalian species have been studied and, while the LC-PUFA biosynthesis pathway is present in rodents and humans (Cook and McMaster, 2004), it is not complete in all mammals with carnivores such as cats unable to produce ARA from LNA due to a lack of Δ6 and, likely, Δ5 desaturase activities (Rivers et al., 1975; Hassam et al., 1977).

In contrast, the situation in fish is complicated and becoming more so as more species are investigated (Monroig et al., 2018). The interest in the LC-PUFA biosynthesis pathways in fish has been driven by aquaculture as a consequence of the development of sustainable feeds, which has essentially been achieved by the replacement of finite and limiting marine ingredients, FM and FO, by ingredients derived from terrestrial agriculture, plant meals, and VO, that do not contain LC-PUFA (Tocher, 2009; Ytrestøyl et al., 2015; Aas et al., 2019). Therefore, sustainable feeds with high levels of FM and FO replaced by plant ingredients have much lower levels of LC-PUFA, especially EPA and DHA, and so fish reared on these feeds also have reduced levels of n-3 LC-PUFA (Turchini et al., 2011c; Henriques et al., 2014; Sprague et al., 2016). This has prompted considerable research into LC-PUFA biosynthesis pathways in fish in the 20 years since the last edition of "Fish Nutrition" in order to optimize and/or maximize endogenous production from VO-derived C_{18} PUFA, as well as providing insight into the qualitative EFA requirements of new species with farming potential. The molecular components of the LC-PUFA biosynthesis pathways, the fatty acyl desaturase and elongase genes, transcripts, and enzymes, have been the subject of several reviews (Tocher, 2010; Monroig et al., 2011b; NRC, 2011; Torstensen and Tocher, 2011; Castro et al., 2012; Castro et al., 2016c; Monroig et al., 2016; Monroig et al., 2018) that have comprehensively covered their fatty acid specificities and activities, as well as species and tissue distributions, and so the present will only focus on the aspects of most relevance to fish nutrition.

5.6.3.1 Desaturases

Similar to mammals, Elasmobranchs have *fads1* and *fads2* genes that were shown in the small-spotted catshark *Scyliorhynus canicula* to code for functional Δ5 and Δ6 desaturases (Castro et al., 2012). Recently, a genome scan study has shown that within the Elopomorpha, one of the oldest teleost lineages that includes tarpons, bone fish, and eels, some species also have both *fads1* and *fads2* genes (Lopes-Marques et al., 2018). Functional characterization of the Fads1 and Fads2 of Japanese eel confirmed that they code for functional Δ5 and Δ6 Fads, respectively (Lopes-Marques et al., 2018). However, it appears that the *fads1* gene has been lost during evolution in the vast majority of teleosts, and some species such as pufferfish appear to have no *fads* genes at all (Leaver et al., 2008a; Castro et al., 2012). The number of *fads2* genes in teleosts can vary from 1 (e.g., zebrafish) to 4 (e.g., salmonids) possibly related to WGD and duplication-loss events, which have been followed by sub- (acquisition of additional activities) and neo-functionalization (acquisition and/or substitution of new activities) such that Fads2 in teleosts can have a range of discrete or combined desaturase activities (Castro et al., 2012; Monroig et al., 2018). Most commonly, *fads2* genes in teleosts code for discrete Δ6 desaturases (Monroig et al., 2018), but other *fads2* genes code for discrete Δ5 desaturases such as in Atlantic salmon and rainbow trout (Hastings et al., 2004; Abdul Hamid et al., 2016). While the functional diversification of teleost Fads2 fatty acyl desaturases was initially thought to relate to trophic level, this has since been shown to be not the case (Garrido et al., 2019).

However, teleost species are also characterized by having desaturase activities so far not found in other vertebrates. Therefore, some *fads2* genes in teleosts code for a discrete Δ4 desaturase such as in rabbitfish *Signaus canaliculatus* (Li et al., 2010c), Senegalese sole (Morais et al., 2012a), pike silverside *Chirostoma estor* (Fonseca-Madrigal et al., 2014), and striped snakehead *Channa striata* (Kuah et al., 2015). Subsequently, the presence of Δ4 Fads2 was identified in a further 11 species and activity as Δ4 desaturases confirmed for Fads2 of medaka *Oryzias latipes* and Nile tilapia (Oboh et al., 2017a). This desaturase activity enables the more direct production of DHA from EPA via a single elongation to 22:5n-3 (DPA) followed by Δ4 desaturation of DPA to produce DHA. Other teleost Fads2 desaturases are bifunctional, enabling them to introduce two different double bonds into carbon chains. Thus, zebrafish have just one desaturase that has both Δ6 and Δ5 activity (Δ6Δ5 Fads2) enabling production of EPA from ALA (Hastings et al., 2001). This is now known to be a relatively common feature of teleost Fads2 with further bifunctional desaturases reported in rabbitfish, Nile tilapia, pike silverside, African catfish (*Clarias gariepinus*), and striped snakehead (Li et al., 2010c; Tanomman et al., 2013; Fonseca-Madrigal et al., 2014; Kuah et al., 2016; Oboh et al., 2016).

Many teleost Fads2 Δ6 desaturases can also have Δ8 desaturation activity and, therefore, the ability to desaturate 20:3n-3 and 20:2n-6 to 20:4n-3 and 20:3n-6, respectively (Monroig et al., 2011a). However, it is important to note that this is not regarded as bifunctionality. This is because the bond introduced into the fatty acyl chain is actually the same one and the position of the desaturation as Δ6 or Δ8 is simply dependent upon whether the desaturation occurs before (Δ6 activity; ALA to 18:4n-3) or after (Δ8 activity; 20:3n-3 to 20:4n-3) elongation of the chain. Similarly, the production of 24:6n-3 from 24:5n-3 in the Sprecher shunt is actually desaturation at the Δ6 position and so is catalyzed by Δ6 FADS2 desaturase in mammals (de Antueno et al., 2001; D'Andrea et al., 2002) and, similarly, Δ6 Fads2 in zebrafish and salmon have been demonstrated to be able to desaturate 24:5n-3 to 24:6n-3 (Tocher, 2003). While Fads are expressed in most tissues in teleost fish, highest expression is found in liver and intestine in salmon and other, often freshwater, species whereas in marine teleost species, brain and eye are also tissues with high expression (Monroig et al., 2018).

5.6.3.2 Elongases

Similar to mammals, Elovl5 and Elovl2 are the primary elongases in the LC-PUFA biosynthesis pathway in fish. Elovl5 elongase appears almost ubiquitous in teleosts having been cloned and functionally characterized in numerous species (Monroig et al., 2016, 2018). In contrast, Elovl2 elongase has only been isolated and functionally characterized in a few species including Atlantic salmon (Morais et al., 2009), rainbow trout (Gregory and James, 2014), zebrafish (Monroig et al., 2009), and African catfish (Oboh et al., 2016). Both these elongases have chain length specificities similar to mammalian ELOVLs. Thus, Elovl5 has primary specificity for C_{18} PUFA with lower activity toward C_{20} PUFA, whereas Elovl2 has specificity for predominantly C_{20} PUFA and, to a lesser extent, C_{22} PUFA (Monroig et al., 2016, 2018). However, as is the case with desaturases, the LC-PUFA elongase repertoire in teleosts is more complicated than in mammals. Specifically, two Elovl4 elongases (a and b) exist in teleost fish compared to only one in mammals (Castro et al., 2016c). Both teleost isoforms are primarily involved in the production of saturated and unsaturated very-long chain fatty acids (VLC-FA) up to C_{36} and have tissue expression patterns reflecting the tissues where VLC-FA have functional roles such as retina/eye, whereas elongases 2 and 5 have a more widespread distribution in most tissues (Monroig et al., 2010, 2011c; Carmona-Antoñanzas et al., 2011; Kabeya et al., 2015; Li et al., 2017; Oboh et al., 2017a). In addition, however, teleost Elovl4, particularly Elovl4b, also has a role in the biosynthesis of DHA from EPA whereas ELOVL4 in mammals does not operate on EPA or DPA and so is not involved in the biosynthesis of DHA (Monroig et al., 2018).

5.6.3.3 Biosynthesis of EPA from ALA

The biosynthesis of LC-PUFA in animals/vertebrates is commonly reported as a single pathway or pathway options whereby ALA is converted to DHA (or LNA to 22:5n-6). In contrast, in fish it is more appropriate to investigate and discuss LC-PUFA biosynthesis as two separate pathways, the EPA pathway from ALA to EPA, and the DHA pathway from EPA to DHA, with EPA being the key component common to both pathways.

The EPA pathway requires two desaturations via Δ6/Δ8 (depending upon whether ALA is first desaturated or elongated) and Δ5 desaturases and Elovl5 elongase (Cook and McMaster, 2004), and is often incomplete in teleost fish, primarily due to a lack of Δ5 desaturase, and so synthesis of EPA from ALA is not possible in many, particularly marine, carnivorous species (Monroig et al., 2018). It is therefore the EPA pathway, specifically the presence or absence of the ability to make EPA from ALA that is the major difference between fish species that require dietary EPA and DHA (and ARA) to satisfy EFA requirements and species whose EFA requirements can be satisfied by ALA (and LNA) (Tocher, 2010). Furthermore, in contrast to mammalian LC-PUFA biosynthesis where Δ6 desaturation is regarded as rate-limiting (Guillou et al., 2010), at least for the production of ARA from LNA, the lack of *fads1* in teleost genomes dictates that the distinctive trait of fish EPA and ARA biosynthesis is that it is limited by Δ5 desaturation (Monroig et al., 2018).

The evolutionary history of the FADS gene family and their respective Δ5 and Δ6 activities in vertebrates has been studied in recent years (Castro et al., 2012, 2016c; Lopes-Marques et al., 2018). This showed that Fads1 and Fads2 genes with distinct, separate Δ5 and Δ6 activities, respectively, emerged in gnathostome ancestry, since the cartilaginous basal gnathostome small-spotted catshark, formerly commonly known as the lesser-spotted dogfish, has functional orthologues of both gene families, as described above (Castro et al., 2012). However, while events of Fads1 gene expansion have taken place in other vertebrate groups including birds and reptiles, the Δ5 Fads1 gene was lost in the vast majority of teleosts other than the oldest lineages such as Elopomorpha (Lopes-Marques et al., 2018). Therefore, it is clear that the acquisition of the capability for Δ5 desaturation, whether as a mono- or bifunctional Fads2 enzyme, presents an advantage to these teleost species as it compensates this loss of Δ5 Fads1 during evolution (Castro et al., 2012). This process may have been driven by the contrasting levels of LC-PUFA in different environments (e.g., lower in freshwater and higher in marine) leading to high or low evolutionary pressure to retain LC-PUFA biosynthesis and is now reflected in the differing qualitative EFA requirements of freshwater/omnivorous versus marine/carnivorous fish species (Leaver et al., 2008a; Carmona-Antoñanzas et al., 2013). In summary, Fads2 enzymes enabling both Δ6 and Δ5 desaturations as well as Elovl5 are required to enable C_{18} PUFA, ALA, and LNA, to be converted into the physiologically important C_{20} LC-PUFA, EPA, and ARA (Castro et al., 2016c).

5.6.3.4 *Biosynthesis of DHA from EPA*

The DHA pathway, the production of DHA from EPA, can proceed in fish via two possible mechanisms, either directly through elongation of EPA to DPA followed by Δ4 desaturation, or via the "Sprecher shunt" that involves elongation to 24:5n-3, Δ6 desaturation (by the same Δ6 desaturase as in the EPA pathway), and peroxisomal chain shortening (Oboh et al., 2017b). Phylogenetic analyses has suggested that the Sprecher pathway is likely present in the majority of teleost species, although the level of activity of the pathway may be generally higher in species of early diverging teleosts, while the Δ4 desaturase has hitherto been reported only in late-diverging teleosts, more specifically species belong to the superorder Acanthopterygii (Oboh et al., 2017a; Lopes-Marques et al., 2018; Machado et al., 2018). Irrespective of precise biochemical mechanism, the DHA pathway from EPA is likely functional in most teleost fish, including marine species, at least in tissues where DHA is particularly critical like brain and eye (Tocher et al., 2006; Zheng et al., 2009; Monroig et al., 2011b), highlighting the importance of endogenous DHA biosynthesis for neural development and function (Zheng et al., 2009). Thus, DHA production from EPA is more conserved than EPA production from ALA consistent with an evolutionary adaptation reflecting the important role of DHA in neural tissues. The absence of an *elovl2* gene in many teleosts suggests that the DHA pathway may also have an impact on fish EFA requirements as a lack of Elovl2 activity could limit DHA production from EPA, although this may be partly mitigated in many species by the presence of Elovl4 that could compensate for the lack of *elovl2* genes (Monroig et al., 2011c).

In species, such as many freshwater species and salmonids, with all the molecular (desaturase and elongase) components necessary for the biosynthesis of LC-PUFA (Tocher, 2010;

Monroig et al., 2011b), there is no physiological driver or evolutionary pressure to synthesize excess EPA and DHA just to be deposited in TAG and stored (Leaver et al., 2008a; Castro et al., 2012). While efforts have been made to enhance endogenous LC-PUFA production through dietary supplements including micronutrients (Senadheera et al., 2012a, 2012b; Lewis et al., 2013; Giri et al., 2016) and biochemicals such as conjugated LNA, 3-thia acids, petroselinic acid, fibrates, and sesamin (Ruyter et al., 1997b; Moya-Falcón et al., 2004; Kennedy et al., 2006; Trattner et al., 2008a, 2008b; Randall et al., 2013), any increased biosynthetic activity has been small and insufficient to fully compensate for the lack of dietary LC-PUFA. Therefore, it appears that endogenous biosynthesis can only produce the relatively low levels of EPA and DHA required to satisfy the normal physiological requirements of fish (Sprague et al., 2019), or little more (Mock et al., 2019), and there appears limited prospect that biosynthesis in fish can be enhanced beyond these levels (Tocher, 2015).

5.6.3.5 *Molecular mechanisms of regulation*

In recent years, the molecular regulation of the LC-PUFA biosynthesis pathway has received increasing interest. In particular, there has been significant research into the mechanisms of transcriptional regulation, particularly in the marine teleost rabbitfish that is regarded as a potential representative model species due to it having both EPA and DHA pathways and thus able to produce DHA from ALA. These studies have revealed roles for several transcription factors including hepatocyte nuclear factor 4α (Hnf4α) (Dong et al., 2016; Wang et al., 2018c), Lxr and Srebp1 (Zhang et al., 2016b), and Pparγ (Li et al., 2019c; Sun et al., 2021). The transcription factor Sp1 has also been implicated as a positive regulator of LC-PUFA biosynthesis in rabbitfish (Li et al., 2019d) and orange-spotted grouper (Xie et al., 2018) by directly targeting Δ6/Δ5 Fads2 and Elovl5. Further studies in rabbitfish have also investigated microRNAs (miR) and recently reported potential roles for miR-33 (Zhang et al., 2016b; Sun et al., 2019), miR-24 (Chen et al., 2019), miR-15/16 cluster (Sun et al., 2021), and miR-26a (Chen et al., 2021) in the regulation of LC-PUFA biosynthesis. The mechanisms of miR regulation appear to involve specific transcription factors including Srebp1 (miR-33 and miR-26a) (Sun et al., 2019; Chen et al., 2020), Pparα (miR-15/16) (Sun et al., 2021), and Lxr (miR-26a) (Chen et al., 2020).

5.6.4 Dietary impacts

The activity of the LC-PUFA biosynthetic pathway(s) can be significantly impacted by diet (Leaver et al., 2008a; Bell and Tocher, 2009a; Torstensen and Tocher, 2011). Indeed in recent years, it has been changes to aquaculture feed formulations, specifically the development of more sustainable feeds with reduced inclusion of FM and FO and, thus, much lower levels of the LC-PUFA, EPA, DHA, and ARA, that have stimulated much of the research into EFA requirements of fish and the role of endogenous LC-PUFA production via the biosynthetic pathways (Tocher, 2015). The impact of sustainable feeds on LC-PUFA production in different fish species depends upon whether the fish has complete pathways for the production of EPA and/or DHA, which, in turn, depends on the genes and enzyme activities that are present (Monroig et al., 2018). Various aspects of these impacts have been reviewed previously (Leaver et al., 2008a; Bell and Tocher, 2009b; NRC, 2011; Torstensen and Tocher, 2011) and, therefore, are summarized only briefly below.

The nutritional regulation of LC-PUFA biosynthesis appears to be largely transcriptional, with the expression of Δ6 and Δ5 *fads* and some *elovl* genes being upregulated by around 2–3–fold in fish fed more sustainable feeds formulated with high levels of VO compared to fish fed traditional FO-based feeds (Vagner et al., 2007; Leaver et al., 2008a, 2008c; Torstensen and Tocher, 2011). In salmonids, the increased gene expression translates into similar fold increases in LC-PUFA biosynthesis in liver and intestine (Zheng et al., 2004, 2005; Torstensen and Tocher, 2011). The molecular mechanism is that *fads* and *elovl* gene expression is suppressed by dietary LC-PUFA, specifically DHA, in fish fed diets rich in FO, and this suppression is reduced by the lower level of DHA in diets formulated with high levels of VO (Leaver et al., 2008c; Thomassen et al., 2012). Some studies suggest that increased substrate concentration (i.e., ALA) may also stimulate enzyme activities (Thanuthong et al., 2011), and *elovl5*, but not *fads*, gene expression (Gregory et al., 2016). The precise mechanisms whereby dietary PUFA exert effects on *fads* and/or *elovl* gene expression is likely through their role as ligands of transcription factors important in fatty acid and lipid biosynthesis including Srebp (Minghetti et al., 2011; Carmona-Antoñanzas et al., 2014). However, it is important to stress what this metabolic upregulation represents in terms of endogenous LC-PUFA production. Thus, metabolic pathways such as LC-PUFA biosynthesis have evolved to satisfy nutrient, specifically EFA, requirements of the fish and, as such, are only sufficient to produce the relatively small amounts of EFA required to satisfy the physiological demands of the fish (see above). The upregulation of LC-PUFA biosynthesis is not sufficient to maintain tissue EPA and DHA at the same level as in fish fed traditional diets based on FO and, therefore, cannot prevent farmed fish fed sustainable feeds formulated with VO having lower levels of EPA and DHA (Bell and Tocher, 2009a; Tocher et al., 2011). In light of current knowledge, it appears that the most likely, if not the only, way to have high levels of n-3 LC-PUFA in farmed fish is to include these fatty acids in the feed. Accordingly, alternative sources of EPA and DHA have also been the subject of considerable research activity in the last decade, as will be discussed in Section 5.13.

5.7 Lipid catabolism and energy production

A major role of lipids, especially TAG, is to store energy, although the majority of the energy is actually held in the constituent fatty acids and released through β-oxidation. Lipid catabolism, as a fundamental metabolic process, represents the breakdown of lipids for further usage, including energy provision. Therefore, one important role of dietary lipid is to provide energy to support fish growth and other physiological processes. Intact TAG cannot enter cells (Whitehead, 1909) and, therefore, require hydrolysis, specifically termed lipolysis, to provide and maintain the supply of energy-rich fatty acids. Overall, production of energy from TAG requires three steps: (1) release of fatty acids from TAG, (2) entry of fatty acids into mitochondria or peroxisomes, and (3) generation of metabolic energy in the form of ATP via the β-oxidation of fatty acids. In recent years, these cellular steps have been verified in fish, and the detailed biological processes and some molecular regulatory mechanisms have been elucidated. Briefly, the release of FFA by degradation of TAG in fish can occur via two metabolic routes: neutral lipid hydrolase (lipase)-mediated lipolysis (Wang et al.,

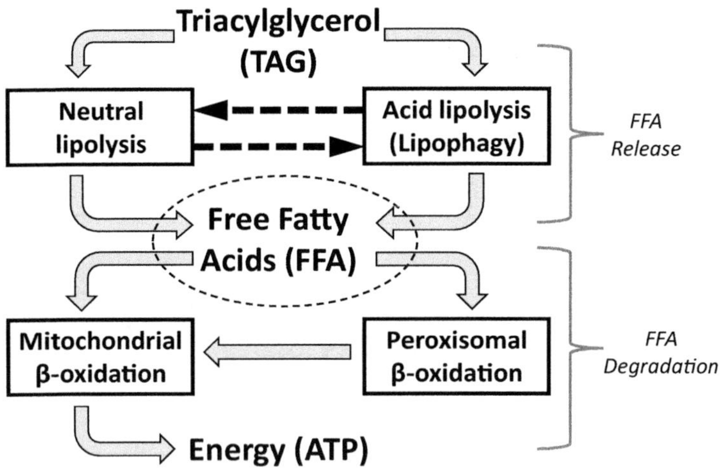

FIGURE 5.12 The cellular triacylglycerol (TAG) catabolism-energy production system. Energy production from TAG consists of two main processes: (1) free fatty acid (FFA) release as a result of neutral and acid lipolysis; (2) FFA degradation by mitochondrial or peroxisomal β-oxidation. Interactions exist between neutral and acid lipolysis, and between mitochondrial and peroxisomal β-oxidation.

2012b; Sun et al., 2018a; Zhang et al., 2019b) and lysosomal acid lipase-mediated lipolysis (lipophagy) (Wang et al., 2018b, 2019b). Moreover, the biochemical reactions of mitochondrial and peroxisomal fatty acid β-oxidation have been described in several fish species (Frøyland et al., 2000b; Tocher, 2003; Du et al., 2006a). Therefore, the entire cellular TAG catabolism-energy production system has been established in fish (Fig. 5.12).

5.7.1 Lipolysis

Lipolysis in animals can be either extracellular or intracellular. Extracellular lipolysis, for example in stomach and intestine, is a key step in lipid digestion, and as such, has been covered in Section 5.4. The present section focusses on intracellular lipolysis that can occur by either neutral lipase (around pH 7) or acid lipase (between pH 4 and 5) (Schweiger et al., 2006).

5.7.1.1 Neutral lipolysis and its regulation

In mammals, neutral lipolysis accounts for more than 90% of total TAG hydrolysis (Schweiger et al., 2006). Three main neutral lipases, adipose TAG lipase (ATGL), hormone-sensitive lipase (HSL), and MAG lipase (MGL), are required for the hydrolysis of fatty acids from TAG (Fig. 5.13). The ATGL catalyzes the first step of TAG hydrolysis to provide fatty acid and DAG. While HSL can also hydrolyze TAG to DAG, it is mainly responsible for the hydrolysis of DAG to MAG. The MGL, as its name implies, hydrolyzes MAG to FFA and glycerol. Although studies on lipases go back to the 19th century (Balser, 1882; Langerhans, 1890), it was the middle of the 20th century before HSL and MGL were described (Vaughan et al., 1964), with ATGL first described in 2004 (Jenkins et al., 2004; Zimmermann et al., 2004).

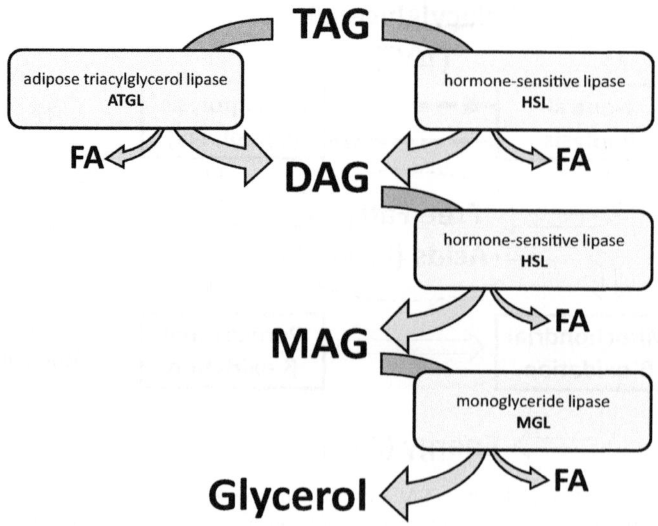

FIGURE 5.13 The lipase system and biochemical process in triacylglycerol (TAG) neutral lipolysis. During this process, three main neutral lipases, ATGL, HSL, and MGL, are required and play specific roles in degrading TAG to diacylglycerol (DAG), then to monoacylglycerol (MAG) and finally glycerol and free fatty acids.

Although studies are limited, neutral lipolysis in fish can be stimulated by several conditions, such as fasting, as in mammals. For example, 15-day fasting stimulated *atgl* and *hsl* expression in European sea bass (Rimoldi et al., 2016), similar to fasted Nile tilapia (Wang et al., 2017e), zebrafish (Lu et al., 2019), and grass carp (Sun et al., 2016). The *atgl* cDNA has been cloned and its structure intensively analyzed in several fish species (Wang et al., 2012b; Sun et al., 2016; Dai et al., 2018). Interestingly, its structure contains the "patatin" domain and "GXSXG" motif highly similar to mammals, which is not surprising as TAG hydrolysis is a common biochemical reaction in all organisms. The *atgl* in large yellow croaker has a similar length and the predicted amino acid sequence showed >93% similarity with mice, pig, chicken, and human *ATGL* (Wang et al., 2012b). Moreover, the N-terminal domain and potential phosphorylation sites in the C-terminal catalytic domain of *atgl* in grass crap was highly conserved among teleosts, humans, and mice (Sun et al., 2016). Although *atgl* has a similar structure in fish as in mammals (Sheridan and Harmon, 2012), *hsl* shows some differences. First, the teleost *hsl* has two isoforms (*hsl1* and *hsl2*), likely reflecting the teleost-specific genome duplication (TSGD) (Kittilson et al., 2011; Chen et al., 2014; Sun et al., 2016; Li et al., 2019b). Moreover, human HSL has five phosphorylation sites at the C-terminal, while grass carp Hsl1 has one potential phosphorylation site and Hsl2 has three (Contreras et al., 1996; Anthonsen et al., 1998; Sun et al., 2016). The isoforms also exhibited differences in mRNA expression, for example *hsl1* showed higher expression than *hsl2* in rainbow trout in response to growth hormone-induced lipolysis (Kittilson et al., 2011; Bergan et al., 2013). The obvious expression difference between *hsl1* and *hsl2* has also been shown in several fish species (Chen et al., 2014; Khieokhajonkhet et al., 2014, 2016; Sun et al., 2016). These differences suggest that neutral lipolysis in fish can be regulated by different pathways compared to mammals as discussed below. Although studies on Mgl in fish are limited, its structure and function in grass carp was similar to mammals (Sun et al., 2016). In general, most studies suggest that neutral lipase enzymes for TAG hydrolysis are similar in fish and mammals and their functions are likely conserved among species.

Neutral lipolysis in mammals is mediated via the cyclic adenosine monophosphate or protein kinase A (cAMP/PKA) pathway (Miyoshi et al., 2007; Shen et al., 2009; Wang et al., 2009a). This signaling pathway likely mediates neutral lipolysis in fish as the potential phosphorylation sites on *atgl* and *hsl* are conserved (Kittilson et al., 2011; Khieokhajonkhet et al., 2014; Sun et al., 2016). Accordingly, Hsl activity was activated 137% in rainbow trout in the presence of cAMP (Michelsen et al., 1994), and the PKA inhibitor H89 reduced neutral lipolysis in yellow catfish (Song et al., 2017). Similarly, the indirect activation of *pka* and *hsl* by forskolin significantly upregulated neutral lipolysis in Nile tilapia (Zhang et al., 2019b). The protein kinase C/mitogen-activated protein kinase (PKC/MAPK) pathway was found to also play a role in neutral lipolysis stimulation in 3T3-L1 cells (Greenberg et al., 2001), and the MAPK inhibitor PD98059 inhibited neutral lipolysis in rainbow trout, suggesting that MAPK may also be involved in fish neutral lipolysis (Albalat et al., 2005b). Moreover, the PKC and MEK/ERK pathways were activated during neutral lipolysis stimulation, and their inhibition largely suppressed neutral lipolysis in rainbow trout hepatocytes (Bergan et al., 2013, 2015). These studies suggest that the regulation of neutral lipolysis is generally conserved in mammals and fish.

Various endogenous and exogenous factors have been demonstrated to influence neutral lipolysis in fish and its regulation in practical aquaculture (Fig. 5.14). Several hormones have been shown to regulate neutral lipolysis in fish. The mechanism of catecholamine-induced neutral lipolysis through the classic regulation pathway has been described in mammals. However, while epinephrine stimulated neutral lipolysis in rainbow trout as in mammals, norepinephrine inhibited this process in fish as opposed to mammals (Magnoni et al., 2008). Similarly, neutral lipolysis was inhibited by catecholamines in different teleost species (Ince and Thorpe, 1975; Minick and Chavin, 1973; Van Raaij et al., 1995; Vianen et al., 2002).

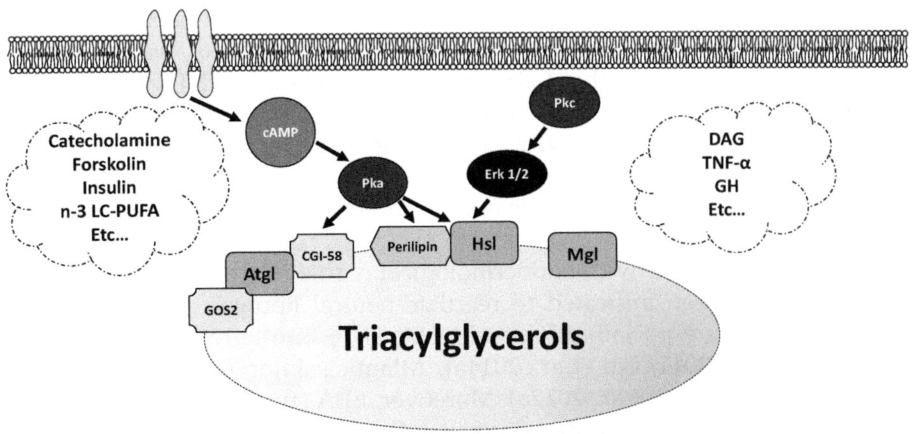

FIGURE 5.14 Regulation of neutral lipolysis. Neutral lipolysis is precisely regulated by hormones, nutrients, and certain metabolites, which play roles through cellular signaling-dependent or signaling-independent processes.

Differences in the effects of catecholamines on neutral lipolysis between fish and mammals may be explained by the different living environments. For example, hypoxia, which increases catecholamine levels, is a common stress for fish (Montpetit and Perry, 1998; Thillart et al., 2002; Perry et al., 2005), and the increase in FFA levels stimulated by catecholamine may cause lipotoxicity and subsequent fish death. Therefore, fish maintain some common regulation of neutral lipolysis as in mammals, but may also require different coping strategies (Vianen et al., 2002).

Another important hormone regulating neutral lipolysis is GH, which upregulates neutral lipolysis in several fish species (Albalat et al., 2005a; Bergan et al., 2013; Vélez et al., 2019). In rainbow trout, GH induced neutral lipolysis through phospholipase C (PLC)/Protein kinase C (PKC) and mitogen-activated protein kinase kinase (MEK)/extracellular signal-regulated kinase (ERK) pathways (Bergan et al., 2012). Insulin and glucagon also showed ability to regulate lipolysis in fish (Harmon and Sheridan, 1992; Albalat et al., 2005a). In mammals, insulin decreased cAMP level and reduced PKA activation, while glucagon is involved downstream of cAMP signaling (Vendrell et al., 2011). In fish, glucagon stimulated phosphorylation of hepatic TAG lipase while insulin depressed phosphorylation by 56% and inhibited glucagon-stimulated neutral lipolysis in trout (Harmon et al., 1993). Leptin, another important hormone in energy homeostasis, increased mRNA expression of *hsl* and *atgl* and thus decreased lipid deposition in grass crap (Lu et al., 2012) and Nile tilapia (Liu et al., 2018a). Leptin may stimulate neutral lipolysis via the JAK-STAT pathway as indicated by a study in grass carp (Lu et al., 2012). Other researchers demonstrated that leptin-induced neutral lipolysis may be an outcome of effects on food intake and energy expenditure (Li et al., 2010a; Won et al., 2012). These studies have indicated that hormones play critical roles in regulating lipid catabolism in fish, as in mammals, although some different phenotypes between fish and mammals suggest that fish may have species-specific endocrinological regulatory mechanisms.

The potential roles of some inflammatory factors in lipid catabolism have attracted interest. Studies have shown that TNFα can promote neutral lipolysis in various fish species, including rainbow trout (Albalat et al., 2005b), large yellow croaker (Wang et al., 2012a), and sea bass (Cruz-Garcia et al., 2009). A recent study indicated that G0S2 (G0/G1 Switch gene 2), which is a specific peptide inhibitor for ATGL (Yang et al., 2010), was downregulated by TNFα in grass crap, suggesting that G0S2-ATGL is involved in TNFα regulated-neutral lipolysis (Sun et al., 2018a). In rainbow trout, the ERK1/2 inhibitor, PD98059, partially inhibited neutral lipolysis caused by TNFα, suggesting that TNFα-induced lipolysis in fish might be mediated partly through the activation of ERK1/2 (Albalat et al., 2005c). Further studies are required to reveal more functions of immune-related cytokines in neutral lipolysis in fish.

Nutrition is closely correlated with endocrinological regulation, and various nutritional/lipid-related factors have been indicated to regulate neutral lipolysis (Bergan et al., 2015). For example, n-3 LC-PUFA, especially EPA and DHA, significantly reduced lipid accumulation in grass carp (Ji et al., 2011; Liu et al., 2014a), Atlantic salmon (Todorčević et al., 2009), and large yellow croaker (Wang et al., 2012a). Moreover, EPA stimulated *atgl* and *hsl1* mRNA expression, but not *hsl2* in grass crap preadipocytes (Lei and Tian, 2017). Meanwhile, DHA suppressed lipid accumulation in grass carp adipocytes by inducing adipocyte apoptosis (Jin et al., 2018a). Diets supplemented with 0.3% ARA reduced fat accumulation in grass carp (Tian et al., 2014, 2017a), but this could be reversed by inhibitors of COX and LOX, indicating that ARA may also regulate neutral lipolysis through LOX and COX pathways (Tian et al., 2016).

In fasted gilthead seabream, neutral lipolysis efficiency correlated with dietary protein content, and moreover, dietary plant protein induced higher neutral lipolysis efficiency in comparison to FM (Albalat et al., 2005a). In addition, a study on rainbow trout also showed that plant proteins can promote fat utilization (Kaushik et al., 1995). As both gilthead seabream and rainbow trout are carnivorous fish, the upregulation of neutral lipolysis caused by plant protein sources may be a compensatory response to increase energy supply given the lower utilization of dietary plant components. Therefore, further studies are required to verify the regulation of neutral lipolysis by dietary protein composition. Knowledge of the impact of micronutrients on neutral lipolysis is also very limited. Dietary inclusion of more than 300 and 600 mg of magnesium (Mg) per kg feed significantly reduced body fat content in grass carp (Wang et al., 2011), and common carp (Dabrowska et al., 1991), respectively. Recently, a study indicated that magnesium ions induced hepatic lipolysis in yellow catfish by regulating PPAR, JAK-STAT, and AMPK pathways (Wei et al., 2017a), revealing that minerals, such as Mg, could also play roles in regulating lipolysis.

Some feed additives such as silymarin, DAG, and α-lipoic acid have been reported to regulate neutral lipolysis in fish. Silymarin is a flavonolignan from the milk thistle (*Silybum marianum*), which has been shown to suppress lipogenic genes and enhance neutral lipolysis in rats (Chou et al., 2012). Dietary silymarin significantly reduced hepatic lipid content and upregulated *hsl* and *cpt1* in grass carp, suggesting it regulated neutral lipolysis and β-oxidation (Xiao et al., 2017). Although the fatty acid composition is similar to TAG, DAG plays a different physiological role in prevention of body fat accumulation and obesity (Umeda et al., 2006; Hue et al., 2009). Indeed, a DAG-diet reduced fat accumulation by inducing neutral lipolysis and β-oxidation in Nile tilapia (Zhang et al., 2019b). α-Lipoic acid exerted its antiobesity effect in rodents by suppression of hypothalamic AMP-activated protein kinase (AMPK) activity (Kim et al., 2004; Stark et al., 2013). In grass carp, dietary α-lipoic acid activated phosphorylation of Ampkα and notably increased Atgl activity, thus promoting lipid catabolism (Shi et al., 2018b). These studies suggested that an evaluation of the potential functions of practical feed additives on lipolysis could be an important new direction for research.

5.7.1.2 Acidic lipolysis (lipophagy)

In cells, TAG are deposited in intracellular lipid droplets (LD) that are dynamic and complex organelles that function in energy metabolism and membrane synthesis by regulating lipid storage and release (Welte and Gould, 2017). Autophagy is a degradation process, present widely in eukaryotic cells, that is important for cell survival, renewal, and homeostasis (Yorimitsu and Klionsky, 2005). While our knowledge of autophagy dates back to the early 1960s, the degradation of LD through the lysosomal pathway was only described in 2009, leading to the term "lipophagy" (Singh et al., 2009; Schulze et al., 2017). As lysosomal lipases work at an acidic pH, lipophagy is also regarded as acidic lipolysis (Zechner et al., 2017). According to mammalian studies, lipophagy involves several cellular processes (Ward et al., 2016): (1) LDs are engulfed by double-membrane vesicles (phagophores: LC3-II positive membranes) to form autophagosomes; (2) autophagosomes fuse with lysosomes to form autolysosomes; (3) acidic lipases hydrolyze lipids to glycerol and FFA (Weidberg

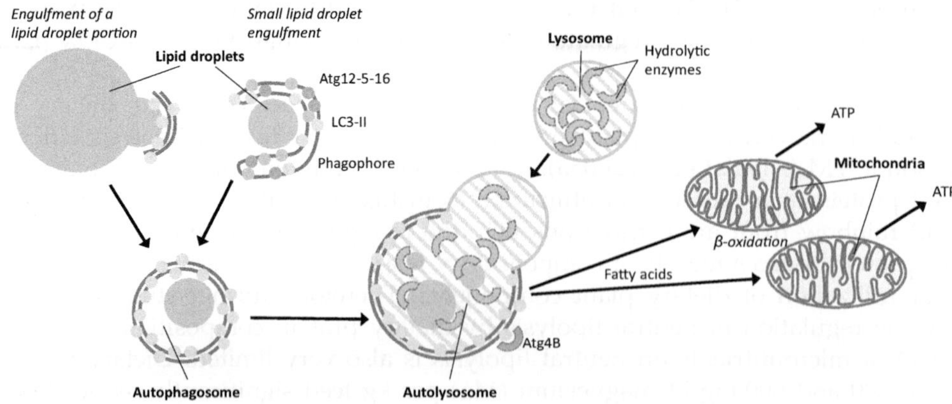

FIGURE 5.15 Cellular processes of lipophagy. Lipophagy is essential in degrading triacylglycerols (TAG) existing in lipid droplets (LD). During this process, LD are firstly engulfed by double-membrane vesicles (phagophores) to form autophagosomes, and then the autophagosomes are fused with lysosomes to form autolysosomes. The lysosome-sourced acidic lipases finally hydrolyze TAG to glycerol and free fatty acids for further energy production.

et al., 2009; Singh and Cuervo, 2012; Wang, 2016; Zhang et al., 2018c). The fatty acids then undergo further metabolism, such as mitochondrial β-oxidation for the provision of energy (Fig. 5.15).

During the process of lipophagy, dozens of autophagy-related proteins (ATG) are sequentially coupled to the vesicle membrane to assemble into preautophagosomal structures (Xia et al., 2019). Among these proteins, Beclin1 interacts with phosphatidylinositol 3-kinase catalytic subunit type 3 (PI3KC3; also known as VPS34) and forms a PI3K core complex (VPS34-Beclin1), which initiates autophagy. The PI3K core complex then phosphorylates phosphatidylinositol (PI) to form phosphatidylinositol 3-phosphate (PI3P) and accelerates the maturation of autophagosomes (Russell et al., 2013). Subsequently, maturation of autophagosomes is followed by transport of ATG proteins (such as ATG3, ATG4, ATG7, ATG12-ATG5-ATG16, ATG10, and LC3) into the nucleation center of the phagophore, where they participate in the initiation and elongation of the phagophore (Molejon et al., 2013). After mature autophagosomes fuse with lysosomes to form autophagolysosomes, the receptor protein SQSTM1/p62 and target cargo (LD) are degraded by lysosomal acid lipase.

In recent years, studies on mammals have revealed that lipophagy plays an important role in maintaining homeostasis in various organisms (Zechner et al., 2017), but similar information is scarce in fish. Nevertheless, the existence of lipophagy in fish was first shown in fasting zebrafish using transmission electron microscopy (TEM) and measuring lipophagy marker proteins (Lc3-I, Lc3-II) (Fig. 5.16). Here it was demonstrated that specific inhibition of different lipophagic steps in vivo and in vitro could cause TAG accumulation and systemic downregulation of lipid metabolism activity, including catabolism and anabolism in both fed and starved states (Wang et al., 2018b, 2019b). This suggests that lipophagy is an important source of FFA as substrates for systemic lipid metabolism, and thus plays an essential role in lipid metabolism. Lipophagy has been found to also play multiple roles in systemic

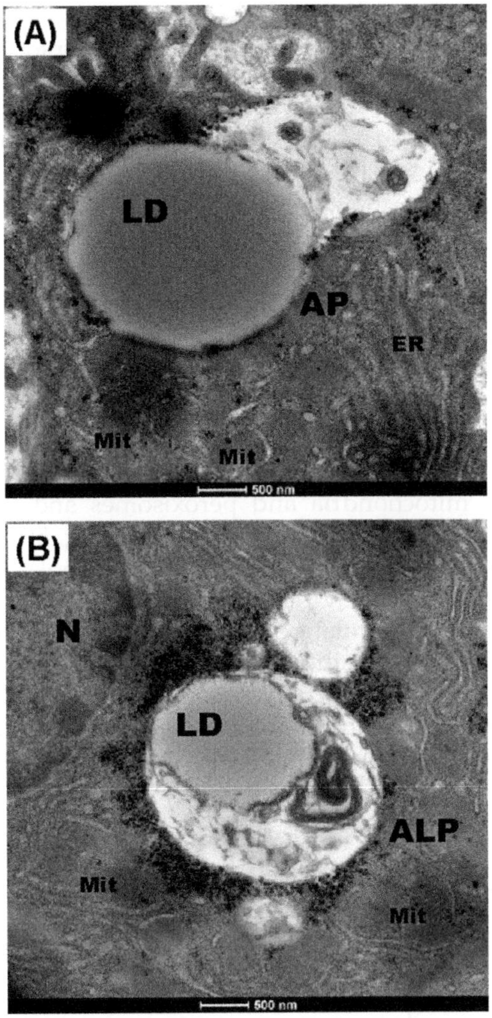

FIGURE 5.16 Key lipophagic processes in liver of fasting zebrafish; electron micrographs. (A): The structure of an autophagosome (AP) starting to enclose a lipid droplet. (B): The structure of an autolipophagosome (ALP) containing a lipid droplet. *ER*, endoplasmic reticulum; *Mit*, mitochondria; *N*, nucleus. *Wang, J., Han, S. L., Li, L. Y., Lu, D. L., Limbu, S. M., Li, D. L., Zhang, M. L. & Du, Z. Y. 2018b. Lipophagy is essential for lipid metabolism in fish. Sci. Bull., 63, 879–882, reprint with permission.*

nutrient metabolism. In yellow catfish, lipophagy was found to be involved in Zn-alleviated hepatocellular lipid accumulation via Zn^{2+}/MTF-1/Pparα and Ca^{2+}/CaMKKβ/AMPK pathways (Wei et al., 2018), and to mediate carbohydrate-induced changes in lipid metabolism via oxidative stress, endoplasmic reticulum (ER) stress, and Chrebp/Pparγ pathways (Zhao et al., 2020a). Moreover, inhibition of lipophagy in Nile tilapia caused lipid accumulation, and lowered antioxidant ability and severe inflammation coupled with disturbed glucose and protein metabolism (Han et al., 2019). In fasted rainbow trout hepatocytes treated with the lipophagy inhibitor Bafilomycine A1, expression levels of gluconeogenesis-related genes were lowered while genes involved in intracellular lipid stores and ER stress were increased (Seite et al., 2019). Therefore, although studies are still limited, lipophagy is definitely an important cellular process in fish metabolism. Considering that lipophagy has been closely connected to nutrient metabolism, immunity, and development in mammals (Yu et al., 2015; Zhang et al., 2018b), it could also be regarded as a promising target for fish nutrition research and application.

5.7.2 Fatty acid β-oxidation

5.7.2.1 Fatty acid β-oxidation and its enzyme system

Fatty acids released from TAG are catabolized to produce energy via β-oxidation, the catabolic process whereby activated fatty acids are broken down to 2-carbon units, acetyl-CoA, with the production of NADH and $FADH_2$. Acetyl-CoA enters the citric acid cycle, producing further NADH and $FADH_2$, which generate ATP via the electron transport chain. Fatty acid β-oxidation occurs in the mitochondria and peroxisomes and is divided into five steps: activation, oxidation, hydration, oxidation, and cleavage (Fig. 5.17).

Normally, a long-chain fatty acid ($>C_{10}$) present in the cytoplasm is first activated by acyl-CoA synthase to form acyl-CoA, which then passes through the mitochondrial bilayer membrane and enters the mitochondrial matrix for β-oxidation facilitated by the "acyl-carnitine shuttle system." During the transportation process, the coenzyme A group is replaced with carnitine to form acyl-carnitine via the action of carnitine-acyltransferase I (CPT1) in the outer membrane of mitochondria. Acyl-carnitine is transported through the mitochondrial inner membrane by carnitine/acyl-carnitine translocase, free carnitine is released into the matrix

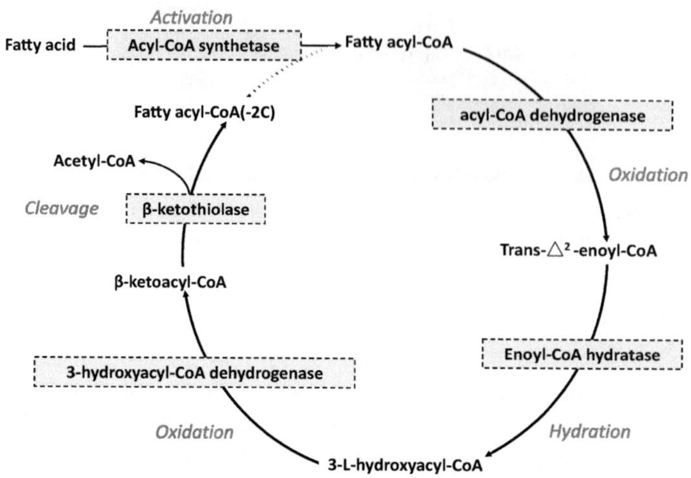

FIGUSRE 5.17 Biochemical reactions of fatty acid β-oxidation. In β-oxidation, fatty acids must first be converted to acyl-CoA (activation), which then go through a repeating cycle of four steps involving oxidation, hydration, oxidation, and cleavage, to be systematically degraded to 2-carbon units, acetyl-CoA, with the production of NADH and $FADH_2$.

and acyl-CoA produced by the action of carnitine-acyltransferase II (CPT2). The released carnitine is transported out of the matrix and acyl-CoA enters the β-oxidation cycle (Fig. 5.17). As the rate-limiting enzyme of fatty acid β-oxidation, CPT1 has been studied in the regulation of lipid metabolism in mammals and fish as an important regulatory target. Very recently, a pioneering study indicated that inhibition of the "acyl-carnitine shuttle system" by either reducing endogenous carnitine concentration or knockout of *cpt-1* not only changed lipid metabolism, but also completely reprogramed energy homeostasis to increase carbohydrate utilization and protein retention in fish (Li et al., 2020a). This highlighted the critical roles of mitochondrial fatty acid β-oxidation in energy metabolism in animals.

Unlike mitochondria, the peroxisome as a monolayer organelle without carnitine acyl transferase is only partially involved in fatty acid β-oxidation. In the peroxisomal matrix, β-oxidation starts via a reaction catalyzed by acyl-CoA oxidase (ACOX), which is a key element of the process. Peroxisomes mainly shorten very long-chain fatty acids including C_{22} fatty acids like DHA (and branched fatty acids etc.) to C_{18}-CoA, which are then transported to mitochondria to complete oxidation and produce ATP (Schrader et al., 2015). Specifically, the initial 2,3-(α,β)-dehydrogenation of DHA, which is the first step in the α-oxidation process of all fatty acids, must be followed in the case of DHA by an NADPH-dependent 2,4-dienoyl"CoA reductase and then by a 3-*cis*"2-*trans*-isomerase in the peroxisome so as to enable oxidation of its β carbon in mammals and fish (Madsen et al., 1999; Sargent et al., 2003). Notably, during peroxisomal β-oxidation, the electrons provided by FAD are transferred to oxygen, thereby leading to the formation of H_2O_2, but not coupled with the respiratory chain for ATP production (Schrader et al., 2015).

Numerous studies have shown that fatty acid catabolism-related enzyme systems and reaction pathways are basically the same in fish and mammals (Henderson and Sargent, 1985; Buzzi et al., 1996; Boukouvala et al., 2010). Presently, *cpt1* cDNA has been cloned in rainbow trout (Gutières et al., 2003), gilthead sea bream (Boukouvala et al., 2010), yellow catfish (Zheng et al., 2013b), blunt snout bream (Lu et al., 2016), and grass carp (Shi et al., 2017b). Nevertheless, *cpt1* has more subtypes in fish compared to mammals because of genome duplication events (Morash et al., 2010; Zheng et al., 2013b). In addition, transcriptional regulation of various isoforms of *cpt1* genes in fish presents diverse mechanisms (Wu et al., 2018). For example, the grass carp *cpt1α1b* possesses the core promoter element Sp1, and NF–Y replaces the classical TATA-box (Wu et al., 2018), which is consistent with Sp1 and NF–Y binding sites in the *CPT1* promoter in rats (Steffen et al., 1999). In zebrafish, *cpt1b* knockout ($cpt1^{-/-}$) resulted in significantly lower mitochondrial fatty acid β-oxidation activity in both liver and muscle (Lu et al., 2019). Moreover, $cpt1^{-/-}$ zebrafish had lower survival rates than wild fish during fasting and low temperature stress, confirming the key roles of *cpt1b* in fish fatty acid β-oxidation and physiology (Lu et al., 2019; Li et al., 2020a). Acyl-CoA dehydrogenase is the first enzyme in β-oxidation, and was regarded as the rate limiting enzyme in grass carp and goldfish (Hamre and Van, 1987). Meanwhile, the activities of β-oxidation-related enzymes vary in different fish species. For example, the activities of Cpt1 in dogfish shark (elasmobranchs) heart and red muscle and hagfish (Myxiniformes) were only 5%–10% of the activities in teleosts, like carp and trout (Moyes et al., 1990). However, the activities of hydroxyacyl-CoA dehydrogenase in those tissues of various bony and cartilaginous fishes were not different (Moyes et al., 1990).

Furthermore, peroxisomal Acox and multifunctional enzymes have been found in rainbow trout, Japanese medaka (Scarano et al., 1994), channel catfish (Gallagher and Di Giulio, 1991), and gilthead seabream (Pedrajas et al., 1996). However, the enzyme necessary for the oxidation of branched chain fatty acids, trihydroxycoprostanoyl-CoA oxidase described in mammalian, avian, and amphibian species, seems to be absent from fish genomes (Bilbao et al., 2009). In Nile tilapia, dietary supplementation with 10,12-tricosadiynoic acid, an ACOX antagonist, significantly decreased peroxisomal β-oxidation in liver, and increased body fat accumulation, verifying that Acox is indispensable for fish peroxisome fatty acid β-oxidation (Liu et al., 2020). In fact, the capacity to oxidize fatty acids through peroxisomal β-oxidation varies depending on fish species. Peroxisomal β-oxidation represents 30% of the total β-oxidation in the liver of the teleost Antarctic fish (*Notothenia gibberifrons*) (Crockett and Sidell, 1993a), but it accounts for 50% of the total β-oxidation in a temperate teleost species like *Myoxocephalus octodecimspinosus* (Crockett and Sidell, 1993b). However, peroxisomal fatty acid β-oxidation in dogfish liver demonstrated similar levels to teleost fish liver (Moyes et al., 1990). In addition, the natural diets of marine fish contain more LC-PUFA (e.g., DHA) requiring peroxisomal β-oxidation (Sargent et al., 2003). In fact, the proportion of Acox, the key long-chain fatty acid catabolic enzyme in peroxisomes, to Cpt, the key enzyme in mitochondria, is not high in fish. For example, the ratio of Acox/Cpt was 0.124 in herring, while in elasmobranch it was about 0.03, while in mice it was greater than 0.3 (Moyes et al., 1991), indicating that the division of mitochondria and peroxisomes in β-oxidation in fish probably is different from mammals.

Furthermore, the activities of Cpt-1, Cpt-2, and Acox vary in different tissues in fish. In salmon, Cpt1 activity in white and red muscle was similar, and higher than in liver (Frøyland et al., 1998). Activity of Cpt-2 was highest in red muscle, whereas liver and white muscle had comparable activity, while Acox activity was higher in liver than muscle tissues (Frøyland et al., 1998). It was concluded that white muscle plays the major role in overall fatty acid β-oxidation capacity based on Cpt-2 activity and total palmitoyl-L-carnitine oxidation (Frøyland et al., 1998). Fatty acid β-oxidation in elasmobranchs differs from other vertebrates, with oxidation of fatty acids and alcohols largely confined to the liver and, to a lesser extent, kidney (Ballantyne and Moon, 1986; Moyes et al., 1990). Carnitine acyltransferase was not detected in the brain (Moon and Mommsen, 1987), heart (Zammit and Newsholme, 1979; Singer and Ballantyne, 1989), red muscle (Zammit and Newsholme, 1979; Moon and Mommsen, 1987; Singer and Ballantyne, 1989; Moyes et al., 1990), and rectal gland (Moon and Mommsen, 1987; Singer and Ballantyne, 1989), indicating the inability of these tissues to oxidize fatty acids as carnitine esters. These studies suggest that elasmobranchs likely rely primarily on ketone bodies or amino acids rather than fatty acids for energy (Speers-Roesch et al., 2006).

5.7.2.2 Preferred substrates for fatty acid β-oxidation in different tissues and organelles

Several previous studies investigated fatty acid substrate preferences for β-oxidation in different cells or tissues, and these data have been summarized in Table 5.3. Many tissues, including liver, muscle, and intestine, have high preference for oxidation of PUFA, especially LNA (Bandyopadhyay et al., 1982; Ruyter and Thomassen, 1999; Lu et al., 2019; Jiao et al., 2020), ALA (Bell et al., 2002; Tocher et al., 2002, 2004; Torstensen and Stubhaug, 2004;

TABLE 5.3 Fatty acid substrate preferences for β-oxidation in mitochondria and peroxisomes in fish.

Fish species	Tissues	Total β-oxidation	Mitochondrial β-oxidation	Peroxisomal β-oxidation	References
Total β-oxidation substrate preference					
Atlantic salmon (*Salmo salar*)	Hepatocytes	ALA	Nd	nd	Torstensen and Stubhaug (2004a)
	Red and white muscle, liver	PAM: red and white muscle > liver			Stubhaug et al. (2005a)
	Liver	LNA > ALA			Ruyter and Thomassen (1999)
	Liver and muscle	ALA > LNA > C22:1 > OLA			Bell et al. (2002)
	Hepatocytes, intestinal enterocytes	ALA			Tocher et al. (2002)
	Skeletal muscle, visceral adipose tissue	C10:0 > OLA: Skeletal muscle; OLA: Visceral adipose tissue (poor)			Denstadli et al. (2011)
	Hepatocytes	Uptake: LNA and EPA (highest), PAM (lowest); Oxidation: ARA > LNA = PAM > OLA > DHA = ALA = EPA			Stubhaug et al. (2005b)
Catfish (*Heteropneustes fossilis* and *Clarias batrachus*)	Liver	LNA > ALA			Bandyopadhyay et al. (1982)
Grass carp (*Ctenopharyngodon idellus*)	Muscle, hepatopancreas	EPA and DHA (poor)			Ji et al. (2011)
Polka dot grouper (*Cromileptes altivelis*)	Whole body	C8:0 > OLA and PAM			Smith et al. (2005)
Rainbow trout (*Oncorhynchus mykiss*)	Hepatocytes, intestinal enterocytes	ALA: Hepatocytes > intestinal enterocytes			Tocher et al. (2004)
Zebrafish (*Danio rerio*)	Liver	LNA > PAM = OLA = ALA > ARA			Lu et al. (2019)

(*Continued*)

TABLE 5.3 Fatty acid substrate preferences for β-oxidation in mitochondria and peroxisomes in fish.—cont'd

Fish species	Tissues	Total β-oxidation	Mitochondrial β-oxidation	Peroxisomal β-oxidation	References
Mitochondrial or peroxisomal β-oxidation substrate preference					
Antarctic fish (*Notothenia gibberifrons*)	Liver	Nd	C16:1, EPA, DHA > ARA, ALA	No obvious preference	Crockett and Sidell (1993b)
Atlantic salmon (*Salmo salar*)	Liver, red and white muscle	FO and VO (poor)	PAM: White muscle	nd	Stubhaug et al. (2007)
	Preadipocytes	Nd	OLA, EPA, DHA	DHA	Todorčević et al. (2008b)
	Liver	Nd	DHA (poor)	DHA	Kjær et al. (2008a)
	Liver, red and white muscle	Nd	PAM: red and white muscle	PAM: liver	Frøyland et al. (2000a)
	Liver, red muscle	Nd	DHA and EPA: liver > red muscle	nd	Stubhaug et al. (2006)
	Skeletal muscle cells	Nd	C8:0 > OLA > EPA	nd	Vegusdal et al. (2004)
Common carp (*Cyprinus carpio*)	Muscle	Nd	C8:0 and C10:0 (poor)	nd	Fontagné et al. (2001)
		Nd	C16:1	nd	Murata and Higashi (1979)
Grass carp (*Ctenopharyngodon idellus*)	Primary hepatocytes	LA > LNA	LNA > ALA	ALA > LNA	Jiao et al. (2020)
	Liver	nd	PAM > DHA, ALA	DHA (poor)	Du et al. (2006a)
	Liver	nd	PAM	PUFA (poor)	Du et al. (2008)
Haddock (*Melanogrammus aeglefinus* L.)	Liver, red and white muscle	nd	PAM: red muscle (91% of total) and white muscle (97% of total)	PA: liver (100% of total)	Nanton et al. (2003)
	Liver	nd	nd	DHA	

Long-horned sculpin (*Myoxocephalus octodecimspinosus*)	Liver	nd	C16:1 > PAM > OLA = LNA > C20:2 > ARA > ALA	PAM = C16:1 > OLA = LNA > C20:2 > ALA = ARA	Crockett and Sidell (1993a)
Nile tilapia (*Oreochromis niloticus*)	Liver, muscle	PA: liver > muscle	PAM: liver > muscle	PAM: liver > muscle	Ning et al. (2016)
	Primaryhepatocytes	LA > LNA	LNA > ALA	ALA > LNA	Jiao et al. (2020)
Orange-spotted grouper (*Epinephelus coioides*)	Liver	nd	C12:0	nd	Tseng and Lin (2020)
Rainbow trout (*Oncorhynchus mykiss*)	Liver	nd	OLA and LNA	DHA and EPA	Du et al. (2004)
	Red muscle	nd	C14:0 = PAM > 16:1n-7 = STA = OLA = 20:1n-9 = 22:1n-9 = DHA > LNA = ALA	nd	Kiessling and Kiessling (1993)
	Liver		18:0 = 18:1 = 18:2 = 18:3	nd	
	Red muscle, liver, adipose tissue	nd	PUFA > SFA	nd	Morash et al. (2009)
	Red muscle, liver	nd	PAM	nd	Morash and McClelland (2011)
	Liver	nd	C12:0 (poor)	nd	Figueiredo-Silva et al. (2011)
	Muscle cells	nd	OLA	nd	Sánchezgurmaches et al. (2010)
Red drum (*Sciaenops ocellatus*)		nd	medium-chain fatty acids (poor)	nd	Davis et al. (1999)
Blunt snout bream (*Megalobrama amblycephala*)	liver	nd	PAM	PAM (poor)	Lu et al. (2014a)

nd, no data.

Trattner et al., 2008a; Jiao et al., 2020), and ARA (Stubhaug et al., 2005b). Furthermore, Atlantic salmon liver showed lower oxidation of 16:0 than muscle (Stubhaug et al., 2005a), while skeletal muscle and visceral adipose tissue showed lower oxidation of OLA than liver and heart (Denstadli et al., 2011). In grass carp, the oxidative capacity for LC-PUFA, especially EPA and DHA, was lower in muscle and liver (Ji et al., 2011) compared to other fatty acids. Thus, fatty acid preferences of β-oxidation vary between tissues and fish species.

Mitochondrial β-oxidation is dominant in all tissues except liver of juvenile Atlantic salmon, which is dominated by peroxisomal β-oxidation (Frøyland et al., 2000b; Kjær et al., 2008a). It has been indicated that fatty acid β-oxidation substrate preferences differ in mitochondria and peroxisomes (Crockett and Sidell, 1993b), which may partly underpin fatty acid preferences for β-oxidation between fish species. Substrate selectivity is broader for peroxisomal β-oxidation than for mitochondrial β-oxidation (Tocher, 2003). Early studies on mitochondrial β-oxidation in fish suggested a preference for SFA and MUFA over PUFA (Kiessling and Kiessling, 1993), whereas peroxisomes preferentially use very long chain fatty acids, PUFA, and branched-chain fatty acids as substrates compared to mitochondria (Crockett and Sidell, 1993b; Tocher, 2003). Most marine and freshwater fish efficiently oxidize LNA and ALA (Bandyopadhyay et al., 1982; Ruyter and Thomassen, 1999; Lu et al., 2019; Jiao et al., 2020) and have similar β-oxidation preference (LNA > ALA), in contrast to mammals (ALA > LNA) (Tran and Christophersen, 2002). Furthermore, LC-PUFA, EPA, and DHA are not preferred substrates in mitochondria of many freshwater fish (Du et al., 2006a), while marine fish, like Atlantic salmon and Antarctic fish, generally have a higher oxidizing activity toward these fatty acids (Crockett and Sidell, 1993b; Kiessling and Kiessling, 1993; Todorčević et al., 2008c). Although mitochondrial β-oxidation is believed to be the main route for fatty acid catabolism in animals, peroxisomes could account for 30%–50% (Crockett and Sidell, 1993a, 1993b) up to 100% (Frøyland et al., 2000b; Nanton et al., 2003) of β-oxidation activity in liver of marine teleosts. However, in the hepatocytes of some freshwater fish, the β-oxidation of C_{18} PUFA mainly occurred in mitochondria (63%–95%), while peroxisomes only contributed 5%–37% of total β-oxidation (Jiao et al., 2020). In addition, peroxisomes from teleosts may catabolize PUFA at lower rates relative to 16:0 than mammals (Frøyland et al., 2000b; Nanton et al., 2003). The fatty acid substrate preferences of mitochondrial and peroxisomal β-oxidation among different fish species and different tissues are summarized in Table 5.3.

5.7.2.3 Biochemical and nutritional regulation of fatty acid β-oxidation

L-carnitine (L-b-hydroxy-c-N, N, N-trimethylaminobutyric acid), which plays a vital role in mitochondrial β-oxidation, is a multifunctional organic nutrient also called vitamin BT that can be synthesized in liver, and subsequently transported to other tissues (Bremer, 1983). However, dietary supplementation of L-carnitine has been reported to improve lipid catabolism and/or alleviate HFD-associated excess lipid accumulation in fish, not only through increasing concentrations of L-carnitine and acetyl-CoA in tissues but also by increasing β-oxidation efficiency and mRNA expression of key β-oxidation-related genes, especially in the liver (Fujisawa et al., 2017; Li et al., 2019b, 2020c). Conversely, dietary mildronate, a carnitine synthesis inhibitor, reduced endogenous carnitine concentration in the liver to around 90% in Nile tilapia and zebrafish, and inhibited mitochondrial β-oxidation efficiency and increased lipid accumulation in tissues, confirming the critical roles of L-carnitine in lipid catabolism (Pan et al., 2017; Li et al., 2018a, 2020b). In fact, dietary L-carnitine supplementation in fish, not only increased lipid catabolism to produce more acetyl-CoA and ATP by stimulating mitochondrial β-oxidation but also increased protein synthesis and deposition by increasing the "protein sparing effect," stimulating the mTOR signaling pathway and saving methionine as an essential amino acid for protein synthesis. These mechanisms are summarized in Fig. 5.18 and have been described in a systemic review of L-carnitine in aquaculture nutrition (Li et al., 2019b). It should be noted that carnitine has two optical isomers, L- and D-carnitine, with L-carnitine confirmed to be physiologically functional in fish, whereas D-carnitine is metabolized as a xenobiotic and may induce metabolic dysfunctions (Li et al., 2019a). Therefore, L-carnitine, but not D-carnitine, is the recommended additive for fish feeds.

Choline, a vitamin-like nutrient, is considered essential for cellular structure and function (Mai et al., 2009). It is a precursor of PC and acetylcholine and required for lipoprotein formation in liver, and thus reduces hepatic fat deposition (Corbin and Zeisel, 2012). Fish are known to possess the ability to synthesize choline, although high energy-dense diets and fast growth in aquaculture means that endogenous choline biosynthesis is insufficient to meet the requirements of cultured fish and, therefore, choline is typically supplemented in diets. In blunt snout bream, yellow catfish, and Nile tilapia, dietary supplementation of choline not only promoted transport of lipid from liver due to the upregulation of *apob*

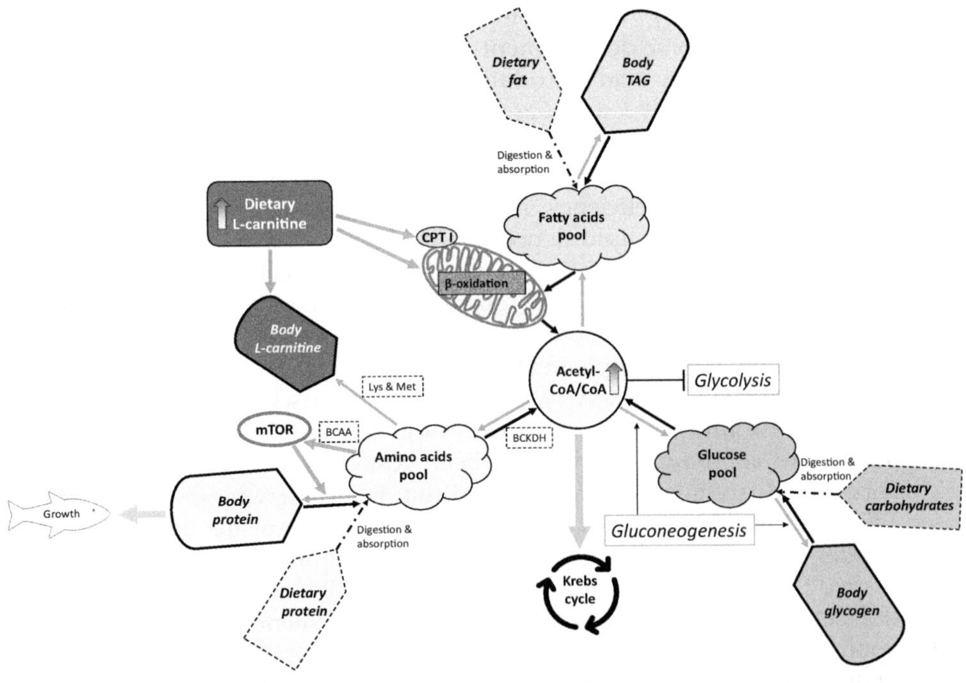

FIGURE 5.18 Systemic effects of L-carnitine supplementation on nutritional metabolism. *BCAA*, branched-chain amino acid; *BCKDH*, branched-chain α-keto acid dehydrogenase complex; *Lys*, lysine; *Met*, methionine; *mTOR*, mechanistic target of rapamycin.

and *mttp* (microsomal TAG transfer protein) gene expression but also decreased lipogenesis and increased Cpt-1 activity, reducing lipid accumulation in the liver (Li et al., 2014; Liu, 2014; Luo et al., 2016). Therefore, choline regulates lipid deposition in fish by increasing lipid transport and enhancing β-oxidation.

Betaine, a naturally occurring tertiary amine (trimethylglycine), exists widely in organisms and participates in the synthesis of carnitine, methionine, PC, and creatine, and, as a methyl donor, is directly involved in methylation (Sheard and Zeisel, 1989). Betaine has been reported to suppress fat accumulation in the liver and muscle via stimulating mitochondrial β-oxidation and lipid transport, and inhibiting lipogenesis (Craig, 2004; Xu et al., 2015b). Similarly, in blunt snout bream fed a high-fat diet, dietary betaine downregulated expression of *fas*, and upregulated expression of *cpt1*, *pparα*, and *mttp*, effectively alleviating severe hepatic fat deposition (Adjoumani et al., 2017), confirming that dietary betaine could contribute to the regulation of β-oxidation.

Traditionally, fish nutritionists have mainly focused on the impact of vitamins on growth performance (Zhang, 2006), but several studies indicate that vitamins A, C, and D, and inositol all affect β-oxidation in fish. For instance, dietary vitamin A enhanced Hsl activity and Fas activity and gene expression, but decreased Cpt-1 activity, and thus increased hepatic lipid content in grouper (*Epinephelus coioides*) (Ding, 2015). Moreover, decreased hepatic vitamin C content in rainbow trout increased long-chain acyl-carnitine and decreased free carnitine contents, suggesting that vitamin C regulated lipid metabolism by affecting carnitine synthesis (Miyasaki et al., 1995). However, size-specific results were obtained in Prussian carp (*Carassius gibelio*), whereby vitamin C increased activity of Cpt-1 during larval development, but not during embryonic development (Jiang et al., 2014a, 2014b). Dietary vitamin D showed similar effects on fatty acid β-oxidation as vitamin C and, thus, dietary vitamin D supplementation decreased Cpt-1 and Fas activities, but increased Hsl activity, reducing hepatic lipid concentration in grouper (Ding, 2015). However, the effects of vitamin D on mitochondrial β-oxidation in fish is dependent on tissue type. For example, depletion of Cyp2r1 resulted in vitamin D deficiency, mitochondrial dysfunction, suppressed mitochondrial β-oxidation, and abnormal lipid deposition in adipose tissue in zebrafish (Peng et al., 2017). Dietary vitamin D supplementation prevented these defects on metabolism in adipose tissue, but not in the liver (Peng et al., 2017), suggesting tissue-specific effects of vitamin D on mitochondrial β-oxidation. Furthermore, tissue-specific effects of vitamins on lipid metabolism were evident in diets supplemented with inositol. For example, dietary inositol supplementation increased Cpt-1 activity in the liver and muscle, and reduced lipid content in muscle, but elevated lipid content in the liver of grass carp (Lin et al., 2018).

Some vitamin-like compounds have also been tested as potential feed additives. For example, α-lipoic acid, a vitamin-like compound that is a coenzyme in the Krebs cycle, has been used widely in health care in mammals and aquatic organisms due to its antioxidant properties (Kütter et al., 2014; Fabiana et al., 2015). The use of α-lipoic acid in obese animals has been reported to affect AMPKα and promote TAG breakdown in different tissues (Kim et al., 2004; Stark et al., 2013). In grass carp, dietary supplementation of α-lipoic acid also promoted lipolysis in the liver and muscle via activation of AMPKα-Atgl pathways, while promoting fatty acid β-oxidation by regulating AMPKα–Cpt-1α pathways (Shi et al., 2018b), showing similar metabolic regulation as in mammals. The above studies show that vitamins and vitamin-like compounds play roles in regulating lipid catabolism; however, more intensive studies are required in future.

While many minerals are essential for fish (NRC 2011), their effects on lipid catabolism have only attracted attention in recent years. Zinc (Zn) is an essential micronutrient, which serves as a catalytic or structural cofactor participating in multiple cellular processes including lipid metabolism (Kambe et al., 2015; King et al., 2015). A lipid-lowing effect of Zn has been reported in fish, with high dietary Zn reducing body lipid content in yellow catfish through increasing tissue Cpt-1, Lpl, and Pparα activities and/or gene expression, while reducing *fas*, *srebp*-1, and *pparγ* gene expression (Zheng et al., 2015c). In yellow catfish, high dietary Zn increased total carnitine and acyl-carnitine contents (Zheng et al., 2014), and enhanced fatty acid β-oxidation and lipophagy via Zn^{2+}/MTF-1/Pparα and Ca^{2+}/CaMKKβ/AMPK pathways (Wei et al., 2018).

Copper (Cu) is an essential micronutrient for vertebrates and has numerous functions including vital roles in cellular respiration and as a cofactor for more than 30 enzymes (Watanabe et al., 1997). Dietary Cu reduced hepatic lipid content in yellow catfish and reduced hepatic free-, acyl-, and total-carnitine contents, Fas and Cpt-1 activities, and mRNA expression of other lipogenic and lipolytic enzymes, such as *6pgd*, *g6pd*, *accα*, *hsl*, and *atgl* (Chen et al., 2015a, 2015b). However, although dietary Cu levels influenced lipid catabolism-related enzyme activities and gene expression in muscle and VAT, it had no effect on the lipid contents of these tissues.

Iron (Fe) is an essential micromineral in all animals including fish (NRC, 2011) that also affects lipid metabolism. Dietary Fe decreased liver lipid content by upregulating *cpt1* and *pparα* mRNA levels in yellow catfish (Luo et al., 2017). Generally, similar results were obtained in javelin goby (*Synechogobius hasta*) whereby waterborne Fe exposure reduced hepatic lipid deposition by downregulating lipogenesis-related genes (*fas*, *srebp-1*, and *pparγ*) and upregulating lipid catabolism-related genes (*cpt-1*, *hsl*, *pparα*) (Chen et al., 2017). Considering many minerals are involved in numerous metabolic progresses, their effects on lipid catabolism are likely more extensive than currently known and could be expected to impact fish health and quality by affecting lipid metabolism.

In recent years, a number of plant extracts have been tested as new feed additives for fish and the beneficial functions of some of these were associated with altered lipid catabolism, including fatty acid β-oxidation. Resveratrol (trans-3,5,4′-trihydroxystilbene) is a natural polyphenol and phytoalexin found in various plants such as grapes, blueberries, raspberries, mulberries, and peanuts and has been studied as a potential health-promoting compound in humans (Baur et al., 2006; Kim et al., 2011; Zhang et al., 2012; Wang et al., 2015). In zebrafish and blunt snout bream fed high-energy diets, resveratrol upregulated *atgl*, *cpt-1a*, and *mttp* mRNA expression in liver, and downregulated hepatic *lpl*, *srebp-1c*, *pparγ*, and *acly* mRNA expression, suggesting resveratrol could improve lipid metabolism by promoting liver lipolysis, β-oxidation, and lipid transport, and suppressing fatty acid uptake and fatty acid synthesis (Ran et al., 2017; Shi et al., 2018a; Zhang et al., 2018b). However, dietary resveratrol reduced body weight through reducing feed intake in Atlantic salmon fed low-FO diets (Menoyo et al., 2019). Therefore, the effect of dietary resveratrol on fish physiology may vary with dietary lipid source.

Green tea extract contains numerous polyphenols known as catechins, particularly epigallocatechin gallate, epicatechin gallate, and gallocatechin gallate. In zebrafish fed HFD, dietary green tea extract supplementation increased hepatic expression of lipid catabolism genes, including *acox1*, *acadm*, *3-had*, and *pparα* (Hasumura et al., 2012; Meguro et al., 2015).

Therefore, the adverse effects of HFD in fish can be modulated by feeding green tea extract via reducing body fat accumulation, accompanied by increased gene expression and activities of fatty acid oxidation-related enzymes in liver and skeletal muscle. Caffeine, a purine and methylxanthine alkaloid, decreased lipid accumulation in overfed fish through inhibiting de novo lipogenesis and ER stress, and enhancing peroxisomal β-oxidation, but not mitochondrial β-oxidation, in liver of zebrafish larvae (Zheng et al., 2015c).

Silymarin is a flavonolignan from the milk thistle that has antilipid peroxidation and anti-inflammatory effects (Metwally et al., 2009; Yao et al., 2011; Chen et al., 2012; Surai, 2015). In grass carp fed HFD, silymarin inhibited lipid deposition in liver via upregulating the expression of genes related to lipolysis and β-oxidation, and positive effects on bile acid synthesis (Xiao et al., 2017). In addition, silibinin, a major bioactive component of silymarin, has been reported to inhibit lipid accumulation in adipocytes and zebrafish by activating AMPKα signaling, which could stimulate lipid catabolism and inhibit lipid synthesis (Suh et al., 2015).

Berberine is an isoquinoline alkaloid found in several plants used in traditional Chinese medicine and has been identified as a new lipid-lowering drug in mammals due to its ability to enhance fatty acid oxidation and inhibit fatty acid synthesis (Lee et al., 2006; Kim et al., 2009). Dietary berberine upregulated mRNA expression of *cpt1*, *aox*, and *pgc-1α*, and down-regulated *fatp*, *lpl*, and *ldlr* expression in blunt snout bream fed HFD (Zhou et al., 2019). These studies suggested that the lipid-lowering effects of berberine were mainly mediated by activating fatty acid β-oxidation and decreasing fatty acid uptake.

Other plant extracts including Yuzu peel, eriocitrin, *Palmaria mollis,* and Rhamnan sulfate have also been demonstrated to reduce body lipid deposition through upregulating fatty acid β-oxidation related enzymic activities and/or gene expression when they were supplemented in diets for fish, mainly zebrafish (Hiramitsu et al., 2014; Zang et al., 2014, 2015; Nakayama et al., 2018). Therefore, plant extracts could be valuable sources of safe and efficient feed additives.

5.7.2.4 Endocrinological regulation of fatty acid β-oxidation

Insulin is a pluripotent anabolic hormone produced by pancreatic β-cells, which has been isolated from several fish species (Mommsen and Plisetskaya, 1991) and performs various functions mediated by insulin receptors, through conserved signal transduction processes such as insulin receptor substrate-phosphatidylinositol-3-kinase (IRS-PI3K) and Akt (Navarro et al., 2002; Caruso and Sheridan, 2011). Similar to mammals, insulin in fish is composed of two units, A and B chains, joined by a disulfide bridge and has been proven to be a hypoglycemic hormone (Navarro et al., 2002). When administered at a physiological dose, piscine insulin exhibits antilipolytic functions by inhibiting fatty acid catabolism including the down-regulation of rate-limiting enzymes involved in both mitochondrial (Cpt-1a, Cpt-1b and Had) and peroxisomal (Acox) β-oxidation pathways (Leaver et al., 2008b). However, a pharmacological dose (10 IU/kg) of bovine insulin did not affect expression of hepatic genes of β-oxidation in rainbow trout (Jin et al., 2014). In contrast, an anabolic effect of insulin was observed in rainbow trout white muscle, as indicated by reduced expression of *cpt1a*, *cpt1b*, and *had* genes, in agreement with earlier reports (Harmon and Sheridan, 1992; Plagnes-Juan et al., 2008; Polakof et al., 2010b).

The precise functions of insulin receptors a and b (Insra and Insrb) on nutrient metabolism have been studied using Insra and/or Insrb knockout zebrafish (Yang et al., 2017a). While

Insra and Insrb exhibit overlapping roles in mediating glucose metabolism through the insulin-signaling pathway, Insrb promoted lipid catabolism, including fatty acid β-oxidation and protein synthesis through activation of the growth hormone-signaling pathway, whereas Insra stimulated lipid synthesis via glucose utilization. This suggests that the roles of insulin and its receptors are complex, with further investigation required.

Leptin is a type-I cytokine hormone primarily secreted by adipocytes in mammals, which is known to regulate metabolism and energy expenditure in vertebrates (Zhang et al., 2013a). In teleosts, liver is considered the primary site for leptin synthesis, while there is almost no leptin expression in VAT (Flynn et al., 2009). However, both liver and SCAT in Nile tilapia showed high expression of leptin (Liu et al., 2018a). Administering leptin to grass carp, yellow catfish and green sunfish (*Lepomis cyanellus*) increased activity and expression of a number of lipid catabolism-related enzymes/genes, including Cpt-1 and Pparα (Song et al., 2015). Furthermore, leptin-induced fatty acid β-oxidation was disrupted by the JAK-STAT inhibitor AG490, suggesting that JAK-STAT signaling pathways were involved in leptin-induced β-oxidation (Lu et al., 2012). In general, although leptin's role in lipid metabolism of fish remains to be clearly characterized, other data suggest that leptin contributes to the starvation-induced transition to lipid-based metabolism by stimulating the mobilization of lipid stores (Copeland et al., 2011; Gorissen and Flik, 2014). Interestingly, a recent study showed that leptin action in nutrient metabolism may vary with nutritional background (Liu et al., 2018a). This study found that long-term feeding of HFD or high-carbohydrate diet (HCD) induced lipid accumulation and leptin resistance in Nile tilapia, although the action of leptin in regulating lipid catabolism was retained in HCD-fed fish, but not in HFD-fed fish. For example, *cpt1a* expression was increased by leptin in control and HCD-fed fish, but not HFD-fed fish, whereas leptin regulation of carbohydrate utilization was retained in HFD-fed fish, but not HCD-fed fish. One week of fasting completely recovered leptin actions in the regulation of lipid and glucose metabolism. This selective leptin action could be an adaptive mechanism in animals to store surplus calories when different types of food are abundant.

Ghrelin is a peptide hormone, mainly synthesized in the gastrointestinal tract, which is subsequently octanoylated by ghrelin O-acyl transferase (Müller et al., 2015) present in several locations including hypothalamic neurons (Al Massadi et al., 2015; Perello and Dickson, 2015). Once octanoylated, ghrelin binds to GH secretagogue receptor R1a, which is also expressed in a panel of fish tissues such as brain (particularly hypothalamus), gastrointestinal tissue, pancreatic cells, and liver (Chan and Cheng, 2004; Kaiya et al., 2009; Cruz et al., 2010). Ghrelin cDNA has been identified in a number of fish species including rainbow trout (Kaiya et al., 2003a), sea bream (Yeung et al., 2006), zebrafish, goldfish, tilapia, and channel catfish (Kojima and Kangawa, 2005; Kaiya et al., 2008; Kang et al., 2011). However, the roles of ghrelin in fish are still unclear although it may be involved in the control of food intake (Jönsson, 2013) and lipid metabolism (Salmerón et al., 2015b). In mammals, the enhanced synthesis of lipids in intracerebroventricular ghrelin-injected animals was accompanied by downregulated oxidative capacity in liver (Sangiao-Alvarellos et al., 2009) and adipose tissue (Claudia et al., 2006). In rainbow trout, intracerebroventricular ghrelin reduced liver oxidative capacity, as evidenced by decreased activities of Cpt1 and Had (Velasco et al., 2016b). Accordingly, mRNA expression of *pparα*, a transcriptional activator of genes encoding enzymes of fatty acid oxidation such as Cpt1a, Mcd and Acox, was decreased in ghrelin-treated fish, consistent with the observed reduced capacity of liver to oxidize fatty acids (Velasco et al., 2016b). Moreover, since expression of uncoupling protein 2 (UCP2) is driven by PPARα activation (Ježek, 2002), decreased *ucp2* mRNA in ghrelin-treated fish was consistent with decreased oxidative capacity induced by central ghrelin treatment. However, in an in vitro study, *pparα* and *cpt1b* gene expression levels were not affected by ghrelin treatment in rainbow trout adipocytes (Salmerón et al., 2015b), suggesting ghrelin regulation of β-oxidation genes is tissue-dependent.

5.7.2.5 Molecular regulation of fatty acid β-oxidation

Many studies have shown that PPAR are major nuclear receptors (NR) controlling lipid metabolism. The PPAR nuclear hormone superfamily has three subtypes, alpha, beta, and gamma, with PPARα playing an important role in promoting fatty acid β-oxidation (Price et al., 2000; Berger and Moller, 2002). The mechanism of PPAR function involves three steps: PPAR are activated by combining with specific ligands, the activated PPAR combines with lipophilic 9-cis-retinoic acid receptor (RXR) to form a heterodimer functional transcription unit, which then binds to PPRE in target gene promoters to activate the transcription of these lipid metabolism-related genes (Kersten et al., 1999; Tan et al., 2005). Physiological PPAR ligands include fatty acids and eicosanoids (such as leukotrienes), and pharmacological ligands include lipid-lowering (e.g., Wy 14,643), and antidiabetic (e.g., thiazolidinedione) drugs (Kliewer et al., 1996). In recent years, *pparα* cDNA have been cloned from a variety of fish species (Boukouvala et al., 2004; Leaver et al., 2005b; Oku and Umino, 2008; Zhao et al., 2011; Dong et al., 2015; Zheng et al., 2015a; Ning et al., 2016), showing similar structures and activator components as in mammals. In zebrafish, *pparα* has two subtypes, *pparα-a* and *pparα-b*, where the latter is the dominant subtype in most tissues, and a deficiency of *pparα-b* significantly lowered fatty acid β-oxidation and elevated lipid deposition, hence verifying the key roles of Pparα in fish metabolism (Li et al., 2021). Fatty acids, eicosanoids like hydroxyeicosatetraenoic acid (HETE) and PGJ_2, and fibrates such as fenofibrate and clofibrate, have been shown to be ligands for Pparα regulating β-oxidation in fish (Ibabe et al., 2005). All three subtypes of Ppar are present in gray mullet (*Mugil cephalus*), with Pparα most expressed in liver, similar to mice (Ibabe et al., 2004).

Different fish species may vary in their sensitivity to Ppar. In gilthead sea bream, *cpt1α* mRNA expression in both muscle and liver was regulated by Pparβ more than Pparα or Pparγ (Boukouvala et al., 2010) but, in Atlantic salmon, Pparγ was suggested to be the key nuclear receptor regulating fatty acid β-oxidation (Ruyter et al., 1997). It was demonstrated in Nile tilapia that Pparα could be activated by fenofibrate and promoted lipid catabolism through increasing mitochondrial proliferation and fatty acid β-oxidation activity (Ning et al., 2016). Liver was an important target organ, but Pparα regulatory activity was shown to be weaker in fish than rodents (Ning et al., 2016). The lipid-lowing regulation of Pparα in Nile tilapia could be activated through increased transcription, protein expression, and changed phosphorylation (Ning et al., 2016); however, protein expression and phosphorylation modification are the main regulatory mechanisms (Tamasi et al., 2008; Ning et al., 2016). In addition, studies in Nile tilapia also found that the lipid-lowering effects of Pparα were positively correlated with dietary lipid and protein levels suggesting that Pparα-induced lipid-lowering effect was dependent on nutritional background and dietary components (Ning et al., 2017, 2019). In short, PPARα regulation of fatty acid catabolism appears conserved in fish and mammals.

Retinoid X receptors (RXR: RXR-α, RXR-β, and RXR-γ), members of the NR superfamily, form homodimers or heterodimers with other nuclear receptors, including retinoic acid receptors, PPAR, LXR, thyroid hormone receptors, and pregnane X receptor (Chawla and A., 2001). The PPAR-RXR transcriptional complex is a key regulator of proteins central to energy metabolism, including lipid transport (lipoprotein metabolism), FFA release (lipases), cellular fatty acid uptake (fatty acid transporters), energy production (β-oxidation), and lipid storage (adipogenesis) (Plutzky, 2011). The RXRα/PPARα heterodimer is required for PPARα regulation of fatty acid oxidation genes such as *ACOX1* and the *P450* system genes (Gorla-Bajszczak et al., 1999). Furthermore, RXR agonist activated PPARα-inducible genes and lowered TAG in hypertriglyceridemic db/db mice (Mukherjee et al., 1998). *rxr* cDNA have been cloned in many finfish species (Jones et al., 1995; Villeneuve et al., 2004; Raingeard et al., 2009a, 2009b), including golden pompano (Yang et al., 2015; Pan et al., 2016). The structures of the cloned Rxr cDNA are highly conserved with respect to amino acid sequence and functions compared to mammalian RXR (Beckett and Petkovich, 1999). However, different from mammals that have three *RXR* genes (α, β and γ), zebrafish has four *rxr* genes: *rxrα*, *rxrβa*, *rxrβb*, and *rxrγ* (Jones et al., 1995; Tallafuss et al., 2006), possibly due to teleost-specific genome duplication events. The RXR have been reported to activate fatty acid β-oxidation similar to PPARα. For example, peroxisome proliferators upregulated palmitoyl-CoA oxidase activity through increasing both *pparα* and *rxrα* mRNA expression levels in European hake (*Merluccius merluccius*) (Raingeard et al., 2009a). However, detailed mechanisms of Rxr regulation of β-oxidation in fish are still unknown and await further investigation.

Farnesoid X receptor (FXR), another NR superfamily, is the primary sensor for endogenous bile acids and plays a crucial role in hepatic TAG homeostasis (Del Bas et al., 2009). Activation of FXR suppressed expression of SREBP-1c, a critical transcription factor that promotes hepatic TAG synthesis (Watanabe et al., 2004). In addition to the suppression of de novo lipogenesis, FXR activation modulates fatty acid oxidation and TAG clearance. Thus, FXR activation inhibited expression of microsomal TAG transfer protein (MTTP) via the small heterodimer partner (Hirokane et al., 2004), which in turn suppressed HNF4α transactivation of the MTTP promoter, and thus decreased hepatic VLDL production (Pan et al., 2010). *fxr* cDNA have been cloned in various fish species, including Japanese medaka (Howarth et al., 2010) and large yellow croaker (Du et al., 2018), and the ligand selectivity profile of green spotted pufferfish (*Tetraodon nigroviridis*) Fxr was found to resemble human FXR (Krasowski et al., 2011).

MicroRNA (miR), small nonprotein-coding RNA molecules generally composed of 18–25 nucleotides, can complement and match with sequences in the 3′UTR region of target gene mRNA to inhibit posttranscriptional expression of the gene and regulate various physiological and metabolic processes. Recently, miR have been shown to regulate lipid metabolism, including fatty acid β-oxidation, in mammals (Moore et al., 2011). Likewise, miR regulation of lipid metabolism has been also found in fish. After feeding in rainbow trout, high expression of mir-33 and mir-122b accompanied by downregulation of liver *cpt1a* and *cpt1b* was found, suggesting a regulatory role of miR homologs in β-oxidation in fish (Mennigen et al., 2012). Similarly, the gene for Had, a key enzyme in β-oxidation, was significantly upregulated in rainbow trout after intraperitoneal injection of miR-122 inhibitor, suggesting this miR has similar functions in fish and mammals (Mennigen et al., 2014). Furthermore, mir-29a influenced the balance of lipid metabolism through the regulation of stearoyl-CoA desaturase (*scd*) expression in tilapia fed a high SFA diet (Qiang et al., 2017). In addition, *scd* and *fas* were identified as potential targets of miR-30c in blunt snout bream (Zhang et al., 2014), while miR-205 acted on *fas* and *scd* genes influencing lipid deposition in yellow catfish exposed to copper (Cui et al., 2018). Studies focusing on tilapia demonstrated that HFD may indirectly reduce the antagonistic effect of malonyl coenzyme A on Cpt1 via miR-205-5p regulation, and promote liver β-oxidation (Tao et al., 2017, 2018). In general, these studies show that miR have similar functions in the regulation of β-oxidation in fish as in mammals, although further studies are required to understand the precise roles of miR in fish nutrition.

5.8 Lipids and the gut microbiota

The microbiota is defined as the collective microbial community inhabiting a specific environment; for example, the gut. In contrast to the human gut where cellular density increases along its length, and the colonic microbiota is the densest and most diverse community in the gut, this may not be the case in fish, as the density and diversity may be similar in both proximal intestine (PrI) and distal intestine (DI). Even though finfish represent the greatest diversity of all vertebrates (Clements et al., 2014), the understanding of the finfish gut microbiota and its significance is less investigated compared to endothermic animals.

In the 1930s, the first studies were published investigating the gut microbiota of fish (Stewart, 1932; Snow and Beard, 1939). Despite the availability of this information and others studies that followed, in the 1970s, controversy existed about the role, and even the existence, of an indigenous gut microbiota in fish (Ringø et al., 2016). This has, however, changed, and today it is accepted that the fish gut microbiota play important roles in mediating and stimulating gastrointestinal (GI) development, aiding digestive function, maintaining mucosal tolerance, enhancing the immune response and providing protection against diseases (Li et al., 2018c; Rowland et al., 2018; Wang et al., 2018a), as well as in the development of metabolic syndrome (Clément, 2011), and vitamin synthesis (Rowland et al., 2018). Therefore, it is of great importance that our knowledge of the gut bacteria that colonize the GI tract is improved, in the context of both enhanced growth performance and fish health.

With respect to fish lipid nutrition, the effects of dietary lipid level and source on the gut microbiota has recently been comprehensively reviewed (Ringø et al., 2016) and, as such, will only briefly be discussed here and presented in Table 5.4.

5.8.1 The effect of dietary lipid level on gut microbiota

Over the last 30 years, several studies have evaluated the effects of dietary lipid level on the gut microbiota of various species, including rainbow trout (Lesel et al., 1989), Arctic charr (Ringø and Olsen, 1999), yellowtail kingfish (*Seriola lalandi*) (Soriano et al., 2018), and Nile tilapia (Ma et al., 2018; Limbu et al., 2019). A general description of the findings revealed in salmonids is presented in Table 5.4.

TABLE 5.4 Studies investigating the effects of dietary lipid on gut microbiota of finfish.

Lipid	Species	Doses/duration	Effects	References
Lipid level	Arctic charr (A)	High and low dietary lipid/40 days	Low dietary lipid modulated the adherent microbiota in proximal (PI) and distal intestine (DI). Dietary manipulation modulated the species composition of carnobacteria. *Carnobacterium* spp. and *C. Mobile*-like strains were only isolated from DI of fish fed low dietary lipid, while *C. Piscicola*-like strains were isolated from PI. In contrast, *C. divergens*-like isolates were associated with the mucus layer of PI and DI of fish fed high dietary lipid. ↑ Yeast in fish fed high dietary lipid.	Ringø and Olsen (1999),[a]
	Rainbow trout (NA)	High (16%) and low (5%) dietary lipid	Modulated fecal gut microbiota. The fecal microbiota of fish fed low lipid consisted of only *Acinetobacter* spp. and Enterobacteria. In contrast, *Acinetobacter* spp., *Aeromonas* spp., Enterobacteria, *Flavobacterium* spp., *Pseudomonas* spp., and coryneforms were isolated from feces of fish fed high lipid.	Lesel et al. (1989),[a]
	Yellowtail kingfish (J)	Three levels of lipid inclusion (7%, 14% and 21%)/30 days	Lipid level modulated the gut microbiota	Soriano et al. (2018)
Lipid source	Arctic charr (A)	Soybean oil, linseed oil and FO/NA	→ Total culturable aerobic bacteria Modulated the adherent hindgut microbiota *Acinetobacter* spp. and *Carnobacterium* spp. were only present in soybean oil-fed fish, while *Micrococcus* spp. were only present inFO-fed charr	Ringø et al. (2002),[a]
	Atlantic salmon (J)	High and low n-3 fatty acid diets	Modulated the allochthonous gut microbiota	Jin et al. (2019)
	Channel catfish (J)	FO and soybean oil/120 days	→ Culturable lactic acid bacteria counts in midintestine by FO and soybean oil feeding	Santerre et al. (2015),[b]
	European sea bass (J)	Replacement of FO by vegetable oil (VO)/90 days	Marginal modulation of the posterior and intestinal digesta microbiota ↓ Adherent *Dermacoccus* in fish fed	Torrecillas et al. (2017)

(*Continued*)

TABLE 5.4 Studies investigating the effects of dietary lipid on gut microbiota of finfish.—cont'd

Lipid	Species	Doses/duration	Effects	References
			15% FO, and *Acinetobacter baumanii* in fish fed 6% FO	
	Gilthead sea bream (NA)	VO/NA	Modulated the gut microbiota	Montero et al. (2006)[a]
	Gilthead sea bream (J)	FO *vs.* VO	Marginally modulated the digesta microbiota with respect to OTU, richness, and diversity VO increased the similarity of bacterial communities	Castro et al. (2019)
	Golden pompano (J)	FO, soybean oil, and linseed oil/8 weeks	Modulated the gut microbiota Firmicutes were dominant in the FO group, while *Tenericutes* were most dominant in the soybean oil and LO groups ↓ Bacterial diversity by soybean oil *vs.* FO and linseed oil	You et al. (2019)
	Nile tilapia (J)	Microencapsulated oil	Modulated the gut microbiota ↑ *Bacillus* and *Paenibacillus*, beneficial bacteria ↓ *Pseudoalteromonas* and *Roseovarius*, "harmful" bacteria	Ma et al. (2018)
	Zebrafish (A)	Control, high saturated fat, and high saturated fat + DHA-enriched diets/6 weeks	Inclusion levels modulated the gut microbiota ↑ *Pseudomonas* by high saturated fat feeding	Arias-Jayo et al. (2019)
Polyunsaturated fatty acids	Arctic charr (J)	4% LNA, 4% ALA, 4% PUFA and 7% ALA/90 days	Modulated the adherent lactic acid bacteria in stomach, PI, DI and feces	Ringø et al. (1998)[a]
	Arctic charr (A)	2.5% LNA/90 days	Modulated the fecal microbiota, LNA feeding ↑ *Vibrio* ↓ *Lactobacillus* and coryneforms	Ringø (1993)[a]
	Tiger puffer (J)	ARA	↓ Alpha diversity of hindgut microbiota ↑ Relative abundance of arcobacter	Yu et al. (2019)

[a]*Studies discussed in the review of Ringø et al. (2016).*
[b]*Study discussed in the review of Zhou et al. (2018c).*
↑, positive effect; ↓, negative effect; →, no effect; *A*, adult; *J*, juvenile; *NA*, no information available.

A recent study revealed that rearing yellowtail kingfish juveniles at 20 and 26°C, and fed three levels of lipid inclusion of 7%, 14%, and 21%, modulated the fecal microbiota (Soriano et al., 2018). Fecal, allochthonous (transient) microbiota, of control fish revealed Alphaproteobacteria, Flavobacteria, and Gammaproteobacteria. In contrast, fish fed the 7% lipid treatment, displayed operational taxonomic units (OTU) belonging to Alphaproteobacteria, Gammaproteobacteria, and Actinobacteria, while in fish fed 14% lipid, significant statistical differences were revealed only at 26°C, where the bacteria belonged to Gammaproteobacteria: Xanthomonadales, Pseudomonadales, and Enterobacteriales. An interesting finding was the difference in the fecal microbiota of fish fed 21% lipid, as they belonged to Alpha Beta and Gammaproteobacteria: Enterobacteriales, Caulobacteriales, Xanthomonadales, Burkholderiales, and Rhodospirillales.

Studies have revealed that HFD can impair gut health by impairing intestinal structures and the intestinal microbiota in endothermic animals (Yang et al., 2020) and finfish. The microencapsulation of oils has been widely applied in the food industry and the dietary microencapsulation of organic acids and essential oils have been applied successfully to endothermic animals, improving growth performance and increasing levels of the beneficial fecal bacteria, *Lactobacillus*. However, in aquaculture, less information is available regarding lipid microencapsulation. A recent study revealed that high fat microencapsulation increased the content of unculturable allochthonous midgut microbiota in Nile tilapia by increasing the level of beneficial bacteria, *Bacillus* and *Paenibacillus*, and decreasing *Pseudoalteromonas* and *Roseovarius* (Ma et al., 2018). While the results are of interest, the autochthonous (resident) bacterial community should also be investigated.

In consideration of the known interactions of an HFD and antibiotic administration on microbial imbalance (dysbiosis) of the gut microbiota, Limbu et al. (2019) evaluated the influence of an HFD (120 g/kg) with or without oxytetracycline on intestinal microbiota and intestinal health of juvenile Nile tilapia. HFD and antibiotic administration decreased *Cytophagales* species, a gliding bacteria with the ability to produce a protein-deamidating enzyme, and increased *Macelibacteroides*. Notably, *Porphyromonas*, an obligate anaerobic, heterotrophic, nonspore-forming, nonmotile, Gram-negative rod that ferments sugars, also decreased. Based on their findings, the authors concluded that HFD and antibiotic administration may affect Nile tilapia health.

The intestinal microbiota of genera *Bifidobacterium*, *Enterococcus*, *Lactobacillus*, *Lactococcus*, *Leuconostoc*, *Streptococcus*, *Microbacteriaceae*, *Nocardia*, and *Ralstonia* decreased in Nile tilapia fed an HFD in comparison to control fish (Zhang et al., 2020a). Furthermore, administration of bacterium strain S1, with high similarity to *Citrobacter freundii*, to HFD increased the proportion of intestinal *Bifidobacterium*, *Enterococcus*, *Lactobacillus*, *Lactococcus*, *Leuconostoc*, *Streptococcus*, and *Ralstonia*. These findings merit further investigation, as they may affect fish health and improve disease resistance against pathogens.

5.8.2 The effect of dietary lipid sources on gut microbiota

A study on the effects of soybean-, linseed- and marine oils on the culturable hindgut microbiota of Arctic charr revealed clear differences in the hindgut microbiota of fish fed different oils, prior and post challenge with *Aeromonas salmonicida* subsp. *salmonicida* (Ringø et al., 2002). As the ability of carnobacteria to inhibit in vitro growth of *A. salmonicida* ssp. *salmonicida* was highest in strains isolated after challenge, these results may be of interest as improved survival after challenge by dietary soybean oil has been demonstrated (Lødemel et al., 2001).

Ringø, Sperstad, and Olsen (unpublished study) fed Atlantic salmon and rainbow trout extruded diets consisting of lipid-free protein meal supplemented with either 30% VO containing either linseed oil, rapeseed oil, sunflower oil or 30% FO, and a commercial diet as a control. Some differences in the culturable bacterial composition, identified by 16S rRNA, between the rearing groups were revealed. In DI of fish fed the commercial diet, *Psychrobacter* and *Staphylococcus* were the dominant bacterial genera. In fish fed the sunflower oil diet, *P. jessenii/fragi*—like strains were dominant followed by *Staphylococcus* spp. However, in fish fed rapeseed oil, *P. submarinus/marincola*—like strains and *Staphylococcus* spp. were prevalent. In the DI of fish fed the linseed oil diet, *Enterococcus pseudoavium*—like strains were the dominant bacterial species, followed by *Acinetobacter* spp. and *Staphylococcus* spp. In contrast to these results, *P. submarinus / marincola*—like strains were dominant in the GI tract of fish fed FO. Furthermore, dietary manipulation modulated the presence of beneficial lactic acid bacteria as 13.6% and 9% of the strains isolated in the hindgut of fish fed rapeseed oil and FO belonged to *Carnobacterium mobile*. However, very few (~2%) *C. mobile* strains were isolated from the GI tract of the other dietary groups.

In a similar experiment, Ringø, Olsen, and Sperstad (unpublished data) revealed that the culturable DI microbiota, identified by 16S rRNA, of rainbow trout fed dietary VO was dominated by *P. submarinus/marincola*—like strains and *Staphylococcus linens/equorum*. However, some changes in microbiota composition were apparent between the dietary groups. For example, *Enterococcus* strains were not isolated from fish fed dietary rapeseed oil, while *Marinilactibacillus psychrotolerans*—like strains were only isolated from this dietary group. A difference in gut microbiota between fish fed VO and FO was observed as *Shewanella putrefaciens* and *Serratia* spp. were only isolated from fish fed FO. On the other hand, the clearest differences in gut microbiota were detected between dietary groups fed VO and FO compared to fish fed the commercial diet prior to the start of the experiment, as the hindgut microbiota of fish fed the commercial diet comprised of *Acinetobacter johnsonii*, *Pseudomonas jessenii*, *Psychrobacter submarinus*, *Carnobacterium* spp., and *Staphylococcus* spp. The results of this study indicated that carnobacteria are present in the alimentary tract of rainbow trout, but generally at relatively low population levels and they seem readily affected by dietary manipulation.

An early study investigated the effect of VO on the gut microbiota of gilthead sea bream after intraperitoneal challenge with *Vibrio alginolyticus* (Montero et al., 2006). This revealed that culturable levels of total Gram-positive bacteria and *Vibrio* spp. were present in fish fed a diet in which 60% of FO was replaced with a blend of rapeseed oil, linseed oil, and palm oil (15:60:25) compared to fish fed 100% FO or 100% VO blend. The controversial hypothesis that the gut microbiota modulation was due to modified fatty acid composition of the intestinal epithelial cell membranes was subsequently put forward. However, these preliminary data should be viewed with caution as the topic requires further evaluation.

A 120-day feeding trial in juvenile channel catfish investigated the effects of FO or soybean oil and supplementation of a commercial probiotic containing *Bacillus*, *Lactobacillus*, and *Saccharomyces* (Santerre et al., 2015). Bacteriological evaluation of the culturable microbiota in midintestine indicated no modulation in intestinal lactic acid bacteria counts by dietary manipulation.

The effects of FO replacement by VO on posterior mucosal gut microbiota and digesta were evaluated by denaturing gradient gel electrophoresis in a 90-day feeding study in

European sea bass (Torrecillas et al., 2017). Marginal differences in pairwise similarity coefficient (Cs) of the digesta bacterial community were revealed. In contrast, for the mucosa bacterial community, significant differences were revealed for Cs between the dietary treatments. With regard to OTU, 37 OTU were observed for the digesta, while 78 were revealed for the mucosa. Some differences were noticed at the genus level including *Acitenobacter baumanii* that can be an opportunistic pathogen in humans, *Dermacoccus* that is normally reported on human skin, and *Clostridium, Corynebacterium, Staphylococcus,* and *Roseburia faecis* that were first isolated from human feces.

A 6-week study with adult zebrafish assessed the effect of a high saturated fat diet versus a high saturated fat DHA-enriched diet on intestinal microbiota (Arias-Jayo et al., 2019). No differences in the total number of bacteria were revealed among groups. At the phylum level, the most abundant phyla in all samples were *Fusobacteria* and *Proteobacteria,* while *Cyanobacteria* and *Tenericutes* were revealed in lesser amounts. At family and genus levels significant differences were revealed among the diets. *Pseudomonas* increased in high-saturated fat diet samples, in contrast to *Tenericutes ck-1c4-19,* which decreased. While the results are interesting, no specific information was presented with regard to whether the autochthonous or allochthonous microbiota or a combination of the two were investigated, or which gut segments were sampled, as the data were presented as total DNA from intestinal samples.

In gilthead seabream juveniles fed VO- and FO-based diets, two distinct clusters of intestinal microbiota were revealed. Notably, the bacterial communities obtained from fish fed the VO diets were more closely related (similarity between 50% and 80%) than those recovered from fish fed the FO diets, which appeared to diverge more (similarity between 50% and 70%) (Castro et al., 2019). Despite this, the average number of OTU, microbial richness, and diversity indices were not significantly affected by dietary manipulation. However, SIMPER similarity indices were significantly higher in fish fed the VO diets compared to fish fed the FO diets. Sequence analysis of selected denaturing gradient gel electrophoresis bands revealed that the dominant bacteria in the intestines of gilthead seabream juveniles were most closely related to the genera *Enterococcus, Bacillus, Pseudomonas, Cronobacter, Burkholderia, Corynebacterium,* and *Bifidobacterium.*

An analysis of the intestinal microbiota of juvenile golden pompano fed FO, soybean oil and linseed oil diets for 8 weeks showed that the soybean and linseed oil diets significantly increased the abundance of potentially pathogenic intestinal bacteria (*Mycoplasma* and *Vibrio*), while the proportions of beneficial *Bacillus* and *Lactococcus* bacteria decreased, especially in the soybean oil group (Yu et al., 2019). However, MetaStat analysis revealed that administration of soybean and linseed oils had no significant effects on taxonomic distribution of intestinal microbiota communities. Correlation analysis of the individual dietary fatty acid contents and predominant intestinal microbial genera revealed that dietary ARA, EPA, and DHA contents had a significant positive correlation with *Leuconostoc, Globicatella, Lactococcus,* and *Bacillus,* while a significant negative correlation was observed with *Mycoplasma.*

For migratory fish that move from fresh to saltwater, there are major dietary changes with respect to the availability of n-3 LC-PUFA, which are present in lower abundance in freshwater systems. In a recent study, Atlantic salmon were fed high or low n-3 LC-PUFA diets in for 210 days (Jin et al., 2019). The allochthonous intestinal microbiota of fish fed the low n-3 LC-PUFA diet in freshwater was dominated by *Firmicutes,* while *Proteobacteria* dominated in the gut of fish fed the high n-3 LC-PUFA diet. After feeding in freshwater, fish were transferred to saltwater and fed the high n-3 LC-PUFA diet for 100 days. Fish previously fed the low n-3 LC-PUFA diet in freshwater revealed a significantly higher *Mycoplasma* content compared to fish fed the high n-3 LC-PUFA diet in freshwater, indicating that the diet in freshwater influenced the establishment of the gut microbiota in saltwater reared fish.

In a study with juvenile tiger puffer, three inclusion levels of ARA; 0.33 (control), 2.45 (L-ARA), and 23.41 (H-ARA) $g.kg^{-1}$ dry matter were administered for 74 days (You et al., 2019). While the dominant genera in all treatment groups were *Arcobacter* and *Mycoplasma*, inclusion of H-ARA decreased the alpha diversity of hindgut microbiota, but increased the relative abundance of *Arcobacter*, a potential pathogenic genus, in contrast to that revealed by L-ARA administration. Moreover, the relative abundance of *Mycoplasma* was significantly increased by L-ARA compared to H-ARA and control treatments.

Less information is available on the effects of C_{18} PUFA on gut microbiota in finfish. An early study revealed that LNA administration modulated the culturable fecal microbiota of Arctic charr by increasing *Vibrio* counts, while *Lactobacillus* and *Coryneforms* counts decreased (Ringø, 1993). In a later study, Arctic charr fed different PUFA showed modulated adherent lactic acid bacteria in stomach, PrI and DI and feces (Ringø et al., 1998).

5.8.3 The effects of gut microbiota on lipid metabolism

The GI tract is inhabited by a wide cluster of microorganisms that have protective, structural, and metabolic roles for the intestinal mucosa. The gut microbiota is involved in a host of functions, including barrier function and the maintenance of its homeostasis, nutrient provision, signaling, the regulation of epithelial development, and immune function. There is great interindividual variability in the number, type, and function of microorganisms in the GI tract. Considering the ability of microbiota to respond to homeostatic and physiological changes, some researchers propose that it can be seen as an endocrine organ. Increasing evidence suggests that both internal and external factors modulate gut microbiota. These changes can be both quantitative and qualitative, resulting in variations of the composition and metabolic activity of the gut microbiota, which subsequently affects health and different disease processes.

In order to provide a satisfactory overview on the interactions between the gut microbiota and lipid metabolism, some information from endothermic animals is included. It is well established that the gut microbiota plays a pivotal role in metabolism, including lipid metabolism (Bäckhed et al., 2004, 2007), as well as the role of the gut microbiota in influencing food intake. The gut microbiota also interact in physiological or pathological conditions, and its composition helps discriminate between healthy and unhealthy individuals (Del Chierico et al., 2012). The gut microbiota plays a central role in maintaining the integrity of the intestinal tract and in regulating multiple host metabolic pathways. Furthermore, several reports highlighted the ability of the gut microbiota to interact with the host's tissue, controlling its energy metabolism, contributing to variations in body weight, fat distribution, insulin sensitivity, and lipid metabolism (Cho and Kim, 2013; Zhang and Zhang, 2013; Falcinelli et al., 2015; Kim et al., 2018). A review based on 30 randomized, controlled trials and devoted to the effect of probiotic products containing live bacteria, mainly lactic acid bacteria, reported improved lipid metabolism by decreased total cholesterol and LDL cholesterol with no significant effect on HDL cholesterol and TAG (Cho and Kim, 2013).

The potential effect of the gut microbiome on metabolism has been examined in a few fish species (Butt and Volkoff, 2019). Administration of *Bacillus clausii* to olive flounder revealed improved growth performance compared to fish fed control diets (Ye et al., 2011), likely attributable to increased food intake and improved nutrient digestibility, suggesting an influence of the microbiota on fish metabolism. In zebrafish, the colonization of the gut by *Firmicutes* promoted lipid bioavailability in intestinal epithelium and liver (Semova et al., 2012; Sullam et al., 2012). In grass carp fed a ryegrass diet, lipid metabolism was significantly improved by changes in the gut microbiota (Ni et al., 2014), suggesting that the metabolic role played by the gut microbiome in grass carp can be affected by feeding. Interestingly, zebrafish with intact microbiota have increased lipid accumulation in the intestinal epithelium and increased expression of genes related to lipid metabolism (e.g., *apoa4*, *cox15*, and *hsl*) compared to germ-free zebrafish that lack microbiota (Sheng et al., 2018). Evaluating the hindgut microbiota of grass carp in response to an extreme dietary shift from FM to Sudan grass (*Sorghum sudanense*) revealed a rapid shift within 11 days (Hao et al., 2017). However, the dietary shift did not significantly change genes associated with lipid transport and metabolism.

α-Lipoic acid is used as an antioxidant in the detoxification of activated oxygen species (Navari-Izzo et al., 2002). Investigation of the effects of α-lipoic acid on lipid metabolism of GIFT tilapia fed HFD revealed that the adipose triglyceride lipase gene (*atgl*) was significantly upregulated in fish fed diets supplemented with 2.4 mg kg^{-1} α-lipoic acid, while the *dgat2* gene was downregulated (Xu et al., 2019a). In addition, a significant upregulation of liver *fabp* gene was displayed as a result of α-lipoic acid administration. Whist this study did not investigate the modulatory effect of α-lipoic on gut microbiota, studies investigating α-lipoic acid and an HFD in mice suggest that this is a likely scenario (Qiao et al., 2013).

The allochthonous microbiota in the hindgut of Western Cascade rainbow trout showed significant enrichment of lipid and fatty acid metabolism genes compared to that of eastern fish, even though Eastern Cascade fish had increased microbial diversity (Yildirimer and Brown, 2018). Kyoto Encyclopedia of Genes and Genomes (KEGG) analysis revealed that 8 out of 19 functional pathways were significantly enhanced in the hindgut microbiota of Western Cascade rainbow trout.

The metabolic functions of the microbial communities of adult zebrafish were analyzed by PICRUst, a bioinformatics software package designed to predict metagenomic functional content from marker genes (Arias-Jayo et al., 2019). In animals fed a high saturated fat diet, pathways involved in amino acid metabolism and transport were dominant, while in fish fed a high saturated fat diet enriched with DHA, pathways concerning transporters and lipid metabolism, including LNA and ARA, were more prominent.

Allochthonous intestinal microbiota of juvenile common carp has been shown to be modulated by waterborne copper exposure (Meng et al., 2018). Furthermore, from a microbial point of view, intestinal Roseburia sequences were positively associated with lipogenic enzymes, total protein and TAG, but negatively associated with lipolysis enzymes.

An early study revealed differences in the gut microbiota of fast and slow growing orange-spotted grouper (Sun et al., 2009). For example, *Bacillus pumilus*, *Bacillus clausii* and *Psychrobacter* sp. were only isolated and dominated in the gut of fast-growing fish, in contrast to slow growing fish, where *Vibrio* was dominant. A later study using fast-growing transgenic common carp suggested that the low *Bacteroides*: *Firmicutes* ratio of transgenic carp could be one of the factors contributing to the fast growth (Li et al., 2013).

When discussing how lipid metabolism can influence the host's health, there is evidence that gut microbiota play a key regulatory role in lipid metabolism, with increasing evidence for interplay between bile acids and lipid metabolism (Xiong et al., 2018; Zhou et al., 2018a). The dietary addition of a bacterial strain, affiliated to *Citrobacter*, to an HFD fed to Nile tilapia modulated the intestinal microbial composition and enhanced HFD-induced lipid accumulation in mesenteric adipose tissue. This observation was concomitant with increased TAG absorption efficiency and reesterification and increased intestinal permeability (Zhang et al., 2020a).

5.8.4 Practical considerations of probiotics, prebiotics, and short-chain fatty acids in aquafeeds

5.8.4.1 *Probiotics*

A review devoted to probiotics and lipid metabolism stated that, within the context of obesity and chronic metabolic diseases of modern societies, "probiotic is a better prevention and treatment strategy for regulating lipid homeostasis" (Zhang and Zhang, 2013), and 5 years later, Falcinelli et al. (2018) stated "probiotics can be used to positively alter gut microbiota, and their ability to improve metabolism." Diseases associated with lipid metabolism disturbances are increasing worldwide, and probiotics, especially *Lactobacillus* and *Bifidobacterium* have been shown to lower serum cholesterol levels (Kumar et al., 2012). (Falcinelli et al., 2015) reported modulation of the zebrafish gut microbiota by *Lactobacillus rhamnosus* IMC 501 administration, showing significantly attenuated weight gain and decreased transcription of genes involved in cholesterol metabolism (*hnf4α* and *npc1l1*) and *fit2* and TAG metabolism (*mgll*), coupled with lowered cholesterol and TAG contents.

A recent study investigated how photoperiod manipulation affected genes related to lipid metabolism and the roles of the intestinal microbiota in zebrafish larvae (Basili et al., 2020). As expression levels of lipid-related genes, *hnf4α*, *npc1l1*, *pparγ*, *srebp1*, *agpat4*, and *fitm2*, were significantly overexpressed in dark-exposed larvae, the authors suggested that the impairment in lipid metabolism was due to the modulation of the gut microbiota by probiotics.

In juvenile olive flounder, the supplementation of multi-strain probiotics, consisting of *Bacillus*, *Lactobacillus*, and *Saccharomyces*, to a low FM diet resulted in an enrichment of *Bacillus* and *Lactobacillus* in the midgut content, and gene functions related to fatty acid and bile acid biosynthesis, and glycerolipid metabolism (Niu et al., 2019).

5.8.4.2 *Prebiotics*

Prebiotics are selectively fermented by specific-health promoting bacteria such as *lactobacilli* and *bifidobacteria*, leading to a decrease of pathogenic bacteria in the GI tract and/or the production of fermentation end products, mainly short-chain fatty acids (SCFA), which can modulate host glucose and lipid metabolism (Gibson and Roberfroid, 1995; Delzenne and Cani, 2008; Delzenne et al., 2008). In comparison to endothermic animals, there is little information concerning the effects of prebiotics on aquatic animals and, despite the availability of numerous original and review articles (Merrifield et al., 2010; Ringø et al., 2010, 2014; Guerreiro et al., 2018), the effects of prebiotics on lipid metabolism have scarcely been investigated in fish.

Profiling mice gut microbiota identified a catalog of putative bacterial targets that may affect host metabolism in obesity and diabetes (Everard et al., 2011). Significant correlations between the genus *Anaerotruncus* and several metabolic parameters, such as plasma TAG content and muscle lipid content were established. In contrast, *Clostridium lactifermentans* revealed no effect on plasma TAG content. Galactooligosaccharides and fructooligosaccharides can have significant health benefits, particularly in relation to their putative anticancer properties, influence on mineral absorption, lipid metabolism, anti-inflammatory, and other immune effects such as atopic disease. In many instances, prebiotics seem to be more effective when used as part of a synbiotic combination. Short-chain fructooligosaccharides lowered plasma cholesterol LDL in gilthead sea bream fed a diet containing 0.25% short-chain fructooligosaccharides compared to a control diet (Guerreiro et al., 2015a). European sea bass fed diets supplemented with short-chain fructooligosaccharides and xylooligosaccharides displayed lower lipogenic enzyme activities (Me, Fas, G6pd) compared to groups fed unsupplemented diets (Guerreiro et al., 2015b). The effect of fructooligosaccharide on lipid metabolism in zebrafish using RNA sequencing revealed that lipid content, total cholesterol, TAG, FFA, and LDL were significantly lower in liver of fish fed a 0.4% fructooligosaccharide diet in comparison to control fed fish (Zhang et al., 2019a). In contrast, fish fed the 0.4% fructooligosaccharide diet had significantly higher HDL than fish fed the control diet. Me and Fas activities were significantly reduced by fructooligosaccharide administration. KEGG pathway analysis showed that the pathways of steroid hormone biosynthesis and steroid biosynthesis were significantly enriched. The profile of genes in these two pathways were affected by fructooligosaccharide, with the authors suggesting new mechanisms underlying the effect of fructooligosaccharide on lipid metabolism.

5.8.4.3 Short chain fatty acids

Organic acids are produced through the microbial fermentation of carbohydrates by various bacterial species under different metabolic pathways and conditions. SCFA, for example acetic, propionic and butyric acids, are produced through carbohydrate fermentation by the gut microbiota of endothermic animals (Cummings et al., 1987; Macfarlane and Macfarlane, 2003) as well as fish microbiota (Hoseinifar et al., 2017; Ng and Koh, 2017).

Several studies have demonstrated that the inclusion of organic acids, their salts, and/or mixtures to aquafeeds influences the bacterial community of the GI tract in aquatic animals. An early study reported evidence of an appreciable reduction of cultivable microbiota in feces and gut surfaces of fish fed organic acid–supplemented diets (Ng et al., 2009). Later, the antimicrobial efficacy of dietary organic acids on gut microbial community was reported in olive flounder (Park et al., 2011), red hybrid tilapia (Koh et al., 2016), and gilthead sea bream (Rimoldi et al., 2018). The inclusion of organic acids in finfish diets has also been shown to reduce the population of potentially deleterious bacteria such as *Vibrio* spp. in the intestinal tract (Park et al., 2011).

Other than a reduction in cultivable bacterial populations, changes in the population composition of the GI microbiota have also been associated with organic acid supplementation. Liu et al. (2014b) observed slight alterations of the adherent microbiota within gut segments of common carp fed a microencapsulated Na-butyrate diet (0.3 g/kg). In contrast, an earlier study reported that Na-butyrate supplementation did not significantly affect the gut bacterial community of African catfish (Owen et al., 2006).

Changes in the population and composition of gut microbiota can have significant impacts on cultured aquatic animals' growth performance, nutrient utilization, immune response, and resistance to pathogens. Further focused research is needed to elucidate the effects of dietary organic acids as mediated through changes in gut microbiota.

5.9 Dietary sterols

Despite no documented reports of a requirement in any fish species, cholesterol is a fundamental component of cell membranes in all animals, where it comprises between 25% and 30% of total membrane lipids. Distributed widely, but not homogeneously, across all cells, cholesterol plays an integral structural role, condensing and rigidifying the cell membrane, dictating fluidity, and ultimately controlling the permeability and subsequent movement of neutral solutes, hydrogen ions, and sodium ions in and out of the cell. In addition to its structural role, cholesterol also plays functional roles in intracellular transport, cell signaling, endocytosis, and nerve conduction (Dietschy and Turley, 2004). Likewise, it is well known as a precursor molecule for calcium metabolism, the synthesis of vitamin D, and the production of sex hormones and bile acids (Blomberg Jensen, 2012; Chay et al., 2013).

As all vertebrates can synthesize cholesterol de novo from acetate, it is not considered an essential nutrient in teleost nutrition. Therefore, until recently, there has been minimal research effort into this important molecule. In addition to biosynthesis, the once abundant concentrations of cholesterol in aquaculture feeds, supplied by the historical high inclusion rates of dietary FM and FO, have further circumvented this need, suppling upwards of 1g of cholesterol per kg of feed (Tocher et al., 2008). However, in an age of rising prices, resource sustainability considerations and a strong desire from the global aquaculture industry to improve its social licence to operate, FM and FO are now relatively minor components of compounded aquafeeds. In the place of these marine-derived raw materials, the industry is now heavily reliant on the inclusion of various alternative meals and oils which, for the most part, are vegetable based. While the compositional differences in the fatty acid composition of these raw materials have been well documented and investigated (Turchini et al., 2009), the marked differences relating to cholesterol composition and the potential impacts on fish lipid metabolism have been given minor consideration (Norambuena et al., 2013). Notably, in comparison to animal-derived meals and oils (e.g., poultry by-product meal, meat and bone meal, tallow etc.) plant meals and VO contain low concentrations of cholesterol, but are abundant in phytosterols, of which β-sitosterol, campesterol, and stigmasterol are present in the highest concentrations (Thomas, 2007). Although structurally similar, phytosterols are known to reduce cholesterol absorption in both fish and mammals via a disruption to the solubilization of intestinal cholesterol into micelles (Brown et al., 2010a). In fish, it has been hypothesized that this may lead to increased TAG accumulation in the liver (Liland et al., 2013a); however, observations that both support and refute this are apparent (Deng et al., 2013; Liland et al., 2013a; Sissener et al., 2017; van Spankeren et al., 2019). For example, in a study investigating cholesterol uptake in zebrafish, van Spankeren et al. (2019) reported no impact on plasma cholesterol concentrations up to a phytosterol: cholesterol ratio of 2.2:1, whereas the opposite effect was evident in juveniles of sea bass and Atlantic salmon (Couto et al., 2015; Sissener et al., 2018).

Targeted studies have demonstrated active cholesterol synthesis in fish following the administration of feeds with "deficient" or low levels of cholesterol (Norambuena et al., 2013). For example, significant elevation of mRNA expression of cholesterol synthesis genes, such as 3-hydroxy-3-methyl-glutaryl-coenzyme. A reductase (*hmgcr*), in yellowtail, Atlantic salmon and rainbow trout fed a non-FM diet indicated that cholesterol synthesis was activated (Maita et al., 2006; Zhu et al., 2018a). Likewise, the expression of cholesterol biosynthetic genes such as mevalonate kinase (*mk*) increased after dietary substitution of FO by VO in Atlantic salmon (Leaver et al., 2008c). However, no significant impacts on fish performance or survival have been reported across a range of species, including Atlantic salmon, rainbow trout, and European seabass (Norambuena et al., 2013; Kortner et al., 2014; Couto et al., 2015). However, there are documented cases of improved performance following cholesterol fortification of diets for channel catfish, Japanese olive flounder, and juvenile turbot (Twibell and Wilson, 2004; Den et al., 2010; Yun et al., 2011). In rainbow trout, growth performance and cholesterol metabolism were improved by cholesterol supplementation of soybean meal-based diets (Deng et al., 2013), and dietary cholesterol supplementation to a plant-based diet suppressed cholesterol synthesis and induced bile acid production in Atlantic salmon (Kortner et al., 2014), with similar results obtained in turbot (Yun et al., 2012; Zhu et al., 2018b). It has been speculated that improved performance in fish fed diets supplemented with cholesterol is linked to a reduction in energy expenditure for cholesterol synthesis (Norambuena et al., 2013), although it is important to highlight that levels above 1% resulted in growth impairment in juvenile turbot, thus suggesting a need for further investigation into the mechanisms underpinning these observations. This need is further highlighted by the contrasting results concerning the effects of cholesterol supplementation on fatty acid metabolism, with an apparent reduction in liver Δ6 Fads activity reported for Atlantic salmon (Sissener et al., 2018) on one hand, and an increase in the expression of both Δ6 *fads2* and *elovl5* genes in rainbow trout reported on the other (Norambuena et al., 2013). In summary, fish can synthesize cholesterol and, when dietary cholesterol is insufficient, endogenous biosynthesis is increased. However, studies in many fish have found that increased plant-derived components in feed can suggest cholesterol "deficiency" and affect fish growth and this can be alleviated by adding cholesterol to feed.

The provision of dietary cholesterol takes on an altogether different role in crustacean nutrition given the inability for de novo synthesis. Thus, while this review is focused on finfish, a few notes about cholesterol in crustaceans are hereto reported. In crustaceans, a dietary supply of cholesterol is essential for maintenance, optimal growth/moulting frequency and survival (Teshima, 1997; Holme et al., 2006). In addition to its roles in fish nutrition, in crustaceans, cholesterol is intimately linked with the production of ecdysone. Synthesized from cholesterol by steriodogenic enzymes, circulating levels of ecdysone mediate the moulting events that facilitate crustacean growth (Kumar et al., 2018). This energy-intensive process has been reported to significantly draw down the stored reserves of cholesterol in the hemolymph of pink shrimp (*Farfantepenaeus duorarum*), presumably for the assembly of the new cell membranes (Bonilla-Gómez et al., 2012). Dietary requirements, typically reported as a

% of the dry diet, range between 0.2% and 2.0% with values established for an array of commercially important species from both marine and freshwater environments, including shrimp, lobster, and crab (NRC, 2011). Notably, it has been suggested that phytosterols can be used to partially, or totally replace dietary cholesterol in some crustacean species in light of (a) their natural omnivorous feeding habits featuring vegetable matter naturally rich in phytosterols (Mondy et al., 2014), or (b) an ability to convert β-sitosterol and ergosterol to cholesterol (Teshima, 1971). The type of phytosterol utilized also appears to exert a major influence on the success of potential cholesterol sparing, where ergosterol successfully replaced cholesterol in larval *Penaeus japonicus* (Teshima et al., 1983), while sitosterol and β-sitosterol were unsuccessful as replacements for cholesterol in juvenile *Homarus* sp. (D'Abramo et al., 1984) and *P. japonicus* (Teshima et al., 1989).

5.10 Dietary phospholipids

As described above, phosphoglycerides, or phospholipids (PL) as they are more commonly referred to, are a diverse range of lipids that play a range of roles in the nutritional physiology of aquatic animals (Tocher et al., 2008). It has long been established that their dietary inclusion can impart a series of performance benefits in marine and freshwater species, but they are perhaps best recognized for their contributions to crustacean and larval/early juvenile fish nutrition, where in the latter, de novo synthesis, if present, is insufficient to meet the requirements dictated by the rapid growth and development of this life-history stage. As outlined above, these compounds are unified by the presence of phosphorous and commonly divided into four major classes (PC, PE, PS, and PI) based on the esterification of choline, ethanolamine, serine, and inositol to a phosphatidic acid backbone (Tocher et al., 2008). In addition to their well-defined roles as integral components of cellular membranes and transporters of essential neutral lipids from the intestinal mucosal cells into serum as chylomicrons and VLDL (Hochachka and Mommsen, 1995; Tocher et al., 2008), PL are well documented as precursors to a series of highly biologically active mediators, including inositol phosphates, platelet-activating factors, DAG, and eicosanoids (Tocher, 2003). In addition, given the amphipathicity of these compounds (i.e., having both hydrophilic and hydrophobic properties) their inclusion has been shown to improve overall diet quality/acceptance via enhanced palatability (Cahu et al., 2009). In particular, PL addition is thought to improve the stability of feeds via a reduction in the leaching of water-soluble vitamins and minerals (Coutteau et al., 1997).

While TAG are widely acknowledged as the key lipid class for storage and subsequent energy production in fish, PLs are known to play important roles in energy provision during the early life history stages of some species (Coutteau et al., 1997). This is particularly evident in fish originating from eggs void of an oil globule (e.g., Atlantic cod and halibut) where, of the lipid that is present, PL (mainly PC) dominate the lipid composition (Finn et al., 1995; Rønnestad et al., 1995). However, there are exceptions to this, where eggs of species which do contain an oil globule (e.g., yellowtail kingfish) can be characterized by high PL concentrations (Hilton et al., 2008). Available evidence suggests that ~20% of marine fish produce eggs without an oil globule, while this figure is as high as 60% in the freshwater realm to assist with egg buoyancy (Baras et al., 2018).

With respect to larval fish species, the efficacy of dietary PL is generally assessed using survival, growth, development, malformation, and, to a lesser extent, stress resistance as the criteria of success. Historically, enriched live feeds were used as the most effective vehicle of administration to assess the impacts and requirements of PL on larval development (Kanazawa et al., 1981). Indeed, although expensive and labor-intensive to produce, this form of feeding invariably yields superior performance under hatchery conditions. However, despite representing the prey of choice for larval fish, the naturally occurring background levels of PL in live feeds means that a deficient diet is difficult to realize (Nordgreen et al., 2007). Likewise, the potential for the remodeling of the intended PL and fatty acid composition further complicates this approach for the establishment of PL requirements using live feeds (NRC, 2011). The onset of microparticulate diets over 30 years ago facilitated the development of deficient basal diets to estimate requirements, albeit at the cost of palatability and fish acceptance considering the need to use PL-free raw materials (e.g., casein) in experimental formulations (Tocher et al., 2008). Observations following the implementation of microparticulate diets have since demonstrated the essential nature of PL across a broad range of larval fish species from both marine and freshwater environments. Notably, this section does not intend to provide a detailed overview of the various reported requirements, and readers are thus directed to one of the major reviews of this topic (Tocher et al., 2008; Cahu et al., 2009).

Of the larval species studied, growth and survival have been the dominant criteria for assessment, demonstrating requirement levels ranging from 2% to 3% (% of dry diet) in common carp, Dojo loach (*Misgurnus anguillicaudatus*) and ayu (*Plecoglossus altivelus*) (Kanazawa et al., 1981; Geurden et al., 1995), 5%–7% in Japanese olive flounder, knife jaw (*Oplegnathus fasciatus*), large yellow croaker, and red sea bream (Kanazawa, 1993; Zhao et al., 2013; Gao et al., 2014) and upwards of 9%–12% in European seabass, pikeperch (*Sander lucioperca*), and gilthead seabream (Cahu et al., 2003; Seiliez et al., 2006; Hamza et al., 2008). In addition to growth, PL inclusion has also been demonstrated to minimize the accumulation of lipid droplets (steatosis) in the intestinal tract of rainbow trout (Dapra et al., 2011), while enhancing fatty acid transport and uptake in large yellow croaker (Cai et al., 2016), confirming the importance of this class in lipid absorption and transport as reported previously (Olsen et al., 2003; Tocher et al., 2008). The requirement for dietary PL in fish larvae is likely due to a limited ability to biosynthesize PL de novo for chylomicron formation and lipid transport out of the enterocytes (Tocher et al., 2008; Cahu et al., 2009). However, a transcriptomic microarray study performed indicated there was no clear genetic constraint on de novo PL synthesis in Atlantic cod larvae that appeared to have relatively high capability for PL biosynthesis at 1 dph (Li et al., 2015). However, few genes involved in de novo synthesis of PL were significantly regulated throughout the live feeding period and up to 60 dph. In contrast, the expression of many genes involved in PL synthesis in the intestine of Atlantic salmon were higher in parr compared to that of fry, suggesting higher capability of PL synthesis in the later developmental stages (Tocher et al., 2008; Cahu et al., 2009) although PL requirements declined as the fish grew (Tocher et al., 2008; Cahu et al., 2009; Feng et al., 2017). Consequently, the requirement for PL in juveniles appears somewhat lower than those observed in larvae, generally ranging between 1.5% and 7% of the dry diet (Kanazawa et al., 1981; Poston, 1990; Takeuchi et al., 1992; Kanazawa, 1993; Geurden et al., 1995). Interestingly, in Atlantic salmon, a stepwise reduction in requirement from 6% to 0% (% of dietary lecithin, not PL) has been reported in fish ranging from 180 mg to 7.5g (Poston, 1990, 1991). Subsequently, the low/nonexistent requirement for PL in more advanced juvenile stages (>2.5g) was reconfirmed (Taylor et al., 2015), who reported that PL inclusion (1.5%–4.2% of the dry diet) did not evoke any additional benefits from a fish growth and survival perspective when salmon were grown to the point of smolting (50–60g). Likely attributable to the cost-prohibitive nature of formulating experimental feeds to assess requirement levels, there is a lack of information pertaining to the essentiality of PL in larger fish. However, while the aquafeed industry continues to incorporate marine ingredients into its feed formulations, this is likely to remain unchanged considering fish receiving an artificial feed would be receiving intact PL from FM and FO sources, therefore improving the probability that any PL requirement would be met. Naturally, if the industry was ever to move away from marine ingredients altogether, it may be necessary to investigate this life stage further.

Despite a wide range of PL preparations being used to assess requirements experimentally (e.g., bonito polar lipid, corn polar lipid, chicken egg lecithin, sunflower PL, rapeseed polar lipid, soybean PL etc.), from a practical viewpoint these can be distilled down to egg lecithins, plant lecithins, and marine PLs, of which, plant lecithins (mainly derived from soybeans), are by far and large the most commercially available. In most cases, studies have used these "complex" PL sources, which tend to vary considerably with respect to the total concentration of PL, the proportions of the classes (e.g., PC, PE, PS, PI) contained within, not to mention the individual fatty acids esterified to the phosphatidic acid backbone. For example, soybean lecithin can contain anywhere between 50% and 86% PL, split evenly between PC, PE, and PI at approximately 20%–25% of total PL (Tocher et al., 2008). As such, when comparing requirement levels between studies using different PL sources, or even the same source, it is difficult to make any definitive statements concerning true quantitative requirements, or the benefit of one PL class over another. In the case of the latter, this can only be realized via the implementation of semi-purified PL sources where, of the limited studies that have been carried out, there is an apparent benefit of PC > PI > PE > PS when assessed from the perspective of growth (Geurden et al., 1998; Tocher et al., 2008). This result is perhaps not surprising given the abundance of PC in the cellular membranes of eukaryotes (Kent, 2005), alongside the well documented role of this PL in the structure of lipoproteins. Alternatively, with respect to larval development and the incidence of malformation, evidence suggests a greater importance of PI. In early juvenile Atlantic cod, a microparticulate diet containing PI at 2.57% of total lipid significantly reduced the incidence of skeletal deformities and increased survival rate in comparison to the control diet containing 1.52% of total lipid as PI (Hansen et al., 2011). However, there is an apparent fine line with respect to the optimal inclusion level of this class in dietary formulations. For example, in juvenile gilthead seabream a reduction in dietary PI increased the incidence of cranial deformities, while an increase in the overall ratio of PC/PI resulted in an increase in the occurrence of skeletal deformities (Sandel et al., 2010).

5.11 Peroxidation and pigments

5.11.1 Lipid peroxidation

Many carnivorous species, including salmonids and marine fish, depend on relatively high levels of dietary LC-PUFA. The challenge with highly unsaturated diets is that they are extremely prone to peroxidation. This is a sequence of processes normally initiated with the generation of a free radical (unpaired electron) by hydrogen abstraction from a lipid molecule (Fig. 5.19). Free radicals rapidly react with oxygen to produce a peroxyl radical, which is an extreme electrophile. Peroxy radicals will therefore abstract an electron (hydrogen) from a lipid compound producing another free radical and a lipoperoxide. The lipoperoxide (analogous to hydrogen peroxide) is unstable and will be further reduced and decomposed to produce a large array of secondary products. Lipid peroxidation is generally autocatalytic in the presence of oxygen, as one free radical gives rise to at least two free radicals (Fig. 5.19). Without oxygen present, the process will not progress and, under normal circumstances, the propagation step can also be prevented by the addition of antioxidants. Antioxidants include hydrogen donors that will eliminate free radicals and prevent the accumulation of peroxides. Donor antioxidants generally give rise to low-reactive free radicals that are subsequently reduced to stable compounds through, for example, dimerization or regeneration. Common antioxidants include vitamins E and C, glutathione, and synthetic phenolic compounds like butylated hydroxyanisole (BHA) or butylated hydroxytoluene (BHT). A

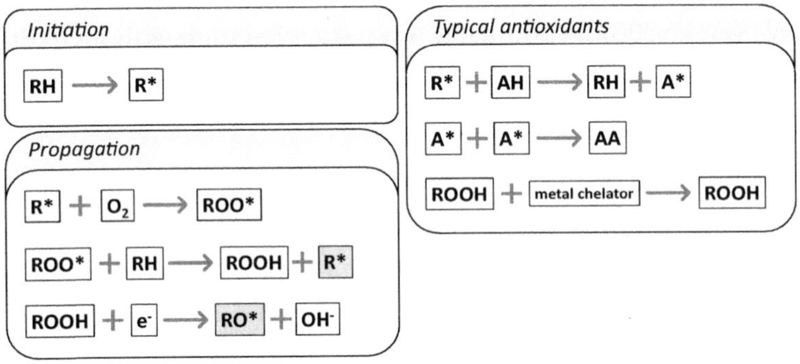

FIGURE 5.19 Schematic progress of lipid peroxidation and antioxidant function. Initiation: generation of a free radical (unpaired electron,*) by hydrogen abstraction from a lipid molecule (RH). Propagation: Free radicals (R*) react with oxygen (O_2) to produce a peroxyl radical (ROO*). Peroxyl radicals abstract an electron from lipid compound producing another free radical (R*, *gray shaded*) and a lipoperoxide (ROOH). The lipoperoxide is unstable and will be reduced and decomposed to produce a large array of secondary products (RO*). Typical antioxidants: hydrogen donors (AH) eliminate free radicals (R*) and prevent the accumulation of peroxides. Donor antioxidants give rise to low-reactive free radicals (A*) that are reduced to stable compounds (AAs). Metal chelators reduce the availability of electron donors which are required as catalyst for lipoperoxide degradation.

crucial point of propagation is lipoperoxide degradation, a process that typically has to be catalyzed by good electron donors such as divalent irons or copper ions. In biological systems, transport of these ions is highly regulated where, for example, iron is transported by transferrin in its oxidized Fe^{3+} form. Iron chelators in food systems include ethylenediaminetetraacetic acid (EDTA).

Lipid peroxidation not only destroys the fatty acids themselves, but intermediate products of free radicals (i.e., aldehydes and ketones) will also attack amino acids and vitamins, eventually ruining the nutritional value of the diet. Many of the free radical products will be absorbed or, when generated in the body, produce similar components, generating oxidative stress and increasing the requirement for nutritional antioxidants. Other compounds, particularly those generated from n-3 LC-PUFA can lead to the production of eicosanoid-like compounds such as isoprostanes and neuroprostanes that can have direct biological functions. The detailed mechanisms of lipid peroxidation in lipid and food systems have been the subject of several books and reviews and will not be discussed in detail here, and the reader is advised to confer with other literature for further details (Hatcher et al., 2009; Decker et al., 2010; Frankel, 2012b; Johnson and Decker, 2015).

The main reason why marine LC-PUFA are extremely prone to peroxidation can be found in the methylene-interrupted double bond structure of unsaturated fatty acids. With the introduction of a double bond, electrons are distributed to the adjacent bis-allylic (α-carbon), lowering its bond dissociation energy from 98 kcal/mol to 89 kcal/mol (Johnson and Decker, 2015) and making it more prone to electron abstraction (Fig. 5.20A). With the introduction of a second methylene-interrupted double bond, both will contribute electrons to the central α-carbon (α*) which reduces the bond dissociation energy further to 80 kcal/mol (Johnson and Decker, 2015) (Fig. 5.20B). Introducing more double bonds will increase the number of methylene-interrupted carbons and, therefore, the oxidizability of fatty acids. The rate of

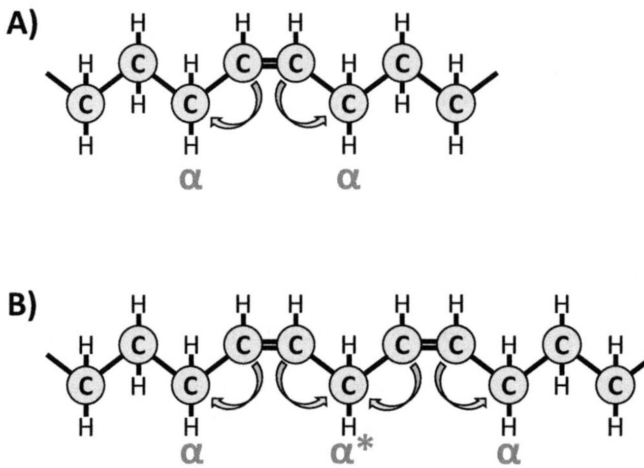

FIGURE 5.20 Electron distribution from double bonds to the adjacent α-carbon in a fatty acid. (A) single double bond, (B) two methylene-interrupted double bonds. Note the contribution to the central α*-carbon by both double bonds.

autooxidation of fatty acids tends to increase linearly and approximately twofold with the number of bis-allylic groups above one double bond. For example, if the peroxidation rate of 18:1 (nonmethylene interrupted hence higher dissociation energy) is set to 1, then 18:2, which has one methylene-interrupted carbon, autooxidizes 41-times faster. Increasing to two methylene-interrupted groups in 18:3 increases peroxidation rate approximately twofold that of 18:2 (98-times that of 18:1), while 20:4 with three methylene interrupted groups, is oxidized with a rate twofold that of 18:3 (195 times faster than that of 18:1). DHA appears to peroxidize at rates around 1.8-times that of 20:4 (Frankel, 2012a; Johnson and Decker, 2015).

Marine oils are generally extracted from conserved raw materials of varying quality under high temperature, potentially leading to an increase in the rate of peroxidation. These products will contain numerous initiators of lipid peroxidation such as iron and copper. Even in more controllable systems, like many plant oils, there are sufficient amounts of metal ions to initiate peroxidation (Johnson and Decker, 2015). It is therefore interesting to note that a majority of the n-3 LC-PUFA oil products on the human nutraceutical market can be characterized as oxidized (Mason and Sherratt, 2017; Wang et al., 2017d) further emphasizing the problems of stabilizing marine-type oils. To stabilize oils containing LC-PUFA it is therefore essential to add antioxidants and to keep oxygen tension as low as possible. In the past, oxidation of FM could generate so much heat that they could spontaneously combust (de Koning, 2002). Due to its excellent antioxidant efficiency, ethoxyquin became a standard ingredient to stabilize meals and oils destined for fish feeds. However, accumulating evidence over many years of the possible toxicity of ethoxyquin (and some of its impurities) in fish and the environment led to several concerns. The effects included increased mortality, deformities, and neuronal damage (Egloff and Pietsch, 2018; Pradhan et al., 2020). As a result, the use of ethoxyquin as an antioxidant in fish and all animal species was banned in the European Union in 2017 (Anonymous, 2017), although the regulation will be reviewed by the end of 2022. There is currently considerable research into evaluating safe levels, and future use may be possible. For the marine oil industry, the lack of good stabilizers has created concerns as there are few good alternatives. While BHT, BHA and other organic antioxidants are used, their efficiency is limited, and FM is now packed in nitrogen atmospheres. Ironically, the elimination of ethoxyquin due to health-issues may just have created other health issues.

While lipid peroxidation in FO was a particular problem in the 1980s, it is still a subject of much research as many of the available oils are of poor quality. There is a plethora of consequences of feeding oxidized oils to fish, although the sensitivity may depend on species or life stage. Common effects include reduced growth, intestinal and tissue damage, increased in vivo peroxidation with generation of oxidation products like malondialdehyde (MDA) or thiobarbituric acid-reactive substances (TBARS), reduced membrane LC-PUFA, and upregulation of inflammation and antioxidant pathways (Hamre et al., 2001; Zhang et al., 2017; Fontagné-Dicharry et al., 2018; Fatima et al., 2019; Liang et al., 2019a; Yin et al., 2019; Xie et al., 2020). As vitamin E is an important in vivo antioxidant, it will be exhausted in fish fed oxidized oils and, therefore, supplementing vitamin E can reduce or mitigate the damage and may also increase growth at moderate levels (Lewis-McCrea and Lall, 2007; Fontagné-Dicharry et al., 2018; Fatima et al., 2019). Increased focus on alternative and natural antioxidants could help protect fish against oxidative stress as synthetic compounds currently used may end up being banned. For example, emodin (6-methyl-1,3,8-trihydroxyanthraquinone), a natural herbal medicine antioxidant has been reported to reverse growth depression, intestinal damage, and mitochondrial damage in Wuchang bream (*Megalobrama amblycephala*) fed oxidized oils partly through Ppar and Nrf2-Keap1 signaling, a master regulator of antioxidant response (Song et al., 2019).

In conclusion, it is well know that feeding high levels of PUFA, especially LC-PUFA, increases oxidative stress in fish (Olsen and Henderson, 1997), which leads to an increased requirement for stabilizing dietary antioxidants like vitamin E (Hamre, 2011). Ironically, the tendency to reduce the inclusion of FO in fish diets may actually lead to improved fish welfare in the long-term as both feed peroxidation and in vivo oxidative stress will be reduced.

5.11.2 Pigments

Carotenoids are produced by all photosynthetic organisms and some heterotrophic fungi and bacteria. They comprise a large group of lipid-soluble pigments with over 750 naturally occurring molecules, of which more than 250 are of marine origin. Carotenoids are tetraterpenes based on a 40-carbon polyene chain with conjugated double bonds that differ in (i) the presence and number of oxygen atoms, with hydrocarbons being carotenes and oxygenated carotenes being xanthophylls, (ii) hydrogenation of the polyene chain, (iii) the cyclization at one or both ends (ε-ione or β-ionone rings), and (iv) the length of the chromophore. The double bond structure gives the different carotenoids distinctive colors and, in animals, have physiological roles in photoprotection, camouflage, and breeding, while color is also important in aquaculture for consumer perception and sales. The best known carotenoids are astaxanthin (red), canthaxanthin (orange-red), β-carotene (orange), tunaxanthin (yellow), doradexanthins (yellow), zeaxanthin (red), and echinone (red) (Amaya and Nickell, 2015; Esteban et al., 2015). As most heterotrophic organisms are incapable of synthesizing carotenoids, any accumulation must be obtained through the diet and, depending on dietary origin, they may be in the free form, or xanthophylls as mono or diesters.

As carotenoids are lipid-soluble compounds with poor solubility in the aqueous environment of the gastrointestinal tract, they partition with lipid in the intestine into the primary lipid emulsions containing neutral lipid (TAG/WE), bile salts, PL and cholesterol. When lipases act upon the emulsions to digest TAG/WE, carotenoids transfer with the hydrolyzed products into more polar mixed micelles dominated by FFA, partial acylglycerols and cholesterol. Carotenoid esters will also be hydrolyzed by nonspecific lipases during the process. Partitioning into mixed micelles is a significant determinant of bioavailability, as these will traverse the unstirred water layer to deliver the compounds to enterocyte surfaces with the rate of transfer to a large extent determined by the polarity of the carotenoid. In experimental systems, only 0.5% of the nonpolar carotene lycopene (noncyclic) will partition from primary emulsions into mixed micelles, compared to 19.8% for the polar xanthophyll astaxanthin (Tyssandier et al., 2001). This is a major factor behind the generally low digestibility of carotenoids, with lower bioavailabilities for carotenes than xanthophylls (Reboul, 2019). In Atlantic salmon, the ratio of astaxanthin to lutein (less polar) was 2.5 in the diet but 25 in blood, suggesting very low bioavailability of lutein (Olsen and Baker, 2006). The typical digestibility of astaxanthin in Atlantic salmon is generally 50% or less (Ytrestøyl et al., 2005). Salmonids also generally preferentially absorb astaxanthin rather than the less polar canthaxanthin, zeaxanthin or carotene, although there are species differences with rainbow trout appearing more selective than Atlantic salmon (Lim et al., 2018).

The bioavailability and digestibility of carotenoids can also be regulated by emulsion droplet size and altering the polarity of the mixed micelles. Reduction of droplet sizes increase lipolytic efficiency and produces mixed micelles with higher FFA content, which can incorporate larger amounts of carotenoids in both artificial (Salvia-Trujillo et al., 2017) and in vivo human studies (Affandi et al., 2012). Although some of these processes can be affected by species differences in endogenous flow of bile salts, PL, and cholesterol, diet is probably the main factor affecting mixed micellar structure and delivery to the enterocyte surface. However, there are relatively few dietary studies focusing on mixed micelles and membrane uptake mechanisms in fish. Accordingly, blood levels of carotenoids or, in the case of salmonids, muscle astaxanthin deposition, are often taken as indicators of carotenoid absorption, which could be challenging, and partially misleading, as they do not take into account factors such as enterocyte efflux or pigment degradation. Bile salts have been shown to increase blood xanthophyll (astaxanthin) levels of Atlantic salmon (Olsen et al., 2005) and tissues of European seabass (Sallam et al., 2017), clearly indicating advantages of altering intestinal micellar structures. Cholesterol also increases blood level and muscle deposition of astaxanthin in Atlantic salmon (Chimsung et al., 2014). There is a general assumption that high dietary lipid levels will enhance astaxanthin absorption, but results are variable and inconsistent (Lim et al., 2018). While early studies in both rainbow trout (Nickell and Bromage, 1998) and Atlantic salmon (Bjerkeng et al., 1999) showed increased muscle pigmentation with high lipid diets, Hamre et al. (2004) found no effect of dietary lipid level on muscle astaxanthin in Atlantic salmon despite the fact that digestibility was probably increased as blood levels were elevated. Oil source may also affect carotenoid uptake as measured by blood level, where replacing FO with less unsaturated olive oil has been shown to enhance blood astaxanthin in rainbow trout (Choubert et al., 2006). Reducing the unsaturation further using lard also increased blood levels of astaxanthin in Atlantic salmon compared to fish fed FO (Olsen et al., 2005). In contrast, dietary soybean lecithin reduced blood astaxanthin levels in Atlantic salmon (Olsen et al., 2005). It is, however, important to note that variations in blood levels do not necessarily lead to enhanced flesh pigmentation. In fact, neither rainbow trout fed olive oil (Choubert et al., 2006), nor Atlantic salmon fed a range of diets with varying unsaturation, showed any difference in muscle astaxanthin deposition (Torstensen et al., 2005; Lim et al., 2018).

The absorption of carotenoids into fish enterocytes was originally thought to proceed through passive diffusion. However, data from mammalian research suggested that both enterocyte absorption and transport to lipoproteins may, to some extent, be carrier-mediated. While at present there is no evidence of a dedicated transporter, the most probable candidates implicated in carotenoid transport include class B scavenger receptors, SR-B1 and SR-B2 (CD36), and NPC1-like transporter 1 (NPC1L1) (Reboul, 2019). SR-B1 is particularly interesting as its low substrate specificity enables it to transport a wide array of components including carotenes, xanthophylls, and vitamins in addition to more common lipids like cholesterol/esters, PL, and FFA. In Atlantic salmon, Sr-b1 is highly expressed in proximal intestine and a paralog, Sr-b1-2, has been suggested as a QTL for flesh pigmentation (Sundvold et al., 2011). Interestingly, SR-B1 is required for carotenoid uptake and coloration of canary birds (Toomey et al., 2017). In addition to these transporters, several ATP-binding cassette (ABC) subfamily members, generally biochemical efflux pumps mediating cellular transport of a wide array of compounds, may be involved in carotenoid absorption. In fish, all the main families have been verified including Abca, Abcb, and Abcg and, in some species, Abch. They often localize to the apical side of polarized membranes and ABCB, ABCC, and some ABCG subfamilies are known drug transporters aiding the elimination of xenobiotics and other compounds from cells (Luckenbach et al., 2014). In mammals, ABCB1 has received considerable interest as a putative carotenoid transporter. Normally ABCB1 pumps lipid-soluble xenobiotics back into the intestinal lumen, but it has also been shown to be involved in exocytosis of carotenoids (Kotake-Nara and Nagao, 2011), thus reducing the apparent digestibility. It is normally highly expressed in fish intestines along with other representatives like ABCG2 (Luckenbach et al., 2014). Some of the ABCG subfamily groups including ABCG5, ABCG8 are sterol transporters involved in phytosterol and bile transport and they have also been suggested as putative carotenoid exporters in mammals (Kotake-Nara and Nagao, 2011).

Once absorbed, carotenoids can either be degraded or embedded into lipoproteins. No intracellular transport proteins have been identified, but several candidates have been suggested in mammals including IFABP, SR-B1, SR-B2, and NPC1L1 (Reboul, 2019). A large part of the absorbed carotenoid is likely metabolized and degraded in enterocytes and other tissues with two of the main enzymes being β-carotene 15,15′ oxygenase 1 (BCO1) and β-carotene 9′,10′ oxygenase 2 (BCO2). BCO1 cleaves symmetrically at the 15,15 double bond. With respect to β-carotene, this results in two molecules of all-trans retinal and, therefore, important for vitamin A synthesis. On the other hand, BCO2 catalyzes eccentric cleavage of a large array of provitamin A and nonprovitamin A carotenoids. The family of genes in teleosts includes five members, two *bco1* and three *bco2* genes, and alignment studies have verified the genes to be common in most fish species studied. In Atlantic salmon, intestinal *bco1*-like and *bco2*-like genes are highly expressed, although lower expression levels are also found for *bco1*, *bco2a*, and *bco2b* (Helgeland et al., 2014).

In mammals, carotenoids and retinyl esters are postprandially packed into chylomicrons through the apoB-dependent pathway. More polar carotenoids such as xanthophylls (e.g., astaxanthin) are likely transported by the apoB-independent HDL pathway mediated by the ABCA1 transporter (Reboul, 2019). Although reports vary with fish, the general view is that carotenoids are distributed in most dense fractions following ultracentrifugation with some fractions carrying more than others. In spawning upstream migrating chum salmon, *Oncorhynchus keta*, astaxanthin was transported in the HDL fraction (Nakamura and Hata, 1985) and, in immature rainbow trout, 66% of astaxanthin was found in the HDL fraction with the remaining distributed in LDL > VHDL > VLDL (Choubert et al., 1994). Chimsung et al. (2013) also found astaxanthin in all lipoprotein fractions, but predominantly in the protein-rich (possibly albumin) and very light VLDL fractions (density < 1.033 mf/g) particularly when cofed cholesterol. Studies using radiolabeled astaxanthin showed most astaxanthin (58%) associated with the protein-rich lipoprotein deficient fraction followed by HDL (36%) and LDL (6%) with some in a very light fraction termed chylomicron (Aas et al., 1999).

Carotenoids are generally stored in the integument and gonads in fish, but some species, particularly in the Salmonidae family, also store carotenoids in the muscle (Lim et al., 2018). In most species, except catfish, integument carotenoids are stored as mono- or di-esters. Fish have some, but often limited, capacity to metabolize carotenoids, with cyprinids able to convert lutein or zeaxanthin to astaxanthin, whereas salmonids cannot (Maoka, 2011; Amaya and Nickell, 2015). Some carotenoids, such as α- and β-carotene and β-cryptoxanthin, have essential properties as vitamin-A precursors (Weber and Grune, 2012). In Atlantic salmon, astaxanthin and canthaxanthin have provitamin A activity while lutein, zeaxanthin, and astaxanthin are precursors for vitamin A_2 in some freshwater fish (Maoka, 2011; Amaya and Nickell, 2015; Lim et al., 2018).

In living organisms, carotenoids are often associated with cellular membranes. Nonpolar carotenoids (e.g., lycopene and β-carotene) are embedded in the bilayer while oxygenated carotenoids (e.g., astaxanthin) span the membrane with their polar rings exposed to the aqueous environment (McNulty et al., 2008). A large part of the biological functions of carotenoids appear to be related to their excellent antioxidative abilities, including free radical and singlet oxygen quenching. The antioxidative ability varies with compounds, but the presence of oxy functional groups gives good antioxidation function (Ambati et al., 2014) probably related to their membrane expanding properties (McNulty et al., 2008). In mammalian systems, physiological effects include anti-inflammatory properties, upregulation of antioxidant capacity (catalase and glutathione peroxidase), reduction in proinflammatory cytokines (e.g., IL-1, IL-6, and TNF-α) and a general suppression of free radical species (Ambati et al., 2014; Milani et al., 2017). In fish and many crustaceans, astaxanthin and other carotenoids have been shown to improve reproductive success, growth rates, survival, stress tolerance, and disease resistance in addition to the more common effects on skin and flesh coloration in salmonids [for review see Lim et al. (2018)]. Many of these effects can be related to the antioxidative potential, but some appear to rely on other pathways.

5.12 Lipids in aquafeeds and planetary boundaries

Since the beginning of the last century, with the expansion of industrialization and the global population, there has been growing interest and focus on sustainability and, in particular, sustainable agriculture. The idea that our planet has limits was formed during the second half of the century and, in 2009, the framework concept of planetary boundaries (PB) was proposed, with these boundaries defined as the safe operating space for humanity, with respect to the planet's biophysical subsystems or processes (Rockstrom et al., 2009). At the United Nations General Assembly in 2015, a collection of 17 global Sustainable Development Goals and 169 targets, which were reported to be "integrated and indivisible," were defined. The aim of these was to balance economic, social, and environmental sustainability, collectively establishing a new universal agenda which continued to build upon the previous Millennium Development Goals (UN, 2015). Within human activities affecting global sustainability, agriculture, farming, and food production systems were identified as playing major roles in many of the Sustainable Development Goals, as major drivers of changes in land use, freshwater depletion, climate change, and pollution of ecosystems (Springmann et al., 2018).

Accordingly, aquaculture is not immune, nor exempt, from sustainability considerations, with aquaculture–environment two-way interactions being increasingly important in shaping the sector. Specifically, the aquafeed sector has directly faced the reality of PB based on the realization that the maximum sustainable limits concerning the extraction and utilization of marine-derived raw materials were reached. Aquafeeds, which had been historically formulated around these two raw materials as primary sources of dietary proteins and lipids, had to rapidly change and evolve to allow increased production as required to support a rapidly growing global aquaculture sector (Turchini et al., 2019). Though the limits of natural resource availability are evident to any aquafeed formulator (in the form of price volatility and unpredictable availability of FM and FO), the concepts and related research efforts underpinning the PB framework have, thus far, focused primarily on terrestrial systems, with limited attention on marine ecosystems (Nash et al., 2017). With the realization that global fisheries have reached their maximum sustainable landings, any possible increase in seafood supply has to come from aquaculture, and the future environmental footprint of global seafood production will, therefore, depend on the aquaculture technologies and approaches implemented, from the species farmed and the location of the farms, to what the cultured aquatic animals eat (Willett et al., 2019). Interestingly, whereas much attention has been given to the (true, speculated or likely) environmental impacts and sustainability considerations of the use of marine-derived raw materials in feed (Naylor et al., 2009; Turchini, 2013; Han et al., 2018; Kok et al., 2020; Zhang et al., 2020b), the environmental impacts and sustainability considerations of current, modern aquafeeds containing relatively low levels of FM and FO (Turchini et al., 2019), such as land use stemming from plant-based ingredients, are scarce (Willett et al., 2019). However, despite this, the aquafeed industry is now challenged by media and public attention on the sustainability of using some terrestrial raw materials of agricultural origin, such as soybean originating from the Amazon (Berge, 2019).

A great diversity of aquafeed formulations and raw materials are currently utilized worldwide. These vary not only in relation to the producer, target species, developmental stage, and farming methods but also in relation to markets, thus having great geographical and temporal variability. However, when it comes to dietary lipid sources (oils and fats), modern aquafeeds generally contain relatively small amounts of FO, typically from 0% to 30% of the total added lipid, with the remainder consisting of lipids of terrestrial origin, such as VO or land animal fats (Turchini et al., 2010). For example, a dramatic change in the utilization and inclusion of FO in favor of terrestrial VO has been documented for the Norwegian Atlantic salmon aquafeed sector. In 2000, FO accounted for 100% of the added dietary lipid in commercial compound aquafeed, equating to over 31% of total feed weight however, by 2016, two-thirds of the added dietary lipid was plant oils (~20% of feed), and marine oils contributed only around 10% of feed (Aas et al., 2019). Within this context, a peculiarity of lipid nutrition in aquaculture, compared to other branches of this scientific sector, is that the choice of oil or fat used for aquafeed formulation will not only affect fish performance, health, and production quantity and cost but also product quality. Indeed, the final fatty acid profile of cultured aquatic animals reflects that of the diet provided (Sargent et al., 2002). For this reason, the n-3 LC-PUFA content of oils used in aquafeeds, and thus the amount of health promoting n-3 LC-PUFA delivered to consumers via aquaculture, is the topic of intense global research and development effort.

Aquaculture currently plays a central role in what has been referred to as the long-chain omega-3 fatty acid gap (Tocher et al., 2019). The importance of aquaculture as both a supplier and a user of n-3 LC-PUFA, and its fundamental role in delivering these health promoting and highly sought after nutrients to humans has been described (Tocher, 2015). Unfed aquaculture, such as filter feeding molluscs and low-trophic finfish species cultured in extensive systems, can be net suppliers of n-3 LC-PUFA (Tan et al., 2020). Additionally, the possibility of transforming intensive aquaculture systems for those species able to biosynthesize n-3 LC-PUFA from dietary ALA that can be provided by VO has been documented (Turchini et al., 2011a). Nevertheless, with current technology it is considered unlikely that aquaculture has the ability to fill the existing gap between supply and demand for n-3 LC-PUFA. Accordingly, a multipronged approach is needed as "we cannot rely on the finite and limited marine resources that we can sustainably harvest or efficiently recycle. We need to produce n-3 LC-PUFA de novo and all possible options should be considered" (Tocher, 2015). Recent studies echoed these conclusions, highlighting the need for increased efficiency of recovery of n-3 LC-PUFA via improved by-product utilization and reduced wastage (Hamilton et al., 2020), and the development of economically viable, larger-scale technologies capable of delivering alternative sources n-3 LC-PUFA, such as microalgae and genetically modified higher plants (Tocher et al., 2019).

With aquaculture production projected to increase to 140 Mt by 2050, and the concomitant need for increased aquafeed supply, a key emerging constraint is the competition for feed resources (Willett et al., 2019). Independent of, and in addition to, the existing gap in sourcing oils containing n-3 LC-PUFA, the aquaculture sector is using large quantities of terrestrial oils and fats to fulfill the energetic requirements of cultured stocks (Turchini et al., 2010). This dependence on agriculturally produced oils and fats is expected to keep rising with the expansion of the sector, further exacerbating the food versus feed versus energy debate (Tilman et al., 2009; Godfray et al., 2010). For example, the freshwater footprint of aquaculture, originating from the agronomic practices for the production of raw materials used in aquafeed has been noted (Troell et al., 2014). Similarly, it is important to quantify and monitor all the other indirect impacts of different raw materials used in aquafeed formulation within the context of sustainability and PB. However, as noted above, these aspects have thus far attracted relatively little research attention.

Several different oils and fats of terrestrial origin are used by the global aquafeed sector. Rapeseed/canola, soybean and palm oils are the most abundantly used VO, and poultry by-product oil the most commonly used animal fat. The use of different oils or fats is driven by a combination of technical and nutritional considerations and other, often predominant considerations, such as availability, price, and other market forces, including legislation and consumer expectation/acceptance. Therefore, the oils used in aquafeed formulations are highly variable geographically and over time. For example, the most abundantly used oil to complement FO and provide the required energy source for Atlantic salmon in Europe is rapeseed (Aas et al., 2019), whereas in America and Australia it is poultry by-product oil (Emery et al., 2014; Aas et al., 2019). Readers interested in deepening their knowledge into the different aspects of oil and fat utilization in aquafeed are suggested to refer to specialized literature (Turchini et al., 2010). Nevertheless, in the following sections a succinct account of specific benefits and challenges provided by different oils commonly used and their environmental interactions are provided.

5.13 n-3 LC-PUFA expectations

5.13.1 The problem

While it is well known that "Omega-3," specifically the n-3 LC-PUFA, EPA, and DHA, are essential components of a healthy, balanced diet in humans, providing beneficial effects on neural development (Innis, 2007; Campoy et al., 2012) and in mitigating several pathological conditions (Delgado-Lista et al., 2012; Calder, 2014, 2018), it is not generally appreciated that these nutrients are also in short supply (Tocher, 2015; Tocher et al., 2019). Based on evidence from randomized controlled trials, epidemiological and laboratory studies (Poudyal et al., 2011; Gil et al., 2012; Calder, 2018), recommendations for EPA and DHA intake for a healthy human diet have been produced by over 50 global and national health agencies and associations, and government bodies (Aranceta and Perez-Rodrigo, 2012; GOED, 2014). Using the most commonly recommended dose for cardiac health of 500 mg EPA+DHA/person/day, the physiological demand of the global population for n-3 LC-PUFA was calculated to be over 1.25 million tonnes, whereas total supply was estimated at around 0.8 million tonnes indicating a shortfall of more than 0.4 million tonnes (Tocher, 2015). While there are different estimates of the precise gap between supply and demand (Hamilton et al., 2020), the fact that it exists is not in question and the ever-expanding global population, which will continually increase demand, means the gap will not lessen (Tocher et al., 2019). The fundamental global lack of n-3 LC-PUFA to supply human needs for a healthy diet were highlighted in a recent worldwide survey of human populations, which concluded that blood levels of EPA + DHA were low to very low for most of the world and this may increase global risk for chronic disease (Stark et al., 2016).

The vast majority of n-3 LC-PUFA is produced by microbes, mainly microalgae, at the base of aquatic food webs (Harwood and Gushina, 2009), whereas terrestrial plants do not produce EPA or DHA (Harwood, 1996). Therefore, the limitation in supply of n-3 LC-PUFA is linked to their aquatic, predominantly marine, origin with fish and seafood, traditionally from capture fisheries, being the source of almost all the EPA and DHA in human diets. With global capture fisheries being at or beyond sustainable limits, there is no prospect of this supply increasing in the future (FAO, 2020), and so the demand for fish has been increasingly supplied by aquaculture with over half of all fish and seafood now being farmed (FAO, 2020). However, essentially all the EPA and DHA in fish is derived from their diet, which in the case of farmed fish, means the manufactured artificial feeds (Tocher, 2015). Until recently, the only way to supply EPA and DHA was to base aquafeed formulations on large amounts of marine-derived raw ingredients, FO and FM, themselves derived from reduction/feed fisheries that, similar to all fisheries, are finite and limiting (Shepherd and Bachis, 2014; FAO, 2020). Thus, while aquaculture has become the major source of EPA and DHA in the human diet, it has also become the greatest consumer of the world's available supply of n-3 LC-PUFA.

The limitation in the supply of FM and FO has been addressed by aquaculture largely by a strategy of substitution with FM being replaced by plant meals, and FO being largely replaced by VO that are devoid of n-3 LC-PUFA and often rich in n-6 PUFA (Turchini et al., 2009, 2011c). This has successfully enabled production to be increased to support the demand for seafood and, simultaneously, increased sustainability through feeds with lower dependence upon pressurized marine ingredients (Ytrestøyl et al., 2015; Aas et al., 2019). However, the lower levels of n-3 LC-PUFA in feeds was reflected in the farmed fish, with reduced levels of EPA and DHA and the consequent impact on the nutritional quality of farmed products (Sprague et al., 2016; de Roos et al., 2017). In addition, as EPA and DHA are also essential nutrients for fish (Tocher, 2010), their lower level in aquafeeds has potential consequences for the health of farmed fish (Montero and Izquierdo, 2011; Tocher and Glencross, 2015). Indeed, replacing n-3 LC-PUFA with n-6 PUFA in aquafeeds is simply reflecting the current situation in human diets with greatly increased n-6: n-3 PUFA ratio, which underpins the beneficial effects of dietary EPA and DHA largely through redressing this imbalance (Simopoulos, 2011; Tocher, 2015). Therefore, further reduction in n-3 LC-PUFA levels in aquafeeds could significantly impact the health of farmed fish and is, therefore, unsustainable (Tocher, 2015). Consequently, while research on lipid nutrition in aquaculture initially focused on mitigating the negative impacts of replacing FM and FO with plant proteins and VO (Turchini et al., 2009, 2011c), the aim of recent research has been to find a long-term solution to the supply of EPA and DHA for aquaculture (Tocher, 2015; Sprague et al., 2017).

5.13.2 Alternative marine sources

Considering that EPA and DHA are essentially only produced in the marine environment, it is unsurprising that alternative marine sources have been one area of focus. Recent reviews have covered the not inconsiderable efforts to investigate sources from lower trophic levels, such as zooplankton including krill and copepods, or other unexploited fisheries, such as mesopelagic fish, as potential sources of EPA and DHA (Olsen et al., 2011). In summary, while there are commercial zooplankton fisheries in the southern hemisphere for Antarctic krill (*Euphausia superba*) providing both krill meal and oil, and in the northern hemisphere for copepods (*Calanus finmarchicus*) providing Calanus oil, they are relatively small-scale and neither are currently able to provide a solution to the problem of n-3 LC-PUFA supply for aquaculture (Olsen et al., 2011). Mesopelagic fish have potential for catch based on their population biomass and the fact that they can be found in reasonable densities (Olsen et al., 2011). Furthermore, mesopelagic species can also be lipid-rich and thus a potential source of EPA and DHA; however, similar to Calanus oil, the lipid often has a high content of WE, rendering them unsuitable for use in feeds for many species of farmed fish (Gjøsæter and Kawaguchi, 1980; Neighbors, 1988; de Koning and Evans, 1991). Recycling, including extracting oils from by-products of aquaculture and wild capture fisheries, is already occurring and, while highly desirable, is simply increasing efficiency in the use of existing, marine-derived sources (Tocher et al., 2019; Hamilton et al., 2020).

5.13.3 New de novo sources of EPA and DHA

All the initiatives described above are, ultimately, simply "harvesting" n-3 LC-PUFA, and a different approach is now required to supply new, entirely de novo supplies of EPA and DHA. This new approach has focused on microalgae as the main primary producers of EPA and DHA and has exploited modern technological and molecular solutions to either directly "farm" the algae themselves or utilize the genes from the algae to produce EPA and DHA in heterologous biological systems. Thus, advances in industrial technologies are enabling scale-up of fermentation of heterotrophic microalgae and bioreactor-like culture of phototrophic algae (Winwood, 2013), while molecular technologies are being utilized to introduce microalgal genes for the production of EPA and DHA into oilseed crops (Haslam et al., 2015; Napier et al., 2015).

5.13.3.1 Microalgae

The microalgal sources that are currently being used in aquaculture are almost exclusively derived from species that can be produced heterotrophically by fermentation such as Thraustochytrids like *Schizochytrium* and, to a lesser extent, dinoflagellates like *Crypthecodinium cohnii* (Tocher et al., 2019). These microalgal species are generally characterized as being rich in DHA but have little or no EPA and, initially, they were only available as biomasses. A variety of algal biomasses have been tested in recent years as ingredients in feeds with generally positive results (Tocher et al., 2019). However, biomasses have a major limitation as, in addition to oil, they contain protein and, especially carbohydrate (largely nonstarch polysaccharide, NSP) and, therefore, cannot directly replace the oil component of feeds. Being a mix of components makes their incorporation into balanced feeds more complicated and, in the case of carnivorous species (salmonids and marine fish), the high carbohydrate/NSP contents limit their inclusion to relatively low levels (~5%–6% of diet) to prevent compromising growth (Kousalaki, 2015). Herbivorous/omnivorous fish species such as carps, tilapia and basa (*Pangasius bocourti*), may have better capacity to digest carbohydrate and cell walls (Krogdahl et al., 2005) and, therefore, may be able to more efficiently digest algal biomasses (Castillo and Gatlin, 2015), which could enable higher inclusion levels in feeds (Sarker et al., 2016). The fact that algal biomasses only supply DHA has also been an issue that has prompted considerable interest into the relative value of DHA and EPA as essential fatty acids (NRC, 2011). Despite these issues, algal biomasses have and are being used in commercial aquafeeds. A number of commercial feeds produced by Alltech Coppens between 2016 and 2019 including feeds for salmonids and marine species as well as broodstock diets for freshwater species included the algal product, ForPlus (Alltech-Coppens, 2018). Similarly, Corbion have supplied AlgaPrime DHA (Corbion, 2019) to the BioMar Group, resulting in over 350,000 tons of fish feed with this algal biomass being delivered to Norwegian salmon farmers between 2016 and 2018 (Aquafeed, 2019). A further algal biomass, DHA Natur, is produced by Archer Daniels Midland [ADM] Animal Nutrition (ADM, 2019).

A recent, possibly game changing, innovation in the microalgal field is Veramaris, a joint venture of DSM and Evonik, which is producing a microalgal oil utilizing a *Schizochytrium* strain that produces both EPA and DHA (Veramaris, 2018). Being an oil will greatly facilitate its inclusion in aquafeeds as it can be used as a direct replacement for the oils currently used (Santigosa et al., 2018, 2020). From October 2018, Norwegian salmon producer, Lingalaks, fed 50% of their production a diet including this microalgal oil containing EPA and DHA, with a German retailer marketing the new salmon (Business, 2019) enabling Veramaris to win the recent F3 Fish Oil Challenge (Feed, 2019). A new fermentation facility was opened in July 2019 in Blair, Nebraska to produce sufficient DHA+EPA algal oil to provide 15% of total current annual demand of the salmon farming industry for Omega-3 (Veramaris, 2019).

5.13.3.2 Genetically modified oilseed crops

The concept of using genetic modification (GM) as a solution to the need for new n-3 LC-PUFA is to take the trait of microalgae for LC-PUFA production and to combine it with the trait of other organisms that produce and store high amounts of lipid, thus producing a transgenic organism that produces a large amount of oil with high proportions of EPA and DHA (Napier and Sayanova, 2005). The oleaginous yeast *Yarrowia lipolytica* was used by DuPont as a platform for EPA production after transformation by a suite of algal genes (Chuang et al., 2010). Two commercial products resulted including New Harvest EPA, an encapsulated oil for direct human consumption, and Verlasso salmon, produced by AquaChile in 2016 using feed containing *Y. lipolytica* biomass (Ledesma-Amaro and Nicaud, 2016), although Verlasso were later produced using microalgal biomass, rather than GM yeast. Recently, myxobacterial PUFA synthases have also been used to transform *Y. lipolytica* for the production of LC-PUFA (Gemperlein et al., 2019). Despite this advance, oilseed crops may be the preferred platform for GM solutions to n-3 LC-PUFA supply. Proof of the GM concept in plants was demonstrated some years ago, initially in model plants like *Arabidopsis*, and then in oilseed crops including soybean, Canola and *Camelina sativa* (Robert et al., 2005; Petrie et al., 2012; Ruiz-Lopez et al., 2014). While scalability is a major challenge with microalgal culture, this is the major advantage of GM oilseed crops as production is readily scalable with infrastructure for the cultivation, harvest, processing, distribution, marketing, and utilization of VO being highly organized and well established (Salunkhe et al., 1992). The GM oilseed products have another potential advantage as seed meal by-products containing residual n-3 LC-PUFA provide added value as they can be used in animal and, possibly, fish feeds at levels within the normal limits for the plant meals based on their nutritional/antinutritional properties (Gatlin et al., 2007; Hixson et al., 2014; Bullerwell et al., 2016).

A stearidonic acid (18:4n-3)-rich oil from GM soybean was developed by Monsanto and was tested successfully as an ingredient in aquafeed for trout, with no negative effects related to GM (Nanton et al., 2012). While this oil proved the general concept it was never fully commercialized and was superseded by oils from GM oilseed crops that produced the LC-PUFA, EPA, and/or DHA (Napier et al., 2015, 2019). The most complete data on GM oilseed crops have come from studies on genetically engineered *Camelina sativa* that have resulted in oils containing 20% EPA alone and up to 25% n-3 LC-PUFA (EPA, DHA, and DPA) which have been extensively tested as ingredients in feeds for Atlantic salmon (Betancor et al., 2015a, 2015b, 2016c, 2017, 2018), gilthead seabream (Betancor et al., 2016b), and European seabass (Betancor et al., 2021). In these studies, the modified camelina oils replaced 100% of the added oil in the feeds without any apparent negative impacts on fish growth or health. These studies culminated with Betancor et al. (2018) showing that replacing the current FO and rapeseed oil blend in salmon feed with camelina oil containing >25% n-3 LC-PUFA doubled the EPA+DHA level in flesh compared to salmon fed standard commercial feed, which suggested the GM oilseed solution could restore levels in farmed salmon to those commonly found before the development of sustainable feeds with high replacement of the traditional marine ingredients (Sprague et al., 2016).

Two GM Canola (rapeseed) products are currently being commercialized with large-scale production likely in 2020/21. One GM Canola oil, originally developed by BASF, contains about 10%–11% n-3 LC-PUFA, predominantly EPA and is being marked by Cargill as

Latitude, "a sustainable, plant-based alternative source of omega-3 for fish feed applications" (Cargill, 2019). This is being grown in the United States with salmon farming in Chile and Canada as target markets. In Australia, a further GM Canola oil that is also around 10%–11% n-3 LC-PUFA, but predominantly DHA (Napier et al., 2019), was developed initially by CSIRO and is being produced and marketed by Nuseed as two products, Aquaterra and Nutriterra targeting aquaculture and direct human sectors, respectively, that have been approved for use in Canada (Anonymous, 2020). While data on trials investigating their use as ingredients in fish feeds are more elusive than for the engineered *Camelina*, reports indicate that both GM Canola oils have been used successfully in trials replacing FO in feeds for Atlantic salmon (Iassonova et al., 2018; Ruyter et al., 2019; Petrie et al., 2020).

5.13.4 Feeding strategies and genetics

Research into the development of sustainable feeds has focused predominantly on formulations with the aim of determining the choice of ingredients (raw materials) to support growth and feed efficiency while mitigating any potential negative effects of the replacement of marine ingredients (Gatlin et al., 2007; Turchini et al., 2011c). However, feeding strategies and genetic approaches to specifically limit impacts on, and meet expectations of, EPA+DHA levels have also been investigated. One such strategy is the use of finishing diets where fish are fed highly substituted, low marine ingredient feeds for most of the growth cycle and then fed a diet with higher levels of FO prior to harvesting (Glencross and Turchini, 2011). In oily species like salmon, the finishing period required could be 3–6 months depending upon the level of substitution and the size of fish and their oil content at harvest, whereas for marine fish like sea bream and sea bass with lower flesh fat content, the finishing period can be much shorter (Bell and Waagbø, 2008). A finishing diet period could successfully restore n-3 LC-PUFA in salmon flesh to around 80% of the level in fish fed marine-based diets until harvest, while at the same time the overall strategy consumed 50% less FO (Bell et al., 2003, 2004). In this respect it has also been demonstrated that oils with relatively high MUFA and/or SFA can reduce the catabolism of dietary n-3 LC-PUFA, thus promoting greater retention in fillets and having an "omega-3 sparing effect" in different species (Turchini et al., 2011b; Emery et al., 2014; Rombenso et al., 2015; Bowzer et al., 2016). A study in Atlantic salmon showed that the right balance of dietary EPA and DHA in both absolute and relative amounts can promote n−3 LC-PUFA retention/sparing (Codabaccus et al., 2012). More recently, potential n-3 LC-PUFA sparing was investigated in a further study that employed a sliding protein to lipid ratio in diets for juvenile Atlantic salmon although results were inconclusive (Francis and Turchini, 2016).

The precise daily feeding regime is another area that has attracted attention, as understanding the natural feeding habits of fish and reflecting that in the feeding regime applied on the farm could improve feed efficiency, reduce waste, and minimize cost. It was known that in nature fish display feeding rhythms related to food availability and that they continue to display these even when provided with a continuous supply of feed as is the situation in farming (López-Olmeda and Sánchez-Vázquez, 2010). A first pioneering study attempting to investigate the possible use of biological rhythms toward improving the overall lipid utilization in farmed fish was published a decade ago (Brown et al., 2010b). These studies in fish resulted in considerable interest in daily rhythms controlled by the endogenous circadian or central "body clock," regulated by various clock genes and their corresponding proteins (Vatine et al., 2011). However, it has been appreciated that there are also local or "peripheral" clock systems that may provide targeted rhythmic control of specific physiological processes. Among these, rhythmic expression within the hepatic transcriptome has been an area of particular interest in teleosts (Vera et al., 2013; Betancor et al., 2014; Paredes et al., 2014, 2015).The rhythmicity of the core clock genes at both central and peripheral levels is impacted by external entrainers and, while photoperiod is the most researched, feeding regime has also been studied (Vera et al., 2013; Betancor et al., 2014; Paredes et al., 2014, 2015). Although research into this area is still in its infancy, some early results have indicated a potential role for the circadian clock in the regulation of lipid metabolism in teleost fish (Brown et al., 2010b; Betancor et al., 2014; Eroldogan et al., 2018).

Another recent strategy has been the application of "nutritional programming" in fish with a view of influencing growth, feed efficiency and nutrient retention. The basic concept is that a nutritional stimulus during critical plastic periods in development can induce permanent changes in animal metabolism and physiology as a result of cellular, molecular and biochemical adaptations that can prepare the animal to respond better to the same nutritional environment when experienced later in life (Patel and Srinivasan, 2002; Gluckman et al., 2005). Underlying mechanisms may include changes in gene expression and epigenetic changes in DNA methylation or histone modifications at regulatory regions of key genes involved in pathways of intermediary metabolism (Symonds et al., 2009; Gut and Verdin, 2013). Some studies have shown that nutritional programming in fish can specifically impact PUFA metabolism, potentially enhancing LC-PUFA biosynthesis pathways in European seabass (Vagner et al., 2007, 2009), and increasing EPA and DHA retention in Atlantic salmon (Clarkson et al., 2017; Vera et al., 2017). Other studies have explored the potential of nutritional programming to improve the acceptance and utilization of plant-based diets in general. Early exposure of rainbow trout to a plant-based diet improved growth and feed utilization when the fish were fed the same diet later in life (Geurden et al., 2013). Nutritional programming improved the acceptance of plant-derived protein in zebrafish (Perera and Yufera, 2016) and gut health of seabass fed a plant-based diet (Torrecillas et al., 2017), carbohydrate utilization in zebrafish (Rocha et al., 2015), Siberian sturgeon (Gong et al., 2015) and gilthead seabream (Rocha et al., 2016), and muscle catabolic capacities in rainbow trout (Panserat et al., 2017). In addition, nutritional programming of broodstock by feeding a low FM and FO diet improved the subsequent performance of the progeny when they were fed plant-based diets (Izquierdo et al., 2015).

While the above areas of research focus on tailoring feeds and feeding regimes to the fish, an alternative approach is to tailor the fish to the feeds (Tocher, 2015). While it is possible to enhance endogenous production of n-3 LC-PUFA in fish via genetic modification, specifically transformation with desaturase and elongase genes (Alimuddin et al., 2005, 2007), the arguably less contentious strategy is the use of selective breeding (Gjedrem, 2000). In an early study, the n—3 LC-PUFA level of flesh was shown to be a heritable trait in Atlantic salmon (Leaver et al., 2011) although differences in levels of n-3 LC-PUFA between different families/strains were only 2%–3% of total fatty acids (Morais et al., 2012c). Genetic variation in muscle EPA and DHA content that could be exploited using selective breeding was shown recently in salmon (Horn et al., 2018), confirming that selection for this trait to optimize the efficiency of n-3 LC-PUFA metabolism and flesh levels was a possible approach (Tocher, 2015; Betancor et al., 2016a). Subsequent studies have attempted to determine which metabolic processes were associated with variation in EPA and DHA contents by investigating liver and muscle gene expression (Horn et al., 2019) or genome-wide association to identify genetic variants associated with n-3 LC-PUFA composition of fillets (Horn et al., 2020b), and to investigate how this information and estimating breeding values can be applied practically to enable selection for n-3 LC-PUFA content in Atlantic salmon (Horn et al., 2020a).

Irrespective of whether new sources of n-3 LC-PUFA become more widely used, EPA and DHA will always be precious essential nutrients and, therefore, it is incumbent on aquaculture to always use them efficiently. Therefore, all options to potentially increase the efficiency of use of n-3 LC-PUFA are important for aquaculture to no longer be criticized as a "user" of omega-3 and be more accepted as a major provider of EPA and DHA to human consumers.

5.14 Dietary lipids and the environment

5.14.1 The rise of "adaptive nutrition"

Since the last iteration of this chapter, the prominence of human-induced climate change and its impacts on global aquaculture practices have moved further to the forefront of discussions concerning the growth and viability of the sector (Galappathi et al., 2020). As the most diverse of all the primary production sectors both in terms of the number of farming systems and species farmed, aquaculture is particularly vulnerable given the poikilothermic nature of its taxa and the susceptibility of the aquatic environment to changes in temperature (de Silva, 2012). Whilst the challenges faced will differ depending on the climatic region and the thermal tolerance of individual species, generalizations regarding impacts from a nutritional point of view include, but are not limited to, changes in appetite, growth, digestion and bioenergetics, increased stress and a subsequent mobilization of the nutrients involved in stress response, and broad ranging changes in physiology through increased metabolic demands (Neubauer and Andersen, 2019). Energy use for basal functioning, standard metabolic rate, is known to be influenced strongly by temperature, interacting with mass specific metabolic demand, which decreases as fish grow (Oldham et al., 2019). Standard metabolic rate scales exponentially with temperature, where a small change in temperature can have a dramatic effect particularly when a species is nearing its upper thermal limit (Gillooly et al., 2001).

Changes in metabolizable energy use (deposition and catabolism) in a commercial setting, particularly where seasonal temperatures exceed the thermal niche of the species, is of high industry relevance. This is of particular importance in species where high-quality ingredients are implemented as a means of minimizing nutrient waste, maximizing performance and final product quality while supplying consumers with products rich in n-3 LC PUFA. An adaptation to temperature, or homeoviscous adaptation as it is more often referred to, varies among species and exerts an influence on fatty acid requirements considering their indispensable roles in regulating cell membrane fluidity (Tocher and Sargent, 1990). Differences are also apparent within the same species group as water temperature changes (Calabretti et al., 2003; Jobling and Bendiksen, 2003), necessitating the need for "adaptive nutrition" approaches with a view of providing farmed species with the appropriate nutritional provisions to mitigate the impacts of a warming rearing environment and the secondary challenges that this brings. It is well documented across several prominent aquaculture species that the retention of n-3 LC PUFA is impacted by temperature, whereby higher retention is observed at lower temperatures (Miller et al., 2006; Skalli et al., 2006; Bowyer et al., 2012). Again, this is intimately linked to differences in the fluidity of individual fatty acids where, when exposed to low temperatures, fish increase the degree of unsaturation and, hence, more malleable fatty acids in their cell membranes. Alternatively, at higher temperatures the opposite trend is apparent where, in rainbow trout for example, concentrations of n-3 LC PUFA decrease, concomitant with an increase in the deposition of less rigid SFA (Hazel, 1979). In recognition of the importance of optimally functioning membranes in the scheme of farmed fish health and performance in the face of climate warming, there has been a significant research focus on this topic in recent years, with an overarching aim of better understanding the relationship between the temperature of the rearing environment and the potential impacts on lipid/fatty acid metabolism (Ng et al., 2004; Guerreiro et al., 2012; Norambuena et al., 2012, 2015, 2016; Huguet et al., 2015). Among many notable results (e.g., changes in lipid digestibility, fatty acid β-oxidation, retention), this work has highlighted a greater requirement for fatty acids such as ARA at elevated temperatures, evidenced by an increase in voluntary intake (Norambuena et al., 2012), while stressing the importance of balanced fatty acid ratios (e.g., ARA/EPA) for maximal performance (Norambuena et al., 2016). Ultimately, this underscores the importance of renewed investigations into nutrient requirements in light of high rates of marine-derived ingredient replacement. In particular, evidence suggests potentially greater importance for n-6 LC-PUFA in species such as Atlantic salmon when reared in unavoidable high-water temperature environments, as supported by an increased accumulation of ARA in wild-caught conspecifics during warm summer months (Ackman and Takeuchi, 1986).

5.14.2 Fatty acids "in" the environment

Alongside the recognition of the issues that our warming climate is placing on aquaculture production, aquaculture itself is recognized for a series of negative impacts on the surrounding environment. The advent of intensification in the aquaculture industry has seen the production of unprecedented levels of aquafeeds, which, combined with the subsequent focus on the substitution of marine-derived raw materials, has resulted in elevated concentrations of terrestrially derived nutrients making their way into the marine environment with potential for ecological consequences from a food web perspective (White et al., 2016). From the point of view of fatty acids, this has resulted in the introduction of higher proportions of OLA, LNA, and ALA being discharged at sites adjacent to the aquaculture farm via uneaten feed, feces, and mortalities. As a result, this has sparked a series of investigations globally researching the apparent trophic subsidy derived by aquaculture waste products in the vicinity of fish farms. Fish sampled from these areas can be characterized by increased concentrations of LNA and OLA and, like those fish in the pens, demonstrate an active reflection of the aquaculture waste products that they are consuming. Interestingly, fish derived from artisanal fisheries caught at a distance of up 10 km from fish farms have returned elevated levels of LNA in their fillets (Izquierdo-Gómez et al., 2015). In a study attempting to evaluate the potential impacts of terrestrially derived n-6 PUFA in the marine environment, female sea urchins (*Heliocidaris erythrogramma*) fed a high n-6 PUFA diet produced larvae with lower survival rates and males with no viable sperm (White et al., 2016). However, labeled a "futureaquafeed," this experimental treatment was formulated to contain ~45% of the dietary lipid source as LNA, therefore presenting a somewhat unrealistic nutritional scenario considering the likelihood of continued FO utilization, the availability of other alternative lipid sources with more favorable fatty acid profiles, and the inevitable widespread availability of oil sources rich in n-3 LC-PUFA. Similar studies conducted in an on-farm environment have documented an increase in the proportions of OLA and LNA in rock crab (*Cancer irroratus*) with concomitant reductions in EPA and DHA (Sardenne et al., 2020). While the potential consequences on reproductive performance was highlighted, the occurrence of this has yet to be adequately assessed. Given the well-documented "reflection" of key dietary fatty acid groups in the tissues of organisms receiving them, these occurrences are undeniable; however, the seriousness and extent and of the nutritional impact, thus far, appear minimal and localized.

5.15 Prospects

In the last 2 decades, since the publication of the previous version of this chapter, an incredible global effort in fish lipid nutrition has been achieved, with our knowledge being substantially expanded. Nevertheless, this new knowledge is also responsible for an increasing number, and depth, of scientific and technical questions. This, coupled with a sector which is becoming progressively more marginal and challenged by many factors, such as resource utilization, price and sustainability, and ever-growing consumer expectations, is fueling a continuous research impetus, which is expected to last many decades into the future.

As there is parallel, and dramatically extensive research in mammalian species, it is natural to interpret data in fish in light of knowledge derived from mammals. However, in recent years it has become evident that these transfer values may not always be correct, with WGD being one possible reason. Mammalian genes have gone through one WGD leading to two potential versions of similar proteins or enzymes. In many cases, these have taken on different and distinct functions in cells and specific analysis is used to describe these different functions. Ray-finned fish have since gone through several WGD events including one teleost-specific event, while salmonids have gone through a fourth WGD. This massively increases the number of potential versions of proteins in fish. This also implies possibilities that further specialization of functions may have taken place. Earlier work using methodology like QPCR did not take into account these differences, and primers could fit one, or several of the variants. In recent years, the specialization of functions of isoforms has become more of a focus, and analysis increasingly takes this into account. This means that many earlier studies may be difficult to interpret in view of recent knowledge.

In addition to the focus on n-3 LC-PUFA metabolism in cultured aquaculture species, a series of other important considerations relative to fish lipid nutrition can also be made. The microbiota, for example, and its interactions with dietary lipids, has thus far attracted proportionally limited research focus. The evaluation of the autochthonous bacterial community, those that adhere and colonize the intestine, merits investigation, as this bacterial community is likely of higher importance than the allochthonous bacterial community. However, the microbial community of the finfish GI tract needs to be characterized with greater precision using more sophisticated methodology in order to provide deeper insight into the underlying mechanisms. Furthermore, the effects of n-6 to n-3 PUFA ratio on gut microbiota has not been investigated in fish, a topic that merits investigation. Knowledge of gut microbiota and lipid metabolism, and the regulation of the genes involved in endothermic animals and finfish has increased in recent years, but several issues remain unanswered. In future studies on the interactions between lipid, gut microbiota and metabolism, it is recommended that teleosts are used as model organisms to understand these interactions. Endothermic studies have revealed that DHA and EPA can reverse intestinal microbial dysbiosis by increasing beneficial bacteria species, including *Lactobacillus*, *Bifidobacterium*, and butyrate-producing bacteria, such as *Roseburia* and *Coprococcus*. In addition, n-3 PUFA decrease the proportions of LPS-producing and mucolytic bacteria in the gut and they can reduce inflammation as well as oxidative stress. As the GI tract of finfish are much more diverse than endothermic animals (e.g., with or without a stomach, nil up to 1000 pyloric caeca, short and long intestine) (Ray and Ringø, 2014), there is a need to clarify these questions, and this will hopefully bring a great step forward in understanding the interactions between dietary lipid and the gut microbiota in finfish.

Research into lipid synthesis and catabolism in fish has also substantially increased our knowledge. For example, the cytosolic degradation of TAG has been clarified in fish to contain two metabolic routes: neutral lipid hydrolase (lipase)-mediated lipolysis and lysosomal acid lipase-mediated lipolysis (lipophagy). Therefore, combining this with previously established knowledge concerning mitochondrial and peroxisomal β-oxidation, a more complete understanding of the cellular TAG degradation system has been achieved. A number of lipid metabolism-related genes, proteins and signaling pathways have been identified and verified. Accordingly, several important molecular regulatory mechanisms of lipid metabolism in fish have been illustrated, such as Ampk/Pparα/Rxr/Pgc-mediated lipid catabolism regulation, and rebpor Pparγ-mediated lipogenesis and lipid deposition regulation. The functions and molecular mechanisms of a number of lipid metabolism-cofactors such as carnitine, choline, lipoic acid, Zn, Cu, n-3 LC-PUFA, and vitamin D etc., have been demonstrated in detail. The regulation of endocrinological factors such as estrogens, leptin, insulin, myostatin etc., on lipid metabolism and fat accumulation in fish has been highlighted.

A number of new technological approaches such as genetic editing and "-omics" have been widely applied in fish or fish cells, and this has significantly increased the speed of advancement. Substantial advances have been achieved as, for example, with respect to lipid metabolism and fat accumulation in farmed fish, as well as the complete understanding of the causes and development of fatty liver in fish. However, many gaps still remain and a series of questions are still unanswered, such as those relative to (i) the cellular biology and molecular regulatory mechanisms of lipid catabolism and synthesis, especially in PL and cholesterol; (ii) the nutritional functions of different dietary fatty acids and their individual or complex regulation on lipid metabolism in fish; (iii) the roles of sensory perceptions (feed palatability) and postingestive signals (nutritional regulation) in determining modulation of voluntary feed intake in fish.

In conclusion, and zooming out from microscopic intracellular mechanisms to macroscopic considerations, the current situation in lipid nutrition in aquaculture remains essentially a strategy of substitution with FO being replaced by a variety of alternative oils, largely VO including rapeseed oil (e.g., salmon feeds in Norway and Scotland), but also others based on both global supply (e.g., soybean and palm oils) or local availability (e.g., sunflower, cottonseed). Likewise, this also includes the utilization of some animal oils, such as poultry oil, especially in Australia and North and South America. Unavoidably, all of these negatively impact n-3 LC-PUFA levels in farmed fish products. The novel microalgal and GM plant omega-3 LC-PUFA products that have been developed are currently being used sparingly, primarily to ensure levels of EPA and DHA are maintained at current levels and do not decline further. How these products will be used in the future will likely depend more on socioeconomic and sociopolitical considerations than on scientific or nutritional grounds. The relatively high cost of production of microalgal sources means that price may be a limiting factor in their use, at least in the short term, while consumer perception and acceptance are already major issues with the use of GM-derived oils, at least in some countries, especially in Europe. While physiological demand for n-3 LC-PUFA can be estimated, actual consumer demand is more difficult to quantify, but it is highly likely that aquaculture will remain the sector driving future demand with use dependent upon availability of supply and cost. Ideally, the new de novo nonmarine sources of EPA and DHA will become available in greater volume and lower cost, enabling increased inclusion in feeds that could restore n-3 LC-PUFA in farmed fish to levels attained prior to the development and application of sustainable feed formulations.

What appears to be sure is that, wherever n-3 LC-PUFA will come from, they will have to be used judiciously and parsimoniously. A greater focus on "return on investment" is needed, in the sense of maximizing the output of n-3 LC-PUFA into the edible products generated by aquaculture and simultaneously reducing the input of these nutrients required by the sector. Aquaculture is, and will increasingly be, the major supplier of n-3 LC-PUFA for a growing global population. The good news is that, as scientists working in this space, we have ample opportunities to engage in these globally relevant and needed R&D activities, and we should aim at always doing so, by "having fun in fish nutrition" (Halver, J.E., personal communication).

References

Aas, G.H., Bjerkeng, B., Storebakken, T., Ruyter, B., 1999. Blood appearance, metabolic transformation and plasma transport proteins of (14)C-astaxanthin in Atlantic salmon (*Salmo salar* L.). Fish Physiol. Biochem. 21, 325–334.

Aas, T.S., Ytrestoyl, T., Asgard, T., 2019. Utilization of feed resources in the production of Atlantic salmon (*Salmo salar*) in Norway: an update for 2016. Aquacult. Rep. 15, 100216.

Abdul Hamid, N.K., Carmona-Antonanzas, G., Monroig, O., Tocher, D.R., Turchini, G.M., Donald, J.A., 2016. Isolation and functional characterisation of a fads2 in rainbow trout (*Oncorhynchus mykiss*) with Delta5 desaturase activity. PLoS One 11, e0150770.

Abu-Elheiga, L., Wu, H., Gu, Z., Bressler, R., Wakil, S.J., 2012. Acetyl-CoA carboxylase $2^{-/-}$ mutant mice are protected against fatty liver under high-fat, high-carbohydrate dietary and de novo lipogenic conditions. J. Biol. Chem. 287, 12578–12588.

Acar, Ü., Kesbiç, O.S., Yılmaz, S., Karabayir, A., 2018. Growth performance, haematological and serum biochemical profiles in rainbow trout (*Oncorhynchus mykiss*) fed diets with varying levels of lupin (*Lupinus albus*) meal. Aquacult. Res. 49, 2579–2586.

Ackman, R.G., 1967. The influence of lipids on fish quality. Int. J. Food Sci. Technol. 2, 169–181.

Ackman, R.G., Takeuchi, T., 1986. Comparison of fatty acids and lipids of smolting hatchery-fed and wild Atlantic salmon *Salmo salar*. Lipids 21, 117–120.

Adjoumani, J.J.Y., Wang, K.Z., Zhou, M., Liu, W.B., Zhang, D.D., 2017. Effect of dietary betaine on growth performance, antioxidant capacity and lipid metabolism in blunt snout bream fed a high-fat diet. Fish Physiol. Biochem. 43, 1733–1745.

ADM, 2019. https://www.admanimalnutrition.com/webcenter/portal/ADMAnimalNutrition/pages_ingredients/sipet/dhanatur1;https://www.youtube.com/watch?v=FGsrKfvnsRM.

Affandi, M.M.R.M.M., Julianto, T., Majeed, A.B.A., 2012. Enhanced oral bioavailability of astaxanthin with droplet size reduction. Food Sci. Technol. Res. 18, 549–554.

Aguilar, A.J., Conde-Sieira, M., López-Patiño, M.A., Míguez, J.M., Soengas, J.L., 2011. In vitro leptin treatment of rainbow trout hypothalamus and hindbrain affects glucosensing and gene expression of neuropeptides involved in food intake regulation. Peptides 32, 232–240.

Al Massadi, O., López, M., Fernø, J., Diéguez, C., Nogueiras, R., 2015. What is the real relevance of endogenous ghrelin? Peptides 70, 1–6.

Albalat, A., Gómez-Requeni, P., Rojas, P., Médale, F., Kaushik, S., Vianen, G.J., Van Den Thillart, G., Gutiérrez, J., Pérez-Sánchez, J., Navarro, I., 2005a. Nutritional and hormonal control of lipolysis in isolated gilthead seabream (*Sparus aurata*) adipocytes. Am. J. Physiol. Regul. Integr. Compar. Physiol. 289, R259–R265.

Albalat, A., Gutiérrez, J., Navarro, I., 2005b. Regulation of lipolysis in isolated adipocytes of rainbow trout (*Oncorhynchus mykiss*): the role of insulin and glucagon. Compar. Biochem. Physiol. Mol. Integr. Physiol. 142, 347–354.

Albalat, A., Liarte, C., Mackenzie, S., Tort, L., Planas, J.V., Navarro, I., 2005c. Control of adipose tissue lipid metabolism by tumor necrosis factor-α in rainbow trout (*Oncorhynchus mykiss*). J. Endocrinol. 184, 527.

Ali, M., Jauncey, K.M., 2004. Evaluation of mixed feeding schedules with respect to compensatory growth and body composition in African catfish *Clarias gariepinus*. Aquacult. Nutr. 10, 39–45.

Ali, A., AL-Ogaily, S., AL-Asgah, N., Goddard, J., Ahmed, S., 2007. Effect of feeding different protein to energy (P/E) ratios on the growth performance and body composition of. J. Appl. Ichthyol. 24, 31–37.

Alimuddin, Y.G., Kiron, V., Satoh, S., Takeuchi, T., 2005. Enhancement of EPA and DHA biosynthesis by over-expression of masu salmon delta 6-desaturase-like gene in zebrafish. Transgenic Res. 14, 159–165.

Alimuddin, Y.G., Kiron, V., Satoh, S., Takeuchi, T., 2007. Expression of masu salmon delta5-desaturase-like gene elevated EPA and DHA biosynthesis in zebrafish. Mar. Biotechnol. 9, 92–100.

Alltech-Coppens, 2018. Industrial Fish Feed. https://www.alltechcoppens.com/en/industrial.

Almaida-Pagan, P.F., Hernandez, M.D., Madrid, J.A., De Costa, J., Mendiola, P., 2011. Effect of masking organoleptic properties of fat on diet self-selection by the sparid Diplodus puntazzo. Aquacult. Nutr. 17, E48–E55.

Alvarez, M.J., Diez, A., Lopez-Bote, C., Gallego, M., Bautista, J.M., 2000. Short-term modulation of lipogenesis by macronutrients in rainbow trout (*Oncorhynchus mykiss*) hepatocytes. Br. J. Nutr. 84, 619–628.

Amaya, E., Nickell, D., 2015. Using feed to enhance the color quality of fish and crustaceans. In: Davis, D.A. (Ed.), Feed and Feeding Practices in Aquaculture. Woodhead Publishing, Oxford.

Ambati, R.R., Phang, S.M., Ravi, S., Aswathanarayana, R.G., 2014. Astaxanthin: sources, extraction, stability, biological activities and its commercial applications–a review. Mar. Drugs 12, 128–152.

Ameer, F., Scandiuzzi, L., Hasnain, S., Kalbacher, H., Zaidi, N., 2014. De novo lipogenesis in health and disease. Metabolism 63, 895–902.

Ando, S., Tachibana, A., Yamada, S., Kishimura, H., 2005. Apolipoprotein complexity in Japanese eel *Anguilla japonica*: truncated apolipoprotein A-I and apolipoprotein A-I-like protein in plasma lipoproteins. Biosci. Biotechnol. Biochem. 69, 2258–2262.

Andreeva, A.M., 2019. The strategies of organization of the fish plasma proteome: with and without albumin. Russ. J. Mar. Biol. 45, 263–274.

Anonymous, 2017. Commission Implementing Regulation (EU) 2017/962 of 7 June 2017 suspending the authorisation of ethoxyquin as a feed additive for all animal species and categories. Off. J. Euro. Union 60.

Anonymous, 2020. Nuseed secures Canadian feed approvals for omega-3 canola oil. Aquafeed.com 20. Available at: http://www.aquafeed.com/af-article/9794/Nuseed-secures-Canadian-feed-approvals-for-omega3-canola-oil/?utm_source=Aquafeed+English+Newsletter&utm_campaign=64230e63ca-EMAIL_CAMPAIGN_2019_06_26_19_COPY_01&utm_medium=email&utm_term=0_0e7f7c0399-64230e63ca-78769.

Anthonsen, M.W., Rönnstrand, L., Wernstedt, C., Degerman, E., Holm, C., 1998. Identification of novel phosphorylation sites in hormone-sensitive lipase that are phosphorylated in response to isoproterenol and govern activation properties in vitro. J. Biol. Chem. 273, 215–221.

Aoyama, T., Fukui, K., Takamatsu, K., Hashimoto, Y., Yamamoto, T., 2000. Soy protein isolate and its hydrolysate reduce body fat of dietary obese rats and genetically obese mice (yellow KK). Nutrition 16, 349–354.

Aquafeed, 2019. http://www.aquafeed.com/news/headline-news-article/8557/Corbions-AlgaPrimeTM-DHA-and-BioMar-improving-omega3-content-of-farmed-fish/.

Aranceta, J., Perez-Rodrigo, C., 2012. Recommended dietary reference intakes, nutritional goals and dietary guidelines for fat and fatty acids: a systematic review. Br. J. Nutr. 107 (Suppl. 2), S8–S22.

Ardail, D., Popa, I., Bodennec, J., Louisot, P., Schmitt, D., Portoukalian, J., 2003. The mitochondria-associated endoplasmic-reticulum subcompartment (MAM fraction) of rat liver contains highly active sphingolipid-specific glycosyltransferases. Biochem. J. 371, 1013–1019.

Ardiyanti, A., Abe, T., Tameoka, N., Kobayashi, E., Shoji, N., Ohtani, Y., Suzuki, K., Roh, S.-G., Katoh, K., 2012. Effects of growth hormone gene polymorphism on lipogenic gene expression levels in diaphragm tissues of Japanese black heifers. Asian-Australas. J. Anim. Sci. 25, 1055–1062.

Arias-Jayo, N., Abecia, L., Lavín, J.L., Tueros, I., Arranz, S., Ramírez-García, A., Pardo, M.A., 2019. Host-microbiome interactions in response to a high-saturated fat diet and fish-oil supplementation in zebrafish adult. J. Funct. Foods 69.

Arishima, T., Tachibana, N., Kojima, M., Takamatsu, K., Imaizumi, K., 2009. Screening of resistant triacylglycerols to the pancreatic lipase and their potentialities as a digestive retardant. J. Food Lipids 16, 72–88.

Arnold Reed, D.E., Bentley, P.J., Phan, C.T., Redgrave, T.G., 1997. The clearance of lipids from the plasma of a teleost fish, the black bream (*Acanthopagrus butcheri*). Compar. Biochem. Physiol. Mol. Integr. Physiol. 116, 167–172.

Aslaksen, M.A., Kraugerud, O.F., Penn, M., Svihus, B., Denstadli, V., Jørgensen, H.Y., Hillestad, M., Krogdahl, Å., Storebakken, T., 2007. Screening of nutrient digestibilities and intestinal pathologies in Atlantic salmon, *Salmo salar*, fed diets with legumes, oilseeds, or cereals. Aquaculture 272, 541–555.

Aubourg, S.P., Losada, V., Prego, R., 2010. Distribution of lipids and trace minerals in different muscle sites of farmed and wild turbot (*Psetta maxima*). Int. J. Food Sci. Technol. 42, 1456–1464.

Aursand, M., Bleivik, B., Rainuzzo, J.R., Leif, J., Mohr, V., 1994. Lipid distribution and composition of commercially farmed Atlantic salmon (*Salmo-Salar*). J. Sci. Food Agric. 64, 239–248.

Babin, P.J., Vernier, J.-M., 1989. Plasma lipoproteins in fish. J. Lipid Res. 30, 467.

Bäckhed, F., Hao, D., Ting, W., Hooper, L.V., Gou Young, K., Nagy, A., Semenkovich, C.F., Gordon, J.I., 2004. The gut microbiota as an environmental factor that regulates fat storage. Proc. Natl. Acad. Sci. U. S. A. 101, 15718–15723.

Bäckhed, F., Manchester, J.K., Semenkovich, C.F., Gordon, J.I., 2007. Mechanisms underlying the resistance to diet-induced obesity in germ-free mice. Proc. Natl. Acad. Sci. U. S. A. 104, 979–984.

Baffy, G., 2015. MicroRNAs in nonalcoholic fatty liver disease. J. Clin. Med. 4, 1977–1988.

Ballantyne, J.S., Moon, T.W., 1986. The effects of urea, trimethylamine oxide and ionic strength on the oxidation of acyl carnitines by mitochondria isolated from the liver of the Little Skate Raja erinacea. J. Compar. Physiol. B Biochem. Syst. Environ. Physiol. 156, 845–851.

Balser, W., 1882. Ueber Fettnekrose, eine zuweilen tödtliche Krankheit des Menschen. Virchows Arch. 90, 520–535.

Bandyopadhyay, G.K., Dutta, J., Ghosh, S., 1982. Preferential oxidation of linolenic acid compared to linoleic acid in the liver of catfish (*Heteropneustes fossilis* and *Clarias batrachus*). Lipids 17, 733–740.

Baras, E., Arifin, O.Z., Slembrouck, J., Subagja, J., Krisanto, A.H., Legendre, M., 2018. Oil globule size in fish eggs: a matter of biome and reproductive strategy. Fish Fish. 19 (6), 996–1002. https://doi.org/10.1111/faf.12307.

Basili, D., Lutfi, E., Falcinelli, S., Balbuena-Pecino, S., Navarro, I., Bertolucci, C., Capilla, E., Carnevali, O., 2020. Photoperiod manipulation affects transcriptional profile of genes related to lipid metabolism and apoptosis in zebrafish (*Danio rerio*) larvae: potential roles of gut microbiota. Microb. Ecol. 79, 933–946.

Batista-Pinto, C., Rodrigues, P., Rocha, E., Lobo-Da-Cunha, A., 2005. Identification and organ expression of peroxisome proliferator activated receptors in brown trout (*Salmo trutta* F. Fario). Biochim. Biophys. Acta 1731, 88–94.

Baur, J., Pearson, K., Price, N., Jamieson, H., Lerin, C., Kalra, A., Prabhu, V., Allard, J., Lopez-Lluch, G., Lewis, K., 2006. Resveratrol improves health and survival of mice on a high-calorie diet. Nature 444, 337–342.

Bayir, M., Bayir, A., Wright, J.M., 2015. Divergent spatial regulation of duplicated fatty acid-binding protein (fabp) genes in rainbow trout (*Oncorhynchus mykiss*). Compar. Biochem. Physiol. Genom. Proteom. 14, 26–32.

Beckett, B.R., Petkovich, M., 1999. Evolutionary conservation in retinoid signalling and metabolism. Am. Zool. 39, 783–795.

Bell, J.G., Koppe, W., 2011. Lipids in aquafeeds. In: Turchini, G.M., Ng, W.K., Tocher, D. (Eds.), Fish Oil Replacement and Alternative Lipid Sources in Aquaculture Feeds. CRC Press, Taylor & Francis Group, Boca Raton, FL, USA.

Bell, J.G., Tocher, D.R., 2009a. Farmed Fish: the impact of diet on fatty acid compositions. In: Rossell, B. (Ed.), Oils and Fats Handbook, Fish Oils, vol. 4. Leatherhead Food International, Leatherhead.

Bell, M.V., Tocher, D.R., 2009b. Biosynthesis of fatty acids; general principles and new directions. In: Arts, M.T., Brett, M.T., Kainz, M.J. (Eds.), Lipids in Aquatic Ecosystems. Springer-Verlag, New York, USA.

Bell, J.G., Waagbø, R., 2008. Safe and nutritious aquaculture produce: benefits and risks of alternative sustainable aquafeeds. In: Holmer, M., Black, K.D., Duarte, C.M., Marba, N., Karakassis, I. (Eds.), Aquaculture in the Ecosystem. Springer Verlag BV, London, UK.

Bell, J.G., Dick, J.R., Mcvicar, A.H., Sargent, J.R., Thompson, K.D., 1993. Dietary sunflower, linseed and fish oils affect phospholipid fatty-acid composition, development of cardiac lesions, phospholipase-activity and eicosanoid production in Atlantic salmon (*Salmo salar*). Prostaglandins Leukot. Essent. Fatty Acids 49, 665–673.

Bell, J.G., Henderson, R.J., Tocher, D.R., Mcghee, F., Dick, J.R., Porter, A., Smullen, R.P., Sargent, J.R., 2002. Substituting fish oil with crude palm oil in the diet of Atlantic salmon (*Salmo salar*) affects muscle fatty acid composition and hepatic fatty acid metabolism. J. Nutr. 132, 222–230.

Bell, J.G., Tocher, D.R., Henderson, R.J., Dick, J.R., Crampton, V.O., 2003. Altered fatty acid compositions in Atlantic salmon (*Salmo salar*) fed diets containing linseed and rapeseed oils can be partially restored by a subsequent fish oil finishing diet. J. Nutr. 133, 2793–2801.

Bell, J.G., Henderson, R.J., Tocher, D.R., Sargent, J.R., 2004. Replacement of dietary fish oil with increasing levels of linseed oil; modification of flesh fatty acid compositions in Atlantic salmon (*Salmo salar*) using a fish oil finishing diet. Lipids 39, 223–232.

Bergan, H.E., Kittilson, J.D., Sheridan, M.A., 2012. Nutrition-regulated lipolysis in rainbow trout (*Oncorhynchus mykiss*) is associated with alterations in the Erk, Pi3K-Akt, JAK-STAT, and PKC signaling pathways. Gen. Compar. Endocrinol. 176, 367–376.

Bergan, H.E., Kittilson, J.D., Sheridan, M.A., 2013. PKC and ERK mediate GH-stimulated lipolysis. J. Mol. Endocrinol. 51, 213.

Bergan, H.E., Kittilson, J.D., Sheridan, M.A., 2015. Nutritional state modulates growth hormone-stimulated lipolysis. Gen. Compar. Endocrinol. 217–218, 1–9.

Berge, A., 2019. Brazilian soy: taking ownership of a problem. SalmonBusiness. Available from: https://salmonbusiness.com/brazillian-soy-taking-ownership-of-a-problem/.

Berger, J., Moller, D.E., 2002. The mechanisms of action of PPARs. Annu. Rev. Med. 53, 409–435.

Betancor, M.B., Mcstay, E., Minghetti, M., Migaud, H., Tocher, D.R., Davie, A., 2014. Daily rhythms in expression of genes of hepatic lipid metabolism in Atlantic salmon (*Salmo salar* L.). PLoS One 9, e106739.

Betancor, M.B., Sprague, M., Sayanova, O., Usher, S., Campbell, P.J., Napier, J.A., Tocher, D.R., 2015a. Evaluation of a high-EPA oil from transgenic *Camelina sativa* in feeds for Atlantic salmon (*Salmo salar* L.): effects on tissue fatty acid composition, histology and gene expression. Aquaculture 444, 1–12.

Betancor, M.B., Sprague, M., Usher, S., Sayanova, O., Campbell, P.J., Napier, J.A., Tocher, D.R., 2015b. A nutritionally-enhanced oil from transgenic *Camelina sativa* effectively replaces fish oil as a source of eicosapentaenoic acid for fish. Sci. Rep. 5, 8104.

Betancor, M.B., Olsen, R.E., Solstorm, D., Skulstad, O.F., Tocher, D.R., 2016a. Assessment of a land-locked Atlantic salmon (*Salmo salar* L.) population as a genetic resource with a focus on long-chain polyunsaturated fatty acid biosynthesis. Biochim. Biophys. Acta 1861, 227–238.

Betancor, M.B., Sprague, M., Montero, D., Usher, S., Sayanova, O., Campbell, P.J., Napier, J.A., Caballero, M.J., Izquierdo, M., Tocher, D.R., 2016b. Replacement of marine fish oil with de novo omega-3 oils from transgenic Camelina sativa in feeds for gilthead sea bream (*Sparus aurata* L.). Lipids 51, 1171–1191.

Betancor, M.B., Sprague, M., Sayanova, O., Usher, S., Metochis, C., Campbell, P.J., Napier, J.A., Tocher, D.R., 2016c. Nutritional evaluation of an EPA-DHA oil from transgenic Camelina sativa in feeds for post-smolt Atlantic salmon (*Salmo salar* L.). PLoS One 11, e0159934.

Betancor, M.B., Li, K., Sprague, M., Sayanova, O., Usher, S., Måsøval, K., Torrissen, O., Napier, J.A., Tocher, D.R., Olsen, R.E., 2017. An oil containing EPA and DHA from transgenic Camelina sativa to replace marine fish oil in feeds for Atlantic salmon (*Salmo salar* L.): effects on intestinal transcriptome, histology, tissue fatty acid profiles and plasma biochemistry. PLoS One 12, e0175415.

Betancor, M.B., Li, K., Bucerzan, V.S., Sprague, M., Sayanova, O., Usher, S., Han, L., Norambuena, F., Torrissen, O., Napier, J.A., Tocher, D.R., Olsen, R.E., 2018. Oil from transgenic Camelina sativa containing over 25 % n-3 long-chain PUFA as the major lipid source in feed for Atlantic salmon (*Salmo salar*). Br. J. Nutr. 119, 1378–1392.

Betancor, M.B., Macewan, A., Sprague, M., Gong, X., Montero, D., Han, L., Napier, J.A., Norambuena, F., Izquierdo, M., Tocher, D.R., 2021. Oils from transgenic Camelina sativa as a source of EPA and DHA in feeds for European sea bass (*Dicentrarchus labrax* L.). Aquaculture (in press).

Bilbao, C., Miren, P., Cancio, Ibon, 2009. Cloning and expression pattern of peroxisomal β-oxidation genes palmitoyl-CoA oxidase, multifunctional protein and 3-ketoacyl-CoA thiolase in mussel *Mytilus galloprovincialis* and thicklip grey mullet Chelon labrosus. Gene 443, 132–142.

Bjerkeng, B., Hamre, K., Hatlen, B., Wathne, E., 1999. Astaxanthin deposition in fillets of Atlantic salmon *Salmo salar* L- fed two dietary levels of astaxanthin in combination with three levels of alpha-tocopheryl acetate. Aquacult. Res. 30, 637–646.

Björnsson, B.T., Johansson, V., Benedet, S., Einarsdottir, I.E., Hildahl, J., Agustsson, T., Jönsson, E., 2004. Growth hormone endocrinology of salmonids: regulatory mechanisms and mode of action. Fish Physiol. Biochem. 27, 227–242.

Blomberg Jensen, M., 2012. Vitamin D metabolism, sex hormones, and male reproductive function. Reproduction 144, 135–152.

Bogevik, A.S., Oxley, A., Olsen, R.E., 2008a. Hydrolysis of acyl-homogeneous and fish oil triacylglycerols using desalted midgut extract from Atlantic salmon, *Salmo salar*. Lipids 43, 655–662.

Bogevik, A.S., Tocher, D.R., Waagbø, R., Olsen, R.E., 2008b. Triacylglycerol-, wax ester- and sterol ester-hydrolases in midgut of Atlantic salmon (*Salmo salar*). Aquacult. Nutr. 14, 93–98.

Bogevik, A.S., Henderson, R.J., Mundheim, H., Waagbø, R., Tocher, D.R., Olsen, R.E., 2010. The influence of temperature on the apparent lipid digestibility in Atlantic salmon (*Salmo salar*) fed *Calanus finmarchicus* oil at two dietary levels. Aquaculture 309, 143–151.

Bonacic, K., Estévez, A., Bellot, O., Conde-Sieira, M., Gisbert, E., Morais, S., 2016. Dietary fatty acid metabolism is affected more by lipid level than source in Senegalese sole juveniles: interactions for optimal dietary formulation. Lipids 51, 105–122.

Bonacic, K., Martinez, A., Gisbert, E., Estevez, A., Morais, S., 2017. Effect of alternative oil sources at different dietary inclusion levels on food intake and appetite regulation via enteroendocrine and central factors in juvenile Solea senegalensis (Kaup, 1858). Aquaculture 470, 169–181.

Bonilla-Gómez, J.L., Chiappa-Carrara, X., Galindo, C., Jeronimo, G., Cuzon, G., Gaxiola, G., 2012. Physiological and biochemical changes of wild and cultivated juvenile pink shrimp Farfantepenaeus duorarum (crustacea: penaeidae) during molt cycle. J. Crust. Biol. 4, 597–606.

Borlongan, I.G., 1990. Studies on the digestive lipases of milkfish, *Chanos chanos*. Aquaculture 89, 315–325.

Bou, M., Todorčević, M., Torgersen, J., Škugor, S., Navarro, I., Ruyter, B., 2016. De novo lipogenesis in Atlantic salmon adipocytes. Biochim. Biophys. Acta 1860, 86–96.

Bou, M., Berge, G.M., Baeverfjord, G., Sigholt, T., Østbye, T.K., Romarheim, O.H., Hatlen, B., Leeuwis, R., Venegas, C., Ruyter, B., 2017. Requirements of n-3 very long-chain PUFA in Atlantic salmon (*Salmo salar* L): effects of different dietary levels of EPA and DHA on fish performance and tissue composition and integrity. Br. J. Nutr. 117, 30–47.

Boukouvala, E., Antonopoulou, E., Favre-Krey, L., Diez, A., Bautista, J.M., Leaver, M.J., Tocher, D.R., Krey, G., 2004. Molecular characterization of three peroxisome proliferator-activated receptors from the sea bass (*Dicentrarchus labrax*). Lipids 39, 1085–1092.

Boukouvala, E., Leaver, M.J., Favre-Krey, L., Theodoridou, M., Krey, G., 2010. Molecular characterization of a gilthead sea bream *(Sparus aurata)* muscle tissue cDNA for carnitine palmitoyltransferase 1B (CPT1B). Compar. Biochem. Physiol. B Biochem. Mol. Biol. 157, 189–197.

Bowyer, J.N., Qin, J.G., Smullen, R.P., Stone, D.A.J., 2012. Replacement of fish oil by poultry oil and canola oil in yellowtail kingfish (*Seriola lalandi*) at optimal and suboptimal temperatures. Aquaculture 356–357.

Bowzer, J., Jackson, C., Trushenski, J.T., 2016. Hybrid Striped Bass feeds based on fish oil, beef tallow, and eicosapentaenoic acid/docosahexaenoic acid supplements: insight regarding fish oil sparing and demand for n-3 long-chain polyunsaturated fatty acids. J. Anim. Sci. 94, 978–988.

Bransden, M.P., Cobcroft, J.M., Battaglene, S.C., Morehead, D.T., Dunstan, G.A., Nichols, P.D., Kolkovski, S., 2005. Dietary 22:6 n-3 alters gut and liver structure and behaviour in larval striped trumpeter (*Latris lineata*). Aquaculture 248, 0–285.

Brauge, C., Medale, F., Corraze, G., 1994. Effect of dietary carbohydrate levels on growth, body composition and glycaemia in rainbow trout, *Oncorhynchus mykiss*, reared in seawater. Aquaculture 123, 109–120.

Bremer, J., 1983. Carnitine-metabolism and functions. Physiol. Rev. 63, 1420–1480.

Brett, M.T., Muller-Navarra, D.C., 1997. The role of highly unsaturated fatty acids in aquatic food web processes. Freshw. Biol. 38, 483–499.

Brockerhoff, H., Hoyle, R.J., 1965. Hydrolysis of triglycerides by the pancreatic lipase of a skate. Biochim. Biophys. Acta 98, 435–436.

Brown, A.W., Hang, J., Dussault, P.H., Carr, T.P., 2010a. Phytosterol ester constituents affect micellar cholesterol solubility in model bile. Lipids 45, 855–862.

Brown, T., Francis, D.S., Turchini, G.M., 2010b. Can dietary lipid source circadian alternation improve omega-3 deposition in rainbow trout? Aquaculture 300, 148–155.

Brunham, L.R., Kruit, J.K., Iqbal, J., Fievet, C., Timmins, J.M., Pape, T.D., Coburn, B.A., Bissada, N., Staels, B., Groen, A.K., Hussain, M.M., Parks, J.S., Kuipers, F., Hayden, M.R., 2006. Intestinal ABCA1 directly contributes to HDL biogenesis in vivo. J. Clin. Invest. 116, 1052–1062.

Bullerwell, C.N., Collins, S.A., Lall, S.P., Anderson, D.M., 2016. Growth performance, proximate and histological analysis of rainbow trout fed diets containing Camelina sativa seeds, meal (high-oil and solvent-extracted) and oil. Aquaculture 452, 342–350.

Bureau, D.P., Hua, K., Harris, A.M., 2008. The effect of dietary lipid and long-chain n-3 Pufa levels on growth, energy utilization, carcass quality, and immune function of rainbow trout, *Oncorhynchus mykiss*. J. World Aquacult. Soc. 39, 1–21.

Bureau, D.P., 2006. Rendered products in fish aquaculture feeds. In: Meekeer, D.L. (Ed.), Essential Rendering. National Rendering Association, Arlington, Virginia.

Burtle, G.J., Lovell, R.T., 1989. Lack of response of channel catfish (*Ictalurus punctatus*) to dietary myo-inositol. Can. J. Fish. Aquat. Sci. 46, 218–222.

Business S, 2019. https://salmonbusiness.com/veramaris-will-cover-salmons-omega-3-needs-with-marine-algae-from-the-american-midwest/.

Butt, R., Volkoff, H., 2019. Gut microbiota and energy homeostasis in fish. Front. Endocrinol. 10, 9.

Buzzi, M., Henderson, R.J., Sargent, J.R., 1996. The desaturation and elongation of linolenic acid and eicosapentaenoic acid by hepatocytes and liver microsomes from rainbow trout (*Oncorhynchus mykiss*) fed diets containing fish oil or olive oil. Biochim. Biophys. Acta 1299, 235–244.

Cahu, C., Zambonino Infante, J., Barbosa, V., 2003. Effect of dietary phospholipid level and phospholipid:neutral lipid value on the development of sea bass (*Dicentrarchus labrax*) larvae fed a compound diet. Br. J. Nutr. 90, 21–28.

Cahu, C.L., Gisbert, E., Villeneuve, L.A., Morais, S., Hamza, N., Wold, P.A., Zambonino Infante, J.L., 2009. Influence of dietary phospholipids on early ontogenesis of fish. Aquacult. Res. 40, 989–999.

Cai, Z., Feng, S., Xiang, X., Mai, K., Ai, Q., 2016. Effects of dietary phospholipid on lipase activity, antioxidant capacity and lipid metabolism-related gene expression in large yellow croaker larvae (*Larimichthys crocea*). Compar. Biochem. Physiol. B Biochem. Mol. Biol. 201, 46–52.

Cai, Z., Mai, K., Ai, Q., 2017. Regulation of hepatic lipid deposition by phospholipid in large yellow croaker. Br. J. Nutr. 118, 1–11.

Calabretti, A., Cateni, F., Procida, G., Favretto, L.G., 2003. Influence of environmental temperature on composition of lipids in edible flesh of rainbow trout (*Oncorhynchus mykiss*). J. Sci. Food Agric. 1497–1498.

Calder, P.C., 2014. Very long chain omega-3 (n-3) fatty acids and human health. Eur. J. Lipid Sci. Technol. 116, 1280–1300.

Calder, P.C., 2018. Very long-chain n-3 fatty acids and human health: fact, fiction and the future. Proc. Nutr. Soc. 77, 52–72.

Campoy, C., Escolano-Margarit, V., Anjos, T., Szajewska, H., Uauy, R., 2012. Omega 3 fatty acids on child growth, visual acuity and neurodevelopment. Br. J. Nutr. 107, S85–S106.

Cao, J.S., Cheng, L., Shi, Y.G., 2007. Catalytic properties of MGAT3, a putative triacylglycerol synthase. J. Lipid Res. 48, 583–591.

Cargill, 2019. https://www.cargill.com/2018/cargill-launches-latitude.

Carlberg, H., Cheng, K., Lundh, T., Brannas, E., 2015. Using self-selection to evaluate the acceptance of a new diet formulation by farmed fish. Appl. Anim. Behav. Sci. 171, 226–232.

Carmona-Antoñanzas, G., Monroig, Ó., Dick, J.R., Davie, A., Tocher, D.R., 2011. Biosynthesis of very long-chain fatty acids ($C \geq 26$) in Atlantic salmon: cloning, functional characterisation, and tissue distribution of an Elovl4. Compar. Biochem. Physiol. B Biochem. Mol. Biol. 159B, 122–129.

Carmona-Antoñanzas, G., Tocher, D.R., Taggart, J.B., Leaver, M.J., 2013. An evolutionary perspective on Elovl5 fatty acid elongase: comparison of Northern pike and duplicated paralogs from Atlantic salmon. BMC Evol. Biol. 13, 85.

Carmona-Antoñanzas, G., Tocher, D.R., Martinez-Rubio, L., Leaver, M.J., 2014. Conservation of lipid metabolic gene transcriptional regulatory networks in fish and mammals. Gene 534, 1–9.

Carmona-Antoñanzas, G., Taylor, J.F., Martinez-Rubio, L., Tocher, D.R., 2015. Molecular mechanism of dietary phospholipid requirement of Atlantic salmon, *Salmo salar*, fry. Biochim. Biophys. Acta 1851, 1428–1441.

Carreiro, A.L., Buhman, K.K., 2019. Absorption of dietary fat and its metabolism in enterocytes. In: Patel, V.B. (Ed.), The Molecular Nutrition of Fats. Academic Press.

Caruso, M.A., Sheridan, M.A., 2011. New insights into the signaling system and function of insulin in fish. Gen. Compar. Endocrinol. 173, 227–247.

Castillo, S., Gatlin, D.M., 2015. Dietary supplementation of exogenous carbohydrase enzymes in fish nutrition: a review. Aquaculture 435, 286–292.

Castro, L.F.C., Wilson, J.M., Gonçalves, O., Galante-Oliveira, S., Rocha, E., Cunha, I., 2011. The evolutionary history of the stearoyl-CoA desaturase gene family in vertebrates. BMC Evol. Biol. 11, 132.

Castro, L.F., Monroig, O., Leaver, M.J., Wilson, J., Cunha, I., Tocher, D.R., 2012. Functional desaturase Fads1 (Delta5) and Fads2 (Delta6) orthologues evolved before the origin of jawed vertebrates. PLoS One 7, e31950.

Castro, C., Corraze, G., Panserat, S., Oliva-Teles, A., 2015a. Effects of fish oil replacement by a vegetable oil blend on digestibility, postprandial serum metabolite profile, lipid and glucose metabolism of European sea bass (*Dicentrarchus labrax*) juveniles. Aquacult. Nutr. 21, 592–603.

Castro, C., Corraze, G., Pérez-Jiménez, A., Larroquet, L., Cluzeaud, M., Panserat, S., Oliva-Teles, A., 2015b. Dietary carbohydrate and lipid source affect cholesterol metabolism of European sea bass (*Dicentrarchus labrax*) juveniles. Br. J. Nutr. 114, 1143–1156.

Castro, C., Corraze, G., Basto, A., Larroquet, L., Panserat, S., Oliva-Teles, A., 2016a. Dietary lipid and carbohydrate interactions: implications on lipid and glucose absorption, transport in gilthead sea bream (*Sparus aurata*) juveniles. Lipids 51, 743–755.

Castro, C., Corraze, G., Diógenes, A., Larroquet, L., Panserat, S., Oliva-Teles, A., 2016b. Regulation of glucose and lipid metabolism by dietary carbohydrate levels and lipid sources in gilthead sea bream juveniles. Br. J. Nutr. 116, 19–34.

Castro, L.F., Tocher, D.R., Monroig, O., 2016c. Long-chain polyunsaturated fatty acid biosynthesis in chordates: insights into the evolution of Fads and Elovl gene repertoire. Prog. Lipid Res. 62, 25–40.

Castro, C., Couto, A., Diógenes, A.F., Corraze, G., Panserat, S., Serra, C.R., Oliva-Teles, A., 2019. Vegetable oil and carbohydrate-rich diets marginally affected intestine histomorphology, digestive enzymes activities, and gut microbiota of gilthead juveniles. Fish Physiol. Biochem. 45, 681–695.

Chan, C.-B., Cheng, C.H., 2004. Identification and functional characterization of two alternatively spliced growth hormone secretagogue receptor transcripts from the pituitary of black seabream *Acanthopagrus schlegeli*. Mol. Cell. Endocrinol. 214, 81–95.

Chang, B.E., Hsieh, S.L., Kuo, C.M., 2001. Molecular cloning of full-length cDNA encoding delta-9 desaturase through PCR strategies and its genomic organization and expression in grass carp (*Ctenopharyngodon idella*). Mol. Reprod. Dev. 58, 245–254.

Chawla, A., 2001. Nuclear receptors and lipid physiology: opening the X-files. Science 294, 1866–1870.

Chay, W., Cooney, R.V., Franke, A.A., Bostick, R.M., 2013. Effects of calcium and vitamin D supplementation on blood pressure and serum lipids and carotenoids: a randomized, double-blind, placebo-controlled, clinical trial. Ann. Epidemiol. 23, 564–570.

Chen, I.S., Chen, Y.C., Chou, C.H., Chuang, R.F., Sheen, L.Y., Chiu, C.H., 2012. Hepatoprotection of silymarin against thioacetamide-induced chronic liver fibrosis. J. Sci. Food Agric. 92, 1441–1447.

Chen, Q.L., Luo, Z., Liu, X., Song, Y.F., Liu, C.X., Zheng, J.L., Zhao, Y.H., 2013a. Effects of waterborne chronic copper exposure on hepatic lipid metabolism and metal-element composition in *Synechogobius hasta*. Arch. Environ. Contam. Toxicol. 64, 301–315.

Chen, Q.L., Luo, Z., Pan, Y.X., Zheng, J.L., Zhu, Q.L., Sun, L.D., Zhuo, M.Q., Hu, W., 2013b. Differential induction of enzymes and genes involved in lipid metabolism in liver and visceral adipose tissue of juvenile yellow catfish *Pelteobagrus fulvidraco* exposed to copper. Aquat. Toxicol. 136, 72–78.

Chen, Q.-L., Luo, Z., Song, Y.-F., Wu, K., Huang, C., Pan, Y.-X., Zhu, Q.-L., 2014. Hormone-sensitive lipase in yellow catfish *Pelteobagrus fulvidraco*: molecular characterization, mRNA tissue expression and transcriptional regulation by leptin in vivo and in vitro. Gen. Compar. Endocrinol. 206, 130–138.

Chen, Q.L., Luo, Z., Liu, C.X., Zheng, J.L., 2015a. Differential effects of dietary Cu deficiency and excess on carnitine status, kinetics and expression of CPT I in liver and muscle of yellow catfish *Pelteobagrus fulvidraco*. Compar. Biochem. Physiol. B Biochem. Mol. Biol. 188, 24–30.

Chen, Q.L., Luo, Z., Wu, K., Huang, C., Zhuo, M.Q., Song, Y.F., Hu, W., 2015b. Differential effects of dietary copper deficiency and excess on lipid metabolism in yellow catfish *Pelteobagrus fulvidraco*. Compar. Biochem. Physiol. B Biochem. Mol. Biol. 184, 19–28.

Chen, M.H., Sun, Y.Y., Kong, C.M., Huijuan, T., Gan, L., 2016. Effect of dietary phosphorus levels on growth and body composition of crucian carp, *Carassius auratus* under indoor and outdoor experiments. Aquacult. Nutr. 23, 702–709.

Chen, G.H., Luo, Z., Chen, F., Shi, X., Song, Y.F., You, W.J., Liu, X., 2017. PPARα, PPARγ and SREBP-1 pathways mediated waterborne iron (Fe)-induced reduction in hepatic lipid deposition of javelin goby *Synechogobius hasta*. Compar. Biochem. Physiol. C Toxicol. Pharmacol. 197, 8–18.

Chen, G.-H., Luo, Z., Hogstrand, C., Wu, K., Ling, S.-C., 2018. SREBP1, PPARG and AMPK pathways mediated the Cu-induced change in intestinal lipogenesis and lipid transport of yellow catfish *Pelteobagrus fulvidraco*. Food Chem. 269, 595–602.

Chen, C., Wang, S., Zhang, M., Chen, B., You, C., Xie, D., Liu, Y., Monroig, O., Tocher, D.R., Waiho, K., Li, Y., 2019. miR-24 is involved in vertebrate LC-PUFA biosynthesis as demonstrated in marine teleost *Siganus canaliculatus*. Biochim. Biophys. Acta 1864, 619–628.

Chen, C., Wang, S., Hu, Y., Zhang, M., He, X., You, C., Wen, X., Monroig, Ó., Tocher, D.R., Li, Y., 2020. miR-26a mediates LC-PUFA biosynthesis by targeting the Lxrα-Srebp1 pathway in the marine teleost Siganus canaliculatus. J. Biol. Chem. 295, 13875–13886.

Chen, C., Wang, S., Hu, Y., Zhang, M., He, X., You, C., Wen, X., Monroig, Ó., Tocher, D.R., Li, Y., 2021. miR-26a mediates LC-PUFA biosynthesis by targeting the Lxrα-Srebp1 pathway in the marine teleost *Siganus canaliculatus*. J. Biol. Chem. (in press).

Cheng, J.B., Russell, D.W., 2004. Mammalian wax biosynthesis II. Expression cloning of wax synthase cDNAs encoding a member of the acyltransferase enzyme family. J. Biol. Chem. 279, 37798–37807.

Cheng, H., Ji, N., Peng, Y., Shen, X., Xu, J., Dong, Z., Wu, C., 2011. Molecular characterization and tissue-specific expression of the acetyl-CoA carboxylase α gene from Grass carp, *Ctenopharyngodon idella*. Gene 487, 46–51.

Chimsung, N., Lall, S.P., Tantikitti, C., Verlhac-Trichet, V., Milley, J.E., 2013. Effects of dietary cholesterol on astaxanthin transport in plasma of Atlantic salmon (*Salmo salar*). Compar. Biochem. Physiol. B Biochem. Mol. Biol. 165, 73–81.

Chimsung, N., Tantikitti, C., Milley, J.E., Verlhac-Trichet, V., Lall, S.P., 2014. Effects of various dietary factors on astaxanthin absorption in Atlantic salmon (*Salmo salar*). Aquacult. Res. 45, 1611–1620.

Cho, Y.A., Kim, J., 2013. Effect of probiotics on blood lipid concentrations. Medicine 94, e1714.

Cho, Y.W., Hong, S., Jin, Q., Wang, L., Lee, J.E., Gavrilova, O., Ge, K., 2009. Histone methylation regulator PTIP is required for PPARγ and C/EBPα expression and adipogenesis. Cell Metabol. 10, 27–39.

Chou, C.-H., Chen, Y.-C., Hsu, M.-C., Tsai, W.-L., Chang, C.-Y., Chiu, C.-H., 2012. Effect of silymarin on lipid and alcohol metabolism in mice following long-term alcohol consumption. J. Food Biochem. 36, 369–377.

Choubert, G., Milicua, J.C.G., Gomez, R., 1994. The transport of astaxanthin in immature rainbow trout *Oncorhynchus mykiss* serum. Compar. Biochem. Physiol. Mol. Integr. Physiol. 108, 245–248.

Choubert, G., Mendes-Pinto, M.M., Morais, R., 2006. Pigmenting efficacy of astaxanthin fed to rainbow trout *Oncorhynchus mykiss*: effect of dietary astaxanthin and lipid sources. Aquaculture 257, 429–436.

Chow, C.K., 2008. Fatty Acids in Foods and Their Health Implications. CRC Press, Taylor and Francis Group, Boca Raton, FL, USA.

Christie, W.W., Han, X., 2010. Lipid Analysis. Isolation, Separation, Identification and Lipidomic Analysis. The Oil Press, PJ Barnes & Associates, Bridgwater, England, UK.

Chuang, L.T., Chen, D.C., Nicaud, J.M., Madzak, C., Chen, Y.H., Huang, Y.S., 2010. Co-expression of heterologous desaturase genes in *Yarrowia lipolytica*. Nat. Biotechnol. 27, 277–282.

Cifarelli, V., Abumrad, N.A., 2018. Enterocyte fatty acid handling proteins and chylomicron formation. In: Said, H.M. (Ed.), Physiology of the Gastrointestinal Tract, sixth ed. Academic Press.

Clarkson, M., Migaud, H., Metochis, C., Vera, L.M., Leeming, D., Tocher, D.R., Taylor, J.F., 2017. Early nutritional intervention can improve utilisation of vegetable-based diets in diploid and triploid Atlantic salmon (*Salmo salar*). Br. J. Nutr. 118, 17–29.

Claudia, T.-C., Petra, W., Philippe, C.-R., Ruben, N., Diego, P.-T., Paul, P., Tamara, R.C., Patrick, M., Annette, S., Ildiko, S., 2006. Ghrelin action in the brain controls adipocyte metabolism. J. Clin. Invest. 116, 1983–1993.

Clément, K., 2011. Bariatric surgery, adipose tissue and gut microbiota. Int. J. Obes. 35, S7–S15.

Clements, K.D., Angert, E.R., Montgomery, W., Howard, C.J., 2014. Intestinal microbiota in fishes: what's known and what's not. Mol. Ecol. 23, 1891–1898.

Codabaccus, B.M., Carter, C.G., Bridle, A.R., Nichols, P.D., 2012. The "n—3 LC-PUFA sparing effect" of modified dietary n—3 LC-PUFA content and DHA to EPA ratio in Atlantic salmon smolt. Aquaculture 356–357, 135–140.

Colombo, S.M., Wacker, A., Parrish, C.C., Kainz, M.J., Arts, M.T., 2017. A fundamental dichotomy in long-chain polyunsaturated fatty acid abundance between and within marine and terrestrial ecosystems. Environ. Rev. 25, 163–174.

Colombo-Hixson, S.M., Olsen, R.E., Milley, J.E., Lall, S.P., 2011. Lipid and fatty acid digestibility in Calanus copepod and krill oil by Atlantic halibut (*Hippoglossus hippoglossus* L.). Aquaculture 313, 115–122.

Conde-Sieira, M., Soengas, J.L., 2017. Nutrient sensing systems in fish: impact on food intake regulation and energy homeostasis. Front. Neurosci. 10.

Conde-Sieira, M., Bonacic, K., Velasco, C., Valente, L.M.P., Morais, S., Soengas, J.L., 2015. Hypothalamic fatty acid sensing in Senegalese sole (*Solea senegalensis*): response to long-chain saturated, monounsaturated, and polyunsaturated (n-3) fatty acids. Am. J. Physiol. Regul. Integr. Compar. Physiol. 309, R1521–R1531.

Contreras, J.A., Karlsson, M., Østerlund, T., Laurell, H., Svensson, A., Holm, C., 1996. Hormone-sensitive lipase is structurally related to acetylcholinesterase, bile salt-stimulated lipase, and several fungal lipases: building of a three-dimensional model for the catalytic domain of hormone-sensitive lipase. J. Biol. Chem. 271, 31426–31430.

Cook, H.W., Mcmaster, R.C.R., 2004. Fatty acid desaturation and chain elongation in eukaryotes. In: Vance, D.E., Vance, J.E. (Eds.), Biochemistry of Lipids, Lipoproteins and Membranes. Elsevier, Amsterdam.

Copeland, D.L., Duff, R.J., Liu, Q., Prokop, J., Londraville, R.L., 2011. Leptin in teleost fishes: an argument for comparative study. Front. Physiol. 2, 26.

Corbin, K., Zeisel, S., 2012. Choline metabolism provides novel insights into nonalcoholic fatty liver disease and its progression. Curr. Opin. Gastroenterol. 28, 159–165.

Corbion, 2019. http://algaprime.com.

Cordier, M., Brichon, G., Weber, J.M., Zwingelstein, G., 2002. Changes in the fatty acid composition of phospholipids in tissues of farmed sea bass (*Dicentrarchus labrax*) during an annual cycle. Roles of environmental temperature and salinity. Compar. Biochem. Physiol. B Biochem. Mol. Biol. 133, 281–288.

Couto, A., Kortner, T.M., Penn, M., Bakke, A.M., Krogdahl, Å., Oliva-Teles, A., 2015. Dietary saponins and phytosterols do not affect growth, intestinal morphology and immune response of on-growing European sea bass (*Dicentrarchus labrax*). Aquacult. Nutr. 21, 970–982.

Coutteau, P., Geurden, I., Camara, M.R., Bergot, P., Sorgeloos, P., 1997. Review on the dietary effects of phospholipids in fish and crustacean larviculture. Aquaculture 155, 149–164.

Craig, S.A., 2004. Betaine in human nutrition. Am. J. Clin. Nutr. 80, 539–549.

Crockett, E.L., Sidell, B.D., 1993a. Peroxisomal beta-oxidation is a significant pathway for catabolism of fatty acids in a marine teleost. Am. J. Physiol. 264, R1004–R1009.

Crockett, E.L., Sidell, B.D., 1993b. Substrate selectivities differ for hepatic mitochondrial and peroxisomal beta-oxidation in an Antarctic fish, *Notothenia gibberifrons*. Biochem. J. 289, 427–433.

Crovetto, C.A., Córdoba, O.L., 2016. Structural and biochemical characterization and evolutionary relationships of the fatty acid-binding protein 10 (Fabp10) of hake (*Merluccius hubbsi*). Fish Physiol. Biochem. 42, 149–165.

Cruz, S.A., Tseng, Y.-C., Kaiya, H., Hwang, P.P., 2010. Ghrelin affects carbohydrate-glycogen metabolism via insulin inhibition and glucagon stimulation in the zebrafish (*Danio rerio*) brain. Compar. Biochem. Physiol. Mol. Integr. Physiol. 156, 190–200.

Cruz-Garcia, L., Saera-Vila, A., Navarro, I., Calduch-Giner, J., Pérez-Sánchez, J., 2009. Targets for TNFα-induced lipolysis in gilthead sea bream (*Sparus aurata* L.) adipocytes isolated from lean and fat juvenile fish. J. Exp. Biol. 212, 2254.

Cui, H.-Y., Chen, Q.-L., Tan, X.-Y., Zhang, D.-G., Ling, S.-C., Chen, G.-H., Luo, Z., 2018. MiR-205 mediated Cu-induced lipid accumulation in yellow catfish *Pelteobagrus fulvidraco*. Int. J. Mol. Sci. 19, 2980.

Cummings, J.H., Pomare, E.W., Branch, W.J., Naylor, C.P., Macfarlane, G.T., 1987. Short chain fatty acids in human large intestine, portal, hepatic and venous blood. Gut 28, 1221–1227.

D'Abramo, L.R., Bordner, C.E., Conklin, D.E., Baum, N.A., 1984. Sterol requirement of juvenile lobsters, *Homarus* sp. Aquaculture 42, 13–25.

D'andrea, S., Guillou, H., Jan, S., Catheline, D., Thibault, J.N., Bouriel, M., Rioux, V., Legrand, P., 2002. The same rat Delta6-desaturase not only acts on 18- but also on 24-carbon fatty acids in very-long-chain polyunsaturated fatty acid biosynthesis. Biochem. J. 364, 49–55.

D'aquila, T., Hung, Y.H., Carreiro, A., Buhman, K.K., 2016. Recent discoveries on absorption of dietary fat: presence, synthesis, and metabolism of cytoplasmic lipid droplets within enterocytes. Biochim. Biophys. Acta 1861, 730–747.

Dabrowska, H., Meyer-Burgdorff, K., Gunther, K.-D., 1991. Magnesium status in freshwater fish, common carp (*Cyprinus carpio*, L.) and the dietary protein-magnesium interaction. Fish Physiol. Biochem. 9, 165–172.

Dai, W., Panserat, S., Mennigen, J.A., Terrier, F., Dias, K., Seiliez, I., Skiba-Cassy, S., 2013. Post-prandial regulation of hepatic glucokinase and lipogenesis requires the activation of TORC1 signalling in rainbow trout (*Oncorhynchus mykiss*). J. Exp. Biol. 216, 4483–4492.

Dai, W., Panserat, S., Plagnes-Juan, E., Seiliez, I., Skiba-Cassy, S., 2015. Amino acids attenuate insulin action on gluconeogenesis and promote fatty acid biosynthesis via mTORC1 signaling pathway in trout hepatocytes. Cell. Physiol. Biochem. 36, 1084–1100.

Dai, W., Panserat, S., Kaushik, S., Terrier, F., Plagnes-Juan, E., Seiliez, I., Skiba-Cassy, S., 2016. Hepatic fatty acid biosynthesis is more responsive to protein than carbohydrate in rainbow trout during acute stimulations. Am. J. Physiol. Regul. Integr. Compar. Physiol. 310, R74–R86.

Dai, Y., Liu, W.-B., Li, X.-F., Zhou, M., Xu, C., Qian, Y., Guang-Zhen, J., 2018. Molecular cloning of adipose triglyceride lipase (ATGL) gene from blunt snout bream and its expression after LPS-induced TNF-α factor. Fish Physiol. Biochem. 44, 1–15.

Damsteegt, E.L., Davie, A., Lokman, P.M., 2018. The evolution of apolipoprotein B and its mRNA editing complex. Does the lack of editing contribute to hypertriglyceridemia? Gene 641, 46–54.

Dapra, F., Geurden, I., Corraze, G., Bazin, D., Zambonino-Infante, J.L., Fontagné-Dicharry, S., 2011. Physiological and molecular responses to dietary phospholipids vary between fry and early juvenile stages of rainbow trout (*Oncorhynchus mykiss*). Aquaculture 319, 377–384.

Darias, M.J., Boglino, A., Manchado, M., Ortiz-Delgado, J.B., Estévez, A., Andree, K.B., Gisbert, E., 2012. Molecular regulation of both dietary vitamin A and fatty acid absorption and metabolism associated with larval morphogenesis of Senegalese sole (*Solea senegalensis*). Compar. Biochem. Physiol. Mol. Integr. Physiol. 161, 130–139.

Davis, D.A., Lazo, J.P., Arnold, C.R., 1999. Response of juvenile red drum (*Sciaenops ocellatus*) to practical diets supplemented with medium chain triglycerides. Fish Physiol. Biochem. 21, 235–248.

Dawood, M.A.O., Koshio, S., 2020. Application of fermentation strategy in aquafeed for sustainable aquaculture. Rev. Aquacult. 12, 987–1002.

De Antueno, R.J., Knickle, L.C., Smith, H., Elliot, M.L., Allen, S.J., Nwaka, S., Winther, M.D., 2001. Activity of human Delta5 and Delta6 desaturases on multiple n-3 and n-6 polyunsaturated fatty acids. FEBS Lett. 509, 77–80.

De Koning, A.J., 2002. The antioxidant ethoxyquin and its analogues: a review. Int. J. Food Prop. 5, 451–461.

De Koning, A.J., Evans, A.A., 1991. Phospholipids of marine origin. The lantern fish (*Lampanyctodes hectoris*). J. Sci. Food Agric. 56, 503–510.

De Roos, B., Sneddon, A.A., Sprague, M., Horgan, G.W., Brouwer, I.A., 2017. The potential impact of compositional changes in farmed fish on its health-giving properties: is it time to reconsider current dietary recommendations? Publ. Health Nutr. 20, 2042–2049.

De Santis, C., Taylor, J.F., Martinez-Rubio, L., Boltana, S., Tocher, D.R., 2015. Influence of development and dietary phospholipid content and composition on intestinal transcriptome of Atlantic salmon (*Salmo salar*). PLoS One 10, e0140964.

de Silva, S.S., 2012. Climate change impacts: challenges for aquaculture. In: Subasinghe, R.P., Arthur, J.R., Bartley, D.M., de Silva, S.S., Halwart, M., Hishamunda, N., Mohan, C.V., Sorgeloos, P. (Eds.), Farming the Waters for People and Food. Proceedings of the Global Conference on Aquaculture 2010, Phuket, Thailand. FAO and NACA, Namgkok, Thailand, pp. 22–25.

De Smet, H., Blust, R., Moens, L., 1998. Absence of albumin in the plasma of the common carp *Cyprinus carpio*: binding of fatty acids to high density lipoprotein. Fish Physiol. Biochem. 19, 71–81.

Deamer, D., 2017. The role of lipid membranes in life's origin. Life-Basel 7.

Decker, E.A., Elias, R.J., Mcclements, D.J., 2010. Oxidation in Foods and Beverages and Antioxidant Applications. Woodhead Publishing Limited.

Del Bas, J.M., Ricketts, M.-L., Vaqué, M., Sala, E., Quesada, H., Ardevol, A., Salvadó, M.J., Blay, M., Arola, L., Moore, D.D., Pujadas, G., Fernandez-Larrea, J., Bladé, C., 2009. Dietary procyanidins enhance transcriptional activity of bile acid-activated FXR in vitro and reduce triglyceridemia in vivo in a FXR-dependent manner. Mol. Nutr. Food Res. 53, 805–814.

Del Chierico, F., Vernocchi, P., Bonizzi, L., Carsette, R., Castellazzi, A.M., Dallapiccola, B., de Vos, W., Guerzoni, M.E., Manco, M., Marseqlia, G.L., Muraca, M., Roncada, P., Salvatori, G., Signore, F., Urbani, A., Putignani, L., 2012. Early-life gut microbiota under physiological and pathological conditions: the central role of combined meta-omics-based approaches. J. Proteom. 75, 4580–4587.

Delgado-Lista, J., Perez-Martinez, P., Lopez-Miranda, J., Perez-Jimenez, F., 2012. Long chain omega-3 fatty acids and cardiovascular disease: a systematic review. Br. J. Nutr. 107, S201–S213.

Delzenne, N.M., Cani, P.D., 2008. Gut microflora is a key player in host homeostasis. Med. Sci. 24, 505–510.

Delzenne, N.M., Cani, P.D., Neyrinck, A.M., 2008. Prebiotics and lipid metabolism. In: Versalovic, J., Wilson, M. (Eds.), Therapeutic Microbiology: Probiotics and Related Strategies. ASM Press, Washington, DC, USA.

Den, J., Mai, K., Ai, Q., Zhang, W., Wang, X., Tan, B., Xu, W., Liufu, Z., Ma, H., 2010. Interactive effects of dietary cholesterol and protein sources on growth performance and cholesterol metabolism of Japanese flounder (*Paralichthys olivaceus*). Aquacult. Nutr. 16, 419–429.

Deng, S.X., Tian, L.X., Liu, F.J., Jin, S.J., Liang, G.Y., Yang, H.J., Du, Z.Y., Liu, Y.J., 2010. Toxic effects and residue of aflatoxin B1 in tilapia (*Oreochromis niloticus*×*O. aureus*) during long-term dietary exposure. Aquaculture 307, 233–240.

Deng, J., Bi, B., Kang, B., Kong, L., Wang, Q., Zhang, X., 2013. Improving the growth performance and cholesterol metabolism of rainbow trout (*Oncorhynchus mykiss*) fed soyabean meal-based diets using dietary cholesterol supplementation. Br. J. Nutr. 110, 29–39.

Dennis, E.A., Norris, P.C., 2015. Eicosanoid storm in infection and inflammation. Nat. Rev. Immunol. 15, 511–523.

Denstadli, V., Bakke, A.M., Berge, G.M., Krogdahl, Å., Hillestad, M., Holm, H., Ruyter, B., 2011. Medium-chain and long-chain fatty acids have different postabsorptive fates in Atlantic salmon. J. Nutr. 141, 1618–1628.

Denver, R.J., Bonett, R.M., Boorse, G.C., 2011. Evolution of leptin structure and function. Neuroendocrinology 94, 21–38.

Desai, A.S., Singh, R.K., 2009. The effects of water temperature and ration size on growth and body composition of fry of common carp, *Cyprinus carpio*. J. Therm. Biol. 34, 276–280.

Dey, I., Buda, C., Wiik, T., Halver, J.E., Farkas, T., 1993. Molecular and structural composition of phospholipid membranes in livers of marine and freshwater fish in relation to temperature. Proc. Natl. Acad. Sci. U. S. A. 90, 7498–7502.

Dias, J., Alvarez, M.J., Diez, A., Arzel, J., Corraze, G., Bautista, J.M., Kaushik, S.J., 1998. Regulation of hepatic lipogenesis by dietary protein/energy in juvenile European seabass (*Dicentrarchus labrax*). Aquaculture 161, 169–186.

Dias, J., Alvarez, M.J., Arzel, J., Corraze, G., Diez, A., Bautista, J.M., Kaushik, S.J., 2005. Dietary protein source affects lipid metabolism in the European seabass (*Dicentrarchus labrax*). Compar. Biochem. Physiol. Mol. Integr. Physiol. 142, 19–31.

Dietschy, J.M., Turley, S.D., 2004. Thematic review series: brain Lipids. Cholesterol metabolism in the central nervous system during early development and in the mature animal. J. Lipid Res. 45, 1375–1397.

Ding, M.Y., 2015. Effect of VA and VD Different Levels in Dietary, Respectively, on Growth, Feed Utilization, Immunity, Fat Metabolism Enzyme Activities and Fas, Hl mRNA Expression of Grouper (*Epinephelus coioides*) at Two Stages. Guangdong Ocean University (in Chinese).

Ding, Q., Zhang, Z., Ran, C., He, S., Yang, Y., Du, Z., Zhang, J., Zhou, Z., 2018. The hepatotoxicity of palmitic acid in zebrafish involves the intestinal microbiota. J. Nutr. 148, 1217–1228.

Dinicolantonio, J.J., O'keefe, J.H., 2018. Effects of dietary fats on blood lipids: a review of direct comparison trials. Open Heart 5, e000871.

Don, A.S., Lim, X.Y., Couttas, T.A., 2014. Re-configuration of sphingolipid metabolism by oncogenic transformation. Biomolecules 4, 315–353.

Dong, X., Xu, H., Mai, K., Xu, W., Zhang, Y., Ai, Q., 2015. Cloning and characterization of SREBP-1 and PPAR-α in Japanese seabass *Lateolabrax japonicus*, and their gene expressions in response to different dietary fatty acid profiles. Compar. Biochem. Physiol. B Biochem. Mol. Biol. 180, 48–56.

Dong, Y., Wang, S., Chen, J., Zhang, Q., Liu, Y., You, C., Monroig, O., Tocher, D.R., Li, Y., 2016. Hepatocyte nuclear factor 4α (HNF4α) is a transcription factor of vertebrate fatty acyl desaturase gene as identified in marine teleost *Siganus canaliculatus*. PLoS One 11, 1–19.

Douros, J.D., Baltzegar, D.A., Breves, J.P., Lerner, D.T., Seale, A.P., Grau, E.G., Borski, R.J., 2014. Prolactin is a major inhibitor of hepatic leptin A synthesis and secretion: studies utilizing a homologous leptin A ELISA in the tilapia. Gen. Compar. Endocrinol. 207, 86–93.

Drew, M.D., Borgeson, T.L., Thiessen, D.L., 2007. A review of processing of feed ingredients to enhance diet digestibility in finfish. Anim. Feed Sci. Technol. 138, 118–136.

Du, Z.Y., 2014. Causes of fatty liver in farmed fish: a review and new perspectives. J. Fish. Sci. China 38, 1628–1638.

Du, Z.Y., Demizieux, L., Degrace, P., Gresti, J., Moindrot, B., Liu, Y.J., Tian, L.X., Cao, J.M., Clouet, P., 2004. Alteration of 20:5n-3 and 22:6n-3 fat contents and liver peroxisomal activities in fenofibrate-treated rainbow trout. Lipids 39, 849–855.

Du, Z.Y., Liu, Y.J., Tian, L.X., Wang, J.T., Wang, Y., Liang, G.Y., 2005. Effect of dietary lipid level on growth, feed utilization and body composition by juvenile grass carp (*Ctenopharyngodon idella*). Aquacult. Nutr. 11, 139–146.

Du, Z.Y., Clouet, P., Zheng, W.H., Degrace, P., Tian, L.X., Liu, Y.J., 2006a. Biochemical hepatic alterations and body lipid composition in the herbivorous grass carp (*Ctenopharyngodon idella*) fed high-fat diets. Br. J. Nutr. 95, 905–915.

Du, Z.Y., Liu, Y.J., Tian, L.X., He, J.G., Cao, J.M., Liang, G.Y., 2006b. The influence of feeding rate on growth, feed efficiency and body composition of juvenile grass carp (*Ctenopharyngodon idella*). Aquacult. Int. 14, 247–257.

Du, Z.Y., Clouet, P., Huang, L.M., Degrace, P., Zheng, W.H., He, J.G., Tian, L.X., Liu, Y.J., 2008. Utilization of different dietary lipid sources at high level in herbivorous grass carp (*Ctenopharyngodon idella*): mechanism related to hepatic fatty acid oxidation. Aquacult. Nutr. 14, 77–92.

Du, J., Xu, H., Li, S., Cai, Z., Mai, K., Ai, Q., 2017. Effects of dietary chenodeoxycholic acid on growth performance, body composition and related gene expression in large yellow croaker (*Larimichthys crocea*) fed diets with high replacement of fish oil with soybean oil. Aquaculture 479, 584–590.

Du, J., Xiang, X., Li, Y., Ji, R., Xu, H., Mai, K., Ai, Q., 2018. Molecular cloning and characterization of farnesoid X receptor from large yellow croaker (*Larimichthys crocea*) and the effect of dietary CDCA on the expression of inflammatory genes in intestine and spleen. Compar. Biochem. Physiol. B Biochem. Mol. Biol. 216, 10–17.

Dumas, A., Raggi, T., Barkhouse, J., Lewis, E., Weltzien, E., 2018. The oil fraction and partially defatted meal of black soldier fly larvae (*Hermetia illucens*) affect differently growth performance, feed efficiency, nutrient deposition, blood glucose and lipid digestibility of rainbow trout (*Oncorhynchus mykiss*). Aquaculture 492, 24–34.

Dutchak, P.A., Katafuchi, T., Bookout, A.L., Choi, J.H., Ruth, T.Y., Mangelsdorf, D.J., Kliewer, S.A., 2012. Fibroblast growth factor-21 regulates PPARγ activity and the antidiabetic actions of thiazolidinediones. Cell 148, 556–567.

Egloff, S., Pietsch, C., 2018. Ethoxyquin: a feed additive that poses a risk for aquatic life. Dis. Aquat. Org. 131, 39–48.

Ekmann, K.S., Dalsgaard, J., Holm, J., Campbell, P.J., Skov, P.V., 2013. Effects of dietary energy density and digestible protein: energy ratio on de novo lipid synthesis from dietary protein in gilthead sea bream (*Sparus aurata*) quantified with stable isotopes. Br. J. Nutr. 110, 1771–1781.

Emery, J.A., Smullen, R.P., Turchini, G.M., 2014. Tallow in Atlantic salmon feed. Aquaculture 422, 98–108.

Emery, J.A., Smullen, R., Keast, R.S.J., Turchini, G.M., 2016. Viability of tallow inclusion in Atlantic salmon diet, as assessed by an on-farm grow out trial. Aquaculture 451, 289–297.

Erkkilä, A., Lankinen, M., 2016. Fish and fish oil and lipoprotein particle number and size. In: Raatz, S.K., Bibus, D.M. (Eds.), Fish and Fish Oil in Health and Disease Prevention. Academic Press, San Diego.

Eroldogan, O.T., Elsabagh, M., Emre, Y., Turchini, G.M., Yilmaz, H.A., Eraslan, D., Emre, N., Evliyaoglu, E., 2018. Circadian feeding schedules in gilthead sea bream (*Sparus aurata*) and European sea bass (*Dicentrarchus labrax*): a comparative approach towards improving dietary fish oil utilization and n-3 LC-PUFA metabolism. Aquaculture 495, 806–814.

Espe, M., Rathore, R.M., Du, Z.Y., Liaset, B., El-Mowafi, A., 2010. Methionine limitation results in increased hepatic FAS activity, higher liver 18: 1 to 18: 0 fatty acid ratio and hepatic TAG accumulation in Atlantic salmon, *Salmo salar*. Amino Acids 39, 449–460.

Esteban, R., Moran, J.F., Becerril, J.M., García-Plazaola, J.I., 2015. Versatility of carotenoids: an integrated view on diversity, evolution, functional roles and environmental interactions. Environ. Exp. Bot. 119, 63–75.

Evans, H., De Tomaso, T., Quail, M., Rogers, J., Gracey, A.Y., Cossins, A.R., Berenbrink, M., 2008. Ancient and modern duplication events and the evolution of stearoyl-CoA desaturases in teleost fishes. Physiol. Genom. 35, 18–29.

Everard, A., Lazarevic, V., Derrien, M., Girard, M., Muccioli, G.G., Neyrinck, A.M., Possemiers, S., Van Holle, A., Francois, P., De Vos, W.M., Delzenne, N.M., Schrenzel, J., Cani, P.D., 2011. Responses of gut microbiota and glucose and lipid metabolism to prebiotics in genetic obese and diet-induced leptin-resistant mice. Diabetes 60, 2775–2786.

Fabiana, A.M., Kivia, Q.D.A., Juliana, C.F.D.S., Marilia, O.F.G., 2015. Lipoic acid: its antioxidant and anti-inflammatory role and clinical applications. Curr. Top. Med. Chem. 15, 458–483.

Falahatkar, B., Soltani, M., Abtahi, B., Kalbassi, M.R., Pourkazemi, M., 2006. Effects of dietary vitamin C supplementation on performance, tissue chemical composition and alkaline phosphatase activity in great sturgeon (*Huso huso*). J. Appl. Ichthyol. 22, 283–286.

Falcinelli, S., Picchietti, S., Rodiles, A., Cossignani, L., Merrifield, D.L., Taddei, A.R., Maradonna, F., Olivotto, I., Gioacchini, G., Carnevali, O., 2015. *Lactobacillus rhamnosus* lowers zebrafish lipid content by changing gut microbiota and host transcription of genes involved in lipid metabolism. Sci. Rep. 5, 9336.

Falcinelli, S., Rodiles, A., Hatef, A., Picchietti, S., Cossignani, L., Merrifield, D.L., Unniappan, S., Carnevali, O., 2018. Influence of probiotics administration on gut microbiota core: a review on the effects on appetite control, glucose, and lipid metabolism. J. Clin. Gastroenterol. 52 (Suppl. 1). Proceedings from the 9th Probiotics, Prebiotics and New Foods, Nutraceuticals and Botanicals for Nutrition & Human and Microbiota Health Meeting, held in Rome, Italy from September 10 to 12, 2017, S50–S56.

FAO, 2020. The State of World Fisheries and Aquaculture (SOFIA). Food and Agriculture Organization of the United Nations, Rome, Italy.

Farese, R.V., Veniant, M.M., Cham, C.M., Flynn, L.M., Pierotti, V., Loring, J.F., Traber, M., Ruland, S., Stokowski, R.S., Huszar, D., Young, S.G., 1996. Phenotypic analysis of mice expressing exclusively apolipoprotein B48 or apolipoprotein B100. Proc. Natl. Acad. Sci. U. S. A. 93, 6393–6398.

Fatima, M., Afzal, M., Shah, S.Z.H., 2019. Effect of dietary oxidized oil and vitamin E on growth performance, lipid peroxidation and fatty acid profile of Labeo rohita fingerlings. Aquacult. Nutr. 25, 281–291.

Faust, I.M., Johnson, P.R., Hirsch, J., 1980. Long-term effects of early nutritional experience on the development of obesity in the rat. J. Nutr. 110, 2027–2034.

Feed, F. F. O. F, 2019. https://oil.f3challenge.org.

Feng, S., Cai, Z., Zuo, R., Mai, K., Ai, Q., 2017. Effects of dietary phospholipids on growth performance and expression of key genes involved in phosphatidylcholine metabolism in larval and juvenile large yellow croaker, *Larimichthys crocea*. Aquaculture 469, 59–66.

Fernández, I., Gisbert, E., 2011. The effect of vitamin A on flatfish development and skeletogenesis: a review. Aquaculture 315, 34–48.

Ferreira, G.F., Ríos Pinto, L.F., Maciel Filho, R., Fregolente, L.V., 2019. A review on lipid production from microalgae: association between cultivation using waste streams and fatty acid profiles. Renew. Sustain. Energy Rev. 109, 448–466.

Figueiredo-Silva, A.C., Corraze, G., Rema, P., Sanchez-Gurmaches, J., Gutiérrez, J., Valente, L.M., 2009. Blackspot seabream (*Pagellus bogaraveo*) lipogenic and glycolytic pathways appear to be more related to dietary protein level than dietary starch type. Aquaculture 291, 101–110.

Figueiredo-Silva, A.C., Kaushik, S., Terrier, F., Schrama, J.W., Médale, F., Geurden, I., 2011. Link between lipid metabolism and voluntary food intake in rainbow trout fed coconut oil rich in medium-chain TAG. Br. J. Nutr. 107, 1714–1725.

Fink, I.R., Benard, E.L., Hermsen, T., Meijer, A.H., Forlenza, M., Wiegertjes, G.F., 2015. Molecular and functional characterization of the scavenger receptor CD36 in zebrafish and common carp. Mol. Immunol. 63, 381–393.

Finn, R.N., Henderson, R.J., Fyhn, H.J., 1995. Physiological energetics of developing embryos and yolk-sac larvae of Atlantic cod (*Gadus morhua*). 2. Lipid metabolism and enthalpy balance. Mar. Biol. 124, 371–379.

Flynn, E.J., Trent, C.M., Rawls, J.F., 2009. Ontogeny and nutritional control of adipogenesis in zebrafish (*Danio rerio*). J. Lipid Res. 50, 1641–1652.

Fonseca-Madrigal, J., Navarro, J.C., Hontoria, F., Tocher, D.R., Martínez-Palacios, C., Monroig, Ó., 2014. Diversification of substrate specificities in teleostei Fads2 fatty acyl desaturases: cloning and functional characterisation of Δ4 and bifunctional Δ6Δ5 desaturases in the freshwater atherinopsid Chirostoma estor. J. Lipid Res. 1408–1419.

Fontagne, S., Geurden, I., Escaffre, A.-M., Bergot, P., 1998. Histological changes induced by dietary phospholipids in intestine and liver of common carp (*Cyprinus carpio* L.) larvae. Aquaculture 161, 213–223.

Fontagné, S., Corraze, G., Bergot, P., 2001. Response of common carp (*Cyprinus carpio*) larvae to different dietary levels and forms of supply of medium-chain fatty acids. Aquat. Living Resour. 13, 429–437.

Fontagné-Dicharry, S., Larroquet, L., Dias, K., Cluzeaud, M., Heraud, C., Corlay, D., 2018. Effects of dietary oxidized fish oil supplementation on oxidative stress and antioxidant defense system in juvenile rainbow trout (*Oncorhynchus mykiss*). Fish Shellfish Immunol. 74, 43–51.

Francis, D.S., Turchini, G.M., 2016. Retro-engineering the protein sparing effect to preserve n-3 LC-PUFA from catabolism and optimise fish oil utilisation: a preliminary case study on juvenile Atlantic salmon. Aquaculture 468, 184–192.

Francis, G., Makkar, H.P.S., Becker, K., 2001. Antinutritional factors present in plant-derived alternate fish feed ingredients and their effects in fish. Aquaculture 199, 197–227.

Francis, D.S., Turchini, G.M., Jones, P.L., de Silva, S.S., 2007. Dietary lipid source modulates in vivo fatty acid metabolism in the freshwater fish, Murray cod (*Maccullochella peelii peelii*). J. Agric. Food Chem. 55, 1582–1591.

Francis, D.S., Thanuthong, T., Senadheera, S.P.S.D., Paolucci, M., Coccia, E., de Silva, S.S., Turchini, G.M., 2014. n-3 LC-PUFA deposition efficiency and appetite-regulating hormones are modulated by the dietary lipid source during rainbow trout grow-out and finishing periods. Fish Physiol. Biochem. 40, 577–593.

Frankel, E.N., 2012a. Free radical oxidation. In: Frankel, E.N. (Ed.), Lipid Oxidation, second ed. Woodhead Publishing.

Frankel, E.N., 2012b. Lipid oxidation. In: Frankel, E.N. (Ed.), Lipid Oxidation, second ed. Woodhead Publishing.

Freudenberg, A., Petzke, K.J., Klaus, S., 2013. Dietary L-leucine and L-alanine supplementation have similar acute effects in the prevention of high-fat diet-induced obesity. Amino Acids 44, 519–528.

Friedman-Einat, M., Cogburn, L.A., Yosefi, S., Hen, G., Shinder, D.S.,A., Seroussi, E., 2014. Discovery and characterization of the first genuine avian leptin gene in the rock dove (*Columba livia*). Endocrinology 155, 3376–3384.

Frøyland, L., Berge, 2000. Mitochondrial and peroxisomal β-oxidation capacities in various tissues from Atlantic salmon *Salmo salar*. Aquacult. Nutr. 6, 85–89.

Frøyland, L., Madsen, L., Eckhoff, K., Lie, O., Berge, R., 1998. Carnitine palmitoyltransferase I, carnitine palmitoyltransferase II, and acyl-CoA oxidase activities in Atlantic salmon (*Salmo salar*). Lipids 33, 923.

Frøyland, L., Lie, O., Berge, R.K., 2000. Mitochondrial and peroxisomal β-oxidation capacities in various tissues from Atlantic salmon *Salmo salar*. Aquacult. Nutr. 6, 85–89.

Fujisawa, K., Takami, T., Matsuzaki, A., Matsumoto, T., Yamamoto, N., Terai, S., Sakaida, I., 2017. Evaluation of the effects of L-carnitine on medaka (*Oryzias latipes*) fatty liver. Sci. Rep. 7, 2749.

Galappathi, E.K., Ichien, S.T., Hyman, A.A., Aubrac, C.J., Ford, J.D., 2020. Climate change adaptation in aquaculture. Rev. Aquacult. 12 (4), 2160–2176.

Gallagher, E.P., di Giulio, R.T., 1991. Effects of 2, 4-dichlorophenoxyacetic acid and picloram on biotransformation, peroxisomal and serum enzyme activities in channel catfish (*Ictalurus punctatus*). Toxicol. Lett. 57, 65–72.

Ganga, R., Bell, J.G., Montero, D., Robaina, L., Caballero, M.J., Izquierdo, M.S., 2005. Effect of dietary lipids on plasma fatty acid profiles and prostaglandin and leptin production in gilthead seabream (*Sparus aurata*). Compar. Biochem. Physiol. B Biochem. Mol. Biol. 142, 410–418.

Gao, J., Koshio, S., Wang, W., Huang, S., Cao, X., 2014. Effects of dietary phospholipid levels on growth performance, fatty acid composition and antioxidant responses of Dojo loach *Misgurnus anguillicaudatus* larvae. Aquaculture 426, 304–309.

Garrido, D., Kabeya, N., Betancor, M.B., Perez, J.A., Acosta, N.G., Tocher, D.R., Rodriguez, C., Monroig, O., 2019. Functional diversification of teleost Fads2 fatty acyl desaturases occurs independently of the trophic level. Sci. Rep. 9, 11199.

Gatlin, D.M., Barrows, F.T., Brown, P., Dabrowski, K., Gibson, G.T., Hardy, R.W., Elliot, H., Hu, G., Krogdahl, A., Nelson, R., Overturf, K., Rust, M., Sealey, W., Skonberg, D., Souza, E.J., Stone, D., Wilson, R., Wurtele, E., 2007. Expanding the utilization of sustainable plant products in aquafeeds: a review. Aquacult. Res. 38, 551–579.

Gemperlein, K., Dietrich, D., Kohlstedt, M., Zipf, G., Bernauer, H.S., Wittmann, C., Wenzel, S.C., Muller, R., 2019. Polyunsaturated fatty acid production by Yarrowia lipolytica employing designed myxobacterial PUFA synthases. Nat. Commun. 10, 4055.

Geng, J.J., Li, H., Liu, J.P., Yang, Y., Jin, Z.L., Zhang, Y.N., Zhang, M.L., Chen, L.Q., Du, Z.Y., 2015. Nutrients and contaminants in tissues of five fish species obtained from Shanghai markets: risk−benefit evaluation from human health perspectives. Sci. Total Environ. 536, 933−945.

Geurden, I., Radünz-Neto, J., Bergot, P., 1995. Essentiality of dietary phospholipids for carp (*Cyprinus carpio* L.) larvae. Aquaculture 131, 303−314.

Geurden, I., Marion, D., Charlon, N., Coutteau, P., Bergot, P., 1998. Comparison of different soybean phospholipidic fractions as dietary supplements for common carp, *Cyprinus carpio*, larvae. Aquaculture 161, 225−235.

Geurden, I., Cuvier, A., Gondouin, E., Olsen, R.E., Ruohonen, K., Kaushik, S., Boujard, T., 2005. Rainbow trout can discriminate between feeds with different oil sources. Physiol. Behav. 85, 107−114.

Geurden, I., Corraze, G., Boujard, T., 2007. Self-feeding behaviour of rainbow trout, *Oncorhynchus mykiss*, offered diets with distinct feed oils. Appl. Anim. Behav. Sci. 108, 313−326.

Geurden, I., Borchert, P., Balasubramanian, M.N., Schrama, J.W., Dupont-Nivet, M., Quillet, E., Kaushik, S.J., Panserat, S., Medale, F., 2013. The positive impact of the early-feeding of a plant-based diet on its future acceptance and utilisation in rainbow trout. PLoS One 8, e83162.

Gibson, G.R., Roberfroid, M.B., 1995. Dietary modulation of the human colonie microbiota: introducing the concept of prebiotics. J. Nutr. 125, 1401−1412.

Gil, A., Serra-Majem, L., Calder, P.C., Uauy, R., 2012. Systematic reviews of the role of omega-3 fatty acids in the prevention and treatment of disease. Br. J. Nutr. 107 (Suppl. 2), S1−S2.

Gillooly, J.F., Brown, J.H., West, G.B., Savage, V.M., Charnov, E.L., 2001. Effects of size and temperature on metabolic rate. Science 293, 2248−2251.

Giri, S.S., Graham, J., Hamid, N.K.A., Donald, J.A., Turchini, G.M., 2016. Dietary micronutrients and in vivo n-3 LC-PUFA biosynthesis in Atlantic salmon. Aquaculture 452, 416−425.

Gjedrem, T., 2000. Genetic improvement of cold-water fish species. Aquacult. Res. 31, 25−33.

Gjellesvik, D.R., Raae, A.J., Walther, B.T., 1989. Partial purification and characterization of a triglyceride lipase from cod (*Gadus morhua*). Aquaculture 79, 177−184.

Gjellesvik, D.R., Lombardo, D., Walther, B.T., 1992. Pancreatic bile salt dependent lipase from cod (*Gadus morhua*): purification and properties. Biochim. Biophys. Acta 1124, 123−134.

Gjøsæter, J., Kawaguchi, K., 1980. A Review of the World Resources of Mesopelagic Fish. FAO Fisheries Technical Paper, Rome, Italy.

Glatz, J.F.C., Luiken, J.J.F.P., 2017. From fat to FAT (CD36/SR-B2): understanding the regulation of cellular fatty acid uptake. Biochimie 136, 21−26.

Glencross, B.D., 2009. Exploring the nutritional demand for essential fatty acids by aquaculture species. Rev. Aquacult. 1, 71−124.

Glencross, B.D., Turchini, G.M., 2011. Fish oil replacement in starter, grow-out and finishing feeds for farmed aquatic animals. In: Turchini, G.M., Ng, W.K., Tocher, D.R. (Eds.), Fish Oil Replacement and Alternative Lipid Sources in Aquaculture Feeds. CRC Press, Taylor & Francis Group, Boca Raton, FL, USA.

Glencross, B.D., Carter, C.G., Duijster, N., Evans, D.R., Dods, K., Mccafferty, P., Hawkins, W.E., Maas, R., Sipsas, S., 2004. A comparison of the digestibility of a range of lupin and soybean protein products when fed to either Atlantic salmon (*Salmo salar*) or rainbow trout (*Oncorhynchus mykiss*). Aquaculture 237, 333−346.

Glencross, B.D., Booth, M., Allan, G.L., 2007. A feed is only as good as its ingredients - a review of ingredient evaluation strategies for aquaculture feeds. Aquacult. Nutr. 13, 17−34.

Gluckman, P.D., Hanson, M.A., Spencer, H.G., 2005. Predictive adaptive responses and human evolution. Trends Ecol. Evol. 20, 527−533.

Godfray, H.C.J., Beddington, J.R., Crute, I.R., Haddad, L., Lawrence, D., Muir, J.F., Pretty, J., Robinson, S., Thomas, S.M., Toulmin, C., 2010. Food security: the challenge of feeding 9 billion people. Science 327, 812−818.

GOED, 2014. Global Organisation for EPA and DHA (GOED), Global Recommendations for EPA and DHA Intake. Available at: www.goedomega3.com/index.php/files/download/304.

Goedeke, L., Fernández-Hernando, C., 2012. Regulation of cholesterol homeostasis. Cell. Mol. Life Sci. 69, 915−930.

Gomez-Abellan, V., Sepulcre, M.P., 2016. The role of prostaglandins in the regulation of fish immunity. Mol. Immunol. 69, 139−145.

Gong, G., Xue, M., Wang, J., Wu, X.F., Zheng, Y.H., Han, F., Liang, X.F., Su, X.O., 2015. The regulation of gluconeogenesis in the Siberian sturgeon (*Acipenser baerii*) affected later in life by a short-term high-glucose programming during early life. Aquaculture 463, 127–136.

Gong, Y., Chen, W., Han, D., Zhu, X., Yang, Y., Jin, J., Liu, H., Xie, S., 2017. Effects of food restriction on growth, body composition and gene expression related in regulation of lipid metabolism and food intake in grass carp. Aquaculture 469, 28–35.

Gong, Y., Sørensen, S.L., Dahle, D., Nadanasabesan, N., Dias, J., Valente, L.M.P., Sørensen, M., Kiron, V., 2020. Approaches to improve utilization of Nannochloropsis oceanica in plant-based feeds for Atlantic salmon. Aquaculture 522, 735122.

Gorissen, M., Flik, G., 2014. Leptin in teleostean fish, towards the origins of leptin physiology. J. Chem. Neuroanat. 61, 200–206.

Gorla-Bajszczak, A., Juge-Aubry, C., Pernin, A., Burger, A.G., Meier, C.A., 1999. Conserved amino acids in the ligand-binding and τi domains of the peroxisome proliferator-activated receptor α are necessary for heterodimerization with RXR. Mol. Cell. Endocrinol. 147, 37–47.

Gøttsche, J.R., Nielsen, N.S., Nielsen, H.H., Mu, H.L., 2005. Lipolysis of different oils using crude enzyme isolate from the intestinal tract of rainbow trout, *Oncorhynchus mykiss*. Lipids 40, 1273–1279.

Greenberg, A.S., Shen, W.-J., Muliro, K., Patel, S., Souza, S.C., Roth, R.A., Kraemer, F.B., 2001. Stimulation of lipolysis and hormone-sensitive lipase via the extracellular signal-regulated kinase pathway. J. Biol. Chem. 276, 45456–45461.

Gregory, M.K., James, M.J., 2014. Rainbow trout (*Oncorhynchus mykiss*) Elovl5 and Elovl2 differ in selectivity for elongation of omega-3 docosapentaenoic acid. Biochim. Biophys. Acta 1841, 1656–1660.

Gregory, M.K., Collins, R.O., Tocher, D.R., James, M.J., Turchini, G.M., 2016. Nutritional regulation of long-chain PUFA biosynthetic genes in rainbow trout (*Oncorhynchus mykiss*). Br. J. Nutr. 115, 1721–1729.

Gu, M., Kortner, T.M., Penn, M., Hansen, A.K., Krogdahl, Å., 2014. Effects of dietary plant meal and soya-saponin supplementation on intestinal and hepatic lipid droplet accumulation and lipoprotein and sterol metabolism in Atlantic salmon (*Salmo salar* L.). Br. J. Nutr. 111, 432–444.

Guay, V., Lamarche, B., Charest, A., Tremblay, A.J., Couture, P., 2012. Effect of short-term low- and high-fat diets on low-density lipoprotein particle size in normolipidemic subjects. Metabolism 61, 76–83.

Guerreiro, I., Peres, H., Castro-Cunha, M., Oliva-Teles, A., 2012. Effect of temperature and dietary protein/lipid ratio on growth performance and nutrient utilization of juvenile Senegalese sole (*Solea senegalensis*). Aquacult. Nutr. 18, 98–106.

Guerreiro, I., Enes, P., Oliva-Teles, A., 2015a. Effects of short-chain fructooligosaccharides (scFOS) and rearing temperature on growth performance and hepatic intermediary metabolism in gilthead sea bream (*Sparus aurata*) juveniles. Fish Physiol. Biochem. 41, 1333–1344.

Guerreiro, I., Oliva-Teles, A., Enes, P., 2015b. Improved glucose and lipid metabolism in European sea bass (*Dicentrarchus labrax*) fed short-chain fructooligosaccharides and xylooligosaccharides. Aquaculture 441, 57–63.

Guerreiro, I., Oliva-Teles, A., Enes, P., 2018. Prebiotics as functional ingredients: focus on mediterranean fish aquaculture. Rev. Aquacult. 10, 800–832.

Guillou, H., Zadravec, D., Martin, P.G.P., Jacobsson, A., 2010. The key roles of elongases and desaturases in mammalian fatty acid metabolism: insights from transgenic mice. Prog. Lipid Res. 49, 186–199.

Gut, P., Verdin, E., 2013. The nexus of chromatin regulation and intermediary metabolism. Nature 502, 489–498.

Gutiéres, S., Damon, M., Panserat, S., Kaushik, S., Médale, F., 2003. Cloning and tissue distribution of a carnitine palmitoyltransferase I gene in rainbow trout (*Oncorhynchus mykiss*). Compar. Biochem. Physiol. B Biochem. Mol. Biol. 135, 139–151.

Hamilton, H.A., Newton, R., Auchterlonie, N.A., Muller, D.B., 2020. Systems approach to quantify the global omega-3 fatty acid cycle. Nat. Food 1, 59–62.

Hamre, K., 2011. Metabolism, interactions, requirements and functions of vitamin E in fish. Aquacult. Nutr. 17, 98–115.

Hamre, K., Van, D.T.G., 1987. Mitochondrial Acyl-CoA dehydrogenase activity of fish red muscle and pig liver. Compar. Biochem. Physiol. B Biochem. Mol. Biol. 88, 19–22.

Hamre, K., Kolås, K., Sandnes, K., Julshamn, K., Kiessling, A., 2001. Feed intake and absorption of lipid oxidation products in Atlantic salmon (*Salmo salar*) fed diets coated with oxidised fish oil. Fish Physiol. Biochem. 25, 209–219.

Hamre, K., Christiansen, R., Waagbø, R., Maage, A., Torstensen, B.E., Lygren, B., Lie, Ø., Wathne, E., Albrektsen, S., 2004. Antioxidant vitamins, minerals and lipid levels in diets for Atlantic salmon (*Salmo salar*, L.): effects on growth performance and fillet quality. Aquacult. Nutr. 10, 113–123.

Hamza, N., Mhetli, M., Khemis, I.B., Cahu, C., Kestemont, P., 2008. Effect of dietary phospholipids levels on performance, enzyme activities and fatty acid composition of pikeperch (*Sander lucioperca*) larvae. Aquaculture 275, 274–282.

Han, C.Y., Zheng, Q.M., Feng, L.N., 2013. Effects of total replacement of dietary fish oil on growth performance and fatty acid compositions of hybrid tilapia (*Oreochromis niloticus* × *O. aureus*). Aquacult. Int. 21, 1209–1217.

Han, D., Shan, X.J., Zhang, W.B., Chen, Y.S., Wang, Q.Y., Li, Z.J., Zhang, G.F., Xu, P., Li, J.L., Xie, S.Q., Mai, K.S., Tang, Q.S., de Silva, S.S., 2018. A revisit to fishmeal usage and associated consequences in Chinese aquaculture. Rev. Aquacult. 10, 493–507.

Han, S.L., Wang, J., Zhang, Y.X., Qiao, F., Chen, L.Q., Zhang, M.L., Du, Z.Y., 2019. Inhibited autophagy impairs systemic nutrient metabolism in *Nile tilapia*. Compar. Biochem. Physiol. Mol. Integr. Physiol. 236, 110521.

Han, S.L., Wang, J., Li, L.Y., Lu, D.L., Chen, L.Q., Zhang, M.L., Du, Z.Y., 2020. The regulation of rapamycin on nutrient metabolism in Nile tilapia fed with high-energy diet. Aquaculture 520, 734975.

Hannun, Y.A., Luberto, C., Argraves, K.M., 2001. Enzymes of sphingolipid metabolism: from modular to integrative signaling. Biochemistry 40, 4893–4903.

Hansen, J.O., Berge, G.M., Hillestad, M., Krogdahl, Å., Galloway, T.F., Holm, H., Holm, J., Ruyter, B., 2001. Apparent digestion and apparent retention of lipid and fatty acids in Atlantic cod (*Gadus morhua*) fed increasing dietary lipid levels. Aquaculture 284, 159–166.

Hansen, Ø.,J., Puvanendran, V., Jostensen, J.P., Ous, C., 2011. Effects of dietary levels and ratio of phosphatidylcholine and phosphatidylinositol on the growth, survival and deformity levels of Atlantic cod larvae and early juveniles. Aquacult. Res. 42, 1026–1033.

Hansen, A.K.G., Kortner, T.M., Krasnov, A., Björkhem, I., Penn, M., Krogdahl, Å., 2020. Choline supplementation prevents diet induced gut mucosa lipid accumulation in post-smolt Atlantic salmon (*Salmo salar* L.). BMC Vet. Res. 16, 32.

Hao, Y.T., Wu, S.G., Xiong, F., Tran, N.T., Jakovlic, I., Zou, H., Li, W.X., Wang, G.T., 2017. Succession and fermentation products of grass carp (*Ctenopharyngodon idellus*) hindgut microbiota in response to an extreme dietary shift. Front. Microbiol. 8, 1585.

Harmon, J.S., Sheridan, M.A., 1992. Effects of nutritional state, insulin, and glucagon on lipid mobilization in rainbow trout, *Oncorhynchus mykiss*. Gen. Compar. Endocrinol. 87, 214–221.

Harmon, J.S., Rieniets, L.M., Sheridan, M.A., 1993. Glucagon and insulin regulate lipolysis in trout liver by altering phosphorylation of triacylglycerol lipase. Am. J. Physiol. Regul. Integr. Compar. Physiol. 265, R255–R260.

Harwood, J.L., 1996. Recent advances in the biochemistry of plant fatty acids. Biochim. Biophys. Acta 1301, 7–56.

Harwood, J.L., 2019. Algae: critical sources of very long-chain polyunsaturated fatty acids. Biomolecules 9, 708.

Harwood, J.L., Gushina, I.A., 2009. The versatility of algae and their lipid metabolism. Biochimie 91, 679–684.

Haslam, R.P., Usher, S., Sayanova, O., Napier, J.A., Betancor, M.B., Tocher, D.R., 2015. The supply of fish oil to aquaculture: a role for transgenic oilseed crops? World Agric. 5.

Hassam, A.G., Rivers, J.P., Crawford, M.A., 1977. The failure of the cat to desaturate linoleic acid; its nutritional implications. Nutr. Metab. 21, 321–328.

Hastings, N., Agaba, M., Tocher, D.R., Leaver, M.J., Dick, J.R., Sargent, J.R., Teale, A.J., 2001. A vertebrate fatty acid desaturase with Delta 5 and Delta 6 activities. Proc. Natl. Acad. Sci. U. S. A. 98, 14304–14309.

Hastings, N., Agaba, M.K., Tocher, D.R., Zheng, X., Dickson, C.A., Dick, J.R., Teale, A.J., 2004. Molecular cloning and functional characterization of fatty acyl desaturase and elongase cDNAs involved in the production of eicosapentaenoic and docosahexaenoic acids from alpha-linolenic acid in Atlantic salmon (*Salmo salar*). Mar. Biotechnol. 6, 463–474.

Hasumura, T., Shimada, Y., Kuroyanagi, J., Nishimura, Y., Meguro, S., Takema, Y., Tanaka, T., 2012. Green tea extract suppresses adiposity and affects the expression of lipid metabolism genes in diet-induced obese zebrafish. Nutr. Metab. 9, 73.

Hatcher, H.C., Singh, R.N., Torti, F.M., Torti, S.V., 2009. Synthetic and natural iron chelators: therapeutic potential and clinical use. Future Med. Chem. 1, 1643–1670.

Haynes, R.R., 1988. Reproductive-biology of selected aquatic plants. Ann. Mo. Bot. Gard. 75, 805–810.

Hazel, J.R., 1979. Influence of thermal acclimation on membrane lipid composition of rainbow trout liver. Am. J. Physiol. 236, R91–R101.

Hazel, J.R., 1984. Effects of temperature on the structure and metabolism of cell membranes in fish. Am. J. Physiol. Regul. Integr. Compar. Physiol. 246, R460–R470.

Hazel, J.R., 1990. Adaptation to temperature: phospholipid synthesis in hepatocytes of rainbow trout. Am. J. Physiol. Regul. Integr. Compar. Physiol. 258, R1495–R1501.

Hazel, J.R., Williams, E.E., 1990. The role of alterations in membrane lipid composition in enabling physiological adaptation of organisms to their physical environment. Prog. Lipid Res. 29, 167–227.

Hazra, A.K., Ghosh, S., Banerjee, S., Mukherjee, B., 1998. Studies on lipid and fatty acid compositions of puffer livers from Indian coastal waters with seasonal variation. J. Am. Oil Chem. Soc. 75, 1673–1678.

He, A.Y., Liu, C.Z., Chen, L.Q., Ning, L.J., Qin, J.G., Li, J.M., Zhang, M.L., Du, Z.Y., 2015a. Molecular characterization, transcriptional activity and nutritional regulation of peroxisome proliferator activated receptor gamma in Nile tilapia (*Oreochromis niloticus*). Gen. Compar. Endocrinol. 223, 139–147.

He, A.Y., Ning, L.J., Chen, L.Q., Chen, Y.L., Xing, Q., Li, J.M., Qiao, F., Li, D.L., Zhang, M.L., Du, Z.Y., 2015b. Systemic adaptation of lipid metabolism in response to low-and high-fat diet in Nile tilapia (*Oreochromis niloticus*). Physiol. Rep. 3, e12485.

Helgeland, H., Sandve, S.R., Torgersen, J.S., Halle, M.K., Sundvold, H., Omholt, S., Våge, D.I., 2014. The evolution and functional divergence of the beta-carotene oxygenase gene family in teleost fish—exemplified by Atlantic salmon. Gene 543, 268–274.

Hellerstein, M.K., Schwarz, J.M., Neese, R.A., 1996. Regulation of hepatic de novo lipogenesis in humans. Annu. Rev. Nutr. 16, 523–557.

Hemre, G.I., Storebakken, T., 2000. Tissue and organ distribution of 14C-activity in dextrin-adapted Atlantic salmon after oral administration of radiolabelled 14C1-glucose. Aquacult. Nutr. 6, 229–234.

Hemre, G.I., Mommsen, T.P., Krogdahl, A., 2002. Carbohydrates in fish nutrition: effects on growth, glucose metabolism and hepatic enzymes. Aquacult. Nutr. 8, 175–194.

Henderson, R.J., 1996. Fatty acid metabolism in freshwater fish with particular reference to polyunsaturated fatty acids. Arch. Anim. Nutr. 49, 5.

Henderson, R.J., Sargent, J.R., 1985. Chain-length specificities of mitochondrial and peroxisimal β-oxidation of fatty acids in livers of rainbow trout (*Salmo gairdneri*). Compar. Biochem. Physiol. B Biochem. Mol. Biol. 82, 79–85.

Henriques, J., Dick, J.R., Tocher, D.R., Bell, J.G., 2014. Nutritional quality of salmon products available from major retailers in the UK: content and composition of n-3 long-chain PUFA. Br. J. Nutr. 112, 964–975.

Henry, M., Gasco, L., Piccolo, G., Fountoulaki, E., 2015. Review on the use of insects in the diet of farmed fish: past and future. Anim. Feed Sci. Technol. 203, 1–22.

Her, G.M., Pai, W.Y., Lai, C.Y., Hsieh, Y.W., Pang, H.W., 2013. Ubiquitous transcription factor YY1 promotes zebrafish liver steatosis and lipotoxicity by inhibiting CHOP-10 expression. Biochim. Biophys. Acta 1831, 1037–1051.

Hernández-Sánchez, C., Mansilla, A., De La Rosa, E.J., De Pablo, F., 2006. Proinsulin in development: new roles for an ancient prohormone. Diabetologia 49, 1142.

Hill, J.A., Kiessling, A., Devlin, R.H., 2000. Coho salmon (*Oncorhynchus kisutch*) transgenic for a growth hormone gene construct exhibit increased rates of muscle hyperplasia and detectable levels of differential gene expression. Can. J. Fish. Aquat. Sci. 57, 939–950.

Hilton, Z., Poortenaar, C.W., Sewell, M.A., 2008. Lipid and protein utilization during early development of yellowtail kingfish (*Seriola lalandi*). Mar. Biol. 154, 855–865.

Hinsdale, M.E., Sullivan, P.M., Mezdour, H., Maeda, N., 2002. ApoB-48 and apoB-100 differentially influence the expression of type-III hyperlipoproteinemia in APOE*2 mice. J. Lipid Res. 43, 1520–1528.

Hiramitsu, M., Shimada, Y., Kuroyanagi, J., Inoue, T., Katagiri, T., Zang, L., Nishimura, Y., Nishimura, N., Tanaka, T., 2014. Eriocitrin ameliorates diet-induced hepatic steatosis with activation of mitochondrial biogenesis. Sci. Rep. 4, 3708.

Hirokane, H., Nakahara, M., Tachibana, S., Shimizu, M., Sato, R., 2004. Bile acid reduces the secretion of very low density lipoprotein by repressing microsomal triglyceride transfer protein gene expression mediated by hepatocyte nuclear factor-4. J. Biol. Chem. 279, 45685–45692.

Hixson, S.M., Parrish, C.C., Anderson, D.M., 2014. Full substitution of fish oil with camelina (*Camelina sativa*) oil, with partial substitution of fish meal with camelina meal, in diets for farmed Atlantic salmon (*Salmo salar*) and its effect on tissue lipids and sensory quality. Food Chem. 157, 51–61.

Hixson, S.M., Sharma, B., Kainz, M.J., Wacker, A., Arts, M.T., 2015. Production, distribution, and abundance of long-chain omega-3 polyunsaturated fatty acids: a fundamental dichotomy between freshwater and terrestrial ecosystems. Environ. Rev. 23, 414–424.

Ho, S.-Y., Lorent, K., Pack, M., Farber, S.A., 2006. Zebrafish fat-free is required for intestinal lipid absorption and Golgi apparatus structure. Cell Metabol. 3, 289–300.

Hochachka, P.W., Mommsen, T.P., 1995. Biochemistry and Molecular Biology of Fishes. In: Environmental and Ecological Biochemistry, vol. 5. Elsevier Press, Amsterdam.

Hodges, P., Scott, J., 1992. Apolipoprotein B mRNA editing: a new tier for the control of gene expression. Trends Biochem. Sci. 17, 77–81.

Hodsdon, M.E., Ponder, J.W., Cistola, D.P., 1996. The NMR solution structure of intestinal fatty acid-binding protein complexed with palmitate: application of a novel distance geometry algorithm. J. Mol. Biol. 264, 585–602.

Holdway, D.A., Beamish, F.W.H., 1984. Specific growth rate and proximate body composition of Atlantic cod (*Gadus morhua* L.). J. Exp. Mar. Biol. Ecol. 81, 147–170.

Holme, M.H., Zeng, C., Southgate, P.C., 2006. The effects of supplemental dietary cholesterol on growth, development and survival of mud crab, *Scylla serrata*, megalopa fed semi-purified diets. Aquaculture 269, 427–435.

Holmes, R.S., 2010. Comparative genomics and proteomics of vertebrate diacylglycerol acyltransferase (DGAT), acyl CoA wax alcohol acyltransferase (AWAT) and monoacylglycerol acyltransferase (MGAT). Compar. Biochem. Physiol. D Genom. Proteonom. 5, 45–54.

Holub, B.J., Nilsson, K., Piekarski, J., Slinger, S.J., 1975. Biosynthesis of lecithin by the CDP-choline pathway in liver microsomes of rainbow trout, Salmo gairdneri. J. Fish. Res. Board Can. 32, 1633–1637.

Hong, S., Tjonahen, E., Morgan, E.L., Lu, Y., Serhan, C.N., Rowley, A.F., 2005. Rainbow trout (*Oncorhynchus mykiss*) brain cells biosynthesize novel docosahexaenoic acid-derived resolvins and protectins-Mediator lipidomic analysis. Prostag. Other Lipid Mediat. 78, 107–116.

Hong, S., Lu, Y., Yang, R., Gotlinger, K.H., Petasis, N.A., Serhan, C.N., 2007. Resolvin D1, protectin D1, and related docosahexaenoic acid-derived products: analysis via electrospray/low energy tandem mass spectrometry based on spectra and fragmentation mechanisms. J. Am. Soc. Mass Spectrom. 18, 128–144.

Hong, H., Zhou, Y., Wu, H., Luo, Y., Shen, H., 2014. Lipid content and fatty acid profile of muscle, brain and eyes of seven freshwater fish: a comparative study. J. Am. Oil Chem. Soc. 91, 795–804.

Horn, S.S., Ruyter, B., Meuwissen, T.H.E., Moghadam, H., Hillestad, B., Sonesson, A.K., 2020a. Accuracy of selection for omega-3 fatty acid content in Atlantic salmon fillets. Aquaculture 519, 734764 (in press).

Horn, S.S., Ruyter, B., Meuwissen, T.H.E., Moghadam, H., Hillestad, B., Sonesson, A.K., 2020b. GWAS identifies genetic variants associated with omega-3 fatty acid composition of Atlantic salmon fillets. Aquaculture 514, 734494 (in press).

Horn, S.S., Ruyter, B., Meuwissen, T.H.E., Hillestad, B., Sonesson, A.K., 2018. Genetic effects of fatty acid composition in muscle of Atlantic salmon. Genet. Sel. Evol. 50, 23.

Horn, S.S., Sonesson, A.K., Krasnov, A., Moghadam, H., Hillestad, B., Meuwissen, T.H.E., Ruyter, B., 2019. Individual differences in EPA and DHA content of Atlantic salmon are associated with gene expression of key metabolic processes. Sci. Rep. 9, 3889.

Horton, J.D., Goldstein, J.L., Brown, M.S., 2002. SREBPs: activators of the complete program of cholesterol and fatty acid synthesis in the liver. J. Clin. Invest. 109, 1125–1131.

Hoseinifar, S.H., Sun, Y.Z., Caipang, C.M., 2017. Short-chain fatty acids as feed supplements for sustainable aquaculture: an updated view. Aquacult. Res. 48, 1380–1391.

Houlihan, D., Boujard, T., Jobling, M., 2001. Food Intake in Fish. Blackwell Science Ltd, Bodmin, Cornwall, UK.

Howarth, D.L., Hagey, L.R., Law, S.H.W., Ai, N., Krasowski, M.D., Ekins, S., Moore, J.T., Kollitz, E.M., Hinton, D.E., Kullman, S.W., 2010. Two farnesoid X receptor alpha isoforms in Japanese medaka (*Oryzias latipes*) are differentially activated in vitro. Aquat. Toxicol. 98, 245–255.

Hsieh, S.L., Liao, W.L., Kuo, C.M., 2001. Molecular cloning and sequence analysis of stearoyl-CoA desaturase in milkfish, *Chanos chanos*. Compar. Biochem. Physiol. B Biochem. Mol. Biol. 130, 467–477.

Hsieh, S.L., Liu, R.W., Wu, C.H., Cheng, W.T., Kuo, C.-M., 2003. cDNA nucleotide sequence coding for stearoyl-CoA desaturase and its expression in the zebrafish (*Danio rerio*) embryo. Mol. Reprod. Dev. 66, 325–333.

Hsieh, S.L., Chang, H.T., Wu, C.H., Kuo, C.M., 2004. Cloning, tissue distribution and hormonal regulation of stearoyl-CoA desaturase in tilapia, *Oreochromis mossambicus*. Aquaculture 230, 527–546.

Hsu, C.C., Lai, C.Y., Lin, C.Y., Yeh, K.Y., Her, G.M., 2018. MicroRNA-27b depletion enhances endotrophic and intravascular lipid accumulation and induces adipocyte hyperplasia in zebrafish. Int. J. Mol. Sci. 19, 93.

Hua, K., Bureau, D.P., 2009. Development of a model to estimate digestible lipid content of salmonid fish feeds. Aquaculture 286, 271–276.

Hua, K., Cobcroft, J.M., Cole, A., Condon, K., Jerry, D.R., Mangott, A., Praeger, C., Vucko, M.J., Zeng, C., Zenger, K., Strugnell, J.M., 2019. The future of aquatic protein: implications for protein sources in aquaculture diets. One Earth 1, 316–329.

Huang, Y., Cai, S., 2019. Molecular characterisation and expression analysis of scavenger receptor class B member 1 (SR-B1) gene in Nile tilapia *Oreochromis niloticus* (Linnaeus, 1758). Indian J. Fish. 66.

Huang, H.Y., Xi, L., Mei, L., Song, T.J., He, Q., Ma, C.G., Tang, Q.Q., 2008. Transcription factor YY1 promotes adipogenesis via inhibiting CHOP-10 expression. Biochem. Biophys. Res. Commun. 375, 496–500.

Huang, C., Chen, Q.L., Luo, Z., Shi, X., Pan, Y.X., Song, Y.F., Zhuo, M.Q., Wu, K., 2014. Time-dependent effects of waterborne copper exposure influencing hepatic lipid deposition and metabolism in javelin goby *Synechogobius hasta* and their mechanism. Aquat. Toxicol. 155, 291–300.

Hue, J.-J., Lee, K.N., Jeong, J.-H., Lee, S.-H., Lee, Y.H., Jeong, S.-W., Nam, S.Y., Yun, Y.W., Lee, B.J., 2009. Anti-obesity activity of diglyceride containing conjugated linoleic acid in C57BL/6J ob/ob mice. J. Vet. Sci. 10, 189–195.

Huguet, C.T., Norambuena, F., Emery, J.A., Hermon, K., Turchini, G.M., 2015. Dietary n-6/n-3 LC-PUFA ratio, temperature and time interactions on nutrients and fatty acids digestibility in Atlantic salmon. Aquaculture 436, 160–166.

Iassonova, D., Arriagada, P., Capdeville, A., 2018. An alternative source of long chain omega-3 fatty acids from novel canola oil for salmonid farming. Feeding trial report. In: 18th International Symposium on Fish Nutrition & Feeding (ISFNF), 3–7 June 2018 Las Palmas de Gran Canaria, Spain.

Ibabe, A., Grabenbauer, M., Baumgart, E., Fahimi, D.H., Cajaraville, M.P., 2002. Expression of peroxisome proliferator-activated receptors in zebrafish (*Danio rerio*). Histochem. Cell Biol. 118, 231–239.

Ibabe, A., Grabenbauer, M., Baumgart, E., V?lkl, A., Fahimi, H.D., Cajaraville, M.P., 2004. Expression of peroxisome proliferator-activated receptors in the liver of gray mullet (*Mugil cephalus*). Acta Histochem. 106, 11–19.

Ibabe, A., Herrero, A., Cajaraville, M.P., 2005. Modulation of peroxisome proliferator-activated receptors (PPARs) by PPARα- and PPARγ-specific ligands and by 17β-estradiol in isolated zebrafish hepatocytes. Toxicol. Vitro 19, 725–735.

Ikonen, E., 2008. Cellular cholesterol trafficking and compartmentalization. Nat. Rev. Mol. Cell Biol. 9, 125–138.

Imrie, D., Sadler, K.C., 2010. White adipose tissue development in zebrafish is regulated by both developmental time and fish size. Dev. Dynam. 239, 3013–3023.

Inaba, M., Murota, K., Nikawadori, M., Kishino, E., Matusda, R., Takagi, M., Ohkubo, T., Tanaka, T., Terao, J., Tokumura, A., 2014. Extracellular metabolism-dependent uptake of lysolipids through cultured monolayer of differentiated Caco-2 cells. Biochim. Biophys. Acta 1841, 121–131.

Ince, B.W., Thorpe, A., 1975. Hormonal and metabolite effects on plasma free fatty acids in the northern pike, *Esox lucius* L. Gen. Compar. Endocrinol. 27, 144–152.

Innis, S.M., 2007. Dietary (n-3) fatty acids and brain development. J. Nutr. 137, 855–859.

Izquierdo, M.S., Socorro, J., Arantzamendi, L., Hernandez-Cruz, C.M., 2000. Recent advances in lipid nutrition in fish larvae. Fish Physiol. Biochem. 22, 97–107.

Izquierdo, M.S., Turkmen, S., Montero, D., Zamorano, M.J., Alfonso, J.M., Karalazos, V., Fernández-Palacios, H., 2015. Nutritional programming through broodstock diets to improve utilization of very low fishmeal and fish oil diets in gilthead sea bream. Aquaculture 449, 18–26.

Izquierdo-Gómez, D., González-Silvera, D., Arechavala-López, P., López-Jiménez, J., Bayle-Sempere, J., Sánchez-Jerez, P., 2015. Exportation of excess feed from mediterranean fish farms to local fisheries through different targeted fish species. ICES J. Mar. Sci. 72, 930–938.

Jakobsson, A., Westerberg, R., Jacobsson, A., 2006. Fatty acid elongases in mammals: their regulation and roles in metabolism. Prog. Lipid Res. 45, 237–249.

Jaromowska, M., Górski, J., 1985. Effect of fasting on skeletal muscle triglyceride content. Experientia 41, 357–358.

Jenkins, C.M., Mancuso, D.J., Yan, W., Sims, H.F., Gibson, B., Gross, R.W., 2004. Identification, cloning, expression, and purification of three novel human calcium-independent phospholipase A2 family members possessing triacylglycerol lipase and acylglycerol transacylase activities. J. Biol. Chem. 279, 48968–48975.

Jeukendrup, A.E., 2010. Regulation of fat metabolism in skeletal muscle. Ann. N. Y. Acad. Sci. 967, 217–235.

JEŽek, P., 2002. Possible physiological roles of mitochondrial uncoupling proteins—UCPn. Int. J. Biochem. Cell Biol. 34, 1190–1206.

Ji, H., Li, J., Liu, P., 2011. Regulation of growth performance and lipid metabolism by dietary n-3 highly unsaturated fatty acids in juvenile grass carp, *Ctenopharyngodon idellus*. Compar. Biochem. Physiol. B Biochem. Mol. Biol. 159, 49–56.

Jiang, Z.Y., Xiong, H.L., Yao, J.J., 2014a. Effects of glucose and vitamin C on activities of acc, Fas and CPT1 during Yolk-sac larva development of of *Carassius auratus* Gibelio. Chin. J. Zool. 49, 904–912.

Jiang, Z.Y., Yao, J.J., An, M., Xiong, H.L., Zhu, Z.S., 2014b. Effects of glucose and vitamin C inhabitation on activities of acetyl-CoA carboxylase, fatty acid synthase and carnitine palmitoyltransferases I during embryo development of *Carassius auratus* gibelio. Chin. J. Animal Nutr. 26, 3510–3516.

Jiao, J.-G., Liu, Y., Zhang, H., Li, L.-Y., Qiao, F., Chen, L.-Q., Zhang, M.-L., Du, Z.-Y., 2020. Metabolism of linoleic and linolenic acids in hepatocytes of two freshwater fish with different n-3 or n-6 fatty acid requirements. Aquaculture 515, 734595.

Jin, J., Panserat, S., Kamalam, B.S., Aguirre, P., Véron, V., Médale, F., 2014. Insulin regulates lipid and glucose metabolism similarly in two lines of rainbow trout divergently selected for muscle fat content. Gen. Compar. Endocrinol. 204, 49–59.

Jin, A., Shi, X.C., Liu, Y., Sun, J., Ji, H., 2018a. Docosahexaenoic acid induces PPARγ-dependent preadipocytes apoptosis in grass carp Ctenopharyngodon idella. Gen. Compar. Endocrinol. 266, 211–219.

Jin, A., Lei, C.X., Tian, J.J., Sun, J., Ji, H., 2018b. Dietary docosahexaenoic acid decreased lipid accumulation via inducing adipocytes apoptosis of grass carp, Ctenopharygodon idella. Fish Physiol. Biochem. 44, 197–207.

Jin, Y., Olsen, R.E., Gillard, G.B., Østensen, M.-A., Korsvoll, S.A., Santi, N., Vik, J.O., Sandve, S.R., Olsen, Y., 2018c. A systemic study of lipid metabolism regulation in salmon fingerlings and early juveniles fed plant oil. Br. J. Nutr. 120, 653–664.

Jin, Y., Olsen, R.E., Østensen, M.-A., Gillard, G.B., Korsvoll, S.A., Santi, N., Gjuvsland, A.B., Vik, J.O., Torgersen, J.S., Sandve, S.R., Olsen, Y., 2018d. Transcriptional development of phospholipid and lipoprotein metabolism in different intestinal regions of Atlantic salmon (*Salmo salar*) fry. BMC Genom. 19, 253.

Jin, Y., Angell, I., RØD Sandve, S., Snipen, L.G., Olsen, Y., Rudi, K., 2019. Atlantic salmon raised with diets low in long-chain polyunsaturated n-3 fatty acids in freshwater have a Mycoplasma-dominated gut microbiota at sea. Aquacult. Environ. Interact. 11, 31–39.

Jobling, M., Bendiksen, E.A., 2003. Dietary lipids and temperature interact to influence tissue fatty acid compositions of Atlantic salmon, *Salmo salar* L., parr. Aquacult. Res. 3, 1423–1441.

Johansson, L., Kiessling, A., Kiessling, K.H., Berglund, L., 2000. Effects of altered ration levels on sensory characteristics, lipid content and fatty acid composition of rainbow trout (*Oncorhynchus mykiss*). Food Qual. Prefer. 11, 247–254.

Johnson, D.R., Decker, E.A., 2015. The role of oxygen in lipid oxidation reactions: a review. Ann. Rev. Food Sci. Technol. 6, 171–190.

Jones, B.B., Ohno, C.K., Allenby, G., Boffa, M.B., Levin, A.A., Grippo, J.F., Petkovich, M., 1995. New retinoid X receptor subtypes in zebra fish (*Danio rerio*) differentially modulate transcription and do not bind 9-cis retinoic acid. Mol. Cell Biol. 15, 5226–5234.

Jones, S.W., Karpol, A., Friedman, S., Maru, B.T., Tracy, B.P., 2020. Recent advances in single cell protein use as a feed ingredient in aquaculture. Curr. Opin. Biotechnol. 61, 189–197.

Jönsson, E., 2013. The role of ghrelin in energy balance regulation in fish. Gen. Compar. Endocrinol. 187, 79–85.

Jönsson, E., Forsman, A., Einarsdottir, I.E., Kaiya, H., Ruohonen, K., Björnsson, B.T., 2007. Plasma ghrelin levels in rainbow trout in response to fasting, feeding and food composition, and effects of ghrelin on voluntary food intake. Compar. Biochem. Physiol. Mol. Integr. Physiol. 147, 1116–1124.

Jönsson, E., Kaiya, H., Björnsson, B.T., 2010. Ghrelin decreases food intake in juvenile rainbow trout (*Oncorhynchus mykiss*) through the central anorexigenic corticotropin-releasing factor system. Gen. Compar. Endocrinol. 166, 39–46.

Jordal, A.E.O., Lie, Ø., Torstensen, B.E., 2007. Complete replacement of dietary fish oil with a vegetable oil blend affect liver lipid and plasma lipoprotein levels in Atlantic salmon (*Salmo salar* L.). Aquacult. Nutr. 13, 114–130.

Jøsrgensen, E., Johansen, S.J., Jobling, M., 2005. Seasonal patterns of lipid deposition and lipid depletion in anadromous Arctic charr. J. Fish. Biol. 51, 312–326.

Kabeya, N., Yamamoto, Y., Cummins, S.F., Elizur, A., Yazawa, R., Takeuchi, Y., Haga, Y., Satoh, S., Yoshizaki, G., 2015. Polyunsaturated fatty acid metabolism in a marine teleost, Nibe croaker *Nibea mitsukurii*: functional characterization of Fads2 desaturase and Elovl5 and Elovl4 elongases. Compar. Biochem. Physiol. B Biochem. Mol. Biol. 188B, 37–45.

Kaiya, H., Kojima, M., Hosoda, H., Moriyama, S., Takahashi, A., Kawauchi, H., Kangawa, K., 2003a. Peptide purification, complementary deoxyribonucleic acid (DNA) and genomic DNA cloning, and functional characterization of ghrelin in rainbow trout. Endocrinology 144, 5215–5226.

Kaiya, H., Kojima, M., Hosoda, H., Riley, L.G., Hirano, T., Grau, E.G., Kangawa, K., 2003b. Amidated fish ghrelin: purification, cDNA cloning in the Japanese eel and its biological activity. J. Endocrinol. 176, 415–424.

Kaiya, H., Miyazato, M., Kangawa, K., Peter, R.E., Unniappan, S., 2008. Ghrelin: a multifunctional hormone in non-mammalian vertebrates. Compar. Biochem. Physiol. Mol. Integr. Physiol. 149, 109–128.

Kaiya, H., Mori, T., Miyazato, M., Kangawa, K., 2009. Ghrelin receptor (GHS-R)-like receptor and its genomic organisation in rainbow trout, *Oncorhynchus mykiss*. Compar. Biochem. Physiol. Mol. Integr. Physiol. 153, 438–450.

Kamalam, B.S., Medale, F., Kaushik, S., Polakof, S., Skiba-Cassy, S., Panserat, S., 2012. Regulation of metabolism by dietary carbohydrates in two lines of rainbow trout divergently selected for muscle fat content. J. Exp. Biol. 215, 2567–2578.

Kamalam, B.S., Panserat, S., Aguirre, P., Geurden, I., Fontagné-Dicharry, S., Médale, F., 2013. Selection for high muscle fat in rainbow trout induces potentially higher chylomicron synthesis and PUFA biosynthesis in the intestine. Compar. Biochem. Physiol. A Mol. Integr. Physiol. 164, 417–427.

Kambe, T., Tsuji, T., Hashimoto, A., Itsumura, N., 2015. The physiological, biochemical, and molecular roles of zinc transporters in zinc homeostasis and metabolism. Physiol. Rev. 95, 749–784.

Kanazawa, A., 1993. Essential phospholipids of fish and crustaceans. In: Kaushik, S.J., Luquet, P. (Eds.), Fish Nutrition in Practice. IV International Symposium on Fish Nutrition and Feeding. National Institute for Agricultural Research, Inra, Francs, pp. 519–530.

Kanazawa, A., Teshima, S., Inamori, S., Iwhashita, T., Nagao, A., 1981. Effect of phospholipids on survival rate and incidence of malformation in the larval ayu. Memory Facult. Fish. Kagoshima Univ. 30, 301–309.

Kang, K.S., Yahashi, S., Matsuda, K., 2011. Central and peripheral effects of ghrelin on energy balance, food intake and lipid metabolism in teleost fish. Peptides 32, 2242–2247.

Karanth, S., Lall, S.P., Denovan-Wright, E.M., Wright, J.M., 2009. Differential transcriptional modulation of duplicated fatty acid-binding protein genes by dietary fatty acids in zebrafish (*Danio rerio*): evidence for subfunctionalization or neofunctionalization of duplicated genes. BMC Evol. Biol. 9, 219.

Karapanagiotidis, I.T., Psofakis, P., Mente, E., Malandrakis, E., Golomazou, E., 2019. Effect of fishmeal replacement by poultry by-product meal on growth performance, proximate composition, digestive enzyme activity, haematological parameters and gene expression of gilthead seabream (*Sparus aurata*). Aquacult. Nutr. 25, 3–14.

Karbiener, M., Fischer, C., Nowitsch, S., Opriessnig, P., Papak, C., Ailhaud, G., Dani, C., Amri, E.Z., Scheideler, M., 2009. microRNA miR-27b impairs human adipocyte differentiation and targets PPARγ. Biochem. Biophys. Res. Commun. 390, 247–251.

Karlsen, Ø., Amlund, H., Ringø, E., Berg, A., Olsen, R.E., 2010. The effect of dietary chitin on growth and nutrient digestibility in farmed Atlantic cod (*Gadus morhua* L.) and Atlantic salmon (*Salmo salar* L.). Aquacult. Nutr. (submitted).

Kasumyan, A.O., Doving, K.B., 2003. Taste preferences in fishes. Fish Fish. 4, 289–347.

Katikou, P., Hughes, S.I., Robb, D.H.F., 2001. Lipid distribution within Atlantic salmon (*Salmo salar*) fillets. Aquaculture 202, 89–99.

Kaushik, S.J., Cravedi, J.P., Lalles, J.P., Sumpter, J., Fauconneau, B., Laroche, M., 1995. Partial or total replacement of fish meal by soybean protein on growth, protein utilization, potential estrogenic or antigenic effects, cholesterolemia and flesh quality in rainbow trout, *Oncorhynchus mykiss*. Aquaculture 133, 257–274.

Kaushik, S.J., Coves, D., Dutto, G., Blanc, D., 2004. Almost total replacement of fish meal by plant protein sources in the diet of a marine teleost, the European seabass, *Dicentrarchus labrax*. Aquaculture 230, 391–404.

Kayama, M., Mankura, M., Ikeda, Y., 1979. Hydrolysis and synthesis of wax esters by different systems of carp hepatopancreas preparation. J. Biochem. 85, 1–6.

Kennedy, S.R., Leaver, M.J., Campbell, P.J., Zheng, X., Dick, J.R., Tocher, D.R., 2006. Influence of dietary oil content and conjugated linoleic acid (CLA) on lipid metabolism enzyme activities and gene expression in tissues of Atlantic salmon (*Salmo salar* L.). Lipids 41, 423–436.

Kennedy, S.R., Bickerdike, R., Berge, R.K., Dick, J.R., Tocher, D.R., 2007. Influence of conjugated linoleic acid (CLA) or tetradecylthioacetic acid (TTA) on growth, lipid composition, fatty acid metabolism and lipid gene expression of rainbow trout (*Oncorhynchus mykiss* L.). Aquaculture 272, 489–501.

Kent, C., 2005. Regulatory enzymes of phosphatidylcholine biosynthesis: a personal perspective. Biochim. Biophys. Acta 1733, 53–66.

Kersten, S., Seydoux, J., Peters, J.M., Gonzalez, F.J., Desvergne, B., Wahli, W., 1999. Peroxisome proliferator-activated receptor alpha mediates the adaptive response to fasting. J. Clin. Invest. 103, 1489–1498.

Khieokhajonkhet, A., Kaneko, G., Ohara, K., Shirakami, H., Ushio, H., 2014. Hormone-sensitive lipase in Japanese flounder *Paralichthys olivaceus*: the potential function of the inclinator muscle of fin as a lipid storage site. Fish. Sci. 80, 341–351.

Khieokhajonkhet, A., Kaneko, G., Hirano, Y., Wang, L., Ushio, H., 2016. Different effects of growth hormone and fasting on the induction patterns of two hormone-sensitive lipase genes in red seabream *Pagrus major*. Gen. Compar. Endocrinol. 236, 121–130.

Kiessling, K.H., Kiessling, A., 1993. Selective utilization of fatty acids in rainbow trout (*Oncorhynchus mykiss* Walbaum) red muscle mitochondria. Can. J. Zool. 71, 248–251.

Kikuchi, K., Furuta, T., Iwata, N., Onuki, K., Noguchi, T., 2009. Effect of dietary lipid levels on the growth, feed utilization, body composition and blood characteristics of tiger puffer *Takifugu rubripes*. Aquaculture 298, 111–117.

Kim, M.-S., Park, J.-Y., Namkoong, C., Jang, P.-G., Ryu, J.-W., Song, H.-S., Yun, J.-Y., Namgoong, I.S., Ha, J., Park, I.-S., Lee, I.-K., Viollet, B., Youn, J., Lee, H.K., Lee, K.-U., 2004. Anti-obesity effects of alpha-lipoic acid mediated by suppression of hypothalamic AMP-activated protein kinase. Nat. Med. 10, 727–733.

Kim, W.S., Lee, Y.S., Cha, S.H., Jeong, H.W., Choe, S.S., Lee, M.-R., Oh, G.T., Park, H.-S., Lee, K.-U., Lane, M.D., 2009. Berberine improves lipid dysregulation in obesity by controlling central and peripheral AMPK activity. Am. J. Physiol. 296, E812–E819.

Kim, S., Jin, Y., Choi, Y., Park, T., 2011. Resveratrol exerts anti-obesity effects via mechanisms involving down-regulation of adipogenic and inflammatory processes in mice. Biochem. Pharmacol. 81, 1343–1351.

Kim, Y., Keogh, J.B., Clifton, P.M., 2018. Probiotics, prebiotics, synbiotics and insulin sensitivity. Nutr. Res. Rev. 17, 1–17.

King, J.C., Brown, K.H., Gibson, R.S., Krebs, N.F., Lowe, N.M., Siekmann, J.H., Raiten, D.J., 2015. Biomarkers of nutrition for development (BOND)—zinc review. J. Nutr. 146, 858S–885S.

Kitamura, T., Kahn, C.R., Accili, D., 2003. Insulin receptor knockout mice. Annu. Rev. Physiol. 65, 313–332.

Kitatani, K., Idkowiak-Baldys, J., Hannun, Y.A., 2008. The sphingolipid salvage pathway in ceramide metabolism and signaling. Cell. Signal. 20, 1010–1018.

Kittilson, J.D., Reindl, K.M., Sheridan, M.A., 2011. Rainbow trout (*Oncorhynchus mykiss*) possess two hormone-sensitive lipase-encoding mRNAs that are differentially expressed and independently regulated by nutritional state. Compar. Biochem. Physiol. Mol. Integr. Physiol. 158, 52–60.

Kjær, M.A., Todorčević, M., Torstensen, B.E., Vegusdal, A., Ruyter, B., 2008a. Dietary n-3 HUFA affects mitochondrial fatty acid β-oxidation capacity and susceptibility to oxidative stress in Atlantic salmon. Lipids 43, 813–827.

Kjær, M.A., Vegusdal, A., Gjøen, T., Rustan, A.C., Todorčević, M., Ruyter, B., 2008b. Effect of rapeseed oil and dietary n-3 fatty acids on triacylglycerol synthesis and secretion in Atlantic salmon hepatocytes. Biochim. Biophys. Acta 1781, 112–122.

Kjær, M.A., Vegusdal, A., Berge, G.M., Galloway, T.F., Hillestad, M., Krogdahl, Å., Holm, H., Ruyter, B., 2009. Characterisation of lipid transport in Atlantic cod (*Gadus morhua*) when fasted and fed high or low fat diets. Aquaculture 288, 325–336.

Kleveland, E.J., Syvertsen, B.L., Ruyter, B., Vegusdal, A., Jørgensen, S.M., Gjøen, T., 2006. Characterization of scavenger receptor class B, type I in Atlantic salmon (*Salmo salar* L.). Lipids 41, 1017–1027.

Kliewer, S.A., Lenhard, J.M., Willson, T.M., Patel, I., Morris, D.C., Lehmann, J.M., 1996. A prostaglandin J2 metabolite binds peroxisome proliferator-activated receptor γ and promotes adipocyte differentiation. Cell 83, 813–819.

Knight, J., Holland, J.W., Bowden, L.A., Halliday, K., Rowley, A.F., 1995. Eicosanoid generating capacities of different tissues from the rainbow trout, *Oncorhynchus mykiss*. Lipids 30, 451–458.

Koh, C.B., Romano, N., Siti-Zahrah, A., Ng, W.K., 2016. Effects of a dietary organic acids blend and oxytetracycline on the growth, nutrient utilization and total cultivable gut microbiota of the hybrid tilapia, *Oreochromis* sp., and resistance to *Streptococcus agalactiae*. Aquacult. Res. 47, 357–369.

Kohyama-Koganeya, A., Sasamura, T., Oshima, E., Suzuki, E., Nishihara, S., Ueda, R., Hirabayashi, Y., 2004. Drosophila glucosylceramide synthase A negative regulator of cell death mediated by proapoptotic factors. J. Biol. Chem. 279, 35995–36002.

Kojima, M., Kangawa, K., 2005. Ghrelin: structure and function. Physiol. Rev. 85, 495–522.

Kok, B., Malcorps, W., Tlusty, M.F., Eltholth, M.M., Auchterlonie, N.A., Little, D.C., Harmsen, R., Newton, R.W., Davies, S.J., 2020. Fish as feed: using economic allocation to quantify the fish in:fish out ratio of major fed aquaculture species. Aquaculture 528, 735474 (in press).

Kondo, H., Morinaga, K., Misaki, R., Nakaya, M., Watabe, S., 2005. Characterization of the pufferfish *Takifugu rubripes* apolipoprotein multigene family. Gene 346, 257–266.

Kortner, T.M., Bjorkhem, I., Krasnov, A., Timmerhaus, G., Krogdahl, A., 2014. Dietary cholesterol supplementation to a plant-based diet suppresses the complete pathway of cholesterol synthesis and induces bile acid production in Atlantic salmon (*Salmo salar* L.). Br. J. Nutr. 111, 2089–2103.

Kota, B.P., Huang, T.H.W., Roufogalis, B.D., 2005. An overview on biological mechanisms of PPARs. Pharmacol. Res. 51, 85–94.

Kotake-Nara, E., Nagao, A., 2011. Absorption and metabolism of xanthophylls. Mar. Drugs 9, 1024–1037.

Kousalaki, K., 2015. Recent developments and future perspectives of using marine microalgae in fish farming. Aquacult. Euro. 40, 5–11.

Kousoulaki, K., Marit, B.G., Mørkøre, T., Krasnov, A., Baeverfjord, G., Ytrestøyl, T., Carlehög, M., Sweetman, J., Ruyter, B., 2020. Microalgal *Schizochytrium limacinum* biomass improves growth and filet quality when used long-term as a replacement for fish oil. In: M. Salmon Diets. Frontiers in Marine Science, vol. 7.

Koven, W.M., Henderson, R.J., Sargent, J.R., 1997. Lipid digestion in turbot (*Scophthalmus maximus*): in-vivo and in-vitro studies of the lipolytic activity in various segments of the digestive tract. Aquaculture 151, 155–171.

Krasowski, M.D., Ai, N., Hagey, L.R., Kollitz, E.M., Kullman, S.W., Reschly, E.J., Ekins, S., 2011. The evolution of farnesoid X, vitamin D, and pregnane X receptors: insights from the green-spotted pufferfish (Tetraodon nigroviridis) and other non-mammalian species. BMC Biochem. 12, 5.

Kraugerud, O.F., Penn, M., Storebakken, T., Refstie, S., Krogdahl, Å., Svihus, B., 2007. Nutrient digestibilities and gut function in Atlantic salmon (*Salmo salar*) fed diets with cellulose or non-starch polysaccharides from soy. Aquaculture 273, 96–107.

Krogdahl, Å., Bakke, A.M., 2015. Antinutrients. In: Lee, C.-S., Lim, C., Gatlin III, D.M., Webster, C.D. (Eds.), Dietary Nutrients, Additives, and Fish Health.

Krogdahl, Å., Bakke-Mckellep, A., Baeverfjord, G., 2003. Effects of graded levels of standard soybean meal on intestinal structure, mucosal enzyme activities, and pancreatic response in Atlantic salmon (*Salmo salar* L.). Aquacult. Nutr. 9, 361–371.

Krogdahl, Å., Hemre, G.I., Mommsen, T.P., 2005. Carbohydrates in fish nutrition: digestion and absorption in postlarval stages. Aquacult. Nutr. 11, 103–122.

Krogdahl, Å., Kortner, T.M., Jaramillo-Torres, A., Gamil, A.A.A., Chikwati, E., Li, Y., Schmidt, M., Herman, E., Hymowitz, T., Teimouri, S., Storebakken, T., 2020. Removal of three proteinaceous antinutrients from soybean does not mitigate soybean-induced enteritis in Atlantic salmon (*Salmo salar*, L). Aquaculture 514, 734495.

Król, E., Douglas, A., Tocher, D.R., Crampton, V.O., Speakman, J.R., Secombes, C.J., Martin, S.A.M., 2016. Differential responses of the gut transcriptome to plant protein diets in farmed Atlantic salmon. BMC Genom. 17, 156.

Kuah, M.K., Jaya-Ram, A., Shu-Chien, A.C., 2015. The capacity for long-chain polyunsaturated fatty acid synthesis in a carnivorous vertebrate: functional characterisation and nutritional regulation of a Fads2 fatty acyl desaturase with Delta4 activity and an Elovl5 elongase in striped snakehead (*Channa striata*). Biochim. Biophys. Acta 1851, 248–260.

Kuah, M.K., Jaya-Ram, A., Shu-Chien, A.C., 2016. A fatty acyl desaturase (fads2) with dual Delta6 and Delta5 activities from the freshwater carnivorous striped snakehead *Channa striata*. Compar. Biochem. Physiol. Mol. Integr. Physiol. 201, 146–155.

Kumar, M., Nagpal, R., Kumar, R., Hemalatha, R., Verma, V., Kumar, A., Chakraborty, C., Singh, B., Marotta, F., Jain, S., Yadav, H., 2012. Colesterol-lowering probiotics as potential biotherapeutics for metabolic diseases. Exp. Diab. Res. 2012, 901917.

Kumar, V., Sinha, A., Romano, N., Allen, K.M., Bowman, B.A., Thompson, K.R., Tidwell, J.H., 2018. Metabolism and nutritive role of cholesterol in the growth, gonadal development, and reproduction of crustaceans. Rev. Fish. Sci. Aquacult. 26, 254–273.

Kurokawa, T., Suzuki, T., Ohta, H., Kagawa, H., Tanaka, H., Unuma, T., 2002. Expression of pancreatic enzyme genes during the early larval stage of Japanese eel *Anguilla japonica*. Fish. Sci. 68, 736–744.

Kurtovic, I., Marshall, S.N., Zhao, X., Simpson, B.K., 2009. Lipases from mammals and fishes. Rev. Fish. Sci. 17, 18–40.

Kurtovic, I., Marshall, S.N., Zhao, X., Simpson, B.K., 2010. Purification and properties of digestive lipases from Chinook salmon (*Oncorhynchus tshawytscha*) and New Zealand hoki (*Macruronus novaezelandiae*). Fish Physiol. Biochem. 36, 1041–1060.

Kütter, M., Romano, L., Ventura-Lima, J., Tesser, M., Monserrat, J., 2014. Antioxidant and toxicological effects elicited by alpha-lipoic acid in aquatic organisms. Compar. Biochem. Physiol. C Toxicol. Pharmacol. 162, 70–76.

Lall, S.P., 2002. The minerals. In: Halver, J.E., Hardy, R.W. (Eds.), Fish Nutrition, third ed. Academic Press, San Diego, CA.

Lands, B., 2014. Historical perspectives on the impact of n-3 and n-6 nutrients on health. Prog. Lipid Res. 55, 17–29.

Langerhans, R., 1890. Ueber multiple fettgewebsnekrose. Arch. für Pathol. Anat. Physiol. für Klin. Med. 122, 252–270.

Lansard, M., Panserat, S., Plagnes-Juan, E., Seiliez, I., Skiba-Cassy, S., 2010a. Integration of insulin and amino acid signals that regulate hepatic metabolism-related gene expression in rainbow trout: role of TOR. Amino Acids 39, 801–810.

Lansard, M., Panserat, S., Plagnes-Juan, E., Dias, K., Seiliez, I., Skiba-Cassy, S., 2010b. L-leucine, L-methionine, and L-lysine are involved in the regulation of intermediary metabolism-related gene expression in rainbow trout hepatocytes. J. Nutr. 141, 75–80.

Laprairie, R.B., Denovan-Wright, E.M., Wright, J.M., 2017. Differential regulation of the duplicated fabp7, fabp10 and fabp11 genes of zebrafish by peroxisome proliferator activated receptors. Compar. Biochem. Physiol. B Biochem. Mol. Biol. 213, 81–90.

Leatherland, J., Nuti, R., 2006. Effects of bovine growth hormone on plasma FFA concentrations and liver, muscle and carcass lipid content in rainbow trout, *Salmo gairdneri* Richardson. J. Fish. Biol. 19, 487–498.

Leaver, M.J., Evridiki, B., Efthimia, A., Amalia, D., Laurence, F.K., Tariq, M.E., Bautista, J.M., Tocher, D.R., Grigorios, K., 2005a. Three peroxisome proliferator-activated receptor isotypes from each of two species of marine fish. Endocrinology 146, 3150–3162.

Leaver, M.J., Evridiki, B., Efthimia, A., Amalia, D., Laurence, F.K., Tariq, E.M., Bautista, J.M., Tocher, D.R., Grigorios, K., 2005b. Three peroxisome proliferator-activated receptor isotypes from each of two species of marine fish. Endocrinology 146, 3150–3162.

Leaver, M.J., Bautista, J.M., Bjornsson, B.T., Jonsson, E., Krey, G., Tocher, D.R., Torstensen, B.E., 2008a. Towards fish lipid nutrigenomics: current state and prospects for fin-fish aquaculture. Rev. Fish. Sci. 16, 73–94.

Leaver, M.J., Bautista, J.M., Björnsson, B.T., Jönsson, E., Krey, G., Tocher, D.R., Torstensen, B.E., 2008b. Towards fish lipid nutrigenomics: current state and prospects for fin-fish aquaculture. Rev. Fish. Sci. 16, 73–94.

Leaver, M.J., Villeneuve, L.A., Obach, A., Jensen, L., Bron, J.E., Tocher, D.R., Taggart, J.B., 2008c. Functional genomics reveals increases in cholesterol biosynthetic genes and highly unsaturated fatty acid biosynthesis after dietary substitution of fish oil with vegetable oils in Atlantic salmon (*Salmo salar*). BMC Genom. 9, 299.

Leaver, M.J., Taggart, J.B., Villeneuve, L., Bron, J.E., Guy, D.R., Bishop, S.C., Houston, R.D., Matika, O., Tocher, D.R., 2011. Heritability and mechanisms of n-3 long chain polyunsaturated fatty acid deposition in the flesh of Atlantic salmon. Compar. Biochem. Physiol. D Genom. Proteonom. 6, 62–69.

Ledesma-Amaro, R., Nicaud, J.-M., 2016. Yarrowia lipolytica as a biotechnological chassis to produce usual and unusual fatty acids. Prog. Lipid Res. 61, 40–50.

Lee, Y.S., Kim, W.S., Kim, K.H., Yoon, M.J., Cho, H.J., Shen, Y., Ye, J.M., Lee, C.H., Oh, W.K., Kim, C.T., 2006. Berberine, a natural plant product, activates AMP-activated protein kinase with beneficial metabolic effects in diabetic and insulin-resistant states. Diabetes 55, 2256–2264.

Lee, J.S., Chi, W.J., Hong, S.K., Yang, J.W., Chang, Y.K., 2013. Bioethanol production by heterologous expression of Pdc and AdhII in Streptomyces lividans. Appl. Microbiol. Biotechnol. 97, 6089–6097.

Lee, J.H., Moniruzzaman, M., Yun, H., Lee, S., Park, Y., Bai, S.C., 2016. Dietary vitamin C reduced mercury contents in the tissues of juvenile olive flounder (*Paralichthys olivaceus*) exposed with and without mercury. Environ. Toxicol. Pharmacol. 45, 8–14.

Leggatt, R.A., Raven, P.A., Mommsen, T.P., Sakhrani, D., Higgs, D., Devlin, R.H., 2009. Growth hormone transgenesis influences carbohydrate, lipid and protein metabolism capacity for energy production in coho salmon (*Oncorhynchus kisutch*). Compar. Biochem. Physiol. B Biochem. Mol. Biol. 154, 121–133.

Lehner, R., Kuksis, A., 1992. Utilization of 2-monoacylglycerols for phosphatidylcholine biosynthesis in the intestine. Biochim. Biophys. Acta 1125, 171–179.

Lehner, R., Kuksis, A., 1996. Biosynthesis of triacylglycerols. Prog. Lipid Res. 35, 169–201.

Lei, C., Tian, J.-J., 2017. Stimulation of glycerol kinase in grass carp preadipocytes by EPA. Fish Physiol. Biochem. 43.

Leng, X.J., Wu, X.F., Tian, J., Li, X.Q., Guan, L., Weng, D.C., 2012. Molecular cloning of fatty acid synthase from grass carp (*Ctenopharyngodon idella*) and the regulation of its expression by dietary fat level. Aquacult. Nutr. 18, 551–558.

Lesel, R., De La Noüe, J., Choubert, G., 1989. Fecal bacterial flora of rainbow trout under antibiotic treatment: effect of the number of pyloric caeca and the lipid content of food. In: De Pauw, N., Jaspers, E., Ackerfors, H., Wilkins, N. (Eds.), Aquaculture – A Biotechnology in Progress, vol. 2. European Aquaculture Society, Bredane, Belgium.

Leslie, J.M., Buckley, J.T., 1976. Phospholipid composition of goldfish (*Carassius auratus* L.) liver and brain and temperature-dependence of phosphatidyl choline synthesis. Compar. Biochem. Physiol. B Biochem. Mol. Biol. 53, 335–337.

Levy, B.D., Clish, C.B., Schmidt, B., Gronert, K., Serhan, C.N., 2001. Lipid mediator class switching during acute inflammation: signals in resolution. Nat. Immunol. 2, 612–619.

Lewis, M.J., Hamid, N.K.A., Alhazzaa, R., Hermon, K., Donald, J.A., Sinclair, A.J., Turchini, G.M., 2013. Targeted dietary micronutrient fortification modulates n-3 LC-PUFA pathway activity in rainbow trout (*Oncorhynchus mykiss*). Aquaculture 412–413, 215–222.

Lewis-Mccrea, L.M., Lall, S.P., 2007. Effects of moderately oxidized dietary lipid and the role of vitamin E on the development of skeletal abnormalities in juvenile Atlantic halibut (*Hippoglossus hippoglossus*). Aquaculture 262, 142–155.

Li, K.S., Olsen, R.E., 2017. Metabolism of sn-1(3)-monoacylglycerol and sn-2-monoacylglycerol in caecal enterocytes and hepatocytes of Brown trout (*Salmo trutta*). Lipids 52, 61–71.

Li, X., Huang, H.Y., Chen, J.G., Jiang, L., Liu, H.L., Liu, D.G., Song, T.J., He, Q., Ma, C.G., Ma, D., 2006. Lactacystin inhibits 3T3-L1 adipocyte differentiation through induction of CHOP-10 expression. Biochem. Biophys. Res. Commun. 350, 1–6.

Li, G.-G., Liang, X.-F., Xie, Q., Li, G., Yu, Y., Lai, K., 2010a. Gene structure, recombinant expression and functional characterization of grass carp leptin. Gen. Compar. Endocrinol. 166, 117–127.

Li, X., Liu, W., Jiang, Y., Zhu, H., Ge, X., 2010b. Effects of dietary protein and lipid levels in practical diets on growth performance and body composition of blunt snout bream (*Megalobrama amblycephala*) fingerlings. Aquaculture 303, 65–70.

Li, Y., Monroig, O., Zhang, L., Wang, S., Zheng, X., Dick, J.R., You, C., Tocher, D.R., 2010c. Vertebrate fatty acyl desaturase with Delta4 activity. Proc. Natl. Acad. Sci. U. S. A. 107, 16840–16845.

Li, X., Yan, Q., Xie, S., Hu, W., Yu, Y., Hu, Z., 2013. Gut microbiota contributes to the growth of fast-growing transgenic common carp (*Cyprinus carpio* L.). PLoS One 8, e64577.

Li, J.Y., Zhang, D.D., Xu, W., Jiang, G.Z., Zhang, C.N., Li, X.F., Liu, W.B., 2014. Effects of dietary choline supplementation on growth performance and hepatic lipid transport in blunt snout bream (*Megalobrama amblycephala*) fed high-fat diets. Aquaculture 434, 340–347.

Li, K., Østensen, M.A., Attramadal, K., Winge, P., Sparstad, T., Bones, A.M., Vadstein, O., Kjørsvik, E., Olsen, Y., 2015. Gene regulation of lipid and phospholipid metabolism in Atlantic cod (*Gadus morhua*) larvae. Compar. Biochem. Physiol. B Biochem. Mol. Biol. 190, 16–26.

Li, A., Yuan, X., Liang, X., Liu, L., Li, J., Li, B., Fang, J., Li, J., He, S., Xue, M., 2016a. Adaptations of lipid metabolism and food intake in response to low and high fat diets in juvenile grass carp (*Ctenopharyngodon idellus*). Aquaculture 457, 43–49.

Li, K., Olsen, R.E., Østensen, M.A., Altin, D., Kjørsvik, E., Olsen, Y., 2016b. Atlantic cod (*Gadus morhua*) larvae can biosynthesis phospholipid de novo from 2-oleoyl-glycerol and glycerol precursors. Fish Physiol. Biochem. 42, 137–147.

Li, S., Monroig, Ó., Wang, T., Yuan, Y., Navarro, J.C., Hontoria, F., Liao, K., Tocher, D.R., Mai, K., Xu, W., Ai, Q., 2017. Functional characterization and differential nutritional regulation of putative Elovl5 and Elovl4 elongases in large yellow croaker (*Larimichthys crocea*). Sci. Rep. 7, 2303.

Li, J.M., Li, L.Y., Qin, X., Degrace, P., Demizieux, L., Limbu, S.M., Wang, X., Zhang, M.L., Li, D.L., Du, Z.Y., 2018a. Inhibited carnitine synthesis causes systemic alteration of nutrient metabolism in zebrafish. Front. Physiol. 9, 509.

Li, K., Egelandsdal, B., Olsen, R.E., 2018b. Hydrolysis activity of pyloric cecal enterocytes of Brown trout (*Salmo trutta*) toward monoacylglycerol and lysophosphatidylcholine. Lipids 53, 615–625.

Li, X., Ringø, E., Hoseinifar, S.H., Lauzon, H.L., Birkbeck, H., Yang, D., 2018c. Adherence and colonisation of microorganisms in the fish gastrointestinal tract. Rev. Aquacult. 11, 603–618.

Li, J.M., Li, L.Y., Zhang, Y.X., Jiang, Z.Y., Limbu, S.M., Qiao, F., Degrace, P., Chen, L.Q., Zhang, M.L., Du, Z.Y., 2019a. Functional differences between l-and d-carnitine in metabolic regulation evaluated using a low-carnitine Nile tilapia model. Br. J. Nutr. 122, 625–638.

Li, L.-Y., Limbu, S.M., Ma, Q., Chen, L.-Q., Zhang, M.-L., Du, Z.-Y., 2019b. The metabolic regulation of dietary L-carnitine in aquaculture nutrition: present status and future research strategies. Rev. Aquacult. 11, 1228–1257.

Li, Y., Yin, Z., Dong, Y., Wang, S., Monroig, O., Tocher, D.R., You, C., 2019c. PPARgamma is involved in the transcriptional regulation of liver LC-PUFA biosynthesis by targeting the Delta6Delta5 fatty acyl desaturase gene in the marine teleost *Siganus canaliculatus*. Mar. Biotechnol. 21, 19–29.

Li, Y., Zhao, J., Dong, Y., Yin, Z., Li, Y., Liu, Y., You, C., Monroig, O., Tocher, D.R., Wang, S., 2019d. Sp1 is involved in vertebrate LC-PUFA biosynthesis by upregulating the expression of liver desaturase and elongase genes. Int. J. Mol. Sci. 20, 5066.

Li, L.Y., Li, J.M., Ning, L.J., Lu, D.L., Luo, Y., Ma, Q., Limbu, S.M., Li, D.L., Chen, L.Q., Lodhi, I., Degrace, P., Zhang, M.L., Du, Z.Y., 2020a. Mitochondrial fatty acid β-oxidation inhibition promotes carbohydrate catabolism and protein deposition through energy homeostasis remodeling in fish. J. Nutr. 150, 2322–2335.

Li, L.Y., Li, J.M., Qiao, F., Chen, L.Q., Zhang, M.L., Du, Z.Y., 2020b. Inhibited carnitine synthesis impairs adaptation to high-fat diet in Nile tilapia (*Oreochromis niloticus*). Aquacult. Rep. 16, 100249.

Li, L.Y., Lu, D.L., Jiang, Z.Y., Limbu, S.M., Qiao, F., Chen, L.Q., Zhang, M.L., Du, Z.Y., 2020c. Dietary L-carnitine improves glycogen and protein accumulation in Nile tilapia via increasing lipid-sourced energy supply: an isotope-based metabolic tracking. Aquacult. Rep. 17, 100302.

Li, Y., Kortner, T.M., Chikwati, E.M., Belghit, I., Lock, E.-J., Krogdahl, Å., 2020d. Total replacement of fish meal with black soldier fly (*Hermetia illucens*) larvae meal does not compromise the gut health of Atlantic salmon (*Salmo salar*). Aquaculture 520, 734967.

Li, L.Y., Lv, H.B., Jiang, Z.Y., Qiao, F., Chen, L.Q., Zhang, M.L., Du, Z.Y., 2021. Peroxisomal proliferator-activated receptor α-b deficiency induces the reprogramming of nutrient metabolism in zebrafish. J. Physiol. (2021).

Liang, J.J., Tian, L.X., Liu, Y.J., Yang, H.J., Liang, G.Y., 2012. Dietary magnesium requirement and effects on growth and tissue magnesium content of juvenile grass carp (*Ctenopharyngodon idella*). Aquacult. Nutr. 18, 56–64.

Liang, H., Habte-Tsion, H.M., Ge, X., Ren, M., Xie, J., Miao, L., Zhou, Q., Lin, Y., Pan, W., 2017a. Dietary arginine affects the insulin signaling pathway, glucose metabolism and lipogenesis in juvenile blunt snout bream *Megalobrama amblycephala*. Sci. Rep. 7, 7864.

Liang, X.F., Hu, L., Dong, Y.C., Wu, X.F., Qin, Y.C., Zheng, Y.H., Shi, D.D., Xue, M., Liang, X.F., 2017b. Substitution of fish meal by fermented soybean meal affects the growth performance and flesh quality of Japanese seabass (*Lateolabrax japonicus*). Anim. Feed Sci. Technol. 229, 1–12.

Liang, G., Li, X., Yang, H., Huan, D., Xu, Z., Leng, X., Gao, Q., 2019a. Dietary oxidized oils decreased growth, antioxidative capacity, and negatively affected skin color of channel catfish, *Ictalurus punctatus*. J. World Aquacult. Soc. 50, 692–706.

Liang, H., Mokrani, A., Chisomo-Kasiya, H., Ji, K., Ge, X., Ren, M., Liu, B., Xi, B., Sun, A., 2019b. Dietary leucine affects glucose metabolism and lipogenesis involved in TOR/PI3K/Akt signaling pathway for juvenile blunt snout bream *Megalobrama amblycephala*. Fish Physiol. Biochem. 45, 719–732.

Libran-Perez, M., Otero-Rodino, C., LOPEZ-Patino, M.A., Miguez, J.M., Soengas, J.L., 2014. Central administration of oleate or octanoate activates. hypothalamic fatty acid sensing and inhibits food intake in rainbow trout. Physiol. Behav. 129, 272–279.

Lie, S., Morrison, J.L., Williams-Wyss, O., Suter, C.M., Humphreys, D.T., Ozanne, S.E., Zhang, S., Maclaughlin, S.M., Kleemann, D.O., Walker, S.K., 2016. Impact of maternal undernutrition around the time of conception on factors regulating hepatic lipid metabolism and microRNAs in singleton and twin fetuses. Am. J. Physiol. 310, E148–E159.

Liland, N.S., Espe, M., Rosenlund, G., Waagbo, R., Hjelle, J.I., Lie, O., Fontanillas, R., Torstensen, B.E., 2013a. High levels of dietary phytosterols affect lipid metabolism and increase liver and plasma TAG in Atlantic salmon (*Salmo salar* L.). Br. J. Nutr. 110, 1958–1967.

Liland, N.S., Rosenlund, G., Berntssen, M.H.G., Brattelid, T., Madsen, L., Torstensen, B.E., 2013b. Net production of Atlantic salmon (FIFO, fish in fish out < 1) with dietary plant proteins and vegetable oils. Aquacult. Nutr. 19, 289–300.

Liland, N.S., Johnsen, E.N., Hellberg, H., Waagbo, R., Sissener, N.H., Torstensen, B.E., Saele, O., 2018. Effects of dietary vegetable oils and varying dietary EPA and DHA levels on intestinal lipid accumulations in Atlantic salmon. Aquacult. Nutr. 24, 1599–1610.

Lim, P.K., Boey, P.L., Ng, W.K., 2001. Dietary palm oil level affects growth performance, protein retention and tissue vitamin E concentration of African catfish, *Clarias gariepinus*. Aquaculture 202, 101–112.

Lim, K.C., Yusoff, F.M., Shariff, M., Kamarudin, M.S., 2018. Astaxanthin as feed supplement in aquatic animals. Rev. Aquacult. 10, 738–773.

Limbu, S.M., Ma, Q., Zhang, M.L., Du, Z.Y., 2019. High fat diet worsens the adverse effects of antibiotic on intestinal health in juvenile Nile tilapia (*Oreochromis niloticus*). Sci. Total Environ. 680, 169–180.

Lin, K., Feng, J., Yang, H., Chen, Y., Chen, X., Luo, H., Huang, W., Wang, H., Luo, L., 2018. Effects of inositol supplementation to practical dietary on growth performance, lipid metabolism and antioxidant activity of *Ctenopharyngodon idella*. J. Fish. China 42, 1428–1437.

Linton, M.F., Farese Jr, R.V., Chiesa, G., Grass, D.S., Chin, P., Hammer, R.E., Hobbs, H.H., Young, S.G., 1993. Transgenic mice expressing high plasma concentrations of human apolipoprotein B100 and lipoprotein(a). J. Clin. Invest. 92, 3029–3037.

Liu, K., 2014. Effects of Dietary Fat and Choline Levels on Growth Performance and Physiological Function of Juvenile tilapia, *Oreochromis niloticus*. Guangxi University (in Chinese).

Liu, X.J., Luo, Z., Xiong, B.X., Liu, X., Zhao, Y.H., Hu, G.F., Lv, G.J., 2010. Effect of waterborne copper exposure on growth, hepatic enzymatic activities and histology in *Synechogobius hasta*. Ecotoxicol. Environ. Saf. 73, 1286–1291.

Liu, P., Li, C., Huang, J., 2014a. Regulation of adipocytes lipolysis by n-3 HUFA in grass carp (*Ctenopharyngodon idellus*) in vitro and in vivo. Fish Physiol. Biochem. 40.

Liu, W., Yang, Y., Zhang, J., Gatlin, D.M., Ringø, E., Zhou, Z., 2014b. Effects of dietary microencapsulated sodium butyrate on growth, intestinal mucosal morphology, immune response and adhesive bacteria in juvenile common carp (*Cyprinus carpio*) pre-fed with or without oxidised oil. Br. J. Nutr. 112, 15–29.

Liu, G.X., Jiang, G.Z., Lu, K.L., Li, X.F., Zhou, M., Zhang, D.D., Liu, W.B., 2016. Effects of dietary selenium on the growth, selenium status, antioxidant activities, muscle composition and meat quality of blunt snout bream, *Megalobrama amblycephala*. Aquacult. Nutr. 23, 777–787.

Liu, C., He, A., Ning, L.-J., Luo, Y., Li, D.-L., Zhang, M.-L., Du, Z.-Y., 2018a. Leptin selectively regulates nutrients metabolism in *Nile tilapia* fed on high carbohydrate or high fat diet. Front. Endocrinol. 9.

Liu, C.Z., He, A.Y., Ning, L.J., Luo, Y., Li, D.L., Zhang, M.L., Chen, L.Q., Du, Z.Y., 2018b. Leptin selectively regulates nutrients metabolism in *Nile tilapia* fed on high carbohydrate or high fat diet. Front. Endocrinol. 9, 574.

Liu, D., Pang, Q., Han, Q., Shi, Q., Zhang, Q., Yu, H., 2019a. Wnt10b participates in regulating fatty acid synthesis in the muscle of zebrafish. Cells 8, 1011.

Liu, J.-Q., Li, W.-X., Zheng, J.-J., Tian, Q.-N., Huang, J.-F., Dai, S.-X., 2019b. Gain and loss events in the evolution of the apolipoprotein family in vertebrata. BMC Evol. Biol. 19, 209.

Liu, Y., Jiao, J.G., Gao, S., Ning, L.J., Mchele Limbu, S., Qiao, F., Chen, L.Q., Zhang, M.L., Du, Z.Y., 2019c. Dietary oils modify lipid molecules and nutritional value of fillet in *Nile tilapia*: a deep lipidomics analysis. Food Chem. 277, 515–523.

Liu, Y., Han, S., Luo, Y., Li, L., Chen, L., Zhang, M., Du, Z., 2020. Impaired peroxisomal fat oxidation induces hepatic lipid accumulation and oxidative damage in *Nile tilapia*. Fish Physiol. Biochem. https://doi.org/10.1007/s10695-020-00785-w.

Ljubojević, D., Radosavljević, V., Puvača, N., Živkov Baloš, M., Đorđević, V., Jovanović, R., Ćirković, M., 2015. Interactive effects of dietary protein level and oil source on proximate composition and fatty acid composition in common carp (*Cyprinus carpio* L.). J. Food Compos. Anal. 37, 44–50.

Lødemel, J.B., Mayhew, T.M., Myklebust, R., Olsen, R.E., Espelid, S., Ringø, E., 2001. Effect of three dietary oils on disease susceptibility in Arctic charr (*Salvelinus alpinus* L.) during cohabitant challenge with *Aeromonas salmonicida* ssp. *salmonicida*. Aquacult. Res. 32, 935–945.

Londraville, R.L., Macotela, Y., Duff, R.J., Easterling, M.R., Liu, Q., Crespi, E.J., 2014. Comparative endocrinology of leptin: assessing function in a phylogenetic context. Gen. Compar. Endocrinol. 203, 146–157.

Lopes-Marques, M., Kabeya, N., Qian, Y., Ruivo, R., Santos, M.M., Venkatesh, B., Tocher, D.R., Castro, L.F.C., Monroig, O., 2018. Retention of fatty acyl desaturase 1 (fads1) in Elopomorpha and Cyclostomata provides novel insights into the evolution of long-chain polyunsaturated fatty acid biosynthesis in vertebrates. BMC Evol. Biol. 18, 157.

López-Olmeda, J.F., Sánchez-Vázquez, F.J., 2010. Feeding rhythms in fish: from behavioural to molecular approach. In: Kulczykowska, E., Popek, W., Kapoor, B.G. (Eds.), Biological Clock in Fish. CRC Press, Taylor and Francis Group, Boca Raton, FL, USA.

Lozano, A.R., Borges, P., Robaina, L., Betancor, M., Hernandez-Cruz, C.M., García, J.R., Caballero, M.J., Vergara, J.M., Izquierdo, M., 2017. Effect of different dietary vitamin E levels on growth, fish composition, fillet quality and liver histology of meagre (*Argyrosomus regius*). Aquaculture 468, 175–183.

Lu, R.H., Liang, X.F., Wang, M., Zhou, Y., Bai, X.L., He, Y., 2012. The role of leptin in lipid metabolism in fatty degenerated hepatocytes of the grass carp *Ctenopharyngodon idellus*. Fish Physiol. Biochem. 38, 1759–1774.

Lu, K.-L., Xu, W.-N., Wang, L.-N., Zhang, D.-D., Zhang, C.-N., Liu, W.-B., 2014a. Hepatic β-oxidation and regulation of carnitine palmitoyltransferase (CPT) I in blunt snout bream *Megalobrama amblycephala* fed a high fat diet. PLoS One 9, e93135.

Lu, K.L., Li, X.F., Wang, L.N., Zhang, C.N., Liu, W.B., 2014b. Effects of dietary nutrient composition on compensatory growth of juvenile blunt snout bream *Megalobrama amblycephala*. North Am. J. Aquacult. 76, 55–60.

Lu, K.L., Xu, W.N., Liu, W.B., Wang, L.N., Zhang, C.N., Li, X.F., 2014c. Association of mitochondrial dysfunction with oxidative stress and immune suppression in Blunt Snout Bream *Megalobrama amblycephala* fed a high-fat diet. J. Aquat. Anim. Health 26, 100–112.

Lu, K.L., Zhang, D.D., Wang, L.N., Xu, W.N., Liu, W.B., 2016. Molecular characterization of carnitine palmitoyltransferase IA in *Megalobrama amblycephala* and effects on its expression of feeding status and dietary lipid and berberine. Compar. Biochem. Physiol. B Biochem. Mol. Biol. 191, 20.

Lu, D.L., Ma, Q., Wang, J., Li, L.Y., Han, S.L., Limbu, S.M., Li, D.L., Chen, L.Q., Zhang, M.L., Du, Z.Y., 2019. Fasting enhances cold resistance in fish through stimulating lipid catabolism and autophagy. J. Physiol. 597, 1585–1603.

Luckenbach, T., Fischer, S., Sturm, A., 2014. Current advances on ABC drug transporters in fish. Compar. Biochem. Physiol. C Toxicol. Pharmacol. 165, 28–52.

Luo, Z., Wei, C.C., Ye, H.M., Zhao, H.P., Song, Y.F., Wu, K., 2016. Effect of dietary choline levels on growth performance, lipid deposition and metabolism in juvenile yellow catfish *Pelteobagrus fulvidraco*. Compar. Biochem. Physiol. B Biochem. Mol. Biol. 202, 1–7.

Luo, Z., Zou, G., Gao, Y., Ye, H., Xi, W., Liu, X., 2017. Effect of dietary iron (Fe) levels on growth performance, hepatic lipid metabolism and antioxidant responses in juvenile yellow catfish *Pelteobagrus fulvidraco*. Aquacult. Nutr. 23, 1475–1482.

Luz, R.K., Boluda-Navarro, D., Sanchez-Vazquez, F.J., Fortes-Silva, R., Portella, M.C., Navarro, S.Z., Madrid, J.A., 2018. European seabass (*Dicentrarchus labrax*) ability to discriminate between diets made with different dietary fat sources. Aquacult. Nutr. 24, 65–73.

Lykidis, A., 2007. Comparative genomics and evolution of eukaryotic phospholipid biosynthesis. Prog. Lipid Res. 46, 171–199.

Ma, Q., Li, L.Y., Le, J.Y., Lu, D.L., Qiao, F., Zhang, M.L., Du, Z.Y., Li, D.L., 2018. Dietary microencapsulated oil improve immune function and intestinal health in Nile tilapia fed with high-fat diet. Aquaculture 496, 19–29.

Macfarlane, S., Macfarlane, G.T., 2003. Regulation of short-chain fatty acids production. Proc. Natl. Acad. Sci. U. S. A. 62, 67–72.

Machado, A.M., Tørresen, O.K., Kabeya, N., Couto, A., Petersen, B., Felício, M., Campos, P.F., Fonseca, E., Bandarra, N., Lopes-Marques, M., Ferraz, R., Ruivo, R., Fonseca, M.M., Jentoft, S., Monroig, Ó., Da Fonseca, R.R., Castro, L.F.C., 2018. "Out of the can": a draft genome assembly, liver transcriptome, and nutrigenomics of the European sardine, Sardina pilchardus. Genes 9, 485.

Madsen, L., Rustan, A.C., Vaagenes, H., Berge, K., Dyrøy, E., Berge, R.K., 1999. Eicosapentaenoic and docosahexaenoic acid affect mitochondrial and peroxisomal fatty acid oxidation in relation to substrate preference. Lipids 34, 951–963.

Magalhães, R., Guerreiro, I., Coutinho, F., Moutinho, S., Sousa, S., Delerue-Matos, C., Domingues, V.F., Olsen, R.E., Peres, H., Oliva-Teles, A., 2020. Effect of dietary ARA/EPA/DHA ratios on growth performance and intermediary metabolism of gilthead sea bream (*Sparus aurata*) juveniles. Aquaculture 516, 734644.

Maglich, J.M., Caravella, J.A., Lambert, M.H., Willson, T.M., Moore, J.T., Ramamurthy, L., 2003. The first completed genome sequence from a teleost fish (*Fugu rubripes*) adds significant diversity to the nuclear receptor superfamily. Nucleic Acids Res. 31, 4051–4058.

Magnoni, L., Vaillancourt, E., Weber, J.-M., 2008. In vivo regulation of rainbow trout lipolysis by catecholamines. J. Exp. Biol. 211, 2460.

Mai, K.S., Xiao, L.D., Ai, Q.H., Wang, X.J., Xu, W., Zhang, W.B., Liufu, Z.G., Ren, M.C., 2009. Dietary choline requirement for juvenile cobia, *Rachycentron canadum*. Aquaculture 289, 124–128.

Maita, M., Maekawa, J., Satoh, K.I., Futami, K., Satoh, S., 2006. Disease resistance and hypocholesterolemia in yellowtail *Seriola quinqueradiata* fed a non-fishmeal diet. Fish. Sci. 72, 513–519.

Maoka, T., 2011. Carotenoids in marine animals. Mar. Drugs 9, 278–293.

Martins, D.A., Valente, L.M.P., Lall, S.P., 2009. Apparent digestibility of lipid and fatty acids in fish oil, poultry fat and vegetable oil diets by Atlantic halibut, *Hippoglossus hippoglossus* L. Aquaculture 294, 132–137.

Mason, R.P., Sherratt, S.C.R., 2017. Omega-3 fatty acid fish oil dietary supplements contain saturated fats and oxidized lipids that may interfere with their intended biological benefits. Biochem. Biophys. Res. Commun. 483, 425–429.

Massart, J., Katayama, M., Krook, A., 2016. Micromanaging glucose and lipid metabolism in skeletal muscle: role of microRNAs. Biochim. Biophys. Acta 1861, 2130–2138.

Matter, F., Peganova, S., Eder, K., 2004. Lipid concentrations of fillets, liver, plasma and lipoproteins of African catfish, *Clarias gariepinus* (Burchell 1822), fed diets with varying protein concentrations. J. Anim. Physiol. Anim. Nutr. 88, 275–287.

Mazzeo, I., Borrell Pichs, Y., Gallego, V., Fernández, C., Sánchez, J., Blanco, G., Pérez, L., Asturiano, J., 2013. Larval width as indicator of growth rate and effect of larval classification on final body composition and flesh quality in cultured gilthead seabream (*Sparus aurata*, L.). J. Appl. Ichthyol. 30, 300–306.

Mcnulty, H., Jacob, R.F., Mason, R.P., 2008. Biologic activity of carotenoids related to distinct membrane physicochemical interactions. Am. J. Cardiol. 101, S20–S29.

Meguro, S., Hasumura, T., Hase, T., 2015. Body fat accumulation in zebrafish is induced by a diet rich in fat and reduced by supplementation with green tea extract. PLoS One 10, e0120142.

Mendelson, K., Pandey, S., Hisano, Y., Carellini, F., Das, B.C., Hla, T., Evans, T., 2017. The ceramide synthase 2b gene mediates genomic sensing and regulation of sphingosine levels during zebrafish embryogenesis. eLife 6, e21992.

Meng, X.L., Li, S., Qin, C.B., Zhu, Z.X., Hu, W.P., Yang, L.P., Lu, R.H., Li, W.J., Nie, G.X., 2018. Intestinal microbiota and lipid metabolism responses in the common carp (*Cyprinus carpio* L.) following copper exposure. Ecotoxicol. Environ. Saf. 160, 257–264.

Mennigen, J.A., Panserat, S., Larquier, M., Plagnes-Juan, E., Medale, F., Seiliez, I., Skiba-Cassy, S., 2012. Postprandial regulation of hepatic microRNAs predicted to target the insulin pathway in Rainbow Trout. PLoS One 7, e38604.

Mennigen, J.A., Martyniuk, C.J., Seiliez, I., Panserat, S., Skiba-Cassy, S., 2014. Metabolic consequences of microRNA-122 inhibition in rainbow trout,*Oncorhynchus mykiss*. BMC Genom. 15, 70.

Menoyo, D., Lopez-Bote, C.J., Bautista, J.M., Obach, A., 2003. Growth, digestibility and fatty acid utilization in large Atlantic salmon (*Salmo salar*) fed varying levels of n-3 and saturated fatty acids. Aquaculture 225, 295–307.

Menoyo, D., Izquierdo, M., Robaina, L., Ginés, R., Lopez-Bote, C., Bautista, J.M., 2004. Adaptation of lipid metabolism, tissue composition and flesh quality in gilthead sea bream (*Sparus aurata*) to the replacement of dietary fish oil by linseed and soyabean oils. Br. J. Nutr. 92, 41–52.

Menoyo, D., Kühn, G., Ruiz-Lopez, N., Pallauf, K., Stubhaug, I., Pastor, J., Ipharraguerre, I., Rimbach, G., 2019. Dietary resveratrol impairs body weight gain due to reduction of feed intake without affecting fatty acid composition in Atlantic salmon. Animal 13, 25–32.

Merrifield, D.L., Dimitroglou, A., Foey, A., Davies, S.J., Baker, R.R., Bøgwald, J., Castex, M., Ringø, E., 2010. The current status and future focus of probiotic and prebiotic applications for salmonids. Aquaculture 302, 1–18.

Merrill Jr., A.H., 2011. Sphingolipid and glycosphingolipid metabolic pathways in the era of sphingolipidomics. Chem. Rev. 111, 6387–6422.

Metcalf, V.J., Gemmell, N.J., 2005. Fatty acid transport in cartilaginous fish: absence of albumin and possible utilization of lipoproteins. Fish Physiol. Biochem. 31, 55–64.

Metwally, M., El-Gellal, A., El-Sawaisi, S., 2009. Effects of silymarin on lipid metabolism in rats. World Appl. Sci. J. 6, 1634–1637.

Meyerhof, W., Korsching, S., 2009. Chemosensory Systems in Mammals, Fishes, and Insects. Springer-Verlag, Berlin, Germany.

Michelsen, K.G., Harmon, J.S., Sheridan, M.A., 1994. Adipose tissue lipolysis in rainbow trout, *Oncorhynchus mykiss*, is modulated by phosphorylation of triacylglycerol lipase. Compar. Biochem. Physiol. B Biochem. Mol. Biol. 107, 509–513.

Milani, A., Basirnejad, M., Shahbazi, S., Bolhassani, A., 2017. Carotenoids: biochemistry, pharmacology and treatment. Br. J. Pharmacol. 174, 1290–1324.

Miller, M.R., Nichols, P.D., Barnes, J., Davies, N.W., Peacock, E.J., Carter, C.G., 2006. Regiospecificity profiles of storage and membrane lipids from the gill and muscle tissue of Atlantic salmon (*Salmo salar* L.) growth at elevated temperature. Lipids 41, 865–876.

Minghetti, M., Leaver, M.J., Tocher, D.R., 2011. Transcriptional control mechanisms of genes of lipid and fatty acid metabolism in the Atlantic salmon (*Salmo salar* L.) established cell line, SHK-1. Biochim. Biophys. Acta 1811, 194–202.

Minick, M.C., Chavin, W., 1973. Effects of catecholamines upon serum FFA levels in normal and diabetic goldfish, *Carassius auratus* L. Compar. Biochem. Physiol. Mol. Integr. Physiol. 44, 1003–1008.

Miyasaki, T., Sato, M., Yoshinaka, R., Sakaguchi, M., 1995. Effect of vitamin C on lipid and carnitine metabolism in rainbow trout. Fish. Sci. 61, 501–506.

Miyoshi, H., Perfield, J.W., Souza, S.C., Shen, W.-J., Zhang, H.-H., Stancheva, Z.S., Kraemer, F.B., Obin, M.S., Greenberg, A.S., 2007. Control of adipose triglyceride lipase action by serine 517 of perilipin A globally regulates protein kinase A-stimulated lipolysis in adipocytes. J. Biol. Chem. 282, 996–1002.

Moazenzadeh, K., Rajabi Islami, H., Zamini, A., Soltani, M., 2018. Effects of dietary zinc level on performance, zinc status, tissue composition and enzyme activities of juvenile Siberian sturgeon, *Acipenser baerii* (Brandt 1869). Aquacult. Nutr. 24, 1330–1339.

Mock, T.S., 2018. Omega-3 Fatty Acids in Atlantic Salmon: Dietary and Environmental Effects. Ph. D. Deakin University.

Mock, T.S., Francis, D.S., Matthew, K.J., Glencross, B.D., Smullen, R.P., Turchini, G.M., 2019. Endogenous biosynthesis of n-3 LC PUFA in Atlantic salmon. Br. J. Nutr. 121, 1108–1123.

Mohanta, K.N., Mohanty, S.N., Jena, J., Sahu, N.P., Patro, B., 2009. Carbohydrate level in the diet of silver barb, *Puntius gonionotus* (Bleeker) fingerlings: effect on growth, nutrient utilization and whole body composition. Aquacult. Res. 40, 927–937.

Mohseni, M., Pourkazemi, M., Bai, S., 2014. Effects of dietary inorganic copper on growth performance and immune responses of juvenile beluga, Huso huso. Aquacult. Nutr. 20, 547–556.

Molejon, M.I., Ropolo, A., Re, A.L., Boggio, V., Vaccaro, M.I., 2013. The vmp1-beclin1 interaction regulates autophagy induction. FASEB J. 27, 4.

Mommsen, T.P., Plisetskaya, E.M., 1991. Insulin in fishes and agnathans: history, structure and metabolic regulation. Rev. Aqualt. Sci. 4, 225–259.

Mondy, N., Grossi, V., Cathalan, E., Delbecque, J.P., Mermillod-Blondin, F., Douady, C.J., 2014. Sterols and steroids in a freshwater crustacean (*Proasellus meridianus*): hormonal response to nutritional input. Invertebr. Biol. 133, 99–107.

Monroig, Ó., Rotllant, J., Sánchez, E., Cerdá-Reverter, J.M., Tocher, D.R., 2009. Expression of long-chain polyunsaturated fatty acid (LC-PUFA) biosynthesis genes during zebrafish *Danio rerio* early embryogenesis. Biochim. Biophys. Acta 1791, 1093–1101.

Monroig, Ó., Rotllant, J., Cerdá-Reverter, J.M., Dick, J.R., Figueras, A., Tocher, D.R., 2010. Expression and role of Elovl4 elongases in biosynthesis of very long-chain fatty acids during zebrafish *Danio rerio* early embryonic development. Biochim. Biophys. Acta 1801, 1145–1154.

Monroig, Ó., Li, Y., Tocher, D.R., 2011a. Delta-8 desaturation activity varies among fatty acyl desaturases of teleost fish: high activity in delta-6 desaturases of marine species. Compar. Biochem. Physiol. B Biochem. Mol. Biol. 159, 206–213.

Monroig, O., Navarro, J.C., Tocher, D.R., 2011b. Long-chain polyunsaturated fatty acids in fish: recent advances on desaturases and elongases involved in their biosynthesis. In: CRUZ-Suarez, L.E., Ricque-Marie, D., Tapia-Salazar, M., Nieto-López, M.G., Villarreal-Cavazos, D.A., Gamboa-Delgado, J., Hernández-Hernández, L.H. (Eds.), Proceedings of the XI International Symposium on Aquaculture Nutrition, 23–25 November 2011, San Nicolás de los Garza, N.L. Universidad Autónoma de Nuevo León, Monterrey, Nuevo León, México. Monterrey, Nuevo León, México: Universidad Autónoma de Nuevo León.

Monroig, Ó., Webb, K., Ibarra-Castro, L., Holt, G.J., Tocher, D.R., 2011c. Biosynthesis of long chain polyunsaturated fatty acids in marine fish: characterization of an Elovl4-like elongase from cobia *Rachycentron canadum* and activation of the pathway during early life stages. Aquaculture 312, 145–153.

Monroig, O., LOPES-Marques, M., Navarro, J.C., Hontoria, F., Ruivo, R., Santos, M.M., Venkatesh, B., Tocher, D.R., Castro, L.F., 2016. Evolutionary functional elaboration of the Elovl2/5 gene family in chordates. Sci. Rep. 6, 20510.

Monroig, O., Tocher, D.R., Castro, L.F.C., 2018. Polyunsaturated fatty acid biosynthesis and metabolism in fish. In: Burdge, G.C. (Ed.), Polyunsaturated Fatty Acid Metabolism in Animals: Applications for Human Health and Research. Elsevier, San Diego, CA, USA.

Monteiro, M., Matos, E., Ramos, R., Campos, I., Valente, L.M.P., 2018. A blend of land animal fats can replace up to 75% fish oil without affecting growth and nutrient utilization of European seabass. Aquaculture 487, 22–31.

Montero, D., Izquierdo, M., 2011. Welfare and health of fish fed vegetable oils as alternative lipid sources to fish oil. In: Turchini, G.M., Ng, W.-K., Tocher, D.R. (Eds.), Fish Oil Replacement and Alternative Lipid Sources in Aquaculture Feeds. CRC Press, Taylor & Francis, Group, Boca Raton, FL, USA.

Montero, D., Caballero, M.J., Gijon, D., Real, F., Robaina, L.E., Izquierdo, M.S., 2006. Influence of dietary blend of vegetable oils on disease resistance and intestinal microbiota of gilthead seabream (*Sparus aurata*). In: XII International Symposium on Fish Nutrition and Feeding, May 28th – June 1st, 2006 Biarritz, France.

Montoya, A., Martins, D.A., Yufera, M., Sanchez-Vazquez, F.J., 2011. Self-selection of diets with different oil oxidation levels in gilthead seabream (*Sparus aurata*). Aquaculture 314, 282–284.

Montoya, A., Zamora, S., Sanchez-Vazquez, F.J., 2012. Dietary selection by gilthead sea bream (*Sparus aurata*) provided with unbalanced mixed-macronutrient feeds dispensed from self-feeders. Aquaculture 358, 35–40.

Montpetit, C.J., Perry, S.F., 1998. The effects of chronic hypoxia on the acute adrenergic stress response in the rainbow trout (*Oncorhynchus mykiss*). Physiol. Zool. 71, 377–386.

Moon, T.W., Mommsen, T.P., 1987. Enzymes of intermediary metabolism in tissue of the little skate, Raja erinacea. J. Exp. Zool. 244, 9–15.

Moore, K.J., Rayner, K.J., Suárez, Y., Fernández-Hernando, C., 2011. The role of microRNAs in cholesterol efflux and hepatic lipid metabolism. Annu. Rev. Nutr. 31, 49–63.

Morais, S., 2017. The physiology of taste in fish: potential implications for feeding stimulation and gut chemical sensing. Rev. Fish. Sci. Aquacult. 25, 133–149.

Morais, S., Monroig, O., Zheng, X., Leaver, M.J., Tocher, D.R., 2009. Highly unsaturated fatty acid synthesis in Atlantic salmon: characterisation of ELOVL5- and ELOVL2-like elongases. Mar. Biotechnol. 11, 627–639.

Morais, S., Pratoomyot, J., Torstensen, B.E., Taggart, J.B., Guy, D.R., Gordon Bell, J., Tocher, D.R., 2011. Diet × genotype interactions in hepatic cholesterol and lipoprotein metabolism in Atlantic salmon (*Salmo salar*) in response to replacement of dietary fish oil with vegetable oil. Br. J. Nutr. 106, 1457–1469.

Morais, S., Castanheira, F., Martinez-Rubio, L., Conceicao, L.E., Tocher, D.R., 2012a. Long chain polyunsaturated fatty acid synthesis in a marine vertebrate: ontogenetic and nutritional regulation of a fatty acyl desaturase with Delta4 activity. Biochim. Biophys. Acta 1821, 660–671.

Morais, S., Edvardsen, R.B., Tocher, D.R., Bell, J.G., 2012b. Transcriptomic analyses of intestinal gene expression of juvenile Atlantic cod (*Gadus morhua*) fed diets with Camelina oil as replacement for fish oil. Compar. Biochem. Physiol. B Biochem. Mol. Biol. 161, 283–293.

Morais, S., Taggart, J.B., Guy, D.E., Bell, J.G., Tocher, D.R., 2012c. Hepatic transcriptome analysis of inter-family variability in flesh n-3 long-chain polyunsaturated fatty acid content in Atlantic salmon. BMC Genom. 13, 410.

Morash, A.J., Mcclelland, G.B., 2011. Regulation of carnitine palmitoyltransferase (CPT) I during fasting in rainbow trout (*Oncorhynchus mykiss*) promotes increased mitochondrial fatty acid oxidation. Physiol. Biochem. Zool. 84, 625–633.

Morash, A.J., Bureau, D.P., Mcclelland, G.B., 2009. Effects of dietary fatty acid composition on the regulation of carnitine palmitoyltransferase (CPT) I in rainbow trout (*Oncorhynchus mykiss*). Compar. Biochem. Physiol. B Biochem. Mol. Biol. 152, 85–93.

Morash, A.J., Moine, C.M.R.L., Mcclelland, G.B., 2010. Genome duplication events have led to a diversification in the CPT I gene family in fish. Am. J. Physiol. Regul. Integr. Compar. Physiol. 299, R579.

Morishita, H., Zhao, Y.G., Tamura, N., Nishimura, T., Kanda, Y., Sakamaki, Y., Okazaki, M., Li, D., Mizushima, N., 2019. A critical role of VMP1 in lipoprotein secretion. eLife 8, e48834.

Moya-Falcón, C., Hvattum, E., Dyrøy, E., Skorve, J., Stefansson, S.O., Thomassen, M.S., Jakobsen, J.V., Berge, R.K., Ruyter, B., 2004. Effects of 3-thia fatty acids on feed intake, growth, tissue fatty acid composition, β-oxidation and N+, K+-ATPase activity in Atlantic salmon. Compar. Biochem. Physiol. B Biochem. Mol. Biol. 139B, 657–668.

Moya-Falcón, C., Thomassen, M.S., Jakobsen, J.V., Ruyter, B., 2005. Effects of dietary supplementation of rapeseed oil on metabolism of [1-14C]18:1n-9, [1-14C]20:3n-6, and [1-14C]20:4n-3 in Atlantic salmon hepatocytes. Lipids 40, 709–717.

Moyes, C., Buck, L., Hochachka, P., 1990. Mitochondrial and peroxisomal fatty acid oxidation in elasmobranchs. Am. J. Physiol. Regul. Integr. Compar. Physiol. 258, R756–R762.

Moyes, C.D., Suarez, R.K., Brown, G.S., Hochachka, P.W., 1991. Peroxisomal β-oxidation: insights from comparative biochemistry. J. Exp. Zool. 260, 267–273.

Mozanzadeh, M.T., Agh, N., Yavari, V., Marammazi, J.G., Mohammadian, T., Gisbert, E., 2016. Partial or total replacement of dietary fish oil with alternative lipid sources in silvery-black porgy (*Sparidentex hasta*). Aquaculture 451, 232–240.

Mukherjee, R., Strasser, J., Jow, L., Hoener, P., Paterniti, J.R., Heyman, R.A., 1998. RXR agonists activate PPARalpha-inducible genes, lower triglycerides, and raise HDL levels in vivo. Arterioscler. Thromb. Vasc. Biol. 18, 272–276.

Müller, T., Nogueiras, R., Andermann, M., Andrews, Z., Anker, S., Argente, J., Tschöp, M., 2015. Ghrelin. Mol. Metab. 4, 437–460.

Murata, H., Higashi, T., 1979. Studies on the metabolism of fatty acid in fish-IV rate of fatty acid decrease based on β-oxidation in carp dark muscle mitochondria. Nippon Suisan Gakkaishi 45, 211–217.

Murray, H.M., Gallant, J.W., Perez-Casanova, J.C., Johnson, S.C., Douglas, S.E., 2003. Ontogeny of lipase expression in winter flounder. J. Fish. Biol. 62, 816–833.

Musso, G., Gambino, R., Cassader, M., 2009. Recent insights into hepatic lipid metabolism in non-alcoholic fatty liver disease (NAFLD). Prog. Lipid Res. 48, 1–26.

Nakamura, K., Hata, M., 1985. A study on astaxanthin in salmon *Oncorhynchus keta* serum. Bull. Jpn. Soc. Sci. Fish. 51, 979–983.

Nakayama, H., Shimada, Y., Zang, L., Terasawa, M., Nishiura, K., Matsuda, K., Toombs, C., Langdon, C., Nishimura, N., 2018. Novel anti-obesity properties of *Palmaria mollis* in zebrafish and mouse models. Nutrients 10, 1–16.

Nanton, D.A., Lall, S.P., Ross, N.W., Mcniven, M.A., 2003. Effect of dietary lipid level on fatty acid β-oxidation and lipid composition in various tissues of haddock, *Melanogrammus aeglefinus* L. Compar. Biochem. Physiol. B Biochem. Mol. Biol. 135, 95–108.

Nanton, D.A., Vegusdal, A., Rørå, A.M.B., Ruyter, B., Baeverfjord, G., Torstensen, B.E., 2007. Muscle lipid storage pattern, composition, and adipocyte distribution in different parts of Atlantic salmon (*Salmo salar*) fed fish oil and vegetable oil. Aquaculture 265, 230–243.

Nanton, D.A., Ruohonen, K., Robb, D.H.F., El-Mowafi, A., Hartnell, G.F., 2012. Effect of soy oil containing stearidonic acid on growth performance and fillet fatty acid composition of Atlantic salmon. Aquacult. Nutr. 18, 640–650.

Napier, J.A., Sayanova, O., 2005. The production of very-long-chain PUFA biosynthesis in transgenic plants: towards a sustainable source of fish oils. Proc. Nutr. Soc. 64, 387–393.

Napier, J.A., Michaelson, L.V., Stobart, A.K., 1999. Plant desaturases: harvesting the fat of the land. Curr. Opin. Plant Biol. 2, 123–127.

Napier, J.A., Usher, S., Haslam, R.P., Ruiz-Lopez, N., Sayanova, O., 2015. Transgenic plants as a sustainable, terrestrial source of fish oils. Eur. J. Lipid Sci. Technol. 117, 1317–1324.

Napier, J.A., Olsen, R.E., Tocher, D.R., 2019. Update on GM canola crops as novel sources of omega-3 fish oils. Plant Biotechnol. 17, 703–705.

Nash, K.L., Cvitanovic, C., Fulton, E.A., Halpern, B.S., Milner-Gulland, E.J., Watson, R.A., Blanchard, J.L., 2017. Planetary boundaries for a blue planet. Nat. Ecol. Evol. 1, 1625–1634.

Nasopoulou, C., Zabetakis, I., 2012. Benefits of fish oil replacement by plant originated oils in compounded fish feeds. A review. LWT Food Sci. Technol. 47, 217–224.

Natalia, Y., Hashim, R., Ali, A., Chong, A., 2004. Characterization of digestive enzymes in a carnivorous ornamental fish, the Asian bony tongue Scleropages formosus (Osteoglossidae). Aquaculture 233, 305–320.

Navari-Izzo, F., Quartacci, M.F., Sgherri, C., 2002. Lipoic acid: a unique antioxidant in detoxification of activated oxygen species. Plant Physiol. Biochem. 40, 463–470.

Navarro, I., Rojas, P., Capilla, E., Albalat, A., Castillo, J., Montserrat, N., Codina, M., Gutiérrez, J., 2002. Insights into insulin and glucagon responses in fish. Fish Physiol. Biochem. 27, 205–216.

Naylor, R.L., Hardy, R.W., Bureau, D.P., Chiu, A., Elliott, M., Farrell, A.P., Forster, I., Gatlin, D.M., Goldburg, R.J., Hua, K., Nichols, P.D., 2009. Feeding aquaculture in an era of finite resources. Proc. Natl. Acad. Sci. U. S. A. 106, 15103–15110.

Neighbors, M.A., 1988. Triacylglycerols and wax esters in the lipids of deep midwater teleost fishes of the southern-California bight. Mar. Biol. 98, 15–22.

Neubauer, P., Andersen, K.H., 2019. Thermal performance of fish is explained by an interplay between physiology, behaviour and ecology. Conserv. Physiol. 7, coz025.

Ng, W.K., Koh, C.B., 2017. The utilization and mode of action of organic acids in the feeds of culture aquatic animals. Rev. Aquacult. 9, 342–368.

Ng, W.K., Sigholt, T., Bell, J.G., 2004. The influence of environmental temperature on the apparent nutrient and fatty acid digestibility in Atlantic salmon (*Salmo salar* L.) fed finishing diets containing different blends of fish oil, rapeseed oil and palm oil. Aquacult. Res. 35, 1228–1237.

Ng, W.K., Chik-Boon, K., Sudesh, K., Siti-Zahrah, A., 2009. Effects of dietary organic acids on growth, nutrient digestibility and gut microflora of red hybrid tilapia, Oreochromis sp., and subsequent survival during a challenge test with Streptococcus agalactiae. Aquacult. Res. 40, 1490–1500.

Ni, J., Yan, Q., Yu, Y., Zhang, T., 2014. Factors influencing the grass carp microbiome and its effect on metabolism. FEMS Microbiol. Ecol. 87, 704–714.

Nichols, J.W., Ozarowski, J., 1990. Sizing of lecithin-bile salt mixed micelles by size-exclusion high-performance liquid chromatography. Biochemistry 29, 4600–4606.

Nickell, D.C., Bromage, N.R., 1998. The effect of dietary lipid level on variation of flesh pigmentation in rainbow trout (*Oncorhynchus mykiss*). Aquaculture 161, 237–251.

Ning, L.J., He, A.Y., Li, J.M., Lu, D.L., Jiao, J.G., Li, L.Y., Li, D.L., Zhang, M.L., Chen, L.Q., Du, Z.Y., 2016. Mechanisms and metabolic regulation of PPARα activation in Nile tilapia (*Oreochromis niloticus*). Biochim. Biophys. Acta 1861, 1036–1048.

Ning, L.J., He, A.-Y., Lu, D.-L., Li, J.-M., Qiao, F., Li, D.-L., Zhang, M.-L., Chen, L.-Q., Du, Z.-Y., 2017. Nutritional background changes the hypolipidemic effects of fenofibrate in Nile tilapia (*Oreochromis niloticus*). Sci. Rep. 7, 41706.

Ning, L.J., Liu, Y., Wang, W.X., Li, Y.Y., Chen, L.Q., Du, Z.Y., 2019. The metabolic regulation of fenofibrate is dependent on dietary protein content in male juveniles of Nile tilapia (*Oreochromis niloticus*). Br. J. Nutr. 122, 648–656.

Niu, K.M., Khosravi, S., Kothari, D., Lee, W.D., Lim, J.M., Lee, B.J., Kim, K.W., Lim, S.G., Lee, S.M., Kim, S.K., 2019. Effects of dietary multi-strain probiotics supplementation in a low fishmeal diet on growth performance, nutrient utilization, proximate composition, immune parameters, and gut microbiota of juvenile olive flounder (*Paralichthys olivaceus*). Fish Shellfish Immunol. 93, 258–268.

Nogales-Mérida, S., Gobbi, P., Józefiak, D., Mazurkiewicz, J., Dudek, K., Rawski, M., Kierończyk, B., Józefiak, A., 2019. Insect meals in fish nutrition. Rev. Aquacult. 11, 1080–1103.

Norambuena, F., Estevez, A., Sanchez-Vazquez, F.J., Carazo, I., Duncan, N., 2012. Self-selection of diets with different contents of arachidonic acid by Senegalese sole (*Solea senegalensis*) broodstock. Aquaculture 364, 198–205.

Norambuena, F., Lewis, M., Hamid, N.K.A., Hermon, K., Donald, J.A., Turchini, G.M., 2013. Fish oil replacement in current aquaculture feed: is cholesterol a hidden treasure for fish nutrition? PLoS One 8, 1–9.

Norambuena, F., Sofia, M., Emery, J.A., Turchini, G.M., 2015. Arachidonic acid and eicosapentaenoic acid metabolism in juvenile Atlantic salmon as affected by water temperature. PLoS One 10 (11), e0143622.

Norambuena, F., Rombenso, A., Tuchini, G., 2016. Towards the optimization of performance of Atlantic salmon reared at different water temperatures via the manipulation of dietary ARA/EPA ratio. Aquaculture 450, 48–57.

Nordgarden, U., Hemre, G.I., Hansen, T., 2002. Growth and body composition of Atlantic salmon (*Salmo salar L.*) parr and smolt fed diets varying in protein and lipid contents. Aquaculture 207, 65–78.

Nordgreen, A., Hamre, K., Langdon, C., 2007. Development of lipid microbeads for delivery of lipid and water-soluble materials to Artemia. Aquaculture 273, 614–623.

Nørrelund, H., 2005. The metabolic role of growth hormone in humans with particular reference to fasting. Growth Hormone IGF Res. 2, 95–122.

NRC, 2011. Nutrient Requirements of Fish and Shrimp. National Research Council, The National Academies Press, Washington DC, USA.

Oboh, A., Betancor, M.B., Tocher, D.R., Monroig, O., 2016. Biosynthesis of long-chain polyunsaturated fatty acids in the African catfish *Clarias gariepinus*: molecular cloning and functional characterisation of fatty acyl desaturase (fads2) and elongase (elovl2) cDNAs7. Aquaculture 462, 70–79.

Oboh, A., Kabeya, N., Carmona-Antoñanzas, G., Castro, L.F.C., Dick, J.R., Tocher, D.R., Monroig, O., 2017a. Two alternative pathways for docosahexaenoic acid (DHA 22:6n-3) biosynthesis are widespread among teleost fish. Sci. Rep. 7, 3889.

Oboh, A., Navarro, J.C., Tocher, D.R., Monroig, Ó., 2017b. Elongation of very long-chain (> C24) fatty acids in *Clarias gariepinus*: cloning, functional characterisation and tissue expression of Elovl4 elongases. Lipids 52, 837–848.

Ogino, C., Takashima, F.L., Chiou, J.Y., 1978. Requirement of rainbow trout for dietary magnesium. Bull. Jpn. Soc. Sci. Fish. 44, 1105–1108.

Oh, S.F., Vickery, T.W., Serhan, C.N., 2011. Chiral lipidomics of E-series resolvins: aspirin and the biosynthesis of novel mediators. Biochim. Biophys. Acta 1811, 737–747.

Oku, H., Umino, T., 2008. Molecular characterization of peroxisome proliferator-activated receptors (PPARs) and their gene expression in the differentiating adipocytes of red sea bream *Pagrus major*. Compar. Biochem. Physiol. B Biochem. Mol. Biol. 151, 268–277.

Oldham, T., Nowak, B., Hvas, M., Oppedal, F., 2019. Metabolic and functional impacts of hypoxia vary with size in Atlantic salmon. Compar. Biochem. Physiol. A Mol. Integr. Physiol. 231, 30–38.

Olsen, R.E., Baker, R.T.M., 2006. Lutein does not influence flesh astaxanthin pigmentation in the Atlantic salmon (*Salmo salar* L.). Aquaculture 258, 558–564.

Olsen, R.E., Henderson, R.J., 1997. Muscle fatty acid composition and oxidative stress indices of Arctic charr, *Salvelinus alpinus* (L.), in relation to dietary polyunsaturated fatty acid levels and temperature. Aquacult. Nutr. 3, 227–238.

Olsen, R.E., Ringø, E., 1997. Lipid digestibility in fish: a review. Recent Res. Dev. Lipid Res. 1, 199–265.

Olsen, R.E., Myklebust, R., Kaino, T., Ringø, E., 1999. Lipid digestibility and ultrastructural changes in the enterocytes of Arctic char (*Salvelinus alpinus* L.) fed linseed oil and soybean lecithin. Fish Physiol. Biochem. 21, 35–44.

Olsen, R.E., Myklebust, R., Ringø, E., Mayhew, T.M., 2000. The influences of dietary linseed oil and saturated fatty acids on caecal enterocytes in Arctic char (*Salvelinus alpinus* L.): a quantitative ultrastructural study. Fish Physiol. Biochem. 22, 207–216.

Olsen, R.E., Dragnes, B.T., Myklebust, R., Ringø, E., 2003. Effect of soybean oil and soybean lecithin on intestinal lipid composition and lipid droplet accumulation of rainbow trout, *Oncorhynchus mykiss* Walbaum. Fish Physiol. Biochem. 29, 181–192.

Olsen, R.E., Henderson, R.J., Sountama, J., Hemre, G.-I., Ringø, E., Melle, W., Tocher, D.R., 2004. Atlantic salmon, *Salmo salar*, utilizes wax ester-rich oil from *Calanus finmarchicus* effectively. Aquaculture 240, 433–449.

Olsen, R.E., Kiessling, A., Milley, J.E., Ross, N.W., Lall, S.P., 2005. Effect of lipid source and bile salts in diet of Atlantic salmon, *Salmo salar* L., on astaxanthin blood levels. Aquaculture 250, 804–812.

Olsen, R.E., Suontama, J., Langmyhr, E., Mundheim, H., Ringø, E., Melle, W., Malde, M.K., Hemre, G.-I., 2006. The replacement of fish meal with Antarctic krill, *Euphausia superba* in diets for Atlantic salmon, *Salmo salar*. Aquacult. Nutr. 12, 280–290.

Olsen, R.E., Waagbø, R., Melle, W., Ringø, E., Lall, S.P., 2011. Alternative marine sources. In: Turchini, G.M., Ng, W.-K., Tocher, D.R. (Eds.), Fish Oil Replacement and Alternative Lipid Sources in Aquaculture Feeds. CRC Press, Florida, USA.

Olsen, Y., Evjemo, J.O., Kjørsvik, E., Larssen, H., Li, K., Overrein, I., Rainuzzo, J., 2014. DHA content in dietary phospholipids affects DHA content in phospholipids of cod larvae and larval performance. Aquaculture 428, 203–214.

Ou, M., Huang, R., Luo, Q., Xiong, L., Chen, K., Wang, Y., 2019. Characterisation of scavenger receptor class B type 1 in rare minnow (*Gobiocypris rarus*). Fish Shellfish Immunol. 89, 614–622.

Owen, M.A.G., Bradley, G., Davies, S.J., 2006. The effect of dietary supplementation of sodium butyrate on growth and the microflora of *Clarias gariepinus* (Burchell 1822). In: XII International Symposium on Fish Nutrition & Feeding, May 28th–June 1st 2006 Biarritz, France, p. 149.

Oxley, A., Tocher, R.D., Torstensen, E.B., Olsen, E.R., 2005a. Fatty acid utilisation and metabolism in caecal enterocytes of rainbow trout (*Oncorhynchus mykiss*) fed dietary fish or copepod oil. Biochem. Biophys. Acta 1737, 119–129.

Oxley, A., Torstensen, B.E., Rustan, A.C., Olsen, R.E., 2005b. Enzyme activities of intestinal triacylglycerol and phosphatidylcholine biosynthesis in Atlantic salmon (*Salmo salar* L.). Compar. Biochem. Physiol. B Biochem. Mol. Biol. 141, 77–87.

Oxley, A., Jutfelt, F., Sundell, K., Olsen, R.E., 2007. Sn-2-monoacylglycerol, not glycerol, is preferentially utilised for triacylglycerol and phosphatidylcholine biosynthesis in Atlantic salmon (*Salmo salar* L.) intestine. Compar. Biochem. Physiol. B Biochem. Mol. Biol. 146, 115–123.

Oxley, A., Bogevik, A.S., Henderson, J.R., Waagbø, R., Tocher, D.R., Olsen, R.E., 2009. Digestibility of *Calanus finmarchicus* wax esters in Atlantic salmon (*Salmo salar*) freshwater and seawater post-smolts maintained at constant water temperature. Aquacult. Nutr. 15, 459–469.

Pan, X., Zhang, Y., Wang, L., Hussain, M.M., 2010. Diurnal regulation of MTP and plasma triglyceride by CLOCK is mediated by SHP. Cell Metabol. 12, 174–186.

Pan, Y.-X., Luo, Z., Wu, K., Zhang, L.-H., Xu, Y.-H., Chen, Q.-L., 2016. Cloning, mRNA expression and transcriptional regulation of five retinoid X receptor subtypes in yellow catfish *Pelteobagrus fulvidraco* by insulin. Gen. Compar. Endocrinol. 225, 133–141.

Pan, H., Li, L., Li, J.M., Wang, W.L., Limbu, S.M., Degrace, P., Li, D.L., Du, Z.Y., 2017. Inhibited fatty acid β-oxidation impairs stress resistance ability in Nile tilapia (*Oreochromis niloticus*). Fish Shellfish Immunol. 68, 500–508.

Panserat, S., Kolditz, C., Richard, N., Plagnes-Juan, E., Piumi, F., Esquerré, D., Médale, F., Corraze, G., Kaushik, S., 2008. Hepatic gene expression profiles in juvenile rainbow trout (*Oncorhynchus mykiss*) fed fishmeal or fish oil-free diets. Br. J. Nutr. 100, 953–967.

Panserat, S., Marandel, L., Geurden, I., Veron, V., Dias, K., Plagnes-Juan, E., Pegourié, G., Arbenoits, E., Santigosa, E., Weber, G., Verlhac Trichet, V., 2017. Muscle catabolic capacities and global hepatic epigenome are modified in juvenile rainbow trout fed different vitamin levels at first feeding. Aquaculture 468, 515–523.

Paredes, J.F., Vera, L.M., Navarro, I., Martínez-López, F.J., Sánchez-Vázquez, F.J., 2014. Circadian rhythms of gene expression of lipid metabolism in gilthead sea bream liver: synchronization to light and feeding time. Chronobiol. Int. 31, 613–626.

Paredes, J.F., López-Olmeda, J.F., Martínez, F.J., Sánchez-Vázquez, F.J., 2015. Daily rhythms of lipid metabolic gene expression in zebrafish liver: response to light/dark and feeding cycles. Chronobiol. Int. 32, 1438–1448.

Park, G.H., Lee, J.H., Yun, H.H., Browdy, C.L., Bharadwaj, A.S., Bai, S.C.C., 2011. Effects of two different organic blends in olive flounder. Kor. J. Organ. Agric. 19, 39–42.

Park, K.H., Ye, Z.W., Zhang, J., Hammad, S.M., Townsend, D.M., Rockey, D.C., Kim, S.H., 2019. 3-ketodihydrosphingosine reductase mutation induces steatosis and hepatic injury in zebrafish. Sci. Rep. 9, 1–13.

Parra, A.M., Rosas, A., Lazo, J.P., Viana, M.T., 2007. Partial characterization of the digestive enzymes of Pacific bluefin tuna *Thunnus orientalis* under culture conditions. Fish Physiol. Biochem. 33, 223–231.

Patel, M.S., Srinivasan, M., 2002. Metabolic programming: causes and consequences. J. Biol. Chem. 277, 1629–1632.

Patton, J.S., Nevenzel, J.C., 1974. Fat digestion in fish. J. Am. Chem. Soc. 51, 521A.

Patton, J.S., Quinn, J.G., 1973. Studies on the digestive lipase of the surf clam *Spisula solidissima*. Mar. Biol. 21, 59–69.

Patton, J.S., Nevenzel, J.C., Benson, A.A., 1975. Specificity of digestive lipases in hydrolysis of wax esters and triglycerides in anchovy and other selected species. Lipids 10, 575–583.

Patton, J.S., Warner, T.G., Benson, A.A., 1977. Partial characterization of the bile salt-dependent triacylglycerol lipase from the leopard shark pancreas. Biochim. Biophys. Acta 486, 322–330.

Pedrajas, J.R., López-Barea, J., Peinado, J., 1996. Dieldrin induces peroxisomal enzymes in fish (*Sparus aurata*) liver. Compar. Biochem. Physiol. C Toxicol. Pharmacol. 115, 0–131.

Peng, S., Chen, L., Qin, J., Hou, J., Yu, N., Long, Z., Li, E., Ye, J., 2009. Effects of dietary vitamin E supplementation on growth performance, lipid peroxidation and tissue fatty acid composition of black sea bream (*Acanthopagrus schlegeli*) fed oxidized fish oil. Aquacult. Nutr. 15, 329–337.

Peng, X., Shang, G., Wang, W., Chen, X., Lou, Q., Zhai, G., Li, D., Du, Z., Ye, Y., Jin, X., 2017. Fatty acid oxidation in zebrafish adipose tissue is promoted by 1α, 25 (OH) 2D3. Cell Rep. 19, 1444–1455.

Penn, A.H., Schmid-Schönbein, G.W., 2008. The intestine as source of cytotoxic mediators in shock: free fatty acids and degradation of lipid-binding proteins. Am. J. Physiol. 294, H1779–H1792.

Perello, M., Dickson, S., 2015. Ghrelin signalling on food reward: a salient link between the gut and the mesolimbic system. J. Neuroendocrinol. 27, 424–434.

Perera, E., Yufera, M., 2016. Soybean meal and soy protein concentrate in early diet elicit different nutritional programming effects on juvenile zebrafish. Zebrafish 13, 61–69.

Peres, H., Lim, C., Klesius, P.H., 2004. Growth, chemical composition and resistance to Streptococcus iniae challenge of juvenile Nile tilapia (*Oreochromis niloticus*) fed graded levels of dietary inositol. Aquaculture 235, 423–432.

Pérez-Jiménez, A., Peres, H., Rubio, V.C., Oliva-Teles, A., 2013. Effects of diet supplementation with white tea and methionine on lipid metabolism of gilthead sea bream juveniles (*Sparus aurata*). Fish Physiol. Biochem. 39, 661–670.

Perry, S.F., Gilmour, K.M., Vulesevic, B., Mcneill, B., Chew, S.F., Ip, Y.K., 2005. Circulating catecholamines and cardiorespiratory responses in hypoxic lungfish (*Protopterus dolloi*): a comparison of aquatic and aerial hypoxia. Physiol. Biochem. Zool. 78, 325–334.

Petrie, J.R., Shrestha, P., Zhou, X.-R., Mansour, M.P., Liu, Q., Belide, S., Nichols, P.D., Singh, S.R., 2012. Metabolic engineering plant seeds with fish oil-like levels of DHA. PLoS One 7, e49165.

Petrie, J.R., Zhou, X.R., Leonforte, A., Mcallister, J., Shrestha, P., Kennedy, Y., Belide, S., Buzza, G., Gororo, N., Gao, W., Lester, G., Mansour, M.P., Mulder, R.J., Liu, Q., Tian, L., Silva, C., Cogan, N.O.I., Nichols, P.D., Green, A.G., De Feyter, R., Devine, M.D., Singh, S.P., 2020. Development of a *Brassica napus* (canola) crop containing fish oil-like levels of DHA in the seed oil. Front. Plant Sci. 11, 727.

Pettersson, A., Johnsson, L., Brannas, E., Pickova, J., 2009. Effects of rapeseed oil replacement in fish feed on lipid composition and self-selection by rainbow trout (*Oncorhynchus mykiss*). Aquacult. Nutr. 15, 577–586.

Pichon, L., Huneau, J.F.O., Fromentin, G., Tomé, D., 2006. A high-protein, high-fat, carbohydrate-free diet reduces energy intake, hepatic lipogenesis, and adiposity in rats. J. Nutr. 136, 1256–1260.

Plagnes-Juan, E., Lansard, M., Seiliez, I., Médale, F., Corraze, G., Kaushik, S., Panserat, S., Skiba-Cassy, S., 2008. Insulin regulates the expression of several metabolism-related genes in the liver and primary hepatocytes of rainbow trout (*Oncorhynchus mykiss*). J. Exp. Biol. 211, 2510–2518.

Plutzky, J., 2011. The PPAR-RXR transcriptional complex in the vasculature: energy in the balance. Circ. Res. 108, 1002–1016.

Polakof, S., Médale, F., Skiba-Cassy, S., Corraze, G., Panserat, S., 2010a. Molecular regulation of lipid metabolism in liver and muscle of rainbow trout subjected to acute and chronic insulin treatments. Domest. Anim. Endocrinol. 39, 26–33.

Polakof, S., Moon, T., Aguirre, P., Skiba-Cassy, S., Panserat, S., 2010b. Effects of insulin infusion on glucose homeostasis and glucose metabolism in rainbow trout fed a high-carbohydrate diet. J. Exp. Biol. 213, 4151–4157.

Polakof, S., Medale, F., Larroquet, L., Vachot, C., Corraze, G., Panserat, S., 2011. Regulation of de novo hepatic lipogenesis by insulin infusion in rainbow trout fed a high-carbohydrate diet. J. Anim. Sci. 89, 3079–3088.

Polakof, S., Panserat, S., Soengas, J.L., Moon, T.W., 2012. Glucose metabolism in fish: a review. J. Compar. Physiol. B Biochem. Syst. Environ. Physiol. 182, 1015–1045.

Polley, S.D., Tiku, P.E., Trueman, R.T., Caddick, M.X., Morozov, I.Y., Cossins, A.R., 2003. Differential expression of cold- and diet-specific genes encoding two carp liver Δ9-acyl-CoA desaturase isoforms. Am. J. Physiol. Regul. Integr. Compar. Physiol. 284, R41–R50.

Poston, H.A., 1990. Effect of body size on growth, survival and chemical composition of Atlantic salmon fed soy lecithin and choline. Progress. Fish Cult. 52, 226–230.

Poston, H.A., 1991. Response of Atlantic salmon fry to feed-grade lecithin and choline. Progress. Fish Cult. 53, 224–228.

Potter, S.M., 1995. Overview of proposed mechanisms for the hypocholesterolemic effect of soy. J. Nutr. 125, 606S–611S.

Poudyal, H., Panchal, S.K., Diwan, V., Brown, L., 2011. Omega-3 fatty acids and metabolic syndrome: effects and emerging mechanisms of action. Prog. Lipid Res. 50, 372–387.

Pradhan, A., Bereketoglu, C., Martin, L., Duhagon, J., Olsson, P.-E., 2020. The food preservative ethoxyquin impairs zebrafish development, behavior and alters gene expression profile. Food Chem. Toxicol. 135, 110926.

Price, P., Nelson, C., Clarke, S., 2000. Omega-3 polyunsaturated fatty acid regulation of gene expression. Curr. Opin. Lipidol. 11, 3–7.

Qi, C., Xie, C., Tang, R., Qin, X., Wang, D., Li, D., 2016. Effect of stocking density on growth, physiological responses, and body composition of juvenile blunt snout bream, *Megalobrama amblycephala*. J. World Aquacult. Soc. 47, 358–368.

Qiang, J., Yi, F.T., Jie, H., Yi, L.S., Xu, P., 2017. miR-29a modulates SCD expression and is regulated in response to a saturated fatty acid diet in juvenile genetically improved farmed tilapia (*Oreochromis niloticus*). J. Exp. Biol. 220, 1481–1489.

Qiao, Y., Sun, J., Ding, Y., Le, G., Shi, Y., 2013. Alterations of the gut microbiota in high-fat diet mice is strongly linked to oxidative stress. Appl. Microbiol. Biotechnol. 97, 1689–1697.

Quinlivan, V.H., Farber, S.A., 2017. Lipid uptake, metabolism, and transport in the larval zebrafish. Front. Endocrinol. 8.

Raatz, S.K., Golovko, M.Y., Brose, S.A., Rosenberger, T.A., Burr, G.S., Wolters, W.R., Picklo, M.J., 2011. Baking reduces prostaglandin, resolvin, and hydroxy-fatty acid content of farm-raised Atlantic salmon (*Salmo salar*). J. Agric. Food Chem. 59, 11278–11286.

Raingeard, D., Bilbao, E., Sáez-Morquecho, C., Cerio, O.D.D., Orbea, A., Cancio, I., Cajaraville, M.P., 2009a. Cloning and transcription of nuclear receptors and other toxicologically relevant genes, and exposure biomarkers in European hake (*Merluccius merluccius*) after the Prestige oil spill. Mar. Genom. 2, 201–213.

Raingeard, D., Cancio, I., Cajaraville, M.P., 2009b. Cloning and expression pattern of peroxisome proliferator-activated receptors, estrogen receptor α and retinoid X receptor α in the thicklip grey mullet *Chelon labrosus*. Compar. Biochem. Physiol. C Toxicol. Pharmacol. 149, 26–35.

Ran, G., Ying, L., Li, L., Yan, Q., Yi, W., Ying, C., Wu, H., Ye, X., 2017. Resveratrol ameliorates diet-induced dysregulation of lipid metabolism in zebrafish (*Danio rerio*). PLoS One 12, e0180865.

Randall, K.M., Drew, M.D., Øverland, M., Østbye, T.K., Bjerke, M., Vogt, G., Ruyter, B., 2013. Effects of dietary supplementation of coriander oil, in canola oil diets, on the metabolism of [1-14C] 18:3n-3 and [1-14C] 18:2n-6 in rainbow trout hepatocytes. Compar. Biochem. Physiol. B Biochem. Mol. Biol. 166B, 65–72.

Rawles, S.D., Smith, S.B., Gatlin Iii, D.M., 2008. Hepatic glucose utilization and lipogenesis of hybrid striped bass (*Morone chrysops* × *Morone saxatilis*) in response to dietary carbohydrate level and complexity. Aquacult. Nutr. 14, 40–50.

Ray, A.K., Ringø, E., 2014. The gastrointestinal tract of fish. In: Merrifield, D., Ringø, E. (Eds.), Aquaculture Nutrition: Gut Health, Probiotics and Prebiotics. Wiley-Blackwell Publishing, Oxford, UK.

Reboul, E., 2019. Mechanisms of carotenoid intestinal absorption: where do we stand? Nutrients 11, 838.

Refstie, S., Sahlstrom, S., Brathen, E., Baeverfjord, G., Krogedal, P., 2005. Lactic acid fermentation eliminates indigestible carbohydrates and antinutritional factors in soybean meal for Atlantic salmon (*Salmo salar*). Aquaculture 246, 331–345.

Regost, C., Arzel, J., Robin, J., Rosenlund, G., Kaushik, S.J., 2003. Total replacement of fish oil by soybean or linseed oil with a return to fish oil in turbot (Psetta maxima): 1. Growth performance, flesh fatty acid profile, and lipid metabolism. Aquaculture 217, 465–482.

Ren, M., Ai, Q., Mai, K., Ma, H., Wang, X., 2011. Effect of dietary carbohydrate level on growth performance, body composition, apparent digestibility coefficient and digestive enzyme activities of juvenile cobia, *Rachycentron canadum* L. Aquacult. Res. 42, 1467–1475.

Ren, M., Habte-Tsion, H.M., Xie, J., Liu, B., Zhou, Q., Ge, X., Pan, L., Chen, R., 2015. Effects of dietary carbohydrate source on growth performance, diet digestibility and liver glucose enzyme activity in blunt snout bream, *Megalobrama amblycephala*. Aquaculture 438, 75–81.

Ren, W., Li, J., Tan, P., Cai, Z., Xu, W., Zhang, Y., Nian, R., Macq, B., Ai, Q., 2017. Lipid deposition patterns among different sizes of three commercial fish species. Aquacult. Res. 49, 1046–1052.

Repa, J.J., Mangelsdorf, D.J., 2000. The role of orphan nuclear receptors in the regulation of cholesterol homeostasis. Annu. Rev. Cell Dev. Biol. 16, 459–481.

Ricardo, A., Szostak, J.W., 2009. Life on earth. Sci. Am. 301, 54–61.

Richieri, G.V., Ogata, R.T., Kleinfeld, A.M., 1994. Equilibrium constants for the binding of fatty acids with fatty acid-binding proteins from adipocyte, intestine, heart, and liver measured with the fluorescent probe ADIFAB. J. Biol. Chem. 269, 23918–23930.

Rietman, A., Schwarz, J., Blokker, B.A., Siebelink, E., Kok, F.J., Afman, L.A., Tomé, D., Mensink, M., 2014. Increasing protein intake modulates lipid metabolism in healthy young men and women consuming a high-fat hypercaloric diet. J. Nutr. 144, 1174–1180.

Riley, L.G., Fox, B.K., Kaiya, H., Hirano, T., Grau, E.G., 2005. Long-term treatment of ghrelin stimulates feeding, fat deposition, and alters the GH/IGF-I axis in the tilapia, *Oreochromis mossambicus*. Gen. Compar. Endocrinol. 142, 234–240.

Rimoldi, S., Benedito-Palos, L., Terova, G., Pérez-Sánchez, J., 2016. Wide-targeted gene expression infers tissue-specific molecular signatures of lipid metabolism in fed and fasted fish. Rev. Fish Biol. Fish. 26, 93–108.

Rimoldi, S., Gliozheni, E., Ascione, C., Gini, E., Terova, G., 2018. Effect of a specific composition of short- and medium-chain fatty acid 1- monoglycerides on growth performances and gut microbiota of gilthead sea bream (*Sparus aurata*). PeerJ 6, e5355.

Ringø, E., 1993. Does dietary linoleic acid affect intestinal microflora in Arctic charr, *Salvelinus alpinus* (L.)? Aquacult. Fish. Manag. 24.

Ringø, E., Olsen, R.E., 1999. The effect of diet on aerobic bacterial flora associated with intestine of Arctic charr (*Salvelinus alpinus* L.). J. Appl. Microbiol. 86, 22–28.

Ringø, E., Bendiksen, H.R., Gausen, S., Sundsfjord, A., Olsen, R.E., 1998. The effect of dietary fatty acids on lactic acid bacteria associated with the epithelial mucosa and from faecalia of Arctic charr, *Salvelinus alpinus* (L.). J. Appl. Microbiol. 85, 855–864.

Ringø, E., Lødemel, J.B., Myklebust, R., Jensen, L., Lund, V., Mayhew, T.M., Olsen, R.E., 2002. Effect of soybean, linseed and marine oils on aerobic gut microbiota of Arctic charr *Salvelinus alpinus* L. before and after challenge with *Aeromonas salmonicida* ssp. *salmonicida*. Aquacult. Res. 33, 591–606.

Ringø, E., Olsen, R.E., Gifstad, T.Ø., Dalmo, R.A., Amlund, H., Hemre, G.I., Bakke, A.M., 2010. Prebiotics in aquaculture: a review. Aquacult. Nutr. 16, 117–136.

Ringø, E., Dimitroglou, A., Hoseinifar, S.H., Davies, S.J., 2014. Prebiotics in finfish: an update. In: Merrifield, D., Ringø, E. (Eds.), Aquaculture Nutrition: Gut Health, Probiotics and Prebiotics. Wiley-Blackwell Publishing, Oxford, UK.

Ringø, E., Zhou, Z., Gonzalez Vecino, J.L., Wadsworth, S., Romero, J., Krogdahl, Å., Olsen, R.E., Dimitroglou, A., Foey, A., Davies, S., Owen, M., Lauzon, H.L., Løvmo Martinsen, L., De Schryver, P., Bossier, P., Sperstad, S., Merrifield, D.L., 2016. Effects of dietary components on the gut microbiota of aquatic animals: a never-ending story? Aquacult. Nutr. 22, 219–282.

Rivers, J.P., Sinclair, A.J., Craqford, M.A., 1975. Inability of the cat to desaturate essential fatty acids. Nature 258, 171–173.

Robb, D.H.F., Kestin, S.C., Warriss, P.D., Nute, G.R., 2002. Muscle lipid content determines the eating quality of smoked and cooked Atlantic salmon (*Salmo salar*). Aquaculture 205, 345–358.

Robert, S., Singh, S., Zhou, X., Petrie, J., Blackburn, S., Mansour, M., Nichols, P., Liu, Q., Green, S., 2005. Metabolic engineering of Arabidopsis to produce nutritionally important DHA in seed oil. Funct. Plant Biol. 32, 473–479.

Rocha, F., Dias, J., Engrola, S., Gavaia, P., Geurden, I., Dinis, M.T., Panserat, S., 2015. Glucose metabolism and gene expression in juvenile zebrafish (*Danio rerio*) challenged with a high carbohydrate diet: effects of an acute glucose stimulus during late embryonic life. Br. J. Nutr. 113, 403–413.

Rocha, F., Dias, J., Geurden, I., Dinis, M.T., Panserat, S., Engrola, S., 2016. High-glucose feeding of gilthead seabream (*Sparus aurata*) larvae: effects on molecular and metabolic pathways. Aquaculture 451, 241–253.

Roche, H.M., Gibney, M.J., 2000. The impact of postprandial lipemia in accelerating atherothrombosis. J. Cardiovasc. Risk 7, 317–324.

Rockstrom, J., Steffen, W., Noone, K., Persson, A., Chapin, F.S., Lambin, E.F., Lenton, T.M., Scheffer, M., Folke, C., Schellnhuber, H.J., Nykvist, B., de Wit, C.A., Hughes, T., Van Der Leeuw, S., Rodhe, H., Sorlin, S., Snyder, P.K., Costanza, R., Svedin, U., Falkenmark, M., Karlberg, L., Corell, R.W., Fabry, V.J., Hansen, J., Walker, B., Liverman, D., Richardson, K., Crutzen, P., Foley, J.A., 2009. A safe operating space for humanity. Nature 461, 472–475.

Rodriguez Sawicki, L., Bottasso Arias, N.M., Scaglia, N., Falomir Lockhart, L.J., Franchini, G.R., Storch, J., Córsico, B., 2017. FABP1 knockdown in human enterocytes impairs proliferation and alters lipid metabolism. Biochim. Biophys. Acta 1862, 1587–1594.

Rombenso, A.N., Trushenski, J.T., Jirsa, D., M., D, 2015. Successful fish oil sparing in white seabass feeds using saturated fatty acid-rich soybean oil and 22:6n-3 (DHA) supplementation. Aquaculture 448, 176–185.

Rønnestad, I., Finn, R.N., Lein, I., Lie, Ø., 1995. Compartmental changes in the contents of total lipid, lipid classes and their associated fatty acids in developing yolk-sac larvae of Atlantic halibut, *Hippoglossus hippoglossus* (L.). Aquacult. Nutr. 1, 119–130.

Rønnestad, I., Nilsen, T.O., Murashita, K., Angotzi, A.R., Moen, A.G.G., Stefansson, S.O., Kling, P., Björnsson, B.T., Kurokawa, T., 2010. Leptin and leptin receptor genes in Atlantic salmon: cloning, phylogeny, tissue distribution and expression correlated to long-term feeding status. Gen. Compar. Endocrinol. 168, 55–70.

Rosen, E.D., Spiegelman, B.M., 2001. PPARγ: a nuclear regulator of metabolism, differentiation, and cell growth. J. Biol. Chem. 276, 37731–37734.

Rowland, I., Gibson, G., Heinken, A., Scott, K., Swann, J., Thiele, I., Tuohy, K., 2018. Gut microbiota functions: metabolism of nutrient and other food components. Eur. J. Nutr. 57, 1–24.

Roy, S.S., Pal, R., 2015. Microalgae in aquaculture: a review with special references to nutritional value and fish dietetics. Proc. Zool. Soc. 68, 1–8.

Roy, J., Mercier, Y., Tonnet, L., Burel, C., Lanuque, A., Surget, A., Larroquet, L., Corraze, G., Terrier, F., Panserat, S., Skiba, S., 2020. Rainbow trout prefer diets rich in omega 3 long chain polyunsaturated fatty acids DHA and EPA. Physiol. Behav. 213, 112692.

Rueda-Jasso, R., Conceiçao, L.E., Dias, J., de Coen, W., Gomes, E., Rees, J.F., Soares, F., Dinis, M.T., Sorgeloos, P., 2004. Effect of dietary non-protein energy levels on condition and oxidative status of Senegalese sole (*Solea senegalensis*) juveniles. Aquaculture 231, 417–433.

Ruiz-Lopez, N., Haslam, R.P., Napier, J.A., Sayanova, O., 2014. Successful high-level accumulation of fish oil omega-3 long-chain polyunsaturated fatty acids in a transgenic oilseed crop. Plant J. 77, 198–208.

Ruiz-Lopez, N., Stubhaug, I., Ipharraguerre, I., Rimbach, G., Menoyo, D., 2015. Positional distribution of fatty acids in triacylglycerols and phospholipids from fillets of Atlantic Salmon (*Salmo salar*) fed vegetable and fish oil blends. Mar. Drugs 13, 4255–4269.

Russell, R.C., Tian, Y., Yuan, H., Park, H.W., Chang, Y.Y., Kim, J., Haerin, K., Neufeld, T.P., Andrew, D., Kun-Liang, G., 2013. ULK1 induces autophagy by phosphorylating Beclin-1 and activating VPS34 lipid kinase. Nat. Cell Biol. 15, 741–750.

Ruyter, B., Thomassen, M.S., 1999. Metabolism of n-3 and n-6 fatty acids in Atlantic salmon liver: stimulation by essential fatty acid deficiency. Lipids 34, 1167–1176.

Ruyter, B., Andersen, Ø., Dehli, A., Östlund Farrants, A.-K., Gjøen, T., Thomassen, M.S., 1997. Peroxisome proliferator activated receptors in Atlantic salmon (*Salmo salar*): effects on PPAR transcription and acyl-CoA oxidase activity in hepatocytes by peroxisome proliferators and fatty acids. Biochim. Biophys. Acta 1348, 331–338.

Ruyter, B., Sissener, N.H., Østbye, T.-K., Simon, C.J., Krasnov, A., Bou, M., Sanden, M., Nichols, P.D., Lutfi, E., Berge, G.M., 2019. n-3 Canola oil effectively replaces fish oil as a new safe dietary source of DHA in feed for juvenile Atlantic salmon. Br. J. Nutr. 122, 1329–1345.

Sahlmann, C., Sutherland, B.J.G., Kortner, T.M., Koop, B.F., Krogdahl, Å., Bakke, A.M., 2013. Early response of gene expression in the distal intestine of Atlantic salmon (*Salmo salar* L.) during the development of soybean meal induced enteritis. Fish Shellfish Immunol. 34, 599–609.

Sæle, Ø., Nordgreen, A., Olsvik, P.A., Hamre, K., 2010. Characterization and expression of digestive neutral lipases during ontogeny of Atlantic cod (*Gadus morhua*). Compar. Biochem. Physiol. A Mol. Integr. Physiol. 157, 252–259.

Sæle, Ø., Rød, K.E.L., Quinlivan, V.H., Li, S., Farber, S.A., 2018. A novel system to quantify intestinal lipid digestion and transport. Biochim. Biophys. Acta 1863, 948–957.

Sales, J., 2009. Linear models to predict the digestible lipid content of fish diets. Aquacult. Nutr. 15, 537–549.

Salhi, M., Hernandez Cruz, C.M., Bessonart, M., Izquierdo, M.S., Fernandez Palacios, H., 1999. Effect of different dietary polar lipid levels and different n-3 HUFA content in polar lipids on gut and liver histological structure of gilthead seabream (*Sparus aurata*) larvae. Aquaculture 179, 253–263.

Sallam, A.E., Mansour, A.T., Srour, T.M., Goda, A.M.A., 2017. Effects of different carotenoid supplementation sources with or without sodium taurocholate on growth, feed utilization, carotenoid content and antioxidant status in fry of the European seabass, *Dicentrarchus labrax*. Aquacult. Res. 48, 3848–3858.

Salmerón, C., Acerete, L., Gutiérrez, J., Navarro, I., Capilla, E., 2013. Characterization and endocrine regulation of proliferation and differentiation of primary cultured preadipocytes from gilthead sea bream (*Sparus aurata*). Domest. Anim. Endocrinol. 45, 1–10.

Salmerón, C., Johansson, M., Angotzi, A.R., Rønnestad, I., Jönsson, E., Björnsson, B.T., Gutiérrez, J., Navarro, I., Capilla, E., 2015a. Effects of nutritional status on plasma leptin levels and in vitro regulation of adipocyte leptin expression and secretion in rainbow trout. Gen. Compar. Endocrinol. 210, 114–123.

Salmerón, C., Johansson, M., Asaad, M., Angotzi, A.R., Rønnestad, I., Stefansson, S.O., Jönsson, E., Björnsson, B.T., Gutiérrez, J., Navarro, I., 2015b. Roles of leptin and ghrelin in adipogenesis and lipid metabolism of rainbow trout adipocytes in vitro. Compar. Biochem. Physiol. Mol. Integr. Physiol. 188, 40–48.

Saltiel, A.R., Kahn, C.R., 2001. Insulin signalling and the regulation of glucose and lipid metabolism. Nature 414, 799.

Salunkhe, D.K., Adsule, R.N., Chavan, J.K., Kadam, S.S., 1992. World Oilseeds: Chemistry, Technology and Utilization. Van Nostrand Reinhold, New York, NY, USA.

Salvia-Trujillo, L., Verkempinck, S.H.E., Sun, L., Van Loey, A.M., Grauwet, T., Hendrickx, M.E., 2017. Lipid digestion, micelle formation and carotenoid bioaccessibility kinetics: influence of emulsion droplet size. Food Chem. 229, 653–662.

Sánchez-Gurmaches, J., Cruz-Garcia, L., Gutiérrez, J., Navarro, I., 2012. mRNA expression of fatty acid transporters in rainbow trout: in vivo and in vitro regulation by insulin, fasting and inflammation and infection mediators. Compar. Biochem. Physiol. A Mol. Integr. Physiol. 163, 177–188.

Sánchezgurmaches, J., Cruzgarcia, L., Gutiérrez, J., Navarro, I., 2010. Endocrine control of oleic acid and glucose metabolism in rainbow trout (*Oncorhynchus mykiss*) muscle cells in culture. Am. J. Physiol. Regul. Integr. Compar. Physiol. 299, R562–R572.

Sandel, E., Nixon, O., Lutzky, S., Ginsbourg, B., Tandler, A., Uni, Z., Koven, W., 2010. The effect of dietary phosphatidylcholine/phosphatidylinositol ratio on malformation in larvae and juvenile gilthead sea bream (*Sparus aurata*). Aquaculture 304, 42–48.

Sanders, F.W.B., Griffin, J.L., 2016. De novo lipogenesis in the liver in health and disease: more than just a shunting yard for glucose. Biol. Rev. Camb. Phil. Soc. 91, 452–468.

Sangiao-Alvarellos, S., Vazquez, M.J., Varela, L., Nogueiras, R., Saha, A.K., Cordido, F., López, M., Dieguez, C., 2009. Central ghrelin regulates peripheral lipid metabolism in a growth hormone-independent fashion. Endocrinology 150, 4562–4574.

Santerre, A., Tellez-Banuelos, M.C., Casas-Solis, J., Castro-Felix, P., Huizar-Lopez, M.R., Zaitseva, G.P., Horta-Fernandez, J.L., Trujillo-Garcia, E.A., De La Mora-Sherer, D., Palafox-Luna, J.A., Juarez-Carrillo, E., 2015. Delta6-fatty acid desaturase and fatty acid elongase mRNA expression, phagocytic activity and weight-to-length relationships in channel catfish (*Ictalurus punctatus*) fed alternative diets with soy oil and a probiotic. Genet. Mol. Res. 14, 11222–11234.

Santigosa, E., Verlhac-Trichet, V., Olsen, R.E., Figuereido-Silva, C., 2018. A microalgal oil containing EPA+DHA can be an effective source of omega 3 for Atlantic salmon post-smolts. In: 18th International Symposium on Fish Nutrition & Feeding (ISFNF) 3–7 June 2018 Las Palmas de Gran Canaria, Spain.

Santigosa, E., Constant, D., Prudence, D., Wahli, T., Verlhac-Trichet, V., 2020. A novel marine algal oil containing both EPA and DHA is an effective source of omega-3 fatty acids for rainbow trout (*Oncorhynchus mykiss*). J. World Aquacult. Soc. 51, 649–665.

Santos, J.D., Burkow, I.C., Jobling, M., 1993. Patterns of growth and lipid deposition in cod (*Gadus morhua* L.) fed natural prey and fish-based feeds. Aquaculture 110, 173–189.

Saravanan, S., Schrama, J.W., Figueiredo-Silva, A.C., Kaushik, S.J., Verreth, J.A.J., Geurden, I., 2012. Constraints on energy intake in fish: the link between diet composition, energy metabolism, and energy intake in rainbow trout. PLoS One 7.

Sardenne, F., Simard, M., Robinson, S.M.C., Mckindsey, C.W., 2020. Consumption of organic wastes from coastal salmon aquaculture by wild decapods. Sci. Total Environ. 711, 134863.

Sargent, J.R., Henderson, R.J., Tocher, D.R., 1989. The lipids. In: Halver, J.E. (Ed.), Fish Nutrition, second ed. Academic Press, Inc, San Diego, CA, USA.

Sargent, J.R., Bell, J.G., Bell, M.V., Henderson, R.J., Tocher, D.R., 1995. Requirement criteria for essential fatty acids. J. Appl. Ichthyol. 11, 183–198.

Sargent, J., Bell, G., Mcevoy, L., Tocher, D., Estevez, A., 1999. Recent developments in the essential fatty acid nutrition of fish. Aquaculture 177, 191–199.

Sargent, J.R., Tocher, D.R., Bell, J.G., 2002. The lipids. In: Halver, J.E., Hardy, R.W. (Eds.), Fish Nutrition, third ed. Academic Press, San Diego, CA, USA.

Sargent, J.R., Tocher, D.R., Bell, J.G., 2003. The lipids. In: Halver, J.E., Hardy, R.W. (Eds.), Fish Nutrition, third ed. Academic Press, San Diego.

Sarker, P.K., Kapuscinski, A.R., Lanois, A., Livesey, E.D., Bernhard, K.P., Coley, M.L., 2016. Towards sustainable aquafeeds: complete substitution of fish oil with marine microalga Schizochytrium sp. improves growth and fatty acid deposition in juvenile Nile tilapia (*Oreochromis niloticus*). PLoS One 11, e0156684.

Scarano, L.J., Calabrese, E.J., Kostecki, P.T., Baldwin, L.A., Leonard, D.A., 1994. Evaluation of a rodent peroxisome proliferator in two species of freshwater fish: rainbow trout (*Oncorhynchus mykiss*) and Japanese medaka (*Oryzias latipes*). Ecotoxicol. Environ. Saf. 29, 13–19.

Schmitz, G., Ecker, J., 2008. The opposing effects of *n*-3 and *n*-6 fatty acids. Prog. Lipid Res. 47, 147–155.

Schrader, M., Costello, J., Godinho, L.F., Islinger, M., 2015. Peroxisome-mitochondria interplay and disease. J. Inherit. Metab. Dis. 38, 681–702.

Schulze, R.J., Sathyanarayan, A., Mashek, D.G., 2017. Breaking fat: the regulation and mechanisms of lipophagy. Biochim. Biophys. Acta 1862, 1178–1187.

Schutz, Y., 2004. Dietary fat, lipogenesis and energy balance. Physiol. Behav. 83, 557–564.

Schweiger, M., Schreiber, R., Haemmerle, G., Lass, A., Fledelius, C., Jacobsen, P., Tornqvist, H., Zechner, R., Zimmermann, R., 2006. Adipose triglyceride lipase and hormone-sensitive lipase are the major enzymes in adipose tissue triacylglycerol catabolism. J. Biol. Chem. 281, 40236–40241.

Seiliez, I., Bruant, J.S., Zambonino-Infante, J., Kaushik, S., Bergot, P., 2006. Effect of dietary phospholipid level on the development of gilthead sea bream (*Sparus aurata*) larvae fed a compound diet. Aquacult. Nutr. 12, 372–378.

Seiliez, I., Panserat, S., Lansard, M., Polakof, S., Plagnes-Juan, E., Surget, A., Dias, K., Larquier, M., Kaushik, S., Skiba-Cassy, S., 2011. Dietary carbohydrate-to-protein ratio affects TOR signaling and metabolism-related gene expression in the liver and muscle of rainbow trout after a single meal. Am. J. Physiol. Regul. Integr. Compar. Physiol. 300, R733–R743.

Seite, S., Pioche, T., Ory, N., Plagnesjuan, E., Panserat, S., Seiliez, I., 2019. The autophagic flux inhibitor bafilomycine A1 affects the expression of intermediary metabolism-related genes in trout hepatocytes. Front. Physiol. 10, 263.

Semova, I., Carten, J.D., Stombaugh, J., Mackey, L.C., Knight, R., Farber, S.A., Rawls, J.F., 2012. Microbiota regulate intestinal absorption and metabolism of fatty acids in zebrafish. Cell Host Microbe 12, 277–288.

Senadheera, S.P.S.D., Turchini, G.M., Thanuthong, T., Francis, D.S., 2012a. Effects of dietary iron supplementation on growth performance, fatty acid composition and fatty acid metabolism in rainbow trout (*Oncorhynchus mykiss*) fed vegetable oil based diets. Aquaculture 342/343, 80–88.

Senadheera, S.P.S.D., Turchini, G.M., Thanuthong, T., Francis, D.S., 2012b. Effects of dietary vitamin B6 supplementation on fillet fatty acid composition and fatty acid metabolism of rainbow trout fed vegetable oil based diets. J. Agric. Food Chem. 60, 2343–2353.

Seo, H.S., Endo, Y., Moku, M., Kawaguchi, K., Fujimoto, K., 2001. Wax ester biosynthesis in the liver of myctophid fishes. Lipids 36, 389–393.

Serhan, C.N., 2014. Pro-resolving lipid mediators are leads for resolution physiology. Nature 510, 92–101.

Serhan, C.N., Levy, B.D., 2018. Resolvins in inflammation: emergence of the pro-resolving superfamily of mediators. J. Clin. Invest. 128, 2657–2669.

Serhan, C.N., Chiang, N., Dalli, J., 2015. The resolution code of acute inflammation: novel pro-resolving lipid mediators in resolution. Semin. Immunol. 27, 200–215.

Seroussi, E., Cinnamon, Y., Yosefi, S., Genin, O., Smith, J.G., Rafati, N., Bornelёv, S., Andersson, L., Friedman-Einat, M., 2015. Identification of the long-sought leptin in chicken and duck: expression pattern of the highly GC-rich avian leptin fits an autocrine/paracrine rather than endocrine function. Endocrinology 157, 737–751.

Shah, M.R., Lutzu, G.A., Alam, A., Sarker, P., Kabir Chowdhury, M.A., Parsaeimehr, A., Liang, Y., Daroch, M., 2018. Microalgae in aquafeeds for a sustainable aquaculture industry. J. Appl. Phycol. 30, 197–213.

Shahpar, Z., Johari, S.A., 2019. Effects of dietary organic, inorganic, and nanoparticulate zinc on rainbow trout, *oncorhynchus mykiss* Larvae. Biol. Trace Elem. Res. 190, 535–540.

Shao, L., Zhu, X., Yang, Y., Jin, J., Liu, H., Han, D., Xie, S., 2016. Effects of dietary vitamin A on growth, hematology, digestion and lipometabolism of on-growing gibel carp (*Carassius auratus* gibelio var. CAS III). Aquaculture 460, 83–89.

Shaw, B.J., Handy, R.D., 2006. Dietary copper exposure and recovery in Nile tilapia, *Oreochromis niloticus*. Aquat. Toxicol. 76, 111–121.

Sheard, N., Zeisel, S., 1989. Choline: an essential dietary nutrient? Nutrition 5, 1–5.

Shen, W.-J., Patel, S., Miyoshi, H., Greenberg, A.S., Kraemer, F.B., 2009. Functional interaction of hormone-sensitive lipase and perilipin in lipolysis. J. Lipid Res. 50, 2306–2313.

Shen, L., Zhang, Y., Du, J., Chen, L., Luo, J., Li, X., Li, M., Tang, G., Zhang, S., Zhu, L., 2015. MicroRNA-23a regulates 3T3-L1 adipocyte differentiation. Gene 575, 761–764.

Sheng, Y., Ren, H., Limbu, S.M., Sun, Y., Qiao, F., Zhai, W., Du, Z.Y., Zhang, M., 2018. The presence or absence of intestinal microbiota affects lipid deposition and related genes expression in zebrafish (*Dania rerio*). Front. Microbiol. 9, 1124.

Shepherd, J., Bachis, E., 2014. Changing supply and demand for fish oil. Aquacult. Econ. Manag. 18, 395–416.

Sheridan, M.A., Harmon, J.S., 2012. Adipose tissue. Biochem. Mol. Biol. Fishes 3, 305–311.

Shi, X., Luo, Z., Chen, F., Huang, C., Zhu, X., Liu, X., 2016. Effects of dietary cellulase addition on growth performance, nutrient digestibility and digestive enzyme activities of juvenile crucian carp *Carassius auratus*. Aquacult. Nutr. 23, 618–628.

Shi, X., Luo, Z., Chen, G.H., Chen, F., Zhang, L.H., Zhu, X., Liu, X., 2017a. Replacement of fishmeal by a mixture of soybean meal and chlorella meal in practical diets for juvenile crucian carp, carassius auratus. J. World Aquacult. Soc. 48, 770–781.

Shi, X.C., Sun, J., Yang, Z., Li, X.X., Ji, H., Li, Y., Chang, Z.G., Du, Z.Y., Chen, L.Q., 2017b. Molecular characterization and nutritional regulation of carnitine palmitoyltransferase (CPT) family in grass carp (*Ctenopharyngodon idellus*). Compar. Biochem. Physiol. B Biochem. Mol. Biol. 203, 11–19.

Shi, H.J., Xu, C., Liu, M.Y., Wang, B.K., Liu, W.B., Chen, D.H., Zhang, L., Xu, C.Y., Li, X.F., 2018a. Resveratrol improves the energy sensing and glycolipid metabolism of Blunt Snout Bream *Megalobrama amblycephala* fed high-carbohydrate diets by activating the AMPK–SIRT1–PGC-1α network. Front. Physiol. 9, 1258.

Shi, X., Jin, A., Sun, J., Tian, J., Ji, H., Chen, L., Du, Z.-Y., 2018b. The protein-sparing effect of α-lipoic acid in juvenile grass carp, *Ctenopharyngodon idellus*: effects on lipolysis, fatty acid β-oxidation and protein synthesis. Br. J. Nutr. 120, 977–987.

Shikata, T., Shimeno, S., 1994. Metabolic response to dietary stearic acid, linoleic acid, and highly unsaturated fatty acid in carp. Fish. Sci. 60, 735–739.

Shimeno, S., Shikata, T., Hosokawa, H., Masumoto, T., Kheyyali, D., 1997. Metabolic response to feeding rates in common carp, *Cyprinus carpio*. Aquaculture 151, 371–377.

Shirai, N., Suzuki, H., Toukairin, S., Wada, S., 2001. Spawning and season affect lipid content and fatty acid composition of ovary and liver in Japanese catfish (*Silurus asotus*). Compar. Biochem. Physiol. B Biochem. Mol. Biol. 129, 185–195.

Siddhuraju, P., Becker, K., 2002. Effect of phenolic nonprotein amino acid L-dopa (L-3,4-dihydroxyphenylalanine) on growth performance, metabolic rates and feed nutrient utilization of common carp (*Cyprinus carpio* L.). Aquacult. Nutr. 8, 69–77.

Siddhuraju, P., Becker, K., 2003. Comparative nutritional evaluation of differentially processed mucuna seeds [*Mucuna pruriens* (L.) DC. var. utilis (Wall ex Wight) Baker ex Burck] on growth performance, feed utilization and body composition in Nile tilapia (*Oreochromis niloticus* L.). Aquacult. Res. 34, 487–500.

Simons, K., Ikonen, E., 1997. Functional rafts in cell membranes. Nature 387, 569–572.

Simopoulos, A.P., 2008. The importance of the omega-6/omega-3 fatty acid ratio in cardiovascular disease and other chronic diseases. Exp. Biol. Med. 233, 674–688.

Simopoulos, A.P., 2011. Evolutionary aspects of diet: the omega-6/omega-3 ratio and the brain. Mol. Neurobiol. 44, 203–215.

Sinclair, A.J., 2019. Docosahexaenoic acid and the brain - what is its role? Asia Pac. J. Clin. Nutr. 28, 675–688.

Singer, T.D., Ballantyne, J.S., 1989. Absence of extrahepatic lipid oxidation in a freshwater elasmobranch, the dwarf stingray *Potamotrygon magdalenae*: evidence from enzyme activities. J. Exp. Zool. 251, 355–360.

Singh, R., Cuervo, A.M., 2012. Lipophagy: connecting autophagy and lipid metabolism. Int. J. Cell Biol. 2012.

Singh, R., Kaushik, S., Wang, Y., Xiang, Y., Novak, I., Komatsu, M., Tanaka, K., Cuervo, A.M., Czaja, M.J., 2009. Autophagy regulates lipid metabolism. Nature 458, 1131–1135.

Sissener, N.H., Hemre, G.I., Espe, M., Sanden, M., Torstensen, B.E., Hevroy, E.M., 2013. Effects of plant-based diets on glucose and amino acid metabolism, leptin, ghrelin and GH-IGF system regulation in Atlantic salmon (*Salmo salar* L.). Aquacult. Nutr. 19, 399–412.

Sissener, N.H., Liland, N.S., Holen, E., Stubhaug, I., Torstensen, B.E., Rosenlund, G., 2017. Phytosterols are not involved in the development of fatty liver in plant oil fed Atlantic salmon (*Salmo salar*) at high or low water temperature. Aquaculture 480, 123–134.

Sissener, N.H., Rosenlund, G., Stubhaug, I., Liland, N.S., 2018. Tissue sterol composition in Atlantic salmon (*Salmo salar* L.) depends on the dietary cholesterol content and on the dietary phytosterol:cholesterol ratio, but not on the dietary phytosterol content. Br. J. Nutr. 119, 599–609.

Sitjà-Bobadilla, A., Peña-Llopis, S., Gómez-Requeni, P., Médale, F., Kaushik, S., Pérez-Sánchez, J., 2005. Effect of fish meal replacement by plant protein sources on non-specific defence mechanisms and oxidative stress in gilthead sea bream (*Sparus aurata*). Aquaculture 249, 387–400.

Skalli, A., Robin, J.H., Le Bayon, N., Le Delliou, H., Person-Le Ruyet, J., 2006. Impact of essential fatty acid deficiency and temperature on tissues' fatty acid composition of European sea bass (*Dicentrarchus labrax*). Aquaculture 255, 223–232.

Smith, D.M., Williams, I.H., Williams, K.C., Barclay, M.C., Venables, W.N., 2005. Oxidation of medium-chain and long-chain fatty acids by polka dot grouper *Cromileptes altivelis*. Aquacult. Nutr. 11, 41–48.

Snow, J.E., Beard, P.J., 1939. Studies on the bacterial flora of North Pacific salmon. Food Res. 4, 563–585.

Soengas, J.L., Cerda-Reverter, J.M., Delgado, M.J., 2018. Central regulation of food intake in fish: an evolutionary perspective. J. Mol. Endocrinol. 60, R171–R199.

Song, G., Xu, G., Ji, C., Shi, C., Shen, Y., Chen, L., Zhu, L., Yang, L., Zhao, Y., Guo, X., 2014. The role of microRNA-26b in human adipocyte differentiation and proliferation. Gene 533, 481–487.

Song, Y.F., Wu, K., Tan, X.Y., Zhang, L.H., Zhuo, M.Q., Pan, Y.X., Chen, Q.L., 2015. Effects of recombinant human leptin administration on hepatic lipid metabolism in yellow catfish *Pelteobagrus fulvidraco*: in vivo and in vitro studies. Gen. Compar. Endocrinol. 212, 92–99.

Song, J.Y., Zhang, C.X., Wang, L., Song, K., Hu, S.C., Zhang, L.F., 2016. Effects of dietary calcium levels on growth and tissue mineralization in Japanese seabass, *Lateolabrax japonicus*. Aquacult. Nutr. 23, 637–648.

Song, Y.F., Hogstrand, C., Wei, C.-C., Wu, K., Pan, Y.X., Luo, Z., 2017. Endoplasmic reticulum (ER) stress and cAMP/PKA pathway mediated Zn-induced hepatic lipolysis. Environ. Pollut. 228, 256–264.

Song, X., Marandel, L., Skiba-Cassy, S., Corraze, G., Dupont-Nivet, M., Quillet, E., Geurden, I., Panserat, S., 2018a. Regulation by dietary carbohydrates of intermediary metabolism in liver and muscle of two isogenic lines of rainbow trout. Front. Physiol. 9, 1579.

Song, Z., Xiaoli, A., Yang, F., 2018b. Regulation and metabolic significance of de novo lipogenesis in adipose tissues. Nutrients 10, 1383.

Song, C., Liu, B., Xu, P., Ge, X., Zhang, H., 2019. Emodin ameliorates metabolic and antioxidant capacity inhibited by dietary oxidized fish oil through PPARs and Nrf2-Keap1 signaling in Wuchang bream (*Megalobrama amblycephala*). Fish Shellfish Immunol. 94, 842–851.

Soriano, E.L., Ramirez, D.T., Araujo, D.R., Gomez-Gil, B., Castro, L.I., Sanchez, C.G., 2018. Effect of temperature and dietary lipid proportion on gut microbiota in yellowtail kingfish *Seriola lalandi* juveniles. Aquaculture 497, 269–277.

Speers-Roesch, B., Ip, Y.K., Ballantyne, J.S., 2006. Metabolic organization of freshwater, euryhaline, and marine elasmobranchs: implications for the evolution of energy metabolism in sharks and rays. J. Exp. Biol. 209, 2495–2508.

Spiegelman, B., 1998. PPAR-gamma: adipogenic regulator and thiazolidinedione receptor. Diabetes 47, 507–514.

Sprague, M., Dick, J.R., Tocher, D.R., 2016. Impact of sustainable feeds on omega-3 long-chain fatty acid levels in farmed Atlantic salmon, 2006–2015. Sci. Rep. 6, 21892.

Sprague, M., Betancor, M.B., Tocher, D.R., 2017. Microbial and genetically engineered oils as replacements for fish oil in aquaculture feeds. Biotechnol. Lett. 39, 1599–1609.

Sprague, M., Gong, X., Betancor, M.B., Olsen, R.E., Torrissen, O., Glencross, B.D., Tocher, D.R., 2019. Endogenous production of n-3 long-chain polyunsaturated fatty acids and the influence of dietary linoleic acid and the alpha-linolenic:linoleic ratio in Atlantic salmon (*Salmo salar*). Br. J. Nutr. 122, 1091–1102.

Sprecher, H., 2000. Metabolism of highly unsaturated n-3 and n-6 fatty acids. Biochim. Biophys. Acta 1486, 219–231.

Springmann, M., Clark, M., Mason-D'croz, D., Wiebe, K., Bodirsky, B.L., Lassaletta, L., de Vries, W., Vermeulen, S.J., Herrero, M., Carlson, K.M., Jonell, M., Troell, M., Declerck, F., Gordon, L.J., Zurayk, R., Scarborough, P., Rayner, M., Loken, B., Fanzo, J., Godfray, H.C.J., Tilman, D., Rockstrom, J., Willett, W., 2018. Options for keeping the food system within environmental limits. Nature 562, 519–+.

Stark, R., Ashley, S.E., Andrews, Z.B., 2013. AMPK and the neuroendocrine regulation of appetite and energy expenditure. Mol. Cell. Endocrinol. 366, 215–223.

Stark, K.D., Van Elswyk, M.E., Higgins, M.R., Weatherford, C.A., Salem, N., 2016. Global survey of the omega-3 fatty acids, docosahexaenoic acid and eicosapentaenoic acid in the blood stream of healthy adults. Prog. Lipid Res. 63, 132–152.

Steffen, M.L., Harrison, W.R., Elder, F.F.B., Cook, G.A., Park, E.A., 1999. Expression of the rat liver carnitine palmitoyltransferase I (CPT-Iα) gene is regulated by Sp1 and nuclear factor Y: chromosomal localization and promoter characterization. Biochem. J. 340, 425–432.

Sternby, B., Larsson, A., Borgström, B., 1983. Evolutionary studies on pancreatic colipase. Biochim. Biophys. Acta 750, 340–345.

Stewart, M.M., 1932. The bacterial flora of the slime and intestinal contents of the haddock (*Gadus aeglefinus*). J. Mar. Biol. Assoc. U. K. 18, 35–50.

Storch, J., Corsico, B., 2008. The emerging functions and mechanisms of mammalian fatty acid-binding proteins. Annu. Rev. Nutr. 28, 73–95.

Stubhaug, I., Lie, Ø., Torstensen, B.E., 2006. β-Oxidation capacity in liver increases during parr-smolt transformation of Atlantic salmon fed vegetable oil and fish oil. J. Fish. Biol. 69, 504–517.

Stubhaug, I., Lie, Ø., Torstensen, B.E., 2007. Fatty acid productive value and β-oxidation capacity in Atlantic salmon (*Salmo salar* L.) fed on different lipid sources along the whole growth period. Aquacult. Nutr. 13, 145–155.

Stubhaug, I., Frøyland, L., Torstensen, B.E., 2005a. β-oxidation capacity of red and white muscle and liver in Atlantic salmon (*Salmo salar* L.)—effects of increasing dietary rapeseed oil and olive oil to replace capelin oil. Lipids 40, 39.

Stubhaug, I., Tocher, D.R., Bell, J.G., Dick, J.R., Torstensen, B.E., 2005b. Fatty acid metabolism in Atlantic salmon (*Salmo salar* L.) hepatocytes and influence of dietary vegetable oil. Biochim. Biophys. Acta Mol. Cell Biol. Lipids 1734, 277–288.

Suh, H.J., Cho, S.Y., Kim, E.Y., Choi, H.-S., 2015. Blockade of lipid accumulation by silibinin in adipocytes and zebrafish. Chem. Biol. Interact. 227, 53–62.

Sullam, K.E., Essinger, S.D., Lozupone, C.A., O'connor, M.P., Rosen, G.L., Knight, R.O.B., Kilham, S.S., Russell, J.A., 2012. Environmental and ecological factors that shape the gut bacterial communities of fish: a meta-analysis. Mol. Ecol. 21, 3363–3378.

Sun, Y., Yang, H., Ling, Z., Chang, J., Ye, J., 2009. Gut microbiota of fast and slow growing grouper (*Epinephelus coioides*). Afr. J. Microbiol. Res. 3, 713–720.

Sun, S., Ye, J., Wang, Y., Chen, L., 2010. Effect of dietary fish oil replacement by rapeseed oil on the growth, fatty acid composition and serum non-specific immunity response of fingerling black carp, *Mylopharyngodon piceus*. Aquacult. Nutr. 17, 441–450.

Sun, J., Ji, H., Li, X.-X., Shi, X.-C., Du, Z.-Y., Chen, L.-Q., 2016. Lipolytic enzymes involving lipolysis in Teleost: synteny, structure, tissue distribution, and expression in grass carp (*Ctenopharyngodon idella*). Compar. Biochem. Physiol. B Biochem. Mol. Biol. 198, 110–118.

Sun, J., Yang, Z., Shi, X.-C., Ji, H., Du, Z.-Y., Chen, L.-Q., 2018a. G0S2a1 (G0/G1 switch gene 2a1) is downregulated by TNF-? in grass carp (*Ctenopharyngodon idellus*) hepatocytes through PPARα inhibition. Gene 641, 1–7.

Sun, Y., Li, B., Zhang, X., Chen, M., Tang, H., Yu, X., 2018b. Dietary available phosphorus requirement of crucian carp, *Carassius auratus*. Aquacult. Nutr. 24, 1494–1501.

Sun, J., Zheng, L., Chen, C., Zhang, J., You, C., Zhang, Q., Monroig, Ó., Tocher, D.R., Wang, S., Li, Y., 2019. MicroRNA involved in the regulation of LC-PUFA biosynthesis in teleosts: miR-33 enhances LC-PUFA biosynthesis in *Siganus canaliculatus* by targeting insig1 which in turn up-regulates srebp1. Mar. Biotechnol. 21, 475–487.

Sun, S.X., Wu, J.L., Lv, H.B., Zhang, H.Y., Zhang, J., Limbu, S.M., Qiao, F., Chen, L.Q., Yang, Y., Zhang, M.L., Du, Z.Y., 2020. Environmental estrogen exposure converts lipid metabolism in male fish to a female pattern mediated by AMPK and mTOR signaling pathways. J. Hazard Mater. 394, 122537.

Sun, J., Chen, C., You, C., Liu, Y., Ma, H., Monroig, Ó., Tocher, D.R., Wang, S., Li, Y., 2021. The miR-15/16 cluster is involved in the regulation of vertebrate LC-PUFA biosynthesis by targeting ppargamma as demonstrated in rabbitfish *Siganus canaliculatus*. Mar. Biotechnol. (in press).

Sundvold, H., Helgeland, H., Baranski, M., Omholt, S.W., Våge, D.I., 2011. Characterisation of a novel paralog of scavenger receptor class B member I (SCARB1) in Atlantic salmon (*Salmo salar*). BMC Genet. 12, 52.

Surai, P.F., 2015. Silymarin as a natural antioxidant: an overview of the current evidence and perspectives. Antioxidants 4, 204–247.

Symonds, M.E., Sebert, S.P., Hyatt, M.A., Budge, H., 2009. Nutritional programming of the metabolic syndrome. Nat. Rev. Immunol. 5, 604–610.

Takeuchi, T., Arakawa, T., Satoh, S., Watanabe, T., 1992. Supplemental effect of phospholipids and requirement of eicosapentaenoic acid and docosahexaenoic acid of juvenile striped jack. Nippon Suisan Gakkaishi 58, 707–713.

Takeuchi, T., Lu, J., Yoshizaki, G., Satoh, S., 2008. Effect on the growth and body composition of juvenile tilapia *Oreochromis niloticus* fed raw Spirulina. Fish. Sci. 68, 34–40.

Tallafuss, A., Hale, L.A., Yan, Y.-L., Dudley, L., Eisen, J.S., Postlethwait, J.H., 2006. Characterization of retinoid-X receptor genes rxra, rxrba, rxrbb and rxrg during zebrafish development. Gene Expr. Patterns 6, 556–565.

Tamasi, V., Miller, K.K.M., Ripp, S.L., Vila, E., Prough, R.A., 2008. Modulation of receptor phosphorylation contributes to activation of peroxisome proliferator activated receptor alpha by dehydroepiandrosterone and other peroxisome proliferators. Mol. Pharmacol. 73, 968–976.

Tan, N.S., Michalik, L., Desvergne, B., Wahli, W., 2005. Multiple expression control mechanisms of peroxisome proliferator-activated receptors and their target genes. J. Steroid Biochem. Mol. Biol. 93, 99–105.

Tan, K., Ma, H.Y., Li, S.K., Zheng, H.P., 2020. Bivalves as future source of sustainable natural omega-3 polyunsaturated fatty acids. Food Chem. 311.

Tang, Q.Q., Lane, M.D., 2000. Role of C/EBP homologous protein (CHOP-10) in the programmed activation of CCAAT/enhancer-binding protein-beta during adipogenesis. Proc. Natl. Acad. Sci. U. S. A. 97, 12446–12450.

Tang, Z., Sun, C., Yan, A., Wu, S., Qin, C., Zhang, Y., Li, W., 2013. Genes involved in fatty acid metabolism: molecular characterization and hypothalamic mRNA response to energy status and neuropeptide Y treatment in the orange-spotted grouper *Epinephelus coioides*. Mol. Cell. Endocrinol. 376, 114–124.

Tang, T., Hu, Y., Peng, M., Chu, W., Hu, Y., Zhong, L., 2019. Effects of high-fat diet on growth performance, lipid accumulation and lipid metabolism-related microRNA/gene expression in the liver of grass carp (*Ctenopharyngodon idella*). Compar. Biochem. Physiol. B Biochem. Mol. Biol. 234, 34–40.

Tanomman, S., Ketudat-Cairns, M., Jangprai, A., Boonanuntanasarn, S., 2013. Characterization of fatty acid delta-6 desaturase gene in Nile tilapia and heterogenous expression in *Saccharomyces cerevisiae*. Compar. Biochem. Physiol. B Biochem. Mol. Biol. 166, 148–156.

Tao, Y.F., Qiang, J., Yin, G.-J., Xu, P., Shi, Q., Bao, J.-W., 2017. Identification and characterization of lipid metabolism-related microRNAs in the liver of genetically improved farmed tilapia (Gift, *Oreochromis niloticus*) by deep sequencing. Fish Shellfish Immunol. 69, 227–235.

Tao, Y.F., Qiang, J., Bao, J.-W., Li, H.-X., Yin, G.-J., Xu, P., Chen, D.-J., 2018. miR-205-5p negatively regulates hepatic acetyl-CoA carboxylase β mRNA in lipid metabolism of *Oreochromis niloticus*. Gene 660, 1–7.

Tasbozan, O., Emre, Y., Gokce, M., Erbas, C., Özcan, F., Kıvrak, E., 2016. The effects of different cycles of starvation and re-feeding on growth and body composition in rainbow trout (*Oncorhynchus mykiss*, Walbaum, 1792). J. Appl. Ichthyol. 32, 583–588.

Taylor, J.F., Martinez-Rubio, L., del Pozo, J., Walton, J.M., Tinch, A.E., Migaud, H., Tocher, D.R., 2015. Influence of dietary phospholipid on early development and performance of Atlantic salmon (*Salmo salar*). Aquaculture 448, 262–272.

Terasawa, F., Hirano, Y., Wada, M., Takita, T., Nakamura, K., Innami, S., 1994. Effects of dietary casein and soy-protein on metabolic conversion of eicosapentaenoic acid to docosahexaenoic acid in the liver of rat. J. Nutr. Sci. Vitaminol. 40, 353–362.

Teshima, S.I., 1971. Bioconversion of β-sitosterol and 24-methylcholesterol to cholesterol in marine crustacea. Compar. Biochem. Physiol. B Biochem. Mol. Biol. 39, 815–822.

Teshima, S., 1997. Phospholipids and sterols. In: D'abramo, L.R. (Ed.), Crustacean Nutrition. Advances in World Aquaculture. World Aquaculture Society, Baton Rouse, Louisiana, USA.

Teshima, S., Kanazawa, A., Sasada, J., 1983. Nutritional value of dietary cholesterol and other sterols to larval prawn, *Penaeus japonicus*. Aquaculture 31, 159–167.

Teshima, S., Kanazawa, A., Koshio, S., Kondo, N., 1989. Nutritive value of sitosterol for the prawn Penaeus japonicus. Nippon Suisan Gakkaishi 54, 1115–1122.

Thanuthong, T., Francis, D.S., Senadheera, S.P.S.D., Paul, L.J., Turchini, G.M., 2011. LC-PUFA biosynthesis in rainbow trout is substrate limited: use of the whole body fatty acid balance method and different 18:3n-3/18:2n-6 ratios. Lipids 46.

Thillart, G., Vianen, G., Zaagsma, J., 2002. Adrenergic regulation of lipid mobilization in fishes; a possible role in hypoxia survival: fish growth and metabolism (Guest Editor: Plisetskaya, E. M.). In: Environmental, Nutritional and Hormonal Regulation. Fish Physiology and Biochemistry, vol. 27.

Thomas, A., 2007. Fats and fatty oils. In: Ullmann's Encyclopedia of Industrial Chemistry, seventh ed. Wiley. https://doi.org/10.1002/14356007.a10_173.

Thomassen, M.S., Berge, R.D.,M.,G., Østbye, T.K., Ruyter, B., 2012. High dietary EPA does not inhibit Δ5 and Δ6 desaturases in Atlantic salmon (*Salmo salar* L.) fed rapeseed oil diets. Aquaculture 360–361, 78–85.

Tian, J., Ji, H., Oku, H., Zhou, J., 2014. Effects of dietary arachidonic acid (ARA) on lipid metabolism and health status of juvenile grass carp, *Ctenopharyngodon idellus*. Aquaculture 430, 57–65.

Tian, J.-J., Lei, C.-X., Chen, L.-Q., Du, Z.-Y., 2016. Dietary arachidonic acid has a time-dependent differential impact on adipogenesis modulated via COX and LOX pathways in grass carp *Ctenopharyngodon idellus*. Lipids 51.

Tian, J.-J., Lei, C.-X., Kaneko, G., Zhou, J.-S., Yu, H., Li, Y., Yu, E.-M., Xie, J., 2017a. Comparative analysis of effects of dietary arachidonic acid and EPA on growth, tissue fatty acid composition, antioxidant response and lipid metabolism in juvenile grass carp, *Ctenopharyngodon idellus*. Br. J. Nutr. 118, 411–422.

Tian, J., Liu, W., Gao, W., Wu, F., Yu, L., Lu, X., Yang, C.-G., Jiang, M., Wen, H., 2017b. Molecular cloning and gene/protein expression of FAT/CD36 from grass carp (Ctenopharyngodon idella) and the regulation of its expression by dietary energy. Fish Physiol. Biochem. 43, 875–888.

Tibbetts, S.M., Mann, J., Dumas, A., 2017. Apparent digestibility of nutrients, energy, essential amino acids and fatty acids of juvenile Atlantic salmon (*Salmo salar* L.) diets containing whole-cell or cell-ruptured *Chlorella vulgaris* meals at five dietary inclusion levels. Aquaculture 481, 25–39.

Tibbetts, S.M., Scaife, M.A., Armenta, R.E., 2020. Apparent digestibility of proximate nutrients, energy and fatty acids in nutritionally-balanced diets with partial or complete replacement of dietary fish oil with microbial oil from a novel *Schizochytrium* sp. (T18) by juvenile Atlantic salmon (*Salmo salar* L.). Aquaculture 520, 735003.

Tilman, D., Socolow, R., Foley, J.A., Hill, J., Larson, E., Lynd, L., Pacala, S., Reilly, J., Searchinger, T., Somerville, C., Williams, R., 2009. Beneficial biofuels-the food, energy, and environment trilemma. Science 325, 270–271.

Tocher, D.R., 1995. Glycerophospholipid Metabolism. In: Biochemistry and Molecular Biology of Fishes: Metabolic and Adaptational Biochemistry. Elsevier Press, Amsterdam.

Tocher, D.R., 2003. Metabolism and functions of lipids and fatty acids in teleost fish. Rev. Fish. Sci. 11, 107–184.

Tocher, D.R., 2009. Issues surrounding fish as a source of omega-3 long-chain polyunsaturated fatty acids. Lipid Technol. 21, 13–16.

Tocher, D.R., 2010. Fatty acid requirements in ontogeny of marine and freshwater fish. Aquacult. Res. 41, 717–732.

Tocher, D.R., 2015. Omega-3 long-chain polyunsaturated fatty acids and aquaculture in perspective. Aquaculture 449, 94–107.

Tocher, D.R., Bendiksen, E.Å., Campbell, P.J., Bell, J.G., 2008. The role of phospholipids in nutrition and metabolism of teleost fish. Aquaculture 280, 21–34.

Tocher, D.R., Glencross, B.D., 2015. Lipids and fatty acids. In: Lee, C.S., Lim, C., Webster, C., Gatlin Iii, D.M. (Eds.), Dietary Nutrients, Additives, and Fish Health. Wiley-Blackwell.

Tocher, D.R., Sargent, J.R., 1990. Effect of temperature on the incorporation into phospholipid classes and metabolism via desaturation and elongation of n-3 and n-6 polyunsaturated fatty acids in fish cells in culture. Lipids 25, 435–442.

Tocher, D.R., Leaver, M.J., Hodgson, P.A., 1998. Recent advances in the biochemistry and molecular biology of fatty acyl desaturases. Prog. Lipid Res. 37, 73–117.

Tocher, D.R., Fonseca-Madrigal, J., Bell, J.G., Dick, J.R., Henderson, R.J., Sargent, J.R., 2002. Effects of diets containing linseed oil on fatty acid desaturation and oxidation in hepatocytes and intestinal enterocytes in Atlantic salmon (*Salmo salar*). Fish Physiol. Biochem. 26, 157–170.

Tocher, D.R., Fonseca-Madrigal, J., Dick, J.R., Ng, W.-K., Bell, J.G., Campbell, P.J., 2004. Effects of water temperature and diets containing palm oil on fatty acid desaturation and oxidation in hepatocytes and intestinal enterocytes of rainbow trout (*Oncorhynchus mykiss*). Compar. Biochem. Physiol. B Biochem. Mol. Biol. 137, 49–63.

Tocher, D.R., Zheng, X., Schlechtriem, C., Hastings, N., Dick, J.R., Teale, A.J., 2006. Highly unsaturated fatty acid synthesis in marine fish: cloning, functional characterization, and nutritional regulation of fatty acyl delta 6 desaturase of Atlantic cod (*Gadus morhua* L.). Lipids 41.

Tocher, D.R., Francis, D.S., Coupland, K., 2011. n-3 Polyunsaturated fatty acid-rich vegetable oils and blends. In: Turchini, G.M., Ng, W.-K., Tocher, D.R. (Eds.), Fish Oil Replacement and Alternative Lipid Sources in Aquaculture Feeds. CRC Press, Taylor & Francis Group, Boca Raton, FL, USA.

Tocher, D.R., Betancor, M.B., Sprague, M., Olsen, R.E., Napier, J.A., 2019. Omega-3 long-chain polyunsaturated fatty acids, EPA and DHA: bridging the gap between supply and demand. Nutrients 11.

Todorčević, M., Vegusdal, A., Gjøen, T., Sundvold, H., Torstensen, B.E., Kjær, M.A., Ruyter, B., 2008a. Changes in fatty acids metabolism during differentiation of Atlantic salmon preadipocytes; effects of n-3 and n-9 fatty acids. Biochim. Biophys. Acta 1781, 326–335.

Todorčević, M., Vegusdal, A., Gjøen, T., Sundvold, H., Torstensen, B.E., Kjær, M.A., Ruyter, B., 2008b. Changes in fatty acids metabolism during differentiation of Atlantic salmon preadipocytes; effects of n-3 and n-9 fatty acids. Biochim. Biophys. Acta Mol. Cell Biol. Lipids 1781, 326–335.

Todorčević, M., Kjær, M.A., Djaković, N., Vegusdal, A., Torstensen, B.E., Ruyter, B., 2009. N-3 HUFAs affect fat deposition, susceptibility to oxidative stress, and apoptosis in Atlantic salmon visceral adipose tissue. Compar. Biochem. Physiol. B Biochem. Mol. Biol. 152, 135–143.

Todorčević, M., Škugor, S., Ruyter, B., 2010. Alterations in oxidative stress status modulate terminal differentiation in Atlantic salmon adipocytes cultivated in media rich in n-3 fatty acids. Compar. Biochem. Physiol. B Biochem. Mol. Biol. 156, 0–318.

Toomey, M.B., Lopes, R.J., Araújo, P.M., Johnson, J.D., Gazda, M.A., Afonso, S., Mota, P.G., Koch, R.E., Hill, G.E., Corbo, J.C., Carneiro, M., 2017. High-density lipoprotein receptor SCARB1 is required for carotenoid coloration in birds. Proc. Natl. Acad. Sci. U. S. A. 114, 5219–5224.

Torrecillas, S., Mompel, D., Caballero, M.J., Montero, D., Merrifield, D.L., Rodiles, A., Robaina, L., Zamorano, M.J., Karalazos, V., Kaushik, S., Izquierdo, M.S., 2017. Effect of fishmeal and fish oil replacement by vegetable meals and oils on gut health of European sea bass (*Dicentrarchus labrax*). Aquaculture 468, 386–398.

Torstensen, B.E., Lie, Ø., Frøyland, L., 2000. Lipid metabolism and tissue composition in Atlantic salmon (*Salmo salar* L.) - effects of capelin oil, palm oil, and oleic acid-enriched sunflower oil as dietary lipid sources. Lipids 35, 653–664.

Torstensen, B.E., Lie, Ø., Hamre, K., 2001. A factorial experimental design for investigation of effects of dietary lipid content and pro- and antioxidants on lipid composition in Atlantic salmon (*Salmo salar*) tissues and lipoproteins. Aquacult. Nutr. 7, 265–276.

Torstensen, B.E., Stubhaug, I., 2004. Beta-oxidation of 18:3n-3 in Atlantic salmon (*Salmo salar* L.) hepatocytes treated with different fatty acids. Lipids 39, 153–160.

Torstensen, B.E., Tocher, D.R., 2011. The Effects of fish oil replacement on lipid metabolism of fish. In: Turchini, G.M., Ng, W.K., Tocher, D.R. (Eds.), Fish Oil Replacement and Alternative Lipid Sources in Aquaculture Feeds. CRC Press, Taylor & Francis Group, Boca Raton, FL, USA.

Torstensen, B.E., Frøyland, L., Lie, Ø., 2004. Replacing dietary fish oil with increasing levels of rapeseed oil and olive oil—effects on Atlantic salmon (*Salmo salar* L.) tissue and lipoprotein lipid composition and lipogenic enzyme activities. Aquacult. Nutr. 10, 175–192.

Torstensen, B.E., Bell, J.G., Rosenlund, G., Henderson, R.J., Graff, I.E., Tocher, D.R., Lie, Ø., Sargent, J.R., 2005. Tailoring of Atlantic salmon (*Salmo salar* L.) flesh lipid composition and sensory quality by replacing fish oil with a vegetable oil blend. J. Agric. Food Chem. 53, 10166–10178.

Tran, T.N., Christophersen, B.O., 2002. Partitioning of polyunsaturated fatty acid oxidation between mitochondria and peroxisomes in isolated rat hepatocytes studied by HPLC separation of oxidation products. Biochim. Biophys. Acta 1583, 195–204.

Trattner, S., Kamal-Eldin, A., Brännäs, E., Moazzami, A., Zlabek, V., Larsson, P., Ruyter, B., Gjøen, T., Pickova, J., 2008a. Sesamin supplementation increases white muscle docosahexaenoic acid (DHA) levels in rainbow trout (*Oncorhynchus mykiss*) fed high alpha-linolenic acid (ALA) containing vegetable oil: metabolic actions. Lipids 43, 989–997.

Trattner, S., Ruyter, B., Østbye, T.K., Gjøen, T., Zlabek, V., Kamal-Eldin, A., Pickova, J., 2008b. Sesamin increases alpha-linolenic acid conversion to docosahexaenoic acid in Atlantic salmon (*Salmo salar* L.) hepatocytes: role of altered gene expression. Lipids 43, 999–1008.

Troell, M., Metian, M., Beveridge, M., Verdegem, M., Deutsch, L., 2014. Comment on 'Water footprint of marine protein consumption-aquaculture's link to agriculture'. Environ. Res. Lett. 9.

Trullàs, C., Fontanillas, R., Tres, A., Sala, R., 2015. Vegetable re-esterified oils in diets for rainbow trout: effects on fatty acid digestibility. Aquaculture 444, 28–35.

Trullàs, C., Fontanillas, R., Tres, A., Barroeta, A.C., Sala, R., 2016. Acid and re-esterified rapeseed oils as alternative vegetable oils for rainbow trout diets: effects on lipid digestibility and growth. Aquaculture 451, 186–194.

Trullàs, C., Fontanillas, R., Tres, A., Sala, R., 2017. Fatty acid digestibility in gilthead sea bream fed diets containing native, re-esterified or acid vegetable oils. Aquacult. Nutr. 23, 537–547.

Tschöp, M., Smiley, D.L., Heiman, M.L., 2000. Ghrelin induces adiposity in rodents. Nature 407, 908.

Tseng, Y., Lin, Y.-H., 2020. Effects of dietary supplementation with coconut oil on the growth, fatty acid profiles and some lipid metabolism relative gene expressions of orange-spotted grouper *Epinephelus coioides*. Aquacult. Nutr. 26, 201–210.

Tsukahara, T., 2013. PPARγ networks in cell signaling: update and impact of cyclic phosphatidic acid. J. Lipid. 2013, 246597.

Turan, F., Akyurt, I., 2005. Effects of red clover extract on growth performance and body composition of African catfish *Clarias gariepinus*. Fish. Sci. 71, 618–620.

Turchini, G.M., 2013. Fish oils, misconceptions and the environment. Am. J. Publ. Health 103, E4.

Turchini, G.M., Mentasti, T., Frøyland, L., Orban, E., Caprino, F., Moretti, V.M., Valfré, F., 2003. Effects of alternative dietary lipid sources on performance, tissue chemical composition, mitochondrial fatty acid oxidation capabilities and sensory characteristics in brown trout (*Salmo trutta* L.). Aquaculture 225, 251–267.

Turchini, G.M., Torstensen, B.E., Ng, W.K., 2009. Fish oil replacement in finfish nutrition. Rev. Aquacult. 1, 10–57.

Turchini, G.M., Ng, W.K., Tocher, D.R., 2010. Fish Oil Replacement and Alternative Lipid Sources in Aquaculture Feeds. CRC Press, Taylor & Francis Group, Boca Raton, FL, USA.

Turchini, G.M., Francis, D.S., Keast, R.S.J., Sinclair, A.J., 2011a. Transforming salmonid aquaculture from a consumer to a producer of long chain omega-3 fatty acids. Food Chem. 124, 609–614.

Turchini, G.M., Francis, D.S., Senadheera, S.P.S.D., Thanuthong, T., de Silva, S.S., 2011b. Fish oil replacement with different vegetable oils in Murray cod: evidence of an "omega-3 sparing effect" by other dietary fatty acids. Aquaculture 315, 250–259.

Turchini, G.M., Ng, W.K., Tocher, D.R., 2011c. Fish Oil Replacement and Alternative Lipid Sources in Aquaculture Feeds. CRC Press, Taylor & Francis Group, Boca Raton, FL, USA.

Turchini, G.M., Nichols, P.D., Barrow, C., Sinclair, A.J., 2012. Jumping on the omega-3 bandwagon: distinguishing the role of long-chain and short-chain omega-3 fatty acids. Crit. Rev. Food Sci. Nutr. 52, 795–803.

Turchini, G.M., Trushenski, J.T., Glencross, B.D., 2019. Thoughts for the future of aquaculture nutrition: realigning perspectives to reflect contemporary issues related to judicious use of marine resources in aquafeeds. N. Am. J. Aquacult. 81, 13–39.

Twibell, R.G., Wilson, R.P., 2004. Preliminary evidence that cholesterol improves growth and feed intake of soybean meal-based diets in aquaria studies with juvenile channel catfish, *Ictalurus punctatus*. Aquaculture 236, 539–546.

Tyssandier, V., Lyan, B., Borel, P., 2001. Main factors governing the transfer of carotenoids from emulsion lipid droplets to micelles. Biochim. Biophys. Acta 1533, 285–292.

Uebanso, T., Taketani, Y., Fukaya, M., Sato, K., Takei, Y., Sato, T., Sawada, N., Amo, K., Harada, N., Arai, H., 2009. Hypocaloric high-protein diet improves fatty liver and hypertriglyceridemia in sucrose-fed obese rats via two pathways. Am. J. Physiol. 297, E76–E84.

Umeda, T., Bauer, J.E., Otsuji, K., 2006. Weight loss effect of dietary diacylglycerol in obese dogs. J. Anim. Physiol. Anim. Nutr. 90, 208–215.

UN, 2015. Transforming Our World: The 2030 Agenda for Sustainable Development. Resolution Adopted by the General Assembly on 25 September 2015. United Nations.

Vagner, M., Zambonino Infante, J.L., Robin, J.H., Person-Le Ruyet, J., 2007. Is it possible to influence European sea bass (*Dicentrarchus labrax*) juvenile metabolism by a nutritional conditioning during larval stage? Aquaculture 267, 165–174.

Vagner, M., Robin, J.H., Zambonino Infante, J.L., Tocher, D.R., Person-Le Ruyet, J., 2009. Ontogenic effects of early feeding of sea bass (*Dicentrarchus labrax*) larvae with a range of dietary n-3 highly unsaturated fatty acid levels on the functioning of polyunsaturated fatty acid desaturation pathways. Br. J. Nutr. 101, 1452–1462.

Van Raaij, M.T.M., Van Den, T.G.E.E.J.M., Hallemeesch, M., Balm, P.H.M., Steffens, A.B., 1995. Effect of arterially infused catecholamines and insulin on plasma glucose and free fatty acids in carp. Am. J. Physiol. Regul. Integr. Compar. Physiol. 268, R1163–R1170.

Van Spankeren, M., Sibinga, N., Reinshol, Ø., Torstensen, B., Sæle, Ø., Liland, N., 2019. A single meal containing phytosterols does not affect the uptake of tissue distribution of cholesterol in zebrafish (*Danio rerio*). Front. Mar. Sci. 6.

Vatine, G., Vallone, D., Gothilf, Y., Foulkes, N.S., 2011. It's time to swim! Zebrafish and the circadian clock. FEBS Lett. 585, 1485–1494.

Vaughan, M., Berger, J.E., Steinberg, D., 1964. Hormone-sensitive lipase and monoglyceride lipase activities in adipose tissue. J. Biol. Chem. 239, 401–409.

Vegusdal, A., Østbye, T.K., Tran, T.N., Gjøen, T., Ruyter, B., 2004. β-oxidation, esterification, and secretion of radiolabeled fatty acids in cultivated atlantic salmon skeletal muscle cells. Lipids 39, 649–658.

Vegusdal, A., Gjøen, T., Berge, R.K., Thomassen, M.S., Ruyter, B., 2005. Effect of 18:1n-9, 20:5n-3, and 22:6n-3 on lipid accumulation and secretion by Atlantic salmon hepatocytes. Lipids 40, 477–486.

Velasco, C., Libran-Perez, M., Otero-Rodino, C., Lopez-Patino, M.A., Miguez, J.M., Cerda-Reverter, J.M., Soengas, J.L., 2016a. Ghrelin modulates hypothalamic fatty acid-sensing and control of food intake in rainbow trout. J. Endocrinol. 228, 25–37.

Velasco, C., Librán-Pérez, M., Otero-Rodiño, C., López-Patiño, M.A., Míguez, J.M., Soengas, J.L., 2016b. Intracerebroventricular ghrelin treatment affects lipid metabolism in liver of rainbow trout (*Oncorhynchus mykiss*). Gen. Compar. Endocrinol. 228, 33–39.

Vélez, E.J., Balbuena-Pecino, S., Capilla, E., Navarro, I., Gutiérrez, J., Riera-Codina, M., 2019. Effects of β2-adrenoceptor agonists on gilthead sea bream (*Sparus aurata*) cultured muscle cells. Compar. Biochem. Physiol. Mol. Integr. Physiol. 227, 179–193.

Vendrell, J., El Bekay, R., Peral, B., García-Fuentes, E., Megia, A., Macias-Gonzalez, M., Real, J.F., Jimenez-Gomez, Y., Escoté, X., Pachón, G., Simó, R., Selva, D.M., Malagón, M.M., Tinahones, F.J., 2011. Study of the potential association of adipose tissue GLP-1 receptor with obesity and insulin resistance. Endocrinology 152, 4072–4079.

Venkatachalam, A.B., Sawler, D.L., Wright, J.M., 2013. Tissue-specific transcriptional modulation of fatty acid-binding protein genes, fabp2, fabp3 and fabp6, by fatty acids and the peroxisome proliferator, clofibrate, in zebrafish (*Danio rerio*). Gene 520, 14–21.

Venkatachalam, A.B., Parmar, M.B., Wright, J.M., 2017. Evolution of the duplicated intracellular lipid-binding protein genes of teleost fishes. Mol. Genet. Genom. 292, 699–727.

Venkatachalam, A.B., Fontenot, Q., Farrara, A., Wright, J.M., 2018. Fatty acid-binding protein genes of the ancient, air-breathing, ray-finned fish, spotted gar (*Lepisosteus oculatus*). Compar. Biochem. Physiol. D Genom. Proteom. 25, 19–25.

Vera, L.M., Negrini, P., Zagatti, C., Frigato, E., Sánchez-Vázquez, F.J., Bertolucci, C., 2013. Light and feeding entrainment of the molecular circadian clock in a marine teleost (*Sparus aurata*). Chronobiol. Int. 30, 649–661.

Vera, L.M., Metochis, C., Taylor, J.F., Clarkson, M., Skjærven, K.H., Migaud, H., Tocher, D.R., 2017. Early nutritional programming affects liver transcriptome in diploid and triploid Atlantic salmon (*Salmo salar*). BMC Genom. 18, 886.

Veramaris, 2018. Harnessing Nature. Enabling Growth. Available at: https://www.veramaris.com/files/assets/downloads/publications/Imagebr_Veramaris_04-2018.pdf.

Veramaris, 2019. https://www.veramaris.com/home.html.

Vianen, G.J., Obels, P.P., Thillart, G.E.E.J.M.V.D., Zaagsma, J., 2002. β-Adrenoceptors mediate inhibition of lipolysis in adipocytes of tilapia (*Oreochromis mossambicus*). Am. J. Physiol. 282, E318–E325.

Villeneuve, L., Gisbert, E., Cahu, C.L., Gall, M.-M.L., Zambonino-Infante, J.L., 2004. Expression and localization of some retinoid receptors during European sea bass (*Dicentrarchus labrax*) larvae development. Aquaculture 242, 537–551.

Volkoff, H., Hoskins, L.J., Tuziak, S.M., 2010. Influence of intrinsic signals and environmental cues on the endocrine control of feeding in fish: potential application in aquaculture. Gen. Compar. Endocrinol. 167, 352–359.

Wang, C.W., 2016. Lipid droplets, lipophagy, and beyond. Biochim. Biophys. Acta 1861, 793–805.

Wang, H., Hu, L., Dalen, K., Dorward, H., Marcinkiewicz, A., Russell, D., Gong, D., Londos, C., Yamaguchi, T., Holm, C., Rizzo, M.A., Brasaemle, D., Sztalryd, C., 2009a. Activation of hormone-sensitive lipase requires two steps, protein phosphorylation and binding to the PAT-1 domain of lipid droplet coat proteins. J. Biol. Chem. 284, 32116–32125.

Wang, W., Mai, K., Zhang, W., Ai, Q., Yao, C., Li, H., Liufu, Z., 2009b. Effects of dietary copper on survival, growth and immune response of juvenile abalone, *Haliotis discus* Hannai Ino. Aquaculture 297, 122–127.

Wang, A., Han, G., Wei, X., Liu, B., Lv, F., Feng, G., Qi, Z., Wang, T., Xu, P., Yang, Z., 2010. Molecular cloning of fatty acid synthase from gift tilapia (*Oreochromis niloticus*): response of its expression to refeeding and different lipid levels in diet. J. Fish. Sci. China 34, 1113–1120.

Wang, F.-B., Luo, L., Lin, S.-M., Li, Y., Chen, S., Wang, Y.-G., Wen, H., Hu, C.-J., 2011. Dietary magnesium requirements of juvenile grass carp, *Ctenopharyngodon idella*. Aquacult. Nutr. 17, e691–e700.

Wang, X., Huang, M., Wang, Y., 2012a. The effect of insulin, TNFα and DHA on the proliferation, differentiation and lipolysis of preadipocytes isolated from large yellow croaker (*Pseudosciaena crocea* R.). PLoS One 7, e48069.

Wang, X., Wang, Y., Li, Y., 2012b. Adipose triglyceride lipase (ATGL) clone, expression pattern, and regulation by different lipid sources and lipid levels in large yellow croaker (*Pseudosciaena crocea* R.). Mar. Biotechnol. 15.

Wang, S., Zhu, M., Du, M., 2015. Prevention of obesity by dietary resveratrol: how strong is the evidence? Expert Rev. Endocrinol. Metabol. 10, 561–564.

Wang, J., Li, X., Han, T., Yang, Y., Jiang, Y., Yang, M., Xu, Y., Harpaz, S., 2016. Effects of different dietary carbohydrate levels on growth, feed utilization and body composition of juvenile grouper *Epinephelus akaara*. Aquaculture 459, 143–147.

Wang, C.Y., Li, Z.B., Sun, Y.Z., Chen, Q., Li, W.J., Huang, Y.C., Lu, J., 2017a. Effects of Chinese herbal medicines mixture on growth performance digestive enzyme activity immune response of juvenile Japanese seabass, *Lateolabrax japonicus*. Aquacult. Nutr. 24, 683–693.

Wang, J.T., Jiang, Y.D., Han, T., Li, X.Y., Wang, Y., Liu, Y.J., 2017b. Effects of dietary carbohydrate-to-lipid ratios on growth and body composition of orange-spotted grouper epinephelus coioides. N. Am. J. Aquacult. 79, 1–7.

Wang, P., Zhu, J., Feng, J., He, J., Lou, Y., Zhou, Q., 2017c. Effects of dietary soy protein concentrate meal on growth, immunity, enzyme activity and protein metabolism in relation to gene expression in large yellow croaker *Larimichthys crocea*. Aquaculture 477, 15–22.

Wang, W., Yang, H., Johnson, D., Gensler, C., Decker, E., Zhang, G., 2017d. Chemistry and biology of omega-3 PUFA peroxidation-derived compounds. Prostag. Other Lipid Mediat. 132, 84–91.

Wang, Y., Zhang, J., Jiao, J., Du, X., Limbu, S., Chen, L., Qiao, F., Li, D., Zhang, M., Du, Z., 2017e. Physiological and metabolic differences between visceral and subcutaneous adipose tissues in Nile tilapia (*Oreochromis niloticus*). Am. J. Physiol. Regul. Integr. Compar. Physiol. 313, R608–R619.

Wang, Y.W., Zhang, J.L., Jiao, J.G., Du, X.X., Limbu, S.M., Chen, L.Q., Qiao, F., Li, D.L., Zhang, M.L., Du, Z.Y., 2017f. Physiological and metabolic differences between visceral and subcutaneous adipose tissues in Nile tilapia (*Oreochromis niloticus*). Am. J. Physiol. Regul. Integr. Compar. Physiol. 313, R608–R619.

Wang, A.R., Ran, C., Ringø, E., Zhou, Z., 2018a. Progress in fish gastrointestinal microbiota research. Rev. Aquacult. 10, 626–640.

Wang, J., Han, S.L., Li, L.Y., Lu, D.L., Limbu, S.M., Li, D.L., Zhang, M.L., Du, Z.Y., 2018b. Lipophagy is essential for lipid metabolism in fish. Sci. Bull. 63, 879–882.

Wang, S., Chen, J., Jiang, D., Zhang, Q., You, C., Tocher, D.R., Monroig, O., Dong, Y., Li, Y., 2018c. Hnf4alpha is involved in the regulation of vertebrate LC-PUFA biosynthesis: insights into the regulatory role of Hnf4alpha on expression of liver fatty acyl desaturases in the marine teleost *Siganus canaliculatus*. Fish Physiol. Biochem. 44, 805–815.

Wang, C., Hu, W., Wang, L., Qiao, H., Wu, H., Xu, Z., 2019a. Effects of dietary supplementation with Sargassum horneri meal on growth performance, body composition, and immune response of juvenile turbot. J. Appl. Phycol. 31, 771–778.

Wang, J., Han, S.L., Li, L.Y., Lu, D.L., Limbu, S.M., Li, D.L., Zhang, M.L., Du, Z.Y., 2019b. Inhibited lipophagy suppresses lipid metabolism in zebrafish liver cells. Front. Physiol. 10.

Wang, Y., Meng, R., Xu, X., Liao, K., Ran, Z., Xu, J., Cao, J., Wang, Y., Wang, D., Xu, S., Yan, X., 2019c. Effects of nutritional status and diet composition on fatty acid transporters expression in zebrafish (*Danio rerio*). Aquacult. Res. 50, 904–914.

Ward, C., Martinez-Lopez, N., Otten, E.G., Carroll, B., Maetzel, D., Singh, R., Sarkar, S., Korolchuk, V.I., 2016. Autophagy, lipophagy and lysosomal lipid storage disorders. Biochim. Biophys. Acta 1861, 269–284.

Watanabe, T., Kiron, V., Satoh, S., 1997. Trace minerals in fish nutrition. Aquaculture 151, 185–207.

Watanabe, M., Houten, S.M., Wang, L., Moschetta, A., Auwerx, J., 2004. Bile acids lower triglyceride levels via a pathway involving Fxr, Shp, and SREBP-1c. J. Clin. Invest. 113, 1408–1418.

Weber, D., Grune, T., 2012. The contribution of β-carotene to vitamin A supply of humans. Mol. Nutr. Food Res. 56, 251–258.

Wei, C.C., Wu, K., Gao, Y., Zhang, L.H., Li, D.D., Luo, Z., 2017a. Magnesium reduces hepatic lipid accumulation in yellow catfish (*Pelteobagrus fulvidraco*) and modulates lipogenesis and lipolysis via ppara, jak-STAT, and AMPK pathways in hepatocytes. J. Nutr. 147.

Wei, C.C., Wu, K., Gao, Y., Zhang, L.H., Li, D.D., Luo, Z., 2017b. Magnesium reduces hepatic lipid accumulation in yellow catfish (*Pelteobagrus fulvidraco*) and modulates lipogenesis and lipolysis via ppara, Jak-STAT, and AMPK pathways in hepatocytes. J. Nutr. 147, 1070–1078.

Wei, C.C., Luo, Z., Hogstrand, C., Xu, Y.H., Wu, L.X., Chen, G.H., Pan, Y.X., Song, Y., 2018. Zinc reduces hepatic lipid deposition and activates lipophagy via Zn2+/MTF-1/PPARα and Ca2+/CaMKKβ/AMPK pathways. FASEB J. 32, 6666–6680.

Weidberg, H., Shvets, E., Elazar, Z., 2009. Lipophagy: selective catabolism designed for lipids. Dev. Cell 16, 628–630.

Weinrauch, A.M., Schaefer, C.M., Goss, G.G., 2019. Activity and post-prandial regulation of digestive enzyme activity along the Pacific hagfish (*Eptatretus stoutii*) alimentary canal. PLoS One 14, e0215027.

Welte, M.A., Gould, A.P., 2017. Lipid droplet functions beyond energy storage. Biochim. Biophys. Acta 1862, 1260–1272.

White, C.A., Dworjanyn, S.A., Nichols, P.D., Mos, B., Dempster, T., 2016. Future aquafeeds may compromise reproductive fitness in a marine invertebrate. Mar. Environ. Res. 122, 67–74.

Whitehead, R.H., 1909. A note on the absorption of fat. Am. J. Physiol. 24, 294–296.

Willett, W., Rockstrom, J., Loken, B., Springmann, M., Lang, T., Vermeulen, S., Garnett, T., Tilman, D., Declerck, F., Wood, A., Jonell, M., Clark, M., Gordon, L.J., Fanzo, J., Hawkes, C., Zurayk, R., Rivera, J.A.W., De Vries, Sibanda, L.M., Afshin, A., Chaudhary, A., Herrero, M., Agustina, R., Branca, F., Lartey, A., Fan, S.G., Crona, B., Fox, E., Bignet, V., Troell, M., Lindahl, T., Singh, S., Cornell, S.E., Reddy, K.S., Narain, S., Nishtar, S., Murray, C.J.L., 2019. Food in the Anthropocene: the EAT-Lancet Commission on healthy diets from sustainable food systems. Lancet 393, 447–492.

Wilson, A.C., Williamson, I.P., 1970. Fatty acid synthetase from the liver of the plaice, *Pleuronectes platessa*. Biochem. J. 117, 26P–27P.

Winwood, R.J., 2013. Recent Developments in the Commercial Production of DHA and EPA Rich Oils from Microalgae. OCL20.

Won, E.T., Baltzegar, D.A., Picha, M.E., Borski, R.J., 2012. Cloning and characterization of leptin in a Perciform fish, the striped bass (*Morone saxatilis*): control of feeding and regulation by nutritional state. Gen. Compar. Endocrinol. 178, 98–107.

Wood, J.D., Richardson, R.I., Nute, G.R., Fisher, A.V., Campo, M.M., Kasapidou, E., Sheard, P.R., Enser, M., 2004. Effects of fatty acids on meat quality: a review. Meat Sci. 66, 21–32.

Wu, J.L., Zhang, J.L., Du, X.X., Shen, Y.J., Lao, X., Zhang, M.L., Chen, L.Q., Du, Z.Y., 2015. Evaluation of the distribution of adipose tissues in fish using magnetic resonance imaging (MRI). Aquaculture 448, 112–122.

Wu, C., Ye, J., Gao, J.E., Yang, X., Zhang, Y., 2016. Effect of varying carbohydrate fractions on growth, body composition, metabolic, and hormonal indices in juvenile Black carp, *Mylopharyngodon piceus*. J. World Aquacult. Soc. 47, 435–449.

Wu, K., Tan, X.Y., Xu, Y.H., Chen, G.H., Zhuo, M.Q., 2018. Functional analysis of promoters of genes in lipid metabolism and their transcriptional response to STAT3 under leptin signals. Genes 9, 334.

Xia, X., Wang, X., Qin, W., Jiang, J., Cheng, L., 2019. Emerging regulatory mechanisms and functions of autophagy in fish. Aquaculture 511, 734212.

Xiao, P., Ji, H., Ye, Y., Zhang, B., Chen, Y., Tian, J., Liu, P., Chen, L., Du, Z., 2017. Dietary silymarin supplementation promotes growth performance and improves lipid metabolism and health status in grass carp (*Ctenopharyngodon idellus*) fed diets with elevated lipid levels. Fish Physiol. Biochem. 43, 245–263.

Xie, D., Fu, Z., Wang, S., You, C., Monroig, O., Tocher, D.R., Li, Y., 2018. Characteristics of the fads2 gene promoter in marine teleost *Epinephelus coioides* and role of Sp1-binding site in determining promoter activity. Sci. Rep. 8, 5305.

Xie, S., Yin, P., Tian, L., Liu, Y., Niu, J., 2020. Lipid metabolism and plasma metabolomics of juvenile largemouth bass *Micropterus salmoides* were affected by dietary oxidized fish oil. Aquaculture 522, 735158.

Xiong, D.M., Xie, C.X., Zhang, H.J., Liu, H.P., 2011. Digestive enzymes along digestive tract of a carnivorous fish *Glyptosternum maculatum* (Sisoridae, Siluriformes). J. Anim. Physiol. Anim. Nutr. 95, 56–64.

Xiong, F., Wu, S.G., Zhang, J., Jakovlic, I., Li, W.X., Zou, H., Li, M., Wang, G.T., 2018. Dietary bile salts types influence the biliary bile acids and gut microbiota in grass cap. Front. Microbiol. 9, 2209.

Xu, J.H., Qin, J., Yan, B.L., Zhu, M., Luo, G., 2011. Effects of dietary lipid levels on growth performance, feed utilization and fatty acid composition of juvenile Japanese seabass (*Lateolabrax japonicus*) reared in seawater. Aquacult. Int. 19, 79–89.

Xu, H.Y., Wang, Y., Wang, J.W., 2012. Advances in animal fatty acid transmembrane transport proteins FATP1 and FATP4. J. Anim. Vet. Adv. 11, 2064–2069.

Xu, H., Zhang, D.L., Yu, D.H., Lv, C.H., Luo, H.Y., Wang, Z.Y., 2015a. Molecular cloning and expression analysis of scd1 gene from large yellow croaker *Larimichthys crocea* under cold stress. Gene 568, 100–108.

Xu, L., Huang, D.P., Hu, Q.L., Wu, J., Wang, Y.Z., Feng, J., 2015b. Betaine alleviates hepatic lipid accumulation via enhancing hepatic lipid export and fatty acid oxidation in rats fed with a high-fat diet. Br. J. Nutr. 113, 1835–1843.

Xu, H., Du, J., Li, S., Mai, K., Xu, W., Ai, Q., 2017a. Effects of dietary n-3 long-chain unsaturated fatty acid on growth performance, lipid deposition, hepatic fatty acid composition and health-related serum enzyme activity of juvenile Japanese seabass *Lateolabrax japonicus*. Aquacult. Nutr. 23, 1449–1457.

Xu, H., Zhang, Y., Wang, C., Wei, Y., Zheng, K., Liang, M., 2017b. Cloning and characterization of fatty acid-binding proteins (fabps) from Japanese seabass (*Lateolabrax japonicus*) liver, and their gene expressions in response to dietary arachidonic acid (ARA). Compar. Biochem. Physiol. B Biochem. Mol. Biol. 204, 27–34.

Xu, H., Zhang, Y., Wang, C., Wei, Y., Zheng, K., Liang, M., 2017c. Cloning and characterization of fatty acid transport proteins in Japanese seabass *Lateolabrax japonicus*, and their gene expressions in response to dietary arachidonic acid. Aquacult. Res. 48, 5718–5728.

Xu, F., Xu, C., Xiao, S., Lu, M., Limbu, S.M., Wang, X., Du, Z.Y., Qin, J.G., Chen, L.Q., 2019a. Effects of α-lipoic acid on growth performance, body composition, antioxidant profile and lipid metabolism of the GIFT tilapia (*Oreochromis niloticus*) fed high-fat diets. Aquacult. Nutr. 25, 585–596.

Xu, H., Liao, Z., Zhang, Q., Wei, Y., Liang, M., 2019b. A moderately high level of dietary lipid inhibited the protein secretion function of liver in juvenile tiger puffer *Takifugu rubripes*. Aquaculture 498, 17–27.

Xue, M., Luo, L., Wu, X., Ren, Z., Gao, P., Yu, Y., Pearl, G., 2006. Effects of six alternative lipid sources on growth and tissue fatty acid composition in Japanese sea bass (*Lateolabrax japonicus*). Aquaculture 260, 206–214.

Yabu, T., Shimuzu, A., Yamashita, M., 2009. A novel mitochondrial sphingomyelinase in zebrafish cells. J. Biol. Chem. 284, 20349–20363.

Yaghoubi, M., Mozanzadeh, M.T., Marammazi, J.G., Safari, O., Gisbert, E., 2016. Dietary replacement of fish meal by soy products (soybean meal and isolated soy protein) in silvery-black porgy juveniles (*Sparidentex hasta*). Aquaculture 464, 50–59.

Yang, X., Lu, X., Lombès, M., Rha, G.B., Chi, Y.-I., Guerin, T.M., Smart, E.J., Liu, J., 2010. The G0/G1 switch gene 2 regulates adipose lipolysis through association with adipose triglyceride lipase. Cell Metabol. 11, 194–205.

Yang, Q., Zheng, P., Ma, Z., Li, T., Jiang, S., Qin, J.G., 2015. Molecular cloning and expression analysis of the retinoid X receptor (RXR) gene in golden pompano Trachinotus ovatus fed Artemia nauplii with different enrichments. Fish Physiol. Biochem. 41, 1449–1461.

Yang, B.-Y., Zhai, G., Gong, Y.-L., Su, J.-Z., Peng, X.-Y., Shang, G.-H., Han, D., Jin, J.-Y., Liu, H.-K., Du, Z.-Y., 2017a. Different physiological roles of insulin receptors in mediating nutrient metabolism in zebrafish. Am. J. Physiol. 315, E38–E51.

Yang, Q., Ding, M., Tan, B., Dong, X., Chi, S., Zhang, S., Liu, H., 2017b. Effects of dietary vitamin A on growth, feed utilization, lipid metabolism enzyme activities, and fatty acid synthase and hepatic lipase mRNA expression levels in the liver of juvenile orange spotted grouper, *Epinephelus coioides*. Aquaculture 479, 501–507.

Yang, Q., Liang, Q., Balakrishnan, B., Belobrajdic, D.P., Feng, Q.J., Zhang, W., 2020. Role of dietary nutrients in the modulation of gut microbiota: a narrative review. Nutrients 12.

Yao, J., Zhi, M., Minhu, C., 2011. Effect of silybin on high-fat-induced fatty liver in rats. Braz. J. Med. Biol. Res. 44, 652–659.

Yasushi, H., Hisayo, E., Shunsuke, T., Ryuji, H., 2010. Novel acyl-coenzyme A: monoacylglycerol acyltransferase plays an important role in hepatic triacylglycerol secretion. J. Lipid Res. 51, 1424–1431.

Ye, J.D., Wang, K., Li, F.D., Sun, Y.Z., 2011. Single or combined effects of fructo- and mannan oligosaccharide supplements and Bacillus clausii on the growth, feed utilization, body composition, digestive enzyme activity, innate immune response and lipid metabolism of Japanese flounder *Paralichthys olivaceus*. Aquacult. Nutr. 17, e902–e911.

Ye, H., Xu, M., Liu, Q., Sun, Z., Zou, C., Chen, L., Su, N., Ye, C., 2019. Effects of replacing fish meal with soybean meal on growth performance, feed utilization and physiological status of juvenile obscure puffer, *Takifugu obscurus*. Compar. Biochem. Physiol. C Toxicol. Pharmacol. 216, 75–81.

Yeh, Y.-H., Hwang, D.-F., 2001. High-performance liquid chromatographic determination for bile components in fish, chicken and duck. J. Chromatogr. B Biomed. Sci. Appl. 751, 1–8.

Yen, C.L., Nelson, D.W., Yen, M.I., 2015. Intestinal triacylglycerol synthesis in fat absorption and systemic energy metabolism. J. Lipid Res. 56, 489–501.

Yeung, C.M., Chan, C.B., Woo, N.Y.S., Cheng, C.H.K., 2006. Seabream ghrelin: cDNA cloning, genomic organization and promoter studies. J. Endocrinol. 189, 365–379.

Yildirimer, C.C., Brown, K.H., 2018. Intestinal microbiota lipid metabolism varies across rainbow trout (*Oncorhynchus mykiss*) phylogeographic divide. J. Appl. Microbiol. 125, 1614–1625.

Yin, B., Liu, H., Tan, B., Dong, X., Chi, S., Yang, Q., Zhang, S., Chen, L., 2018. Cottonseed protein concentrate (CPC) suppresses immune function in different intestinal segments of hybrid grouper♀ *Epinephelus fuscoguttatus* ×♂ *Epinephelus lanceolatu* via TLR-2/MyD88 signaling pathways. Fish Shellfish Immunol. 81, 318–328.

Yin, P., Xie, S., Huo, Y., Guo, T., Fang, H., Zhang, Y., Liu, Y., Tian, L., Niu, J., 2019. Effects of dietary oxidized fish oil on growth performance, antioxidant defense system, apoptosis and mitochondrial function of juvenile largemouth bass (*Micropterus salmoides*). Aquaculture 500, 347–358.

Yorimitsu, T., Klionsky, D.J., 2005. Autophagy: molecular machinery for self-eating. Cell Death Differ. 12, 1542–1552.

You, C., Chen, B., Wang, M., Wang, S., Zhang, M., Sun, Z., Juventus, A.J., Ma, H., Li, Y., 2019. Effects of dietary lipid sources on the intestinal microbiome and health of golden pompano (*Trachinotus ovatus*). Fish Shellfish Immunol. 89, 187–197.

Ytrestøyl, T., Struksnæs, G., Koppe, W., Bjerkeng, B., 2005. Effects of temperature and feed intake on astaxanthin digestibility and metabolism in Atlantic salmon, *Salmo salar*. Compar. Biochem. Physiol. B Biochem. Mol. Biol. 142, 445–455.

Ytrestøyl, T., Aas, T.S., Asgard, T., 2015. Utilisation of feed resources in production of Atlantic salmon (*Salmo salar*) in Norway. Aquaculture 448, 365–374.

Yu, Y., Fang, L., Zhang, Y., Sheng, H.X., Fang, W.H., 2015. VgrG2 of type VI secretion system 2 of Vibrio parahaemolyticus induces autophagy in macrophages. Front. Microbiol. 6, 168.

Yu, G., Qu, W., Liao, Z., Xu, H.Y., Liang, M., Zhang, Y., Mai, K., 2019. Intestinal homeostasis of juvenile tiger puffer *Takifugu rubripes* was sensitive to dietary arachidonic acid on mucosal microbiota. Aquaculture 502, 97–106.

Yuan, X., Liang, X.F., Liu, L., Fang, J., Li, J., Li, A., Cai, W., Xue, M., Wang, J., Wang, Q., 2016. Fat deposition pattern and mechanism in response to dietary lipid levels in grass carp, *Ctenopharyngodon idellus*. Fish Physiol. Biochem. 42, 1–13.

Yun, B.A., Mai, K.S., Zhang, W.B., Xu, W., 2011. Effects of dietary cholesterol on growth performance, feed intake and cholesterol metabolism in juvenile turbot (*Scophthalmus maximus* L.) fed high plant protein diets. Aquaculture 319, 105–110.

Yun, B., Ai, Q., Mai, K., Xu, W., Qi, G., Luo, Y., 2012. Synergistic effects of dietary cholesterol and taurine on growth performance and cholesterol metabolism in juvenile turbot (*Scophthalmus maximus* L.) fed high plant protein diets. Aquaculture 324, 85–91.

Yun, C., Xiao, L., Cao, X., Jian, G., 2018. Molecular characterization of peroxisome proliferator activated receptor gamma (PPARγ) in loach *Misgurnus anguillicaudatus* and its potential roles in fatty acid metabolism in vitro. Process Biochem. 66, 205–211.

Zammit, V.A., Newsholme, E.A., 1979. Activities of enzymes of fat and ketone-body metabolism and effects of starvation on blood concentrations of glucose and fat fuels in teleost and elasmobranch fish. Biochem. J. 184, 313.

Zang, L., Shimada, Y., Kawajiri, J., Tanaka, T., Nishimura, N., 2014. Effects of Yuzu (*Citrus junos* Siebold ex Tanaka) peel on the diet-induced obesity in a zebrafish model. J. Funct. Foods 10, 499–510.

Zechner, R., Madeo, F., Kratky, D., 2017. Cytosolic lipolysis and lipophagy: two sides of the same coin. Nat. Rev. Mol. Cell Biol. 18, 671–684.

Zeng, D., Ai, Q., Milley, J.E., Lall, S.P., 2010. Lipid and fatty acid compositions of cod (*Gadus morhua*), haddock (*Melanogrammus aeglefinus*) and halibut (*Hippoglossus hippoglossus*). J. Ocean Univ. China 9, 381–388.

Zhang, L., 2006. Nutritional Physiology of Several Vitamins and Protein Sources Development for Japanese Seabass, *Lateolabrax japonicus* and Large Yellow Croaker, *Pseudosciaena crocea*. Ocean University of China (in Chinese).

Zhang, Y., Zhang, H., 2013. The effects of probiotics on lipid metabolism. In: Baez, R.V. (Ed.), Lipid Metabolism. InTech.

Zhang, Y., Guo, K., Leblanc, R.E., Loh, D., Schwartz, G.J., Yu, Y.H., 2007. Increasing dietary leucine intake reduces diet-induced obesity and improves glucose and cholesterol metabolism in mice via multimechanisms. Diabetes 56, 1647–1654.

Zhang, J., Chen, L., Zheng, J., Zeng, T., Li, H., Xiao, H., Deng, X., Hu, X., 2012. The protective effect of resveratrol on islet insulin secretion and morphology in mice on a high-fat diet. Diab. Res. Clin. Pract. 97, 474–482.

Zhang, H., Chen, H., Zhang, Y., Li, S., Lu, D., Zhang, H., Meng, Z., Liu, X., Lin, H., 2013a. Molecular cloning, characterization and expression profiles of multiple leptin genes and a leptin receptor gene in orange-spotted grouper (*Epinephelus coioides*). Gen. Compar. Endocrinol. 181, 295–305.

Zhang, M., Xu, Y., Li, L., Wei, S., Zhang, S., Liu, Z., 2013b. Identification, evolution and expression of a CD36 homolog in the basal chordate amphioxus Branchiostoma japonicum. Fish Shellfish Immunol. 34, 546–555.

Zhang, D., Lu, K., Dong, Z., Jiang, G., Xu, W., Liu, W., 2014. The effect of exposure to a high-fat diet on microRNA expression in the liver of blunt snout bream (*Megalobrama amblycephala*). PLoS One 9, e96132.

Zhang, C.X., Huang, F., Li, J., Wang, L., Song, K., Mai, K.S., 2016a. Interactive effects of dietary magnesium and vitamin E on growth performance, body composition, blood parameters and antioxidant status in Japanese seabass (*Lateolabrax japonicus*) fed oxidized oil. Aquacult. Nutr. 22, 708–722.

Zhang, Q., You, C., Wang, S., Dong, Y., Monroig, O., Tocher, D.R., Li, Y., 2016b. The miR-33 gene is identified in a marine teleost: a potential role in regulation of LC-PUFA biosynthesis in *Siganus canaliculatus*. Sci. Rep. 6, 32909.

Zhang, Y., Li, Y., Liang, X., Cao, X., Huang, L., Yan, J., Wei, Y., Gao, J., 2017. Hepatic transcriptome analysis and identification of differentially expressed genes response to dietary oxidized fish oil in loach *Misgurnus anguillicaudatus*. PLoS One 12, e0172386.

Zhang, J., Hu, Y., Ai, Q., Mao, P., Tian, Q., Zhong, L., Xiao, T., Chu, W., 2018a. Effect of dietary taurine supplementation on growth performance, digestive enzyme activities and antioxidant status of juvenile black carp (*Mylopharyngodon piceus*) fed with low fish meal diet. Aquacult. Res. 49, 3187–3195.

Zhang, Y., Li, S., Li, J., Han, L., He, Q., Wang, R., Wang, X., Liu, K., 2018b. Developmental toxicity induced by PM2.5 through endoplasmic reticulum stress and autophagy pathway in zebrafish embryos. Chemosphere 197, 611–621.

Zhang, Z., Yao, Z., Chen, Y., Qian, L., Jiang, S., Zhou, J., Shao, J., Chen, A., Zhang, F., Zheng, S., 2018c. Lipophagy and liver disease: new perspectives to better understanding and therapy. Biomed. Pharmacother. 97, 339–348.

Zhang, C.N., Wand, J.H., Huang, M.X., Liu, M., Zhang, J., 2019a. Evaluation of growth performance and lipid metabolism in zebrafish fed fructooligosaccharide using RNA sequencing. Aquacult. Nutr. 25, 1194–1206.

Zhang, H., Luo, Y., Lu, D.-L., Jiao, J.-G., Li, L.-Y., Qin, J.-G., Du, Z.-Y., Chen, L.-Q., 2019b. Diacylglycerol oil reduces fat accumulation and increases protein content by inducing lipid catabolism and protein metabolism in Nile tilapia (*Oreochromis niloticus*). Aquaculture 510, 90–99.

Zhang, Y., Cao, X., Gao, J., 2019c. Cloning of fatty acid-binding protein 2 (fabp2) in loach (*Misgurnus anguillicaudatus*) and its expression in response to dietary oxidized fish oil. Compar. Biochem. Physiol. B Biochem. Mol. Biol. 229, 26–33.

Zhang, M.L., Li, M., Sheng, Y., Tan, F., Chen, L.Q., Cann, I., Du, Z.Y., 2020a. Citrobacter species increase energy harvest by modulating intestinal microbiota in fish: nondominant species play important functions. mSystems 5, e00303–e00320.

Zhang, W.B., Liu, M., De Mitcheson, Y.S., Cao, L., Leadbitter, D., Newton, R., Little, D.C., Li, S.L., Yang, Y., Chen, X., Zhou, W., 2020b. Fishing for feed in China: facts, impacts and implications. Fish Fish. 21, 47–62.

Zhao, Y., Gul, Y., Li, S., Wang, W., 2011. Cloning, identification and accurate normalization expression analysis of PPARα gene by GeNorm in *Megalobrama amblycephala*. Fish Shellfish Immunol. 31, 462–468.

Zhao, J., Ai, Q., Mai, K., Zuo, R., Luo, Y., 2013. Effects of dietary phospholipids on survival, growth, digestive enzymes and stress resistance of large yellow croaker, *Larmichthys crocea* larvae. Aquaculture 410–411, 122–128.

Zhao, T., Wu, K., Hogstrand, C., Xu, Y.H., Chen, G.H., Wei, C.C., Luo, Z., 2020a. Lipophagy mediated carbohydrate-induced changes of lipid metabolism via oxidative stress, endoplasmic reticulum (ER) stress and ChREBP/PPARγ pathways. Cell. Mol. Life Sci. 77, 1987–2003.

Zhao, Y., Cao, X., Fu, L., Gao, J., 2020b. n-3 PUFA reduction caused by fabp2 deletion interferes with triacylglycerol metabolism and cholesterolhomeostasis in fish. Appl. Microbiol. Biotechnol. 104, 2149–2161.

Zheng, X., Tocher, D.R., Dickson, C.A., Bell, J.G., Teale, A.J., 2004. Effects of diets containing vegetable oil on expression of genes involved in highly unsaturated fatty acid biosynthesis in liver of Atlantic salmon (*Salmo salar*). Aquaculture 236, 467–483.

Zheng, X., Torstensen, B.E., Tocher, D.R., Dick, J.R., Henderson, R.J., Bell, J.G., 2005. Environmental and dietary influences on highly unsaturated fatty acid biosynthesis and expression of fatty acyl desaturase and elongase genes in liver of Atlantic salmon (*Salmo salar*). Biochim. Biophys. Acta 1734, 13–24.

Zheng, X., Ding, Z., Xu, Y., Monroig, O., Morais, S., Tocher, D.R., 2009. Physiological roles of fatty acyl desaturases and elongases in marine fish: characterisation of cDNAs of fatty acyl Δ6 desaturase and elovl5 elongase of cobia (*Rachycentron canadum*). Aquaculture 290, 122–131.

Zheng, J.L., Luo, Z., Liu, C.X., Chen, Q.L., Tan, X.Y., Zhu, Q.L., Gong, Y., 2013a. Differential effects of acute and chronic zinc (Zn) exposure on hepatic lipid deposition and metabolism in yellow catfish *Pelteobagrus fulvidraco*. Aquat. Toxicol. 132, 173–181.

Zheng, J.L., Luo, Z., Zhu, Q.L., Chen, Q.L., Gong, Y., 2013b. Molecular characterization, tissue distribution and kinetic analysis of carnitine palmitoyltransferase I in juvenile yellow catfish *Pelteobagrus fulvidraco*. Genomics 101, 195–203.

Zheng, J.L., Luo, Z., Zhu, Q.L., Tan, X.Y., Chen, Q.L., Sun, L.D., Hu, W., 2013c. Molecular cloning and expression pattern of 11 genes involved in lipid metabolism in yellow catfish *Pelteobagrus fulvidraco*. Gene 531, 53–63.

Zheng, J.L., Luo, Z., Hu, W., Liu, C.X., Chen, Q.L., 2014. Differential effects of dietary Zn deficiency and excess on carnitine status, kinetics and expression of CPT I in yellow catfish *Pelteobagrus fulvidraco*. Aquaculture 420, 10–17.

Zheng, J.L., Luo, Z., Hu, W., Liu, C.X., Chen, Q.L., Zhu, Q.L., Gong, Y., 2015a. Different effects of dietary Zn deficiency and excess on lipid metabolism in yellow catfish *Pelteobagrus fulvidraco*. Aquaculture 435, 10–17.

Zheng, J.L., Zhuo, M.Q., Luo, Z., Song, Y.F., Pan, Y.X., Huang, C., Hu, W., Chen, Q.L., 2015b. Peroxisome proliferator-activated receptor alpha1 in yellow catfish *Pelteobagrus fulvidraco*: molecular characterization, mRNA tissue expression and transcriptional regulation by insulin in vivo and in vitro. Compar. Biochem. Physiol. B Biochem. Mol. Biol. 183, 58–66.

Zheng, X., Dai, W., Chen, X., Wang, K., Zhang, W., Liu, L., Hou, J., 2015c. Caffeine reduces hepatic lipid accumulation through regulation of lipogenesis and ER stress in zebrafish larvae. J. Biomed. Sci. 1, 1–12.

Zheng, J.L., Zhu, Q.L., Shen, B., Zeng, L., Zhu, A.Y., Wu, C.W., 2016. Effects of starvation on lipid accumulation and antioxidant response in the right and left lobes of liver in large yellow croaker *Pseudosciaena crocea*. Ecol. Indicat. 66, 269–274.

Zhou, Q.C., Liu, Y., Mai, K., Tian, L., 2007. Effect of dietary phosphorus levels on growth, body composition, muscle and bone mineral concentrations for orange-spotted grouper *epinephelus coioides* reared in floating cages. J. World Aquacult. Soc. 35, 427–435.

Zhou, J.S., Chen, H.J., Shi, X.C., Li, X.X., Chen, L.Q., Du, Z.Y., Yu, H., 2017. Effect of dietary bile acids on growth, body composition, lipid metabolism and microbiota in grass carp (*Ctenopharyngodon idella*). Aquacult. Nutr. 24, 802–813.

Zhou, J.S., Chen, H.J., Ji, H., Shi, X.C., Li, X.X., Chen, L.Q., Du, Z.Y., Yu, H.B., 2018a. Effect of dietary bile acids on growth, body composition, lipid metabolism and microbiota in grass carp (*Ctenopharyngodon idella*). Aquacult. Nutr. 24, 802–813.

Zhou, J.S., Li, R.W., Lin, Y.Q., Wang, J.G., Li, Y., Yang, Y.H., Gisbert, E., Yu, H.B., Ji, H., 2018b. LCFA Uptake and FAT/CD36: molecular cloning, tissue expression and mRNA expression responses to dietary oil sources in grass carp (*Ctenopharyngodon idellus*). J. Appl. Anim. Res. 46, 572–582.

Zhou, Z., Ringø, E., Olsen, R.E., Song, S.K., 2018c. Dietary effects of soybean products on gut microbiota and immunity of aquatic animals: a review. Aquacult. Nutr. 24, 644–665.

Zhou, W., Rahimnejad, S., Lu, K., Wang, L., Liu, W., 2019. Effects of berberine on growth, liver histology, and expression of lipid-related genes in blunt snout bream (*Megalobrama amblycephala*) fed high-fat diets. Fish Physiol. Biochem. 45, 83–91.

Zhu, T., Corraze, G., Plagnes-Juan, E., Quillet, E., Dupont-Nivet, M., Skiba-Cassy, S., 2018a. Regulation of genes related to cholesterol metabolism in rainbow trout (*Oncorhynchus mykiss*) fed a plant-based diet. Am. J. Physiol. Regul. Integr. Compar. Physiol. 314, R58–R70.

Zhu, T., Mai, K., Xu, W., Ai, Q., 2018b. Effect of dietary cholesterol and phospholipids on feed intake, growth performance and cholesterol metabolism in juvenile turbot (*Scophthalmus maximus* L.). Aquaculture 495, 443–451.

Zhuo, M.Q., Luo, Z., Wu, K., Zhu, Q.L., Zheng, J.L., Zhang, L.H., Chen, Q.L., 2014. Regulation of insulin on lipid metabolism in freshly isolated hepatocytes from yellow catfish (*Pelteobagrus fulvidraco*). Compar. Biochem. Physiol. B Biochem. Mol. Biol. 177, 21–28.

Zimmermann, R., Strauss, J.G., Haemmerle, G., Schoiswohl, G., Birner-Gruenberger, R., Riederer, M., Lass, A., Neuberger, G., Eisenhaber, F., Hermetter, A., Zechner, R., 2004. Fat mobilization in adipose tissue is promoted by adipose triglyceride lipase. Science 306, 1383.

Zou, B., Ge, Z., Zhu, W., Xu, Z., Li, C., 2015. Persimmon tannin represses 3T3-L1 preadipocyte differentiation via up-regulating expression of miR-27 and down-regulating expression of peroxisome proliferator-activated receptor-γ in the early phase of adipogenesis. Eur. J. Nutr. 54, 1333–1343.

CHAPTER

6

The minerals

Santosh P. Lall

Retired from National Research Council of Canada, Halifax, NS, Canada

6.1 Introduction

Minerals are naturally occurring inorganic elements (metals, metalloids, and nonmetals) required by all forms of aquatic animals in small amounts to maintain essential physiological functions necessary for normal life processes. Unlike most terrestrial animals, fish have the ability to absorb some inorganic elements not only from their diets but also from their external environment in both freshwater (FW) and seawater (SW). The exchange of ions from the aquatic environment across gills and skin of fish complicates the determination of the quantitative dietary requirements of minerals (Lall, 2002). Many essential elements are required in such small quantities that it is difficult to formulate diets and maintain an environment that is sufficiently low in minerals to demonstrate a mineral deficiency. The biochemical mechanisms of mineral metabolism in fish are generally similar to those of terrestrial animals with the exception of mineral uptake from water and maintenance of osmotic balance between body fluids and the aquatic environment, which can vary in salinity from 0‰ to 35‰ in FW, brackish water (BW), and SW. Despite new developments in mineral nutrition of fish, most research on inorganic elements has been focused on toxicity, osmoregulation, and related physiological functions. This chapter examines the mineral requirements of finfish and compares mineral metabolism of fish with that of other animals. Although significant progress has been made in the mineral nutrition of fish in the past 3 decades, environmental concerns of air and water pollution have prompted more research on the metabolism and functional role of toxic elements. The ability to regulate abnormal high concentrations of minerals varies among aquatic organisms. Certain fish are able to excrete high proportions of excessive metal intake and consequently regulate the concentration in their body at relatively normal levels (Wood, 2012).

It is well recognized that all living organisms contain most naturally occurring elements in the periodic table. At present, 29 of the 94 naturally occurring elements are known to be essential for animal life. The greater proportion of living matter consists of six basic structural elements carbon, hydrogen, nitrogen, oxygen, phosphorus, and sulfur. For historical reasons, the elements are divided into two groups, macro- and microelements (Table 6.1).

Fish Nutrition, Fourth Edition
https://doi.org/10.1016/B978-0-12-819587-1.00005-7

TABLE 6.1 Essential minerals for animals and fish.

Essentiality established by biochemical mechanism(s)		Essentiality suggested by physiological impairment	
Macroelements			
Calcium	Sodium		
Phosphorus	Potassium		
Magnesium	Chloride		
Microelements			
Iron	Selenium	Chromium[b]	Fluorine[b]
Copper	Iodine	Arsenic	Boron[b]
Manganese	Cobalt	Lithium[b]	Lead[b]
Zinc	Molybdenum[a]	Nickel[b]	Silicon[b]
		Vanadium[b]	

[a]*No published information for their essentiality for fish.*
[b]*Only weak or no evidence for physiological impairment reported in fish.*

Six macroelements, calcium (Ca), phosphorus (P), magnesium (Mg), sodium (Na), potassium (K), and chloride (Cl), are required in gram quantities. The requirement for the 17 microelements is in the range of micrograms to milligrams per kilogram. Initial difficulties in the accurate determination of low levels of many microelements inevitably led to their description as "trace elements" and "ultratrace elements." Today modern analytical and molecular techniques permit the accurate analysis of these elements and the study of their unique functions in many metabolic processes.

By the simplest definition, an essential element is one required for the maintenance of life; its absence results in death of the organism. The criterion for an essential element has been described by Frieden (1984): "An element is considered essential when a deficient intake produces an impairment of function when restoration of physiological levels of the element prevents or relieves the deficiency. The organism can neither grow nor complete its life cycle without the element in question. The element should have a direct influence on the organism and be involved in metabolism. The effect of an essential element cannot be wholly replaced by any other element." In a later review of the scientific basis for establishing the essentiality of trace elements, the following definition was adapted (Mertz, 1998): "An element is considered essential to an organism when reduction of its exposure below a certain limit results consistently in a reduction in a physiologically important function, or when the element is an integral part of an organic structure performing a vital function in the organism." This definition omits a previous assumption that the mechanism of action of an essential trace element should be well defined.

Seventeen trace elements are considered to be essential for animals. Among these the essentiality of cobalt (Co), copper (Cu), iodine (I), iron (Fe), manganese (Mn), molybdenum (Mo), selenium (Se), and zinc (Zn) has been confirmed by biochemical mechanisms.

Essentiality of the remaining nine trace elements, fluoride (Fl), chromium (Cr), vanadium (V), silicon (Si), nickel (Ni), arsenic (As), lithium (Li), lead (Pb), and boron (B), has been based on physiological impairment. Four trace elements, Cu, Fe, Mn, and Zn, are required between 1 and 100 mg/kg of diet. The term "ultratrace element" is used to define essential elements required by animals in amounts between 50 and 1000 μg/kg of diet for growth and maintenance, which include Se, Mo, and I and for some species Co, Cr, Ni, B, Li, and V. Although deficiencies of Ni, V, Si, and As have been demonstrated in an ultraclean environment, with the exception of Si, the physiological function of these trace elements has not been clearly demonstrated. Other elements, including cadmium (Cd), bromine (Br), and rubidium (Rb), have also been claimed to be essential, but their essentiality remains to be confirmed. Most of these trace elements have been detected in fish (Lall, 1995); however, the essentiality of only a few of these trace elements (Cu, Co, Fe, I, Mn, Se, and Zn) has been reported NRC (2011). Boron is required to complete the life cycle of zebrafish (Eckhert et al., 1999).

The main functions of essential elements include four main body functions: (1) formation of skeletal structure, (2) physiological, (3) catalytic, and (4) regulatory functions. Certain minerals form structural components of body organs and tissues such as calcium, phosphorus, and magnesium; silicon in bones, fins, and scales; and phosphorus and sulfur in muscle proteins. Minerals such as zinc and phosphorus can also contribute structural stability to molecules and tissue membranes. Inorganic elements in body fluids and tissues, as electrolytes function in the maintenance of acid–base balance, osmotic pressure, membrane permeability, and transmission of nerve impulses. Sodium, K, Cl, Ca, and Mg, along with phosphates and bicarbonates in the blood, cerebrospinal fluid, and gastric juice, perform these functions. Trace elements are important components of hormones, enzymes, and activators of enzymes. Several trace elements (Fe, Mn, Cu, Co, Zn, Mo, Se) are firmly associated with a specific protein in metalloenzymes, which produce a unique catalytic function. Certain minerals, such as Ca, Mg, and Mn, are of particular significance as enzyme activators. A number of different metalloenzymes and coenzymes have been identified (Table 6.2), and new developments in molecular biology have increased our knowledge of their specific roles in metabolism. Minerals also regulate cell replication and differentiation, e.g., Ca ions influence signal transduction, and selenocysteine plays a role in gene transcription. The nonmetal, I, is necessary for biosynthesis of thyroid hormones, which in turn greatly affects development and metabolism in all vertebrates. The essential metabolic role of thyroxine has been attributed to the influence of triiodothyronine on gene transcription (Bassett et al., 2003). Some biologically important compounds also contain minerals as an inherent part of their structure, e.g., hemoglobin and vitamin B_{12}.

6.1.1 Difficulties in determining mineral requirements of fish

Inorganic elements, as nutrients, are difficult to study, particularly trace elements. Many trace elements are required in such small amounts that it is difficult to formulate purified diets low in minerals and maintain water sufficiently free of the test element. Despite advances in instrumental analysis of trace elements making lower detection limits possible, there are still many problems associated with their accurate measurement in fish tissue. A critical factor in the determination of ultratrace elements, such as vanadium and chromium, is the need for

TABLE 6.2 Trace elements in certain essential metalloenzymes and metalloproteins and their functions.

Trace elements	Metalloenzyme and metalloproteins	Function
Copper	Cytochrome *c* oxidase	Catalyzes the reduction of molecular oxygen to water
	Lysyl oxidase	Catalyzes the oxidative deamination of lysine and hydroxylysine residues in collagens and elastin
	Caeruloplasmin	Copper transport in blood
	Superoxide dismutase	Dismutation of superoxide radical (O_2^-) to molecular oxygen and H_2O_2
Iron	Succinate dehydrogenase	Participates in both citric acid cycle and electron transport chain
	Catalase	Promotes the reduction of H_2O_2 to H_2O and O_2
	Hemoglobin	Oxygen transport in blood
	Hepcidin	Central regulator of Fe homeostasis
Manganese	Pyruvate carboxylase	Pyruvate metabolism
	Superoxide dismutase	Antioxidant by removing O_2^-
	Glycosyl aminotransferases	Metabolism of polysaccharides, saponins, glycolipids, glycoproteins, proteoglycans, peptidoglycans
Selenium	Glutathione peroxidases	Removal of H_2O_2 and hydroperoxides
	Type I, II, and III deiodinases	Synthesis of the active form of thyroid hormone
Zinc	Carbonic anhydrase	Formation of carbon dioxide
	Alkaline phosphatase	Hydrolysis of phosphate esters
	Carboxypeptidase	Hydrolysis of peptide bonds
	Aminopeptidase	Cleavage of amino acids
	Phospholipase A_2	Hydrolysis of phosphatidylcholine
	RNA and DNA polymerase	Synthesis of RNA and DNA

meticulous sample preparation. Techniques that involve the use of high-purity reagents, acid-cleaned glassware, and clean-room facilities should be employed to avoid contamination. Often normal values of trace elements in fish tissue vary widely in reports from laboratory to laboratory. The use of certified reference materials (CRMs) is also essential to assure differences arise from the fish tissues being examined and not the analytical techniques employed.

Although a wide range of functions have been established for the essential elements for domestic animals and humans, such information available for fish is fragmentary and incomplete. Relatively little is known about the uptake, cellular metabolism, and biological availability of many trace elements. Environmental concerns of air and water pollution have prompted more research on the metabolism and functional role of toxic elements. To measure the nutritional status of trace elements, information is also required on age, sex, health, and physiological conditions (e.g., smoltification, sexual maturation) of the animal. Nutrient requirements of most farmed animals are clearly defined from birth/hatching to market size (Suttle, 2010); however, such information is lacking or limited for fish (NRC, 2011). The most commonly used measure of nutritional status is growth and the level of trace elements in the whole body, blood, liver, muscle, and bone, and specific enzymes linked to their metabolism. For many essential elements, there are a range of tissue levels compatible with optimum growth and function. A reduced level of mineral intake influences tissue concentration, which causes a gradual decline in the function of an organ until clinical toxicity and deficiency occur. However, it is the intermediate stage that is difficult to establish. More desirable measures of mineral nutritional status are those which test some metabolic function of the element rather than tissue analysis alone. For iodine, there is a well-defined functional test (thyroxine binding) to assess its nutritional status. However, for other elements, such functional tests are not clearly established. Attempts should be made to express body and mineral composition data on a fat-free basis. The amount of lipid in the body of fish particularly salmonids during growth makes it difficult to estimate and compare results between and within the same fish species. It is also difficult to compare results obtained on mineral requirements of fish measured in flow-through and recirculation systems because of the uptake of minerals from water and retention by fish. The water exchange rate in the system and mineral accumulation are known to affect acquisition, retention, and requirement of certain minerals (Antony Jesu Prabhu et al., 2017; Martins et al., 2009; Lall, 2002). Therefore, it is important that experiments designed to determine nutrient requirements contain information about the available nutrient contents in experimental diets, that the main determinants of nutrient requirements be characterized, and that standardized research methodologies and laboratory procedures be used.

Elucidation of techniques that allow for the identification of subclinical, pathological changes in the assessment of nutritional status remains a challenging area for mineral nutrition research. Application of molecular techniques continues to provide new information related to mineral interactions and their metabolism at the cellular level to diagnose diseases and improve health of animals and humans and to a limited extent fish. Some of the novel transcriptomics, proteomics, and metabolomics techniques available now offer opportunities to resolve some of the complex problems of mineral metabolism and requirements, characterize specific trace element deficiencies, and improve nutrient balance.

6.1.2 Nature and distribution of elements in fish

The concentration of minerals in the body of aquatic organisms depends on food source, environment, species, stage of development, and physiological status of the animal. Most organisms accumulate and retain minerals from the aquatic environment; however, their incorporation is highly selective. Early studies conducted in fish caught from FW lakes showed that quantities of Mn, Fe, Cu, and Zn in the muscle and liver reflect the geochemical characteristics of the earth surrounding the lake (Love, 1980). In marine food chains, the mechanism of trace metal transfer at various trophic levels has been reported (Bernhard and Andreae, 1984). Most trace metals show the highest concentration increase at the first trophic level (SW phytoplankton). In zooplankton feeding on phytoplankton, mainly Cd, Cu, and Zn concentrations increase. However, plankton-feeding fish show higher Cu and Zn levels than fish preying on invertebrates and fish feeding on invertebrates and fish. Generally, the concentration of Cd, Cu, and Zn increases along the food chain at the lower levels, reaching a maximum with crustaceans and then decreasing in fish. Limited information is available on the transport of other essential elements in the marine food chain. Fertilization of fish ponds is common practice in pond culture of warm water fish such as catfish, tilapia, and carp. It has been demonstrated that fertilization of ponds with inorganic fertilizers and manures augments the production of natural food organisms, and these organisms supply essential trace elements for growth and reproduction (Mischke, 2012). Several aspects of certain essential trace elements and heavy metals transferred from the abiotic environment to living organisms that accumulate in biota at different trophic levels and are known to contaminate the aquatic ecosystems have been subject of intensive research and modeling (Ali and Khan, 2018; Reinfelder et al., 2008; Hamilton, 2004) and are not discussed here.

Unlike some soft tissues of fish such as liver, muscle, and kidney, skeletal tissues, otolith, and lenses retain metals for a longer duration, and their concentration has been used to detect environmental pollution (Dove and Kingsford, 1998). The distribution of many essential and toxic elements in several fish, crustaceans, and mollusks has been reported; however, the complete elemental composition is available for only a few fish species and is restricted mainly to edible parts. The mineral composition of rainbow trout varies with fish size, stage of life cycle, and reproductive status. The concentrations of Ca, Cu, Fe, K, Mg, Mn, Na, P, Sr, and Zn were higher in juvenile fish than in adults (Shearer, 1984). A decrease in somatic Mn, Fe, and Zn was found during gonadal maturation in female but not male fish. Disease state, water quality, and other environmental factors may also modify tissue mineral distribution. With some exceptions, mineral element concentrations in skeletal tissues are less susceptible to dietary and environmental changes than those of soft tissue and body fluids. Fish eggs also absorb certain minerals from water. Zeitoun et al. (1976) demonstrated absorption of Na, Fe, and Zn at hatching by eggs in rainbow trout and the uptake of Ca, Na, K, Fe, and Zn from the water after yolk absorption. They concluded that Cu and P found in the larvae probably originated from an unfertilized oocyte. A remarkable increase in Mn content of Atlantic salmon and Japanese rice fish (*Oryzias latipes*) eggs after fertilization has been reported (Hori and Ivasaki, 1976; Lall and Hines, 1985 unpublished). However, the uptake of heavy metals by eggs and developing embryos shows some differences. The uptake of Cd by trout eggs is restricted by chorion (Beattie and Pascoe, 1978). Waterborne selenium accumulation in embryos increases with the development of gills (Hodgson et al., 1986).

6.1.3 Osmoregulation and saltwater adaptations

The mechanism of osmotic and ionic regulation in fish is a complex physiological process and the subject of extensive research and comprehensive reviews (Evans and Claiborne, 2009; McCormick, 2012a). Briefly, fish maintain the ionic composition and osmolarity of their body fluids at levels significantly different from those of the surrounding water. Although large areas of body surface, i.e., the gills serving respiratory function, remain exposed to the external environment, the blood of FW fish is hypertonic and that of marine fish is hypotonic compared with their external medium. Maintenance of body fluid balance principally involves gills, gastrointestinal tract, and kidney. Gills directly sense osmotic changes and actively absorb monovalent ions (Na^+, K^+, Cl^-) maintaining plasma osmolality within a narrow range (Fig. 6.1). In marine fish, ingested SW containing Na and Cl is absorbed by the intestine; however, the intestine plays a minor role in osmoregulation. They regulate their internal salt concentrations and osmotic pressure at levels well below that of SW and lose water osmotically across the large, permeable surface area of the gills. To replace the water, fish drink SW. Secretion of bicarbonate (HCO_3) into the gut

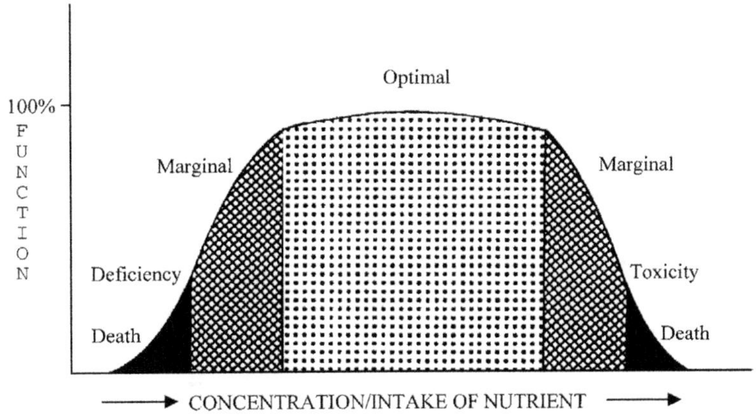

FIGURE 6.1 Biological dose–response curve. Dependence of animal function on intake of an essential nutrient according to Mertz (1986).

lumen raises the pH of the ingested fluid and precipitates most Ca^{2+} and some Mg^{2+} (Wilson et al., 2002), which significantly reduces the osmotic pressure of the gut fluid without the absorption of these divalent salts. However, over 90% of Na^+, Cl^-, and K^+ from drinking are absorbed across the gut (Kirschner, 2004). The major proportion of the Na and Cl may enter passively, but some must be actively transported across the gut epithelium, a process driven by Na/K-ATPase. To replace water lost by osmosis across the gills, fish incur a sizable salt load across the gut, which combines with any salts taken up across the branchial epithelium. Kidney is a critical organ for secretion of divalent ions (Ca^{2+}, Mg^{2+}, SO_4^-) in marine fish and actively secretes large amount of dilute urine to counter copious water influx across the body surfaces (Fig. 6.2).

Salt water adaptation and osmotic and ionic regulation in euryhaline fish, particularly salmon, trout, and eels, are well documented (Boeuf, 1987; Steffanson et al., 2008; McCormick, 2012a, 2012b). Euryhaline fish withstand abrupt changes in environment salinity with several physiological adjustments. These include modification of branchial chloride cell number, drinking rate, urine production rate, gill epithelium permeability, and ion efflux. Smoltification of migratory salmonids is associated with marked changes in morphology, chemical composition, physiological and biochemical function, and behavior (Steffanson et al., 2008). The underlying mechanisms for these changes have not been fully elucidated. It is widely accepted that gill Na–K-activated adenosine triphosphatase (Na^+ K^+-ATPase) plays an important role in SW adaptation. The activity of this enzyme increased in smolts prior to SW entry. Several other physiological and biochemical changes occur in muscle during smolt development that may relate to swimming capacity. White muscle buffering capacity was higher in masu salmon (*Oncorhynchus masou*) smolts relative to parr, which showed minimum pH disturbances associated with anaerobic metabolism (Ogata et al., 1998). Mizuno et al. (2007) reported lower levels of hematocrit, hemoglobin, and burst swimming speed in hatchery masu salmon compared with wild fish, all of which was minimized by dietary Fe supplementation. Salt water adaptation and osmotic and ionic regulation are also stimulated by several hormones including growth hormone, insulin-like growth factor-1, cortisol, and thyroid hormones, whereas the effect of prolactin is generally inhibitory (McCormick, 2012b).

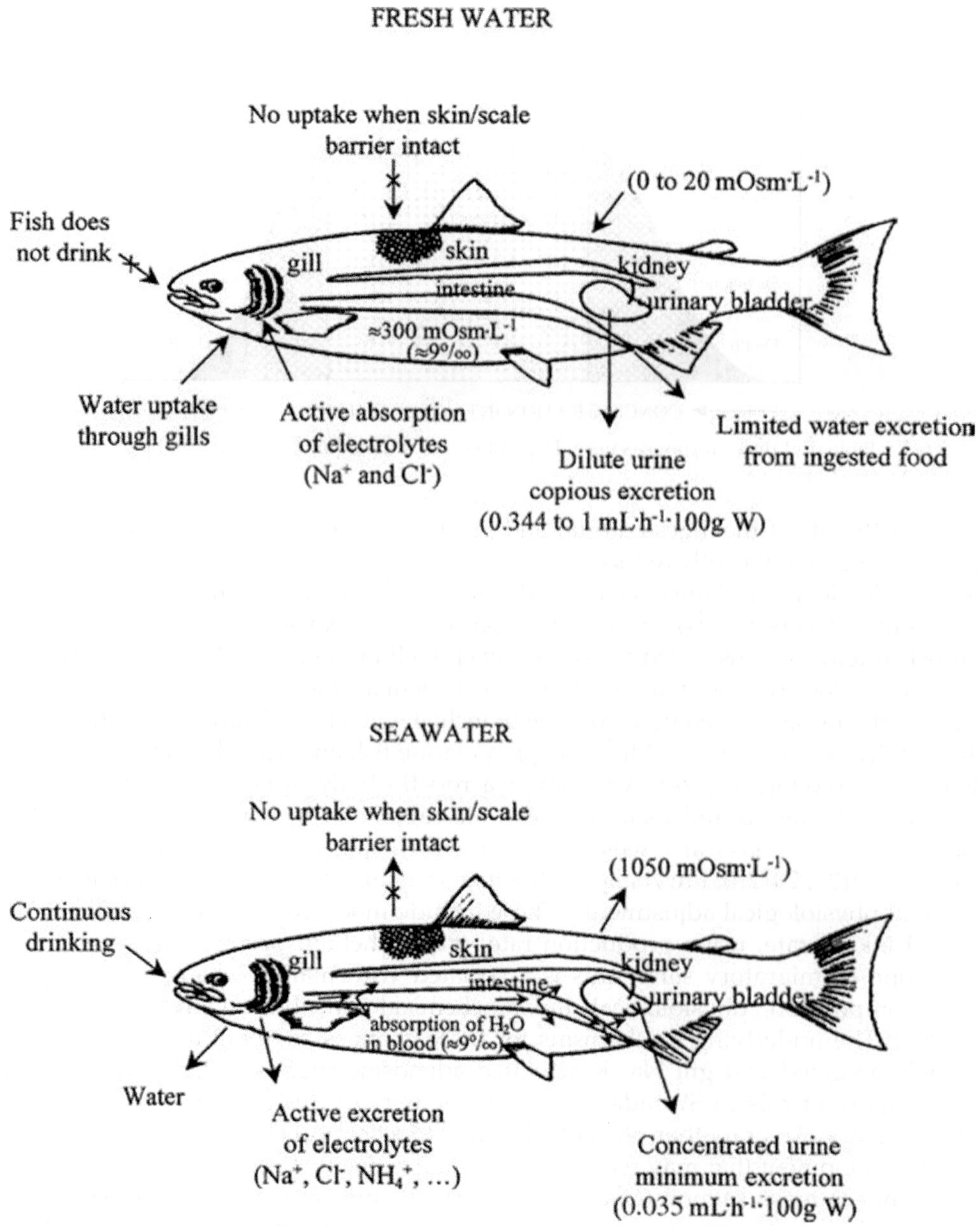

FIGURE 6.2 Osmoregulation by rainbow trout in freshwater (0‰) and seawater (35‰). *From Boeuf, G., 1987. Bases physiologiques de la salmoniculture: osmoregulation et adapatation à l'eau de mer. La Pisciculture Francaise, 87, 28–41 with permission.*

Evidence indicates that nutritional status influences adaptation of euryhaline fish to SW. Feeding a diet high in salt content enables trout and salmon to adapt to SW more easily (Zaugg and McLain, 1969; Basulto, 1976; Salman and Eddy, 1990). Furthermore, starved rainbow trout are unable to maintain their gill Na^+/K^+-ATPase activity at the same level as fed animals (Jurss et al., 1983). Free amino acids, apart from their metabolic function, play an important role in osmotic regulation mechanism during salinity changes in rainbow trout (Jurss, 1980; Jurss et al., 1983). The concentration of several amino acids in the muscle of rainbow trout, held in SW, was higher than in animals starved in FW (Jurss, 1980).

6.1.4 Mineral interactions

All dietary nutrients are interrelated to some degree, and there is an optimal concentration for each nutrient to achieve the most efficient performance and health of fish. The term "interaction" was defined by O'Dell (1997) as "interrelationships among mineral elements as revealed by physiological or biochemical consequences." Several nutrient interactions and interrelationships are well known (e.g., Zn—Cu and Ca—P—vitamin D and K) in terrestrial animals, fish, and human nutrition (O'Dell, 1997; Lall, 2002). Minerals may interact more than other nutrients because they are labile and tend to form chemical bonds and elicit mineral—mineral or mineral—vitamin interactions. Specific mineral interactions may include competition for binding sites on transport proteins, chemical bonding to form insoluble complexes, which affect the availability of certain minerals, and in some cases mineral cofactor enzymes, which synthesize mineral containing macromolecules. In addition, there are mineral—vitamin interactions as a result of some minerals and vitamins having complementary biochemical functions. A synergism between dietary Se and vitamin E is known. Zinc is required for the metabolism of vitamin A. The strong redox potential of vitamin C may alter the valence of copper and iron and thus reduce or enhance absorption. A wide range of potential mineral—mineral and mineral—vitamin interactions in fish have also been reported (Hilton, 1989). Specific mineral interactions are discussed in appropriate individual inorganic element sections.

Nutrient interactions are broadly classified as positive or synergistic, or negative or antagonistic. Synergistic relationships in which one element enhances the role of another, e.g., Fe and Cu, are difficult to explain. There may be direct interaction between elements in structural processes such as requirement of Cu and Fe for hemoglobin formation, Ca, P, and Mg for formation of bone hydroxyapatite and an interaction of Mn with Zn for proper conformational shape of RNA molecules in the liver. The complex interrelationship between Cu, Zn, Fe, and Ca, as well as that of Cu, Mo, and S, is of practical concern (O'Dell, 1997) but beyond the scope of this chapter.

Antagonistic relationships occur when elements with a similar electronic configuration and ionic radius compete for binding sites, e.g., Zn and Cd in metallothionein, and Mg/Mn substitutions at enzyme active sites. There are several ways in which antagonistic relationships can occur within the gastrointestinal tract. The simplest mechanism involves a chemical reaction forming an insoluble complex between minerals such as Cu and S to form copper sulfide or a mineral and another dietary component such as Zn combined with phytic acid to form phytate. Mineral elements can also be adsorbed onto the surface of colloidal particles, such as Mn and Fe onto the surface of Mg or Al salts. Competition between minerals such as Co and Fe for carriers in the intestinal wall has also been observed. Finally, some ions, including B and Pb, have inhibitory effects on processes such as oxidative phosphorylation in the intestinal wall or on the activity of some enzymes, which interfere with the breakdown of feed ingredients and liberation of inorganic ions for absorption. At the cellular level, there can also be antagonistic processes resulting in formation of insoluble complexes like the formation of copper thiomolybdates from Cu, Mo, and S. There can be a competition among ions for the active centers in enzyme systems as occurs with Mg and Mn in alkaline phosphatase. Competition for binding sites on carrier proteins in blood has been observed with Zn competing with Fe for the transport protein, transferrin. Antagonistic effects have been reported whereby one ion will activate an enzyme and another will cause inhibition as with Mg and Ca with ATPase.

Minerals may work in competition by activating enzyme systems with opposite effects. Copper activates ascorbate oxidase, which oxidizes ascorbic acid, versus activation by Zn and Mn of lactonases, which promote synthesis of this vitamin. Reciprocal interrelationships are also found in which both elements influence the metabolism of the other. The negative interactions that exist between Zn and Fe would be an example. One popular theory of mineral interactions based on chemical and physical properties of the elements was proposed by Hill and Matrone (1970). Ions with similar ionic size or electron configurations of their outer orbitals are likely to interact competitively.

Another type of interrelationship involves the interaction between the elements themselves. Selenium has a high affinity for certain toxic elements such as mercury (Hg) and silver (Ag); hence, Se exerts a protective effect against the toxicity of these metals by forming complexes in vitro, resulting in a decrease in the biological availability of both Se and the heavy metal (NRC, 2005). Minerals can also interact with other nutrients. Some mineral interactions determined with experimental animals may not be directly applicable to fish because of differences in mineral uptake from food and water in fish (Hilton, 1989). However, the application of techniques to assess mineral interactions and bioavailability is extremely useful.

Fish can also accumulate excessive minerals from water, which can interact with as well as inhibit proteins that facilitate essential ion transport (Wood, 2012). Metal-binding reactions include competition at the biotic ligand with other cations (e.g., Na^+, H^+, Ca^{2+}) and competing complexation reactions for the metal by other ligands in solution including inorganic anions (e.g., chloride, hydroxide, sulfide) as well as organic ligands such as dissolved organic carbon (Brix et al., 2017). For example, accumulation of Cu or Ag at the fish gill interferes with Na^+ uptake, thereby disrupting ionic balance in the organism and leading to toxicity (Grosell, 2012). Similarly, Cd and Zn interfere with Ca uptake at the fish gill, leading to hypocalcemia and toxicity (Hogstrand, 2012). These interactions are beyond the scope of this chapter.

6.1.5 Absorption and bioavailability of minerals

The gastrointestinal tract is the major site for absorption of minerals in fish in FW and SW. The general mechanisms for absorption of minerals from both the gastrointestinal tract and gills involve simple diffusion through epithelium, metal-specific carriers, and nutrient ion transporters (Bakke et al., 2010). Certain inorganic elements (e.g., Zn) may bind to amino acids (e.g., histidine and cystine) in the gut of fish and are transported by amino acid transporters (Glover et al., 2013). After absorption, minerals are transported through blood and delivered to target organs where essential inorganic elements contribute to normal metabolic functions, and both essential and nonessential minerals may be detoxified, transformed, exert beneficial or toxic effects, and are excreted. Physiological mechanisms of mineral absorption and transport are beyond the scope of this chapter: however, and certain aspects of this topic are covered in individual mineral sections.

The term "bioavailability" is widely used to describe the absorption and utilization of nutrients. Certain minerals in the diet found in either free or bound ligands can be absorbed and are present in body fluids and tissues but may not be utilized by fish to perform specific functions. To determine how efficiently the body utilizes dietary mineral elements, one must know the availability of that element from a feed ingredient or complete diet. Generally, chemical analysis does not indicate the level of biologically effectiveness of a nutrient. Many mineral bioavailability problems are being increasingly recognized in fish, human, and animal nutrition (Ammerman, 1995; Lall and Milley, 2008b). Several methods have been used to predict the bioavailability of trace elements (Ammerman, 1995; O'Dell, 1997) including (1) apparent absorption, (2) true absorption, (3) mineral retention or balance, (4) plasma or tissue mineral concentrations, (5) growth, (6) biologically active form of mineral (e.g., enzyme activity, hemoglobin level), and (7) prevention of deficiency signs. Many of these methods do not directly measure the efficiency of mineral utilization; however, they provide information that can be useful to estimate bioavailability. Generally, it is considered desirable to measure the bioavailability of minerals to evaluate at least three concentrations of a standard source and test minerals within a range of concentrations that will give a linear response for the bioavailability criteria measured. The ratio between the slopes of the dose–response curve can then be used to estimate bioavailability of the test mineral to standard source.

It has been shown that the bioavailability of an element can differ markedly when supplied from different feedstuffs and with the same element from feed in different diets. Many factors can influence the bioavailability of minerals including intake level of the nutrient, its chemical form, digestibility of the diet that supplies the element, particle size, interaction with other nutrients, chelators, inhibitors, physiological and pathological states of the animal, water chemistry, type of feed processing, and the species of animal being tested (Lall. 2002). Trace element biological availability in a diet depends on the various molecular forms in which the element is present, its valence states, and the ligands present when ingested from different diets. Mechanisms that involve the formation of insoluble and nonabsorbable substances in the gut may either hinder or facilitate mucosal uptake, transport, and metabolism of an element in the body (Ammerman, 1995). Zinc, Cu, and Mn have been supplemented in fish feeds using inorganic salts such as oxides, chloride, and sulfates. However, trace mineral salts may dissociate in the low pH environment of the upper gastrointestinal tract, leaving the minerals susceptible to various nutrient and ingredient antagonisms that impair absorption and thus reduce bioavailability. Antagonisms can occur between minerals, e.g., high levels of Zn reduce the availability of Cu, and the opposite is also true (O'Dell, 1997). In addition, phytic acid is able to form complexes with trace minerals that are very stable and highly insoluble, rendering the minerals unavailable for absorption (Section 6.2.4).

In recent years, organic chelated trace elements have been introduced that may have the advantage of potential binding with organic ligand(s) and forming a complex that is likely to provide stability in the upper gastrointestinal tract, thereby minimizing mineral losses to antagonists and allowing the complex to be delivered to the absorptive epithelium of the small intestine for mineral uptake. However, it has been shown with other animals that all organic trace minerals are not equally stable at low pH and therefore will not necessarily increase the bioavailability of a given mineral to the same extent (Guo et al., 2001; Suttle, 2010). Limited research has been conducted on the bioavailability of inorganic and organic trace elements in fish, and differences exist between fish species grown in FW and SW as discussed in specific mineral sections.

Certain inorganic elements may compete with the test element for important binding sites during these processes, and factors influencing each trace element are discussed in later sections. Factors that influence the bioavailability of Zn and Fe in terrestrial animals are well documented (O'Dell, 1969; Forbes and Erdman, 1983; Henry and Miller, 1995). In fish, Zn is better absorbed from animal protein supplements than from plant protein sources. Cereals and other plant feedstuffs contain a number of substances, particularly phytate, which can bind Zn, making it unavailable for absorption. The bioavailability of iron is influenced not only by its chemical form but also by interaction between Fe and other dietary components. Apart from these dietary components, several environmental factors can also influence mineral bioavailability.

6.1.6 Minerals and skeletal tissue metabolism

Bone and cartilage are the major mineralized skeletal tissues in fish. Several nutrients are considered essential for growth and development of these tissues. Adequate supplies of energy, amino acids, and major bone-forming minerals (Ca, P, Mg, Mn, Zn) and vitamins (e.g., vitamin C, D, K) are involved in cartilage and bone metabolism. Skeletal tissues are comprised of a mixture of minerals deposited around a protein matrix, which provide strength and flexibility to vertebrae. Approximately 65% of bone tissue is inorganic mineral that provides the hardness of bone. The remaining 35% of bone tissue is an organic matrix with a major proportion as collagen fibers. In most vertebrates, the skeleton represents a reservoir of calcium, phosphorous, and other ions that are in a state of continual exchange with electrolytes found in blood and extracellular fluids. There are two types of fish bones, cellular and acellular. Cellular bones are confined to only a few groups of fish, e.g., Salmonidae, Cyprinadae, and Clupeidae. The skeletal tissue of higher orders of teleost fish such as Perciformes is unique among vertebrates in that it lacks osteocyte cells. Acellular bones are formed from osteoblast cells, which move away from the site of mineralization as bone deposition occurs, and therefore are incapable of extensive modeling (Moss, 1963). Skeletal physiology and skeletogenesis processes in fish have been comprehensively covered in earlier reviews (Simkiss, 1974; Boglione et al., 2013; Hall, 2015; Boglione, 2020).

The bone matrix is mostly comprised of collagen and hydroxyapatite, a hydroxylated polymer of calcium phosphate [$Ca_{10}(PO_4)_6(OH)_2$]; however, cartilage consists of cells in an extracellular matrix, which may or may not be mineralized depending on the cartilage type (Hall, 2015). In bone and scales of fish, hydroxyapatite salts are embedded in a matrix of type I collagen fibers. Magnesium, Na, K, and citrate ions are also present, conjugated to hydroxyapatite crystals. Other constituents of vertebrae include Sr, Pb, citrate, carbonate, fluoride, hydroxide, and sulfate. The chemical compositions of bone may vary because of substitution of Mg, Ba, or Sr for Ca^+; HPO_4^{2-} or CO_3^{2-} for PO_4^{2-}; and HPO_4^{2-}, CO_3^{2-} , Cl^-, or F^- for OH^- (Giraud-Guille, 2014). Bone also contains small amounts of trace elements, which are either essential or toxic to the health of skeletal tissues depending on the concentrations.

The collagen fibers of the organic bone matrix provide the interior scaffolding upon which minerals are deposited. The other components of organic matrix are noncollagenous proteins including γ-carboxyglutamic acid (Gla)-containing proteins that may serve as anchor points for cells involved in bone remodeling (Heaney, 2012). The development of bone requires continuous remodeling, a process that involves three types of bony cells: osteoblasts (bone forming cells), osteocytes (entrapped inside the bone matrix), and osteoclasts (multinucleated bone resorbing cells). Osteocytes are thought to be involved in the maintenance of bone substances and the exchange of ions from body fluids. Thus, the skeleton of most vertebrae exerts a massive buffering effect on changes in plasma electrolyte values. Among fish, a variety of different skeletal systems perform these functions.

Among the trace elements, low concentrations of Mn, Cu, Fe, Zn, Cr, and Co have been detected in fish bone; however, limited information is available on their role in bone metabolism including the mechanisms and interaction between the elements (Lall and Lewis-McCrea, 2007; Lall, 2010; Baeverfjord et al., 2019). Copper deficiency affects lysyl oxidase activity preventing cross-linking between collagen and elastin, reducing the strength of the bone organic matrix. Iron deficiency may decrease the serum 1, α-dihydroxycholecalciferol, insulin-like growth factors, and osteocalcin concentration. Manganese is a cofactor in numerous enzymes (glycosyl transferase, xylosyl transferases, phosphohydrolases, phosphotransferases) that directly affect bone metabolism. Many proteins, enzymes, zinc finger proteins, and bone markers are closely related and sensitive to Zn homeostasis (metallothionein, Zn transporters, alkaline phosphatase, collagenase, and aminoacyl-tRNA synthetase) and are linked to Zn in bone metabolism. Selenium plays an important role in neutralization of reactive oxygen species, and selenoproteins regulate miRNAs that influence bone development. Many elements are ubiquitous in the aquatic environment and may impact bone and other skeletal tissue metabolism. Toxicity of P, F, Cd, and As may also affect bone metabolism and health of fish (Dermience et al., 2015).

Marginal nutrient deficiency at early stages of the life cycle of fish may impact later bone growth of fish and increase their susceptibility to malformation (Boglione et al., 2013). The elemental concentration of skeletal tissues varies with age in both cellular and acellular bones. Any change in metabolism and bioavailability of minerals and certain nutrients in the body affects bone metabolism, involving specific changes to the bone's cellular system. Several types of vertebral and spinal malformations such as kyphosis (humpback, hunchback), lordosis (saddleback, swayback), scoliosis (lateral curvature with rotation of the vertebrae), and platyspondyly (short-tail, compressed vertebrae) have been reported in fish. Often these deformities may be a combination of several deformities; however, neck, vertebral, and spinal disorders are most prevalent and are often linked to dietary factors (Baeverfjord et al., 2019; Lall and Lewis-McCrea et al., 2007). Information available on the pathogenesis of bone disorders linked to specific nutrient deficiencies in fish is limited except for phosphorus. A high prevalence of skeletal deformities in triploid rather than diploid Atlantic salmon has been shown to be due to a higher P requirement for triploids (Smedley et al., 2018). In farmed fish, several skeletal disorders have also been linked not only to nutrients but also to interactions between the stage of development, environment, genetic, and dietary factors. The relationship between these factors is complex and poorly understood.

The effects of P deficiency on skeletal development and deformity of vertebral bodies have been extensively studied and are discussed in more detail later. Early work on deficiency of Zn and Mn in rainbow trout showed that these elements caused short body dwarfism and skull abnormalities (Lall, 2002); however, histomorphic changes associated with these trace elements have not been characterized. Several vitamins are also involved in skeletal tissue metabolism (ascorbic acid, vitamins A, K, and D) particularly ascorbic acid–related collagen defects reflected to pathogenesis of scoliosis and lordosis in fish (Lall, 2010; Lewis-McCrea and Lall, 2010; Lock et al., 2010; Hamre, 2011; Fernández et al., 2018). Another question related to the high amounts of lipid found in skeletal tissues, the impact of eicosanoids production and cellular metabolism, and susceptibility of skeletal tissues to lipid peroxidation by prooxidants (Lewis-McCrea and Lall, 2010) and their biochemical function remain to be examined. Phosphorus deficiency is known to increase lipid accumulation in fish tissues, and dietary phospholipid reduces the prevalence of malformation in larval fish (Cahu et al., 2003; Ammerman, 1995). The metabolism of P in bone metabolism is discussed in more detail in Section 6.2.1.2.

6.1.7 Uptake of toxic elements from water and impact of dietary minerals on the environment

It is widely recognized that fish and aquatic organisms are able to accumulate and retain trace elements drawn from their environment. Metals enter the hydrosphere from either natural processes or pollution through activities such as mining operations, burning of fossil fuels, agriculture, and urbanization. The solubility of trace metals in natural waters is mainly controlled by pH, type and concentration of ligands and chelating agents, oxidation state of the mineral components, and redox environment of the system. Soluble forms are usually ions (simple or complex) or unionized organometallic chelates or complexes. The ability of fish to regulate abnormal concentrations varies with species. Certain fish are able to excrete high proportions of excessive metal intake and thus regulate the concentration in the body at relatively normal levels (Bury et al., 2003). Generally, the major excretion route is feces and urine. Newly hatched fry and fingerlings are poorer regulators of these metals than nonessential minerals such as Hg, Cd, and Pb. Experimental studies on the sublethal effects of several metals conducted under laboratory conditions show morphological, physiological (growth, swimming performance, respiration, and reproduction), and behavioral changes in several aquatic organisms. The toxicity mechanisms of metal ions include blocking essential biological functional groups of enzymes, displacing the essential metal ion in the biomolecule (enzyme or protein), and modifying the active conformation of the biomolecule (Handy, 1996). The physiological mechanisms and toxicology of several inorganic elements (Co, Cr, Cu, Fe, Mo, Ni, Se, and Zn) in fish have been comprehensively documented by Wood et al. (2012).

Fish feeds require mineral supplementation due to low availability of trace elements and P from feed ingredients. Feeds of carnivorous fish may contain relatively high proportions of fishmeal and other marine by-products. These feed ingredients contain a high concentration of minerals particularly Ca, P, and Fe. Bioavailability studies have shown that supplementation of fishmeal-based diets with Zn and Mn is necessary for optimum growth, bone mineralization, and prevention of nutritional diseases in fish (Lall, 2010). Inorganic and organic compounds of Zn, Fe, Cu, Mn, and Se are commonly used as supplements in feeds. The discharge of minerals from uneaten food and excretion of urine and feces by fish has a direct influence on the enrichment of the aquatic environment. Minerals excreted in soluble and particulate forms affect the water quality directly, whereas the particulate form can settle to the bottom of the tank or accumulate in the sediment under fish cages or ponds. Soluble P is readily available as a nutrient for the growth of algae leading to algal blooms and fish kills in lakes and coastal areas. Indigestible P settles in sediments and is gradually released as the soluble form during anaerobic or other related biological processes. It appears that the form of minerals in feeds consumed by fish will affect both the amount of soluble and particulate mineral excreted. The breakdown of organically bound minerals from feed and feces varies considerably due to differences in the chemical characteristics of feed ingredients and type of mineral supplements used, the environmental conditions (e.g., temperature, oxygen, pH and salinity, water current), and the type of microorganisms in natural waters (Lall and Milley, 2008a).

Elevated levels of Zn and Cu have been found in sediments under sea cages of salmon farms (Dean et al., 2007). Experimental studies have demonstrated that the concentration of these trace elements in sediments decreases to background levels after chemical remediation (Brooks et al., 2003). These studies indicate that metals complexed with free sulfides, particularly Cu and Zn, are released to the water column during chemical remediation process. Studies conducted in several countries have shown that Zn sediment levels were elevated in the surface 20 cm, below which the sediment concentration was not affected (Dean et al., 2007).

Copper, Zn, and Cd accumulated from feed and feces may be incorporated into the sediment; however, these trace elements can also be associated with naturally occurring organic debris. The decomposition of organic material during early stages of diagenesis in the oxidized surface layers influences the pore water concentration of Cu, Zn, and Cd and results in elevated pore water levels. Organic particles—associated metals are remobilized into the dissolved phase, and soluble Cu, Zn, and Cd form complexes with organic ligands in pore waters (Petersen et al., 1995). The bioavailability of sediment metals is determined by the abundance of organic compounds such as sulfides and organic material. The bioavailability of some metals may be low around organically enriched sediments of highly productive farm sites. It appears that Zn and Cu enrichment of sediment under marine farm sites affects recolonization by benthic organisms. Generally, fish feeds are not supplemented with Cd; however, fishery by-products and other ingredients may contribute to minor amounts of Cd in finished feed. Cadmium is relatively less soluble in reducing pore waters than toxic overlying waters (Sundby et al., 2004). These studies indicate that the difference in solubility may cause a flux of soluble oxidized species from the overlying water into the anoxic sediment and the formation of insoluble species, resulting in authigenic enrichment of Cd in sediments.

6.2 Calcium and phosphorus

Calcium and P are considered together because they are closely related to the development and maintenance of the skeletal system and participate in several physiological processes including the maintenance of acid–base equilibrium. The major proportion of Ca and P are located in bones and scales, where they are combined with other minerals as a Ca and P complex, hydroxyapatite. In pure state hydroxyapatite, Ca:P ratio is 1:66, which may be somewhat modified with chloride or fluoride ions substituting for hydroxyl groups or Sr, replacing some Ca (Simkiss and Wilbur, 1989). Carbonate may also replace PO_4 to a limited extent. Hydroxyapatite provides rigidity and strength in bone and thus structural strength to the body and protection for brain and other organs. Fish scales are an important site of Ca metabolism and Ca and P deposition (Schönbörner et al., 1979). The calcium exchange rate of fish scales is three times that in bones (Berg, 1968). The inner ear of all vertebrates contains otolith or calcium deposits of several types. The ratio of Ca to P has been used to estimate the bone mineralization of fish during their life cycle and, to some extent, the amount of these minerals in the diet of some farm animals and certain fish at various stages of their growth. The proportion of cartilage and bone differ from hatching to adult stage and also vary among different fish species, which may reflect differences in the Ca and P concentration in the body. The ratio of Ca to P in scales and bones ranges from 1.3 to 2.1 in several fish species (Lall, 2002). The ratio of Ca to P in the whole body of several farmed fish ranges from 0.7 to 1.6 (Lall, 1991). In Atlantic salmon, the Ca:P ratio in FW phase was 0.2–0.4 after hatching, increased to 1.04 at the presmolt stage and declined after smoltification and transfer to SW to 0.6, and ranged between 0.7 and 0.8 at the adult stage (Beaversfjord et al., 2019). During intensive growth and skeletal mineralization, the concentration of P may decrease. With regard to adult animals of a given species, the individual differences are caused by the extent of fat deposition, but if the P content is calculated on a fat-free basis, the differences are likely to be relatively small. From applied nutritional perspectives, it is important to determine the bioavailability of Ca and P, which must be retained and used for metabolism and/or in bone formation and mineralization.

6.2.1 Functions and metabolism

6.2.1.1 Calcium

Calcium is one of the most abundant cations in the body of a fish. Despite the importance of Ca in bone formation and maintenance of skeletal tissues, calcium ions are widely distributed in soft tissues. Other functions of Ca include muscle contraction, blood clot formation, nerve transmission, maintenance of cell membrane integrity, and activation of several important enzymes. In the cell membrane, calcium is closely bound to phospholipid, where it controls the permeability of membrane and thus regulates the uptake of nutrients by the cell. In blood, Ca is either complexed with protein and other compounds or present as free ion. The free Ca is a physiologically important fraction. Fish regulate their ionic plasma Ca level more strictly than their protein-bound Ca level, and any minor disruptions in ionic Ca concentrations may lead to severe stress and disturbance of Ca balance (Flik et al., 1995). Unlike terrestrial vertebrates, which depend solely on the diet as their Ca source, fish live in an environment with a readily available source of Ca.

In an aquatic environment with variable Ca levels in FW, BW, and SW, fish maintain the homeostasis of this element in the body through an efficient Ca regulation mechanism. Fish obtain Ca by waterborne exposure via gills and integument as well as absorption from their diet from the gastrointestinal tract or drinking SW (Berg, 1968; Simkiss, 1974; Perry and Flick, 1988; McCormick et al., 1992). In FW fish, the major organ for ion regulation in fish is gills, which are responsible for over 95% of Ca uptake from water (Flik et al., 1995). They maintain Ca homeostasis by active branchial and intestinal Ca^{2+} absorption by a basolateral ATP-dependent Ca^{2+} pump and a Na^{+}/Ca^{2+} exchanger. The latter derives its force from a Na^{+}/K^{+}-ATPase-settled Na^{+} gradient. Seawater has a calcium concentration of ~10 mmol/L, whereas the total plasma Ca concentration of marine fish ranges from 2 to 3 mmol/L; thus, marine fish live in a hypercalcic environment and face an inward gradient of Ca. As Ca availability in the aquatic environment varies, fish have developed Ca regulatory systems that can react rapidly to changes in environmental Ca concentrations (Wendelaar Bonga and Pang, 1991; Bjornsson et al., 1999). Drinking in SW may also provide a route for Ca influx via the intestine; however, the magnitude of intestinal Ca influx from imbibed water appears to vary depending on species (Schoenmakers et al., 1993).

Calcium ion transport across the gills appears to be more efficient in FW. The skin serves as the main organ for ion regulation at early developmental stages of fish, before the gills are fully developed (Hwang et al., 2011). Calcium absorption in these organs occurs mainly through two pathways in specific epithelial cells: (1) paracellular transport, a passive diffusion of Ca, and (2) transepithelial transport, an active transport through Ca transporters (Lin and Hwang, 2016). A general model for Ca^{2+} uptake has been developed whereby Ca^{2+} entry across the apical membrane of specific epithelial cells, ionocytes, occurs via facilitated diffusion mediated by Ca^{2+} channels (ECaC), while basolateral movement of Ca^{2+} from the cytosol to the blood is facilitated by high-affinity plasma membrane Ca^{2+}-ATPases and/or Na^{+}/Ca^{2+} - exchangers. In summary, it is believed that Ca^{2+} entry across the apical membrane is the rate-limiting step to transepithelial Ca^{2+} uptake in fish. Some minerals (e.g., Cd, Cu, Mg, Sr, Zn) may reduce Ca absorption from gills or gastrointestinal tract (Berntssen et al., 2003; Baldisserotto et al., 2006; NRC, 2005; Zimmer et al., 2019).

The endocrine control of Ca metabolism in fish has been investigated for more than five - decades and has been the subject of several reviews (Copp, 1970; Wendelaar Bonga and Pang, 1991; Flik et al., 1995; Bjornsson et al., 1999, 2011; Lin and Hwang, 2016; Takei and Hwang, 2016). There are major differences in the hormonal mechanisms that regulate Ca and P balance in fish and terrestrial animals. Several hormones (i.e., parathyroid hormone, PTH, and vitamin D) have been well known in the maintenance of Ca homeostasis in mammals (Renkema et al., 2008). The function of other hormones (e.g., calcitonin, isotocin, insulin-like growth factor, and stanniocalcin-1 (STC1)) on Ca regulation has also been investigated, but their mechanisms are not completely conclusive. Endocrine control of Ca metabolism in fish is regulated by both hyper- and hypocalcemic hormones. The major hypocalcemic hormone in fish is stanniocalcin (STC), secreted by the corpuscle of stannius. In response to rising Ca level, STC 1 is secreted by corpuscle of stannius to inhibit gill and intestinal Ca transport and to promote phosphate resorption in the kidney, and in this manner is able to maintain the normal physiological limit of serum Ca and phosphate (Wagner et al., 1998). The level of dissolved Ca in water also acts as stimulus for hormone synthesis and secretion. The effect of Ca concentration in water is mediated by its diffusion across the gills and skin, as well as absorption via the gut as marine fish ingest large amounts of SW. Thus, STC cells in marine fish are metabolically more active and secrete more hormones than FW fish.

The role of another hypocalcemic hormone, calcitonin, secreted by ultimobranchial gland, is not clearly established. Prolactin and parathyroid hormone-related protein (PTHrP) are considered the major hypercalcemic hormones (Flik et al., 1995; Fuentes et al., 2006; Guerreiro et al., 2007). Moreover, hormones such as growth hormone, vitamin D, and cortisol are also known to be involved in the Ca balance of fish and estradiol in specific stages of the life cycle (Wendelaar Bonga and Pang, 1991; Bjornsson et al., 1999; Lin and Hwang, 2016).

In vertebrates, vitamin D regulates transcellular Ca uptake (via a luminal calcium channel) and induces cytosolic Ca transport (calbindinD-9k) (Bronner, 2012). Fish have a vitamin D endocrine system, which has similar functions to the system in mammals (Fraser, 2018). They have similar vitamin D metabolites as terrestrial mammals (Sundell et al., 1992; Graff et al., 1999; Fraser, 2018). The function of vitamin D and endocrine control of Ca metabolism is beyond the scope of this chapter. Other factors that promote changes in osmoregulatory, acid–base, and metabolic parameters have indirect actions over Ca^{2+} balance.

6.2.1.2 *Phosphorus*

Phosphorus exists naturally as inorganic phosphate, denoted as P_i, and as organic phosphate, bound in molecules such as sugars, proteins, or other components of the cell. Phosphorus in hydroxyapatite has the important role of being the principal crystalline material of bone. In addition, phosphorus is located in every cell of the body. Phosphate occurs in equilibrium with H_3PO_4, $H_2PO_4^-$, HPO_4^{2-}, and PO_4^{3-}. The predominant form at neutral pH is HPO_4^{2-}, whereas phosphoric acid (H_3PO_4) is the major form at low pH. In soft tissues, phosphate plays different role as (1) a structural component, (2) a factor in intermediate metabolism, and (3) a component of genetic material. Phospholipids (e.g., phosphatidyl choline) are a major constituent of cell membranes and intracellular organelles. In DNA and RNA, phosphate is an essential part of nucleic acids. Release of high energy phosphate by hydrolysis of adenosine phosphate (ATP) provides the main source of energy for metabolic processes and for muscle contraction. A number of enzymes, hormones, and cell-signaling molecules depend on phosphorylation for their activation. Thus, P plays an important role in carbohydrate, lipid, and amino acid metabolism and in muscle and nervous tissue metabolism. Phosphorus also helps to maintain normal acid–base balance by acting as the body's most important buffer. In addition, the P-containing molecule 2,3-diphosphate binds to hemoglobin in red cells and regulates oxygen delivery to tissues.

The regulation of phosphate is considered more critical than that of Ca because fish must effectively absorb, store, mobilize, and conserve phosphate in both FW and SW environments. Food is the main source of P for fish because FW and SW are low in phosphate (Lall, 2002). Normal physiological levels of serum phosphate and calcium are maintained by the interplay of intestinal absorption from diet, branchial absorption from surrounding water, kidney reabsorption, and mobilization/mineralization in skeleton. The mechanism of P absorption and transport in fish has not been well studied. In higher vertebrates, intestinal P absorption is dependent upon a Na gradient caused by the active transport of

Na, and thus the transport of P from the lumen into the cells through the brush-border membrane is secondary to that of Na. Phosphate is taken up from the circulation into cells via type II and type III sodium phosphate cotransporters (Bergwitz and Jüppner, 2010).

Dietary P concentration is a major regulator of P metabolism in fish. The amount of phosphate absorbed from the food is affected by the level of phosphate in the blood (Lall, 1991). Phosphorus absorption increases at low dietary P concentration and decreases at high dietary P concentration through the dynamic adjustment of the number of NaPi-IIb transporters in the apical membrane of the intestine or the acute adaptation of NaPi-IIb transporters in the intestine and renal proximal tubules (Hattenhauer et al., 1999; Avila et al., 2000). The absorbed P accumulates in soft tissues (heart, liver, kidney, muscle, and blood), and deposition in skeletal tissues is relatively low. When fish were exposed to water containing ^{32}P and then transferred to water free of this isotope, there was a rapid loss of' ^{32}P from soft tissues; hard tissue ^{32}P remained unaffected (Tomiyama et al., 1956; Asano and Ito, 1957). High concentrations of dietary Ca and P interfere with the absorption and retention of certain trace elements and Mg. This will be discussed in subsequent sections related to specific elements.

Approximately 90% of the phosphate lost from the body is excreted renally (Tomiyama et al., 1956). To replace this loss and to supply the various structural and physiological needs of the body, phosphate must be absorbed from both food and water. Urinary loss of phosphate is much higher in FW than SW fish because of the large volume of urine produced in FW fish. In SW-adapted eel, P excretion was 40 times lower than those held in FW (Chester Jones et al., 1969). The uptake of ^{32}P from water has been well documented (Lall, 1991, 2002). In carp, the P uptake increases with a rise in water temperature and from any decrease in the Ca content of water (Solomalina and Arsan, 1979). The level of blood Ca is similar in bony and cartilaginous fish; however, the blood phosphate level is much lower in cartilaginous fish (Dean et al., 2015). It is possible that the mechanism for regulating phosphate is similar in both FW and SW because the concentration of this ion is low in both environments. Both Ca and P are distributed in inorganic, organic, and free forms in blood and present some analytical challenges to accurately measure either free or bound form of P to properly interpret and investigate different biochemical mechanisms involved in metabolism and excretion of these minerals (Dean et al., 2015). The Ca level is high in the urine of SW fish, whereas the phosphate concentration is low. Phosphorus deficiency lowers the plasma and urine phosphate concentrations; however, the total body P is significantly depleted when fish are fed a P-deficient diet for a relatively long period (Vielma and Lall, 1998a).

It appears that metabolism of P in fish may be similar to humans and other vertebrates, but the endocrine control may be different. Like Ca homeostasis, the maintenance of adequate phosphate levels seems to be linked to the coordinated interrelationship of several systemic and related local factors eliciting their control more in intestine and kidney than skeletal tissues. Extensive studies conducted on terrestrial vertebrates for many decades show that inadequate Ca intake causes lower blood Ca levels that stimulate increased PTH secretion, and rapidly decreases urinary Ca excretion but increases urinary P excretion, which stimulates bone resorption (Fleet, 2011). PTH also stimulates conversion of vitamin D to its active form (1,25-dihydroxyvitamin D) in the kidneys. Increased circulating 1,25-dihydroxyvitamin D in turn stimulates increased intestinal absorption of both Cs and P. In recent years, characterization of fibroblast growth factor 23 (FGF23), stanniocalcin, and other neuroendocrine factors has revealed the novel regulatory axis of Ca and P metabolism (Bergwitz and Jüppner 2010). FGF-23 plays a central role in P homeostasis, which is secreted by bone-forming cells (osteoblasts/osteocytes) in response to increases in P intake. In a negative feedback loop, FGF-23 inhibits the production and stimulates the degradation of 1,25-dihydroxyvitamin D, as well as promotes an increase in urinary P excretion independently of PTH and 1,25-dihydroxyvitamin D.

In contrast to the endocrine regulation of Ca homeostasis in fish, relatively little is known about the regulation of phosphate homeostasis. Several endocrine factors and local paracrine and autocrine signaling pathways and negative feedback mechanisms have been implicated in maintaining P homeostasis in fish (Suarez-Bregua et al., 2017). The hormones involved in phosphate regulation include stanniocalcin (STC), prolactin, and parathyroid like hormones (Pth1h). With the rise in serum Ca, STC is secreted by the corpuscles of stannius to inhibit gill and intestinal Ca transport and to promote P reabsorption in the kidney to maintain normal physiological serum Ca and P levels (Wagner et al., 1993). It is also not clear whether the effect is mediated by the vitamin D metabolites as happens in terrestrial vertebrates (Wendelaar Bonga and Flick, 1995). Although intraperitoneal injection of vitamin D metabolites influences P homeostasis (Fenwick and Vermette, 1989), dietary intake of cholecalciferol had no clear effect on P absorption and retention in rainbow trout (Vielma et al., 1999).

The removal of pituitary hormones caused a net decrease in phosphate absorption in FW eel (Nakamura and Hirano, 1986). However, the pituitary plays a limited role in the regulation of plasma phosphate in fish. Parathyroid hormone–like hormones, Pth3 and Pth4, play an important role in functions related to P and bone mineral homeostasis (Potts 2005; Guerreiro et al., 2007; Suarez-Bregua et al., 2017). It has also been observed that Pth1a is involved in the maintenance of Ca homeostasis through scales resorption in goldfish *Carassius auratus* (Suzuki et al., 2011) and by promoting the differentiation of ionocytes in low water calcium (Kwong and Perry 2015). This hypercalcemic effect was not detected in gilthead sea bream scales; however, Pth1b caused ion mobilization from postcranial dermal skeleton (Canario et al., 2006). FGF23 signaling pathway has been suggested in zebrafish; however, the underlying mechanism remains to be determined (Mangos et al., 2012).

6.2.2 Requirement

6.2.2.1 Calcium

The Ca requirement of fish is affected by several dietary, physiological, and environmental factors including uptake of Ca from FW, BW, and SW, Ca bioavailability from diet, P level, and species differences (NRC, 2011; Hossain and Yoshimatsu, 2014). In large part, the requirement of most fish is met by absorption of Ca through gills and skin in FW and by drinking SW. A low concentration of calcium (0.34% or less) is required in the diet of carp, striped bass, tilapia, catfish, and Chum salmon (Ogino and Takeda, 1976; Lovell, 1978; Nakamura, 1980; Robinson et al., 1984; Dougall et al., 1996) for optimum growth. Catfish and tilapia reared in water with a low calcium concentration (<1 mg Ca/L) required 0.45% and 0.7% Ca in the diet, respectively (Robinson et al., 1986, 1987). Atlantic salmon absorbs Ca from SW, thus making dietary supplementation unnecessary (Lall and Bishop, 1977). The uptake of Ca from SW is sufficient to meet the dietary Ca requirement of red sea bream (Sakamoto and Yone, 1973, 1976a). Calcium requirements of several fish species are listed in Table 6.4.

TABLE 6.3 Mineral deficiency symptoms in certain fish.[a-s]

Mineral	Deficiency symptoms
Calcium	Reduced growth and low feed conversion
Phosphorus	Reduced growth, low feed conversion, poor bone mineralization,[a-s] skeletal deformity,[a,g] cranial deformity,[g] increased body or visceral fat[g,j]
Magnesium	Reduced growth and feed intake, sluggishness,[d,g,p] nephrocalcinosis,[p] convulsions,[g] cataracts,[g,p] degeneration of muscle fibers and epithelial cells of pyloric cecum and gill filaments,[p] skeletal deformity,[p] reduced bone ash,[p] and body, vertebrae, and scale Mg concentration,[a,d,g,i,l] high mortality,[d,g,p] low MnSOD, and GPx activity[i,l]
Iron	Reduced growth and low feed conversion, hypochromic microcytic anemia,[b,f,l,p] low hematocrit and hemoglobin levels,[a,d,p] reduced plasma iron and iron transferrin saturation[d]
Zinc	Reduced growth, anorexia,[d,g] short body dwarfism,[p] cataracts,[g,p] fin erosion,[g,p] skin erosion,[g] reduced bone calcium and zinc concentrations,[d,h,i,k,l,n,o,r] low serum zinc level,[d] high mortality,[d,g,p] low SOD activity[k,l,r]
Manganese	Reduced growth, short body dwarfism,[g,p] cataract,[g,p] reduced body and/or vertebrae Mn concentration,[a,c,e,g,h,k,n,r,s] poor hatchability of eggs,[a,c,p] abnormal tail growth,[p] low MnSOD activity[e,h,k,o,r,s]
Copper	Reduced growth, cataracts,[f] and reduced activities of liver Cu, Zn SOD,[a,r] heart cytochrome *c* oxidase,[a,d] and serum ceruloplasmin[q]
Selenium	Reduced growth, anemia,[g] cataracts,[g] muscular dystrophy,[a] exudative diathesis,[p] reduced tissue GPx activity[a,c,d,e,h,n,p]
Iodine	Thyroid hyperplasia[c,f,p]

[a]*Atlantic salmon* (Salmo salar).
[b]*Black sea bream* (Acanthopagrus schlegelii).
[c]*Brook trout* (Salvelinus fontinalis).
[d]*Channel catfish* (Ictalurus punctatus).
[e]*Cobia* (Rachycentron canadum).
[f]*Chinook salmon* (Oncorhynchus tshawytscha).
[g]*Common carp* (Cyprinus carpio).
[h]*Gibel carp* (Carassius auratus gibelio).
[i]*Grass carp* (Cichlasoma urophthalmus).
[j]*Haddock* (Melanogrammus aeglefinus).
[k]*Hybrid tilapia* (Oreochromis niloticus O. aureus).
[l]*Indian major carp* (Labeo rohita).
[m]*Japanese eel* (Anguilla japonica).
[n]*Malabar grouper* (Epinephelus malabaricus).
[o]*Nile tilapia* (Tilapia mossambica).
[p]*Rainbow trout* (Oncorhynchus mykiss).
[q]*Red sea bream* (Chrysophrys major).
[r]*Yellow catfish* (Pelteobagrus fulvidrac).
[s]*Yellow croaker* (Pseudosciaena crocea).
Note: (1) Gross deficiency signs appear when specific mineral intake is restricted for longer term. Most studies have been conducted for a relatively short-term duration. (2) Different stages of development of deficiency signs: Reduced growth and feed utilization >> depletion of nutrients in tissues (e.g., liver, skeletal tissues etc.) and reduced specific enzyme activity >> clinical symptoms (e.g., anemia, cataracts etc.; histopathological changes) >> mortality (advanced stages).

TABLE 6.4 Calcium requirement of certain fish.

Fish species	Requirement[a] (g/100 g)	P in diet (g/100 g)	Ca in water (mg/L)	References
Freshwater fish				
Channel catfish	0.45	0.5	Ca free	Robinson et al. (1986)
	0.05	0.45	14	Lovell (1978)
	1.5	0.8	56	Andrews et al. (1973)
Nile tilapia	0.7	0.5	Ca free	Robinson et al. (1987)
Blue tilapia	0.17–0.65		Ca free	Robinson et al. (1984)
Hybrid tilapia	0.35–0.42[b]	0.9	27.1–33.3	Shiau and Tseng (2007)
Common carp	0.1	0.6–0.7	5	Nakamura (1982)
	NR	0.6–0.7	20	Ogino and Takeda (1976)
Grass carp	1.04[b]	1.02	27.8–43.2	Liang et al. (2012c)
American cichlid	0.18	0.15	86	Chavez-Sanchez et al. (2000)
Marine or anadromous fish				
Atlantic salmon	NR[c]	0.3		Lall and Bishop (1977)
Japanese flounder	0.1–0.25	0.9		Hossain and Furuichi (1999a)
Black sea bream	NR	0.9		Hossain and Furuichi (1999b)
Red sea bream	NR	0.9		Hossain and Furuichi (1999c) Sakamoto and Yone (1976b)
Tiger pufferfish	0.1–0.20	0.9		Hossain and Furuichi (1999d)
	0.25	0.9		Furuichi et al. (1997)
Giant croaker	0.10	0.9		Hossain and Furuichi (1999e)

[a]*Unless specified flow-through system used.*
[b]*Recirculation system used.*
[c]*Not required.*

6.2.2.2 *Phosphorus*

The P requirements of various fish species determined in FW, BW, and SW are listed in Table 6.5. Generally, these requirement estimates are solely based on juvenile fish. Research on the P requirement of fish has been the subject of numerous investigations for three decades; however, the minimum requirement for this nutrient has shown a wide range of values not only for different fish species but also within some species. In some instances, the differences in requirement values may be associated with differences in experimental conditions, methods used to estimate P bioavailability from feed ingredients and inorganic P supplements, response criteria to estimate requirement, stage of development, and other physiological factors that affect P metabolism and retention in body and skeletal tissues. A metaanalytical approach used to quantify the dietary P requirement of fish based on several

TABLE 6.5 Phosphorus requirement reported for certain fish.[a]

Fish species	Requirement[b] (g/100 g)	References
Atlantic salmon (*Salmo salar*)	0.6	Ketola (1975)
	0.83–0.93	Vielma and Lall, 1998b
	1.0	Asgaard and Shearer, 1997
	0.6[d]	Lall and Bishop (1977)
Rainbow trout (*Oncorhynchus mykiss*)	0.7–0.8	Ogino and Takeda (1976)
	0.24–0.59	Rodehutscord (1996)
	0.34–0.54	Ketola and Richmond (1994)
Chum salmon (*Oncorhynchus keta*)	0.5–0.6	Watanabe et al. (1980)
Channel catfish (*Ictalurus punctatus*)	0.8	Andrews et al. (1973)
	0.45	Lovell et al. (1978)
	0.33	Wilson et al. (1982)
Milkfish (*Chanos chanos*)	0.85	Borlongan and Saito (2001)
Blue tilapia (*Oreochromis aureus*)	0.5	Robinson et al. (1987)
Nile tilapia (*Oreochromis niloticus*)	0.86	Yao et al. (2014)
	0.65	Carvalho et al. (2018)
Common carp (*Cyprinus carpio*)	0.6–0.7	Ogino and Takeda (1976)
Gibel carp (*Carassius auratus gibelio*)	0.67–1.07	Xie et al. (2017)
Grass carp (*Ctenopharyngodon idella*)	0.85	Liang et al. (2012d)
Crucian carp (*Carassius auratus*)	0.78–0.83	Sun et al. (2018)
Stinging catfish (*Heteropneustes fossilis*)	0.9–1.1	Zafar and Khan (2018)
African giant catfish (*Heterobranchus bidorsalis*)	1.23	Nwanna et al. (2009)
Walking catfish (*Clarias leather*)	0.58–0.75	Yu et al., 2012
Chinese sucker (*Myxocyprinus carpio*)	0.83–0.86	Yuan et al. (2011)
Hybrid striped bass (*Morone chrysops* × *Morone saxatile)*	0.5	Brown et al. (1993)
Snakehead (*Channa argus* × *Channa maculata*)	0.96	Shen et al. (2016)
Tambaqui (*Colossoma macropomum*)	0.7	Araujo et al. (2016)
Japanese eel (*Anguilla japonica*)	0.29	Arai et al. (1974)
Red drum (*Sciaenops ocellatus*)	0.86[c]	Davis and Robinson (1987)
Red sea bream (*Chrysophyrs major*)	0.68[d]	Sakamoto and Yone (1978a)
Gilthead sea bream (*Sparus auratus*)	0.75[d]	Pimental-Rodrigues and Olivia-Teles (2001)
Black sea bream (*Acanthopagrus schlegelii*)	0.55[d]	Shao et al. (2008)
Haddock (*Melanogrammus aeglefinus*)	0.96[d]	Roy and Lall (2003)

(Continued)

TABLE 6.5 Phosphorus requirement reported for certain fish.[a]—cont'd

Fish species	Requirement[b] (g/100 g)	References
Japanese sea bass (*Lateolabrax japonicus)*	0.86-0.90[d]	Zhang et al. (2006)
European seabass (*Dicentrarchus labrax*)	0.65[d]	Olivia-Teles and Pimental-Rodrigues (2004)
Orange spotted grouper (*Epinephelus coioides*)	1.09[d]	Ye et al. (2006)
Yellow croaker (*Pseudosciaena crocea*)	0.89–0.91[d]	Mai et al., 2006
Japanese flounder (*Paralichthys olivaceus*)	0.6–1.5[d]	Choi et al. (2005); Wang et al. (2005); Uyan et al. (2007)

[a]*NRC recommended requirement of P (g/100 g): Atlantic salmon, 0.8; rainbow trout, 0.7; Pacific salmon, 0.6; channel catfish, 0.33; common carp, 0.7;* Tilapia *sp., 0.4; hybrid striped bass, 0.5; red drum, 0.8; European sea bass, 0.65; Japanese flounder, 0.6.*
[b]*Unless marked requirement determined in freshwater.*
[c]*Brackish water (5‰–6‰).*
[d]*Seawater.*

published studies suggested that among several response criteria analyzed, weight gain was more reliable and whole-body P was the most stringent in determining the P requirement of fish (Antony Jesu Prabhu et al., 2013).

Many studies with monogastric animals have shown that an optimum Ca-to-P ratio is important. Increasing the Ca-to-P ratio of the diet interferes with the absorption of P, and conversely, a high P-to-Ca ratio may restrict Ca absorption. Although the magnitude of effect changes with the species and forms of Ca and P present in the diet, studies on Ca and P ratio in fish diets are limited. Optimum Ca:P ratio is important in the diet of red sea bream, 1:2 (Sakamoto and Yone, 1973), and eel, 1:1 (Nose and Arai, 1979).

6.2.3 Deficiency

6.2.3.1 Calcium

Clinical signs of Ca deficiency have not been reported in fish (Table 6.3). Growth and feed utilization were reduced in rainbow trout, Japanese eel, red sea bream, tilapia, olive flounder, and red mullet. Calcium deficiency has not been detected in carp and catfish in FW (Andrews et al., 1973; Ogino and Takeda, 1976) or in Atlantic salmon in SW (Lall and Bishop, 1977) fed low Ca diet. As mentioned earlier, absorption of Ca from water varies from species to species, and a low requirement for certain fish has been demonstrated (Hossain and Yoshimatsu, 2014). Therefore, clinical Ca deficiency is difficult to detect in fish. Generally, Ca from water and the feed ingredients in practical diets supply sufficient Ca to meet the requirements of most finfish with some exceptions, which is discussed in Section 6.14.

6.2.3.2 Phosphorus

Most studies conducted on the P requirement of 29 fish species have shown reduced weight gain, feed utilization, and bone mineralization when P content of the diets was low, with some exceptions, gibel carp (Xie et al., 2016), red drum (Davis and Robinson, 1987), gilthead bream (Pimental-Rodrigues and Olivia-Teles, 2001), red sea bream (Sakamoto and Yone, 1978a) and European sea bass (Olivia-Teles and Pimental-Rodrigues, 2004). Studies with red drum, European sea bass, and gilthead sea bream were conducted in SW recirculation systems, and it is likely that low uptake of P from the water did not produce P deficiency as other fish species. Other P deficiency signs experimentally produced in several fish species included increases in liver or body fat, reduced blood phosphate levels, and poor mineralization of scales. There is evidence that P deficiencies primarily trigger scale resorption (Metz et al., 2014) and also that wild Atlantic salmon resorb scales during upstream spawning (Persson et al., 1998; Shearer, 1992). Fish fed diet containing either mineral deficient or low P diet mobilize minerals including Ca and P to maintain normal physiological functions (Lall, 2002), and fish scales appear to be the most sensitive indicator of P deficiency in young fish (Skonberg et al., 1997; Vielma and Lall, 1998b). The cause of low dietary P-induced skeletal deformities in salmonids and certain marine fish has been the subject of intensive research, and several reviews have been published in this area (Sugiura et al., 2004; Lall and Lewis-McCrea, 2007; Lall, 2010; Baeverfjord et al., 2019). Common skeletal deformities with low intake of P include curved spines and soft bones in Atlantic salmon (Baeverford et al., 1998), cephalic deformities in the frontal bones of common carp (Ogino and Takeda, 1976), and compressed vertebral bodies resulting in scoliosis in haddock (Roy and Lall, 2003) and halibut (Lall and Lewis-McCrea,2007). Atlantic halibut and haddock juveniles fed low P diets possessed twisted neural and hemal spines in the midsection of the fish (Roy and Lall, 2003; Lall and Lewis-McCrea, 2007). These bones are commonly affected, as they are soft and brittle due to the reduction of mineral content (Sakamoto and Yone, 1978a), and with muscular action, the bones become twisted. Histological and histochemical examination of P-deficient haddock showed an initial increase in bone resorption, which was subsequently followed by a decrease in bone mineralization and reduced bone formation (Roy and Lall, 2003).

Severe P deficiency can only be induced using semipurified test diets containing recommended levels of other nutrients without any P supplement (Table 6.3). Recently, histological studies of scale and bone tissues have revealed that mobilization of minerals does not fully explain the biochemical mechanism involved in the demineralization process (Witten et al., 2015, 2019). A decrease in dietary P content in practical diet of larger Atlantic salmon smolts showed vertebrae without any significant changes in organic matrix and demineralization of bone (Fjelldal et al., 2007; Witten et al., 2019). Early work on salmon had also showed that salmon fed low P (0.3%) purified diets had normal growth and bone development, but vertebrae ash and P content were significantly reduced in P-deficient fish (Lall and Bishop, 1977). Several nutrients support skeletal growth including bone and scale matrix mineralization, and interactions between P and other nutrients in skeletal tissue biomineralization are poorly understood (Section 6.1.4). It is widely recognized that mineralization of skeletal tissues occurs throughout the life cycle of fish, and it is affected by several factors including environment, genetics, and interaction with other nutrients and triploidy. Certain genes have been identified that respond to early P deficiency by increasing or decreasing their expression at mRNA level in rainbow trout fed practical diets (Kirchner et al., 2007; Lake et al., 2010).

6.2.4 Sources and bioavailability

Most salts of alkali metals present in foods are found in a freely soluble ionized form. However, the salts of alkali earth metals, such as Ca and Mg, are less soluble and form complex organic salts with varying degrees of solubility in the digestive tract. Certain naturally occurring calcium salts such as Ca oxalate are more readily degraded during the digestion process. Feedstuffs of animal origin show the highest concentrations of Ca and P. Among common feedstuffs used in fish feed formulation, fishmeal (1.5%–3.2% P) and meat and bone meal (3.5%–5.5% P) are the richest sources of Ca and P. Skeletal tissue contributes a significant proportion of P in fishmeal and meat meal. Phosphorus composition of meat and bone meal varies according to the type of material used for processing and bone-to-meat ratio. The major proportion of P in these feedstuffs is found in an inorganic form, whereas the remainder is present as phosphate complexes of protein, lipid, and carbohydrate. In these forms, P is more readily available to fish than from forms present in plant protein supplements.

The bioavailability of dietary P may differ markedly among feed ingredients and inorganic P supplements (Lall, 2002; NRC, 2011). Digestibility and growth measurements have frequently been used to determine the bioavailability of P to fish from selected feed ingredients. To some extent, apparent P digestibility data are of limited use unless a correction is made for endogenous P excretion in the feces. A model to estimate the digestible P fraction in fish feeds has been developed (Hua and Bureau, 2006). Phosphorus retention in the body and/or vertebrae is considered the most sensitive practical criterion for evaluating dietary P utilization and can provide a better estimate than values based on body weight or apparent digestibility. Plasma alkaline phosphatase assays are the least sensitive method for predicting P bioavailability.

Bioavailability of P is influenced by several factors including chemical form, digestibility of diet, particle size, interaction with other nutrients, feed processing, and water chemistry. There are significant differences in the availability of P from a variety of inorganic salts: the more soluble the salt, the higher the availability of P, thus P is more readily available from mono- or dicalcium phosphates than from tricalcium phosphate (NRC National Research Council, 1993, 2011). Salmonids utilize P present in fishmeal more efficiently than do carp, tilapia, and channel catfish (Lovell, 1978; Ogino et al., 1979; Watanabe et al., 1988; Köprücü and Özdemir, 2005). Phosphorus in fishmeal is mainly in the form of an insoluble Ca–P complex, hydroxyapatite, which varies from the source of fish used, and its bioavailability varies among fish species (NRC National Research Council, 1993, 2011). Acid hydrolysis of fish bone by-products appears to increase P availability of the hydroxyapatite in bone (Albrektsen et al., 2017).

Phosphorus in cereal grains and vegetable protein concentrate may range 0.3%–0.4% and 0.5%–1.4%, respectively. Phytic acid (myoinositol-1,2,3,4,5,6-hexakisphosphate) is the storage form of P in seeds, representing 60%–90% of seed total P. The term phytin implies a Ca–Mg salt of phytic acid, and feedstuffs of plant origin primarily contain phytin. The concentration of phytate and phytase in feedstuffs varies considerably. Phytate constitutes between 0.5% and 5% of most cereal grains and oilseeds (Erdman 1979; Eeckhout and Paepe, 1994). The oilseed-derived proteins such as soybean meal, rapeseed meal, and sesame meal contain 1.0%–1.5%, 5.0%–7.5%, and 2.4% phytate, respectively (Erdman, 1979; Francis et al., 2001).

The remaining P is present as soluble inorganic phosphate and cellular P (phosphorus bound in nucleic acids, phosphorylated proteins, phospholipids, phosphosugar). In soybean meal, phytic acid, phosphatides, and inorganic P constitute about 75%, 12%, and 4% of total P, respectively. Distillers' dried corn with solubles contains a lower percentage of P in the phytate form because of a breakdown of P during the fermentation process. In contrast to other feed ingredients of plant origin, alfalfa meal P and meals produced from macromicroalgal are nonphytate in nature. Phytate P is unavailable to fish and reduces the digestibility of other nutrients and the performance of fish due to its antinutritional effects.

Several reviews have considered various aspects of phytates in fish nutrition and potential strategies to improve the bioavailability of phytates in feed ingredients of plant origin (Cao et al., 2007; Kumar et al., 2012; Dersjant-Li et al., 2015; Lemos and Tacon, 2017). The major findings are briefly summarized in this section. In addition to unavailability of phytate P in plant ingredients, the main concerns about phytate include the uptake of certain minerals and lower digestibility of protein and amino acids. In the gastrointestinal tract, phytate chelates with positively charged cations such as Ca^{2+}, Mg^{2+}, Zn^{2+}, and Fe^{2+}, and thus reduces the bioavailability of these minerals (Erdman, 1979). A decrease in availability of these minerals has also been observed in several fish species. The reduction in Ca and Zn availability appears to be higher than Mg, Fe, or Cu because of interaction of these ions with phytate. The formation of insoluble metal cation–phytate complexes at physiological pH values is considered one of the main reasons for poor mineral availability because these complexes are essentially not absorbed from the gastrointestinal tract. Phytate depresses protein and amino acid digestibility and utilization efficiency in fish and other higher animals. Phytate interactions with proteins are also pH dependent (Cheryan, 1980). At pH values below the isoelectric point of the protein, the anionic phosphate groups of phytate bind strongly to the cationic groups of the protein to form insoluble complexes that dissolve only below pH 3.5. The α-NH_2 terminal group, the ε-NH_2 of lysine, the imidazole group of histidine, and the guanidyl group of arginine have been implicated as protein-binding sites for phytate at low pH values. In acidic environments of Nile tilapia stomach (pH 1–2), nearly half of the phosphate moiety is negatively charged, which favors binding to soluble proteins at amino groups of lysine, imidazole groups of histidine, and guanidyl groups of arginine. However, in the alkaline environment of tilapia intestine (pH 8.5–8.8), ternary complexes are formed. Both complexes appear to be resistant to proteolytic digestion. Phytate also forms complexes with lipid and derivatives along with other nutrients (Vohra and Satanarayana, 2003), but this aspect of chelate formation has not been investigated in fish.

The hydrolysis of phytic acid can be achieved by enzymatic and nonenzymatic degradation. Development of phytases and optimization of their catalytic features has been a promising strategy for efficient reduction of phytate in animal feeds. Phytases (myoinositol hexakisphosphate phosphohydrolase; EC 3.1.3.8/EC 3.1.3.26) are a group of enzymes that catalyze the sequential hydrolysis of phytate in a stepwise manner to lower inositol phosphate, myoinositol, and inorganic phosphate (Greiner and Konietzny, 2002). Several factors affect the efficiency of phytate hydrolysis in the gastrointestinal tract of fish including the origin of phytase (e.g., 3- or 6-phytase, bacterial or fungal) incorporated into feeds, dietary factors (type of feed ingredients, phytate, Ca, P, and Na content, feed processing), species differences (cold water vs. warm water, gastrointestinal tract), and other physiological factors (e.g., gut pH, intrinsic phytase, age, etc.). To eliminate the antinutritional effects of phytate of feed ingredients, it must be hydrolyzed by phytase in the upper part of the gut. Most of these factors have been comprehensively considered in the aforementioned reviews. Although phytase-producing bacterial floras in the foregut and hindgut regions of the gastrointestinal tracts of several FW fish have been demonstrated, Nile tilapia digests more phytate than other fish. Low digestibility (1%–5%) of dietary phytate P has been reported for most farmed fish including cold water rainbow trout (Kumar et al., 2012). A wide range of microbial phytase products for fish are now commercially available. The most commonly used microbial phytases are derived from fungi (*Aspergillus niger*) and bacteria (*Escherichia coli*). The catalytic properties of many phytases and hydrolysis pathway of phytate by phytate-degrading enzymes are summarized by Greiner and Konietzny (2002). These enzymes have a potential for improving the biological availability of several plant proteins in fish feeds; however, their stability during feed processing must be maintained. The preferable method of phytase application in feeds is by coating a liquid form of the enzyme after extrusion and drying, which prevents loss of enzyme activity during processing (Verlhac-Trichet et al., 2014).

6.2.5 Phosphorus excretion and strategies to minimize P output from aquaculture operations

The form of P, which is excreted in feces and urine by fish, has a direct influence on the enrichment of the aquatic environment and the growth of algae. The breakdown of organically bound P from uneaten feed and feces varies considerably due to differences in the chemical characteristics of feed ingredients used, the environmental conditions (temperature, oxygen, pH, and salinity), and the type of microorganisms in natural waters. Generally, P is excreted in soluble and particulate forms: the soluble forms of P consisting of organic P and PO_4^{3-} affect water quality directly, whereas the particulate form settles to the bottom of tank or accumulates in the sediment. Soluble P is readily available as a nutrient for plant growth. The amount of particulate P excreted by fish maintained in hatchery or fish farm conditions ranges between 30% and 64% of the total excreted P. A significant amount of the phosphate fraction consists of free inorganic orthophosphate. Under hatchery and farm conditions, excreted P settles in sediment and is gradually released as the soluble form during anaerobic or other related biological processes. Obviously, the form of P consumed by fish will affect both the amount of soluble and particulate P excreted and the quantity of P that could subsequently be biologically degraded from sediment. In the past 3 decades, significant effort has been directed to minimize the adverse effects of solid and soluble P wastes as well as other organic matter in effluent water through changes in feed formulation, better nutrition and feeding strategies, improved aquaculture practices, and stringent environmental regulations.

The following strategies have been recommended to reduce P output from aquaculture operations: (1) estimation of correct P balance, (2) reduction in P levels in feeds without affecting growth, feed efficiency, health and reproduction, etc., (3) selection of feed ingredients and P supplements with high P bioavailability, (4) selection of P supplements with high P absorption and low solubility in water, (5) use of phytase to increase the digestibility of plant protein containing phytic acid, (6) development of diets with better feed efficiency, and (7) minimizing feed wastage. It is important that P content of effluents from land-based fish farms, fish hatcheries, and sea cages be minimized. For farmed fish, mass balance models have been developed for nitrogen and P, indicating that 50% of the nitrogen and 28% P supplied with the food is wasted in the dissolved form. Several organic (bone, organic, phytate, etc.) and inorganic (mono calcium, sodium and potassium phosphate, or di- and tricalcium phosphate) forms of P may be present in fish feeds that originate from feed ingredients and feed supplements. A model developed for salmonid fish to predict the digestible P content is based on the following formula: digestible P = 0.68 bone P + 0 phytate P + 0.84 organic P + 0.89 Ca monobasic/Na/K Pi supplement + 0.64 Ca dibasic Pi supplement + 0.51 phytase/phytate − 0.02 $(\text{phytase/phytate})^2$ − 0.03 $(\text{bone P})^2$ − 0.14 bone P (Hua and Bureau, 2006). A bioenergetic approach to manage production and control phosphorus discharge from fish hatcheries has been developed that takes into the account growth, feed consumption, and P mass balance (Canale et al., 2016). A significant amount of bound P in fishmeals is not digested and settles into the sediments. Maximum nutrient release can be estimated from the hydrographic conditions, such as water volume, tidal water exchange, and currents, in the immediate vicinity of the farm. In recent years, better site management and production levels, improvements in feeding efficiency, development of highly digestible nutrient-dense diets, and feed quality have significantly reduced aquaculture waste and consequently the environmental impact.

6.3 Magnesium

6.3.1 Functions and metabolism

Magnesium is a major intracellular cation that is an essential cofactor in more than 300 enzymatic reactions in intermediary metabolism. These include those that transfer phosphate groups (phosphokinase), hydrolyze phosphate and pyrophosphate groups (phosphokinase and pyrophosphatase) and fatty acid oxidation involving acetyl coenzyme A (thiokinase), and activate amino acid synthesis (amino acid synthetase). Magnesium is involved in energy production, nucleic acid and protein synthesis, ion transport, and cell signaling and also has structural functions. Extracellular Mg is vital to normal nerve conduction, muscle function, and skeletal tissue metabolism. It plays an important role in the respiratory adaptation of FW fish (Houston, 1985). The major proportion (50%–70%) of Mg in the body of fish is located in skeletal tissues and scales (Bijwelds et al., 1998). The remainder is found within the cells of soft tissues. The red blood cells of fish contain significantly higher levels of Mg and lower levels of Ca than found in humans. The Mg pools of bones and scales are mobilized when dietary Mg intake is low (Cowey et al., 1977). In muscle, it comprises approximately 20% of the total body Mg pool (Knox et al., 1981).

The absorption of Mg from the gastrointestinal tract has received limited attention. Generally, dietary Mg is solubilized within the acidic environment of the stomach, and a large proportion remains available for absorption in the intestine. FW fish absorb Mg ions by active uptake either from the aquatic environment or from dietary sources

(Bijvelds et al., 1998); however, dietary Mg is the main source for growth and development of fish (Lall, 2002). There is some evidence that part of the Mg requirement can be met by uptake from water via gills when dietary Mg concentration is low (Shearer and Asgard, 1992; Dabrowska et al., 1991), but for most fish species studied, Mg uptake from FW was insufficient to meet their dietary Mg requirement (Lall, 2002; NRC, 2011). In FW, excess Mg is excreted renally (Oikari and Rankin, 1985). In SW, fish obtain Mg by drinking and excrete exclusively by the kidneys (Hickman and Trump, 1969). The major part of the Mg requirement of rainbow trout and other fish could be met by absorption from drinking SW (Lall and Bishop, 1977).

6.3.2 Deficiency

Magnesium deficiency in carp (Ogino and Chiou, 1976; Satoh et al., 1983a), catfish (Gatlin et al., 1982), eel (Nose and Arai, 1979), rainbow trout (Cowey et al., 1977; Ogino et al., 1978; Knox et al., 1981, 1983), Atlantic salmon (El-Mowafi and Maage, 1998), hybrid tilapia (Lin et al., 2013), Indian major carp (Musharaff and Khan, 2018), grass carp (Wang et al., 2011; Liang et al.,2012a, 2012b, 2012c), and grouper (Ye et al., 2010) generally includes one or more of the following deficiency signs (Table 6.3): anorexia, reduced growth, sluggishness, high mortality and reduced body, vertebra, and scale Mg content. In rainbow trout, Mg deficiency also causes calcinosis of kidney, vertebrae deformity, and degeneration of muscle fibers and epithelial cells of the pyloric cecum and gill filaments (Cowey et al., 1977; Ogino et al., 1978). Catfish and rainbow trout fed Mg-deficient diets showed flaccid appearance of their muscle (Knox et al., 1981; Gatlin et al., 1982). Common carp fed a low-Mg diet also developed convulsions and cataracts (Ogino and Chiou, 1976). Low Mg intake caused high mortality in rainbow trout, channel catfish, and common carp and reduced Mn SOD and GPx activities of carps (Musharaff and Khan, 2018; Wang et al., 2011). The concentration of Mg in eggs and water may affect the hatchability of eggs and survival of offspring. A low concentration of Mg in water (<0.0.01 mmol/L) resulted in a decrease in Mg uptake by eggs and caused higher mortality, deformities, embolism, tissue necrosis, and higher mortality in carp (Van der Valden et al., 1991).

Atlantic salmon and red sea bream do not show signs of Mg deficiency in the SW environment, where Mg concentration is much higher than in FW where they obtain Mg by drinking SW. However, it is not uncommon to find nephrocalcinosis in rainbow trout reared in SW. Poor water quality particularly low oxygen and high carbon dioxide levels during the FW rearing period of salmonids and other factors may induce early signs of nephrocalcinosis, but the clinical signs develop after SW transfer. Cowey et al. (1977) observed that Mg-deficient diet caused renal calcification at Ca levels of 2.6% (Ca:P ratio = 1:1) or more. A lower dietary Ca level (1.4%) or Mg supplementation (0.1%) produced trout with normal Ca levels. Thus, dietary Ca levels of 2.6% were not detrimental to trout provided sufficient Mg was present in the diet. An interaction between dietary protein and Mg level has been demonstrated in tilapia (Dabrowska et al., 1989a), where excess Mg (0.32%) in a low-protein diet (24%) produced toxicity signs and Mg deficiency in a high-protein diet (44%) caused whole-body hypercalcinosis.

6.3.3 Requirement

The quantitative dietary Mg requirements of several fish species are listed in Table 6.6. Magnesium requirements of most farmed fish species have ranged from 0.4 to 0.6 g/kg diet NRC (2011). Hybrid tilapia showed the Mg requirement of 0.2% and 0.02% in FW and

TABLE 6.6 Magnesium and manganese requirement of certain fish.

	Magnesium[a]		Manganese[b,d]	
Fish species	**g/kg**	**References**	**mg/kg**	**References**
Atlantic salmon	0.33	El-Mowafi and Maage, 1998	15 7.5–10.5	Lorentzen et al. (1996) Maage et al. (2000)
Rainbow trout	0.5 0.6 0.5 0.33	Knox et al. (1981) Shearer (1989) Satoh et al. (1991) Shearer and Asgaard (1992)	12–13	Ogino and Yang (1980)
Channel catfish	0.4	Gatlin et al. (1982)	2.4	Gatlin and Wilson (1984c)
Yellow catfish			5.5–6.4	Tan et al. (2012)
Common carp	0.6 0.4–0.5	Dabrowska et al. (1991) Ogino and Chiou (1976)	12–13	Ogino and Yang (1980)
Gibel carp			13.8	Pan et al., 2008
Grass carp	0.5–0.7 0.6–0.7	Wang et al. (2011) Liang et al. (2012d)		
Indian major carp	0.44–0.45	Musharraf and Khan (2018)		
Hybrid tilapia	0.2 0.02[f]	Lin et al. (2013) Lin et al. (2013)	7	Lin et al., 2008a
Blue tilapia	0.5	Reigh et al. (1991)		
Nile tilapia	0.6–0.77	Dabrowska et al. (1989)		
Japanese eel	0.4	Nose and Arai (1979)		
Grouper[c]	NR	Ye et al., 2010	15	Ye et al., 2009
Yellow croaker	NR	El-Zibdeh et al. (1996)	16.4	Zhang et al. (2016)
Cobia			21.7–24.9 10.5–15.4[e]	Liu et al., 2012 Nie et al. (2016)

[a] *NRC (2011) recommended requirement of Mg (g/kg): Atlantic salmon, 0.4; rainbow trout, 0.5; channel catfish, 0.4; common carp, 0.5;* Tilapia *sp., 0.6.*
[b] *NRC (2011) recommended requirement of Mn (mg/kg): Atlantic salmon, 10; rainbow trout, 12; channel catfish, 2.4; common carp, 12;* Tilapia *sp., 7.*
[c] *Orange spotted grouper (*Epinephelus coioides*).*
[d] *Unless specified $MnSO_4 \cdot H_2O$ used as Mn supplement.*
[e] *$MnSO_4 \cdot H_2O$, manganese glycine, and manganese 2-hydroxy-4-(methylthio) butyrate showed Mn requirement of 15.4, 11.2, and 10.5 mg/kg, respectively.*
[f] *Magnesium requirement in seawater.*

SW, respectively. No requirement of Mg for yellow croaker and red mullet has been reported. A metaanalysis of Mg requirement reported for several fish species showed relatively close estimates for the Mg requirement (g/kg diet) based on the following parameters: weight gain, 0.34; whole body Mg, 0.49; vertebrae Mg, 0.42; and plasma Mg, 0.5 (Antony Jesu Prabhu et al., 2016a, 2016b). Generally, Mg concentration of water may show a wide range of values, but it is generally considered low in FW (1−3 mg/L) to meet the metabolic demands of fish. In the marine environment where Mg concentration is high (1350 mg/L) and fish blood contains a lower level of this element, Mg supplementation of diet may not be necessary (Lall and Bishop, 1977; Sakamoto and Yone, 1979a). It is important to consider concentration and bioavailability of Mg from water and diet for feed formulation to provide sufficient amounts of Mg to meet the requirement of this element. The Mg requirement of rainbow trout is not influenced by an increase in dietary Ca and P levels (Knox et al., 1981). In terrestrial animals, Mg requirement is affected by Ca as well as P level in the diet.

6.3.4 Sources

The Mg concentration of FW is approximately 200 μM. In SW, it is the third most abundant ion with a concentration of more than 50 mM (Rankin and Davenport, 1981). Among food sources, cereal grains contain less Mg (0.15%−2%) than other common feedstuffs. In cereals, most of the Mg is found in the bran fraction. Plant proteins may contain 0.4%−0.6% Mg. The skeletal tissues present in meat and fishmeals contribute to the variable amount of Mg in these rich sources of Mg. In plant products, Mg is present as the chelated metal in the porphyrin moiety of chlorophyll. Little is known about the form of Mg in other feedstuffs. Magnesium feed supplements include magnesium sulfate, magnesium chloride, magnesium oxide, and magnesium acetate. Magnesium sulfate is more water soluble than magnesium oxide and therefore more available for absorption. Magnesium present as Mg acetate is more efficiently used by tilapia than either Mg oxide or sulfate (Dabrowska et al., 1989).

6.4 Sodium, potassium, and chloride

6.4.1 Functions and metabolism

Sodium, potassium, and chloride are the most abundant electrolytes in the body of living organisms. Sodium and chloride are the major cation and anion, respectively, of extracellular fluids of the body, whereas K and Mg are the major intracellular cations. They serve a vital function in controlling osmotic pressures, ion regulation, and acid−base equilibrium (Marshall and Grosell, 2006). Potassium concentrations are about 30 times higher inside than outside the cells, whereas Na concentrations are more than 10 times lower inside the cells. Sodium and K serve a vital function in controlling osmotic pressures and acid−base equilibrium. The concentration differences between Na and K across cell membranes create an electrochemical gradient known as membrane potential. A cell's membrane potential is maintained by ion pumps in the cell membrane, particularly Na/K-ATPase pumps. These pumps use energy (ATP) to pump Na out of the cell in exchange for K. A tight control of cell membrane potential is critical for muscle contraction, cardiac function, and nerve impulse transmission. The osmotic pressure of the intracellular and extracellular fluids is rigidly controlled, largely through energy-dependent regulatory mechanisms that determine the rate of absorption of sodium ions and water by the epithelial membranes of gill, gut, integument, and kidney. The mechanisms of Na uptake by fish are complex and the subject of comprehensive investigation for FW fish (Kumai and Perry, 2012). Three models have been proposed for the molecular mechanisms of Na uptake of FW fish: (1) direct exchange of Na^+ and H^+ via Na^+/H^+ exchangers, (2) H^+-ATPase-assisted Na^+ uptake, and (3) Na^+/Cl^- cotransporter. The physiological regulation of Na by fish in FW, SW, and BW water is beyond the scope of this chapter.

Chloride ions remain mainly extracellular, as intracellular fluid already contains sufficient anions (HPO_4^{2-}, protein, etc.) to balance the K^+ and Mg^+. To maintain osmotic balance across the cell membrane, divalent ions within the cell with intracellular protein particles balance the monovalent electrolytes extracellularly. Chloride is the major anion of gastric juice and blood. In gastric juice, it is accompanied by hydrogen rather than sodium ion, as in plasma and other extracellular fluids of the body. Another specific function of the chloride ion is the "chloride shift" in blood during the transport of carbon dioxide and carbonate.

Feeding sodium chloride-enriched diets to salmonids (Zaugg and McLain, 1969; Basulto, 1976; Zaugg et al., 1983; Salmon and Eddy, 1990; Staurnes and Finstad, 2000) and Nile tilapia (Fontaínhas-Fernandes et al., 2000) for a short period prior to SW transfer facilitates acclimatization from FW to higher salinity. Dietary sodium chloride supplementation increased the gill Na/K-ATPase activity, elevated plasma chloride and Na concentrations, and increased the ability of fish to maintain electrolyte balance. The specific activity of this enzyme is greater in the gills of SW fish than in those of FW fish, and this difference has been attributed to the increased load of Na^+ in marine forms, which must be pumped across the gills from blood to the external environment. In euryhaline fish, red drum (Gatlin et al., 1992), European sea bass (Eroldagan et al., 2005), and Asian sea bass (Harpaz et al., 2005) grown in FW, dietary sodium chloride supplementation increased growth and feed utilization. One explanation for the positive response was an increase in amino acid absorption. Dietary sodium chloride supplementation has also been shown to reduce toxicity to water-borne cations, such as Cu in rainbow trout (Kamunde et al., 2005) and nitrite toxicity in catfish (Welker et al., 2011). Potassium enrichment of water was necessary for barramundi reared in hyperosmotic salinity (45‰) as compared with near isosmotic (15‰) or hyposmotic (5‰) salinities to meet the physiological requirement of K (Partridge and Lymberry, 2008).

6.4.2 Deficiency

Deficiencies of Na, K, and chloride are difficult to produce because fish derive these elements from the aquatic environment. In terrestrial animals, sodium deficiency causes growth retardation, softening of bones, corneal keratinization, gonadal inactivity, adrenal hypertrophy, changes in cellular function, and impairment in food utilization. Potassium deficiency causes overall muscle weakness, resulting in intestinal distention, weakness of cardiac and respiratory muscles, and their ultimate failures (Lall, 2002). Chloride deficiency resembles that of Na deficiency.

6.4.3 Requirement

Studies on the quantitative requirements of Na, K, and chloride have been the subject of limited investigations because fish readily absorb these elements from the surrounding water. The abundance of these three elements in feedstuffs makes their supplementation unnecessary in practical diets containing natural feed ingredients for many farmed fish. The quantitative requirement (g/100 g) for K has been reported for four species of fish, Chinook salmon, 0.8 (Shearer, 1988); channel catfish, 0.26 Wilson and Naggar (1992); hybrid tilapia, 0.2–0.3 (Shiau and Hsieh, 2001); grass carp, 0.95–1.0 (Liang et al., 2014), 0.46–0.6 (Zhu et al., 2014), and 0.4–0.74 (Chen et al., 2016); and Indian major carp, 0.36 (Musharraf and Khan, 2019a). For Chinook salmon, whole-body K saturation was reached between 0.6% and 1.2% of the diet (Shearer, 1988). Some of the differences in K requirement values of grass carp may be due to the concentration of K and other minerals in the water and size differences of fish. Studies on carp and hybrid tilapia also measured gill Na/K-ATPase activity in addition to growth parameters to determine the K requirement of fish. Fish reared in SW do not require K supplementation (Lall and Bishop, 1977; Lall, unpublished data; Sakamoto and Yone, 1978b), where the K concentration is much higher than FW.

6.4.4 Sources

Most animal protein concentrates (particularly fishmeal) are rich in sodium (0.5%–1%), whereas plant protein concentrates contain relatively low levels (0.01%–0.07%). The major proportion of Na in common feedstuffs occurs in a free ionizable form as chloride, sulfate, or phosphate or as ionizable salts of carboxylic acids in plant and animal tissues. Thus Na, in most feed ingredients, is freely soluble and assumed to be highly available. The availability of Na in bone and meat meal varies with the species because of differences in the rate of solubilization in the digestive tract. The distribution of K in feed ingredients shows a significantly different pattern than that of Na. Of the protein supplements, corn gluten meal and fishmeal are relatively poor sources of K, while soybean meal, cottonseed meal, dehydrated alfalfa meal, and cane molasses are good sources, containing 2% or more of this element. Whole cereal grains contain between 0.1% and 0.5% K. Like Na, the greater proportion of K in feed ingredients is present in a freely soluble ionic form and considered to be available biologically. Chloride distribution in feedstuffs shows a similar pattern to Na and appears to be equally available.

6.5 Copper

6.5.1 Functions and metabolism

Copper is an essential trace element for all living organisms. It is a vital component of several enzymes known as cuproenzymes that are involved in oxidation–reduction reactions and occurs tightly bound to proteins in the cell rather than as free ions. Copper metalloenzymes are involved in iron metabolism, cellular energy production (cytochrome *c* oxidase), protection of cells from free radical damage (superoxide dismutase), brain neurotransmitters (dopamine hydroxylase and peptidyl alpha amidating monooxygenase), collagen synthesis (lysyl oxidase), and melanin production (tyrosinase). Four Cu-containing enzymes, known as multicopper oxidases (MCO) or ferroxidase, oxidize Fe^{2+} to Fe^{3+} ion, the form of iron that can be incorporated onto the protein transferrin for transport to the site of red blood cell formation. The MCO family comprises the circulating and membrane-bound ceruloplasmin and two other proteins (Prohaska, 2011). Apart from the enzymatic function, Cu proteins and chelates are also involved in other metabolic roles. In the body, Cu shifts from cuprous (Cu^{2+}) to cupric (Cu^{3+}) forms, though the major proportion of body's copper is in Cu^{2+} form. The ability of Cu to easily accept and donate electrons explains its important role in oxidation—reduction (redox) reactions and in scavenging free radicals.

Uptake of Cu from water has been extensively studied particularly as it relates to toxicology (McDonald and Wood 1993; Grosell, 2012). Although fish absorb Cu via the gills and digestive tract, diet is considered a major source of Cu for optimum growth and development. The gill may contribute to a significant amount of Cu uptake depending upon the Cu concentration of the water (Taylor et al., 2003). The relative contribution of waterborne Cu to whole-body Cu content typically accounts for about 10% and increases up to 60% at a dietary Cu concentration of 0.8 mg/kg (Kamunde et al., 2002). The interaction between dietary uptake and waterborne Cu has been covered by Grosell et al. (2012). Branchial epithelia play an important role in Cu uptake by fish (Bury et al., 2003). The mechanism of Cu transport across the gill appears to involve apical transport via sodium-sensitive pathways that may involve epithelial sodium channel and a sodium insensitive mechanism (Grosell and Wood, 2002). Basolateral transfer is considered carrier mediated with involvement of a P-type ATPase (Campbell et al., 1999). In gastrointestinal Cu transport, apical Cu uptake appears to be passive with basolateral transport to be active and rate limiting (Clearwater et al., 2000; Handy et al., 2000). Unlike terrestrial vertebrates, intestinal Cu uptake in fish primarily occurs in the midposterior region of the intestine. Excessive amounts of Cu supplied in the diet do not enter the body; however, it is retained in gut tissue by metallothionein. This metallothionein-bound Cu may then be excreted into the feces through sloughing off of the epithelial membrane (Clearwater et al., 2000). The uptake of Cu from FW and SW appears to be different and is discussed in Section 6.5.4. Marine invertebrates, especially mollusks, accumulate large quantities of Cu. Hemocyanin, a blue-colored Cu-containing complex, is found in the hemolymph of mollusks and crustaceans. Copper serves as an oxygen carrier in the hemolymph of these organisms.

In terrestrial vertebrates, the liver is the main organ involved in Cu homeostasis and metabolism (Cousins, 1985). Most studies on Cu homeostasis in fish have been conducted in FW fish (Grosell, 2012) and are beyond the scope of this section. After gill uptake and absorption from the intestine, Cu is cleared from blood by the liver and incorporated into ceruloplasmin for transport to extrahepatic organs, stored in Cu—protein complexes or excreted via the bile. Ceruloplasmin in the cell and plasma is involved in iron utilization. It transports Cu as well as helps convert iron to a form that can be transported to other tissues. The ceruloplasmin differs from the iron protein, transferrin, in that it shows oxidative activity, especially toward p-phenylenediamine. The function of ceruloplasmin appears to be enzymatic, but its specific role continues to be emerging (Linder, 2016).

6.5.2 Deficiency

Gross deficiency signs of Cu have not yet been reported for fish. Copper is widely distributed in feeds and water, and therefore, its deficiency would only occur in fish under extreme conditions. Among the 14 fish species cultured in FW and SW, several biological hematological and biochemical response criteria have been used to determine the requirement and to characterize deficiency and adequacy of Cu (Table 6.3). A decrease in growth, body and liver Cu concentration, and ceruloplasmin and enzyme activities (liver CuZn SOD, heart cytochrome *c* oxidase) have been reported in several species. In a long-term study, carp fed diets containing white fishmeal without Cu supplement showed reduced growth and cataract formation (Satoh et al., 1983a). Copper deficiency is well recognized in domestic animals (Suttle, 2010). Symptoms include anemia, bone disorders, depigmentation of hair, fur, wool, or feathers, demyelination of the spinal cord, fibrosis of the myocardium, gastrointestinal disturbances, and aortic rupture. Copper toxicity arising from pollutants in natural water or produced experimentally may cause severe damage to the gills and necrotic changes in the liver and kidneys (Grosell, 2012).

6.5.3 Requirement

Since fish absorb Cu via gills and the intestine, Cu is available from both water and dietary sources. However, Cu in the rearing water alone cannot meet the requirements, and oral administration of Cu is essential for aquatic animals (NRC, 2011). The dietary Cu requirement has been reported for several fish species (Table 6.7). Copper requirement would also depend on the physiological state of the animal, Cu concentration in the water, and probably the level of elements that are metabolic antagonists of copper, such as Fe, Zn, Cd, and Mo. A metaanalysis of published information on Cu requirement of several fish species showed relatively close estimates for the Cu requirement (mg/kg diet) based on the following parameters: weight gain, 5; liver Cu concentration, 6; liver CuZn SOD activity, 4 (Antony Jesu Prabhu et al., 2016a). In rainbow trout, the antagonistic effects of Zn and Cu have not been observed (Knox et al., 1982).

6.5.4 Toxicity

Oral toxicity of Cu has been investigated in rainbow trout, Atlantic salmon, rockfish (*Sebastes schlegelii*), Nile tilapia, and African walking catfish (*Clarias gariepinus*), and the tolerance of Cu varies among fish species (Lanno et al., 1985; Berntssen et al., 1999a, 1999b; Kim and Kang, 2004; Hoyle et al., 2007; Damasceno et al., 2016). In addition to reduced growth and feed utilization, changes in hematological parameters and tissue lipid peroxidation have also been observed. Dietary Cu toxicity measured as growth inhibition in rainbow trout was approximately 664–730 mg Cu/kg feed corresponding to 44 mg Cu/kg body weight per day (Lanno et al., 1985; Clearwater et al., 2002). The maximum tolerable levels recommended by NRC (2005) for Atlantic salmon and rainbow trout were 100 and 500 mg Cu/kg diet, respectively. Relatively low concentrations of Cu (34 mg/kg) in diet caused tissue lipid

TABLE 6.7 Copper and iron requirement of certain fish.

	Copper[a,c]		Iron[b,f]	
Fish species	**mg/kg**	**References**	**mg/kg**	**References**
Atlantic salmon	5–10	Lorentzen et al. (1998)	60–100 60	Andersen et al. (1996) Naser (2000)
Rainbow trout	3	Ogino and Yang (1980)		
Channel catfish	5	Gatlin and Wilson (1986b)	30	Gatlin and Wilson, 1986a
Yellow catfish	3.1–4.2	Tan et al. (2011b)	55.7	Luo et al. (2017)
Common carp	3[d]	Ogino and Yang (1980)	147.4	Ling et al. (2010)
Gibel carp			202	Pan et al. (2009)
Grass carp	4.7–5	Tang et al., 2013		
Hybrid tilapia	4	Shiau and Ning (2003)	150–160[f] 85[g]	Shiau and Su (2003)
Japanese eel			170	Nose and Arai (1979)
Asian stinging catfish	5.2–5.7	Zafar and Khan (2020)		
Russian sturgeon	7–8	Wang et al. (2016)		
Red sea bream			150	Sakamoto and Yone (1978d)
Tongue sole	11–12	Wang et al. (2015)		
Malabar grouper	4–6 2–3[e]	Lin et al. (2008a, 2008b) Lin et al. (2010a)	100[h]	Ye et al. (2007)
Yellow croaker	3.4–7	Cao et al., 2014		
Cobia			80.5–94.7[f] 71.3–75.1[i]	Qiao et al. (2013)

[a] *NRC (2011) recommended requirement of copper (mg/kg): Atlantic salmon, 5; rainbow trout, 3; channel catfish, 5; common carp, 3;* Tilapia *sp., 5; grouper* (Epinephelus *sp.), 5.*

[b] *NRC (2011) recommended requirement of iron (mg/kg): Atlantic salmon, 30–60; channel catfish, 30; common carp, 150;* Tilapia *sp., 85.*

[c] *Unless specified* $CuSO_4 \cdot 5H_2O$ *used as Cu supplement.*

[d] $CuCl_2$ *used as Cu supplement.*

[e] *Copper peptide used as Cu supplement.*

[f] *Unless specified* $FeSO_4 \cdot 6H_2O$ *used as Fe supplement.*

[g] *Ferric citrate used as Fe supplement.*

[h] *Requirement for orange- spotted grouper* (Epinephelus coioides).

[i] *Iron methionine used as Fe supplement.*

peroxidation in Atlantic salmon (Berntssen et al., 2000). The threshold of diet-borne Cu toxicity for Atlantic salmon parr and fry was in the range of 15–17 mg Cu/kg body weight per day based on organ Cu burden and reduced growth (Clearwater et al., 2002). The current EU commission authorized maximum content of Cu in complete feed for salmonid and other fish species is 25 mg/kg.

Waterborne Cu toxicity in fish and other aquatic organisms has been subject to extensive research (Grosell, 2012; Malhotra et al., 2020). Copper absorbed through the gills accumulates in gill, intestine, liver, and other organs and causes ion regulatory disturbances. Copper toxicity in fish varies with water hardness, pH, anions, and dissolved organic carbon. Copper toxicity is more lethal in soft compared with hard waters rich in cations (e.g., Ca^{2+} and Mg^{2+}) as cations reduce bioavailability of Cu and thus toxic effects (Pagenkopf 1983). Copper is more toxic under acidic conditions ($pH < 6$). Anions and dissolved organic carbon bind to Cu and form compounds, thus reducing the toxic effects (Niyogi and Wood, 2004). The mechanism of Cu toxicity differs in FW and SW. In SW, waterborne Cu affects not only gills but also the intestine, since SW fish drink to compensate for water loss into the surrounding aquatic environment. An exposure to higher levels of Cu in water can also affect the performance of the whole organism by decreasing metabolic rate and causing oxidative stress (Beaumont, 2000; McDonald and Wood, 1993). The cause of reduced oxidative metabolism appears to be gill damage, manifested as either disruption of branchial structure, secretion of mucus that binds metals and impedes the rates of diffusion, inhibition of respiratory enzymes, and damage to gill oxygen receptors (Grosell, 2012). Oxidative damage from Cu exposure has been observed as changes in biomarkers (e.g., protein carbonyls, lipid peroxidation, and DNA damage products) in the gill, liver, and intestine of numerous fish species (McDonald and Wood, 1993). Salinity appears to protect liver and gill tissues of kill fish (*Fundulus heteroclitus*) against bioaccumulation of Cu, but whether it also protects against Cu-induced toxicity on oxidative stress and metabolic response is not clear (Ransberry et al., 2015).

6.5.5 Sources and bioavailability

Processed feed ingredients of plant and animal origin show a variable Cu content resulting from contaminants. Moisture during processing of materials rich in protein provides a favorable condition for the uptake of Cu by the protein fraction. High concentrations of Cu (>85 mg/kg) have been detected in condensed fish solubles and whey products. The Cu content of cereal grain ranges from 5 to 20 mg/kg. Most animal and vegetable protein supplements contain 5–30 mg Cu/kg. While it is easy to measure total levels of Cu in major ingredients, ascertaining the bioavailability of this element is a complex procedure.

Several Cu compounds are commercially available as feed additives to supplement fish and animal feeds including cupric sulfate pentahydrate; cupric chloride, dihydrate; cupric carbonate, monohydrate; cupric acetate, monohydrate; cupric methionate; cupric oxide; cupric chelate of amino acid hydrate; copper lysine sulfate; cupric chelate of glycine hydrate; copper chelate of hydroxy analog of methionine; dicopper chloride trihydroxide (TBCC); and copper bislysinate. Efforts are now being made to minimize the supply of Cu and other minerals in feeds as a means of reducing excretion in the aquatic environment. Organic forms of trace elements show potential to reduce Cu excretion; however, limited effort has been made to measure the bioavailability of Cu from inorganic and organic forms or the residual Cu present in feed ingredients. Studies conducted on bioavailability from different organic Cu sources in animal feeds show similar bioavailability, while others show higher bioavailability relative to cupric sulfate (NRC, 2005).

6.6 Iron

6.6.1 Functions and metabolism

Iron is an essential element for fish and other vertebrates. It is an integral component of proteins and enzymes that support essential biological functions, such as oxygen transport, energy production, immune function, and DNA synthesis. Iron is found in the body mainly in complex forms bound to protein such as heme compounds (hemoglobin and myoglobin), several heme enzymes, and nonheme compounds (transferrin, ferritin, and iron-containing flavoproteins). Hemoglobin and myoglobin are involved in oxygen transport and storage. Heme enzymes are of fundamental importance in vital respiratory processes, namely cytochrome system (e.g., cytochromes *a, b, f*; cytochrome *c* oxidase) and/or with oxidases activity (e.g., sulfite oxidase, cytochrome P450 oxidases, myeloperoxidase, peroxidases, catalase, endothelial nitric oxide synthase, cyclooxygenase). Iron-sulfur cluster proteins with oxidoreductase activities are involved in energy metabolism (e.g., succinate dehydrogenase, isocitrate dehydrogenase, NADH dehydrogenase, aconitase, xanthine oxidase, ferredoxin-1) or involved in DNA replication and repair (DNA polymerases, DNA helicases). As a transition metal, Fe also has useful ligand-binding and redox properties. The interconversion of Fe^{2+} and Fe^{3+} (ferrous to ferric) in the presence of oxygen is an important chemical property of this element in the biological redox processes. However, the formation of the reduced form of iron, Fe^{2+}, can produce highly reactive hydroxyl and lipid radicals, which can damage lipid membranes, nucleic acids, and proteins. Therefore, Fe can be both toxic and beneficial to organisms, and Fe status in the body must be carefully regulated to provide sufficient Fe for biological functions, while limiting excess Fe^{2+} which can lead to oxidative stress.

Iron absorption from the digestive tract in teleost fish is considered similar to that in mammals (Bury et al., 2003; Bury, 2012). In the acidic environment of the stomach, ferric ion is released from ingested food materials and binds to mucin, which may facilitate metal solubility in the small intestine of fish (Whitehead et al., 1996). Gills of teleost fish also play an important role in Fe acquisition (Bury and Grosell, 2003), especially in developing fish without functional digestive tracts (Andersen et al., 1997). Iron uptake across gills is considered less than uptake across the intestine, which suggests the latter is the predominant route of Fe absorption (Cooper et al., 2006). The presence of reducing substances in the diet (e.g., ascorbic acid) can enhance the ability of fish to absorb Fe. The addition of ferrous sulfate to water may also enhance growth and hemoglobin level in certain warm-water fish (*Xiphophorus helleri* and *Xiphophorus maculatus*) (Roeder and Roeder, 1966). The amount and chemical form of Fe ingested, and the amount and proportion of both inorganic and organic components present in the diet can also affect the absorption of Fe in fish. Absorption of Fe is influenced by the age, state of health, iron status, and conditions within the gastrointestinal tract of the fish.

Transferrin is the transport form of body Fe and performs the role of regulating Fe distribution within the body. Ferritin is an intracellular protein and a primary storage form of body Fe that can contain up to 20% Fe. Absorbed Fe is transported in the blood by transferrin, which has been identified in several fish species (Ikeda et al., 1973; Hershberger, 1981). Hagfish transferrin has a similar structure to humans (Aisen and Leibman, 1972). Molecular and physiological studies indicate that apical Fe transport is achieved in the Fe^{2+} state, via a proton-dependent divalent metal transporter (DMT1, also known as Nramp2 (SLC11A2) (Kwong et al., 2009, 2013). Iron-regulated transporter ferroportin (SLC40A1) is involved in basolateral transport (Cooper et al., 2006). One of the key regulators of systematic Fe homeostasis is hepcidin, a small cysteine-rich antimicrobial peptide synthesized by liver cells in vertebrates (Neves et al., 2016); however, hepcidin homologs, hamp 1 and hamp 2, have been found in several fish species (Hilton and Lambert, 2008; Neves et al., 2016). Studies on sea bass showed that after experimental induction of anemia, the liver assumes the main role of Fe metabolism through hepcidin, whereas spleen and the kidney increase their erythropoietic activity, and the intestine enhances Fe absorption (Neves et al., 2016). Each hepcidin type appears to play a diverse role in response to anemia, with different transcriptional responses. Hamp1 assumes a role as the major regulator of Fe homeostasis when enhanced erythropoiesis is required, with a possible secondary role for hamp2 in severe anemia.

6.6.2 Deficiency and toxicity

Iron deficiency is caused from insufficient Fe intake, poor iron absorption, transport or storage, significant blood loss, and prevalence of certain infectious diseases (Lall, 2010). Anemia is the most common blood disorder resulting from an abnormally low level of red blood cells or hemoglobin. It causes a decrease in blood oxygen carrying capacity, which may eventually lead to lower energy production, cell proliferation and repair, and muscular activity. Severe oxygen deprivation in the brain may cause death. Signs of Fe deficiency have been experimentally produced in certain fish fed low-iron diets (Table 6.3). Iron deficiency causes characteristic microcytic anemia or low hemoglobin levels in brook trout (Kawatsu, 1972), rainbow trout (Desjardins et al., 1987), Atlantic salmon (Bjornvic and Maage, 1993; Andersen et al., 1996), red sea bream (Sakamoto and Yone, 1976a; Sakamoto and Yone, 1978d), yellowtail (Ikeda et al., 1973), eels (Nose and Arai, 1979), and carp (Sakamoto and Yone, 1978c). Ferrous sulfate, ferric chloride, ferric citrate, ferric choline citrate, and ferric ammonium citrate have been effective in preventing Fe deficiency anemia in fish and animals. Iron compounds with low solubility, such as ferric oxide, were ineffective for animals (Henry and Miller, 1995). In most cases, the growth of fish was not influenced by Fe deficiency. The normal liver color changed to yellowish-white during Fe deficiency in carp (Sakamoto and Yone, 1978b). In catfish, Fe deficiency suppresses hemotocrit, hemoglobin, and plasma iron levels and causes transferrin saturation (Gatlin and Wilson, 1986a).

Iron toxicity can occur from excessive Fe exposure of fish in water and high levels of dietary Fe intake, which may interfere with Fe homeostatic regulation by causing Fe overload in tissues (Bury et al., 2012). Dietary Fe toxicity signs develop in rainbow trout fed greater than 1380 mg Fe/kg (Desjardins et al., 1987). The major effects of Fe toxicity include reduced growth, poor feed utilization, feed refusal, increased mortality, diarrhea, and histopathological damage to liver cells. In aqueous environments, iron naturally exists as a soluble ferrous (Fe^{2+}) and an insoluble ferric particulate iron (Fe^{3+}). In oxygenated waters, soluble Fe^{2+}-oxidizes to Fe^{3+}, and in circumneutral waters ($pH > 6.5$), Fe^{3+} ions are insoluble and rapidly precipitate as hydroxides and oxyhydroxides. Iron toxicity in water is closely related to Fe speciation and the interaction of Fe with body and gill surfaces. Excess iron in the water is known to cause respiratory disorders in the gills. Dietary Fe toxicity signs develop in rainbow trout fed higher than 1380 mg Fe/kg (Desjardins et al., 1987). Major effects of Fe toxicity include reduced growth, poor feed utilization, feed refusal, increased mortality, diarrhea, and histopathological damage to liver cells.

6.6.3 Requirement

A wide range of Fe requirement values (30–245 mg/kg diet) have been reported for several fish species (Table 6.7). Response criteria used to estimate Fe requirement were weight gain, whole body, liver and plasma Fe concentration, and hematological parameters (hemoglobin, hematocrit, mean corpuscular volume, mean corpuscular hemoglobin). A metaanalysis of published information on Fe requirement of several fish species showed estimates for Fe requirement ranging from 58.8 to 166.4 mg/kg with a wide variation for all parameters tested (Antony Jesu Prabhu et al., 2016a). It should be emphasized that weight gain alone may not provide a complete estimate of Fe requirement. Most of the Fe body pool is in the form of hemoglobin in red blood cells. At later stages of growth and maturity, Fe requirement is likely to change as the blood volume declines and a lower rate of iron deposition occurs in tissues.

6.6.4 Sources and bioavailability

Food is considered the major source of Fe for metabolic purposes. The amount and chemical form of Fe ingested, and the amount and proportion of both inorganic and organic components present in the diet can also affect Fe absorption in fish. The presence of reducing substances in the diet (e.g., ascorbic acid) can enhance the ability of fish to absorb Fe. Iron absorption is influenced by the age, state of health, Fe status, and conditions within the gastrointestinal tract of fish. The concentration of Fe in common feedstuffs is highly variable and greatly influenced by the degree of contamination from ferrous metal during processing. Feeds of animal origin, other than milk by-products, are rich sources of Fe. Fishmeal and meat meal contain approximately 150–800 mg Fe/kg. The Fe content of cereal grain varies from 30 to 60 mg/kg, whereas oilseed protein may contain 100–200 mg Fe/kg. Certain feed-grade calcium phosphates and limestones may contain 2000 mg Fe/kg.

Little is known about the form of Fe found in common feedstuffs and their bioavailability to fish. In animal protein supplements, Fe may be present as iron porphyrin, myoglobin, and hemoglobin. In cereal grains, a small proportion of Fe occurs as an iron–phytin complex. Utilization of ferrous sulfate and ferric chloride was essentially the same. In red sea bream, ferrous and ferric chloride are more efficiently utilized than ferric citrate (Sakamoto and Yone, 1979b). Biological availability of Fe measured by a hemoglobin regeneration assay in Atlantic salmon showed that the relative availability of Fe from ferric chloride, ferric oxide, blood meal, and herring meal was 98.8%, 17.8%, 52.3%, and 47.1%, respectively (Naser, 2000). However, Andersen et al. (1997) observed that heme Fe from blood meal was more efficiently utilized than ferrous sulfate in Atlantic salmon using liver Fe concentration and hemoglobin slope ratio methods. For grouper, bioavailability of iron based on hematological index and catalase activity was higher for Ferric hydroxy methionine analog than ferrous sulfate and Ferric citrate (Huang et al., 2018). Ferric methionine also showed higher bioavailability based on catalase activity than ferrous sulfate in cobia (Qiao et al., 2013).

It is well known that in animals, bioavailability of Fe is affected by the composition of test diets, processing methods used to produce blood meal, Fe concentration in the diet, age and species of test animals, intake of Ferelative to the need, chemical form in which Fe is supplied from other dietary components, and the amount and proportion of other dietary components with which Fe interacts metabolically (Forbes and Erdman, 1983).

6.7 Manganese

6.7.1 Function and metabolism

Manganese is an essential element, but its function and metabolism in fish is poorly understood. Physiological functions, metabolism, and toxicity of Mn in animals and humans are well studied (Keen et al., 1999; Aschner and Aschner, 2005). The highest concentration of Mn is found in bone, and significant amounts are present in liver, muscle, kidney, gonadal tissues, and skin. Manganese is a cofactor for enzymes involved in protein and energy metabolism, antioxidant action, and mucopolysaccharide synthesis. These enzymes include metalloenzymes, such as Mn-dependent superoxide dismutase, pyruvate carboxylase, and arginase, and the Mn-activated enzymes, such as glycosyl transferase, glutamine synthetase, farnesyl pyrophosphate synthetase, and phosphoenolpyruvate carboxykinase (Leach et al., 1997). Manganese can activate numerous other enzymes such as kinases, transferases, hydrolases, lyases, ligases, and decarboxylases, which are also activated by Mg and other metals. Through involvement of these metalloproteins, Mn plays an important role in growth, development, reproduction, antioxidant defense, energy production, immune response, and regulation of brain functions.

Uptake of Mn from FW has been demonstrated from gills and gastrointestinal tract (Miller et al., 1980; Rouleau et al., 1995); however, the mechanism of Mn uptake from either gills or gut is not clearly defined. Homeostatic regulation of Mn in animal tissues is primarily through biliary excretion, which may constitute the main regulatory role of this trace element (Keen et al., 1999). Excretion of Mn in bile and feces has also been observed in Atlantic salmon (Antony Jesu Prabhu et al., 2018a). Urinary excretion of Mn is relatively small and plays a negligible role in Mn homeostasis. The absorption from gastrointestinal tract and gills is considered an important factor in Mn homeostasis in fish. Divalent Mn ion entering in the circulation system is removed rapidly by the liver. It is well recognized that Mn is efficiently absorbed from diet, but the absorption may be reduced by high levels of Ca and P (Lall, 2002). High concentrations of Mn in an aquatic environment induce sodium imbalance, reduce the absorption of Ca and P, affect carbohydrate metabolism, and impair the immunological functions of fish (Ye et al., 2009; Partridge and Lymbery, 2008). It also causes oxidative stress, tissue damage, inflammation, neurodegeneration, and disruption of homeostasis of other metals in fish (Vieira et al., 2012).

6.7.2 Deficiency

Manganese deficiency (Table 6.3) causes reduced growth and skeletal abnormalities in rainbow trout, carp, and tilapia (Ishak and Dollar, 1968; Ogino and Yang, 1980; Yamamoto et al., 1983). Other deficiency signs in rainbow trout and carp include cataracts and dwarfism associated with disturbances in bone metabolism (Satoh et al., 1992). In several fish species (Atlantic salmon, rainbow trout, tilapia, cobia, gibel carp, yellow catfish), low Mn intake decreases the activities of Mn-dependent SOD in either liver or cardiac muscle, and it also suppressed Mn concentration of vertebrae. Liver Mn-SOD activity was not influenced in catfish fed a diet containing 2.4 mg Mn/kg (Gatlin and Wilson, 1984a). The reduction in Mn-SOD activities has been correlated with high levels of mitochondrial lipid peroxidation in animals (Zidenberg-Carr et al., 1983). In broodstock rainbow and brook trout, a fishmeal-based diet without Mn supplement caused poor hatchability and low Mn levels of eggs (Takeuchi et al., 1981). High concentrations of dietary Mn supplementation (1 g/kg) caused changes in feeding behavior, decrease in body Fe concentration, and elevation of Zn concentration in vertebrae and body of grouper (Ye et al., 2006). Manganese is considered to be one of the least toxic of the essential elements (NRC, 2005). There are no important studies reporting toxic effects of an oral dose of Mn in fish, but exposure studies to inorganic Mn in water have shown hematological and histopathological changes in certain warm water fish (NRC, 2005).

6.7.3 Requirement

Manganese requirements (mg/kg diet) of certain FW, and marine fish species have been determined using semipurified experimental diets for most juvenile fish (Table 6.6). The Mn requirement of marine fish species (grouper, cobia, turbot, yellow croaker) has ranged between 12 and 25 mg/kg and Atlantic salmon reared in SW, 14–22 mg/kg of diet. When three different forms of Mn were used to investigate the Mn requirement of cobia, the requirements were 15.4, 11.2, and 10.5 mg Mn/kg for manganese sulfate, Mn-glycine, and Mn-2-hydroxy-4-(methylthio) butyrate, respectively. The response criteria used to determine Mn requirement of fish in various studies include growth, feed efficiency, deficiency signs, vertebrae Mn concentration, and Mn-dependent SOD activities in liver and cardiac muscle. The supplementation of Mn in either purified or fishmeal diet has not shown consistent positive growth responses in all fish species. In addition to growth and biochemical parameters, vertebrae ash content must be considered to estimate quantitative Mn requirement of fish. A metaanalysis of Mn requirement determined for several fish species based on estimates of weight gain, whole body Mn, and vertebrae Mn content were 10.7, 13.4, and 18.4 mg/kg, respectively (Antony Jesu Prabhu et al., 2016a). Although the Mn requirement of juvenile fish may range between 2 and 25 mg/kg diet, broodstock fish may require higher amounts of this element (Lall, 2002). Higher amounts of P in the diet may also lower Mn absorption and thus increase the need for supplemental Mn for fish.

6.7.4 Sources and bioavailability

Protein supplements of animal origin are generally low in Mn. Whole herring and capelin meal are particularly low (4–12 mg/kg). Manganese levels of 4–38 mg/kg are common in other fishmeals. Wheat middling, rice bran, alfalfa meal, and corn distillers' dried soluble are the richest plant sources of Mn. Corn is particularly low in Mn (4–11 mg/kg), whereas other cereal grains have contents ranging from 8 to 50 mg/kg.

Manganese availability differs in various inorganic salt supplements. Manganese in manganous oxide is poorly utilized by rainbow trout and Atlantic salmon (Watanabe et al., 1997; Lall, 2002). The most commonly used source of Mn in fish feeds is manganous sulfate, monohydrate ($MnSO_4 \cdot H_2O$). Other forms of Mn supplements include manganous chloride, tetrahydrate ($MnCl_2 \cdot 4H_2O$), manganous oxide (MnO), and manganese chelates of amino acids or glycine. The availability of Mn is low in manganous carbonate for carp (Satoh et al., 1987a). Mn-glycine was effectively utilized by cobia, turbot, and Atlantic salmon reared in SW (Ma et al., 2015; Nie et al., 2016; Antony Jesu Prabhu et al., 2019a). Mn-glycine was better utilized by cobia and turbot than manganous sulfate (Ma et al., 2015; Nie et al., 2016). Bioavailability of Mn-methionine and Mn-2-hydroxy-4-(methylthio) butyrate was relatively high for turbot (Ma et al., 2015) and cobia (Nie et al., 2016), respectively. Some challenges in evaluation of inorganic and organic Mn supplements in fish diets have been recently investigated (Antony Prabhu Jesu et al., 2019b; Silva et al., 2019, 2020). Phytic acid like other divalent ions (Zn^{2+}, Cu^{2+}, and Fe^{2+}) binds Mn. Dietary supplementation of 15 mg/kg of Mn-Gly caused lower digestibility of Zn and Cu (Antony Jesu Prabhu et al., 2019a; Silva et al., 2019). The maximum total content of Mn approved for complete fish feed in Europe by the European Food Safety Authority is 100 $mgkg^{-1}$. The upper limit of Mn may vary in other countries.

6.8 Selenium

6.8.1 Functions and metabolism

Selenium is an essential trace element for fish and other vertebrates that are chemically similar to sulfur and can replace S in cysteine in certain enzymes. Selenium exists in four natural oxidation states (−2, 0, +6, +4). At physiological pH, the primary form found in tissues is Se^{2+} in selenocysteine. In nature, inorganic Se is present in four oxidation states: selenate, selenite, elemental Se, and selenide. These forms are converted by all biological systems into more bioavailable organic forms, mainly as the two seleno-amino acids selenocysteine (SeC) and selenomethionine (SeMet). The biochemical role of Se puzzled researchers until the discovery that selenium is an integral part of glutathione peroxidase (Rostruck et al., 1973).

Several physiological roles and health benefits of Se have been made since the discovery that the amino acid residue in proteins, selenocysteine (SeC), is a major form of Se in the cell and that Sec is encoded by the UGA codon (Papp et al., 2007). A comprehensive study for the identification and comparative analysis of vertebrate selenoproteomes has shown more than 45 selenoproteins in mammals and, among bony fish, 38 selenoproteins in zebrafish (Mariotti et al., 2012). Expression of selenoenzyme mRNA and activity is largely regulated by dietary Se level or body Se status. Selenium is incorporated into functional and structural proteins as SeC. Several of these proteins are enzymes that provide cellular antioxidant protection. Among the selenoproteins, five glutathione peroxidases, three thioredoxin reductases, three iodothyronine deiodinases, and one methionine sulfoxide reductase B1 are among the best characterized selenoproteins with known functions (Labunsky et al., 2014; Burk and Hill, 2015). Selenium as an integral part of glutathione peroxidase (GPx) and thioredoxin reductase interacts with other nutrients (e.g., α-tocopherol) that affect redox status (i.e., prooxidant and antioxidant balance). Glutathione peroxidase destroys hydrogen peroxide and hydroperoxides by using reducing equivalents from glutathione, thereby protecting cells and membranes against oxidative damage. Other trace elements that are critical components of antioxidant enzymes include Cu (as SOD), Zn (as SOD), and Fe (catalase). At the metabolic functional level, Se is closely associated with vitamins E and C, polyunsaturated fatty acids, Fe, S-containing amino acids, and iodine. In the past decade, significant progress has been made on the role of selenoproteins in immune function, metabolism, and pathogenesis of certain diseases and gene expression under different physiological conditions. The function of Se in these areas in fish is poorly understood.

Although fish can absorb limited amounts of Se from the environment via the gills and skin under certain conditions, the gastrointestinal tract is the primary site for Se absorption. Low concentrations of selenium are found extensively in aquatic ecosystems. The uptake of Se as selenite across gills is very efficient at low waterborne concentrations (Hodson et al., 1986). Selenocysteine and selenomethionine are likely to be absorbed by an active amino acid transport mechanism, whereas selenite is absorbed by simple diffusion and selenate by a sodium-mediated carrier shared with sulfate (Burk and Hill, 2015). Absorbed Se is associated with proteins in the plasma and transported to tissues. Fish and other vertebrates excrete Se via feces, but urine is the primary excretion route and likely plays a quantitatively important role in Se homeostasis (Janz, 2012; Burk and Hill, 2015).

Beneficial effects and toxicity of dietary Se supplementation in fish and animals are well documented (NRC, 2005, 2011; Suttle, 2010; Hosnedlova et al., 2017). The effects of dietary Se also show an increase in expression of selenoprotein P in rainbow trout and zebrafish (Fontagné-Dicharry et al., 2015; Pacitti et al., 2015; Penglase et al., 2015). In Se-adequate animals, the kidney and liver have the highest Se content. Muscle has moderate levels of Se content but accounts for the largest pool of body Se. Retention and distribution of Se in tissues are affected by the body Se status and chemical form of Se (Berntssen et al., 2017). Selenium-deficient animals retain Se more efficiently than Se-adequate animals. Dietary Se supplementation showed an increase in expression of selenoprotein P in rainbow trout and zebrafish (Fontagné-Dicharry et al., 2015; Pacitti et al., 2015; Penglase et al., 2015).

Dietary SeMet supplementation increased muscle and serum Se levels to a higher level than inorganic selenite due to differences in the metabolism of inorganic and organic forms of Se in fish (Kücükbay et al., 2009; Rider et al., 2010; Fontagne-Dicharry et al., 2015; Godin et al., 2015). Selenomethionine can be directly and not specifically incorporated into any protein containing methionine, while selenite must react with glutathione (GSH) to form hydrogen selenide (H_2Se) before incorporation into specific selenoproteins (Suzuki, 2005; Godin et al., 2015). Selenium concentrations in different tissues can be highly variable among species (Janz et al., 2012). Catabolism of SeMet and selenocysteine releases reduced form of inorganic Se as selenide (HSe^-), which can be reincorporated into selenoproteins, or can be methylated to form the excretory forms methyl selenol, or CH_3SeH, dimethyl selenide, or $(CH_3)_2Se$, trimethyl selenium ion, or $(CH_3)_3Se^+$, and 1β-methylseleno-N-acetyl-D-galactosamine ($(CH_3)_2SeGa1N$). With excess intake, both SeMet and hydrogen selenite were excreted into urine after methylation in the liver using s-adenosylmethionine (SAM) as methyl donor (Suzuki, 2005).

6.8.2 Deficiency

Limited research has been undertaken to characterize Se deficiency in fish (Table 6.3). Selenium deficiency causes growth depression in rainbow trout (Hilton et al., 1980), carp (Satoh et al., 1983b), and catfish (Gatlin and Wilson, 1984c), but Se deprivation alone did not produce any pathological signs in these fish. Both Se and vitamin E are required to prevent muscular dystrophy in Atlantic salmon (Poston et al., 1976) and exudative diathesis in rainbow trout (Bell et al., 1985). Glutathione peroxidase activity in liver and serum or plasma decreases during Se deficiency (Poston et al., 1976; Hilton et al., 1980; Gatlin and Wilson, 1984a; Bell et al., 1985). In animals, it is very difficult to produce or distinguish symptoms of Se deficiency alone from that of a Se-vitamin E deficiency (Suttle, 2010).

Although effects of selenium deficiency on reproduction of animals are well documented (Suttle, 2010; Hosnedlova et al., 2017), the results obtained on reproductive performance of female and milt quality of fish are not clear. Recently, Wischhusen et al. (2019) found that Se-supplementation of diets based on a major proportion of plant ingredients enhanced the total number of spawning fish and higher levels of hydroxy methionine supplementation led to earlier spawning. There was no evidence of transfer of maternal Se to progeny, and supplementation of organic Se increased activity of GPx activity and mRNA expression of other proteins involved in antioxidant protection at the cellular levels as well as elevated tissue vitamin C and E concentrations.

6.8.3 Requirement

The minimum selenium requirement of fish varies with the form of Se ingested, Se availability in the diet, form of Se (inorganic or organic), vitamin E content of the diet, and concentrations of waterborne selenium. Selenium requirements (mg Se/kg diet) of several fish species determined on the basis of optimum growth and maximal liver and serum or plasma glutathione peroxidase activity are summarized (Table 6.8). The estimated Se requirement based on weight gain, whole-body Se retention, and liver GSH-Px activities have shown different values for some fish species. A metaanalysis of published information on Se requirements of several fish species showed different estimates for the Se requirement (mg/kg) based on different parameters: weight gain, 0.35; enzyme activities of liver GPx, 0.78; serum GPx, 0.43; and liver GR activity, 0.41 (Antony Jesu Prabhu et al., 2016a). In a recent study, the estimated requirement of Atlantic salmon smolts based on available Se in plant ingredient diet was much lower (0.27 mg/kg). Supplementation of SeMet as compared with sodium selenite has shown lower Se requirements for Atlantic salmon smolts (Antony Jesu Prabhu et al., 2020) and gibel carp (Zhu et al., 2016). In addition to differences between the bioavailability of Se from the experimental diets, Se uptake from water, age, dietary vitamin E levels and Se bioavailability, and differences among fish species in the utilization of Se should be considered when published requirement values are applied in feed formulation.

6.8.4 Toxicity

Prior to the discovery of the essential role of Se, its toxic properties were well recognized. Certain geographic areas are seleniferous and produce plants with high selenium content.

TABLE 6.8 Zinc and selenium requirement of certain fish.

	Zinc[a,c]		Selenium[b,d]	
Fish species	mg/kg	References	mg/kg	References
Atlantic salmon	37–67	Maage and Julshamn (1993)	0.27[e]	Antony Jesu Prabhu et al. (2020)
Rainbow trout	15–30 30.1	Ogino and Yang (1978) Welker et al. (2016)	0.15–0.38	Hilton et al. (1980)
Channel catfish	20	Gatlin and Wilson (1983)	0.25 0.2–0.3	Gatlin and Wilson (1984c) Wang and Lovell (1997)
Yellow catfish	17.1–20.9	Luo et al. (2011)		
Common carp	15	Ogino and Yang (1979)		
Jian carp	43.2–48.7	Tan et al. (2011a)		
Grass carp	55.1	Liang et al. (2012d)		
Gibel carp			1.18[f] 0.73–1.19[f]	Han et al. (2011) Zhu et al. (2016)
Indian major carp	47.8–52.9	Musharraf and Khan (2019b)		
Hybrid tilapia	26–29 105–115	Lin et al. (2008c) Li and Huang (2015)		
Blue tilapia	20	McClain and Gatlin (1988)		
Nile tilapia	30 37.2	Eid and Ghonim (1994) Huang et al. (2015)		
Largemouth bass			1.60–1.85	Zhu et al., 2011
Russian sturgeon	28.2–34.6	Moazenzadeh et al. (2017)		
Red drum	20	Gatlin et al. (1991)		
Blunt snout sea bream	52.1–86.2	Jiang et al. (2016)		
Gilthead sea bream			0.94	Dominguez et al. (2019)
Black sea bream			0.86[g]	Wang et al. (2019)
Malabar grouper	28.9–33.7	Houng-Yung et al., 2014	0.7[f]	Lin and Shiau (2005)
Cobia	42.9	Xu et al. (2007)	0.8[f]	Liu et al. (2010a)

[a]*NRC (2011) recommended requirement of Zn (mg/kg): Atlantic salmon, 37; rainbow trout, 15; channel catfish, 20; common carp, 15;* Tilapia *sp., 20; hybrid striped bass, 37; grouper* (Epinephelus *sp.), 20.*

[b]*NRC (2011) recommended requirement of Se (mg/kg): rainbow trout, 0.15; channel catfish, 0.25; hybrid striped bass, 0.35; grouper* (Epinephelus *sp.), 0.7.*

[c]*Unless specified $ZnSO_4 \cdot 5H_2O$ used as Zn supplement.*

[d]*Unless specified Na_2SeO_3 used as Se supplement.*

[e]*Based on available Se.*

[f]*Se-methionine used as Se supplement.*

[g]*Se-polysaccharide used as Se supplement.*

Animals grazing on these alkali pastures show several toxicity syndromes such as blind staggers. Signs of Se toxicity are extensively described (NRC, 2005; Suttle, 2010). Three possible mechanisms of Se toxicity have been proposed: (1) substitution of Se for sulfur in important biochemical reactions and structures (e.g., disulfide bonds may disrupt normal function and cell integrity); (2) reaction between selenite and glutathione depletes cellular-free and protein-bound thiol levels, thus affecting the activities of certain enzymes; (3) free radicals such as superoxide anions produced by the reactions of certain forms of Se with tissue thiols may cause oxidative injuries to tissues. Any single model may not fully explain all modes of action due to the diverse chemical properties and metabolism of various Se compounds. High levels of selenium exert their toxic effects in animals probably through competition with sulfur compounds or a strong affinity for sulfur in the formation of S–Se complexes. However, the precise biochemical mechanism involved in Se toxicity remains to be established. Diets high in S or protein provide some protection against selenium toxicity.

The toxicity of both dietary and waterborne selenium has been experimentally produced in fish (Hodgson and Hilton, 1983). An important aspect of Se toxicity is the relatively low narrow range between dietary requirement and toxicity (Hamilton, 2004; Berntssen et al., 2018). A detailed account of Se toxicity in fish related to aquatic environment has been comprehensively reviewed (Hamilton, 2004; Janz et al., 2012). Selenium toxicity occurs in rainbow trout and catfish when dietary Se exceeds 13 and 15 mg/kg dry feed, respectively (Hilton et al., 1980; Gatlin and Wilson, 1984c). Reduced growth, poor feed efficiency, and high mortality were the major effects of selenium toxicity in juvenile fish. Other adverse effects of feeding higher level of selenite and SeMet in fish diets include reduced growth, decrease in energy retention, lower egg viability, reduced neurological and immunological functions, pathological changes in liver, kidney, and ovaries, and skeletal deformities. For the assessment of early sublethal adverse effects of selenite and SeMet toxicity, several biomarkers such as tissue lipid peroxidation, reduced glutathione (oxidative stress marker), and changes in lipid composition have also been proposed (Berntssen et al., 2017). Elevated Se levels for rainbow trout, Chinook salmon, fathead minnow, striped bass, bluegill, and razorback sucker ranged from 2.4 to 70 mg Se/kg in feed and 47 to 472 μg/L of water (Hamilton, 2004). In most cases, reduced growth or survival occurred at dietary Se levels close to 3 mg Se/kg. Atlantic salmon tolerated either 1–2 or 3 mg Se/kg of selenite and SeMet supplementation, respectively, in a feed based on a high proportion of plant ingredients that contained 0.45 mg Se/kg (Berntssen et al., 2018). Relative toxicity of different Se supplements may be largely related to their solubility in water and nutrient bioavailability and can be modulated by dietary factors such as protein, sulfate, vitamin E, and a number of trace elements including As, Cu, and Hg. The maximum limit for total Se in animal feeds, including fish, has been set at 0.5 mg Se/kg feed by the U.S. Food and Drug Administration and European Union.

Susceptibility of different fish species to Se toxicity varies greatly. Coho salmon are more sensitive than Chinook salmon to inorganic Se (Hamilton and Buhl, 1990). The elimination of fish species from contaminated aquatic communities is highly selective (Hamilton, 2004). As in the case of trace element bioavailability, relative toxicity of a given Se compound is affected by its chemical form and solubility. Highly insoluble elemental Se is much less toxic to many species than other more soluble forms such as selenite and selenate (NRC, 2005; Janz, 2012). Seleno-methionine was less toxic to Chinook salmon and coho salmon than selenite or selenate (Hamilton and Buhl, 1990). However, the opposite was true in other studies (Hamilton, 2004). Liver and kidneys play an important role in the excretory process in trout, in which major excretory routes appear to be gills and urine. The route, amount, and time course of Se excretion are affected by levels of Se intake, form of Se, body Se status, and dietary levels of Cu, Pb, Hg, and As (NRC, 2005).

Oxidative stress has been identified as a main cause for excess dietary Se exposures in fish. In Atlantic salmon fed high levels of selenite and SeMet yeast, oxidative stress was a main driver for Se toxicity (Berntssen et al., 2017); however, white sturgeon and brown trout fed high levels of organic Se did not show oxidative stress (Knight et al., 2016; Zee et al., 2016). Altered liver lipid synthesis and metabolism have been shown as the central mechanism in dietary organic Se toxicity in rainbow trout (Knight et al., 2016; Pacitti et al., 2016). Recent wide-scope pathway assessments by the application of metabolomics techniques show that disturbance in lipid metabolism is an important factor in inorganic and organic Se toxicity (Berntssen et al., 2017).

Several trace elements including antimony (Ab), As, bismuth (Bi), Cd, Cu, germanium (Ge), Hg, silver (Ag), and tungsten (W) affect Se utilization and thus attenuate Se toxicity in fish, mammals, and birds (Hamilton, 2004; Janz, 2012). The protection of As compounds against Se toxicity has been shown in various species fed different forms of Se (Hamilton, 2004). A possible mechanism is the enhanced biliary excretion of Se by As. A correlation between liver Se and Cu has been observed in rainbow trout and Atlantic salmon (Hilton, 1989). Selenium and Cu interaction reduces metabolically active forms of these elements by formation of a Se–Cu complex. High levels of Cu reduced Se retention in tissues of fish but had no biological effect (Hamilton, 2004).

6.8.5 Sources and bioavailability

The chemical form of Se distributed in feed ingredients, foods, and Se feed supplements (organic and inorganic) affects the bioavailability and metabolism of this element, which has been extensively studied in animals (Henry and Ammerman, 1995) and humans (Thiry et al., 2012). Most of the Se in feed ingredients occurs largely as seleno-amino acids with SeMet being the most predominant form. Inorganic forms of Se, selenite (SeO_3^{2-}) or selenite (SeO_4^{2-}), and organic forms (selenized yeast and SeMet) are used to supplement fish feed. Selenium content of feedstuffs from plant origin varies widely due to the level and uptake of Se in the soil of various geographical locations. In cereal grains, it may range from <0.1 to >0.8 mg/kg. Fishmeals are the best natural sources of Se among the common feedstuffs with concentrations ranging from 1 to 2.4 mg/kg with the exception of tuna and mackerel meals where concentrations may exceed 5 mg/kg. Selenium is widely distributed in small concentrations in FW (0.1–0.3 μg/L) and SW (0.052–0.2 μg/L) (NRC, 2005).

Several criteria have been used to determine the bioavailability of Se including GSH-Px activity, tissue Se concentration, and prevention of Se deficiency. Activity of GSH-Px in plasma, red blood cells, and a number of tissues respond in a manner to dietary Se concentration. Organic sources of Se supplements have shown a higher bioavailability as compared with inorganic forms for fish (Wang and Lovell, 1997; Rider et al., 2010; Sele et al., 2018; Le and Fotedar, 2014; Fontagné Dicharry et al., 2015; Antony Jesu Prabhu et al., 2020).

Selenium-enriched yeast showed higher bioavailability than sodium selenite (Godin et al., 2015). Early work of Bell and Cowey (1989) showed higher bioavailability in Atlantic salmon, and it follows the order from highest to lowest: selenomethionine > selenite > selenocysteine > fishmeal (Bell and Cowey, 1989). Certain fishmeals, e.g., tuna, may have poor biological availability because of heavy metal complexing of Se. A wide variation in Se availability (38.5%–60%) of different batches of capelin meal measured using GSH-Px activity in a chick bioassay was observed (Gabrielson and Opstvedt, 1980). Selenium supplementation of fish diets appears to be the most effective method to meet its requirement.

6.9 Zinc

6.9.1 Functions and metabolism

The essentiality of Zn for fish and other vertebrates is related to the important role of this element as an integral part of a number of metalloenzymes (e.g., alkaline phosphatase, RNA nucleotide transferases, carbonic anhydrases) as well as a catalyst for regulating the activity of specific Zn-dependent enzymes in lipid, carbohydrate, and protein metabolism. The zinc ion is a component of at least 300 metalloenzymes isolated from multiple species that catalyze >50 different physiologic reactions (Maret, 2013). Zinc is required for structural and functional integrity of more than 2000 transcriptional factors involved in expression of many genes. The homeostasis of Zn is tightly controlled at the whole body, tissue, cellular, and subcellular levels by a number of proteins. Two types of Zn transporters that maintain Zn homeostasis, ZnTs (zinc transporters) and ZIPs (Zrt-, Irt-like proteins) are particularly important in characterizing diseases related to dysfunction of Zn homeostasis (Kambe et al., 2015). Biochemical roles for Zn transporters and regulation of specific genes have been studied in zebrafish and are beyond the scope of this chapter (Feeney et al., 2005; Zhao et al., 2014). Although Zn has no redox potential, it can influence oxidative processes, especially those involving Fe possibly Cu, by protecting sulfhydryl groups from oxidation. Cu–Zn SOD is an example of the structural role of Zn metalloenzyme where Cu serves as the catalytic site, while Zn serves a role in structures. Most of the knowledge related to the biochemical function of Zn has emerged from research on other vertebrates including humans, and this area is wide open for research from fish perspectives.

Fish accumulate Zn from both water and dietary sources. The major routes of zinc uptake in fish are through the gill and intestinal tract both in FW and SW (Hogstrand, 2012). It is generally agreed, however, that Zn is more efficiently absorbed from the diet (Lall, 2002). In FW, even when dietary Zn levels are adequate, there is an active uptake of waterborne Zn (Spry et al., 1988). In winter flounder, the entire digestive tract is capable of absorbing Zn, but the uppermost portion of the intestine has the highest capacity and the stomach has the lowest (Shears and Fletcher, 1983). The gills in rainbow trout also play a major role in excretion of dietary Zn (Hardy et al., 1987). The Zn status of fish is tightly controlled, and surplus Zn is excreted via bile, the sloughing of intestinal mucosa in feces and through gills (Handy, 1996). Mucus, gut pH, and luminal ionic concentration of gastrointestinal (GI) tract appear to have a significant role in absorption of Zn as well as certain other minerals. In the intestine, dietary Zn binds to the mucus of the intestinal epithelium and is transported into the epithelial cells either as Zn ions or these ions bound to amino acids (Bury et al., 2003; Hogstrand et al., 2012). Recent development of enterocyte models provides new opportunities to investigate specific mechanisms involved in absorption of Zn from fish diets, which contain different forms of organic and inorganic Zn (Antony Jesu Prabhu et al., 2018b).

Calcium has been shown to have an inhibitory or a stimulatory effect on Zn uptake depending on the concentration (Glover and Hogstrand, 2003; Ojo et al., 2009). Waterborne Ca is a competitive inhibitor of branchial Zn influx. Zinc inhibits the influx of Ca across gills; thus, Ca in water protects fish against Zn toxicity. The accumulation of Zn in gills is also regulated through alteration in Zn uptake mechanisms limiting its excessive uptake (reviewed by Bury et al., 2003). Zinc chelation with certain amino acids, such as histidine and cysteine that have a high affinity for binding this element, may enhance Zn absorption and distribution in various tissues of fish (Glover and Hogstrand, 2002). In vitro studies have also demonstrated that amino acid taurine with limited zinc-binding activity also influences Zn uptake (Glover and Hogstrand, 2002). High intakes of phytate (Section 6.2.4) and iron reduce the absorption of Zn. It appears that excretory mechanisms and control of gastrointestinal tract uptake may play important roles in maintaining Zn homeostasis.

6.9.2 Deficiency

Although essentiality of Zn for various physiological functions is widely recognized, few studies have reported clinical signs of Zn deficiency in fish (Table 6.3). In terrestrial animals, Zn deficiency is associated with a loss of appetite, growth depression, skeletal and skin disorders, and impaired immune function and reproduction (Suttle, 2010). The ubiquity of Zn in the aquatic environment and in feed ingredients makes it difficult to induce deficiency of this element in short-term experiments. In rainbow trout, Zn deficiency causes growth depression, high mortality, lens cataracts, erosion of fins and skin (Ogino and Yang, 1978), and short body dwarfism (Satoh et al., 1983b). Caudal fin Zn concentration is a good indicator of Zn status in rainbow trout (Wekell et al., 1986). In catfish, diets low in Zn cause reduced growth, appetite, bone Zn and Ca levels, and serum Zn concentrations (Gatlin and Wilson, 1983). Broodstock diets low in Zn reduced egg production and hatchability of eggs (Takeuchi et al., 1981). In the early 1970s, widespread occurrence of cataracts in salmonids fed practical diets in US hatcheries ultimately led to the realization that Zn was unavailable in white fishmeal (Ketola, 1979). Excess Ca and P in fishmeal affected Zn bioavailability resulting in lens cataracts. Zinc and multiple dietary, genetic, and environmental factors may be involved in the pathogenesis of cataracts (Lall, 2010).

6.9.3 Requirement

A dietary requirement for Zn (mg/kg) has been reported for several fish species fed semi-purified diets (Table 6.8). The minimum Zn requirement varies with age, sexual maturity, composition of diet, water temperature, and water quality. The response criteria used to determine Zn requirement of fish in various studies include growth, feed efficiency, deficiency signs, whole-body Zn concentration and retention, serum or plasma levels, and enzyme activities. Most of the above criteria show response to increasing dietary concentration; the major proportion of dietary Zn is retained in skeletal tissues particularly vertebrae. A metaanalysis of published information on Zn requirement of several fish species showed estimates ranging from 33.5 to 64.6 mg/kg on the basis of different parameters tested (weight gain, 36; whole body Zn, 33.5; vertebrae Zn, 64.6; serum Zn, 53.4; serum ALP activity, 47) (Antony Jesu Prabhu et al., 2016a).

6.9.4 Toxicity

Rainbow trout and carp can tolerate 1700–1900 mg Zn/kg in diet without any apparent signs of toxicity (Wekell et al., 1983). Common carp accumulate much higher concentrations of Zn in their tissues, particularly in the viscera, than other fish studied, without any apparent toxicity signs (Jeng and Sun, 1981). Elevated levels of dietary Zn (500–1000 mg Zn/kg) caused reduced hemoglobin, hematocrit, and hepatic Cu concentrations in rainbow trout (Knox et al., 1982). However, Cu status of catfish was not impaired by diets containing 200 mg Zn/kg (Gatlin et al., 1989b).

Environmental pollution and heavy metal contamination of fish has resulted in many investigations into accumulation and distribution of Zn in aquatic organisms (Alsop and Wood, 1999). Bioavailability and toxicity of zinc and other metals are known to vary, depending on an organism's physiology and osmoregulatory mechanisms (Alsop et al., 1999; Bielmyer et al., 2012). Most research studies on sublethal physiological effects of Zn have been conducted in FW and demonstrate that the principal toxic effect of Zn toxicity is hypocalcemia caused by Zn interfering with active Zn uptake at the gills (Hoagstrand et al., 2012). Studies conducted on killifish in FW and SW suggest that sublethal Zn exposure causes pathological changes in both Ca and Na homeostasis, and increase in salinity exerts protective effects against sublethal and lethal Zn toxicities (Loro et al., 2014). Mineral sensitivity is highest during the larval stage, compared with other times in the life history of a fish. Zinc content of scales reflects environmental metal concentrations.

6.9.5 Sources and bioavailability

Among feedstuffs, common cereal grains contain 15–30 mg Zn/kg. Most of the Zn is found in the bran and germ fraction of grain; thus a major proportion of Zn is lost during the milling of wheat. Typical plant protein concentrates and oilseed meals may contain 30–120 mg Zn/kg. Zinc concentration in fishmeal depends on species as well as whole or partial fish parts used for processing. Levels of 80–130 mg Zn/kg are more common in fishmeal. Certain feed ingredients of animal origin (meat and bone meal, poultry feather meal) contain high zinc levels (90–140 mg Zn/kg). Egg albumin, because of its low Zn content (<3 mg/kg), is generally used in low-Zn experimental diets.

Bioavailability of Zn is the fraction of Zn intake that is retained and used for physiological functions. Zinc absorption is determined by three factors: the animal's Zn status, total Zn content of the diet, and availability of soluble Zn from dietary components (Lonnerdal, 2000). Among dietary factors, Ca and P levels, phytic acid, protein source, and form of Zn and Ca content affect Zn absorption and retention in fish (Gatlin and Wilson, 1984b;

Richardson et al., 1985; Wekell et al., 1986; Hardy et al., 1987; Satoh et al., 1987a, 1987b, 1989; McClain and Gatlin, 1988; Apines et al., 2004; Silva et al., 2019). Phytate in plant products, especially cereals and legumes, irreversibly binds zinc in the intestinal lumen and accounts for the lower efficiency of absorption (see Section 6.2.4). Sometimes fiber is implied as having a negative effect on Zn absorption, but this is usually because of high fiber plant feedstuffs that are often high in phytates. The binding of Zn to low-molecular-weight ligands or chelators that can be absorbed also has a positive effect on zinc absorption because its solubility is increased. Certain chelators, e.g., EDTA and amino acids (histidine or methionine) enhance Zn absorption.

Common zinc supplements available to provide adequate Zn in diets include zinc sulfate, zinc chloride, zinc nitrate, zinc oxide, zinc acetate, zinc carbonate, zinc citrate, zinc picolinate, zinc–amino acid complexes (methionine, glycine, or histidine), and zinc proteinates. Zinc sulfate and zinc nitrate were effectively utilized by rainbow trout as Zn supplements (40 mg/kg) in diets containing white fishmeal, and they also alleviated dwarfism and cataract problems (Satoh et al., 1987b). The bioavailability of Zn oxide is low in Atlantic salmon and other fish species probably due to lower solubility of this compound. Although Zn oxide is widely used in monogastric nutrition, recent studies show high variability of zinc oxide in color, texture, and Zn content, and origin and manufacture techniques affect its bioavailability. Amino acid chelates of Zn appear to be more readily available than inorganic sources in rainbow trout (Apines et al., 2004). No apparent differences in the bioavailability of zinc sulfate and zinc methionine were noted in catfish (Li and Robinson, 1996). Bioavailability of Zn in fishmeal is affected by the hydroxyapatite (mainly tricalcium phosphate) content (Satoh et al., 1987b). Higher levels of supplemental Zn should be included in practical feeds to compensate for reduced Zn bioavailability caused by dietary inhibitors; however, excess Zn has adverse effects on aquatic environments. A higher limit in fish feed has been set for salmonid (150 mg kg^{-1} of diet) and other fish (100 mg kg^{-1} of diet) is set by the European Food Safety Authority.

6.10 Iodine

6.10.1 Functions and metabolism

Iodine, a nonmetallic element of the halogen group, is volatile at ambient temperature and pressure. It can exist in various oxidation states with valences of $^{-}1$, $^{+}1$, $^{+}3$, $^{+}5$, or $^{+}7$, but the main oxidation state is $^{-}1$ as iodide. The term iodide refers to its ionic form I^{-}. Iodate refers to IO^{3-}. Iodine is an essential element for fish and other animal species because it is an integral component of the thyroid hormones, 3,5,3′-triiodo-L-thyronine (T3) and 3,5,3′, 5′-tetraiodo-L-thyronine (thyroxine, T4). Thyroid hormones regulate cell activity and growth in virtually all tissues and are therefore essential for cellular oxidation, hematopoiesis, circulation, reproduction, neuromuscular functioning, and metabolism of major nutrients. Recently, Eales (2019) has reviewed the relationships between thyroid hormones, thyroid homeostasis, and iodine metabolism in fish and humans.

In FW, food and gill uptake are the main iodine sources for fish. Gill iodide uptake alone should satisfy most thyroid demands, as gills have access to a high water flow (Eales, 2019). Marine fish drink SW ensuring an adequate intestinal I supply, supplemented by I in food and possibly I uptake via other body surfaces (Moren et al., 2008). Rainbow trout derive 80% of their iodide supply from water, 19% from diet, and less than 1% from recycling iodide originating from thyroid hormone degradation (Hunt and Eales, 1979). When dietary uptake is low or absent, uptake of ambient iodide, iodide binding to plasma proteins, and iodide mobilization from tissue stores maintain the plasma iodide level. Approximately 5% of the iodide intake is utilized by the thyroid. Iodide trapped in the thyroid gland is oxidized to I by a process probably mediated by peroxide enzyme, which is followed by iodinization of tyrosine to form monoiodotyrosine and diiodotyrosine (DIT). The coupling of two iodotyrosines to form iodotyronines involves two possible routes. The two iodotyrosine molecules combine to form one molecule of thyroxine (T4), or one mono and one diiodotyrosine combine to form one molecule of triiodotyronine (T3). T3 is the predominant hormone secreted by the thyroid gland and is regarded as an active precursor for T4. Both T4 and T3 occur in blood. In rainbow trout, T3 binds more strongly to plasma proteins than T4, and consequently, T3 turnover is slower in plasma than T4. Major differences exist between fish and mammals in the handling of iodine and extrathyroidal metabolism of T4 and T3 (Eales, 2019).

Unlike T3, orally administered T4 is unable to enhance fish growth (Higgs et al., 1982). It was found later that while orally administered T3 raised plasma T3 levels, T4 at comparable doses did not increase systemic plasma levels of either T4 or T3 in three taxonomically diverse teleosts, rainbow trout, red drum, and Nile tilapia (Eales, 2019). However, by administering T4 by injection or addition to water, T4 level increased in rainbow trout and Atlantic salmon (Morin et al., 1995). Several factors could explain this difference, but it is likely that T4 might be degraded to a greater extent than T3 by gastrointestinal deiodination. The role of thyroid and thyroid hormone metabolism may differ during different stages of development (Blanton and Specker, 2007), parr-smolt transformation (Specker et al., 2000), interactions with sex-related hormones during reproduction (Cyr and Eales, 1996), and metamorphosis (Power, 2008). Females transfer considerable amounts of thyroid hormones to developing ova (Blanton and Specker, 2007), and the offspring may rely on parental iodine storage to complete their early development. Some of these physiological changes may affect the iodine metabolism and requirement at different stages of the life cycle of fish (Lall, 2002; Higgs et al., 1982). Both T4 and T3 are excreted extensively in bile; however, other routes (kidney and gills) may also be involved. Several factors other than dietary and water iodide levels influence blood iodide levels, including elevated water temperature that causes greater iodide excretion, sexual maturation, and differences in the capacity of fish to bind iodide to plasma proteins (Eales, 2019).

6.10.2 Deficiency

Insufficient iodine in the diet is the most common cause of hypothyroidism in vertebrates. More than a century ago, the first element to prevent a specific mineral deficiency, goiter, in salmonid fish was iodine (Marine, 1914). It was later confirmed that this disease could be controlled by administration of iodine. Experimental thyroid hormone deficiency created by glucosinolates present in the diet, by the administration of thiocarbamide derivatives such as thiourea and propylthiouracil, and by radiothyroidectemy resulted in growth depression of salmonids (Higgs and Eales, 1978). Accumulation of ^{131}I by the thyroid gland is significantly reduced in the scorbutic snakehead (Agarwal and Mahajan, 1981).

6.10.3 Requirement

The minimum dietary iodine requirement of most fish species has not been established. The requirement is likely to be influenced by growth, sex, age, physiological status, environmental stress, disease, and iodine content of the water. Goitrogenic substances present in the diet also influence iodine requirement. Woodall and LaRoche (1964) reported a higher iodine requirement for advanced parr compared with fingerlings due to increased thyroid activity during smoltification. Lall et al. (1985) observed that 4.5 mg I kg^{-1} of diet was essential to protect Atlantic salmon from bacterial kidney disease infections.

6.10.4 Sources

Iodine is widely distributed in nature. It is found in relative abundance in marine plants and animals, in deposits of organic origin such as coal, in certain natural mineral waters, in sedimentary phosphate rock, and particularly in large nitrite deposits of Chile. This Chilean ore contains nearly 0.1% iodine and serves as the chief commercial source of this element. The concentration of iodine in common feedstuffs is highly variable. Protein sources of animal origin other than fishery by-products contain nutritionally insignificant quantities of this element. Atlantic white fishmeal may contain 60–90 mg I kg^{-1}, whereas herring and capelin meal contain only 5–10 mg I kg^{-1}. Substantial amounts of iodine are lost during fishmeal processing. Iodine content of protein concentrates from plant origin ranges from 30 to 100 μg kg^{-1}. Oilseed proteins or their concentrates contain 100–300 μg I kg^{-1}, and common cereal grains contain 40–100 μg I kg^{-1}. Certain seaweeds also contain high levels of iodine (Teas et al., 2004). Marine fish have much higher iodine contents than FW fish (Lall, 1995).

Iodine sources permitted as feed additives include calcium iodate, sodium iodate, potassium iodate, potassium iodide, sodium iodide, ethylenediamine dihydroiodide (EDDI), calcium iodobehenate, cuprous iodide, 3,5-diiodosalicylic acid, pentacalcium orthoperiodate, and thymol iodide. Iodine occurs in foods and feeds largely as inorganic iodide. In this form, it is almost completely absorbed from the gastrointestinal tract (Suttle, 2010).

6.11 Chromium

Chromium is an essential element for animals and humans (Mertz, 1998), but its essentiality has not yet been established for fish. The essentiality of Cr for both animals and humans has been questioned by the European Commission. Chromium exists in nature in three oxidation states, 2^+, 3^+, and 6^+. Trivalent chromium (Cr^{3+}) is the most stable oxidation state in which Cr is found in living organisms. It has a low reactivity, which is the most significant biological feature distinguishing it from hexavalent chromium (Cr^{6+}). Trivalent Cr forms relatively inert complexes with proteins and nucleic acids and is required for normal carbohydrate, protein, and lipid metabolism. No Cr-dependent enzymes have been identified. The biological function of Cr is closely related to that of insulin. It is biologically active as part of an oligopeptide, chromodulin, which potentiates the effect of insulin by facilitating insulin binding to receptors at the cell surface (Vincent, 2000). This oligopeptide, comprised primarily of glycine, cysteine, aspartate, and glutamate, has been isolated from the liver or kidney of a number of animal species. In response to insulin release, Cr from the blood is transported to insulin-sensitive cells where it binds to apochromodulin. When insulin concentrations decrease, chromodulin is believed to be released from cells and excreted in the urine. Urine is the major route of excretion for absorbed Cr. Consumption of high carbohydrate diets that increase blood insulin concentrations increases urinary excretion of chromium (Pechova and Pavlata, 2007).

Hexavalent Cr is more soluble than Cr^{3+} and is at least five times toxic than Cr^{3+} (Barceloux, 1999b). Uptake of Cr from water as it relates to physiology and toxicology has been reviewed by Reid (2012). Chromium is absorbed across the gills and transported via blood to tissues, but mechanisms of absorption from gills and the gastrointestinal tract, and excretion are unknown. The toxic Cr^{6+} readily passes cellular membranes and is then reduced to the trivalent form. This Cr^{3+}combines with several macromolecules including genetic material inside the cytosol and ultimately exposes the toxic and mutagenic alterations of Cr toxicity. Although the mechanism of Cr toxicity in fish has received limited attention, higher levels of Cr in diet and water caused histological changes in intestine, gills, liver, and kidney (Reid, 2012; Bakshi and Panigrahi, 2018).

Limited research has been conducted on Cr metabolism in fish and its effect on glucose utilization. Chromium supplementation in carp diet improved glucose utilization probably by modulation of endogenous insulin activity (Hertz et al., 1989); however, glucose utilization was not affected by dietary Cr in channel catfish (Ng and Wilson, 1997). In rainbow trout, supplementation of a practical diet with chelated Cr (0.5 mg/kg) increased blood glucose clearance but did not affect growth, or protein and energy retention (Bureau et al., 1995). Supplementation of tilapia diet containing glucose with inorganic Cr improved growth, energy retention, and liver glycogen deposition of hybrid tilapia (Shiau and Chen, 1993; Shiau and Liang, 1995; Shiau and Shy, 1998). Studies conducted on hybrid tilapia, rohu (*Labeo rohita*), and golden pompano (*Trachinotus ovatus*) fed high carbohydrate (starch or glucose) diet showed that growth and feed utilization and certain biochemical parameters were improved by organic Cr (Cr-picolinate or Cr-nicotinate) supplementation (Pan et al., 2003; Giri et al., 2014; Asad et al., 2018; Wang et al., 2019).

The National Research Council (United States) does not list Cr requirement values for any major farm animal species including fish. An early study with fish on the effects of dietary Cr^{3+} did not show any gross deficiency signs or a change in tissue distribution of rainbow trout fed a low Cr purified diet (Tacon and Beveridge, 1982). Chromium is ubiquitous in aquatic environments and feedstuffs, which make it difficult to formulate diet low enough to induce its deficiency signs. Reduced growth in fish fed diets supplemented with relatively high levels of Cr has been observed in croaker (Wang et al., 2014), tilapia (Shiau and Liang, 1995; Liu et al. 2010b), and rainbow trout (Tacon and Beveridge, 1982). In most of these studies, decreased growth and feed efficiency were likely due to a toxic effect of Cr that occurs at relatively high levels. Common sources of Cr for fish and animal feeds include chromic chloride, chromium propionate, high Cr-yeast, Cr-nicotinate, and Cr-picolinate. Research studies on animals show that organic forms of dietary chromium such as Cr-picolinate, Cr- nicotinate, and Cr-enriched yeast are absorbed more efficiently (25%–30%) than inorganic forms such as chromic chloride ($CrCl_3$) or chromic oxide (Cr_2O_3), which are poorly absorbed (1%–3%) regardless of dose or dietary chromium status (NRC, 2005; Suttle, 2010).

6.12 Other minerals

Several minerals have shown biochemical functions in animals and humans (As, B, F, Li, Ni, Pb, Si, and V), but their status as essential nutrients determined by physiological impairment has not been confirmed. Limited information is available on most of the aforementioned minerals for their biochemical functions as well as physiological impairment for fish. Some circumstantial evidence for the essentiality of Al, Br, Li, and Sr exist for some animals (Nielsen, 2000), but they are not considered here. The information available on As, B, Co, Ni, F, Si, V, and sulfate in fish is briefly discussed. Essentiality based on physiological impairment or any disorders at a low intake of As, F, Li, Mo, Pb, Si, and V has not been observed. Although there may be some lower limit of exposure at which clinical signs of toxicity may not be observed, it is possible to detect adverse effects at the molecular level. The physiology and toxicology of As, Pb, and Cd in terms of their uptake from water have been the subject of extensive research, which provides some fundamental knowledge on their absorption from gills and gastrointestinal tract and interactions with other essential minerals.

6.12.1 Arsenic

Arsenic is generally not considered an essential nutrient for higher animals. The toxicity and physiological aspects of As in fish have received significant attention (McIntyre and Linton, 2012), but its role as an essential nutrient remains to be investigated. Arsenic affects the methylation of molecules and has a function in methionine metabolism and gene expression (Nielsen, 1991; Uthus, 2003). Some organic arsenic compounds, e.g., roxarsone, have been used extensively as growth promoters for pigs and poultry because of their antimicrobial and anticoccidial properties. Incorporation of As in the diet of rainbow trout has shown reduced growth and As accumulation in several tissues (Cockell et al., 1991; Erickson et al., 2011). A maximum tolerable limit for fish may be in the range of 5 mg As/kg diet (NRC, 2005). Seafood and fish have been identified as major sources of arsenic in the human diet, and it is present predominantly in the organic forms of arsenobetaine and arsenocholine, which are virtually nontoxic (NRC, 2005).

6.12.2 Boron

Boron has been found to be essential for zebrafish (Eckhert and Rowe, 1999) and frogs (Fort et al., 1999) to complete their life cycle. It also stimulates embryonic growth in trout depending on the dose (Eckhert, 1998). According to Nielsen (1998, 2002), there is substantial evidence from experiments comparing low intake (<100 μg/kg) diet for animals and 0.25 mg/d for humans to suggest that B is needed for optimal bone health, brain function, and immune function in higher animals and humans (Nielsen, 1998, 2008). However, B is not currently accepted as an essential nutrient for animals, humans, and fish. More evidence related to clearly defined specific biochemical function is needed for wide acceptance of B essentiality for vertebrates. Acute toxicity of B to fry of Chinook salmon and coho salmon has been reported (Hamilton and Buhl, 1990).

6.12.3 Cadmium

Cadmium is not considered an essential nutrient for animals and fish. However, a number of studies with animals have reported increased weight gain when low levels of Cd were added to the diets (Bokori and Fekete, 1995). Cadmium is known to affect Ca homeostasis in both fish and animals (Giles 1984; Verbost et al., 1989). Acute Cd toxicity causes disruption of ion homeostasis in fish, mainly Ca but also Na and Mg (McGeer et al., 2012). The toxicity of Cd is reduced by complexation of Cd with inorganic and organic anions and through competitive interactions between Ca and Cd for uptake sites. Hypocalcemia after waterborne and dietary Cd exposure has been observed in fresh and BW water fish (Hwang and Yang 1997). Cadmium appears to enter the gill epithelium via the same pathway as Ca^{2+} (apical Ca^{2+} channel of the chloride cells) and inhibits the basolateral Ca^{2+}ATPase, thereby blocking active Ca^{2+} uptake (Verbost et al., 1989). In Atlantic salmon, inhibition of ATP-dependent Ca uptake measured as $Ca^{2+}$and Na^+/K^+-ATPase occurred in the intestine of salmon fed 25 mg Cd/kg (Berntssen et al., 2003). The accumulation of dietary Cd in salmon tissues was in the following decreasing order: intestine > kidney > gill > bone. The concentration of Cd in most uncontaminated feed ingredients is relatively low (Adamse et al., 2017). Dietary phosphate sources are likely sources of Cd contamination, which can contribute significantly to the total diet Cd concentration (NRC, 2005).

6.12.4 Cobalt

The biological function of Co relates to its role as a component of vitamin B_{12} (cyanocobalamin). Approximately 4.5% of the molecular weight of vitamin B_{12} is contributed by elemental Co. Vitamin B_{12} is a cofactor for the enzymes methylmalonyl CoA mutase and methionine synthase. Methylmalonyl CoA mutase is responsible for conversion of methylmalonyl CoA to succinyl CoA and is important in propionate metabolism. Methionine synthase is involved in the regeneration of methionine following loss of its methyl group and in the maintenance of biologically active folate concentrations in tissues. Uptake of Co in fish can occur by either uptake directly from water or absorption from food (Baudin et al., 1997, 2000). In carp, Co uptake occurs mainly via water, while food is a minor route (Baudin and Fritsch, 1989). Some uptake of Co occurred in rainbow trout eggs during embryonic development (Kuenze et al., 1978). Most Co detected in maturing Atlantic salmon ovaries is present as part of vitamin B_{12} (Julshamn and Bræcken, 1975). The homeostasis, other physiological aspects, and toxicity of Co have been reviewed (Blust, 2012).

Bacteria and algae synthesize vitamin B_{12} from inorganic cobalt. Most animals and carnivorous fish require the intake of vitamin B_{12} because they lack the ability to synthesize this vitamin from dietary Co in significant amounts by microbiota in their digestive tract. Cobalt is of particular significance in ruminant nutrition, where it is used solely in the synthesis of vitamin B_{12} by rumen microflora. Kashiwada et al. (1970) isolated bacteria from the intestinal tract of carp that produced vitamin B_{12} in vitro. Removal of Co from the diet of catfish significantly reduced intestinal synthesis of vitamin B_{12} (Limsuwan and Lovell, 1981).

Addition of cobalt chloride and/or cobalt nitrate to feed or cobalt chloride to water of fish ponds enhances growth and hemoglobin formation in carp. Lin et al. (2010b) found that dietary Co promoted gastrointestinal bacterial synthesis of vitamin B_{12} in sufficient quantity to meet the requirement of this vitamin for grouper. The optimum requirement for growth was 10 mg Co/kg of diet, and no vitamin B_{12} supplementation was needed. Cobalt chloride supplementation resulted in higher growth in tilapia (Anadu et al., 1990). Dietary feed ingredients may supply Co, but fish feeds are often supplemented with vitamin B_{12} to ensure that the requirement of this micronutrient is met.

6.12.5 Fluorine

Fluorine is considered an essential element, although specific deficiency signs have not been observed in animals. It is difficult to produce an environment sufficiently devoid of this element that survival of an animal will be virtually threatened. The biochemical basis of acute fluorine toxicity causes enzyme inhibition, calcium complex formation, and specific organ injury (Hodge and Smith, 1981). Fluoride inhibits a large number of metalloenzymes containing Cu, Mn, Zn, Ni, and F by binding to the metal on the active site (NRC, 2005). Bone has a great affinity to incorporate fluoride into hydroxyapatite to form fluorapatite. Aquatic organisms absorb fluorine from their diet and surrounding water and accumulate it in skeletal tissues. Toxicity data are available for terrestrial and aquatic animals; however, oral maximum tolerable levels for fish are not established (NRC, 2005). The primary sources of dietary fluorides for farmed animals are P supplements and feed ingredients of animal origin. Phosphorus supplements vary greatly in their fluoride content, depending on origin and manufacturing processes.

6.12.6 Lead

Lead is not known to be an essential nutrient for animals and does not participate in any known important biochemical functions. However, several studies have shown that the addition of lead to the diet improves growth rates and lipid metabolism of animals (Reichlmayr-Lais and Kirchgessner, 1981; NRC, 2005). Calcium, Fe, and Zn deficiencies enhance Pb toxicity by increasing absorption of toxic metals and by exchanging with the essential cation on biochemically active sites, including receptor proteins in the brain (Goyer, 1997). The other mechanism at the gut level involves Pb competing with Ca for binding sites on Ca-binding proteins. Mechanisms of Pb uptake from water and its toxicity in fish have been comprehensively reviewed (Mager, 2012).

6.12.7 Nickel

Nickel generally is not considered an essential nutrient for higher animals, humans, and fish because biochemical functions of this element are not clearly defined. It is essential for some lower forms of life where it participates in hydrolysis and redox reactions, regulates gene expression, and stabilizes certain cellular structures (NRC, 2005). The function of Ni may be associated with vitamin B_{12}, because lack of this vitamin inhibits the response to Ni supplementation when dietary Ni concentration is low (Nielsen et al., 1998). Nickel can alleviate vitamin B_{12} deficiency in higher animals. Nickel deprivation also affects distribution and proper functioning of other nutrients including Ca, Fe, and Zn (Nielsen, 1991). The effect of dietary Ni toxicity on reproduction of zebrafish has been observed (Alsop et al., 2014). An antagonism exists between Ni and Mg in fish and other organisms (Pyle and Couture, 2012). Nickel also inhibits Fe absorption in rainbow trout and other fish (Kwong and Niyogi, 2009). The physiology and toxicology aspects of Ni have been reviewed by Pyle and Couture (2012). The nickel content of feed ingredients varies with the uptake of Ni from soil, and some reported values include wheat (0.08–0.3 mg kg^{-1}), linseed meal (5.24 mg kg^{-1}), soybean meal (7.91 mg kg^{-1}), and sunflower meal (7.78 mg kg^{-1}). Nickel in animal by-products is low; however, in fishmeal or fish protein concentrate, its level is variable and may range between 0.5 and 3.5 mg Ni kg^{-1}.

6.12.8 Silicon

Silicon, a tetravalent trace element, has a strong affinity for oxygen. The chemistry of Si is similar to that of carbon, its sister element. Silicon forms silicon–silicon, silicon–hydrogen, silicon–oxygen, silicon–nitrogen, and silicon–carbon bonds. Silicon is essential for some lower forms of life and is an essential element for diatoms (Round et al., 1990). They are the primary producers in the high-production areas of the ocean, and Si is a key factor in the transfer of energy to higher tropic levels. Silicon has an essential role in collagen and glycosaminoglycan formation or function and thus influences bone formation, wound healing, and calcification in rats (Seaborn and Nielsen, 2002). A possible mechanism of action for the beneficial effects of Si is the binding of hydroxyl group of polyols in a manner that influences the formation/or utilization of glycosaminoglycans, mucopolysaccharides, and collagen in connective tissues and bone (Neilsen, 2014). In addition, silicon may affect the absorption, retention, or action of other minerals (Al, Cu, and Mg). To date, silicon is not accepted as an essential nutrient for higher animals and fish because of the lack of a clearly defined specific biochemical function. However, new evidence strongly suggests that Si is an important biological element in humans and that its body distribution is regulated, in part, via mammalian transporters. Cereal grains and their by- products high in fiber are the major sources of Si for animals (Pennington, 1991). Silicon, present in plants as silica and soluble silicates, and in organic combinations, is bound to the cellulosic cell structure.

6.12.9 Sulfur

Sulfur is primarily incorporated into a number of organic molecules that are vital to life, making it an essential nutrient. Sulfur-containing compounds include sulfur-containing amino acids and their metabolites (methionine, cystine, cysteine, homocysteine, taurine, cysteic acid, and cystathionine), thiamine, biotin, lipoic acid, coenzyme A, glutathione, chondroitin sulfate, cartilage mucopolysaccharides, and fibrinogen. Most of these compounds are found in nearly all body tissues. Many of these sulfur-containing compounds required by the body can be synthesized from methionine. It appears that vertebrates including fish are not capable of producing methionine from inorganic sulfur in the diet. Therefore, the organic sulfur form of sulfur as methionine is an essential nutrient in the diets of fish and most vertebrate animals. Inorganic sulfate failed to promote growth or provide the requirements of sulfur amino acids in rainbow trout. Sulfur amino acid metabolism is discussed in Chapter 4.

Water-soluble sulfur-containing vitamins, thiamin, and biotin are not synthesized from methionine. Certain sulfate forms of minerals (Ca, Mg, K, Zn, Fe, Mn, Cu etc.) used in mineral premix also supply sulfate in the diet.

6.12.10 Vanadium

Essentiality of V to fish has not been reported. Circumstantial evidence suggests that V is considered an essential element for rats and goats. However, because a defined biochemical function has not been identified in higher animals, V is currently not considered an essential element. There is a low risk of animals suffering from a deficiency in V; however, V toxicity can occur and in certain species at fairly low intake levels (NRC, 2005). Vanadium binds to nucleic acids, amino acids, phosphates, phospholipids, glutathione, oxalate, citrate, and lactate (Nechay, 1984). It inhibits a large number of enzyme systems in vitro including ATPases, phosphates, and phosphoric transfer enzymes. The vanadate ion is a potent inhibitor of the Na^+/K^+ATPase pump (Barceloux, 1999a). However, specific activation or inhibition of an enzyme by V has not been found (Nielsen, 1995). Incorporation of sodium orthovanadate in the diet (>10 mg V kg^{-1} kg) of rainbow trout caused reduced growth and feed utilization, and levels of 493 mg kg^{-1} showed liver lipid peroxidation and high mortality (Hilton and Bettger, 1988). One of the main concerns for V in feedstuffs is the concentration of this element in P mineral sources, as V concentrations in other feedstuffs are low (NRC, 2005).

6.13 Use of alternate feed ingredient sources to fishmeal and criteria for their mineral supplementation

Fishmeals have constituted the major source of protein in fish feeds since the inception of fish feed development in the late 1960s. In addition to high-quality protein, it was considered an important source of minerals particularly P. An outbreak of cataracts in the early 1970s in salmonid hatcheries in North America led to research on trace element bioavailability in fishmeals. The cause of cataracts was linked to low levels of Zn and high amounts of Ca and other minerals in diets containing white fishmeal (Ketola, 1979). Later, Watanabe (1988) also found that omission of trace mineral supplementation in fishmeal diet was responsible for short body dwarfism and lens cataracts and other deficiency signs in chum salmon. Since then, it has been recognized that higher Zn and Mn supplementation is necessary in fishmeal-based diets to overcome the problems associated with low bioavailability of these elements. In certain fishmeals, high concentrations of Fe (>300 mg/kg) were also recognized (Soevik et al., 1981). Technological changes in handling and processing of raw material during processing minimized Fe contamination during fishmeal production (Bimbo and Cowther, 1992).

Fishmeals are produced from whole fish (e.g., herring, capelin, anchovy, horse mackerel, menhaden, capelin, sand eel, blue whiting), by-catches from fisheries, trimmings, offal, and other discards left from fish processed for human consumption. The proportion of bone and skeletal tissues in these raw materials contribute to variable amounts of total ash/minerals in different fishmeals. Generally, ash content ranges from 10% to 20%; some fishmeal may contain higher amounts (>25%) depending on the source of material

(Hertramp and Piedad-Pascal, 2000; https://www.feedipedia.org). Fishmeal contains significant amounts of minerals particularly Ca (1.9%–8%), P (1.5%–4.8%), and trace elements (Fe, 150–450; Zn, 90–120; Mn, 2.4–40; Cu, 5–12; Se, 1.5–2.8; I, 0–2.4 mg/kg) (Julshamn et al., 1978; Barlow and Windsor, 1984; Bimbo and Cowther, 1992). In the past 2 decades, with the increasing demands of the world's aquaculture production on the finite quantity of this high-quality protein source, alternative sources of plant and/or animal origin or their mixture with fishmeal now increasingly comprise the major constituents of fish feeds (NRC, 2011). Numerous studies have investigated the potential of alternate plant proteins, particularly proteins from oilseed (soybean and rapeseed/canola meal or their concentrates), grains (wheat and corn gluten), and legumes (pea and lupin). In addition, better quality meals from animal by-products (meat and bone, poultry), fishery by-products (fish concentrate and fish bone meal), and insect and algae products have been developed. Applied research conducted in the past two decades has established the maximum amounts of some plant protein sources that carnivorous fish can tolerate and the adverse effects of their antinutritional factors or toxicants (e.g., protease inhibitors, lectins, phytic acid, saponins, phytoestrogens, alkaloids, tannins, cyanogens, glucosinolates, etc.) that adversely affect digestion, absorption, and physiological utilization of nutrients (Krogdahl et al., 2010, Chapter 11). The levels of phytate vary widely depending on plant protein sources, thereby affecting mineral availability in fish (Section 6.2.4). Phosphorus availability from different feed ingredients and fishmeal is also discussed earlier.

Despite significant amounts of trace elements in fishmeal, trace element supplementation is considered essential for fishmeal as well as diets based on a major proportion of alternate protein sources. To determine the amounts of mineral supplements in practical diets, the basic principles of feed formulation are applied (Lall and Dumas, 2015). The following information is needed to determine adequate levels of minerals: (1) mineral content of feed ingredients, (2) requirement values for specific species, (3) stage of development (juvenile, adult, and broodstock), (4) bioavailability of minerals, and (5) impact of mineral excretion on the aquatic environment. The composition of fishmeals is well documented; however, the current trend is to blend fishmeals for the aquaculture feed market. Although it is relatively easy to determine the mineral concentrations of plant feed ingredients, their bioavailability is essential to be used in conjunction with the NRC-recommended requirement values to estimate the amounts of trace elements to be incorporated into feeds to support normal performance. However, these values do not take into account safety factors required to account for the variability that exists in the mineral content of feed ingredients or the variability that exists between fish species or within the same species of fish. The mineral content of oilseeds and legumes, such as soybeans, is generally high, but the bioavailability is poor due to the presence of phytate, which is a main inhibitor of Fe and Zn absorption (Section 6.2.4). It is apparent that higher than NRC trace element recommendations (NRC, 2011) are necessary for salmonid and marine fish (Read et al., 2014; Antony Jesu Prabhu et al., 2016a, 2016b, 2019b, 2020; Welker et al., 2016, 2017). Among them, Zn and Se supplementation is considered more critical. Fishmeal contains high concentrations of Se in organic form (Lund et al., 1973) with relatively high bioavailability (Bell and Cowey, 1989; Antony Jesu Prabhu, 2020). An insurance (extrasupplementation with trace elements beyond NRC or published values) to provide sufficient minerals in feeds may be necessary for optimum performance of fish.

Trace metal supplements are essential in broodstock diets. Takeuchi et al. (1981) demonstrated the effects of maternal trace element deficiency on rainbow trout egg quality. The trace element-deficient group showed a reduction in egg production, hatchability of eggs, Mn content of eggs, and female body weight. Fishmeal-based practical diets without any inorganic Mn supplement for Atlantic salmon and brook trout caused a reduction in hatchability of eggs and a decrease in Mn concentration of seminal fluid and eggs (Lall, 1995, unpublished data). The bioavailability and effects of inorganic and organic (amino acid chelate, yeast complexes, etc.) trace element supplements are discussed in individual sections.

6.14 Concluding remarks

In the past two decades, significant progress has been made in the determination of the mineral requirements of certain fish species; however, there are significant gaps in knowledge related to quantitative dietary requirements of most minerals at various stages of the life cycle. Many new fish species are now farmed under different culture and environmental conditions with limited knowledge of their mineral digestion, absorption, metabolism, and excretory products from practical feeds. It is increasingly recognized that dietary minerals play a pivotal role in immune function, health, disease prevention, and reproduction of fish. The physiological mechanisms underlying observed acute toxicity in fish are relatively well understood for some minerals (Cu, Zn, Mn, Se, Cr, Cd, etc.), and more information generated in these areas would be useful to better understand the relative uptake of these essential minerals from feed as well as water. Further research based on a better understanding of the biochemical basis of minerals in reducing the detrimental effects of free radicals and toxic metabolites on immune response is needed. Advances in molecular biology, microscopy, immunology, and analytical chemistry now provide many new opportunities to explore trace element metabolism and regulation in fish.

The bioavailability of a mineral is a key factor in assessing mineral requirements of fish. However, a better understanding of interactions between trace elements and other dietary components, and uptake from water remains a bottleneck to better estimate the requirements for essential trace elements. While it is clear that the GI tract is an important site of mineral absorption in fish, no clear picture has emerged as to the regional uptake capacity from the GI tract of P and other trace elements in a single fish species. The greater use of alternate plant proteins to replace fishmeal has put more emphasis on increasing the utilization of P from phytates and the bioavailability of trace elements. The new generation of phytase enzymes offers opportunities to improve phytic acid utilization in plant proteins, the pH specificity for different species, and optimum physiological conditions necessary for their utilization. The environmental constraints to reduce P discharge in aquaculture effluents and the need to improve the bioavailability of P and reduce excretory material in water in flow through and recirculation systems remain major challenges. Limited studies have been conducted on some organic trace elements that show higher bioavailability and performance than inorganic sources. As excessive amounts of mineral supplementation not only cause adverse environmental impacts but also increase feed cost, their margin of safety in feed formulation should be carefully assessed.

NRC requirements are guidelines that nutritionists use as a starting point to establish recommendation allowances for fish. They provide estimates of minimal requirements necessary to support normal performance. However, caution is necessary to use these values as safety factors designed to account for the variability that exists in the nutrient content of feed ingredients or the variability that exists between fish species or within the same species of fish. They are not intended to necessarily be recommended levels for all stages of fish (larvae, fingerlings, juveniles to adult or broodstock). Sophisticated modeling and metaanalysis approaches are now being applied to estimate mineral requirements and metabolism of farmed animals and fish; however, additional reliable new data on mineral requirements for major farmed fish species are needed to effectively utilize these methods.

References

Adamse, P., Van der Fels-Klerx, H.J., de Jong, J., 2017. Cadmium, lead, mercury and arsenic in animal feed and feed materials – trend analysis of monitoring results. Food Addit. Contam. 34 (8), 1298–1311.

Agrawal, N.K., Mahajan, C.L., 1981. Effect of ascorbic acid deficiency on the uptake of iodine by thyroid and non-thyroid tissues of an air-breathing freshwater fish *Channa punctatus* Bloch. J. Fish. Biol. 18, 411–416.

Aisen, P., Leibmen, A., Sia, C., 1972. Molecular weight and subunit structure of hagfish transferrin. Biochemistry 11, 3461–3464.

Albrektsen, S., Lock, E.-J., Baeverfjord, G., Pedersen, M., Krasnov, A., Takle, H., Veiseth-Kent, E., Veiseth-Kent, E., Waagbø, R., Ytteborg, E., 2017. Utilization of $H_2SO_4^-$ hydrolysed phosphorus from herring bone by-products in feed for Atlantic salmon (*Salmo salar*) 0+ postsmolt. Aquacult. Nutr. 24, 348–365.

Ali, H., Khan, E., 2018. Trophic transfer, bioaccumulation, and biomagnification of non-essential hazardous heavy metals and metalloids in food chains/webs—concepts and implications for wildlife and human health. Human Ecol. Risk Assess. Int. J. 25 (6), 1353–1376.

Alsop, D.H., Wood, C.M., 1999. Influence of waterborne cations on zinc uptake and toxicity in rainbow trout, *Oncorhynchus mykiss*. Can. J. Fish. Aquat. Sci. 56, 2112–2119.

Alsop, D., Lall, S.P., Wood, C.M., 2014. Reproductive impacts and physiological adaptations of zebrafish to elevated dietary nickel. Comp. Biochem. Physiol. C 165, 67–75.

Ammerman, C.B., 1995. Methods for estimation of mineral bioavailability. In: Ammerman, C.B., Baker, D.J., Lewis, A.J. (Eds.), Bioavailability of Nutrients for Animals: Amino Acids, Minerals and Vitamins. Academic Press, San Diego, pp. 83–94.

Anadu, D.I., Anozie, O.C., Anthony, A.D., 1990. Growth responses of *Tilapia zillii* fed diets containing various levels of ascorbic acid and cobalt chloride. Aquaculture 88, 329–336.

Andersen, F., Maage, A., Julshamn, K., 1996. An estimation of iron requirement of Atlantic salmon *Salmo salar* L., parr. Aquacult. Nutr. 2, 41–47.

Andersen, F., Lorentzen, M., Waagbø, R., Maage, A., 1997. Bioavailability and interactions with other micronutrients of three dietary iron sources in Atlantic salmon, *Salmo salar*, smolts. Aquacult. Nutr. 3, 239–246.

Andrews, J.W., Murai, T., Campbell, C., 1973. Effects of dietary calcium and phosphorus on growth, food conversion, bone ash and hematocrit levels of catfish. J. Nutr. 103, 766–771.

Antony Jesu Prabhu, P., Schrama, J.W., Kaushik, S.J., 2013. Quantifying the dietary phosphorus requirement of fish – a meta-analytic approach. Aquacult. Nutr. 19, 243–249.

Antony Jesu Prabhu, P., Schrama, J.W., Kaushik, S.J., 2016a. Mineral requirements of fish: a systematic review. Rev. Aquacult. 8, 172–219.

Antony Jesu Prabhu, P., Geurden, I., Fontagné-Dicharry, S., Vernon, V., Larroquet, Marijouis, L.C., Schrama, J.W., Kaushik, S.J., 2016b. Responses in micro-mineral metabolism in rainbow trout to change in dietary ingredient composition and inclusion of a micro-mineral premix. PLoS One 1 (2), e0149378.

Antony Jesu Prabhu, P., Kaushik, S., Geurden, I., Stouten, T., Fontagne-Dicharry, S., Veron, V., Mariojouls, C., Verreth, J., Eding, E., Schrama, J., 2017. Water exchange rate in RAS and dietary inclusion of micro-minerals influence growth, body composition and mineral metabolism in common carp. Aquaculture 471, 8–18.

Antony Jesu Prabhu, P., Schrama, J., Fontagné-Dicharry, S., Mariojouls, C., Surget, A., Bueno, M., Geurden, I., Kaushik, S.J., 2018a. Evaluating dietary supply of microminerals as a premix in a complete plant ingredient-based diet to juvenile rainbow trout (*Oncorhynchus mykiss*). Aquacult. Nutr. 24, 539–547.

Antony Jesu Prabhu, P., Stewart, T., Silva, M., Amlund, H., Ørnsrud, R., Lock, E.-J., Waagbo, R., Hogstrand, C., 2018b. Zinc uptake in fish intestinal epithelial model RTgutGC: impact of media ion composition and methionine chelation. J. Trace Elem. Med. Biol. 50, 377–383.

Antony Jesu Prabhu, P., Silva, M.S., Kröeckel, S., Holme, M., Ørnsrud, R., Amlund, H., Erik-Jan Lock, E.-J., Waagbø, R., 2019a. Effect of levels and sources of dietary manganese on growth and mineral composition of post-smolt Atlantic salmon fed low fish meal, plant-based ingredients. Aquaculture 512, 734287.

Antony Jesu Prabhu, P., Fountoulaki, E., Maas, R., Heinsbroek, L.T.N., Eding, E.H., Kaushik, S.J., Schrama, J.W., 2019b. Dietary ingredient composition alters faecal characteristics and waste production in common carp reared in recirculation system. Aquaculture 512, 734357.

Antony Jesu Prabhu, P., Holen, E., Espe, M., Silva, M.S., Holme, M.-H., Hamre, K., Lock, E., Waagbø, R., 2020. Dietary selenium required to achieve body homeostasis and attenuate pro-inflammatory responses in Atlantic salmon post-smolt exceeds the present EU legal limit. Aquaculture 526, 735413. https://doi.org/10.1016/j.aquaculture.2020.735413.

Apines-Amar, M.J.S., Satoh, S., Caipang, C.M.A., Kiron, V., Watanabe, T., Aoki, T., 2004. Amino acid-chelate: a better source of Zn, Mn and Cu for rainbow trout, *Oncorhynchus mykiss*. Aquaculture 240, 345–358.

Arai, S., Nose, T., Kawatsu, H., 1974. Effect of minerals supplemented to the fish meal diet growth of eel, *Anguilla japonica*. Bull. Freshw. Fish. Res. Lab. 24, 95–99.

Araújo, J.G., Guimarães, I.G., Mota, C.S., de Paula, F.G., Café, M.B., Pádua, D.M.C., 2016. Dietary available phosphorus requirement for tambaqui, *Colossoma macropomum* , juveniles based on growth, haematology and bone mineralization. Aquacult. Nutr. 23, 822–832.

Åsgård, T., Shearer, K.D., 1997. Dietary phosphorus requirement of juvenile Atlantic salmon, *Salmo salar* L. Aquacult. Nutr. 3, 17–23.

Asad, F., Mubarik, M.S., Ali, T., Zahoor, M.K., Ashrad, R., Qamer, S., 2018. Effect of organic and in-organic chromium supplementation on growth performance and genotoxicity of *Labeo rohita*. Saudi J. Biol. Sci. 26, 1140–1145. https://doi.org/10.1016/j.sjbs.2018.12.015.

Asano, M., Ito, M., 1957. Comparative biochemical studies on aquatic animals. II. Phosphorus turnover in fish and shellfish. Tohoku J. Agric. Res. 7, 291–302.

Aschner, J.L., Aschner, M., 2005. Nutritional aspects of manganese homeostasis. Mol. Aspect. Med. 26, 353–362.

Avila, E.M., Tu, H., Basantes, S., Ferraris, R.P., 2000. Dietary phosphorus regulates intestinal transport and plasma concentrations of phosphate in rainbow trout. J. Comp. Physiol. B Biochem. Syst. Environ. Physiol. 170, 201–209.

Baeverfjord, G., Asgard, T., Shearer, K., 1998. Development and detection of phosphorus deficiency in Atlantic salmon, *Salmo salar* L., parr and post-smolts. Aquacult. Nutr. 4, 1–11.

Baeverfjord, G., Antony Jesu Prabhu, P., Fjelldal, P.G., Albrektsen, S., Hatlen, B., Denstadli, V., Waagbø, R., 2019. Mineral nutrition and bone health in salmonids. Rev. Aquacult. 11, 740–765.

Bakke, A.M., Glover, C., Krogdahl, Å., 2010. Feeding, digestion and absorption of nutrients. In: Grosell, M., Farrell, A.P., Brauner, C.J. (Eds.), Multifunctional Gut of Fish, Fish Physiology, vol. 30. Academic Press, UK, pp. 57–110.

Bakshi, A., Panigrahi, A.K., 2018. A comprehensive review on chromium induced alterations in fresh water fishes. Tox. Rep. 5, 440–447.

Baldisserotto, B., Chowdhury, M.J., Wood, C.M., 2006. In vitro analysis of intestinal absorption of cadmium and calcium in rainbow trout fed with calcium- and cadmium-supplemented diets. J. Fish. Biol. 69, 658–667.

Barceloux, D.G., 1999a. Chromium. Clin. Toxicol. 37, 173–194.

Barceloux, D.G., 1999b. Vanadium. Clin. Toxicol. 37, 265–278.

Barlow, K., Windsor, M., 1984. Fishery By-Products, International Association of Fish Meal Manufacturers, Technical Bulletin No. 19. St Albans, Hertfordshire, UK.

Basulto, S., 1976. Induced saltwater tolerance in connection with inorganic salts in the feeding of Atlantic salmon (*Salmo salar* L.). Aquaculture 8, 45–55.

Bassett, J.H., Williams, G.R., 2003. The molecular actions of thyroid hormone in bone. Trends Endocrinol. Metabol. 14, 356–364.

Baudin, J.P., Fritsch, A.F., 1989. Relative contributions of food and water in the accumulation of ^{60}Co by a freshwater fish. Water Res. 23, 817–823.

Baudin, J.P., Veran, M.P., Adam, C., Garnier-LaPlace, J., 1997. ^{60}C transfer from water to the rainbow trout (*Oncorhynchus mykiss* Walbaum). Arch. Environ. Contam. Toxicol. 33, 230–237.

Baudin, J.P., Veran, M.P., Adam, C., Garnier-LaPlace, J., 2000. Dietary uptake, retention and tissue distribution of ^{54}Mn, ^{60}Co and and ^{137}Cs in rainbow trout (*Oncorhynchus mykiss* Walbaum). Water Res. 34, 2869–2878.

Beattie, J.H., Pascoe, D., 1978. Cadmium uptake by rainbow trout (*Salmo gairdneri*) eggs and alevins. J. Fish. Biol. 13, 631–637.

Beaumont, M., 2000. Exposure of brown trout, *Salmo trutta*, to a sub-lethal concentration of copper in soft acidic water: effects upon muscle metabolism and membrane potential. Aquat. Toxicol. 51, 259–272.

Bell, J.G., Cowey, C.B., 1989. Digestibility and bioavailability of dietary selenium from fishmeal, selenite, selenomethionine and selenocystine in Atlantic salmon (*Salmo salar*). Aquaculture 81, 61–68.

Bell, J.G., Cowey, C.B., Adron, J.W., Shanks, A.M., 1985. Some effects of vitamin E and selenium deprivation on tissue enzyme levels and indices of tissue peroxidation in rainbow trout (*Salmo gairdneri*). Br. J. Nutr. 53, 149–157.

Berg, A., 1968. Studies on the metabolism of calcium and strontium in freshwater fish. 1. Relative contribution of direct and intestinal absorption. Mem. Ist. Ital. Idrobiol. Dott. Maroco Marchi 23, 161–196.

Bergwitz, C., Jüppner, H., 2010. Regulation of phosphate homeostasis by PTH, vitamin D, and FGF23. Annu. Rev. Med. 61, 91–104.

Bernhard, M., Andreae, M.O., 1984. Transport of trace metals in marine food chains. Life Sci. Res. Rep. 28, 143–168.

Berntssen, M.H.G., Lundebye, A.-K., Hamre, K., 2000. Tissue lipid peroxidative responses in Atlantic salmon (*Salmo salar* L.) parr fed high levels of dietary copper and cadmium. Fish Physiol. Biochem. 21, 35–48.

Berntssen, M.H.G., Lundebye, A.K., Maage, A., 1999a. Effects of elevated dietary copper concentrations on growth, feed utilisation and nutritional status of Atlantic salmon (*Salmo salar* L.) fry. Aquaculture 174, 167–181.

Berntssen, M.H.G., Waagbø, R., Toften, H., Lundebye, A.K., 1999b. Effects of dietary cadmium on calcium homeostasis, Ca mobilization and bone deformities in Atlantic salmon (*Salmo salar* L.) parr. Aquacult. Nutr. 9, 175–183.

Berntssen, M.H.G., Waagbo, R., Toften, H., Lundebye, A.-K., 2003. Effects of dietary cadmium on calcium homeostasis, Ca mobilization and bone deformities in Atlantic salmon (*Salmo salar* L.) parr. Aquacult. Nutr. 9, 175–183.

Berntssen, M.H.G., Sundal, T.K., Olsvik, P.A., Amlund, H., Rasinger, J.D., Sele, V., Hamre, K., Hillsted, M., Buttle, L., Ørnsrud, R., 2017. Sensitivity and toxic mode of action of dietary organic and inorganic selenium in Atlantic salmon (*Salmo salar*). Aquat. Toxicol. 192, 116–126.

Berntssen, M.H.G., Betancor, M., Caballero, M.J., Hillestad, M., Rasinger, J., Hamre, K., Sele, V., Amlund, H., Ørnsrud, R., 2018. Safe limits of selenomethionine and selenite supplementation to plant-based Atlantic salmon feeds. Aquaculture 495, 617–630.

Bielmyer, G.K., Bullington, J.B., DeCarlo, C.A., Chalk, S.J., Smith, K., 2012. The effects of salinity on acute toxicity of zinc to two euryhaline species of fish, *Fundulus heteroclitus* and *Kryptolebias marmoratus*. Integr. Comp. Biol. 52, 753–760.

Bijvelds, M., Velden, J., Kolar, Z.I., Flik, G., 1998. Magnesium transport in freshwater teleosts. J. Exp. Biol. 201, 1981–1990.

Bimbo, A.P., Crowther, J.B., 1992. Fish meal and oil: current uses. J. Am. Oil Chem. Soc. 69, 221–227.

Bjornsson, B.T., Persson, P., Larsson, D., Johannsson, S.H., Sundell, K., 1999. Calcium balance in teleost fish: transport and endocrine control mechanism. In: Danks, J., Dacke, C., Flik, G., Gay, D. (Eds.), Calcium Metabolism: Comparative Endocrinology. Bioscientifica Ltd, Bristol, UK, pp. 29–38.

Bjørnevik, M., Maage, A., 1993. Effects of dietary iron supplementation on tissue iron concentration and haematology in Atlantic salmon (*Salmo salar*). Fisk. Dir. Skr. Ser. Ernaer. 6, 35–45.

Björnsson, B.T., Stefansson, S.O., McCormick, S.D., 2011. Environmental endocrinology of salmon smoltification. Gen. Comp. Endocrinol. 170, 290–298.

Blanton, M.L., Specker, J.L., 2007. The hypothalamic-pituitary-thyroid (HPT) axis in fish and its role in fish development and reproduction. Crit. Rev. Toxicol. 37, 97–115.

Blust, R., 2012. In: Wood, C.M., Farrel, A.P., Brauner, C.J. (Eds.), Cobalt. Homeostasis and toxicology of essential metals, Fish physiology 31A. Elsevier, Amsterdam, pp. 291–326.

Boeuf, G., 1987. Bases physiologiques de la salmoniculture: osmoregulation et adapatation à l'eau de mer. La Pisciculture Francaise 87, 28–41.

Boglione, C., 2020. Skeletal abnormalities. In: Woo, P.T.K., Iwama, G. (Eds.), Climate Change and Non-infectious Fish Disorders. CABI, Wallingford, UK, pp. 54–79.

Boglione, C., Gavaia, P., Koumoundouros, G., Gisbert, E., Moren, M., Fontagne, S., Witten, P.E., 2013. A review on skeletal anomalies in reared European fish larvae and juveniles. Part 1: normal and anomalous skeletogenic processes. Rev. Aquacult. 5 (Suppl. 1), S99–S120.

Bokori, J., Fekete, S., 1995. Complex study of the physiological role of cadmium. I. Cadmium and its physiological role. Acta Vet. Hung. 43, 3–43.

Borlongan, I., Satoh, S., 2001. Dietary phosphorus requirement of juvenile milkfish, *Chanos chanos* (Forsskal). Aquacult. Res. 32, 26–32.

Brix, K.V., Tellis, M.S., Crémazy, A., Wood, C.M., 2017. Characterization of the effects of binary metal mixtures on short-term uptake of Cd, Pb, and Zn by rainbow trout (*Oncorhynchus mykiss*). Aquat. Toxicol. 193, 217–227.

Bronner, F., 2012. Recent developments in intestinal calcium absorption. Nutr. Rev. 67, 109–113.

Brooks, K.M., Stierns, A.R., Mahnken, C.V.W., Blackburn, D.B., 2003. Chemical and biological remediation of the benthos near Atlantic salmon farms. Aquaculture 219, 355–377.

Brown, M.L., Jaramillo Jr., F., Gatlin III, D.M., 1993. Dietary phosphorus requirement of juvenile sunshine bass, *Moronechrysops* × *M. saxatilis*. Aquaculture 113, 355–363.

Bureau, D.P., Kirkland, J.B., Cho, C.Y., 1995. The effects of dietary chromium supplementation on performance, carcass yield and blood glucose of rainbow trout *Oncorhynchus mykiss* fed two practical diets. J. Anim. Sci. 73, 194.

Burk, R.F., Hill, K.E., 2015. Regulation of selenium metabolism and transport. Annu. Rev. Nutr. 35, 109–134.

Bury, N., Grosell, M., 2003. Iron acquisition by teleost fish. Comp. Biochem. Physiol. C Toxicol. Pharmacol. 135, 97–105.

Bury, N.R., Walker, P.A., Glover, C.N., 2003. Nutritive metal uptake in teleost fish. J. Exp. Biol. 206, 11–23.

Bury, N.R., Boyle, D., Chen Cooper, C.A., 2012. Iron. In: Wood, C.M., Farrell, A.P., Brauner, C.J. (Eds.), Homeostasis and Toxicology of Essential Metals. Academic Press, San Diego, pp. 201–251.

Cao, J., Miao, X., Xu, W., Li, J., Zhang, W., Mai, K., 2014. Dietary copper requirements of juvenile large yellow croaker *Larimichthys croceus*. Aquaculture 432, 346–350.

Cao, L., Wang, W., Yang, C., Yang, Y., Diana, J., Yakupitiyage, A., Luo, Z., Li, D., 2007. Review: application of microbial phytase in fish feed. Enzym. Microb. Technol. 40, 497–507.

Cahu, C.L., Zambonino Infante, J.L., Barbosa, V., 2003. Phospholipid level in dietary lipid fraction is determining for sea bass (*Dicentrarchus labrax*) larval development. Br. J. Nutr. 90, 21–28.

Campbell, H.A., Handy, R.D., Nimmo, M., 1999. Copper uptake kinetics across the gill of rainbow trout (*Oncorhynchus mykiss*) measured using an improved isolated perfused head technique. Aquat. Toxicol. 46, 177–190.

Canale, R.P., Whelan, G., Switzer, A., Eisch, E., 2016. A bioenergetic approach to manage production and control phosphorus discharges from a salmonid hatchery. Aquaculture 451, 137–146.

Canario, A.V., Rotllant, J., Fuentes, J., Guerreiro, P.M., Teodósio, H., Power, D.M., Clark, M.S., 2006. Novel bioactive parathyroid hormone and related peptides in teleost fish. FEBS Lett. 580, 291–299.

Carvalho, P.L.P.F., Koch, J.F.A., Cintra, F.T., Fernandes Jr., A.C., Sartori, M.M.P., Barros, M., Padihla, P.D., Pezzato, L.L., 2018. Available phosphorus as a reproductive performance enhancer for female Nile tilapia. Aquaculture 486, 202–209.

Chavez-Sanchez, C., Martinez-Palacios, C.A., Martinez-Perez, G., Ross, L.G., 2000. Phosphorus and calcium requirements in the diet of the American cichlid *Cichlasoma urophthalmus* (Gunther). Aquacult. Nutr. 6, 1–9.

Chen, C., Zhu, W., Wu, F., Liu, M., Tan, Q., Han, D., Xie, S., Liang, X., 2016. Quantifying the dietary potassium requirement of subadult grass carp (*Ctenopharyngodon idellus*). Aquacult. Nutr. 22, 541–549.

Cheryan, M., Rackis, J.J., 1980. Phytic acid interactions in food systems. Crit. Rev. Food Sci. Nutr. 13, 297–335.

Chester Jones, I., Chan, D.K.O., Rankin, J.C., 1969. Renal function in the European eel (*Anguilla anguilla* L.). Effect of caudal neurosecretory system, corpuscles of stannius, neurohypophysial peptides and vasoactive substances. J. Endocrinol. 43, 21–31.

Choi, S.M., Kim, K.W., Kang, Y.J., Wang, X.J., Kim, J.W., Yoo, G.Y., Bai, S.C., 2005. Re-evaluation of the phosphorus requirement of juvenile olive flounder *Paralichthys olivaceus* and the bioavailability of various inorganic phosphorus sources. J. World Aquacult. Soc. 36, 217–222.

Clearwater, S.J., Baskin, S.J., Wood, C.M., McDonald, D.G., 2000. Gastrointestinal uptake and distribution of copper in rainbow trout. J. Exp. Biol. 203, 2433–2466.

Clearwater, S.J., Farag, A., Meyer, J., 2002. Bioavailability and toxicity of diet borne copper and zinc to fish. Comp. Biochem. Physiol., C 132 (132), 269–313.

Cockell, K.A., Hilton, J.W., Bettger, W.J., 1991. Chronic toxicity of dietary disodium arsenate heptahydrate to juvenile rainbow trout (*Oncorhynchus mykiss*). Arch. Environ. Contam. Toxicol. 21, 518–527.

Cooper, C.A., Bury, N.R., Grosell, M., 2006. The effects of pH and the iron redox state on iron uptake in the intestine of a marine teleost fish, gulf toadfish (*Opsanus beta*). Comp. Biochem. Physiol. A Mol. Integ. Physiol. 143, 292–298.

Copp, D.H., 1970. Endocrine regulation of calcium homeostasis. Ann. Rev. Physiol. 32, 61–87.

Cousins, R.J., 1985. Absorption, transport, and hepatic metabolism of copper and zinc: special reference to metallothionein and ceruloplasmin. Physiol. Rev. 65, 238–309.

Cowey, C.B., Knox, D., Adron, J.W., George, S., Pirie, B., 1977. The production of renal calcinosis by magnesium deficiency in rainbow trout (*Salmo gairdneri*). Br. J. Nutr. 38, 127–135.

Cyr, D.G., Eales, J.G., 1996. Interrelationships between thyroidal and reproductive systems in fish. Rev. Fish Biol. Fish. 6, 165–200.

Dabrowska, H., Meyer-Burgdorff, K., Gunther, K.D., 1989. Interaction between dietary protein and magnesium level in tilapia (*Oreochromis niloticus*). Aquaculture 76, 277–291.

Dabrowska, H., Meyer-Burgdorff, K., Gunther, K.D., 1991. Magnesium status in freshwater fish, common carp (*Cyprinus carpio*, L.) and the dietary protein-magnesium interaction. Fish Physiol. Biochem. 9, 165–172.

Damasceno, F.M., Fleuri, L.F., Sartori, M.M.P., Amorim, R.L., Pezzato, L.E., da Silva, R.L., Carvalho, P.L.P.F., Barros, M.M., 2016. Effect of dietary inorganic copper on growth performance and hematological profile of Nile tilapia subjected to heat-induced stress. Aquaculture 454, 257–264.

Davis, D.A., Robinson, E.H., 1987. Dietary phosphorus requirement of juvenile red drum *Sciaenops ocellatus*. J. World Aquacult. Soc. 18, 129–136.

Dean, M.N., Ekstrom, L., Monsonego-Ornan, E., Ballantyne, J., Witten, P.E., Riley, C., Habrakena, W., Omelon, S., 2015. Mineral homeostasis and regulation of mineralization processes in the skeletons of sharks, rays and relatives (Elasmobranchii). Semin. Cell Dev. Biol. 46, 51–67.

Dersjant-Li, Awati, A., Schulze, H., Partridge, G., 2015. Phytase in non-ruminant animal nutrition: a critical review on phytase activities in the gastrointestinal tract and influencing factors. J. Sci. Food Agric. 95, 878–896.

Dean, R.J., Shimfield, T.M., Black, K.D., 2007. Copper, zinc and cadmium in marine cage fish farm sediments: an extensive survey. Environ. Poll 145, 84–95.

Dermience, M., Lognay, G., Mathieu, F., Goyens, P., 2015. Effects of thirty elements on bone metabolism. J. Trace Elem. Med. Biol. 32, 86–106.

Desjardins, L.M., Hicks, B.D., Hilton, J.W., 1987. Iron catalysed oxidation of trout diets and its effect on growth and physiological response of rainbow trout. Fish Physiol. Biochem. 3, 173–182.

Domínguez, D., Sehnine, Z., Castro, P., Robaina, L., Fontanillas, R., Prabhu, P.A.J., Izquierdo, M., 2019. Optimum selenium levels in diets high in plant-based feedstuffs for gilthead sea bream (*Sparus aurata*) fingerlings. Aquacult. Nutr. 26, 579–589.

Dougall, D.S., Woods, L.C., Douglass, L.W., Soares, J.H., 1996. Dietary phosphorus requirement of juvenile striped bass *Morone saxatilis*. J. World Aquacult. Soc. 27, 82–91.

Dove, S.G., Kingsford, M.J., 1998. Use of otoliths and eye lenses for measuring trace-metal incorporation in fishes: a biogeographic study. Mar. Biol. 130, 377–387.

Eales, J.G., 2019. The relationship between ingested thyroid hormones, thyroid homeostasis and iodine metabolism in humans and teleost fish. Gen. Comp. Endocrinol. 280, 62–72.

Eckhert, C.D., 1998. Boron stimulates embryonic trout growth. J. Nutr. 128, 2488–2493.

Eckhert, C.D., Rowe, R.I., 1999. Embryonic dysplasia and adult retinal dystrophy in boron-deficient zebrafish. J. Trace Elem. Exp. Med. 12, 213–219.

Eid, A.E., Ghonim, S.I., 1994. Dietary zinc requirement of fingerling *Oreochromis niloticus*. Aquaculture 119, 259–264.

Eeckhout, W., Paepe, M., 1994. Total phosphorus, phytate-phosphorus and phytase activity in plant feedstuffs. An. Feed Sci. Tech. 47, 19–29.

El-Mowafi, A., Maage, A., 1998. Magnesium requirement of Atlantic salmon (*Salmo salar* L.) parr in seawater-treated fresh water. Aquacult. Nutr. 4, 31–38.

El-Zibdeh, M., Ide, K., Furuichi, M., 1996. Effects of the deletion of Mg or Fe from semi purified diets on growth and efficiency of feed utilization of yellow croaker *Nibea albiflora*. J. Facult. Agr. Kyushu Univ. 40, 391–397.

Erdman Jr., J.W., 1979. Oilseeds phytates: nutritional implications. J. Am. Oil Chem. Soc. 66, 736–741.

Erickson, R.J., Mount, D.R., Highland, T.L., Russell Hockett, J., Jenson, C.T., 2011. The relative importance of waterborne and dietborne arsenic exposure on survival and growth of juvenile rainbow trout. Aquat. Toxicol. 104, 108–115.

Eroldagan, O.T., Kumlu, M., Kir, M., Kiris, G.A., 2005. Enhancement of growth and feed utilization of the European sea bass *Dicentrarchus labrax* fed supplementary dietary salt in freshwater. Aquacult. Res. 36, 361–369.

Evans, D.H., Claiborne, J.B., 2009. Osmotic and ionic regulation n fishes. In: Evans, D.H. (Ed.), Osmotic and Ionic Regulation: Cells and Animals. Taylor and Francis Group, Boca Raton, pp. 295–366.

Feeney, G.P., Zheng, D., Kille, P., Hogstrand, C., 2005. The phylogeny of teleost ZIP and ZnT zinc transporters and their tissue specific expression and response to zinc in zebrafish. Biochim. Biophys. Acta Gene Struct. Expr. 1732, 88–95.

Fenwick, J.C., Vermette, M.G., 1989. Vitamin D3 and the renal handling of phosphate in American eels. Fish Physiol. Biochem. 7, 351–358.

Fernández, I., Gavaia, P., Darias, M.J., Gisbert, E., 2018. In: Yufera, M. (Ed.), Fat-soluble Vitamins in Fish: A Transcriptional Tissue-Specific Crosstalk that Remains to Be Unveiled and Characterized, Emerging Issues in Fish Larvae Research. Springer International Publishing AG, Switzerland, pp. 159–208.

Fleet, J., 2011. Molecular regulation of calcium/phosphate absorption. In: Feldman, D., Adams, J., Pike, W. (Eds.), Vitamin D, third ed. Elsevier, San Diego, pp. 349–362.

Flik, G., Verbost, P.M., Bonga, S.E.W., 1995. Calcium transport processes in fishes. Fish Physiol. 12, 317–342.

Fjelldal, P.G., Nordgarden, U., Hansen, T., 2007. The mineral content affects vertebral morphology in underyearling smolt of Atlantic salmon (*Salmo salar* L.). Aquaculture 270, 231–239.

Fontegné-Dicharry, S., Godin, S., Liu, H., Antony Jesu Prabhu, P., Bouyssière, B., Bueno, M., Tacon, P., Médale, F., Kaushik, S.J., 2015. Influence of the forms and levels of dietary selenium on antioxidant status and oxidative stress-related parameters in rainbow trout (*Oncorhynchus mykiss*) fry. Br. J. Nutr. 113, 1876–1887.

Fontaínhas-Fernandes, A., Russell-Pinto, F., Gomes, E., Reis-Henriques, M.A., Coimbra, J., 2000. The effect of dietary sodium chloride on some osmoregulatory parameters of the teleost, *Oreochromis niloticus*, after transfer from freshwater to seawater. Fish Physiol. Biochem. 23, 307–316.

Forbes, R.M., Erdman Jr., J.W., 1983. Bioavailability of trace mineral elements. Ann. Rev. Nutr. 3, 213–221.

Fort, D.J., Stover, E.L., Strong, P.L., Murray, F.J., 1999. Effect of boron deprivation on reproductive parameters in *Xenopus laevis*. J. Trace Elem. Exp. Med. 12, 187–204.

Francis, G., Makkar, H.P.S., Becker, K., 2001. Antinutritional factors present in plant-derived alternate feed ingredients and their effects in fish. Aquaculture 199, 197–227.

Fraser, D.R., 2018. Evolutionary biology: mysteries of vitamin D in fish. In: Feldman, D., Pike, J.W., Bouillon, R., Giovannucci, E., Goizman, D., Hewison, M. (Eds.), Vitamin D, Biochemistry, Physiology and Diagnostics, vol. 1. Academic Press, San Diego, pp. 13–27.

Frieden, E., 1984. A survey of the essential biochemical elements. In: Frieden, E. (Ed.), Biochemistry of the Essential Ultratrace Elements. Plenum, New York, pp. 1–15.

Fuentes, J., Figueiredo, J., 2006. Parathyroid hormone-related protein regulates intestinal calcium transport in sea bream (*Sparus auratus*). Am. J. Physiol. Regul. Integr. Comp. Physiol. 291, 1499–1506.

Furuichi, M., Furusho, Y., Hossain, M.A., Matsui, S., Azuma, R., 1997. Essentiality of calcium supplement to white fish meal diet for tiger puffer. J. Fac. Agric. Kyushu Univ. 42, 69–76.

Gabrielsen, B.O., Opstvedt, J., 1980. Availability of selenium in fish meal in comparison with soybean meal, corn gluten meal and selenomethionine relative to selenium in sodiumselenite for restoring glutathione peroxidase activity in selenium-depleted chicks. J. Nutr. 110, 1096–1100.

Gatlin, D.M., O'Connell, J.P., Scarpa, J., 1991. Dietary zinc requirement of the red drum, *Sciaenops ocellatus*. Aquaculture 92, 259–265.

Gatlin, D.M., Phillips, H.F., 1989. Dietary calcium, phytate and zinc interactions in channel catfish. Aquaculture 79, 259–266.

Gatlin, D.M., Robinson, E.H., Poe, W.E., Wilson, R.P., 1982. Magnesium requirement of fingerling channel catfish and signs of magnesium deficiency. J. Nutr. 112, 1182.

Gatlin, D.M., Wilson, R.P., 1983. Dietary zinc requirement of fingerling channel catfish. J. Nutr. 113, 630–635.

Gatlin, D.M., Wilson, R.P., 1984a. Zinc supplementation of practical channel catfish diets. Aquaculture 41, 31–36.

Gatlin, D.M., Wilson, R.P., 1984b. Studies on the manganese requirement of fingerling channel catfish. Aquaculture 41, 85–92.

Gatlin, D.M., Wilson, R.P., 1984c. Dietary selenium requirement of fingerling channel catfish. J. Nutr. 114, 627–633.

Gatlin, D.M., Wilson, R.P., 1986a. Characterization of iron deficiency and the dietary iron requirement of fingerling channel catfish. Aquaculture 52, 191–198.

Gatlin, D.M., Wilson, R.P., 1986b. Dietary copper requirement of fingerling channel catfish. Aquaculture 54, 277–285.

Gatlin, D.M., Mackezie, D.S., Craig, S.R., Neill, W.H., 1992. Effects of dietary sodium chloride on red drum juveniles in waters of various salinities. Prog. Fish-Cult. 54, 220–227.

Giles, M.A., 1984. Electrolyte and water balance in plasma and urine of rainbow trout (*Salmo gairdneri*) during chronic exposure to cadmium. Can. J. Fish. Aquat. Sci. 41, 1678–1685.

Giraud-Guille, M.-M., Belamie, E., Mosser, G., 2014. Organic and mineral networks in carapaces, bones and biomimetic materials. Comptes Rendus Palevol 3, 503–513.

Glover, C.N., Hogstrand, C., 2003. Effects of dissolved metals and other hydrominerals on in vivo intestinal zinc uptake in freshwater rainbow trout. Aquat. Toxicol. 62, 281–293.

Giri, A.F., Sahu, N.P., Saharan, N., Dash, G., 2014. Effect of dietary supplementation of chromium on growth and biochemical parameters of *Labeo rohita* (Hamilton) fingerlings. Indian J. Fish. 61, 73–81.

Glover, C.N., Bucking, C., Wood, C.M., 2013. The skin of fish as a transport epithelium: a review. J. Comp. Physiol. B 183, 877–891.

Glover, C.N., Hogstrand, C., 2009. In vivo characterization of intestinal zinc uptake in freshwater rainbow trout. J. Exp. Biol. 205, 141–150.

Godin, S., Fontagné-Dicharry, S., Bueno, M., Tacon, P., Antony Jesu Prabhu, P., Kaushik, S., Médale, F., Bouyssiere, B., 2015. Influence of dietary selenium species on selenoamino acid levels in rainbow trout. J. Agric. Food Chem. 63, 6484–6492.

Goyer, R.A., 1993. Lead toxicity: current concerns. Environ. Health Perspect. 100, 177–187.

Graff, I.E., Aksnes, L., Lie, O., 1999. In vitro hydroxylation of vitamin D3 and 25-hydroxy vitamin D3 in tissues of Atlantic salmon *Salmo salar*, Atlantic mackerel *Scomber scombrus*, Atlantic halibut *Hippoglossus hippoglossus* and Atlantic cod *Gadus morhua*. Aquacult. Nutr. 5, 23–32.

Greiner, R., Konietzny, U., 2002. Molecular and catalytic properties of phytate-degrading enzymes (phytases). Int. J. Food Sci. Tech. 37, 791–812.

Grosell, M., 2012. Copper. In: Wood, C.M., Farrell, A.P., Brauner, C.J. (Eds.), Homeostasis and Toxicology of Essential Metals. Academic Press, Elsevier, San Diego, pp. 53–133.

Grosell, M., Wood, C.M., 2002. Copper uptake across rainbow trout gills: mechanisms of apical entry. J. Exp. Biol. 205, 1179–1188.

Guerreiro, P.M., Renfro, J.L., Power, D.M., Canario, A.V., 2007. The parathyroid hormone family of peptides: structure, tissue distribution, regulation, and potential functional roles in calcium and phosphate balance in fish. Am. J. Physiol. Regul. Integr. Comp. Physiol. 292, R679–R696.

Guo, R., Henry, P.R., Holwerda, R.A., Cao, J., Littell, R.C., Miles, R.D., Ammerman, C.B., 2001. Chemical characteristics and relative bioavailability of supplemental organic copper sources for poultry. J. Anim. Sci. 79, 1132–1141.

Hall, B.K., 2015. Bones and Cartilage: Developmental Skeletal Biology. Academic Press, London.

Hamilton, S.J., 2004. Review of selenium toxicity in the aquatic food chain. Sci. Total Environ. 326, 1–31.

Hamilton, S.J., Buhl, K.J., 1990. Acute toxicity of boron molybdenum and selenium to fry of Chinook salmon and coho salmon. Arch. Environ. Contam. Toxicol. 19, 366–373.

Hamre, K., 2011. Metabolism, interactions, requirements and functions of vitamin E in fish. Aquacult. Nutr. 17, 98–115.

Han, D., Xie, S., Liu, M., Xiao, X., Liu, H., Zhu, X., Yang, Y., 2011. The effects of dietary selenium on growth performances, oxidative stress and tissue selenium concentration of gibel carp (*Carassius auratus* Gibelio). Aquacult. Nutr. 17, e741–e749.

Handy, R.D., 1996. Dietary exposure to trace metals in fish. In: Taylor, E.W. (Ed.), Toxicology of Aquatic Pollution. Cambridge University Press, Cambridge, pp. 29–60.

Handy, R.D., Musonda, M.M., Phillips, C., Falla, S.J., 2000. Mechanisms of gastrointestinal copper absorption in the African walking catfish: copper dose-effects and a novel anion-dependent pathway in the intestine. J. Exp. Biol. 203, 2365–3237.

Hardy, R.W., Sullivan, C.V., Koziol, A.M., 1987. Absorption, body distribution, and excretion of dietary zinc by rainbow trout (*Salmo gairdneri*). Fish Physiol. Biochem. 3, 133–143.

Harpaz, S., Hakim, Y., Slosman, T., Eroldogan, O.T., 2005. Effect of adding salt to the diet of Asian sea bass *Lates calcarifer* reared in fresh or salt water recirculating tanks on growth and brush border enzyme activities. Aquaculture 248, 315–324.

Hattenhauer, O., Traebert, M., Murer, H., Biber, J., 1999. Regulation of small intestinal Na-P(i) type IIb cotransporter by dietary phosphate intake. Am. J. Physiol. 277, G756–G762.

Heaney, R.P., 2012. Phosphorus. In: Erdman Jr., J.F., Macdonald, I.A., Zeisel, S.H. (Eds.), Present Knowledge of Nutrition. John Wiley & Sons Inc., Ames, pp. 447–458.

Henry, P.R., Ammerman, C.B., 1995. Selenium bioavailability. In: Ammerman, C.B., Baker, D.H., Lewis, A.J. (Eds.), Bioavailability of Nutrients for Animal: Amino Acids, Minerals, and Vitamins. Academic Press, San Diego, pp. 303–336.

Henry, P.R., Miller, E.R., 1995. Iron bioavailability. In: Ammerman, C.B., Baker, D.H., Lewis, A.J. (Eds.), Bioavailability of Nutrients for Animal: Amino Acids, Minerals, and Vitamins. Academic Press, San Diego, pp. 169–199.

Hershberger, W.K., Pratschner, G.A., 1981. Iron binding capacities of six transferrin phenotypes of Coho salmon and potential fish health applications. Prog. Fish-Cult. 43, 27–31.

Hertramp, J.W., Piedad-Pascal, F., 2000. Handbook on Ingredients for Aquaculture Feeds. Kluwer Academic Publishers, Dordrecht, p. 573.

Hertz, Y., Madar, Z., Hepher, B., Gertler, A., 1989. Glucose metabolism in the common carp (*Cyprinus carpio* L.): the effects of cobalt and chromium. Aquaculture 76, 255–267.

Hickman, C.P., Trump, B.F., 1969. The kidney. In: Hoar, W.S., Randall, D.J. (Eds.), Fish Physiology, vol. 1. Academic Press, New York, pp. 91–239.

Higgs, D.A., Eales, J.G., 1978. Radiothyroxine kinetics in yearling brook trout, *Salvelinus fontinalis* (Mitchill), on different levels of dietary intake. Can. J. Zool. 56, 80–85.

Higgs, D.A., Fagerlund, U.H.M., Eales, J.G., McBride, J.R., 1982. Application of thyroid and steroid hormones as anabolic agents in fish culture. Comp. Biochem. Physiol. 73B, 143–176.

Hill, C.H., Matrone, G., 1970. Chemical parameters in the study of in vivo and in vitro interactions of transition elements. Fed. Proc. 29, 1474–1481.

Hilton, J.W., 1989. The interaction of vitamins, minerals and diet composition in the diet of fish. Aquaculture 79, 223–244.

Hilton, J.W., Bettger, W.J., 1988. Dietary vanadium toxicity in juvenile rainbow trout: a preliminary study. Aquat. Toxicol. 12, 63–71.

Hilton, K.B., Lambert, L.A., 2008. Molecular evolution and characterization of hepcidin gene products in vertebrates. Gene 415, 40–48.

Hilton, J.W., Hodson, P.V., Slinger, S.J., 1980. The requirement and toxicity of selenium in rainbow trout (*Salmo gairdneri*). J. Nutr. 110, 2527–2535.

Hodge, H.C., Smith, F.A., 1981. Fluoride. In: Bronner, F., Coburn, J.W. (Eds.), Disorders of Mineral Metabolism, vol. 1. Academic Press, New York, pp. 439–483.

Hodson, P.V., Hilton, J.W., 1983. The nutritional requirements and toxicity to fish of dietary and waterborne selenium. Ecol. Bull. 35, 335–340.

Hodson, P.V., Hilton, J.W., Slinger, S.J., 1986. Accumulation of waterborne selenium by rainbow trout (*Salmo gairdneri*), eggs, fry and juveniles. Fish Physiol. Biochem. 1, 187–196.

Hogstrand, C., 2012. Zinc. In: Wood, C.M., Farrell, A.P., Brauner, C.J. (Eds.), Homeostasis and Toxicology of Essential Metals. Academic Press, Elsevier, San Diego, pp. 135–200.

Hori, R., Iwasaki, S., 1976. On the manganese content of the egg of *Oryzias latipes* and its changes during early development. Protoplasma 87, 403–407.

Hosnedlova, B., Kepinska, M., Skalickova, S., Fernandez, C., Ruttkay-Nedecky, B., Malevu, T.D., Sochor, J.S., Baron, M., Melcova, M., Zidkova, J., Kizek, R., 2017. A summary of new findings on the biological effects of selenium in selected animal species—a critical review. Int. J. Mol. Sci. 18, 2209.

Hossain, M.A., Furuichi, M., 1999a. Requirement of Japanese flounder for dietary calcium. Aquacult. Sci. 47, 567–571.

Hossain, M.A., Furuichi, M., 1999b. Necessity of dietary calcium supplement in black sea bream. Fish. Sci. 65, 893–897.

Hossain, M.A., Furuichi, M., 1999c. Effect of deletion of calcium supplement from purified diet on growth and bone mineralization in red sea bream. J. Fac. Agric. Kyushu Univ. 44, 91–97.

Hossain, M.A., Furuichi, M., 1999d. Calcium requirement of tiger puffer fed a semi-purified diet. Aquacult. Int. 7, 287–293.

Hossain, M.A., Furuichi, M., 1999e. Dietary calcium requirement of giant croaker Nibea japonica. J. Fac. Agric. Kyushu Univ. 44, 99–104.

Hossain, M.A., Yoshimatsu, T., 2014. Dietary calcium requirements of fishes. Aquacult. Nutr. 20, 1–11.

Houng-Yung, C., Yu-Chun, C., Li-Chi, H., Meng-Hsien, C., 2014. Dietary zinc requirements of juvenile grouper, *Epinephelus malabaricus*. Aquaculture 432, 360–364.

Houston, A., 1985. Erythrocytic magnesium in freshwater fishes. Magnesium 4, 106–128.

Hoyle, I., Shaw, B.J., Handy, R.D., 2007. Dietary copper exposure in the African walking catfish, *Clarias gariepinus*: transient osmoregulatory disturbances and oxidative stress. Aquat. Toxicol. 83, 62–72.

Hua, K., Bureau, D.P., 2006. Modelling digestible phosphorus content of salmonid fish feeds. Aquaculture 254, 455–465.

Huang, F., Jiang, M., Wen, H., Wu, F., Liu, W., Tian, J., Yang, C., 2015. Dietary zinc requirement of adult Nile tilapia (*Oreochromis niloticus*) fed semi-purified diets, and effects on tissue mineral composition and antioxidant responses. Aquaculture 439, 53–59.

Huang, Q.-C., Kwaku, A., Wang, E.-L., Dong, X.H., 2018. Iron bioavailability of different sources in juvenile grouper *Epinephelus coioides*. Aquacult. Res. 49, 2799–2807.

Hwang, P., Yang, C., 1997. Modulation of calcium uptake in cadmium-pretreated tilapia (*Oreochromis mossambicus*) larvae. Fish Physiol. Biochem. 16, 403–410.

Hunt, D.W.C., Eales, J.G., 1979. Iodine balance in rainbow trout (*Salmo gairdneri*) and effects of testosterone propionate. J. Fish. Res. Board Can. 36, 282–285.

Hwang, P.-P., Lee, T.-H., Lin, L.-Y., 2011. Ion regulation in fish gills: recent progress in the cellular and molecular mechanisms. Am. J. Physiol. Regul. Integr. Comp. Physiol. 301, R28–R47.

Ikeda, Y., Ozaki, H., Vematsu, K., 1973. Effect of enriched diet with iron in culture of yellow-tail. J. Tokyo Univ. Fish 59, 91–99.

Ishak, I.M., Dollar, A.M., 1968. Studies on manganese uptake in *Tilapia mossambica* and *Salmo gairdneri*. I. Growth and survival of *Tilapia mossambica* in response to manganese. Hydrobiologia 31, 572–584.

Janz, D.M., 2012. Selenium. In: Wood, C.M., Farrell, A.P., Brauner, C.J. (Eds.), Homeostasis and Toxicology of Essential Metals. Academic Press, San Diego, USA, pp. 329–375.

Jeng, S.S., Sun, L.T., 1981. Effects of dietary zinc levels on zinc concentrations in tissues of common carp. J. Nutr. 111, 134–140.

Jiang, M., Wu, F., Huang, F., Wen, H., Liu, W., Tian, J., Changgeng, Y., Wang, W., 2016. Effects of dietary Zn on growth performance, antioxidant responses, and sperm motility of adult blunt snout bream, *Megalobrama amblycephala*. Aquaculture 464, 121–128.

Julshamn, K., Braekkan, O.R., 1975. The relation between total cobalt and cobalt in vitamin B_{12} during the maturation of ovaries in salmon (*Salmo salar*). Comp. Biochem. Physiol. 52B, 381–382.

Julshamn, K., Haugnes, J., Utne, F., 1978. The contents of 14 major and minor elements (minerals) in Norwegian fish species and fish by-products, determined by atomic absorption spectrophotometry. Fisk. Dir. Skr. Ser. Ernaer. 1, 117–135.

Jürss, K., 1980. The effect of changes in external salinity on the free amino acids and two aminotransferases of white muscle from fasted *Salmo gairdneri,* Richardson. Comp. Biochem. Physiol. 65A, 501–504.

Jürss, K., Bittorf, T., Vokler, T., Wacke, R., 1983. Influence of nutrition on biochemical sea water adaptation of the rainbow trout (*Salmo gairdneri*, Richardson). Comp. Biochem. Physiol. 75B, 713–717.

Kambe, T., Tsuji, T., Hashimoto, A., Itsumura, N., 2015. The physiological, biochemical, and molecular roles of zinc transporters in zinc homeostasis and metabolism. Physiol. Rev. 95, 749–784.

Kamunde, K., Niyogi, S., Wood, C.M., 2005. Interaction of dietary sodium chloride and waterborne copper in rainbow trout: sodium and chloride homeostasis, copper homeostasis, and chronic copper toxicity. Can. J. Fish. Aquat. Sci. 62, 390–399.

Kamunde, C., Grosell, M., Higgs, D., Wood, C.M., 2002. Copper metabolism in actively growing rainbow trout (*Oncorhynchus mykiss*): interactions between dietary and waterborne copper uptake. J. Exp. Biol. 205, 279–290.

Kashiwada, K., Teshima, S., Kanazawa, A., 1970. Studies on production of B vitamins by intestinal bacteria of fish 5. Evidence of production of vitamin B_{12} by microorganisms in the intestinal canal of carp. Bull. Jap. Soc. Sci. Fish. 36, 421–424.

Kawatsu, H., 1972. Studies on the anemia of fish. 5. Dietary iron deficient anemia in brook trout (*Salvelinus fontinalis*). Bull. Freshw. Fish. Res. Lab. 22, 59–67.

Keen, C.L., Ensunsa, J.L., Clegg, M.S., 1999. Manganese metabolism in animals and humans including the toxicity of manganese. Met. Ions Biol. Syst. 37, 89–121.

Ketola, H.G., 1975. Requirement of Atlantic salmon for dietary phosphorus. Trans. Am. Fish. Soc. 104, 548–551.

Ketola, H.G., 1979. Influence of dietary zinc on cataracts in rainbow trout (Salmo gairdneri). J. Nutr. 105, 965–969.

Ketola, H.G., Richmond, M.E., 1994. Requirement of rainbow trout for dietary phosphorus and its relationship to the amount discharged in hatchery effluents. Trans. Am. Fish. Soc. 123, 587–594.

Kim, S.G., Kang, J.C., 2004. Effect of dietary copper exposure on accumulation, growth and hematological parameters of the juvenile rockfish, *Sebastes schlegeli*. Mar. Environ. Res. 58, 65–82.

Kirchner, S., McDaniel, N.K., Sugiura, S.H., Soteropoulos, P., Tian, B., Fletcher, J.W., Ferraris, R.P., 2007. Salmonid microarrays identify intestinal genes that reliably monitor P deficiency in rainbow trout aquaculture. Anim. Genet. 38, 319–331.

Kirchner, S., McDaniel, N.K., Sugiura, S.H., Soteropoulos, P., Tian, B., Fletcher, J.W., Ferraris, R., et al., 2007. Salmonid microarrays identify intestinal genes that reliably monitor P deficiency in rainbow trout aquaculture. Anim. Genet. 38, 319–331.

Kirschner, L.B., 2004. The mechanism of sodium chloride uptake in hyperregulating aquatic animals. J. Exp. Biol. 207, 1439–1452.

Knight, R., Marlatt, V.L., Baker, J.A., Lo, B.P., Debruyn, A.M.H., Elphick, J.R., Martyniuk, C.J., 2016. Dietary selenium disrupts hepatic triglyceride stores and transcriptional networks associated with growth and Notch signaling in juvenile rainbow trout. Aquat. Toxicol. 180, 103–114.

Knox, D., Cowey, C.B., Adron, J.W., 1981. Studies on the nutrition of salmonid fish. The magnesium requirement of rainbow trout (*Salmo gairdneri*). Br. J. Nutr. 45, 137–148.

Knox, D., Cowey, C.B., Adron, J.W., 1982. Effects of dietary copper and copper: zinc ratio on rainbow trout (*Salmo gairdneri*). Aquaculture 27, 111–119.

Knox, D., Cowey, C.B., Adron, J.W., 1983. Studies on the nutrition of rainbow trout (*Salmo gairdneri*). Magnesium deficiency: the effect of feeding with a Mg-supplemented diet. Br. J. Nutr. 50, 121–127.

Köprücü, K., Özdemir, Y., 2005. Apparent digestibility of selected feed ingredients for Nile tilapia (*Oreochromis niloticus*). Aquaculture 250, 308–316.

Krogdahl, A., Penn, M., Thorsen, J., Refstie, S., Bakke, A.M., 2010. Important anti-nutrients in plant feedstuffs for aquaculture: an update on recent findings regarding responses in salmonids. Aquacult. Res. 41, 333–344.

Küçükbay, F., Yazlak, H., Karaca, I., Sahin, N., Tuzcu, M., Cakmak, M., Sahin, K., 2009. The effects of dietary organic or inorganic selenium in rainbow trout (*Oncorhynchus mykiss*) undercrowding conditions. Aquacult. Nutr. 15 (6), 569–576.

Kuenze, J., Buhringer, H., Harms, V., 1978. Accumulation of cobalt during embryonic development of rainbow trout (*Salmo gairdneri* Rich.). Aquaculture 13, 61–66.

Kumai, Y., Perry, S.F., 2012. Mechanisms and regulation of Na+ uptake by freshwater fish. Respir. Physiol. Neurobiol. 184, 249–256.

Kumar, V., Sinha, A.K., Makkar, H.P.S., De Boeck, G., Becker, K., 2012. Phytate and phytase in fish nutrition. J. Anim. Physiol. Anim. Nutr. 96, 335–364.

Kwong, R.W., Hamilton, C.D., Niyogi, S., 2013. Effects of elevated dietary iron on the gastrointestinal expression of Nramp genes and iron homeostasis in rainbow trout (*Oncorhynchus mykiss*). Fish Physiol. Biochem. 39, 363–372.

Kwong, R.W., Perry, S.F., 2015. An essential role for parathyroid hormone in gill formation and differentiation of ion-transporting cells in developing zebrafish. Endocrinology 156, 2384–2394.

Kwong, R.W.M., Niyogi, S., 2009. The interactions of iron with other divalent metals in the intestinal tract of a freshwater teleost, rainbow trout (*Oncorhynchus mykiss*). Comp. Biochem. Physiol. C Toxicol. Pharm. 150, 442–449.

Labunskyy, V.M., Hatfield, D.L., Gladyshev, V.N., 2014. Selenoproteins: molecular pathways and physiological roles. Physiol. Rev. 94, 739–777.

Lake, J., Gravel, C., Koko, G.K.D., Robert, C., Vandenberg, G.W., 2010. Combining suppressive subtractive hybridization and cDNA microarrays to identify dietary phosphorus responsive genes of the rainbow trout (*Oncorhynchus mykiss*) kidney. Comp. Biochem. Physiol. D 5, 24–35.

Lall, S.P., 1991. Digestibility, metabolism and excretion of dietary phosphorous in fish. In: Cowey, C.B., Cho, C.Y. (Eds.), Nutritional Strategies and Aquaculture Waste. University of Guelph, Guelph, ON, Canada, pp. 21–36.

Lall, S.P., 1995. Macro and trace elements in fish and shellfish. In: Ruiter, A. (Ed.), Fish and Fishery Products. CAB International, Wallingford, UK, pp. 187–213.

Lall, S.P., 2002. The minerals. In: Halver, J.E., Hardy, R.W. (Eds.), Fish Nutrition, third ed. Academic Press, San Diego, pp. 259–308.

Lall, S.P., 2010. Disorders of nutrition and metabolism. In: Leatherland, J.F., Woo, P.T.K. (Eds.), Fish Diseases and Disorders, Non-infectious Disorders, vol. 2. CABI, Wallingford, pp. 202–237.

Lall, S.P., Bishop, F.J., 1977. Studies on mineral and protein utilization by Atlantic salmon (*Salmo salar*) grown in sea water. Fish. Mar. Serv. Tech. Rep. 688, 1–16.

Lall, S.P., Dumas, A., 2015. Nutritional requirements of cultured fish: formulating nutritionally adequate feeds. In: Davis, D.A. (Ed.), Feed and Feeding Practices in Aquaculture. Woodhead Publishing, Cambridge, UK, pp. 53–109.

Lall, S.P., Lewis-McCrea, L.M., 2007. Role of nutrients in skeletal metabolism and pathology in fish-an overview. Aquaculture 267, 3–19.

Lall, S.P., Milley, J.E., 2008a. Impact of aquaculture on aquatic environment: trace minerals discharge. In: Schlegel, P., Durosoy, S., Jongbloed, A.W. (Eds.), Trace Elements in Animal Production Systems. Wageningen Academic Publishers, The Netherlands, pp. 77–87.

Lall, S.P., Milley, J.E., 2008b. Trace mineral requirements of fish and crustaceans. In: Schlegel, P., Durosoy, S., Jongbloed, A.W. (Eds.), Trace Elements in Animal Production Systems. Wageningen Academic Publishers, The Netherlands, pp. 203–214.

Lall, S.P., Paterson, W.D., Hines, J.A., Adams, N.J., 1985. Control of bacterial kidney disease in Atlantic salmon *Salmo salar* L. by dietary modification. J. Fish. Dis. 8, 113–124.

Lanno, R.P., Slinger, S.J., Hilton, J.W., 1985. Maximum tolerable and toxicity levels of dietary copper in rainbow trout (*Salmo gairdneri* R.). Aquaculture 49, 257–268.

Le, K.T., Fotedar, R., 2014. Bioavailability of selenium from different dietary sources in yellowtail kingfish (*Seriola lalandi*). Aquaculture 420–421, 57–62.

Leach, R.M., Harris, E.D., 1997. Manganese. In: O'Dell, B.L., Sunde, R.A. (Eds.), Handbook of Nutritionally Essential Minerals. Marcel Dekker, New York, pp. 335–355.

Lemos, D.E.L., Tacon, A.G.J., 2017. Use of phytases in fish and shrimp feeds: a review. Rev. Aquacult. 9, 266–282.

Li, M.H., Robinson, E.H., 1996. Comparison of chelated zinc and zinc sulfate as zinc sources for growth and bone mineralization of channel catfish (*Ictalurus punctatus*) fed practical diets. Aquaculture 146, 237–243.

Lewis-McCrea, L.M., Lall, S.P., 2010. Effects of phosphorus and vitamin C deficiency, vitamin A toxicity, and lipid peroxidation on skeletal abnormalities in Atlantic halibut (*Hippoglossus hippoglossus*). J. Appl. Ichthy 26, 334–343.

Li, M.-R., Huang, C.-H., 2015. Effect of dietary zinc level on growth, enzyme activity and body trace elements of hybrid tilapia, *Oreochromis niloticus* × *O. aureus*, fed soya bean meal-based diets. Aquacult. Nutr. 22, 1320–1327.

Liang, J.J., Tian, L.X., Liu, Y.J., Yang, H.J., Liang, G.Y., 2012a. Dietary magnesium requirement and effects on growth and tissue magnesium content of juvenile grass carp (*Ctenopharyngodon idella*). Aquacult. Nutr. 18, 56–64.

Liang, J.-J., Yang, H.-J., Liu, Y.-J., Tian, L.-X., Liang, G.-Y., 2012b. Dietary zinc requirement of juvenile grass carp (*Ctenopharyngodon idella*) based on growth and mineralization. Aquacult. Nutr. 18, 380–387.

Liang, J.-J., Liu, Y.-J., Tian, L.-X., Yang, H.-J., Liang, G.-Y., 2012d. Dietary available phosphorus requirement of juvenile grass carp (Ctenopharyngodon idella). Aquacul. Nutr. 18, 181–188.

Liang, J.-J., Liu, Y.-J., Yang, Z.-N., Tian, L.-X., Yang, H.-J., LIang, G.-Y., 2012c. Dietary calcium requirement and effects on growth and tissue calcium content of juvenile grass carp (*Ctenopharyngodon idella*). Aquacult. Nutr. 18, 544–550.

Liang, J.J., Yang, H.J., Liu, Y.J., Tian, L.X., 2014. Dietary potassium requirement of juvenile grass carp (*Ctenopharyngodon idella*, Val.) based on growth and tissue potassium content. Aquacult. Res. 45, 701–708.

Limsuwan, T., Lovell, R.T., 1981. Intestinal synthesis and absorption of vitamin B_{12} in channel catfish. J. Nutr. 111, 2125–2132.

Lin, C.H., Hwang, P.P., 2016. The control of calcium metabolism in zebrafish (*Danio rerio*). Int. J. Mol. Sci. 17, 1783. https://doi.org/10.3390/ijms17111783.

Lin, Y.H., Shiau, S.Y., 2005. Dietary selenium requirements of juvenile grouper, *Epinephelus malabaricus*. Aquaculture 250, 356–363.

Lin, Y.H., Lin, S.M., Shiau, S.Y., 2008a. Dietary manganese requirements of juvenile tilapia, *Oreochromis niloticus* x *O. aureus*. Aquaculture 284, 207–210.

Lin, Y.H., Shie, Y.-Y., Shiau, S.-Y., 2008b. Dietary copper requirements of juvenile grouper, *Epinephelus malabaricus*. Aquaculture 274, 161–165.

Lin, Y.H., Jiang, L.-C., Shiau, S.-Y., 2008c. Dietary zinc requirements of juvenile tilapia, *Oreochromis niloticus* x *O. aureus*. J. Fish. Soc. Taiwan 35, 117–125.

Lin, Y.H., Shih, C.-C., Kent, M., Shiau, S.-Y., 2010a. Dietary copper requirement re-evaluation for juvenile grouper, *Epinephelus malabaricus*, with an organic copper source. Aquaculture 310, 173–177.

Lin, Y.-H., Wu, J.-Y., Shiau, S.-Y., 2010b. Dietary cobalt can promote gastrointestinal bacterial production of vitamin B_{12} in sufficient amounts to supply growth requirements of grouper, *Epinephelus malabaricus*. Aquaculture 302, 89–93.

Lin, Y.-H., Ku, C.-Y., Shiau, S.-Y., 2013. Estimation of dietary magnesium requirements of juvenile tilapia, *Oreochromis niloticus*×*Oreochromis aureus*, reared in freshwater and seawater. Aquaculture 380–383, 47–51.

Linder, M.C., 2016. Ceruloplasmin and other copper binding components of blood plasma and their functions: an update. Metallomics 8, 887–905.

Ling, J., Feng, L., Liu, Y., Jiang, J., Jiang, W.D., Hu, K., Li, S.H., Zhou, X.Q., 2010. Effect of dietary iron level on growth, body composition and intestinal enzyme activities of juvenile Jian carp (*Cyprinus carpio* Var. Jian). Aquacult. Nutr. 16, 616–624.

Liu, K., Ai, Q.H., Mai, K.S., Zhang, W.B., Zhang, L., Zheng, S.X., 2012. Dietary manganese requirement for juvenile cobia, *Rachycentron canadum* L. Aquacult. Nutr. 19, 461–467.

Liu, K., Wang, X.J., Ai, Q., Mai, K., Zhang, W., 2010a. Dietary selenium requirement for juvenile cobia, *Rachycentron canadum* L. Aquacult. Res. 41, e594–e601.

Liu, T., Wen, H., Jiang, M., Yuan, D., Gao, P., Zhao, Y., Wu, F., Liu, W., 2010b. Effect of dietary chromium picolinate on growth performance and blood parameters in grass carp fingerling, *Ctenopharyngodon idellus*. Fish Physiol. Biochem. 36, 565–572.

Lock, E.-J., Waagbø, R., Wendelaar Bonga, S., Flik, G., 2010. The significance of vitamin D for fish: a review. Aquacult. Nutr. 16, 100–116.

Lönnerdal, B., 2000. Dietary factors influencing zinc absorption. J. Nutr. 130, 1378S–1383S.
Lorentzen, M., Maage, A., Julshamn, K., 1996. Manganese supplementation of a practical, fish meal based diet for Atlantic salmon parr. Aquacult. Nutr. 2, 121–125.
Lorentzen, M., Maage, A., Julshamn, K., 1998. Supplementing copper to a fish meal based diet fed to Atlantic salmon parr affects liver copper and selenium concentrations. Aquacult. Nutr. 4, 67–72.
Loro, V.L., Nogueira, L., Nadella, S.R., Wood, C.M., 2014. Zinc bioaccumulation and ionoregulatory impacts in *Fundulus heteroclitus* exposed to sublethal waterborne zinc at different salinities. Comp. Biochem. Physiol. Toxic. Pharm. 166 C, 96–104.
Love, R.M., 1980. In: Love, R.M. (Ed.), The Chemical Biology of Fishes. Academic Press, London, pp. 133–151.
Lovell, R.T., 1978. Dietary phosphorus requirement of channel catfish (*Ictalurus punctatus*). Trans. Am. Fish. Soc. 107, 617–621.
Lunde, G., 1973. Trace metal contents of fish meal and of the lipid phase extracted from fish meal. J. Food Sci. Agr. 24, 413–419.
Luo, Z., Guo, G.-Y., Gao, Y., Ye, H.-M., Xi, W.-Q., Liu, X., 2017. Effect of dietary iron (Fe) levels on growth performance, hepatic lipid metabolism and antioxidant responses in juvenile yellow catfish *Pelteobagrus fulvidraco*. Aquacul. Nutr. 23, 1475–1482.
Luo, Z., Tan, X.-Y., Zheng, J.-L., Chen, Q.-L., Liu, C.-X., 2011. Quantitative dietary zinc requirement of juvenile yellow catfish *Pelteobagrus fulvidraco*, and effects on hepatic intermediary metabolism and antioxidant responses. Aquaculture 319, 150–155.
Ma, R., Hou, H., Mai, K., Bharadwaj, A.S., Ji, F., Zhang, W., 2015. Comparative study on the effects of chelated or inorganic manganese in diets containing tricalcium phosphate and phytate on the growth performance and physiological responses of turbot *Scophthalmus maximus*. Aquacult. Nutr. 21, 780–787.
Maage, A., Julshamn, K., 1993. Assessment of zinc status in juvenile Atlantic salmon (*Salmo salar*) by measurement of whole body and tissue levels of zinc. Aquaculture 117, 179–191.
Maage, A., Lygren, B., Ali El-Mowafi, A.F., 2000. Manganese requirement of Atlantic salmon (*Salmo salar*) fry. Fish. Sci. 66, 1–8.
Mai, K., Zhang, C., Ai, Q., Duan, Q., Xu, W., Zhang, L., Liufu, Z., Tan, B., 2006. Dietary phosphorus requirement of large yellow croaker, *Pseudosciaena crocea* R. Aquaculture 251, 346–353.
Malhotra, N., Ger, T.-R., Uapipatanakul, B., Huang, J.-C., Chen, K.-C., Hsiao, C.-D., 2020. Review of copper and copper nanoparticle toxicity in fish. Nanomaterials 10, 1126.
Mangos, S., Amaral, A.P., Faul, C., Jüppner, H., Jüppner, 2012. Expression of fgf23 and a klotho in developing embryonic tissues and adult kidney of the zebrafish, *Danio rerio*. Nephrol. Dial. Transplant. 27, 4314–4322.
Mager, E.M., 2012. Lead. In: Wood, C.M., Farrell, A.P., Brauner, C.J. (Eds.), Homeostasis and Toxicology of Non-essential Metals. Academic Press, San Diego, pp. 185–236.
Maret, W., 2013. Zinc Biochemistry: from a single zinc enzyme to a key element of life. Adv. Nutr. 4, 82–91.
Marine, D., 1914. Further observations and experiments on goiter in brook trout. J. Exp. Med. 19, 70–75.
Mariotti, M., Ridge, P.G., Zhang, Y., Lobanov, A.V., Pringle, T.H., Guigo, R., Hatfield, D.L., Gladyshev, V.N., 2012. Composition and evolution of the vertebrate and mammalian selenoproteomes. PLoS One 7 (3), e33066.
Martins, C.I.M., Pistrin, M.G., Ende, S.S.W., Eding, E.H., Verreth, J.A.J., 2009. The accumulation of substances in Recirculating Aquaculture Systems (RAS) affects embryonic and larval development in common carp *Cyprinus carpio*. Aquaculture 291, 65–73.
Marshall, W.S., Grosell, M., 2006. Ion transport, osmoregulation acid-base balance. In: Evans, D.H., Claiborne, J.B. (Eds.), Physiology of Fishes. CRC Press, Boca Raton, pp. 170–230.
McClain, W., Gatlin, D.M., 1988. Dietary zinc requirement of *Oreochromis aureus* and effects of dietary calcium and phytate on zinc bioavailability. J. World Aquacult. Soc. 19, 103–108.
McCormick, S.D., Hasegawa, S., Hirano, T., 1992. Calcium uptake in the skin of a freshwater teleost. Proc. Natl. Acad. Sci. U. S. A. 89, 3635–3638.
McCormick, S.D., 2012a. In: McCormick, S.D., Farrell, A.P., Brauner, C.J. (Eds.), Smolt Physiology and Endocrinology, Fish Physiology: Euryhaline Fishes, vol. 32. Academic Press, San Diego, pp. 199–251.

McCormick, S.D., 2012b. The hormonal control of osmoregulation in teleost fish. In: Farrell, A.P. (Ed.), Encyclopedia of Fish Physiology: From Genome to Environment, vol. 2. Academic Press, San Diego, pp. 1466–1473.

McDonald, D.G., Wood, C.M., 1993. Branchial mechanism of acclimation to metals in freshwater fish. Fish Ecophysiol. Fish Fish. Ser. 9, 297–321.

McGeer, J.C., Niyogi, S., Scott Smith, D., 2012. Cadmium. In: Wood, C.M., Farrell, A.P., Brauner, C.J. (Eds.), Homeostasis and Toxicology of Non-essential Metals. Academic Press, San Diego, pp. 125–184.

McIntyre, D.O., Linton, T.K., 2012. Arsenic. In: Wood, C.M., Farrell, A.P., Brauner, C.J. (Eds.), Homeostasis and Toxicology of Non-essential Metals. Academic Press, San Diego, pp. 297–349.

Mertz, W., 1986. Trace elements in human and animal nutrition, 5th ed., vol.1. Academic Press, Orlando, FL.

Mertz, W., 1998. Review of the scientific basis for establishing the essentiality of trace elements. Biol. Trace Elem. Res. 66, 185–191.

Metz, J.R., Leeuwis, R.H.J., Zethof, J., Flik, G., 2014. Zebrafish (*Danio rerio*) in calcium-poor water mobilise calcium and phosphorus from scale. J. Appl. Ichthyol. 30, 671–677.

Miller, D.W., Vetter, R.J., Atchison, G.J., 1980. Effect of temperature and dissolved oxygen on uptake and retention of ^{54}Mn in fish. Health Phys. 38, 221–225.

Mischke, C.M., 2012. Aquaculture Pond Fertilization: Impacts of Nutrient Input on Production. Wiley-Blackwell.

Mizuno, S., Misaka, N., Ando, D., Torao, M., Urabe, H., Kitamura, T., 2007. Effects of diets supplemented with iron citrate on some physiological parameters and on burst swimming velocity in smoltifying hatchery-reared masu salmon (*Oncorhynchus masou*). Aquaculture 273, 284–297.

Moazenzadeh, K., Islami, H.R., Zamini, A., Soltani, M., 2017. Dietary zinc requirement of Siberian sturgeon (*Acipenser baerii*, Brandt 1869) juveniles, based on the growth performance and blood parameters. Int. Aquat. Res. 9, 25–35.

Moren, M., Sloth, J.J., Hamre, K., 2008. Uptake of iodide from water in Atlantic halibut larvae (*Hippoglossus hippoglossus* L.). Aquaculture 285, 174–178.

Morin, P.P., Hara, T.J., Eales, J.G., 1995. T4 depresses olfactory responses to L-alanine and plasma T3 and T4 production in smoltifying Atlantic salmon. Am. J. Physiol. 269, R1434–R1440.

Moss, M.L., 1963. The biology of acellular bone. Ann. N. Y. Acad. Sci. 74, 337–350.

Musharraf, M., Khan, M.A., 2018. Dietary magnesium requirement for fingerlings of Rohu (*Labeo rohita*). Aquaculture 496, 96–104.

Musharraf, M., Khan, M.A., 2019a. Dietary potassium requirement of fingerling Indian major carp, *Labeo rohita* (Hamilton), based on growth parameters, gill Na^+-K^+ATPase activity, tissue mineralization and antioxidant activities. Aquacult. Nutr. 25, 271–280.

Musharraf, M., Khan, M.A., 2019b. Dietary zinc requirement of fingerling Indian major carp, *Labeo rohita* (Hamilton). Aquaculture 503, 489–498.

Nakamura, Y., Yamada, J., 1980. Effects of dietary calcium levels. Ca/P ratios, and calcium components on calcium absorption rate in carp. Bull. Fac. Fish. Hokkaido Univ. 31, 277–282.

Nakamura, Y., 1982. Effects of dietary phosphorous and calcium contents on the absorption of phosphorous in the digestive tract of carp. Fish. Sci. 51, 605–608.

Nakamura, Y., Hirano, T., 1986. Effects of hypophysectomy on absorption of inorganic phosphate by the eel intestine. Comp. Biochem. Physiol. 84A, 595–599.

Naser, N., 2000. Role of Iron in Atlantic Salmon (*Salmo salar*) Nutrition: Requirement, Bioavailability, Disease Resistance and Immune Response. Ph.D. thesis. Dalhousie University, Halifax, Nova Scotia, p. 282.

NRC (National Research Council), 1993. Nutrient Requirements of Fish. The National Academies Press, Washington, DC.

NRC (National Research Council), 2005. Mineral Tolerance of Animals. The National Academies Press, Washington, DC.

NRC (National Research Council), 2011. Nutrient Requirements of Fish and Shrimp. National Research Council, The National Academies Press, Washington, DC.

Nechay, B.R., 1984. Mechanisms of action of vanadium. Annu. Rev. Pharmacol. Toxicol. 24, 501–524.

Neves, J.V., Caldas, C., Ramos, M.F., Rodrigues, P.N.S., 2016. Hepcidin-dependent regulation of erythropoiesis during anemia in a teleost fish, *Dicentrarchus labrax*. PLoS One 11 (4), e0153940.

Ng, W.-K., Wilson, R.P., 1997. Chromic oxide inclusion in the diet does not affect glucose utilization or chromium retention by channel catfish, *Ictalurus punctatus*. J. Nutr. 127, 2357–2362.

Nie, J.-Q., Dong, X.-H., Tan, B.-P., Chi, S.-Y., Yang, Q.-H., Liu, H.-Y., Shuang, Z., 2016. Effects of dietary manganese sources and levels on growth performance, relative manganese bioavailability, antioxidant activities and tissue mineral content of juvenile cobia (*Rachycentron canadum* L). Aquacult. Res. 47, 1402–1412.

Nielsen, F.H., 1991. Nutritional requirements for boron, silicon, vanadium, nickel, and arsenic: current knowledge and speculation. Faseb. J. 5, 2661–2667.

Nielsen, F.H., 1995. Vanadium in mammalian physiology and nutrition. Met. Ions Biol. Syst. 31, 543–573.

Nielsen, F.H., 1998. Ultratrace elements in nutrition: current knowledge and speculation. J. Trace Elem. Exp. Med. 11, 251–274.

Nielsen, F.H., 2000. Importance of making dietary recommendations for elements designated as nutritionally beneficial, pharmacologically beneficial, or conditionally essential. J. Trace Elem. Exp. Med. 13, 113–129.

Nielsen, F.H., 2002. The nutritional importance and pharmacological potential of boron for higher animals and human. In: Goldbach, H.E., Rerkasem, B., Wimmer, M.A., Brown, P.H., Thellier, M., Bell, R.W. (Eds.), Boron in Plant and Animal Nutrition. Kluwer Academic/Plenum Publishers, New York, pp. 37–49.

Nielsen, F.H., 2008. Is boron nutritionally relevant? Nutr. Rev. 66, 183–191.

Nielsen, F.H., 2014. Update on the possible nutritional importance of silicon. J. Trace Elem. Exp. Med. 28, 379–382.

Niyogi, S., Wood, C.M., 2004. Biotic ligand model, a flexible tool for developing site-specific water quality guidelines for metals. Environ. Sci. Technol. 38, 6177–6192.

Nose, T., Arai, S., 1979. Recent advances on studies on mineral nutrition of fish in Japan. In: Pillay, V.R., Dill, W.A. (Eds.), Advances in Aquaculture. Fishing News, Farnam, England, pp. 584–590.

Nwanna, L.C., Adebayo, I.A., Omitoyin, B.O., 2009. Phosphorus requirements of African catfish, *Clarias gariepinus*, based on broken-line regression analysis methods. Sci. Asia 35, 227–233.

Ogino, C., Chiou, J.Y., 1976. Mineral requirements in fish. 2. Magnesium requirements of carp. Bull. Jpn. Soc. Sci. Fish. 42, 71–75.

Ogino, C., Takeda, H., 1976. Mineral requirements in fish. 3. Calcium and phosphorus requirements of carp. Bull. Jpn. Soc. Sci. Fish. 42, 793–799.

Ogino, O., Yang, G.Y., 1978. Requirement of rainbow trout for dietary zinc. Bull. Jpn. Soc. Sci. Fish. 44, 1015–1018.

Ogino, C., Takashima, F., Chiou, J.Y., 1978. Requirement of rainbow trout for dietary magnesium. Bull. Jpn. Soc. Sci. Fish. 44, 1105–1108.

Ogino, C., Yang, G.Y., 1979. Requirement of carp for dietary zinc. Bull. Jpn. Soc. Sci. Fish. 45, 967–969.

Ogino, C., Takeuchi, L., Takeda, H., Watanabe, T., 1979. Availability of dietary phosphorus in carp and rainbow trout. Bull. Jpn. Soc. Sci. Fish. 45, 1527–1532.

Ogino, C., Yang, G., 1980. Requirements of carp and rainbow trout for dietary manganese and copper. Bull. Jpn. Soc. Sci. Fish. 46, 455–458.

O'Dell, B.L., 1969. Effect of dietary components upon zinc availability. A review with original data. Am. J. Clin. Nutr. 22, 1315–1322.

O'Dell, B.L., 1997. Mineral-ion interaction as assessed by bioavailability and ion-channel function. In: O'Dell, B.L., Sunde, R.A. (Eds.), Handbook of Nutritionally Essential Element. Marcel Dekker, Inc., New York, NY, pp. 641–659.

Ogata, H.Y., Konno, S., Silverstein, J.T., 1998. Muscular buffering capacity of the parr and smolts in *Oncorhynchus masou*. Aquaculture 168, 303–310.

Oikari, A.O.J., Rankin, J.C., 1985. Renal excretion of magnesium in a freshwater teleost, *Salmo gairdneri*. J. Exp. Biol. 117, 319–333.

Ojo, A.A., Nadella, S.R., Wood, C.M., 2009. In Vitro examination of interactions between copper and zinc uptake via the gastrointestinal tract of the rainbow trout (oncorhynchus mykiss). Arch. Environ. Contam. Toxicol. 56, 244–252.

Olivia-Teles, A., Pimentel-Rodrigues, A., 2004. Phosphorus requirement of European sea bass (*Dicentrarchus labrax* L.) juveniles. Aquacult. Res. 35, 636–642.

Pacitti, D., Lawan, M.M., Feldmann, J., Sweetman, J., Wang, T., Martin, S.A.M., Secombes, C.J., 2016. Impact of selenium supplementation on fish antiviral responses: a whole transcriptomic analysis in rainbow trout (*Oncorhynchus mykiss*) fed supra nutritional levels of Sel-Plex®. BMC Genom. 17, 1–26.

Pacitti, D., Lawan, M.M., Sweetman, J., Martin, S.A.M., Feldmann, J., Secombes, C.J., 2015. Selenium supplementation in fish: a combined chemical and biomolecular study to understand sel-plex assimilation and impact on selenoproteome expression in rainbow trout (*Oncorhynchus mykiss*). PLoS One 10, e0127041.

Pagenkopf, G.K., 1983. Gill surface interaction model for trace-metal toxicity to fishes: role of complexation, pH, and water hardness. Environ. Sci. Technol. 17, 342–347.

Pan, Q.S., Liu, Y., Tan, Bi, Y., 2003. The effect of chromium picolinate on growth and carbohydrate utilization in tilapia, *Oreochromis niloticus* and *Oreochromis aureus*. Aquaculture 225, 421–429.

Pan, L., Xie, S., Lei, W., Han, D., Yang, Y., 2009. The effect of different dietary iron levels on growth and hepatic iron concentration in juvenile gibel carp (*Carassius auratus gibelio*). J. Appl. Ichthyol. 25, 428–431.

Pan, L., Zhu, X., Xie, S., Lei, W., Han, D., Yang, Y., 2008. Effects of dietary manganese on growth and tissue manganese concentrations of juvenile gibel carp, *Carassius auratus* gibelio. Aquacult. Nutr. 14, 459–463.

Papp, L.V., Lu, J., Holmgren, A., Khanna, K.K., 2007. From selenium to selenoproteins: synthesis, identity, and their role in human health. Antioxidants Redox Signal. 9, 775–806.

Partridge, G.J., Lymbery, A.J., 2008. The effect of salinity on the requirement for potassium by barramundi (*Lates calcarifer*) in saline groundwater. Aquaculture 278, 164–170.

Pechova, A., Pavlata, L., 2007. Chromium as an essential nutrient: a review. Vet. Med-CZECH. 52, 1–18.

Penglase, S., Hamre, K., Ellingsen, S., 2015. The selenium content of SEPP1 versus selenium requirements in vertebrates. PeerJ 3, e1244.

Pennington, G.A.T., 1991. Silicon in foods and diets. Food Addit. Contam. 8, 97–118.

Perry, S.F., Flick, G., 1988. Characterization of branchial transepithelial calcium fluxes in the freshwater trout (*Salmo gairdneri*). Am. J. Phys. 254, R491–R498.

Persson, P., Sundell, K., Björnsson, B.T., Lundqvist, H., 1998. Calcium metabolism and osmoregulation during sexual maturation of river running Atlantic salmon. J. Fish. Biol. 52, 334–349.

Petersen, W., Wallmann, K., Li, P.L., Schroeder, F., Knauth, H.D., 1995. Exchange of trace elements at the sediment-water interface during early diagenesis processes. Mar. Fresh. Res. 46, 19–26.

Pimentel Rodrigues, A., Oliva Teles, A., 2001. Phosphorus requirements of gilthead sea bream (*Sparus aurata* L.) juveniles. Aquacult. Res. 32, 157–161.

Poston, H.A., Combs, G.F., Leibovitz, L., 1976. Vitamin E and selenium interrelations in the diet of Atlantic salmon (*Salmo salar*): gross, histological and biochemical deficiency signs. J. Nutr. 106, 892–904.

Potts, J.T., 2005. Parathyroid hormone: past and present. J. Endocrinol. 187, 311–325.

Power, D.M., Einarsdottir, I.E., Pittman, K., Sweeney, G.E., Hildahl, J., Camphino, M.A., Silva, N., Saele, O., Galay-Burgos, M., Smaradottir, H., Bjornsson, B.T., 2008. The molecular and endocrine basis of flatfish metamorphosis. Rev. Fish. Sci. 16, 95–111.

Prohaska, J.R., 2011. Impact of copper limitation on expression and function of multicopper oxidases (ferroxidases). Adv. Nutr. 2, 89–95.

Pyle, G., Couture, P., 2012. Nickel. In: Wood, C.M., Farrell, A.P., Brauner, C.J. (Eds.), Homeostasis and Toxicology of Non-essential Metals. Academic Press, San Diego, pp. 253–289.

Qiao, Y.-G., Mai, K.-S., Ai, Q.-H., Zhang, W.-D., Xu, W., 2013. Evaluation of iron methionine and iron sulphate as dietary iron sources for juvenile cobia (*Rachycentron canadum*). Aquacult. Nutr. 19, 721–730.

Rankin, J.C., Davenport, D., 1981. Animal osmoregulation. Blackie, Glasgow, pp. 83–100.

Ransberry, V.E., Morash, A.J., Blewett, T.A., Wood, C.M., McClelland, G.B., 2015. Oxidative stress and metabolic responses to copper in freshwater- and seawater-acclimated killifish, *Fundulus heteroclitus*. Aquat. Toxicol. 161, 242–252.

Read, E.S., Barrows, F.T., Gaylord, T.G., Paterson, J., Petersen, M.K., Sealey, W.M., 2014. Investigation of the effects of dietary protein source on copper and zinc bioavailability in fishmeal and plant-based diets for rainbow trout. Aquaculture 432, 97–105.

Reichlmayr-Lais, A.M., Kirchgessner, M., 1981. Essentiality of lead for growth and metabolism. Z. für Tierphysiol. Tierernaehrung Futtermittelkd. 46, 1–8.

Reid, S.D., 2012. Molybdenum and chromium. In: Wood, C.M., Farrell, A.P., Brauner, C.J. (Eds.), Homeostasis and Toxicology of Essential Metals, Fish Physiology, vol. 31A. Academic Press, San Diego, pp. 376–415.

Reigh, R.C., Robinson, E.H., Brown, P.B., 1991. Effects of dietary magnesium on growth and tissue magnesium content of blue tilapia *Oreochromis aureus*. J. World Aquacult. Soc. 22, 192–200.

Reinfelder, J.R., Fisher, N.S., Luoma, S.N., Nichols, J.W., Wang, W.-X., 2008. Trace element trophic transfer in aquatic organisms: a critique of the kinetic model approach. Sci. Total Env. 219, 117–135.

Renkema, K.Y., Alexander, R.T., Bindels, R.J., Hoenderop, J.G., 2008. Calcium and phosphate homeostasis: concerted interplay of new regulators. Ann. Med. 40, 82–91.

Richardson, N.L., Higgs, D.A., Beames, R.M., McBride, J.R., 1985. Influence of dietary calcium, phosphorus, zinc and sodium phytate level on cataract incidence, growth and histopathology in juvenile chinook salmon (*Oncorhynchus tshawytscha*). J. Nutr. 115, 553–567.

Rider, S.A., Davies, S.J., Jha, A.N., Clough, R., Sweetman, J.W., 2010. Bioavailability of co-supplemented organic and inorganic zinc and selenium sources in a white fishmeal-based rainbow trout (*Oncorhynchus mykiss*) diet. J. Anim. Physiol. Anim. Nutr. 94, 99–110.

Robinson, E., Rawles, S.D., Yette, H.E., Greene, L.W., 1984. An estimate of the dietary calcium requirement of fingerling Tilapia aurea reared in calcium-free water. Aquaculture 41, 389–393.

Robinson, E.H., Rawles, S.D., Yette, H.E., Greene, L.W., 1986. Dietary calcium requirement of channel catfish (*Ictalurus punctatus*), reared in calcium-free water. Aquaculture 53, 263–270.

Robinson, E.H., LaBomascus, D., Brown, P.B., Linton, T.L., 1987. Dietary calcium and phosphorus requirements of *Oreochromis aureus* reared in calcium-free water. Aquaculture 64, 267–276.

Rodehutscord, M., 1996. Response of rainbow trout (*Oncorhynchus mykiss*) growing from 50 to 200 g to supplements of dibasic sodium phosphate in a semipurified diet. J. Nutr. 126, 324–331.

Roeder, M., Roeder, R.H., 1966. Effect of iron on the growth rate of fishes. J. Nutr. 90, 86–90.

Rostruck, J.T., Pope, A.L., Ganther, H.E., Swanson, A.B., Hafeman, D.G., Hoekstra, W.G., 1973. Selenium: biochemical role as a component of glutathione peroxidase. Science 179, 588–590.

Rouleau, C., Tjalve, H., Gottofrey, J., Pelletier, E., 1995. Uptake, distribution, and elimination of ^{54}Mn(II) in brown trout (*Salmo trutta*). Environ. Toxicol. Chem. 14, 483–490.

Round, E.E., Crawford, R.M., Mann, D.G., 1990. The Diatoms. Biology & Morphology of the Genera. Cambridge University Press, Cambridge, 477 pp.

Roy, P.K., Lall, S.P., 2003. Dietary phosphorus requirement of juvenile haddock (*Melanogrammus aeglefinus* L.). Aquaculture 221, 451–468.

Sakamoto, S., Yone, Y., 1973. Effect of dietary calcium/phosphorus ratio upon growth, feed efficiency and blood serum Ca and P level in red sea bream. Bull. Jpn. Soc. Sci. Fish. 39, 343–348.

Sakamoto, S., Yone, Y., 1976a. Requirement of red sea bream for dietary Fe - 1. Rep. Fish. Res. Lab. Kyushu Univ. 3, 53–58.

Sakamoto, S., Yone, Y., 1976b. Requirement of red sea bream for dietary Ca. Rep. Fish. Res. Lab. Kyushu Univ. 3, 59–64.

Sakamoto, S., Yone, T., 1978a. Effect of dietary phosphorus level on chemical composition of red sea bream. Bull. Jpn. Soc. Sci. Fish. 44, 227–229.

Sakamoto, S., Yone, Y., 1978b. Requirement of red sea bream for dietary Na and K. J. Fac. Agr. Kyushu Univ. 23, 79–84.

Sakamoto, S., Yone, Y., 1978c. Iron deficiency symptoms of carp. Bull. Jap. Soc. Sci. Fish. 44, 157–160.

Sakamoto, S., Yone, Y., 1978d. Requirement of red sea bream for dietary iron II. Bull. Jap. Soc. Sci. Fish. 44, 223–225.

Sakamoto, S., Yone, Y., 1979a. Requirement of red sea bream for dietary Mg. Bull. Jap. Soc. Fish. 45, 57–60.

Sakamoto, S., Yone, Y., 1979b. Availabilities of three iron compounds as dietary iron sources for red sea bream. Bull. Jpn. Soc. Sci. Fish. 45, 231–235.

Salman, N.A., Eddy, F.B., 1990. Increased sea-water adaptability of non-smolting rainbow trout by salt feeding. Aquaculture 86, 259–270.

Satoh, S., Takeuchi, T., Watanabe, T., 1987a. Availability to carp of manganese in white fish meal and of various manganese compounds. Nippon Suisan Gakkaishi 53, 825–832.

Satoh, S., Takeuchi, T., Watanabe, T., 1991. Availability of manganese and magnesium contained in white fish meal to rainbow trout *Oncorhynchus mykiss*. Nippon Suisan Gakkaishi 57, 99–104.

Satoh, S., Yamamoto, H., Takeuchi, T., Watanabe, T., 1983a. Effects on growth and mineral composition of carp of deletion of trace elements or magnesium from fish meal diet. Nippon Suisan Gakkaishi 49, 431–435.

Satoh, S., Takeuchi, T., Narabe, Y., Watanabe, T., 1983b. Effects of deletion of several trace elements from fish meal diets on growth and mineral composition of rainbow trout fingerlings. Nippon Suisan Gakkaishi 49, 1909–1916.

Satoh, S., Tabata, K., Izume, K., Takeuchi, T., Watanabe, T., 1987b. Effect of dietary tricalcium phosphate on availability of zinc to rainbow trout. Nippon Suisan Gakkaishi 53, 1199–1205.

Satoh, S., Poe, W.E., Wilson, R.P., 1989. Effect of supplemental phytate and/or tricalcium phosphate on weight gain, feed efficiency and zinc content in vertebrae of channel catfish. Aquaculture 80, 155–161.

Satoh, S., Izume, K., Takeuchi, T., Watanabe, T., 1992. Effect of supplemental tricalcium phosphate on zinc and manganese availability to common carp. Nippon Suisan Gakkaishi 58, 539–545.

Sele, V., Ørnsrud, R., Sloth, J.J., Berntssen, M.H.G., Amlund, H., 2018. Selenium and selenium species in feeds and muscle tissue of Atlantic salmon. J. Trace Elem. Med. Biol. 47, 124–133.

Shao, Q., Ma, J., Xu, Z., Hu, W., Xu, J., Xie, S., 2008. Dietary phosphorus requirement of juvenile black seabream, *Sparus macrocephalus*. Aquaculture 277, 92–100.

Shears, M.A., Fletcher, G.L., 1983. Regulation of Zn super(2+) uptake from the gastrointestinal tract of a marine teleost, the winter flounder (*Pseudopleuronectes americanus*). Can. J. Fish. Aquat. Sci. 40, 197–205.

Shiau, S.Y., Chen, M.J., 1993. Carbohydrate utilisation by tilapia (*Oreochromis niloticus* x *Oreochromis aureus*) as influenced by different chromium sources. J. Nutr. 123, 1747–1753.

Shiau, S.Y., Hsieh, J.F., 2001. Quantifying the dietary potassium requirement of juvenile hybrid tilapia (*Oreochromis niloticus* x *O. aureus*). Br. J. Nutr. 85, 213–218.

Shiau, S.Y., Liang, H.S., 1995. Carbohydrate utilisation and digestibility by tilapia (*Oreochromis niloticus* x *O. aureus*) are affected by chromium oxide inclusion in the diet. J. Nutr. 125, 976–982.

Shiau, S., Ning, Y., 2003. Estimation of dietary copper requirements of juvenile tilapia, *Oreochromis niloticus* x *O. aureus*. Anim. Sci. 77, 287–292.

Shiau, S.Y., Shy, S.M., 1998. Dietary chromic oxide inclusion level required to maximize glucose utilization in hybrid tilapia (*Oreochromis niloticus* x *O. aureus*). Aquaculture 161, 357–364.

Shiau, S.Y., Su, L.W., 2003. Ferric citrate is half as effective as ferrous sulfate in meeting the iron requirement of juvenile tilapia, *Oreochromis niloticus* × *O. aureus*. J. Nutr. 133, 483–488.

Shiau, S.Y., Tseng, H.C., 2007. Dietary calcium requirements of juvenile tilapia, *Oreochromis niloticus* x *O. aureus* reared in fresh water. Aquacult. Nutr. 13, 298–303.

Schönbörner, A.A., Boivin, G., Baud, C.V., 1979. The mineralization process in teleost fish scales. Cell Tissue Res. 202, 203–212.

Schoenmakers, T.J.M., Verbost, P.M., Flik, G., Wendelaar Bonga, S.E., 1993. Transcellular intestinal calcium transport in freshwater and seawater fish and its dependence on sodium/calcium exchange. J. Exp. Biol. 176, 195–206.

Seaborn, C.D., Nielsen, F.H., 2002. Silicon deprivation decreases collagen formation in wounds and bone, and ornithine transaminase enzyme activity in liver. Biol. Trace Elem. Res. 89, 251–261.

Shen, H.M., Chen, X.R., Chen, W.Y., Lin, S.M., Chen, Y.J., Zhang, L., Luo, L., 2016. Influence of dietary phosphorus levels on growth, body composition, metabolic response and antioxidant capacity of juvenile snakehead (*Channa argus* × *Channa maculata*). Aquacult. Nutr. 23, 662–670.

Shearer, K.D., 1984. Changes in elemental composition of hatchery reared rainbow trout, Salmo gairdneri, associated with growth and reproduction. Can. J. Fish. Aquat. Sci. 41, 1592–1600.

Shearer, K.D., 1988. Dietary potassium requirement of juvenile chinook salmon. Aquaculture 73, 119–129.

Shearer, K.D., 1989. Whole body magnesium concentration as an indicator of magnesium status in rainbow trout (*Salmo gairdneri*). Aquaculture 77, 201–210.

Shearer, K.D., Åsgård, T., 1992. The effect of water-borne magnesium on the dietary magnesium requirement of the rainbow trout (*Oncorhynchus mykiss*). Fish Physiol. Biochem. 9, 387–392.

Silva, M.S., Kröckel, S., Jesu Prabhu, P.A., Koppe, W., Ørnsrud, R., Waagbø, R., Araujo, P., Amlund, H., 2019. Apparent availability of zinc, selenium and manganese as inorganic metal salts or organic forms in plant-based diets for Atlantic salmon (*Salmo salar*). Aquaculture 503, 562–570.

Silva, M.S., Prabhu, P.A.J., Ornsrud, R., Sele, V., Krockel, S., Sloth, J.J., Amlund, H., 2020. In vitro digestion method to evaluate solubility of dietary zinc, selenium and manganese in salmonid diets. J. Trace Elem. Med. Biol. 57, 126418.

Simkiss, K., 1974. Calcium metabolism of fish in relation to ageing. In: Bagenal, T.B. (Ed.), The Proceedings of an International Symposium on the Ageing Fish. Unwin Brothers, Old Woking, UK, pp. 1–12.

Skonberg, D.I., Yogev, L., Hardy, R.W., Dong, F.M., 1997. Metabolic response to dietary phosphorus intake in rainbow trout (*Oncorhynchus mykiss*). Aquaculture 157, 11–24.

Solomalina, V.D., Arson, O.M., 1979. Effect of phosphorus in water on certain metabolic links in carp. Vopt. Ikhtiologii 19, 724–729.

Simkiss, K., Wilbur, K.M., 1989. Biomineralization. Cell Biology and Mineral Deposition. Academic Press, Inc., San Diego.

Smedley, M.A., Migaud, H., McStay, E.L., Clarkson, M., Bozzolla, P., Campbell, P., Taylor, J.F., 2018. Impact of dietary phosphorous in diploid and triploid Atlantic salmon (*Salmo salar* L.) with reference to early skeletal development in freshwater. Aquaculture 490, 329–343.

Soevik, T., Opstvedt, J., Braekkan, O.R., 1981. Biological availability of iron found in fish-meals. J. Sci. Food Agric. 32, 1063–1068.

Specker, J.L., Eales, J.G., Tagawa, M., Tyler III, W.A., 2000. Parr-smolt transformation in Atlantic salmon: thyroid hormone deiodination in liver and brain and endocrine correlates of change in rheotactic behavior. Can. J. Zool. 78, 696–705.

Spry, D.I., Hodson, P.V., Wood, C.M., 1988. Relative contributions of dietary and waterborne zinc in the rainbow trout, *Salmo gairdneri*. Can. J. Fish. Aquat. Sci. 45, 32–41.

Stefansson, S.O., Björnsson, B.T., Ebbesson, L.O.E., McCormick, S.D., 2008. Smoltification. In: Finn, R.N., Kapoor, B.G. (Eds.), Fish Larval Physiology. Science Publishers Enfield, NH, USA, pp. 639–681.

Staurnes, M., Finstad, B., 2000. The effects of dietary NaCl supplement on hypo-osmoregulatory ability and sea water performance of Arctic charr (*Salvelinus alpinus* L.) smolts. Aquacult. Res. 31, 737–743.

Suarez-Bregua, P., Cal, L., Cañestro, C., Rotllant, J., 2017. PTH reloaded: a new evolutionary perspective. Front. Physiol. 8, 776. https://doi.org/10.3389/fphys.2017.00776.

Sugiura, S.H., Hardy, R.W., Roberts, R.J., 2004. The pathology of phosphorus deficiency in fish – a review. J. Fish. Dis. 27, 255–265.

Sun, Y., Li, B., Zhang, X., Chen, M., Tang, H., Yu, X., 2018. Dietary available phosphorus requirement of crucian carp, *Carassius auratus*. Aquacult. Nutr. 24, 1494–1501.

Sundby, B., Martinez, P., Gobeil, C., 2004. Comparative geochemistry of cadmium, rhenium, uranium, and molybdenum in continental marine sediments. Geochem. Cosmochim. Acta 68, 2485–2493.

Sundell, K., Bishop, J.E., Bjornsson, B.T., Norman, A.W., 1992. 1,25-Dihydroxyvitamin-D_3 in the Atlantic cod -plasma-levels, a plasma binding component, and organ distribution of a high affinity receptor. Endocrinology 131, 2279–2286.

Suttle, N., 2010. Mineral Nutrition of Livestock, fourth ed. Commonwealth Agricultural Bureaux International, Oxfordshire, UK.

Suzuki, K.T., 2005. Metabolomics of selenium: Se metabolites based on speciation studies. J. Health Sci. 51 (2), 107–114.

Suzuki, N., Danks, J.A., Maruyama, Y., Ikegame, M., Sasayama, Y., Hattori, A., et al., 2011. Parathyroid hormone 1 (1-34) acts on the scales and involves calcium metabolism in goldfish. Bone 48, 1186–1193.

Tacon, A.G.J., Beveridge, M.M., 1982. Effects of dietary trivalent chromium on rainbow trout. Nutr. Rep. Int. 25, 49–56.

Takei, Y., Hwang, P.-P., 2016. Homeostatic responses to osmotic stress. Biol. Stress Fish Fish Physiol. 35, 207–249.

Takeuchi, T., Watanabe, T., Ogino, C., Saito, M., Nishimura, K., Nose, T., 1981. Effects of low protein-high calorie diets and deletion of trace elements from a fish meal diet on reproduction of rainbow trout. Bull. Jpn. Soc. Sci. Fish. 47, 645–654.

Tan, L.-N., Feng, Liu, Y., Jiang, J., Jiang, W.-D., Hu, K., Li, S.-H., Zhou, X.-Q., 2011a. Growth, body composition and intestinal enzyme activities of juvenile Jian carp (*Cyprinus carpio* var. Jian) fed graded levels of dietary zinc. Aquacult. Nutr. 17, 338–345.

Tan, X.Y., Luo, Z., Liu, X., Xie, C.X., 2011b. Dietary copper requirement of juvenile yellow catfish *Pelteobagrus fulvidraco*. Aquacult. Nutr. 17, 170–176.

Tan, X.Y., Xie, P., Luo, Z., Lin, H.Z., Zhao, Y.H., Xi, W.Q., 2012. Dietary manganese requirement of juvenile yellow catfish *Pelteobagrus fulvidraco*, and effects on whole body mineral composition and hepatic intermediary metabolism. Aquaculture 326–329, 68–73.

Tang, Q.Q., Feng, L., Jiang, W.D., Liu, Y., jiang, J., Li, S.H., Kuang, S.,Y., Tang, L., Zhu, X.Q., 2013. Effects of dietary copper on growth, digestive, and brush border enzyme activities and antioxidant defense of hepatopancreas and intestine for young grass carp (*Ctenopharyngodon idella*). Biol. Trace Elem. Res. 155, 370–380.

Taylor, L.N., Wood, C.M., McDonald, D.G., 2003. An evaluation of sodium loss and gill metal binding properties in rainbow trout and yellow perch to explain species differences in copper tolerance. J. Exp. Biol. 22, 2159–2166.

Teas, J., Pino, S., Critchley, A., Braverman, L.E., 2004. Variability of iodine content in common commercially available edible seaweeds. Thyroid 14, 836–841.

Thiry, C., Ruttens, A., De Temmerman, L., Schneider, Y.-J., Pussemier, L., 2012. Current knowledge in species-related bioavailability of selenium in food. Food Chem. 130, 767–784.

Tomiyama, T., Kobayashi, K., Ishio, S., 1956. Distribution and excretion of intramuscularly administered $^{32}PO_4$ by carp. Res. Eff. Inf. Nucl. Atom Bomb Exp.Tokyo 1201–1203.

Uyan, O., Koshio, S., Ishikawa, M., Uyan, S., Ren, T., Yokoyama, S., Komilus, F.C., Michael, F.R., 2007. Effects of dietary phosphorus and phospholipid level on growth, and phosphorus deficiency signs in juvenile Japanese flounder, *Paralichthys olivaceus*. Aquaculture 267, 44–54.

Ura, K., Onodera, Y., Fukuda, H., Misaka, N., Hara, A., Adachi, S., Yamauchi, K., 2001. Changes in transcript levels of gill cortisol receptor during smoltification in wild masu salmon, *Oncorhynchus masou*. Zool. Sci. 18, 863–860.

Uthus, E.O., 2003. Arsenic essentiality: a role affecting methionine metabolism. J. Trace Elem. Exp. Med. 16, 345–355.

Van Der Velden, J.A., Kolar, Z.I., Flik, G., 1991. Intake of magnesium from water by freshwater tilapia fed on a low-Mg diet. Comp. Biochem. Physiol. 99 A, 103–105.

Vieira, M.C., Torronteras, R., Córdoba, F., Canalejo, A., 2012. Acute toxicity of manganese in goldfish *Carassius auratus* is associated with oxidative stress and organ specific antioxidant responses. Ecotoxicol. Environ. Saf. 78, 212–217.

Verbost, P.M., van Rooij, J., Flik, G., Lock, R.A.C., Wendelaar Bonga, S.E., 1989. The movement of cadmium through freshwater trout branchial epithelium and its interference with calcium transport. J. Exp. Biol. 145, 185–197.

Verlhac-Trichet, V., Vielma, J., Dias, J., Rema, P., Santigosa, E., Wahli, T., Vogl, K., 2014. The efficacy of a novel microbial 6-phytase expressed in Aspergillus oryzae on the performance and phosphorus utilization of cold- and warm-water fish: rainbow trout, *Oncorhynchus mykiss*, and Nile tilapia, *Oreochromis niloticus*. J. World Aquacult. Soc. 45, 367–379.

Vielma, J., Lall, S.P., 1998a. Control of phosphorus homeostasis of Atlantic salmon (*Salmo salar*) in fresh water. Fish Physiol. Biochem. 19, 83–93.

Vielma, J., Lall, S.P., 1998b. Phosphorus utilization by Atlantic salmon (*Salmo salar*) reared in freshwater is not influenced by higher dietary calcium intake. Aquaculture 160, 117–128.

Vielma, J., Lall, S., Koskela, J., Mattila, P., 1999. Influence of low dietary cholecalciferol intake on phosphorus and trace element metabolism by rainbow trout (*Oncorhynchus mykiss*, Walbaum). Comp. Biochem. Physiol. Mol. Integr. Physiol. 122B, 117–125.

Vincent, J.B., 2000. The biochemistry of chromium. J. Nutr. 130, 715–718.

Vohra, A., Satanarayana, T., 2003. Phytases: microbial sources, production, purification, and potential biotechnological applications. Crit. Rev. Biotechnol. 23, 29–60.

Wagner, G.F., 1993. Stanniocalcin: structure, function, and regulation. In: Hochachka, Mommsen, T. (Eds.), The Biochemistry and Molecular Biology of Fishes. Elsevier Science, Amsterdam, pp. 419–434.

Wagner, G.F., Jaworski, E.M., Haddad, M., 1998. Stanniocalcin in the seawater salmon: structure, function, and regulation. Am. J. Physiol. 274, R117–R118.

Wang, C., Lovell, R.T., 1997. Organic selenium sources, selenomethionine and selenoyeast, have higher bioavailability than an inorganic selenium source, sodium selenite, in diets for channel catfish (*Ictalurus punctatus*). Aquaculture 152, 223–234.

Wang, F.B., Luo, L., Lin, S.M., Li, Y., Chen, S., Wang, Y.G., Wen, H., Hu, C.J., 2011. Dietary magnesium requirements of juvenile grass carp, *Ctenopharyngodon idella*. Aquacult. Nutr. 17, 691–700.

Wang, J., Ai, Q., Mai, K., Xu, H., Zuo, R., 2014. Dietary chromium polynicotinate enhanced growth performance, feed utilization, and resistance to Cryptocaryon irritans in juvenile large yellow croaker (*Larmichthys crocea*). Aquaculture 432, 321–326.

Wang, Q., Zhang, Y., Bai, D., Chen, C., Guo, Y., Xing, K., 2015. Estimation of dietary copper (Cu) requirement of Cynoglossus semilaevis Günther. In: Zhang, T.C., Nakajima, M. (Eds.), Adv. Appl. Biotech, Lecture Notes in Electrical Engineering, vol. 332. Springer, Berlin, Heidelberg.

Wang, H., Li, E., Zhu, H., Du, Z., Qin, J., Chen, L., 2016. Dietary copper requirement of juvenile Russian sturgeon *Acipenser gueldenstaedtii*. Aquaculture 454, 118–124.

Wang, L., Xiao, J.-X., Hua, Y., Xiang, X.-W., Zhou, Y.-F., Ye, L., Shao, Q.-J., 2019. Effects of dietary selenium polysaccharide on growth performance, oxidative stress and tissue selenium accumulation of juvenile black sea bream, *Acanthopagrus schlegelii*. Aquaculture 503, 389–395.

Wang, X.J., Choi, S., Park, S., Yoo, G., Kim, K., Kang, J.C., Bai, S.C., 2005. Optimum dietary phosphorus level of juvenile Japanese flounder *Paralichthys olivaceus* reared in the recirculating system. Fish. Sci. 71, 168–173.

Wang, J., Gatlin, D.M., Li, L., Wang, Y., Jin, N., Lin, H., Zhou, C., Huang, Z., Yu, W., Guo, Y., 2019. Dietary chromium polynicotinate improves growth performance and feed utilization of juvenile golden pompano (*Trachinotus ovatus*) with starch as the carbohydrate. Aquaculture 505, 405–411.

Watanabe, T., Kiron, V., Satoh, S., 1997. Trace minerals in fish nutrition. Aquaculture 151, 185–207.

Watanabe, T., Murakami, A., Takeuchi, L., Nose, T., Ogino, C., 1980. Mineral requirements in fish 9. Requirement of chum salmon held in fresh-water for dietary phosphorus. Bull. Jpn. Soc. Sci. Fish. 46, 361–367.

Watanabe, K., Satoh, S., Takeuchi, T., 1988. Availability of minerals in fish meal. Asian Fish. Soc. 1, 175–195.

Wekell, J.C., Shearer, K.D., Gauglitz Jr., E.J., 1986. Zinc supplementation of trout diets: tissue indicators of body zinc status. Prog. Fish-Cult. 48, 205–212.

Wekell, J.C., Shearer, K.D., Houle, C.R., 1983. High zinc supplementation of rainbow trout diets. Prog. Fish-Cult. 45, 144–147.

Welker, T.L., Lim, C., Yildirim-Aksoy, M., Klesius, P.H., 2011. Susceptibility of channel catfish (*Ictalurus punctatus*) fed with dietary sodium chloride to nitrite toxicity. Aquacult. Nutr. 17, 892–901.

Welker, T., Barrows, F., Overturf, K., Gaylord, G., Sealey, W., 2016. Optimizing zinc supplementation levels of rainbow trout (*Oncorhynchus mykiss*) fed practical type fishmeal- and plant-based diets. Aquacult. Nutr. 22, 91–108.

Welker, T.L., Overturf, K., Abernathy, J., Barrows, F.T., Gaylord, G., 2017. Optimization of dietary manganese fo rainbow trout, *Oncorhynchus mykiss*, fed a plant-based diet. J. World Aquacult. Soc. 49, 71–82.

Wendalaar Bonga, S.E., Flik, G., 1995. Calcium regulation in fish. In: Lahlou, B., Vitiello, P. (Eds.), Aquaculture: Fundamental and Applied Research. American Geophysical Union, Washington, DC, pp. 47–59.

Wendelaar Bonga, S.E., Pang, P.K.T., 1991. Control of calcium regulation hormones in the vertebrates: parathyroid hormone, calcitonin, prolactin and stanniocalcin. Int. Rev. Cytol. 128, 139–213.

Whitehead, M.W., Thompson, R.P.H., Powell, J.J., 1996. Regulation of metal absorption in the gastrointestinal tract. Gut 39, 625–628.

Wilson, R.P., Naggar, G.E., 1992. Potassium requirement of fingerling channel catfish, Ictalurus punctatus. Aquaculture 108, 169–175.

Wilson, R.P., Rpbinson, E.H., Gatlin III, D.M., Poe, W.E., 1982. Dietary phosphorus requirement of channel catfish. J. Nutr. 112, 1197–2002.

Wilson, R.W., Wilson, J.M., Grosell, M., 2002. Intestinal bicarbonate secretion by marine teleost fish—why and how? Biochim. Biophys. Acta 1566, 182–193.

Wischhusen, P., Parailloux, M., Geraert, P.-A., Briens, M., Bueno, M., Mounicou Bouyssiere, B., Antony Jesu Prabhu, P., Kaushik, S.J., Fauconneau, B., Fontagné-Dicharry, S., 2019. Effect of dietary Se in rainbow trout (*Oncorhynchus mykiss*) broodstock on antioxidant status, its parental transfer and oxidative status in the progeny. Aquaculture 507, 126–138.

Witten, P.E., Fjelldal, P.G., Huysseune, A., McGurk, C., Obach, A., Owen, M.A.G., 2019. Bone without minerals and its secondary mineralization in Atlantic salmon (*Salmo salar*): the recovery from phosphorus deficiency. J. Exp. Biol. https://doi.org/10.1242/jeb.188763.

Witten, P.E., Owen, M.A.G., Fontanillas, R., Soenens, M., McGurk, C., Obach, A., 2015. A primary phosphorus-deficient skeletal phenotype in juvenile Atlantic salmon *Salmo salar*: the uncoupling of bone formation and mineralization. J. Fish Biol. 88, 690–708.

Wood, C.M., 2012. An introduction to metals in fish physiology and toxicology: basic principles. Fish Physiol. 31A, 1–53.

Wood, C.M., Farrell, A.M., Brauner, C.J., 2012. Homeostasis and Toxicology of Non-essential Metals, vol. 31B, 531 pp.

Woodall, A.N., LaRoche, G., 1964. Nutrition of salmonid fishes. 11. Iodide requirements of chinook salmon. J. Nutr. 82, 475–482.

Xie, D., Han, D., Zhu, X., Yang, Y., Jin, J., Liu, H., Xie, S., 2016. Dietary available phosphorus requirement for on-growing gibel carp (*Carassius auratus* gibelio var. CAS III). Aquacult. Nutr. 48, 28411–32851.

Xie, D., Han, D., Zhu, X., Yang, Y., Jin, J., Liu, H., Xie, S., 2017. Dietary available phosphorus requirement for on-growing gibel carp (*Carassius auratus gibelio* var. CAS III). Aquacul. Nutr. 23, 1104–1112.

Xu, Z., Dong, X., Liu, C., 2007. Dietary zinc requirement of juvenile cobia (*Rachycentron canadum*). Fish. Sci. 26, 138–141.

Yamamoto, H., Satoh, S., Takeuchi, T., Watanabe, T., 1983. Effects on rainbow trout of deletion of manganese or trace elements from fish meal diet. Nippon Suisan Gakkaishi 49, 287–293.

Yao, Y.F., Jiang, M., Wen, H., Wu, F., Liu, W., Tian, J., Yang, C.G., 2014. Dietary phosphorus requirement of GIFT strain of Nile tilapia *Oreochromis niloticus* reared in fresh water. Aquacult. Nutr. 20, 273–280.

Ye, C.-X., Liu, Y.-J., Mai, K.-S., Tian, L.-X., Yang, H.-J., Niu, J., Huang, J.-W., 2007. Effect of dietary iron supplement on growth, haematology and microelements of juvenile grouper, *Epinephelus coioides*. Aquacult. Nutr. 13, 471–477.

Ye, C.X., Liu, Y.J., Tian, L.X., Mai, K.S., Du, Z.Y., Yang, H.J., Niu, J., 2006. Effect of dietary calcium and phosphorus on growth, feed efficiency, mineral content and body composition of juvenile grouper, *Epinephelus coioides*. Aquaculture 255, 263–271.

Ye, C.X., Tian, L.X., Mai, K.S., Yang, H.J., Niu, J., Liu, Y.J., 2010. Dietary magnesium did not affect calcium and phosphorus content in juvenile grouper *Epinephelus coioides*. Aquac. Nutr. 16, 378–384.

Ye, C.X., Tian, L.X., Yang, H.J., Liang, J.J., Niu, J., Liu, Y.J., 2009. Growth performance and tissue mineral content of juvenile grouper (*Epinephelus coioides*) fed diets supplemented with various levels of manganese. Aquacult. Nutr. 15, 608–614.

Yu, H.-R., Zhang, Q., Xiong, D.-M., Huang, G.-Q., Li, W.-Z., Liu, S.-W., 2012. Dietary available phosphorus requirement of juvenile walking catfish, *Clarias leather*. Aquacult. Nutr. 19, 483–490.

Yuan, Y.C., Yang, H.J., Gong, S.Y., Luo, Z., Yu, D.H., Yan, J.L., Yang, X.F., 2011. Dietary phosphorus requirement of juvenile Chinese sucker *Myxocyprinus asiaticus*. Aquacult. Nutr. 17, 159–169.

Zafar, N., Khan, M.A., 2018. Determination of dietary phosphorus requirement of stinging catfish *Heteropneustes fossilis* based on feed conversion, growth, vertebrae phosphorus, whole body phosphorus, haematology and antioxidant status. Aquacult. Nutr. 24, 1577–1586.

Zafar, N., Khan, M.A., 2020. Dietary copper requirement of fingerling *Heteropneustes fossilis* for formulating copper-balanced commercial feeds. Aquacult. Nutr. 26, 248–260.

Zaugg, W.S., McLain, L.R., 1969. Inorganic salt effects on growth, salt water adaptation, and gill ATPase of Pacific salmon. In: Neuhaus, W.W., Halver, J.E. (Eds.), Fish in Research. Academic Press, New York, pp. 293–306.

Zaugg, W.S., Roley, D.D., Prentice, E.F., Gores, K.X., Waknitz, F.W., 1983. Increased seawater survival and contribution to fishery of chinook salmon (*Oncorhynchus tshawytscha*) by supplemented dietary salt. Aquaculture 32 (2), 15–22 17.

Zee, J., Patterson, S., Wiseman, S., Hecker, M., 2016. Is hepatic oxidative stress a main driver of dietary selenium toxicity in white sturgeon (*Acipenser transmontanus*)? Ecotoxicol. Environ. Saf. 133, 334–340.

Zeitoun, I.H., Ullrey, D.E., Bergen, W.G., Magee, W.T., 1976. Mineral metabolism during ontogenesis of rainbow trout (*Salmo gairdneri*). J. Fish. Res. Board Can. 33, 2587–2591.

Zidenberg-Carr, S., Keen, C.L., Lonnerdal, B., Hurley, L.S., 1983. Superoxide dismutase activity and lipid peroxidation in the rat: correlation affected by manganese deficiency. J. Nutr. 113, 2498–2504.

Zimmer, A.M., Brix, K.V., Wood, C.M., 2019. Mechanisms of Ca^{2+} uptake in freshwater and seawater-acclimated killifish, *Fundulus heteroclitus*, and their response to acute salinity transfer. J. Comp. Physiol. B 189, 47–60.

Zhao, L., Xia, Z., Wang, F., 2014. Zebrafish in the sea of mineral (iron, zinc, and copper) metabolism. Front. Pharmacol. 5 https://doi.org/10.3389/fphar.2014.00033.

Zhang, C., Mai, K., Ai, Q., Zhang, W., Duan, Q., Tan, B., Ma, H., Xu, W., Linfu, Z., Wang, X., 2006. Dietary phosphorus requirement of juvenile Japanese seabass, *Lateolabrax japonicus*. Aquaculture 255, 201–209.

Zhang, H., Sun, R., Xu, W., Zhou, H., Zhang, W., Mai, K., 2016. Dietary manganese requirement of juvenile large yellow croaker *Larimichthys crocea* (Richardson, 1846). Aquaculture 450, 74–79.

Zhu, Y., Chen, Y., Liu, Y., Yang, H., Liang, G., Tian, L., 2011. Effect of dietary selenium level on growth performance, body composition and hepatic glutathione peroxidase activities of largemouth bass Micropterus salmoide. Aquacult. Res. 43, 1660–1668.

Zhu, L., Han, D., Zhu, X., Yang, Y., Jin, J., Liu, H., Xie, S., 2016. Dietary selenium requirement for on-growing gibel carp (*Carassius auratus* gibelio var. CAS III). Aquacult. Res. 48, 2841–2851.

Zhu, W., Liu, M., Chen, C., Wu, F., Yang, J., Tan, Q., Xie, S., Liang, X., 2014. Quantifying the dietary potassium requirement of juvenile grass carp (*Ctenopharyngodon idellus*). Aquaculture 430, 218–223.

CHAPTER

7

Carbohydrates

Sadasivam J. Kaushik[1], *Stéphane Panserat*[1] *and Johan W. Schrama*[2]

[1]INRAe, Univ Pau and Pays de l'Adour, Nutrition, Metabolism and Aquaculture Unit, Saint Pée sur Nivelle, France; [2]Aquaculture and Fisheries Group, Wageningen University, Wageningen, The Netherlands

7.1 Introduction

Carbohydrates form the most widespread organic compounds in the biosphere and are major constituents of most fruits, vegetables, legumes, and cereal grains. Carbohydrates are also present in animals but in smaller amounts. The term "carbohydrates" covers a large number of neutral chemical compounds, formed of carbon, oxygen, and hydrogen atoms in proportions, which almost always correspond to the empirical formula $(CH_20)n$, where n is 3 or more. The carbohydrates as a group contain sugars, polyhydroxy aldehydes, ketones, alcohols and acids, their simple derivatives, and any compound that may be hydrolyzed to these. The carbohydrates can be broadly classified as sugars and nonsugars. The simplest sugars are monosaccharides, which are divided into subgroups—trioses ($C_3H_6O_3$), tetroses ($C_4H_8O_4$), pentoses ($C_5H_{10}O_5$), hexoses ($C_6H_{12}O_6$), and heptoses ($C_7H_{14}O_7$) depending upon the number of carbon atoms present in the molecule. Polysaccharides, also called glycans, are polymers of monosaccharides linked together, with the elimination of one molecule of water at each linkage, to produce di-, tri-, tetra-, or polysaccharides, containing, respectively, 2, 3, 4, or even up to 10 monosaccharide units linked by glycoside bonds. When more than 10 monosaccharide units are linked together by glycosidic bonds, then these are also called as nonsugars, examples of which are glycans, cellulose, glycogen, pectin, and lignin. The polysaccharides can have either straight or branched chains and are classified into two groups, the homoglycans, which contain only a single type of monosaccharide unit, and the heteroglycans, which on hydrolysis yield mixtures of monosaccharides and derivatives. There are also some compounds containing carbohydrates in combination with noncarbohydrate molecules such as glycolipids and glycoproteins.

Fish Nutrition, Fourth Edition
https://doi.org/10.1016/B978-0-12-819587-1.00008-2

Although there are a large number of "carbohydrates," only a few provide nutritional value in animal nutrition. These involve mainly hexoses (glucose), disaccharides, such as maltose, sucrose, or lactose, formed of two hexose molecules and some homopolysaccharides, including starch and glycogen. Other carbohydrates can play a secondary nutritional role, or even a negative one: This involves fermentable carbohydrates and fibers (Fig. 7.1).

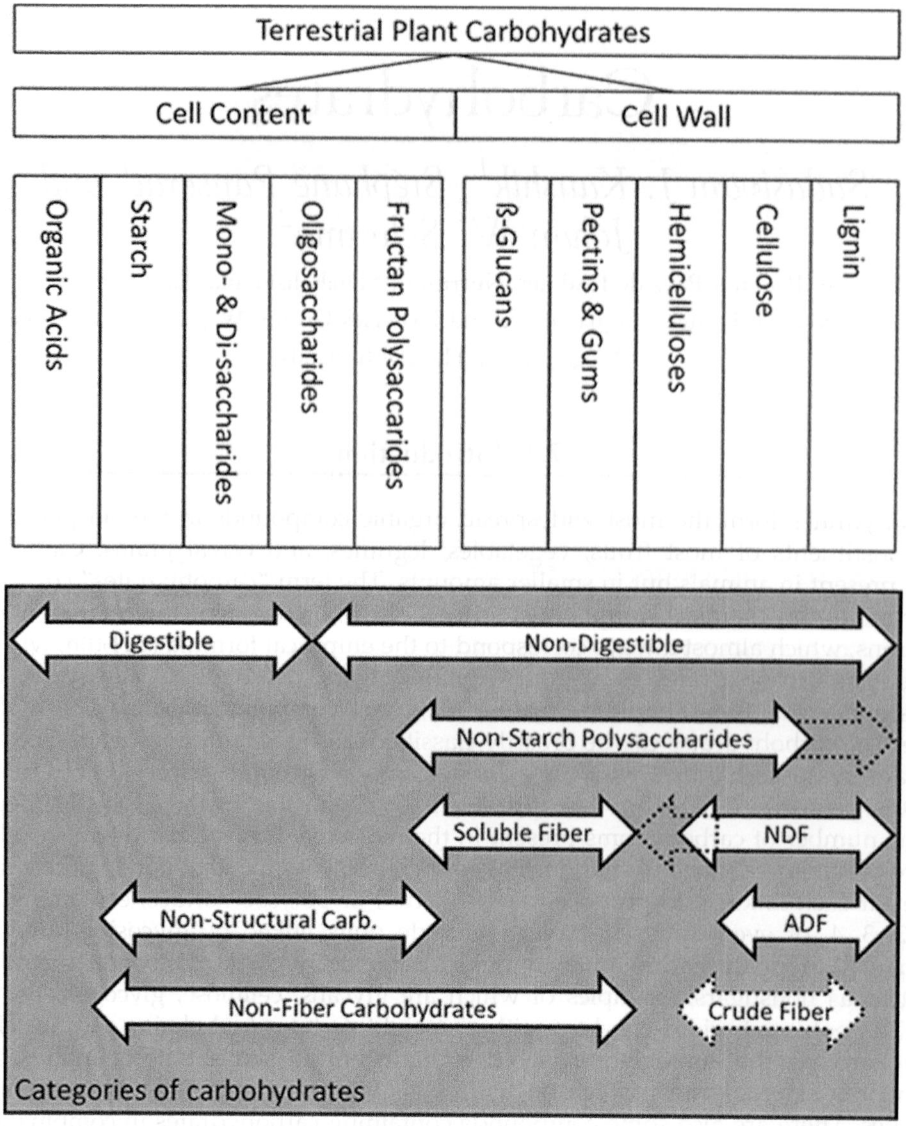

FIGURE 7.1 Classification/categories of dietary carbohydrates of terrestrial plant origin. NDF is neutral detergent fiber; ADF is acid detergent fiber; Carb. is carbohydrates. *Dotted arrows* indicate that variability exists in which type of carbohydrates is included into the various categories, which can be due to differences in definition (e.g., is lignin a carbohydrate or not) and analytical method. Adapted from National Research Council (NRC) 2011.

"Starch" is a glucan and is present in many plants as a reserve carbohydrate and energy storage. It is the most abundant in seeds, fruits, tubers, and roots. Starches occur naturally in the form of granules, built up in concentric layers, whose size and shape vary in different plants. Although glucan is the main component of the starch granules, they can also contain some other minor constituents such as protein, fatty acids, and phosphorus compounds. Starches differ in their chemical composition and, except in rare instances, are mixtures of two structurally different polysaccharides, amylose and amylopectin, the proportions of which can vary depending upon the source. Amylose is mainly linear in structure, formed of long, unbranched chains of glucose units linked between carbon atom 1 of one molecule and carbon atom 4 of the adjacent molecule ($\alpha 1 \rightarrow 4$ linkages). Amylopectin is highly branched containing $\alpha 1 \rightarrow 4$ bonds as well as a large number of $\alpha 1 \rightarrow 6$ bonds. In most starches, amylopectin is the main component, amounting to about 70%–80% of the total. *Cellulose* is the most abundant carbohydrate in the plant kingdom, forming the fundamental structure of plant cell walls, where cellulose is closely associated with other components such as hemicelluloses and lignin. Cellulose is distinguished from the starches by their $\beta 1 \rightarrow 4$ bonds. Other structural polyosides such as peptidoglycan in bacteria and chitin in the animal kingdom or in the fungi are similar to cellulose in their $\beta 1 \rightarrow 4$ linkages. Some special cases are the laminarines from algae, where the bonds are of the $\beta 1 \rightarrow 3$ type.

In animal cells, the main form of carbohydrate storage is "glycogen," found in liver, muscle, and some other tissues, and plays a major role in energy metabolism. Glycogen is in fact a generic term used to describe a group of highly branched polysaccharides similar to amylopectins in structure. The molecular weights of glycogen molecules can vary much depending upon the species, the tissue, and the physiological state of the animal. Glycogen can be mobilized rapidly within an animal under conditions requiring the mobilization of glucose, such as exercise and stress.

The distinction between α and β allows the classification of glucose polymers into two main categories—available carbohydrates and fibers—but there are also digestible disaccharides with $\beta 1 \rightarrow 3$ links such as lactose or other indigestible oligosaccharides such as the α-galactosides (Table 7.1). Sometimes termed as cellulose compounds, plant cell wall compounds, "fibers" include carbohydrates of varying degrees of complexity but have never been satisfactorily defined. Resistant starch is defined as the sum of starch and the products of starch degradation not absorbed in the small intestine (Asp et al., 1996). Available carbohydrates are those that are hydrolyzed by enzymes of the gastrointestinal tract to monosaccharides that are absorbed in the small intestine and enter the pathways of carbohydrate metabolism. Indigestible carbohydrates are not hydrolyzed by digestive enzymes but may be partially fermented by bacteria in the intestine, forming short-chain fatty acids that may be absorbed and contribute to the body's energy needs. Amylose and amylopectin molecules are hydrolyzed into maltose by α-amylase and the residues of hydrolysis such as maltose and short-chain dextrins (which are intermediate products of the hydrolysis of either starch or of glycogen and are water soluble) are hydrolyzed by maltase and sucrase-isomaltase.

Terrestrial animals consume large quantities of starch as part of cereal grains, cereal by-products, and tubers. In the case of aquatic organisms, natural food for fish or shrimp is low or sometimes even completely devoid of carbohydrates of plant origin but can contain

TABLE 7.1 Classification of different types of carbohydrates (those in bold are recognized as having a nutritional role).

Monosaccharides	
Trioses ($C_3 H_6O_3$)	Glyceraldehyde, dihydroxyacetone
Tetroses ($C_4H_6O_4$)	Erythrose
Pentoses ($C_5H_{10}O_5$)	Ribose. arabinose., xylose ...
Hexoses ($C_6H_{12}O_6$)	**Glucose, galactose, mannose, fructose**
Oligosaccharides	
Disaccharides ($C_{12}H_{22}Ou$)	**Maltose (2 glucose)**
	Saccharose or sucrose (glucose + fructose)
	Lactose (glucose + galactose)[a]
Trisaccharides ($C_{18} H_{32}O_{16}$)	Raffinose (galactose + glucose + fructose)
Tetrasaccharides ($C_{24}H_{42}O_{21}$)	Stachyose (2-galactose + glucose + fructose)
Pentasaccharides ($C_{30}H_{52}O_{26}$)	Verbascose (3 galactose + glucose + fructose)
Polysaccharides	
Homopolysaccharides	Pentosanes[a]
	Hexosanes: starch, glycogen, cellulose[a], laminarine[a]
	Fructosanes: inulin[b]
	Galactanes
	Mannanes
Heteropolysaccharides	Pectin, hemicellulose[a], mucopolysaccharides
Others	
Chitin	(Polymer of *N*-acetyl-glucosamine)

[a]*Compound with at least one beta-type bond.*
[b]*Polysaccharides also containing glucose units.*

chitin found in crustaceans. With regard to carbohydrates in fish nutrition, given the trophic level (>2.5) of all farmed fish and crustaceans (Duarte et al., 2009), it is in fact difficult to state whether a given species has "carnivorous, herbivorous, or omnivorous" feeding habits. From an etymological point of view as well as some of the definitions of carnivory implying reliance on diet composed of at least 70% animal tissue (Schermerhorn, 2013), we can hardly say any species of fish or shrimp is a carnivore. Under farming conditions, although a dietary supply of carbohydrates/sugars is not considered indispensable to fish or shrimp, they contribute in a significant manner as a source of digestible energy (DE). In the absence of carbohydrates, there is generally an increased utilization of other macronutrients as an energy source.

7.2 Carbohydrate digestion

7.2.1 Amylases

Fish exhibit large variations in the anatomical structures and functions of the gastrointestinal tract, and such differences have also a clear impact on the ability of different species to digest carbohydrates (Krogdahl et al., 2005). The presence of digestive carbohydrases has long been reported in fish (Kitamikado and Tachino, 1960; Chiu and Benitez, 1981; Kawai and Ikeda, 1971; Krogdahl et al., 2005) and that they are present from early larval stages (Cahu and Zambonino-Infante, 1994; Yúfera et al., 2018). Carbohydrate digestion involves two categories of enzymes: α-amylase of pancreatic origin and disaccharidases at the brush border membrane level in the intestinal epithelial cells (Hidalgo et al., 1999; Hemre et al., 2002; Krogdahl et al., 2005). In general, species known to have planktivorous or omnivorous feeding habits in the wild are considered to exhibit higher levels of amylase activities compared with those with insectivorous or piscivorous feeding habits, which have no or low carbohydrate contents in their natural food. Very early, comparative studies have shown that the amylase activity in the digestive system of the common carp (*Cyprinus carpio*) was to be about 10–30 times higher than that of rainbow trout, *Oncorhynchus mykiss* (Hofer and Strumbauer, 1985), and about 80 times higher than that in the yellowtail, *Seriola quinqueradiata* (Shimeno et al., 1977). Karasov et al. (2011) suggested that the activities of carbohydrases correlate positively with dietary carbohydrate intake or the trophic levels of fish. But a direct relationship between the activities of α-amylase and the feeding habits in the wild are not clearly established as was found in different species living under different habitats. Chakrabarti et al. (1995) measured the activities of α-amylase among other enzymes in 11 species of finfish and found no relationship between enzyme activities and feeding habits. It appears that the activity of α-amylase follows a pattern influenced more by phylogeny than by diet (Chan et al., 2004). Similar observations were made by Lopez-Vasquez et al. (2009) in different species of Amazonian fish having varied feeding habits. The total carbohydrase activity in farmed fish such as the Atlantic salmon (*Salmo salar*), rainbow trout, European sea bass (*Dicentrarchus labrax*), and gilthead sea bream (*Sparus aurata*) is considerably lower than those in tilapia (Papoutsoglou and Lyndon, 2005).

Under farming conditions, it has also not been clearly demonstrated that there is a direct relation between dietary carbohydrate intake and amylase secretion or activities. The ability to adapt or regulate amylase secretion and activities to match dietary carbohydrate intake is, however, considered to be restricted more to species having omnivorous feeding habits, while in those having piscivorous feeding habits, this would not be the case (Krogdahl et al., 2005). In Atlantic salmon, for instance, amylase secretion is not upregulated by dietary carbohydrate content, similar to the situation encountered in strictly carnivorous mammals (Bakke et al., 2010). The activity of α-amylase has been found to increase in the larvae of marine fish occurring as early as when the mouth opening occurs (Zambonino Infante and Cahu, 2001), suggesting that these activities were not induced by dietary factors. As mentioned by Krogdahl et al. (2005), not much is known on the regulation of amylase activity, secretion, or biosynthesis in farmed fish under different feeding regimes with varying levels and forms of carbohydrates. Although the main endogenous source of amylase in fish is of the exocrine pancreatic origin, there have been suggestions that amylase can also be due to the potential action of intestinal microbiota (Sugita et al., 1997).

At the molecular level, fish amylases appear to be closely related to those in mammals with comparable characteristics (Krogdahl et al., 2005; Froystad et al., 2006; Xie et al., 2016). The presence of multiple amylases was reported in winter flounder, *Pleuronectes americanus* (Douglas et al., 2000) similar to what was reported also in the Pacific white shrimp, *Penaeus vannamei*, by Van Wormhoudt and Sellos (1996). In this regard, specific analysis undertaken by German et al. (2016) led them to conclude that the variations in amylase activity in the guts of fish are not necessarily linked to variations in amylase gene copy numbers and that they all contribute toward the digestion of starch to varying degrees.

7.2.2 Disaccharidases

The enzymes responsible for this hydrolysis of a number of low-molecular-weight saccharides including maltose, sucrose, galactosides, and trehalose are present in the brush border membrane (Krogdahl et al., 2004). The activity of disaccharidases can vary much between species with higher values generally reported in species known to have "herbivorous" feeding habits than in others of a higher trophic level. In some studies, disaccharidase activities did not respond to changes in dietary carbohydrate levels (Buddington and Hilton, 1987; Krogdahl et al., 1999; Shiau and Liang, 1995), whereas in some other studies, increases have been noted, such as in salmonids (Krogdahl et al., 2004) and Amazonian fishes (Correa et al., 2007). Moreover, food deprivation decreases enzyme activities, whereas a rapid increase in enzyme activities appears to occur upon refeeding (Hakim et al., 2009; Krogdahl et al., 1999, 2004).

High maltase activities and much lower activities of sucrase and lactase were detected in white sturgeon, *Acipenser transmontanus* (Hung et al., 1989) albeit not affected by dietary carbohydrate sources. Similar observations of lack of regulation of maltase activities have also been made in tilapia fed different mono-, di-, or polysaccharides (Shiau and Chuang, 1995). The expression of maltase gene was found to be induced by high levels of dietary dextrin in rainbow trout (Geurden et al., 2007) although they did not measure the activity of the enzyme.

7.2.3 Nonstarch polysaccharides

As regards nonstarch polysaccharides (NSPs), the levels can vary much, ranging from 100 to more than 300 g/kg dry matter, in some of the plant-derived feedstuffs of potential interest for use in fish or shrimp feeds (Sinha et al., 2011). Cereal brans, which are commonly used in carp, tilapia, or shrimp feeds, can have an NSP content up to 500 g/kg dry matter. Endogenous production of NSP-degrading enzymes, such as cellulase, beta-xylanase, and beta-glucanase, is thought to be absent in fish (Maas et al., 2020b) like in terrestrial monogastric animals. Precise quantitative data on digestive enzymes, which can hydrolyze the β-glycosidic bonds in the NSPs, are scarce. Due to the lack of specific NSP-degrading digestive enzymes, these carbohydrates are often considered mostly unavailable as an energy source.

However, this differs between fish species. In Nile tilapia, it was observed that NSP digestibility ranges between 0 up to 70% depending on the type of NSP and can contribute up to 17% of the total DE intake (Maas et al., 2020b; Haidar et al., 2016). Low levels of chitinase and cellulase activities have been reported in some species (Clark et al., 1984; Lindsay et al., 1984; Chakrabarti et al., 1995) as well as in shrimp (Clark et al., 1993). However, it is not clear whether chitinolytic or cellulolytic activities are due to endogenous digestive enzymes or due to microbial actions in the digestive tract. Under pond culture conditions, ingestion of NSP-degrading enzymes and NSP-producing bacteria present in detritus may occur (German and Bittong, 2009). The digestibility of polyholosides of aquatic origin, such as the alginates and carrageenans, has been little studied in fish.

7.2.4 Chitinase

Chitin, which is a major heteropolysaccharide composed of the hexosamine β-(1−4)-linked *N*-acetylglucosamine molecules, is considered to be the second most abundant compound after cellulose in living organisms and is widely found in the exoskeleton of insects and aquatic invertebrates. Gastric chitinase splits chitin into n-acetyl-D-glucosamine, which may be further broken down by chitobiase (exo-*N*-acetyl-hexosaminidase) in the pyloric caecae or the intestine. Studies with some marine as well as freshwater fish have suggested the presence of bacterial strains in the gastrointestinal tract capable of hydrolyzing chitin (Lindsay and Gooday, 1985; Sugita et al., 1999) although the relative importance of microbial action in chitinolysis is not quantified. In fact, many species of fish are reported to have endogenous chitinase activities mainly in the stomach (Jeuniaux, 1961; Ikeda et al., 2017). Based on analysis of some marine elasmobranchs and one bony fish, Fange et al. (1979) observed that chitinase activities can be correlated to feeding habits involving dietary chitin intake. But, Lindsay et al. (1984) found relatively high levels of chitinase activity in the stomachs and of chitobiase in the intestines of rainbow trout fed different diets with or without chitin. Gastric chitinase activity was detected even from the late yolk sac stage and feeding or the presence of dietary chitin was not required for the development of chitinase activity (Lindsay, 1985). More recent work has shown that teleosts have several endogenous chitinases, which can be classified into specific groups having different substrate specificities, suggesting that they can degrade ingested chitinous substances efficiently (Ikeda et al., 2017).

7.2.5 Cellulase

Very early studies by Stickney and Shumway (1974) suggested the presence of cellulase activity in the digestive tracts of many species of fish from the wild as well as in some farmed fish. They also showed that channel catfish exposed to an antibiotic lost their cellulase activity while untreated, starved controls did not. Bergot (1981) undertook a specific study with purified cellulose in the diets of rainbow trout and common carp and found that cellulose and hemicellulose were not digested by either species and also that volatile fatty acid concentrations in the different segments of the digestive tract were low. Cellulase activity is often considered to be associated with the microflora in the digestive tract. Lesel et al. (1986) found that the cellulolytic population was always very low ($<10^3$ bacterial bodies/g) in the grass carp (*Ctenopharyngodon idella*) and even absent in the goldfish (*Carassius auratus*) and that the role of bacterial flora to ensure cellulolytic function would be negligible, if at all in both the species. In the milkfish (*Chanos chanos*) consuming naturally occurring algae in the ponds, no cellulase activity was detected in any region of the digestive tract (Chiu and Benitez, 1981).

The possible action of gut microflora in the digestion of cellulose has been proposed by many. Both endogenous and bacterial cellulase were found in grass carp (Das and Tripathy, 1991). In rohu (*Labeo rohita*), an Indian major carp, a diet-dependent cellulase activity was reported by Saha and Ray (1988) who also found that there was possibly a significant contribution of microflora in this species. Cellulase-producing bacteria have been isolated from the digestive tracts of tilapia (*Oreochromis niloticus*) as well as grass carp by Saha et al. (2006). The cell walls of marine algae may contain other more complex polysaccharides such as carrageenan, agarose, laminaria, and alginate. There are indications that at least some tropical marine fish appear to degrade these carbohydrates although it is not clear whether this is due to endogenous digestive enzymes or due to microbial actions (Skea et al., 2007). To assess the contribution of such intestinal microflora to digestion of complex carbohydrates requires information on the relative importance of exogenous enzymes produced by the gastrointestinal flora and endogenous digestive enzymes secreted by the host fish themselves. Despite a relatively large number of reports on microbial communities in the guts of several fishes (Wang et al., 2018) reportedly having "herbivorous" feeding habits in nature, there is still a lack of baseline data from wild fish of different trophic levels. Our understanding of the possible contribution of these symbioses in digestion and their functional significance lags what is known in land animals (Clements, 1997; Clements et al., 2014).

7.3 Digestibility of dietary carbohydrates

Digestion of carbohydrates requires hydrolysis to liberate monosaccharides followed by their absorption by enterocytes. One of the earliest studies on the measurement of digestibility of starch was undertaken by Singh and Nose (1967) with rainbow trout fed purified diets containing graded levels (20%–60% of the diet) of different carbohydrate sources such as glucose, sucrose, lactose, dextrin, and native potato starch. They found that the digestibility (predominantly absorption) of the mono- and disaccharides was very high (>95%) and was unaffected by the inclusion levels, whereas the digestibility of carbohydrates from dextrin and potato starch was low and decreased with increasing levels of incorporation. A number of studies have been undertaken with various species of interest from the global fish production point of view: cyprinids such as the common carp (Kirchgessner et al., 1986; Hernandes et al., 1994; Degani et al., 1997), grass carp (Law, 1986; Takeuchi et al., 1994; Tan et al., 2006), Indian major carps (Erfanullah and Jafri, 1998), cichlids such as tilapia (Takeuchi et al., 1994; Shiau and Liang, 1995; Kaushik et al., 1995; Maas et al., 2020b), salmonids such as the rainbow trout (Bergot and Breque, 1983; Kim and Kaushik, 1992; Bergot, 1993; Brauge et al., 1994; Pfeffer, 1995; Storebakken et al., 1998) or the Atlantic salmon (Arnesen and Krogdahl, 1993; Hemre et al., 1995; Hillestad et al., 2001), sturgeon (Hung et al., 1989; Herold et al., 1995), catfish (Wilson and Poe, 1987; Hung et al., 2003); marine fish such as the European seabass (Dias et al., 1998; Peres and Oliva-Teles, 2002; Enes et al., 2006), the Asian seabass, *Lates calcarifer* (Glencross et al., 2012), the gilthead seabream (Lupatsch et al., 1997; Venou et al., 2003; Fountoulaki et al., 2005; Enes et al., 2011), the turbot, *Scophthalmus maximus* (Jollivet et al., 1988; Burel et al., 2000), the Atlantic cod, *Gadus morhua* (Hemre et al., 1989), or the cobia, *Rachycentron canadum* (Ren et al., 2011). Davis and Arnold (1993) found the digestibility of carbohydrates to vary with source and thermal treatments. Cousin et al. (1996) found that native starches were as digestible as pretreated cereals or starches in shrimp. Summarized data on the digestibility of carbohydrates from various feedstuffs in different fish have been made available by some authors (Stone, 2003; Krogdahl et al., 2005; Enes et al., 2011; Maas et al., 2020b).

Generalizing, the major factors affecting the digestibility of carbohydrates in fish are species (trophic level), inclusion level, composition of carbohydrates (mono- and disaccharides and starch vs. pectins, hemicellulose, cellulose, etc.; Fig. 7.1), and technological treatment, which alters the gelatinization degree of starch (e.g., steam pelleting vs. extrusion, thermal treatments, etc.). Some examples of these major factors on measured digestibility of total carbohydrates, starch, and NSP of different ingredients are given in Table 7.2.

In many species, the digestibility of starch is found to decrease with increasing dietary inclusion levels (Bergot, 1979a,b; Bergot and Breque, 1983; Kirchgessner et al., 1986; Hemre et al., 1989; Krogdahl et al., 2004; Fountoulaki et al., 2005; Booth et al., 2006; Moreira et al., 2008; Enes et al., 2011). Irrespective of the species, the digestibility of carbohydrates is also linked to the complexity of the molecule: simple sugars such as glucose or sucrose having a higher digestibility than dextrin or raw starch. A clear inverse relation between dietary amylopectin levels and starch digestibility was observed by Glencross et al. (2012) in the Asian seabass. A systematic review of carbohydrate utilization by tilapia was made by Maas et al. (2020b), who found a positive linear relationship between dietary starch level and carbohydrate digestibility, whereas the digestibility was adversely affected by the dietary crude fiber levels.

Data on the digestibility of NSPs in fish are scarce. Some data on NSP digestibility in Nile tilapia and African catfish are given in Table 7.2, showing differences between species and ingredients. Leenhouwers et al. (2007a,b) showed that the digestibility differed between types of NSP: soluble NSP having a higher digestibility than insoluble NSP. They also reported the presence of significant amount of volatile fatty acids in the chyme especially in the distal intestine, an indication of occurrence of fermentation (Amirkolea et al., 2006; Leenhouwers et al., 2007a,b). As regards cellulose, even in the grass carp and in the goldfish, Lesel et al. (1986) found that the digestibility of cellulose was nil. Law (1986) found that the digestibility of carbohydrates from grass meals, copra, or rice bran was found to be nil in the same species.

In almost all species of fish and shrimp studied so far, preliminary hydrothermal treatment (cooking-extrusion, flaking, toasting) improves the digestibility of complex carbohydrates, thus increasing the dietary DE supply. Even different thermal treatments of feeds can have an impact on the ADC of starch as was reviewed by Maas et al. (2020b) that in tilapia, the ADC of starch for steam pelleted feeds was 91% versus 99% for extruded feeds. This treatment is all the more efficient in species with a low capacity for digestion of complex carbohydrates and has been demonstrated in both freshwater and marine fish. The use of cereals (wheat, maize, and triticale) or pulses (peas) as an energy source in fish feeds

TABLE 7.2 Some examples of studies showing the impact of factors affecting the apparent digestibility (ADC) of total carbohydrates (Tot. Carb.), starch, and nonstarch polysaccharides (NSPs) of different ingredients in different fish species.

			Dietary content (g/kg DM)			ADC (%)					
								Soluble	Insoluble		
Fish species	Ingredient	Type of diet	Tot. Carb.	Starch	NSP	Tot. Carb.	Starch	NSP	NSP	NSP	Sources
Salmon	Gelatinized corn flour	Extrusion	79a	70	–	–	86.4	–	–	–	Krogdahl et al. (2004)
Salmon	Gelatinized corn flour	Extrusion	252a	236	–	–	58.1	–	–	–	Krogdahl et al. (2004)
Trout	Gelatinized corn flour	Extrusion	79a	70	–	–	99.1	–	–	–	Krogdahl et al. (2004)
Trout	Gelatinized corn flour	Extrusion	252a	236	–	–	97.5	–	–	–	Krogdahl et al. (2004)
Nile tilapia	Gelatinized corn flour	Steam pelleting	258a	181	–	–	99.2	–	–	–	Amirkolaie et al. (2006)
Nile tilapia	Native corn flour	Steam pelleting	260a	177	–	–	93.8	–	–	–	Amirkolaie et al. (2006)
Nile tilapia	Gelatinized corn flour	Steam pelleting	397a	309	–	–	99.5	–	–	–	Amirkolaie et al. (2006)
Nile tilapia	Native corn flour	Steam pelleting	393a	296	–	–	93.9	–	–	–	Amirkolaie et al. (2006)
Nile tilapia	Maize	Steam pelleting	320b	268	52	–	87.4	neg.	21.6	neg.	Leenhouwers et al. (2007a)
Nile tilapia	Wheat	Steam pelleting	321b	259	66	–	91.0	13.6	60.1	neg.	Leenhouwers et al. (2007a)
Nile tilapia	Barley	Steam pelleting	346b	257	89	–	90.6	23.9	55.4	neg.	Leenhouwers et al. (2007a)
Nile tilapia	Rye	Steam pelleting	327b	242	85	–	92.8	17.9	53.6	neg.	Leenhouwers et al. (2007a)
African catfish	Maize	Steam pelleting	322b	273	49	–	89.7	48.4	60.7	44.1	Leenhouwers et al. (2007b)
African catfish	Wheat	Steam pelleting	314b	273	41	–	87.4	4.0	7.5	2.1	Leenhouwers et al. (2007b)
African catfish	Barley	Steam pelleting	290b	208	82	–	80.3	22.2	59.6	neg.	Leenhouwers et al. (2007b)
African catfish	Rye	Steam pelleting	292b	224	68	–	96.4	21.0	25.7	17.6	Leenhouwers et al. (2007b)
Nile tilapia	*Nannochloropsis gaditana*	Extrusion	140	2	138	21.6	–	46.0	–	–	Teuling et al. (2017)
African catfish	*N. gaditana*	Extrusion	140	2	138	46.9	–	70.2	–	–	Teuling et al. (2017)
Nile tilapia	*Arthrospira maxima*	Extrusion	138	33	105	68.2	85.3	77.5	–	–	Teuling et al. (2017)
African catfish	*A. maxima*	Extrusion	138	33	105	66.3	103.5	70.1	–	–	Teuling et al. (2017)
Nile tilapia	*Chlorella vulgaris*	Extrusion	164	31	134	70.4	108.2	97.8	–	–	Teuling et al. (2017)
African catfish	*C. vulgaris*	Extrusion	164	31	134	84.6	103.7	114.8	–	–	Teuling et al. (2017)
Nile tilapia	*Scenedesmus dimorphus*	Extrusion	204	81	123	56.9	96.9	60.2	–	–	Teuling et al. (2017)
African catfish	*S. dimorphus*	Extrusion	204	81	123	62.3	99.7	66.9	–	–	Teuling et al. (2017)

thus involves technological pretreatments. A level of gelatinization above 70% appears to be required to maximize carbohydrate digestibility in most fish. Such carbohydrate sources can thus contribute to protein sparing (Chapter 2, *Bioenergetics*).

In conclusion, carbohydrates are not inert for digestion by fish and shrimp and contribute partially to their energy requirements. However, the extent of carbohydrate digestion differs between fish species (trophic level) and type of carbohydrate. In general, in increasing order of digestibility, the general classification of polysaccharides can be as follows: lignin < cellulose < hemicelluloses < pectins < native starch < gelatinized starch.

7.3.1 Gastrointestinal transit

The extent of hydrolysis of macronutrients by endogenous enzymes and subsequent absorption is dependent on the passage rate of digesta throughout the gastrointestinal tract. Most studies in fish on intestinal transit time have focused on gastric emptying. Gastric evacuation rates and duration can be affected by the forms and levels of dietary starch. The type and level of inclusion of dietary fiber, be it soluble or insoluble, can have a major influence on gastrointestinal transit and even the physical characteristics of feces. Depending on the level of inclusion, dietary fiber levels in the feed have been found to affect the rate of passage of the chyme through the intestine, thus affecting the time available for absorption in several species (Hilton et al., 1981, 1982; Spannhof and Plantikow, 1983; Hernandez et al., 1994; Venou et al., 2003; Amirkolae et al., 2006; Bonvini et al., 2018; Le Hoang et al., 2019).

In African catfish (Harter et al., 2015), the stomach evacuation rate was reduced with an increase in dietary starch content. Gelatinized starch sources had a lower stomach evacuation rate than raw starch sources in sea bream (Venou et al., 2003). Extrusion of feeds as compared with steam pelleting was shown to reduce the stomach passage rate in rainbow trout (Hilton et al., 1981), which is most likely related to the gelatinization degree of starch. As regards the effects of nonstarch polysaccharides on the passage rate of digesta, the effect seems to be different between soluble and insoluble NSP. Insoluble NSP seems to increase the passage rate of digesta in fish. Carboxymethylcellulose (an insoluble NSP) increased the stomach emptying rate in tilapia (Shia et al., 1988) and increasing the dietary inclusion of wheat bran (mainly insoluble NSP) resulted in a quicker stomach emptying in rainbow trout (Hilton and Slinger, 1983). In contrast to these fish studies, gastric emptying rates were reduced with higher inclusion levels of α-cellulose in the freshwater prawn, *Macrobrachium rosenbergii* (del Carmen Gonzalez-Pena et al., 2002). Soluble NSPs seem to delay the gastric emptying in fish. Increasing dietary inclusion levels of guar gum (high in soluble NSP) in rainbow trout (Storebakken, 1985) and carob seed germ meal (rich in soluble NPS) in European sea bass (Nikolopoulou et al., 2011) reduced the gastric emptying rate. The extent to which both soluble and insoluble NSPs affect the digestive passage rate most likely differs between fish species. For instance, carob seed germ meal affected the stomach emptying in seabass, but the same diets did not have an impact in sea bream (Nikolopoulou et al., 2011). The reduced passage rate induced by soluble NSP is likely related to its ability to increase the viscosity of digesta. Like in most monogastric land animals, the increased digesta passage rate by insoluble NSP might be related to their role as filler of bulk (dilution of dietary nutrients).

7.4 Absorption/uptake and transport

As mentioned earlier, despite big differences in the structural and functional anatomy of the digestive tract (Chapter 8, *Nutritional Physiology*), all fish possess the necessary enzymes for hydrolysis and absorption of simple and more complex carbohydrates (Krogdahl et al., 2005; Enes et al., 2011). After digestion of carbohydrates, the main monosaccharides such as D-glucose, D-galactose, or D-fructose are then absorbed along the intestinal tract and transported to the different metabolic organs via the circulatory system through specific transporters in the enterocyte membranes (Buddington et al., 1987; Reshkin and Ahearn, 1987; Collie and Ferraris, 1995). Currently, it is recognized that these transporters comprise two structurally and functionally distinct groups, such as the Na^+-dependent glucose cotransporters (SGLTs) in the brush border/apical membrane and the facilitative Na^+-independent glucose/sugar transporters (GLUT family) in the basolateral membrane, very much similar to those found in land animals (Wood and Trayhurn, 2003). The SGLT has been detected in almost all freshwater and marine fish studied so far (Kamalam et al., 2017; Deng et al., 2020; Drai et al., 1990; Enes et al., 2011; Sala-Rabanal et al., 2004). Several members of glucose transporters have been identified in fish, and the different glucose transporters appear to be fairly well conserved within the vertebrate groups (Krogdahl et al., 2005; Bakke et al., 2010; Marin-Juez et al., 2014). The active uptake and transport of glucose occurs all along the digestive tract in fish, but the majority of glucose uptake occurs in the anterior intestine, and in species having pyloric caecae, these tiny evaginations also contribute significantly to the uptake of glucose. The existence of an alternate metabolic pathway for the export of glucose to the bloodstream dependent on phosphorylation of glucose by hexokinase (HK) and further dephosphorylation by glucose-6-phosphatase (G6Pase) is also established in fish (Polakof et al., 2010).

In crustaceans, D-glucose transport processes occurring at the hepatopancreatic epithelium have been detected in the lobster (*Homarus americanus*) as well as in the kuruma prawn (*Penaeus japonicus*) (Verri et al., 2001; Martínez-Quintana and Yepiz-Plascencia, 2012). Likely et al. (2015) analyzed mucosal to serosal transport of labeled glucose, sucrose, and fructose and found that there was transepithelial transport of sucrose, suggesting that disaccharide sugars can be transported intact across the intestine of lobster possibly through a functional disaccharide membrane transporter. In some marine shrimp, carrier systems able to use both Na^+- and K^+-dependent glucose transport processes on D-glucose transport have been identified (Obi et al., 2013). As regards the different GLUT proteins, our knowledge in crustaceans is still limited (Martínez-Quintana and Yepiz-Plascencia, 2012; Wang et al., 2016).

The rate of uptake and transport of glucose across the digestive barrier in fish is considered to be lower in fish than in mammals with differences also between species of fish (Collie and Ferraris, 1995; Bakke et al., 2010). Boge et al. (1982) observed that the time course of D-glucose uptake by intestinal brush border membranes has specific requirements for Na^+ as in mammals and that marine fish have different absorption kinetics. Buddington et al. (1987) fed different species of fish with the same feed to demonstrate differences in the regulation of intestinal amino acid (AA) and glucose uptake. Santigosa et al. (2011) showed that D-glucose influx was much higher in the pyloric caecae in the rainbow trout than in the proximal or distal intestines, whereas in the gilthead seabream, influx at the distal intestine was much higher than in the proximal intestine. Based on in vitro studies, Subramaniam et al. (2019, 2020) found that tilapia had a higher electrogenic glucose absorption than rainbow trout with different kinetic systems and that fish differ from mammals in that electrogenic glucose absorption occurs even in the hind gut of fish, a phenomenon already noted by Collie and Ferraris (1995). As regards dietary factors, the capacity for modulation of glucose transport in tune with the dietary supply has been demonstrated in many species of fish studied so far Soengas and Moon (1998), Bakke-McKellep et al. (2000), Kirchner et al. (2008), Polakof et al. (2010), Santigosa et al. (2011), Li et al. (2018a,b).

7.5 Postprandial glycemia and control

More than 50 years ago, Chavin and Young (1970) analyzed blood sugar levels in different species of fish and found them to vary considerably, to range from 0 to 3.8 g/L (>21 mmol/L). A comparison of data on basal glycemia values from different fish species shows that they can vary greatly (Polakof et al., 2012; Biju Sam Kamalam, 2017) but are generally lower in teleosts than in other vertebrate groups (Polakof et al., 2012). Glucose turnover rates in fish are reported to be lower than in mammals and birds (Garin et al., 1987; Polakof et al., 2012). In metabolic terms, monitoring postprandial glycemia is considered a good indicator of the utilization of dietary carbohydrates. Measurements of changes in glycemia have been made in fish subjected to an oral glucose challenge, an intravenous glucose infusion or intraperitoneal injections in different species (Palmer and Ryman, 1972; Furuichi and Yone, 1981; Hung, 1991; Legate et al., 2001; Hemre et al., 2002; Gisbert et al., 2003; Polakof et al., 2011, 2012, Li et al., 2018a,b). Early studies (Palmer and Ryman, 1972) suggested that some fish are "intolerant" to an oral glucose challenge. Furuichi and Yone (1981) compared the pattern of change in plasma glucose and insulin levels when subjected to a glucose tolerance test in three species of fish, common carp *Cyprinus carpio*, red sea bream *Chrysophrys major*, and yellowtail. Their data showed that the three species exhibited different levels and patterns of changes in plasma glucose and insulin levels after the oral glucose challenge.

After feed intake, there is a marked increase in plasma glucose levels, and the hyperglycemia can be long-lasting compared with terrestrial vertebrates, exceeding more than 12 h in most teleosts (Fig. 7.2). The peak postabsorptive plasma glucose concentrations can reach very high levels (>20 $mmol^{-1}$ or 300 $mg \cdot 100\ mL^{-1}$), depending on the amount and type of dietary carbohydrate sources (Bergot and Breque, 1983; Kaushik et al., 1989; Bergot, 1993; Peres et al., 1999; Deng et al., 2001; Figueiredo-Silva et al., 2012), water temperature, and body size, which together affect overall intestinal transit and metabolic rates. In comparison, the postprandial peak in man or other land animals is normally less than 7 mmols and lasts only for a few hours (Polakof et al., 2012). Even in shrimp, such postprandial glycemia levels appear to follow similar patterns as found in fish. Abdel Rahman et al. (1979) reported that the level of glucose in the hemolymph of kuruma shrimp increased rapidly after they were fed a diet containing glucose and remained at high levels for 24 h. Hall and Van Ham (1998) found that the mean blood glucose levels varied between 0.77 and 1.39 mmol/L, in the tiger shrimp, *Penaeus monodon*, depending on culture conditions. Pronounced and significant increases in blood glucose levels occurred within 20 min of feeding, with peak levels

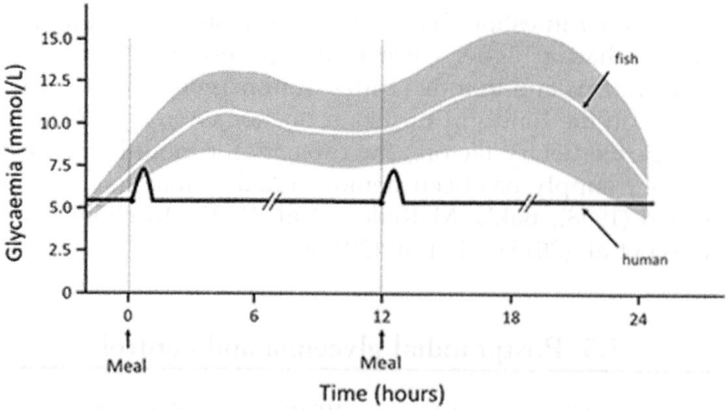

FIGURE 7.2 Comparison of glycemia in fish and man. The shaded area corresponds to fluctuation in glycemia levels due to variation in digestible carbohydrate input.

after 100 min. The prolonged hyperglycemia observed in fish either after glucose administration or after a carbohydrate-rich meal reflects a lack of control of glycemia and, often associated as an "intolerance" to high dietary carbohydrates in teleosts, has been well analyzed (Moon, 2001). The term "glucose intolerance" is a clinical term describing persistent hyperglycemia after an oral glucose load in mammals. The prolonged hyperglycemia seen in teleosts after a glucose challenge or after a meal with high carbohydrate levels is, however, transient. Besides, teleost fish generally have high plasma insulin levels compared with mammals, and thus, an insulin deficiency cannot be a suitable explanation for this persistent hyperglycemia but has more to do with peripheral utilization of glucose. The range of change in hemolymph glucose concentration in crustaceans (0.2–2 mM) is smaller than that in fish (Wang et al., 2016).

Initially, the postprandial hyperglycemia in fish fed diets rich in carbohydrates was attributed to an absence of insulin secretion and led some authors to consider fish as "naturally diabetic" animals. The development of specific assays (radioimmunoassay) for fish insulin has demonstrated that they do secrete insulin (Mommsen and Plisetskaya, 1991; Plisetskaya, 1998), with the values ranging from 1 to 3 ng/mL in starved fish. After feeding, the plasma insulin levels much higher, ranging from 5 to 50 ng/mL. It is now clear that in most fish, insulin secretion is not a limiting factor. Dietary carbohydrate supply and consequent glucose uptake stimulates insulin secretion in almost all species (Polakof et al., 2012). In addition, insulin administration lowers glycemia in trout (Kaushik et al., 1989) just as effectively as in other species (Furuichi and Yone, 1981). The respective roles of insulin and of glucagon seem to be the same in fish as they are in terrestrial animals (Plisetskaya and Mommsen, 1996). Insulin receptors are present in several fish tissues, but also differences between species in terms of tissue receptivity and tyrosine kinase activity. As regards urinary glucose excretion, there is little information on the renal threshold for plasma glucose in fish. Specific techniques such as those developed by Deng et al. (2000) can be of use to assess the glycemic and glycosuric response of fish under practical conditions. The hormones involved in glucose regulation in crustaceans are different from those in fish. Besides insulin-like peptides and insulin-like growth factor, the main hormone involved in the regulation of hemolymph glucose in shrimp is the crustacean hyperglycemic hormone produced in the eyestalk (Wang et al., 2016).

7.6 Glucose metabolism

The gastrointestinal tract is the first organ confronted with diets and the nutrients contained therein. It is also a highly active tissue with high protein turnover rates (Houlihan, 1991). As mentioned earlier, in response to dietary carbohydrate supply, the intestine of fish is shown to increase glucose uptake (Blasco et al., 1996, 2001) and accumulate glycogen (Polakof et al., 2010). Glucose utilization by the intestinal tract itself has been measured and found to be significant, being dependent on the amount available from dietary or endogenous origins and the rates of transport and uptake mechanisms involved (Soengas and Moon, 1998; Soengas et al., 2006; Polakof et al., 2010). Glucose metabolism by the intestine itself depends on glucose transport rates, and glucose utilization rates although significant are lower than those of transport, indicating that gastrointestinal tract cells metabolize only a fraction of the transported glucose. The intestine can regulate its own glucose metabolism, exerting control over glycogenolysis (Mommsen and Busby, 2006; Polakof et al., 2010) and decreasing glucose oxidation (Soengas and Moon, 1998). The fish gut has been shown to be a major glycolytic tissue (Polakof et al., 2010); hence the balance between glycolysis and gluconeogenesis in the intestinal tissue has a major role in glucose homeostasis at the tissue or the whole body level. Few studies have examined the time course of changes in intestinal glucose metabolism (transport, glycolysis, glycogenesis, and gluconeogenesis) at the biochemical and molecular levels. The anterior intestine plays a major role in the regulation of postprandial glucose homeostasis with significant glycolytic potential and glucose storage (Polakof et al., 2010; Chen et al., 2017). A lack of control of gluconeogenesis at the digestive tract level is one factor contributing possibly to the persistent postprandial hyperglycemia at least in some species such as the rainbow trout (Polakof et al., 2010).

7.6.1 Glycolysis

Glycolytic reactions transform glucose into pyruvate with ATP formation and produces oxidizable substrates for the tricarboxylic acid cycle (under aerobic conditions) or to lactate under anaerobic conditions, prevailing for instance, in the white muscle. All enzymes involved in the glycolytic pathway(s) (Fig. 7.3; Polakof et al., 2012) have been detected in all the species of fish studied thus far. As in other animals, all glycolytic reactions are reversible and can thus function in the direction of gluconeogenesis, with the exception of the three stages catalyzed by HK (formation of glucose-6-P), phosphofructokinase (formation of fructose-1,6-diphosphate), and pyruvate kinase (transformation of phosphoenolpyruvate into pyruvate), respectively (Fig. 7.3). The regulation of the enzymes involved in glycolysis (HKs, 6-phosphofructo-1-kinase, and pyruvate kinase) by dietary factors, especially of dietary carbohydrate supply, has been demonstrated in almost all teleosts studied so far (Enes et al., 2009; Biju Sam Kamalam et al., 2017). Much less is known in shrimp (Wang et al., 2016).

Glucose phosphorylation: The conversion of glucose into glucose- 6-phosphate (glucose phosphorylation) is the first step of its metabolic utilization (Fig. 7.3) and is carried out by the HKs. The HKs catalyze the phosphorylation of hexoses in general and found in all cells involved in glucose metabolism. HKs have low Km and hence high affinity to glucose, even at low concentrations. Over the years, at least four isoforms of HKs (I to IV) with different Kms have been identified in various species. Glucokinase (GK, HK-IV) is specific to glucose

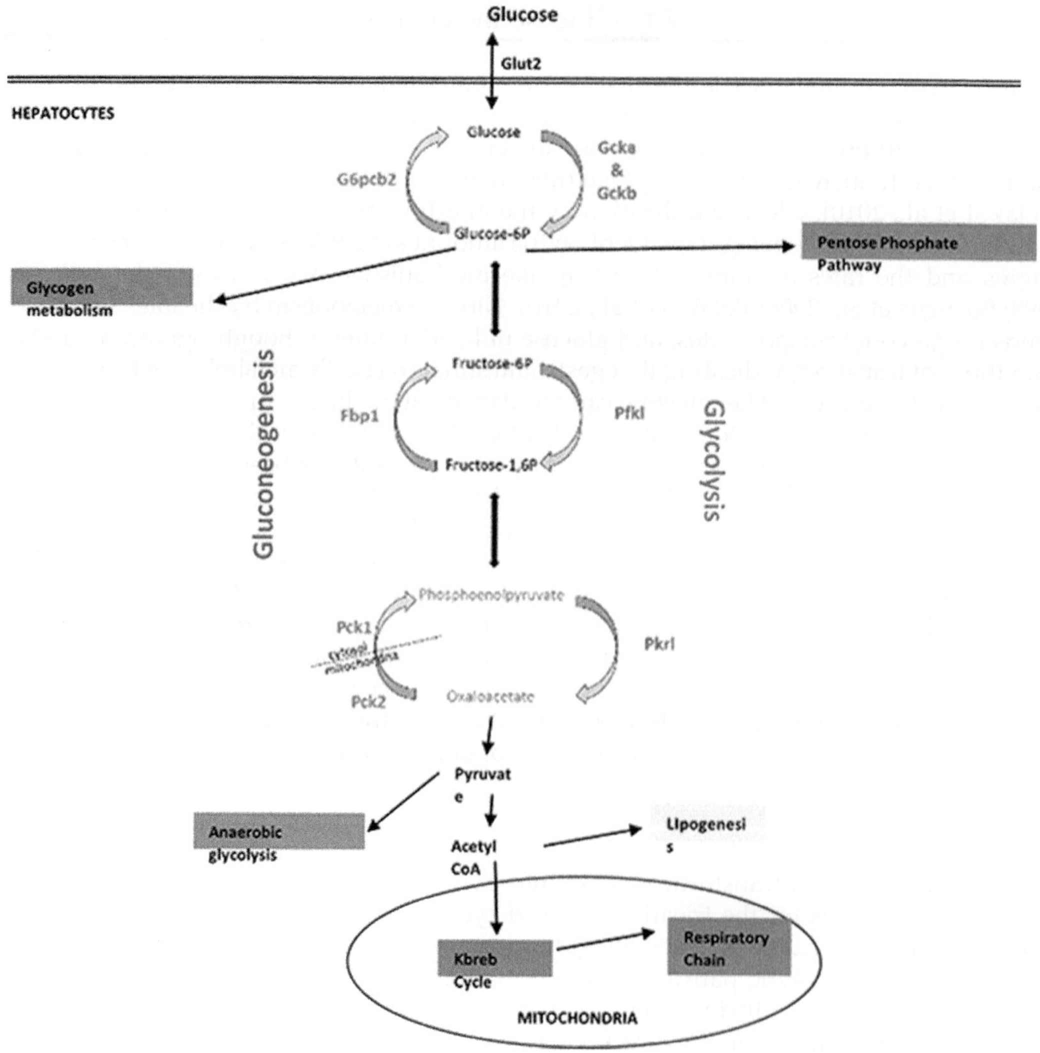

FIGURE 7.3 Schematic representation of pathways of glucose phosphorylation (glycolysis) and glucose production (gluconeogenesis) and their links with other metabolic pathways involved in the storage of excess glucose (glycogenesis and lipogenesis) and energy metabolism (pentose phosphate pathway, citric acid cycle, and respiratory chain) in liver.

with a high Km or low affinity to glucose and responds to high dietary glucose inputs. It was initially suggested that the prolonged hyperglycemia observed in fish after challenge with glucose or when fed diets with high levels of carbohydrates might be due to the low level of glucose phosphorylation by HKs in the liver and to the absence of an inducible hepatic GK expression (Walton and Cowey, 1982). Subsequent studies have demonstrated the presence of GK-like enzymes in the Atlantic salmon (Borrebaek et al., 1993; Tranulis et al., 1996). Work undertaken by Panserat et al. (2000a,b, 2001) also clearly showed that GK is very much present in different species such as the common carp, the rainbow trout, and the gilthead seabream. Although liver is the main organ where the expression and activities of HK(s) are found, they have also been detected in a number of other tissues (Polakof et al., 2007).

With respect to nutritional regulation, the activities of the low Km HKs (HK I–III) do not seem to be under nutritional regulation, not much affected by dietary carbohydrate levels (Panserat et al., 2000; Kirchner et al., 2008; Enes et al., 2009), whereas GK (HK IV) activity and expression levels respond to dietary digestible carbohydrate supply and the consequent changes in blood glucose levels in different species of fish (Enes et al., 2009; Panserat et al., 2014; Nie et al., 2015; Song et al., 2018) (Fig. 7.4). However, the potential significance/role of the adaptive response of hepatic GK to high carbohydrate intake in species such as the rainbow trout remains an intriguing issue (Polakof and Panserat, 2016).

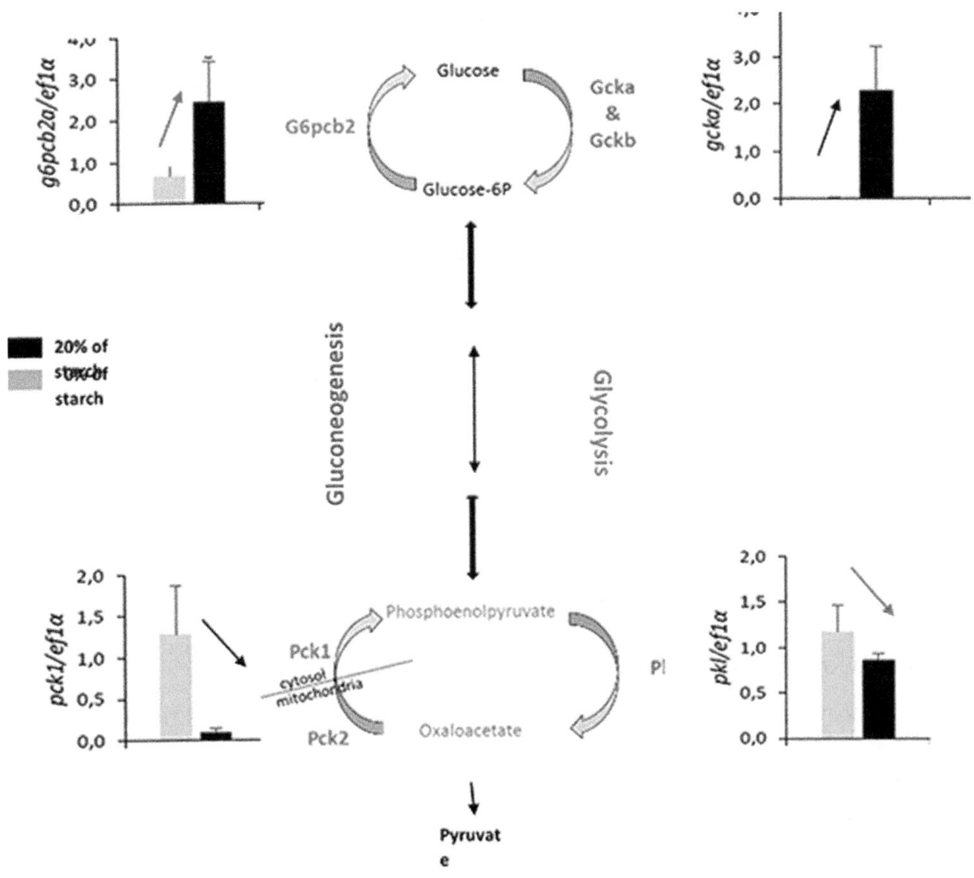

FIGURE 7.4 Example of molecular regulations by dietary carbohydrates of hepatic gene expression in rainbow trout. The upregulation of the first enzyme of glycolysis (gck) and the downregulation of the first gluconeogenic enzyme (pck1) by dietary carbohydrates (*black box*) are expected. By contrast, the downexpression of the last glycolytic enzyme (pkl) and the upregulation of the last gluconeogenic enzyme are surprising and different to mammals. *G6pcb2*, glucose-6-phosphatase; *Gck*, glucokinase; *pck1*, phosphoenolpyruvate carboxykinase; *pkl*, pyruvate kinase. *Data extracted from Song, X., Marandel, L., Skiba-Cassy, S., Corraze, G., Dupont-Nivet, M., Quillet, E., Geurden, I., Panserat, S., 2018. Regulation by dietary carbohydrates of intermediary metabolism in liver and muscle of two isogenic lines of rainbow trout. Front. Physiol. 9. doi: 10.3389/fphys.2018.01579.*

6-Phosphofructo-1-kinase (PFK) catalyzes the phosphorylation of fructose-6-phosphate into fructose-1,6-bisphosphate. The activity of PFK activity has been detected in several tissues such as the liver, red, and white muscle, kidney, gills, heart, and brain in cod, plaice, and rainbow trout (Knox et al., 1980). Fish PFK resembles that of mammals, being activated by fructose-1,6-bisphosphate and AMP and being inhibited by ATP (Cowey and Walton, 1989). Dietary carbohydrate levels have been shown to modulate hepatic PFK-1 activity in some fish (Fideu et al., 1983; Walton, 1986; Fernández et al., 2007), but the response appears to be less than that of GK (Enes et al., 2009; Song et al., 2018).

Pyruvate kinase (PK) catalyzes the last step in glycolysis for the conversion of phosphoenolpyruvate (PEP) to pyruvate. PK activities are reported in several tissues with high activities detected in the skeletal muscle, heart, and brain in different species (Knox et al., 1980), besides the liver, intestine, and kidney in rainbow trout (Panserat et al., 2001; Enes et al., 2009). Dietary carbohydrate intake does not seem to induce a direct and immediate increase in PK activities or expression in salmonids (Hilton and Atkinson, 1982; Fideu et al., 1983; Panserat et al., 2001; Enes et al., 2009; Song et al., 2018) (Fig. 7.4) but appears to be upregulated by high dietary carbohydrate levels in the grass carp (Yuan et al., 2013).

7.6.2 Pentose pathway

The pentose phosphate pathway also called the phosphogluconate pathway or shunt leads to the synthesis of ribose 5 phosphate from glucose 6 phosphate, with NADPH generation. This metabolite is an energy transporter with powerful reducing potential and is required for synthesis of fatty acids and steroids. The pentose pathway also provides the ribose necessary for the synthesis of nucleic acids. In fish, the enzymes for this oxidation pathway are mainly found in the liver, and their synthesis is temperature sensitive. Excess dietary carbohydrates induce lipogenesis increased activities/expression of enzymes involved in de novo lipogenesis in all species of fish (Lin and Shiau, 1995; Dias 1999; Hemre et al., 2002, Chapter 5, *Lipids*).

7.7 Endogenous glucose production, storage, and mobilization

7.7.1 Gluconeogenesis

Gluconeogenesis is the pathway responsible for the de novo synthesis of glucose, possibly also of glycogen (called the indirect pathway of glycogen synthesis), from different glucogenic precursors such as lactate, AAs, glycerol, or fructose (Walton and Cowey, 1982; Suarez and Mommsen, 1987; Moon, 1988). The metabolic pathways for gluconeogenesis in fish are similar to those found in other animal species. The liver and kidney are the main sites concerned. Gluconeogenesis involves four successive key enzymes, pyruvate carboxylase (PC), phosphoenolpyruvate carboxykinase (PEPCK; pck genes), fructose-1,6-bisphosphatase (FBPase; fpb gene), and glucose-6-phosphatase (G6Pase; G6pc genes), which have been reported in almost all animals (Fig. 7.3).

PC catalyzes the carboxylation of pyruvate to oxaloacetate and is not exclusively related to gluconeogenesis alone. As in other animals, it is found mainly in the liver with low levels in white muscle. PEPCK catalyzes the reversible conversion of oxaloacetate to PEP and is considered the rate-limiting step of gluconeogenesis. Glucose-6-phosphatase (G6Pase) catalyzes the hydrolysis of glucose-6-phosphate, produced from glycogen or gluconeogenic precursors, into inorganic phosphate and glucose, having thus a central role in glucose homeostasis by completing the final steps of gluconeogenesis as well as glycogenolysis. It is mainly found in the liver and widely distributed in all species of fish and shrimp studied so far. The activities of these enzymes are generally well correlated with glucose production in vivo (Mommsen, 1986).

Available information on the nutritional regulation of the enzymes involved in gluconeogenesis have also been assessed by several authors (Knox et al., 1980; Suarez and Mommsen, 1987; Enes et al., 2009; Polakof et al., 2012; Biju Sam Kamalam et al., 2017). The activities of G6Pase, FDPase, and PEPCK increase when fish are fed a high protein/low carbohydrate diet as compared with when fed a low protein/high carbohydrate diet. Protein intake has been shown to be the main nutritional factor, which regulates hepatic gluconeogenesis in different species of fish (Cowey et al., 1977, 1981; Walton, 1986; Kirchner et al., 2003; Enes et al., 2006; Guerreiro et al., 2014). AAs play a large part in gluconeogenesis in most teleosts, with glucogenic AAs such as alanine, serine, or glycine being preferred substrates. A dietary excess of some dispensable AAs such as alanine, aspartic acid, or glutamic acid does not, however, affect gluconeogenesis (Kirchner et al., 2003). Dietary carbohydrate supply does not seem to affect the activities/expression of gluconeogenic enzymes neither in the intestine (Kirchner et al., 2005, 2008; Polakof et al., 2010) nor in the liver (Lin and Shiau, 1995; Panserat et al., 2000, 2002; Figueiredo-Silva et al., 2012; Marandel et al., 2015, 2017). The persistent gluconeogenesis independent of carbohydrate intake explains the lack of control of glycemia and glucose homeostasis at the whole organism level. Even in fasted or starving fish, the contribution of body proteins and fats for the provision of energy is greater than that of glucose provided by the mobilization and hydrolysis of glycogen (Navarro and Gutierrez, 1995). The specific demands of different tissues such as the brain and nervous tissues for glucose are met by gluconeogenesis rather than by glycogenolysis, thus maintaining the blood glucose levels. There is indirect evidence that gluconeogenesis plays a significant role in maintaining glucose homeostasis in migratory fish undergoing starvation, which manages to maintain high hepatic glycogen levels (Chang and Idler, 1960), suggesting that glucose production by gluconeogenesis can even exceed glucose demand, the excess being stored as glycogen.

7.7.2 Fate of pyruvate resulting from glycolysis

Lactate formation: Under anaerobic conditions, such as during intense muscular exercise, pyruvate is converted into lactate via a lactate dehydrogenase (Fig. 7.3). This reversible reaction occurring in the cytosol generates NAD+, which allows the maintenance of the cells' redox potential and stimulates the formation of oxaloacetate. This pathway for pyruvate utilization is preponderant in comparison with the oxidative pathway in the white muscle of fish. As in mammals, lactate dehydrogenase is present in the form of several isozymes, which differ between tissues. However, fish isozymes have particular characteristics. The temperature required to obtain maximal activity is much lower than that required by the mammalian enzyme owing to a different enzyme–substrate link. In carp, muscle lactate dehydrogenase is only slightly sensitive to pH and maintains a high activity level even at low pH, which can be considered as an adaptation to severe hypoxic conditions. The activity of this enzyme in the white muscle is variable from one species to another, being higher in fast-swimming fish (Somero and Childres, 1990; Moon and Foster, 1995).

Formation of acetyl-CoA: Pyruvate is converted by decarboxylation to acetyl-CoA, which can then be oxidized via the TCA cycle (Fig. 7.2). The decarboxylation stage is carried out in the mitochondrial matrix by the pyruvate dehydrogenase complex, to thiamine pyrophosphate that coexists in the inactive phosphorylated form and the active dephosphorylated form. The activity of the enzyme complex is regulated by the concentration of mitochondrial ATP in the oxidative tissues. In white muscle, citrate inhibits the appearance of the active form.

7.7.3 Glycogen reserves and mobilization

As most animals, teleost fish also do have carbohydrate stores in the form of glycogen. In the fish, hepatic glycogen consists of 10,000 to 14,000 units of glucose, depending on species. There is significant variation between tissues: the brain, liver, and red muscle contain more than white muscle (Moon and Forster, 1995). In the brain, concentrations can reach $350\ \mathrm{mg \cdot 100\ g^{-1}}$. In the white muscle, these vary between 40 and $200\ \mathrm{mg \cdot 100\ g^{-1}}$ and, in the red muscle, between 300 and $900\ \mathrm{mg \cdot 100\ g^{-1}}$. Values in the liver vary from 1% to 12% of fresh tissue, in direct proportion to dietary carbohydrate supply. In crucian carp (*Carassius carassius*), for instance, glycogen content in the liver can be even as high as 20% of liver weight (Vornanen et al., 2011), and the capacity of massive glycogen stores in different tissues ensures survival of such species during the long anoxic periods in winter. Glycogen content is in general significantly higher in liver ranging from 65 to 252 glucosyl units/g, than in all other tissues for most teleosts studied so far (Short and Driedzik, 2018).

Breakdown of glycogen is carried out by the action of glycogen phosphorylase and has been observed in many species, localized in the white muscle but also in the liver. Glycogenolysis is very rapid in the white muscle compared with red muscle or to the liver. This mobilization of glycogen reserves from the white muscle, noticed under conditions of stress (hypoxia, exercise), leads to an increase in lactate concentrations. In contrast, there does not appear to be any relationship between glycogen catabolism and phosphorylase activity, the function of which in fish is the subject of controversy. For glycogen, as for the fatty acids, there is also a lysosomal degradation pathway (glycophagy), which appears to be preponderant in the muscle.

In general, glycogen as an energy substrate is used very little in fish tissues; glycogen content and its mobilization depend on the previous nutritional history (Navarro and Gutierrez, 1995). In fish fed with diets rich in digestible carbohydrates, hepatic glycogen reserves are high but are rapidly depleted during starvation. Conversely, in fish fed low-carbohydrate diets, mobilization during starvation is lower. In the wild, even after long periods of starvation, as observed in salmons that starve in synchrony with migration, or in carp during the winter months, glycogen reserves undergo only slight changes (French et al., 1983). This is very different from the situation encountered in higher vertebrates in which mobilization of hepatic glycogen is very rapid and that of muscle glycogen is slow. Despite this fact, hypoglycemia is rarely observed in starved fish, probably because there is a decrease in metabolic activity in general (and consequently in glucose utilization) and activation of gluconeogenesis.

Glycolysis and glycogenolysis, the main catabolic pathways for carbohydrates, appear to function at low levels in fish. Increasing carbohydrate supply results in an increase in enzyme activity involved in both the glycolysis and the pentose pathway (notably glucose-6-phosphate dehydrogenase) and hence lipogenesis. An inverse relationship is often observed with increasing lipid supply (Chapter 5).

7.8 Efficiency of utilization of digested carbohydrates for growth

In the aforementioned sections, it clear that a portion of dietary carbohydrates can be digested, especially starch when properly gelatinized, and provide a substantial part of the daily DE intake. For example, in Nile tilapia, Haidar et al. (2016) measured that about 45% of DE intake originated from digested carbohydrate. The aforementioned sections indicate that some fish species have limited capacity to metabolize absorbed glucose. This should then be reflected in a lower energetic utilization efficiency of DE for energy gain when fish are fed diets with a high digestible carbohydrate content.

The gross energy content of glucose is 15.9 kJ/g and of starch and cellulose is 17.6 kJ/g. The (mean) value for gross energy content of carbohydrates used by NRC (2011) is 17.2 kJ/g (for glucose and 17.6 for starch). Information about the net energy value (NE, the potential for energy retention) of digested carbohydrates is limited in fish. One of the earliest studies on NE values of dietary carbohydrates in fish was undertaken by Hilton et al. (1987) with rainbow trout. They determined NE values for glucose and starch to be, respectively, 2.2 kJ/g and 4 kJ/g, representing between 13% and 25% of gross energy values of these carbohydrates. Analysis of available data on the effects of dietary macronutrient composition on the efficiency of utilization of DE for growth by fish (Schrama et al., 2012) shows that the relationships of energy utilization with trophic level of fish and with growth composition were predominantly induced by dietary macronutrient composition. Efficiency of utilization of DE was found to decrease with an increasing ratio of dietary digestible carbohydrate over digestible fat. NE values for digested protein, fat, and carbohydrates have been estimated only for a few species such as the Nile tilapia and rainbow trout (Schrama et al., 2018), common carp and Asian seabass (Phan et al., 2019), and snakehead, *Channa striata* (Phan et al., 2021). The largest variability in NE value between fish species was observed for digested carbohydrates ranging between 0.9 and 12.1 kJ/g compared with protein and fat. Though carbohydrates were digested, Fig. 7.5 shows that the contribution of digested carbohydrates to NE is minimal in Asian seabass and snakehead. This is in contrast to rainbow trout, common carp, and Nile tilapia, which appear only slightly less efficient in utilizing digested carbohydrates for growth (10.2–12.1 kJ/g) compared with values used for pigs (12.1–14.4 kJ/g; CVB, 1993; Noblet et al., 1994). In the rainbow trout, the NE values were curvilinearly related to digestible carbohydrate intake (Schrama et al., 2018).

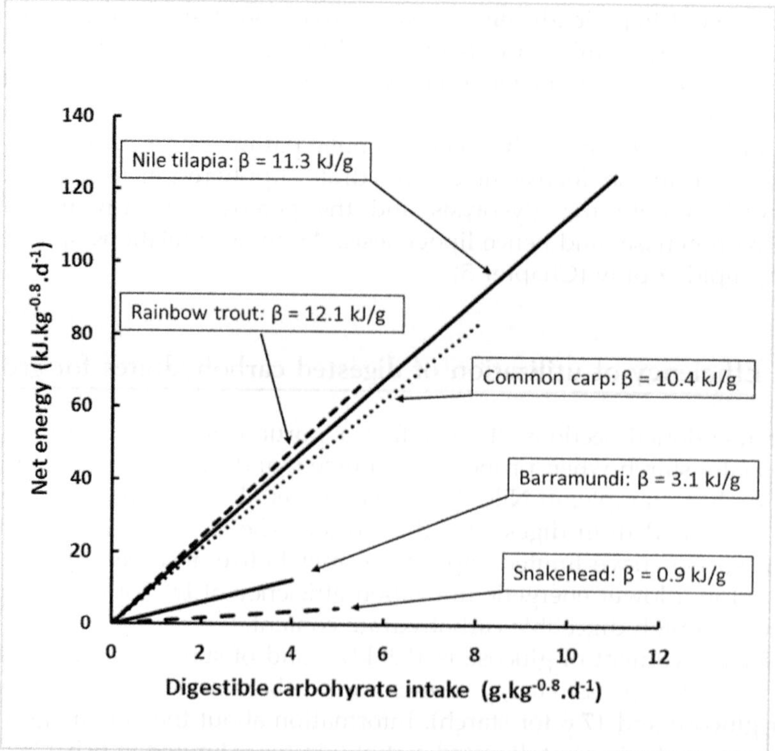

FIGURE 7.5 The relationships between net energy and digestible carbohydrate intake in different species of fish. Net energy in this graph is the sum of energy retention and fasting heat production (i.e., net energy for maintenance). For species comparison, net energy and digestible carbohydrate intake are expressed per unit of metabolic body weight ($kg^{0.8}$). The lines are only expressed for the range of digestible carbohydrate intake applied for the different species. *These relationships were based on data derived from Schrama, J.W., Haidar, M.N., Geurden, I., Heinsbroek, L.T.N., Kaushik, S.J., 2018. Energy efficiency of digestible protein, fat and carbohydrate utilisation for growth in rainbow trout and Nile tilapia. Br. J. Nutr. 119, 782–791, Phan, L.T.T., Groot, R., Konnert, G.D.P., Masagounder, K., Figueiredo-Silva, A.C., Glencross, B.D., Schrama, J.W., 2019. Differences in energy utilisation efficiencies of digestible macronutrients in common carp (*Cyprinus carpio*) and barramundi (Lates calcarifer). Aquaculture 511, 734238, and Phan, L.T.T., Masagounder, K., Mas-Munoz, J., Schrama, J.W., 2021. Differences in energy utilization efficiency of digested protein, fat and carbohydrates in snakehead (*Channa striata*). Aquaculture 532, 736066.*

7.9 Dietary levels of incorporation

In practical feed formulation, a minimal constraint for "carbohydrates" is seldom incorporated into the optimization given the absence of an established requirement for carbohydrates by fish or shrimp. However, under cooking-extrusion feed manufacturing, which is commonly used to produce feeds for many species of fish and shrimp, a minimal level of inclusion of starch at 60 g/kg is used as a technical constraint for ensuring proper physical properties (Sorensen, 2012). Consequently, the actual carbohydrate content in aquafeeds is constrained by levels of protein, AAs, fat, fatty acids, and energy content set during feed formulation. In marine fish species like salmon, common practice is to formulate high nutrient density diets by setting constraints (minimum levels) for high dietary protein and fat contents. This limits remaining space in the formulation, and consequently, such nutrient dense feeds have minimal/low carbohydrate levels. In shrimp, tilapia, and pangasius feed, the nutrient density is lower compared with salmon feeds due to lower minimum levels set for fat and/or energy in the formulation of these diets. In shrimp diets, which until recently were mainly produced as compressed (steam) pellets, high fat inclusion level (>10%) limits compression during pellet production and therefor water stability of the pellets. As a consequence, such feed formulas require a filler, which results in higher carbohydrate levels. Often this leads to the inclusion of inexpensive carbohydrates sources such as rice bran and wheat bran in feeds for shrimp, tilapia, pangasius, and other freshwater fish feeds, thus introducing substantial amounts on NSP in such low nutrient dense feeds.

The question remains whether specific fish species have a maximal or optimal dietary carbohydrate need. Research on this issue in fish has mainly focused on the digestible carbohydrate fraction (mono- and disaccharides and starch). Very early works undertaken with salmonids dealing with the effect of different carbohydrate sources (Buhler and Halver, 1961) showed no significant differences in fish growth or health when fed diets containing from zero up to 48% of dextrin replacing equivalent amounts of alpha-cellulose. They also observed that protein utilization was the best at a level of inclusion of 20% dextrin, which is most likely due to a protein sparing effect as the dietary nonprotein energy content increased with the inclusion of carbohydrates (dextrin). In fish fed diets containing glucose, maltose, dextrin, and potato starch as carbohydrate sources, growth rate was found to decrease with increasing complexity of carbohydrates (Bergot and Breque, 1979; Wilson and Poe, 1987; Hung et al., 1989). Wilson and Poe (1987) fed channel catfish (*Ictalurus punctatus*) diets containing similar amounts of glucose, maltose, fructose, sucrose, corn starch, and dextrin and found that growth and nutrient/energy utilization were higher in fish fed starch or dextrin than when fed mono or disaccharides. Hung et al. (1989) analyzed the effects of different dietary carbohydrate sources such as glucose, fructose, maltose, sucrose, lactose, dextrin, raw corn starch, or cellulose in the diets of white sturgeon (*Acipenser transmontanus*) and found that maltose and glucose were better utilized than other sources. Kaushik et al. (1989) found that relatively high levels of extruded or pregelatinized starches or even extruded whole corn could be used in the diets of Siberian sturgeon (*Acipenser baeri*). Even in the grass carp, Tian et al. (2010) observed that glucose was superior to starch as the carbohydrate source. Grass carp also appears to show higher tolerance to dietary starch than gibel carp, *Carassius gibelio*, where levels of corn starch above 300 g/kg lead to lipid accumulation in the liver and in the whole body (Li et al., 2016). In shrimp, it has been shown that even at dietary inclusion levels of 25%–35% of wheat flour, there were no adverse effects on carbohydrate digestibility or growth of tiger shrimp (Catacutan, 1991; Shiau et al., 1991). Available data on maximum inclusion levels of various carbohydrate sources in the feeds for different species of fish ranges from 12% to 48% (Kamalam et al., 2017).

An analysis of global carbohydrate production through terrestrial agriculture and the digestible fraction of these based on starch and sugar contents was made by Biju Sam Kamalam et al. (2017). Looking at global feed usage in aquaculture (Tacon and Metian, 2015; FAO, 2020), the contribution of terrestrial plant products would represent more than to 50% of feed composition, irrespective of the type of farmed fish or shrimp. But as mentioned earlier, starches differ depending on their botanical origins. In some fish species, raw wheat starch is better utilized than that of maize, triticale, or potato. The presence of amylase inhibitors in wheat does not appear to have any significantly harmful effect, at least in species such as the common carp. The protein sparing effect of digestible carbohydrates is also subject to much variability. However, the supply of DE by carbohydrates appears to have an overall beneficial effect in terms of improving growth and protein utilization in most species.

7.10 Other beneficial or detrimental effects of dietary carbohydrates

Besides having potential nutritional value (energy, glucose source), carbohydrates can interfere with the nutritional value of other nutrients. Both digestible and indigestible carbohydrates can influence the digestibility of other nutrients. Dilution of diets by the inclusion of 30% gelatinized corn starch in common carp (Phan et al., 2019) and the inclusion of 30% wheat flour in snakehead (Phan et al., 2021) decreased the digestibility of protein and fat. Krogdahl et al. (2004) showed that 20% inclusion of gelatinized corn flour reduced protein and fat digestion in salmon but had no impact in rainbow trout. Also the degree of starch gelatinization in nonextruded feeds can influence the digestibility of other nutrients. Replacement of gelatinized corn flour by native maize flour starch reduced protein and fat digestibility in fish fed steam pelleted feeds (Amirkolaie et al., 2006). This impact of gelatinization degree is important when comparing digestibility values in literature in relation to the method of feed production (steam pelleting, extrusion, etc.).

Regarding the impact of NSPs on the digestibility of other nutrients, a distinction should be made between soluble and insoluble NSPs. Soluble NSPs have often the capacity to increase the viscosity of digesta and feces, which is shown in Nile tilapia (Amirkolaie et al., 2005), African catfish (Leenhouwers et al., 2006), rainbow trout (Brinker, 2007), and striped catfish (Tran-Tu et al., 2019). In most of these studies, increasing digesta viscosity paralleled with a decrease in digestibility of all macronutrients, except for the study on rainbow trout by Brinker (2007). Increased digesta viscosity is hypothesized to reduce mixing of endogenous enzymes and digesta and hamper the diffusion and consequently absorption of nutrients by enterocytes.

Insoluble NSPs can also hamper digestion of other macronutrients, which can be attributed to increasing bulkiness of feeds but also to specific properties such as bile acid binding. As in land animals, there is a negative correlation between dietary fiber content and energy digestibility and consequently feed efficiency, mainly due to a dilution of digestible macronutrients. Inclusion of fiber affects rate of passage of foodstuffs through the digestive tract and possibly leads to an increased feed/DE intake, as described earlier. Many other effects of fiber, both beneficial and detrimental, have been described or are still the subject of studies with higher vertebrates (desquamation of intestinal cells, increased elimination of cholesterol and minerals). Most of these are still poorly understood or unknown in fish. Even the incorporation limits in carnivorous or omnivorous fish are impossible to determine precisely; the value of 10% which is sometimes used appears to be rather arbitrary. In experimental diets, cellulose can be used conveniently to dilute diets without altering nutrient utilization, on the assumption that cellulose is inert. However, dietary cellulose (insoluble NSP) inclusion decreases lipid digestibility in Atlantic salmon (Aslaksen et al., 2007) and Nile tilapia (Amirkolaie et al., 2005). Staessen et al. (2020) found that bile acid losses via feces in rainbow trout were increased when fed diets with high NSP levels, which correlated with a reduction in fat digestion.

Due to their low to indigestible nature, carbohydrates and especially NSP form the major fraction in fish feces, i.e., fecal waste production as shown in rainbow trout by Meriac et al. (2014). Therefore, dietary carbohydrates strongly determine the chemical composition and characteristics of fecal waste. For example, some binding agents such as agar, guar gum, alginates, carrageenans, carboxymethylcellulose, or even cereal brans can improve feces consistency (Brinker, 2007). The impact of dietary NSP content of fecal stability and fecal removal efficiency from the water depends on fish species. In rainbow trout, guar gum supplementation at a low dose improved feces stability and the feces removal efficiency from the water (Brinker et al., 2005). However, in Nile tilapia (Amirkolaie et al., 2005, 2006) and striped catfish (Tran-Tu et al., 2020), guar gum negatively affects fecal stability and recovery efficiency in addition to an increased amount of fecal waste produced (i.e., lower digestibility).

7.11 Improving carbohydrate utilization through exogenous enzymes

Given the recognition of the need for reducing reliance of the aquaculture feed industry on capture fishery-based feedstuffs, primarily fishmeal or fish oil, and given the increasing use of plant protein sources in feeds for fish or shrimp, there is due concern as to the possibility of increasing levels of indigestible/nonavailable carbohydrates including NSPs in the feeds. The nutritional value of starch can be increased by processing technologies, such as cooking-extrusion, which increases starch digestibility and eliminates its impact on protein and fat digestion. Regarding NSPs, it has been suggested that exogenous enzymes may improve nutrient digestibility of plant-based fish diets. Nevertheless, there are a number of issues such as maintaining enzyme activity during processing and efficiency of such exogenous enzymes when supplemented to fish or shrimp feeds (Castillo and Gatlin, 2015). Few studies on exogenous enzymes targeting NSPs have been undertaken with fish or shrimp. The majority have used cocktails of different enzymes, i.e., combinations of proteases, phytases, and carbohydrases. Few studies in fish have evaluated the use of single NSP-degrading enzymes. In Nile tilapia, exogenous xylanase improved carbohydrate digestibility and reduced the negative impact of NSPs on fat and protein digestion (Maas et al., 2018). This study also showed that combining enzymes, in this case phytase and xylanase, had a synergic effect on growth, protein retention, and protein efficiency. The efficacy of enzymes is dependent on the ingredient composition of the diets. Xylanase supplementation in Nile tilapia leads to a greater improvement of NSP digestibility in an NSP-rich diet compared with a normal tilapia diet (Maas et al., 2020a). Next to dietary factors, water temperature and species-related factors such as stomach pH will determine the potential of exogenous enzymes.

7.12 Analytical issues on measurement and characterization of dietary starches

In feed and food characterization, numerous analytical methods are applied to quantify different carbohydrate fractions as classified given in Fig. 7.1. Examples include the Weende crude fiber analysis and the van Soest method to quantify neutral detergent fiber (NDF) and acid detergent fiber (ADF). However, most food composition databases and most other sources presenting total carbohydrate data to do so by calculating "carbohydrate" level "by difference" as "nitrogen-free extract" (NFE). According to the original Weende method, NFE is defined as dry matter minus crude protein minus crude ash minus crude fat minus crude fiber. However, it is not uncommon that the total carbohydrate content is presented with the inclusion of the crude fiber fraction. In addition to the variability in calculation method of total carbohydrates, this method should be used with caution because of the possible accumulation of the analytical errors involved in proximate analysis. For instance, the factor 6.25 applied to calculate the crude protein content from the measured kjeldahl N content is based on the average nitrogen content of animal tissue. Using this factor for other feed ingredients, such as plant protein concentrates, yields inaccurate values due to differences in AA contents of the proteins with concurrent differences in the percentage of molecular weight assigned to nitrogen (Mariotti et al., 2008), hence affecting the carbohydrate fraction calculated by difference. Ingredients containing chitin, such as shrimp meal or insect meal, will have an overestimation of the crude protein and an underestimation of the total carbohydrate fraction since chitin is a nitrogen containing polysaccharide and therefore contributes to crude protein content. By calculating the total carbohydrates as the remaining fraction, substantial amount of noncarbohydrate components can be included into this fraction like lignin (a polyphenol), part of the nucleotides, polyhydroxyalkanoate etc. Numerous bacteria store energy in the form of polyhydroxyalkanoate, which is chemically classed as fat but not included in the standard fat analysis. This issue is of concern as feeds increasingly use novel ingredients, specifically single-cell ingredients (bacterial, algal, and fungal) and insect meals. Under these conditions, using values derived by calculating total carbohydrates "by difference" can be a source of significant error in feed formulation and research studies.

7.13 Conclusions

In general, fish do not have as great a capacity for utilizing dietary carbohydrates as terrestrial higher vertebrates. This difference is true for both available carbohydrates and the carbohydrates that fall into the category of fibers. The amounts of starch currently incorporated into fish feeds far exceed the limits imposed two decades ago. For certain species, it is possible to supply an even larger amount of carbohydrate energy when the economic context is favorable. Before doing this, however, it is advisable to target certain species and study their physiological characteristics, which are still far from being known in full detail. Compared with those in fish, few studies have examined the nutrition or metabolism of crustacean species. Analysis of the relative importance of and the interplay between gluconeogenetic and glycolytic pathways in dietary carbohydrate utilization and glucose homeostasis needs to be strengthened from a comparative perspective between species and for exploiting the potential of all sources of digestible carbohydrates.

References

Abdel-Rahman, S.H., Kanazawa, A., Teshima, S.-I., 1979. Effects of dietary carbohydrate on the growth and the levels of the hepatopancreatic glycogen and serum glucose of prawn. Bull. Japanese Soc. Sci. Ser. 45 (12), 1491–1494.

Amirkolaie, A.K., Leenhouwers, J.I., Verreth, J.A.J., Schrama, J.W., 2005. Type of dietary fibre (soluble versus insoluble) influences digestion, faeces characteristics and faecal waste production in Nile tilapia (*Oreochromis niloticus* L.). Aquac. Res. 36, 1157–1166.

Amirkolaie, A.K., Verreth, J.A.J., Schrama, J.W., 2006. Effect of gelatinization degree and inclusion level of dietary starch on the characteristics of digesta and faeces in Nile tilapia (*Oreochromis niloticus* (L.)). Aquaculture 260, 194–205.

Arnesen, P., Krogdahl, A., 1993. Crude and pre-extruded products of wheat as nutrient sources in extruded diets for Atlantic salmon (*Salmo salar* L) grown in sea water. Aquaculture 118, 105–117.

Aslaksen, M.A., Kraugerud, O.F., Penn, M., Svihus, B., Denstadli, V., Jorgensen, H.Y., Hillestad, M., Krogdahl, A., Storebakken, T., 2007. Screening of nutrient digestibilities and intestinal pathologies in Atlantic salmon, *Salmo salar*, fed diets with legumes, oilseeds, or cereals. Aquaculture 272, 541–555.

Asp, N.G., van Amelsvoort, J.M.M., Hautvast, J.G.A.J., 1996. Nutritional implications of resistant starch. Nutr. Res. Rev. 9, 1–31.

Bakke, A.M., Glover, C., Krogdahl, Å., 2010. Feeding, digestion and absorption of nutrients. In: Grosell, M., Farrell, A.P., Brauner, C.J. (Eds.), The Multifunctional Gut of Fish. Fish Physiology, vol. 30. Academic Press/Elsevier, Amsterdam, pp. 57–110.

Bakke McKellep, A.M., Nordrum, S., Krogdahl, A., Buddington, R.K., 2000. Absorption of glucose, amino acids, and dipeptides by the intestines of Atlantic salmon (*Salmo salar* L.). Fish Physiol. Biochem. 22, 33–44.

Bergot, F., 1979a. Carbohydrate in rainbow trout diets: effects of the level and source of carbohydrate and the number of meals on growth and body composition. Aquaculture 18, 157–167.

Bergot, F., 1979b. Effects of dietary carbohydrates and or their mode of distribution on glycaemia in rainbow trout (Salmo gairdneri Richardson). Comp. Biochem. Physiol. 64A, 543–547.

Bergot, F., 1981. Etude de l'utilisation digestive d'une cellulose purifiée chez la truite arc-en-ciel (Salmo gairdneri) et la carpe commune (*Cyprinus carpio*). Reprod. Nutr. Dev. 21, 83–93.

Bergot, F., 1993. Digestibility of native starches of various botanical origins by rainbow trout. (*Oncorhynchus mykiss*). In: Luquet, P., Kaushik, S. (Eds.), Fish Nutrition in Practice. INRA (France), pp. 857–865.

Bergot, F., Breque, J., 1983. Digestibility of starch by rainbow trout: effects of the physical state of starch and of the intake level. Aquaculture 34, 203–212.

Blasco, J., Fernandez-Borras, J., Marimon, I., Requena, A., 1996. Plasma glucose kinetics and tissue uptake in brown trout in vivo: effect of an intravascular glucose load. J. Comp. Physiol. B 165, 534–541.

Blasco, J., Marimon, I., Viaplana, I., Fernandez Borras, J., 2001. Fate of plasma glucose in tissues of brown trout in vivo: effects of fasting and glucose loading. Fish Physiol. Biochem. 24, 247–258.

Bogé, G., Rigal, A., Pérès, G., 1982. The use of intestinal brush border membrane vesicles for comparative studies of glucose and 2-amino isobutyric acid transport by four species of marine teleost. Comp. Biochem. Physiol. Physiol. 72, 85–89.

Bonvini, E., Bonaldo, A., Parma, L., Mandrioli, L., Sirri, R., Grandi, M., Fontanillas, R., Viroli, C., Gatta, P.P., 2018. Feeding European sea bass with increasing dietary fibre levels: impact on growth, blood biochemistry, gut histology, gut evacuation. Aquaculture. 494, 1–9.

Booth, M.A., Andersen, A.J., Allan, G.L., 2006. Investigation of the nutritional requirements of Australian snapper *Pagrus auratus* (Bloch & Schneider 1801): digestibility of gelatinized wheat starch and clearance of an intraperitoneal injection of D-glucose. Aquac. Res. 37, 975–985.

Borrebaek, B., Waagbo, R., Christophersen, B., Tranulis, M.A., Hemre, G.I., 1993. Adaptable hexokinase with low affinity for glucose in the liver of Atlantic salmon (*Salmo salar*). Comp. Biochem. Physiol. B Biochem. Mol. Biol. 106, 833–836.

Brauge, C., Medale, F., Corraze, G., 1994. Effect of dietary carbohydrate levels on growth, body composition and glycaemia in rainbow trout, *Oncorhynchus mykiss*, reared in seawater. Aquaculture 123, 109–120.

Brinker, A., 2007. Guar gum in rainbow trout (*Oncorhynchus mykiss*) feed: the influence of quality and dose on stabilisation of faecal solids. Aquaculture 267, 315–327.

Brinker, A., Koppe, W., Rösch, R., 2005. Optimised effluent treatment by stabilised trout faeces. Aquaculture 249, 125–144.

Buddington, R.K., Hilton, J.W., 1987. Intestinal adaptations of rainbow trout to changes in dietary carbohydrate. Am. J. Physiol. Gastrointest. Liver Physiol. 253, G489–G496.

Buddington, R.K., Chen, J.W., Diamond, J., 1987. Genetic and phenotypic adaptation of intestinal nutrient transport to diet in fish. J. Physiol. (London) 393, 261–281.

Buhler, D.R., Halver, J.E., 1961. Nutrition of salmonid fishes. Carbohydrate requirements of Chinook salmon. J. Nutr. 74, 307–318.

Burel, C., Boujard, T., Tulli, F., Kaushik, S.J., 2000. Digestibility of extruded peas, extruded lupin, and rapeseed meal in rainbow trout (*Oncorhynchus mykiss*) and turbot (*Psetta maxima*). Aquaculture 188, 285–298.

Cahu, C.L., Zambonino Infante, J.L., 1994. Early weaning of seabass (*Dicentrarchus labrax*) larvae with a compound diet: effect on digestive enzymes. Comp. Biochem. Physiol. 109A, 213–222.

Castillo, S., Gatlin, D.M., 2015. Dietary supplementation of exogenous carbohydrase enzymes in fish nutrition: a review. Aquaculture 435, 286–292.

Catacutan, M.R., 1991. Apparent digestibility of diets with various carbohydrate levels and the growth response of *Penaeus monodon*. Aquaculture 95, 89–96.

Chakrabarti, I., Gani, M.A., Chaki, K.K., Sur, R., Misra, K.K., 1995. Digestive enzymes in 11 freshwater teleost fish species in relation to food habit and niche segregation. Comp. Biochem. Physiol. Physiol. 112, 167–177.

Chan, A.S., Horn, M.H., Dickson, K.A., Gawlicka, A., 2004. Digestive enzyme activities in carnivores and herbivores: comparisons among four closely related prickleback fishes (Teleostei: Stichaeldae) from a California rocky intertidal habitat. J. Fish. Biol. 65, 848–858.

Chang, V.M., Idler, D., 1960. Biochemical studies on sockeye salmon during spawning migration: xii. Liver glycogen. Can. J. Biochem. Physiol. 38, 553–558.

Chavin, W., Young, J.E., 1970. Factors in the determination of normal serum glucose levels of goldfish, *Carassius auratus* L. Comp. Biochem. Physiol. 33, 629–653.

Chen, Y.-J., Zhang, T.-Y., Chen, H.-Y., Lin, S.-M., Luo, L., Wang, D.-S., 2017. Simultaneous stimulation of glycolysis and gluconeogenesis by feeding in the anterior intestine of the omnivorous GIFT tilapia, *Oreochromis niloticus*. Biol. Open 6, 818–824.

Chiu, Y.N., Benitez, L.V., 1981. Studies on the carbohydrases in the digestive tract of the milkfish *Chanos chanos*. Mar. Biol. 61, 247–254.

Clark, J., McNaughton, J.E., Stark, J.R., 1984. Metabolism in marine flatfish. 1. Carbohydrate digestion in Dover sole (*Solea solea* L.). Comp. Biochem. Physiol. B 77B, 821–827.

Clark, D., Lawrence, A., Swakon, D., 1993. Apparent chitin digestibility in penaeid shrimp. Aquaculture 109, 51–57.

Clements, K.D., 1997. Fermentation and gastrointestinal microorganisms in fishes. Gastrointest. Microbiol. 156 https://doi.org/10.1007/978-1-4615-4111-0_6.

Clements, K.D., Angert, E.R., Montgomery, W.L., Choat, J.H., 2014. Intestinal microbiota in fishes: what's known and what's not. Mol. Ecol. 23, 1891–1898.

Collie, N.L., Ferraris, R.P., 1995. Nutrient fluxes and regulation in fish intestine. In: Mommsen, H.a. (Ed.), Biochemistry and Molecular Biology of Fishes, vol. 4. Elsevier Science, pp. 221–239.

Correa, C.F., de Aguiar, L.H., Lundstedt, L.M., Moraes, G., 2007. Responses of digestive enzymes of tambaqui (*Colossoma macropomum*) to dietary cornstarch changes and metabolic inferences. Comp. Biochem. Physiol. A Mol. Integr. Physiol. 147, 857–862.

Cousin, M., Cuzon, G., Guillaume, J., 1996. Digestibility of starch in *Penaeus vannamei*: in vivo and in vitro study on eight samples of various origin. Aquaculture 140, 361–372.

Cowey, C.B., Cooke, D.J., Matty, A.J., Adron, J.W., 1981. Efects of quantity and quality of dietary protein on certain enzyme activities in rainbow trout. J. Nutr. 111 (1), 8–15.

Cowey, C.B., de la Higuera, M., Adron, J.W., 1977. The effect of dietary composition and of insulin on gluconeogenesis in rainbow trout (*Salmo gairdneri*). Br. J. Nutr. 38, 385–395.

Cowey, C.B., Walton, M.J., 1989. Intermediary metabolism. In: Halver, J.E.H. (Ed.), Fish Nutrition, 2nd. Academic Press, New York, pp. 259–329.

CVB, 1993. Centraal Veevoederbureau Veevoedertabel (Animal Feedstuffs Table) (Lelystad).

Das, K.M., Tripathi, S.D., 1991. Studies on the digestive enzymes of grass carp, *Ctenopharyngodon idella* (Val.). Aquaculture. 92, 21–32.

Davis, D.A., Arnold, C.R., 1993. Evaluation of five carbohydrate sources for *Penaeus vannamei*. Aquaculture 114, 285–292.

Degani, G., Yehuda, Y., Viola, S., 1997. The digestibility of nutrient sources for common carp, *Cyprinus carpio* Linnaeus. Aquac. Res. 28, 575–580.

del Carmen Gonzalez-Pena, M., Gomes, S.Z., Moreira, G.S., 2002. Effects of dietary fiber on growth and gastric emptying time of the freshwater prawn *Macrobrachium rosenbergii* (De Man, 1879). J. World Aquac. Soc. 33, 441–447.

Deng, D.F., Refstie, S., Hemre, G.I., Crocker, C.E., Chen, H.Y., Cech Jr., J.J., Hung, S.S.O., 2000. A new technique of feeding, repeated sampling of blood and continuous collection of urine in white sturgeon. Fish Physiol. Biochem. 22, 191–197.

Deng, D.F., Refstie, S., Hung, S.S.O., 2001. Glycemic and glycosuric responses in white sturgeon (*Acipenser transmontanus*) after oral administration of simple and complex carbohydrates. Aquaculture 199, 107–117.

Deng, D., Yan, X., Zhao, W., Qin, C., Yang, G., Nie, G., 2020. Glucose transporter 2 in common carp (*Cyprinus carpio* L.): molecular cloning, tissue expression, and the responsiveness to glucose, insulin, and glucagon. Fish Physiol. Biochem. 46, 1207–1218.

Dias, J., 1999. Lipid Deposition in Rainbow Trout (*Oncorhynchus mykiss*) and European Sea Bass (*Dicentrarchus labrax*): Nutritional Control of Hepatic Lipogenesis. PhD thesis. Univ. Porto, Portugal & Univ, Bordeaux I, France, 190p +annexes.

Dias, J., Alvarez, M.J., Diez, A., Arzel, J., Corraze, G., Bautista, J.M., Kaushik, S.J., 1998. Regulation of hepatic lipogenesis by dietary protein/energy in juvenile European seabass (*Dicentrarchus labrax*). Aquaculture 161, 169–186.

Douglas, S.E., Mandla, S., Gallant, J.W., 2000. Molecular analysis of the amylase gene and its expression during development in the winter flounder, Pleuronectes americanus. Aquaculture 190, 247–260.

Drai, P., Albertini-Berhaut, J., Lafaurie, M., Sudaka, P., Giudicelli, J., 1990. Simultaneous preparation of basolateral and brush-border membrane vesicles from sea bass intestinal epithelium. Biochim. Biophys. Acta Biomembr. 1022, 251–259.

Duarte, C.M., Holmer, M., Olsen, Y., Soto, D., Marbà, N., Guiu, J., Black, K., Karakassis, I., 2009. Will the oceans help feed humanity? Bioscience 59, 967–976.

Enes, P., Panserat, S., Kaushik, S., Oliva-Teles, A., 2006. Effect of normal and waxy maize starch on growth, food utilization and hepatic glucose metabolism in European sea bass (*Dicentrarchus labrax*) juveniles. Comp. Biochem. Physiol. A Mol. Integr. Physiol. 143, 89–96.

Enes, P., Panserat, S., Kaushik, S., Oliva-Teles, A., 2009. Nutritional regulation of hepatic glucose metabolism in fish. Fish Physiol. Biochem. 35, 519–539. https://doi.org/10.1007/s10695-008-9259-5.

Enes, P., Panserat, S., Kaushik, S., Oliva-Teles, A., 2011. Dietary carbohydrate utilization by European sea bass (*Dicentrarchus labrax* L.) and gilthead sea bream (*Sparus aurata* L.) juveniles. Rev. Fish. Sci. 19, 201–215. https://doi.org/10.1080/10641262.2011.579363.

Erfanullah Jafri, A.K., 1998. Evaluation of digestibility coefficients of some carbohydrate-rich feedstuffs for Indian major carp fingerlings. Aquac. Res. 29, 511–519.

Fange, R., Lundblad, G., Lind, J., Slettengren, K., 1979. Chitinolytic enzymes in the digestive-system of marine fishes. Mar. Biol. 53 (4), 317–321.

FAO, 2020. The State of World Fisheries and Aquaculture 2020. Sustainability in Action. FAO, Rome, p. 224.

Fernández, F., Miquel, A.G., Córdoba, M., Varas, M., Metón, I., Caseras, A., Baanante, I.V., 2007. Effects of diets with distinct protein-to-carbohydrate ratios on nutrient digestibility, growth performance, body composition and liver intermediary enzyme activities in gilthead sea bream (*Sparus aurata*, L.) fingerlings. J. Exp. Mar. Biol. Ecol. 343, 1–10.

Fideu, M.D., Soler, G., Ruiz-Amil, M., 1983. Nutritional regulation of glycolysis in rainbow trout (*Salmo gairdneri* R.). Comp. Biochem. Physiol. 74B, 795–799.

Figueiredo-Silva, A.C., Panserat, S., Kaushik, S., Geurden, I., Polakof, S., 2012. High levels of dietary fat impair glucose homeostasis in rainbow trout. J. Exp. Biol. 215, 169–178.

Fountoulaki, E., Alexis, M.N., Nengas, I., Venou, B., 2005. Effect of diet composition on nutrient digestibility and digestive enzyme levels of gilthead sea bream (*Sparus aurata* L.). Aquac. Res. 36, 1243–1251.

French, C.J., Hochachka, P.W., Mommsen, T.P., 1983. Metabolic organization of liver during spawning migration of sockeye salmon. Am. J. Physiol. Regul. Integr. Comp. Physiol. 245, R827–R830.

Froystad, M.K., Lilleeng, E., Sundby, A., Krogdahl, A., 2006. Cloning and characterization of alpha-amylase from Atlantic salmon (*Salmo salar* L.). Comp. Biochem. Physiol. Mol. Integr. Physiol. 145, 479–492.

Furuichi, M., Yone, Y., 1981. Change of blood sugar and plasma insulin levels of fishes in glucose tolerance test. Nippon Suisan Gakkaishi 47, 761–764.

Garin, D., Rombaut, A., Freminet, A., 1987. Determination of glucose turnover in sea bass *Dicentrarchus labrax*. Comparative aspects of glucose utilization. Comp. Biochem. Physiol. B Biochem. Mol. Biol. 87, 981–988.

German, D.P., Bittong, R.A., 2009. Digestive enzyme activities and gastrointestinal fermentation in wood-eating catfishes. J. Comp. Physiol. B Biochem. Syst. Environ. Physiol. 179, 1025–1042.

German, D.P., Foti, D.M., Heras, J., Amerkhanian, H., Lockwood, B.L., 2016. Elevated gene copy number does not always explain elevated amylase activities in fishes. Physiol. Biochem. Zool. 89, 277–293.

Geurden, I., Aramendi, M., Zambonino-Infante, J., Panserat, S., 2007. Early feeding of carnivorous rainbow trout (*Oncorhynchus mykiss*) with a hyperglucidic diet during a short period: effect on dietary glucose utilization in juveniles. Am. J. Physiol. Regul. Integr. Comp. Physiol. 292, R2275–R2283.

Gisbert, E., Sainz, R.D., Hung, S.S.O., 2003. Glycemic responses in white sturgeon after oral administration of graded doses of D-glucose. Aquaculture 224, 301–312.

Glencross, B., Blyth, D., Tabrett, S., Bourne, N., Irvin, S., Anderson, M., Fox-Smith, T., Smullen, R., 2012. An assessment of cereal grains and other starch sources in diets for barramundi (*Lates calcarifer*)–implications for nutritional and functional qualities of extruded feeds. Aquac. Nutr. 18, 388–399.

Guerreiro, I., Peres, H., Castro, C., Pérez-Jiménez, A., Castro-Cunha, M., Oliva-Teles, A., 2014. Water temperature does not affect protein sparing by dietary carbohydrate in Senegalese sole (*Solea senegalensis*) juveniles. Aquac. Res. 45, 289–298.

Haidar, M.N., Petie, M., Heinsbroek, L.T.N., Verreth, J.A.J., Schrama, J.W., 2016. The effect of type of carbohydrate (starch vs. nonstarch polysaccharides) on nutrients digestibility, energy retention and maintenance requirements in Nile tilapia. Aquaculture 463, 241–247.

Hakim, Y., Harpaz, S., Uni, Z., 2009. Expression of brush border enzymes and transporters in the intestine of European sea bass (*Dicentrarchus labrax*) following food deprivation. Aquaculture 290, 110–115.

Hall, M.R., Van Ham, E.H., 1998. Diurnal variation and effects of feeding on blood glucose in the giant tiger prawn, *Penaeus monodon*. Physiol. Zool. 71, 574–583.

Harter, T.S., Heinsbroek, L.T.N., Schrama, J.W., 2015. The source of dietary non-protein energy affects in vivo protein digestion in African catfish (*Clarias gariepinus*). Aquac. Nutr. 21, 569–577.

Hemre, G.-I., Lie, O., Lied, E., Lambertsen, G., 1989. Starch as an energy source in feed for cod (*Gadus morhua*): digestibility and retention. Aquaculture 80, 261–270.

Hemre, G.I., Sandnes, K., Lie, O., Torrissen, O., Waagboe, R., 1995a. Carbohydrate nutrition in Atlantic salmon, *Salmo salar* L.: growth and feed utilization. Aquac. Res. 26, 149–154.

Hemre, G.I., Torrissen, O., Krogdahl, A., Lie, O., 1995b. Glucose tolerance in Atlantic salmon, *Salmo salar* L., dependence on adaption to dietary starch and water temperature. Aquac. Nutr. 1, 69–75.

Hemre, G.I., Mommsen, T.P., Krogdahl, A., 2002. Carbohydrates in fish nutrition: effects on growth, glucose metabolism and hepatic enzymes. Aquac. Nutr. 8, 175–194.

Hernandez, M., Takeuchi, T., Watanabe, T., 1994. Effect of gelatinized corn meal as a carbohydrate source on growth performance, intestinal evacuation, and starch digestion in common carp. Fish. Sci. 60, 579–582.

Herold, M.A., Hung, S.S.O., Fynn-Aikins, K., 1995. Apparent digestibility coefficients of carbohydrates for white sturgeon. Prog. Fish-Cult. 57, 137–140.

Hidalgo, M.C., Urea, E., Sanz, A., 1999. Comparative study of digestive enzymes in fish with different nutritional habits. Proteolytic and amylase activities. Aquaculture 170, 267–283.

Hillestad, M., Johnsen, F., Asgard, T., 2001. Protein to carbohydrate ratio in high-energy diets for Atlantic salmon (*Salmo salar* L.). Aquac. Res. 32, 517–529.

Hilton, J.W., Atkinson, J.L., 1982. Response of rainbow trout (*Salmo gairdneri*) to increased levels of available carbohydrate in practical trout diets. Br. J. Nutr. 47, 597.

Hilton, J.W., Slinger, S.J., 1983. Effect of wheat bran replacement of wheat middlings in extrusion processed (floating) diets on the growth of juvenile rainbow trout (*Salmo gairdneri*). Aquaculture 35, 201–210.

Hilton, J.W., Cho, C.Y., Slinger, S.J., 1981. Effect of extrusion processing and steam pelleting diets on pellet durability, pellet water absorption, and the physiological response of rainbow trout (*Salmo gairdneri* R.). Aquaculture 25, 185–194.

Hilton, J.W., Atkinson, J.L., Slinger, S.J., 1982a. Effect of increased dietary fiber on the growth of rainbow trout (*Salmo gairdneri*). Can. J. Fish. Aquat. Sci. 40, 81–85.

Hilton, J.W., Atkinson, J.L., Slinger, S.J., 1982b. Maximum tolerable level, digestion, and metabolism of D-glucose (cerelose) in rainbow trout (*Salmo gairdneri*) reared on a practical trout diet. Can. J. Fish. Aquat. Sci. 39, 1229–1234.

Hilton, J.W., Atkinson, J.L., Slinger, S.J., 1987. Evaluation of the net energy value of glucose (cerelose) and maize starch in diets for rainbow trout (*Salmo gairdneri*). Br. J. Nutr. 58, 453–461.

Hofer, R., Sturmbauer, C., 1985. Inhibition of trout and carp alpha-amylase by wheat. Aquaculture 48, 277–283.

Houlihan, D.F., 1991. Protein turnover in ectotherms and its relationship to energetics. In: Gilles, R. (Ed.), Advances in Comparative and Environmental Physiology. Springer-Verlag, New York, pp. 1–43.

Hung, S.S.O., 1991. Carbohydrate utilization by white sturgeon as assessed by oral administration tests. J. Nutr. 121, 1600–1605.

Hung, S.S.O., Fynn-Aikins, K., Lutes, P.B., Xu, R., 1989. Ability of juvenile white sturgeon (*Acipenser transmontanus*) to utilize different carbohydrate sources. J. Nutr. 119, 727–733.

Hung, L.T., Lazard, J., Mariojouls, C., Moreau, Y., 2003. Comparison of starch utilization in fingerlings of two Asian catfishes from the Mekong River (Pangasius bocourti Sauvage, 1880, Pangasius hypophthalmus Sauvage, 1878). Aquac. Nutr. 9, 215–222.

Ikeda, M., Kakizaki, H., Matsumiya, M., 2017. Biochemistry of fish stomach chitinase. Int. J. Biol. Macromol. 104, 1672–1681.

Jeuniaux, C., 1961. Chitinase: an addition to the list of hydrolases in the digestive tract of vertebrates. Nature 192, 135–136.

Jollivet, D., Gabaudan, J., Metailler, R., 1988. Some Effects of Physical State and Dietary Level of Starch, Temperature and Meal Size on Turbot (*Scophthalmus maximus* L.). ICES C.M.1988/F:25.

Kamalam, B.S., Medale, F., Panserat, S., 2017. Utilisation of dietary carbohydrates in farmed fishes: new insights on influencing factors, biological limitations and future strategies. Aquaculture 467, 3–27. https://doi.org/10.1016/j.aquaculture.2016.02.007.

Karasov, W.H., Martinez del Rio, C., Caviedes-Vidal, E., 2011. Ecological physiology of diet and digestive systems. Annu. Rev. Physiol. 73, 69–93.

Kaushik, S., Médale, F., Fauconneau, B., Blanc, D., 1989. Effect of digestible carbohydrates on protein/energy utilization and on glucose metabolism in rainbow trout (*Salmo gairdneri* R.). Aquaculture 79, 63–74.

Kaushik, S.J., Doudet, T., Médale, F., Aguirre, P., Blanc, D., 1995. Protein and energy needs for maintenance and growth of Nile tilapia (*Oreochromis niloticus*). J. Appl. Ichthyol. 11, 290–296.

Kawai, S.-i., Ikeda, S., 1971. Studies on digestive enzymes of fishes-I. Carbohydrases in digestive organs of several fishes. Nippon Suisan Gakkaishi 37, 333–337.

Kim, J.D., Kaushik, S.J., 1992. Contribution of digestible energy from carbohydrates and estimation of protein/energy requirements for growth of rainbow trout (*Oncorhynchus mykiss*). Aquaculture 106, 161–169.

Kirchgessner, M.K.,H., Schwarz, F.J., 1986. Digestibility of crude nutrients in different feeds and estimation of their energy content for carp (*Cyprinus carpio* L. Aquaculture 58, 189–194.

Kirchner, S., Kaushik, S., Panserat, S., 2003. Effect of partial substitution of dietary protein by a single gluconeogenic dispensable amino acid on hepatic glucose metabolism in rainbow trout (*Oncorhynchus mykiss*). Comp. Biochem. Physiol. A Mol. Integr. Physiol. 134, 337–347.

Kirchner, S., Panserat, S., Lim, P.L., Kaushik, S., Ferraris, R.P., 2008. The role of hepatic, renal and intestinal gluconeogenic enzymes in glucose homeostasis of juvenile rainbow trout. J. Comp. Physiol. B Biochem. Syst. Environ. Physiol. 178, 429–438.

Kirchner, S., Seixas, P., Kaushik, S., Panserat, S., 2005. Effects of low protein intake on extra-hepatic gluconeogenic enzyme expression and peripheral glucose phosphorylation in rainbow trout (*Oncorhynchus mykiss).* Comp. Biochem. Physiol. B, Biochem. Mol. Biol. 140, 333–340.

Kitamikado, M., Tachino, S., 1960. Studies on the digestive enzymes of rainbow trout. I. Carbohydrases. Bull. Jap. Soc. Sci. Fish. 26, 679–684.

Knox, D., Walton, M.J., Cowey, C.B., 1980. Distribution of enzymes of glycolysis and gluconeogenesis in fish tissues. Mar. Biol. 56, 7–10. https://doi.org/10.1007/BF00390588.

Krogdahl, A., Nordrum, S., Sorensen, M., Brudeseth, L., Rosjo, C., 1999. Effects of diet composition on apparent nutrient absorption along the intestinal tract and of subsequent fasting on mucosal disaccharidase activities and plasma nutrient concentration in Atlantic salmon *Salmo salar* L. Aquac. Nutr. 5, 121–133.

Krogdahl, Å., Sundby, A., Olli, J.J., 2004. Atlantic salmon (*Salmo salar*) and rainbow trout (*Oncorhynchus mykiss*) digest and metabolize nutrients differently. Effects of water salinity and dietary starch level. Aquaculture 229, 335–360.

Krogdahl, A., Hemre, G.I., Mommsen, T.P., 2005. Carbohydrates in fish nutrition: digestion and absorption in post-larval stages. Aquac. Nutr. 11, 103–122.

Law, A.T., 1986. Digestibility of low-cost ingredients in pelleted feed by grass carp (*Ctenopharyngodon idella* C. et V.). Aquaculture 51, 97103.

Le Hoang, T.M.D., Lie, K.K., Etayo, A., Rønnestad, I., Sæle, Ø., 2019. Nutrient Sensing; Transcriptomic Response and Regulation of Gut Motility in an Agastric Vertebrate. bioRxiv, p. 827659.

Leenhouwers, J.I., Adjei-Boateng, D., Verreth, J.A.J., Schrama, J.W., 2006. Digesta viscosity, nutrient digestibility and organ weights in African catfish (*Clarias gariepinus*) fed diets supplemented with different levels of a soluble non-starch polysaccharide. Aquac. Nutr. 12, 111–116.

Leenhouwers, J.I., Ortega, R.C., Verreth, J.A.J., Schrama, J.W., 2007a. Digesta characteristics in relation to nutrient digestibility and mineral absorption in Nile tilapia (*Oreochromis niloticus* L.) fed cereal grains of increasing viscosity. Aquaculture 273, 556–565.

Leenhouwers, J.I., ter Veld, M., Verreth, J.A.J., Schrama, J.W., 2007b. Digesta characteristiscs and performance of African catfish (*Clarias gariepinus*) fed cereal grains that differ in viscosity. Aquaculture 264, 330–341.

Legate, N.J., Bonen, A., Moon, T.W., 2001. Glucose tolerance and peripheral glucose utilization in rainbow trout (*Oncorhynchus mykiss*), American eel (*Anguilla rostrata*), and black bullhead catfish (*Ameiurus melas*). Gen. Comp. Endocrinol. 122, 48–59.

Li, X., Zhu, X., Han, D., Yang, Y., Jin, J., Xie, S., 2016. Carbohydrate utilization by herbivorous and omnivorous freshwater fish species: a comparative study on gibel carp (*Carassius auratus* gibelio. var CAS III) and grass carp (*Ctenopharyngodon idellus*). Aquac. Res. 47, 128–139.

Lesel, R., Fromageot, C., Lesel, M., 1986. Cellulose digestibility in grass carp, *Ctenopharyngodon idella* and in goldfish, *Carassius auratus*. Aquaculture 54, 11–17.

Li, R.-X., Liu, H.-Y., Chen, Q., Tan, B.-P., Dong, X.-H., Chi, S.-Y., Yang, Q.-H., Zhang, S., Chen, L.-Q., 2018a. Glucose tolerance of grass carp *Ctenopharyngodon idellus* after a long-term adaptation to carbohydrate-to-lipid ratio diets. Aquac. Res. 49, 3881–3888.

Li, R., Liu, H., Dong, X., Chi, S., Yang, Q., Zhang, S., Tan, B., 2018b. Molecular characterization and expression analysis of glucose transporter 1 and hepatic glycolytic enzymes activities from herbivorous fish *Ctenopharyngodon idellus* in respond to a glucose load after the adaptation to dietary carbohydrate levels. Aquaculture 492, 290–299.

Likely, R., Johnson, E., Ahearn, G.A., 2015. Functional characterization of a putative disaccharide membrane transporter in crustacean intestine. J. Comp. Physiol. B 185, 173–183.

Lin, J.S., Shiau, S.Y., 1995. Hepatic enzyme adaptation to different dietary carbohydrates in juvenile tilapia *Oreochromis niloticus* x *O. aureus*. Fish Physiol. Biochem. 14, 165–170.

Lindsay, G.J.H., 1985. Development of gastric chitinase activity in rainbow trout (Salmo gairdneri). Aquaculture 48, 285–290.

Lindsay, G.J.H., Gooday, G.W., 1985. Chitinolytic enzymes and the bacterial microflora in the digestive tract of cod, *Gadus morhua*. J. Fish. Biol. 26, 255–265.

Lindsay, G.J.H., Walton, M.J., Adron, J.W., Fletcher, T.C., Cho, C.Y., Cowey, C.B., 1984. The growth of rainbow trout (*Salmo gairdneri*) given diets containing chitin and its relationship to chitinolitic enzymes and chitin digestibility. Aquaculture 37, 315–334.

Lopez-Vasquez, K., Castro-Perez, C.A., Val, A.L., 2009. Digestive enzymes of eight Amazonian teleosts with different feeding habits. J. Fish. Biol. 74, 1620–1628.

Lupatsch, I., Kissil, G.W., Sklan, D., Pfeffer, E., 1997. Apparent digestibility coefficients of feed ingredients and their predictability in compound diets for gilthead seabream, Sparus aurata L. Aquac. Nutr. 3, 81–89.

Maas, R.M., Verdegem, M.C.J., Dersjant-Li, Y., Schrama, J.W., 2018. The effect of phytase, xylanase and their combination on growth performance and nutrient utilization in Nile tilapia. Aquaculture 487, 7–14.

Maas, R.M., Verdegem, M.C.J., Stevens, T.L., Schrama, J.W., 2020a. Effect of exogenous enzymes (phytase and xylanase) supplementation on nutrient digestibility and growth performance of Nile tilapia (*Oreochromis niloticus*) fed different quality diets. Aquaculture 529, 735723.

Maas, R.M., Verdegem, M.C.J., Wiegertjes, G.F., Schrama, J.W., 2020b. Carbohydrate utilisation by tilapia: a meta-analytical approach. Rev. Aquac. 16p. https://doi.org/10.1111/raq.12413.

Marandel, L., Seiliez, I., Véron, V., Skiba-Cassy, S., Panserat, S., 2015. New insights into the nutritional regulation of gluconeogenesis in carnivorous rainbow trout (*Oncorhynchus mykiss*): a gene duplication trail. Physiol. Genom. 47, 253–263.

Marandel, L., Panserat, S., Plagnes-Juan, E., Arbenoits, E., Soengas, J.L., Bobe, J., 2017. Evolutionary history of glucose-6-phosphatase encoding genes in vertebrate lineages: towards a better understanding of the functions of multiple duplicates. BMC Genom. 18, 342. https://doi.org/10.1186/s12864-017-3727-1.

Marín-Juez, R., Capilla, E., Carvalho-Simoes, F., Camps, M., Planas, J.V., 2014. Structural and functional evolution of glucose transporter 4 (GLUT4): a look at GLUT4 in fish. In: Szablewski, L. (Ed.), Glucose Homeostasis. Intech Open. https://doi.org/10.5772/58094.

Mariotti, F., Tomé, D., Mirand, P.P., 2008. Converting nitrogen into protein-beyond 6.25 and Jones' factors. Crit. Rev. Food Sci. Nutr. 48, 177–184.

Martínez-Quintana, J.A., Yepiz-Plascencia, G., 2012. Glucose and other hexoses transporters in marine invertebrates: a mini review. Electron. J. Biotechnol. 15, 16.

Meriac, A., Eding, E.H., Schrama, J., Kamstra, A., Verreth, J.A.J., 2014. Dietary carbohydrate composition can change waste production and biofilter load in recirculating aquaculture systems. Aquaculture 420–421, 254–261.

Mommsen, T.P., 1986. Comparative gluconeogenesis in hepatocytes from salmonid fishes. Can. J. Zool. 64, 1 1 10–1 1 15.

Mommsen, T.P., Busby, E.R., 2006. Glucagon and friends. In: Reinecke, M., Zaccone, G., Kapoor, B. (Eds.), Fish Endocrinology, 6. Science, Enfield, NH, pp. 223–256.

Mommsen, T.P., Plisetskaya, E.M., 1991. Insulin in fishes and agnathans: history, structure and metabolic regulation. Rev. Aquat. Sci. 4, 225–259.

Moon, T.W., 1988. Adaptation, constraint, and the function of the gluconeogenic pathway. Can. J. Zool. 66, 1059–1068.

Moon, T.W., 2001. Glucose intolerance in teleost fish: fact or fiction? Comp. Biochem. Physiol. B Biochem. Mole. Biol. 129, 243–249.

Moon, T.W., Foster, G.D., 1995. Tissue Carbohydrate Metabolism, Gluconeogenesis and Hormonal and Environmental Influences, Biochemistry and Molecular Biology of Fishes. Elsevier, pp. 65–100.

Moreira, I., Peres, H., Couto, A., Enes, P., Oliva-Teles, A., 2008. Temperature and dietary carbohydrate level effects on performance and metabolic utilisation of diets in European sea bass (*Dicentrarchus labrax*) juveniles. Aquaculture 274, 153–160.

National Research Council (NRC), 2011. Nutrient requirements of fish and shrimp. National Academy of Sciences, Washington DC, USA.

Navarro, I., Gutiérrez, J., 1995. Fasting and starvation. In: Hochachka, Mommsen (Eds.), Biochemistry and Molecular Biology of Fishes, vol. 4. Elsevier Science, pp. 393–434.

Nie, Q., Miao, H., Miao, S., Zhou, H., Zhang, Y., Zhang, W., Mai, K., 2015. Effects of dietary glucose and dextrin on activity and gene expression of glucokinase and fructose-1, 6-bisphosphatase in liver of turbot *Scophthalmus maximus*. Fish Physiol. Biochem. 41, 819–832.

Nikolopoulou, D., Moutou, K.A., Fountoulaki, E., Venou, B., Adamidou, S., Alexis, M.M., 2011. Patterns of gastric evacuation, digesta characteristics and pH changes along the gastrointestinal tract of gilthead seabream (*Sparus aurata* L.) and European sea bass (*Dicentrarchus labrax* L.). Comp. Biochem. Physiol., A 158, 406–414.

Noblet, J., Fortune, H., Shi, X.S., Dubois, S., 1994. Prediction of net energy value of feeds for growing pigs. J. Anim. Sci. 72, 344–354.

NRC, 2011. Nutrient Requirements of Fish and Shrimp. National Academies Press.

Obi, I.E., Sterling, K.M., Ahearn, G.A., 2013. K+-dependent 3 H-d-glucose transport by hepatopancreatic brush border membrane vesicles of a marine shrimp. J. Comp. Physiol. B 183, 61–69.

Palmer, T.N., Ryman, B.E., 1972. Studies on glucose intolerance in fish. J. Fish Bioi. 4. J ll-319.

Panserat, S., Blin, C., Medale, F., Plagnes-Juan, E., Breque, J., Krishnamoorthy, R., Kaushik, S., 2000a. Molecular cloning, tissue distribution and sequence analysis of complete glucokinase cDNAs from gilthead seabream (*Sparus aurata*), rainbow trout (*Oncorhynchus mykiss*) and common carp (*Cyprinus carpio*). Biochim. Biophys. Acta 1474, 61–69. https://doi.org/10.1016/S0304-4165(99)00213-5.

Panserat, S., Medale, F., Blin, C., Breque, J., Vachot, C., Plagnes-Juan, E., Gomes, E., Krishnamoorthy, R., Kaushik, S., 2000b. Hepatic glucokinase is induced by dietary carbohydrates in rainbow trout, gilthead seabream, and common carp. Am. J. Physiol. Regul. Integr. Comp. Physiol. 278, R1164–R1170.

Panserat, S., Medale, F., Breque, J., PlagnesJuan, E., Kaushik, S., 2000c. Lack of significant long-term effect of dietary carbohydrates on hepatic glucose-6-phosphatase expression in rainbow trout (*Oncorhynchus mykiss*). J. Nutr. Biochem. (JNB) 11, 22–29.

Panserat, S., Plagnes-Juan, E., Kaushik, S., 2001. Nutritional regulation and tissue specificity of gene expression for proteins involved in hepatic glucose metabolism in rainbow trout (*Oncorhynchus mykiss*). J. Exp. Biol. 204, 2351–2360.

Panserat, S., Plagnes-Juan, E., Kaushik, S., 2002. Gluconeogenic enzyme gene expression is decreased by dietary carbohydrates in common carp (*Cyprinus carpio*) and gilthead seabream (*Sparus aurata*). Biochim. Biophys. Acta 1579, 35–42. https://doi.org/10.1016/S0167-4781(02)00501-8.

Panserat, S., Rideau, N., Polakof, S., 2014. Nutritional regulation of glucokinase: a cross-species story. Nutr. Res. Rev. 27, 21–47.

Papoutsoglou, E.S., Lyndon, A.R., 2005. Effect of incubation temperature on carbohydrate digestion in important teleosts for aquaculture. Aquac. Res. 36, 1252–1264.

Peres, H., Oliva-Teles, A., 2002. Utilization of raw and gelatinized starch by European sea bass (*Dicentrarchus labrax*) juveniles. Aquaculture 205, 287–299.

Peres, M.H., Goncalves, P., Oliva-Teles, A., 1999. Glucose tolerance in gilthead seabream (*Sparus aurata*) and European seabass (*Dicentrarchus labrax*). Aquaculture 179, 415–423.

Pfeffer, E., 1995. Carbohydrate utilization and its determination. J. Appl. Ichthyol. 11, 175–182.

Phan, L.T.T., Groot, R., Konnert, G.D.P., Masagounder, K., Figueiredo-Silva, A.C., Glencross, B.D., Schrama, J.W., 2019. Differences in energy utilisation efficiencies of digestible macronutrients in common carp (*Cyprinus carpio*) and barramundi (*Lates calcarifer*). Aquaculture 511, 734238.

Phan, L.T.T., Masagounder, K., Mas-Munoz, J., Schrama, J.W., 2021. Differences in energy utilization efficiency of digested protein, fat and carbohydrates in snakehead (*Channa striata*). Aquaculture 532, 736066.

Plisetskaya, E.M., 1998. Some of my not so favorite things about insulin and insulin-like growth factors in fish. Comp. Biochem. Physiol. B Biochem. Mol. Biol. 121, 3–11.

Plisetskaya, E.M., Mommsen, T.P., 1996. Glucagon and glucagon-like peptides in fishes. Int. Rev. Cytol. 168, 187–257.

Polakof, S., Panserat, S., 2016. How Tom Moon's research highlighted the question of glucose tolerance in carnivorous fish. Comp. Biochem. Physiol. B Biochem. Mol. Biol. 199, 43–49.

Polakof, S., Moon, T., Aguirre, P., Skiba-Cassy, S., Panserat, S., 2010. Effects of insulin infusion on glucose homeostasis and glucose metabolism in rainbow trout fed a high-carbohydrate diet. J. Exp. Biol. 213, 4151–4157.

Polakof, S., Ceinos, R.M., Fernandez-Duran, B., Miguez, J.M., Soengas, J.L., 2007. Daily changes in parameters of energy metabolism in brain of rainbow trout: dependence on feeding. Comp. Biochem. Physiol. a-Mol. Integr. Physiol. 146, 265–273.

Polakof, S., Mommsen, T.P., Soengas, J.L., 2011. Glucosensing and glucose homeostasis: from fish to mammals. Comp. Biochem. Physiol. B Biochem. Mol. Biol. 160, 123–149.

Polakof, S., Panserat, S., Soengas, J.L., Moon, T.W., 2012. Glucose metabolism in fish: a review. J. Comp. Physiol. B 182, 1015–1045. https://doi.org/10.1007/s00360-012-0658-7.

Ren, M., Ai, Q., Mai, K., Ma, H., Wang, X., 2011. Effect of dietary carbohydrate level on growth performance, body composition, apparent digestibility coefficient and digestive enzyme activities of juvenile cobia. *Rachycentron canadum* L. Aquac. Res. 42, 1467–1475.

Reshkin, S.J., Ahearn, G.A., 1987. Basolateral glucose transport by intestine of teleost, *Oreochromis mossambicus*. Am. J. Physiol. Regul. Integr. Comp. Physiol. 252, R579–R586.

Saha, A.K., Ray, A.K., 1998. Cellulase activity in rohu fingerlings. Aquac. Int. 6, 281–291.

Saha, S., Roy, R.N., Sen, S.K., Ray, A.K., 2006. Characterization of cellulase-producing bacteria from the digestive tract of tilapia, *Oreochromis mossambica* (Peters) and grass carp, *Ctenopharyngodon idella* (Valenciennes). Aquac. Res. 37, 380–388.

Sala-Rabanal, M., Gallardo, M., Sanchez, J., Planas, J., 2004. Na-dependent D-glucose transport by intestinal brush border membrane vesicles from gilthead sea bream (*Sparus aurata*). J. Membr. Biol. 201, 85–96.

Santigosa, E., García-Meilán, I., Valentin, J.M., Pérez-Sánchez, J., Médale, F., Kaushik, S., Gallardo, M.A., 2011. Modifications of intestinal nutrient absorption in response to dietary fish meal replacement by plant protein sources in sea bream (*Sparus aurata*) and rainbow trout (*Onchorynchus mykiss*). Aquaculture 317, 146–154.

Schermerhorn, T., 2013. Normal glucose metabolism in carnivores overlaps with diabetes pathology in non-carnivores. Front. Endocrinol. 4, 188.

Schrama, J.W., Haidar, M.N., Geurden, I., Heinsbroek, L.T.N., Kaushik, S.J., 2018. Energy efficiency of digestible protein, fat and carbohydrate utilisation for growth in rainbow trout and Nile tilapia. Br. J. Nutr. 119, 782–791.

Schrama, J.W., Saravanan, S., Geurden, I., Heinsbroek, L.T.N., Kaushik, S.J., Verreth, J.A.J., 2012. Dietary nutrient composition affects digestible energy utilisation for growth: a study on Nile tilapia (*Oreochromis niloticus*) and a literature comparison across fish species. Br. J. Nutr. 108, 277–289.

Shiau, S.Y., Chuang, J.C., 1995. Utilization of disaccharides by juvenile tilapia, *Oreochromis niloticus* × *O. aureus*. Aquaculture 133, 249–256.

Shiau, S.Y., Liang, H.S., 1995. Carbohydrate utilization and digestibility by Tilapia, *Oreochromis niloticus* × *O.* aureus, are affected by chromic oxide inclusion in the diet. J. Nutr. 125, 976–982.

Shiau, S.Y., Yu, H.L., Hwa, S., Chen, S.Y., Hsu, S.I., 1988. The influence of carboxymethylcellulose on growth, digestion, gastric emptying time and body composition of Tilapia. Aquaculture 70, 345–354.

Shiau, S.Y., Lin, S.F., Lu, L.T., 1991. Effects of different types of wheat flour in feed for grass prawn *Penaeus monodon*. Nippon Suisan Gakkaishi 57, 705–710.

Shimeno, S., Hosokawa, H., Hirata, H., Takeda, M., 1977. Comparative studies on carbohydrate metabolism of yellowtail and carp. Bull. Jpn. Soc. Sci. Fish. 43, 213–217.

Short, C.E., Driedzic, W.R., 2018. Species-specific low plasma glucose in fish is associated with relatively high tissue glucose content and is inversely correlated with cardiac glycogen content. J. Comp. Physiol. B Biochem. Syst. Environ. Physiol. 188, 809–819.

Singh, R.P., Nose, T., 1967. Digestibility of carbohydrates in young rainbow trout. Bull. Freshw. Fish. Res. Lab. 17, 21–25.

Sinha, A.K., Kumar, V., Makkar, H.P.S., De Boeck, G., Becker, K., 2011. Non-starch polysaccharides and their role in fish nutrition – a review. Food Chem. 127, 1409–1426.

Skea, G.L., Mountfort, D.O., Clements, K.D., 2007. Contrasting digestive strategies in four New Zealand herbivorous fishes as reflected by carbohydrase activity profiles. Comp. Biochem. Physiol. A Mol. Integr. Physiol. 146, 63–70.

Soengas, J.L., Moon, T.W., 1998. Transport and metabolism of glucose in isolated enterocytes of the black bullhead Ictalurus melas: effects of diet and hormones. J. Exp. Biol. 201, 3263–3273.

Soengas, J.L., Polakof, S., Chen, X., Sangiao-Alvarellos, S., Moon, T.W., 2006. Glucokinase and hexokinase expression and activities in rainbow trout tissues: changes with food deprivation and refeeding. Am. J. Physiol. Regul. Integr. Comp. Physiol. 291, R810–R821.

Somero, G., Childress, J., 1990. Scaling of ATP-supplying enzymes, myofibrillar proteins and buffering capacity in fish muscle: relationship to locomotory habit. J. Exp. Biol. 149, 319–333.

Song, X., Marandel, L., Skiba-Cassy, S., Corraze, G., Dupont-Nivet, M., Quillet, E., Geurden, I., Panserat, S., 2018. Regulation by dietary carbohydrates of intermediary metabolism in liver and muscle of two isogenic lines of rainbow trout. Front. Physiol. 9 https://doi.org/10.3389/fphys.2018.01579.

Sorensen, M., 2012. A review of the effects of ingredient composition and processing conditions on the physical qualities of extruded high-energy fish feed as measured by prevailing methods. Aquac. Nutr. 18, 233–248.

Spannhof, L., Plantikow, H., 1983. Studies on carbohydrate digestion in rainbow trout. Aquaculture 30, 95–108.

Staessen, T.W.O., Verdegem, M.C.J., Weththasinghe, P., Schrama, J.W., 2020. The effect of dietary non-starch polysaccharide level and bile acid supplementation on fat digestibility and the bile acid balance in rainbow trout (*Oncorhynchus mykiss*). Aquaculture 523, 735174.

Stickney, R.R., Shumway, S.E., 1974. Occurrence of cellulase activity in the stomachs of fishes. J. Fish BioE. 6, 779–790.

Stone, D.A.J., 2003. Dietary carbohydrate utilization by fish. Rev. Fish. Sci. 11, 337–369.

Storebakken, T., 1985. Binders in fish feeds I. Effect of alginate and guar gum on growth, digestibility, feed intake and passage through the gastrointestinal tract of rainbow trout. Aquaculture 47, 11–26.

Storebakken, T., Shearer, K.D., Refstie, S., Lagocki, S., McCool, J., 1998. Interactions between salinity, dietary carbohydrate source and carbohydrate concentration on the digestibility of macronutrients and energy in rainbow trout (*Oncorhynchus mykiss*). Aquaculture 163, 347–359.

Suarez, R.K., Mommsen, T.P., 1987. Gluconeogenesis in teleost fishes. Can. J. Zool. 65, 1869–1882.

Subramaniam, M., Weber, L.P., Loewen, M.E., 2019. Intestinal electrogenic sodium-dependent glucose absorption in tilapia and trout reveal species differences in SLC5A-associated kinetic segmental segregation. Am. J. Physiol. Regul. Integr. Comp. Physiol. 316, R222–R234.

Subramaniam, M., Enns, C.B., Luu, K., Weber, L.P., Loewen, M.E., 2020. Comparison of intestinal glucose flux and electrogenic current demonstrates two absorptive pathways in pig and one in Nile tilapia and rainbow trout. Am. J. Physiol. Regul. Integr. Comp. Physiol. 318, R245–R255.

Sugita, H., Kawasaki, J., Deguchi, Y., 1997. Production of amylase by the intestinal microflora in cultured freshwater fish. Lett. Appl. Microbiol. 24, 105–108.

Sugita, H., Yamada, S., Konagaya, Y., Deguchi, Y., 1999. Production of beta;-*N*-acetylglucosaminidase and chitinase by *Aeromonas* species isolated from river fish. Fish. Sci. 65, 155–158.

Tacon, A., Metian, M., 2015. Feed matters: satisfying the feed demand of aquaculture. Rev. Fish. Sci. Aquac. 23, 1–10.

Takeuchi, T., Hernandez, M., Watanabe, T., 1994. Nutritive value of gelatinized corn meal as a carbohydrate source to grass carp and hybrid tilapia *Oreochromis niloticus* × *O. aureus*. Fish. Sci. 60, 573–577.

Tan, Q., Xie, S., Zhu, X., Lei, W., Yang, Y., 2006. Effect of dietary carbohydrate sources on growth performance and utilization for gibel carp (*Carassius auratus* gibelio) and Chinese longsnout catfish (Leiocassis longirostris Gunther). Aquac. Nutr. 12, 61–70.

Teuling, E., Schrama, J.W., Gruppen, H., Wierenga, P.A., 2017. Effect of cell wall characteristics on algae nutrient digestibility in Nile tilapia (*Oreochromis niloticus*) and African catfish (*Clarus gariepinus*). Aquaculture 479, 490–500.

Tian, L.X., Liu, Y.J., Hung, S.S.O., Deng, D.F., Yang, H.J., Niu, J., Liang, G.Y., 2010. Effect of feeding strategy and carbohydrate source on carbohydrate utilization by grass carp (*Ctenopharyngodon idella*). Am. J. Agric. Biol. Sci. 5 (2), 135–142.

Tran-Tu, L.C., Bosma, R.H., Verstegen, M.W.A., Schrama, J.W., 2019. Effect of dietary viscosity on digesta characteristics and progression of digestion in different segments of the gastrointestinal tract of striped catfish (*Pangasionodon hypophthalmus*). Aquaculture 504, 114–120.

Tran-Tu, L.C., Nguyen, T.-C., Verreth, J.A.J., Schrama, J.W., 2020. Doses response of dietary viscosity on digestibility and faecal characteristics of striped catfish (*Pangasionodon hypophthalmus*). Aquac. Res. 51, 595–604.

Tranulis, M.A., Dregni, O., Christophersen, B., Krogdahl, A., Borrebaek, B., 1996. A glucokinase-like enzyme in the liver of Atlantic salmon (*Salmo salar*). Comp. Biochem. Physiol. B. 114B, 35–39.

Van Wormhoudt, A., Sellos, D., 1996. Cloning and sequencing analysis of three amylase cDNAs in the shrimp *Penaeus vannamei* (Crustacea Decapoda): evolutionary aspects. J. Mol. Evol. 42, 543–551.

Venou, B., Alexis, M.N., Fountoulaki, E., Nengas, I., Apostolopoulou, M., Castritsi-Cathariou, I., 2003. Effect of extrusion of wheat and corn on gilthead sea bream (*Sparus aurata*) growth, nutrient utilization efficiency, rates of gastric evacuation and digestive enzyme activities. Aquaculture 225, 207–223.

Verri, T., Mandal, A., Zilli, L., Bossa, D., Mandal, P.K., Ingrosso, L., Zonno, V., Vilella, S., Ahearn, G.A., Storelli, C., 2001. D-Glucose transport in decapod crustacean hepatopancreas. Comp. Biochem. Physiol. Mol. Integr. Physiol. 130 (3), 585–606.

Vornanen, M., Asikainen, J., Haverinen, J., 2011. Body mass dependence of glycogen stores in the anoxia-tolerant crucian carp (*Carassius carassius* L.). Naturwissenschaften 98, 225–232.

Walton, M.J., 1986. Metabolic effects of feeding a high protein/low carbohydrate diet as compared to a low protein/high carbohydrate diet to rainbow trout *Salmo gairdneri*. Fish Physiol. Biochem. 1, 7–15.

Walton, M.J., Cowey, C.B., 1982. Aspects of intermediary metabolism in salmonid fish. Comp. Biochem. Physiol. 73B, 59–79.

Wang, X., Li, E., Chen, L., 2016. A review of carbohydrate nutrition and metabolism in crustaceans. N. Am. J. Aquac. 78, 178–187.

Wang, A.R., Ran, C., Ringø, E., Zhou, Z.G., 2018. Progress in fish gastrointestinal microbiota research. Rev. Aquac. 10, 626–640.

Wilson, R.P., Poe, W.E., 1987. Apparent inability of channel catfish to utilize dietary monosaccharides and disaccharides as energy sources. J. Nutr. 117, 280–285.

Wood, I.S., Trayhurn, P., 2003. Glucose transporters (GLUT and SGLT): expanded families of sugar transport proteins. Br. J. Nutr. 89, 3–9.

Xie, D., Xu, S., Wang, S., You, C., Li, Y., 2016. Cloning, tissue expression, and nutritional regulation of the α-amylase gene in the herbivorous marine teleost *Siganus canaliculatus*. Aquaculture 454, 229–236.

Yuan, X., Zhou, Y., Liang, X.-F., Li, J., Liu, L., Li, B., He, Y., Guo, X., Fang, L., 2013. Molecular cloning, expression and activity of pyruvate kinase in grass carp *Ctenopharyngodon idella*: effects of dietary carbohydrate level. Aquaculture 410–411, 32–40.

Yúfera, M., Moyano, F.J., Martínez-Rodríguez, G., 2018. The digestive function in developing fish larvae and fry. From molecular gene expression to enzymatic activity. In: Yufera, M. (Ed.), Emerging Issues in Fish Larvae Research. Springer International Publishing, pp. 51–86. https://doi.org/10.1007/978-3-319-73244-2_3.

Zambonino Infante, J.L., Cahu, C.L., 2001. Ontogeny of the gastrointestinal tract of marine fish larvae. Comp. Biochem. Physiol. C Toxicol. Pharmacol. 130, 477–487.

CHAPTER

8

Nutritional physiology

Brian C. Small

Aquaculture Research Institute, Hagerman Fish Culture Experiment Station, University of Idaho, Hagerman, ID, United States

8.1 Introduction

Nutrition can be defined as the study of food and how it affects the body, while nutritional physiology emphasizes the mechanisms by which nutrients are extracted from food and utilized for maintenance, growth, reproduction, and health of the fish. To understand the nutritional physiology of a fish, food intake, absorption, assimilation, biosynthesis, catabolism, and excretion must all be considered in total. This is a large topic that exceeds the scope of a single book chapter. Fortunately, many texts, from ichthyology to ecological and general fish physiology and nutrition, have reviewed the anatomy and life history of fish classifications. Even so, sufficient basic information is provided here to guide consideration of digestion, nutrient absorption, and gut physiology.

Since the last edition of this book, the fields of animal nutrition and fish nutrition have expanded to include the microbial populations of the intestinal tract. The field of "microbiome" research has exploded. As with most new fields of study, the human microbiome has received the greatest effort, leading to the discovery of clear functional roles of microbial communities in normal and abnormal physiology (e.g., Tremaroli and Bäckhed, 2012; Vila et al., 2018; Budden et al., 2019). While active research efforts to understand the roles and interactions of mucosal microbial populations of fish are ongoing, the diversity of fish of economic importance, lack of general capacity, and insufficient funding limit our understanding. Even so, the nutritional physiology of fish is now incomplete without including some discussion of the intestinal microbiome.

This chapter will also address what is known regarding the relationship and interactions among nutrients entering the gut and gut neuropeptides, as they affect gut function and regulate feeding, whether through paracrine, endocrine, or neuronal signaling. This area of study within nutritional physiology focuses on the gastroenteropancreatic (GEP) system. GEP endocrine cells distributed in the intestinal mucosa and those located in pancreatic tissue can have both direct and indirect actions on appetite regulation, digestion, nutrient utilization, energetics, and growth, sometimes interacting with the enteric nervous system.

Fish Nutrition, Fourth Edition
https://doi.org/10.1016/B978-0-12-819587-1.00007-0

New discoveries in fish nutritional physiology since the previous edition of *Fish Nutrition* have been exciting and worth exploring, but they build upon extant knowledge. Therefore, this chapter will provide an overview of fish anatomy and physiology, as it relates to nutrient intake and utilization, while also providing insight into the broader connections between the gastrointestinal (GI) tract, gut-associated lymphoid tissues (GALT), GI microbial communities, the enteric nervous system, and the hormones of the GEP system within the fish.

8.2 Feeding strategy

Much of the material in this book is focused on the internal processes of metabolism and nutrient utilization. This chapter will also focus on internal processes, specifically those regulating feeding, digestion, and nutrient absorption. However, given the diversity of life histories and highly variable environments to which fish are adapted, one cannot ignore fundamental aspects of feed detection, acquisition, and ingestion, as these processes often modulate internal nutritional processes, which in turn affect fish performance. Further complicating understanding of these processes are the large ontogenetic changes in feeding strategies across life stages observed for many species of fish.

8.2.1 Trophic diversity

In their natural environment, fish vary greatly in the character of the food they consume. Foodstuffs may be benthic, suspended in the water column, or even terrestrial. Given the diversity of aquatic environments, it should not be surprising that the variety of foodstuffs consumed by fish exceed that of any other classes of vertebrates (Nikolsky, 1963). As such, it should also be no surprise that there is great diversity among fish foraging types, GI systems (Fig. 8.1), and the capacity to extract nutrients from foodstuffs.

For classification purposes, it is easiest to group fish into a limited number of foraging types. Feeding guilds, such as planktivore, herbivore, detritivore, omnivore, and carnivore, paint a broad picture of the foodstuffs consumed. Some would also suggest that generalizations can be made regarding freshwater versus marine fish and temperate versus tropical fish. These differences result from the availability of foodstuffs in different ecosystems, e.g., lakes, rivers, estuarine, and marine. For example, fish living in freshwater systems may have access to vascular plants and terrestrial insects, whereas fish in marine systems may have access to coral polyps and thallose algae, and detritus feeders are often abundant in large rivers and estuaries (Hyatt, 1979). As a group, tropical fish have a greater range of trophic diversity, attributed to longer evolutionary history and relatively stable environments (Hyatt, 1979). Many tropical fish have also achieved a higher degree of flexibility and adaptability at the family level, including species that span the range of foraging classifications from extreme specialists to extreme generalists, while temperate fish tend to lack such a range at the family level (Hyatt, 1979).

Regardless of foraging strategy, the ecological theory of optimal foraging states that an organism will preferentially seek out and consume food resources that promote the greatest long-term net nutrient return for the least energy investment and risk on the part of the consumer (Bone and Moore, 2008). While it is true that fish must acquire enough energy to survive and reproduce, foraging strategies often exhibit a trophic ontogeny or shift in feeding

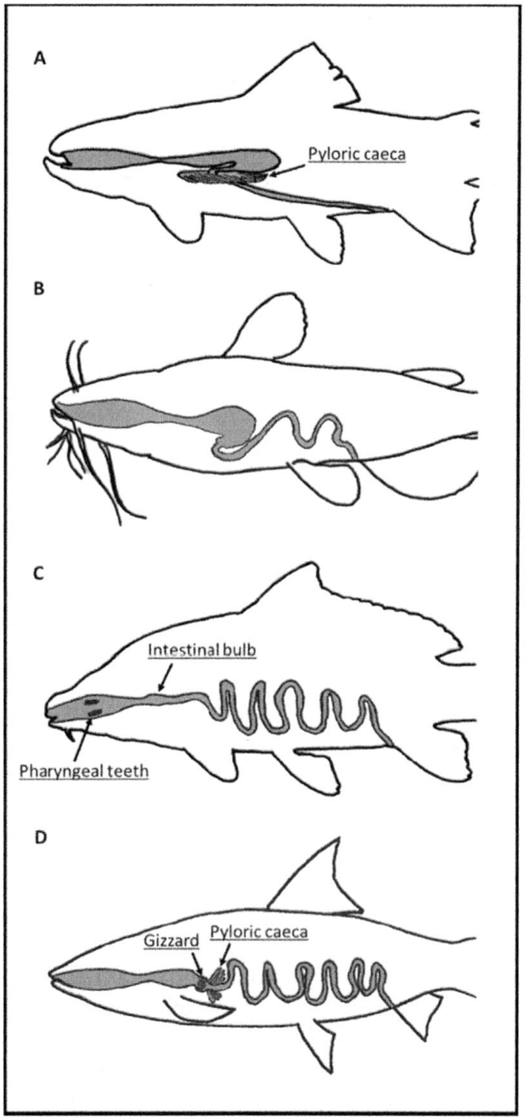

FIGURE 8.1 Simplified diagram of typical fish digestive tract anatomy representing different feeding guilds: (A) carnivore with U-shaped stomach and pyloric caeca (e.g., salmon); (B) omnivore/detritivore emphasizing animal food sources with a sac-like J-shaped stomach (e.g., catfish); (C) omnivore emphasizing plant food sources with no stomach and pharyngeal teeth (e.g., carp); (D) planktivore with a straight gastric stomach and muscular gizzard (e.g., milkfish). *Credit: Brian Small.*

styles, food sizes, and types at different stages in their life cycle (Bone and Moore, 2008). Such shifts in foraging strategy follow changes in fish growth, development, and habitat changes. Throughout these life stages, anatomical, physiological, and behavioral adaptations allow fish to best utilize a variety of foodstuffs and lead to classifications by dietary specialization into feeding guilds.

8.2.1.1 Planktivorous fish

Many fish exhibit a planktonic feeding stage early in development. Species maintaining planktivory are represented by filter feeders, pickers, and particle feeders. Filter feeders must separate plankton from the water. The most common means of separation is by way of mechanical filtration across gill rakers. For species whose diet consists primarily of phytoplankton, an exceedingly long GI tract is necessary to digest plant cell walls and extract nutritional content. Bone and Moore (2008) provide an excellent example of differences in gut morphology between a primary phytoplankton consumer and a primary zooplankton consumer, the menhaden (*Brevoortia tyrannus*) and herring (*Clupea harengus*), respectively. According to their account, the menhaden gut is 10 times longer and has 400 pyloric caeca, compared with 10 pyloric caeca for the herring. In aquaculture, two planktivorous fish are among the most cultivated fish globally, the silver carp (*Hypophthalmichthys molitrix*) and bighead carp (*Aristichthys nobilis*).

8.2.1.2 Herbivorous fish

Only 5% of all fish families contain herbivores (Horn, 1989). Compared with carnivores, herbivores expend more time and energy acquiring food. It has been reported that herbivorous tilapia species spend as much as 14 h/day feeding, compared with 1–3 h/day for carnivorous salmonids (Bone and Moore, 2008). Although herbivorous fish are among the least common in marine and temperate freshwater environments, tropical freshwater environments appear to be ideal for herbivorous species, such as grass carp (*Ctenopharyngodon idella*). Bone and Moore (2008) suggest this is due to seasonal flooding, and greater access to grasses and decaying vegetation, as well as fruits, seeds, and nuts that fall into the water. Herbivorous fish have longer intestinal tracts than omnivorous or carnivorous fish, again providing a greater surface area for nutrient digestion and absorption from difficult to digest plant cells.

8.2.1.3 Detritivorous fish

Because detritivores feed on food items along the benthos, many possess small undershot, inferior mouths, suctorial lips, and barbels (Bone and Moore, 2008), which allow them to vacuum, or suck, and detect buried prey. Others may feed directly on phytoplankton and detritus particles, which they pulverize in the pharyngeal space or a gizzard-like organ. Detritivores can be subclassified as those with a one- or two-part stomach (Bone and Moore, 2008). Those with a one-part stomach have a simple stomach and long intestine, where the sac-like stomach lacks the musculature for mechanical digestion but provides an acidic digestive environment. In addition to the distensible cardiac portion, species with a two-part stomach have a gizzard-like pyloric portion used for grinding or triturating the food (Gerking, 1994).

8.2.1.4 Omnivorous fish

Many species of fish are opportunistic feeders, making omnivory common. Omnivorous species consume a variety of animal and vegetable matter; detritivores are typically also omnivorous. Their GI tract can be thought of as intermediate between strict carnivores and herbivores. Generally, the stomach is sac-like in omnivores with a higher prevalence of animal matter in their diet. Those whose diet tends to be higher in plant matter may have no stomach and a longer intestine.

8.2.1.5 *Carnivorous fish*

Most fish species are predatory, and many are strict carnivores (Bone and Moore, 2008). Carnivores, particularly piscivorous species, often have large terminal or subterminal mouths with well-developed teeth for grasping and biting and/or pharyngeal teeth (along the tongue and top of the mouth). If gill rakers are present, they are often short and blunt. Although GI tract morphology is as diverse as the prey, the GI tract of carnivorous species typically consists of a well-defined muscular stomach, which provides both mechanical and chemical digestion of the food, followed by a short intestine, relative to that of herbivorous fish.

8.2.2 Food acquisition

Nutrient intake begins with food acquisition. Food acquisition, or predation, is a complicated process, as first described by Holling (1966) and later reduced to the basic components of encounter, detection, attack, and capture (Sullivan et al., 2016). Many factors are involved with food acquisition. A fish must first perceive hunger and then detect the presence of food. These two perceptions are sensory and involve the activation of orexigenic (appetite-stimulating) neurons in the brain and stimulation of external sensory neurons, respectively. Finally, a fish must attack and capture the foodstuff.

8.2.2.1 *Hunger and satiety*

In general, hunger drives feeding activity, and hunger is stimulated in response to the nutritional needs of the fish. Hunger itself is a function of a complex network of sensory receptors monitoring environmental, metabolic, neuronal, and endocrine factors, as well as the presence of food and specific nutrients present in the GI tract. As a result, external and internal signals influencing feed intake are received by the brain and processed and communicated as either orexigenic or anorexigenic (appetite inhibiting) signals on neuroendocrine cells in the hypothalamus (Fig. 8.2). The two primary central orexigenic factors are neuropeptide Y (NPY) and agouti-related peptide (AgRP), and the two primary central anorexigenic factors are cocaine and amphetamine-regulated transcript (CART) and corticotropin-releasing factor (CRF) (Volkoff, 2016; Rønnestad et al., 2017).

In the GI tract, a complex array of neuroendocrine factors interacts to regulate feeding behavior and satiety. To date, we only have clear evidence of one orexigenic factor produced in the GI tract, ghrelin (Volkoff, 2019). All others appear to serve as satiety signals and/or modulators of gut function and nutrient assimilation in response to the presence of certain compounds liberated from ingested foodstuffs during digestion. Chemical messenger-secreting cells of the enteric nervous system interact with GEP endocrine cells within the stomach, intestine, and pancreas to monitor and respond to ingested food items. As discussed in more detail later in this chapter, these systems are not only involved in appetite regulation but are also important regulators of digestion, absorption, and metabolism.

8.2.2.2 *Sensory detection and identity of food*

To acquire and process food, fish must first be able to locate it and identify it as food. To accomplish this, a complex of sensory stimuli, including optical, acoustic, tactile, chemical, and electrical, is received through a combination of sensory pathways and interpreted by

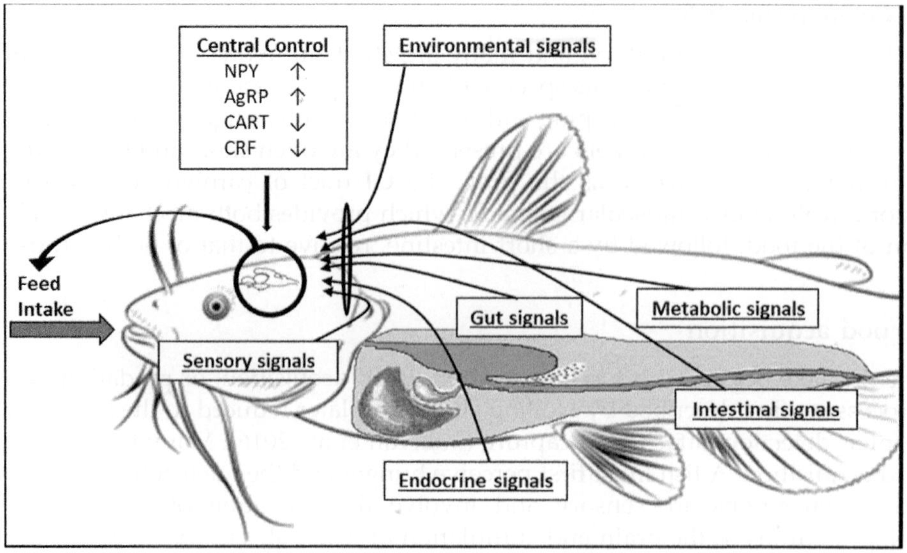

FIGURE 8.2 Simplified diagram of the signaling networks acting on the central control of feed intake via hypothalamic orexigenic factors, neuropeptide Y (NPY) and agouti-related peptide (AgRP), and anorexigenic factors, cocaine and amphetamine-regulated transcript (CART) and corticotropin-releasing factor (CRF). *Credit: Brian Small.*

the brain to identify potential foodstuffs. Even so, many species rely more on one sensory system than others to locate and identify food. Here, the visual and chemical senses used for food recognition are reviewed.

8.2.2.2.1 Vision

Early research on fish feeding provided correlative evidence that in many fish, vision is a predominant factor in locating food. Woodhead (1966) conducted a robust examination of feeding periodicity based on diel changes in stomach contents and observed that in many species, food was only present during daylight hours. Given the vast trophic diversity of fish, one must accept that a heavy reliance on vision is not absolute, but among cultured fish species, vision plays a paramount role in detecting food as part of the larger summation of sensory pathways. The following material has been extracted from Walls (1942), Ali (1975), Nicol and Somiya (1989), Douglas and Djamgoz (1990), Gerking (1994), and Rust (2002).

Fish eyes are often positioned bilaterally, but not always. Other eye positions include unilateral, as seen with flounder, dorsal, as is common for bottom dwelling species, and on eyestalks, observed in certain larval and adult species of fish, although not observed in commonly cultured species. Teleost fish eyes are also often lifted above the body. This provides fish with increased field of view, forward and backward. For fish with eyes angled forward on the head, binocular vision is achieved. On the contrary, fish whose eyes are located far back on the head have only monocular vision.

The retina is an internal photosensitive membrane at the back of the eye, containing photoreceptor cells that translate light into electrical signals transmitted to the brain. The light-sensitive cells of the retina are known as rods and cones, which house the visual pigments. Rods and cones in fish function by typical retinomotor responses to light and dark.

The number and nature of rods primarily dictates visual acuity under different light intensities, with rods dominating at low intensities and cones dominating at higher intensities. As such, rods tend to be more numerous and denser, providing both light sensitivity under low-light conditions and contrast detection. Rods contain a single visual pigment, rhodopsin. Cones are typically more widely spaced than rods and contain between one and four visual pigments with different absorption maxima, allowing for the discrimination of color and extending into the ultraviolet spectrum for some species. Together these two cell types define the resolution of the eye. Different patterns of cell density and composition across the retina have been associated with feeding behaviors (Rust, 2002).

The visual system exhibits significant inter- and intraspecies variation developmentally (Evans, 2004), making it difficult to generalize. The larval stages of marine fish are likely to depend heavily on vision. Helvik et al. (2001) described a functional larval eye at the onset of feeding in Atlantic halibut (*Hippoglossus hippoglossus*), which provides for good vision and object detection under bright light. Later in larval development, double and mosaic cone structure appears, and the structure of the larval retina is complete, allowing the larvae to see at low light and to detect movement (Rønnestad et al., 2013). Given early eye development and functionality in larval marine fishes, determination of optimum light intensity, color, distribution, and daylength is an important aspect of larval nutrition. The detection of movement also appears to be a crucial aspect of feeding marine larvae, and, as such, live feeds remain obligatory for first feeding of many marine species.

Nutrition-related visual disorders and pathologies have been described for a number of fish species. Hughes (1985) described six visual impairments in salmonids having their basis in nutrition. In salmonids, nutritional deficiencies of riboflavin (vitamin B_2), sulfur amino acids (methionine and cystine), tryptophan, and zinc have all been individually linked to cataract formation, while thiamin (vitamin B) deficiency causes corneal inflammation and opacification of the cornea, and insufficient vitamin A not only affects the cornea but also results in exophthalmia and retinal degeneration (Hughes, 1985). More recently, dietary histidine has been identified as crucial for the prevention of cataracts in Atlantic salmon (Breck et al., 2005; Waagbø et al., 2010; Taylor et al., 2015). The importance of vitamin A in larval fish retinal development has also been described (Marsh-Armstrong et al., 1994; Rønnestad et al., 1999), as has the critical role of docosahexaenoic acid (DHA) in rod formation (Bell and Dick, 1993; Shields et al., 1999).

8.2.2.2.2 Chemoreception

Regarding feeding, olfaction (smell) and gustation (taste) are the primary chemosensory mechanisms in fish. Chemoreception in fish differs from terrestrial animals in that the compounds detected by these two sensory systems in fish are dissolved in water, while in terrestrial animals, soluble compounds are tasted, and volatile (airborne) compounds are smelled. The morphology and physiology of chemoreception in fish have been extensively reviewed (Hara 1971, 1992; Hansen and Reutter, 2004; Reutter and Kapoor, 2005).

8.2.2.2.2.1 Olfaction Briefly, the morphology of peripheral olfactory organs in fish is variable, reflecting the diversity of species, environmental conditions, and evolutionary pathways (Zeiske et al., 1992). Generally, however, olfactory receptors are of one of two types, ciliated or microvillar (Hara, 1992). These sensory neurons are the only neurons in vertebrates whose axons carry external information directly to the brain, suggesting a high level of importance to the survival of the animal. The olfactory bulb in fish is similar to higher vertebrates and segregated into lateral and medial tracts, as reviewed by Satou (1992), with stimulatory responses differing according to distinct chemical classes.

Hara (1992) makes the point that chemical stimuli have three distinct properties. First, chemicals have distinct molecular structures and stereoisomeric configurations, which allow for extreme specificity in detection. Second, chemical signals persist in the environment beyond when they were produced, unlike visual and other stimuli. Third, chemical stimuli are nondirectional, with distribution and gradients determined by diffusion characteristics and water turbulence. Even so, many studies, both in the laboratory and in the field, have observed fish using chemical stimuli to orient precisely to food items. In fact, fish can detect the presence of chemical substances that other vertebrates cannot and often at extremely low thresholds (Bone and Moore, 2008). Such compounds include amino acids, steroids, and prostaglandins (Hara, 1992).

Of importance to feeding, much research has been conducted on amino acids as olfactory stimuli for fish, and many reviews exist (Kleerekoper, 1969; Hara, 1982, 1986, 1992, 2005; Mackie, 1985; Caprio, 1984; Carr and Derby, 1986; Carr, 1988). Numerous stimulatory hierarchies have been developed for various species using electrophysiological and cardiac responses. For channel catfish (*Ictalurus punctatus*), Little (1981) reported cysteine > threonine > phenylalanine > asparagine > serine = alanine > glutamine > methionine > arginine = control. Based upon electrophysiological studies, stimulatory thresholds for amino acids are reportedly as low as 10^{-8} to 10^{-9} M with a response range that covers six to seven log units of concentrations (Hara, 1992; Bone and Moore, 2008).

8.2.2.2.2.2 Gustation Many texts suggest olfaction is generally the more sensitive chemoreceptive sense, but it is likely species dependent, as olfaction in salmonids appears to be an important aspect of food search behavior, but channel catfish appear to be driven primarily by gustation (Hara, 2006). In fact, some fish have been observed to have very low gustatory thresholds of detection. Early research demonstrated sugar thresholds in minnows 512 times (4×10^{-4} M) more sensitive than that of humans and salt thresholds 184 times (2×10^{-5} M) more sensitive (Krinner, 1935). More recent research on the physiology of taste in fish suggests even greater sensitivity ($<10^{-9}$ M) for the most potent substances (Kasumyan and Døving, 2003). Numerous electrophysiology studies have provided extensive characterizations of a variety of compounds, including amino acids (AAs), nucleic acids, organic and inorganic acids, fatty acids, alkaloids, and salts on several species (Morais, 2017). Together with new knowledge on fish-specific gustatory receptors and transduction pathways, unique avenues for improving feed palatability may be discerned for cultured fish.

As previously stated, the morphology and physiology of chemoreception have been extensively reviewed (Hara 1971, 1992; Hansen and Reutter, 2004; Reutter and Kapoor, 2005), as have, more specifically, taste preferences in fish (Kasumyan and Døving, 2003). Taste buds, their innervations, and central organization are described as being similar to mammals (Morais, 2017). Where the large differences occur between fish and mammals is the number and placement of taste buds. Fish are generally portrayed as having the most taste buds of any animal and have them distributed not only in the oropharyngeal space but also extraorally. Many species may have taste buds located in the lips, gill rakers, oral cavity, pharynx, esophagus, body surface, barbels, and/or fins (Ishimaru et al., 2005). Different roles have been proposed for internal and external taste buds, with those in close proximity to the head and mouth being used to detect food and with oral taste buds determining acceptance or rejection of the food item (Kasumyan and Døving, 2003; Hansen and Reutter, 2004). Often in aquaculture, fish are observed to ingest feed particles, only to reject and expel them rapidly from their mouths. Although mechanosensory receptors in the oropharyngeal area likely also play a role in feed rejection, the extreme sensitivity to chemical signals in the feed likely dominates.

To better understand taste preferences among fish, both electrophysiological and behavioral studies have been implemented. Together, these types of studies have provided extensive information on not only the classical taste modalities, as reviewed by Kasumyan and Døving (2003), but also numerous compounds with attractive and aversive taste modalities (reviewed by Morais, 2017). Once again, amino acids have been best described, but not all amino acids are attractive and not all fish respond the same. Hara (1994) grouped fish into two classifications based upon amino acid responsiveness: (1) fish responding to a wide range of amino acids (e.g., channel catfish, Japanese bream, mullet, tilapia), and fish with a limited response to only a few amino acids (e.g., amberjack, cyprinids, eel, pufferfish, salmonids, yellowtail). Even so, a review of amino acid stimulatory effects across 21 fish species suggests the most frequently stimulatory are L-alanine > L-cysteine > L-serine > L-glutamine = L-glycine = L-glutamic acid = L-tyrosine (Kasumyan and Døving, 2003). These rankings, however, represent individual amino acids, but certain mixtures of amino acids have been shown to elicit synergistic responses not observed with the individual component amino acids (Adams and Johnsen, 1986; Ogawa and Caprio, 2000). In their review, Morais (2017) emphasized the need for further characterization of these types of interactions and the activation of different taste receptors, as they pertain to aquaculture feeds, especially with regard to palatability issues with medicated feeds and alternative ingredients to fishmeal and fish oil.

8.2.2.3 *Capture and ingestion*

Taste and texture may determine whether food is retained or rejected, but the process of capture and swallowing is dictated by differences in the morphology and mechanics of the mouth, jaws, gill arches, and teeth, each playing important roles in ingestion. There are many different mouth morphologies, teeth types, and locations. Together with knowledge of natural habitats and life history, the mouth can provide important clues regarding what type of formulated feed a species might accept. Rust (2002) suggested that by examining the mouth, one might predict whether to feed a hard or soft, floating or sinking pellet. For example, fish having the structural capacity to feed on shellfish should have no problem consuming a hard pellet and that bottom feeders would likely prefer a sinking pellet (Rust, 2002). Furthermore, mouth structure and location can provide clues about the type of diet a fish feeds on, which is often a useful starting point for developing feeds for newly cultured species.

In aquaculture, jawed fish dominate, but there is a large diversity of teeth types and locations among them. Teeth located in the jaw may be found on the premaxillary, maxillary, and mandible bones. These teeth are used for biting, tearing, grasping, rasping, and scraping. Many species only have small, pointed teeth used to immobilize prey and rely on suction forces in the mouth to finish ingestion and swallow the food item (Bone and Moore, 2008). While immobilization of prey is not often an issue in aquaculture, it is important to realize that many species do not chew, but instead, simply ingest feed by suction. Thus, the physical attributes of the pellet and ingredients must be such that they are easily digested to make the nutrients available for absorption.

Other common teeth arrangements include those on the tongue (palatine) or roof of the mouth (vomerine), which help to hold food and direct it to the back of the mouth, and pharyngeal teeth, those found on, above or below the gill arches. Pharyngeal teeth are important for shearing or crushing prey and often found in fish without jaw teeth, such as carps, minnows, and suckers, but may also be found in fish having jaw teeth, such as cichlids and drums (Bone and Moore, 2008). Pharyngeal teeth provide a means of mechanical digestion of food for many stomachless fish, passing masticated food through the esophagus into the intestine. Ingestion is completed as food enters and leaves the esophagus.

8.3 Digestive system

8.3.1 The alimentary canal

The alimentary canal is divided into the mouth, buccal cavity, pharynx, esophagus, stomach, intestine, anus, and associated organs. There is much diversity in the morphology of the digestive tract among fish (Fig. 8.1). For some, the tract may be straight from the mouth to the anus, while for others, it may be a complex series of twists and turns, divided structurally and functionally, sometimes separated by valves and sphincters. As discussed previously, generalizations have been made regarding GI tract morphology, as it relates to the natural diet and trophic position of fish. However, modifications are often superimposed on the basic GI morphology, and many in-depth reviews on the morphology and histology of the alimentary canal of fish exist, including more recent reviews by Wilson and Castro (2010) and Ray and Ringø (2014).

8.3.1.1 Esophagus

As food leaves the pharynx, it passes into a short muscular tube, the esophagus, which is often highly elastic to allow expansion and accommodation of large food items. The esophagus connects the oropharynx with the stomach in gastric fish (those with a stomach) or the intestine in stomachless fish. The process is facilitated by the production of mucus. The mucosal epithelium is dominated by goblet cells (mucus cells) in teleost fish and has a striated musculature. Fish do not appear to secrete digestive enzymes with mucus, and as such, this secretion functions only to lubricate (Smith, 1989). The esophagus may or may not terminate in a cardiac sphincter. It is more common in gastric freshwater fish, but not observed in marine fish. The latter is likely due to their need to continuously drink water for osmoregulation, while the opposite is true for freshwater fish, which must minimize water intake (Guillaume et al., 1999).

8.3.1.2 Stomach and intestinal bulb

The most basic digestive tract distinction separating fish is the presence or absence of a stomach. The stomach can be defined as an elastic muscular organ, which receives food from the esophagus and secretes acid and enzymes that digest food. In microphagous detritivores or herbivores, the posterior of the stomach may be modified into a gizzard-like structure for trituration and mixing. Fish lacking a stomach, such as many herbivorous fish, often have a sac-like bulge in the anterior intestine, which receives food from the esophagus. This structure, referred to as the intestinal bulb, provides temporary storage of ingested food but provides no digestive function. Histological examination of the intestinal bulb suggests that the mucosa lacks the capacity to secrete acid and digestive enzymes (Horn et al., 2006; Manjakasy et al., 2009).

In gastric fish, food exits the esophagus and enters the stomach anteriorly, and digesta exits posteriorly through the pyloric sphincter into the intestine. The stomach produces hydrochloric acid, which initiates digestion by denaturing protein and converting inactive zymogen (proenzyme) pepsinogen into the active proteolytic enzyme pepsin, making it the primary site of protein digestion. An acid environment is necessary for the activation of pepsinogen to pepsin, which functions at an optimum pH between 1 and 5 (Bakke et al., 2011). Fish stomachs have been classified into categories based upon their morphological appearance, which has been related to the duration between meals and diet type (Suyehiro, 1942). Suyehiro's stomach classifications include (1) straight tube (Pleuronectidae), (2) U-shape (Salmonids), (3) V-shape (Plecoglossidae, Mugilidae, Salmonidae, Sparidae), (4) Y-shape (Mugilidae, Clupeidae), and (5) I-shape (Carangidae, Gadidae, Scrombidae, Serranidae). Other classifications have since been added to the literature, such as J-shape (Ictaluridae; Sis et al., 1979) and C-shape (Pseudopimelodidae; Mello et al., 2019. Fig. 8.3).

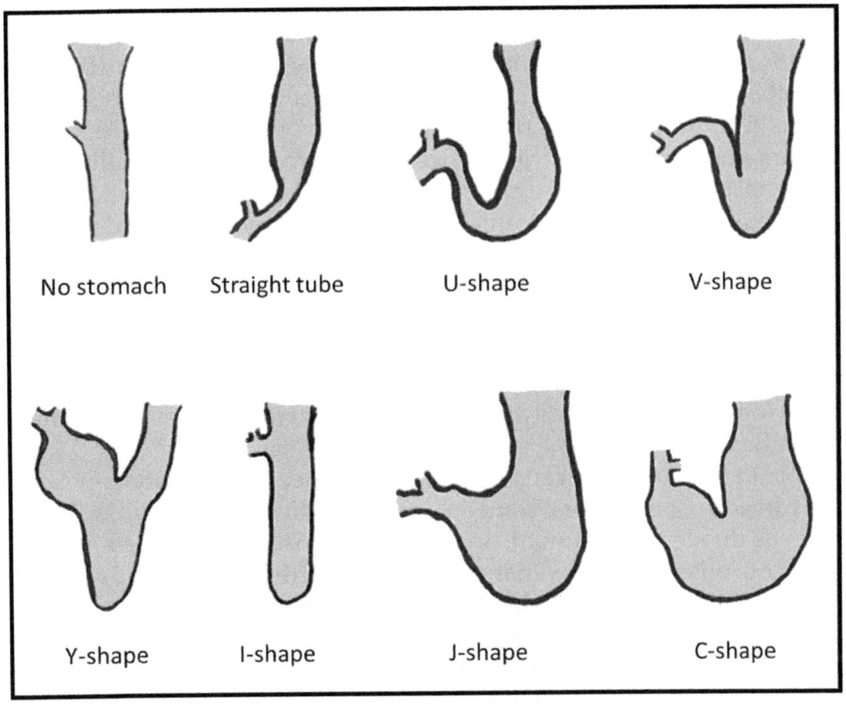

FIGURE 8.3 Simplified stomach classifications based upon morphological appearance according to Suyehiro (1942), Sis et al. (1979), and Mello et al. (2019) and modified from Bertin (1958). *Credit: Brian Small.*

8.3.1.3 Pyloric caeca

Many gastric teleost fish have blind tubes, termed pyloric caeca, connected with an opening into the anterior intestine in the region of the pylorus. Pyloric caeca vary in length, width, and number, with as few as one to nearly a thousand in certain members of the Gadidae family. Pyloric caeca resemble the intestine in structure, having a well-developed muscularis and inner epithelium containing goblet cells for mucus production. There is no evidence of secretory cells producing digestive enzymes in the epithelium. As such, it is often proposed that the pyloric caeca simply provide increased absorptive surface area. Yet observations of enteroendocrine cells and enterocytes capable of facilitating digestion and absorption have been reported (Hibiya, 1982). In truth, the function of pyloric caeca is poorly understood. It is clear, however, that pancreatic enzymes make their way into the caeca. How this occurs is unclear. Pancreatic exocrine cells are dispersed within the connective tissue between the caeca. Pancreatic enzymes are either secreted through several small ducts directly into the lumen of the pyloric caeca or released into ductules and then into a common pancreatic duct emptying into the intestinal lumen proximal to the pyloric caeca. The presence of enteroendocrine cells with cholecystokinin (CCK)- and gastrin-like immunoreactivity in the pyloric caeca epithelium, as observed in of the flounder (Yoshida et al., 1983), suggests that these blind appendages of the proximal intestine may function as an endocrine organ, stimulating the release of digestive enzymes from the stomach and pancreas.

8.3.1.4 Intestine

As stated earlier, the intestine of fish is highly variable in shape and length. It may be straight, bent, twisted, folded, coiled, helical, and more. The length varies from 1/5th to 20× body length and is longest in microphagous and herbivorous fish (Bryan, 1975). As a generalization, the ratio of gut length to body length is often greater than 3 in microphagous and herbivorous fish, from 1 to 3 in omnivores, and less than 1 in carnivores.

Although numerous and distinct differences exist across species, Guillaume et al. (1999) identified five constant features of fish intestines. First, structurally and functionally, the fish intestine resembles the small intestine of mammals, having some degree of successive differentiation, reminiscent of the duodenum, jejunum, and ileum of mammals. Fish do not have a true colon. Second, the intestinal wall is composed of only three layers: mucous, muscular, and serous. Third, intestinal villi are poorly differentiated and appear more like pseudo-villi or folds. Fourth, some enterocytes retain the endocytic function, capable of engulfing and transporting proteins into enterocytes, throughout life and not, as in mammals, just during early development. Finally, the suite of digestive enzymes found in fish is similar to that found in higher vertebrates.

The sections of the fish intestine can be divided into three. The terminology used to describe these sections varies among authors, sometimes referred to as anterior, middle, and posterior, and other times as duodenum, jejunum, and ileum (after mammalian terminology). Some authors simply refer to anterior or proximal and posterior or distal, apparently resulting from the difficultly in differentiating sections due to a lack of external differences. Here they will be referred to as proximal, medial, and distal, as is common to the more recent literature.

In stomachless fish, the proximal intestine begins after the esophagus and typically has an intestinal bulb (previously described) at the anterior end. In these fish, the bile duct connects to the intestine, anterior to the intestinal bulb, and the pH is never acidic, ranging from 7 to 9. The proximal intestine of gastric fish begins after the pyloric sphincter separating it from the stomach. In these fish, the bile duct connects to the proximal intestine below the stomach and above pyloric caeca, if they are present. Histologically, the intestinal mucosa is similar to higher vertebrates. As described by Guillaume et al. (1999), the proximal intestine is lined by a simple columnar epithelium, which possesses pseudovilli increasing the surface area by approximately 20-fold. Mucus-secreting (goblet) cells and endocrine cells are scattered among the epithelial cells. Enterocytes line the luminal surface and themselves are covered in microvilli on their luminal surface, further increasing surface area. Bound to the enterocyte cell membranes are enzymes, transporter proteins, and ion channels.

The medial intestine, while anatomically difficult to differentiate, is histologically distinct with apparent endocytic function (Guillaume et al., 1999). Enterocytes possess several invaginations between the microvilli, and large supranuclear vacuoles appear in the intercellular space. Through the process of endocytosis, protein molecules can be absorbed and intracellularly digested.

The distal intestine in fish is typically short but may be elongated in some species, widening and differentiating as the "rectum," sometimes separated by the ileo-rectal valve, largely formed by smooth muscles. The enterocytes of the distal intestine differ from the proximal and medial intestine by having very short microvilli and lack many of the characteristics associated with absorbent cells. They do, however, have numerous mitochondria and are thought to play a role in osmoregulation through selective absorption of minerals (Guillaume et al., 1999). The mucosa of the distal intestine (and rectum) is replete with goblet cells for mucus production.

8.3.1.5 Exocrine pancreas

The exocrine pancreas of teleost fish is typically diffuse but in rare instances can be discretely located. Whether discrete or diffuse, the exocrine pancreas contains zymogen granules that produce and store digestive enzymes. Pancreatic enzymes, or their activity, have been found in fish, corresponding to lipase, colipase, phospholipase, α-amylase, trypsin, chymotrypsin, elastase, carboxypeptidases A and B, and DNAase and RNAase (reviewed by Bakke et al., 2011). Diffuse pancreatic tissue forms sheaths around blood vessels, and numerous small ducts release the enzymes into the lumen of the proximal intestine and pyloric caeca, although in some species a single duct may join the common bile duct. In cyprinids, labrids, and certain other species, the pancreatic tissue is found within the portal canals of the liver, likely releasing enzymes into the bile ducts (Fänge and Grove, 1979). For species having a discrete pancreas, the distinct organ is embedded in the connective tissue on the surface of the proximal intestine and releases enzymes through a common pancreatic duct into the intestine.

8.3.1.6 Liver and gallbladder

The following material has been extracted from Hinton (1994), Kruslé and Anadon (1996), Rust (2002), and Bakke et al. (2011). The liver lies in the anterior portion of the peritoneal cavity and may be arranged in two or more distinct lobes or as a single organ in fish. The liver receives blood from the hepatic artery and from one or more portal veins that transport blood from the gastric and intestinal mucosa, swim bladder, spleen, and pancreas. The rich vascularization gives the liver a reddish-brown color, although large depots of lipid may cause the liver to appear pale or yellow, and disruptions of bile production can cause a green appearance.

The morphology of the fish liver resembles that of other vertebrate livers, except for not being organized into discrete lobules or acinar units. The liver parenchyma is a complex structure made up of anastomosing epithelial lamellae separating numerous vascular sinusoids and biliary canaliculi. It consists of two cellular plates surrounded by sinusoids, with each plate having polarized hepatocytes showing a sinusoidal face for absorption and biliary face for excretion. The bile ducts consist of a simple cubic epithelium with a brush border, underlain by a basal lamina and a wall of collagen and muscle fibers. The biliary tree of fish livers resembles that of higher vertebrates. For many teleost fish, it originates as an intercellular canaliculus formed by close opposition of two hepatocytes. Bile secretory activity is evident by the presence of cytoplasmic Golgi vesicles and lysosomes concentrated near the canaliculi. Bile flow through the bile ducts is regulated by numerous contractile microfilaments located concentrically around the canaliculi.

The bile duct opens into the anterior of the proximal intestine. In most fish, bile (the exocrine secretion of the liver) is stored in the gallbladder and is released into the intestinal lumen when chyme enters the proximal intestine from the stomach. Contraction of the gallbladder occurs in response to the presence of nutrients and is under neuroendocrine control. Cholecystokinin (CCK), secreted via neural signals from neuroendocrine cells lining the intestine, is responsible for both gall bladder and pancreatic secretion.

Bile acids are hydroxylated steroids, synthesized in the liver from cholesterol (Fig. 8.4), which aid in the emulsification of dietary lipids to allow for the efficient action of lipases and the formation of micelles. Specifically, bile acids are cholic and chenodeoxycholic acids.

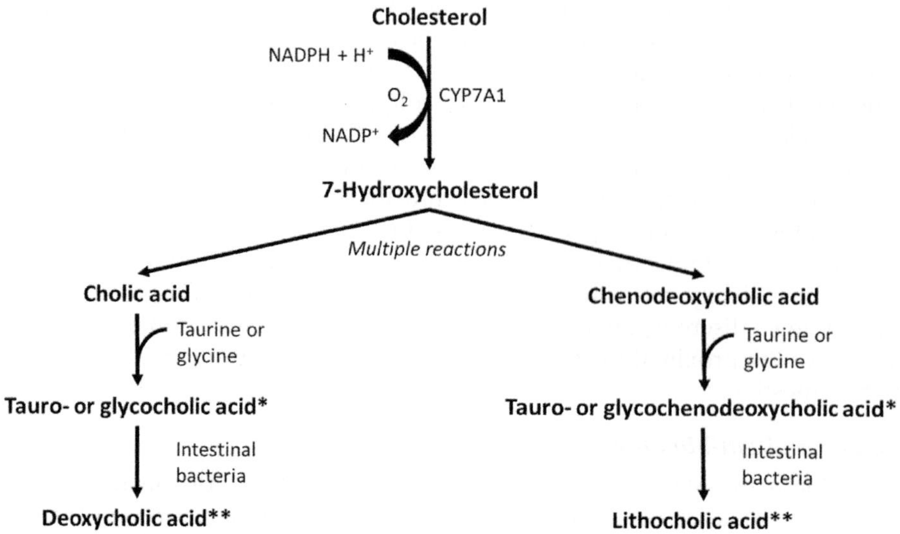

FIGURE 8.4 Bile acid synthesis. *Primary bile acids; these are synthesized in the liver and for most fish are conjugated to taurine. **Secondary bile acids; these are synthesized by bacteria in the distal intestine. *Credit: Brian Small.*

In most fish, the majority of bile acids are conjugated to taurine to form taurocholic acid and taurochenodeoxycholic acid, while some conjugate to glycine to form glycocholic acid. Bile acids decrease in the chyme as the intestinal contents move distally. Bile acids that are not absorbed in the proximal intestine reach the distal intestine and are metabolized by the microbiota into secondary bile acids and ultimately reabsorbed or excreted. The resorption of bile acids is referred to as enterohepatic recycling, as observed in higher vertebrates; however, the extent to which this occurs in fish is not known.

8.3.2 Digestion

Knowledge of the physiological and biochemical specifics of digestion remains limited for most fish. The digestive enzyme apparatus of fish appears to be qualitatively similar to higher vertebrates, having many of the same enzymes present in the digestive tract and with similar substrate specificity. However, fish enzymes often show higher specific activity and affinity for their substrates than in mammals. This is said to be an adaption to poikilothermy, allowing for enzymes to function at lower temperatures (Bakke et al., 2011). Much of what we know about digestive physiology in fish relates to gastric species, and because stomachless fish are defined more by what is absent (a stomach), there is a paucity of knowledge into form or function of the alimentary canal.

Digestion of foodstuffs is a process that for some fish begins in the pharynx with the physical mastication of ingesta and for others begins in the stomach (gastric fish) or proximal intestine (stomachless fish). Digestion continues through a series of chemical and enzymatic processes to break down food into substances that can be used by the body. The following material has been extracted from Rust (2002), Kuz'mina (2008), and Bakke et al. (2011) and supports topical nutrient discussions in Chapter 3, Vitamins; Chapter 4, Amino acids; Chapter 5, Lipids; Chapter 6, Minerals; and Chapter 7, Carbohydrates.

8.3.2.1 *Protein digestion*

In gastric fish, the stomach provides an essential role in denaturing and hydrolyzing protein (Fig. 8.5). This process begins with the production of hydrochloric acid in the gastric fluid and the resultant denaturing of most proteins, opening the proteins for easier access by the proteolytic enzyme pepsin. In many fish, the pH of the stomach decreases after food intake. Although pepsin is undoubtedly the major acid protease in the stomach, the gastric fluid contains several forms of pepsin, which function at different pH optima (between 1 and 5), allowing for different activation rates, each having specific activities and substrate specificities (Bakke et al., 2011).

Pepsins are secreted from the gastric gland cells as a zymogen called pepsinogen. Although inactive as a zymogen, the acidic stomach environment converts pepsinogen to the active proteolytic enzyme pepsin. Pepsins are endopeptidases, hydrolyzing peptide linkages formed by amino groups of aromatic amino acids, such as phenylalanine, tryptophan, and tyrosine. The hydrolysis of protein increases solubility and makes them more available for further processing by intestinal proteases.

As proteins leave the stomach and enter the proximal intestine, they become solubilized and diluted in alkaline solutions from the liver, pancreas, and intestine. The actions of pancreatic proteases secreted into the intestine and caeca (if present) result in a mixture of free

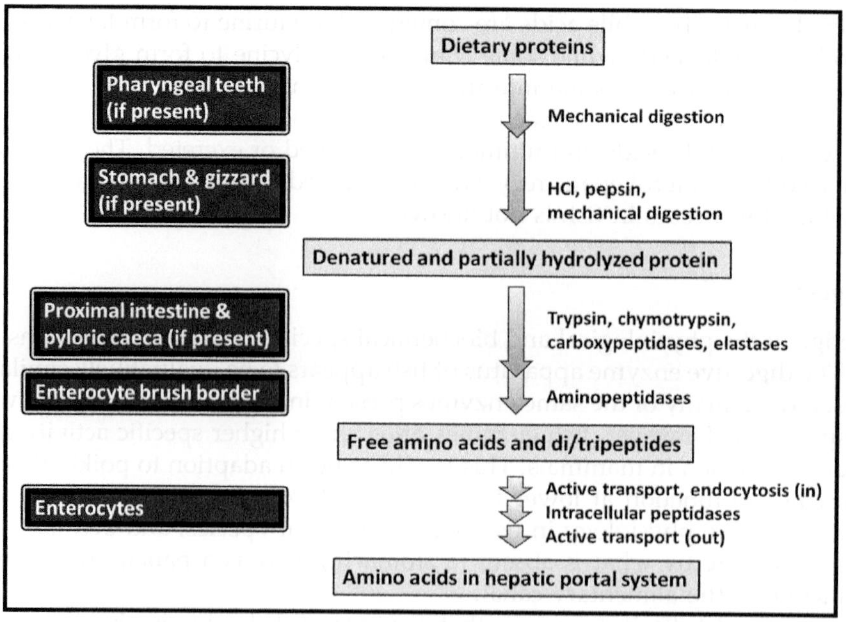

FIGURE 8.5 Simplified diagram of reported mechanisms of protein digestion and absorption in the fish gastrointestinal tract. *Credit: Brian Small.*

amino acids and small peptides. Trypsin, chymotrypsin, carboxypeptidase, and elastase are stored as zymogen granules. The presence of chyme in the intestine stimulates pancreatic secretion. Trypsinogen is then transformed into trypsin by enterokinases produced by intestinal mucosal cells. Trypsin in turn activates other pancreatic zymogens, converting them into functional enzymes.

Trypsin is an endopeptidase catalyzing the hydrolysis of peptide bonds, having an optimal pH around 7. It functions to cleave peptide chains at carboxyl groups coming from arginine or lysine. The resultant smaller peptide products are then further hydrolyzed into amino acids and small peptides via other proteases. Chymotrypsin is formed by the action of trypsin on chymotrypsinogen. Chymotrypsin is also an endopeptidase, selectively cleaving peptide bonds formed by aromatic side chains coming from tyrosine, tryptophan, and phenylalanine. Other pancreatic endopeptidases include elastases I and II, again activated by the action of trypsin on the zymogen proelastase. Elastases are especially active on peptide bonds in the protein elastin, an extracellular matrix protein present in connective tissue. Two exopeptidases round out the common pancreatic peptidases, carboxypeptidase A and B. Again, activated by trypsin, these exopeptidases hydrolyze the terminal peptide bonds of their substrates. The resulting mixture of amino acids and small peptides produced by the actions of these pancreatic proteases is further hydrolyzed at the brush border of the enterocytes by aminopeptidases and intracellular peptidases once transported across the membrane. However, some proteins and peptides may resist proteolysis and reach the distal intestine relatively intact. There appears to be some capacity to assimilate protein and peptides directly by endocytosis and intracellular digestion throughout the life of a fish; pinocytosis is the primary means of protein assimilation in larvae.

8.3.2.2 *Lipid digestion*

The primary mechanism for lipid digestion in fish is extracellular hydrolysis and requires emulsifiers for efficient enzyme activity (Fig. 8.6). Exogenous emulsifiers, mainly proteins and phospholipids, come from the food, while endogenous emulsifiers include bile acids and phospholipids. When chyme enters the proximal intestine and is sensed by enteroendocrine cells, cholecystokinin is released and stimulates gallbladder contraction and secretion of the bile into the intestine. The emulsification of dietary lipids is essential for efficient digestion and absorption of fatty acids, cholesterol, fat-soluble vitamins, and other hydrophobic components of the diet. As amphipathic molecules with hydrophobic and hydrophilic regions, conjugated bile acids have detergent action on particles of dietary fat, which causes fat globules to break down or be emulsified into minute microscopic droplets. This action substantially increases the surface area of the lipid for access and digestion by lipases. Pancreatic lipase, along with phospholipase A2 and bile salt-activated lipase, is the principal enzyme hydrolyzing triglyceride into monoglyceride and free fatty acids (Tocher, 2003). Bile acids also play important roles after monoglycerides and fatty acids are liberated by complexing with them and other hydrophobic substances (e.g., phospholipids and fat-soluble vitamins) to form structures called micelles suspended within the ingesta.

Pancreatic lipases are the primary source of lipolytic activity in fish, and two predominant forms exist. In freshwater fish, the main form is likely colipase-dependent pancreatic lipase (PL), and in marine fish, bile acid–dependent carboxyl ester lipase (CEL) seems to dominate (Kurtovic et al., 2009). PL has a higher specificity and efficiency for digesting triglycerides. CEL is said to hydrolyze a broader range of lipids, including wax esters, while also having

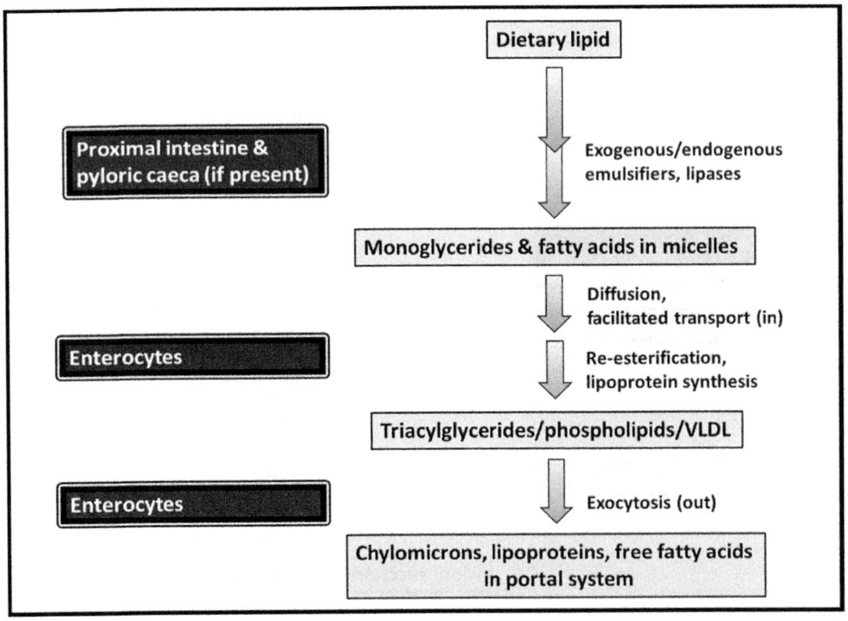

FIGURE 8.6 Simplified diagram of reported mechanisms of lipid digestion and absorption in the fish gastrointestinal tract. *Credit: Brian Small.*

a strong affinity for glycerides. Phospholipids are also efficiently hydrolyzed by fish. Phospholipase and other lipolytic enzymes activity have been detected, but the general nonspecific action of lipases makes it difficult to confirm the source, and it is unclear whether a specific phospholipase is important to lipid digestion in fish. Hydrolysis of lipids by lipases can result in the production of several breakdown products depending on the substrate; however, free fatty acids and monoglycerides appear to be the major products. In fish, monoglycerides may be the preferential substrate for the synthesis of triglycerides in the enterocytes (Oxley et al., 2007). Free fatty acids are presumed to be packaged in mixed micelles along with fat-soluble vitamins and cholesterol esters. Mixed micelles encounter the brush border of intestinal enterocytes and dissociate as a result of the lower pH at the mucosal surface. The released contents of the micelle are then taken up into the intestinal epithelial cells.

8.3.2.3 *Carbohydrate digestion*

Fish vary greatly in their capacity to digest and absorb carbohydrates. Carbohydrate hydrolysis can occur in the proximal intestine and pyloric caeca, as well as at the brush-border membrane of the intestinal epithelial cells (Fig. 8.7). Pancreatic α-amylase is secreted along with the other pancreatic enzymes discussed previously. Amylase is an α-1-4-glucosidase, which can hydrolyze α-1 → 4 bonds in amylase or linear fragments of amylopectin or glycogen. Pancreatic α-amylase is found in all fish, including carnivorous species that never encounter starch in their natural diet, suggesting a main role in the hydrolysis of glycogen in prey animals. Of course, formulated fish feeds are likely to contain some starch, and the presence of α-amylase across fish ensures some capacity for its digestion.

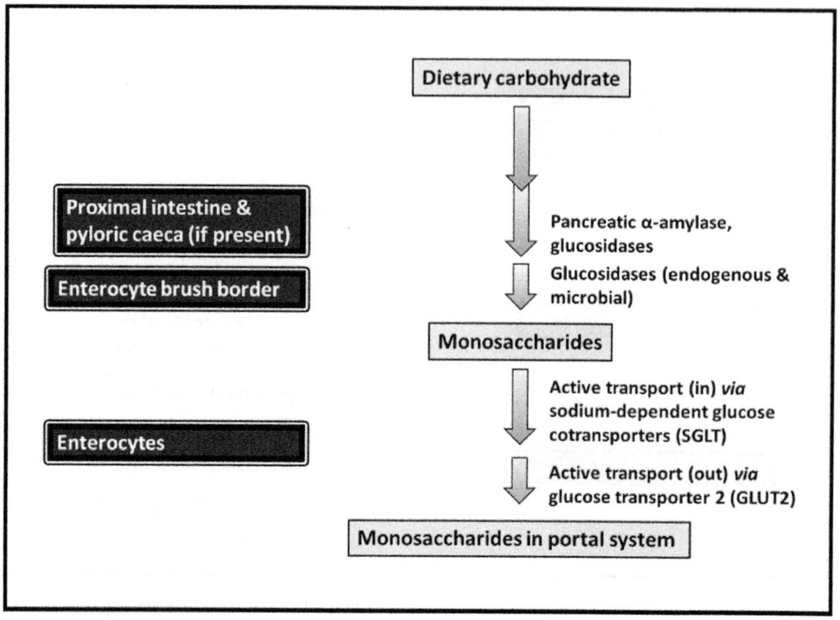

FIGURE 8.7 Simplified diagram of reported mechanisms of carbohydrate digestion and absorption in the fish gastrointestinal tract. *Credit: Brian Small.*

Even so, differences in amylase activity may be more than 100-fold higher in herbivorous fish compared with carnivorous fish (Kuz'mina, 2008). Most amylases have two pH optima, between 4 and 9, and different isoforms may exist (Krogdahl et al., 2005). There is also evidence of maltase and sucrase activity (Goel, 1975; Agrawal et al., 1975; Weinrauch et al., 2019), suggesting that certain oligosaccharides and disaccharides are digested by specific enzymes at the brush-border membrane.

The activities of two other glucosidases have been observed in fish intestine, albeit at low activities: cellulase (β-1-4-glucosidase) and laminarinase (β-1-3-glucosidase) (Guillaume et al., 1999). Cellulase activity has been measured in the guts of several species, and both intestinal and hepatopancreatic cellulase activity in grass carp increased in response to dietary cellulose (Das and Tripathi, 1991; German and Bittong, 2009). Undoubtedly, some of the cellulase activity found in the fish gut is of microbial origin, linked to microbial fermentation of cellulose. Laminarinase activity has been found in the gut of fish consuming algae (Guillaume et al., 1999). It allows the partial hydrolysis of laminarin, a structural protein in algae. Although it is not surprising to find this enzyme in algae-eating fish, there is little evidence of endogenous production.

Chitinase is another enzyme that hydrolyzes complex β-link carbohydrates and is often found in carnivorous species, especially those naturally consuming insects and crustaceans (Guillaume et al., 1999). The polysaccharide chitin is a primary component of cell walls in fungi, the exoskeletons of arthropods, the radulae of mollusks, cephalopod beaks, and the scales of fish and amphibians. The source of endogenous chitinase appears to vary among fish and may include secretions from the gastric and/or intestinal mucosa (Fange, 1979; Divakaran et al., 1999; Gutowska et al., 2004; Ikeda et al., 2009). Like cellulase and laminase, the contribution of exogenous chitinase from the diet is also a likely source.

8.3.2.4 Enzymatic digestion in stomachless fish

There is relatively little information concerning digestion in stomachless fish. Stomachless fish cover the entire trophic spectrum, including herbivory, omnivory, detritivory and carnivory (Logothetis et al., 2001; Crossman et al., 2005; Horn et al., 2006; German, 2009). Cyprinids are the most studied of stomachless fish due to their prevalence in nature and importance as aquaculture species. Generally, stomachless herbivores are thought to rely on a robust pharyngeal jaw apparatus to triturate their food. This may not be the case for carnivores; stomachless needlefish (Belonidae) apparently use their pharyngeal jaws to grasp and manipulate prey into a "head first" position, moving it directly into the posterior intestine where digestion and absorption occur (Manjakasy et al., 2009).

In a study comparing two herbivorous and two carnivorous stomachless fish, Day et al. (2011) observed not only some expected trends based on trophic preference but also some unexpected results. In two herbivorous halfbeak species (Hemiramphidae), α-amylase activity was predictably high; however, the activities of trypsin and lipase were similar or greater than those of the two carnivorous needlefish species. The digestive enzyme activity profile of the two carnivorous needlefish species emphasized protein digestion, as evidenced by the dominant role of trypsin and high levels of aminopeptidase; however, lipase activity was not as high as predicted for carnivores. The herbivorous halfbeaks generally displayed an overall pattern distally decreasing digestive enzyme activity, and the carnivorous needlefish generally had the greatest enzymatic activity in the medial and distal intestine.

Day et al. (2011) concluded that the herbivorous stomachless halfbeak "employs an aggressive, 'shotgun' approach to nutrient assimilation, targeting a wide range of nutrients available from its narrow dietary range of plant material that has been rendered by the pharyngeal jaw apparatus into a high surface area to volume bolus," and the carnivorous, stomachless needlefish "were more specific in their 'rifle' approach to enzyme production, with activity levels modulated to specific nutrient levels, characterized by high trypsin activity, directed at the abundant protein available from its range of large, low surface area to volume prey items."

Relatively few digestibility studies have been conducted in stomachless fish. In one study evaluating protein digestion and amino acid absorption along the intestine of the common carp (*Cyprinus carpio*), Dabrowski (1986) observed that, except for methionine and histidine, the essential free amino acid concentration of digesta decreased significantly along the intestine, suggesting absorption occurs primarily in the proximal to the medial intestine. In a study evaluating the intestinal function of the ballan wrasse (*Labrus bergylta*), Le et al. (2019) demonstrated that macronutrients were digested and absorbed along the whole intestinal tract but mainly in the proximal portion. They also observed the apparent digestibility of crude protein to be comparable to many gastric species (ADC of 90% for ~50%–61% protein feed). From this, they concluded that stomach conditions of pepsin and low pH are not a definitive requirement for the efficient digestion of protein.

8.3.3 Absorption

Compared with mammals, knowledge of the detailed mechanisms of absorptive processes in fish has advanced slowly. Following digestion, smaller, solubilized nutrients released from ingested food are transported to the brush-border membrane of the enterocytes lining the intestine. These nutrients must be absorbed by the enterocytes and make their way to the basolateral membrane, where they will exit and enter into circulation. Transport into the enterocyte from the intestinal lumen occurs by simple diffusion, active transport, ion exchange, or endocytosis. The latter, occurring at all life stages in fish, likely plays a primary role in recycling the components of digestive secretions (e.g., enzymes, bile salts).

Like the higher substrate affinity observed among digestive enzymes in fish compared with mammals, nutrient transporters have higher affinity in fish intestine than in mammalian intestine. Higher transport affinity may be a function of temperature, lower nutrient content in the intestinal lumen, transporter density, and relative total intestinal surface area in fish (Karasov and Diamond, 1981; Ferraris et al., 1989; Collie and Ferraris, 1995). Not surprisingly, species, environmental differences, and trophic preferences affect nutrient absorption among fish. One commonality appears to be the capacity for active transport across the entire length of the intestine.

8.3.3.1 Protein absorption

While it is generally agreed that amino acids and small peptides liberated by digestion or ingested as components of the diet enter the intestinal enterocytes against concentration gradients coupled to the transport of inorganic ions, it has recently been suggested that dietary di/tripeptides are more efficiently absorbed, likely transported via one or more peptide transporters (PepT). Amino acid transport systems have been reviewed by Bakke et al. (2011).

In fish, at least four distinct Na^+-dependent amino acid transport systems have been identified (Storelli et al., 1989; reviewed by Bakke et al., 2011). These systems transport acid, base, and neutral amino acids, respectively, and are complemented by a system for transporting proline and N-methylated amino acids. Na^+-independent amino acid transport systems have also been identified for alanine, glycine, and lysine (Storelli et al., 1989). In addition to these systems, histidine transport in the rainbow trout (*Oncorhynchus mykiss*) intestine has been shown to be facilitated by a high affinity transporter (Glover and Wood, 2008). The absorption of free amino acids and small peptides appears to occur earlier in the intestine after a single meal than intact dietary proteins (Ambardekar et al., 2009). More recently, it was shown in trout that synchronization of amino acid absorption is important to optimize protein utilization (Brezas and Hardy, 2020). This may prove significant when feeding formulated feeds supplemented with free amino acids, especially in farmed fish fed only once per day.

Several authors have also reported the presence of a low-affinity/high capacity H^+-dependent peptide transporter (PepT1) and a high-affinity peptide transporter (PepT2) in fish (Thamotharan et al., 1996a,b; Maffia et al., 1997, 2003; Verri et al., 2000, 2003; Romano et al., 2006; Goncalves et al., 2007; Hakim et al., 2009; Sangaletti et al., 2009; Terova et al., 2009), with a focus on PepT1. PepT1 function is conserved in vertebrates, and it is highly expressed in the brush-border membrane of intestinal epithelium. Although functional studies are limited in fish, PepT1 gene characterization and expression have been reported for several fish species (reviewed by Wang et al., 2017). In their review, Wang et al. (2017) provide evidence of pH- and salinity-dependent regulation of intestinal PepT1 regulation. As in mammals, dietary protein source and concentration may affect PepT1 expression in fish (Frøystad-Saugen et al., 2009; Bakke et al., 2010; Rimoldi et al., 2015); however, contradictory results are also reported (Gonçalves et al., 2007; Amberg et al., 2008), suggesting potential differences among species while reinforcing a need to functionally characterize the peptide transporter (PepT) family. As many as five peptide transporters have been identified in other vertebrates (Wang et al., 2017).

Vacuolated enterocytes, similar to those observed in the ileum of neonatal mammals, have also been observed in larval fish (Ng et al., 2005; Wallace et al., 2005; Park et al., 2019) and appear to persist throughout development into adulthood in at least some fish (McLean and Ash, 1987; McLean and Donaldson, 1990). In their review of intestinal protein absorption in fish, Sire et al. (1992) describe the absorption of proteins by endocytosis followed by intracellular digestion attributed to lysosomal acidity in the vacuole system. More recently, Park et al. (2019) described specialized intestinal cells in larval zebrafish, which they referred to as "lysosome-rich enterocytes (LREs)." These LREs were observed to internalize dietary proteins by receptor-mediated endocytosis.

8.3.3.2 Lipid absorption

Intestinal lipid absorption in fish is thought to follow mechanisms that have been well described in mammals (Iqbal and Hussain, 2009); however, direct knowledge of lipid absorption in the fish intestine and circulatory transport is limited. Bakke et al. (2011) reported that released micelle contents, including monoglycerides and fatty acids, are presumed to be taken up into the intestinal epithelial cells of the proximal intestine and pyloric caeca by diffusion or facilitated transport, the latter likely involving intestinal fatty acid–binding proteins (I-FABPs) (Andre et al., 2000; Concha et al., 2002). Characterization of I-FABPs in fish is limited mostly to zebrafish (Andre et al., 2000; Pierce et al., 2000; Sharma et al., 2004) and carp (Concha et al., 2002; Chen et al., 2012), with more recent interest in changes in gene expression during diet-induced intestinal enteritis in salmonids (Venold et al., 2012, 2013). While present and presumably affected by intestinal inflammation in salmonids, the precise function of I-FABP in fatty acid transport has not been well characterized. Regardless of the transport mechanism, it appears that microbes in the brush-border membrane play an important role in facilitating fatty acid uptake and lipid droplet formation in the intestinal epithelium (Semova et al., 2012). Several potential mechanisms were proposed by Semova et al. (2012), but once again, functional studies are needed to understand this relationship. Once in the enterocytes, lipids accumulate in the supranuclear space where they are reesterified, predominantly by glycerol-3-phosphate, into glycerophospholipids and triacylglycerides (Babin and Vernier, 1989). Triacylglycerides are packed into chylomicrons and presumed to exit the enterocyte by exocytosis (Rogie and Skinner, 1985; Hernandez-Blazquez et al., 2006). The direct passage of unesterified fatty acids from the intestinal lumen into the circulatory system has also been described in fish (Robinson and Mead, 1973; Kayama and Iijima, 1976).

8.3.3.3 Carbohydrate absorption

Absorption of carbohydrates occurs by active transport mechanisms and can take place against considerable concentration gradients. The sodium-dependent glucose cotransporters (SGLTs) are responsible for the transport of D-glucose across the brush-border membrane in fish, simultaneously transporting sodium. First identified in mammals, the SGLTs are generally characterized as either high-affinity (Ha)/low-capacity (Lc) transporters or low-affinity (La)/high-capacity (Hc) transporters. The SGLT1 transporter, functioning similarly to the same Ha/Lc transporter as in mammals, has been identified in several species of fish (Storelli et al., 1986; Buddington et al., 1987; Reshkin and Ahearn, 1987a,b; Ahearn et al., 1992; Collie and Ferraris, 1995; Maffia et al., 1996; Soengas and Moon, 1998; Sala-Rabanal et al., 2004; Geurden et al., 2007; Goncalves et al., 2007). More recently, expression of SGLT2 transcripts was identified in trout and characterized as a La/Hc transporter, functioning similarly to the mammalian counterpart (Subramaniam et al., 2019). In the same study, the researchers identified a single Ha/Hc Na^+-dependent glucose transport system along the entirety of the tilapia intestinal tract and a low capacity, three-kinetic heterogeneous system of Na^+-dependent glucose transport in trout, with trout showing different transport kinetics in the pyloric caeca, midgut, and hindgut intestinal segments. Glucose transporter 2 (GLUT2) transcripts have also been reported in the intestine of certain fish (Krasnov et al., 2001; Panserat et al., 2001; Hall et al., 2006). In mammals, it is generally stated that SGLTs transport glucose from the intestinal lumen into the enterocytes, whereas GLUT2 transports glucose from cytosol to the blood. This is also assumed to be the case in fish, although the precise location and characteristics of GLUT2 within the enterocytes have yet to be investigated.

8.3.3.4 Vitamin absorption

As reviewed by Bakke et al. (2011), very little information exists on the absorption of vitamins in the fish intestine. Of the fat-soluble vitamins, only retinol (vitamin A) absorption has been described. In a study using hybrid striped bass (*Morone chrysops* × *Morone saxatilis*) proximal intestinal tissue, in vitro absorption was found to be by a saturable, transporter-mediated process (Buddington et al., 2002), much like in mammals. For many of the water-soluble vitamins studied, transport-mediated processes have been suggested, while for nicotinamide, a nonsaturable process has been proposed (Vilella et al., 1989; Rose and Choi, 1990; Buddington et al., 1993; Casirola et al., 1995; Casirola and Ferraris, 1997). Folic acid in catfish, however, appeared to be absorbed by facilitated diffusion (Casirola et al., 1995).

8.3.3.5 *Mineral absorption*

Minerals play critical roles in many metabolic systems and are important structural components of the skeletal and dermal (i.e., scales) systems in fish. However, there is a paucity of information on dietary absorption, in part due to the ability of fish to efficiently absorb several minerals across gill epithelium from the aquatic environment in which they live. As such, much of what we know about mineral physiology comes from research of the osmoregulatory systems in fish. Chloride, potassium, sodium, and calcium are readily absorbed from the water across gill epithelia; however, calcium requirements are high, and dietary contributions are often necessary. As such, the macrominerals, calcium, phosphorus, and magnesium, and a handful of microminerals are discussed here.

8.3.3.5.1 Calcium

The calcium requirement of fish is primarily met by uptake across the gill epithelium (Flik and Verbost, 1993), with some calcium transport occurring across the skin (Marshall et al., 1992; McCormick et al., 1992), and intestinal absorption is likely playing a small role (reviewed by Hossain and Yoshimatsu, 2014). In their review, Bakke et al. (2011) describe several types of putative calcium transporters and channels: a P2 purinoceptor-responsive transporter operating well below saturation (Klaren et al., 1997), an L-type calcium channel (Larsson et al., 1998), and an epithelial calcium channel (ECaC) (Qiu and Hogstrand, 2004; Pan et al., 2005). Generally, it is believed that in fish, dietary intestinal calcium transport is under primary control of stanniocalcin, a hypocalcemic hormone released from the corpuscles of Stannius (Flik and Verbost, 1993). In fact, several endocrine factors have been implicated in the control of intestinal calcium uptake also in fish. Stanniocalcin (Sundell et al., 1992), vitamin D (Larsson et al., 2002), parathyroid hormone-like protein (Fuentes et al., 2006), vitamin A in concert with vitamin D (Ørnsrud et al., 2009), and prolactin (Wongdee and Charoenphandhu, 2013) have all been described as having regulatory effects on intestinal calcium absorption. However, a clear mechanistic understanding of intestinal calcium uptake in fish is missing, and species differences likely exist (Flik and Verbost, 1993).

8.3.3.5.2 Phosphorus

Unlike calcium, the phosphorus requirement of fish is mainly provided through the diet because of low concentrations and low absorption rates in natural waters. Consequently, intestinal absorption is the major route of uptake and has been well characterized. Phosphorus is mainly absorbed in the proximal intestine and appears to rely on the function of a type II sodium phosphate cotransporter (NaP_i-II; Scl34) in trout (Sugiura et al., 2003). The NaP_i-II cotransporter was first cloned in winter flounder kidney (Werner et al., 1994) and later observed on the apical surface of flounder enterocytes (Kohl et al., 1996). Furthermore, NaP_i-II is also located on the apical and basolateral side of renal tubular cells

(Elger et al., 1998), supporting kidney excretion and reabsorption as crucial to phosphorus homeostasis in fish.

In more recent years, NaP_i-II expression has been found in the pyloric caeca and intestine of rainbow trout, correlating with areas of high phosphorus uptake (Avila et al., 2000; Sugiura and Ferraris, 2004). Interestingly, transporter activity appears to differ in the two sections. The intestinal form of NaP_i-II responds to dietary phosphorus content, but the pyloric caeca form appears to be a high affinity, low-capacity transporter (Sugiura and Ferraris, 2004). Zebrafish also express high- and low-affinity forms, Na-Pi-II transporters (Nalbant et al., 1999; Graham et al., 2003). The functional characterization of the zebrafish NaP_i-II system shows considerable differences between the two forms with regard to intracellular regulation (Nalbant et al., 1999). Vitamin D_3 likely plays a role in intestinal uptake and phosphorus homeostasis in fish (Sugiura and Ferraris, 2004), and in yellow catfish (*Pelteobagrus fulvidraco*), Chen et al. (2016) demonstrated vitamin D_3−regulated expression of slc34 genes isoforms encoding NaP_i-II transporters. The two primary transcript isoforms identified in zebrafish (Slc34a2a and Scl34a2b) appear to encode NaP_i-II cotransporters with different affinities and capacity for phosphorus, NaP_i-IIb1 (low affinity/high capacity) and NaP_i-IIb2 high affinity/low capacity (reviewed by Verri and Werner, 2019).

8.3.3.5.3 Magnesium

Much of what is known regarding magnesium transport in fish comes from early work conducted with tilapia (*Oreochromis mossambicus*) and was summarized by Bijvelds et al. (1998). From their observation, a hypothetical 2 Na^+/Mg^{2+}exchanger (NME) localized to the basolateral membrane was proposed to play a role in intestinal transport (Van der Velden et al., 1990; Bijvelds et al., 1998). More recently, solute carrier 41a1 (SLC41a1), previously identified in the kidney of torafugu (*Takifugu rubripes*) and involved in magnesium homeostasis (Islam et al., 2013), was identified in goldfish (*Carassius auratus*) intestine and was responsive to dietary and environmental magnesium manipulations (Kodzhahinchev et al., 2017).

8.3.3.5.4 Microminerals

Except for copper and zinc, there is insufficient information on the intestinal absorption of microminerals; what little is known has been summarized by Bakke et al. (2011). In their review, Bakke et al. (2011) report evidence of a sodium−iodine symporter in the gut of larval fish (Moren et al., 2008), the presence of a divalent metal ion transporter in fish similar to the divalent metal transporter-1 (DMT-1) in mammalian intestines (Conrad and Umbreit, 2002), and basolateral transfer of iron by ferroportin in zebrafish (Donovan et al., 2000).

Copper absorption across the fish gut has been well studied in trout. Nadella et al. (2006) observed that the concentration of dissolved Cu was several-fold higher in stomach digesta compared with chyme sampled from various parts of the intestine. Their results support the stomach and medial intestine as the two major sites of net Cu absorption. Further studies examined the mechanisms of intestinal transport, leading them to propose novel Na^+/H^+ exchange (NHE) mechanisms at the apical membrane, which subsequently acidify the apical microenvironment for the effective transport of Cu via metal-specific carriers such as copper proton symporters DMT-1 and/or copper transporter-1 (CTR1) on the apical membrane and Cu-ATPase on the basolateral membrane (Nadella et al., 2006, 2007). More recently, Nadella et al. (2011) concluded that while Cu uptake is linked to Na^+ gradients in the intestine, probably moving through DMT1, this is not the case in the stomach. In their study, they identified and characterized CTR1 transcript in the trout stomach and concluded Cu is most probably transported by CTR1 in the stomach.

Intestinal uptake of Zn in fish is also believed to involve solute carrier families Slc30 (ZNT) and Slc39 (ZIP) and potentially L-type calcium channel (LTCC) and divalent metal transporter-1 (DMT1) (Glover and Hogstrand, 2003; Feeney et al., 2005). In the intestine, dietary Zn binds to the mucus of the intestinal epithelium and is transported into the epithelial cells either as the Zn^{2+} ion or bound to amino acids (Bury et al., 2003). The relative efficiency and functional importance of ZIP-mediated Zn^{2+} uptake versus amino acid–facilitated Zn uptake in GI tract remain to be characterized. Investigation of intestinal ZIP genes in zebrafish intestine showed ZIP3, ZIP7 and ZIP8 were downregulated in response to higher aquatic and food zinc concentrations, while putative ZnT4 and ZnT5 transporters were upregulated in response to increased levels of zinc. Physiological studies on Zn uptake in fish using in vivo perfusion (Glover and Hogstrand, 2002) and in vitro brush border vesicle models (Glover et al., 2003) suggest that amino acids with high binding affinity for Zn^{2+} can improve Zn uptake.

8.4 Chemical and cellular signaling

The GI tract is arguably the largest chemical signaling system in the body. In addition to the sensory and digestive signaling mechanisms discussed already, the GI tract and associated tissues support a robust immune system, an interactive microbial community, and neurological and hormonal signaling pathways. These systems are complex and not fully elucidated in any vertebrate. Furthermore, they are interactive and often interdependent. As physiologists work to better understand the complex signaling of the GI tract, nutritionists work to manipulate these systems through nutritive and nonnutritive supplements, pushing fish nutrition toward the development of "functional" (a.k.a. "performance" or "precision") feeds that target improved nutrient utilization and enhanced growth and robustness, as discussed in Chapter 11, Nutritional supplements and fish health.

8.4.1 Gut-associated lymphoid tissues

Like other vertebrates, the gut of fish is lined with mucus, which is secreted by goblet cells. The mucosal layer is a critical line of defense against environmental threats and creates not only a physical barrier, but chemical and cellular barriers as well. Mucus in the gut is much like that of other mucosal tissues, but with added roles in digestion and nutrient absorption. Although said to be less developed than in mammals (Montalban-Arques et al., 2015), intestinal immunity in fish relies on both innate and adaptive responses to threats. Mucosal-associated lymphoid tissue (MALT) plays an important function in immunity. A competent of the MALT, the GALT, in concert with chemical compounds found in the mucus, functions to prevent and remove pathogens from the gut.

In addition to producing mucus, goblet cells of the MALT have been observed to secrete cytokines, chemokines, and antimicrobial peptides, said to demonstrate functions in innate immunity beyond barrier maintenance (Knoop and Newberry, 2018). Specifically in fish, lectins (Itami et al., 1993), complement (Dalmo et al., 1997), proteases (Aranishi and Mano, 2000), cytokines (Linderstrøm et al., 2003), lysozyme (Fernandes et al., 2004), and other chemical substances associated with innate immune function have been identified (reviewed by Cain and Swan, 2010). Like the other MALTs, the GALT includes diffuse B and T cells that respond to mucosal infection; however, the GALT is said to demonstrate immune tolerance to avoid constant immune responses against nonharmful antigens present from consumed substances (i.e., food and water) or microbiota (Salinas, 2015). For more information on gut immune function and the mucosal immune system in fishes, see Cain and Swan (2010), Salinas (2015), Salinas and Magadan (2017), Cabillon and Lazado (2019), and Dawood (2020).

Gut immunity is important to overall fish performance and health, impacting feed digestibility and absorption, osmoregulation, pathogen susceptibility, and the gut microbiome. The investigation and the use of functional feed additives in aquafeeds are rapidly evolving. While some of these additives target enhanced nutrient utilization (e.g., exogenous enzymes), a substantial focus is on improving intestinal health, pathogen resistance, and stress tolerance. Additives, such as prebiotics, probiotics, and immune stimulants, target the GALT and commensal microbial communities to enhance the performance of the fish.

8.4.2 Gut microbiome

Early studies on the microbial communities of teleost intestines were conducted using culture-dependent techniques and often were unsuccessful in identifying even a single culturable microbe (Margolis, 1953; Obst, 1919). Such results led to a prolonged disregard of commensal gut microbes in teleost fish, assuming them to be of little relevance to the physiology of fish. A resurgence in interest came about in the 1970s, and with better culture methods, researchers not only began identifying culturable bacteria in the fish gut but also began characterizing the influence that host and environmental factors have on gut microbial composition (Yoshimizu and Kimura, 1976). More recently, the use of culture-independent molecular techniques, including next-generation sequencing and advanced bioinformatics, has substantially advanced our knowledge of the gut microbiome, as it is now termed, and its interactions with the environment and physiology of the fish host. The gut microbiome of more than 145 teleost species, including predominate aquaculture species, model species (i.e., zebrafish), and numerous wild species (marine and freshwater), have now been characterized using next-generation sequencing, with important sources of variation identified under select conditions (Perry et al., 2020).

Our understanding of the influence of the microbiome on nutritional physiology of fish is certainly not complete, though it is now clear that the gut microbiota has a significant impact on nutrient digestion, absorption, and metabolism in fish. In terms of nutrient breakdown, fish gut bacteria are known to produce digestive enzymes that are often useful to the host, such as amylases (Sugita et al., 1996), cellulases (Ni et al., 2014), proteases, lipases, phytases, tannases (i.e., tannin acyl hydrolase), xylanases, and chitinases (Ray et al., 2012). Depending on the fish species and its diet, the microbiota may provide novel digestive capabilities to the host. Genomic comparisons of the grass carp and its intestinal bacteria showed that microbiota provided cellulases, xylanases, pectinases, and peptidase, which are not present in the host genome, despite being crucial to the digestion of the host's herbivorous diet (Hao et al., 2017). Along with these digestive enzymes, certain gut microbes can anaerobically ferment nondigestible carbohydrates present in many fish diets and in turn produce short-chain fatty acids (SCFAs). These SCFAs not only serve as an energy source to the host, particularly the intestinal enterocytes but are also known to play a role in enteroendocrine signaling involved in cellular proliferation, inflammation, and metabolism (Piazzon et al., 2017; Butt and Volkoff, 2019). Vitamin production is yet another of the nutritional attributes offered to finfish by their microbiota, as many bacteria commonly detected in the fish gut are capable of synthesizing cobalamin (vitamin B_{12}) and other vitamin precursors (LeBlanc et al., 2013; Sugita et al., 1991).

Gut microbes not only impact the intralumenal digestion and biochemical conversion of nutrients but also have the capacity to affect nutrient absorption and metabolism, directly and indirectly, at the local and systemic levels. Specific groups of gut bacteria have been found to increase absorption and accumulation of lipids in teleost enterocytes, resulting in downstream increases in extraintestinal accumulation and metabolism of dietary fatty acids in fish (Semova et al., 2012; Sheng et al., 2018). The gut microbiota also has been implicated in indirect modulation of host systemic metabolism. One example is through the production of specific bacterial proteins that stimulate proliferation of insulinogenic beta cells in the pancreatic islets of the host, which are required for proper maintenance of glucose homeostasis (Hill et al., 2016). Additional research also suggests that microbiota have the potential to modulate systemic regulation of fish appetite and energetics, both peripherally by influencing intestinal serotonin levels and centrally through effects on circulating metabolites and direct vagal nerve stimulation (Volkoff, 2019).

Gut health and the nutritional functions of fish gut microbiota are of particular importance, as commercial aquaculture looks toward increasing the use of alternative dietary ingredients to increase economic and environmental sustainability. Some alternative ingredients, particularly plant-based ingredients, show reduced digestibility or may be immunogenic in certain fish species due to the presence of antigens and antinutritional factors (ANFs) (Krogdahl et al., 2010). This has led to an interest in better understanding diet–microbiota interactions within teleost species with the aim of overcoming any negative impacts of alternative dietary ingredients. Microbial diversity in the gut has been repeatedly shown to be correlated with the level of host herbivory (Li et al., 2018), and the ability of dietary changes to shift the gut microbiota composition is well documented (Desai et al., 2012; Schmidt et al., 2016; Lyons et al., 2017; Antonopoulou et al., 2019). While many studies have described how these alternative dietary formulations alter gut microbiota, recent efforts also show promise in using gut microbiota manipulations to increase nutrient digestibility and reduce the negative impacts of alternative dietary ingredients (Piazzon et al., 2017; Santos et al., 2020). Together these studies highlight the bidirectional relationship among host-nutrition and the gut microbiota, as each can have a positive or negative influence on the other.

Pre- and probiotic supplementation is the most common method of attempting to modulate the gut microbiota toward improved nutrient metabolism, growth, and gut health. Prebiotics are dietary compounds that are indigestible to the host but serve to stimulate the proliferation of certain members of the gut microbiota that produce beneficial outcomes in the host, while probiotic supplementation involves the direct dietary inoculation of live microorganisms with desirable attributes to the fish gut (Caipang and Lazado, 2015). A combination of these two products, often termed synbiotic supplementation, is also a common strategy aimed at improving the nutritional utility of gut microbiota. Such supplements have been shown to improve overall growth rates; increase lipase, protease, and carbohydrate activity; and improve apparent lipid, protein, and carbohydrate digestibility under certain situations (Hoseinifar et al., 2017). However, each of these categories of supplements encompasses a broad range of products that produce varied outcomes depending on fish species and environment, and in many cases, results from studies on pre-, pro-, and synbiotics have contradictory or inconclusive results. The inconsistency of results from these functional dietary supplements suggests that host–diet–microbiota interactions are indeed species and environment specific, and a better understanding of these dynamics is still needed to achieve reliable improvements to the nutritional physiology of teleost fish through gut microbiota manipulation.

Although the study of the functional contributions of fish gut microbial communities is still somewhat in its infancy, the field is rapidly expanding, and future prospects from this knowledge should allow for improved gut health, nutrient assimilation, and precision feed formulation for cultured species. As a quickly evolving field, many aspects related to the gut microbiome are beyond the scope of this chapter, and the reader is referred to the following reviews for more detail: Hoseinifar et al. (2017), Legrand et al. (2020), Perry et al. (2020), and Yukgehnaish et al. (2020).

8.4.3 Enteric nervous system

Much of what we know regarding the enteric nervous system has been learned from pharmacological studies mainly conducted with Atlantic cod (*Gadus morhua*) and rainbow trout, which have been reviewed by Olsson and Holmgren (2011). In many regards, the enteric nervous system of teleost fish has a similar arrangement to other vertebrates (Olsson, 2009). Part of the autonomic nervous system, the enteric nervous system can act independently from the central nervous system or the rest of the autonomic nervous system (Olsson, 2009). Even so, extrinsic innervation of the GI tract still plays important roles in coordinating gut activity and relaying information such as hunger and pain (Olsson, 2009). The vagal nervous system provides the primary extrinsic innervation of the GI tract, innervating the stomach and proximal intestine of gastric fish (Burnstock, 1959; Nilsson, 1983) and the entire tract in some stomachless fish (Olsson, 2009). Spinal nerves innervate the distal intestine and rectum in gastric species (Burnstock, 1959; Nilsson, 1983).

Nerve cells of the enteric nervous system can be found spread over the myenteric plexus, located entirely within the gut wall (Olsson, 2009; Olsson and Holmgren, 2011). Reflex pathways of different nerve types respond to chemical (presence of nutrients) and mechanical (stretching of the gut wall) stimuli. In doing so, sensory nerves regulate excitatory motor nerves, which may target smooth muscle, blood vessels, or glands (Olsson, 2009). Each of these reflex actions affects the efficiency of digestion and absorption. Especially important is the peristaltic reflex, which controls gut motility.

Comprehensive studies of the mechanisms controlling gut motility in cod, trout, and zebrafish have been reviewed by Olsson and Holmgren (2011). From those studies, gut contractions in fish appear to be regulated similarly to other vertebrates. Acetylcholine and tachykinins stimulate muscle contraction, while VIP/PACAP and nitric oxide are inhibitory (Olsson and Holmgren, 2011). The exact mechanisms of action, however, remain unclear. In their review, Olsson and Holmgren (2011) suggest serotonergic interneurons may stimulate cholinergic motor neurons, with tachykinins possibly having direct or indirect effects, while interneurons stimulating VIP/PACAP or nitric oxide might release serotonin and/or acetylcholine. In addition to VIP/PACAP, other neuropeptide hormones of GEP system (e.g., GRP; Holmgren and Jönsson, 1988) likely play roles in gut motility.

8.4.4 Gastroenteropancreatic endocrine system

The GEP endocrine system comprises the chemical messenger–secreting cells of the stomach, intestine, and endocrine pancreas. Although correctly referred as the enteropancreatic endocrine system in stomachless fish species, for simplicity, the term GEP is broadly applied for all fish. Secretory neuroendocrine cells of the GEP are diffusely arranged throughout the gut mucosa and found as aggregates of cells within the pancreatic islets. Within the gut mucosa, the GEP cells commonly span the full epithelial layer, acquiring sensory input from the gut lumen while anchored to the basement membrane where chemical messenger (neurohormone) secretion occurs (Takei and Loretz, 2010).

Hormones of the GEP are numerous and play important roles in appetite regulation, enzyme secretion, gall bladder contraction, metabolism, gut motility, and osmoregulation (Takei and Loretz, 2010). As described briefly earlier, GEP hormones are central in the coordination of foraging and feeding in fish (Smith, 1999). Once ingested, sensory signals initiate a network of responses stimulating digestion, metabolic processes, and appetite regulation. Food entering the stomach and intestine interacts with GEP cells in the mucosa that in turn secrete hormones that (1) coordinate the breakdown of food, (2) stimulate the release of other GEP hormones, and (3) circulate through the bloodstream to act on other target organs and physiological processes (Fig. 8.8, Nelson and Sheridan, 2006). In an effort to

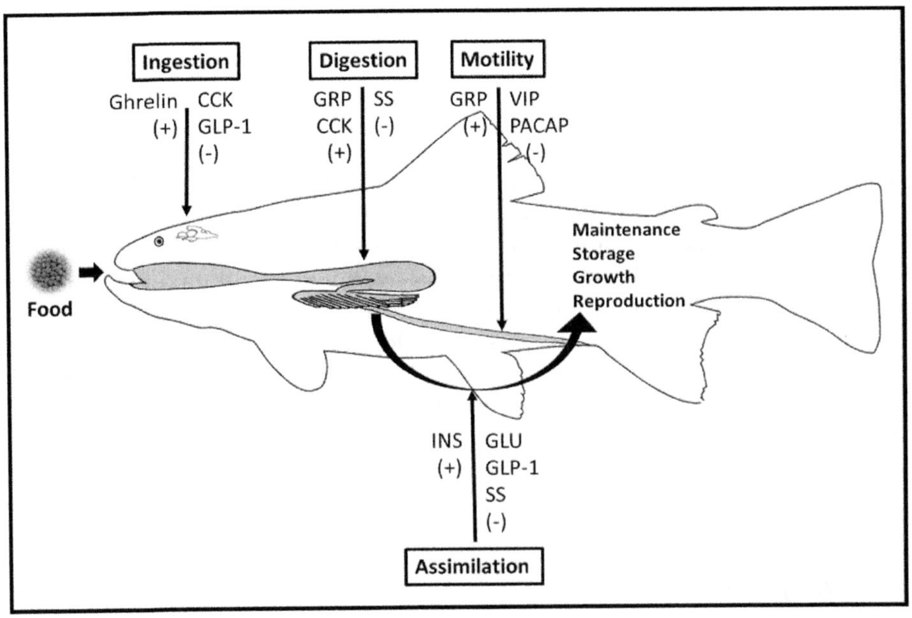

FIGURE 8.8 Simplified diagram of known gastroenteropancreatic (GEP) hormone action on the nutritional physiology of fish. Hormones in a column either stimulate (+) or inhibit (−) the labeled function. *CCK*, cholecystokinin; *GLP-1*, glucagon-like peptide-1; *GLU*, glucagon; *GRP*, gastrin-releasing peptide; *INS*, insulin; *PACAP*, pituitary adenylate cyclase-activating peptide; *SS*, somatostatins; *VIP*, vasoactive intestinal peptide. *Credit: Brian Small.*

efficiently provide an overview of these systems in the context of fish nutrition, this discussion will focus on major GEP hormone families, rather than particular processes (e.g., gut motility), following the approach of Nelson and Sheridan (2006). For more systematic reviews, see Takei and Loretz (2010), Jönsson (2013), Volkoff (2016), Rønnestad et al. (2017), and Bertucci et al. (2019). Several GEP hormones are primarily involved in osmoregulation and will not be discussed here. The reader is referred to the book *Fish Osmoregulation* edited by Baldisserotto et al. (2007) and the review by Takei and Loretz (2010) to better understand the role of the gut in osmoregulation.

8.4.4.1 Gastrin family

Gastrin and cholecystokinin (CCK) are structurally related peptides with a common C-terminal tetrapeptide sequence. CCK/gastrin-like immunoreactivity has been shown in the gut and nervous system of several fish species (Volkoff et al., 2005); however, the existence of gastrin in fish has not been demonstrated (Jensen et al., 2001). Research in trout supports a similar role for CCK in fish as in mammals (Jönsson et al., 2006). CCK has been demonstrated to stimulate gall bladder contraction and slow gastric emptying in fish (Olsson et al., 1999), as well as significantly decrease food intake (Volkoff et al., 2005). In trout, CCK was found in the pyloric caeca and was released into circulation following food intake (Jönsson et al., 2006). In the same study, plasma CCK levels were not affected by the amount of food consumed, but rather by the ratio of dietary protein to lipid content, with higher lipid content eliciting an increase in CCK release and peaking 4 and 6 h postprandial. In Atlantic cod, CCK inhibited gastric secretion (Holstein, 1983), and in the stomachless killifish, CCK stimulated lipase activity (Honkanen et al., 1988).

8.4.4.2 Bombesin/gastrin-releasing peptide family

The bombesin/gastrin-releasing peptide family of neuropeptides includes bombesin, gastrin-releasing peptide (GRP), and neuromedin B (NMB). Several immunohistochemical studies have established that fish GI tissues contain peptides with gastrin-releasing peptide (GRP)/bombesin-like immunoreactivity (Bjenning and Holmgren, 1988), and GRP sequences are available for goldfish (Volkoff and Peter, 2000) and zebrafish (Chen et al., 2017). Bombesins arise from independent genes (Nagalla et al., 1996), while the other BBS/GRP-like peptides are encoded by single genes and posttranscriptionally modified (Nagalla et al., 1994; Spindel, 1986, 1990). In mammals, GRP reduces food intake, increases in plasma gastrin and gastric acid secretion, and causes the secretion of various digestive enzymes (Gajjar and Patel, 2017); however, the roles of GRP in fish are less clear. In studies with cod, exogenous injections of bombesin and GRP stimulate acid gastric secretion and motility (Holstein and Humphrey, 1980; Holmgren and Jönsson, 1988). Exogenous injections of GRP/bombesin also inhibit food intake in goldfish (Himick and Peter, 1994) and stimulate growth hormone (GH) secretion at least in part by suppressing somatostatin production (Canosa and Peter, 2004). Mechanistic studies are needed to better understand the role GRP plays in fish gut function.

8.4.4.3 Secretin family

The secretin family of neuropeptides is comprised of secretin, glucagon-like peptide (GLP), gastric inhibitory peptide (GIP), enteroglucagon, vasoactive intestinal peptide (VIP), growth hormone–releasing hormone (GHRH), pituitary adenylate cyclase-activating peptide (PACAP), and oxyntomodulin. Little is known about the function of most of these hormones in fish. Generally, this family of hormones exerts effects of a wide range of metabolic, appetite, and growth-related processes. In teleosts, secretin has not yet been isolated, or the gene identified. Glucagon activity and gene function have been best studied in fish, while some information exists for GIP, VIP, and PACAP. The glucagon genes in fish are expressed in the endocrine pancreas and intestine (Plisetskaya and Mommsen, 1996). As in mammals, fish glucagon antagonizes the actions of insulin by promoting cellular glucose release and increasing plasma glucose concentrations (Plisetskaya and Mommsen, 1996). Glucagon also has been shown to mobilize lipids and increase plasma fatty acid levels in fish (Harmon and Sheridan, 1992; Sheridan, 1994). Fish GLP-1 has been shown to promote glycogenolysis and gluconeogenesis, while also inhibiting feed intake (Silverstein et al., 2001).

Less is known regarding the function of the other members of this family in fish. Known to stimulate insulin production in mammals, gastric inhibitory peptide (GIP)–immunoreactive cells have been identified in both the intestinal tract and pancreatic islets (Pan et al., 2000; Wagner and McKeown, 1981). Various functions have been shown for VIP, PACAP, secretin, and oxyntomodulin, including the inhibition of gut motility (VIP, PACAP), pancreatic stimulation in coordination with CCK (secretin), and glucagon-like actions (oxyntomodulin) (Plisetskaya and Mommsen, 1996; Pollock et al., 1988; Olsson and Holmgren, 2001, 2011; Wong et al., 2000).

8.4.4.4 *Insulin family*

A rather well-characterized family in fish, the insulin family of neuropeptides, includes insulin and the insulin-like growth factors (IGF-I, IGF-II). Important growth and metabolic regulators, the IGFs, primarily produced in hepatic and muscle cells, have been well studied and reviewed in fish (see Reinecke et al., 2006). As in mammals, insulin has been found in the pancreatic islets and promotes a metabolic shift from catabolism to anabolism (Nelson and Sheridan, 2006). Generally, insulin treatment in fish stimulates glycogen synthesis, activates lipogenesis, and promotes protein synthesis and fish growth (Planas et al., 2000; Plisetskaya and Dunguay, 1993). Stimulation of insulin secretion and responsiveness to increases in circulating insulin varies greatly among fish species, with insulin receptor number, nutritional history, and foraging strategy all likely serving as mediators (Nelson and Sheridan, 2006).

8.4.4.5 *Somatostatin family*

Somatostatins are important inhibitors of catabolic processes and feed intake in fish (Sheridan and Kittilson, 2004; Volkoff et al., 2005). Produced in the pancreatic islets, throughout the GI tract, and other body tissues, somatostatins are often colocalized with CCK and NPY (Patel, 1999; Nelson and Sheridan, 2005). Somatostatins directly and indirectly alter growth. Reviewed by Klein and Sheridan (2008), somatostatins inhibit the release of GH, decrease hepatic GH sensitivity, lower plasma IGF-I levels, stimulate the breakdown of energy stores, and influence digestion, food intake, nutrient absorption, and food conversion both directly and through the modulation of other hormonal systems.

8.4.4.6 *Ghrelin*

Ghrelin is the only known GI tract orexigenic factor in mammals (Higgins et al., 2007) and was first identified in fish as a potent stimulator of GH in fish (Sheperd et al., 2000; Kaiya et al., 2005) and later as a stimulator of feed intake (Unniappan et al., 2002). Fish ghrelin expression is highest in the stomach (Unniappan et al., 2002; Kaiya et al., 2003a,b,c; Kaiya et al., 2005). Ghrelin receptors are expressed highest in the pituitary and pancreatic islets in channel catfish, suggesting, beyond its role in pituitary GH stimulation, that ghrelin may play a role in the modulation of pancreatic hormones (Small et al., 2009). Discovery of a novel ghrelin receptor, highly expressed throughout the brain of channel catfish, suggests receptor-mediated control of ghrelin's activities on appetite stimulation, GH release, and putative pancreatic function; however, mechanistic studies have not been conducted (Small et al., 2019).

8.4.4.7 Other gastroenteropancreatic hormones

As shown in Table 8.1, there are a multitude of hormones produced by GEP system. A better understanding of the mechanisms of how food intake, nutrient status, life history, and so

TABLE 8.1 Hormones of the gastroenteropancreatic system found among fishes as described by Takei and Loretz (2010).

GI tract	Endocrine pancreas
Ghrelin	Insulin
Motilin	Glucagon
Cholecystokinin	Glucagon-like peptide
Gastrin	Somatostatins
Insulin	Peptide tyrosine-tyrosine
Galanin	Pancreatic polypeptide
Galanin-like peptide	Peptide Y
Glucagon	Amylin
Glucagon-like peptide	
Glucagon-like peptide	
Glucose-dependent insulinotropic peptide	
Vasoactive intestinal polypeptide	
Pituitary adenylate cyclase-activating polypeptide	
Somatostatins	
Peptide Y	
Guanylin	
Uroguanylin	
Natriuretic peptides	
Calcitonin gene-related peptide	
Adrenomedullin 2/intermedin	
Adrenomedullin 5	
Tachykinins	
Neuromedin U	
Gastrin-releasing peptide	

on affects these hormones and, in return, how these hormones affect fish feeding, digestion, metabolism, and growth is important to the improvement of feed utilization and the development of precision feeds and feeding regimes for cultured species. There is still much to learn about chemical signaling within and between the gut and associated organs.

8.5 Impacts of terrestrial plant-based ingredients on fish physiology

Over the past decade, there has been a tremendous effort to replace fishmeal and fish oil in the diets of cultured fish due to perceptions of unsustainability and the high costs of these ingredients. As a result, plant protein meals have been a major focus, with soybean meal and its derivatives (e.g., soy protein concentrate) leading the way. Soybean meal is not only the leading oilseed crop produced globally, but it is also economical and nutritious. Soybean meal contains an average of 48% crude protein and has an amino acid profile suitable for fish meal replacement, although often requiring supplementation with feed-grade methionine and sometimes lysine and threonine, depending on the species' requirements and diet formulation. Numerous other plant protein feedstuffs have been evaluated across several species, each with certain limitations.

Trophic preferences and alimentary tract differences have a substantial impact on the capacity of a species to digest and absorb the nutrients from a given feedstuff, as discussed earlier. However, beyond these basic differences relating to fish morphology and digestive function, all plant-derived proteins contain antinutritional factors or antigenic compounds. While many antinutritional factors are at low enough concentrations in the complete diet to be unlikely to affect fish performance, others can have significant impacts on fish physiology and performance. Some of the common antinutritional factors found in plant proteins and their effects on the nutritional physiology of fish are discussed in the following; however, for more thorough reviews, see Francis et al. (2001) and Chapter 12, Adventitious toxins and antinutrients.

8.5.1 Antinutritional factors

ANFs are substances in feed ingredients that interfere with food utilization and affect fish health and production performance, by themselves or through their metabolic products (Makkar, 1993). Numerous ANFs have been reported for commonly used plant protein feedstuffs. Soybean meal contains protease inhibitors, lectins, phytic acid, saponins, phytoestrogens, antivitamins, and allergens. While many of the same ANFs are found in other plant protein sources, glucosinolates (rapeseed meal, mustard oil cake), tannins (pea seed meal, mustard oil cake), cyanogens (pea seed meal), alkaloids (lupin seed meal), arginase inhibitor (sunflower oil cake), gossypol (cottonseed meal), cyclopropenoic acid (cottonseed meal), and mimosine (Leucaena leaf meal) may also be present (Francis et al., 2001). Broad classifications of ANFs proposed by Francis et al. (2001) include (1) factors affecting protein utilization and digestion, such as protease inhibitors, tannins, and lectins; (2) factors affecting mineral utilization, which include phytates, gossypol, and glucosinolates; (3) antivitamins; and (4) miscellaneous substances such as mycotoxins, mimosine, cyanogens, alkaloids, phytoestrogens, and saponins. Across the groupings, several ANFs irritate the intestinal epithelia or otherwise cause excess mucus production, thus impairing the function of the intestine and altering cellular and chemical signaling.

Fortunately, many of the ANFs found in terrestrial plant protein sources can be removed or reduced by heat treatment or further processing. Rumsey (1993) suggested categorizing ANFs into heat-labile and heat-stable groups. Heat-labile ANFs include protease inhibitors, phytates, lectins, goitrogens, and antivitamins, whereas heat-stable ANFs include saponins, nonstarch polysaccharides, antigenic proteins (allergens), phytoestrogens, and some phenolic compounds. Among the heat-stable ANFs, saponins and antigenic proteins are among the most problematic for fish, especially salmonids and have been associated with distal enteritis and digestive dysfunction.

8.5.2 Distal enteritis and digestive disfunction

Distal enteritis was first observed when feeding soybean meal to Atlantic salmon (*Salmo salar*; Baeverfjord and Krogdahl, 1996) and has since been well studied, albeit precise molecular mechanisms remain largely unknown. Soybean meal–induced enteritis (SBMIE) in salmonids can be defined as a shortening and widening of the distal intestinal mucosal folds, loss of supranuclear vacuoles in the absorptive enterocytes, and an infiltration of inflammatory cells (Baeverfjord and Krogdahl, 1996). This condition has since been observed in a rainbow trout (Escaffre et al., 2007), chinook salmon (Booman et al., 2018), and a number of carnivorous marine species, such as Japanese flounder (*Paralichthys olivaceus*; Chen, et al., 2011), yellow kingfish (*Seriola lalandi*; Bansemer et al., 2015), turbot (*Scophthalmus maximus*; Gu et al., 2017), and pearl gentian grouper (*Epinephelus lanceolatus*♂ × *Epinephelus fuscoguttatus*♀; Zhang et al., 2019). While SBMIE appears to be prevalent among salmonids and marine carnivores, it has also been observed in common carp (Urán et al., 2008) and Nile tilapia fry (*Oreochromis niloticus*; Obirikorang et al., 2020) and larval zebrafish (Hedrera et al., 2013; Coronado et al., 2019).

More recently, researchers have attempted to identify the causative ANFs, which induced SBMIE. A large focus has been on soybean saponins, with a smaller effort to address antigenic proteins. These studies and modern approaches to understanding SBMIE in fish, such as transcriptomics, metabolomics, and microbiomics, have provided greater insight into this condition and suggestions for mitigation. ANFs in the alcohol-soluble fraction of soybean meal were shown early on to be a major contributor to SBMIE (Van den Ingh et al., 1996; Bureau et al., 1998). This information led to further research focusing specifically on alcohol-soluble saponins (Knudsen et al., 2007, 2008; Iwashita et al., 2008; Chikwati et al., 2012). As a heat-stable ANF in soybean meal, these ANFs are not removed during processing. These studies demonstrated that saponins are a major player in the induction of SBMIE of Atlantic salmon and also cause an increase in intestinal permeability (Knudsen et al., 2007). The latter impact on the distal intestine has also been observed in turbot, albeit not as a result of saponins, but following enteritis induced by the soybean antigen β-conglycinin (Zheng et al., 2020). In their study, they also report an increase in proinflammatory cytokines in the first 3 weeks of feeding and a decrease in the gene expression of tight junction proteins, providing insight into the mechanisms of inflammation and increased intestinal permeability.

Other studies into the mechanisms of SBMIE indicate effects of soybean meal on mucosal digestive enzymes of the distal intestine, fecal trypsin, and pancreatic insulin production (Krogdahl et al., 2003). Distal intestine mucosal enzymes were the most impacted with dose-dependent decreases in alkaline phosphatase, leucine aminopeptidase, maltase, isomaltase, lactase, and sucrase. Increasing levels of soybean meal also decrease pancreatic lipase in

Atlantic salmon (Chikwati et al., 2013). More recently, nutrigenomics studies in rainbow trout and pearl gentian grouper (*E. lanceolatus* ♂ × *E. fuscoguttatus* ♀) have identified potential biomarkers for SBM. Using metabolomics, Zhang et al. (2019) identified 17 potential metabolites related to and possibly influencing SBMIE in grouper. Using transcriptomics, Abernathy et al. (2017) identified risk loci in rainbow trout associated with inflammatory bowel disease in humans, specifically IL17REL, a major risk factor in ulcerative colitis. Together with studies on the impact of soybean meal on the microbiome of the distal intestine (Desai et al., 2012; Wang et al., 2015; Blaufuss et al., 2020), "-omic" studies such as these are beginning to define the molecular mechanisms and interactions associated with SBMIE. As more of these studies are published and the data set of omics information grows, it should become possible to develop a comprehensive model of SBMIE and improved strategies for mitigation in sensitive species.

8.5.3 Bile salts and enterohepatic circulation

The impacts of terrestrial plant-based ingredients on fish physiology extend beyond distal enteritis and general reductions in diet utilization and fish performance. ANFs, particularly those associated with soybean meal, have also been linked to dysfunctions in bile acid metabolism and enterohepatic circulation in fish (reviewed by Romano et al., 2020). Unfortunately, the regulation of bile acid synthesis and recycling is not completely understood, even in humans (Russell, 2009; Chiang, 2017). What is known from the human literature is that bile acid synthesis is tightly regulated and involves complex feedback mechanisms to maintain homeostasis of both the liver and the intestine (Chiang, 2017).

As previously discussed, bile acids play important roles in lipid digestion and the absorption of fat-soluble nutrients through their emulsification properties. When plant proteins are used at significant levels in feeds, however, lipid digestibility often decreases. While several possible explanations exist (e.g., intestinal enteritis, excess mucus production, binding of nutrients to ANFs or nonstarch polysaccharides, GEP imbalances), impaired bile acid function has been identified as a potential cause (Krogdahl, 1989), likely occurring through either changes in bile acid reabsorption (enterohepatic circulation) or bile acid synthesis.

Impairments in enterohepatic circulation resulting in reduced reabsorption of bile acids have been attributed to ANF-induced distal enteritis that reduces their absorptive function (Romano et al., 2020). Although SBMIE has been demonstrated to decrease carrier-mediated transport and increase intestinal permeability (Nordrum et al., 2000), the impact on bile acid reabsorption has not been clearly demonstrated. It is likely that ANFs, such as nonstarch polysaccharides and hydrophobic peptides, bind to bile acids, preventing reabsorption and resulting in losses through fecal elimination. Choi et al. (2002) identified a bile acid–binding region on soy glycinin capable of binding to cholesterol and/or bile acids and purported to increase fecal excretion. Plant-derived antinutritional factors, such as saponins, are known to form large micelles with bile acids, in turn, also increasing fecal excretion (Francis et al., 2002). Romano et al. (2020) suggest that other ANFs may also be responsible. Interactive effects should also not be ruled out, as definitive mechanisms remain unclear.

Romano et al. (2020) suggested that decreased bile acid production could be a result of one of two factors: (1) lower amounts of dietary cholesterol in plant-based diets for bile acid synthesis (Yun et al., 2011); and/or (2) reduced activity of the hepatic 7-alpha-hydroxylase (CYP7A1), a rate-limiting enzyme in the bile acid biosynthetic pathway (Murashita et al., 2018).

Upon further consideration, one might broaden these to simply state (1) reduced bile acid component compounds and (2) reduced synthetic efficiency. Cholesterol and taurine are critical components of bile acid production in fish. As previously discussed, most bile acids in fish are conjugated to taurine to form taurocholic acid, taurolithocholic acid, or taurochenodeoxycholic acid (Fig. 8.4). Disruptions in bile acid recycling can reduce the pool of these taurine-conjugated bile acids coming from enterohepatic circulation, thus creating a requirement for greater bile acid synthetic capacity and a need for more taurine and cholesterol.

The ability of fish to synthesize taurine varies by species and developmental stage (Yokoyama et al., 2001; Kim et al., 2003, 2005). This, and observations that plant-based feeds are deficient in taurine and can create a taurine limitation through reduced enterohepatic circulation (Krogdahl, 1989), may cause a conditional requirement for this nutrient. Taurine is synthesized from methionine by way of cysteine (Oja and Kontro, 1983), and limitations in de novo synthesis have been linked to varying activity of L-cysteine sulfinate decarboxylase, a key enzyme for the conversion of cysteine to taurine (de la Rosa and Stipanuk, 1985; Takeuchi et al., 2001; Yokoyama et al., 2001). Taurine may also play a role in stimulating the activity of CYP7A1, as shown in cod (Nandhini et al., 2002). As such, limitations in bile acid reabsorption and/or taurine synthesis and availability can also lead to reduced bile acid production resulting from substrate and enzymatic limitations, in turn potentially disrupting the efficiency of downstream lipid digestion and absorption. Salze and Davis (2015), however, remind us that many interacting factors contribute to lipid absorption, and it is difficult at this time to draw any definitive conclusions regarding the role of plant-derived ANFs or taurine on disruptions to bile acid metabolism or lipid absorption. As fish nutrition and associated research advances, a greater understanding of the complex systems involved in feeding, digestion, absorption, and utilization will aid in the mitigation of functional impairments resulting from increased use of alternative feed ingredients.

8.6 Concluding remarks

The foundation of fish nutritional physiology lies in a basic understanding of trophic systems, fish biology, and the alimentary tract. Since the last edition of *Fish Nutrition*, new research tools (e.g., -omics and advanced bioinformatics) have provided greater insight into the mechanisms associated with digestion, nutrient absorption, and the impact of cellular and chemical signaling on fish nutrition from immune, microbiome, and neuroendocrine systems. Historically studied in isolation, research has begun to model how these systems interact and function together to drive changes in fish performance. For long-studied species (e.g., trout, salmon, catfish), feeds have become fairly efficient, in part because formulations/ingredients are now being matched with fish physiology to optimize nutritional capacity. But for more recently cultured species, particularly marine species, there is still much to learn. The future of nutritional physiology as it applies to feeds and feeding lies in characterizing the impacts of feeds, nutrients, environment, and genetics on fish performance to fine-tune nutrient utilization in long-studied species and match efficiency with more recently cultured species. There will continue to be an emphasis on diet sustainability and cost. Because of this, alternate proteins and oils, as well as functional dietary supplements, will make up larger proportions of future fish feeds. Increased knowledge of nutritional physiology will be needed to maintain and improve fish performance under these and future constraints.

References

Abernathy, J., Brezas, A., Snekvik, K.R., Hardy, R.W., Overturf, K., 2017. Integrative functional analyses using rainbow trout selected for tolerance to plant diets reveal nutrigenomic signatures for soy utilization without the concurrence of enteritis. PloS One 12, e0180972.

Adams, M.A., Johnsen, P.B., 1986. Chemical control of feeding in herbivorous and carnivorous fish. In: Chemical Signals in Vertebrates, vol. 4. Springer, pp. 45–61.

Agrawal, V., Sastry, K., Kaushab, S., 1975. Digestive enzymes of three teleost fishes. Acta Physiol. Acad. Sci. Hungar. 46, 93.

Ahearn, G.A., Behnke, R.D., Zonno, V., Storelli, C., 1992. Kinetic heterogeneity of Na-D-glucose cotransport in teleost gastrointestinal tract. Am. J. Physiol. Regul. Integr. Comp. Physiol. 263, R1018–R1023.

Ali, M.A., 1975. Vision in Fishes: New Approaches in Research. Plenum Press, New York.

Ambardekar, A.A., Reigh, R.C., Williams, M.B., 2009. Absorption of amino acids from intact dietary proteins and purified amino acid supplements follows different time-courses in channel catfish (*Ictalurus punctatus*). Aquaculture 291, 179–187.

Amberg, J., Myr, C., Kamisaka, Y., Jordal, A.-E., Rust, M., Hardy, R., Koedijk, R., Rønnestad, I., 2008. Expression of the oligopeptide transporter, PepT1, in larval Atlantic cod (*Gadus morhua*). Comp. Biochem. Physiol. B Biochem. Mol. Biol. 150, 177–182.

Andre, M., Ando, S., Ballagny, C., Durliat, M., Poupard, G., Briancon, C., Babin, P.J., 2000. Intestinal fatty acid binding protein gene expression reveals the cephalocaudal patterning during zebrafish gut morphogenesis. Int. J. Dev. Biol. 44, 249–252.

Antonopoulou, E., Nikouli, E., Piccolo, G., Gasco, L., Gai, F., Chatzifotis, S., Mente, E., Kormas, K.A., 2019. Reshaping gut bacterial communities after dietary *Tenebrio molitor* larvae meal supplementation in three fish species. Aquaculture 503, 628–635.

Aranishi, F., Mano, N., 2000. Antibacterial cathepsins in different types of ambicoloured Japanese flounder skin. Fish Shellfish Immunol. 10, 87–89.

Avila, E., Tu, H., Basantes, S., Ferraris, R., 2000. Dietary phosphorus regulates intestinal transport and plasma concentrations of phosphate in rainbow trout. J. Comp. Physiol. B 170, 201–209.

Baeverfjord, G., Krogdahl, Å., 1996. Development and regression of soybean meal induced enteritis in Atlantic salmon, *Salmo salar* L., distal intestine: a comparison with the intestines of fasted fish. J. Fish. Dis. 19, 375–387.

Babin, P.J., Vernier, J., 1989. Plasma lipoproteins in fish. JLR (J. Lipid Res.) 30, 467–489.

Bakke, A.M., Glover, C., Krogdahl, Å., 2010. Feeding, digestion and absorption of nutrients. In: Fish Physiology, vol. 30. Elsevier, pp. 57–110.

Bakke, A., Glover, C., Krogdahl, Å., 2011. In: Grosell, M., Farrell, A.P., Brauner, C.J. (Eds.), The Multifunctional Gut of Fish. Academic Press, pp. 57–110 (Ch. 2).

Baldisserotto, B., Mancera Romero, J.M., Kapoor, B.G., 2007. Fish Osmoregulation. Science Publishers, Enfield, N.H.

Bansemer, M., Forder, R., Howarth, G., Suitor, G., Bowyer, J., Stone, D., 2015. The effect of dietary soybean meal and soy protein concentrate on the intestinal mucus layer and development of subacute enteritis in Yellowtail Kingfish (*Seriola lalandi*) at suboptimal water temperature. Aquacult. Nutr. 21, 300–310.

Bell, M., Dick, J., 1993. The appearance of rods in the eyes of herring and increased di-docosahexaenoyl molecular species of phospholipids. J. Mar. Biol. Assoc. U. K. 73, 679–688.

Bertin, L., 1958. Appareil digestif. Traité de Zoologie 13, 1249–1301.

Bertucci, J.I., Blanco, A.M., Sundarrajan, L., Rajeswari, J.J., Velasco, C., Unniappan, S., 2019. Nutrient regulation of endocrine factors influencing feeding and growth in fish. Front. Endocrinol. 10, 83.

Bijvelds, M., Velden, J., Kolar, Z.I., Flik, G., 1998. Magnesium transport in freshwater teleosts. J. Exp. Biol. 201, 1981–1990.

Bjenning, C., Holmgren, S., 1988. Neuropeptides in the fish gut. Histochemistry 88, 155–163.

Blaufuss, P.C., Bledsoe, J.W., Gaylord, T.G., Sealey, W.M., Overturf, K.E., Powell, M.S., 2020. Selection on a plant-based diet reveals changes in oral tolerance, microbiota and growth in rainbow trout (Oncorhynchus mykiss) when fed a high soy diet. Aquaculture 735287.

Bone, Q., Moore, R., 2008. Biology of Fishes. Taylor & Francis.

Booman, M., Forster, I., Vederas, J.C., Groman, D.B., Jones, S.R., 2018. Soybean meal-induced enteritis in Atlantic salmon (*Salmo salar*) and Chinook salmon (*Oncorhynchus tshawytscha*) but not in pink salmon (*O. gorbuscha*). Aquaculture 483, 238–243.

Breck, O., Bjerkås, E., Campbell, P., Rhodes, J.D., Sanderson, J., Waagbø, R., 2005. Histidine nutrition and genotype affect cataract development in Atlantic salmon, Salmo salar L. J. Fish Dis. 28, 357–371.

Brezas, A., Hardy, R.W., 2020. Improved performance of a rainbow trout selected strain is associated with protein digestion rates and synchronization of amino acid absorption. Sci. Rep. 10, 1–12.

Buslé, J., Anadon, G.G., 1996. The structure and function of fish liver. Fish Morphol. 76, 545–551.

Bryan, P.G., 1975. Food habits, functional digestive morphology, and assimilation efficiency of the rabbitfish *Siganus spinus* (Pisces, Siganidae) on Guam. Pac. Sci. 29 (3), 269–277.

Budden, K.F., Shukla, S.D., Rehman, S.F., Bowerman, K.L., Keely, S., Hugenholtz, P., Armstrong-James, D.P., Adcock, I.M., Chotirmall, S.H., Chung, K.F., 2019. Functional effects of the microbiota in chronic respiratory disease. Lancet Respir. Med. 7, 907–920.

Buddington, R.K., Chen, J.W., Diamond, J., 1987. Genetic and phenotypic adaptation of intestinal nutrient transport to diet in fish. J. Physiol. 393, 261–281.

Buddington, R.K., Puchal, A.A., Houpe, K.L., Diehl III, W.J., 1993. Hydrolysis and absorption of two monophosphate derivatives of ascorbic acid by channel catfish *Ictalurus punctatus* intestine. Aquaculture 114, 317–326.

Buddington, R.K., Buddington, K.K., Deng, D.-F., Hemre, G.-I., Wilson, R.P., 2002. A high retinol dietary intake increases its apical absorption by the proximal small intestine of juvenile sunshine bass (*Morone chrysops*×*M. saxatilis*). J. Nutr. 132, 2713–2716.

Bureau, D.P., Harris, A.M., Cho, C.Y., 1998. The effects of purified alcohol extracts from soy products on feed intake and growth of chinook salmon (*Oncorhynchus tshawytscha*) and rainbow trout (*Oncorhynchus mykiss*). Aquaculture 161, 27–43.

Burnstock, G., 1959. The innervation of the gut of the brown trout (*Salmo trutta*). J. Cell Sci. 3, 199–220.

Bury, N.R., Walker, P.A., Glover, C.N., 2003. Nutritive metal uptake in teleost fish. J. Exp. Biol. 206, 11–23.

Butt, R.L., Volkoff, H., 2019. Gut microbiota and energy homeostasis in fish. Front. Endocrinol. 10, 9.

Cabillon, N.A.R., Lazado, C.C., 2019. Mucosal barrier functions of fish under changing environmental conditions. Fishes 4, 2.

Cain, K., Swan, C., 2010. Barrier function and immunology. In: Fish Physiology, vol. 30. Elsevier, pp. 111–134.

Caipang, C.M.A., Lazado, C.C., 2015. Nutritional impacts on fish mucosa: immunostimulants, pre-and probiotics. Mucosal Health Aquac. 211–272 (Elsevier).

Canosa, L.F., Peter, R.E., 2004. Effects of cholecystokinin and bombesin on the expression of preprosomatostatin-encoding genes in goldfish forebrain. Regul. Pept. 121, 99–105.

Caprio, J., 1984. Olfaction and taste in fish. Comp. Physiol. Sens. Syst. 257–283.

Carr, W.E., 1988. The molecular nature of chemical stimuli in the aquatic environment. In: Sensory Biology of Aquatic Animals. Springer, pp. 3–27.

Carr, W.E., Derby, C.D., 1986. Chemically stimulated feeding behavior in marine animals. J. Chem. Ecol. 12, 989–1011.

Casirola, D.M., Ferraris, R.P., 1997. Intestinal absorption of water-soluble vitamins in rainbow trout (*Onchorhynchus mykiss*). Comp. Biochem. Physiol. Physiol. 116, 273–279.

Casirola, D., Vinnakota, R., Ferraris, R., 1995. Intestinal absorption of water-soluble vitamins in channel catfish (*Ictalurus punctatus*). Am. J. Physiol. Regul. Integr. Comp. Physiol. 269, R490–R496.

Chen, X., Jiang, S., Shi, Z., 2012. Identification and expression analysis of fabp2 gene from common carp *Cyprinus carpio*. J. Fish. Biol. 80, 679–691.

Chen, P., Tang, Q., Wang, C., 2016. Characterizing and evaluating the expression of the type IIb sodium-dependent phosphate cotransporter (slc34a2) gene and its potential influence on phosphorus utilization efficiency in yellow catfish (*Pelteobagrus fulvidraco*). Fish Physiol. Biochem. 42, 51–64.

Chen, S., Reichert, S., Singh, C., Oikonomou, G., Rihel, J., Prober, D.A., 2017. Light-dependent regulation of sleep and wake states by prokineticin 2 in zebrafish. Neuron 95, 153–168 e6.

Chiang, J.Y., 2017. Recent advances in understanding bile acid homeostasis. F1000Res. 6.

Chikwati, E., Sahlmann, C., Holm, H., Penn, M., Krogdahl, Å., Bakke, A., 2013. Alterations in digestive enzyme activities during the development of diet-induced enteritis in Atlantic salmon, *Salmo salar* L. Aquaculture 402, 28–37.

Chikwati, E.M., Venold, F.F., Penn, M.H., Rohloff, J., Refstie, S., Guttvik, A., Hillestad, M., Krogdahl, Å., 2012. Interaction of soyasaponins with plant ingredients in diets for Atlantic salmon, *Salmo salar* L. Br. J. Nutr. 107, 1570–1590.

Choi, S.-K., Adachi, M., Utsumi, S., 2002. Identification of the bile acid-binding region in the soy glycinin A1aB1b subunit. Biosci. Biotechnol. Biochem. 66, 2395–2401.

Collie, N.L., Ferraris, R.P., 1995. Nutrient fluxes and regulation in fish intestine. In: Biochemistry and Molecular Biology of Fishes, vol. 4. Elsevier, pp. 221–239.

Concha, M.I., Santander, C., Villanueva, J., Amthauer, R., 2002. Specific binding of the endocytosis tracer horseradish peroxidase to intestinal fatty acid-binding protein (I-FABP) in apical membranes of carp enterocytes. J. Exp. Zool. 293, 541–550.

Conrad, M.E., Umbreit, J.N., 2002. Pathways of iron absorption. Blood Cell Mol. Dis. 29, 336–355.

Coronado, M., Solis, C.J., Hernandez, P.P., Feijóo, C.G., 2019. Soybean meal-induced intestinal inflammation in zebrafish is T cell-dependent and has a Th17 cytokine profile. Front. Immunol. 10, 610.

Crossman, D.J., Choat, J.H., Clements, K.D., 2005. Nutritional ecology of nominally herbivorous fishes on coral reefs. Mar. Ecol. Prog. Ser. 296, 129–142.

Dabrowski, K.R., 1986. Ontogenetical aspects of nutritional requirements in fish. Comp. Biochem. Physiol. Physiol. 85, 639–655.

Dalmo, R., Ingebrigtsen, K., Bøgwald, J., 1997. Non-specific defence mechanisms in fish, with particular reference to the reticuloendothelial system (RES). J. Fish. Dis. 20, 241–273.

Das, K., Tripathi, S., 1991. Studies on the digestive enzymes of grass carp, Ctenopharyngodon idella (Val.). Aquaculture 92, 21–32.

Dawood, M.A., 2020. Nutritional immunity of fish intestines: important insights for sustainable aquaculture. Rev. Aquac.

Day, R.D., German, D.P., Manjakasy, J.M., Farr, I., Hansen, M.J., Tibbetts, I.R., 2011. Enzymatic digestion in stomachless fishes: how a simple gut accommodates both herbivory and carnivory. J. Comp. Physiol. B 181, 603–613.

de la Rosa, J., Stipanuk, M.H., 1985. Evidence for a rate-limiting role of cysteinesulfinate decarboxylase activity in taurine biosynthesis in vivo. Comp. Biochem. & Physiol. B Comp. Biochem. 81, 565.

Desai, A.R., Links, M.G., Collins, S.A., Mansfield, G.S., Drew, M.D., Van Kessel, A.G., Hill, J.E., 2012. Effects of plant-based diets on the distal gut microbiome of rainbow trout (*Oncorhynchus mykiss*). Aquaculture 350, 134–142.

Divakaran, S., Kim, B., Ostrowski, A., 1999. Digestive enzymes present in Pacific threadfin *Polydactylus sexfilis* (Bloch & Schneider 1801) and bluefin trevally *Caranx melampygus* (Cuvier 1833). Aquacult. Res. 30, 781–787.

Donovan, A., Brownlie, A., Zhou, Y., Shepard, J., Pratt, S.J., Moynihan, J., Paw, B.H., Drejer, A., Barut, B., Zapata, A., 2000. Positional cloning of zebrafish ferroportin1 identifies a conserved vertebrate iron exporter. Nature 403, 776–781.

Douglas, R.D., Djamgoz, M., 1990. The Visual System of Fish. Springer, Dordrecht.

Elger, M., Werner, A., Herter, P., Kohl, B., Kinne, R., Hentschel, H., 1998. Na-Pi cotransport sites in proximal tubule and collecting tubule of winter flounder (*Pleuronectes americanus*). Am. J. Physiol. Ren. Physiol. 274, F374–F383.

Escaffre, A.-M., Kaushik, S., Mambrini, M., 2007. Morphometric evaluation of changes in the digestive tract of rainbow trout (*Oncorhynchus mykiss*) due to fish meal replacement with soy protein concentrate. Aquaculture 273, 127–138.

Evans, B.I., 2004. A fish's eye view of habitat change. In: The Senses of Fish. Springer, pp. 1–30.

Fange, R., 1979. Digestion. In: Fish Physiology, pp. 161–260.

Fange, R., Grove, D., 1979. Digestion. In: Hoar, W.A., Randoll, D.J., Brett, J.R. (Eds.), Fish Physiology, vol. 8. Academic Press, New York.

Feeney, G.P., Zheng, D., Kille, P., Hogstrand, C., 2005. The phylogeny of teleost ZIP and ZnT zinc transporters and their tissue specific expression and response to zinc in zebrafish. Biochim. Biophys. Acta Gene Struct. Expr. 1732, 88–95.

Fernandes, J., Kemp, G., Smith, V., 2004. Two novel muramidases from skin mucosa of rainbow trout (*Oncorhynchus mykiss*). Comp. Biochem. Physiol. B Biochem. Mol. Biol. 138, 53–64.

Ferraris, R.P., Lee, P.P., Diamond, J.M., 1989. Origin of regional and species differences in intestinal glucose uptake. Am. J. Physiol. Gastrointest. Liver Physiol. 257, G689–G697.

Flik, G., Verbost, P., 1993. Calcium transport in fish gills and intestine. J. Exp. Biol. 184, 17–29.

Francis, G., Makkar, H.P., Becker, K., 2001. Antinutritional factors present in plant-derived alternate fish feed ingredients and their effects in fish. Aquaculture 199, 197–227.

Francis, G., Kerem, Z., Makkar, H.P., Becker, K., 2002. The biological action of saponins in animal systems: a review. Br. J. Nutr. 88, 587–605.

Frøystad-Saugen, M., Lilleeng, E., Bakke-McKellep, A., Vekterud, K., Valen, E., Hemre, G.I., Krogdahl, Å., 2009. Distal intestinal gene expression in Atlantic salmon (*Salmo salar* L.) fed genetically modified maize. Aquacult. Nutr. 15, 104–115.

Fuentes, J., Figueiredo, J., Power, D.M., Canário, A.V., 2006. Parathyroid hormone-related protein regulates intestinal calcium transport in sea bream (Sparus auratus). Am. J. Physiol. Regul. Integr. Comp. Physiol. 291, R1499–R1506.

Gajjar, S., Patel, B.M., 2017. Neuromedin: an insight into its types, receptors and therapeutic opportunities. Pharmacol. Rep. 69, 438–447.

Gerking, S.D., 1994. Feeding Ecology of Fish. Academic Press, San Diego.

German, D.P., 2009. Do herbivorous minnows have "plug-flow reactor" guts? Evidence from digestive enzyme activities, gastrointestinal fermentation, and luminal nutrient concentrations. J. Comp. Physiol. B 179, 759–771.

German, D.P., Bittong, R.A., 2009. Digestive enzyme activities and gastrointestinal fermentation in wood-eating catfishes. J. Comp. Physiol. B 179, 1025–1042.

Geurden, I., Aramendi, M., Zambonino-Infante, J., Panserat, S., 2007. Early feeding of carnivorous rainbow trout (*Oncorhynchus mykiss*) with a hyperglucidic diet during a short period: effect on dietary glucose utilization in juveniles. Am. J. Physiol. Regul. Integr. Comp. Physiol. 292, R2275–R2283.

Goel, K., 1975. Carbohydrase activity in the digestive system of some teleost fishes. Acta Physiol. Acad. Sci. Hungar. 46, 191–196.

Glover, C.N., Hogstrand, C., 2002. Amino acid modulation of in vivo intestinal zinc absorption in freshwater rainbow trout. J. Exp. Biol. 205, 151–158.

Glover, C.N., Hogstrand, C., 2003. Effects of dissolved metals and other hydrominerals on in vivo intestinal zinc uptake in freshwater rainbow trout. Aquat. Toxicol. 62, 281–293.

Glover, C.N., Wood, C.M., 2008. Absorption of copper and copper–histidine complexes across the apical surface of freshwater rainbow trout intestine. J. Comp. Physiol. B 178, 101–109.

Glover, C.N., Bury, N.R., Hogstrand, C., 2003. Zinc uptake across the apical membrane of freshwater rainbow trout intestine is mediated by high affinity, low affinity, and histidine-facilitated pathways. Biochim. Biophys. Acta Biomembr. 1614, 211–219.

Gonçalves, A.F., Castro, L.F.C., Pereira-Wilson, C., Coimbra, J., Wilson, J.M., 2007. Is there a compromise between nutrient uptake and gas exchange in the gut of *Misgurnus anguillicaudatus*, an intestinal air-breathing fish? Comp. Biochem. Physiol. Genom. Proteom. 2, 345–355.

Graham, C., Nalbant, P., Scholermann, B., Hentschel, H., Kinne, R.K., Werner, A., 2003. Characterization of a type IIb sodium-phosphate cotransporter from zebrafish (*Danio rerio*) kidney. Am. J. Physiol. Ren. Physiol. 284, F727–F736.

Gu, M., Bai, N., Xu, B., Xu, X., Jia, Q., Zhang, Z., 2017. Protective effect of glutamine and arginine against soybean meal-induced enteritis in the juvenile turbot (*Scophthalmus maximus*). Fish Shellfish Immunol. 70, 95–105.

Guillaume, J., Kaushik, S., Bergot, P., Metailler, R., 1999. Nutrition and Feeding of Fish and Crustaceans. INRA, IFRAMER. Springer, UK.

Gutowska, M.A., Drazen, J.C., Robison, B.H., 2004. Digestive chitinolytic activity in marine fishes of Monterey Bay, California. Comp. Biochem. Physiol. Mol. Integr. Physiol. 139, 351–358.

Hakim, Y., Harpaz, S., Uni, Z., 2009. Expression of brush border enzymes and transporters in the intestine of European sea bass (*Dicentrarchus labrax*) following food deprivation. Aquaculture 290, 110–115.

Hall, J.R., Short, C.E., Driedzic, W.R., 2006. Sequence of Atlantic cod (*Gadus morhua*) GLUT4, GLUT2 and GPDH: developmental stage expression, tissue expression and relationship to starvation-induced changes in blood glucose. J. Exp. Biol. 209, 4490–4502.

Hansen, A., Reutter, K., 2004. Chemosensory systems in fish: structural, functional and ecological aspects. In: The Senses of Fish. Springer, pp. 55–89.

Hao, Y.T., Wu, S.G., Xiong, F., Tran, N.T., Jakovlić, I., Zou, H., Li, W.X., Wang, G.T., 2017. Succession and fermentation products of grass carp (*Ctenopharyngodon idellus*) hindgut microbiota in response to an extreme dietary shift. Front. Microbiol. 8, 1585.

Hara, T.J., 1971. 4 Chemoreception. Fish Physiol. 5, 79–120 (Elsevier).

Hara, T., 1982. Structure-activity relationships of amino acids as olfactory stimuli. Chemorecept. Fishes 135–157.

Hara, T.J., 1986. Role of olfaction in fish behaviour. In: The Behaviour of Teleost Fishes. Springer, pp. 152–176.

Hara, T.J., 1992. Fish Chemoreception. Springer, Netherlands.

Hara, T., 1992. Overview and introduction. In: Fish Chemoreception. Springer, pp. 1–12.

Hara, T.J., 1994. Olfaction and gustation in fish: an overview. Acta Physiol. Scand. 152, 207–217.

Hara, T.J., 2005. Olfactory responses to amino. In: Fish Chemosenses, vol. 31.

Hara, T.J., 2006. Gustation. Fish Physiol. 25, 45–96.

Harmon, J.S., Sheridan, M.A., 1992. Effects of nutritional state, insulin, and glucagon on lipid mobilization in rainbow trout, *Oncorhynchus mykiss*. Gen. Comp. Endocrinol. 87, 214–221.

Hedrera, M.I., Galdames, J.A., Jimenez-Reyes, M.F., Reyes, A.E., Avendaño-Herrera, R., Romero, J., Feijóo, C.G., 2013. Soybean meal induces intestinal inflammation in zebrafish larvae. PloS One 8, e69983.

Helvik, J.V., Drivenes, Ø., Harboe, T., Seo, H.-C., 2001. Topography of different photoreceptor cell types in the larval retina of Atlantic halibut (*Hippoglossus hippoglossus*). J. Exp. Biol. 204, 2553–2559.

Hernandez-Blazquez, F.J., Guerra, R.R., Kfoury, J.R., Bombonato, P.P., Cogliati, B., Da Silva, J.R.M.C., 2006. Fat absorptive processes in the intestine of the Antarctic fish *Notothenia coriiceps* (Richardson, 1844). Polar Biol. 29, 831–836.

Hibiya, T., 1982. Atlas of Fish Histology. Kodansha.

Higgins, S.C., Gueorguiev, M., Korbonits, M., 2007. Ghrelin, the peripheral hunger hormone. Ann. Med. 39, 116–136.

Hill, J.H., Franzosa, E.A., Huttenhower, C., Guillemin, K., 2016. A conserved bacterial protein induces pancreatic beta cell expansion during zebrafish development. Elife 5, e20145.

Himick, B.A., Peter, R.E., 1994. CCK/gastrin-like immunoreactivity in brain and gut, and CCK suppression of feeding in goldfish. Am. J. Physiol. Regul. Integr. Comp. Physiol. 267, R841–R851.

Hinton, D.E., 1994. Cells, cellular responses, and their markers in chronic toxicity of fishes. In: Aquatic Toxicology: Molecular, Biochemical and Cellular Perspectives, pp. 207–239.

Holling, C.S., 1966. The functional response of invertebrate predators to prey density. Mem. Entomol. Soc. Can. 98, 5–86.

Holmgren, S., Jönsson, A.-C., 1988. Occurrence and effects on motility of bombesin related peptides in the gastrointestinal tract of the Atlantic cod, *Gadus morhua*. Comp. Biochem. Physiol. C Comp. Pharmacol. 89, 249–256.

Holstein, B., 1983. Effect of vasoactive intestinal polypeptide on gastric acid secretion and mucosal blood flow in the Atlantic cod, *Gadus morhua*. Gen. Comp. Endocrinol. 52, 471–473.

Holstein, B., Humphrey, C., 1980. Stimulation of gastric acid secretion and suppression of VIP-like immunoreactivity by bombesin in the Atlantic codfish, *Gadus morhua*. Acta Physiol. Scand. 109, 217–223.

Honkanen, R., Crim, J., Patton, J., 1988. Effects of cholecystokinin peptides on digestive enzymes in killifish in vivo. Comp. Biochem. & Physiol. A, Comp. Physiol. 89, 655–660.

Horn, M.H., 1989. Biology of marine herbivorous fishes. Oceanogr. Mar. Biol. 27, 167–272.

Horn, M., Gawlicka, A., German, D., Logothetis, E., Cavanagh, J., Boyle, K., 2006. Structure and function of the stomachless digestive system in three related species of New World silverside fishes (Atherinopsidae) representing herbivory, omnivory, and carnivory. Mar. Biol. 149, 1237–1245.

Hoseinifar, S.H., Dadar, M., Ringø, E., 2017. Modulation of nutrient digestibility and digestive enzyme activities in aquatic animals: the functional feed additives scenario. Aquacult. Res. 48, 3987–4000.

Hossain, M., Yoshimatsu, T., 2014. Dietary calcium requirement in fishes. Aquacult. Nutr. 20, 1–11.

Hughes, S.G., 1985. Nutritional eye diseases in salmonids: a review. Progress. Fish Cult. 47, 81–85.

Hyatt, K.D., 1979. Feeding strategy. Fish Physiol. 8, 71–119.

Ikeda, M., Miyauchi, K., Mochizuki, A., Matsumiya, M., 2009. Purification and characterization of chitinase from the stomach of silver croaker *Pennahia argentatus*. Protein Expr. Purif. 65, 214–222.

Iqbal, J., Hussain, M.M., 2009. Intestinal lipid absorption. Am. J. Physiol. Endocrinol. Metabol. 296, E1183–E1194.

Ishimaru, Y., Okada, S., Naito, H., Nagai, T., Yasuoka, A., Matsumoto, I., Abe, K., 2005. Two families of candidate taste receptors in fishes. Mech. Dev. 122, 1310–1321.

Islam, Z., Hayashi, N., Yamamoto, Y., Doi, H., Romero, M.F., Hirose, S., Kato, A., 2013. Identification and proximal tubular localization of the Mg^{2+} transporter, Slc41a1, in a seawater fish. Am. J. Physiol. Regul. Integr. Comp. Physiol. 305, R385–R396.

Itami, T., Ishida, Y., Endo, F., Kawazoe, N., Takahashi, Y., 1993. Hemagglutinins in the skin mucus of ayu. Fish Pathol. 28, 41–47.

Iwashita, Y., Yamamoto, T., Furuita, H., Sugita, T., Suzuki, N., 2008. Influence of certain soybean antinutritional factors supplemented to a casein-based semipurified diet on intestinal and liver morphology in fingerling rainbow trout *Oncorhynchus mykiss*. Fish. Sci. 74, 1075–1082.

Jensen, H., Rourke, I.J., Møller, M., Jønson, L., Johnsen, A.H., 2001. Identification and distribution of CCK-related peptides and mRNAs in the rainbow trout, *Oncorhynchus mykiss*. Biochim. Biophys. Acta Gene Struct. Expr. 1517, 190–201.

Jönsson, E., 2013. The role of ghrelin in energy balance regulation in fish. Gen. Comp. Endocrinol. 187, 79–85.

Jönsson, E., Forsman, A., Einarsdottir, I.E., Egnér, B., Ruohonen, K., Björnsson, B.T., 2006. Circulating levels of cholecystokinin and gastrin-releasing peptide in rainbow trout fed different diets. Gen. Comp. Endocrinol. 148, 187–194.

Kaiya, H., Kojima, M., Hosoda, H., Riley, L.G., Hirano, T., Grau, E.G., Kangawa, K., 2003a. Identification of tilapia ghrelin and its effects on growth hormone and prolactin release in the tilapia, *Oreochromis mossambicus*. Comp. Biochem. Physiol. B Biochem. Mol. Biol. 135, 421–429.

Kaiya, H., Kojima, M., Hosoda, H., Riley, L., Hirano, T., Grau, E., Kangawa, K., 2003b. Amidated fish ghrelin: purification, cDNA cloning in the Japanese eel and its biological activity. J. Endocrinol. 176, 415–424.

Kaiya, H., Kojima, M., Hosoda, H., Moriyama, S., Takahashi, A., Kawauchi, H., Kangawa, K., 2003c. Peptide purification, complementary deoxyribonucleic acid (DNA) and genomic DNA cloning, and functional characterization of ghrelin in rainbow trout. Endocrinology 144, 5215–5226.

Kaiya, H., Small, B.C., Bilodeau, A.L., Shepherd, B.S., Kojima, M., Hosoda, H., Kangawa, K., 2005. Purification, cDNA cloning, and characterization of ghrelin in channel catfish, *Ictalurus punctatus*. Gen. Comp. Endocrinol. 143, 201–210.

Karasov, W., Diamond, J., 1981. Intestinal glucose-transport varies with taxa and diet. Am. Zool. 21, 1030-1030.

Kasumyan, A.O., DÖving, K.B., 2003. Taste preferences in fishes. Fish Fish. 4, 289–347.

Kayama, M., Iijima, N., 1976. Studies on lipid transport mechanism in the fish. Bull. Jpn. Soc. Sci. Fish. 42, 987–996.

Kim, S.-K., Takeuchi, T., Yokoyama, M., Murata, Y., 2003. Effect of dietary supplementation with taurine, β-alanine and GABA on the growth of juvenile and fingerling Japanese flounder *Paralichthys olivaceus*. Fish. Sci. 69, 242–248.

Kim, S.-K., Takeuchi, T., Yokoyama, M., Murata, Y., Kaneniwa, M., Sakakura, Y., 2005. Effect of dietary taurine levels on growth and feeding behavior of juvenile Japanese flounder *Paralichthys olivaceus*. Aquaculture 250, 765–774.

Klaren, P.H., Wendelaar bonga, S.E., Flik, G., 1997. Evidence for P2-purinoceptor-mediated uptake of Ca^{2+} across a fish (*Oreochromis mossambicus*) intestinal brush border membrane. Biochem. J. 322, 129–134.

Kleerekoper, H., 1969. Olfaction in Fishes. Indiana University Press, Bloomington.

Klein, S.E., Sheridan, M.A., 2008. Somatostatin signaling and the regulation of growth and metabolism in fish. Mol. Cell. Endocrinol. 286, 148–154.

Knoop, K.A., Newberry, R.D., 2018. Goblet cells: multifaceted players in immunity at mucosal surfaces. Mucosal Immunol. 11, 1551–1557.

Knudsen, D., Urán, P., Arnous, A., Koppe, W., Frøkiær, H., 2007. Saponin-containing subfractions of soybean molasses induce enteritis in the distal intestine of Atlantic salmon. J. Agric. Food Chem. 55, 2261–2267.

Knudsen, D., Jutfelt, F., Sundh, H., Sundell, K., Koppe, W., Frøkiær, H., 2008. Dietary soya saponins increase gut permeability and play a key role in the onset of soyabean-induced enteritis in Atlantic salmon (*Salmo salar* L.). Br. J. Nutr. 100, 120–129.

Kodzhahinchev, V., Kovacevic, D., Bucking, C., 2017. Identification of the putative goldfish (*Carassius auratus*) magnesium transporter SLC41a1 and functional regulation in the gill, kidney, and intestine in response to dietary and environmental manipulations. Comp. Biochem. Physiol. Mol. Integr. Physiol. 206, 69–81.

Kohl, B., Herter, P., Hulseweh, B., Elger, M., Hentschel, H., Kinne, R., Werner, A., 1996. Na-Pi cotransport in flounder: same transport system in kidney and intestine. Am. J. Physiol. Ren. Physiol. 270, F937–F944.

Krasnov, A., Teerijoki, H., Mölsä, H., 2001. Rainbow trout (*Onchorhynchus mykiss*) hepatic glucose transporter. Biochim. Biophys. Acta Gene Struct. Expr. 1520, 174–178.

Krinner, M., 1935. Über die Geschmacksempfindlichkeit der Elritzen. Z. für Vgl. Politikwiss. (ZfVP) 21, 317–342.

Krogdahl, Å., 1989. Alternative protein sources from plants contain antinutrients affecting digestion in salmonids. Proc. III. Feed. & Nutr. Fish 253–261 (Toba, Japan).

Krogdahl, Å., Bakke-McKellep, A., Baeverfjord, G., 2003. Effects of graded levels of standard soybean meal on intestinal structure, mucosal enzyme activities, and pancreatic response in Atlantic salmon (*Salmo salar* L.). Aquacult. Nutr. 9, 361–371.

Krogdahl, Å., Hemre, G.I., Mommsen, T., 2005. Carbohydrates in fish nutrition: digestion and absorption in postlarval stages. Aquacult. Nutr. 11, 103–122.

Krogdahl, Å., Penn, M., Thorsen, J., Refstie, S., Bakke, A.M., 2010. Important antinutrients in plant feedstuffs for aquaculture: an update on recent findings regarding responses in salmonids. Aquacult. Res. 41, 333–344.

Kurtovic, I., Marshall, S.N., Zhao, X., Simpson, B.K., 2009. Lipases from mammals and fishes. Rev. Fish. Sci. 17, 18–40.

Kuz'mina, V.V., 2008. Classical and modern concepts in fish digestion. In: Feeding and Digestive Functions of Fishes, pp. 85–104.

Larsson, D., Aksnes, L., Björnsson, B.T., Larsson, B., Lundgren, T., 2002. Antagonistic effects of 24R, 25-dihydroxyvitamin D_3 and 25-hydroxyvitamin D_3 on L-type Ca^2 channels and Na/Ca^2 exchange in enterocytes from Atlantic cod (*Gadus morhua*). J. Mol. Endocrinol. 28, 53–68.

Larsson, D., Lundgren, T., Sundell, K., 1998. Ca 2+ uptake through voltage-gated L-type Ca 2+ channels by polarized enterocytes from Atlantic cod Gadus morhua. J. Membr. Biol. 164, 229–237.

Le, H.T., Shao, X., Krogdahl, Å., Kortner, T.M., Lein, I., Kousoulaki, K., Lie, K.K., Sæle, Ø., 2019. Intestinal function of the stomachless fish, ballan wrasse (*Labrus bergylta*). Front. Mar. Sci. 6, 140.

LeBlanc, J.G., Milani, C., De Giori, G.S., Sesma, F., Van Sinderen, D., Ventura, M., 2013. Bacteria as vitamin suppliers to their host: a gut microbiota perspective. Curr. Opin. Biotechnol. 24, 160–168.

Legrand, T.P., Wynne, J.W., Weyrich, L.S., Oxley, A.P., 2020. A microbial sea of possibilities: current knowledge and prospects for an improved understanding of the fish microbiome. Rev. Aquacult. 12, 1101–1134.

Li, W., Liu, J., Tan, H., Yang, C., Ren, L., Liu, Q., Wang, S., Hu, F., Xiao, J., Zhao, R., 2018. Genetic effects on the gut microbiota assemblages of hybrid fish from parents with different feeding habits. Front. Microbiol. 9, 2972.

Lindenstrøm, T., Buchmann, K., Secombes, C., 2003. *Gyrodactylus derjavini* infection elicits IL-1β expression in rainbow trout skin. Fish Shellfish Immunol. 15, 107–115.

Little, E.E., 1981. Conditioned cardiac response to the olfactory stimuli of amino acids in the channel catfish, *Ictalurus punctatus*. Physiol. Behav. 27, 691–697.

Logothetis, E., Horn, M., Dickson, K., 2001. Gut morphology and function in *Atherinops affinis* (Teleostei: Atherinopsidae), a stomachless omnivore feeding on macroalgae. J. Fish. Biol. 59, 1298–1312.

Lyons, P.P., Turnbull, J.F., Dawson, K.A., Crumlish, M., 2017. Effects of low-level dietary microalgae supplementation on the distal intestinal microbiome of farmed rainbow trout *Oncorhynchus mykiss* (Walbaum). Aquacult. Res. 48, 2438–2452.

Mackie, A., 1985. Identification of gustatory feeding stimulants for fish-applications in aquaculture. In: Nutrition and Feeding in Fish. Academic Press, pp. 177–189.

Maffia, M., Acierno, R., Cillo, E., Storelli, C., 1996. Na (+)-D-glucose cotransport by intestinal BBMVs of the Antarctic fish *Trematomus bernacchii*. Am. J. Physiol. Regul. Integr. Comp. Physiol. 271, R1576–R1583.

Maffia, M., Verri, T., Danieli, A., Thamotharan, M., Pastore, M., Ahearn, G., Storelli, C., 1997. H (+)-glycyl-L-proline cotransport in brush-border membrane vesicles of eel (*Anguilla anguilla*) intestine. Am. J. Physiol. Regul. Integr. Comp. Physiol. 272, R217–R225.

Maffia, M., Rizzello, A., Acierno, R., Verri, T., Rollo, M., Danieli, A., Döring, F., Daniel, H., Storelli, C., 2003. Characterisation of intestinal peptide transporter of the Antarctic haemoglobinless teleost *Chionodraco hamatus*. J. Exp. Biol. 206, 705–714.

Makkar, H., 1993. Antinutritional factors in foods for livestock. BSAP Occas. Pub. 16, 69–85.

Manjakasy, J.M., Day, R.D., Kemp, A., Tibbetts, I.R., 2009. Functional morphology of digestion in the stomachless, piscivorous needlefishes *Tylosurus gavialoides* and *Strongylura leiura ferox* (Teleostei: Beloniformes). J. Morphol. 270, 1155–1165.

Margolis, L., 1953. The effect of fasting on the bacterial flora of the intestine of fish. J. Fish. Board Can. 10, 62–63.

Marshall, W.S., Bryson, S.E., Wood, C.M., 1992. Calcium transport by isolated skin of rainbow trout. J. Exp. Biol. 166, 297–316.

Marsh-Armstrong, N., McCaffery, P., Gilbert, W., Dowling, J.E., Dräger, U.C., 1994. Retinoic acid is necessary for development of the ventral retina in zebrafish. Proc. Natl. Acad. Sci. 91, 7286–7290.

McCormick, S., Hasegawa, S., Hirano, T., 1992. Calcium uptake in the skin of a freshwater teleost. Proc. Natl. Acad. Sci. U. S. A. 89, 3635–3638.

McLean, E., Ash, R., 1987. Intact protein (antigen) absorption in fishes: mechanism and physiological significance. J. Fish. Biol. 31, 219–223.

McLean, E., Donaldson, E.M., 1990. Absorption of bioactive proteins by the gastrointestinal tract of fish: a review. J. Aquat. Anim. Health 2, 1–11.

Mello, G.C., Santos, M.L., Arantes, F.P., Pessali, T.C., Brito, M.F., Santos, J.E., 2019. Morphological characterisation of the digestive tract of the catfish *Lophiosilurus alexandri* Steindachner, 1876 (Siluriformes, Pseudopimelodidae). Acta Zool. 100, 14–23.

Montalban-Arques, A., De Schryver, P., Bossier, P., Gorkiewicz, G., Mulero, V., Gatlin III, D.M., Galindo-Villegas, J., 2015. Selective manipulation of the gut microbiota improves immune status in vertebrates. Front. Immunol. 6, 512.

Morais, S., 2017. The physiology of taste in fish: potential implications for feeding stimulation and gut chemical sensing. Rev. Fish. Sci. & Aquac. 25, 133–149.

Moren, M., Sloth, J.J., Hamre, K., 2008. Uptake of iodide from water in Atlantic halibut larvae (*Hippoglossus hippoglossus* L.). Aquaculture 285, 174–178.

Murashita, K., Rønnestad, I., Furuita, H., Matsunari, H., Oku, H., Yamamoto, T., 2018. Effects of dietary soybean meal on the bile physiology in rainbow trout, *Oncorhynchus mykiss*. Aquaculture 490, 303–310.

Nadella, S.R., Grosell, M., Wood, C.M., 2006. Physical characterization of high-affinity gastrointestinal Cu transport in vitro in freshwater rainbow trout *Oncorhynchus mykiss*. J. Comp. Physiol. B 176, 793–806.

Nadella, S.R., Grosell, M., Wood, C.M., 2007. Mechanisms of dietary Cu uptake in freshwater rainbow trout: evidence for Na-assisted Cu transport and a specific metal carrier in the intestine. J. Comp. Physiol. B 177, 433–446.

Nadella, S.R., Hung, C.C., Wood, C.M., 2011. Mechanistic characterization of gastric copper transport in rainbow trout. J. Comp. Physiol. B 181, 27–41.

Nagalla, S.R., Barry, B.J., Spindel, E.R., 1994. Cloning of complementary DNAs encoding the amphibian bombesin-like peptides Phe8 and Leu8 phyllolitorin from *Phyllomedusa sauvagei*: potential role of U to C RNA editing in generating neuropeptide diversity. Mol. Endocrinol. 8, 943–951.

Nagalla, S.R., Barry, B.J., Falick, A.M., Gibson, B.W., Taylor, J.E., Dong, J.Z., Spindel, E.R., 1996. There are three distinct forms of bombesin Identification of [Leu] bombesin,[Phe] bombesin, and [Ser3, Arg, Phe] bombesin in the frog *Bombina orientalis*. J. Biol. Chem. 271, 7731–7737.

Nalbant, P., Boehmer, C., Dehmelt, L., Wehner, F., Werner, A., 1999. Functional characterization of a Na+–phosphate cotransporter (NaPi-II) from zebrafish and identification of related transcripts. J. Physiol. 520, 79–89.

Nandhini, A.A., Balakrishnan, S., Anuradha, C., 2002. Taurine improves lipid profile in rats fed a high fructose-diet. Nutr. Res. 22, 343–354.

Nelson, L.E., Sheridan, M.A., 2005. Regulation of somatostatins and their receptors in fish. Gen. Comp. Endocrinol. 142, 117–133.

Nelson, L.E., Sheridan, M.A., 2006. Gastroenteropancreatic hormones and metabolism in fish. Gen. Comp. Endocrinol. 148, 116–124.

Ng, A.N., de Jong-Curtain, T.A., Mawdsley, D.J., White, S.J., Shin, J., Appel, B., Dong, P.D.S., Stainier, D.Y., Heath, J.K., 2005. Formation of the digestive system in zebrafish: III. Intestinal epithelium morphogenesis. Dev. Biol. 286, 114–135.

Ni, J., Yan, Q., Yu, Y., Zhang, T., 2014. Factors influencing the grass carp gut microbiome and its effect on metabolism. FEMS Microbiol. Ecol. 87, 704–714.

Nicol, J.A.C., Somiya, H., 1989. The Eyes of Fishes. Oxford University Press.

Nikolsky, G., 1963. In: Birkett, T.L. (Ed.), The Ecology of Fishes. Academic Press, London.

Nilsson, S., 1983. Autonomic Nerve Function in the Vertebrates. Springer Verlag.

Nordrum, S., Bakke-McKellep, A., Krogdahl, Å., Buddington, R., 2000. Effects of soybean meal and salinity on intestinal transport of nutrients in Atlantic salmon (*Salmo salar* L.) and rainbow trout (*Oncorhynchus mykiss*). Comp. Biochem. Physiol. B Biochem. Mol. Biol. 125, 317–335.

Obirikorang, K.A., Gyamfi, S., Goode, M.E., Amisah, S., Edziyie, R.E., Quagrainie, K., Egna, H., Frimpong, E., 2020. Effect of soybean meal diets on the growth performance, ammonia excretion rates, gut histology and feed cost of Nile tilapia (*Oreochromis niloticus*) fry. Aquacult. Res. 51, 3520–3532.

Obst, M.M., 1919. A bacteriologic study of sardines. J. Infect. Dis. 158–169.

Ogawa, K., Caprio, J., 2000. Glossopharyngeal taste responses of the channel catfish to binary mixtures of amino acids. Chem. Senses 25, 501–506.

Oja, S.a., Kontro, P., 1983. Taurine. In: Metabolism in the Nervous System. Springer, pp. 501–533.

Olsson, C., 2009. Autonomic innervation of the fish gut. Acta Histochem. 111, 185–195.

Olsson, C., Aldman, G., Larsson, A., Holmgren, S., 1999. Cholecystokinin affects gastric emptying and stomach motility in the rainbow trout *Oncorhynchus mykiss*. J. Exp. Biol. 202, 161–170.

Olsson, C., Holmgren, S., 2001. The control of gut motility. Comp. Biochem. Physiol. Mol. Integr. Physiol. 128, 479–501.

Olsson, C., Holmgren, S., 2011. Autonomic control of gut motility: a comparative view. Auton. Neurosci. 165, 80–101.

Ørnsrud, R., Lock, E., Glover, C., Flik, G., 2009. Retinoic acid cross-talk with calcitriol activity in Atlantic salmon (*Salmo salar*). J. Endocrinol. 202, 473.

Oxley, A., Jutfelt, F., Sundell, K., Olsen, R.E., 2007. Sn-2-monoacylglycerol, not glycerol, is preferentially utilised for triacylglycerol and phosphatidylcholine biosynthesis in Atlantic salmon (*Salmo salar* L.) intestine. Comp. Biochem. Physiol. B Biochem. Mol. Biol. 146, 115–123.

Pan, Q.-S., Fang, Z.-P., Zhao, Y.-X., 2000. Immunocytochemical identification and localization of APUD cells in the gut of seven stomachless teleost fishes. World J. Gastroenterol. 6, 96.

Pan, T.-C., Liao, B.-K., Huang, C.-J., Lin, L.-Y., Hwang, P.-P., 2005. Epithelial Ca^{2+} channel expression and Ca^{2+} uptake in developing zebrafish. Am. J. Physiol. Regul. Integr. Comp. Physiol. 289, R1202–R1211.

Panserat, S., Plagnes-Juan, E., Kaushik, S., 2001. Nutritional regulation and tissue specificity of gene expression for proteins involved in hepatic glucose metabolism in rainbow trout (*Oncorhynchus mykiss*). J. Exp. Biol. 204, 2351–2360.

Park, J., Levic, D.S., Sumigray, K.D., Bagwell, J., Eroglu, O., Block, C.L., Eroglu, C., Barry, R., Lickwar, C.R., Rawls, J.F., 2019. Lysosome-rich enterocytes mediate protein absorption in the vertebrate gut. Dev. Cell 51, 7–20 e6.

Patel, Y.C., 1999. Somatostatin and its receptor family. Front. Neuroendocrinol. 20, 157–198.

Perry, W.B., Lindsay, E., Payne, C.J., Brodie, C., Kazlauskaite, R., 2020. The role of the gut microbiome in sustainable teleost aquaculture. Proc. R. Soc. B 287, 20200184.

Piazzon, M.C., Calduch-Giner, J.A., Fouz, B., Estensoro, I., Simó-Mirabet, P., Puyalto, M., Karalazos, V., Palenzuela, O., Sitjà-Bobadilla, A., Pérez-Sánchez, J., 2017. Under control: how a dietary additive can restore the gut microbiome and proteomic profile, and improve disease resilience in a marine teleostean fish fed vegetable diets. Microbiome 5, 164.

Pierce, M., Wang, Y., Denovan-Wright, E.M., Wright, J.M., 2000. Nucleotide sequence of a cDNA clone coding for an intestinal-type fatty acid binding protein and its tissue-specific expression in zebrafish (*Danio rerio*). Biochim. Biophys. Acta Gene Struct. Expr. 1490, 175–183.

Planas, J.V., Méndez, E., Baños, N., Capilla, E., Castillo, J., Navarro, I., Gutiérrez, J., 2000. Fish insulin, IGF-I and IGF-II receptors: a phylogenetic approach. Am. Zool. 40, 223–233.

Plisetskaya, E.M., Dunguay, S.J., 1993. Pancreatic hormones and metabolism in ectotherm vertebrates: current views. In: Schreibman, M.P., Scanes, C.G., Pang, P.K.T. (Eds.), The Endocrinology of Growth, Development, and Metabolism in Vertebrates. Academic Press, pp. 265–287.

Plisetskaya, E.M., Mommsen, T.P., 1996. Glucagon and glucagon-like peptides in fishes. In: International Review of Cytology, vol. 168. Elsevier, pp. 187–257.

Pollock, H., Kimmel, J., Ebner, K., Hamilton, J., Rouse, J., Lance, V., Rawitch, A., 1988. Isolation of alligator gar (*Lepisosteus spatula*) glucagon, oxyntomodulin, and glucagon-like peptide: amino acid sequences of oxyntomodulin and glucagon-like peptide. Gen. Comp. Endocrinol. 69, 133–140.

Qiu, A., Hogstrand, C., 2004. Functional characterisation and genomic analysis of an epithelial calcium channel (ECaC) from pufferfish, *Fugu rubripes*. Gene 342, 113–123.

Ray, A.K., Ringø, E., 2014. The gastrointestinal tract of fish. In: Aquaculture Nutrition: Gut Health, Probiotics and Prebiotics. Wiley-Blackwell, p. 488.

Ray, A., Ghosh, K., Ringø, E., 2012. Enzyme-producing bacteria isolated from fish gut: a review. Aquacult. Nutr. 18, 465–492.

Reinecke, M., Zaccone, G., Kapoor, B.G., 2006. Fish Endocrinology. Science Publishers, Enfield, NH.

Reshkin, S.J., Ahearn, G.A., 1987a. Basolateral glucose transport by intestine of teleost, *Oreochromis mossambicus*. Am. J. Physiol. Regul. Integr. Comp. Physiol. 252, R579–R586.

Reshkin, S.J., Ahearn, G.A., 1987b. Intestinal glucose transport and salinity adaptation in a euryhaline teleost. Am. J. Physiol. Regul. Integr. Comp. Physiol. 252, R567–R578.

Reutter, K., Kapoor, B.G., 2005. Fish Chemosenses. Science Publishers, Enfield, NH, USA.

Rimoldi, S., Bossi, E., Harpaz, S., Cattaneo, A.G., Bernardini, G., Saroglia, M., Terova, G., 2015. Intestinal B0AT1 (SLC6A19) and PEPT1 (SLC15A1) mRNA levels in European sea bass (*Dicentrarchus labrax*) reared in fresh water and fed fish and plant protein sources. J. Nutr. Sci. 4, e21.

Robinson, J.S., Mead, J.F., 1973. Lipid absorption and deposition in rainbow trout (*Salmo gairdnerii*). Can. J. Biochem. 51, 1050–1058.

Rogie, A., Skinner, E.R., 1985. The roles of the intestine and liver in the biosynthesis of plasma lipoproteins in the rainbow trout, *Salmo gairdnerii* Richardson. Comp. Biochem. Physiol. A Part B: Comp. Biochem. 81, 285–289.

Romano, N., Kumar, V., Yang, G., Kajbaf, K., Rubio, M.B., Overturf, K., Brezas, A., Hardy, R., 2020. Bile acid metabolism in fish: disturbances caused by fishmeal alternatives and some mitigating effects from dietary bile inclusions. Rev. Aquac.

Romano, A., Kottra, G., Barca, A., Tiso, N., Maffia, M., Argenton, F., Daniel, H., Storelli, C., Verri, T., 2006. High-affinity peptide transporter PEPT2 (SLC15A2) of the zebrafish *Danio rerio*: functional properties, genomic organization, and expression analysis. Physiol. Genom. 24, 207–217.

Rønnestad, I., Hamre, K., Lie, Ø., Waagbø, R., 1999. Ascorbic acid and α-tocopherol levels in larvae of Atlantic halibut before and after exogenous feeding. J. Fish. Biol. 55, 720–731.

Rønnestad, I., Yúfera, M., Ueberschär, B., Ribeiro, L., Sæle, Ø., Boglione, C., 2013. Feeding behaviour and digestive physiology in larval fish: current knowledge, and gaps and bottlenecks in research. Rev. Aquacult. 5, S59–S98.

Rønnestad, I., Gomes, A.S., Murashita, K., Angotzi, R., Jönsson, E., Volkoff, H., 2017. Appetite-controlling endocrine systems in teleosts. Front. Endocrinol. 8, 73.

Rose, R.C., Choi, J., 1990. Intestinal absorption and metabolism of ascorbic acid in rainbow trout. Am. J. Physiol. Regul. Integr. Comp. Physiol. 258, R1238–R1241.

Rumsey, G.L., 1993. Fish meal and alternate sources of protein in fish feeds update 1993. Fisheries 18, 14–19.

Russell, D.W., 2009. Fifty years of advances in bile acid synthesis and metabolism. JLR (J. Lipid Res.) 50, S120–S125.

Rust, M.B., 2002. Nutritional physiology. In: Fish Nutrition. Academic Press, Amsterdam, pp. 367–452.

Sala-Rabanal, M., Gallardo, M., Sanchez, J., Planas, J., 2004. Na-dependent D-glucose transport by intestinal brush border membrane vesicles from gilthead sea bream (*Sparus aurata*). J. Membr. Biol. 201, 85–96.

Salinas, I., 2015. The mucosal immune system of teleost fish. Biology 4, 525–539.

Salinas, I., Magadán, S., 2017. Omics in fish mucosal immunity. Dev. Comp. Immunol. 75, 99–108.

Salze, G.P., Davis, D.A., 2015. Taurine: a critical nutrient for future fish feeds. Aquaculture 437, 215–229.

Sangaletti, R., Terova, G., Peres, A., Bossi, E., Corà, S., Saroglia, M., 2009. Functional expression of the oligopeptide transporter PepT1 from the sea bass (*Dicentrarchus labrax*). Pflueg. Arch. Eur. J. Physiol. 459, 47.

Santos, K.O., Costa-Filho, J., Spagnol, K.L., Nornberg, B.F., Lopes, F.M., Tesser, M.B., Marins, L.F., 2020. The inclusion of a transgenic probiotic expressing recombinant phytase in a diet with a high content of vegetable matter markedly improves growth performance and the expression of growth-related genes and other selected genes in zebrafish. Aquaculture 519, 734878.

Satou, M., 1992. Synaptic organization of the olfactory bulb and its central projection. In: Fish Chemoreception. Springer, pp. 40–59.

Schmidt, V., Amaral-Zettler, L., Davidson, J., Summerfelt, S., Good, C., 2016. Influence of fishmeal-free diets on microbial communities in Atlantic salmon (*Salmo salar*) recirculation aquaculture systems. Appl. Environ. Microbiol. 82, 4470–4481.

Semova, I., Carten, J.D., Stombaugh, J., Mackey, L.C., Knight, R., Farber, S.A., Rawls, J.F., 2012. Microbiota regulate intestinal absorption and metabolism of fatty acids in the zebrafish. Cell Host & Microbe 12, 277–288.

Sharma, M.K., Denovan-Wright, E.M., Degrave, A., Thisse, C., Thisse, B., Wright, J.M., 2004. Sequence, linkage mapping and early developmental expression of the intestinal-type fatty acid-binding protein gene (fabp2) from zebrafish (*Danio rerio*). Comp. Biochem. Physiol. B Biochem. Mol. Biol. 138, 391–398.

Sheng, Y., Ren, H., Limbu, S.M., Sun, Y., Qiao, F., Zhai, W., Du, Z.-Y., Zhang, M., 2018. The presence or absence of intestinal microbiota affects lipid deposition and related genes expression in zebrafish (*Danio rerio*). Front. Microbiol. 9, 1124.

Sheperd, B., Eckert, S., Parhar, I., Vijayan, M., Wakabayashi, I., Hirano, T., Grau, E., Chen, T., 2000. The hexapeptide KP-102 (D-Ala-D-Nal-Ala-Trp-D-Phe-Lys-NH2) stimulates growth hormone release in a cichlid fish (*Oreochromis mossambicus*). J. Endocrinol. 167, R7–R10.

Sheridan, M.A., 1994. Regulation of lipid metabolism in poikilothermic vertebrates. Comp. Biochem. Physiol. A Part B: Comp. Biochem. 107, 495–508.

Sheridan, M.A., Kittilson, J.D., 2004. The role of somatostatins in the regulation of metabolism in fish. Comp. Biochem. Physiol. B Biochem. Mol. Biol. 138, 323–330.

Shields, R.J., Bell, J.G., Luizi, F.S., Gara, B., Bromage, N.R., Sargent, J.R., 1999. Natural copepods are superior to enriched Artemia nauplii as feed for halibut larvae (*Hippoglossus hippoglossus*) in terms of survival, pigmentation and retinal morphology: relation to dietary essential fatty acids. J. Nutr. 129, 1186–1194.

Silverstein, J.T., Bondareva, V.M., Leonard, J.B., Plisetskaya, E.M., 2001. Neuropeptide regulation of feeding in catfish, *Ictalurus punctatus*: a role for glucagon-like peptide-1 (GLP-1)? Comp. Biochem. Physiol. B Biochem. Mol. Biol. 129, 623–631.

Sire, M., Dorin, D., Vernier, J., 1992. Intestinal absorption of macromolecular proteins in rainbow trout. Aquaculture 100, 234–235.

Sis, R., Ives, P., Jones, D., Lewis, D., Haensly, W., 1979. The microscopic anatomy of the oesophagus, stomach and intestine of the channel catfish, *Ictalurus punctatus*. J. Fish. Biol. 14, 179–186.

Small, B.C., Quiniou, S.M., Kaiya, H., 2009. Sequence, genomic organization and expression of two channel catfish, *Ictalurus punctatus*, ghrelin receptors. Comp. Biochem. Physiol. Mol. Integr. Physiol. 154, 451–464.

Small, B.C., Quiniou, S.M., Kaiya, H., Bledsoe, J.W., Musungu, B., 2019. Characterization of a third ghrelin receptor, GHS-R3a, in channel catfish reveals novel expression patterns and a high affinity for homologous ligand. Comp. Biochem. Physiol. Mol. Integr. Physiol. 229, 1–9.

Smith, L.S., 1989. Digestive functions in teleost fishes. In: Halver, J.E. (Ed.), Fish Nutrition, second ed. Academic Press Inc., San Diego, USA, pp. 331–421.

Smith, G., 1999. Introduction to the reviews on peptides and the control of food intake and body weight. Neuropeptides 33, 323–328.

Soengas, J.L., Moon, T.W., 1998. Transport and metabolism of glucose in isolated enterocytes of the black bullhead *Ictalurus melas*: effects of diet and hormones. J. Exp. Biol. 201, 3263–3273.

Spindel, E., 1986. Mammalian bombesin-like peptides. Trends Neurosci. 9, 130–133.

Spindel, E., Giladi, E., Brehm, P., Goodman, R., Segerson, T., 1990. Cloning and functional characterization of a complementary DNA encoding the murine fibroblast bombesin/gastrin-releasing peptide receptor. Mol. Endocrinol. 4, 1956–1963.

Storelli, C., Vilella, S., Cassano, G., 1986. Na-dependent D-glucose and L-alanine transport in eel intestinal brush border membrane vesicles. Am. J. Physiol. Regul. Integr. Comp. Physiol. 251, R463–R469.

Storelli, C., Vilella, S., Romano, M.P., Maffia, M., Cassano, G., 1989. Brush-border amino acid transport mechanisms in carnivorous eel intestine. Am. J. Physiol. Regul. Integr. Comp. Physiol. 257, R506–R510.

Subramaniam, M., Weber, L.P., Loewen, M.E., 2019. Intestinal electrogenic sodium-dependent glucose absorption in tilapia and trout reveal species differences in SLC5A-associated kinetic segmental segregation. Am. J. Physiol. Regul. Integr. Comp. Physiol. 316, R222–R234.

Sugita, H., Kawasaki, J., Kumazawa, J., Deguchi, Y., 1996. Production of amylase by the intestinal bacteria of Japanese coastal animals. Lett. Appl. Microbiol. 23, 174–178.

Sugita, H., Miyajima, C., Deguchi, Y., 1991. The vitamin B12-producing ability of the intestinal microflora of freshwater fish. Aquaculture 92, 267–276.

Sugiura, S.H., Ferraris, R.P., 2004. Contributions of different NaPi cotransporter isoforms to dietary regulation of P transport in the pyloric caeca and intestine of rainbow trout. J. Exp. Biol. 207, 2055–2064.

Sugiura, S.H., McDaniel, N.K., Ferraris, R.P., 2003. In vivo fractional Pi absorption and NaPi-II mRNA expression in rainbow trout are upregulated by dietary P restriction. Am. J. Physiol. Regul. Integr. Comp. Physiol. 285, R770–R781.

Sullivan, L.J., Ignoffo, T.R., Baskerville-Bridges, B., Ostrach, D.J., Kimmerer, W.J., 2016. Prey selection of larval and juvenile planktivorous fish: impacts of introduced prey. Environ. Biol. Fish. 99, 633–646.

Sundell, K., Björnsson, B.T., Itoh, H., Kawauchi, H., 1992. Chum salmon (*Oncorhynchus keta*) stanniocalcin inhibitis in vitro intestinal calcium uptake in Atlantic cod (*Gadus morhua*). J. Comp. Physiol. B 162, 489–495.

Suyehiro, Y., 1942. A study on the digestive system and feeding habits of fish. Jpn. J. Zool. 10, 1–303.

Takei, Y., Loretz, C.A., 2010. The gastrointestinal tract as an endocrine/neuroendocrine/paracrine organ: organization, chemical messengers and physiological targets. Fish Physiol. 30, 261–317 (Elsevier).

Takeuchi, T., Park, G.S., Seikai, T., Yokoyama, M., 2001. Taurine content in Japanese flounder *Paralichthys olivaceus* T. & S. and red sea bream *Pagrus major* T. & S. during the period of seed production. Aquacult. Res. 32, 244–248.

Taylor, J.F., Waagbø, R., Diez-Padrisa, M., Campbell, P., Walton, J., Hunter, D., Matthew, C., Migaud, H., 2015. Adult triploid Atlantic salmon (Salmo salar) have higher dietary histidine requirements to prevent cataract development in seawater. Aquac. Nutr. 21, 18–32.

Terova, G., Corà, S., Verri, T., Rimoldi, S., Bernardini, G., Saroglia, M., 2009. Impact of feed availability on PepT1 mRNA expression levels in sea bass (*Dicentrarchus labrax*). Aquaculture 294, 288–299.

Thamotharan, M., Zonno, V., Storelli, C., Ahearn, G., 1996a. Basolateral dipeptide transport by the intestine of the teleost *Oreochromis mossambicus*. Am. J. Physiol. Regul. Integr. Comp. Physiol. 270, R948–R954.

Thamotharan, M., Gomme, J., Zonno, V., Maffia, M., Storelli, C., Ahearn, G., 1996b. Electrogenic, proton-coupled, intestinal dipeptide transport in herbivorous and carnivorous teleosts. Am. J. Physiol. Regul. Integr. Comp. Physiol. 270, R939–R947.

Tocher, D.R., 2003. Metabolism and functions of lipids and fatty acids in teleost fish. Rev. Fish. Sci. 11, 107–184.

Tremaroli, V., Bäckhed, F., 2012. Functional interactions between the gut microbiota and host metabolism. Nature 489, 242–249.

Unniappan, S., Lin, X., Cervini, L., Rivier, J., Kaiya, H., Kangawa, K., Peter, R.E., 2002. Goldfish ghrelin: molecular characterization of the complementary deoxyribonucleic acid, partial gene structure and evidence for its stimulatory role in food intake. Endocrinology 143, 4143–4146.

Urán, P., Gonçalves, A., Taverne-Thiele, J., Schrama, J., Verreth, J., Rombout, J., 2008. Soybean meal induces intestinal inflammation in common carp (*Cyprinus carpio* L.). Fish Shellfish Immunol. 25, 751–760.

Van den Ingh, T., Olli, J., Krogdahl, Å., 1996. Alcohol-soluble components in soybeans cause morphological changes in the distal intestine of Atlantic salmon, *Salmo salar* L. J. Fish. Dis. 19, 47–53.

Van der Velden, J., Groot, J., Flik, G., Polak, P., Kolar, Z., 1990. Magnesium transport in fish intestine. J. Exp. Biol. 152, 587–592.

Venold, F.F., Penn, M.H., Krogdahl, Å., Overturf, K., 2012. Severity of soybean meal induced distal intestinal inflammation, enterocyte proliferation rate, and fatty acid binding protein (Fabp2) level differ between strains of rainbow trout (*Oncorhynchus mykiss*). Aquaculture 364, 281–292.

Venold, F.F., Penn, M.H., Thorsen, J., Gu, J., Kortner, T.M., Krogdahl, Å., Bakke, A.M., 2013. Intestinal fatty acid binding protein (fabp2) in Atlantic salmon (*Salmo salar*): localization and alteration of expression during development of diet induced enteritis. Comp. Biochem. Physiol. Mol. Integr. Physiol. 164, 229–240.

Verri, T., Werner, A., 2019. Type II Na+-phosphate cotransporters and phosphate balance in teleost fish. Pflueg. Arch. Eur. J. Physiol. 471, 193–212.

Verri, T., Maffia, M., Danieli, A., Herget, M., Wenzel, U., Daniel, H., Storelli, C., 2000. Characterisation of the H(+)/peptide cotransporter of eel intestinal brush-border membranes. J. Exp. Biol. 203, 2991–3001.

Verri, T., Kottra, G., Romano, A., Tiso, N., Peric, M., Maffia, M., Boll, M., Argenton, F., Daniel, H., Storelli, C., 2003. Molecular and functional characterisation of the zebrafish (*Danio rerio*) PEPT1-type peptide transporter 1. FEBS Lett. 549, 115–122.

Vila, A.V., Imhann, F., Collij, V., Jankipersadsing, S.A., Gurry, T., Mujagic, Z., Kurilshikov, A., Bonder, M.J., Jiang, X., Tigchelaar, E.F., 2018. Gut microbiota composition and functional changes in inflammatory bowel disease and irritable bowel syndrome. Sci. Transl. Med. 10.

Vilella, S., Reshkin, S.J., Storelli, C., Ahearn, G.A., 1989. Brush-border inositol transport by intestines of carnivorous and herbivorous teleosts. Am. J. Physiol. Gastrointest. Liver Physiol. 256, G501–G508.

Volkoff, H., 2016. The neuroendocrine regulation of food intake in fish: a review of current knowledge. Front. Neurosci. 10, 540.

Volkoff, H., 2019. Fish as models for understanding the vertebrate endocrine regulation of feeding and weight. Mol. Cell. Endocrinol. 497, 110437.

Volkoff, H., Peter, R.E., 2000. Effects of CART peptides on food consumption, feeding and associated behaviors in the goldfish, *Carassius auratus*: actions on neuropeptide Y-and orexin A-induced feeding. Brain Res. 887, 125–133.

Volkoff, H., Canosa, L., Unniappan, S., Cerda-Reverter, J., Bernier, N., Kelly, S., Peter, R., 2005. Neuropeptides and the control of food intake in fish. Gen. Comp. Endocrinol. 142, 3–19.

Waagbø, R., Tröße, C., Koppe, W., Fontanillas, R., Breck, O., 2010. Dietary histidine supplementation prevents cataract development in adult Atlantic salmon, Salmo salar L., in seawater. Br. J. Nutr. 104, 1460–1470.

Wagner, G., McKeown, B., 1981. Immunocytochemical localization of hormone-producing cells within the pancreatic islets of the rainbow trout (*Salmo gairdneri*). Cell Tissue Res. 221, 181–192.

Wallace, K.N., Akhter, S., Smith, E.M., Lorent, K., Pack, M., 2005. Intestinal growth and differentiation in zebrafish. Mech. Dev. 122, 157–173.

Walls, G.L., 1942. The Vertebrate Eye and its Adaptive Radiation. Cranbrook Institute of Science. Cranbrook Press.

Wang, J., Yan, X., Lu, R., Meng, X., Nie, G., 2017. Peptide transporter 1 (PepT1) in fish: a review. Aquac. & Fish. 2, 193–206.

Wang, C., Zhu, X., Han, D., Jin, J., Yang, Y., Xie, S., 2015. Responses to fishmeal and soybean meal-based diets by three kinds of larval carps of different food habits. Aquacult. Nutr. 21, 552–568.

Weinrauch, A.M., Schaefer, C.M., Goss, G.G., 2019. Activity and post-prandial regulation of digestive enzyme activity along the Pacific hagfish (*Eptatretus stoutii*) alimentary canal. PloS One 14, e0215027.

Werner, A., Murer, H., Kinne, R., 1994. Cloning and expression of a renal Na-Pi cotransport system from flounder. Am. J. Physiol. Ren. Physiol. 267, F311–F317.

Wilson, J., Castro, L., 2010. Morphological diversity of the gastrointestinal tract in fishes. Fish Physiol. 30, 1–55 (Elsevier).

Wong, A.O., Li, W.S., Lee, E.K., Leung, M.Y., Tse, L.Y., Chow, B.K., Lin, H.R., Chang, J.P., 2000. Pituitary adenylate cyclase activating polypeptide as a novel hypophysiotropic factor in fish. Biochem. Cell. Biol. 78, 329–343.

Wongdee, K., Charoenphandhu, N., 2013. Regulation of epithelial calcium transport by prolactin: from fish to mammals. Gen. Comp. Endocrinol. 181, 235–240.

Woodhead, P., 1966. The behaviour of fish in relation to light in the sea. Oceanogr. Mar. Biol. Annu. Rev. 4, 337–403.

Yokoyama, M., Takeuchi, T., Park, G., Nakazoe, J., 2001. Hepatic cysteinesulphinate decarboxylase activity in fish. Aquacult. Res. 32, 216–220.

Yoshida, K., Iwanaga, T., Fujita, T., 1983. Gastro-entero-pancreatic (GEP) endocrine system of the flatfish, *Paralichtys olivaceus*: an immunocytochemical study. Arch. Histol. Jpn. 46, 259–266.

Yoshimizu, M., Kimura, T., 1976. Study on the intestinal microflora of salmonids. Fish Pathol. 10, 243–259.

Yukgehnaish, K., Kumar, P., Sivachandran, P., Marimuthu, K., Arshad, A., Paray, B.A., Arockiaraj, J., 2020. Gut microbiota metagenomics in aquaculture: factors influencing gut microbiome and its physiological role in fish. Rev. Aquac.

Yun, B., Mai, K., Zhang, W., Xu, W., 2011. Effects of dietary cholesterol on growth performance, feed intake and cholesterol metabolism in juvenile turbot (*Scophthalmus maximus* L.) fed high plant protein diets. Aquaculture 319, 105–110.

Zeiske, E., Theisen, B., Breucker, H., 1992. Structure, development, and evolutionary aspects of the peripheral olfactory system. In: Fish Chemoreception. Springer, pp. 13–39.

Zhang, W., Tan, B., Ye, G., Wang, J., Dong, X., Yang, Q., Chi, S., Liu, H., Zhang, S., Zhang, H., 2019. Identification of potential biomarkers for soybean meal-induced enteritis in juvenile pearl gentian grouper, *Epinephelus lanceolatus*♂×*Epinephelus fuscoguttatus*♀. Aquaculture 512, 734337.

Zheng, J., Yang, P., Dai, J., Yu, G., Ou, W., Xu, W., Mai, K., Zhang, Y., 2020. Dynamics of intestinal inflammatory cytokines and tight junction proteins of turbot (*Scophthalmus maximus* L.) during the development and recovery of enteritis induced by dietary β-conglycinin. Front. Mar. Sci. https://doi.org/10.3389/fmars.2020.00198.

CHAPTER

9

Diet formulation and manufacture

Ronald W. Hardy[3] *and Andreas Brezas*[1,2]

[1]Hagerman Fish Culture Experiment Station, University of Idaho, Hagerman, ID, United States; [2]CJ Europe GmbH, Main Airport Center, Frankfurt am Main, Germany; [3]Aquaculture Research Institute, Hagerman Fish Culture Experiment Station, University of Idaho, Hagerman, ID, United States

9.1 Introduction

Diet formulation and diet manufacture are processes anchored in practical reality rather than on ideal concepts. Together, diet formulation and manufacture involves the selection and combination of feed ingredient to form products that deliver the nutrients needed to meet production goals in animal and fish husbandry. Formulation and manufacture are linked by compromises that each impose on the other. Feed formulations that are nutritionally complete and formulated to least-cost specifications are relatively simple to develop, especially when using feed formulation software. However, formulations based on nutrient content alone may be impossible to manufacture into pellets that possess characteristics necessary for their use, such as water stability or a specific density and buoyancy. Similarly, formulations designed to optimize the manufacturing process may not be the best mixture to support rapid, economical fish growth. The goal of both formulation and manufacture of feeds is to maximize delivery and utilization of the nutrients in the mixture to the animal or fish for which it has been produced. Diet formulation and manufacture, thus, require a combination of nutritional knowledge and production experience that must be gained by both study and practice.

Feeds are formulated and manufactured to meet specific production goals in aquaculture. Production goals for fish larvae or fry are rapid growth and high survival. Both depend on maintaining the quality of rearing water, so producing feed particles that are water stable yet highly attractive and digestible to fish is a high priority. In contrast, production goals for postjuvenile fish are related to supporting efficient and economical fish growth. As a result, feeds for these distinct life stages of fish will differ both in formulation and manufacturing processes. Similarly, feeds for broodstock fish may be formulated to supply extra nutrients at key periods during gonadal maturation such that fecundity and egg quality are maximized.

Fish Nutrition, Fourth Edition
https://doi.org/10.1016/B978-0-12-819587-1.00002-1

In freshwater trout farming using flow-through water systems, feeds might be formulated to minimize release of enriching nutrients, e.g., phosphorus or suspended solids, to the aquatic environment, making it necessary to alter feed formulations by adding certain ingredients or limiting levels of others such that the cost of feed increases. However, regulatory restrictions on farm effluent nutrient content often make such changes necessary. Feed used in recirculating aquaculture systems (RAS) may differ from those used in flow-through rearing systems to minimize nitrogen or phosphorus excretion as well as accumulation of certain minerals such as copper and zinc that are required dietary nutrients but also toxic if water levels get too high. Placing restrictions on formulations for RAS feeds will increase feed costs but not necessarily increase the cost of fish production. Feeds used in research are designed to achieve specific goals, such as induction of a vitamin deficiency or establishment of a minimum dietary requirement. Highly refined ingredients containing extremely low levels of the nutrient(s) being studied must be used in research feeds, so that the nutrient being studied can be added to the experimental diet in stepwise increments. Regular feed ingredients cannot be used in such feeds because they supply an array of essential nutrients that complicate nutritional requirement studies. As a result, experimental feeds used in research studies are expensive. The preceding examples demonstrate that formulation and manufacturing practices depend on the intended purpose of the feed or the goal of production.

Production diets must be economical to manufacture, ship, store, and deliver to the fish. Manufacturing high-quality pellets requires that ingredients are ground to a particle size suitable for pelleting and mixed properly to produce a homogenous blend. The mixture of ingredients must contain sufficient binding capacity to form stable pellets that can withstand being conveyed in the feed mill and being handled during bagging or bulk transport to the end user, the fish farmer, without fracturing or disintegrating, producing fines. Storage stability, both physical and chemical, is another important attribute of fish feed.

Clearly feed formulation and feed manufacturing are not independent of each other but rather synergistic elements of feed production. When feeds are formulated, the physical characteristics of ingredient mixtures are just as important as the nutritional content of mixtures. Some aquaculture species consume pellets within seconds after feed is introduced to the water, but for other species, pellets must remain intact in water for up to several minutes until they are consumed. Water stability is a crucial property of pellets for such situations to minimize loss of nutrients through leaching and also to prevent disintegrating pellets from reducing water quality. Clearly many factors are involved in fish feed formulation and manufacturing, not simply the cost of the feed.

During the last 30 years, fish feed formulations and manufacturing practices have changed greatly, driven primarily by the demands of the aquaculture industry. New methods of feed and ingredient production, changes in the availability and quality of ingredients used in feeds, development of new feed ingredients, cultivation of new fish species, and advances in our knowledge of fish nutritional requirements have all influenced fish feed formulation and manufacture. Concerns about the effects of fish farming on aquatic environments along with long-standing concerns about the dependence of fish feeds on fish meal produced from fully utilized or overutilized marine resources have led to changes in fish feed formulations across species, mainly associated with reductions in the percentage of fish meal and fish oil used in formulations. An increasing number of retail seafood marketers now require that seafood is produced according to specific standards designed to enhance seafood quality, safety, and sustainable production. Certification programs include standards for sourcing feed ingredients and operating feed manufacturing facilities. Demand for aquaculture products is predicted to increase substantially in the coming years, and this will lead to higher fish feed production. Industry-driven advances in fish feeds over the past decade have significantly improved the efficiency of aquaculture, especially salmon farming. Over the next decade, further improvements are necessary to permit aquaculture to evolve into the socially beneficial enterprise that it ought to be.

9.1.1 Brief history of diet preparation

Prior to the beginning of the 20th century, farmed fish were produced using extensive farming systems, defined as those in which nutrients for fish were supplied by natural food production, often enhanced by pond fertilization. Early studies with trout showed that feed containing combinations of animal, fish, shrimp, and vegetable meals was not capable of supporting normal growth and health (Embody, 1918). However, when 15% of the feed consisted of fresh liver or kidney, fish growth and health were maintained. Vitamins had not yet been discovered and, therefore, were not supplemented in early fish diets. Fresh liver and kidney supplied the necessary vitamins to such feeds.

In the early years of the 20th century, biologists investigated the natural diet of trout, which varied somewhat with location and season, enumerating the species of aquatic creatures consumed, along with their relative proportions in the total diet. They analyzed the proximate composition of the natural diet and used this information as a guide to arrive at an appropriate proximate composition of trout feed (Embody and Gordon, 1924). The proximate composition of the natural diets was remarkably similar to the proximate composition of feeds fed to young salmon and trout today, i.e., 49% crude protein, 15%–16% fat, 8% crude fiber, and 10% ash, on a dry weight basis.

9.1.1.1 *Wet foods*

During the 1920s and 1930s, salmon and trout were fed a variety of farm or hatchery-made feeds produced using locally available ingredients. Typical ingredients included salmon eggs, fresh, canned, or frozen fish, oilseed meals, and brewer's yeast, combined with beef liver, spleen, horse meat, chicken eggs, and cottage cheese. Mixtures such as one-third beef liver, one-third hog liver, and one-third salmon viscera were chopped and mixed at the hatchery, 2% salt was added to congeal the mixture, and it was delivered to the fish by a hatchery worker using a spoon or ladle. This process was labor intensive, and since the mixture was wet and runny, water quality suffered. Feed for small fish was dropped through a fan to reduce particle size. These feeds contained about 60% moisture with a texture similar to moist sawdust (Hastings and Dickie, 2002). The feed cost per kg of fish weight gain ranged from $0.07 to $1.25 depending on the mixture (Donaldson and Foster, 1937).

9.1.1.2 *Wet-dry feed*

In the 1940s, the cost of fresh animal organs used in wet fish feeds increased due to expanded fish hatchery production and demand from other users. To extend the traditional ingredients, meat–meal mixtures for feeding salmon and trout were developed. These feeds were blends of slaughterhouse by-products and dry, commercially available feed ingredients. Cortland dry feed mixture No. 6, consisting of 24% each of dry skim milk, cottonseed meal, whitefish meal, and wheat middlings, with 4% salt added, was a typical dry feed mixture

(Phillips et al., 1940). This mixture was combined with an equal weight of beef liver and hog spleen, water was added, and the mixed material was squeezed through a potato ricer to feed small fish in troughs. Versions of the potato ricer were adopted to use compressed air for large-scale applications to feed fish in large raceways under production conditions (Pelnar, 1947). Feeding trials using brook trout at the Cortland Hatchery in New York indicated that growth rates equivalent to those obtained using a meat diet were achieved using the meat-dry meal mixtures. Feed cost per unit of production was reduced by about 50% (Tunison et al., 1941).

Today, in many areas of the world, farmed fish are fed farm-made feeds that combine fresh fish, fish, or shellfish processing by-products with meal mixtures similar to how salmon and trout feeds were made over 70 years ago. In some situations, fish processing by-products are preserved with acid to produce fish silage which is then combined with dry meal mixes to form moist feeds. The dry mixes contain fish meal, cooked starch, vitamin and mineral premixes, fish oil, and binders, such as seaweed or alginates. These are high moisture feeds that must be fed within several days. For many years in Japan, wet diets consisting mainly of groundfish, sometimes combined with binders (90:10 ratio), were the primary feed for yellowtail and salmon, but farmers have now transitioned to extruded pellets. Generally, farm-made feeds support excellent fish growth, but they are only practical when the availability and cost of fresh fish or fish processing waste and the cost of food preparation (labor) warrant their use.

9.1.1.3 Semimoist feeds

During the late 1940s and 1950s, the Oregon Fish Commission and the Seafood Laboratory of Oregon State University developed a wet mixture–dry mixture combination feed for juvenile salmon. At the time, a disease caused by *Mycobacterium piscium*, called fish tuberculosis, was a major problem in salmon hatcheries. This disease was not fatal to juvenile salmon, but it had serious consequences when fish infected as juveniles returned to hatcheries as maturing fish, namely incomplete gonadal development. Salmon hatcheries used maturing fish that returned to hatcheries as the source of eggs for the next generation. Salmon infected with fish tuberculosis did not contribute eggs to hatcheries, resulting in lower production of smolts released later to support commercial, recreational, and tribal fishing. After salmon were spawned at hatcheries, their carcasses were frozen and used as components of hatchery feeds fed to their offspring. Researchers discovered that chopped salmon carcasses or salmon viscera were the primary vector by which juvenile salmon were infected with *Mycobacterium piscium* (Wood, 1979). The pathogen was killed by heat treatment, so salmon carcasses and processing by-products had to be cooked (pasteurized) before being used as salmon feed ingredients. This eliminated vertical transmission of *Mycobacterium piscium*, and fish tuberculosis was no longer a problem in salmon hatcheries. Due to the necessity to introduce thermal processing step in salmon feed production, feeds were no longer made at salmon hatcheries but rather shifted to commercial companies according to specifications developed by the Oregon Fish Commission. The new feed was named the Oregon Moist Pellet (OMP). The OMP was cold pelleted using commercial pasta machines with external knives to cut pellets to appropriate lengths. Pellets were immediately quick-frozen, sacked, and stored frozen until they were fed at hatcheries (Crawford et al., 1973).

The OMP formula was refined over time, leading to three formulations, a mash for fry and two formulations for larger fish (Table 9.1). Commercial companies produced the OMP for state, federal, and tribal salmon hatcheries until 1985, when the State of Oregon ended its fish feed development program. Commercial companies then developed closed-formula, semimoist pellet feeds for salmon hatchery use. It is important to note that the OMP was developed for salmon hatcheries where fish were reared to the smolt stage for release to enhance commercial and recreational fisheries. The goal of hatcheries is to produce robust, healthy smolts, not to produce fish at a profit for consumption. Although the OMP was used in early stages of the salmon aquaculture industry in coastal states of the United States and Canada, its cost and the complex logistics associated with storage of frozen pellets rapidly led to its being replaced with dry feeds by salmon farms.

9.1.1.4 Compressed (steam) pellets

Grassl published the first study describing the successful use of dry pellets for trout rearing (Grassl, 1956). He reported that trout hatchery production increased by 60% and feed costs were reduced by 40% when compressed pellets replaced hatchery-made feeds. Grassl's formulas did not include a supplemental vitamin premix, so it was necessary to feed beef liver to the trout once every 3 weeks. The 10-cm fish reached stocking size of 18 cm after 7 months of feeding, a slow growth rate by today's standards.

Addition of a vitamin mixture to the dry pellet formulation made it possible to rear trout to spawning without supplemental feeding of liver or other organ meats. Open feed

TABLE 9.1 Oregon moist pellet specifications.

	Percent in diet		
Ingredient	**Oregon mash (OM-3)**	**OP-4**	**OP-2 (1/8 and larger pellets)**
Herring meal (or anchovy or hake meals up to $^{1}/_{2}$ fish meals) except mash	49.9	47.5	28.0
Wheat germ meal	10.0	Remainder	Remainder
Cottonseed meal (48.5%)	–	–	15.0
Dried whey product or dried whey	8.0	4.0	5.0
Corn distillers dried solids	–	–	4.0
Sodium bentonite	–	3.0	–
Trace mineral premix	0.1	0.1	0.1
Vitamin premix	1.5	1.5	1.5
Choline chloride (70%)	0.5	0.5	0.5
Wet fish	20.0	30.0	30.0
Fish oil	10.0	6.5–7.0	6.0–6.75

formulations for trout (Phillips et al., 1964) and salmon (Fowler and Burrows, 1971) were developed for use in government-operated hatcheries and were also adopted by commercial trout farms in the United States. Their use resulted in rapid growth of trout production through the 1960s and 1970s, eliminating the need for trout farmers to make their own feed. These formulations were the basis for feeds used in salmon farming in the state of Washington and British Columbia, Canada, although compressed pellets were subsequently replaced by pellets produced using cooking-extrusion technology. Specifications for several of these diets are shown in Table 9.2.

Dry, compressed pellets have obvious advantages over moist pellets. First, dry pellets do not require frozen storage; room temperature storage is sufficient for at least 90 days after manufacture. Second, dry pellets can be used in inexpensive, platform-type demand feeders. Third, dry pellets are less expensive to produce than moist pellets, especially when the cost of moist pellets is expressed on a dry weight basis. However, dry, compressed pellets have several disadvantages. First, some species of young Pacific salmon start feeding less readily on dry feed than on semimoist feed, particularly in cold water (<7°C). Second, dry, compressed pellets are dense and sink rapidly in water. Third, the upper limit of lipid content

TABLE 9.2 Open-formula diet specifications for dry salmon diets (Fowler, 1980).

	Percent in diet		
Ingredient	**Starter S8-2**	**Crumbles (18-2)**	**Pellets (19-2)**
Herring meal	58	55	50
Dried whey	5	5	5
Blood flour (or meal)	10	10	10
Condensed fish solubles	3	3	3
(or)			
Poultry by-product meal	1.5	1.5	1.5
Wheat germ meal	–	5	5
Wheat middlings mill run or shorts	Remainder	Remainder	Remainder
Vitamin premix	1.5	1.5	1.5
Choline chloride (60%)	0.58	0.58	0.58
Ascorbic acid	0.1	0.1	0.1
Trace mineral mixture	0.05	0.1	0.1
Lignin sulfonate pellet binder	2.0	2.0	2.0
Fish oil	12	9	9
(or)			
Soybean lecithin (max. 2%)			

of the feed mixture being pelleted must be restricted to enable pellets to be compressed sufficiently for durability. High-lipid content in the feed mixture lubricates the die through which the mixture is compressed, limiting pellet compression and density. After pelleting, additional lipid can be top-dressed to dried pellets, but there is an upper limit to the amount of lipid-compressed pellets will absorb due to pellet density. It is difficult to produce compressed pellets with total lipid levels above 18%–20%.

9.1.1.5 Extruded pellets

Extruded pellets are now the most common feed type used in aquaculture. Extruded pellets are made by forcing a heated, moist feed mixture through a barrel containing one or two screws that have decreasing distance between flights along the course of the screw within the barrel. As the feed mixture moves down the barrel, it is compressed to a high pressure that keeps feed mixture moisture in its liquid state as small water droplets. When the mixture exits suddenly through the die, pressure is released and water droplets instantly turn into vapor (steam), creating tiny air pockets throughout the pellet. This causes pellets to expand, and this reduces pellet density. After drying, pellets are top-dressed with lipid, often using vacuum chambers. Extruded pellets float, an obvious advantage in pond culture. Extruded pellets were adopted by the catfish industry in the United States and later by the Atlantic salmon industry in Scotland and Norway when formulations shifted to high-energy diets. Trout farmers were late adopters of extruded pellets because of problems that arose when extruded pellets were first introduced into the feed market in the United States in the 1970s. At the time, extruded feeds used formulations designed for compressed pellets that contained over 20% starch (Hilton and Slinger, 1984). Because cooking extrusion increases the percentage of starch gelatinization compared to compression pelleting, the resulting extruded pellets contained high levels of digestible carbohydrates. This overwhelmed the capacity of trout to metabolize glucose, causing liver abnormalities and high mortality in trout. Trout farmers blamed the extrusion process, not the formulation, for this problem. Extrusion technology was finally adopted by trout feed producers over 20 years later after researchers worked out the relationship between available carbohydrate levels in feeds and liver problems in salmonids and reduced the starch content of extruded feed formulations.

The advantages of extrusion pelleting over compression pelleting include lower pellet density, higher capacity to absorb top-dressed lipid, higher durability and, therefore, lower amounts of fines, and less starch needed as a nutritional binder due to higher gelatinization. Pellet density can be controlled by adjusting conditions of pelleting to produce floating, slowly sinking, or sinking pellets that are sufficiently durable to be used in automatic and demand feeders. For these reasons, the advantages of extruded pellets outweigh the higher cost of pelleting compared to compressed pellets.

9.1.1.6 Semipurified diets

The adaptation of semipurified research diets used in nutritional studies with rats or other lab animals for trout (McLaren et al., 1947) (Wolf, 1951), salmon (Halver, 1957), and catfish (Dupree, 1966) was the breakthrough that allowed researchers to estimate nutritional requirements of many species of cultivated fish. Semipurified diets contain highly refined ingredients with extremely low levels of vitamins and essential minerals required by fish. Essential vitamins and minerals must be added to the diet to support optimum growth and health, and by deleting one nutrient at a time and adding it back at increasing levels in experimental diets, the dietary requirements of fish for vitamins and some minerals were established. By substituting a mixture of crystalline amino acids for a portion of the casein and gelatin, semipurified diets were adapted to determine the amino acid requirements of fish (Dupree and Halver, 1970). Formulations of semipurified diets continue to be updated as new information on requirements for essential nutrients becomes available. There are many examples of semipurified diets developed for studies of different essential nutrient requirements, e.g., arginine, phosphorus, zinc, and others. Semipurified diet formulations are also modified to be palatable to species of fish other than salmonids or catfish. Examples of several are shown in Table 9.3.

9.1.1.7 Microdiets

The most challenging types of fish feed to produce are microdiets to feed larvae. Larval feed particles must be extremely small, homogenous in terms of nutrient content, water-

TABLE 9.3 Semipurified diet formulations for salmonids (g/kg diet).[a]

Ingredient	H-440	Oregon test diet	Guelph test diet
Vitamin-free casein	380	495	400
Gelatin	120	87	40
Dextrin	280	156	90
Wheat starch	–	–	115
Alpha-cellulose	–	77	30
Carboxymethyl cellulose	–	13	–
Marine fish oil	90	100	150
Vitamin E	–	2	In premix
Choline chloride	–	10	In premix
DL-Methionine	–	–	5
L-Arginine	–	–	10
Mineral mix	40[b]	40[d]	80[a]
Vitamin premix	90[c]	20[e]	30[a]

[a] *H-440 and Oregon test diet (NRC, 1973); Guelph test diet (Cho and Bureau, 1997; Hardy et al., 1984; Jacobs et al., 2002).*

[b] *Mineral mix contains the following (g/kg premix): calcium biophosphate, 135.7; calcium lactate, 326.9; ferric citrate, 29.7; magnesium sulfate, 132; potassium phosphate (dibasic), 239.8; sodium biophosphate, 87.2; sodium chloride, 43.5; $A1C1_3.6H_20$, 0.15; KI, 0.15; $CuC1_2$, 0.1; $MnSO_4.H_2O$, 0.8; $CoCl_2.H_2O$, 1.0; $ZnSO_4.H_2O$, 3.0.*

[c] *Vitamin premix contains the following (g/kg premix): alpha-cellulose, 893; choline chloride, 56; inositol, 22; ascorbic acid, 11; niacin, 8.4; Ca-pantothenate, 5.6; riboflavin, 2.2; menadione, 0.45; pyridoxine.HC1, 0.56; thiamin.HC1, 0.56; folic acid, 0.17; biotin, 0.06; vitamin B_{12}, 10 g.*

[d] *Mineral mix contains the following (g/kg premix): $CaCO_3$, 21; $CaHPO_4.2H_2O$, 735; K_2HPO_4, 81; K_2SO_4, 68; NaCl, 30.6; $Na_2H\text{-}PO_4.6H_2O$, 21.4; MnO, 25.0; $FEC_6H_5O_7.3H_2O$, 5.58; $MnCO_3$, 4.18; $2CuCO_2Cu(OH)_2$, 0.34; $ZnCO_3$, 0.81; KI, 0.01; NaF, 0.02; $CoCl_2$, 0.2; citric acid, 6.88.*

[e] *Vitamin premix contains the following (g/kg premix): thiamin.HCl, 3.2; riboflavin, 7.2; niacinamide, 25.6; biotin, 0.08; Ca-pantothenate, 14.4; pyridoxine.HCl, 2.4; folic acid, 0.96; menadione, 0.8; vitamin B_{12}, 2.67; inositol, 125; ascorbic acid, 60.0; para-amino benzoic acid, 20; vitamin D_2, 0.4; BHA, 0.75; celite, 735.8.*

stable, resistant to nutrient leaching, and must contain easily digestible ingredients for larval fish having immature and incompletely developed digestive systems. Some of these characteristics are mutually exclusive, e.g., water-stable and easily digestible. Larval diets are often called microdiets. Flaked diets, prepared by drying a feed ingredient slurry on a steam-heated, double-drum dryer, have been produced for years to rear ornamental fish for the hobbyist industry, and larval feeds can be produced by pulverizing flakes and screening to recover small particles. Microencapsulation is another technology to produce feed for larval fish due to the water stability of the capsule and the nutritional completeness of the feed particles. Methods of manufacturing larval diets are discussed later in this chapter.

9.2 Aims of diet preparation

Diet preparation is a complicated task that involves many considerations beyond formulation. Anyone with a computer and formulation software can formulate a diet that contains all of the known required nutrients at levels that meet or exceed the requirements of fish. However, practical matters such as ingredient prices, quality and availability, pelleting characteristics of a formulation, storage stability, durability of pellets, antinutrient content, palatability, pellet density, and nutrient digestibility are all critically important, and all must be considered when formulating feeds, especially when using least-cost formulation software. Some of these practical considerations are relatively easy to address but others are not, in part because the knowledge needed to address them can only be gained by experience.

The goal of feed production for commercial aquaculture is to produce a feed that supports maximum fish performance or production at the lowest cost. This is usually not the same as producing the lowest cost feed. Aquaculture has many different production aims, such as high hatchery survival of larvae and fry, economical grow-out of juvenile or postjuvenile fish, producing attractive and robust fish for stocking public or private waters, producing viable eggs, or propagation of attractive ornamental fish. Even when one species of fish is being reared with one goal in mind, different formulations are typically used to rear fish at different stages of growth or production. For example, a "finishing diet" might be fed during the final months before harvest to produce specific organoleptic properties, such as flesh pigmentation in salmonids or to enhance the omega-3 content of the flesh leading to an added value product (salmon).

9.3 Feed ingredients

Many of the ingredients used in animal and fish feeds are by-products of food processing to obtain high-value food items, such as meat products, edible oils, or flour, from a crop, livestock, poultry, or fish. After high-value products are removed, the remaining material is further processed and used in feeds to produce additional high-value products, such as livestock, poultry, or fish. In a sense, animal and fish production is supported by converting materials not consumed directly by humans into feed ingredients. For example, soybean meal and soy protein concentrate are important ingredients in swine, poultry, and fish feeds. Both are by-products of the material remaining after soy oil (the primary product) is extracted from crushed soybeans. The remaining material, called press cake, is further processed by drying to make soybean meal or by extracting soluble carbohydrates and then drying to make soy protein concentrate. Similarly, corn gluten meal is a by-product of corn oil and starch extraction and wheat gluten from flour milling. Feeds often contain whole grain meals, such as ground wheat, ground corn, and broken rice that are not by-products of food processing industries but rather grown specifically for use in livestock and fish feeds. Seafood processing by-products are commonly recovered and processed into fish meal and fish oil, and now account for over 30% of the fish meal and fish oil used in fish feeds worldwide (Stevens et al., 2018; Tacon, 2019). Most fish meal and fish oil are produced from fish species for which there is little or no market demand for direct human consumption. Food processing wastes and other materials are used to raise insects that are deoiled and dried to make insect meal. Single-cell protein ingredients can be produced by growing bacteria or yeast using methane, methanol, wood waste, and other carbon sources as substrates. In summary, most feed ingredients are products of other processes that are recycled into feeds to produce livestock, poultry, and fish.

9.3.1 Feed nomenclature

Feed ingredient nomenclature and classification began in Germany in the early 19th century when methods of determining chemical composition of livestock ingredients were developed, mainly to evaluate forages for ruminant animals. Feed ingredients were classified based on nitrogen (protein) content and digestible nutrient content. At the time, monogastric farm animals, e.g., swine and poultry, were raised by small-scale farmers to feed their families using locally available feed ingredients and kitchen leftovers, with the assumption that the animals would obtain a significant portion of their nutritional needs by foraging in barnyards and fields. Fish were not cultivated in most parts of the world, at least on an intensive scale, and fish consumption was dominated by wild-caught species.

Nomenclature was originally based on common names of ingredients, but as the number of by-products from single, raw ingredients increased, the use of common names, such as wheat, became confusing. To further complicate nomenclature, the same ingredient sometimes had different common names depending on location. A systematic investigation during the development of today's nomenclature system revealed that more than 20% of the common names in use for feed ingredients were simply different names for the same product (Harris, 1980). Today's system of nomenclature is called the International Feed Vocabulary (IFV), and it is used worldwide (Harris, 1963; Harris et al., 1968). Each ingredient is assigned a comprehensive name and number based on assignment of the ingredient to one of eight feed classes (Table 9.4) plus numbers based on one or more of six categories. The categories are: (1) origin, which includes the scientific and common names for specific plants and animals, poultry, fish, cereals, grass, mineral, chemical product, and drug, or other names for nonspecific materials; (2) part fed to animal as affected by processing; (3) process(es) and treatment(s) to which the feed ingredient was subjected; (4) stages of maturity and development; (5) cutting (for forage crops); and (6) grade. Using this system, herring meal is described as fish herring, *Clupea harengus*, meal, mechanically extracted, International Feed Number 5-02-000. The last five numbers of the International Feed Number are assigned to

TABLE 9.4 Classes of feed ingredients used to assign international feed numbers.

1. Dry forages and roughages
2. Pasture, range plants, and forages fed green
3. Silages (ensiled forages only)
4. Energy feeds – less than 20% protein (dry basis) less than 18% crude fiber (dry basis)
5. Protein supplements—more than 20% protein (dry basis)
6. Mineral supplements
7. Vitamin supplements
8. Additives—includes antibiotics, coloring materials, flavors, hormones, and medicants

each ingredient name, the first number being a code for the feed class, a protein supplement in the case of herring meal. Wheat mill run is described as wheat, *Triticum aestivum*, flour by-product, less than 9.5% fiber, International Feed Number 4-05-205. Over 18,000 feed ingredients were assigned numbers using this system (Harris, 1980).

The International Network of Feed Information Centers (INFIC) was formed to catalog information on feed ingredients and disseminate this information throughout the world. One of the participating centers, the Feed Composition Data Bank, National Agriculture Library, Beltsville, Maryland, US, provides the data used in feed composition tables, such as those found in the United States–Canadian Tables of Feed Composition, available as a download from https://www.nap.edu/catalog/1713. Another source of feed information is INRA-CIRAD-AZF at https://www.feedtables.com/. These databases are the sources of feed composition information found in National Research Council bulletins on the nutrient requirements of domestic animals and fish, such as the Nutrient Requirements of Fish and Shrimp (NRC, 2011). Databases used in many feed formulation programs also use values from these sources.

9.3.2 Protein supplements

As indicated in Table 9.4, protein supplements are feed ingredients having crude protein contents above 20% on a dry weight basis. There are three general groups of protein supplements. The first group consists of ingredients with protein levels of 20%–30% containing materials of plant origin that are by-products of the brewing and distilling industries, or extraction of primary products, such as flour to produce wheat germ meal and corn oil to produce corn gluten feed. The second group is composed of ingredients with protein level of 30%–50% and includes the oilseed meals, by-product meals of crustacean processing and ethanol production, and dried milk products. The third group comprised ingredients having over 50% protein and includes protein meals from marine resources, land animal proteins, yeast and bacterial protein products, insect meals, and algae proteins.

9.3.2.1 *Fish products and by-products*

Fish meal is the most important and nutritionally useful protein supplement in formulated diets of salmonids, other carnivorous fishes, and fry diets for omnivorous fishes, such as tilapia, catfish, and carp. Fish meals are produced from numerous fish species in different areas of the world. As a result, there is considerable variability in proximate and nutrient content of fish meals. Common fish meals include anchovy, capelin, menhaden, herring, whitefish, and salmon meals. Fish meals are produced from whole fish, as is the case with anchovy and capelin, or from processing residue, as is the case with menhaden, herring, and salmon. White fish meal is produced from trimmings of pollock, cod, and haddock processing, and often does not contain fish solubles, giving it physical characteristics desirable for specific feed applications, such as eel feed. Brown fish meal is an archaic term for products made from menhaden, anchovies, and other whole fish. These meals contain fish solubles, giving the meal a brown color. Fish meals are rich sources of essential amino acids. Their fat contents range from 4% to 20%, and ash contents are highly variable, ranging from around 10% in herring meal to over 23% in some white fish meal made from filleting wastes. The ash content of white fish meals can be reduced by removing large bones from dried meal by screening prior to grinding to produce the final product (Rathbone et al., 2001). Typical quality standards for fish meal are shown in Table 9.5.

By-products from crab, shrimp, and squid are also recovered and processed to produce crab, shrimp, and squid liver meals. These products contain about 32%–45% protein, respectively, while squid meal contains over 75% crude protein. The ash content of crab and shrimp meals is high because of the shell that remains after processing but not so for squid meal. Shrimp and crab meals are good sources of trace elements and carotenoid pigments. All of these meals are attractants and/or palatability enhancers in feeds for many farmed fish species.

9.3.2.1.1 Wet fish products (hydrolysates)

Fish processing by-products can be preserved by means other than drying, such as acid preservation. Acid-preserved processing waste is called fish silage. Mineral or organic acids

TABLE 9.5 Quality standards for fish meal required for salmonid diets (Cho and Bureau, 1997).

Component	Level
Crude protein (N × 6.25)	>68%
Lipid	<10%
Ash	<13%
Salt (NaCl)	<3%
Moisture	<10%
Ammonia-N	<0.2%
Antioxidant	<200 μg/g

are added to groundfish trimmings to lower the pH of the material to about 4.0. Antifungal agents, such as propionic acid, are also added to prevent mold growth. During acid preservation, proteolytic enzymes in the fish processing material hydrolyze the protein, resulting in a liquid product composed of peptide fragments and free amino acids. The moisture content of fish silage is similar to that of whole fish, between 75% and 80%, and fish silage can be used as is, concentrated, or dried with other ingredients (Hardy et al., 1984). Raw fish by-products may also be heated to 40–50°C with added proteolytic enzymes and then heated to 85°C for 15 min to stop enzymatic activity and pasteurize the product. The resulting product can be used directly as an ingredient in semimoist diets or dried after removing bones to produce hydrolyzed fish protein meal. Hydrolyzed fish protein is used in feeds for larvae and fry because proteins are partially hydrolyzed, facilitating digestion and absorption by young fish. It is also added to feeds as an attractant.

9.3.2.1.2 Other marine products

Krill meal, produced from *Euphausia superba* and *E. pacifica*, is a specialty marine protein source for fish feeds. Krill meal contains 33%–55% protein, 15%–20% fat, and 15%–28% ash. Besides having a well-balanced amino acid profile, krill meal contains astaxanthin and imparts color to skin of many species of fish and shrimp, plus muscle tissue of salmon and trout. Krill meal is also considered an attractant in feeds. Research to assess meal produced from the Arctic amphipod, *Themisto libellula*, as a protein source for salmon and cod has demonstrated its value as protein source, but high concentrations of copper and cadmium in meal limit its use (Moren et al., 2006).

9.3.2.2 Land animal by-products

Land animal by-products are dried meals produced from livestock and poultry processing industries. The protein content of these products after drying ranges from 50% to over 85%, and there are established standards for the quality of the proteins, generally a minimum pepsin digestibility value. The essential amino acid composition of animal by-products meal is similar to that of whole-egg protein, which is the standard by which protein quality is judged. Animal by-product meals are excellent sources of essential amino acids for fish.

9.3.2.2.1 Meat meal and meat and bone meal

These products are defined as dried mammalian tissues, exclusive of hair, hooves, horn, hide trimmings, manure, and stomach contents. Their protein content is about 50%, and fat levels average 9%–10%. The principal difference between the two products is the phosphorus content, with meat meal having, by definition, less than 4.4% phosphorus. The calcium content of meat and bone meal ranges from 8.8% to 12%, while meat meal generally contains about 3% less calcium. Both ingredients have relatively high ash contents, about 27% and 31% for meat meal and meat and bone meal, respectively.

9.3.2.2.2 Blood meal

Blood meal is a dry product made from clean, fresh animal blood, exclusive of all extraneous materials. Blood meal, spray-drying after an initial low-temperature vacuum evaporation had reduced the moisture content to about 70%, is the most common blood meal product used in fish feeds. Other processes of drying blood include flash drying and drum drying.

Blood meal has a minimum protein content of 85%, while is considered to be rich in lysine and histidine. Blood from animal processing can be refined further by separating blood cells from plasma to produce higher value products.

9.3.2.2.3 Feather meal

Feather meal is made from poultry feathers that are hydrolyzed under pressure in the presence of $Ca(OH)_2$ and dried. It has a protein content of 80%–85%, and not less than 75% of the protein must be digestible by the pepsin digestibility method. Digestibility of feather meal products can vary depending on processing conditions, and the crude protein apparent digestibility coefficient of individual feather meal products dictates appropriate levels used in fish feeds.

9.3.2.2.4 Poultry by-product meal

Poultry by-product meal is made from by-products of poultry processing, exclusive of feathers and the contents of the gizzard and intestine. The material remaining after chickens are dressed is rendered and dried. The ash content cannot be higher than 16%, of which not more than 4% can be acid-insoluble ash. Poultry by-product meals are available in regular, low-ash, and pet food grades. Protein level and price increase, and ash level decreases with grade. An extra quality index is considered to be the cysteine content. If cysteine levels are much higher than usual, poultry by-product meal may contain feather meal. This is easily confirmed by feed microscopy.

9.3.2.2.5 Milk by-products

Milk by-products, e.g., dried whey, dried whey product, casein, and dried skim milk, are sometimes used in feed formulation for fish but mainly in specialized feeds. Dried whey product is the residue obtained when a portion of the lactose has been removed, while dried whey contains a minimum of 65% lactose. The protein content of whey products is relatively low (13%–17%), yet these products are classified as protein supplements. Dried skim milk is sometime used in starter diets, primarily due to its excellent digestibility and amino acid balance. Its protein content is about 34%. Casein is the residue obtained by acid or rennet coagulation of defatted milk. By definition, it contains at least 80% protein. Vitamin-free casein is the standard protein used in semipurified diets.

9.3.2.2.6 Gelatin

Gelatin is an important ingredient in most semipurified feed formulations. It is a glutinous material obtained by the partial hydrolysis of collagen from animal skin, tendons, and ligaments. It is hard and brittle when dried but dissolves in hot water and forms a gel when cooled. It contains 88%–92% protein but lacks tryptophan. It serves as protein source and binding agent in semipurified diets.

9.3.2.3 *Plant protein supplements*

The most important plant protein supplements in fish feeds are oilseed meals, produced from press cake remaining after edible oil has been extracted from soybeans, cottonseed, rapeseed or canola, peanuts, sunflower seeds, safflower seeds, and linseed or palm seed. Oils are commonly extracted using solvents, leaving about 1.5%–2% residual oil in the meals.

Soybean meal is the most common oilseed protein supplement used in livestock, poultry, and fish feeds. Dehulled soybean meal contains a minimum of 47% protein, whereas nondehulled meals average 44%. Dehulled meals are used in feeds for fish, poultry, and monogastric livestock. Full-fat soybean meal is obtained by processing raw soybeans from which no oil has been removed and is sometimes used in feeds for Asian catfish and tilapia, and it can be used in feeds for salmonids trout. Cottonseed meal was used in the OMP formulation to supply protein and act as a pellet binder. Rapeseed meal has become a common constituent of salmon feeds. Other oilseed meal levels are limited in feeds for carnivorous species due to their fiber content but are used in feed for omnivorous species when availability and prices are favorable.

Soybeans contain antinutritional factors that affect feed intake, nutrient utilization, and intestinal morphology. Some antinutritional factors, notably trypsin inhibitors, are destroyed or inactivated by thermal processing (heating or drying), while others, such as phytic acid, are unaffected by processing but can be reduced using enzymatic treatment (phytase) prior to drying. Phytic acid in fish feeds can reduce zinc availability and protein digestibility in salmonid diets (Spinelli et al., 1983; Richardson et al., 1985; Sajjadi and Carter, 2004; Denstadli et al., 2006). Cottonseeds and rapeseeds contain antinutritional factors that affect fish performance (see Chapter 11).

9.3.2.3.1 Protein concentrates

Soy protein concentrate, rapeseed protein concentrate, wheat gluten, and corn gluten are important protein sources for fish feeds. Soy and rapeseed protein concentrates are produced from press cake remaining after oil is removed. Press cake is washed with water/alcohol mixtures to remove soluble carbohydrates, leaving protein, fiber, and some nonsoluble carbohydrates. Wheat gluten is the separation product from wheat starch by the wet milling process and has a minimum guaranteed protein content of 75% (Apper-Bossard et al., 2013). Wheat gluten is considered one of the best plant protein concentrates to use in fish feed due to its high protein digestibility and the fact that it lacks antinutritional factors, with the exception of phytic acid (Gatlin et al., 2007). Corn gluten meal is a high-protein (60%) by-product of corn processing to remove oil, starch, and bran. High-protein corn gluten meal is commonly used in feeds and is similar to corn gluten meal except that nonsoluble residues remaining after refining are not included in the product.

Brewery and distiller's by-products are another major plant protein supplement category. These products are dried materials remaining after brewing or distilling corn, rye, rice, and other grains to make beer and ethanol. Most of the starch in the grains is fermented to alcohol, which is distilled, thereby concentrating the protein and fiber fractions. Brewer's dried grains and brewer's yeast are by-products of brewing that are sometimes used in fish feed formulation. The protein and fiber content of brewer's dried grains are 27% and 13%, respectively. Brewer's dried yeast contains about 44% protein and only 3% fiber. Brewer's grains have limited use in fish diets because of their bulk and the difficulty in grinding it to a consistent particle size. Brewer's yeast acts as a binder in some formulations, but its use is limited because of its relatively high cost. Three main distilling by-products are available as feed ingredients: distillers' dried solubles, which is the dried stillage fraction; distillers' dried grains, which is the fraction remaining after removal of the stillage; and distillers' dried grains with solubles, which is a combination of stillage and grain sediments. The protein content of these products ranges from 26% to 28%, while the crude fiber level is higher in the dried grains (9%–11%) than in the dried solubles (4%–5%).

A fourth group of plant protein supplements includes concentrates from peas, beans, and lentils. Seeds from peas, beans, and lentils contain 20%–30% protein, but the protein content can be increased considerably by using air classification to reduce hulls and starch.

9.3.2.4 Other protein sources

The need to replace fish meal in fish feeds has renewed interest in microbial protein, single-cell proteins derived from yeast or bacteria grown on waste material such as wood products or by-products, municipal waste, crop processing wastes, or methane/methanol. The microorganisms are then harvested and dried, producing dry powdery meals contain 60%–70% protein. The economics of producing microbial proteins for fish feed use has become more favorable, and several commercial products have performed well in feeding trials with salmonids and other fish species. These products contain relatively high levels of nucleotides and nonprotein nitrogen.

An interesting unconventional ingredient that is rapidly expanding is insect meal produced from black soldier fly (*Hermetia illucens*), yellow mealworm (*Tenebrio molitor*), or common housefly (*Musca domestica*) larvae. Insects are natural food sources for many freshwater fish species and have been used historically as food for farmed fish, such as the use of silkworm pupae in Japan and Asia. More recently, research has demonstrated that insect meal is a suitable ingredient in feeds for a number of farmed fish (Henry et al., 2015). Insect meals have favorable amino acid profiles for fish, and their fatty acid contents are malleable, reflecting the fatty acid profiles of substrates used to grow them. Using inexpensive substrates, such as food processing waste, to grow insect larvae has lowered production costs to the extent that insect meals are being used at low levels in feeds for some high-value species at specific life history stages.

9.3.3 Basal feed ingredients (energy sources)

9.3.3.1 Carbohydrate (starch) sources

Ingredients used in livestock and poultry feeds to provide energy, mainly carbohydrates, have been historically called basal feeds. Farm animals are omnivores, and in feeds for omnivorous fish species, basal or high-energy feed ingredients are important constituents. By convention, the upper protein content of basal feed ingredients is 20%; most are in the 10%–17% range. Most grains, with the exception of oats, contain 68%–72% starch and about 10%–12% protein. Energy in seeds is stored as starch, and livestock and poultry can digest about 95% of the starch in grains. However, starch digestibility varies widely among fishes. Carnivorous species, such as salmonids, derive little energy from unprocessed plant starch. Omnivorous species, such as catfish, and herbivorous species, such as some carps, derive a high amount of energy from starch.

By convention in animal nutrition, energy feeds cannot contain over 18% fiber (Crampton and Harris, 1969). Otherwise, they are classified as roughages. Fiber is undigestible by carnivorous fish, while omnivorous and herbivorous fish are able to utilize fiber in varying degrees, thanks to their microbiome. The fiber content of grains is about 6%, although oats and barley contain higher levels.

A number of by-products of grain processing are used in animal feeds and are classified as basal feed ingredients. Wheat milling, which produces flour as its main product, also produces wheat shorts, wheat bran, wheat mill run, and wheat middlings, which are used in fish feed formulation. By-products of oats, corn, and other grains are also potential fish feed ingredients. After removal of flour, these products contain higher protein, fiber, and fat levels than the grains from which they were derived.

Factors other than nutrient content influence the use of basal feed ingredients in animal diets. The most important factor is cost, which is usually quite low. Thus, after nutrient requirements have been met, basal feeds can be used to fill out animal feed formulations. This function is not important in feed formulations for carnivorous fish species. Basal feeds have excellent binding properties that hold feed pellets together in both dry and semimoist diets. Finally, they are relatively indigestible, and this quality is useful in animal production when a reduced rate of growth at a normal feeding level is desired. Rarely does this situation arise in fish culture.

A large number of feed ingredients other than those listed here are candidates to include in fish feed formulations. As mentioned earlier, over 18,000 individual feed ingredients have been identified throughout the world. Although most of these ingredients will never be used in fish feeds, other ingredients not now in use may have value in the feeds of various species of fish whose potential in intensive aquaculture has not yet been developed, such as Amazonian species. It is up to fish nutritionists to identify potential fish feed ingredients and obtain information on their chemical composition and nutritional value to fish.

9.3.3.2 *Fats and oils*

Fats and oils are included in formulations of high-energy diets for salmonids and many other fishes to reach the desired energy levels in the finished feeds. Fats and oils are high-energy ingredients which are highly digestible by fish, such as rainbow trout (Austreng et al., 1979; Ellis and Smith, 1984; Trushenski and Lochmann, 2009). Even when it is desirable to have low levels in formulations, such as in catfish diets, some fat and oil is added to facilitate pelleting and reduce dustiness of the feed.

Fats and oils are lipid sources, fats being the term used for lipids that are solid at room temperature and oils being the term for lipids that are liquid at room temperature. The melting point of lipid sources dictates whether or not they are solid or liquid at room temperature. Fish oils are liquid at room temperature but become solid at lower temperature whereas poultry fat is a solid at room temperature and must be heated to become liquid. These are important considerations for feed manufacturers who must add fats and oils in their liquid states during feed production. The melting point of a lipid depends upon its fatty acid composition. Lipids such as tallow contain a high proportion of saturated fatty acids, and because they are solid at room temperature (have a high melting point), they are termed fats. Fish oils are the opposite—they contain a high proportion of unsaturated fats. Fats and oils are concentrated energy sources for fish and animals, having 2.25 times as much total energy (joules or calories) per gram as carbohydrates.

There are many commercially available fats and oils suitable as ingredients for fish diets. In the past, fish feeds for salmonids, shrimp, and marine fish contained fish oil, but as aquaculture feed production for these species has increased, demand for fish oil has exceeded annual global production, making it necessary to replace a portion of fish oil with other oils in feeds. Common fish oils used in fish feeds are anchovy, herring, salmon, menhaden, tuna, and capelin. Alternative oils are mainly those from plant sources, e.g., rapeseed, soybean, corn, palm, or cottonseed, or those from rendered animal processing, e.g., poultry fat, tallow, and lard. As long as the essential fatty acid requirements of the fish for which feed is intended are met, alternative oils can provide the necessary dietary energy in fish feeds (Watanabe, 1982; Turchini et al., 2009, 2010). Fats and oils are chosen using the same criteria as for other ingredients: price, availability, nutritional value, and quality as measured by chemical tests. Fats and oils are susceptible to oxidative and hydrolytic degradation during storage, especially under abusive conditions. Therefore, limits are often made on the percent of free fatty acids and products of oxidation, such as malonaldehyde and peroxides, in fats and oils used in fish feeds. Product specifications for fats and oils also include water content, unsaponifiable matter content, and insoluble matter, all of which indicate contamination with other materials. Occasionally fish oils which have been refined to reduce free fatty acid values and nonfat material are available on the market. While these oils are within specifications, they may be low in naturally present antioxidants, shortening the induction period prior to the onset of oxidation. This can be prevented by adding antioxidants. Antioxidants are usually added to fish oils during manufacture to prevent oxidation, but buyers of oil for fish feed use should make certain this has been done.

Another concern with fats and oils is the possible presence of persistent organic pollutants (POPs), such as polychlorinated biphenyls (PCBs) and 1,1-dichloro-2, 2-Bis(p-chlorophenyl) ethylene (DDE) (Jacobs et al., 2002). Methods to remove POPs from fish oil have been developed and are used in Northern European fish meal plants, where environmental contamination has resulted in elevated levels in fish from these areas (Oterhals, 2011). An example of fish oil quality standards is shown in Table 9.6.

A large number of feed ingredients other than those listed here are candidates to include in fish feed formulations. As mentioned earlier, over 18,000 individual feed ingredients have been identified throughout the world. Although most of these ingredients will never be used in fish feeds, other ingredients not now in use may have value in the feeds of various

TABLE 9.6 Quality standards for fish oil required for salmonid diets (Ontario Ministry of Natural Resources, 1998, cited in Hardy, 2002).

Component	Level
Iodine value	Report value
Peroxide value	<5 meq/kg
Anisidine value	<10
Pesticides (total)	<0.4 μg/g
Polychlorinated biphenyls	<0.6 μg/g
Nitrogen	<1.0%
Moisture	<1.0%
Antioxidant	<500 μg/g

species of fish whose potential in intensive aquaculture has not yet been developed, such as Amazonian species. It is up to fish nutritionists to identify potential fish feed ingredients and obtain information on their chemical composition and nutritional value to fish.

9.3.4 Vitamin premixes

Vitamin premixes are concentrates in which stable forms of essential vitamins are mixed with a carrier such as a wheat by-product. Vitamin premixes are formulated to meet or exceed the dietary requirements of fish at recommended levels of inclusion in fish feeds without considering any contributions to vitamin requirements by feed ingredients. Although vitamin premixes are expensive, they are usually included in fish feeds at 0.5%–2% and make up a small proportion of the total cost of feeds. Thus, it is economical and prudent to ignore vitamin contributions from feed ingredients. Choline chloride is generally not included in vitamin premixes. Although it is a stable product, due to its hygroscopic nature it interacts with other vitamins, reducing their activity (Frye, 1994; Gabaudan and Hardy, 2000). Ascorbic acid is included in some vitamin premixes but is added separately to others. In practical salmonid diets, the concentration of vitamins in the premix is sufficient to supply vitamins to finished feeds in excess of recommended levels, providing a safety margin to account for losses during feed manufacturing and storage (NRC, 2011).

9.3.5 Mineral premixes

Mineral premixes are concentrates of essential compounds that provide both macro and micro (trace) minerals in practical fish diets to make up for low levels in the formulation. Although the necessity of mineral supplementation of practical diets varies with diet formulation and mineral content of fish rearing water, mineral fortification is a wise practice, especially when diets contain high levels of plant protein concentrates. Plant proteins contain low amounts of many trace elements and may contain components, such as phytic acid, that reduce the bioavailability of divalent cations, such as zinc. Mineral premixes are inexpensive, so including them in fish feed formulations to prevent possible deficiencies is a sensible practice. Several mineral premixes used in open-formula salmonid diets are shown in Table 9.6.

Mineral fortification is required in formulations of semipurified, experimental fish diets since the highly refined ingredients used in these diets are low in mineral content. Several experimental diet formulations along with mineral premix specifications are shown in National Research Council publications on the nutrient requirements of fish (NRC, 1973, 1983, 1993).

9.3.6 Nonnutritive feed ingredients

Nonnutritive feed ingredients are additives that are included for reasons other than to provide nutrients. For the most part, these compounds have no nutritional value; yet they are of great practical value in a number of ways. Nonnutritive feed ingredients include feed binders, carotenoid supplements, drugs and antibiotics, hormones, antifungals, antioxidants, fiber, flavorings, and water.

9.3.6.1 Pellet binders

There are many feed ingredients that act as binders in fish feeds. Some are by-products of cereal grains or other plants and may provide nutrients to fish when added to the diet. For example, 20% pregelatinized potato starch is added to eel diets to increase water stability of the dough and to provide energy. Other commonly used binders include bentonite, lignin sulfonate, and hemicellulose extract. Bentonite is a naturally occurring mineral consisting mainly of trilayered, aluminum silicate. It is available as either sodium bentonite or calcium bentonite. Sodium bentonite has, by definition, between 1% and 2% available ion content, or sodium exchange. It swells when added to water, while calcium bentonite does not. Both sodium and calcium bentonite can be added to dry, compressed fish feeds at no more than 2% to act as a binding agent and also as a lubricant, increasing pellet mill production rates and pellet mill die life (Reinitz, 1983).

Lignin sulfonate is a product of the wood pulping industry. It aids in pellet binding, reduces fines, and permits the addition of more steam during the manufacture of compressed pellets. Lignin sulfonate is added at up to 4% as a pelleting aid in dry, compressed feeds. Hemicellulose extract is a product made by spray drying the concentrated, soluble by-product of pressed wood manufacture. It is less commonly used than lignin sulfonate.

Moist and semimoist fish feed production requires the use of both nutritive and nonnutritive binder materials. Nutritive binders include oat groats, vital wheat gluten, finely milled wheat bran, cottonseed meal, and pregelatinized starches. Nonnutritive binders include tapioca, carboxymethyl cellulose, alginates, agar, and various gums. Chitosan, carrageenan, and collagen have also been tried (Heinen, 1981). Semimoist feeds, having moisture contents of 25%–35%, can be made into stable pellets by careful selection of feed ingredients. However, when the use of low binding feed ingredients is desirable for some reason, supplementary binders may be needed.

Moist feeds, having moisture contents of 35%–70%, require the addition of binder. For example, semipurified test diets, such as H440 and the Oregon Test Diet (NRC, 1973), include gelatin and carboxymethyl cellulose as binders. Moist diets, which are combinations of wet fish ingredients and dry meal, may contain 0.5%–2.0% alginates as binders. Sodium bentonite has also been used in some formulations. Alginates were found to be better binders than gum, carrageenan, chitosan, collagen, carboxymethyl cellulose, and cornstarch in a 41% moisture diet (Heinen, 1981). Agar was an effective binder, but expensive. Calcium ions and a sequestrant, such as sodium hexametaphosphate, must be added to diets containing alginates as binders to prevent premature alginate activation.

Extruded pellets are the most common form of fish feed used in intensive aquaculture. Extruded pellet manufacturing involves several steps that precondition the feed mixture, increase moisture content, and subject the mixture to high-pressure extrusion. These forces increase starch gelatinization which holds the mixture together as hard, durable pellets after cooling. It is not necessary to add additional feed binders to extruded pellets, but ingredient composition and processing conditions affect the quality of extruded feeds (Sorensen, 2012).

9.3.6.2 Carotenoid supplements

Carotenoid pigments are added to fish feeds to impart color to skin, flesh, and/or eggs (Simpson et al., 1981; Ibrahim et al., 1984; Storebakken et al., 1987; Shahidi and Brown, 1998; Garcia and Lara, 2013), and to feeds for crustacea to improve color and value of farmed products (Yamada et al., 1990) (Wade et al., 2015). Over 300 pigments are found in plants and animals, with xanthophylls and carotenoids being the most important classes in fish. For the most part, xanthophylls are found in plants and carotenoid pigments in crustacea and fish. Carotenoid pigments are produced by algae and bioconcentrated up the aquatic food chain.

Some finfish and shellfish possess the ability to convert certain xanthophyll pigments to carotenoid pigments. Goldfish and common carp can convert the yellow xanthophyll pigment, zeaxanthin, to the red carotenoid pigment, astaxanthin (Hata and Hata, 1973; Hata and Hata, 1976). Similarly, the kuruma prawn (*Marsupenaeus japonicus*) can convert both beta-carotene and zeaxanthin to astaxanthin (Tanaka et al., 1976). Salmon and red sea bream, which normally have pigmented flesh and skin, do not convert xanthophyll pigments to canthaxanthin, and astaxanthin. Rather, they must obtain canthaxanthin and/or astaxanthin from their diets, similar to wild fish.

Carotenoid supplementation of fish diets involves adding natural materials containing the desired carotenoid pigments, carotenoid extracts of natural products, or chemically synthesized pigments. Natural materials used include algae (*Haematococcus pluvialis*, spirulina), certain yeasts or bacterial meals, zooplankton, krill products, and processing waste from shrimp, crab, and crayfish processing. Dietary levels of 20% or more of wet crustacea processing waste are required to reach the desired astaxanthin level in trout and salmon fillets. Concentrated carotenoid extracts of red crab and crayfish are effective dietary supplements for salmonids (Spinelli and Mahnken, 1978; Chen et al., 1984). The amount added to the diet depends on the concentration of carotenoid pigments in the extract, but dietary levels normally range from 3% to 7%, replacing added fats and oils. Synthetic astaxanthin (Carophyll pink, DSM Nutritional Products Ltd) is a commercial product produced from beta-carotene containing a minimum of 10% astaxanthin in beadlets with a cornstarch-coated matrix of lignosulfonate and corn oil. It is added to commercial feeds at 0.05%−0.1% to yield a nominal level of 50−100 mg astaxanthin/kg feed. Astaxanthin is easily oxidized and sensitive to air, heat, light, and humidity. As a result, a portion of the supplemented astaxanthin in feeds is destroyed during feed pelleting and long-term storage of feed, especially under abusive conditions. For ornamental fish species, skin color is a valuable attribute. Carophyll yellow (xanthophyll), carophyll red (canthaxanthin) and carophyll pink are all effective color-enhancing feed additives for ornamental fish species.

9.3.6.3 Medications

Medications are sometimes added to fish feeds in order to treat, cure, or mitigate disease. A number of antibiotics are effective against fish diseases, although in the United States and the European Union (EU), only three are approved as feed additives: sulfamethazine, oxytetracycline (Terramycin), sulfadimethoxine (Romet-30), and florfenicol (Aquaflor). Sulfamerazine is also approved for fish feed use but no longer marketed. In Europe, oxalinic acid is used in feeds as an antimicrobial drug. As with livestock feeds, medicated fish feeds have specific labeling requirements, including a warning to withdraw medicated feed for a proscribed length of time before the fish are marketed. The use of antibiotics to treat fish diseases has largely been replaced by vaccination in salmonid aquaculture.

9.3.6.4 Hormones

Supplementing anabolic steroids to domestic animal feeds is no longer permitted in many parts of the world due to concerns about hormone residues in products destined for human consumption. The same concern exists for fish products, and the addition of steroids and other hormones to fish feeds to enhance growth is not permitted. However, the addition of hormones to the diets of fry to produce monosex populations is allowed. Monosex populations are desirable when one sex grow faster than another. Sexual differentiation occurs during the fry stage for most fish, meaning phenotypic gender is undefined for a period after fish hatch and begin to grow. In tilapia, males grow more rapidly than females, so all-male culture is advantageous to growers. Sex reversal is achieved in tilapia by supplementing feed with 17α-methyltestosterone (MT) at 60 mg MT/kg feed for about 35 days. This converts female fish to phenotypic (XX) males. In trout culture, it is advantageous to rear all-female populations to eliminate the production of precocious males that makes them unfit for the market. Trout fry are fed MT feeds to produce females that are phenotypic males. At maturation, the XX males are crossed with normal females, also XX, producing all-female progeny. Feeding MT to tilapia fry for a short period may pose no risk to humans who subsequently consume the fish after they grow to harvest size. For trout, the progeny of mature fish that sex reversed as fry using MT is the final product to consumers, not fish treated as fry. As such, there is no risk of carryover of MT to consumer products. Detailed reviews of hormonal induction of sex reversal in fish (Pandian and Kirankumar, 2003) and fish gonadogenesis (Penman and Piferrer, 2008) and their applications to aquaculture are available to readers interested in these topics.

9.3.6.5 Antimicrobial agents

Plant-based ingredients are increasingly used in fish feeds at high inclusion rates. Unfortunately, the use of ingredients, such as corn, wheat, and soy, is associated with the potential of mycotoxins contamination from fungal growth that can occur during ingredient storage. In warm regions and especially in tropical and subtropical conditions (high temperature and humidity), this potential is further increased. However, even when mold is seen on ingredients, it does not necessarily mean that mycotoxins are present.

Microbial spoilage can also occur on pelleted feeds. Spoilage microorganisms need a source of unbound water to grow. For that reason, feeds are dried to less than 10% moisture before being shipped to farms. Microbial growth occurs very rapidly in semimoist foods, and many molds produce compounds that are toxic to fish. Mold growth is visible within 3 days in semimoist feeds held at room temperature (22°C), while at refrigerated temperatures (1–3°C), it may not be visible for 10–20 days. During frozen storage, microbial growth is completely stopped; thus, if semimoist diets are fed to fish shortly after manufacture, there is no need to use antimicrobial additives in the feed. If longer storage is required, additives must be used if the material is not frozen.

Over 20 compounds are used in the feed industry to inhibit fungal or microbial growth (Table 9.7). Some have general and some specific applications in feed products. The benzoates and parabens are wide-spectrum antimicrobial agents, which are effective against bacteria, fungi, and yeast. Propionates are used primarily to inhibit yeasts and molds and are also effective against bacteria, fungi, and yeast. Choosing an appropriate antimicrobial agent

TABLE 9.7 Antimicrobial agents used in feeds.

Name	Limit or restriction
Benzoic acid	0.1%
Calcium propionate	None
Calcium sorbate	None
Distearyl thiodipropionate	0.005%
Formic acid	2.5%
Methylparaben	0.1%
Potassium bisulfite	Not for use in B_1 sources
Potassium metabisulfite	Not for use in B_1 sources
Potassium sorbate	None
Propionic acid	None
Propylparaben	0.1%
Sodium benzoate	0.1%
Sodium bisulfite	Not for use in B_1 sources
Sodium metabisulfite	Not for use in B_1 sources
Sodium nitrite	0.002%
Sodium propionate	None
Sodium sorbate	None
Sodium sulfite	Not for use in B_1 sources
Sorbic acid	None

Source: Food Industry Red Book, 1981.

for fish food involves more than picking the least expensive one. An intelligent choice includes a complete study of the specific microbial problem, solubility and ease of application of the agent, pH requirements, empirical information, and acceptability in the feed to the animal being fed.

Control of spoilage in intermediate moisture products (18%–36% moisture), such as semi-moist fish feeds, is not simply a matter of adding antimicrobial compounds to the product. The best results are obtained when a combination of approaches is used. The first factor is the microbial load of the product. It is essential that the feed has a low plate count at the start, with no more than 500,000 organisms per g of material. If the microbial load of the starting material is high, it is difficult to control spoilage with antimicrobial compounds alone. A second means of controlling microbial spoilage in feeds is by the use of special packaging that maintains a controlled atmosphere after manufacture until use of the feed. The aim of this process is to reduce O_2 tension in the feed package during storage and thereby restrict the growth of aerobic microorganisms. Feeds are stored in hermetically sealed pouches or bags that are impermeable to water vapor or atmospheric gases. An inert gas, such as N_2 or CO_2, is introduced into the bag before it is sealed.

A third approach to controlling spoilage in intermediate moisture products is by controlling water activity (Aw). Microorganisms require an abundant supply of water to grow. Reducing the supply of water in a feed is, therefore, an effective way to limit their growth. In intermediate moisture products, sufficient water to permit microbial growth is present. However, water exists in feeds in three forms, as a liquid, as a vapor, and as chemically bound water. Control of water activity in intermediate moisture products involves increasing the proportion of total moisture in the product that is chemically bound, thereby decreasing the proportion present as a liquid. Water activity is defined as the ratio of water vapor in a feed and the vapor pressure of pure water at the same temperature. The water activity of pure water is 1.0. Water in feeds has a value of Aw less than 1.0. Small differences in Aw may result in relatively large differences in the resistance of a product to microbial spoilage. The approximate lower limits of Aw for microorganism growth are 0.91 for bacteria, 0.88 for yeasts, and 0.80 for molds. It is not terribly difficult to achieve Aw values in this range in intermediate moisture products by the addition of compounds such as sugar (sucrose), glycerol, NaCl, and propylene glycol.

Semimoist fish feeds not requiring frozen storage are commercially available. These products contain agents to inhibit microbial growth, and often manufacturers use other means to restrict microbial spoilage. The products are highly palatable to salmonids and marine fish fry.

9.3.6.6 Antioxidants

Antioxidants are chemical compounds added to feed ingredients to prevent oxidation of lipids. However, other feed components, such as carotenoid pigments and tocopherols, can also undergo oxidation. Oxidation can occur in several ways, but the mechanism of most interest in feed manufacturing is autoxidation, also known as atmospheric oxidation, which is the oxidation of moderately unsaturated fatty acids resulting in products that produce off-flavors and off-odors. The rate of autoxidation of fats can be accelerated by ionizing or UV radiation, divalent cation concentration, temperature, and oxygen concentration.

Autoxidation of fats is a process involving three steps. The first step is initiation in which free radicals are formed. Initiation is enhanced by a number of factors, including light, UV radiation, and the presence of divalent cations, such as copper and iron. The second step in autoxidation is the propagation step, in which the free radicals formed in the initiation step react with oxygen to form more free radicals. The final step is termination, in which stable end products are formed. Since the propagation step itself forms free radicals, autoxidation reactions are autocatalytic.

Antioxidants work by chelating prooxidant divalent cations, by acting as free radical acceptors, or by donating hydrogen. The last two functions are considered sacrificial because once an antioxidant molecule reacts, it no longer possesses antioxidant properties and is, therefore, "destroyed" in the process.

Antioxidants added to lipids and feeds to prevent oxidation by reacting with free radicals are phenolics, such as butylated hydroxytoluene (BHT) and butylated hydroxyanisole (BHA), and amines, such as ethoxyquin. BHA and BHT are added to feeds at a level up to 150 ppm while ethoxyquin is added at 0.015%. Other antioxidants in use include dilauryl thiodipropionate, propyl gallate, and thiodipropionate.

Antioxidants that prevent oxidation by chelating metallic prooxidants include ascorbic acid, citric acid, phytic acid, tartaric acid, oxalic acid, and ethylenediaminetetraacetic acid (EDTA). There is a synergistic effect when phenolic or amine antioxidants are combined with an antioxidant that chelates prooxidants.

Many lipid sources contain naturally occurring antioxidants. These compounds effectively inhibit autoxidation of lipids until they are used up, at which time the rate of oxidation reactions increases very rapidly. The period of time during which antioxidant in a lipid or feed prevents oxidation is called the induction time. Induction time can be measured using the Schall oven test (see Chapter 10).

9.3.6.7 *Fiber*

Fiber is the nonnutritive portion of feed ingredients that is classified as crude fiber in proximate analysis. It is undigestible by salmonids and other carnivorous fish, but omnivorous species, such as catfish, carp, and tilapia, have intestinal microflora capable of digesting a portion of dietary fiber (Stickney and Shumway, 1974). Fiber is added to semipurified diets to facilitate binding as well as to increase digestion efficiency (Buhler and Halver, 1961). Generally, fiber is not added to practical diets; rather it is avoided. Upper limits are often specified, thus eliminating many potential fish feed ingredients and restricting the levels of others. In diets for carnivorous fish species, levels of fiber above 3%–5% are not recommended (NRC, 2011). Levels up to 8%–12% fiber can be tolerated by most fish, but higher levels often result in growth depression (Leary and Lovell, 1975; Edwards et al., 1977). Fish fed diets high in undigestible fiber increase their feed consumption and gastric evacuation time, but this compensation is limited (Hilton et al., 1983). Since undigested dietary fiber passes through the fish, diets high in fiber increase fecal waste and, therefore, reduce water quality, an important variable when formulating feeds to be used in RAS.

9.3.6.8 *Flavorings*

As fish feed formulations shift from using high levels of fish meal to using plant protein concentrates, the effects of these changes on feed intake must be considered. Fish are sensitive to certain tastes in their food, a trait that can be advantageous and detrimental in feed production. Some ingredients, such as krill or blood worms, enhance feed intake for salmon and sturgeon, respectively. Shrimp increase feed intake when compounds associated with their natural prey are present (Tantikitt, 2014). However, the main concern with feed palatability is reduced feed intake when certain feed ingredients are used. For example, chinook salmon fry are extremely sensitive to the presence of low levels of dietary soybean meal (Fowler, 1980) and respond to its presence in feeds by reducing feed intake. Pacific salmon reduce intake of feed containing the antibiotic erythromycin (Schreck and Moffitt, 2011).

Flavorings or, more accurately, palatability-enhancing substances, can be added to feeds to restore feed intake. Betaine is an example of a compound that increases intake of plant protein–based feeds (Tiril et al., 2008). Betaine-containing materials increase feed intake of rainbow trout fed feeds containing certain plant protein concentrates, such as canola meal or pea protein concentrate (Thiessen et al., 2003). Other natural products, such as krill hydrolysate or stickwater from fish meal production, stimulate feed intake in salmon (Kousonlaki et al., 2013). From a survey of the scientific literature, it is clear that: (1) certain natural products and combinations of amino acids stimulate feed intake; (2) the effectiveness of palatability-enhancing compounds varies with dietary level; (3) fish and shrimp species differ in their responses to palatability-enhancing compounds; and (4) measuring feed intake over a short period to assess effectiveness of palatability-enhancing compounds is not always correlated with long-term benefits in fish growth. Fish often reduce feed intake when feed formulations are changed but adjust to a new feed after a short period of time. Measuring weight gain over a fixed period of time can indicate that a feed formulation containing an alternative protein ingredient is inferior to a standard diet, but in reality, the growth effect may simply be due to a brief period of feed intake reduction, followed by a restoration of normal growth. New knowledge of the regulatory roles of feeding behavior by neuropeptides (Matsuda, 2009) as well as the signaling properties of certain amino acids (Kaushik and Seilitz, 2010; Habte-Tsion, 2020) suggests that the subject of enhancing feed intake in fish by supplementing feeds with various products requires further investigation.

9.3.6.9 Water

The water content of feeds ranges from 6% to 10% for dry-compressed or extruded pellets to 65%–70% for some wet feeds produced at the farm using trash fish or other such materials. The moisture content of feeds is important because of the potential for microbial growth when feed moisture levels exceed 10% feeds. High moisture, farm-made feeds should be fed soon after being produced. For semimoist feeds, freezing or controlling water activity prevents microbial growth. In the past, there was concern about feeding dry, pelleted feeds to some species, such as salmon or largemouth bass, at the first feeding or fry stage. Semimoist feeds are typically fed to marine larvae and fry of marine species, and some freshwater species. Texture and palatability of semimoist feeds is an advantage for many species at these stages.

9.3.7 Feed ingredient regulations

Feed ingredients used in feeds for livestock, poultry, companion animals, laboratory animals, and fish are regulated by governmental entities in most countries. In the United States, the Association of American Feed Control Officials (AAFCO; https://www.aafco.org) is the entity empowered by law to regulate the sale and distribution of animal feeds and drugs. It oversees processes to define feed ingredients, establish quality standards, and harmonize language of regulatory laws used by state and federal agencies. AAFCO provides information on quality assurance/quality control methods for ingredients and feeds, guides industry on feed labeling, maintains an online database of ingredients, and lists government requirements for feeds. Most other countries with significant animal or fish farming sectors have similar governmental regulatory authorities and structures. For example, the European Food Safety Authority (EFSA; https://efsa.europa.eu) and the European Commission, Animal Feed section (https://ec.europa.eu/food/safety/animal-feed_en) play similar roles in EU countries.

9.4 Diet formulation

Diet formulation appears to be a simple matter, but in practice, it is a complicated process involving many considerations that require substantial knowledge and experience. Feed formulators select dietary ingredients to combine into a mixture that is palatable, pelletable, nutritious, inexpensive, and easy to ship, store, and use. Ingredients are chosen on the basis of cost, availability, chemical composition, nutritional value, and ease of use. Diet formulation involves making choices between designing an inexpensive feed that may be relatively inefficient or a high-performance feed that may be relatively expensive. Most formulators formulate feeds to hit the sweet spot between extremes where high fish performance, such as fry survival, weight gain, or egg production, is obtained at the least cost per unit of performance. Thus, the aim of production is the first factor that must be considered in feed formulation.

Formulating a fish feed involves selecting a combination of ingredients which results in a mixture containing levels of essential nutrients that meet or exceed the minimum requirements of the fish. To do this, one has to know the nutritional requirements of the fish and the nutrient content of the ingredients. A quick examination of nutrient requirement information presented for various species in the NRC (2011) reveals that our knowledge of nutritional requirements of most farmed fish and shrimp species is incomplete. Feeds for fish species for which nutritional requirements are incomplete are formulated to contain nutrient levels similar to those required by related species. In contrast to information on nutritional requirements of fish, nutrient content of feed ingredients is well known and readily available in feed composition tables. However, values listed in such tables are not absolute values. Rather, they are averages of multiple analyses of samples of each ingredient. Individual batches of ingredients routinely vary in nutrient content depending upon the season or stage of maturity at harvest, the procedures used to process the raw material into the feed commodity, and many other factors. Chemical analysis methods are also subject to experimental error, generally minor but present nonetheless. Averaging results of multiple analyses should, in theory, reduce error associated with variability in chemical analysis. As a result, formulating a feed to contain the exact amount of essential, limiting nutrients at required levels is a risky practice. Nutrient requirements listed in various references, such as the NRC (2011), are minimum values determined under laboratory conditions. If nutrient levels in one or more ingredients used in a feed are below average tabled values, the feed may underperform in real-world conditions. To guard against this, feeds are commonly formulated to contain levels of protein, amino acids, energy, and other essential nutrients above minimum requirement levels.

The availability of nutrients in feed ingredients is another aspect that must be considered in feed formulation. Incorporating information on nutrient availability into the feed formulation process is easy, but this information is incomplete. Such factors as the procedures used to manufacture the ingredient and interactions with other dietary ingredients can reduce the availability of a nutrient to fish even though the results of chemical analysis of feeds show that adequate dietary levels are present. An example of a reduction in nutrient availability as a result of processing is the reduction that occurs when protein supplements are dried in the presence of reducing sugars. Amino acid analysis will measure total lysine content, but the level of bioavailable lysine will be lower. An example of reduction in nutrient availability due to interaction with other dietary ingredients is that of zinc availability in fish fed diets containing high-ash meals, elevated levels of calcium and phosphorus, or elevated levels of calcium, phosphorus, and phytic acid (Ketola, 1979; Hardy and Shearer, 1985; Richardson et al., 1985). A dietary zinc level in the range of the requirement for trout (15–30 ppm) will be adequate in a semipurified diet, but insufficient in a practical diet containing high-ash fish meal or in a semipurified diet containing elevated levels of calcium, phosphorus, and/or phytic acid. Determining dietary availability involves feeding experimental diets to fish and measuring such parameters as digestibility, enzyme activity, or tissue nutrient saturation. It is essential that nutrient availability values are considered in the feed formulation process. For nutrients for which availability can vary and for which no reliable availability values exist, prudent practice is to formulate to levels significantly above minimum requirements.

Feed formulation involves making assumptions, and formulators must be aware of these assumptions to avoid surprises in feed performance. The first assumption made in feed formulation is that nutrients in ingredients are interchangeable, that is, a nutrient in one ingredient is interchangeable with the same nutrient in another ingredient. This assumption is not always justified or even recognized. However, it is an essential assumption that allows ingredients in feed formulas to be manipulated as prices and availability change. Feed formulators must be aware that they are making this assumption to avoid peril. A second assumption made in feed formulation is that nutrient levels in ingredients are additive, meaning that the content of nutrients in a feed can be summed up from the amount of each nutrient contributed by each ingredient in the feed. Several problems arise from this assumption. The assumption is based on absence of interactions between or among ingredients when they are combined that may reduce the level of an essential nutrient. Second, feed processing may alter, inactivate, or destroy certain nutrients in feed mixtures at a rate that differs among mixtures. Gaps in knowledge contribute to these uncertainties. Despite these gaps in knowledge, fish feeds are formulated, manufactured, and used successfully to rear fish every day. Successful feed formulations usually contain enough leeway in available nutrient levels to cover any potential shortfalls in meeting dietary requirements of fish. However, as feed formulations for carnivorous species shift to much lower levels of fish meal, gaps in knowledge that were previously not critical are becoming so. Put another way, high levels of fish meal in feeds covered up gaps in fish nutrition knowledge that are now critical to know as feeds shift to higher levels of alternative protein sources, especially plant protein concentrates.

9.5 Feed categories

The general goal of feed formulation is to support the highest performance of the species being fed at the lowest cost. For some feeds, this may involve setting nutrient limits in the feed that are well above estimated nutrient requirements, thereby increasing the cost of the feed. However, this may be necessary to meet the goal of the feed. An example could be the formulation of starter diets for first-feeding fry. Successfully transitioning fish to exogenous feeding is the goal, and feed cost is a minor consideration. Fry feeds are used once fish having a larval stage have passed through metamorphosis and possess functional digestive systems. This is a stage where fish grow rapidly, so providing a high-quality diet that supports rapid growth is the primary goal of formulation. For grow-out fish, feed efficiency is a major determinant of profitability of a farming enterprise. In this case, feeds will be formulated carefully to attain the most efficient protein/energy ratio and amino acid balance. Regulatory constraints on levels of enriching nutrients in farm effluent water are another consideration that affects feed formulation. Feeds in this case must be formulated to ensure maximum retention efficiency of feed nutrients, such as phosphorus, in fish tissues and minimize excretion to meet regulatory standards. Once the function or use of the feed has been defined, the type of feed, its formulation, and conditions of manufacture (grinding, mixing, pelleting) can be selected.

Phase feeding is a term used to describe the selection of feed formulations based upon the life stage of an animal. Feed formulations can be categorized by the intended function or phase of production for which the feed will be used. This facilitates communication between feed producers and fish farmers. The purpose of phase feeding is to minimize input costs (feed) and maximize output value (fry survival, fish weight gain, etc.). Feeds can be formulated to enhance fry survival, growth rates, feed efficiency, fish or fillet quality, or pollution reduction to meet regulations or minimize buildup of nutrients in RAS. The characteristics of feeds for major phases of fish production are described in detail below.

9.5.1 First feed for fry and larvae

Several terms are commonly used to describe the first feeds offered to young fish when they have absorbed most of their yolk and begin to consume exogenous material, namely "larval feeds" and "starter feeds." Although these terms are sometimes used interchangeably, they are distinctly different. Larval feeds are produced for species of fish that are very small at the first-feeding stage and do not possess fully functional digestive systems. For some species, feed particles must be 100–200 μm in diameter to be consumed by larvae, and nearly neutrally buoyant so that they remain in the water column long enough for larvae to locate and ingest them. In addition, they must contain ingredients that are easily digested while, at the same time, be somewhat resistant to nutrient leaching. These characteristics can be mutually exclusive. Starter feeds are used as first feed for species of fish that do not pass through a larval stage but rather initiate exogenous feeding at a relatively large size and with a fully functional digestive system. Catfish, salmon, trout, and tilapia are examples of such fish. These fish can consume feed particles larger than 400 μm in diameter, and they tend to be aggressive surface feeders, making feed buoyancy less of a concern. Unpelleted mash was typically used as a starter feed for the first few days of juvenile trout and salmon feeding, after which fish were fed crumbles, sieved to appropriate particle sizes. Advances in extrusion pelleting technology now make it possible to produce 300 μm diameter pellets, after which fish are fed using high-protein 0.5 and 0.7 mm pellets as they grow.

Larval feeds and starter feeds have several characteristics in common that relate to their intended use as the first feeds consumed by larvae and fry. By the time larvae and fry initiate feeding on exogenous foods, they have nearly used up their yolk nutrient reserves. They must obtain nutrients and energy from exogenous feeding to prevent starvation and death, and they are behaviorally programmed to do so. The goal of feed formulation and manufacture of larval and starter feeds is to produce highly nutritious, palatable, correctly sized, visually attractive, and water-stable particles. The formulation affects feed consumption by influencing the color, texture, taste, and smell of the feed, whereas pellet manufacturing affects consumption by influencing feed particle size, shape, texture, density, and buoyancy. Formulation and pellet manufacturing must be tailored together to produce high-quality larval and starter feeds. If not, poor fish performances including low larval survival and fry growth are likely. For example, formulating a feed using expensive and highly palatable feed ingredients will not yield support to larval or fry survival if the method of feed manufacturing results in a feed particle texture that lowers or prevents feed consumption by fish or shrimp larvae. Likewise, formulations that include ingredients affecting feed taste/smell and lowering feed consumption by larvae cannot be improved by feed manufacturing.

As mentioned, the first-feeding stage of fish life history is a critical one, and fish survival and fish health are primary (and only) considerations. Larval and starter feeds are expensive but only represent a tiny fraction of the total cost of feed consumed throughout the production cycle of fish and shrimp. For that reason, formulating and manufacturing anything less than the best larval and starter feed is not only false economy, but a recipe for failure.

Starter feeds are those fed to first-feeding fry or larvae. First-feeding fry or larvae have nearly used up their yolk material when the urge to feed begins, and it is absolutely essential that they accept their starter feed or they will rapidly starve to death. A starter feed must not only provide complete nourishment to the first-feeding fry or larvae but must also be appealing, so that larvae consume it. Stater feeds are formulated with these qualities foremost in mind; cost is largely irrelevant. Feed particles must be very small; yet they must also be water-stable without being indigestible.

Starter feeds differ in composition and type depending upon the size of the fry or larvae at first feeding. Salmonid fry are relatively large at first feeding and can initiate exogenous feeding when fed a dry powder similar to fish meal in consistency or a moist mash in which the particles are smaller than 0.07 mm (Table 9.8). This particle size corresponds to material which will pass through a US Bureau of Standards #25 screen and not pass through a #30 screen. Fry of intermediate size, such as catfish fry, will also accept feeds of this size. The larvae of some species of fish require particles as small as 50 microns in diameter. Manufacturing feed particles of small sizes that are water-stable, digestible, nutritious, and appealing to the larvae is difficult and requires specialized formulations and novel manufacturing technology.

9.5.1.1 Fry feed

Once fry or larvae are actively feeding and reach a weight of 0.50–0.75 g, they can be switched from a starter feed to a fry feed. Fry feeds formerly were produced by screening crumbled compressed particles to appropriate sizes, but extrusion is now the primary pelleting process to produce fry feeds. The fry stage is notable for being the life history stage where the greatest relative gains in growth can be made for the least total feed investment. Any gain in growth made at the fry stage, however, small in absolute amount, is likely to persist in subsequent stages of production. For example, a 25% increase in weight gain between groups of fry fed different diet formulations may be small difference in actual weight, but the 25% advantage can be carried over to harvest size, assuming appropriate fish culture practices are followed. The gains realized at the fry stage will increase later production. Fry feed

TABLE 9.8 Vitamin premix specifications for several open-formula salmonid diets.[a]

Ingredient (form)	NRC dietary recommended levels	Abernathy salmon diet premix	Ontario Ministry of Natural Resources vitamin concentration per kg premix+	USDA Agriculture Research service premix ARS 702
Vitamin A (palmitate or acetate)	2500 (I.U)	440,000 (6600) (IU)	1.93 g	965,000 (4825 IU)
Vitamin D3 (as cholecalciferol)	2400	29, 398 (441)	16 mg	640,000 (3300 IU)
Vitamin E (as DL-a-tocopheryl acetate)	30	33,510 (502)	15,200 IU	13,200 (66 IU)
Vitamin K (as menadione Na-bisulfate)	10	1858 (28)	278 mg	110 (1)
Vitamin B12 (as cyanocobalamin)	0.02	3.97 (0.06)	3 mg	(0.02)
Ascorbic acid (as ascorbyl polyphosphate)	100	(1000)	15000 mg	Added separately
D-Biotin	1	39.7 (0.6)	18 mg	33 (0.17)
Choline (as chloride)	400	8818 (132)	300,000 mg	Added separately
Folic acid	5	849 (12.7)	500 mg	250 (1.25)
Niacin (nicotinic acid)	150	14, 704 (220)	6700 mg	2180 (10.9)
Pantothenic acid	40	7668 (115)	3200 mg	4.651 (23.25)
Pyridoxine	10	2504 (38)	1500 mg	1370 (7)
Riboflavin	20	3527 (53)	1600 mg	960 (4.80)
Thiamin	10	3117 (47)	1300 mg	910 (4.55)
Carrier (wheat middlings or starch)	–	+	+	
Total premix		1000 g	1000 g	

[a] *All quantities are expressed on the basis of vitamin activity. Values in parentheses are dietary levels (IU or mg/kg) at recommended premix supplementation levels.*

formulation should be made with this in mind. Cost savings at the expense of nutritional quality in fry feed are false economy due to advantages gained in growth rate and to the fact that fry feed is a small percentage of the total feed consumed by production fish.

9.5.1.2 Fingerling feeds

Fish feed formulations for fingerling fish differ from fry feeds mainly in the dietary protein level, which is typically lowered by 12%–15% in feeds for fish once they reach 1–2 g. From this weight until fish reach 10–25 g depending on species, they are fed fingerling diets.

Extruded pellets between 1.0 and 2.5 mm (1/16″ to 3/32″) cover this fish size range. Fish growth rates during this stage are higher than at later life cycle stages. Thus, it is prudent to insure that the growth potential of fish or shrimp is achieved by formulating feeds that are overfortified in critical essential nutrients. Most of the research on fish nutritional requirements has been obtained from studies with fish of this size. Using this information, fingerling feeds can be formulated quite precisely.

9.5.1.3 *Grow-out feeds*

Grow-out feeds are formulated to promote efficient and economical growth of fish up to market size. Fish above 15%–20% g gain weight at a constant rate and appear to have nutritional requirements that are relatively constant. Formulations for these fish contain 15%–25% lees protein that fingerling formulations. The energy content and proportion of the total dietary energy derived from protein must be carefully formulated to insure that dietary protein is used as much as possible for growth rather than as a source or energy. Most of the feed consumed by a production fish during its life will be this phase, so attention to feed cost and/or the cost of feed per unit of production is an important consideration in the formulation process.

9.5.1.4 *Broodstock feeds*

Somatic growth continues uninterrupted in most species of cultured fish, exhibiting indeterminate growth until a combination of external and internal factors initiate sexual maturation. The temporal aspects of the sexual maturation process in fish are variable, proceeding for weeks, or months, depending upon the species. During sexual maturation, somatic growth slows and gonadal growth accelerates until the fish spawns. In many species of cultured fish, feeding ceases during part or all of the maturation process. If the species of fish is one that survives spawning, feeding then begins again soon after spawning.

Broodstock feeds are formulated to meet the nutritional needs of reproducing fish, although relatively little is known about specific nutritional needs of broodstock fish. Generally, formulations contain higher levels of protein, energy, and certain vitamins associated with interconversion of maternal tissue nutrients to egg protein. Supplementing feeds with squid or krill meal is practiced with certain marine species (Xu et al., 2017). Ascorbic acid level is also often increased in broodstock feeds. For salmonids, astaxanthin is needed in broodstock diets to ensure adequate astaxanthin in eggs and support robust fry. For marine species, dietary lipid source affects egg fatty acid levels, egg buoyancy, and larval survival (Watanabe et al., 1982; Izquierdo et al., 2000). More recently, interest has increased in nutritional programming through broodstock nutrition as a means of influencing metabolism of offspring, although the effects differ among species. In rainbow trout, supplementing diets with methionine and choline, both methyl donors, improved performance, presumably through epigenetic mechanisms (Cleveland et al., 2020). Improved tolerance to alternate ingredients in feeds, such as linseed oil, has been demonstrated in gilthead sea bream offspring after feeding broodstock diets containing linseed oil (Izquierdo et al., 2015), but similar studies in trout did not show similar improvement (Lazzarotto et al., 2016), suggesting species differences in responses to nutritional programming.

9.5.1.5 Seawater transfer feeds for salmon

When salmonids migrate from juvenile freshwater rearing habitats to the ocean, they transition from freshwater to seawater in estuaries over a period of time, moving back and forth between partial and full-strength seawater as they adapt to the marine environment. In contrast, transition of farmed salmon from freshwater hatcheries to marine pens is abrupt and results in transitory disruption of osmoregulation and tissue levels of sodium, potassium, and magnesium, all essential for normal functioning of nerves and muscle. This disruption can be mitigated using transfer feeds for several weeks prior to and after seawater transfer. Transfer feeds are typically supplemented with minerals and vitamin C plus immunostimulants. Supplementing transfer feeds with histidine reduces the incidence of cataracts in salmon (Waagbo et al., 2010).

9.5.1.6 Low-pollution feeds

Freshwater fish farms and hatcheries use surface and ground water resources, and release water into the environment containing uneaten feed, fecal waste, metabolic excretions, notably ammonia, and nutrients in feeds that are not retained by growing fish. These wastes enrich receiving water bodies, stimulating algae and aquatic plant growth. Governmental agencies enact and enforce regulations designed to reduce the downstream effects of farms and hatcheries by limiting concentrations of select materials in farm water discharges, the most common materials being phosphorus and suspended solids. Phosphorus is regulated because it is the first limiting nutrient for freshwater algae and macrophytes. In some jurisdictions, feed conversion ratios are limited, but, more commonly, phosphorus and solid levels in water released by farms are routinely measured to calculate the total amount released. Since feeds are the source of phosphorus, solids, and other materials in farm discharge water, farms must employ several strategies to meet regulations, including the use of low-pollution feeds.

Low-pollution feeds for aquaculture are characterized by having low levels of total phosphorus, levels of available phosphorus that meet but do not exceed levels required by fish, and by having high feed efficiency values. For many years, phosphorus levels in salmonid feeds were higher than required levels because fish meal was the primary protein source used in feeds. The phosphorus content of fish meal is relatively high due to the presence of fish bones, scales, and other hard tissues. Other high-phosphorus ingredients used in fish feeds are certain land animal proteins containing bones, e.g., poultry by-product meal, meat and bone meal, and plant protein concentrates containing high levels of phytate phosphorus. Replacing high-phosphorus feed ingredients with low-phosphorus ingredients is the first step in developing low-pollution feeds. This lowers the total phosphorus level in the feed to be closer to the dietary requirement for available phosphorus. The second step in low-pollution feed development is to formulate feeds to contain available phosphorus at or slightly above the nutritional requirement of the fish for which feeds are being formulated. This requires precise estimates of phosphorus requirements for all life stages of fish, not just for fingerlings (Rodehutscord, 1996; Ruohonen et al., 1999; Sugiura and Hardy, 2000). Studies with grow-out stage trout have shown that the dietary phosphorus requirement is lower than that for fingerling trout (Sugiura et al., 2000). The requirement was based on urinary excretion of phosphorus, which remains low until dietary phosphorus levels exceed metabolic and storage needs, at which point it increases proportionally to dietary phosphorus intake. Circulating levels of phosphorus are tightly regulated, and fish maintain a depot of stored phosphorus in bone and scales to ensure metabolic needs are met. If dietary intake is below required levels, fish draw upon stored phosphorus and can maintain growth and health for a period until phosphorus reserves are exhausted (Hardy et al., 1991). At that point, clinical deficiency signs rapidly appear. One strategy to reduce phosphorus discharge levels from trout farms is to feed a phosphorus-deficient diet during the final months of production. Trout draw upon reserves to maintain phosphorus homeostasis during this period, thus maintaining growth and health (Lellis et al., 2004).

Increasing the proportion of total dietary phosphorus that is available to fish is a critical element of developing low-pollution feeds. Low-ash fish meal can be produced by mechanical deboning of seafood processing waste prior to fish meal production or by screening to remove bones after drying before it is ground (Babbitt et al., 1994). Reducing bone content in fish meal increases phosphorus digestibility and lowers calcium levels in products (Sugiura et al., 2000). Plant protein concentrates contain low levels of phosphorus compared to fish meal or land animal proteins, but a high proportion of phosphorus is present as phytate phosphorus, the principle storage form of phosphorus in seeds. Phytate phosphorus is indigestible by monogastric animals, including fish, passing through the gut and being excreted in feces, thus contributing to phosphorus levels in farm or hatchery discharge water. Phytase in a natural enzyme that releases phosphorus from phytic acid as seeds sprout. It is produced commercially and widely used as a feed additive in poultry and swine feeds. Phytase activity increases with temperature, reaching a maximum at 50−55°C. It is active at body temperature of birds and mammals, but in fish its activity depends on rearing water temperature. It is, therefore, more active in warmwater species than in cold-water species. As a result, higher amounts of phytase must be supplemented in feeds for cold-water species to achieve the same degree of phosphorus digestibility as seen in warmwater species. Phytase can also be used to treat plant protein ingredients to release phosphorus prior to their being used in feeds (Cain and Garling, 1995). Several commercially available soy protein products use this approach. A final processing step to release phosphorus from plant proteins involves microbial fermentation. Products treated to release phytate phosphorus exhibit apparent phosphorus digestibility coefficients similar to or higher than fish meal (Sugiura and Hardy, 2000).

Supplementing diets with citric acid to prevent formation of calcium−phytate complexes in the gut increases phosphorus digestibility in grow-out stage rainbow trout (Sugiura et al., 1998; Vielma et al., 1999). However, feed intake is reduced when citric acid is added to feeds for first-feeding trout. Citric acid supplementation may be a component of development of low-pollution feeds for larger fish.

Increasing overall feed digestibility and balancing nutrient and energy content results in higher feed efficiency and lower fecal and metabolic losses. Transitioning to high-energy diets for salmon and trout over the past 2 decades has led to lower feed conversion ratios and higher nutrient retention, especially protein and energy retention. These changes have reduced nutrient losses from farms and are elements of low-pollution feeds. However, not all changes in feed formulations have lowered pollution potential of feeds, especially increasing soybean meal levels in feeds. Soybean meal increases the rate of passage of food through the gut, thus lowering water recovery from feces in the distal intestine leading to watery feces that disperse as suspended solids in the water column. Suspended solids are very difficult to recover from farm rearing water prior to discharge through settling or screening. Thus, using feed additives or formulating feeds to increase the proportion of intact, high-density fecal particles that can be recovered by settling in quiescent zones or screening is another strategy to production of low-pollution feeds (Schumann et al., 2017).

9.5.1.7 Recirculating aquaculture system feeds

Maintaining water quality in RAS is crucial, and since water quality is directly related to feed formulation and nutrient retention, feeds for RAS require some changes from regular commercial grow-out feeds. Feeds for rearing fish using RAS should be formulated with many of the features of low-pollution feeds to minimize the amount of fecal and metabolic excretions by the fish. In addition, feed formulators must be aware of the potential for trace mineral built-up in rearing water from feed mineral premixes, particularly for copper and zinc.

9.5.1.8 Product quality feeds

Product quality feeds are formulated to enhance quality characteristics of market products. These characteristics include sensory attributes, such as color, flavor and texture, color and flavor, nutritional attributes, such as docosahexaenoic acid (DHA) and eicosapentaenoic acid (EPA) content, and shelf life of products. Color is the primary characteristic associated with product quality feeds for salmon, and astaxanthin is the carotenoid responsible for fillet pigmentation. Astaxanthin supplements used in product quality feeds increase feed cost by about 10%, despite being a minor constituent of feed (0.1% on average). Color is also an important attribute, and hence, for shrimp and red sea bream, skin color must meet consumer expectation, so product quality feeds for these species are supplemented with carotenoid pigment sources, both natural and synthesized. Flavor is influenced by odor associated with feed lipid sources and by off-flavors connected to lipid oxidation. Fish oil imparts a fishy flavor to fillets whereas plant oils, such as soy or safflower, impart a neutral flavor. Protein sources in feeds have little or no effect on flavor or texture. However, fillet texture of some species, notably salmon and trout, is affected by fillet lipid level. Reducing the fat level in feeds fed near the end of production cycles may reduce fillet lipid levels slightly. Much research effort has been put into developing finishing feeds intended to increase fillet DHA and EPA levels in salmonids, but results support the dilution hypothesis that fatty acid turnover in tissues is slow, so changing fatty acid contents of feeds only results in overlaying the new fatty acids onto fatty acids already present in tissues (Jobling, 2003). Blending plant and fish lipids in feeds results in similar fillet fatty acid compositions as feeding one or the other for a period of time, then switching. Shelf life of fresh fish products is mainly dictated by sanitation and storage temperature to inhibit microbial spoilage. Shelf life of frozen fish products, however, is associated with prevention of lipid oxidation. Feeds with lower levels of fish oil and elevated levels of alpha-tocopheryl acetate increase tissue levels of vitamin E and reduce indicators of lipid oxidation (malondialdehyde levels) at conventional frozen storage temperatures (Boggio et al., 1985).

9.5.2 Practical considerations

9.5.2.1 Ingredient limits

For the most part, fish feeds are formulated using 40–50 common feed ingredients, a relatively small number. After feed ingredients available to formulate feeds are chosen, the next step is to place upper, lower, or fixed limits on each ingredient. Upper limits are placed on ingredients that contain antinutritional factors or toxicants, or that affect palatability or pelletability. Lower limits are placed on ingredients that are desirable in the formulation despite cost. For example, fish meal is an expensive source of protein in tilapia feed formulations. Nevertheless, a lower limit of 5% fish meal may be made in feed formulations for tilapia fingerlings to guarantee performance. In this case, fish performance is a higher priority than feed cost. Fixed levels are used for vitamin and trace element premixes and for other ingredients for which the level in the formulation is not appropriate to alter. In practical fish feed manufacture, an ingredient occasionally is in the inventory of a mill, and the manufacturer may wish to use that ingredient to remove it from inventory. This situation would call for a lower limit to be placed on the level of the ingredient in a formulation, even though the ingredient might not be the least expensive or best choice at that level. Finally, it is not necessary to place upper and lower on each ingredient if the formulator wishes to see the consequences of allowing the software to formulate without restrictions. Formulations usually require substantial adjustments of limits and recalculations to arrive at the most desirable product.

9.5.2.2 Manufacturing process

The pelleting technology being used for manufacture feeds determines the limits placed on the levels of many ingredients in a formulation. Dry, compressed pellets, for example, must contain minimum levels of wheat by-products to allow the exterior surface of the pellet to gelatinize during pelleting. If the pellets are to be crumbled and screened, ingredients and their levels must be chosen to increase pellet hardness.

Cooking extrusion is now the most common pelleting method for fish feeds. This technology produces hard, durable pellets without the need for supplemental binders in the formulation; ingredients such as starch or wheat gluten present in the formulation are gelatinized by the cooking extrusion process and bind pellets together. Limits are placed on these ingredients to optimize pellet hardness, expansion, and buoyancy.

In moist pellet or semimoist pellet formulations, the ingredients and their limits are chosen with the water content and water activity in mind. Ingredients that act as pellet binders, such as pregelatinized potato starch, must be included at levels that provide sufficient binding capacity to produce a high-quality pellet. In formulations containing moisture levels over 30%–35%, such as those containing liquefied fish products or ground fishery waste, additional binders must be used.

9.5.2.3 Hardness and water solubility

In practical fish farming, pellet hardness and the absence of fines are qualities that fish farmers can observe themselves. To minimize fines, feed manufacturers screen pellets to remove fines before the pellets are sacked, but pellet disintegration can occur in feeds during shipping and handling at the fish farm. Pellet durability can be improved by including certain ingredients in the formulation that might not otherwise be included if least-cost formulation procedures are rigidly followed.

Pellets used for slow-eating fish and crustacea must remain in water for some time without disintegrating until they are located and consumed. This requires special consideration during feed formulation and manufacturing. As an extreme example, eel diets contain pregelatinized potato starch in the formulation to maintain water stability of the dough-like feed. Pregelatinized potato starch would not be included in eel feed formulations based on price and nutrient content alone.

9.5.3 Calculations

9.5.3.1 Best-buy ingredients

Aside from the limitations discussed in previous sections, ingredients should be selected on the basis of value. If several different fish meals are available and suitable for use in a formulation, and the fish meals are equal in protein content and protein availability, the decision about which to select is simple; choose the least expensive. However, if the meals differ in protein content or protein availability and price, calculations are required to determine which is the best value. For example, suppose one can select either Danish fish meal having a protein content of 64% and costing US \$1435/MT, or Icelandic fish meal having a protein content of 70% and costing US \$1530/MT. To compare the two meals, compute the cost per kg protein as follows:

Cost per kg protein

$$\text{(1) Danish fish meal} = \frac{\$1435/\text{MT}}{640\ \text{kg protein/MT}} = \$2.242/\text{kg protein}$$

$$\text{(2) Icelandic fish meal} = \frac{\$1530/\text{MT}}{700\ \text{kg protein/MT}} = \$2.186/\text{kg protein}$$

The results of this example show that Icelandic fish meal, although it is more expensive per MT, is a better value, since the cost per kg protein is lower than for Danish fish meal. Ofter protein supplements are compared on the basis of cost per unit protein or "protein unit." This simple calculation is made by dividing the price per MT by the percentage protein in the product. For example:

$$\text{(1) Peruvian fish meal} = \frac{\$1450/\text{MT}}{64\%\ \text{protein}} = \$22.66\ \text{per protein unit}$$

$$\text{(2) Icelandic fish meal} = \frac{\$1530/\text{MT}}{70\%\ \text{protein}} = \$21.86\ \text{per protein unit}$$

$$\text{(3) Herring meal} = \frac{\$1585/\text{MT}}{72\%\ \text{protein}} = \$22.01\ \text{per protein unit}$$

$$\text{(4) US Meat and bone meal} = \frac{\$520/\text{MT}}{50\%\ \text{protein}} = \$10.4\ \text{per protein unit}$$

Icelandic fish meal is the best value among the fish meals listed in the example above. Meat and bone meal is the best value on a protein unit basis.

This process of comparing feed ingredients on the basis of cost per unit nutrient can be applied to any nutrient of interest and to the energy content. For example, suppose one wishes to determine if it is less expensive to add L-lysine or increase the level of herring meal in a feed formulation to increase the lysine content. One approach is to calculate the cost of lysine in each product as follows:

(1) Lysine HCl costs US $1.54 per kg and contains 98% L-lysine. The cost is:

$$\frac{\$1540/\text{MT}}{98} = \text{U. S. } \$15.71 \text{ per kg lysine}$$

(2) Herring meal costs US $15850/MT and contains 5.36% lysine. The cost per unit is:

$$\frac{\$1585/\text{MT}}{5.36} = \text{U. S. } \$295.71 \text{ per kg lysine}$$

It is less expensive to add lysine-HCl to the formulation rather than add herring meal to increase the lysine content of the formulation.

A similar calculation can be used to compare the costs of adding energy to a fish feed from various ingredients or to compare the costs of obtaining energy from dietary protein or lipid. For example, if herring meal costs US $1585/MT and contains 19.749 MJ DE/kg, the cost per 1000 MJ is US $0.080. Although fish oil contains more MJ per kg than fish meal, the exact amount that is available to fish is uncertain. A close estimate based on lipid digestibility values of 90% given by Ellis and Smith (1984) is 33.49–33.91 MJ DE/kg. Using a value of 33.49 MJl/kg, one can calculate the price below which fish oil must sell to warrant replacing herring meal with fish oil as an energy source.

(1) Price per 1000 MJ DE for herring meal is US $0.080. DE content of fish oil is 33.49 MJ/kg.
(2) $(0.080) (33.49) = 2.679/kg or $2679/MT. This is the price one could afford to pay for fish oil to provide energy to the fish diet as cheaply as energy from herring meal. If the price of fish meal increases, the cost advantage of providing dietary energy from fish oil increases.

9.5.3.2 Using simultaneous equations to formulate feeds

Simultaneous equations can be used to solve simple feed formulations once ingredients have been chosen. This method is preferable to trial and error substitutions, although an experienced formulator can produce a very good formulation in a short time by substitution and recalculation. The process of formulating a fish feed is divided into several steps (Cho et al., 1985). The first step is to define the nutrient levels desired in the feed (Table 9.8). In the example, we wish to formulate a feed containing 45% protein and 16.747 MJ digestible energy per kg feed. We will ignore, at this point, the essential amino acid levels of the formulation, although they can be included if we desired.

The second step in the formulation process is to choose ingredients and to list the protein and digestible energy content of each (Table 9.9). The third step is to list the ingredients which

TABLE 9.9 Mineral premix specifications for several open-formula salmonid diets.

Element (form[a])	U.S. Fish and Wildlife service trace Mineral premix #3 g element/kg premix	Ontario Ministry of natural resources mineral premix MIN-8204 g element/kg premix
Zn (as $ZnSO_4$)	75	14.4
Mn (as $MnSO_4$)	20	8.6
Cu (as $CuSO_4$)	1.54	2.5
I (as KIO_3)	10	0.8
(or)		
I ($C_2H_8N_22HI$)	–	–
Fe (as $FeSO_4$)	–	6.3

[a] *Hydrated forms may be used.*

will be in the feed formulation at fixed levels and to calculate the contribution of each ingredient to the total desired levels of protein and digestible energy in the feed (Table 9.7). These values are then added and the totals subtracted from the desired levels in the finished feed. In the example, the levels of soybean meal, poultry by-product meal, and vitamin and mineral supplements are fixed, while the levels of fish meal, wheat middlings, and fish oil are variable in the formulation. The fourth step is to determine the appropriate levels of the various ingredients in the formulation using simultaneous equations (Table 9.10).

The solution of the simultaneous equations yields the following: fish meal, 43.7 kg/100 kg feed; wheat middlings, 22.4 kg/100 kg feed; and fish oil, 11.22 kg/100 kg feed. The last step is to check the final feed levels of protein and digestible energy to insure the desired levels are present (Table 9.11). The levels of individual essential amino acids can also be calculated at this point to insure that the levels in the feed meet or exceed the dietary requirements of the fish.

9.5.3.3 *Linear programming*

Computers are used in the feed industry to calculate least-cost formulations. The process by which this is done is linear programming, which involves the simultaneous solution of a series of linear equations. Linear programming has been used in animal feed formulation for over 30 years and is used almost exclusively in modern feed formulation in agriculture today. In the past, linear programming required access to a mainframe computer and specialized knowledge in mainframe use, i.e., programming ability. While the actual computer time

TABLE 9.10 Nutrient levels desired in the feed.

Feed (kg)	Protein (kg)	Digestible energy (MJ)
100	45.0	16.747 (4000/kg feed)

TABLE 9.11 Composition of dietary ingredients on a dry matter basis.

Ingredients	Protein (%)	Digestible energy (MJ/kg)
Fish meal (herring)	72.0	18.7987
Poultry by-product meal	58.0	13.900
Soybean meal	48.0	13.498
Wheat middlings	17.0	7.000
Fish oil	–	37.681
Vitamin premix	–	–
Mineral premix	–	–
Choline chloride	–	–
Ascorbic acid	–	–

required to arrive at a least-cost formulation was short, the time required for data input and setting up was lengthy. For simple formulations or for small feed manufacturers, hand calculations were often more practical. The widespread availability of mainframe computers and the use of remote terminals that can be connected to a mainframe by the use of a modem (modulator/demodulator) and commercial telephone line, combined with stored data and programming on the computer, have increased the accessibility of linear programming to virtually every feed mill in North America. Today, one does not need to know how to program a computer or how to set up linear equations to solve least-cost feed programs.

Today, computers using feed formulation programs are routinely used to calculate the combination and levels of ingredients that provide the desired nutrient content of the diet at the least cost. The formulation program must have up-to-date data on the nutrient content of each feed ingredient, obtained from feed composition tables or from analyses at the mill itself. The current price of each ingredient must be entered into the program along with the nutrient specifications of the formulation being calculated. Ingredient limits, i.e., maximum and minimum levels allowed in the formulation, must be imposed. At this point, the software calculates the formulation. Outputs vary with different programs but include the least-cost formula in which ingredients and levels to add to the mixture are listed, listings of calculated levels of nutrients in the formulation and information on the effects of the restrictions are placed on the levels of individual ingredients or desired nutrient levels on the cost of the diet. Information on the effects of restrictions placed on ingredient levels or nutrient levels is useful to the feed formulator. It shows the cost of maintaining the level of a nutritional parameter, such as protein, and the savings that would result from relaxing the restriction, say, from 45% to 44%, in a salmon diet. This output is sometimes called dual activity. Information on the effects of relaxing upper and lower restrictions on the levels of individual ingredients is shown in an output called reduced costs. For example, if the upper limit of soybean meal was set at 10% in a trout diet formulation, and the feed formulation calculated by the software contained the maximum amount of soybean 10% output, a value will be shown in the reduced cost output indicating the savings that would result by relaxing this restriction to allow higher levels of soybean meal to be used. Of course, the decision to limit the soybean meal level to 10% of the formulation may have been made for good reasons, such as availability, pellet texture, palatability concerns, and so on. The nutritionist must decide what to do in such a situation, but the reduced cost output clearly shows what changes can be made in the formulations to lower the cost of the formulation.

To illustrate how the output from linear programming can aid the nutritionist, two cases are presented in which the restrictions placed on soybean meal and canola meal are changed in a trout feed formulation. In Diet 1 (Table 9.12), the levels of wheat middlings are restricted to between 10% and 40% of the diet, soybean meal is fixed at 10%, a maximum level of 10% is placed on blood meal, a minimum level of 25% herring meal is used, herring oil is not restricted, the level of canola meal is fixed at 15%, and the remaining ingredients are fixed at levels that remain constant. In the first column, under Amount, is the number of pounds per ton needed for each ingredient to achieve the least-cost formula with restrictions imposed. In the second column, under Number, is the feed number assigned by the feed company. This is followed by the feed name, the percent of the ingredient in the formula, and the ingredient cost, which is in US dollars per 100 pounds at 1981 prices of ingredients. Next, under Ingredient Minimum/Maximum, are the restrictions placed on ingredient levels in pounds per ton. The next columns are Low Range and High Range. These columns show the low and/or high prices the ingredient could cost before the computer would change the level of the ingredient in the formulation. The last column, Reduced Cost, lists the dollar savings that could be made per ton of feed if the restrictions were relaxed. In Diet 1, the computer shows that an increased cost of US $10.43 per ton of feed is incurred because of the restriction being placed on the level of soybean meal used in the formulation. In other words, if more soybean meal could be used in this formulation, the cost would be lower. The same is true for blood meal and canola meal. For other ingredients listed in this formulation, which have values appearing in the reduced cost, the computer shows a lower cost per ton could be achieved by lowering the minimum restriction placed on each ingredient (Table 9.13).

TABLE 9.12 Contribution of fixed ingredients to nutrient content of the final formulation.

Ingredients	Percent	Protein (kg)	Digestible energy (MJ/kg)
Poultry by-product meal	10	5.8	139
Soybean meal	10	4.8	135
Vitamin premix	2.0	–	–
Mineral premix	0.1	–	–
Choline chloride	0.5	–	–
Ascorbic acid	0.1	–	–
TOTAL—fixed ingredients	22.7	10.6	274
Amount needed from ingredients	77.3	34.4	1400

TABLE 9.13 Least-cost formulation for salmon feed, with minimum dietary protein and lipid set at 45% and 23%, respectively.

Ingredient	Minimum	Actual	Maximum
Fish meal	32.00	32.00	40.00
Corn gluten meal	3.00	3.00	10.00
Meat and bone meal		10.00	10.00
Soybean meal (48 CP)		5.137	20.00
Wheat middlings	5.00	5.00	15.00
Wheat, ground grain	10.00	10.00	
Poultry by-product meal	5.00	17.73	20.00
Fish oil	12.00	15.98	
Ascorbate-2-phos.	0.10	0.10	0.10
Vitamin premix	0.50	0.50	0.50
Trace mineral premix	0.10	0.01	0.10
Choline chloride	0.35	0.35	0.35
Astaxanthin	0.10	0.10	0.10
		Cost is $776.79 per metric ton	

Nutrient analysis—Diet 1			
Category	**Minimum**	**Actual**	**Cost/ton**
Crude protein	41.00	41.01	310.5
Arginine		2.635	
Lysine		2.911	
Methionine		0.793	
Methionine + cystine		1.313	
Tryptophan		0.486	
Histidine		1.217	
Leucine		3.590	
Isoleucine		1.510	
Phenylalanine		1.921	
Phenylalanine + tyrosine		2.886	
ME energy—kcal/kg		3080	

TABLE 9.13 Least-cost formulation for salmon feed, with minimum dietary protein and lipid set at 45% and 23%, respectively.—cont'd

Nutrient analysis—Diet 1			
Category	**Minimum**	**Actual**	**Cost/ton**
Crude fat	10.00	9.999	59.08
Crude fiber		3.607	
Percent fish protein	17.00	17.46	
Calcium		0.854	
Phosphorus		1.009	

In the nutrient analysis output, Table 9.12, the calculated levels of a number of nutrients and categories are shown. Specifications were made for several, including crude protein level, crude fat level, and the percent protein of fish origin, i.e., herring meal, in the formulation. For these items, numbers appearing under Minimum indicate the minimum level specified in the program, and numbers under Cost/Ton refer to the savings per ton that could be achieved if the minimum specification values are removed.

In Diet 2, the same ingredients were used (Table 9.14), but the restrictions on soybean meal and canola meal were changed. Soybean meal was allowed in the formulation up to 25%, and the canola meal upper limit was increased from 10% to 15% with no minimum level being imposed. The solution to the least-cost feed formulation was quite different, as can be seen from the percent of each ingredient in the formula.

Soybean meal and canola meal were used at maximum permitted levels, blood meal and wheat middling levels were decreased, and herring meal was used at a level slightly above the minimum level permitted. The cost per ton for ingredients was reduced by US $7.64. The values are listed in Table 9.15.

Reduced costs show that if the maximum limits on soybean meal and canola meal were futher relaxed, additional savings would be possible.

In Diet 3, the maximum level of canola meal was increased to 20%, with no minimum level entered (Table 9.16). Soybean meal was restricted to 15%–25% of the formulation. As can be seen, the least-cost formulation included 25% soybean meal and 20% canola meal, which replaced wheat middlings and blood meal in the formulation. This change lowered the cost per ton an additional US $3.73 without significantly changing the nutrient analysis values. These diets were fed to groups of rainbow trout, and the high canola meal–high soybean meal diet supported growth equal to or higher than the other diets (Hardy and Sullivan, 1983).

The preceding discussion illustrates the various facets of linear programming methods for feed formulation. To become a competent, linear programmer requires years of study of computer science and higher mathematics and years of practice. Fortunately, that is not necessary since there are many feed formulation software programs suitable for a range of applications.

TABLE 9.14 Simultaneous equations to determine levels of wheat middlings (WMs), fish meal (FM), and fish oil (FO) needed in the formulation.

(1) WM + FM + FO = 77.3 kg feed

(2) 1672 WM + 4490 FM + 9000 FO = 334,560 kcal DE

(3) 0.17 WM + 0.70 FM + 0 FO = 34.4 kg protein

(4) From Eq. (1), let WM = 77.3 − FM − FO
Substitute Eq. (4) in Eqs. (2) and (3) as follows:
(a) 1672 (77.3 −FM − FO) + 4490 FM + 9000 FO = 334,560
(b) 0.17 (77.3 − FM − FO) + 0.70 FM + 0 FO = 34.4

Solve (a) and (b) to make Eqs. (5) and (6)

(5) 2818 FM + 7328 FO = 205,314.4

(6) 0.53 FM − 0.17 FO = 21.259

From Eq. (5), rearrange to make Eq. (7)

(7) FM = 205,314.4 − 7328 FO/2818 or 72.858 − 2.6 FO

Substitute Eq. (7) into Eq. (6)

(8) 0.53 (72.858 − 2.6 FO) − 0.17 FO = 21.259

Solve Eq. (8) for FO to obtain:
FO = 11.2
Substitute 11.2 for FO in Eq. (5)

(9) 2818 FM + 7328 (11.2) = 205,314.4

Solve Eq. (9) for FM to obtain
FM = 43.7
Solve for WM using Eq. (4)
WM = 77.3 −43.7 − 11.22 = 22.4

Solution:
Fish meal = 43.7 kg/100 kg diet
Wheat middlings = 22.4 kg/100 kg diet
Fish oil = 11.2 kg/100 kg diet

TABLE 9.15 Least-cost formulation for salmon feed, with minimum dietary protein, and lipid changed to 44% and 20%, respectively.

Ingredient	Minimum	Actual	Maximum
Fish meal	32.00	32.00	40.00
Corn gluten meal	3.00	3.00	10.00
Meat and bone meal		10.00	10.00
Soybean meal (48 CP)		**16.78**	20.00
Wheat middlings	5.00	5.00	15.00
Wheat, ground grain	10.00	10.00	
Poultry by-product meal	5.00	**7.39**	20.00
Fish oil	12.00	**14.68**	

TABLE 9.15 Least-cost formulation for salmon feed, with minimum dietary protein, and lipid changed to 44% and 20%, respectively.—cont'd

Ingredient	Minimum	Actual	Maximum
Ascorbate-2-phos.	0.10	0.10	0.10
Vitamin premix	0.50	0.50	0.50
Trace mineral premix	0.10	0.01	0.10
Choline chloride	0.35	0.35	0.35
Astaxanthin	0.10	0.10	0.10

Cost = $735.24 per metric ton.

Nutrient analysis—Diet 2

Category	Minimum	Actual	Cost/ton
Crude protein	41.00	41.00	264.70
Arginine		2.820	
Lysine		2.807	
Methionine		0.815	
Methionine + cystine		1.327	
Tryptophan		0.493	
Histidine		1.151	
Leucine		3.344	
Isoleucine		1.761	
Phenylalanine		1.872	
Phenylalanine + tyrosine		2.033	
ME energy—kcal/kg		3236	
Crude fat	10.00	9.979	56.17
Crude fiber		3.652	
Percent fish protein	17.00	17.01	24.34
Calcium		0.839	
Phosphorus		1.026	

Nutrient analysis—Diet 3

Category	Minimum	Actual	Cost/ton
Crude protein	41.00	41.00	234.70
Arginine		2.831	
Lysine		2.751	

(*Continued*)

TABLE 9.15 Least-cost formulation for salmon feed, with minimum dietary protein, and lipid changed to 44% and 20%, respectively.—cont'd

Nutrient analysis—Diet 3			
Category	**Minimum**	**Actual**	**Cost/ton**
Methionine		0.832	
Methionine + cystine		1.335	
Tryptophan		0.490	
Histidine		1.142	
Leucine		3.271	
Isoleucine		1.791	
Phenylalanine		1.841	
Phenylalanine + tyrosine		3.040	
ME energy—kcal/kg		3280	
Crude fat	10.00	9.996	
Crude fiber		4.133	
Percent fish protein	17.00	17.01	
Calcium		0.864	
Phosphorus		1.057	

Changes are in bold.

TABLE 9.16 Recalculation of feed formulation to check nutrient levels.

		Nutrient contributions	
Ingredient	**kg**	**Protein (kg)**	**Digestible energy (MJ)**
Fish meal	43.7	30.6	821.5
Poultry by-product meal	10	5.8	139
Soybean meal	10	4.8	135
Wheat middlings	22.4	3.8	157
Fish oil	11.2	–	422
Vitamin premix	2.0	–	–
Mineral premix	0.1	–	–
Choline chloride	0.5	–	–
Ascorbic acid	0.1	–	–

9.6 Diet manufacture and storage

Diet manufacture and storage is a complex process, and it is beyond the scope of this chapter to cover all phases of the subject in comprehensive detail. The information presented here is intended to provide the reader with an overview of the major steps and processes in diet manufacture. This will help biologists and fish farmers appreciate the complexity and limitations of commercial diet manufacture as well as understand the differences associated with manufacturing various types of fish feeds. The information will inform readers how to best utilize feeds. Those interested in more detail on feed manufacture should consult Schofield (Schofield, 2005). Further, there are several sources that describe best practices for feed manufacturing, such as the US Food and Drug Administration (https://www.fda.gov/animal-veterinary/guidance-industry/feedfood-guidances), Best Aquaculture Practices (https://bapcertification.org/blog/environmental-responsibility/), and the Aquaculture Stewardship Council (https://www.asc-aqua.org/), to name a few.

The basic steps in diet manufacture are mixing, grinding, conditioning, pelleting or extruding, cooling, drying, screening, top-dressing, bagging or transferring to bulk storage bins, and shipping (Fig. 9.1). These steps are discussed in sequence, and where different steps are required to produce different types of pellets, these are noted. Ingredients arrive in sacks and bulk, in both dry and liquid form. Ingredients should be examined for condition on arrival, dated and tested for chemical composition, using near-infrared spectrometry (NIR). Drugs and medications must be segregated from other ingredients in a secure location. Ingredients must be stored to protect against contamination, moisture, and pest damage. Each ingredient should be used in the order of batch delivery to prevent deterioration associated with long storage. In most modern plants, bulk ingredients are cleaned using screens before being conveyed to storage bins.

9.6.1 Grinding

Particle size reduction by mechanical grinding is necessary to ensure that ingredient mixtures do not segregate by particle size during processing and to eliminate large particles that can plug pellet dies. Grinding increases the surface area of ingredients, thus facilitating mixing and pelleting. Some ingredients arrive at the mill in coarse form and must be ground prior to mixing, but, generally, dry ingredients for a formulation batch are first mixed and then ground to uniform particle size. Grinders reduce particle size by impact (hammer mills and attrition mills, Fig. 9.2), cutting (rotary cutters, roller mills, and attrition mills), or crushing (roller mills). Mixtures used to produce small extruded pellets are often ground using a hammer mill or a roller mill for pregrinding and then undergo final grinding with the use of air-swept pulverizers that combine air flow with high-speed impact to move material through the grinder, keep material cool, and prevent screens on grinders from becoming clogged. Choice of the type of grinders for a feed mill depends on the type of pelleting technology being used and the ingredients that will comprise the formulation. Ingredients possessing elevated residual lipid levels are particularly difficult to grind without plugging grinder screens. High-lipid ingredients, such as fish meal, can be combined with low-lipid ingredients such as soybean meal, corn gluten meal, or wheat by-products prior to grinding to reduce problems with plugged grinding screens.

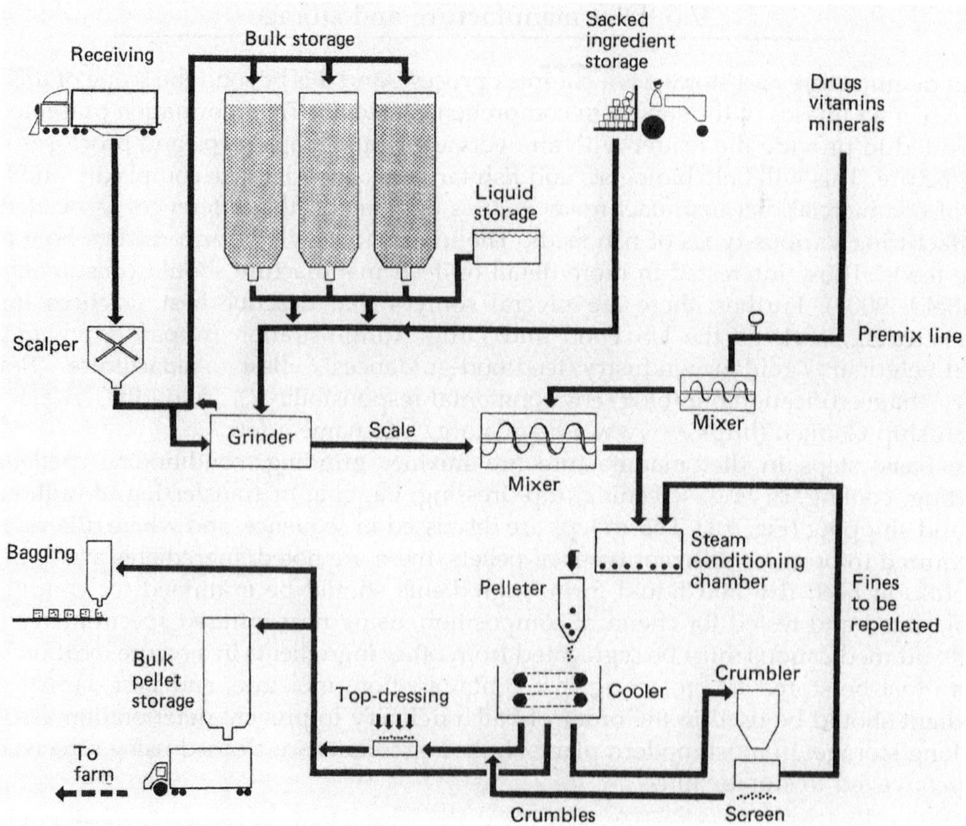

FIGURE 9.1 Simplified flow diagram of compressed pellet manufacturing.

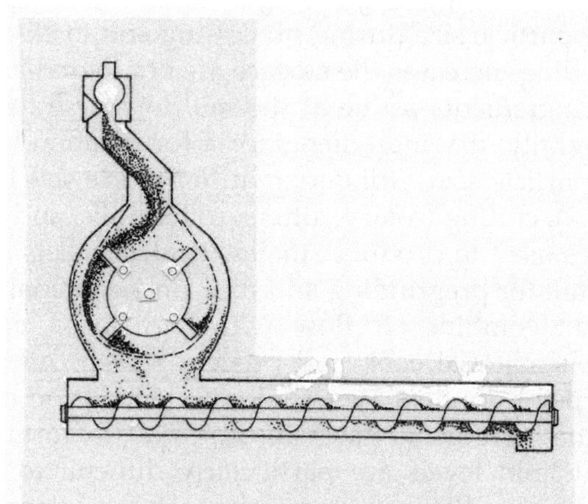

FIGURE 9.2 Hammer mill.

Grinding requires substantial energy and is an expensive part of the feed manufacturing process. A combination of mills, such as roller mill or a hammer mill, is often used as a pre–hammer mill or pulverizer with the aim of: (a) most economical grinding (energy wise); (b) overheating avoidance; and (c) smaller screens used at the final grinding. Overgrinding should be avoided, not only to control costs but also to prevent damage to ingredients associated with heat buildup in the grinding process. The degree of particle size reduction depends on the process by which feed will be pelleted and by the pellet size to be produced. A general rule of thumb for livestock feeds is to reduce particle size of mixtures to 25%–33% of the diameter pellets to be produced. For fish feeds, a common goal is to achieve particle size reduction to 1000 μm, although grinding to 50–100 μm is necessary to produce small extruded pellets for larval and fry feeds.

9.6.2 Mixing

The next step in diet manufacture is mixing the desired proportions of ingredients to produce a homogenous blend. Microingredients (vitamins, minerals, drugs, carotenoid pigments, etc.) are sometimes mixed separately with a diluent (carrier) before being mixed with other dietary ingredients to facilitate distribution within the feed mixture. Done properly, mixing procedures result in a blend from which pellets can be produced which contain the same proportion of ingredients as the formulation. Ingredients are generally added to the mixer in a predetermined sequence to insure adequate mixing. Dry ingredients are first combined, with dry premixes added in the middle of the sequence. If particle size is about the same for each ingredient, particle segregation will be minimized. After sufficient mixing of dry ingredients, liquid ingredients are added (sprayed) and mixing continues.

In the past, mixing was done with shovels on the floor with a set number of turns per batch. Most mixing today is done mechanically with batch or continuous mixers. Both kinds of mixers are cylinders or cones in which the feed ingredients are mixed with moving paddles, augers, or ribbons. Continuous mixers are similar, except that the design of the moving parts is such that material moves through the mixer as it is being mixed. Types of mixers include horizontal ribbon mixers, vertical mixers, Nanta mixers, and turbine mixers (Figs. 9.3 and 9.4). For laboratory use, dough mixers are appropriate. Mixing microingredients may require the use of specially designed mixers, such as twin-shell blenders, to disperse the microingredients in a carrier, such as wheat starch or alpha cellulose.

As mentioned above, the goal of mixing is to produce a homogeneous blend of feed ingredients. Although ingredients are ground prior to mixing, all particles in a mixture will not be identical in size or density. Mixing equipment should be calibrated to determine optimum mixing time to achieve a homogeneous mixture. Undermixing prevents even distribution of particles throughout the batch of ingredients, while overmixing (mixing for longer than necessary) causes particles to segregate based upon particle size or density. This is undesirable because it will result in pellets that vary in ingredient content and nutritional value. Mixers are calibrated by adding a tracer material, such as salt, silica, or an inert tracer to the feed ingredient batch. The mixer is then operated, and samples are removed at timed intervals to check for concentration of the tracer throughout the mixture. Manufacturers of

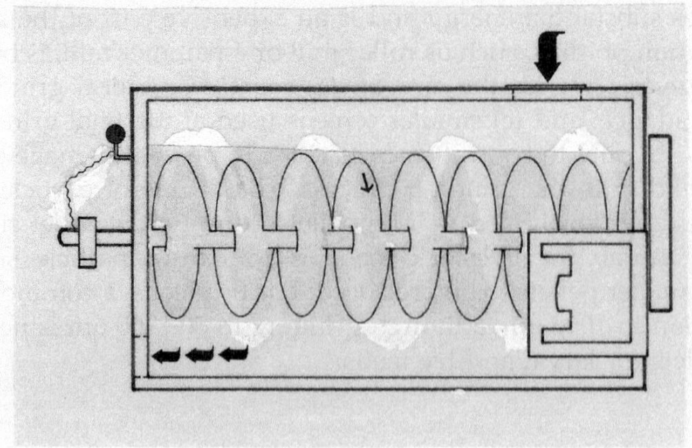

FIGURE 9.3 Horizontal mixer.

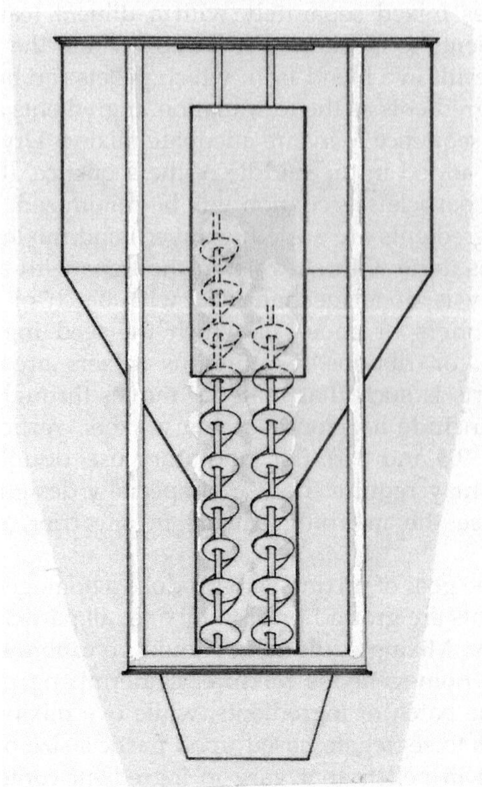

FIGURE 9.4 Vertical mixer.

mixing equipment provide guidelines for optimum mixing time, but feed mills should periodically calibrate their mixers to account for changes in grinding efficiency or feed formulation, as part of best feed mill practices.

Premixes are concentrated mixtures of vitamins and minerals that are added to mixers at low percentages of the total feed blend. They are produced using specialized mixing equipment, such as twin-shell blenders, using a carrier to facilitate mixing and to dilute micronutrient contents.

9.6.3 Pelleting

The objective in pelleting is to convert a homogenous blend of ingredients into durable particles having physical characteristics that make them suitable for feeding fish or shrimp in water. There are three main pelleting technologies used to produce fish feed pellets: compression, cooking extrusion, and cold extrusion.

9.6.3.1 Compressed pellets

Compressed pellets are made by forcing the feed mixture, which has been exposed to dry steam during conditioning for about 5–25 s to increase the temperature to about 85°C and the moisture to about 16%, through holes in a rotating metal die by the action of a static roller located inside the die (Fig. 9.1). The combination of heat, moisture, and pressure compresses the mixture as it passes through the die from the inside out. The force involved generates heat that gelatinizes starch and compresses the mixture into pellets having a bulk density of 0.5–0.6 g/cc. As the feed emerge from the outside surface of the rotating die, it is cut off by a static adjustable knife to achieve desired pellet length, generally 1–2 times pellet diameter.

Compressed pellet quality is influence by several factors, including fat level, moisture, and humidity in the feed mixture. The fat level of the mixture should be no lower than 2%–3% to lubricate the holes in the die and to reduce dustiness, and no higher than 8%–10% to avoid excessive die lubrication causing insufficient compression of the feed mixture. The moisture level is critical in that it affects pellet quality. Either insufficient or excessive moisture in the feed mixture reduces pellet hardness. Pellets made with insufficient moisture are dry and crumbly, while excessive moisture results in soft pellets due to insufficient compression. Changes in atmospheric humidity can change pellet quality, and the pellet machine operator must make adjustments in the amount of steam added to the feed and/or the rate at which the feeds mixture is introduced into the steam chamber to compensate for changes in the humidity. The addition of steam improves pellet quality by wetting and softening the feed particles and thereby improving compression and by increasing gelatinization of raw starch, which helps to bind the ingredients together in the pellet. The lubricating effects of fat and moisture together lower the electrical power required to operate a pellet mill and extend the operating life of the die. Experienced pellet mill operators become familiar with various feed mixtures and manufacturing conditions and make minor adjustments in steam addition and material flow through the die necessary to make high-quality compressed pellets.

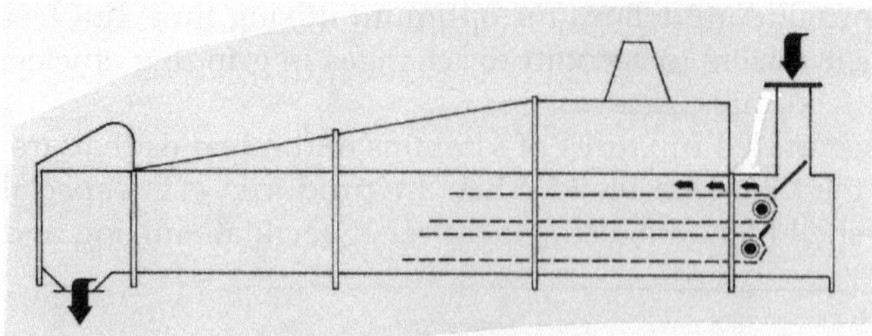

FIGURE 9.5 Horizontal pellet cooler-dryer.

Pellets exit the die at approximately 95°C and are immediately cooled and dried by passing them through a cooler-dryer (Fig. 9.5). Cool air is blown through the pellets, which are spread thinly on a moving belt in a horizontal cooler. A vertical cooler drops the hot pellets through a cooling tower. The latent heat in pellets facilitates drying; typically no additional heating is needed. Passage through the cooler-dryer takes 5–15 min, and when the pellets leave the cooler-dryer, they are at ambient temperature and have about 10% moisture. For laboratory-scale pelleting, cooling and drying can be accomplished by spreading the pellets and blowing air over them with a fan.

9.6.3.2 Extruded dry pellets

Cooking extrusion is now the most common technology used to produce fish feeds. The process involves three main steps: conditioning, extrusion, and cooling/drying (Fig. 9.6). During conditioning, steam is added to a chamber in which the feed mixture is being agitated using counterrotating augers. The feed mixture is heated by the steam and friction associated with mechanical force, reaching 125–150°C with a moisture content of 20%–24%. These conditions initiate starch gelatinization. The mixture is worked to a dough-like consistency and then forced into a long, tapered barrel with steam ports containing an auger with increasingly tight flights. The pressure within the barrel keeps water vapor (steam) in a liquid state until the feed mixture is forced through a stationary plate with tapered holes (the die). The feed is extruded through the die as long noodles that are cut to desired length by a rotating knife on the outside surface of the die. As the feed mixture exits the die, high pressure is released, that instantly converts moisture to water to water vapor, creating small air pockets to form within the pellets. This step is called expansion. Pellets immediately pass through a dryer on a moving belt where they are exposed to forced, heated air. After cooling and drying, pellet density (0.25–0.3 g/cc) is such that pellets float or slowly sink in water, depending on conditions under which the extruder is operated. Proper combinations of ingredients to include starch and to limit fat in the mixture produce water-stable, hard pellets.

9.6.3.3 Cold extrusion

Modified pasta-type extruders are used to manufacture semimoist pellets. By varying the moisture content and by careful selection of ingredients with binding activity, stable moist pellet can be produced. Moist pellet extrusion does not involve heating the mixture with steam or other means before extrusion, but rather the wet mixture is forced through a plate containing tapered holes to compress the feed mixture. This means that sanitation standards

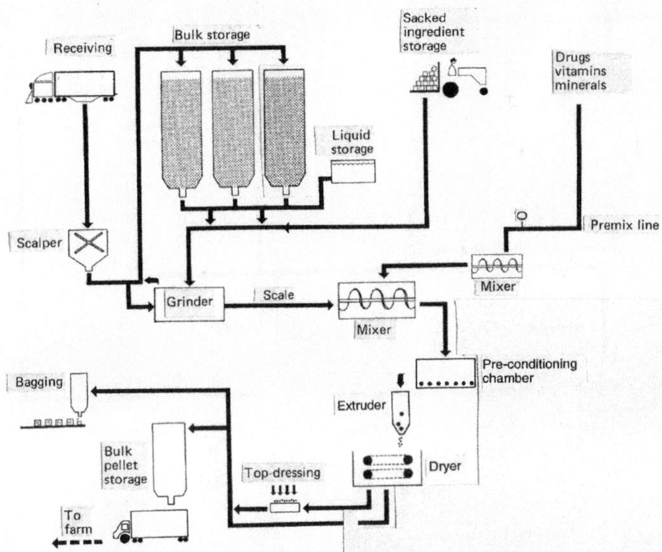

FIGURE 9.6 Simplified flow diagram of extruded pellet manufacturing.

throughout the process must be very high to prevent microbial contamination. The resulting noodles are cut into appropriate lengths by an external rotating knife as they exit the die. After extrusion, the pellets are screened, "quick-frozen," sacked, and stored. Soft moist pellets can be stored at room temperature providing that the feeds contain ingredients that lower water activity and prevent mold growth. Such pellets are typically packed in foil bags that are flushed with nitrogen to remove oxygen and then sealed. This reduced the chance of lipid oxidation in the pellets. The overall process is depicted in Fig. 9.7.

To summarize pelleting methods, each manufacturing process results in pellets having specific physical properties (or more accurately, a range of properties). The properties important to feed producers and fish farmers are shown in Table 9.17.

9.6.4 Cooling and drying

After pelleting, pellets pass through dryers on moving screened belts where warm air cools pellets and removes moisture (Fig. 9.7). Compressed pellets exit the die at approximately 95°C, and the latent heat in the pellets facilitates drying; typically no additional heating is needed. Passage through the cooler-dryer takes 5–15 min, and when the pellets leave the cooler-dryer, they are at ambient temperature and have about 10% moisture. Extruded pellets

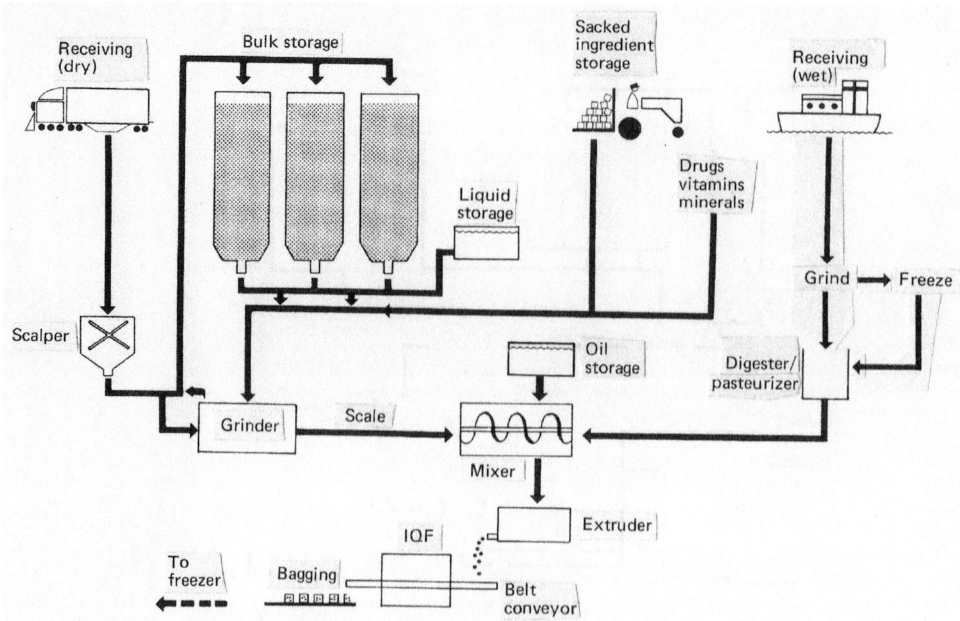

FIGURE 9.7 Simplified flow diagram of moist pellet manufacturing.

TABLE 9.17 Physical qualities and attributes of types of fish feed pellets.

Physical quality	Compressed	Annular gap	Extruded
Density (g/L)[a]	590	680	400–550
Max. temperature (°C)	95	135	150
Time exposed to steam	<1 min.	<1 min.	2–5 min.
Starch gelatinization (%)	<40	65–70	>80
Starch digestibility[b]	Low	High	High
Maximum fines (%)	2–3	<1	<1
Maximum fat level (%)	18	25	38
Pellet buoyancy	Sinking	Sinking	Floating/sinking
Water stability	Low	Low	High
Durability	Low	Medium	High
Nutrient destruction[c]	Low	Low	Medium

[a] *480 g/L is the break point for pellets to float; higher density pellets sink, and lower density pellets float in freshwater.*
[b] *Starch digestibility is a function of the degree of gelatinization.*
[c] *Nutrient destruction during pelleting is caused by high temperature and pressure plus time that feed mixtures are exposed to high temperature. Certain vitamins and the carotenoid pigments are the most sensitive to loss due to excess heat.*

contain more moisture than compressed pellets, and, therefore, dryers for extruded pellets pass heated air through and over the pellets to achieve final moisture levels. For laboratory-scale pelleting, cooling and drying can be accomplished by spreading the pellets and blowing air over them with a fan.

9.6.5 Crumbling and screening

Compressed or extruded pellets are crumbled after cooling and drying by passing them through the corrugated rollers of a crumbler. In a crumbler, one roller is driven by a motor and the other spins freely. One roller rotates faster than the other, and this fractures the pellets into smaller particles (Fig. 9.8). The objective in crumbling is to maintain a high rate of production of smaller particle sizes without producing excessive amounts of fines (dust). This is achieved by adjusting the speeds of the rollers and the distance between them. After passing through the crumbler, particles are screened to appropriate feed sizes using a shaking, vibrating screen that separates particles into different size-class ranges (Fig. 9.9).

9.6.6 Top-dressing

The total lipid level in feed formulations for many fish species is too high for pelleting systems to produce strong durable compressed or extruded pellets. To achieve higher lipid levels, additional lipid is added after pelleting by spraying fat or oil onto the pellet surface, a process known as "top-dressing." As mentioned, fat levels over 8%–10% in feed mixtures result in soft compressed pellets due to overlubrication of the holes in the die of the pellet machine. About 7%–8% fat can be sprayed on compressed pellets with repeated spraying, but the high density and highly gelatinized exterior of compressed pellets limit the amount

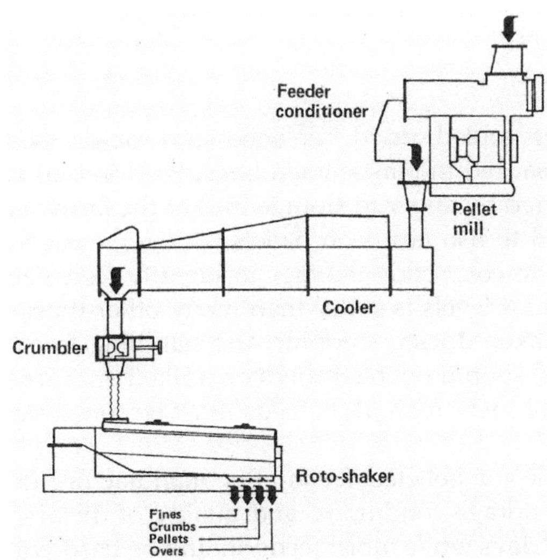

FIGURE 9.8 Action of rollers in crumbler.

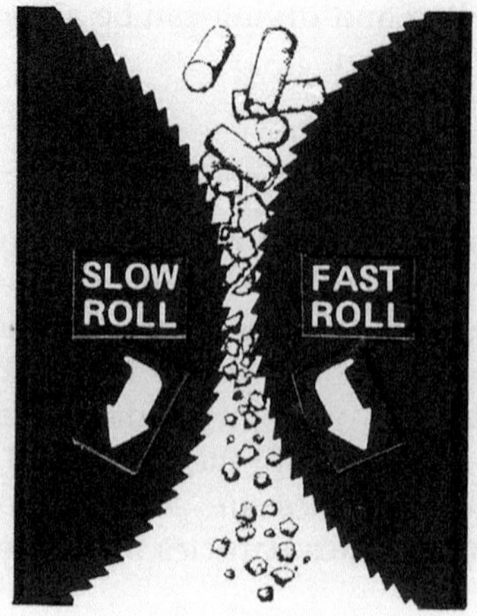

FIGURE 9.9 Equipment used to make pellets and crumbles.

of fat that can be added. In contrast, extruded pellets can absorb higher levels of fat during top-dressing to reach total fat levels over 30% with the use of vacuum chambers. Batches of dried pellets are added to the vacuum chamber, and air is removed. Fat or other liquids are then introduced into the chamber and moved into micropockets in pellets, dispersing fat throughout the pellet. Top-dressing is also used to apply heat-sensitive additives or nutrients, such as probiotics or enzymes, along with fat.

9.6.7 Storage

After pellets have been cooled, dried, screened, and top-dressed, they are transferred to bins for bulk storage or sacked in plastic-lined bags. Feed in bins can be delivered to farms in bulk by truck, transferred by auger to storage bins at the farm and fed. Bagged, or sacked, feed is typically delivered to fish farms on pallets.

It is common practice to color-code the tags to identify feed sizes and formulas. Pelleted fish feed containing high-fat levels is softer than many other types of pelleted animal feeds; therefore, care must be taken during shipping and handling to avoid breaking pellets and creating fines. Sacked feed should not be thrown or walked on. Sacks of feed should be stored on pallets no more than 10 sacks high. Bulk feeds must be conveyed gently to and from bulk feed bins.

Maximum storage time for pelleted feeds after manufacture depends on many factors, including type of pellet, storage conditions, and quality of dietary fat. Generally, dry feeds should be used within 90 days while moist feeds should be used within 60 days. Longer storage times may result in nutrient loss and/or rancidity problems. Storage life of feeds can be reduced by abusive storage conditions. For maximum storage life, dry feeds should be kept in areas where temperatures do no exceed 20–22°C and relative humidity remains below 75%. Dry feeds should not be stored in a freezer because freezing leads to migration of moisture in bags, creating pockets of high moisture where mold can develop after feeds are thawed. Semimoist feeds which require frozen storage should be kept at −20°C and thawed just before use. At the time a feed is used, the user should be alert for any sign of mold in a feed. If mold is found in a sack of feed, the entire sack should be discarded.

9.6.8 Semipurified diets

Semipurified diets were first used by researchers to determine the vitamin requirement of juvenile salmon. Most of the ingredients used in semipurified diets are highly refined and essentially devoid of water-soluble vitamins, making it possible to produce experimental diets containing graded levels of individual vitamins to establish required levels. The exception is fish oil which is rich in vitamins A and D, or plant oils which are rich in tocopherols. The first successful semipurified diet for fish was developed in the early 1950s and was named diet H440 (Halver and Coates, 1957). This formulation contained all of the nutrients needed to successfully raise juvenile salmon, although fish fed this diet grew slowly compared to standard growth rates today. As knowledge accumulated on the nutritional requirements of salmonids, the diet was adapted to study dietary requirements for amino acids, fat-soluble vitamins, and minerals, and the H440 formula was updated and modified. To study amino acid requirements, a mixture of amino acids is used to replace casein and gelatin. Semipurified diets contain vitamin-free casein, gelatin, dextrin, fish oil, carbohydrate binders, a vitamin premix, a mineral premix, and water. To make the diet, the dry ingredients except gelatin are mixed and added to the wet mix, which included gelatin dissolved in hot water, choline chloride, and fish oil. After further mixing, the blend is transferred to pans, cooled, and cut or pelleted. Since the diet contains over 20% water, it must be stored frozen until it is used. Laboratory-scale compression pelleting equipment can also be used to pellet semipurified diets, but conditions of pelleting must be controlled to minimize heat production and plug pelleting dies. Semipurified diets have been developed to conduct research with crustacea using vitamin-free casein (Conklin et al., 1980) or crab protein concentrate (Castell et al., 1989). Details on the use of semipurified diets in crustacean nutritional studies are given by the NRC (2011).

Other protein sources have been used in place of casein and gelatin in semipurified diet formulations to produce amino acid–deficient diets. Whole egg protein, fish protein concentrate, and semipurified fish protein have specific applications in experiments in which casein and gelatin are not appropriate protein sources.

9.6.9 Microdiets: larval feeds

Production of microdiets to feed fish at the larval stage is an active area of research and development. Microdiets are difficult and time-consuming to produce, and, therefore, they are more expensive than conventional feeds used at other stages of fish production. There are several reasons for this. First, microdiets are extremely small particles (5–150 μm for filter feeders, 40–700 μm for particle feeders), and each particle must contain all essential nutrients for the fish. Second, the digestive system of many fish larvae is not fully developed at first feeding. Therefore, selection criteria for ingredients used in larval feeds must include digestibility in the gut of fish lacking fully functional digestive capacity. Third, microdiets must possess properties that allow larval fish to detect and capture feed particles. They must be neutrally buoyant and attract larval fish by being visible and palatable. Finally, microdiets have large surface area to volume ratios. This makes them susceptible to nutrient leaching and disintegration. These characteristics can be controlled, but care must be taken to maintain nutrient availability while maintaining particle integrity and nutrient content, two seemingly mutually exclusive properties. Clearly, larval diet manufacture presents tremendous challenges to producers. There are three main categories of larval diets based on manufacturing method: (1) microbound feeds; (2) microencapsulated feeds; and (3) complex or combined feeds (Barrows and Hardy, 2002).

9.6.9.1 Microbound feeds

Microbound feeds can be separated into two classes based on the processes used to produce each class: crumbled and on-size feeds. Crumbled feeds are produced by fracturing pellets, a dried sheet—like feed or a cake into smaller particles that are then passed through sifting screens to separate particles into several size classes. On-size feeds are small particles produced directly by microextrusion. New developments in microextrusion processes enable the production of pellets in the 200—300 μm diameter range. This eliminates the crumbling and sieving steps, but more importantly, it results in spherical or round particles rather that sharp-sided particles that result from the crumbling process.

9.6.9.1.1 Crumbed feed

Crumbled feeds are generally produced from compressed pellets, but they can also be produced from extruded pellets. Both types have been used for many years as first feeds for salmon, trout, and channel catfish. They differ in buoyancy, water stability, and shape, but since these species are aggressive feeders, these characteristics are not critically important. However, for other farmed fish species, particular marine species, these characteristics are critical. For this reason, crumbled feeds are not effective as first feeds for larval fish.

Flake feeds are another type of crumbled feeds that are commonly used to feed ornamental fish species for hobbyists. Flake feeds are produced using double-drum dryers. These consist of two parallel drums that are heated internally, usually with steam. The distance between the drums is adjustable, and motors rotate the drums slowly in opposite directions. A feed slurry is applied continuously between the top of the drums, and the slurry spreads to thinly coat each drum at their closest point of contact. As the drums rotate, the slurry cooks and dries in approximately 30 s. Knives peel off the dried slurry in sheets or flakes, the thickness of which can be adjusted by varying the distance between the drums. The sheets are allowed to cool and then gently grounded and screened to different size classes.

Crumbled cake feeds are produced from feed mixtures that are formulated and treated to produce cake-type product that is subsequently converted into small particles. Crumbled cake feeds contain nutritional binders to hold feed mixtures together. Alginate, zein, cold-water gelling starches, carrageenan, gelatin, and egg albumin are examples of nutritional binders used to make crumbled cake feeds (Kanazawa et al., 1982; Teshima and Kanazawa, 1983). Activation of the binding ingredient is required to solidify the cake. For example, alginate is activated by calcium ions. Egg albumin is activated by heat. Two or more binders can be used together to improve the water stability and physical integrity of the mixture. Cakes are dried by freeze drying, oven drying, or drum drying. Crumbled cake feeds are particularly useful to feed crustaceans during early life history stages.

9.6.9.1.2 On-size feeds

As mentioned above, on-size feeds are those that are produced directly from a feed mixture into a shaped particle or pellet, rather than a product made from larger pellets that are ground and screened to various sizes. The manufacturing techniques used to produce on-size feeds are briefly described below.

Microextruded Marumerization: Microextrusion marumerization (MEM) is a process used in the pharmaceutical industry to manufacture small, homogenous particles. MEM is a two-step process involving cold extrusion of a moist feed mixture to produce long strings resembling small noodles, followed by a second step that converts the noodles into spherical particles (Barrows et al., 1993). Noodles can be produced with diameters as low as 300 μm. Feed mixtures used in microextrusion must be finely ground to 50–100 μm to avoid plugging the extrusion equipment.

In the second step in the MEM process, the noodles are placed in a cylindrical device having a high-speed rotating plate at the bottom, called a marumerizer. Plates contain grooves, and the depth of the grooves affects the amount of energy that is transferred to the noodles during marumerization. The marumerizer first reshapes the noodles by breaking them into spherical particles approximately the same length as the diameter of the noodles. Particle shaping is affected by feed formulation, moisture level, and other factors. As the spherical particles spin in the marumerizer, centrifugal forces cause water and small ingredient particles to migrate to the particle surface. As a result, the surface of the particle becomes more dense than the interior. Particle size remains within a narrow range, making further sifting to sort particles by size unnecessary.

MEM formulations include binders that are moisture or pressure activated. The MEM process does not involve thermal treatment, but some heat is generated by friction associated with microextrusion. Gums and protein hydrolysates are commonly used binders.

Particle-Assisted Rotational Agglomeration Feeds (PARA): The PARA process produces on-size, microbound feed particles directly from a wet feed mixture using the marumerizer. This process skips the noodle-making step in the MEM process. Briefly, wet mash is placed directly into the marumerizer along with inert particles. Rotation imparts energy to the inert particles which then shape the mash into spheroid particles that vary in size depending on feed formulation and moisture content of the mash. Size distribution of particles and their density are controlled by other process variables. Particles are then dried and sieved into different size classes.

The PARA process involves less equipment and has lower operating costs compared to the MEM process because the extruder step is eliminated. Binders that are effective in the MEM process are equally effective in the PARA process. As mentioned above, particles produced by the PARA process vary in size, making it necessary to sort particles into different size classes. However, PARA particles have lower density than MEM particles and are, therefore, more buoyant. Reducing the sinking rate of particles is an advantage to species of fish that feed throughout the water column.

Spray Beadlets: Spray beadlets are a third on-size, microbound feed particle. They are produced by spraying a feed slurry into a liquid that assists in the formation of particles (Villamar and Langdon, 1993). The process uses alginate and/or gelatin as binders. Particles formed using the spray beadlet process vary in size, making sieving necessary to sort particles into appropriate size classes.

9.6.10 Microencapsulated feeds

Microencapsulation is a process in which a material (the wall) surrounds and coats another substance (the payload). For fish feeds, the payload is the feed mixture. Microencapsulation technology is used in a range of industrial applications, notably in producing products for the pharmaceutical industry, and has been adapted to produce larval fish feeds. Microcapsules are separated into two categories: (1) impermeable and (2) controlled release. The permeability of the wall depends on the its composition, which can be varied to control the release of the payload, an important consideration for pharmaceuticals. For fish feeds, cross-linked proteins (Langdon et al., 1985) and lipid-walled microcapsules (Langdon, 2000; Langdon and Seigfried, 1984) are the most appropriate. Important physical characteristics when evaluating microcapsules include: (1) nutrient retention, (2) digestibility of the capsule, (3) payload capacity, and (4) water stability or durability. An important characteristic of microcapsules is the rate of nutrient leaching after feeding. Varying the fatty acid composition of lipid-wall microcapsules can change the rate at which certain nutrient leach from particles. The molecular weight of the nutrient in the payload determines the type of wall best suited for that nutrient (Langdon, 2000). The digestibility of the wall material to a particular species or to fish at specific life stages is also important to assure nutrients in the payload are available. The wall may be a nutrient source itself. Some types of microcapsules, notably lipid-walled, have low payload capacity, making them most suitable as micronutrient carriers (Villamar and Langdon, 1993). The survivability or durability of a microcapsule determines which feed delivery system it is best suited for and also how microcapsules can be used in complex feeds (next section).

The double-emulsion method can also be sued to produce lipid-walled microcapsules (Langdon et al., 1985). This technique involves emulsifying the aqueous payload in lipid wall material. A secondary emulsion is then formed by placing the primary emulsion into an aqueous solution. Lipid particles that result are removed and sized (Villamar and Langdon, 1993). Different lipids used in the primary emulsion will impart different characteristics to the final particles, such as nutrient stability and particle durability.

9.6.11 Complex feeds

Complex particles are an evolutionary development in larval feeds and allow greater flexibility for development of feeds to more closely match the needs of different fish species and life stages. Larval feed production methods and their products each have advantages and disadvantages. Many of the disadvantages can be overcome by combining different particle types to produce complex particles. These are feeds produced by combining two or more particles produced by different methods into a single particle (Villamar and Langdon, 1993).

For example, a water-stable microencapsulated particle can be embedded in an MEM, PARA, or spray beadlet particle, producing a particle within a particle. Microcoated feeds can be considered as complex particles because there are two processes involved in making them. Microcoated particles are made by coating microbound particles with coating of cholesterol or lecithin (Teshima et al., 1982) or with modified corn gluten or zein (Kanazawa et al., 1982). This is a two-step process, the first to form the particle and the second to coat the particle to reduce nutrient leaching. Microcoated particles are used in commercial production of marine fish larvae.

9.7 Concluding remarks

Diet formulation and manufacturing of fish feeds have undergone continous, incremental improvement over the past 2 decades as knowledge of the nutritional requirements of numerous farmed aquatic species has accumulated, and new feed manufacturing technologies have been developed. However, the most significant driver of change in fish feeds is associated with feed ingredients. First, the need to reduce dependance on fish meal and fish oil produced from marine resources has led to increasing use of other protein and oil commodities used more widely in animal and fish feeds. As is well known, these commodities, mainly grain and oilseed protein concentrates, plant oils, and land animal proteins and oils, are all nutritionally inferior in one or more ways to fish meal and fish oil as fish feed ingredients. The rapid growth of aquaculture feed production has created opportunities to modify these ingredients through by various means to improve their nutritional value to fish. It is likely that further improvements will be made to increase nutrient availability, reduce or eliminate antinutrients, and even modify ingredients through genetic selection or genetic modification in the case of crops, so that they are better suited for use in fish feeds.

In addition to improvements in feed ingredient commodities as fish feed ingredients, new feed ingredients are being developed specifically for use in fish feeds. These include insect meals, single-cell proteins, and algae oils containing high levels of essential fatty acids. Both insect meals and single-cell proteins differ from conventional feed ingredients in that they are not by-products of processes that produce human foods, such as soy proteins produced from press cakes after oil is removed for human consumption. Rather they are produced from novel waste streams, such as food processing waste in the case of insect meals or novel carbon sources in the case of single-cell proteins. If production of such ingredients can be economically scaled up, it will improve the sustainability of aquaculture production and ensure supplies of farmed aquatic products will keep up with global demand.

Glossary

Bulk feed Feed delivered to a user without being sacked.
Compound feed Any formulated feed prepared from conventional feed ingredients.
Crumbles Small particles produced by cracking pellets and screening particles to size ranges.
Diet A prescribed or controlled mixture or combination of feed ingredients or foods.
Dry diet Diet containing less than 13% water.
Farm-made feed Feed prepared from locally available ingredients at the farm rather than at a feed manufacturing facility.

Feed A mixture or preparation used for feeding livestock, poultry, fish, and other animals.
Fines Dust and small particles that result from pelleted disintegration.
Food Any material containing nutrients that can be consumed, absorbed, and used by the body.
Formulation A formula listing the proportions of ingredients for a feed.
Mash An unpelleted feed mixture.
Moist diet Same as wet diet.
Pellet A feed particle mechanically forced into a formed shape by compression or extrusion.
Practical diet A diet composed of nonrefined, commercially available ingredients.
Premix A concentrated mixture of micronutrients, usually vitamins or minerals, with a carrier or diluent that is added to feeds.
Protein unit A value used to compare the cost of protein supplements having different prices and protein contents.
Purified diet A research diet composed of completely refined ingredients.
Ration A share or allotment of feed over a given time period, typically 24 h.
Semimoist diet Diet containing 17%–35% water.
Semipurified diet A research diet containing a combination of refined and nonrefined ingredients.
Supplement A feedstuff rich in a particular nutrient use to balance a livestock diet or supply a combination of biochemically active compounds.
Wet diet Diet containing more than 35% moisture.

References

Apper-Bossard, E., Feneuil, A., Wagner, A., Respondek, F., 2013. Use of vital wheat gluten in aquaculture feeds. Aquat. Biosyst. 9, 21.

Austreng, E., Skrede, A., Eldegard, A., 1979. Effect of dietary fat source on the digestibility of fat and fatty acids in rainbow trout and mink. Acta Agric. Scand. 29, 119–126.

Babbitt, J., Hardy, R., Reppond, K., Scott, T., 1994. Processes for improving the quality of whitefish meal. J. Aquat. Food Tech. 3, 59–68.

Barrows, F., Hardy, R., 2002. Nutrition and feeding. In: Wedemeyer, G. (Ed.), Fish Hatchery Management. American FIsheries Society, Bethesda, MD, pp. 483–558.

Barrows, F., Zitzow, R., Kindschi, G., 1993. Effects of surface water spray, diet, and phase feeding on swim bladder inflation, survival, and cost of production of intensively reared larval walleyes. Prog. Fish-Cult. 55, 224–228.

Boggio, S., Hardy, R., Babbitt, J., Brannon, E., 1985. The influence of dietary lipid source and alpha-tocopheryl acetate level on product quality of rainbow trout (*Salmo gairdneri*). Aquaculture 51, 13–24.

Buhler, D., Halver, J., 1961. Nutrition of salmon fishes. XI. Carbohydrate requirement of Chinook salmon. J. Nutr. 74, 307–318.

Cain, K., Garling, D., 1995. Pretreatment of soybean meal with phytase for salmonid diets to reduce phosphorus concentrations in hatchery effluents. Prog. Fish-Cult. 57, 114–119.

Castell, J., Kean, J., D'Abramo, L., 1989. A standard reference diet for crustacean nutrition research. J. World Aquacult. Soc. 29 (3), 93–99.

Chen, H., Meyers, S., Hardy, R., Biede, S., 1984. Color stability of astaxanthin pigmented rainbow trout under various packaging conditions. J. Food Sci. 49, 1337–1340.

Cho, C.Y., Bureau, D.P., 1997. Reduction of waste output from salmonid aquaculture through feeds and feeding. Prog. Fish-Cult. 59, 155–160.

Cho, C., Cowey, C., Watanabe, T., 1985. Fish Nutrition in Asia. Methodological Approaches to Research and Development. International Development Research Centre, Ottawa.

Cleveland, B., et al., 2020. Supplementing rainbow trout (*Oncorhynchus mykiss*) broodstock diets with choline and methionine improves growth in offspring. J. World Aquacult. Soc. 51, 226–282.

Conklin, D., D'Abramo, L., Bordner, C., Baum, N., 1980. A successful purified diet for the culture of juvenile lobsters: the effect of lecithin. Aquaculture 21, 243–249.

Crampton, E., Harris, L., 1969. Applied Animal Nutrition. W.H. Freeman and Co, San Francisco.

Crawford, D., Law, D., McKee, T., Westgate, J., 1973. Storage and nutritional characteristics of modified Oregon moist rations as an intermediate moisture product. Prog. Fish-Cult. 35, 33–38.

Denstadli, V., et al., 2006. Feed intake, growth, feed conversion, digestibility, enzyme activities and intestinal structure in Atlantic salmon (*Salmo salar* L.) fed graded levels of phytic acid. Aquaculture 256, 365–376.

Donaldson, L., Foster, F., 1937. A summary table of some experimental tales in feeding young salmon and trout. Trans. Am. Fish. Soc. 67, 262–270.

Dupree, H., 1966. Vitamins Essential for the Growth of Channel Catfish. Bureau of Sport Fish and Wildlife, Washington, DC.

Dupree, H., Halver, J., 1970. Amino acids essential for the growth of catfish (*Ictalurus punctatus*). Trans. Am. Fish. Soc. 99, 90–92.

Edwards, D., Austreng, E., Risa, S., Gjedrem, T., 1977. Carbohydrate in rainbow trout diets. I. Growth of fish of different families fed diets containing different proportions of carbohydrates. Aquaculture 11, 31–38.

Ellis, R., Smith, R., 1984. Determining fat digestibilities in trout using a metabolic chamber. Prog. Fish-Cult. 46, 116–119.

Embody, G., 1918. Results of some trout feeding experiments carried on in the experimental hatching station of Cornell University. Trans. Am. Fish. Soc. 48, 26–33.

Embody, G., Gordon, M., 1924. A comparative study of natural and artificial foods of brook trout. Trans. Am. Fish. Soc. 48, 185–200.

Fowler, L., 1980. Substitution of soybean and cottonseed products for fish meal in diets fed to chinook and coho salmon. Prog. Fish-Cult. 42, 87–91.

Fowler, L., Burrows, R., 1971. The Abernathy salmon diet. Prog. Fish-Cult. 33, 67–75.

Frye, T., 1994. The Performance of Vitamins in Multicomponent Premixes. Jefferson, GA.

Gabaudan, J., Hardy, R., 2000. Vitamin sources for fish feeds. In: Stickney, R.R. (Ed.), Encyclopedia of Aquaculture. John Wiley & Sons, Inc., New York, pp. 961–965.

Garcia, C., Lara, F., 2013. The use of carotenoid in aquaculture. Res. J. Fish. Hydrobiol. 8 (2), 38–49.

Gatlin, D.I., et al., 2007. Expanding the utilization of sustainable plant products in aquafeeds - a review. Aquacult. Res. 38, 551–579.

Grassl, E., 1956. Pelleted dry rations for trout propagation in Michigan hatcheries. Trans. Am. Fish. Soc. 86, 307–322.

Habte-Tsion, H., 2020. A review on fish immuno-nutritional response to indispensable amino acids in relation to TOR, NF-κB and Nrf2 signaling pathways: trends and prospects. Comp. Biochem. Physiol. B Biochem. Mol. Biol. 241.

Halver, J., 1957. Nutrition of salmonid fishes. III. Water-soluble vitamin requirements of Chinook salmon. J. Nutr. 62, 225–243.

Halver, J., Coates, J., 1957. A vitamin test diet for long-term feeding studies. Prog. Fish-Cult. 19, 112–119.

Hardy, R., Fairgrieve, W., Scott, T., 1991. Periodic Feeding of Low-Phosphorus Diet and Phosphorus Retention in Rainbow Trout (*Oncorhynchus mykiss*). INRA, Paris, pp. 403–412.

Hardy, R., Shearer, K., 1985. Effect of dietary calcium, phosphate, and zinc supplementation on whole body zinc concentration of rainbow trout (*Salmo gairdneri*). Can. J. Fish. Aquat. Sci. 42, 181–184.

Hardy, R., Shearer, K., Spinelli, J., 1984. The nutritional properties of co-dried fish silage in rainbow trout (Salmo gairdneri) dry diets. Aquac. 38, 35–44.

Hardy, R., Sullivan, C., 1983. Canola meal in rainbow trout (*Salmo gairdneri*) production diets. Can. J. Fish. Aquat. Sci. 40, 281–286.

Harris, L., 1963. A system for naming and describing feeds, energy terminology and the use of such information in calculating diets. J. Anim. Sci. 22, 535–547.

Harris, L., 1980. Feedstuffs. Food and Agriculture Organization of the United Nations/United Nations Development Program, Rome.

Harris, L., Asplund, J., Crampton, E., 1968. An International Feed Nomenclature and Methods for Summarizing and Using Feed Data to Calculate Diets. Utah Agriculture Experiment Station, Logan, UT.

Hastings, W., Dickie, L., 2002. Feed formulation and evaluation. In: Halver, J. (Ed.), Fish Nutrition. Academic Press, New York, pp. 327–374.

Hata, M., Hata, M., 1973. Carotenoid pigments in goldfish. V. Conversion of zeaxanthin to astaxanthin. Bull. Jpn. Soc. Sci. Fish. 39, 339–343.

Hata, M., Hata, M., 1976. Carotenoid pigments in fancy red carp, *Cyprinus carpio*. II. Metabolism of 14C-zeaxanthin. Bull. Jpn. Soc. Sci. Fish. 42, 203–205.

Heinen, J., 1981. Evaluation of some binding agents for crustacean diets. Prog. Fish-Cult. 142–145.

Henry, M., Gasco, L., Piccolo, G., Fountoulaki, E., 2015. Review on the use of insects in the diet of farmed fish: past and future. Anim. Feed Sci. Technol. 203, 1–22.

Hilton, J., Atkinson, J., Slinger, S., 1983. Effect of increased dietary fiber on the growth of rainbow trout (*Salmo gairdneri*). Can. J. Fish. Aquat. Sci. 40, 81–85.

Hilton, J., Slinger, S., 1984. Effect of wheat bran replacement of wheat middlings in extrusion processed (floating) diets on the growth of juvenile rainbow trout (*Salmo gairdneri*). Aquaculture 35, 201–210.

Ibrahim, A., Shimizu, C., Kono, M., 1984. Pigmentation of cultured red sea bream, Chrysophrys major, using astaxanthin from Antarctic krill, Euphausia superba, and mysid, Neomysis sp. Aquaculture 38, 45–57.

Izquierdo, M., Socorro, J., Arantzanmendi, L., Hernandez-Cruz, C., 2000. Recent advances in lipid nutrition in fish larvae. Fish Physiol. Biochem. 22, 97–107.

Izquierdo, M., et al., 2015. Nutritional programming through broodstock diets to improve utilization of very low fishmeal and fish oil diets in glithead bream. Aquaculture 449, 18–26.

Jacobs, M.N., Covaci, A., Schepens, P., 2002. Investigation of selected persistent organic pollutants in farmed Atlantic salmon (Salmo salar), salmon aquaculture feed, and fish oil components of the feed. Environ. Sci. Technol. 36 (13), 2797–2805.

Jobling, M., 2003. Do changes in Atlantic salmon, *Salmo salar* L., fillet fatty acids following a dietary switch represent wash-out or dilution? Test of a dilution model and its application. Aquacult. Res. 34, 1215–1221.

Kanazawa, A., Teshima, S., Sosada, H., Rahman, S., 1982. Culture of the prawn larvae with microparticulate diets. Bull. Jpn. Soc. Sci. Fish. 48, 195–199.

Kaushik, S., Seilitz, I., 2010. Protein and amino acid nutrition and metabolism in fish: current knowledge and future needs. Aquacult. Res. 41, 322–332.

Ketola, H., 1979. Influence of dietary zinc on cataracts in rainbow trout (*Salmo gairdneri*). J. Nutr. 109, 965–969.

Kousonlaki, K., et al., 2013. Krill hydrolysate free amino acids responsible for feed intake stimulation in Atlantic salmon (*Salmo salar*). Aquacult. Nutr. 19, 47–61.

Langdon, C., 2000. Larval diets. In: Stickney, R. (Ed.), Encyclopedia of Aquaculture. John Wiley & Sons, Inc., New York, NY, pp. 529–530.

Langdon, C., Levine, D., Jones, D., 1985. Microparticulate feeds for marine suspension-feeders. J. Microencapsul. 2, 1–11.

Langdon, C., Seigfried, C., 1984. Progress in the development of artificial diets for bivalve feeders. Aquaculture 39, 135–153.

Lazzarotto, V., Corraze, G., Larroquet, L., Mazurais, D., 2016. Does broodstock nutritional history affect the response of progeny to different first-feeding diets? A whole-body transcriptomic study of rainbow trout alevins. Br. J. Nutr. 115, 2079–2092.

Leary, D., Lovell, R., 1975. Value of fiber in production-type diets for channel catfish. Trans. Am. Fish. Soc. 104, 328–332.

Lellis, W., Barrows, F., Hardy, R., 2004. Effects of phase-feeding dietary phosphorus on survival, growth, and processing characteristics of rainbow trout (*Oncorhynchus mykiss*). Aquaculture 242, 607–616.

Matsuda, K., 2009. Recent advances in the regulation of feeding behavior by neuropeptides in fish. Trends in comparative endocrinology and neurobiology. Ann. N. Y. Acad. Sci. 1163, 241–250.

McLaren, B., Keller, E., O'Donnell, D., Ellehjem, C., 1947. The nutrition of trout. I. Studies of vitamin requirements. Arch. Biochem. 15, 169–178.

Moren, M., Suontama, J., Hemre, G.-I., Karlsen, O., Olsen, R.E., Mundheim, H., Julshamn, K., 2006. Element concentrations in meals from krill and amphipods, — Possible alternative protein sources in complete diets for farmed fish. Aquac. 261 (1), 174–181.

NRC, 1973. Nutrient Requirements of Trout, Salmon and Catfish. National Academy Press, Washington DC.

NRC, 1983. Nutrient Requirements of Warm-Water Fishes and Shellfishes. National Academy Press, Washington DC.

NRC, 1993. Nutrient Requirements of Fish. National Academy Press, Washington DC.

NRC, 2011. Nutrient Requirements of Fish and Shrimp. National Academy Press, Washington, DC.

Oterhals, A., 2011. Decontamination of Persistent Organic Pollutants in Fishmeal and Fish Oil. University of Bergen, Bergen.

Pandian, T., Kirankumar, S., 2003. Recent advances in hormonal induction of sex-reversed fish. J. Appl. Aquacult. 13, 205–230.

Pelnar, J., 1947. Compressed air-ash feeding apparatus. Prog. Fish-Cult. 9, 97–98.

Penman, D., Piferrer, F., 2008. Fish gonadogenesis. Part 1: genetic and environmental mechanisms of sex determination. Rev. Fish. Sci. 16, 16–34.

Phillips, A.J., et al., 1964. Dry Concentrates as Complete Fish Foods. New York State Conservation Department, Albany, NY.

Phillips, A., et al., 1940. Feed Mixtures for the Production Hatchery. New York State Conservation Department, Albany.

Rathbone, C., Babbitt, J., Dong, F., Hardy, R., 2001. Performance of juvenile Coho salmon *Oncorhynchus kisutch* fed diets containing meals from fish wastes, deboned fish wastes, or skin-and-bone by-product as the protein ingredient. J. World Aquacult. Soc. 32 (1), 21–29.

Reinitz, G., 1983. Evaluation of sodium bentonite in practical diets for rainbow trout. Prog. Fish-Cult. 45, 100–103.

Richardson, N., Higgs, D., Beames, R., McBride, J., 1985. Influence of dietary calcium, phosphorus, zinc and sodium phytate levels on cataract incidence, growth, and histopathology in juvenile Chinook salmon (*Oncorhynchus tshawytscha*). J. Nutr. 115, 553–567.

Rodehutscord, M., 1996. Response of rainbow trout (*Oncorhynchus mykiss*) growing from 50 to 200 g to supplements of dibasic sodium phosphate in a semi-purified diet. J. Nutr. 126, 324–331.

Ruohonen, K., Vielma, J., Grove, D., 1999. Low-phosphorus supplement increases protein retention and reduces the amount of nitrogen and phosphorus wasted by rainbow trout fed on low-fat herring. Aquacult. Nutr. 5, 83–91.

Sajjadi, M., Carter, C., 2004. Effect of phytic acid and phytase on feed intake, growth, digestibility and trypsin activity in Atlantic salmon (*Salmo salar* L.). Aquacult. Nutr. 10, 135–142.

Schofield, E., 2005. Feed Manufacturing Technology V. American Feed Industry Association, Arlington.

Schreck, J., Moffitt, C., 2011. Palatability of feed containing different concentrations of erythromycin thiocyanate to Chinook salmon. Prog. Fish-Cult. 49, 241–247.

Schumann, M., Unger, J., Brinker, A., 2017. Floating feces: effects on solid removal and particle size distribution in RAS. Aqua. Eng. 78, 75–84.

Shahidi, R., Brown, J., 1998. Carotenoid pigments in seafoods and aquaculture. Crit. Rev. Food Sci. Nutr. 38, 1–67.

Simpson, K., Chichester, C., Katayama, T., 1981. Carotenoids in fish feeds. In: Bauernfeind, J. (Ed.), Carotenoids as Colorants and Vitamin A Precursors: Technical and Nutritional Applications. Academic Press, New York, pp. 463–538.

Sorensen, M., 2012. A review of the effects of ingredient composition and processing conditions on the physical qualities of extruded high-energy fish feed as measured by prevailing methods. Aquacult. Nutr. 18, 233–248.

Spinelli, J., Houle, C., Wekell, J., 1983. The effect of phytates on growth of rainbow trout (*Salmo gairdneri*) fed purified diets containing varying quantities of calcium and magnesium. Aquaculture 13, 71–83.

Spinelli, J., Mahnken, C., 1978. Carotenoid deposition in pen-reared salmonids fed diets containing oil extracts of red crab (*Pleuroncodes planipes*). Aquaculture 13, 213–223.

Stevens, J., Newton, R., Tlusty, M., Little, D., 2018. The rise of aquaculture by-products: increasing food production, value and sustainability through strategic utilization. Mar. Pol. 90, 115–124.

Stickney, R., Shumway, S., 1974. Occurrence of cellulase activity in the stomachs of fishes. J. Fish. Biol. 6, 779–790.

Storebakken, T., Foss, P., Schiedt, E., Austreng, E., 1987. Carotenoids in diets for salmons: IV. Pigmentation of Atlantic salmon with astaxanthin, astaxanthin dipalmitate and canthaxanthin. Aquaculture 65, 279–292.

Sugiura, S., Dong, F., Hardy, R., 2000a. A new approach to estimating the minimum dietary requirement of phosphorus for large rainbow trout based on nonfecal excretions of phosphorus and nitrogen. J. Nutr. 130, 865–872.

Sugiura, S., Dong, F., Hardy, R., 2000b. Primary responses of rainbow trout to dietary phosphorus concentrations. Aquacult. Nutr. 6, 235–245.

Sugiura, S., Dong, F., Rathbone, C., Hardy, R., 1998. Effects of dietary supplements on the availability of minerals in fish meal; preliminary observations. Aquaculture 160, 283–303.

Sugiura, S., Hardy, R., 2000. Environmentally friendly feeds. In: Stickney, R. (Ed.), Encyclopedia of Aquaculture. John Wiley & Sons, Inc., New York, pp. 299–310.

Tacon, A., 2019. Trends in global aquaculture and aquafeed production: 2000-2017. Rev. Fish. Sci. Aquacult. 28, 43–56.

Tanaka, Y., et al., 1976. The biosynthesis of astaxanthin. XVII. The metabolism of carotenoids in the prawn, *Penaeus japonicus* Bate. Bull. Jpn. Soc. Sci. Fish. 42, 197–202.

Tantikitt, C., 2014. Feed palatability and the alternative protein sources in shrimp feed. Songklanakarin J. Sci. Technol. 36, 51–55.

Teshima, S., Kanazawa, A., Sakamoto, M., 1982. Microparticulate diets for the larvae of aquatic animals. Min. Rev. Data File Fish. Res. 2, 67–86.

Teshima, T., Kanazawa, A., 1983. Effects of several factors on growth and survival of the prawn larval reared with microparticulate diets. Bull. Jpn. Soc. Sci. Fish. 49, 1893–1896.

Thiessen, D., Campbell, G., Tyler, R., 2003. Utilization of thin distillers' solubles as a palatability enhancer in rainbow trout (*Oncorhynchus mykiss*) diets containing canola meal or air-classified pea protein. Aquacult. Nutr. 9, 1–10.

Tiril, S., Agagil, F., Yagci, F., Aral, O., 2008. Effects of betaine supplementation in plant protein based diets on feed intake and growth performance in rainbow trout (*Oncorhynchus mykiss*). Israel. J. Aqua. Bamigden 60, 57–64.

Trushenski, J., Lochmann, R., 2009. Potential, implications and solutions regarding the use of rendered animal fats in aquafeeds. Am. J. Anim. Vet. Sci. 4, 108–128.

Tunison, A., et al., 1941. Feeding Trials with Practical Diets. New York State Conservation Department, Albany, NY.

Turchini, G., Torstensen, B., Ng, W.-K., 2009. Fish oil replacement in finfish nutrition. Rev. Aquacult. 1, 10–57.

Turchini, T., Ng, W.-K., Tocher, D., 2010. Fish Oil Replacement and Alternative Lipid Sources in Aquaculture Feeds. CRC Press, Boca Raton.

Vielma, J., Ruohonen, K., Lall, S., 1999. Supplemental citric acid and particle size of fish bone-meal influence the availability of minerals in rainbow trout *Oncorhynchus mykiss* (Walbaum). Aquacult. Nutr. 5, 65–71.

Villamar, D., Langdon, C., 1993. Delivery of dietary components to larval shrimp (*Penaeus vannamei*) by means of complex microcapsules. Mar. Biol. 115, 635–642.

Waagbo, R., et al., 2010. Dietary histidine supplementation prevents cataract development in adult Atlantic salmon, *Salmo salar* L., in seawater. Br. J. Nutr. 104, 1460–1470.

Wade, N., Budd, A., Irvin, S., Glencross, B., 2015. The combined effects of diet, environment and genetics and pigmentation in the giant tiger prawn, *Penaeus monodon*. Aquaculture 449, 78–86.

Watanabe, T., 1982. Lipid nutrition of fish. Comp. Biochem. Physiol. 73B, 3–15.

Watanabe, T., Ohta, M., Kitajima, C., Fujita, S., 1982. Improvement of dietary value of brine shrimp Artemia salina for fish larvae by feeding them on W-3 highly unsaturated fatty acids. Nippon Suisan Gakkaishi 48, 1775–1782.

Wolf, L., 1951. Diet experiments with trout. Prog. Fish-Cult. 13, 17–24.

Wood, J., 1979. Diseases of Pacific Salmon, third ed. Department of Fisheries, Olympia, WA.

Xu, H., Zhao, M.Z.W.Y., Yan, L., Liang, M., 2017. Antarctic krill (*Euphausia superba*) meal in the diets improved the reproductive performance of tongue sole (*Cynoglossus semilaevis*) broodstock. Aquacult. Nutr. 2017 (00), 1–9.

Yamada, S., Tanaka, Y., Sameshima, M., Ito, Y., 1990. Pigmentation of prawn (*Penaeus japonicus*) with carotenoids: I. Effect of dietary astaxanthin, ß-carotene and canthaxanthin on pigmentation. Aquaculture 87, 323–330.

CHAPTER

10

Diet analysis and evaluation

S. Charles Bai[1], *Ronald W. Hardy*[2] *and Ali Hamidoghli*[3]

[1]FAO World Fisheries University Pilot Program; Feeds and Foods Nutrition Research Center, Pukyong National University, Busan, South Gyeongsang Province, Republic of Korea; [2]Aquaculture Research Institute, Hagerman Fish Culture Experiment Station, University of Idaho, Hagerman, ID, United States; [3]Feeds and Foods Nutrition Research Center, Pukyong National University, Busan, South Gyeongsang Province, Republic of Korea

10.1 Introduction

As the global aquaculture industry continues its rapid development, there will be a greater dependence on the commercial feed sector to accommodate this growth. Proper feed formulation and manufacturing requires a deep understanding of the nutritional requirements of aquatic species, which is encompassed by the discipline of nutrition science. Nutritional science relies on several core sciences such as chemistry, biology, physics, mathematics, and the field of marketing. Because of the need to verify the efficacy of feed formulations across these disciplines, diet and ingredient analysis and evaluation are essential to fish feed formulation and manufacturing. It is important not only that the quality and quantity of ingredients meet the nutritive requirements of a farmed fish but also that the processed diets are efficiently metabolized by fish while feed quality control aspects are ensured. In addition, economic feasibility of the feed must ensure the continued sustainability of the aquaculture enterprise itself. These are considerations that have real-world impacts.

Technological advancements during the past decades have opened new doors for diet analysis and evaluation along with the fast development of the fish feed industry. Precise and detailed information on ingredient and diet characteristics are made possible by new chemical analysis methodologies increasing employed by modern facilities. Comparing the official methods of analysis presented by the Association of the Official Analytical Chemists (AOAC) in the 1980s, and more recently in 2019, reveals significant advances in analytical methodologies. Antiquated time-consuming protocols have been replaced by fast analytical techniques that are often performed by fully automated machinery with high accuracy, eliminating variations caused by differences in technical expertise. Furthermore, the biological approaches of diet evaluation have developed beyond weight and length measurements to analysis at molecular levels. Nutrigenomics is a burgeoning field that has been highlighted

Fish Nutrition, Fourth Edition
https://doi.org/10.1016/B978-0-12-819587-1.00010-0

in fish nutrition research to better understand nutrient metabolic pathways and the influence of nutrients on gene expression. User-friendly kits provide full packages for easy and fast genomic studies that are adding powerful dimensions to fish nutrition research. These advancements in knowledge can aid fish nutritionists in providing precision feed formulations to the aquaculture community.

Several approaches are utilized to evaluate feed ingredients, diet formulations, and fish performance on new feeds. The first involves chemical analysis of diets and ingredients. Measuring the proximate composition (protein, fat, moisture, fiber, and ash) and content of micronutrients (amino acids, fatty acids, vitamins, and minerals) in every ingredient used in feeds is essential to formulate appropriate feeds. The importance of chemical analysis has become more important than ever due to the ability of companies to source ingredients globally, leading to more complex feed formulations. As such, variations in the nutritive value of basic feed ingredients can be observed. For example, the nutritive value of soybean meal can vary from region to region due to differences in climate and soil. Moreover, ingredients such as fishmeal are sourced from different species and seasons, which can alter ratios of crude lipid and protein and thereby have a significant impact on the nutritive quality of a feed formulation. To ensure quality control in feed formulations, it is absolutely essential to verify the nutrient profile of ingredients prior to formulation. Diet analysis is therefore an indispensable tool for fish nutritionists that allows them to match up all the nutrients in a complex diet with the requirements of each species at different life stages. Second is quality control assurance of finished feeds. Manufactured feeds should pass quality control tests to identify nutrient losses during processing, oxidation, deformations, and physical properties of diets. Furthermore, antinutritional factors and microbial toxins (e.g., mycotoxins, aflatoxins, etc.) are nonnutrient components that occur naturally in finished feeds containing certain ingredients or as a result of contamination. To minimize the deleterious effects of these compounds on growth, immune function, and body composition of cultured fish, quality control assessments are a vital safeguard. The third approach is biological evaluation of diets which involves understanding the influence of diets on the performance of target species and the retention and loss of nutrients. After short- or long-term feeding trials, growth, feed utilization performance, digestibility, metabolic assessments, and gene expression analyses are necessary to evaluate diets at the biological level. The last but not least criteria is economic evaluation of feed. Without economic evaluation, the highest quality and perfectly balanced feed formulations may not be practical to use in production settings. Fluctuations in prices of common ingredients can affect fish production costs and the sustainability of aquaculture itself. Identification and evaluation of alternative ingredients, by-products, and feed additives can enhance the economic feasibility of feed formulations. Therefore, understanding diet analysis and evaluation is essential to develop high-quality, nutritious, and cost-effective feeds and therefore support the sustainable development of aquaculture.

10.2 Chemical analysis

Chemical analysis of ingredients and feeds is a critical part of fish nutrition research and industrial feed manufacturing. After the selection of ingredients, based on feed formulations, each ingredient should be analyzed to confirm contents of essential nutrients to ensure that formulations containing these ingredients will meet the nutrient requirements of the cultured species and life history stage for which will be used. The complexity of feed formulations in modern aquaculture has elevated the importance of understanding chemical analysis results. Modern aquaculture feeds are designed to minimize the use of ingredients sourced from marine resources, primarily fishmeal and fish oil, replacing them with less expensive alternative ingredients obtained from more sustainable sources, mainly from plant sources and/or by-products of seafood and land animal processing. Fish nutritionists are therefore often investigating novel ingredients in new feed formulations. Chemical analysis is an essential tool to assess the nutritional quality of these feedstuff for development of practical diets that can provide for the nutritional needs of target species.

There are several methods used for chemical evaluation of ingredients and diets. Proximate analysis is commonly used to characterize feed ingredients by partitioning ingredient constituents into categories based on their chemical properties (Table 10.1). Proximate analysis is not a nutrient analysis but rather a useful tool with which to compare ingredients or feeds. The categories in proximate analysis are moisture, crude protein, crude lipid, fiber, ash, and nitrogen-free extract (NFE). Other analytical methods are used to determine amino acid, fatty acid, vitamin, and mineral contents. Although there are standardized

TABLE 10.1 Categories in proximate analysis and their constituents.

Category	Composition
Water	Water
Crude protein	Essential amino acids
	Nonessential amino acids
	Free amino acids, amines, nucleic acids
Ether extract	Triglycerides
	Phospholipids
	Sterols
	Fat-soluble vitamins
	Miscellaneous lipids (waxes, sphingomyelins, etc.)
Crude fiber	Insoluble polysaccharides (cellulose, hemocellulose, chitin)
Nitrogen-free extract	Monosaccharides
	Oligosaccharides
	Soluble saccharides
	Water-soluble vitamins
Ash	Essential elements
	Nonessential elements
	Toxic elements

methodologies for these procedures in references such as AOAC, analytical results from chemical analyses of the ingredients from different laboratories often differ. These differences may be attributed to the ingredient's production location, the season which the ingredient was harvested or made, the source of ingredient, or the processing methods (Glencross et al., 2007). In addition, inner laboratorial specifications and slight differences in procedures may influence the results for chemical analysis of feed. In a publication by the Food and Agriculture Organization (FAO), several principles have been suggested to minimize these differences (Mueller-Harvey, 2004). First, properly validated methods should be used for the measurements. Second, certified reference materials should be used for routine calibration and verification of measurements, assuring the quality of measurements, verifying the correct application of methods, and developing new measurement methods. Third, independent assessment of laboratories should be performed by international proficiency testing schemes. Employing these principles will result in more accurate chemical analyses results that will lessen differences often reported between laboratories. This section will review the most common methods used in chemical analysis of ingredients and feeds used in fish nutrition studies and commercial production.

10.2.1 Water and dry matter

Every ingredient or diet used in fish nutrition contains a certain amount of water. The role of water is not only as a filler or solvent but also to maintain the structure and function of other molecules. If water is removed from feeds or feedstuffs, the remaining dry matter will be comprised of organic and inorganic compounds. Feed ingredients vary in moisture content, so proximate analysis for the other five categories is conducted on dry matter. This standardizes comparisons of the nutritional value of ingredients. Results of proximate analysis can be expressed on a dry-weight basis or on an as-is basis that takes into consideration the moisture content of the ingredient or feed.

The amount of water in feed ingredients significantly influences structure, texture, and physical properties. For example, bulk density and the total weight of ingredients are important aspects of transportation and storage that are dependent on water content. Also, the amount of water in fish feeds influences shelf life and bulk density because microbial growth and enzyme activities are highly dependent on the water content (Pojić et al., 2015). Therefore, in fish nutrition, it is essential to determine the water content of ingredients and feed.

Commercial extruded fish feed typically contain 7%–12% moisture depending on manufacturers' guidelines and policies. There are several methods to determine water content and dry matter in feed and feed ingredients. Among these methods, the best ones are those that require less resources, time, money, and personnel and, at the same time, are accurate. The oldest and most commonly used method is drying the samples in an oven until they reach a constant weight. This is known as the gravimetric method. This method is based on the release (evaporation) of water from the sample and measurement of difference between the initial weight and final dry weight. The temperature used to dry samples depends on the type of sample and the processing time. Although the boiling temperature of water is 100°C, because of the dissolvent of solute, a higher temperature is required for complete evaporation of water. According to the AOAC (2000), the temperature range used for moisture determination should be between 100 and 135°C for 2–16 h in the air oven.

Faster determinations could be possible with higher temperature ranges, although samples dried at high temperatures cannot be used for crude lipid or fatty acid determination because volatile compounds such as fatty acids will evaporate at high temperatures. If the same samples are being used for lipid or fatty acid determination, vacuum-oven drying or temperatures below 70°C can be used. For feed samples containing molasses, high temperatures should be avoided due to potential loss of volatile compounds and caramelization of sugars.

10.2.2 Crude protein and amino acids

Protein is the most important nutrient and most expensive component on a mass basis in fish feed. Proteins are large, complex, and nitrogen-containing compounds and are essential for maintenance of life and growth in fish. Proteins are mostly comprised of carbon (50%–55%), hydrogen (6%–8%), oxygen (20%–23%), nitrogen (15%–18%, depending on the amino acid composition), and sulfur (0%–4%). Common methods for determining the protein content of feed or ingredients are based on measuring the total nitrogen content. Due to the importance of this group of nutrients in fish nutrition, precise determination of their quantity in aquafeed is indispensable (Jobling, 2001). There are several methods to measure protein levels in feeds, but only the most common will be reviewed here.

In 1883, a Danish chemist named Johan Gustav Christoffer Thorsager Kjeldahl introduced a method to determine the nitrogen content of common feed ingredients. Kjeldahl's method includes three main steps: digestion with strong acid, distillation of ammonia nitrogen, and titration. In the digestion step, samples are broken down by boiling the sample in a strong acid (namely sulfuric acid). Carbon dioxide and water are produced by oxidation, and ammonium sulfate is released. In the distillation step, the digested material is cooled, and water and alkali are added to convert ammonium ions to ammonia, which is absorbed in boric acid. In the titration step, an acid is slowly added to determine the amount of generated ammonia in samples, identified by a color change. The protein content is calculated by multiplying the nitrogen value by a conversion factor, typically 6.25. Proteins contain approximately 16% nitrogen so 100/16 is 6.25. One of the most obvious limitations in the Kjeldahl method is the use of 6.25 as a conversion factor. The nitrogen content of common proteins ranges from 15% to 18% (conversion factor 6.667 to 5.556, respectively) differing as a result of the amino acid composition of the protein. A second limitation with the Kjeldahl method is associated with the presence of nonprotein nitrogenous compounds in feed ingredients. Nitrogen associated with inorganic nitrogen, urea, and nucleotides contributes to total nitrogen and is counted as protein. The limitations of Kjeldahl method must be considered based on the type of sample. Despite its limitations and potential errors, the Kjeldahl method is widely used in animal and fish nutrition studies to determine crude protein content in ingredients, feed, and fish tissue. Other methods such as Dumas and radiochemical assays have also been tested for the determination of total nitrogen but are not widely used except in research because they are not standardized and/or practical to use routinely due to high costs (Greaser and Guo, 2015).

In fish nutrition, the requirement for protein is actually the requirement for amino acids, the building blocks of proteins. Among all the amino acids that have been identified (approximately 300), only around 20 of those are constituents of proteins. Ten amino acids cannot be synthesized by fish or other vertebrate animals and are deemed essential dietary nutrients for fish. Therefore, measuring the amino acid content of ingredients and fish feeds is a crucial aspect that must be considered by nutritionists. The two major methods for measuring amino acids are high-performance liquid chromatography (HPLC) and dedicated amino acid analyzers (AAAs). Proteins are hydrolyzed to release amino acids, and samples are injected into ion-exchange columns that separate amino acids based on their characteristics. Amino acids are tagged with a detection compound, either before injection (precolumn) or after they exit the column, after which they pass through a detection chamber. The HPLC method uses reagents such as *o*-phthalaldehyde, whereas AAA method uses ninhydrin solution for detection of amino acids (Teruel, 2002).

10.2.3 Lipids and fatty acids

Lipids are water-insoluble organic compounds and the main source of energy, essential fatty acids, phospholipids, and fat-soluble vitamins in fish feed. They play important roles in the structure of cell membrane and as precursors for some hormones. Lipids are classified into two main groups: polar and neutral lipids. Polar lipids are structural components of cell membranes and play important roles in cell membrane fluidity. Neutral lipids supply metabolic energy and are the stored form of lipids in the body. The most important polar lipids are phospholipids, whereas neutral lipids consist of fatty acids, waxes, alcohols, triglycerides, and sterols. Both contain fatty acids as their main components. Fatty acids are important nutrient groups in fish feed with a carbon backbone structure that is classified as saturated or unsaturated. The main difference between these two classifications is the presence or absence of double bonds between carbon atoms. Fats containing unsaturated fatty acids with double bonds are mainly liquid at room temperature, while fats containing saturated fatty acids with no double bonds, such as animal fats, are solid. Fish typically require unsaturated fatty acids with a high number of double bonds that cannot by synthesized de novo. These fatty acids are referred to as essential, and their levels in feeds must be considered by feed formulators. It is important to monitor the levels of dietary lipid and essential fatty acids in fish diets (Molina-Poveda, 2015).

Crude lipids are measured by extracting ingredient or feed samples with a nonpolar solvent, such as petroleum ether or chloroform. The solvent extraction method is called Soxhlet because of the apparatus used during extraction. Samples are placed in paper or fiber thimbles where evaporated solvent condenses back to liquid and extracts fats, fatty acids, compound lipids, and fat-soluble vitamins from samples where they accumulate in a preweighed flask containing the liquid solvent. When extraction is complete, the solvent is evaporated and difference in weight of the flask is measured and considered as crude lipid. Solvents used in this method do not completely remove all lipids from samples; some lipoproteins and phospholipids are not extracted. The accuracy of this method depends, to a high extent, on the type of samples and their lipid composition. Extraction efficiency of extruded pellets can be improved by subjecting the sample to acid hydrolysis before solvent extraction (Hardy and Barrows, 2002). For the determination of fatty acids, a gas chromatography (GC) is used. The GC method can also be used to determine volatile compounds and nutritionally important lipid sources such as cholesterol. Other important lipid sources such as phospholipids are measured with HPLC coupled to an evaporative light scattering detector.

10.2.4 Vitamins

Vitamins are organic compounds that cannot be synthesized sufficiently in animals and are required in trace amounts in the diet for normal maintenance, health, growth, and reproduction. They are classified into two main groups based on how they were first identified by extracting tissues with different solvents: the water-soluble vitamins including the B vitamins, myo-inositol, choline, and vitamin C; and the fat-soluble vitamins including vitamins A, D, E, and K. The water-soluble vitamins are not stored in excessive amounts in animal tissue, and excess dietary amounts are excreted. In contrast, fat-soluble vitamins are stored in the liver and storage fats after absorption in the digestive tract. Excessive accumulation of fat-soluble vitamins can lead to the production of toxic compounds (Jobling, 2001). The requirements of fish for water and fat-soluble vitamins need to be known to precisely formulate feeds. Analysis of vitamins in feeds and ingredients is an important consideration that needs to be carefully considered.

Before fat-soluble vitamins can be analyzed in feed ingredients, lipoprotein bonds need to be broken. Saponification is needed for extraction of fat-soluble vitamins by nonpolar solvents such as ether. The AOAC suggests HPLC as the standard analytical tool for fat-soluble vitamin analysis. However, fat-soluble vitamins can also be determined with the GC or spectrophotometric methods. For most of the water-soluble vitamins, HPLC is the appropriate analytical instrument. Microbiological assays are another option for vitamin B12, folates, pantothenic acid, and biotin analysis due to their lower price.

10.2.5 Minerals

Minerals are the homogenous and inorganic compounds that are required for normal life processes in animals. Nutritionists classify minerals into two main groups: macro minerals such as calcium (Ca), chloride (Cl), magnesium (Mg), phosphorus (P), potassium (K), sodium (Na), and sulfur (S); and microminerals such as chromium (Cr), cobalt (Co), copper (Cu), fluoride (F), iodine (I), iron (Fe), manganese (Mn), molybdenum (Mo), selenium (Se), and zinc (Zn). Macrominerals are required in the diet of fish for several functions including calcium in skeletal tissues, chloride in osmotic balance, magnesium in enzymes, and phosphorus in phospholipids. Microminerals also take part in various functions such as copper in metalloenzymes, chromium in carbohydrate metabolism, iodine in thyroid hormones, and iron in hemoglobin (NRC, 2011). Minerals can be obtained by fish either from the diet or from the surrounding water, although fish, especially in freshwater environments, obtain the bulk of their mineral requirements from dietary sources (Jobling, 2001). It is important, therefore, to measure the level of minerals in feeds to prevent mineral deficiencies.

Using the appropriate methods for determination of minerals in ingredients and feeds is important in precise formulation. There are several methods for determining the quality or quantity of minerals in feedstuffs. Qualitative methods are relatively easy to perform and do not require expensive equipment or facilities. These methods differ depending on the type of mineral and are generally based on changes in color with chemical reactions. For example, cobalt, copper, and iron in the presence of potassium sodium tartrate and disulfonic acid yield pink, light brown, and green colors, respectively (Teruel, 2002). For quantitative determination of mineral levels, based on the cation or anion properties, methods may differ.

The most common analytical method involves inductively coupled plasma spectroscopy (ICP) that is ideally coupled with mass spectrometry (ICP-MS). Although the ICP method involves expensive laboratory equipment, it is considerably more sensitive and precise than colorimetric methods (Burlingame and Charrondiere, 2015).

10.2.6 Crude fiber

In fish nutrition, fiber refers to compounds that cannot be digested or absorbed and are not required dietary components by fish (Davies, 1985). The main fiber in feeds consists of non-starch polysaccharides (NSPs) that include different types of polysaccharides, excluding starch, such as arabinoxylans, pectin, cellulose, hemicellulose, mucilage, β-glucan, and gums. These are found in plant ingredients such as cereal grains, wheat, corn, and sorghum. There is one major component of the crude fiber fraction that is not contributed by plant material, namely, chitin. Chitin is a polymer of glucosamine that constitutes a large portion of the exoskeleton of crustaceans and insects. This is an important fact to remember in light of the growing use of insect meal in fish feeds. NSPs can be evaluated by three measures including crude fiber (CF), acid detergent fiber (ADF), and neutral detergent fiber (NDF). CF is the remaining material of feed or ingredients after reaction with a weak acid and base and removing the inorganic compounds (ash). Fiber stimulates the digestion process occurring in the gastrointestinal tract. For most of the plant ingredients, crude fiber consists mainly of cellulose, lignin, and insoluble hemicellulose. ADF gives a similar measure as CF with a simpler procedure for the sum of cellulose, lignin, and acid-insoluble hemicellulose (Molina-Poveda, 2015). Cellulose can either be determined by removal with acid and the lignin by oxidation. The limitation is that ADF does not determine the hemicellulose and, therefore, is not suitable for the evaluation of total structural fiber. NDF is a common measure of fiber in nutritional studies that consist of cellulose, lignin, and neutral detergent-insoluble hemicellulose. This method of fiber determination is often used by nutritionists and researchers (Udén et al., 2005).

10.2.7 Ash

Ash is the portion of feed that remains after incinerating samples at 550−600°C in a muffle furnace. Incineration at 550°C burns all organic material in feed or ingredient samples, leaving inorganic minerals that are referred to as ash. This value is an important quality indication for ingredients, particularly FM and some land animal proteins, and for fish feeds as it contains essential, nonessential, and toxic compounds (Hardy and Barrows, 2002).

10.2.8 Nitrogen-free extract

As mentioned, proximate analysis partitions feed constituents into categories based upon their physical and chemical properties. NFE is a fraction that is not actually measured but rather calculated by subtracting all of the other proximate values from 100. NFE includes a complex mixture of compounds including starch, sugars, cellulose, hemicellulose, lignin, pectins, tannins, resins, water soluble vitamins, and organic acids.

10.2.9 Near-Infrared Spectroscopy

Near-infrared spectroscopy (NIRS) is an important method in feed analysis that allows for rapid, accurate, and cost-effective characterization of many important nutrients in a sample. This method is easily repeatable and routinely used in the assessment of crude protein, crude fat, moisture, ash, starch, amino acids, crude fiber, NDF, and ADF content of ingredients received at many feed companies. The accuracy of NIRS has been successfully tested for chemical composition, digestibility and energy content of compound feed (Aufrère et al., 1996). In addition, prediction of free amino acids and nonprotein nitrogen of ingredient (Prevolnik et al., 2011), digestible protein and digestible energy of ingredients (Glencross et al., 2015), and digestible protein and digestible energy of diet (Glencross et al., 2017) was validated through NRIS analysis. The NIRS principle is based on the absorption of reflected light (700–2500 nm wavelength) by organic molecules and contains compositional information that can be converted by the computer to report different analysis values. The NIRS instruments are calibrated with reference to a multivariate mathematical model that is based on the careful analysis of samples with standardized methods (Cozzolino et al., 2002).

10.3 Feed quality assessments

The quality of feeds depends largely on the quality of ingredients that are combined according to a feed formulation. Assuring the expected quality of ingredients received at the mill is of utmost importance for any feed manufacturer. This can be challenging, as ingredients can vary in nutritional profile and nutritional quality depending on origin, processing conditions, storage conditions, or adulterations. For example, soybean meal from India, Argentina, and the United States may differ in crude protein content or essential amino acid composition (Tangendjaja, 2015). The chemical analysis methods described in the previous section can help manufacturers to continuously monitor the composition of ingredients they obtain and feeds that they produce. However, analyzing every ingredient that is received from suppliers for the 40 or so essential nutrients would be time-consuming and expensive. One of the potential strategies to overcome this problem is to integrate information provided by laboratories at a global level. This strategy can help scientists and manufacturers access global databases with valuable information on feed ingredients such as chemical analysis and origin. While this approach is helpful, feed quality assessment cannot be fully guaranteed without the routine analysis of ingredients. These assessments will verify quality standards for each ingredient after receiving, storage, and processing and will continue after the feed is made and stored under different conditions. In this section, protein and lipid quality assessments along with processing effects, antinutritional factors (ANFs) and feed contaminants will be discussed.

10.3.1 Protein quality

10.3.1.1 Main protein ingredient sources

The main protein ingredients used in aquafeed are fishmeal, soybean meal and soy protein concentrate, other plant protein concentrates, and animal by-product meals. The quality of these ingredients dictates the quality of final feed.

10.3.1.1.1 Fishmeal

Fishmeal is one of the most valuable ingredients in aquafeeds because it possesses excellent nutritional properties including a balanced amino acid profile, high digestibility, and low/no antinutritional factors. It is therefore very important to ensure the quality standards for this commonly used ingredient. Fishmeal is produced either from whole fish (low economic value fish, sometimes called trash fish) or from by-products of fish fillet production factories that contain inedible parts, head, skin, fins, and bones. The quality of fishmeal depends largely on where it is produced. Fishmeal made of by-products contains high ash levels and lower protein levels than fishmeal made from whole fish. However, by-products are usually removed from human food-grade fish and are therefore fresh if handled carefully. In contrast, forage fish are used in fishmeal production harvested in bulk from targeted fisheries and not always handled prior to fishmeal production using procedures routinely used for human food-grade fish. This can lead to degradation and spoilage of the fish, causing enzymatic hydrolysis of proteins to amino acids and bacterial spoilage causing amino acids to be broken down further to ammonia and amine compounds. The value of fishmeal is mainly based on the amount of protein determined with the methods described earlier (N × 6.25). In the case of degraded fishmeal, a portion of the nitrogen present will not be derived from protein and therefore not accurately measure the crude protein level in the product. In other words, measuring Kjeldahl nitrogen in fishmeal is not a suitable measure of protein quality or, in some cases, even crude protein content. Additional chemical testing is required to detect protein breakdown products or nitrogen-based adulterants. This can be done by measuring total volatile nitrogen or ammonia nitrogen using Kjeldahl distillation units (Hardy and Barrows, 2002). Other nonprotein nitrogen-based compounds such as urea and melamine can be detected by simple color tests and chromatography methods, respectively. But the main quality aspect of fishmeal is the profile of amino acids, especially essential amino acid levels. In addition, some fishmeals contain high levels of arginine, histidine, lysine, tryptophan, asparagine, and/or glutamine, and these amino acids could give us an overestimated protein content because of high levels of nitrogen.

10.3.1.1.2 Soybean meal

One of the most important plant protein sources used in animal feeds and a valuable ingredient to partially replace fishmeal in fish feeds is soybean meal. The quality of soybean meal depends on processing steps, including crushing and removing the hulls and heat treatment. Dehulled soybean meal contains a higher protein (48%) and lower fiber content (3.5%) than hulled soybean meal and is preferred in fish feeds. Measuring bulk density can partially indicate the quality of soybean meal. Lower than normal bulk density could be due to presence of hulls in soybean meal, and higher bulk density could be due to adulteration with sand. Similar to fishmeal, feed formulators should be mindful of adulteration of soybean meal with nonprotein nitrogen. Performing a urea test is recommended to determine the quality of soybean meal along with amino acid analysis (Tangendjaja, 2015). One of the major problems with plant ingredients such as soybean meal is the presence of antinutritional factors that is discussed in another chapter. Methods of measuring antinutritional factors will be briefly discussed.

10.3.1.1.3 Land-animal proteins

In fish nutrition, commonly used land-animal proteins are meat and bone meal (MBM), poultry by-product meal (PBM), feather meal, and blood meal. These ingredients are made through the rendering process, and therefore, can contain rumen, wool, or hair that can significantly influence the fiber content. MBM and PBM typically contain 50%–58% protein and 10%–11% fat, whereas feather meal and blood meal contain 80%–85% protein and 2%–5% fat (Masagounder et al., 2016). Special care should be given to the storage of MBM and PBM due to considerable amounts of fat present in these ingredients. Antioxidants should be added to MBM and PBM right after production to prevent lipid oxidation. Despite their high nutritional value (protein and phosphorus content) and competitive prices, the nutritional profiles and quality of land animal protein ingredients can be variable depending on the source and the species from which were obtained. This is a complication for feed formulators and makes it prudent to test each batch of ingredients upon arrival at the feed mill to ensure ingredient specifications are being met. The pepsin digestibility test is a common in vitro method used to estimate the digestibility (and therefore quality) of animal by-products. The National Renders Association has suggested a pepsin digestibility of more than 86% for MBM, 88% for PBM, and 75% for feather meal. MBM is sometimes adulterated with added limestone to increase the calcium content. A measurement of calcium to phosphorus ratio can reveal the presence of limestone because limestone (calcium carbonate) does not contain phosphorus.

10.3.1.2 Chemical score and indispensable amino acid index

The following calculated indices are used to estimate protein quality as part of feed quality assessments. Some of the biological tests such as protein efficiency ratio (PER), net protein utilization (NPU), and biological value (BV) will be discussed in the biological evaluation of diet section. Chemical score (CS) and indispensable amino acid index (IAAI) of protein are influenced by the presence or absence of essential amino acids. The CS calculates percentage of the most limiting amino acid in a test protein compared with whole-egg protein (WEP) and is calculated as follows:

$$\begin{aligned} \text{CS} &= \left(\frac{\%\ \text{1st limiting amino acid in test protein}}{\%\ \text{the amino acid in whole egg proten}}\right) \times 100 \\ &= (\%\ \text{1st limiting amino acid in test protein} \\ &\div \%\ \text{amino acid in reference protein, normally taken as whole} - \text{egg protein}) \times 100 \end{aligned}$$

WEP is considered as an ideal reference protein because of its standard amino acid composition. Limiting amino acids are those that are most deficient. For example, the CS of wheat is 32.65, and note that egg protein contains approximately 57% essential amino acids, whereas wheat gluten meal contains only 33.45% (Table 10.2).

CS of wheat is 32.65, and note that egg contains approximately 57% essential amino acids, whereas wheat gluten meal contains only 33.45%

The IAAI is the ratio of the indispensable amino acid in the test protein (TP) divided by the indispensable amino acids in WEP as follows:

TABLE 10.2 Calculation of Chemical Score for wheat gluten meal, 80.7% crude protein (5-05-220).

Essential amino acid	% In egg	% In wheat protein	(Wheat/egg) × 100
Arginine	6.4	3.8	59.40
Histidine	2.1	2.0	95.24
Isoleucine	8.0	3.7	46.25
Leucine	9.2	6.3	68.48
Lysine	7.2	4.9	68.06
Methionine	4.1	1.6	39.02
Phenylalanine	6.3	4.5	71.43
Threonine	4.9	1.6	*32.65
Tryptophan	1.5	1.05	70.00
Valine	7.3	4.0	54.79

Chemical score of wheat is 32.65, and note that egg contains approximately 57% Essential Amino Acids whereas wheat gluten meal contains only 33.45%

$$\text{IAAI} = \left[\frac{\text{ARG(TP)} + \text{HIS(TP)} + \text{ISO(TP)} + \cdots + \text{VAL(TP)}}{\text{ARG(WEP)} + \text{HIS(WEP)} + \text{ISO(WEP)} + \cdots + \text{VAL(WEP)}}\right] \times 100$$

The IAAI is a more complex and accurate calculation than CS as it evaluates all essential amino acids rather than the limiting one. The IAAI of wheat gluten meal (5-05-220) could be calculated using the essential amino acid values from Table 10.2 as follows:

$$\left[\frac{(3.8 + 2.0 + 3.7 + 6.3 + 4.9 + 1.6 + 4.5 + 1.6 + 1.05 + 4.0)}{(6.4 + 2.1 + 8.0 + 9.2 + 7.2 + 4.1 + 6.3 + 4.9 + 1.5 + 7.3)}\right] \times 100 = \left(\frac{33.45}{57}\right) \times 100$$
$$= 58.68(\%)$$

It is important that protein quality calculations be made when protein is limiting growth to maximize efficient utilization of protein (Hardy and Barrows, 2002).

10.3.2 Lipid quality

Lipid is the main source of energy and essential fatty acids in aquafeed that is mainly supplied by fish oil or plant-derived oils. Fish oil can be obtained from freshwater fish or marine fish, and plant oils can be obtained from soybean, rapeseed, palm, corn, or coconut. The inclusion levels of oil in feeds is generally between 0% and 10% depending of the specific requirements of target species and the fatty acid profile of each ingredient. The quality of oil sources is indicated by the fatty acid composition that is analyzed using gas chromatography. More simple methods such as iodine value can be used to measure the number of double bonds in fatty acids, as an indication of the percentage of unsaturated fatty acids. The main concern for oil sources in the fish feed industry is hydrolytic and oxidative rancidity that degrades the quality of these ingredients.

10.3.2.1 *Hydrolytic rancidity*

Enzymatic hydrolysis of fatty acids from triglycerides and phospholipids is called hydrolytic rancidity. Hydrolytic rancidity is also known as acid value and is the measurement of free fatty acids in oil ingredients and a method to estimate oxidation status. Moisture is required for lipases, the enzymes involved in releasing fatty acids, to operate, and lipases can originate from the tissues of fish or from microorganisms living in feed or lipids. Free fatty acids are measured by a titration method (AOAC, 2000) and are expressed as a percentage. In fish oils used in fish feeds, the usual upper limit for fatty acids is 3%.

10.3.2.2 *Oxidative rancidity*

Oxidative rancidity occurs when a free radical reacts with double bonds of fatty acids. Free radicals arise through various mechanisms, including enzymatic activity and radiation, and their production is enhanced by the presence of divalent cations, especially Cu and Fe. The primary products of oxidation are peroxides, dienes, and free fatty acids that can lead to toxic secondary products or oxidized polymers. There are nearly an infinite number of intermediate products of oxidation of a complex lipid such as fish oil, but intermediate products include hydroperoxides, peroxides that in turn produce aldehydes and ketones. Oxidative rancidity is assessed by measuring aldehydes and ketones in oil sources, or by measuring final products of oxidation. Unsaturated fatty acids can take up oxygen or other compounds under abusive moisture, temperature, and light conditions and cause oil to become rancid. Fish oil is more susceptible to oxidation than plant oils due to its concentration of highly unsaturated fatty acids with four, five, and six double bounds. Oxidative stability of oils, as an indication of quality, can be measured using chemical tests involving oxidative stress of an oil source (Choe and Min, 2006). This provides a feed manufacturer with a predictive estimate of the oxidative stability of a batch of oil. Chemical tests for lipid oxidation measure intermediate products that rise as the process of oxidation proceeds.

10.3.2.2.1 Peroxide value

The most common test to measure oxidation is peroxide value that rises quickly during the first stages of oxidation but then decreases, as the intermediate products are converted to final products. Unless peroxide values are measured over time, it is impossible to know whether the lipid being tested is in the initial stages or later stages of oxidation. This problem can be overcome by conducting an accelerated oxidation test that heats the oil sample while oxygen is bubbled through it. Peroxide tests done at the start of the accelerated test and at intervals thereafter will reveal the stage of oxidation of the sample. Peroxide value will be quite low in highly oxidized oils and expressed as milliequivalents of peroxide per kilogram of sample (Eldin, 2010). Determination of peroxide value, as described by AOAC (2000), is based on the reaction of hydroperoxides with iodide ions and titration with thiosulfate. Oils used in fish feeds should have a peroxide value of less than 10; fresh herring oil has a peroxide value of less than 6 (Hardy and Barrows, 2002).

10.3.2.2.2 Thiobarbituric acid–reactive substances test

The thiobarbituric acid–reactive substances (TBARS) test is a method to investigate secondary oxidative products especially in polyunsaturated fatty acids such as malonaldehydes. Malonaldehyde reacts with thiobarbituric acid to produce a red color that can be detected colorimetrically. The TBARS number is expressed as milligrams of malonaldehyde per kilogram of sample. This method can be used in final feed and ingredients although it is better suited for ingredients as it can interact with sugar compounds (Hardy et al., 1983). Fresh herring oil can have about 50 mg/kg of TBARS while oxidized ones can have over 500 mg/kg of TBARS. The heat and pressure of pelleting a fish feed destroy malonaldehyde, making evaluation of freshly pelleted feed by TBARS a less valuable tool than evaluation of the ingredients or mixtures before pelleting (Hardy and Barrows, 2002).

10.3.2.2.3 Anisidine value

The anisidine value test measures the presence of aldehyde (responsible for the rancid smell of oil at the second stage of oxidation) in a sample. Measuring anisidine value, which increases throughout the oxidation process, provides a good understanding of oxidation status. But the anisidine value has its own limitations, as this value can be influenced by other factors such as fatty acid composition, physical state of lipid, and storage condition while being less specific as a sensory quality evaluation method (Yang et al., 2016). It is suggested that the anisidine value test is conducted with the peroxide value test to estimate the aldehydes that are decomposed from hydroperoxides. The Totox value can be calculated by adding the anisidine value to 2 × peroxide value to understand the overall oxidation state (Wai et al., 2009).

10.3.2.2.4 Kreis test

Kreis test was proposed in 1902 and was known as a reliable and rapid test to detect oxidative rancidity in samples. When a sample is mixed with phloroglucinol, the appearance of red or pink colors indicates incipient rancidity. This change of color indicates the presence of an aldehyde, as the test can be used for quantitative or qualitative measurements (Narasimhan et al., 1999).

10.3.2.2.5 Schaal oven test

Frankel (1993) described a method used to determine the induction time remaining for a lipid before oxidation moves from stage 1, initiation, to stage 2, propagation. This method overcomes the problem of trying to estimate the quality of a lipid source from a single measurement of the peroxide value or TBARs; in other words, the problem of not knowing whether the value is going up or going down. By measuring the peroxide value in a lipid sample, and then subjecting the sample to an elevated temperature in an oven for a day or more and retesting, one can determine if the sample is in the early or late stages of initiation and/or propagation, plus estimate the degree of protection provided by antioxidants in the sample.

10.3.3 Processing effects on ingredients

The main goal of processing ingredients is to maximize nutritional value and increase fish production efficiency. Different processing activities such as crushing, mixing, heating, and drying can affect the quality of ingredients and final feed. For example, dehulling of certain grains reduces fiber content or drying blood increases the protein content. Heat treatment is one of the most challenging parts of processing and can improve the oil extraction, inactivate antinutrients, and destroy microbes in feed or ingredients. On the other hand, high temperatures during heat treatment can damage ingredients and reduce their nutritional value.

10.3.3.1 *Amino acid availability*

High temperature exposure can reduce the availability of amino acids, especially lysine, arginine, and cysteine (Masagounder et al., 2016). The ε-amino group of lysine reacts with reducing sugars during the heating process to form a Schiff's base, which is a complex that is indigestible by animals. The process is called the Maillard reaction. Amino acid analysis will show the presence of lysine, but its availability to animals or fish will be reduced. Available lysine can be measured by the fluorodinitrobenzene (FDNB) method that reacts with the protein and attaches to the (free) ε-amino group of lysine. Then the protein is hydrolyzed, and the available lysine is determined by the ratio of FDNB lysine to total lysine. This method is appropriate to assess the quality of protein sources for which heat damage is a concern. In an experiment by Ljøkjel et al. (2000), moist heat treatment of soybean meal at 130°C reduced the content of arginine, lysine, and cysteine. These authors stated that this could have occurred due to the Maillard reaction.

10.3.3.2 *Processed soybean meal quality assessment tests*

Soy press cake is heated to inactivate antinutritional factors (trypsin inhibitors and lectin) and increase nutritional quality. Finding a balance between sufficient heating to inactivate antinutritional factors and excessive heating that reduces protein digestibility is crucial. The color of soybean meal provides a simple assessment tool if the product was subjected to the proper amount of heating. Soybean meal appears dark brown when overcooked and light yellow when undercooked. Vohra and Kratzer (1991) summarized the various chemical tests used to determine the adequacy of heat treatment of soybean meal. They divided the chemical tests into two groups: those that detect underheated soybean meal and those that detect overheated meal. Chemical tests to detect underheated soybean meal were urease activity, trypsin activity, and protein solubility. Urease is an enzyme naturally present in soybeans, which does not have any significant nutritional relevance except that it is heat-sensitive and its activity correlates well with residual trypsin activity in dried soybean meal. It is also relatively easy to measure (AOAC, 2000). Urease activity in commercial soybean meal ranges from a 0.02 to a 0.1 increase in pH (Vohra and Kratzer 1991). Values of over a 0.5 increase in pH indicate insufficient heat treatment of the soybean meal. If no increase in pH is detected with the urease test, this may mean that the soybean meal has been overheated, so some residual urease activity in the meal is preferred, at least for soybean meal intended for use in poultry feeds. Unheated soybean meal has a urease activity of a >2.25 pH rise (Waldroup et al., 1985).

As mentioned, trypsin activity in soybean meal decreases with heat treatment in proportion to urease activity. Unheated, solvent-extracted soybean meal can contain over 21 trypsin inhibitor units/mg sample (Araba and Dale, 1990), but commercial soybean meal subjected to normal heating during the press cake drying process generally contains about half the trypsin inhibitor activity of unheated meal. Additional heating further reduces trypsin inhibitor activity, the amount of reduction depending on the temperature and the duration of heat treatment (Arndt et al., 1999). The heat generated by extrusion cooking is sufficient to lower trypsin inhibitor levels, at least in full-fat soybeans (Wilson, 1992). Wilson (1992) extruded full-fat soybeans having an initial trypsin inhibitor activity of 46.5 trypsin units inhibited/mg sample, reducing the trypsin inhibitor activity to 8.1 trypsin units inhibited/mg sample.

A third method for measuring the extent of heat treatment of soybean meal is the water solubility test, which involves measuring Kjeldahl nitrogen levels in the soybean meal and in a water extract of the soybean meal (Vohra and Kratzer, 1991). The method has been slightly modified by extracting the sample in 0.2%KOH (Araba and Dale, 1990). Heating decreases the percentage of 0.2% KOH-extractable protein, from about 99% in raw soybean meal to about 72% after 20 min of autoclaving, corresponding to a decrease in trypsin inhibitor units from 21.1 to 1.0 (Araba and Dale, 1990).

10.3.3.3 *Pepsin digestibility*

High temperatures for long periods can degrade the quality of fishmeal by making the amino acids indigestible for fish. The pepsin digestibility test has been widely used to determine the effect of heat processing on fishmeal. The pepsin enzyme obtained from pig stomachs is used to perform in vitro digestibility test for fishmeal protein. Fishmeal damaged by thermal treatment has lower values for pepsin digestibility compared with high-quality fishmeal. High-quality fishmeals are produced from fresh whole fish by the wet reduction method and dried at low temperatures. Anderson et al. (1997) compared the quality of 16 types of fishmeals and reported that the dilute-pepsin digestibility is an accurate in vitro test for herring fishmeal quality assessment. Multienzyme pH-stat method is another method to measure the in vitro digestibility of fishmeal using a combination of proteolytic enzymes from land animals while holding the pH of the solution being analyzed constant as amino acids are released. Dimes et al. (1994) refined the pH-stat method to more accurately reflect protein digestibility in fish by using crude enzyme extracts from rainbow trout and Pacific salmon and reported a high correlation between pH-stat results and in vivo digestibility results for fishmeal and land animal proteins, but a lower correlation for soybean meal. The accuracy of this method was confirmed by Rungruangsak-Torrissen et al. (2002) with crude enzyme extracts from the pyloric caeca of Atlantic salmon. To improve accuracy of pH-stat results, the proteolytic enzymes should be obtained from the same species at the age of interest and validated for various ingredients with in vivo digestibility.

10.3.3.4 *Biogenic amines*

An important indicator of fishmeal quality is the content of biogenic amines that are formed during fishmeal production. Processing temperature, freshness of fish, fish species, and storage of raw material are the most important factors in the formation of biogenic amines (Clancy et al., 1995). Cadaverine, putrescine, spermidine, spermine, tyramine, tryptamine, and histamine are well-known biogenic amines produced from decarboxylation of free amino acids in fishmeal (Rawles et al., 1996). For example, histamine results from the decomposition of histidine in fishmeal that combines with lysine and forms gizzerosine. Cadaverine, putrescine, and tyramine are produced by decomposition of lysine, arginine, and tyrosine, respectively. Gizzerosine is a toxic compound that causes gizzard damage in young poultry, leading to black vomit and death of these animals. In Atlantic salmon, significantly higher histamine, putrescine, cadaverine, and tyramine did not influence digestibility nor feed utilization performance (Anderson et al., 1997). Even though swelling and thinning of the stomach wall was reported in rainbow trout due to high histamine levels, the effects of biogenic amines on growth and survival in fish are negligible (Fairgrieve et al., 1994). Generally, high levels of biogenic amines reveal decomposition of raw material prior to manufacturing of fishmeal and, as such, are an indicator of quality. Biogenic amines are determined by HPLC methods because of their heat stability (Pike and Hardy, 1997).

10.3.4 Feed microscopy assessment

Feed microscopy is the determination and evaluation of ingredient and feed using a microscope. This quality control technique is fast, flexible, and cheap and therefore can provide general information on ingredients that can ideally complement the chemical analysis. Feed microscopy can help to identify (1) overheating of some ingredients, especially FM and land-animal meals, like blood meal and (2) adulteration, especially to detect feather meal being added to other land-animal meals, etc. For feeds, microscopy is useful to assess effectiveness of grinding and even contamination. A good feed microscopy expert can almost determine the formulation of a feed by looking at a sample under a microscope. Generally, feed microscopy is divided into two main areas, quantitative and qualitative microscopy. The quantitative microscopy aims at measuring relative proportion of ingredients in final feed or contaminants and adulterants in ingredients or feed. Qualitative microscopy identifies ingredients and foreign materials in samples. Both of these methods can be performed by a stereo and/or compound microscope. The stereo microscope has a lower magnification and can recognize the external appearance and observe the morphological conformation of material, whereas compound microscope has a high magnification and can perform evaluations based on the cell features and histological structure (Klein and Marquard, 2005). The accuracy of feed microscopy as a quality control method in feed industry has been validated by several authors (Kamphues et al., 1998; Sanches et al., 2006; Veys and Baeten, 2018). Although feed microscopy is an accurate, cost-effective, and rapid evaluation tool, it requires an experienced microscopist to identify ingredients, adulterants, and contaminants. One of the methods that can help microscopists to perform routine quality control with feed microscopy is to have standard samples of ingredients and diets. A simple comparison of standard and sample material could help the feed microscopists to identify the quality status of samples. For example, color of soybean meal is an indication of proper/improper heating, or texture of rice bran indicates the amount of rice hull (Tangendjaja, 2015). The information provided in various literature such as "Atlas for the microscopic examination of feed containing vegetable and animal products" can be an asset to microscopists (Klein and Marquard, 2005). The methodology of sample preparation for feed microscopy has also been explained by the Association of Official Analytical Chemists (AOAC, 2000).

10.3.5 Assessment of buoyancy and water stability of pellets

Pelleting characteristics such as buoyancy and water stability are equally important as the nutritional composition of feed. Irrespective of how nutritionally balanced a feed is, if it is not ingested or the essential nutrients are not properly delivered to the fish, it will not be successful. Buoyancy and water stability are basically influenced by the ingredient type and processing technique. Assessment of these characteristics in pellets is a critical part of quality control procedure (Glencross et al., 2007).

10.3.5.1 Assessment of pellet buoyancy

Buoyancy is the ability or tendency of pellets to float in water and an important aspect of aquafeed that needs to be tailored based on the cultured species. For example, shrimp requires sinking pellets, whereas olive flounder cannot ingest the feed after it reaches the bottom surface of the tank (Hamidoghli et al., 2020). Assessment of feed buoyancy is simply performed by placing an adequate number of pellets in a testing water tank and counting the number of sinking pellets at different intervals during 60 min. It is important to adjust the salinity and temperature of testing water tank according to the culture condition of target species. Pellet buoyancy can be presented as percentage and a buoyancy percentage of over 70% after 20 min can be considered ideal for most aquaculture species (Rolfe et al., 2001; Solomon et al., 2011; Orire and Emine, 2019).

$$\%\ \text{buoyancy} = \frac{\text{Number of floating pellets}}{\text{Initial number of pellets}} \times 100$$

10.3.5.2 Assessment of pellet water stability

A properly formulated and processed pellet should be stable in water and retain its nutritional integrity until it is consumed by the fish. Pellets with high stability in the aquatic medium can guarantee the delivery of essential nutrients to the cultured fish. Nutrients that are leached from the feed cannot be ingested by the fish and are washed through the water system causing pollution, poor growth, and low survival. Pellets need to have higher water stability for species such as shrimp, which is a slow consumer, compared with aggressive consumers such as trout or salmon. There have been several methods for determining the water stability of pellets in water. These methods include immersing the pellets in still or agitated water or within a flow through system that demonstrates practical farms (Cuzon et al., 1982; Gu et al., 1990; Zettl et al., 2019). Obaldo et al. (2002) compared three methods of pellet water stability assessment and recommended horizontal and vertical shaking methods of water tanks as the best methods to approximate culture conditions. These authors also concluded that water salinity and temperature are important in performing studies on pellet water stability. Generally, pellets are placed inside the water with different intervals of immersion time, the remains of pellets are removed from the leached water using a filter, and leached feeds are analyzed for dry matter retention.

10.3.6 Antinutritional factors

Plant ingredients are probably one of the most suitable candidates to replace or reduce the amount of marine ingredients such as fishmeal in sustainable feed formulations. But the inclusion of these ingredients is limited due to the presence of ANFs. ANFs are described as endogenous substances in some ingredients that can negatively influence the fish health and growth (NRC, 2011). These biologically active substances are present in plants to protect them from grazing animal as a defense mechanism. Some antinutrients in plants can cause poisoning in animals but, more commonly, long-term ingestion of these products can lower fish performance. There are several types of ANFs present in different plant ingredients (NRC, 2011; Glencross et al., 2020) that are indicators of feed and ingredient quality. Detection and quantification of these compounds in plant ingredients is very important before including them in the feed formulation.

10.3.6.1 Lectins

Lectins are soluble heterogeneous proteins and possess domains that can bind to some carbohydrates. They are also known as hemagglutinins, which are driven from their ability in agglutinating erythrocytes. These ANFs are mostly found in plants including legumes such as soybean meal, as well as some animals and microorganisms. Chemical analysis of lectin in soybean meal and other plant ingredients can be performed by enzyme-linked immunosorbent assay (ELISA, Schulze et al., 1995).

10.3.6.2 Saponins

Saponins are a diverse group of glycosides that contain a hydrophobic steroidal or triterpenoid structure attached to sugar chains. Due to their chemical structure, they are dissolved in alcohol but stable in heat. Different chemical structures of saponins exist in a variety of plant species (such as legumes) that can give a bitter taste to the plant. Therefore, as the first antinutritional effect, saponins can influence feed intake. Saponins have some antioxidant, antifungal, and antiviral properties that can stimulate immune system but at the same time inhibit the proper digestion on protein and absorption of vitamins (Francis et al., 2002). Chemical analysis of saponins is complicated due to the complex structure of these compounds. There are two main steps in the saponin evaluation: the first step is extraction of ingredients that is performed with 80% methanol, and the second step is detection with a nano-HPLC electrospray ionization multistage tandem mass spectrometry (nLC-ESI-MS/MS, Mad et al., 2006).

10.3.6.3 Tannins

Tannins are a group of water-soluble phenolic compounds that can have either hydrolyzable or nonhydrolyzable (condensed) forms (Fig. 10.1). Hydrolyzable tannins (at 2% in diet)

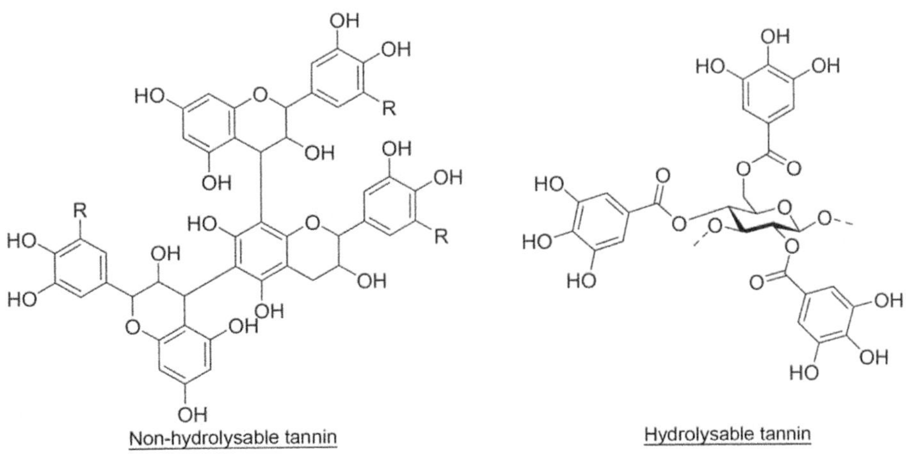

FIGURE 10.1 Demonstration of nonhydrolyzable and hydrolyzable forms of tannins. *Source: Peña, C., De La Caba, K.O.R.O., Eceiza, A., Ruseckaite, R., Mondragon, I., 2010. Enhancing water repellence and mechanical properties of gelatin films by tannin addition. Bioresour. Technol. 101 (17), 6836–6842.*

can reduce the metabolic rate, feed intake, and finally growth in fish (Glencross, 2015). Nonhydrolyzable tannins are present in legume seeds and can bind digestive enzymes or proteins to reduce their digestion. Generally, tannins can reduce the feed intake due to their bitter taste. There are few studies on the effect of tannins on fish, but their inclusion in feed should be limited.

Some previous methods on tannin analysis measured total phenol, which may not indicate the true index of polyphenols in ingredients. Methods that have been used for tannin analysis were spectrophotometric, colorimetric, chromatographic, and nuclear magnetic resonance. These techniques have received criticism due to their unsatisfying results and incapability to be used for different types of feedstuffs (Deshpande et al., 1986). The quantification of tannins should be based on the binding activity, which separates them from the analysis of other polyphenols. Similar to other procedures, tannins should be first isolated from the plant ingredients using conventional extraction methods, microwaves, ultrasonic waves, or subcritical solvents. Then, tannins can be measured by various methods. Generally, it is more difficult to analyze the content of hydrolyzable tannins compared with those of condensed due to sensitivity to hydrolysis. For accurate analysis of tannins (down to 100 ng), chromatography combined with mass spectrometry method can be used (Lochab et al., 2014).

10.3.6.4 *Alkaloids*

Alkaloids are a diverse group of toxic molecules derived from amino acids and present in about 25% of plants. Quinolizidines are predominant alkaloids and the most concern for fish health. Alkaloids are mainly found in legumes such as peas, beans, and lupins at variable concentrations of less than 200 mg/kg to more than 5000 mg/kg. Probably, most species have never been exposed to major alkaloids, but the threshold has been reported less than 100 mg/kg in the diet of rainbow trout (Glencross et al., 2020). Alkaloid analysis in plants could be performed by a comprehensive two-dimensional reverse phase liquid chromatography technique (Wei et al., 2016).

10.3.6.5 *Enzyme inhibitors*

Enzyme inhibitors such as proteinase, amylase, and lipase inhibitors are present in plant seeds to prevent them from being digested. Proteinase inhibitors are protein molecules that inhibit the activity of proteolytic enzymes such as trypsin, chymotrypsin, elastase, and enterokinase in the gastrointestinal tract. Soybean meal is abundant in proteolytic enzymes, especially trypsin inhibitor. Measurement of trypsin inhibitor activities for ingredients such as soybean meal can be done by mixing trypsin with different inhibitor levels and then performing colorimetric reactions. This method was approved by the American Oil Chemists Society and American Association of Cereal Chemists International (Liu, 2019).

10.3.6.6 *Glucosinolates*

Major glucosinolates are goitrin, isothiocyanates, and thiocyanates (Fig. 10.2) that naturally exist in cruciferous plants such as cabbage, kale, broccoli, and green leafy vegetables. Glucosinolates are also found in oilseed crops such as rapeseed and soybean meal. They are water soluble and derived from glucose and amino acids that contain a sulfur group. Glucosinolates can contain different side groups that play important roles in their biological activities. The hydrolysis products of glucosinolates are antithyroid and reduce the level of the thyroid hormones triiodothyronine (T_3) and thyroxine (T_4) in animals. The threshold of

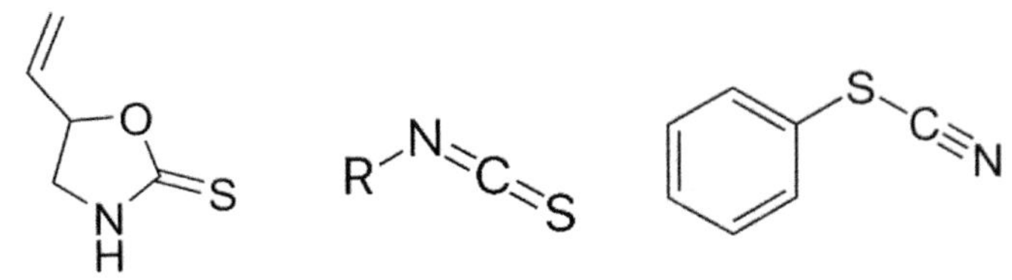

FIGURE 10.2 Structure of goitrin, isothiocyanates, and thiocyanate (Crews and Clarke, 2014).

glucosinolates in fish diets depends on the species, but 1.4 μmol/g of diet is considered a safe for most species (NRC, 2011).

Analysis of intact glucosinolates in various plant ingredients is done by their extraction at high temperatures with methanol—water mixture. This method can disable myrosinase activity that can be purified in an ion-exchange column. Finally, detection is done with an HPLC with a photodiode array or ultraviolet device (Grosser and van Dam, 2017).

10.3.6.7 Phytic acid

Phytic acid (inositol-hexaphosphate) and phytate (Fig. 10.3) are present in plant seeds as the storage forms of inositol and phosphate. They are abundant in cereal products and oilseed meals such as soybean, rapeseed, and cottonseed; rapeseed meal can contain up to 75 g/kg phytate. In high pH environments, phytic acid can bind with cations of Fe, Zn, K, Ca, Mg, Mn, and Ba, and at low pH values, phytic acid can complex with proteins that can reduce their availability to the animal. As a result, phytic acid can negatively influence protein utilization, skeletal development, and growth. Feed formulations should avoid high levels of phytic acid in diets, or alternatively, exogenous phytase can be used to release phosphorus from phytic acid (Glencross et al., 2020). For the analysis of phytic acid in feed, first it should be precipitated with ferric chloride to form sodium phytate. Then sodium phytate should be injected onto a C_{18} reversed-phase column and quantified with HPLC (Camire and Clydesdale, 1982).

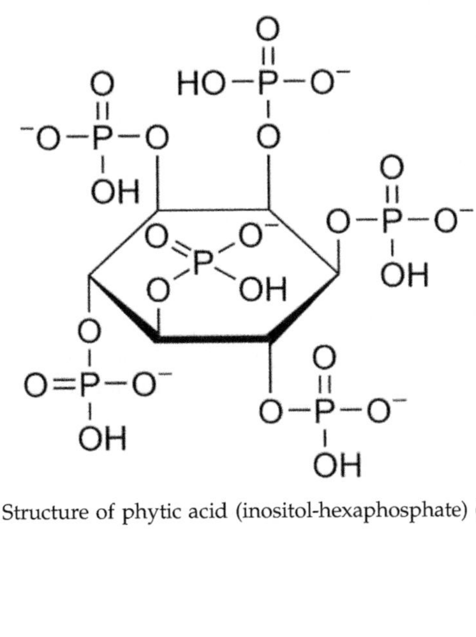

FIGURE 10.3 Structure of phytic acid (inositol-hexaphosphate) (Kumar et al., 2010).

10.3.7 Contaminants

There are several compounds that can contaminate animal and plant ingredients, final feed, cultured fish, and eventually humans. Some of these compounds can be produced by microorganisms such as bacteria and fungi as toxins. Some others can be retained in plant and animal ingredients from the environment such as the environmental persistent organic pollutants.

10.3.7.1 Mycotoxins

Fungi are responsible for the production of mycotoxins that can contaminate ingredients and final feed. The degree of mycotoxins effect in animals highly depends on the species type, age, nutrition, and health status and also the toxin type. For example, a 1.0 mg/kg dose can result in complete mortality of rainbow trout within few days, and 3.0 mg/kg can kill 40% of channel catfish in the same period (Glencross et al., 2020). Mycotoxins have a significant impact on feed intake, digestive performance, immune system, nerve system, and survival in animals. There are different types of mycotoxins that are produced by various types of fungal species. Aflatoxins, ochratoxin, deoxynivalenol, fumonisin, zearalenone, and T-2 toxin produced by *Aspergillus flavus, Aspergillus ochraceus, Fusarium culmorum, Fusarium moniliforme, Fusarium culmorum,* and *Fusarium sporotrichioides*, respectively, are the most important mycotoxins in fish feed. Some mycotoxins can be produced by more than one species of fungi. Feed producers should avoid using ingredients that contain these toxins or fungal contaminations. Measuring these compounds in ingredients and feed is important to ensure the quality of feed (D'Mello, 2004).

Efficient analysis of the type and amount of mycotoxins in feed is important in quality assessment procedures. Common procedures for the evaluation of many mycotoxins include sampling, extraction, cleanup (reducing unwanted compounds), separation, and detection. Cleanup could be done by the immunoaffinity method, separation by HPLC, and detection by UV or fluorescence method. For the qualitative or semiquantitative analysis of samples that are contaminated with mycotoxins, a cheaper alternative is the thin-layer chromatography (TLC). More updated techniques that can provide faster results with less labor-intensive protocols include enzyme-linked immunosorbent assays (ELISA). Another fast and nondestructive screening method is the near-infrared transmittance spectroscopy to detect mycotoxins in feed ingredients. This method requires minimum sample preparation and can be integrated with online monitoring systems. But, as mentioned previously, there are several types of mycotoxins that can contaminate feed. Therefore, methods that can detect several mycotoxins simultaneously are of interest. The latter has been possible by the development of liquid chromatography with tandem mass spectrometry (LC–MS/MS). This method can be used to identify large number of mycotoxins (up to 87) at the same time (Krska et al., 2008).

10.3.7.2 Environmental pollutants

Different types of organic and inorganic pollutants that can contaminate feed ingredients are industrial residues, pesticides, heavy metals, and radioactive compounds (D'Mello, 2004; Glencross et al., 2020).

10.3.7.2.1 Industrial residues

Dioxin and dioxin-like compounds are the main by-products of industrial processes that are potential contaminants of feed ingredients. Dioxins, also known as polyhalogenated dibenzo-*p*-dioxins, are produced by industrial activities or natural forest fires. When high levels of these compounds are present in feeds, they are deposited in lipids of fish. The primary methods for analyzing these compounds are chromatographic methods such as GC or HPLC with different detection techniques such as electron capture detection, mass spectroscopy, or photodiode array. In addition, some biological methods can be used for the detection of dioxin-like compounds such as biomarkers, whole animal exposures (in vivo), and protein-binding assays (Petronijevic et al., 2015).

10.3.7.2.2 Pesticides

Pesticides can originate from organochlorine, organophosphate, and pyrethroid compounds and are one of the most important environmental persistent organic pollutants. Pesticides in animal feeds are toxic, but the main concern in fish is that the residues of pesticides contaminate seafood products. Fish diets with high plant ingredients are more susceptible to pesticide contamination. Analysis of pesticides using GC on capillary columns has a higher separation efficiency and can result in precise detection (Müller and Stan, 1990).

10.3.7.2.3 Heavy metals

Heavy metals are considered as environmental resistant pollutants that can contaminate feed and fish. These compounds can have toxic effects of fish when administered at high levels. The most important metalloids in animal feed and health are mercury, cadmium, lead, and arsenic. The degree of toxicity for these compounds depends on the chemical form. Cadmium may occur in plant ingredients as a result of fertilizer usage during production. Lead from industrial pollution can contaminate many ingredients, whereas mercury is generally a concern in fishmeal. Analysis of heavy metals at very low concentrations and high specificity is possible with high-resolution ICP-MS (Choi et al., 2019).

10.4 Biological evaluation of diets

Biological evaluation of diets in fish nutrition involves measuring using a test fish or shrimp as a model organism to evaluate the performance and nutrient composition of formulated diets. This includes feeding different groups of fish compound diets that are varied in some nutrients and/or ingredients. A thorough knowledge of dietary chemical analysis is necessary prior to beginning any biological evaluation. A chemical analysis will provide information on the composition (dry matter, protein, fat, ash, etc.) of the diet, while a biological evaluation determines the nutritional value of feedstuffs and how they are metabolized by fish (Gizzi and Givens, 2004). The advantage of biological evaluation is that it can measure the direct performance of the diet on the fish while also providing information on palatability, bioavailability, and digestibility. However, the disadvantage is that it requires time-consuming feeding trails and expensive livestock management and is limited to evaluating only a few nutrients at a time. Lastly, results derived from evaluations of individual ingredients are not necessarily reliable when mixed with others in compound diets due to unpredictable interactions among other feed ingredients (Hardy and Barrows, 2002). For this reason, a revaluation must be done with the compound diet to verify its effects.

Designing a proper feeding trail for the biological evaluation of diets requires careful consideration of the experimental objectives. The experimental objectives could be either the investigation of nutrient retention in fish, evaluation of fish performance, or measurement of nutrient losses from excretion. Based on these objectives, experimental diets, feeding management, culture system, species, and period of the experiment will be adjusted (NRC, 2011). For example, experimental diets with nearly identical nutrient compositions but different ingredients are used for biological evaluations of ingredients, or diets with similar types of ingredients but different nutritional levels are used for nutrient requirement studies. It is important to make sure that the experimental diets are made with ingredients that were sourced from similar suppliers, therefore having similar qualities. Also, prepared diets must possess identical physical properties that match the feeding behavior and size of the cultured species. Selection of a species for biological evaluation of a diet depends on a study's objectives, while factors such as source, age, genotype, ploidy level, and nutritional background can play a vital role in the nutrient utilization performances of fish and shrimp. The period of the feeding trial also depends, to a great extent, on the life stage of the cultured species. Larval and juvenile cultured species can show quick responses to nutritional manipulations due to their fast growth, whereas studies on adults and broodstock fish often require long periods to bear out effects. Another factor that is indispensable in the biological evaluation of diets is the culture system. Proper feeding trials should occur in identical tanks with similar water flow rates, temperature, oxygenation, pH, salinity, light intensity, and photoperiod. Although in some experiments, such as actual farm and cage culture feeding trials, tightly controlling such environmental conditions is difficult. Nonetheless, researchers should strive to carry out biological evaluations in controlled environmental conditions (Lovell, 1989), unless environmental conditions are considered as variables and the interaction between the diet and these factors is being investigated (Soriano et al., 2018). Therefore, a sound biological evaluation requires all the aforementioned specifications put together according to the objectives of the experiment. The next step is the measurement of parameters such as growth (weight gain), feed utilization performance, digestibility, metabolic reactions, and gene expression in fish that will be discussed in the following section.

10.4.1 Growth

Growth is the most common and widely used parameter in biological evaluation of diet in fish nutrition science. It is the sensitive, direct, and practical indication of the accuracy of feed formulation and adequacy of essential nutrients (Teruel, 2002). In addition to diet, growth in fish is influenced by other factors such as age, genotype, and rearing conditions. Starting with a similar mean weight (MW) and during a fixed period, different groups of fish are fed with experimental diets.

$$MW = \frac{\text{total biomass of fish}}{\text{number of fish}}$$

By measuring the difference between the mean initial weight (IW) and final weight (FW), percentage of weight gain (WG) can be calculated.

$$WG(\%) = \frac{FW - IW}{IW} \times 100$$

WG is a reliable indicator for growth and new tissue deposition in fish. In some fish species, measurement of differences in final length (FL) and initial length (IL) could demonstrate the influence of feed more clearly. Length gain (LG) can be presented in fish nutrition studies along with WG calculations.

$$LG(\%) = \frac{FL - IL}{IL} \times 100$$

When the interaction between diet, weight, and length is being measured, the condition factor (K) is ideal. For describing the physical and biological responses to feed and indicating the general fish condition, this value is used in fish nutrition studies and at fish farms (Datta et al., 2013).

$$K = \frac{FW}{FL^3} \times 100$$

If the response of feed formulation or nutrients are being measured on a daily basis, percentage of specific growth rate (SGR) can be calculated by measuring gain in weight over time.

$$SGR = \frac{\ln FW - \ln IW}{days} \times 100$$

The SGR can help us to understand the weight increase for the days of experiment and the trend that might occur in future. However, this prediction is only valid if fish grow at a constant weight over the duration of the study. For example, at larval stage which the growth is rather exponential, this measurement can be valid. But during long feeding trials and with changes in water temperature, thermal-unit growth coefficient (TGC) is more appropriate to predict growth rate. Previous studies on different types of trout and salmon have proven the accuracy of this equation (NRC, 2011):

$$TGC = \frac{FW^{1/3} - IW^{1/3}}{\sum (T \times D)} \times 100$$

where T is temperature in ℃ and D is days of experiment.

10.4.2 Feed utilization performance

Measurement of the conversion of feed to tissue and weight gain is either assessed by feed conversion ratio (FCR) or feed efficiency (FE):

$$FCR = \frac{\text{Total feed fed}}{\text{WG}} \quad FE = \frac{\text{WG}}{\text{Total feed fed}} \times 100$$

FCR and FE are used to describe how feed can contribute to the gain in body weight, and therefore, feeding frequency and body weight can influence these values. These models do not describe the nutrient composition of feed or fish body but are useful measures for evaluating the economic aspects of fish culture.

For the comparison of different protein sources in a single experiment and to determine the interaction of weight gain and protein intake, protein efficiency ratio (PER) can be calculated. Therefore, PER can be used to evaluate the quality of protein in diet based on weight gain.

$$PER = \frac{\text{WG}}{\text{Protein intake}}$$

Net protein utilization (NPU) describes the gain of protein by fish in relation to the protein fed over a specific time period. Chemical analysis for protein content of diet, fish tissue at the beginning of feeding trial, and fish tissue at the end of feeding trial along with protein digestibility coefficient is required to calculate NPU.

$$\begin{aligned} NPU &= \frac{\text{protein gain}}{\text{Protein intake}} \\ &= \frac{(\text{protein content of fish at final feeding} - \text{protein content of fish at first feeding})}{\text{Protein in diet} \times \text{protein digestability coefficinet}} \end{aligned}$$

Another method for the evaluation of protein quality is biological value (BV) that deals with measuring the nitrogen retained in fish tissue by the difference in nitrogen absorbed and excreted. The BV could be calculated by

$$\begin{aligned} BV &= \frac{N \text{ retained}}{N \text{ absorbed}} \times 100 \\ &= \frac{(N \text{ intake} - \text{metabolic fecal } N) - (\text{urinary } N - \text{endogenous uninary } N)}{N \text{ itnake} - (\text{fecal } N - \text{metabolic fecal } N)} \times 100 \end{aligned}$$

Generally, in aquaculture, it is more preferred that weight gain consists mostly of protein rather than lipid. Measurement of nutrient retention is important for determination of food fish quality and can indicate how well nutrients were absorbed. This measurement is calculated by calculating proximate composition of fish body before and after the feeding trial in relation to initial and final weight over nutrient intake. This model can be used for the retention percentage of dietary protein, lipid, or energy.

$$\%\text{ retention} = \frac{[(\text{FW} \times \text{final nutrient concentration}) - (\text{IW} \times \text{initial nutrient concentration})] \times 100}{\text{nutrient intake}}$$

10.4.3 Digestibility

The definition of digestion is the breakdown of feed and its incorporated ingredients into simple components that fish are able to absorb. Digestibility is the term used to identify the efficiency of digestion for diets or ingredients. In a dynamic industry such as fish feed, the search for new and sustainably sourced ingredients with high nutritional values is expanding. Feed formulations are being modified, to a high extent, to include new ingredients and formulations are increasingly setting limits based upon digestible nutrient levels rather than total nutrient levels. Therefore, digestibility values for these ingredients and diets are indispensable. Fish nutritionists need to predict the digestibility of their formulations to estimate the nutrient retention and waste management efficiency. Nutrients that are not digested and retained in fish tissues are excreted, affecting water quality. A good example is that of phosphorus, an essential nutrient for fish and also a limiting nutrient for growth of freshwater algae and macrophytes. As a result, phosphorus levels in fish farm effluent water is regulated by many government agencies. Phosphorus digestibility is a crucial parameter in feed evaluation to ensure that the amount of digestible phosphorus meets the nutritional needs of fish and that the levels of indigestible phosphorus are not excessive. Using ingredients with high phosphorus digestibility is essential to promote efficient growth of farmed fish and support the economic feasibility of aquaculture itself.

Digestibility of ingredients and feeds is measured using both in vitro and in vivo assays. The in vitro digestibility assays have been mostly tested for ruminants and humans (Mabjeesh et al., 2000; Butts et al., 2012). The application of these methods in aquaculture has been thoroughly reviewed by Moyano et al. (2015). Stimulating the physical and chemical conditions of gastrointestinal tract that feed undergoes after ingestion by the animal is mostly important in these methods. The simplest procedure for in vitro digestibility determination is based on incubation of protein substrates in presence of enzymes such as pepsin, trypsin, renin, and so on and quantification of final digestive products. The potential benefits of this method are fast determination, reproducibility, applicable for every ingredient, and no life stage limitations. But the major drawbacks of this method are that results are not uniformly correlated with in vivo digestibility values, especially for plant proteins, and do not reliably reflect the complex reactions occurring in different parts of gastrointestinal tract. Future configurations may result in more reliable findings from in vitro digestibility methods in aquatic animals. Further discussion of in vitro digestibility is beyond the scope of this chapter.

In vivo digestibility determination requires feeding a basal diet alone and combined with a feed ingredient to different groups of fish and collecting feces. Collecting feces from aquatic animals presents challenges. There are several methods to collect feces in fish nutrition digestibility experiments, each of which has strengths and weaknesses. Feces collection methods including dissection, striping of feces by gentle pressure, siphoning or netting of feces from tanks, and passive feces collection using specialized tanks, also called settling, have been described and compared by several authors (Windell et al., 1978; Satoh et al., 1992). Each collection method must meet important specifications including inducing minimum stress to the experimental fish, collection of feces with lowest nutrient loss, and being applicable for different fish sizes. Digestibility diets must contain a nontoxic, indigestible, and nonabsorbable marker such as yttrium oxide (0.1%) or chromium oxide (0.5%). The marker content of feeds and feces is then measured. Chromium oxide can be easily measured in most laboratories with spectrophotometric methods, whereas yttrium is measured using inductively coupled atomic absorption spectrophotometry. Using these values, apparent digestibility coefficient (ADC) is calculated using dry weight values as follows:

$$\mathrm{ADC}(\%) = \left(1 - \frac{\text{concentration of marker in feed}}{\text{concentration of marker in feces}}\right) \times 100$$

The same method can be used to determine the ADC of nutrients (e.g., protein and fat) and energy in diets.

$$\mathrm{ADC}(\%) = \left(1 - \frac{\text{concentration of marker in feed}}{\text{concentration of marker in feces}} \times \frac{\text{nutrient concentration in feces}}{\text{nutrient concentration in feed}}\right) \times 100$$

The ADC model does not consider the endogenous excretions that are excreted in feces such as residues of body, mucosal cells, enzymes, secretions from the digestive tract, and the microflora residues. To measure the true ADC, a diet without the nutrient being measured must be included to allow for correction for endogenous losses as shown in the following:

$$\mathrm{ADC}(\%) = \left(1 - \frac{\text{concentration of marker in feed}}{\text{concentration of marker in feces}} \times \frac{\text{nutrient level in feces (experimental diet)} - \text{nutrient level in feces (control diet)}}{\text{nutrient level in feed}}\right) \times 100$$

The ADC of ingredients can be measured with same feces collection methods and markers. But these ingredients cannot be fed individually to the fish and, therefore, should be included in 70:30 ratios with the basal diet. This means 70% basal diet that includes 30% of the tested ingredient (Forster, 1999; NRC, 2011).

$$\text{Ingredient ADC} = \text{diet ADC} + \left[(\text{diet ADC} - \text{Ref. diet ADC}) \times \left(\frac{0.7 \times \mathrm{DR}}{0.3 \times \mathrm{DI}}\right)\right]$$

In this model, DR is the percentage of the tested nutrient or kJ/g energy of the basal diet. DI is the percentage or energy content of the tested ingredient.

10.4.4 Metabolic assessment

A broad spectrum of physiological and metabolic changes in fish is influenced by nutrition. Fish nutritionists should understand the effects of different nutrients on metabolism along with growth. Nutrient sensing mechanisms allow fish to respond to a specific nutrient through different regulatory mechanisms. For example, nutrient availability can regulate cell signaling pathways that modulate basic processes such as secretion of enzymes and hormones. Regulation of endocrine signaling pathways helps fish to rapidly adapt to the diet and the environmental conditions. Pituitary, thyroid, insulin, and sex hormones have important roles in the feed intake and growth of fish. These hormones can directly influence the central nervous system that can eventually influence the gastrointestinal tract (MacKenzie et al., 1998). In addition, the digestive enzyme activities such as amylase, pepsin, and lipase provide information on the fish nutritional physiology. Measurement of these parameters is important in biological evaluation of diet and ingredients.

One of the major properties of a balanced diet is to promote the immunity of cultured fish for maximum survival and welfare. Modern aquaculture emphasizes more on the strong role of nutrition on fish health management to reduce the economic losses related to disease. Studies on the immunological functions of fish in relation to diet have revealed that feed formulations for optimal immunity may differ from those suitable for optimal growth (Pohlenz and Gatlin III, 2014). Fish have a well-developed immune system that consists of innate and adaptive mechanisms. Innate immune response is the first step of defense mechanism after the detection and recognition of invading pathogens. The forefront of innate immunity are skin and mucosal defense barriers that act strongly in protecting the body. Mucus contains lysozyme, antimicrobial peptides, lectins, IgM, pentraxins, and other factors that are elements of the innate immune system. The innate immunity at the cell level involves macrophages, inflammatory cells, cytotoxic cells, dendritic cells, and other phagocytes. For example, neutrophils are the most abundant white blood cells with phagocytic activities that are present in fish blood. In addition, humoral parameters such as transferrin, interferon, and protease inhibitors in fish serum and body fluids are involved in pathogen defense mechanisms. Lysozyme is a bacteriolytic enzyme present in fish tissue and fluids and acts as an important nonspecific immune parameter against pathogens (Magnadóttir, 2006). This enzyme can be detected in tissue or plasma of fish using spectrophotometric methods.

10.4.5 Gene expression

Recent developments in our knowledge regarding the fish genome has introduced new concepts in fish nutrition and new ways to evaluate the effects of nutrition on fish health and performance. The modern field of nutrigenomics focuses on the relationship between diet or nutrients and gene expression. Nutrigenomics has furthered understanding of biochemical and metabolic pathways during digestion, absorption, and utilization of nutrients in fish. Research has demonstrated that regulation of various genes related to digestion, transport, metabolism, and hormonal actions in fish is affected by diet (Panserat and Kaushik, 2010). There are several molecular techniques associated with nutrigenomics in fish nutrition studies including transcriptomics, proteomics, metabolomics, and genomics. The study of DNA is known as genomics, the study of patterns of gene expression is known as transcriptomics, the study of protein synthesis, structure, and patterns is known as proteomics, and the study of functioning metabolites is known as metabolomics. All these methods are used to investigate the influence of nutrients and diet on the body (Rimbach and De Pascual-Teresa, 2005).

Gene expression investigates the amount of transcribed messenger RNA (mRNA) in the tissue and provides information on how genotype gives rise to phenotype. These studies aim at detecting and quantifying mRNA for selected genes. Northern blot was the first method for RNA-based expression that could detect specific RNA molecules among a mixture of RNA. This method was developed in 1977 and was soon replaced by new methods. The most common methods for gene expression are quantitative polymerase chain reaction (qPCR), expression microarray, and next-generation sequencing (RNAseq) for microarray (Segundo-Val and Sanz-Lozano 2016). The PCR technique was developed in 1984 and is based on amplifying segment of DNA associated with the gene of interest to multiple copies (Saiki et al., 1985). In 1992, the qPCR was developed that enabled the PCR technique to perform gene expression studies. The qPCR or real-time PCR method amplifies and detects the increase of fluorescence generated by a dye in response to new strands of DNA by a special camera. The microarray is another common technique that can detect the expression of thousands of genes at the same time. In this method, microscope slides (gene chips) containing tiny spots in specific positions with known genes in each spot are used to detect gene expression. mRNA samples are collected from an experimental group and control and converted into complementary DNA (cDNA). Each sample group is labeled with a different color fluorescent probe, mixed, and allowed to bind in the slide (hybridization). Finally, by scanning the slide, the amount of expression for each sample can be measured by changes in color in each spot. RNAseq is a method to examine the quantity and sequence of RNA by the transcriptome of gene expression patterns. After understanding the gene sequencing in 1964, in 2005 the high-throughput technologies were introduced commercially. These sequencers can measure the transcriptome with different kinds of RNA (mRNA, tRNA, rRNA, snRNA, and ncRNA) in the cell. The expressions of these RNAs are highly related to the nutrition and physiological condition of body (Segundo-Val and Sanz-Lozano, 2016). However, the challenge with gene expression results is that they measure cellular responses at a single point in time, thereby excluding the temporal aspects of metabolic responses to diet or other external influence. For that reason, integrating "omic" results with physiological responses is necessary to develop a mechanistic understanding of fish nutrition.

10.5 Feed formulation and economic evaluation

Feed accounts for approximately over 50% of the operating costs of growing fish. This amount could be varied in different contraries with different production systems and species. A case study comparing 340 Asian fish farms including carp, catfish, prawn, and milkfish revealed that feed can comprise 10%–80% of total production costs (Hasan, 2007). This creates a temptation to focus on the cost of feed, the idea being to purchase a feed that can maximize production and minimize the costs of production. Therefore, economic evaluation of feed is a very important factor determining the profitability of an aquaculture enterprise.

10.5.1 Feed cost

For the precise evaluation of feed cost, both direct and indirect costs involved in the production process of feed should be calculated. Direct costs are those related to ingredients that are used in feed formulation and labor that is directly involved in the production of feed. Indirect costs are those related to management costs, energy costs (gas, oil and electricity), maintenance costs, equipment and buildings, and incidental expenses (Agbayani, 2002). Each of these variables contributes a specific percentage to the final feed price, which varies among manufacturing country and target species. For example, in Vietnam, catfish feed price is mainly dependent on ingredient (86.5%), labor and management (6.3%), and supply (3.2%) expenses. In addition, storage and transportation (2%), feed quality assessments (0.25%), and research and development (0.02%) contribute to the final price of catfish feed in Vietnam (Hasan et al., 2019). Computing the actual feed cost is an essential part of economic evaluation and needs to be performed carefully considering all the variables involved.

10.5.2 Feed economic efficiency

As previously mentioned in Section 10.4, FCR is a measure that can define the efficiency of feed formulation. FCR is the ratio of given feed weight over animal weight gain in a certain period of time. Lower FCR values indicate that a feed is efficiently converted into fish weight gain while overfeeding or underfeeding increases the ratio. But FCR does not directly measure feed economic efficiency. To accomplish this, one must include feed cost in calculations. The economic conversion ratio (ECR, \$/kg fish) provides a measure of feed economic efficiency by multiplying FCR and feeds cost.

$$\text{ECR} = \text{FCR} \times \text{ feed cost } (\$ / \text{kg})$$

Economic profit index (EPI) is another model that can be used to measure the feed economic efficiency. Measurement of EPI is performed by calculating the total value of fish produced minus the total feed costs (Martínez-Llorens et al., 2007):

$$\text{EPI} = \left[\text{fish weight (kg)} \times \text{fish price} \left(\frac{\$}{\text{kg}}\right)\right] - \left[\text{ECR} \left(\frac{\$}{\text{kg}}\right) \times \text{fish weight increase (kg)}\right]$$

These measures can be used along with other biological parameters (e.g., growth, immunity, digestibility, etc.) to evaluate feed formulations at a more practical level. Even though these models are basically integrated, sometimes biological and economic evaluations of feed can result in slightly different conclusions. For example, in a study by Lozano et al. (2007), sunflower meal (at 0%, 12%, 24%, and 36%) was used to replace fishmeal in sea bream diet to reduce the feed price. Results indicated that optimum inclusion of sunflower meal was 10%–12% based on growth and nutrient utilization performance but 14%–15% based on ECR and EPI. Economic evaluation along with biological measurements can be used to establish benchmarks and goals for production and finance in the aquaculture industry. Decisions that increase productivity without increasing production costs can be made with economic evaluation of feed.

References

Agbayani, R.F., 2002. Economics of feeding. In: Millamena, O.M., Coloso, R.M., Pascual, F.P. (Eds.), Nutrition in Tropical Aquaculture: Essentials of Fish Nutrition, Feeds, and Feeding of Tropical Aquatic Species. Southeast Asian Fisheries Development Center: Aquaculture Department, pp. 209–221.

Anderson, J.S., Higgs, D.A., Beames, R.M., Rowshandeli, M., 1997. Fish meal quality assessment for Atlantic salmon (*Salmo salar* L.) reared in sea water. Aquacult. Nutr. 3 (1), 25–38.

Araba, M., Dale, N.M., 1990. Evaluation of protein solubility as an indicator of overprocessing soybean meal. Poultry Sci. 69 (1), 76–83.

Arndt, R.E., Hardy, R.W., Sugiura, S.H., Dong, F.M., 1999. Effects of heat treatment and substitution level on palatability and nutritional value of soy defatted flour in feeds for Coho Salmon, *Oncorhynchus kisutch*. Aquaculture 180 (1), 129–145.

Association of Official Analytical Chemists, AOAC, 2000. Official Methods of Analysis, seventeenth ed. AOAC, Washington, DC.

Aufrère, J., Graviou, D., Demarquilly, C., Perez, J.M., Andrieu, J., 1996. Near infrared reflectance spectroscopy to predict energy value of compound feeds for swine and ruminants. Anim. Feed Sci. Technol. 62 (2), 77–90.

Burlingame, B., Charrondiere, R., 2015. Food composition. In: Lovegrove, J.A., Hodson, L., Sharma, S. (Eds.), Nutrition Research Methodologies. John Wiley & Sons, West Sussex, UK, pp. 71–90.

Butts, C.A., Monro, J.A., Moughan, P.J., 2012. In vitro determination of dietary protein and amino acid digestibility for humans. Br. J. Nutr. 108 (S2), S282–S287.

Camire, A.L., Clydesdale, F.M., 1982. Analysis of phytic acid in foods by HPLC. J. Food Sci. 47 (2), 575–578.

Choi, S.H., Kim, J.Y., Mi Choi, E., Lee, M.Y., Yang, J.Y., Ho Lee, G., Su Kim, K., Yang, J.S., Russo, R.E., Yoo, J.H., Kang, G.J., 2019. Heavy metal determination by inductively coupled plasma–mass spectrometry (ICP-MS) and direct mercury analysis (DMA) and arsenic mapping by femtosecond (fs)–laser ablation (LA) ICP-MS in cereals. Anal. Lett. 52 (3), 496–510.

Clancy, S., Beames, R., Higgs, D., Dosanjh, B., Haard, N., Toy, B., 1995. Influence of spoilage and processing temperature on the quality of marine fish protein sources for salmonids. Aquacult. Nutr. 1 (3), 169–177.

Crews, C., Clarke, D., 2014. Natural toxicants: naturally occurring toxins of plant origin. In: Mortajemi, Y. (Ed.), Encyclopedia of Food Safety, Vol 2. Hazards and Diseases. Elsevier, London, pp. 261–268.

Choe, E., Min, D.B., 2006. Mechanisms and factors for edible oil oxidation. Compr. Rev. Food Sci. Food Saf. 5 (4), 169–186.

Cozzolino, D., Chree, A., Murray, I., Scaife, J.R., 2002. The assessment of the chemical composition of fishmeal by near infrared reflectance spectroscopy. Aquacult. Nutr. 8 (2), 149–155.

Cuzon, G., Hew, M., Cognie, D., Soletchnik, P., 1982. Time lag effect of feeding on growth of juvenile shrimp, *Penaeus japonicus* (Bate). Aquaculture 29, 33–44.

Deshpande, S.S., Cheryan, M., Salunkhe, D.K., Luh, B.S., 1986. Tannin analysis of food products. Crit. Rev. Food Sci. Nutr. 24 (4), 401–449.

D'Mello, J.P.F., 2004. Contaminants and Toxins in Animal Feeds. FAO animal production and health paper, pp. 107–128.

Datta, S.N., Kaur, V.I., Dhawan, A., Jassal, G., 2013. Estimation of length-weight relationship and condition factor of spotted snakehead *Channa punctata* (Bloch) under different feeding regimes. SpringerPlus 2, 436.

Davies, S.J., 1985. The role of dietary fibre in fish nutrition. In: Muir, J.F., Roberts, R.J. (Eds.), Recent Advances in Aquaculture. Westview Press, CO, USA, pp. 219–249.

Dimes, L.E., Haard, N.F., Dong, F.M., Rasco, B.A., Forster, I.P., Fairgrieve, W.T., Arndt, R., Hardy, R.W., Barrows, F.T., Higgs, D.A., 1994. Estimation of protein digestibility—II. In vitro assay of protein in salmonid feeds. Comp. Biochem. Physiol. Physiol. 108 (2–3), 363–370.

Eldin, A.K., 2010. Methods to determine the extent of lipid oxidation in foods. In: Decker, E.A., Elias, R.J., McClements, D.J. (Eds.), Oxidation in Foods and Beverages and Antioxidant Applications, Understanding Mechanisms of Oxidation and Antioxidant Activity, vol. 1. Woodhead Publishing Limited, Cambridge, UK, pp. 181–191.

Fairgrieve, W.T., Myers, M.S., Hardy, R.W., Dong, F.M., 1994. Gastric abnormalities in rainbow trout (*Oncorhynchus mykiss*) fed amine- supplemented diets or chicken gizzard-erosion-positive fishmeal. Aquaculture 127 (2–3), 219–232.

Forster, I., 1999. A note on the method of calculating digestibility coefficients of nutrients provided by single ingredients to feeds of aquatic animals. Aquacult. Nutr. 5, 143–145.

Francis, G., Makkar, H.P.S., Becker, K., 2002. Dietary supplementation with a *Quillaja* saponin mixture improves growth performance and metabolic efficiency in common carp (*Cyprinus carpio* L). Aquaculture 203, 311–320.

Frankel, E.N., 1993. In search of better methods to evaluate natural antioxidants and oxidative stability in food lipids. Trends Food Sci. Technol. 4 (7), 220–225.

Gizzi, G., Givens, D.I., 2004. Variability in Feed Composition and its Impact on Animal Production. FAO Animal Production and Health Paper 160. FAO, United nations, pp. 36–52.

Glencross, B., 2015. Understanding the nutritional and biological constraints of ingredients to optimize their application in aquaculture feeds. In: Nates, S.F. (Ed.), Aquafeed Formulation. Academic Press, New York, pp. 39–42.

Glencross, B.D., Booth, M., Allan, G.L., 2007. A feed is only as good as its ingredients – a review of ingredient evaluation strategies for aquaculture feeds. Aquacult. Nutr. 13, 17–34.

Glencross, B., Bourne, N., Hawkins, W., Karopoulos, M., Evans, D., Rutherford, N., McCafferty, P., Dods, K., Burridge, P., Veitch, C., Sipsas, S., 2015. Using Near Infrared Reflectance Spectroscopy (NIRS) to predict the protein and energy digestibility of lupin kernel meals when fed to rainbow trout, *Oncorhynchus mykiss*. Aquacult. Nutr. 21 (1), 54–62.

Glencross, B., Bourne, N., Irvin, S., Blyth, D., 2017. Using near-infrared reflectance spectroscopy to predict the digestible protein and digestible energy values of diets when fed to barramundi, *Lates calcarifer*. Aquacult. Nutr. 23 (2), 397–405.

Glencross, B.D., Baily, J., Berntssen, M.H., Hardy, R., MacKenzie, S., Tocher, D.R., 2020. Risk assessment of the use of alternative animal and plant raw material resources in aquaculture feeds. Rev. Aquacult. 12 (2), 703–758.

Greaser, M.L., Guo, W., 2015. Proteins. In: Nollet, L.M.L., Toldra, F. (Eds.), Handbook of Food Analysis. CRC Press, Novi Sad, Serbia, pp. 357–366.

Grosser, K., van Dam, N.M., 2017. A straightforward method for glucosinolate extraction and analysis with high-pressure liquid chromatography (HPLC). JoVE 121, e55425.

Gu, W.Y., Zhu, X.S., Behnke, K.C., Posner, E.S., 1990. Evaluation of Wheat and its Milling Fractions for Water Stability Enhancement in Pelleted Shrimp Feeds. Contribution No 90–112-J, Kansas Agricultural Experiment Station. Kansas State University, Manhattan, KS, USA.

Hamidoghli, A., Won, S., Lee, S., Lee, S., Farris, N.W., Bai, S.C., 2020. Nutrition and feeding of olive flounder *Paralichthys olivaceus*: a review. Rev. Fish. Sci. Aquacult. 1–18.

Hardy, R.W., Barrows, F.T., 2002. Diet formulation and manufacture. In: Halver, J.E., Hardy, R.W. (Eds.), Fish Nutrition. Academic Press, San Diego, CA, pp. 505–600.

Hardy, R.W., Mugrditchian, D.S., Iwaoka, W.T., 1983. Storage stability of lipids in a dry salmonid diet. Aquaculture 34 (3–4), 239–246.

Hasan, M.R., 2007. Economics of Aquaculture Feeding Practices in Selected Asian Countries (No. 505). Food & Agriculture Organization (FAO).

Hasan, M.R., Shipton, T.A., Bueno, P.B., 2019. Aquafeed Value Chain Analysis and a Review of Regulatory Framework of Striped Catfish Farming in Viet Nam. FAO Fisheries and Aquaculture Technical Paper No. 648. FAO, Rome.

Jobling, M., 2001. Feed composition and analysis. In: Houlihan, D., Boujard, T., Jobling, M. (Eds.), Food Intake in Fish. Blackwell, Oxford, pp. 1–24.

Kamphues, J., Gies, P., Vohringer, H., Hahn, H., 1998. Feed microscopy—its special relevance for veterinary nutrition. J. Anim. Physiol. Anim. Nutr. 80 (1-5), 207–212.

Klein, H., Marquard, R., 2005. Feed Microscopy: Atlas for the Microscopic Examination of Feed Containing Vegetable and Animal Products. Agrimedia GmbH, Bergen.

Krska, R., Schubert-Ullrich, P., Molinelli, A., Sulyok, M., MacDonald, S., Crews, C., 2008. Mycotoxin analysis: an update. Food Addit. Contam. 25 (2), 152–163.

Kumar, V., Sinha, A.K., Makkar, H.P., Becker, K., 2010. Dietary roles of phytate and phytase in human nutrition: a review. Food Chem. 120 (4), 945–959.

Liu, K., 2019. Soybean trypsin inhibitor assay: further improvement of the standard method approved and reapproved by American oil chemists' society and American association of cereal chemists international. J. Am. Oil Chem. Soc. 96 (6), 635–645.

Ljøkjel, K., Harstad, O.M., Skrede, A., 2000. Effect of heat treatment of soybean meal and fishmeal on amino acid digestibility in mink and dairy cows. Anim. Feed Sci. Technol. 84, 83–95.

Lochab, B., Shukla, S., Varma, I.K., 2014. Naturally occurring phenolic sources: monomers and polymers. RSC Adv. 4 (42), 21712–21752.

Lovell, R.T., 1989. Nutrition and Feeding of Fish. Van Nostrand Reinhold Publishers, New York, YSA.

Lozano, N.B.S., Vidal, A.T., Martínez-Llorens, S., Mérida, S.N., Blanco, J.E., López, A.M., Torres, M.P., Cerdá, M.J., 2007. Growth and economic profit of gilthead sea bream (*Sparus aurata*, L.) fed sunflower meal. Aquaculture 272 (1–4), 528–534.

Mabjeesh, S.J., Cohen, M., Arieli, A., 2000. In vitro methods for measuring the dry matter digestibility of ruminant feedstuffs: comparison of methods and inoculum source. J. Dairy Sci. 83, 2289–2294.

MacKenzie, D.S., VanPutte, C.M., Leiner, K.A., 1998. Nutrient regulation of endocrine function in fish. Aquaculture 161, 3–25.

Mad, T., Sterk, H., Mittelbach, M., Rechberger, G.N., 2006. Tandem mass spectrometric analysis of a complex triterpene saponin mixture of *Chenopodium quinoa*. J. Am. Soc. Mass Spectrom. 17 (6), 795–806.

Magnadóttir, B., 2006. Innate immunity of fish (overview). Fish Shellfish Immunol. 20, 137–151.

Martínez-Llorens, S., Moñino, A.V., Tomás, A., Pla, M., Jover, M., 2007. Soybean meal as partial dietary replacement for fish meal in gilthead sea bream (*Sparus aurata*) diets: effects on growth, nutritive efficiency and body composition. Aquacult. Res. 38, 82–90.

Masagounder, K., Ramos, S., Reimann, I., Channarayapatna, G., 2016. Optimizing nutritional quality of aquafeeds. In: Nates, S.F. (Ed.), Aquafeed Formulation. Academic Press, New York, pp. 239–264.

Molina-Poveda, C., 2015. Nutrient requirements. In: Nates, S.F. (Ed.), Aquafeed Formulation. Academic Press, New York, pp. 75–216.

Moyano, F.J., Saenz Rodrigañes, M.A., Diaz, M., Tacon, A.G.J., 2015. Application of in vitro digestibility methods in aquaculture: constraints and perspectives. Rev. Aquacult. 7, 223–242.

Müller, H.M., Stan, H.J., 1990. Pesticide residue analysis in food with CGC—study of long-term stability by the use of different injection techniques. J. High Resolut. Chromatogr. 13 (10), 697–701.

Mueller-Harvey, I., 2004. Modern Techniques for Feed Analysis. FAO, Assessing Quality and Safety of Animal Feeds. 160 Animal Production and Health Paper, Rome.

Narasimhan, S., Kumar, A.V., Ravi, R., Chand, N., 1999. Optimization of Kreis test for edible oils. J. Food Lipids 6 (2), 107–115.

NRC (National Research Council), 2011. Nutrient Requirements of Fish and Shrimp. The National Academies Press, Washington DC.

Obaldo, L.G., Divakaran, S., Tacon, A.G., 2002. Method for determining the physical stability of shrimp feeds in water. Aquacult. Res. 33 (5), 369–377.

Orire, A.M., Emine, G.I., 2019. Effects of crude protein levels and binders on feed buoyancy. J. Aquatic Res. Devel. 10 (565), 2.

Panserat, S., Kaushik, S.J., 2010. Regulation of gene expression by nutritional factors in fish. Aquacult. Res. 41, 751–762.

Peña, C., De La Caba, K.O.R.O., Eceiza, A., Ruseckaite, R., Mondragon, I., 2010. Enhancing water repellence and mechanical properties of gelatin films by tannin addition. Bioresour. Technol. 101 (17), 6836–6842.

Petronijevic, R., Radicevic, T., Jankovic, V., 2015. Methods for determination of dioxins and dioxin-like compounds—a brief review of recent advances. Procedia Food Sci. 5, 227–230.

Pike, I.H., Hardy, R.W., 1997. Standards for assessing quality of feed ingredients. In: D'Abramo, L.R., Conclin, D.E., Akiyama, D.M. (Eds.), Crustacean Nutrition—Advances in World Aquaculture, vol. 6. World Aquaculture Society, Baton Rouge, LA, pp. 473–492.

Pojić, M., Kravić, S., Stojanović, Z., Nollet, L.M.L., Toldra, F., 2015. Analytical methods for determination of moisture and ash in foodstuffs. In: Nollet, L.M.L., Toldra, F. (Eds.), Handbook of Food Analysis. CRC Press, Novi Sad, Serbia, pp. 275–296.

Pohlenz, C., Gatlin III, D.M., 2014. Interrelationships between fish nutrition and health. Aquaculture 431, 111–117.

Prevolnik, M., Škrlep, M., Janeš, L., Velikonja-Bolta, Š., Škorjanc, D., Čandek-Potokar, M., 2011. Accuracy of near infrared spectroscopy for prediction of chemical composition, salt content and free amino acids in dry-cured ham. Meat Sci. 88 (2), 299–304.

Rawles, D.D., Flick, G.J., Martin, R.E., 1996. Biogenic amines in fish and shellfish. Adv. Food Nutr. Res. 39, 329–365.

Rimbach, G., De Pascual-Teresa, S., 2005. Application of Nutrigenomics Tools to Analyze the Role of Oxidants and Antioxidants in Gene Expression. CRC Press, Taylor & Francis Group, FL, US.

Rolfe, L.A., Huff, H.E., Hsieh, F., 2001. Effects of particle size and processing variables on the properties of an extruded catfish feed. J. Aquat. Food Prod. Technol. 10 (3), 21–34.

Rungruangsak-Torrissen, K., Rustad, A., Sunde, J., Eiane, S.A., Jensen, H.B., Opstvedt, J., Nygård, E., Samuelsen, T.A., Mundheim, H., Luzzana, U., Venturini, G., 2002. In vitro digestibility based on fish crude enzyme extract for prediction of feed quality in growth trials. J. Sci. Food Agric. 82 (6), 644–654.

Saiki, R.K., Scharf, S., Faloona, F., Mullis, K.B., Horn, G.T., Erlich, H.A., Arnheim, N., 1985. Enzymatic amplification of beta-globin genomic sequences and restriction site analysis for diagnosis of sickle cell anemia. Science 230 (4732), 1350–1354.

Sanches, R.L., Alkmin-Filho, J.F., de Souza, S.V., Junqueira, R.G., 2006. In-house validation of a method for detection of animal meals in ruminant feeds by microscopy. Food Control 17 (2), 85–92.

Satoh, S., Young Cho, C., Watanabe, T., 1992. Effects of fecal retrieval timing of digestibility of nutrients in rainbow trout diet with the Guelph and YUF feces collection systems. Nippon Suisan Gakkaishi 58, 1123–1127.

Schulze, H., Saini, H.S., Huisman, J., Hessing, M., Van Den Berg, W., Verstegen, M.W., 1995. Increased nitrogen secretion by inclusion of soya lectin in the diets of pigs. J. Sci. Food Agric. 69 (4), 501–510.

Segundo-Val, I.S., Sanz-Lozano, C.S., 2016. Introduction to the gene expression analysis. In: García, M.I. (Ed.), Molecular Genetics of Asthma. Springer, New York, pp. 29–43.

Solomon, S.G., Ataguba, G.A., Abeje, A., 2011. Water stability and floatation test of fish pellets using local starch sources and yeast (*Saccahromyces cerevisae*). Int. J. Latest Trends Agric. Food Sci. 1 (1), 1–5.

Soriano, E.L., Ramírez, D.T., Re Araujo, D., Gómez-Gil, B., Sánchez, C.G., 2018. Effect of temperature and dietary lipid proportion on gut microbiota in yellowtail kingfish Seri*ola lalandi juv*eniles. Aquaculture *497*, 269–277.

Tangendjaja, B., 2015. Quality control of feed ingredients for aquaculture. In: Davis, D.A. (Ed.), Feed and Feeding Practices in Aquaculture. Woodhead publishing, Cambridge, UK, pp. 141–168.

Teruel, M.B., 2002. Evaluation of feedstuffs and aquafeeds. In: Millamena, O.M., Coloso, R.M., Pascual, F.P. (Eds.), Nutrition in Tropical Aquaculture: Essentials of Fish Nutrition, Feeds, and Feeding of Tropical Aquatic Species. Southeast Asian Fisheries Development Center: Aquaculture Department, pp. 149–168.

Udén, P., Robinson, P.H., Wiseman, J., 2005. Use of detergent system terminology and criteria for submission of manuscripts on new, or revised, analytical methods as well as descriptive information on feed analysis and/or variability. Anim. Feed Sci. Technol. 118, 181–186.

Veys, P., Baeten, V., 2018. Protocol for the isolation of processed animal proteins from insects in feed and their identification by microscopy. Food Contr. 92, 496–504.

Vohra, P., Kratzer, F.H., 1991. Evaluation of soybean meal determines adequacy of heat treatment. Feedstuffs 63 (8), 23–28.

Wai, W.T., Saad, B., Lim, B.P., 2009. Determination of TOTOX value in palm oleins using a FI-potentiometric analyzer. Food Chem. 113 (1), 285–290.

Waldroup, P.W., Ramsey, B.E., Hellwig, H.M., Smith, N.K., 1985. Optimum processing for soybean meal used in broiler diets. Poultry Sci. 64, 2314–2320.

Wei, X., Shen, H., Wang, L., Meng, Q., Liu, W., 2016. Analyses of total alkaloid extract of *Corydalis yanhusuo* by comprehensive RP× RP liquid chromatography with pH difference. J. Anal. Methods Chem. 9752735, 8.

Wilson, T.P., 1992. Full-fat Soybean Meal—An Acceptable, Economical Ingredient in Chinook Salmon Grower Diets. Ph.D. Dissertation. Univ., Washington, Seattle.

Windell, J.T., Foltz, J.W., Sarokon, J.A., 1978. Methods of fecal collection and nutrient leaching in digestibility studies. Progress. Fish Cult. 40, 51–55.

Yang, X., Boyle, R.A., 2016. Sensory evaluation of oils/fats and oil/fat–based foods. In: Hu, M., Jacobsen, C. (Eds.), Oxidative Stability and Shelf Life of Foods Containing Oils and Fats. AOCS Press, Denmark, pp. 157–185.

Zettl, S., Cree, D., Soleimani, M., Tabil, L., 2019. Mechanical properties of aquaculture feed pellets using plant-based proteins. Cogent Food Agricul. 5 (1), 1656917.

CHAPTER

11

Nutritional supplements and fish health

Delbert M. Gatlin, III and Fernando Y. Yamamoto

Department of Ecology and Conservation Biology, Texas A&M University System, College Station, TX, United States

11.1 Introduction

As commercial aquaculture has continued to expand in terms of both global distribution and level of production intensity, more emphasis is being placed on improving nutritional regimes to not only reduce production costs but also increase the health of the cultured organism. As such, a considerable amount of effort has been expended since the last edition of this book was published to evaluate various nutritional supplements to improve fish health. A wide variety of supplements have been evaluated including several different nutrients as well as diverse nonnutritive supplements, which may exert their influences either through the immune system or by various other means. This chapter will highlight some of the most recent information pertaining to these various supplements after reviewing the various factors affecting fish health.

11.2 Factors affecting fish health

11.2.1 Environmental conditions/stressors

The health and well-being of fish can be influenced by a wide variety of abiotic and biotic factors. Particularly under intensive culture conditions, environmental factors such as water quality conditions are typically less than optimal. Such factors including high levels of nitrogenous wastes or low levels of dissolved oxygen can be chronic stressors to fish and adversely affect their physiological and immunological functioning.

In addition, there are various procedures routinely practiced in aquaculture that subject the fish to various other stressors such as crowding, grading, handling, and movement to

Fish Nutrition, Fourth Edition
https://doi.org/10.1016/B978-0-12-819587-1.00004-5

different culture systems. These various stressors are generally considered to be immunosuppressive in that they will impair the ability of fish to mount an immune response. It is also generally believed that exposure to chronic stress will make fish more susceptible to infectious agents.

11.2.2 Infectious agents

A wide variety of disease-causing agents may be encountered in aquaculture including bacteria, fungi, viruses, and parasitic organisms. The prevalence of these disease agents varies with the species of cultured organism and its culture environment. However, most all aquacultured organisms can be adversely affected by the various groups of pathogens, although bacterial and parasites are the ones most commonly treated for cure. Nevertheless, maintaining an optimally functioning immune system can be of great assistance to the cultured organism in being resistant to various types of pathogens.

11.2.3 Immunity and disease resistance

The cells, molecules, and tissues responsible for immunity constitute the immune system, and their collective and coordinated responses upon exposure to a foreign substance or an organism are called the immune response (Abbas et al., 2014; Murphy and Weaver, 2016). The immune system is classically divided into two branches: the innate and adaptive or acquired immune systems. However, with discoveries from the past decades, it has been shown that both branches orchestrate together, where their components shift from one group to another, communicating and having specific roles on both types of immune responses (Castro and Tafalla, 2015; Pohlenz and Gatlin, 2014). The immune responses from the innate and adaptive immune systems also can be subdivided into two groups: humoral and cell-mediated immunity (Table 11.1). Additionally, components from the teleost fish and

TABLE 11.1 Main effector components of fish immune responses.

Immune system branch	Cellular components	Humoral components	Specificity
Innate	Macrophages	Complement proteins	Pathogen-associated molecular patterns
	Polymorphonuclear cells	Lysozyme	Damage-associated molecular patterns
	Dendritic cells	AMP	
	Natural killer–like cells	Lectins	
	Nonspecific cytotoxic cells	Transferrin	
	Thrombocytes	Natural Igs	
Adaptive	B lymphocytes	Ig D, Ig M, IgZ/T	Specific epitopes
	T lymphocytes		

AMPs, antimicrobial peptides; *Ig*, immunoglobulin.
Adapted from Pohlenz, C., Gatlin III, D.M., 2014. Interrelationships between fish nutrition and health. Aquaculture 431, 111–117.

crustacean innate immune systems, such as macrophages and hemocytes, may present a "memory-like" feature called trained immunity, which will be further discussed in this section. The immunological responses are a collectively complex set of mechanisms, which ultimately protect the animals against toxins, endogenous and exogenous components, and a wide variety of pathogens and parasites, as well as reoccurring infections (Hoseinifar et al., 2019).

Under their natural habitat, the teleost fish developed a very elaborate innate immune system, capable of having robust responses against the many pathogens and parasites that naturally occur in the aquatic environment (Foey and Picchietti, 2014). The innate immune system is the earliest line of defense, in a nonspecific manner, recognizing the conserved molecules shared by pathogenic organisms (pattern-associated molecular patterns, PAMPs), by the receptors present in their cell walls (pattern recognition receptors, PRRs). The nonspecific responses of the innate immune system comprise cellular and biochemical defense mechanisms that are in place before infection and are poised to respond rapidly in case of infection (Abbas et al., 2014). Some of the nonspecific immunological responses focused on teleost fish are discussed in the following.

Cytokines constitute a broad category of small proteins that mediate cell signaling within the immune system (Castro and Tafalla, 2015). These proteins are soluble factors that are secreted by cells, mainly leukocytes, and are one of the earliest assets of the innate immune response to infection and tissue damage (Abbas et al., 2014). The outcome of cytokine signaling induces inflammation, chemotaxis of immune cells, respiratory burst, and pathogen killing, as well as antiviral responses to prevent viral replication (Foey and Picchietti, 2014). However, they are also involved in the later adaptive immune response, where they will commence the development, activation, and functionality of adaptive immune responses from B and T cells (Foey and Picchietti, 2014; Wang and Secombes, 2013). Within the cytokine family, the chemokines are responsible for regulating immune cell migration, maturation, and functions of the recruited leukocytes to the site of inflammation. Teleost fish appear to synthesize all proinflammatory cytokines (TNF-α, IL-1β, IL-6, IL-8, and IL-12) and antiinflammatory cytokines (IL-10, TGF- β1) (Secombes, 2016). The regulation of the fish immune responses by cytokines is a complex mechanism and has been extensively studied, especially using zebrafish as an animal model. Comprehensive reviews about fish cytokines can be found in Secombes (2016), Wang and Secombes (2013), and Zou and Secombes (2016).

Antimicrobial peptides (AMPs) are secreted by fish as an innate defense mechanism. They can be expressed in several tissues and organs (e.g., skin, liver, kidney, spleen, gills, eyes, gonads, etc.). The AMPs molecule size are relatively small (less than 100 amino acids) constituting peptides or small proteins, with low molecular weight, and typically with potent, broad-spectrum activity targeting pathogens (viruses, fungi, parasites, and bacteria, or tumor cells) (Noga et al., 2011). They can be classified into five major families based on their molecular structure: β-defensins, cathelicidins, histone-derived peptides, and piscidins (Katzenback, 2015). An AMP binds their positively charged amino acids to the negatively charged molecules or substances in the membrane of a foreign organism, and their mechanism of protection involves disruption (lytic) or pore-forming (ionophoric) actions (Rajanbabu and Chen, 2011).

The complement system consists of a set of proteins and glycoproteins that are synthesized by hepatocytes, tissue macrophages, blood monocytes, and epithelial cells of the gastrointestinal (GI) tract (Castro and Tafalla, 2015). The mammalian complement system is a set of about 30 distinct proteins, and it is a major humoral response that can be part of innate and adaptive immunity (Nakao et al., 2011; Whyte, 2007). It plays an important and versatile role for the immune system because it can promote inflammatory responses, and clearance of apoptotic cells and necrotic cell debris. In addition, it can opsonize, recruit phagocytes, and directly kill pathogens by assembling a membrane attack complex (MAC) (Abbas et al., 2014; Nakao et al., 2011). The complement system comprises three pathways: the classical pathway, the lectin pathway, and the alternative pathway. To be activated, the classical pathway is dependent on the recognition of immune complexes (antigen–antibody complex). The activation of the lectin pathway relies on the interaction of mannose-binding lectin and ficolins with carbohydrates present on the pathogen's surface. Moreover, the alternative pathway is triggered when the C3 protein recognizes certain structures in the surface of the pathogen cell wall (e.g., lipopolysaccharides). All three pathways can ultimately converge to the cytolytic pathway, where the set of proteins attach to the surface of the foreign cell assembling the MAC, which creates small pores and lysis of certain pathogens (Murphy and Weaver, 2016).

Lysozyme is a mucolytic enzyme, produced by leukocytes and possibly by intestinal cells, that disrupts the linkage between the *N*-acetylmuramic acid and the *N*-acetyl-D-glucosamine of gram-positive bacterial cell walls (Castro and Tafalla, 2015; Saurabh and Sahoo, 2008). Gram-negative bacteria are only affected by lysozyme when the outer cell wall is breached by other components of the immune system (e.g., complement proteins) (Saurabh and Sahoo, 2008). This enzyme is an important marker for physiological stress and immune response, and it can be detected in the mucus, eggs, various organs, and blood of fish (Ringø et al., 2011; Saurabh and Sahoo, 2008).

Respiratory burst is an antimicrobial response from the immunological repertoire of phagocytes (i.e., neutrophils, macrophages, monocytes, dendritic cells, and B cells). During phagocytosis or during the recognition of PAMPs by the phagocyte's receptors, a heightened consumption of molecular oxygen takes place, and O_2 is reduced by membrane-associated NADPH oxidase, hence generating superoxide anions (O_2^-), which can be further converted to the bactericidal hydrogen peroxide (H_2O_2) by superoxide dismutase (Belosevic et al., 2009; Chung and Secombes, 1988). Reactive nitrogen species are also an asset of the antimicrobial responses of phagocytes, where N_2 is reduced by the enzyme inducible nitric oxide synthase (iNOS). The conversion of arginine to citrulline is catalyzed by iNOS releasing freely diffusible nitric oxide gas, which can kill bacterial pathogens (Abbas et al., 2014).

Natural antibodies, or natural polyreactive antibodies, are produced in a T cell–independent manner, by the B1 subset of B cells in mammals. They have a low-binding affinity and play a role in early host defense, by opsonizing pathogens and activating the classical pathway of the complement system (Scapigliati et al., 2018). Natural antibodies have the ability to recognize evolutionarily fixed epitopes present in foreign antigens and to recognize autoantigens such as single-stranded DNA, thyroglobulin, and heat shock proteins (Holodick et al., 2017). They mainly confine the infection to enhance further specific antibody response and take part in homeostasis by clearing cell debris and linking the responses of the innate and adaptive immune systems (Foey and Picchietti, 2014).

The relatively new concept of "trained immunity" postulates that an initial stimulation of the innate immune system may confer increased nonspecific resistance to homologous or heterologous pathogens (Rojo-Cebreros et al., 2018). This implies that by priming the innate immune system with PAMPs or an immunostimulant, the monocytes and macrophages will be epigenetically reprogrammed, leading to an increase in cytokine production and a change in the cellular metabolic pathway from oxidative phosphorylation to aerobic glycolysis (Rojo-Cebreros et al., 2018). These modifications will confer the host with ability to elicit a heightened immune response to a secondary infection.

The characteristics proposed for a trained innate immune response are as follows: they are independent of B and T cell adaptive responses and increase resistance of the host to infections, but in a less specific manner, thus providing a cross-protection against other infections. The heightened activation state of trained immunity will have mechanisms involving macrophages and natural killer cells, with improved recognition of pathogens by PRRs and enhanced protective inflammatory response (Netea et al., 2011). Studies using mice and zebrafish with knocked-out genes for the adaptive mechanisms (recombination activating gene, RAG) of B and T cells presented evidence of the trained innate immunity in which animals receiving a sublethal exposure to a pathogen will significantly enhance protection to a lethal exposure (Hohn and Petrie-Hanson, 2012; Quintin et al., 2012). The trained innate immunity concept has allowed investigators dealing with aquacultured organisms the possibility of better elucidating the protective mechanisms of immunostimulants against subsequent infections. More detailed reviews about the subject have been recently published (Petit and Wiegertjes, 2016; Rojo-Cebreros et al., 2018; Zhang et al., 2019).

Teleost fish are the first vertebrates in the evolutionary tree to present a developed adaptive immune system. The key elements in the evolution of the adaptive response are the development of B and T lymphocytes, thymus, the RAG enzymes (which allow the rearrangement of the genes necessary for the diverse repertoire of B and T cell receptors), and the major histocompatibility complex (MHC) molecules associated with specific immune responses and memory (Secombes and Wang, 2012). The remarkable features of the adaptive immune system are specificity against particular epitopes and memory, which provides a long-lasting ability to respond more vigorously to repeated exposures to the same pathogen (Abbas et al., 2014). However, when compared with the prompt response of the innate immune system, the adaptive immune response of fish can take from several days to months to develop. The adaptive immune responses also can branch in two categories: a humoral immunity that is mediated by antibodies secreted in the bloodstream and mucosa by B cells; and cell-mediated immunity mediated by T cells.

Upon the direct recognition of a specific foreign antigen or by the presentation of antigens by antigen-presenting cells (APCs; e.g., macrophages), B cells are activated and stimulated by their receptors and antibodies are secreted to target and bind the specific pathogen that stimulated their production. Antibodies can inactivate viruses, and microbial toxins, and also opsonize pathogens for phagocytes, and for the protein of the classical complement cascade. Unlike mammals, teleost fish have a more primitive adaptive immune response, and only three classes of antibodies have been described for fish: IgM, IgD, and IgT/IgZ. Moreover, fish lack the ability of class switch recombination, meaning that the expression of IgM isotype would block the generation of IgT/IgZ transcripts and vice versa (Mashoof and Criscitiello, 2016; Parra et al., 2015). In most teleost, IgM is mainly expressed as a tetramer. However, IgM monomers also have been reported for different fish species and are the most predominant Ig present in serum and to a lesser extent in the mucosa. IgT/IgZ has been associated with mucosal immunity and is not present in all fish species, and also can be found as tetramers

(Mashoof and Criscitiello, 2016). Teleost fish immunoglobulins are mainly secreted by plasmablasts and plasma cells, which are mainly located within the head kidney (Parra et al., 2015).

An antigen-specific cell-mediated immune response by T cells is activated by the interaction between the T cell receptor (TCR) and the specific peptide presented by the MHCs. T cells can be classified into two populations: cytotoxic T cells, which express the coreceptor $CD8^+$, and will interact with the host cell MHC I, and helper T cells, which express the coreceptor $CD4^+$, and interact with MHC II presented by APCs (Nakanishi et al., 2015). Cytotoxic T cells are an essential asset against injured cells, and phagocytes and other cells infected with bacteria and viruses, by ultimately inducing them to undergo apoptosis, and making the pathogens accessible to circulating antibodies (Smith et al., 2019). Upon recognition of the MHC loaded with specific epitope by APCs, helper T cells can release cytokines and assist B cells and macrophages, by helping B cells to make capable antibodies and to activate the antimicrobial activities of macrophages. T cells usually develop and mature in the thymus and, in teleost fish, are located near the gill cavity (Nakanishi et al., 2015), but in sea bass, T cells were detected developing in the gut-associated lymphoid tissue (GALT) (Smith et al., 2019).

To successfully infect fish, pathogenic organisms must first overcome the physical barriers such as mucus, scales, and the epithelium of the skin, gills, or gut tissues (Castro and Tafalla, 2015). If they manage to surpass these barriers, they must thrive in the mucosal layer, where they will be subjected to antimicrobial factors present in the mucus (e.g., lectins, lysozyme, complement proteins, AMPs, and antibodies) (Castro and Tafalla, 2015). For the context of this chapter, more focus will be given to the infiltration of pathogens through the gut and the implication in the GALT. In the stomach, chemical barriers such as low pH, digestive enzymes, bile salts, and fatty acids also can serve to prevent infection by their microbicidal or bacteriostatic properties (Murphy and Weaver, 2016). When reaching the intestinal tissue, pathogenic microorganisms, for instance, must outcompete the commensal microbiota, colonize the intestinal mucus, and breach the epithelial brush border (Bøgwald and Dalmo, 2014). Several bacterial species that afflict farmed aquatic species (e.g., *Aeromonas hydrophila, Edwardsiella ictaluri, Edwardsiella tarda, Streptococcus agalactiae, Streptococcus iniae, Vibrio anguillarum, Vibrio salmonicida, Yersinia ruckeri*, etc.) have been classified as enteric pathogens, i.e., using the intestinal tissues as their main portal of entry into the host, or to potentially have their pathogenesis in the intestinal tract (Bøgwald and Dalmo, 2014).

The general tendency over the past decade of reducing the inclusion of marine ingredients in aquafeeds with increasing levels of alternative plant protein ingredients along with their antinutritional factors may impair the immunological competency of the fish intestine (Trushenski, 2015). Several studies have reported chronic intestinal inflammation (enteritis), shortening of microvilli height, swelling of lamina propria, increasing number of goblet cells due to infiltration/recruitment of granulocytes, loss of mucosal fold structure, and disruption of intestinal tissue in response to feeding high levels of plant protein ingredients to fish, with more pronounced effects in carnivorous species (Rombout et al., 2011). The replacement of marine ingredients by rendered animal ingredients or plant coproducts can also shift the intestinal microbiota of the fish, which may disturb the intestinal homeostasis. These disorders can directly facilitate the enteric pathogen translocation into the gut epithelia. Other factors that can increase the susceptibility of fish to infections are the numerous stressors that can be imposed on them by rearing conditions during various phases of aquacultural production. High stocking densities, water quality fluctuations, and handling and transport may inflict acute and/or chronic stress, which can ultimately lead to immunosuppression. Furthermore, once a cultured organism becomes infected, the pathogen can quickly spread by horizontal transmission, with the culture water serving as the transmission medium.

The use of feed additives to ameliorate and/or improve fish health can be an important asset to prevent losses and increase disease resistance. They may do so by either providing nutrients that play roles during metabolism of the immune responses, by manipulating the intestinal microbiota, by direct immunostimulation, or by training the innate immune system of the fish prior to a pathogenic insult. The diverse types of feed additives, which have been identified to improve disease resistance in aquaculture and their prospective benefits, are discussed in the following.

11.3 Dietary components influencing fish health

11.3.1 Nutrients

Fish, like other animals, require a dietary supply of at least 40 different nutrients to maintain normal growth, health, and various physiological functions. Some nutrients have more direct effects of fish health than others; therefore, those nutrients demonstrated to influence various aspects of fish health are reviewed in the following section.

11.3.1.1 Amino acids

Amino acids are obtained from the diet or synthesized by organisms to produce proteins with various unique characteristics. Numerous biochemicals required for normal bodily functions are proteins such as enzymes, hormones, immunoglobulins, and structural tissues including skeletal muscle. The distinctive characteristics of various proteins are determined by the number and composition of amino acids, which are genetically controlled. Fish, like other animals, synthesize body proteins from amino acids that are provided in the diet as well as some amino acids that can be synthesized in the body from other precursors. Those that must be provided in the diet due to lack of endogenous synthesis capabilities are called "essential" or "indispensable" amino acids. There are typically 10 indispensable amino acids that have been established for fish as well as many other terrestrial animals. There are also 10 "nonessential" or "dispensable" amino acids that are classified as such because they can be synthesized from other precursors by the body. These dispensable amino acids also may be found in dietary protein and used for synthesizing body proteins by the fish. However, the indispensable amino acids are most critical to provide in the diet because a deficiency of any one can limit protein synthesis, which is often manifested as reduced weight gain as well as other specific deficiency signs including increased susceptibility to disease.

In addition to the immunoglobulins and various enzymes involved in immunological responses, some individual amino acids have been shown to have immune-potentiating effects in various animals including some aquatic species (Li et al., 2009). Arginine, in particular, is one such amino acid, which has been shown to influence various immunological responses due to its role as a precursor of nitric oxide (Li et al., 2009; Pohlenz and Gatlin, 2014).

Glutamine is a dispensable amino acid, which has been shown to influence health aspects of various aquatic species due to its role as a source of energy for enterocytes (Pohlenz and Gatlin, 2014).

11.3.1.2 Minerals

The minerals consist of inorganic elements that are required in the body for various purposes. It has been established that fish require the same minerals as terrestrial animals for tissue formation and other metabolic functions, including osmoregulation (NRC, 2011). However, dissolved minerals in the aquatic environment may contribute to satisfying the metabolic requirements of fish and interact with dietary requirements.

Minerals are typically classified as either macro- or microminerals, based on the quantities required in the diet and stored in the body. Macrominerals consist of calcium, phosphorus, magnesium, chloride, sodium, potassium, and sulfur. Dietary deficiencies of most macrominerals such as calcium generally have been difficult to produce with fish species because of the uptake of waterborne ions by the gills. Phosphorus is not only a major constituent of hard tissues such as bone and scales, but it is also present in various biochemicals. Impaired growth and feed efficiency as well as reduced tissue mineralization and impaired skeletal formation in juvenile fish are common deficiency signs exhibited by various fish species when fed diets deficient in phosphorus (NRC, 2011). However, deficiencies of phosphorus can be readily produced, and supplementation of phosphorus in fish diets is usually most critical because its presence in the water and utilization by fish is limited. Some of the dietary phosphorus contributed by plant feedstuffs is in the form of phytate, which is not appreciably digested by monogastric animals including various aquatic species that produce negligible amounts of the phytase enzyme (Kumar et al., 2012). Due to the fact that excessive phosphorus in the aquatic environment may cause eutrophication, considerable efforts have been made to increase phosphorus availability in aquafeeds by either adding microbial phytase or supplementing more available forms of inorganic phosphorus. Once a species' requirement for available phosphorus is satisfied, additional supplementation to the diet typically will not confer additional health benefits. Thus, considerable advancements have been made in more closely matching the amount of available phosphorus in the diet to the metabolic requirements of various aquatic species so that excessive levels of phosphorus are not excreted and contribute to eutrophication (NRC, 2011).

The microminerals (also known as trace minerals) include cobalt, chromium, copper, iodine, iron, manganese, selenium, and zinc. These various trace minerals have diverse metabolic functions in fish similar to those of terrestrial animals, and quantitative dietary requirements for some fish species have been established (NRC, 2011). In contrast to phosphorus, impaired growth and feed efficiency are not readily induced with deficiencies of microminerals but may occur after an extended period of feeding semipurified diets deficient in one or more of these trace minerals (NRC, 2011). Of all the microminerals, copper, iron, manganese, selenium, and zinc have been demonstrated in some fish species to be most important to supplement in diets due to low levels in practical feedstuffs and/or interactions with other dietary components that may reduce bioavailability. Although supplementation of practical diets with other microminerals has not been shown to be essential in most instances, a relatively inexpensive trace mineral premix has traditionally been added to most nutritionally complete diets to ensure adequacy. Such premixes commonly consist of inorganic salts of copper, iron, manganese, selenium, and zinc. However, it is well established that less soluble forms of inorganic trace mineral salts may have limited bioavailability compared with more soluble forms, which also may be more susceptible to leaching in the aquatic environment. Various types of organic minerals have been developed by combining certain minerals with organic chelates such as amino acids or concentrating specific minerals such as copper or selenium in yeast (NRC, 2011). The bioavailability of chelated organic minerals is generally greater than that of inorganic minerals because they are not susceptible to interactions with natural antinutritional factors such as phytic acid (phytate), which is found in many plant feedstuffs and can bind inorganic mineral salts making them unavailable to the organism being fed. Providing adequate amounts of trace minerals to meet metabolic requirements of aquatic species can be achieved with premixes regardless of whether they are composed of inorganic or organic constituents.

Numerous macro- and microminerals have been evaluated in terms of immunological responses and disease resistance of various aquatic species fed deficient or levels above established requirements (reviewed by Webster and Lim, 2015). Investigations have included magnesium, phosphorus, as well as chromium, copper, iron, iodine, selenium, and zinc. Dietary deficiency of most nutrients including minerals impairs various physiological functions, which may include immunological responses (Webster and Lim, 2015). However, effects of supplementing various minerals above minimum requirement levels on immune responses and disease resistance have been limited and often inconsistent (Webster and Lim, 2015).

11.3.1.3 Vitamins

There are 15 vitamins established as being essential for terrestrial animals as well as for several fish species that have been examined to date (Halver, 2002). These nutrients are characterized as being organic compounds distinct from carbohydrates, protein, and lipid that are required in relatively small concentrations to support specific structural or metabolic functions. Vitamins are divided into two groups based on solubility, with the fat-soluble vitamins including vitamin A (retinol), vitamin D (cholecalciferol), vitamin E (alpha-tocopherol), and vitamin K. These fat-soluble vitamins are metabolized and deposited in association with body lipids such that extended periods without a dietary supply are needed before overt deficiency signs are manifested. In contrast, the water-soluble vitamins including ascorbic acid (vitamin C), biotin, choline, folic acid, inositol, niacin, pantothenic acid, pyridoxine, riboflavin, thiamin, and vitamin B_{12} are not stored in appreciable amounts in the body. As such, deficiency signs will be generally manifested within weeks in young, rapidly growing fish fed a diet deficient in one or more of these vitamins. Most of these water-soluble vitamins serve as components of coenzymes that have specific metabolic functions. Detailed information about the functions of these vitamins and their quantitative dietary requirements have been established for numerous cultured fish (NRC, 2011).

Based on the quantified dietary requirements for these various vitamins, premixes have been developed for inclusion in prepared diets to provide adequate levels of each vitamin independent of levels in dietary ingredients and also provide a margin of safety for losses associated with processing and storage. The stability of vitamins during feed manufacture and storage has been improved over the years with protective coatings and/or chemical modifications. This is particularly evident in the development of various stabilized forms of the very labile ascorbic acid (NRC, 2011). As such, vitamin deficiencies are rarely observed in commercial production.

Dietary deficiencies of any water-soluble or fat-soluble vitamin, if allowed to progress, will be manifested in various aquatic species as reduced growth, increased susceptibility to disease as well as specific deficiency signs (NRC, 2011). Several vitamins including vitamin A, vitamin D, thiamin, pyrodoxine, folic acid, pantothenic acid, and myo-inositol have been shown to affect immune responses and disease resistance of fish. However, dietary supplementation of vitamin C and vitamin E above minimum requirement levels has been most widely studied and consistently shown to enhance various immunological responses and disease resistance of numerous aquatic species (Izquierdo and Betancor, 2015; Trichet et al., 2015). The beneficial effects of those two vitamins have most often been attributed to their antioxidant properties, which can work synergistically. Supplementation of vitamin C and vitamin E at levels above that required for normal growth has most consistently shown positive effects on immunity and disease resistance of aquatic species under conditions in which the immune system is challenged (Trichet et al., 2015). This includes stressful conditions encountered due to changing environmental conditions as well as those associated with common aquacultural practices such as handling, grading, and disease treatment.

11.3.2 Nonnutritive supplements

11.3.2.1 Acidifiers

Acidifiers are a group of compounds consisting of weak acids and their salts, which may be added to the diet to reduce its pH and may also influence the pH of the GI tract as well as the composition of bacterial species present there. Some acidifiers such as propionic acid traditionally have been used as mold inhibitors. These compounds have been used in the diets of livestock and poultry for quite some time and more recently have been evaluated with various aquatic species (Lim et al., 2015). The number of studies conducted on various aquatic species has grown considerably in recent years, although they have yielded some variable results.

Most acidifiers are organic carboxylic acids varying in carbon number from 1 to 6, along with their salts of calcium, magnesium, potassium, or sodium. These compounds of four or fewer carbons are water soluble, but solubility decreases considerably for acids with six or more carbons (Lim et al., 2015). The low pH of these compounds results in bacteriostatic and bactericidal properties when included in diets at concentrations ranging from 1% to 3% by weight. They also may lower the pH of the stomach and intestine as well as limit the growth of gram-negative bacteria. These bacteria are particularly susceptible to organic acids because the acids can pass through the cell membrane, and once in the cytosol, they dissociate and decrease the pH as well as deplete the bacteria of cellular energy in an attempt to maintain homeostasis. In contrast to the gram-negative bacteria that exhibit reduced growth and mortality as pH decreases below 5, lactic acid bacteria are either unaffected or enhanced at low pH.

Citric acid and formic acid have been shown in various species to reduce the pH of the diet and GI tract, which typically results in increased bioavailability of dietary minerals, especially phosphorus. Other acids such as sodium formate, sodium butyrate, or potassium diformate (KDF) have been reported to have more variable effects on reducing GI tract pH and influencing the digestive process. Sodium lactate is another acid, which has been shown in several studies to enhance the growth of Arctic char but not Atlantic salmon (summarized by Lim et al., 2015).

A majority of the studies that have evaluated organic acids and their salts in the diet on various aquatic species have focused on salmonids and tilapia. However, in one study conducted with red drum, the pH of experimental diets was lowered by the addition of 0.75% citric acid (CA) (pH 5.63), 1.5% CA (pH 5.36), and 1.5% potassium diformate (KDF) (pH 5.79) when compared with the basal diet (pH 6.15) and also reduced the pH values of stomach contents, which were correlated with the pH of the experimental diets. Additionally, supplementation of the organic acids at 1.5% of diet tended to improve growth performance of juvenile red drum, although only weight gain in the 1.5% CA treatment (1547%) was significantly higher than the basal treatment (1357%). This enhanced performance appeared to be due in part to an increment in the activity of digestive enzymes pepsin, trypsin, and lipase; leucine-aminopeptidase and phosphatases were higher with the inclusion of organic acids in the diet (Castillo et al., 2014). Results from various other studies with salmonids and tilapia have not always been consistent, but some have reported improved growth performance, nutrient digestibility, and alteration of microbiota in the GI tract and increased disease resistance. Therefore, due to the positive effects that have been observed in some studies, additional research with various acidifiers may be warranted to optimize their inclusion in the diet of various aquatic species.

11.3.2.2 Beta glucans

β-glucans (BGs) are all naturally occurring polysaccharides having glucose as their building blocks linked by beta-linkages (Meena et al., 2013; Raa, 2015). BG molecules are widespread in nature and can be found in plants, yeasts, fungi, algae, protozoa, and bacteria (Dalmo and Bogwald, 2008; Volman et al., 2008; Wismar et al., 2010), but the majority of these molecules (e.g., cellulose, starch) do not directly stimulate the immune system. The interaction and recognition by the PRRs from the fish leukocytes and other cells will depend on the conformation of the BG molecules, such as degrees of branching, polymer charge, molecular mass, and solubility. The early detection of microorganism pathogens is granted by the host PRRs when recognizing PAMPs, and BGs from fungi and bacteria are examples of these patterns (Ringø et al., 2010, 2011). The interaction between the PAMPs and the leukocyte PRRs will trigger a downstream intracellular signaling cascade, leading to the activation of transcription factors (NFAT and NF-κβ), which will induce the production of inflammatory cytokines and respiratory burst and ultimately affect the immune responses of the host (Batbayar et al., 2012; Meena et al., 2013; Petit and Wiegertjes, 2016; Volman et al., 2008).

The extraction process and methods of preparing BGs as supplements for animal feeds will depend on the BG source and the targeted molecular structure, and it can be a multistep process (Zhu et al., 2016). For instance, several protocols patented for BG extraction can rely on hot water and detergent washes, acid and alkaline treatments, enzymatic incubation, and even ultrasound- and microwave-assisted extractions (Zhu et al., 2016). These laborious steps can affect the integrity and purity of BGs, which may influence their immunostimulatory properties (Raa, 2015; Sletmoen and Stokke, 2008). The majority of studies conducted using aquatic organisms as animal models focused on testing BGs either extracted from brewer's yeast (*Saccharomyces cerevisiae*) or using the dried yeast itself as a functional ingredient

(Gannan, 2015). However, providing the brewer's yeast per se, or low-purity BGs extracted from the yeast, not only supplies branched β-1,3/1,6 glucan molecules, but also some nutrients and other compounds, such as mannose oligosaccharides and nucleotides, which can interfere with the immunological responses of the host (Ferreira et al., 2010; Ringø et al., 2011).

The immunostimulation and resistance against pathogenic bacteria conferred by dietary BGs have been evaluated with aquatic farmed animals for quite some time, but the underlying mechanisms for this protection still remain obscure. It has been hypothesized that antigen sampling by intestinal cells could elicit the intestinal inflammation and cytokine production, consequently increasing resistance to infections (Volman et al., 2008). Another line of thought is that the relatively new concept of trained immunity plays a role, whereby the primary stimulus (i.e., the dietary BG stimulus) will prime the innate immunity and lead to epigenetic changes, providing a heightened and long-lasting secondary innate immune response (i.e., when the host is challenged by the pathogen) (Petit and Wiegertjes, 2016; Zhang et al., 2019).

Several studies reported other beneficial effects of supplementing BGs in the diet of teleost fish. Beside the enhancement of immune responses and disease resistance, improved growth performance, modulation of cortisol secretion, and wound healing also have been observed. Comprehensive reviews about the use of BGs in aquaculture have been published in peer-reviewed scientific journals, covering the various benefits of BGs in fish diets (Meena et al., 2013; Petit and Wiegertjes, 2016; Ringø et al., 2011). Studies utilizing BG as a feed additive published since the latest review are compiled in Table 11.2.

One interesting aspect that has been little explored is the impact of dietary BGs in fish diets on the composition of the gut microbiota. It is hypothesized that the supplementation of BGs to fish diets can affect the microbial community by either serving as a substrate for β-glucanase-producing bacteria (Lam and Cheung, 2013) or indirectly reducing the pathogenic microorganisms by the immune enhanced cells in the GI tissues (Dalmo and Bøgwald, 2008). A study using larvae of turbot (*Scophthalmus maximus*) fed rotifers and *Artemia nauplii* enriched with MacroGard reported not only a higher survivability of these fish but also a faster development and maturity of the microbial community of the gut at 24 days posthatch when compared with the control treatment (Miest et al., 2016). Alteration of the gut microbiome also was observed for common carp (*Cyprinus carpio*) fed the same commercial yeast-derived BG, in four independent studies (Harris et al., 2019; Jung-Schroers et al., 2016, 2018; Kühlwein et al., 2013). In contrast to these findings, a study conducted in our laboratory, using the same molecular technique as the aforementioned studies, did not find any differences in the transient microbial community of the red drum (*Sciaenops ocellatus*) intestine, when they were fed diets supplemented with *Euglena gracilis* as a source of BG (Yamamoto et al., 2018). Contrasting these studies, the BG delivery method, the physiology of the fish tested, and/or the nature of the BG may be factors influencing the ability to alter the intestinal microbiota. The bioencapsulation of BG with live feed such as rotifers and the physiology of stomachless fish such as common carp could minimize the deleterious effects caused by acid digestion on the integrity of BG molecules, which could ultimately influence its immunostimulatory properties in the intestine (Kudrenko et al., 2009). Future studies should further investigate the impact of BGs on the gut microbiota using more refined techniques, such as next-generation sequencing, to better portray the impact of dietary BGs on the intestinal bacterial population.

TABLE 11.2 Compilation of recent studies on the effects of oral administration of β-glucan and its physiological implications.

Species used	Type of β-glucan	Dosage	Length of administration	Route of administration	Mechanisms of action/results	References
Red drum (*Sciaenops ocellatus*)	Algamune (dried *Euglena gracilis*)	100 and 200 mg/kg	24 h	In vitro	↑ reactive oxygen species and bactericidal activity in vitro	Yamamoto et al. (2018a)
			8 weeks	Oral	↑ alternative complement pathway → no change in the intestinal microbiota	
Hybrid striped bass (*Morone chrysops* × *Morone saxatilis*)	Paramylon (*E. gracilis*)	50 and 100 mg /g	4 and 8 weeks	Oral	↑ respiratory burst and total immunoglobulins	Yamamoto et al. (2020a)
			7 days	Injection	↑ hemolytic and lysozyme activity	
	Zymosan (*Saccharomyces cerevisiae*)	50 and 100 mg/kg	4 and 8 weeks	Oral	↑ respiratory burst	
			7 days	Injection	↑ respiratory burst and antiprotease activity	
	Paramylon + vitamin C	100 mg/kg and 500 mg/kg of vitamin C	18 h	In vitro	↑ respiratory burst, lysozyme activity	Yamamoto et al. (2020b)
			8 weeks	Oral	↑ respiratory burst, intestinal acid phosphatase, and immunoglobulins	
Nile tilapia (*Oreochromis niloticus*)	Algamune (dried *E. gracilis*)	50 and 100 mg/kg	24 h	in vitro	↑ respiratory burst	Yamamoto et al. (2018b)
			3 weeks	Oral	↑ respiratory burst	
	β-Glucan (daigon do)	500 mg/kg	60 days	Oral	↑ production performance, villi length in the anterior intestine, plasma total protein, lysozyme activity, gene expression IFN-G, TNF-α, IL-1β ↓ feed conversion ratio, plasma cortisol	Dawood et al. (2020a)

(*Continued*)

TABLE 11.2 Compilation of recent studies on the effects of oral administration of β-glucan and its physiological implications.—cont'd

Species used	Type of β-glucan	Dosage	Length of administration	Route of administration	Mechanisms of action/results	References
		500 mg/kg	15 and 30 days	Oral	↑ plasma lysozyme activity, phagocytic index, gene expression of CAT, GPx, IL-8 and IL-1β when challenged with deltamethrin ↓ cortisol and glucose levels in plasma, gene expression of HSP70 and CASP3, when challenged with deltamethrin	Dawood et al. (2020b)
Persian sturgeon (*Acipenser persicus*)	Macrogard	100, 200, and 300 mg/kg	6 weeks	Oral	↑ production performance, lysozyme, and hemolytic activity	Aramli et al. (2015)
Red sea bream (*Pagrus major*)	β-Glucan (daigon do)	250, 500, and 1000 mg/kg	8 weeks	Oral	↑ production performance, bactericidal and peroxidase activity, plasma protein and hematocrit, and mucus production, ↓ glutamic oxaloacetic transaminase	Dawood et al. (2015)
Pompano (*Trachinotus olivaceus*)	Macrogard	500, 1000, 2000 and 4000 mg/kg	8 weeks	Oral	↑ production performance, red blood cells count, leukocyte count, resistance against stress challenge, ↓ intestinal *Vibrio* counts	Huu et al. (2016)
Zebra fish (*Danio rerio*)	Paramylon (*E. gracilis*)	1250 mg/kg	1 week	Oral from parental to eggs and embryos	↑ gene expression of C3 molecules and factor B	Jiang et al. (2016)
	Oat nanoscale	500 μg/mL	5 days	Oral	↑ survival after bacterial challenge, gene expression of TNF-α, IL-1β, β-defensin, lysozyme, IL-10, IL-12, C-rel	Udayangani et al. (2017)

TABLE 11.2 Compilation of recent studies on the effects of oral administration of β-glucan and its physiological implications.—cont'd

Species used	Type of β-glucan	Dosage	Length of administration	Route of administration	Mechanisms of action/results	References
Turbot (*Scophthalmus maximus*)	Macrogard	500 mg/L	24 days	Oral through live feed (rotifers and artemia)	↑ gene expression of trypsin and chymotrypsin, richness of larval microbiota, C3. ↓ HSP 70, TNF-α, IL-1β	Miest et al. (2016)
Matrinxa (*Brycon amazonicus*)	β-glucan (*S. cerevisiae*)	100 mg/kg	15 days	Oral	↑ alternative complement pathway, respiratory burst postinfection, leukocyte count poststress, ↓ lysozyme activity poststress	Montoya et al. (2017)
Pacu (*Piaractus mesopotamicus*)	Macrogard	100 mg/kg	15 days	Oral	↑ leukocyte count, cortisol production after stress challenge, ↓ cortisol production after bacterial challenge, alternative complement pathway after bacterial challenge	de Mello et al. (2019)
	Macrogard™	500 mg/kg	0, 3, 6, and 24 h, and 7 days	Oral	↑ leukocyte respiratory activity after stress and glucose levels, ↓ number of leukocytes 1 week after challenge	Sabioni et al. (2020)
Rainbow trout (*Oncorhynchus mykiss*)	Macrogard	100, 200, and 500 mg/kg	15 and 30 days	Oral	↑ gene expression of TNF-α, IL-1β, IL-10, COX-2, TGF-β, MC2R, and HSP70	Douxfils et al. (2017)
	β-glucan (*S. cerevisiae*)	50, 100, and 200 mg/kg	6 weeks	Oral	↑ production performance, survival after bacterial challenge, superoxide dismutase, peroxidase, catalase activities, heat shock protein 70. ↓ glutamic oxalacetic transaminase and pyruvic transaminase levels	Ji et al. (2017)

(Continued)

TABLE 11.2 Compilation of recent studies on the effects of oral administration of β-glucan and its physiological implications.—cont'd

Species used	Type of β-glucan	Dosage	Length of administration	Route of administration	Mechanisms of action/results	References
Common carp (*Cyprinus carpio*)	Macrogard	1000 mg/kg	0, 7, 21, 35, and 49 days	Oral	↓ gene expression of IL-1β and intestinal microbiota richness	Harris et al. (2019)
Atlantic salmon (*Salmo salar*)	Macrogard	1000 mg/kg and 100 mg/kg of astaxanthin	0, 30, and 37 days	Oral	↑ gene expression for IFN-α1 and mx under acute hypoxia. ↑ gene expression of IL-12 and IFN-γ with astaxanthin. ↑ gene expression of IL-12 and CD4 after vaccination	Rodriguez (2016)
Silver catfish (*Rhamdia quelen*)	Macrogard	0.1 and 0.5 mg/L	1 h	Bath	↓ blood cortisol levels. ↑ wound healing	Voloski et al. (2019)

Another caveat in the BGs investigations is the feed regimen imposed on the animals. During experimental evaluations, the animals are usually tested in short-term trials, where the length of exposure can range from days to a couple of months (Table 11.1). Studies sampling at different time points usually have convoluted results, with high variability and complicated statistical interpretation. It is still questioned if the prolonged exposure to BGs can lead to an overstimulation or exhaustion of the immune system, or if the host will be susceptible to develop tolerance to the dietary BG.

11.3.2.3 Essential oils and herbal extracts

A wide variety of chemical compounds derived from plants, herein referred to as phytochemicals, have been evaluated as supplements with various aquatic species. These include a wide variety of plants including herbs, spices, seaweeds, and traditional Chinese medicines. Many of these compounds have been derived from plants historically known for their medicinal value, and most commonly have been extracted with solvents such as alcohols and administered by injection or more conveniently in the diet (Van Hai, 2015; Jeney et al., 2015). The chemical compositions of the active compounds in such extracts are as diverse as the plants from which they are derived and include various alkaloids, terpenes, phenolic compounds, and acids and amides (Jeney et al., 2015). The lack of purity in some extracts has made it difficult to identify all the bioactive compounds and their specific influences on the target organism. However, some of these purified compounds as well as crude extracts have been noted in some studies to promote growth by possibly stimulating appetite or providing specific nutrients. Others have been shown to reduce the prevalence of microbes as well as prevent or control disease (reviewed by Jeney et al., 2015).

One particular group of phytochemicals, which has received heightened attention over the past decade, is referred to as essential oils (Sutilli et al., 2018). These compounds are naturally produced by various aromatic plants as secondary metabolites to protect the plants from bacterial, viral, and fungal pathogens as well as repel higher organisms that might consume the plant. They are characterized by a very strong odor and being lipid soluble due to being derived from terpenes. Chemically these essential oils are generally classified into two major groups including hydrocarbon terpenes/terpenoids and a second group of oxygenated derivatives of hydrocarbon terpenes including alcohols, aldehydes, ketones, acids, phenols, ethers, and esters. The composition of essential oils derived from a given plant species may vary considerably depending on the plant's stage of development, its culture conditions including environment, and the method of extraction. Essential oils in a given plant typically consist of several different compounds in variable concentrations but typically make up less than 1% of wet plant weight (Carlson and Hammer, 2011). Many different essential oils have been derived from a variety of plants including citrus, cinnamon, and oregano. The recent review by Sutilli et al. (2018) provides a detailed summary of several different kinds of essential oils, which has been evaluated with aquatic species and the numerous responses observed when exposed to those compounds. In several studies, positive effects of the essential oil on growth performance, microbiota of the GI tract, and immune responses have been documented in some of the more prominent warm-water and cold-water fish species (Sutilli et al., 2018). In addition, Souza et al. (2019) recently reviewed potential applications of essential oils as sedatives/anesthetics, as well as their potential use in reducing stress responses of various fish produced in aquaculture. Both Sutilli et al. (2018) and Souza et al. (2019) highlighted the limited stability of essential oils and various means to maintain their integrity when incorporated into aquafeeds. Refinements in extraction procedures and various means of encapsulating essential oils to improve their stability are currently in development.

11.3.2.4 Probiotics, prebiotics and synbiotics

These additives represent a diverse group of components which, are intended to favorably affect the GI tract and thus improve the health and well-being of the cultured organism. As with other diet additives reviewed in this chapter, probiotics, prebiotics, and synbiotics were originally applied in livestock and human nutrition, and thus have been more extensively evaluated in those organisms, but have received heightened attention with various aquatic species in recent years. This is especially due to their potential influence on immunity and disease resistance. Before discussing specific aspects of these various supplements, the distinguishing characteristics of each are described in the following.

Probiotics—The term probiotic was introduced by Parker (1974) as "organisms and substances which contribute to intestinal microbial balance." Fuller (1989) revised the definition as "live microbial feed supplement which beneficially affects the host animal by improving its intestinal microbial balance." Subsequently, Moriarty (1998) proposed that the definition of probiotics be extended to microbial "water additives." Administration of probiotics to the water has been associated with improved water quality such as reductions in the concentrations of nitrogen and phosphorus (Wang et al., 2005). In addition to affecting water quality, probiotics administered to the water or diet also may inhibit the growth of pathogenic microorganisms, contribute digestive enzymes to increase feed utilization, provide other growth-promoting factors, and stimulate the immune response of the host organism.

Recognized probiotics that may influence fish immunity, disease resistance, and other performance indices include gram-positive rods of the genus *Bacillus* and various lactic acid bacteria (*Lactobacillus, Lactococcus, Carnobacterium, Pediococcus, Enterococcus*, and *Streptococcus*) (Kiron, 2015). Viability of the bacteria during storage and processing is critical for probiotics to confer their beneficial effects to the host organism; however, the applications of dead cells, freeze-dried cells, or cell-free extracts or spores have all shown some degree of success (Merrifield et al., 2010). The logistical constraint of culturing live microorganisms under controlled laboratory conditions prior to application on the feed has constrained the use of probiotics at aquaculture facilities, and thus, administration of lyophilized cells or spores may be more practical. Potential applications of probiotics in fish, shrimp, and molluskan aquaculture have been reviewed by Burr et al. (2005), Wang et al. (2008), and Kesarcodi-Watson et al. (2008) and most recently by Ringø et al. (2010).

Prebiotics—Prebiotics are defined as nondigestible food ingredients that beneficially affect the host by selectively stimulating the growth of and/or activating the metabolism of one or a limited number of health-promoting bacteria in the intestinal tract, thus improving the host's intestinal balance (Gibson and Roberfroid, 1995). The health-promoting bacteria most commonly augmented by prebiotics include those of the genus *Lactobacillus* and *Bifidobacter*, which tend to limit the presence of harmful bacteria. Examples of prebiotics include mannanoligosaccharides, lactose, galacto-gluco-mannans, and oligofructose and inulin (Gatlin, 2015). Many of these prebiotics are carbohydrates and primarily short-chain oligosaccharides consisting of 3–10 carbohydrate units, which are derived from various plants or cell wall components of yeast. Other commercial products shown to possess prebiotic properties include the yeast-based product GroBiotic®-A, which is a mixture of partially autolyzed brewer's yeast, dairy ingredient components, and dried fermentation products. The various prebiotic compounds are generally not altered by diet processing and require limited regulatory approval such that their incorporation into diets is much simpler than required for drugs or chemical therapeutic agents.

Synbiotics—This classification is given to components that include both live probiotic bacteria as well as prebiotic substances, which in some instances may interact synergistically with the probiotic and thus augment its effectiveness.

Information concerning the application of probiotics, prebiotics, and synbiotics in aquatic organisms has increased rapidly in recent years and has been summarized in several review articles (Song et al., 2014; Wang et al., 2017; Caipang et al., 2020; Van Doan et al., 2020) and book chapters (Gatlin, 2015; Kiron, 2015). In addition to describing the various kinds or prebiotics and probiotics evaluated with aquatic species, those reviews also have provided detailed information on their modes of action. Summarized in the following are some of major effects exerted by probiotics and prebiotics on aquatic organisms that have been documented to date. As would be expected, most of these effects are mediated by the GI tract of the host organism, which possess bacteria and may provide a defensive barrier and protection against the invasion of bacterial pathogens. Healthy gut microbiota have the ability to prevent pathogenic bacteria from colonizing the intestine and thus causing infection because the autochthonous bacteria of the GI tract present under normal conditions act to competitively exclude pathogens simply by their presence. By taking up space and resources along the mucosal lining of the GI tract, pathogenic bacteria are forced to continue in a transient state where the likelihood of damaging intestinal cells or causing infection is reduced. Autochthonous bacteria also have the capacity to produce antimicrobial substances, which add to their ability to fend off pathogens from attempting to colonize the GI tract. Probiotics help to directly accomplish a favorable balance by providing an increased number of desirable bacteria when ingested; whereas, prebiotics accomplish their goal more indirectly, by acting as a food source, preferentially, to the beneficial bacteria. It also has been shown that some pathogenic bacteria may become bound to certain prebiotics, as opposed to attaching to the mucosal lining of the GI tract and thus may be passed from the GI tract.

The microbiota of the GI tract also may influence a wide variety of metabolic processes via the microbiota stimulating epithelial proliferation and expression of numerous genes. Prominent among these are various physiological, biochemical, and immunological responses that must be maintained or enhanced to improve health status, stress responses, and disease resistance.

The first line of defense within the GI tract is the mucosa that separates the gut microbiota from direct contact with the epithelial cells of the GI tract. It is because of this direct contact with the mucus that the immune system of the GI tract, often referred to as gut-associated lymphoid tissue or GALT, has developed mechanisms to distinguish between potentially pathogenic bacteria and the normal, commensal autochthonous bacteria. Consequently, the GALT can determine whether to mount an attack or tolerate a specific bacteria's presence. In the event that potentially pathogenic bacteria are detected, the cellular and humoral mechanisms of the GALT activate the innate immune system and subsequently the adaptive immune system (via antibodies) to prevent bacteria from causing and/or spreading infection (Gomez and Balcázar, 2008). Components of the innate or nonspecific immune system have been described in detail previously in this chapter. Although greater insights into the effects of prebiotics and probiotics on nonspecific and adaptive immunity of aquatic species are continuing to be generated, additional research in this area is warranted to more fully characterize these responses and understand how they can protect the cultured organism from infectious agents. Disease resistance is one of the most important responses to the aquaculturist as it directly affects the production efficiency and profitability of the enterprise. Numerous studies have been conducted demonstrating that supplementation of prebiotics and probiotics may enhance the ability of various aquatic species to resist disease from numerous bacterial, viral, and protozoan pathogens and have been summarized in several review articles (Gatlin, 2015).

Another group of physiological responses that are influenced by the beneficial bacteria in the GI tract have been associated with improved digestion of dietary nutrients and energy by some fish species. The increased nutrient digestibility associated with prebiotic or probiotic supplementation may be due to the favored microbial community producing enzymes that are either lacking or occurring only at low levels in the host (reviewed by Gatlin, 2015). Other studies have shown that prebiotic supplementation to the diet may increase the absorptive area of the GI tract based on quantitative changes in histological measurements of the GI tract such as intestinal fold height, enterocyte height, and microvillus height (reviewed by Gatlin, 2015). Such changes also could contribute to increased nutrient absorption. Increased nutrient utilization of the cultured organism as well as enhanced metabolism associated with prebiotic and probiotic supplements also may result in increased weight gain and feed efficiency. Such improvements in weight gain and feed efficiency in response to prebiotic supplementation have been most readily observed when organisms are cultured under less than optimal environmental conditions or in the presence of pathogenic organisms. Additional insights concerning endocrine and molecular effects of prebiotic and probiotic supplementation are becoming more readily available as various molecular techniques are being commonly applied in research with aquatic species.

11.3.2.4.1 Practical application of prebiotics, probiotics, and synbiotics

As mentioned previously, maintaining the viability of probiotics during storage and processing is important for them to exert their beneficial effects to the host organism. Therefore, some logistical constraints may be encountered with the cultivation of live microorganisms in conjunction with manufacturing feeds. To ensure probiotic viability, its application to the feed typically must occur after extrusion so the probiotic organism is not exposed to excessive heat and pressure. Administration of probiotics in the form of lyophilized cells or spores may be less demanding. Feed manufacturing constraints are generally of less concern when dealing with prebiotics because they are not living organisms. Although the efficacy of several prebiotics has been shown when incorporated into extrusion-processed feeds, the potential chemical alteration of prebiotic compounds during feed manufacturing has not been studied to any appreciable extent and deserves further consideration.

Administration regimes for specific prebiotics and probiotics also have not been widely studied to date. Although these compounds have immunostimulating effects, it does not appear that long-term administration causes immunosuppression as noted with other potent immunostimulants. Therefore, these diet additives may be administered for extended periods. However, developing more refined administration protocols for individual prebiotics or probiotics should be investigated to optimize their effectiveness. For example, administering probiotics or prebiotics at prescribed times before the culture organism is exposed to a stressful event or at particular times of the year when pathogenic organisms are most prevalent may be developed to most efficiently derive the benefits of these compounds under particular culture regimes.

A number of probiotics and prebiotics are now commercially available and used to various extents. Although there are added costs associated with using these products, improved production efficiency and reduced disease incidence may offset such costs due to the tremendous expense associated with mortality in commercial aquaculture.

11.3.2.5 Nucleotides

Nucleotide supplementation of diets for fish produced in aquaculture has received heightened worldwide attention over the past two decades. Nucleotides are monomers of nucleic acids. The potential benefits of nucleotide supplementation were reported over two decades ago; however, nucleotides are not considered essential given that de novo synthesis and salvage pathways exist in animals and plants (Cosgrove, 1998; Gil, 2002). However, during extraordinary stressful events, sufficient amounts of nucleotides may not be synthesized (Devresse, 2000). It is known in aquacultural production that the cultured organism may undergo stress associated with physiological responses including rapid growth, reproduction, resisting disease, and recovering from injury. In addition, environmental changes, marginal water quality, and stress from confinement and handling are commonly encountered in aquaculture. These stressful conditions may require that the cultured organism be provided with additional nucleotides to be readily available for cell proliferation (Hoffmann, 2007). The essentiality of ribonucleotides is more apparent under these conditions. Also, de novo synthesis and salvage of nucleotides are metabolically costly processes; therefore, a supplemental source of exogenous nucleotides in the diet may enhance the functions of rapidly dividing tissues, principally when growth is rapid (Sanderson and He, 1994).

Exogenous nucleotides are commonly distributed in food especially those containing cellular elements and nucleoproteins such as animal tissues including organ meats and seafood (Kojima, 1974; Mateo et al., 2004). Dietary supplementation of nucleotides was not particularly relevant a few decades ago in aquaculture because most of the aquafeeds at that time had relatively high inclusion of FM, which is a rich source of nucleotides (Mateo et al., 2004). Nonetheless, FM is being increasingly replaced with alternative feedstuffs which have a lower concentration of nucleotides. In comparison with animal tissues, single-cell proteins have up to sevenfold the concentration of nucleic acids, and thus, products such as yeast provide a good source of nucleotides (Ingledew, 1999). Commercial brewer's yeast (*S. cerevisiae*) is inactive yeast remaining after the brewing process, which has been used widely for commercial production of supplemental nucleotides (Ferreira et al., 2010). One of the earliest works on dietary nucleotide supplementation in fish nutrition was reported by Ramadan et al. (1991, 1994) who demonstrated that supplementation of the commercial nucleotides Ascogen P could improve growth performance and survival of hybrid tilapia. A decade after this initial publication, Burrells et al. (2001) reported the inclusion of exogenous nucleotides in salmonid diets increased disease resistance to challenges from viral and rickettsial diseases, as well as ectoparasitic infestation.

Supplementation of nucleotides in fish diets has received considerable amounts of attention in recent years, and the current knowledge of using such additives and their effects on fish has been widely reviewed (Li and Gatlin, 2006). Some of the recent studies have focused on commercial nucleotide products such as AccelerAid and NuPro. Berto et al. (2016) demonstrated that inclusion up to 8% of NuPro in tilapia diets resulted in increased weight gain and feed efficiency. On the other hand, inclusion of 0.05%–0.4% of AccelerAid in tilapia diets did not have any significant effects on growth performance or immunological responses (Barros et al., 2015). A potential limitation of using these commercial products is that the concentrations and types of nucleotides present in such products are typically not disclosed due to trade secret. However, the efficacy of nucleotides is determined by the type and concentration of nucleotides incorporated in the diet (Hossain et al., 2016; Song et al., 2012). Many studies have reported recommended nucleotide dosages between 0.2% and 0.5% of dry weight in the fish diet. Furthermore, more recent studies have indicated that, instead of using combinations of nucleotides, a single nucleotide supplementation may be more effective and sufficient to improve growth responses, stress tolerance, and immune responses (Hossain et al., 2016; Song et al., 2012).

Several feeding trials were recently conducted with juvenile hybrid striped (HSB) to determine potential effects of supplementing purified dietary nucleotides (de Cruz et al., 2020a,b). The basal diet in these trials was formulated to contain approximately 40% protein and 10% lipid. Moreover, the nucleotide supplements were added to the basal diet at the expense of cellulose. Adenosine monophosphate (AMP), uridine monophosphate (UMP), cytidine monophosphate (CMP), guanosine monophosphate (GMP), and inosine monophosphate (IMP) were individually supplemented at 0.5% of diet and evaluated in a 9-week trial. Results showed significant enhancement of weight gain and innate immunity of fish fed the AMP-supplemented diet compared with those fed the basal diet.

Furthermore, the effects of dietary AMP and IMP, each at 0.5% of diet, or combinations of both (equaling 1% of diet) also were evaluated in a 9-week trial. No significant effects of AMP or IMP were observed on growth performance, whole-body composition, or innate immunity; however, fish fed IMP had significant enhancement of lymphocyte proliferation compared with fish fed the basal diet. These dietary treatments were retained and evaluated for immune modulation prior to and after acute-stress challenge. Result indicated some of the dietary nucleotides provided significant enhancement of innate immunity at 0.5 and 12 h poststress challenge, imposed by 1-minute air exposure, compared with those fed the basal diet. Supplementation of AMP at 0.5% of diet provided the greatest enhancement of innate immunity during poststress.

11.3.2.6 Other components

In contrast to the various diet additives discussed in the preceding sections, there are several components that may be found in the diet that can negatively affect the health of various aquatic species. These include adventitious toxins including those produced by various molds and fungi. These toxins may enter the diet through contaminated ingredients, or they may be produced by mold and fungi that become established due to excessive moisture in the diet or poor storage conditions. Detailed information about the various kinds of mycotoxins and their negative effects on immune responses and disease resistance of aquatic species was thoroughly reviewed by Manning (2015). Certain feedstuffs used in aquafeeds may contain antinutritional factors (substances) that may indirectly interfere with nutrient availability to the cultured organism or directly exert adverse effects on them. A detailed review of these antinutritional factors including their structures, sources, and mechanisms of action, as well as their effects on fish growth, immune responses, and disease susceptibility, was provided by Krogdahl and Bakke (2015).

11.4 Integration of diet formulations and feeding practices to optimize fish health

11.4.1 Specialized feeding regimes

It is generally accepted that fish should be fed adequate quantities of a nutritionally complete diet on a daily basis for optimal growth and health (Lim et al., 2015). However, there are certain conditions or circumstances in which alternative feeding regimes have been applied. For example, several fish species have been shown to exhibit compensatory growth in which their weight gain and feed efficiency improve markedly after certain periods of feed deprivation (NRC 2011). Thus, eliciting this compensatory growth response by intermittent periods of feed deprivation is one means of increasing production efficiency. However, other studies have shown that feed deprivation may lower the health condition of fish by causing anemia and other physiological changes (reviewed by Lim et al., 2015). For example, not feeding channel catfish for up to 4 weeks resulted in decreased immune response and increased mortality after infection with *Edwardsiella ictaluri* (Lim et al., 2015). Therefore, the practice of feed deprivation to promote compensatory growth is typically not recommended unless it is realized under typical culture conditions such as pond production of fish in temperate regions.

Under such conditions, it is not uncommon for warm-water fish species such as channel catfish to be exposed to water temperatures during the winter, which may suppress their feeding response for extended periods of time. Under such conditions, a series of studies (reviewed by Lim et al., 2015) showed that not feeding the fish for up to 3 months (December through February) did not result in reduced production if feeding was reinitiated in March. However, channel catfish starved from November 1 through April were immunosuppressed and had reduced resistance to a controlled exposure to *E. ictaluri*. Therefore, based on these results and other observations, feed deprivation for extended periods of time should be avoided, and regularly feeding of nutritionally complete diets should be maintained when fish are willing to actively feed.

11.4.2 Vaccination

The development of vaccines has resulted in significant advancements in controlling various diseases in aquaculture. This was quite evident when they were implemented in Norwegian salmon aquaculture in the late 1980s and early 1990s (Plant and LaPatra, 2011; Brudeseth et al., 2013). The positive effects of vaccines in providing protection from specific bacterial and viral pathogens have resulted in their development for several prominent freshwater and marine fish species. The types of vaccines include those in which the targeted pathogen has been inactivated (killed) or live but attenuated. In addition, DNA-based vaccines have particular use against viruses or intracellular bacteria.

Vaccines are commonly administered by injection, immersion or orally. Injection, in either the peritoneal cavity or musculature, is generally most effective but labor intensive, although various means of automation have improved the process (Plant and LaPatra, 2011). However, vaccination by injection can be stressful and generally not feasible for fish under 20 g. Oral vaccination is more easily applied, but its effectiveness can vary due to variability in gastric degradation or uptake from the GI tract (Brudeseth et al., 2013). Some vaccination regimes have been developed in which fish are vaccinated by immersion, then receive a booster either orally or by bath, and finally are injected with another vaccine (Brudeseth et al., 2013).

Although the efficacy of oral vaccines is generally less than those administered by injection, the application of adjuvants in the diet has some potential to enhance the immune response (Tafalla et al., 2013). Adjuvants are primarily used with inactivated vaccines. Some microparticulate adjuvants that also may be efficacious and applied as dietary immunostimulants include alginates, beta-glucans, chitosan, Ergosan, and various liposomes (Plant and LaPatra, 2011; Ringø et al., 2014). Other types of oral adjuvants may include macroalgae or microalgae such as *Chlamydomonas reinhardtii*, which has been transformed with foreign genes or proteins or *Artemia* containing specific antigens. Some adjuvants of a liquid nature may be injected as can genetic adjuvants, which contain DNA plasmids that encode immunostimulatory proteins (Brudeseth et al., 2013).

The application of specific nutrients to enhance immunization efficiency has been reviewed by Pohlenz and Gatlin (2014) and evaluated according to the different stages of immunization. Several different additives were noted as improving the detection and recognition of pathogens including alginic acid, beta-glucans, fructooligosaccharides, mannanoligosaccharides, lipopolysaccharides, nucleotides, peptidoglycans, and extracts of marine algae and yeast (Pohlenz and Gatlin, 2014). Other nutrients that have been observed to increase pathogen phagocytosis and killing, as well as improve antigen presentation, include arachidonic acid, arginine, astaxanthin, beta-hydroxy-beta-methyl butyrate, glutamine, linoleic acid, n-3 HUFA, polyamines, and vitamins C and E. Most of the nutrients and additives mentioned earlier, except for arachidonic acid, astaxanthin, and linoleic acid, but with the addition of nucleotides and beta-carotene, have been noted to enhance expansion and immunological memory (Pohlenz and Gatlin, 2014). To date, the dietary supplementation of nutrients or additives in concert with vaccination protocols has been evaluated to only a limited extent. However, systematic evaluation of nutritional supplementation prior to, during, and after vaccination may offer opportunities to increase the organism's response to vaccination and thus increase its efficiency.

11.4.3 Medicated feeds

Supplementing antibiotics in feed is one of the most effective means of providing medication to fish for treating bacterial infections, especially in large culture systems. Three commercial antibiotic products, i.e., oxytetracycline, sulfadimethoxine/ormetoprim, and florfenicol, have been approved by the US Food and Drug Administration (FDA) for use in the farming of fish destined for human consumption after prescribed withdrawal periods (http://www.fda.gov/AnimalVeterinary/DevelopmentApprovalProcess/Aquaculture/ucm132954.htm). The drugs and their use are governed by the FDA under a veterinary feed directive, which established requirements related to the distribution and use of drugs and animal feeds containing such drugs. As such, the use of medicated feeds must be under the professional supervision of a licensed veterinarian.

It is critical to control the quantity of antibiotic fed to cultured organisms. Proper feeding rates and withdrawal times must be followed to reduce the deposition of antibiotics into fish tissues or release into the rearing water that may be discharged into the aquatic environment. Specific administration and withdrawal procedures for the various antibiotics and targeted fish species were established during the registration process. Antibiotics may only be added to feeds in the United States by a licensed manufacturer.

11.5 Concluding remarks

This chapter has reviewed a number of different types of diet additives, which may positively affect the growth and health of various aquatic species. Many of these additives are naturally occurring and represent a more natural way to limit the use of antibiotics and other chemotherapeutics in aquacultural production. In addition, several of these compounds have been developed into commercial products, while others have remained at the experimental scale. The effectiveness of these various additives has been shown to vary somewhat from study to study and in some cases from species to species. Administration regimes for some additives have not been widely studied to date and deserve further attention. Developing more refined administration protocols for individual supplements should be investigated to optimize their effectiveness, especially related to certain practices employed in aquaculture or phases of production in which stress and increased disease susceptibility are likely to be encountered.

References

Abbas, A.K., Lichtman, A.H., Pillai, S., 2014. Cellular and Molecular Immunology E-Book. Elsevier Health Sciences.

Aramli, M.S., Kamangar, B., Nazari, R.M., 2015. Effects of dietary β-glucan on the growth and innate immune response of juvenile Persian sturgeon, *Acipenser persicus*. Fish Shellfish Immunol. 47, 606–610.

Barros, M.M., Guimarães, I.G., Pezzato, L.E., de Oliveira Orsi, R., Fernandes Junior, A.C., Teixeira, C.P., Fleuri, L.F., Padovani, C.R., 2015. The effects of dietary nucleotide mixture on growth performance, haematological and immunological parameters of Nile tilapia. Aquacult. Res. 46, 987–993.

Batbayar, S., Lee, D.H., Kim, H.W., 2012. Immunomodulation of fungal β-glucan in host defense signaling by dectin-1. Biomole. Therap. 20, 433.

Belosevic, M., Haddad, G., Walsh, J.G., Grayfer, L., Katzenback, B.A., Hanington, P.C., Neumann, N.F., Stafford, J.L., 2009. Innate immunity of fish: antimicrobial responses of fish macrophages. In: Zaccone, G., Meseguer, J., Garcia-Ayala, A., Kapoor, B.G. (Eds.), Fish Defenses. Science Publishers, Enfield, NH, USA, pp. 145–184.

Berto, R.d.S., Pereira, G.d.V., Mouriño, J.L.P., Martins, M.L., Fracalossi, D.M., 2016. Yeast extract on growth, nutrient utilization and haemato-immunological responses of Nile tilapia. Aquacult. Res. 47, 2650–2660.

Bøgwald, J., Dalmo, R.A., 2014. Gastrointestinal pathogenesis in aquatic animals. In: Merrifield, D., Ringø, E. (Eds.), Aquaculture Nutrition: Gut Health, Probiotics and Prebiotics, pp. 53–74.

Brudeseth, B.E., Wiulsrød, R., Fredriksen, B.N., Lindmo, K., Løkling, K.-E., Bordevik, M., Steine, N., Klevan, A., Gravningen, K., 2013. Status and future perspectives of vaccines for industrialised fin-fish farming. Fish Shellfish Immunol. 35, 1759–1768.

Burr, G., Gatlin III, D., Ricke, S., 2005. Microbial ecology of the gastrointestinal tract of fish and the potential application of prebiotics and probiotics in finfish aquaculture. J. World Aquacult. Soc. 36, 425–436.

Burrells, C., Williams, P., Forno, P., 2001. Dietary nucleotides: a novel supplement in fish feeds: 1. Effects on resistance to disease in salmonids. Aquaculture 199, 159–169.

Caipang, C.M.A., Suharman, I., Avillanosa, A.L., Bargoyo, V.T., 2020. The 8th international and national seminar on fisheries and marine science. IOP Conf. Ser. Earth Environ. Sci. 430.

Carson, C.F., Hammer, K.A., 2011. Chemistry and bioactivity of essential oils. In: Thormar, H. (Ed.), Lipids and Essential Oils as Antimicrobial Agents. Wiley, Chichester, pp. 203–238.

Castillo, S., Rosales, M., Pohlenz, C., Gatlin, D.M., 2014. Effects of organic acids on growth performance and digestive enzyme activities of juvenile red drum Sciaenops ocellatus. Aquaculture 433, 6–12.

Castro, R., Tafalla, C., 2015. Overview of fish immunity. In: Beck, B.H., Peatman, E. (Eds.), Mucosal Health in Aquaculture. Elsevier, pp. 3–54.

Chung, S., Secombes, C., 1988. Analysis of events occurring within teleost macrophages during the respiratory burst. Comp. Biochem. Physiol. Part B: Comp. Biochem. 89, 539–544.

Cosgrove, M., 1998. Perinatal and infant nutrition. Nucleot. Nutrit. 14, 748–751.

Dalmo, R.A., Bogwald, J., 2008. beta-glucans as conductors of immune symphonies. Fish Shellfish Immunol. 25, 384–396.

Dawood, M.A., Koshio, S., Ishikawa, M., Yokoyama, S., 2015. Interaction effects of dietary supplementation of heat-killed *Lactobacillus plantarum* and β-glucan on growth performance, digestibility and immune response of juvenile red sea bream, *Pagrus major*. Fish Shellfish Immunol. 45, 33–42.

Dawood, M.A., Metwally, A.E.-S., El-Sharawy, M.E., Atta, A.M., Elbialy, Z.I., Abdel-Latif, H.M., Paray, B.A., 2020a. The Role of β-glucan in the Growth, Intestinal Morphometry, and Immune-Related Gene and Heat Shock Protein Expressions of Nile tilapia (*Oreochromis niloticus*) under Different Stocking Densities. Aquaculture, p. 735205.

Dawood, M.A., Abdo, S.E., Gewaily, M.S., Moustafa, E.M., SaadAllah, M.S., AbdEl-Kader, M.F., Hamouda, A.H., Omar, A.A., Alwakeel, R.A., 2020b. The influence of dietary β-glucan on immune, transcriptomic, inflammatory and histopathology disorders caused by deltamethrin toxicity in Nile tilapia (*Oreochromis niloticus*). Fish Shellfish Immunol. 98, 301–311.

de Cruz, C.R., Yamamoto, F.Y., Castillo, S., Gatlin III, D.M., 2020a. Establishing the optimal adenosine 5′-monophosphate level for hybrid striped bass *Morone chrysops*× *Morone saxatilis*: effects on growth performance, nutrient digestibility, and immune modulation during acute and chronic stress. Aquaculture 520, 734668.

de Cruz, C.R., Yamamoto, F.Y., Ju, M., Chen, K., Velasquez, A., Gatlin III, D.M., 2020b. Efficacy of purified nucleotide supplements on the growth performance and immunity of hybrid striped bass *Morone chrysops* x *Morone saxatilis*. Fish Shellfish Immunol. 98, 868–874.

de Mello, M.M.M., de Faria, C.D.P., Zanuzzo, F.S., Urbinati, E.C., 2019. beta-glucan modulates cortisol levels in stressed pacu (*Piaractus mesopotarnicus*) inoculated with heat-killed Aeromonas hydrophila. Fish Shellfish Immunol. 93, 1076–1083.

Devresse, B., 2000. Nucleotides—a key nutrient for shrimp immune system. Feed Mix 8, 20–22.

Do Huu, H., Sang, H.M., Thuy, N.T.T., 2016. Dietary β-glucan improved growth performance, Vibrio counts, haematological parameters and stress resistance of pompano fish, *Trachinotus ovatus* Linnaeus, 1758. Fish Shellfish Immunol. 54, 402–410.

Douxfils, J., Fierro-Castro, C., Mandiki, S.N.M., Emile, W., Tort, L., Kestemont, P., 2017. Dietary β-glucans differentially modulate immune and stress-related gene expression in lymphoid organs from healthy and *Aeromonas hydrophila*-infected rainbow trout (*Oncorhynchus mykiss*). Fish Shellfish Immunol. 63, 285–296.

Ferreira, I., Pinho, O., Vieira, E., Tavarela, J., 2010. Brewer's Saccharomyces yeast biomass: characteristics and potential applications. Trends Food Sci. Technol. 21, 77–84.

Foey, A., Picchietti, S., 2014. 2 immune defences of teleost fish. In: Merrifield, D., Ringø, E. (Eds.), Aquaculture Nutrition: Gut Health, Probiotics and Prebiotics, pp. 14–52.

Fuller, R., 1989. Probiotics in man and animals. J. Appl. Bacteriol. 66, 365–378.

Gannan, A.L., 2015. β-Glucans. In: Lee, C.-S., Lim, C., Gatlin III, D.M., Webster, C.D. (Eds.), Dietary Nutrients, Additives and Fish Health. John Wiley & Sons, pp. 111–124.

Gatlin III, D.M., 2015. Prebiotics. In: Lee, C.-S., Lim, C., Gatlin III, D.M., Webster, C.D. (Eds.), Dietary Nutrients, Additives and Fish Health. John Wiley & Sons, pp. 271–281.

Gibson, G.R., Roberfroid, M.B., 1995. Dietary modulation of the human colonic microbiota: introducing the concept of prebiotics. J. Nutr. 125, 1401–1412.

Gil, A., 2002. Modulation of the immune response mediated by dietary nucleotides. Eur. J. Clin. Nutr. 56, S1–S4.

Gómez, G.D., Balcázar, J.L., 2008. A review on the interactions between gut microbiota and innate immunity of fish. FEMS Immunol. Med. Microbiol. 52, 145–154.

Halver, J.E., 2002. The vitamins. In: Halver, J.E., Hardy, R. (Eds.), Fish Nutrition. Elsevier, pp. 61–141.

Harris, S.J., Bray, D.P., Adamek, M., Hulse, D.R., Steinhagen, D., Hoole, D., 2019. Effect of β-1/3, 1/6-glucan upon immune responses and bacteria in the gut of healthy common carp (*Cyprinus carpio*). J. Fish. Biol. 96, 444–455.

Hoffmann, K., 2007. The use of nucleotides in animal feed. Feed Mix 15, 25–27.

Hohn, C., Petrie-Hanson, L., 2012. Rag1−/− mutant zebrafish demonstrate specific protection following bacterial re-exposure. PloS One 7, e44451.

Holodick, N.E., Rodríguez-Zhurbenko, N., Hernández, A.M., 2017. Defining natural antibodies. Front. Immunol. 8, 872.

Hoseinifar, S.H., Van Doan, H., Dadar, M., Ringø, E., Harikrishnan, R., 2019. Feed additives, gut microbiota, and health in finfish aquaculture. In: Derome, N. (Ed.), Microbial Communities in Aquaculture Ecosystems. Springer, pp. 121–142.

Hossain, M.S., Koshio, S., Ishikawa, M., Yokoyama, S., Sony, N.M., 2016. Dietary effects of adenosine monophosphate to enhance growth, digestibility, innate immune responses and stress resistance of juvenile red sea bream, *Pagrus major*. Fish Shellfish Immunol. 56, 523–533.

Ingledew, W., 1999. Yeast-could you base a business on this bug, under the microscope—focal points for the new millennium—biotechnology in the feed industry. In: Proceedings of Alltech's 15th Annual Symposium. Nottingham University Press, Nottingham, UK, pp. 27–47.

Izquierdo, M., Betancor, M. Vitamin E. *In*: Lee, C.-S., Lim, C., Gatlin III, D.M., Webster, C.D. (Eds.), Dietary Nutrients, Additives and Fish Health. John Wiley & Sons, pp. 173-194.

Jeney, G., Wet, L.D., Jeney, Z., Yin, G., 2015. Plant extracts. In: Lee, C.-S., Lim, C., Gatlin III, D.M., Webster, C.D. (Eds.), Dietary Nutrients, Additives and Fish Health. John Wiley & Sons, pp. 321–332.

Ji, L., Sun, G., Li, J., Wang, Y., Du, Y., Li, X., Liu, Y., 2017. Effect of dietary β-glucan on growth, survival and regulation of immune processes in rainbow trout (*Oncorhynchus mykiss*) infected by *Aeromonas salmonicida*. Fish Shellfish Immunol. 64, 56–67.

Jiang, C., Wang, P., Li, M., Liu, S., Zhang, S., 2016. Dietary β-glucan enhances the contents of complement component 3 and factor B in eggs of zebrafish. Dev. Comp. Immunol. 65, 107–113.

Jung-Schroers, V., Adamek, M., Jung, A., Harris, S., Dóza, Ö.S., Baumer, A., Steinhagen, D., 2016. Feeding of β-1, 3/1, 6-glucan increases the diversity of the intestinal microflora of carp (*Cyprinus carpio*). Aquacult. Nutr. 22, 1026–1039.

Jung-Schroers, V., Adamek, M., Harris, S., Syakuri, H., Jung, A., Irnazarow, I., Steinhagen, D., 2018. Response of the intestinal mucosal barrier of carp (*Cyprinus carpio*) to a bacterial challenge by *Aeromonas hydrophila* intubation after feeding with β-1, 3/1, 6-glucan. J. Fish. Dis. 41, 1077–1092.

Katzenback, B.A., 2015. Antimicrobial peptides as mediators of innate immunity in teleosts. Biology 4, 607–639.

Kesarcodi-Watson, A., Kaspar, H., Lategan, M.J., Gibson, L., 2008. Probiotics in aquaculture: the need, principles and mechanisms of action and screening processes. Aquaculture 274, 1–14.

Kiron, V., 2015. Gastrointestinal microorganisms of fish and probiotics. In: Lee, C.-S., Lim, C., Gatlin III, D.M., Webster, C.D. (Eds.), Dietary Nutrients, Additives and Fish Health. John Wiley & Sons, pp. 283–304.

Kojima, K., 1974. Safety evaluation of disodium 5′-inosinate, disodium 5′-guanylate and disodium 5′-ribonucleotide. Toxicology 2, 185–206.

Krogdahl, Å., Bakke, A.M., 2015. Antinutrients. In: Lee, C.-S., Lim, C., Gatlin, D., Webster, C.D. (Eds.), Dietary Nutrients, Additives and Fish Health. Wiley-Blackwell, pp. 211–235.

Kudrenko, B., Snape, N., Barnes, A.C., 2009. Linear and branched ß (1–3) D-glucans activate but do not prime teleost macrophages *in vitro* and are inactivated by dilute acid: implications for dietary immunostimulation. Fish Shellfish Immunol. 26, 443–450.

Kühlwein, H., Emery, M., Rawling, M., Harper, G., Merrifield, D., Davies, S., 2013. Effects of a dietary β-(1, 3)(1, 6)-D-glucan supplementation on intestinal microbial communities and intestinal ultrastructure of mirror carp (*Cyprinus carpio* L.). J. Appl. Microbiol. 115, 1091–1106.

Kumar, V., Sinha, A., Makkar, H., De Boeck, G., Becker, K., 2012. Phytate and phytase in fish nutrition animal physiology and animal. J. Anim. Physiol. Anim. Nutr. 96, 335–364.

Lam, K.-L., Cheung, P.C.-K., 2013. Non-digestible long chain beta-glucans as novel prebiotics. Bioactive Carbohyd. Diet. Fibre 2, 45–64.

Li, P., Gatlin III, D.M., 2006. Nucleotide nutrition in fish: current knowledge and future applications. Aquaculture 251, 141–152.

Li, P., Mai, K., Trushenski, J., Wu, G., 2009. New developments in fish amino acid nutrition: towards functional and environmentally oriented aquafeeds. Amino Acids 37, 43–53.

Lim, C., Webster, C.D., Lee, C.S., 2015. Feeding practices and fish health. In: Lee, C.-S., Lim, C., Gatlin, D., Webster, C.D. (Eds.), Dietary Nutrients, Additives and Fish Health. Wiley-Blackwell, pp. 333–346.

Manning, B.B., 2015. Mycotoxin contamination of fish feeds. In: Lee, C.-S., Lim, C., Gatlin, D., Webster, C.D. (Eds.), Dietary Nutrients, Additives and Fish Health, pp. 237–248.

Mashoof, S., Criscitiello, M.F., 2016. Fish immunoglobulins. Biology 5, 45.

Mateo, C., Dave, R., Stein, H., 2004. Effects of supplemental nucleosides for newly weaned pigs. J. Anim. Sci. 82, 71.

Meena, D.K., Das, P., Kumar, S., Mandal, S.C., Prusty, A.K., Singh, S.K., Akhtar, M.S., Behera, B.K., Kumar, K., Pal, A.K., Mukherjee, S.C., 2013. Beta-glucan: an ideal immunostimulant in aquaculture (a review). Fish Physiol. Biochem. 39, 431–457.

Merrifield, D.L., Dimitroglou, A., Foey, A., Davies, S.J., Baker, R.T., Bøgwald, J., Castex, M., Ringø, E., 2010. The current status and future focus of probiotic and prebiotic applications for salmonids. Aquaculture 302, 1–18.

Miest, J.J., Arndt, C., Adamek, M., Steinhagen, D., Reusch, T.B., 2016. Dietary β-glucan (MacroGard®) enhances survival of first feeding turbot (*Scophthalmus maximus*) larvae by altering immunity, metabolism and microbiota. Fish Shellfish Immunol. 48, 94–104.

Montoya, L.N.F., Martins, T.P., Gimbo, R.Y., Zanuzzo, F.S., Urbinati, E.C., 2017. β-Glucan-induced cortisol levels improve the early immune response in matrinxã (*Brycon amazonicus*). Fish Shellfish Immunol. 60, 197–204.

Moriarty, D., 1998. Control of luminous Vibrio species in penaeid aquaculture ponds. Aquaculture 164, 351–358.

Murphy, K., Weaver, C., 2016. Janeway's Immunobiology, ninth ed. Garland Science/Taylor & Francis Group, LLC, New York, NY.

Nakanishi, T., Shibasaki, Y., Matsuura, Y., 2015. T cells in fish. Biology 4, 640–663.

Nakao, M., Tsujikura, M., Ichiki, S., Vo, T.K., Somamoto, T., 2011. The complement system in teleost fish: progress of post-homolog-hunting researches. Dev. Comp. Immunol. 35, 1296–1308.

Netea, M.G., Quintin, J., Van Der Meer, J.W., 2011. Trained immunity: a memory for innate host defense. Cell Host Microbe 9, 355–361.

Noga, E.J., Ullal, A.J., Corrales, J., Fernandes, J.M., 2011. Application of antimicrobial polypeptide host defenses to aquaculture: exploitation of downregulation and upregulation responses. Comp. Biochem. Physiol. Genom. Proteonomics 6, 44–54.

NRC, 2011. Nutrient Requirement of Fish and Shrimp. The National Academies Press, Washington, D.C.

Parker, R., 1974. Probiotics, the other half of the antibiotic story. Anim. Nutr. Health 29, 4–8.

Parra, D., Reyes-Lopez, F.E., Tort, L., 2015. Mucosal immunity and B cells in teleosts: effect of vaccination and stress. Front. Immunol. 6, 354.

Petit, J., Wiegertjes, G.F., 2016. Long-lived effects of administering β-glucans: indications for trained immunity in fish. Dev. Comp. Immunol. 64, 93–102.

Plant, K.P., LaPatra, S.E., 2011. Advances in fish vaccine delivery. Dev. Comp. Immunol. 35, 1256–1262.

Pohlenz, C., Gatlin III, D.M., 2014. Interrelationships between fish nutrition and health. Aquaculture 431, 111–117.

Quintin, J., Saeed, S., Martens, J.H., Giamarellos-Bourboulis, E.J., Ifrim, D.C., Logie, C., Jacobs, L., Jansen, T., Kullberg, B.-J., Wijmenga, C., 2012. Candida albicans infection affords protection against reinfection via functional reprogramming of monocytes. Cell Host Microbe 12, 223–232.

Raa, J., 2015. Immune modulation by non-digestible and non-absorbable beta-1,3/1,6-glucan. Microb. Ecol. Health Dis. 26, 27824-27824.

Rajanbabu, V., Chen, J.-Y., 2011. Applications of antimicrobial peptides from fish and perspectives for the future. Peptides 32, 415–420.

Ramadan, A., Atef, M., Afifi, N., 1991. Effect of the biogenic performance enhancer (Ascogen" S") on growth rate of tilapia fish. Acta Vet. Scand. 87, S304–S306.

Ramadan, A., Afifi, N.A., Moustafa, M., Samy, A., 1994. The effect of ascogen on the immune response of tilapia fish to *Aeromonas hydrophila* vaccine. Fish Shellfish Immunol. 4, 159–165.

Ringø, E., Olsen, R., Gifstad, T., Dalmo, R., Amlund, H., Hemre, G.I., Bakke, A., 2010. Prebiotics in aquaculture: a review. Aquacult. Nutr. 16, 117–136.

Ringø, E., Erik Olsen, R., Gonzalez Vecino, J., Wadsworth, S., 2011. Use of immunostimulants and nucleotides in aquaculture: a review. J. Mar. Sci. Res. Dev. 2, 1–22.

Ringø, E., Olsen, R.E., Jensen, I., Romero, J., Lauzon, H.L., 2014. Application of vaccines and dietary supplements in aquaculture: possibilities and challenges. Rev. Fish Biol. Fish. 24, 1005–1032.

Rodríguez, F.E., Valenzuela, B., Farías, A., Sandino, A.M., Imarai, M., 2016. β-1, 3/1, 6-Glucan-supplemented diets antagonize immune inhibitory effects of hypoxia and enhance the immune response to a model vaccine. Fish Shellfish Immunol. 59, 36–45.

Rojo-Cebreros, A.H., Ibarra-Castro, L., Martínez-Brown, J.M., 2018. Immunostimulation and trained immunity in marine fish larvae. Fish Shellfish Immunol. 80, 15–21.

Rombout, J.H., Abelli, L., Picchietti, S., Scapigliati, G., Kiron, V., 2011. Teleost intestinal immunology. Fish Shellfish Immunol. 31, 616–626.

Sabioni, R.E., Zanuzzo, F.S., Gimbo, R.Y., Urbinati, E.C., 2020. β-Glucan enhances respiratory activity of leukocytes suppressed by stress and modulates blood glucose levels in pacu (*Piaractus mesopotamicus*). Fish Physiol. Biochem. 46, 629–640.

Sanderson, I.R., He, Y., 1994. Nucleotide uptake and metabolism by intestinal epithelial cells. J. Nutr. 124, 131S–137S.

Saurabh, S., Sahoo, P., 2008. Lysozyme: an important defence molecule of fish innate immune system. Aquacult. Res. 39, 223–239.

Scapigliati, G., Fausto, A.M., Picchietti, S., 2018. Fish lymphocytes: an evolutionary equivalent of mammalian innate-like lymphocytes? Front. Iimmunol. 9, 971.

Secombes, C., Wang, T., 2012. The innate and adaptive immune system of fish. In: Austin, B. (Ed.), Infectious Disease in Aquaculture. Elsevier, pp. 3–68.

Secombes, C.J., 2016. What's new in fish cytokine research. Fish Shellfish Immunol. 53, 1–3.

Sletmoen, M., Stokke, B.T., 2008. Review : higher order structure of (1,3)-beta-D-glucans and its influence on their biological activities and complexation abilities. Biopolymers 89, 310–321.

Smith, N.C., Rise, M.L., Christian, S.L., 2019. A comparison of the innate and adaptive immune systems in cartilaginous fish, ray-finned fish and lobe-finned fish. Front. Immunol. 10, 2292.

Song, J.-W., Lim, S.-J., Lee, K.-J., 2012. Effects of dietary supplementation of inosine monophosphate on growth performance, innate immunity and disease resistance of olive flounder (*Paralichthys olivaceus*). Fish Shellfish Immunol. 33, 1050–1054.

Song, S.K., Beck, B.R., Kim, D., Park, J., Kim, J., Kim, H.D., Ringø, E., 2014. Prebiotics as immunostimulants in aquaculture: a review. Fish Shellfish Immunol. 40, 40–48.

Souza, C., Baldissera, M., Baldisserotto, B., Heinzmann, B., Martos-Sitcha, J.A., Mancera, J.M., 2019. Essential oils as stress-reducing agents for fish aquaculture: a Review. Front. Physiol. 10, 785.

Sutili, F.J., Gatlin III, D.M., Heinzmann, B.M., Baldisserotto, B., 2018. Plant essential oils as fish diet additives: benefits on fish health and stability in feed. Rev. Aquacult. 10, 716–726.

Tafalla, C., Bøgwald, J., Dalmo, R.A., 2013. Adjuvants and immunostimulants in fish vaccines: current knowledge and future perspectives. Fish Shellfish Immunol. 35, 1740–1750.

Trichet, V.V., Santigosa, E., Cochin, E., Gabaudan, J., 2015. The effect of vitamin C on fish health. In: Lee, C.-S., Lim, C., Gatlin III, D.M., Webster, C.D. (Eds.), Dietary Nutrients, Additives and Fish Health. John Wiley & Sons, pp. 151–171.

Trushenski, J., 2015. Nutritional impacts on fish mucosa: dietary considerations. In: Beck, B.H., Peatman, E. (Eds.), Mucosal Health in Aquaculture. Elsevier, pp. 199–209.

Udayangani, R.M.C., Dananjaya, S.H.S., Fronte, B., Kim, C.H., Lee, J., De Zoysa, M., 2017. Feeding of nano scale oats beta-glucan enhances the host resistance against *Edwardsiella tarda* and protective immune modulation in zebrafish larvae. Fish Shellfish Immunol. 60, 72–77.

Van Doan, H., Hoseinifar, S.H., Ringø, E., Ángeles Esteban, M., Dadar, M., Dawood, M.A., Faggio, C., 2020. Host-associated probiotics: a key factor in sustainable aquaculture. Rev. Fish. Sci. Aquaculture. 28, 16–42.

Van Hai, N., 2015. The use of medicinal plants as immunostimulants in aquaculture: a review. Aquaculture 446, 88–96.

Volman, J.J., Ramakers, J.D., Plat, J., 2008. Dietary modulation of immune function by beta-glucans. Physiol. Behav. 94, 276–284.

Voloski, A.P.S., Soveral, L.F., Dazzi, C.C., Sutili, F., Frandoloso, R., Kreutz, L.C., 2019. β-Glucan improves wound healing in silver catfish (*Rhamdia quelen*). Fish Shellfish Immunol. 93, 575–579.

Wang, Y.B., Li, J.R., Lin, J., 2008. Probiotics in aquaculture: challenges and outlook. Aquaculture 281, 1–4.

Wang, T., Secombes, C.J., 2013. The cytokine networks of adaptive immunity in fish. Fish Shellfish Immunol. 35, 1703–1718.

Wang, Y.-B., Xu, Z.-R., Xia, M.-S., 2005. The effectiveness of commercial probiotics in northern white shrimp Penaeus vannamei ponds. Fish. Sci. 71, 1036–1041.

Wang, W., Sun, J., Liu, C., Xue, Z., 2017. Application of immunostimulants in aquaculture: current knowledge and future perspectives. Aquacult. Res. 48, 1–23.

Webster, C.D., Lim, C., 2015. Minerals. In: Lee, C.-S., Lim, C., Gatlin III, D.M., Webster, C.D. (Eds.), Dietary Nutrients, Additives and Fish Health. John Wiley & Sons, pp. 195–210.

Whyte, S.K., 2007. The innate immune response of finfish—a review of current knowledge. Fish Shellfish Immunol. 23, 1127–1151.

Wismar, R., Brix, S., Frokiaer, H., Laerke, H.N., 2010. Dietary fibers as immunoregulatory compounds in health and disease. Food. Health 21st Cent.: Road Map Fut. 1190, 70–85.

Yamamoto, F.Y., Yin, F., Rossi, W., Hume, M., Gatlin, D.M., 2018a. β-1, 3 glucan derived from *Euglena gracilis* and Algamune™ enhances innate immune responses of red drum (*Sciaenops ocellatus* L.). Fish Shellfish Immunol. 77, 273–279.

Yamamoto, F.Y., Sutili, F.J., Hume, M., Gatlin III, D.M., 2018b. The effect of β-1, 3-glucan derived from *Euglena gracilis* (Algamune™) on the innate immunological responses of Nile tilapia (*Oreochromis niloticus* L.). J. Fish. Dis. 41, 1579–1588.

Yamamoto, F.Y., Castillo, S., Gatlin III, D.M., 2020a. Immunomodulatory effects of β-glucans derived from *Euglena gracilis* or *Saccharomyces cerevisiae* for hybrid striped bass (*Morone chrysops* × *M. saxatilis*). Aquacult. Res. 51, 1211–1221.

Yamamoto, F.Y., Castillo, S., de Cruz, C.R., Chen, K., Hume, M.E., Gatlin III, D.M., 2020b. Synergistic Effects of the β-1, 3 Glucan Paramylon and Vitamin C on Immunological Responses of Hybrid Striped Bass (*Morone Chrysops* × *M. saxatilis*) Were Pronounced *in Vitro* but More Moderate *in Vivo*. Aquaculture, p. 735394.

Zhang, Z., Chi, H., Dalmo, R.A., 2019. Trained innate immunity of fish is a viable approach in larval aquaculture. Front. Immunol. 10.

Zhu, F., Du, B., Xu, B., 2016. A critical review on production and industrial applications of beta-glucans. Food Hydrocolloid. 52, 275–288.

Zou, J., Secombes, C.J., 2016. The function of fish cytokines. Biology 5, 23.

CHAPTER

12

Antinutrients and adventitious toxins*

Åshild Krogdahl[1], Trond M. Kortner[1] and Ronald W. Hardy[2]

[1]Faculty of Veterinary Medicine, Norwegian University of Life Sciences, Ås, Norway;
[2]Aquaculture Research Institute, Hagerman Fish Culture Experiment Station, University of Idaho, Hagerman, ID, United States

12.1 Introduction

The range of ingredients used for fish feed production is expanding rapidly, as fish cultivation is escalating worldwide and supply of marine ingredients is decreasing. New plant and animal ingredients are included at increasing level in the diets, as the efforts to produce diets free of fishmeal are continuing. Moreover, new biomasses are utilized as nutrient sources, e.g., agricultural waste, food waste, insects, wood, seaweed, and so on, some after extensive bioconversion (Agboola et al., 2020; Giovanni et al., 2019; Morais et al., 2020). New ingredients often demand new processing technologies, which may lead to the production of new compounds and require new additives. Consequently, the range of dietary compounds, which are alien to fish living in the wild, is widening. Some of these may be detrimental to the fish.

In this chapter, potentially harmful compounds, which may be found in ingredients used in fish feed, are gathered in two groups, antinutritional factors (ANFs) and adventitious toxins (ATs). The ANFs are naturally occurring constituents of feed ingredients, whereas the ATs come from other sources than the plant. The function of an ANF in a plant is, in most cases, to protect the plant against bacteria, fungi, insects, and larger animals (Gustafson and Ryan 1974; Kim et al. 2005, 2006, 2009; Lawrence and Koundal 2002; Ryan and Clary 1964; Ryan 1990). However, some antinutrients are clearly important for internal

* This chapter is an update of Chapter 10. Antinutrients and adventitious toxins by Jerry D. Hendriks published in the previous edition of this book. Important parts of his chapter are included in the present update. Thanks are due to my colleague Nicole Frost Nyquist and her mother Elizabeth Dianne Frost for their help in proof reading and improving the language of this chapter.

Fish Nutrition, Fourth Edition
https://doi.org/10.1016/B978-0-12-819587-1.00001-X

functions of the organism. Many plants with seemingly high nutrient content are not considered as food or feed due to their high content of antinutrients, which are difficult to inactivate or remove. In castor beans (*Ricinus communis*) containing the lectin ricin and rosary peas (*Abrus precatorius*) containing the lectin abrin, the toxic potency of the ANFs in the seeds is so high that they are useless as food.

Possible harmful effects of ANFs include reduced palatability, less efficient nutrient utilization, alteration of nutrient balances of the diets, inhibition of growth, intestinal dysfunction, immune modulation, altered gut microbiota, goitrogenesis, pancreatic hypertrophy, hypoglycemia, or liver and kidney malfunction. Although they may have detrimental effects when present at high level, at low or moderate level some of these compounds may also have beneficial effects, such as being antioxidative, improving immune functions, or serving as prebiotics. Adventitious toxins are compounds not naturally contained in feed ingredients. They are present as a result, e.g., of microbial growth on the feed ingredients while they are growing in the field or after harvest, of the use of disease control chemicals, of additives used to improve processing condition and product quality or the use of by-products of processing, and so on. To be able to produce fish feeds of high nutritional value, which secure good health and welfare, detailed information of the presence and effects of ANFs and AT is required. This chapter intends to give an overview of the most relevant ANFs and ATs regarding structure, mode of action, and possible harmful effects on fish production and health. As the effects of such compounds may vary with species of animal, age or stage of development, size, sex, stress, state of health, and feeding strategy, attention is also given to such aspects to the extent that information is available. The overview is based on information included in the previous version of this chapter (Hendricks, J. 2002), three later reviews (Gatlin et al., 2007; Krogdahl et al., 2010; NRC 2011), as well as more recent, relevant results presented in original scientific literature.

12.2 Antinutrients

12.2.1 Antinutrients in plants

Antinutrients present in plants may, in a nutrition context, be divided in two categories, the heat labile and the heat stable. The heat-labile ANFs of the common feed ingredients comprise proteinaceous compounds such as protease and amylase inhibitors, and lectins (Belitz and Grosch 1999a, 1999b). However, the heat treatment needed to inactivate these antinutrients, at the same time, may reduce availability of and inactivate essential nutrients, such as amino acids and vitamins, resulting in low nutritional value (Anderson 1979; Belitz and Grosch 1999a, 1999b; Krogdahl et al., 2020). For some of the heat-labile ANFs, treatment with a mixture of alcohol and water may inactivate their antinutritive potential (Cantoral et al., 1995) to levels which the animal can handle, e.g., by adjustments of the digestive processes. Regarding the heat-stable ANFs, some may be removed by alcohol extraction, e.g., saponins, estrogens, phytosterols, and oligosaccharides. A combination of heat and alcohol extraction is used for processing of key animal feed ingredients e.g. in the production of soy protein concentrate. Removal of some ANFs may need specific procedures, adapted to the chemical characteristics of the individual antinutrient, such as fermentation or other extraction procedures. Recently, breeding strategies have also been used for removal of antinutrients (Krogdahl et al., 2020; Schmidt et al., 2015).

Indigestible carbohydrates, oligo- as well as polysaccharides, may be categorized under the definition of antinutrients given before. These are, however, not usually included among the antinutrients and are not discussed in this chapter.

12.2.1.1 *Enzyme inhibitors*

Many plants contain enzyme inhibitors, which protect their proteins, starch, and lipids from being exploited by other organisms. This is achieved by inhibiting their corresponding digestive enzymes. Legume seeds are, among the plant ingredients used in animal production, the nutrient sources with the highest number and concentration of such enzyme inhibitors. The proteinaceous enzyme inhibitors comprise inhibitors of trypsin, chymotrypsin, elastase, carboxypeptidase A and B, enterokinase, and amylase (Bhat et al., 1981; Hellinger and Gruber 2019). Proteinaceous lipase inhibitors also exist, but most lipase inhibitors are not proteins or peptides, but compounds such as saponins, terpenes, phenols, and chitosans (reviewed by Singh et al., 2015).

12.2.1.1.1 Protease inhibitors

Protease inhibitors as a group comprise compounds such as amino acids, di- and tripeptides, larger peptides and proteins, as well as polyphenols, terpenes, flavones, saponins, alkaloids, tannins, amino acids, and derivatives of these (Atta-ur-Rahman 2003). Hellinger and Gruber (2019) recently published a comprehensive review of present knowledge on proteins and peptides with antiproteolytic activity, which has served as an important source of information for the following. Protease inhibitors, as a group of molecules with distinct characteristics, have been found to comprise several thousand compounds. The majority of them are found in four plant families: grasses (Poaceae), legumes (Fabaceae), night shade (Solanaceae), e.g., potato and tomato, and cabbage (Brassicaceae). Many members of the legume and night shade families contain several inhibitors, others a few, and some only one (Atanasov et al., 2015; Hellinger and Gruber 2019). Some proteinaceous protease inhibitors bind only one specific enzyme at one site of the molecule, some may be able to bind more than one, e.g., trypsin as well as chymotrypsin at the same site, and others may bind two or more proteases at different sites. Some bind the proteases covalently, and others by competitive inhibition. Inhibition may take place in the active site but also as a result of attachments at other sites of the enzyme. The strength of the inhibition depends on the characteristics of the binding. It is the strongest when covalent binding is involved and weakest when sites of the molecule other than the active site are involved.

Among the many compounds with protease-inhibiting effects, only those of protein nature have been addressed for such characteristics in fish studies. Therefore, in this chapter, only proteinaceous inhibitors are included. Most of them are medium-sized proteins in the range 6–25 kDa, but smaller and larger proteins also exist. They can be categorized, according to the structure of their active sites, into 12 families (Hellinger and Gruber 2019). Most of the protease inhibitors contain several sulfur bridges, which stabilize the molecule against denaturation by heat as well as toward low pH in the stomach. A well-characterized representative is the Bowman–Birk inhibitor, a compact 8000 kDa protein with seven sulfur bridges (Fig. 12.1). It has two active sites, one optimal for trypsin binding and the other for binding of chymotrypsin (Birk 1985; Odani 1972). The enzyme binding is competitive. When the inhibitor binds to the enzyme, a peptide bond is broken. A steric alteration is taking

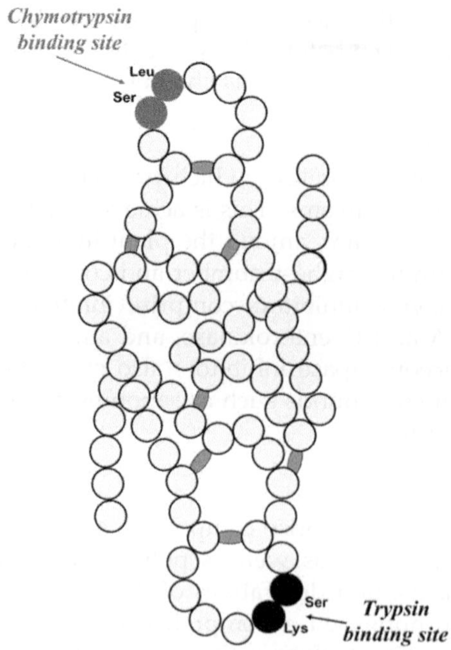

FIGURE 12.1 A schematic presentation of the peptide chain of the Bowman–Birk inhibitor from soybeans. Green ovals indicate the positions of the seven cystin bridges. The amino acids in the active sites, interacting with trypsin (Ser-Lys) and chymotrypsin (Leu-Ser), are indicated. *Adapted from Odani, S., Ikenaka, T., 1972. Studies on soybean trypsin inhibitors. IV. Complete amino acid sequence and the anti-proteinase sites of Bowman-Birk soybean proteinase inhibitor. J. Biochem. 5, 839–848.*

place, which increases the stability of the molecule complex, making the release of the enzymes very slow. Thereby the enzyme is effectively inactivated. Similar mechanisms underlie the action of most protease inhibitors, whether one or more active sites are involved.

In the late 1980s, the salmon feed industry was searching for new nutrient sources to be able to meet the increased demand for feed. As soybean meal was the most interesting ingredient among the available protein sources, a series of experiments were conducted to answer questions regarding possible effects of the soybean protease inhibitors (Berg-Lea et al., 1989; Krogdahl and Holm 1982; Krogdahl et al., 1994; Olli et al. 1989, 1994). From these studies, it became clear that salmonid proteases are highly sensitive to soybean protease inhibitors. A dose–response study, with levels of trypsin inhibitors in the diet from 0 to 1 mg/kg diet, fed to Atlantic salmon for 10 days, showed an accelerating decrease in protein digestibility from 92% to 75% in parallel with a drop in trypsin activity by 80% in the chyme sampled from the pyloric intestine. Also, lipid digestibility was affected, this being most pronounced for C18:0, for which digestibility decreased from 97% to 81%. The results of a similar, longer-term study with Atlantic salmon gave more pronounced results regarding nutrient digestibility. The results for trypsin, both the trypsin protein as such and for the enzyme activity, are illustrated in Fig. 12.2 for samples collected at 12 as well as 31 days of feeding (Olli et al., 1994). They show that longer-term feeding with protease inhibitors may exhaust the trypsin production capacity and indicate that even quite low levels in the diet may affect gut function and supposedly also nutrient utilization.

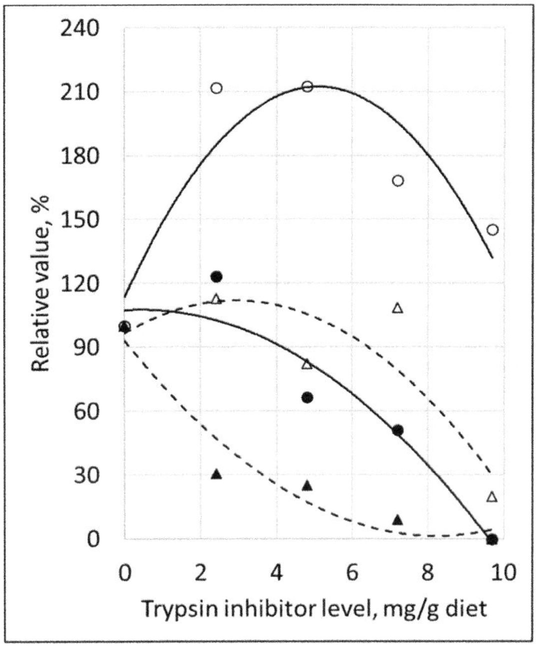

FIGURE 12.2 Results of a dose–response study of effects of a soybean protease inhibitor preparation (Sigma type II, no T9128, containing a mixture of Kunitz and Bowman–Birk inhibitor), included in the diet up to a level comparable to the inhibitor level in a diet with 60% raw soybean meal (Olli et al., 1994). The observed biomarkers were trypsin protein (*open indicators*) and trypsin activity (*filled indicators*) in digesta samples from the pyloric region in Atlantic salmon, collected after 12 (*circles*) and 31 days (*triangles*) of feeding. The results are expressed as percent of the values observed for diets without the inhibitor preparation.

Not only Atlantic salmon but also sea bream trypsin activity may be outcompeted by the presence of soybean trypsin inhibitors in the diet (Santigosa et al., 2010), but the effect was temporary. After 30 days of feeding, the effect disappeared. A temporary effect was observed also on growth (Santigosa et al., 2010). The mechanism behind this temporary effect may be a change with time in secretion of various proteases from the pancreas with varying sensitivity to protease inhibitors. In humans, long-term duodenal infusion of proteases has been shown to alter the proportion of proteases in the pancreatic juice. At the beginning, proteases that were inhibited by the soybean inhibitors dominated, and the protease activity was low. Gradually, the protease activity resumed due to a switch in secretion of proteases to types that were resistant to the inhibitors (Reseland et al., 1995).

12.2.1.1.2 Amylase inhibitors

Plant parts with high levels of starch often contain amylase inhibitors supposedly protecting the starch from being consumed by microbes, insects, or higher animals. Ho et al. (1994) have published a comprehensive review regarding content of amylase inhibitor in plants. They cluster the amylase inhibitors into three groups: (1) the microbial N-containing carbohydrates; (2) the microbial polypeptides; and (3) the higher plant proteins. In this context, the plant inhibitors that are relevant for aquaculture are described. One plant may contain several proteins with amylase inhibitor activity, and some of them may bind more than one amylase molecule, or another, additional enzyme, e.g., trypsin. The different inhibitors may differ in their activity toward insect and animal amylases.

Jaffe et al. (1973) screened 95 cultivars of beans and found that lima beans (*Phaseolus lunatus*), mung beans (*Phaseolus aureus*), and horse gram (*Dolichos biflorus*) had the highest levels of amylase inhibitory activity. Also, grains contain amylase inhibitors. Wheat and rye, but not barley and oats, have shown high amylase inhibitor activity (Granum 1979). The structure of the inhibitor in the beans and grains is not well described. The mechanism of action may be as for the amylase inhibitor produced by *Streptomyces tendae* (Konig et al., 2003). It works by binding to the active site of the amyalse and prevents binding of starch molecules. The available information on effects of amylase inhibitors in fish is very limited. An in vitro study has shown that amylases from carp and rainbow trout are inhibited by amylase inhibitors from wheat (Hofer and Sturmbauer 1986). The results indicate that the sensitivity of the amylase inhibitors may differ between fish species. It seemed to be lower for the rainbow trout than the carp amylase. No reports of feeding studies with fish have been found. A study with rats showed negative effects of amylase inhibitor from kidney beans, when included at 3.3 and 6.6 g/kg in the diet, not only on starch digestibility but also on protein digestibility. The amylase inhibitor also caused increased losses of lipids from the body and retarded growth. This demonstrates the interaction between the digestive processes taking place in the intestine. When one part of the digestive process is disturbed, also other processes are affected (Pusztai et al., 1995). Moreover, the relative weights of the intestines and the pancreas were higher, whereas liver and thymus weights were lower than in control rats. There are good reasons to believe that at least some species of cultivated fish may be affected by amylase inhibitors in a similar way as the rat. However, the Atlantic salmon may be affected differently, as their amylase seems to be defective due to deletion of an amino acid sequence essential for binding of amylose to the active site. It may therefore be unable to bind an inhibitor (Froystad et al., 2006).

12.2.1.1.3 Lipase inhibitors

In the intensive search of the aquaculture industry for new nutrient sources, research is underway to find whether also macroalgae may be of value. A wide range of macroalgae contain heat stable inhibitors of pancreatic lipase. A review by Wan-Loy and Siew-Moi (2016), with a main focus of finding compounds for fighting obesity in humans, gives detailed and comprehensive insight into the presence of lipase inhibitor in macroalgae as well as in other marine sources. The lipase inhibitors are polysaccharides such as alginate and fucoidans, polycyclic compounds such as phlorotannins, and simpler compounds such as the carotenoid fucoxanthin (Wan-Loy and Siew-Moi 2016). They all show antilipolytic activities and cause weight reduction in experimental animals. They may also have beneficial, functional, and medicinal effects (Wan-Loy and Siew-Moi 2016). As the lipase inhibitors in macroalgae are heat stable, they will survive extrusion. It may be questioned whether fish lipases are inhibited by the compounds inhibiting pancreatic lipase in land animals and humans, as lipase from fish appear to work independently of colipase (Smichi et al., 2017). One study investigated the effects of a heat-stable lipase inhibitor supplemented to the Atlantic salmon diet, i.e., effects of Orlistat, a commercial product used for reducing obesity in humans. It is a saturated form of lipstatin, a lipase inhibitor isolated from the bacteria *Streptomyces toxytricini*. Inclusion of Orlistat at a level of 1 g/kg diet reduced lipid digestibility from 97% to 32% in the salmon, indicating that binding of an inhibitor from algae will also bind to fish lipases (Chikwati, 2007). If macroalgae ingredients become important nutrient sources for fish and other production animals, an eye should also be kept on possible effects on lipid digestion.

12.2.1.2 Lectins (hemagglutinins)

Lectins, also called hemagglutinins due to their ability to agglutinate blood cells, i.e., clumping the cells, are proteins that bind to specific carbohydrate structures, mono- or oligosaccharides. Several hundred plant lectins have been isolated and characterized in detail (see review of Van Damme et al., 1997). They are found in all organisms. The carbohydrates they bind may be part of receptors on cell membranes or free-floating complexes. In an animal body, lectins are important in regulation of cell and organ function, and in signaling. They also mediate attachment and binding of bacteria, viruses, and fungi to their intended targets. The review of Ghazarinan et al. (2011) gives a comprehensive overview of glycobiology and the role of lectins. Effects of lectins eaten by an animal will depend on the function of the receptors they bind to in the intestinal mucosa, and possibly after absorption. Some lectins are aggregates of more than one protein molecule, and some of them are heterogeneous. Tetramers are the most common, allowing the lectins to connect to receptors of four cells simultaneously and thereby, efficiently, agglutinate the cells (Fig. 12.3). The lectin of several plants, e.g., the mistletoe, contains in addition to the carbohydrate binding polypeptide, a protein with ribosome inactivating ability of type II (RIP II), which may have anticancerogenic action (Voelter et al., 2005).

Hendriks et al. (1990) studied binding of soybean lectins to enterocytes from Atlantic salmon. The results showed that tissues from both the proximal and the distal intestine bind this lectin. The binding appeared to be stronger for the cells from the distal than the proximal intestine. However, feeding soybean lectin does not appear to impair digestive

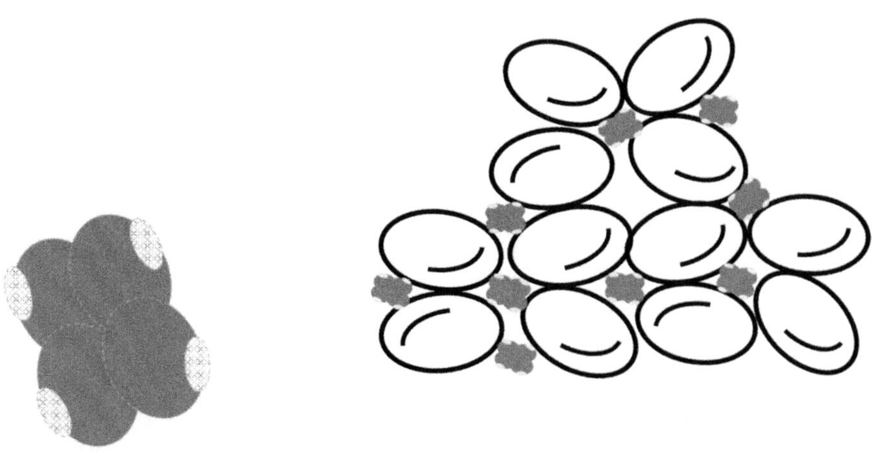

FIGURE 12.3 Illustration of a lectin molecule with four monomers, each with one active site (left), and of blood cells agglutinate by s lectin (right).

and absorptive functions in salmonids (Buttle et al., 2001; Iwashita et al., 2008). As various lectins from different sources are highly specific in their binding, upper levels must be determined for each individual lectin. Legumes used as foods and feeds contain the less potent lectins, but without heating, they disturb intestinal functions and are without value as nutrient sources. The extrusion process used for fish feed today inactivates the lectins to a large extent. However, incomplete extrusion and pelleting may render some lectins active, resulting in impaired gut function and health.

12.2.1.3 Saponins

Saponins are a highly diverse group of triterpenoidal or steroidal glycosides naturally occurring in plants. Over 100 families of plants are known to contain saponins, and considerable heterogeneity of chemical structures exists. Seeds of soy, pea, sunflower, and lupin are among the most important animal feed ingredients of plant origin containing significant amounts of saponins (Shi et al., 2004). Soybeans generally contain higher levels of saponins than other feed ingredients, typically in the range of 5–7 g/kg (Ireland et al., 1986; Knudsen et al., 2006). Saponins are heat-stable amphipathic molecules, consisting of a hydrophobic steroidal or triterpenoid aglycone (sapogenin) to which one or more hydrophilic sugar chains are glycosidically bound (Fig. 12.4).

The sugar sidechains usually contain glucose, arabinose, galactose, glucuronic acid, xylose, rhamnose, or methyl pentose. The complexity of the saponin structure arises from the variability of the aglycone structure, the nature of the sugar side chains, and the position of attachment of these moieties on the aglycone. For example, soybeans contain triterpenoid saponins that can be divided into group A, B, and E soyasaponins based on their aglycone structure (Guclu-Ustundag and Mazza 2007). Saponins are reported to have diverse biological effects including antifungal and antiviral activity, immune stimulation, anticancer effects, antioxidant properties, inhibition of protein digestion and vitamin absorption, and glucocorticoid like effects (Francis et al., 2002; Guclu-Ustundag and Mazza 2007). Many of these effects have been attributed to the amphiphilic structure of saponins. For example, saponins

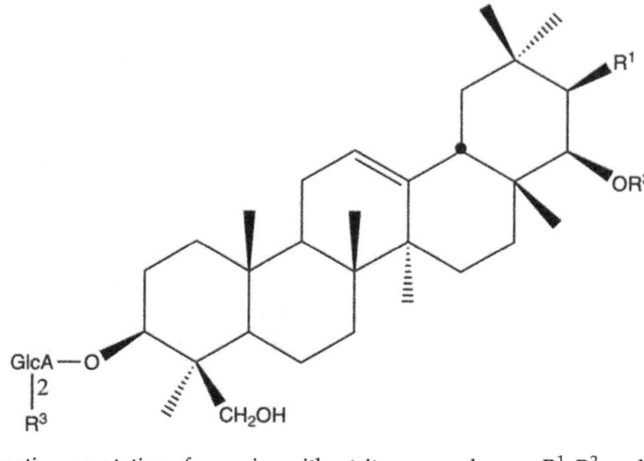

FIGURE 12.4 Schematic presentation of saponins with a triterpene aglycone. R^1, R^2, and R^3 represent a hydrogen atom, a hydroxyl group, and/or glucose, arabinose, galactose, glucuronic acid, xylose, rhamnose, or methyl pentose, alone or in pairs. Saponins may also contain other aglycones such as the steroids spirostanol and furostanol.

can bind to membrane cholesterol, form holes, and thereby increase cellular permeability, which may in turn facilitate intestinal uptake of macromolecules, including allergens and antigens (Francis et al., 2002; Gee et al., 1996). As a result of their detergent and surfactant properties, dietary saponins likely disturb fat emulsification and formation of micelles and absorption of their constituents, i.e., bile salts, fatty acids, fat-soluble vitamins, and other lipid soluble compounds. Most saponins can form complexes with intestinal bile salts and cholesterol, thus decreasing intestinal absorption and reabsorption (Messina 1999). The usefulness of some of the saponins as adjuvants in vaccines is well documented (Oda et al., 2000). Saponins are highly toxic to fish when added to the surrounding water. They are the active ingredient in several traditional fish poisons and act by damaging the respiratory epithelium of the gills by their detergent actions.

The work by Knudsen et al. demonstrated that orally administered soyasaponins are not degraded during gastrointestinal passage in Atlantic salmon (Knudsen et al., 2006). Several studies have reported that low dietary supplementation levels of *Quillaja* or soy saponins can improve growth in fish (Francis and Makkar 2005; Krogdahl et al., 2015; Serrano Jr 2013). This effect may be a result of increased feed intake, as previously demonstrated in oriental clouded yellow larva, *Coliaserate poliographus* (Matsuda et al., 1998). However, it is clear that higher dietary levels of saponin can exert adverse effects in fish, and there are now strong indications that saponins play a key role in the onset and development of the noninfectious distal intestine enteritis induced by full-fat and solvent-extracted soybean meal in salmonids (Baeverfjord and Krogdahl 1996; van den Ingh et al., 1991). The general good correspondence between predicted saponin content in various soy feed ingredients and their negative effects on growth performance, nutrient digestibility, and gut health in salmonids has led to a number of studies using purified or semipurified saponins. These studies investigate whether saponins, alone or in combination with other ANFs, can be the responsible agent for the observed negative effects. Extensive damage to the intestinal mucosa by dietary saponins was first observed in chinook salmon (*Oncorhynchus tshawytscha*) and rainbow trout (*Oncorhynchus mykiss*) fed levels of 0.15%–0.30% *Quillaja* bark saponin (Bureau et al., 1998). Later, a study conducted in Atlantic salmon (Chikwati, E. 2007) found that highly purified (95%) soy saponins added at a level of 0.2%, equivalent to a diet containing about 400 g soybean meal kg^{-1}, decreased brush border membrane enzyme activities and induced mild histological changes in the distal intestine, resembling the inflammatory changes observed in salmon suffering from soybean meal–induced enteritis. Similar exposure studies in sea bass (*Dicentrarchus labrax*) and sea bream (*Sparus aurata*), at juvenile and on-growing stages, have indicated that these species are not responding with inflammation as the Atlantic salmon, when fed purified saponins, although the sea bass juveniles showed some alterations in digestive and immune functions (Couto et al. 2014a, 2014b, 2015a, 2015b). Recently, high levels of purified soy saponins (0.5% and 1.0%) was found to reduce growth performance and feed efficiency and disturb the morphology of the intestinal mucosa in zebrafish (Tian et al., 2018). Waterborne exposure to purified soy saponins in zebrafish larvae further demonstrated interaction effects between saponin-induced immune responses, antibiotic treatment, and microbial composition (Lopez Nadal et al., 2018). Although the practical relevance of such studies for fish nutrition can be questioned, they may provide insights into saponin's mode of action in induction of inflammatory phenotypes.

There are several indications of interaction and/or potentiation effects between saponins and plant feed ingredients. For example, supplementation of a semipurified soy saponin to a fishmeal-based diet (0.17% and 0.26%) did not alter the intestinal morphology of Atlantic salmon, but when the diet contained lupin kernel meal, the saponin supplementation caused histological and pathophysiological changes similar to those that occur in salmonids fed soybean meal and increased epithelial permeability in vitro (Knudsen et al., 2008). Similar interactions/potentiation effects between purified soy saponin and plant feed ingredients were observed in the studies conducted by Chikwati and Kortner (Chikwati et al., 2012; Kortner et al., 2012). Supplementation of 0.2% purified soy saponin to Atlantic salmon diets induced distal intestine enteritis and a wide range of accompanying pathophysiological changes, but only when the basal diet contained high levels of pea protein concentrate. Peas also contain significant levels of saponins and have been found to induce similar histopathological changes in the distal intestine of salmon at high inclusion levels (Penn et al., 2011). Recently, a study with Atlantic salmon by Krogdahl and coworkers described the dose–response relationship between highly purified (95%) soya saponin supplemented at 0, 2, 4, 6, or 10 g/kg to two basal diets, one containing fishmeal as the major protein source and the other 20% lupin meal (Krogdahl et al., 2015). Saponins caused dose-dependent increases in the severity of inflammation independent of the basal diet, with concomitant pathophysiological alterations in digestive, barrier, and immune functions of the distal intestine. However, responses were often stronger in fish fed the saponin-supplemented lupin meal diets than in the corresponding fishmeal diet groups, suggesting potentiation by other legume component(s). A significant interaction between high dietary saponin levels and basal diet composition was also observed on hypocholesterolemic effects and alternations in sterol metabolism (Gu et al., 2014). A similar dose–response study conducted in juvenile turbot (*Scophthalmus maximus*) fed fishmeal-based diets containing graded levels of purified soy saponins (0, 2.5, 7.5, 15 g/kg) also demonstrated that soy saponins induced intestinal inflammation and barrier dysfunction in a dose-dependent manner (Gu et al., 2018). Synergism between dietary saponins and phytosterols on gut structure and function has also been suggested (Chikwati, 2007; Couto et al., 2014a, 2014b, 2015a, 2015b). Caution should therefore be exercised to avoid ingredient combinations giving high saponin levels, or combination of high levels of saponins and other plant ANFs.

12.2.1.4 *Phytoestrogens and phytosterols*

Phytoestrogens (Fig. 12.5) are nonsteroidal compounds of plant origin with structural similarity to 17β-estradiol, the primary female sex hormone. Phytosterols, on the other hand, are plant sterols with a structure similar to that of cholesterol, varying only in sterol side chain and/or double bond structure (Fig. 12.6). Both phytoestrogens and phytosterols are considered ANFs due to their effects on lipid, sterol, and glucose metabolism. Because of the similarity of observed responses in fish, the two compound classes are presented together in this section.

There are four main classes of phytoestrogens: isoflavones, coumestans, prenylflavonoids, and lignans. Phytoestrogens appear in more than 300 plant species, but few of these are consumed by animals or humans (Dixon, 2004). A great number of phytoestrogens have been identified in fruits, vegetables, and whole grains but are most abundant in leguminosae. Soybeans and other oilseeds (such as linseed), clover, and alfalfa sprouts are the most

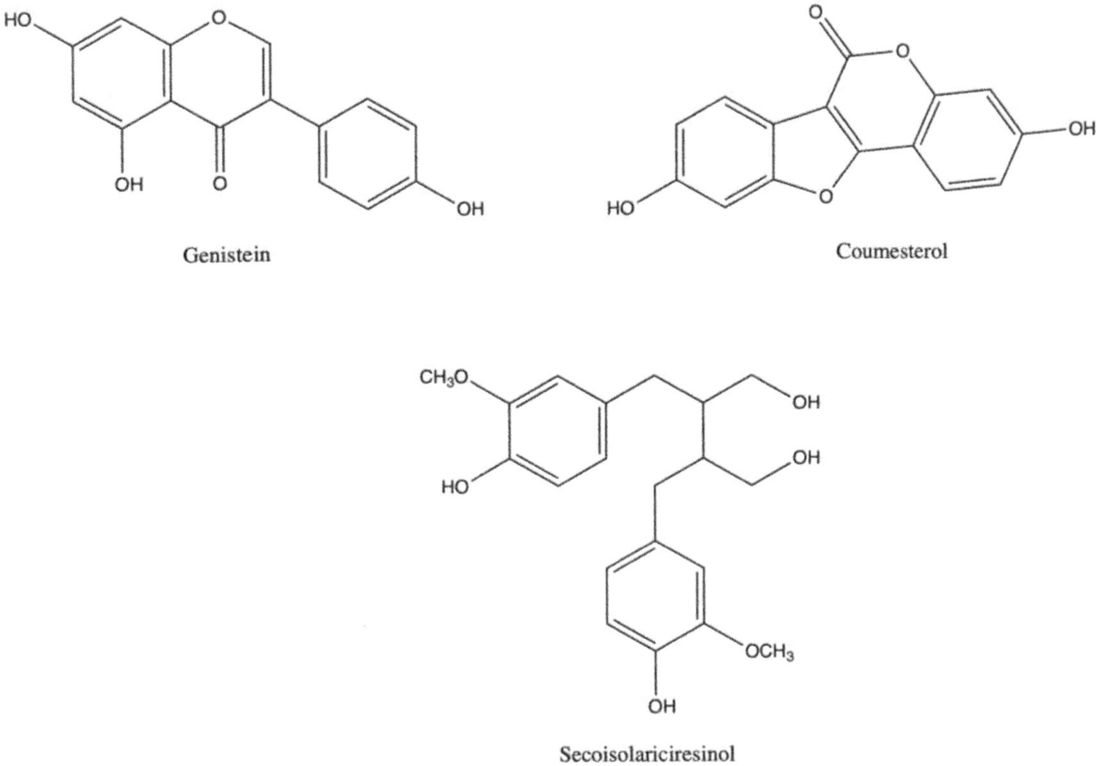

FIGURE 12.5 Schematic presentation of the phytoestrogens genistein, coumesterol, and secoisolariciresinol belonging to different groups of compounds, i.e., isoflavonoids, coumestans, and lignans, respectively.

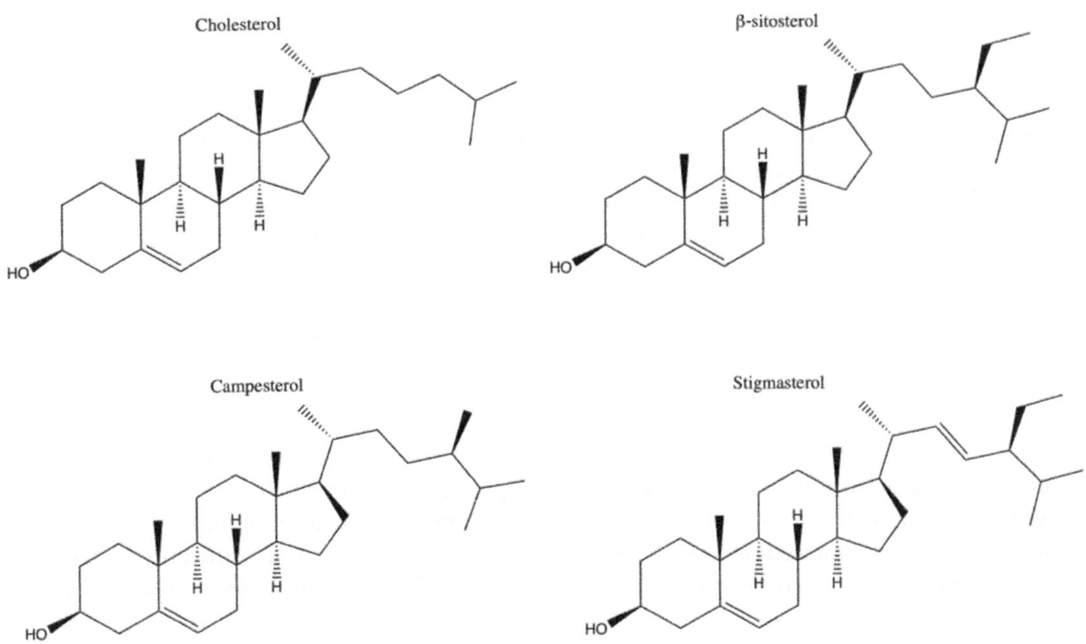

FIGURE 12.6 Chemical structures of cholesterol and various phytosterols.

significant dietary sources of isoflavones, coumestans, and lignans, respectively (Kurzer and Xu 1997). The main soybean isoflavones, i.e., daidzein, genistein, and glycitein, commonly exist as glucosides which are cleaved after ingestion by intestinal glucosidases. The main mechanism of action by which phytoestrogens may exert their estrogen-like biological activities is by binding to the estrogen and other hormone receptors. Dietary phytoestrogens may interfere with estrogen uptake, bind to receptors regulating reproduction, and affect feed intake and lipid, sterol, and glucose homeostasis in humans and animals. In humans, various beneficial health effects have been ascribed to phytoestrogens, such as a lowered risk of menopausal symptoms, cardiovascular disease, obesity, metabolic syndrome and type 2 diabetes, brain function disorders, breast cancer, prostate cancer, and bowel cancer. The estrogenic properties of phytoestrogens have also raised concerns since they might act as endocrine disruptors, indicating a potential to cause adverse health effects (Rietjens et al., 2017).

Phytosterols, including sitosterol, stigmasterol, and campesterol, are the plant equivalents of cholesterol. They are mostly found in plant oils and lipid-rich plant meals. For example, soybean phytosterols consist mainly of β-sitosterol (43%–67%), campesterol (17%–34%), and stigmasterol (10%–30%) (Yamaya et al., 2007). In contrast to the phytoestrogens, most phytosterols do not bind to the human estrogen receptors. They are known to lower plasma total cholesterol and low-density lipoprotein cholesterol levels in mammals, thus offering a prophylactic treatment against cardiovascular disease in humans. Inhibition of intestinal cholesterol absorption due to displacement of cholesterol from micelles is generally accepted as the underlying mechanism (Ling and Jones 1995).

Effects of phytoestrogens, in particular genistein, on growth, health, and reproductive function of fish have been an active area of research during the past decades. Direct evidence for estrogenic activity of the phytoestrogens daidzein, biochanin A, genistein, equol, and coumestrol, as assessed by their ability of inducing vitellogenin synthesis, was first reported in Siberian sturgeon (*Acipenser baeri*) by Pelissero et al. (1991). However, all the phytoestrogens tested were considerably less potent than 17β-estradiol. Later studies have demonstrated increased levels of vitellogenin production after genistein exposure in several different fish species (Bennetau-Pelissero et al., 2001; Inudo et al., 2004; Kausch et al., 2008). Endocrine-disrupting effects of genistein and other soybean isoflavones have also been demonstrated by inhibition of hepatic and renal 17β-estradiol metabolism in salmonid fish (Ng et al., 2006). The affinity of genistein for blood plasma–binding proteins in fish is, however, much lower than endogenous estrogens (Milligan et al., 1998; Tollefsen et al., 2004). The main soybean isoflavones daidzein, genistein, and glycitein were found to be effectively cleared by phase II uridine–diphosphate–glucuronosyltransferase in salmon liver microsomes and primary hepatocytes, suggesting that bioaccumulation of phytoestrogens in farmed salmon is unlikely (Johny et al., 2020).

Contrasting effects of genistein and other phytoestrogens on fish growth performance, nutrient utilization, and health have been reported, and the effects seem to depend on dietary dosage, species, and, possibly, developmental stage. Low dietary supplementation of genistein seemed to improve growth in yellow perch (Ko And et al., 1999) but has been shown not to affect growth in rainbow trout (Bennetau-Pelissero et al., 2001; D'Souza et al., 2005). Significant reductions in feed intake, feed efficiency, and nutrient digestibility were observed in Japanese flounder (*Paralichthys olivaceus*) fed high levels of soy isoflavones (Zhang et al., 2012). Dietary supplementation of genistein also led to reduced feed intake and slightly increased protein utilization in rainbow trout, but did not affect whole body fatty acid composition (Torno et al., 2019). High dietary administration of genistein to southern flounder (*Paralichthys lethostigma*) induced feminizing effects and reduced growth and survival (DiMaggio et al., 2014). Negative effects of soy isoflavones on growth performance, digestive function, and bone formation have also been observed in Atlantic salmon fry (Gu et al., 2015). In a recent study in rainbow trout, no effects of dietary isoflavones were observed on reproduction, growth, and gut health (Pastore et al., 2018). In contrast, dietary administration of soybean isoflavones was reported to improve growth performance, innate immune responses, hepatic antioxidant abilities, and disease resistance of juvenile golden pompano (*Trachinotus ovatus*) (Zhou et al., 2015). Based on available literature, the sensitivity to phytoestrogens seems to vary among species, and caution should be exerted to avoid high dietary levels.

The replacement of fish oil and fishmeal with plant raw materials in feeds for farmed carnivore fish species has caused a shift in dietary sterol profile, with increased phytosterol concentrations and corresponding decreases in cholesterol content. It has been a concern whether this shift in sterol composition could lead to decreased availability of dietary cholesterol and that compensatory mechanisms could lead to unwanted metabolic effects (Liland et al., 2013). A number of studies conducted in recent years have shed light on how dietary phytosterols may affect fish, including effects on sterol absorption and homeostasis, growth performance, and immune response. Similar to observations in humans, dietary phytosterols seem to be poorly absorbed in the fish intestine as compared with cholesterol, but phytosterols accumulate in fish liver and muscle (Miller et al., 2008). Campesterol and brassicasterol appeared to be the phytosterols with the highest intestinal absorption in Atlantic salmon (Sissener et al., 2018), corresponding to observations of higher apparent digestibility and accumulation of campesterol than β-sitosterol (Miller et al., 2008). The very low plasma levels of sitosterol and campesterol found in fish fed diets supplemented with cholesterol are likely to reflect a competition between cholesterol and phytosterols in uptake from the intestine (Kortner et al., 2014). On the other hand, dietary phytosterols do not seem to affect cholesterol absorption or tissue cholesterol or phytosterol levels (Sissener et al., 2018), indicating that total cholesterol level and the phytosterol:cholesterol ratio are the most important dietary factors affecting sterol composition of the body. In accordance, a study in zebrafish, employing radiolabeled cholesterol, showed no effects of dietary phytosterol supplementation on the excretion or tissue distribution of dietary cholesterol 24 h after the meal (Spankeren et al., 2019). The hypocholesterolemic effect of phytosterols is, however, well described in mammals and has been demonstrated in feeding trials with Atlantic salmon fed diets with 0.5% and 1.0% phytosterols (Chikwati, 2007) or vegetable oils with high phytosterol levels (Liland et al., 2013), and in sea bass juveniles fed a diet with 0.5% phytosterols (Couto et al., 2015b). In Chikwati's study, dietary phytosterols lowered lipid digestibility as well as level of bile salts in the intestinal chyme and in blood plasma (), probably resulting from reduced hepatic levels of cholesterol as substrate for primary bile acid synthesis. However, graded levels of phytosterols, cholesterol, and their ratio exhibited few effects on salmon growth performance, lipid metabolism, and health (Sissener et al., 2017). In a series of studies conducted in juvenile and on-growing gilthead sea bream (*Sparus aurata*) and European sea bass (*Dicentrarchus labrax*), Couto and coworkers evaluated the effects of dietary supplementation with purified phytosterols, alone or in combination with soya saponin, on growth performance and feed utilization, gut physiology, and immune responses (Couto et al. 2014a, 2014b, 2015a, 2015b). Overall, the conclusion from these studies was that the relatively high concentrations of the antinutrients tested had little effect on growth performance and animal health. However, certain histological alterations and gastrointestinal disturbances were observed in the trials with the juveniles. A follow-up study with gilthead sea bream, targeting systemic immune responses, showed that dietary phytosterols, alone or in combination with saponins, induced a number of changes indicating induction of inflammation processes (Costas et al., 2014). Whether these changes might affect the outcome of pathogen challenges has not been investigated to date. Anyhow, the sterol profile of the feed should be monitored if high levels of plant feed ingredients are being used.

12.2.1.5 Glucosinolates

Glucosinolates are amino acid–derived sulfur-containing plant metabolites with variable glycosylation (Fig. 12.7). A wide range of glucosinolates have been identified in plants. Most members of the brassica family used for food or feed, including rapeseed, cabbage, broccoli, and cauliflower, mustard seed, and kale, contain glucosinolates (Blažević et al., 2020). The glucosinolates are not particularly harmful themselves, but hydrolysis by the enzyme myrosinase releases toxic compounds. Myrosinase is a constituent of plant cells, which is released from the cell upon stress, or under processing. Examples of relevant toxins are goitrin, sinapine, organic nitrites, isothiocyanates, and thiocyanate ions (Fig. 12.7). All of

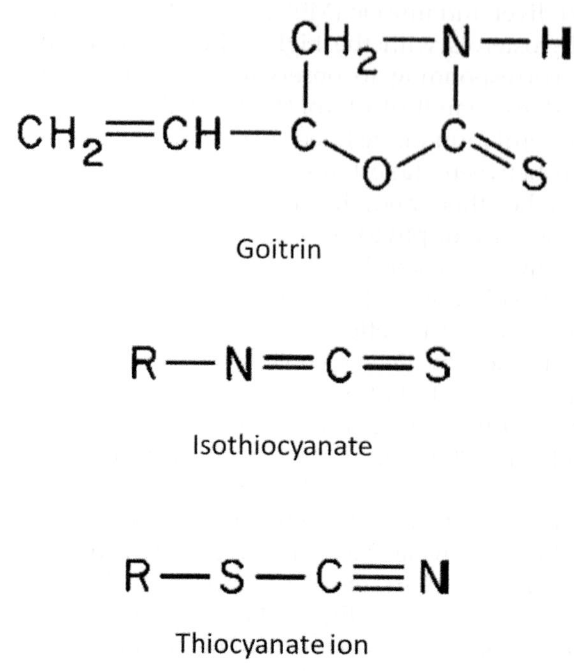

FIGURE 12.7 Schematic presentation of goitrin, one of the glucosinolates found in members of the brassica family, and the key metabolites, isothiocyanate and thiocyanate (Hendricks, 2002a).

the hydrolysis products of glucosinolates are antithyroid substances (see the review by Bala et al., 2015). Goitrin, the most potent, exerts its effects through the inhibition of binding of iodine to the thyroxin hormone. The addition of iodine to the diet does not reverse the thyroid effect. In contrast, thiocyanate acts by lowering iodine uptake by the thyroid, and increased iodine supply will prevent goiter. Less is known about the mode of action of other breakdown products, i.e., organic nitrites and organic isothiocyanates, but they are thought to function as goitrogens through the production of thiocyanate ion and subsequent lowering of iodine uptake by the thyroid (Tookey et al., 1980).

The glucosinolates have become important issues in fish nutrition with the increasing interest in rapeseed meal as a low-cost protein source. Successful efforts have been made by plant breeders to remove the glucosinolates from this crop together with the other antinutrient in rape seeds, i.e., erucic acid, sinapine, and fiber (Hannoufa et al., 2014). The beneficial effects for the rape seed of these antinutrients are insecticidal, bactericidal, or fungicidal properties. The level of glucosinolates in the new rape seed varieties is as low as 10 mmol/kg, corresponding to 3 g/kg goitrin, compared with that of older varieties, which may contain more than 25 g/kg (EFSA, 2008).

Burel et al. (2001) studied effects of a rapeseed meal containing 26 mmol/kg glucosinolates, incorporated at increasing levels from 10% to 30% in the diet of juvenile rainbow trout. Disturbances in the thyroid axis appeared after 14 days of feeding with a rapid decrease of plasma T_4 and T_3 levels. Growth was not affected until after 6 months of feeding. Dietary supplementation with T3 or iodine induced an increase in plasma T_3 levels and reduced the negative effect of the rape seed meal on growth. In a more recent experiment (Hannoufa et al., 2014) with hybrid striped bass (*Morone chrysops*), reduced feed intake was observed with diets containing 6–10 mmol/kg diet of glucosinolates (from the *Brassica carinata*). The fish exhibited lethargic swimming and feeding behaviors, and consequently poor feed utilization and growth. Extrusion of the feed reduced glucosinolates by about 50%. The fish seemed to tolerate up to 2.7 mmol of glucosinolates and up to 2.5 mmol sinapine/kg of diet (0.8 mg/kg) without affecting feed consumption and utilization, growth, and thyroid metabolism. Although literature describing effects of purified glucosinolates is limited, it seems clear that glucosinolate levels should be kept low in fish diets.

12.2.1.6 *Erucic acid*

Erucic acid is a 22-carbon monounsaturated fatty acid (22:1 ω-9; *cis*-13-docosenoic acid) (Fig. 12.8). It is mainly found in oil from rape seeds, an oil that is used extensively in human foods as well as in fish diets. In the early 1970s, years with very high incidence of heart diseases in humans in the western world, this fatty acid was investigated for its potential role in such severe health problems. Rat studies indicated that erucic acid increased lipid accumulation in the heart muscle and that the cause was slower metabolism in the organ (Kako and Peckett 1981). These observations triggered research in many laboratories and with many animal species, with the aim to understand the mechanism of action and

$$CH_3(CH_2)_7HC{=}CH(CH_2)_{11}COOH$$

FIGURE 12.8 Schematic presentation of erucic acid, an antinutrient in rapeseed oil.

consequences for human health. The results are summarized in a risk evaluation from Australia and New Zealand (2003). Although the studies, all in all, indicated that the erucic acid did not play an important role in the severe situation regarding heart diseases in humans at that time, safety levels were developed for humans, which were adopted worldwide. The consequence of the concern regarding negative effects of erucic acid stimulated plant breeders to reduce the level in the oil, and they were successful. Rapeseed oil from traditional breeds contains erucic acid in the range 20%–55% (Slinger 1977), while today oils with less than 1% erucic acid are available.

Knowledge on effects of erucic acid in fish is very limited, but the available information indicates that fish may be more vulnerable than humans. In a study of Parker and Hendricks (unpublished, cited by Hendricks and Bailey, 1989), coho salmons were fed diets with rapeseed oil containing 50% erucic acid, at a level of 6% for 6 months and thereafter at 12% for 4 months. The fish fed the rapeseed oil grew more slowly, had higher mortality, and showed histopathological alterations in the skin, gills, kidney, and heart. Detrimental effects of erucic acid have also been observed in grass carp (*Ctenopharyngodon idella*) in a recent challenge trial with *Aeromonas hydrophila* causing inflammation in various tissues including the intestine (Gan et al., 2020). Diets with 0, 3, 6, 9, 12, and 15 g/kg of erucic acid were fed to the fish for 60 days before they were challenged with an intraperitoneal injection of the bacteria (Kong et al., 2017). Clear negative effects of the erucic acid on severity of the intestinal inflammation were observed. Based on the results regarding enteritis, morbidity, and the activities of lysozyme in proximal, mid-, and distal intestine as response criteria, the maximum tolerance levels of erucic acid for on-growing grass carp were estimated to be in the range 3.6–5.3 g/kg diet, averaging 4.5 g/kg. If the same levels are valid for salmons, the current levels in salmon diets should be well below the tolerance level even at the highest inclusion level of rapeseed oil.

As discussed earlier, erucic acid is a concern not only for cultivated fish but also for the consumer of the fish. The work of Sissener et al. (2018) discusses this issue showing that the level of erucic acid in salmon fish feed in Norway, collected over the years 2012–16, averaged 0.6% in the oils and 1.3 g/kg in fish diets (Sissener et al., 2018). The conclusion of this review was that, with today's fish diets, with low erucic acid oils, the fish contains erucic acid at a level, which implicates that one meal containing 200 g of fatty fish supplies this fatty acid at a level close to the tolerable daily intake defined by EFSA (Knutsen et al., 2016). Rapeseed oil has become invaluable for the aquaculture industry helping to keep the recommended n-3/n-6 balance in the diet. It is, however, necessary to use the new oils with low content of erucic acid.

12.2.1.7 Tannins

Tannins are polyphenolic compounds, which can be divided in three groups based on their monomers. One group is flavone derived (Fig. 12.9), also called condensed tannins, and a second group, derived from gallic acid, is often called hydrolyzable tannins. Both of these are found in plants living on land. A third group based on phloroglucinol and called phlorotannins is found in seaweed. Many feed ingredients contain tannins, in particular legumes and grains. The level is particularly high in sorghum, which may contain up to 5% of flavone-derived tannins. Also beans such as lima and faba beans, as well as sunflower seeds, contain substantial amounts (reviewed by Tacon, 1997). The antinutritional characteristic of tannins is

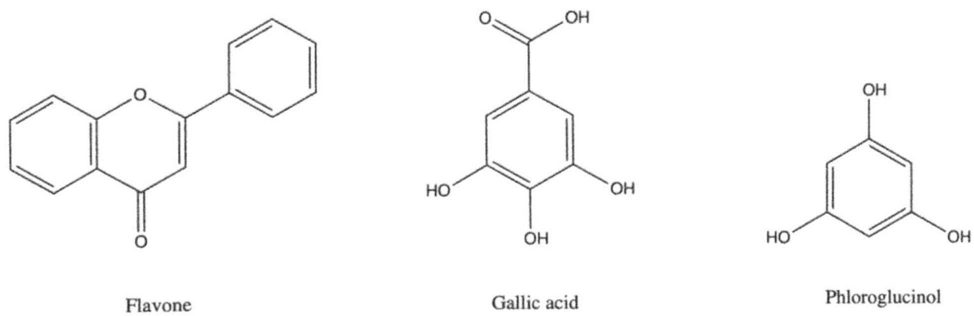

FIGURE 12.9 Schematic presentation of the most common core molecules of tannins. (Left) flavone, in the polymer form called condensed tannins. (Middle) gallic acid, in the polymer form called hydrolyzable tannins. Both the condensed and hydrolyzable tannins are found in plants. A tannin based on phloroglucinol (right), contained in phlorotannins, is found in seaweed.

their ability to precipitate protein and thereby cause reduced protein digestibility and utilization in monogastrics (Duodua et al., 2003; Maclean et al., 1983). Heat treatment increases the binding and the negative effects. In ruminants, on the other hand, this ability has been used to reduce protein degradation in the rumen and thereby increase protein availability to these animals (Naumann et al., 2017). Tannins are claimed to have functional characteristics and are investigated for their antimicrobial, antiparasitic, and immunomodulation properties (Huang et al., 2018). They are efficient antioxidants.

Very limited information is published regarding effects of tannins in fish. One dose–response study of effects of hydrolyzable tannins in grass carp was recently reported by Yao et al. (2019). A purified preparation was used and included at four levels, 0%, 0.75%, 1.25%, and 1.75%, in a semipurified basal diet. The diets were pelleted and thereafter dried at 40°C for 12 h. No effect on growth was observed, but feed intake and feed conversion ratio increased. Nutrient digestibility was not measured. Muscle lipid and viscera weight decreased, indicating an energy cost of inclusion of the tannins. Increasing activity of digestive enzymes was also observed (U/mg protein), suggesting that the tannins challenged the digestive processes. The mechanism may be that the tannins bind to the proteolytic enzymes and act as protease inhibitors. This will trigger the pancreas to produce and secrete more enzymes. However, the diets were processed under very mild conditions in this study. It might be expected that processing at high temperatures, such as used in production of salmon feed, would have resulted in clearer effects of the inclusion of these tannins.

12.2.1.8 Phytic acid

Phytic acid is the hexaphosphate of myo-inositol (Fig. 12.10), which strongly binds divalent ions. A high level in the diet reduces the availability of ions such as Fe^{++}, Ca^{++}, Mg^{++}, and Zn^{++}. The antinutritional effects are mainly related to this characteristic, which reduces the availability of such minerals. Phytic acid is found in all cereal products, as well as in oilseeds such as soybeans, rapeseeds, and cottonseeds (Tahir et al., 2012). Of importance is also the high content of phosphate in phytic acid. The phytic acid phosphate in plant feedstuffs accounts for a large proportion of the phosphate in feeds (Tahir et al., 2012), and as the phytic

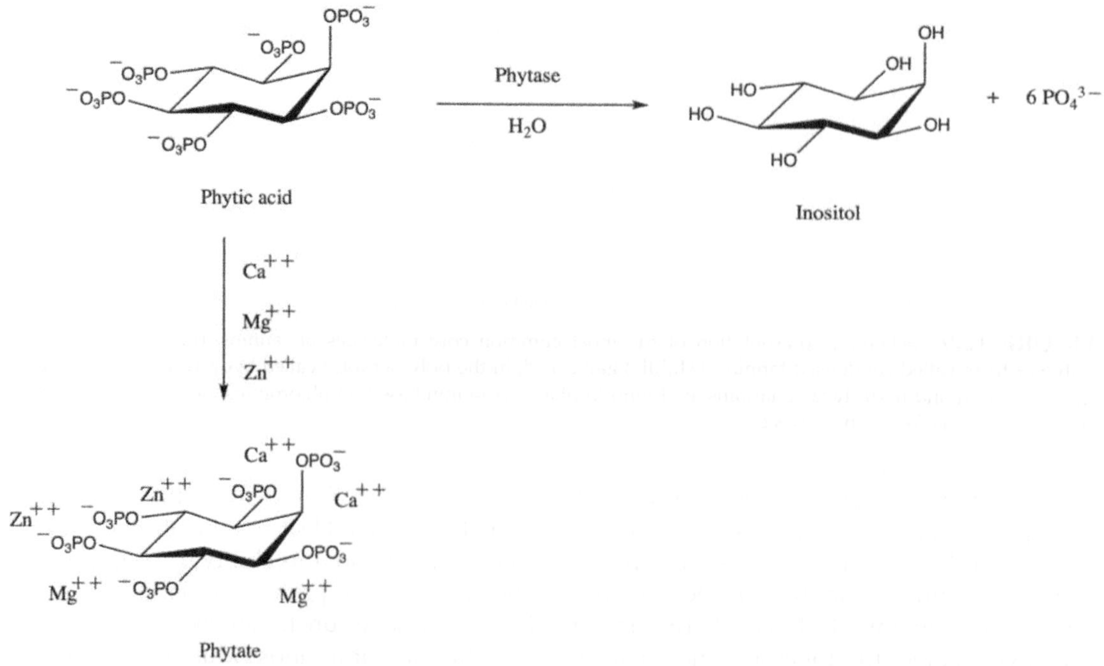

FIGURE 12.10 Schematic presentation of phytic acid with indication of the salt form and hydrolysis by phytase, which releases the phosphate groups as well as divalent ions bound to the molecule.

acid has low digestibility, the phosphate digestibility in plants is low. The consequence of low digestibility of the phytic acid phosphate is high output to the environment, and loss of phosphate, mostly to rivers and lakes where it becomes a pollution problem. It is not uncommon to increase mineral and phosphate availability of today's monogastric animal feeds by inclusion of phytases (Dersjant-Li et al., 2015). Phytases, isolated from bacteria and fungi, are available. They seem to work in the stomach and proximal part of the small intestine before they are inactivated (Yan et al., 2002). However, in cold-water fish species, the activity of the phytases is low, and higher inclusion levels are needed (Morales et al., 2011). Studies of effects of phytase inclusion in fish diets show variable effects on digestibility of phosphate (Jackson et al., 1996; Storebakken et al., 1998; Vielma et al., 1998). Improvements from digestibilities as low as 10% to about 60% have been observed, depending on the digestibility in the basal diet (Verlhac-Trichet et al., 2014; Vielma et al. 1999, 2000). These studies also show concomitant improvement in growth. In a recent study with turbot, negative effects on feed efficiency and protein digestibility were observed when the fish were fed a semipurified diet based on casein, supplemented with 10 g/kg phytic acid, i.e., a level commonly found in the range of today's commercial diets (von Danwitz and Schulz 2020). A series of studies with grass carp (Zhong et al. 2019, 2020) addressed phytic acid effects on growth and feed as well as on immune and inflammatory responses in the intestine, head kidney, spleen and skin, and disease resistance. The fish were fed semipurified diets supplemented with 0, 8, 16, 24, 32, or 40 g phytic acid/kg. After 60 days of feeding, the fish were challenged with *Aeromonas hydrophila* for 6 days. As in the work of von Danwitz and Schulz, the phytic acid reduced growth and feed efficiency. In addition, the study showed clear dose-related dependencies between the level of phytic acid and severity of the enteritis and the skin lesions induced by the *A. hydrophila*. The study gives, based on the biomarkers weight gain, enteritis morbidity, IgM content, and lysozyme activity in the proximal intestine, an estimate of maximum tolerance levels of phytic acid for on-growing grass carp of about 15 g/kg diet. The cause of the observed effects of phytic acid, not discussed in the paper, was most likely mineral imbalances and multiple deficiencies, increasing in severity with increasing dietary level of phytic acid (Richards et al., 2010). The study also revealed accumulation of phytic acid in the head kidney, spleen, and skin, increasing with increasing dietary level, from about 0.04% in the skin of fish fed unsupplemented diet, to 0.2% in fish fed the highest phytic acid level. This new information clearly indicates that phytic acid was absorbed to some extent and was distributed throughout the organism. The negative effects of phytic acid could therefore be a result of its action on mineral availability in both the intestine and the internal organs. Although present information on the antinutritional effect of phytic acid in fish is limited, it is clear that high levels are detrimental for growth and feed utilization and may severely interfere with the function and health of internal organs.

12.2.1.9 Gossypol

Gossypol is a dark yellow, polyphenolic aldehyde (Fig. 12.11) found in all parts of cotton, with the highest concentration in the seeds where it is stored in specific glandular structures appearing as dark dots. Gossypol exists in two forms, free gossypol, which is readily extractable with solvents, and bound gossypol, covalently bound to proteins, e.g., to lysine. The free form is the most toxic. In unprocessed seeds, most of the gossypol is in the free form, but heating may cause binding to lysine in proteins and thereby reduce toxicity. The concentration in conventional seeds varies between breed in the range 8–20 g gossypol/kg seeds (Berardi and Goldblatt 1980). However, new breeds may vary even more in gossypol content. Some breeding programs aim to increase gossypol level to obtain breeds that are more resistant to pests and diseases. Other programs aim to reduce the gossypol level to zero, to obtain breeds that can supply the world with both cotton and protein-rich foods and feeds. In the low gossypol varieties, resistance to pests is, however, reduced to the extent that it reduces

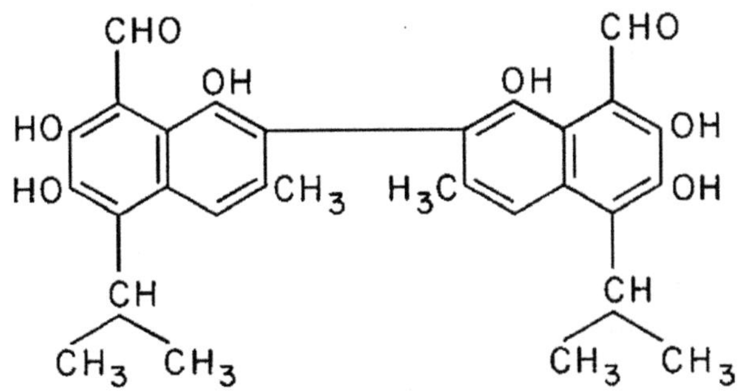

FIGURE 12.11 Schematic presentation of gossypol, the important antinutrient in cotton seeds.

production severely. By use of gene modification, however, varieties which express other antipest compounds, gossypol-free breeds, with resistance to insects as well as insecticides, can be obtained (see reviews by Cai et al., 2020 and Rathore et al., 2020).

Free gossypol is lipid soluble and readily absorbed from the intestine. It is toxic, in particular for monogastric land animals. A comprehensive review of chemical and biological characteristics of gossypol has been compiled by Knutsen et al. (2017). It also reviews knowledge on antinutritional aspects of gossypol in monogastric land animals. The mechanism of action of gossypol is not well described, but iron binding seems to be important. In land animals, intake of gossypol causes anemia, with consequences for the function of internal organs such as gonads, heart and liver (see the review by Rathore et al., 2020).

A large number of studies have been conducted to evaluate the nutritional values of various breeds of cottonseed meal for different fish species. However, as also low gossypol varieties of cotton cause some effects, studies with pure gossypol are necessary to distinguish effects of gossypol from those of other antinutrients (Liu et al., 2020; Shen et al., 2020). The available studies using purified, free gossypol are few, and only channel catfish and rainbow trout have been used as experimental species. Dorsa et al. (1982), fed 200 g channel catfish with diet containing gossypol at 0.3, 0.9, and 1.4 g/kg diet. They found growth depression at 1.4 g/kg, but there was no indication of growth effects at 0.9 g/kg. Observation of accumulation of gossypol in whole body, muscle, liver, and kidney showed increasing levels with increasing gossypol supply. Dorsa et al. concluded that 0.9 g/kg was a safe level for channel catfish. A study with juvenile channel catfish, feeding the fish soybean meal–based diets with doses of gossypol up to 0.8 g/kg, focused on effects on immune mechanisms. It included a challenge test with *Edwardsiella ictaluri* (Yildirim-Aksoy et al., 2004). The fish were fed for 12 weeks. No effect was seen on final weight gain, feed intake, feed efficiency, body composition, or survival after the challenge. The same regarded the many observed immune biomarkers, including antibody production against *E. ictaluri* 21-day postinfection. Evans et al. (2010) did a study with fingerling channel catfish, feeding the fish purified diets containing gossypol at levels of up to 1.5 g/kg diet for 12 weeks. They observed necrosis of gastric glands and significant mild-to-severe pancreatic necrosis at an inclusion level of 0.6 g/kg and above. At 0.9 and 1.5 g/kg, the liver exhibited significantly higher pigment deposition. The authors conclude that dietary gossypol levels should be below 0.6 g/kg diet.

Roehm et al. (1967), working with fingerling rainbow trout, reported no adverse effect on growth when the fish were fed gossypol acetate at 0.3 g/kg for 18 months. They found high concentrations of free and bound gossypol in the liver but low levels in the muscle. Gossypol, which had accumulated in the liver, decreased very slowly when the fish were switched to a gossypol-free diet. The work of Dabrowski et al. (2000) with rainbow trout weighing 250 g and fed for 131 days gave results in line with those of Roehm et al. The fish were fed diets with cottonseed up to 40%, replacing fishmeal. The level of free gossypol in the diets ranged from 0 to 1 g/kg, and total gossypol ranged from 0 to 4.4 g/kg. The results showed high concentration of gossypol in the blood of fish fed the highest level of cottonseed meal. Blood hemoglobin concentration decreased with increasing levels, but there were no signs of alteration in the reproductive organs and sperm. In contrast, in the study with rainbow trout by Herman (1970), effects of various cottonseeds as gossypol sources were investigated. Histological changes in the liver and kidney occurred in fish receiving 0.1 g/kg or more gossypol. Growth suppression occurred on diets containing 0.3 g/kg or more of free gossypol. Fish receiving

0.5 g/kg free gossypol had severe reductions in hematocrit, hemoglobin, and plasma proteins, in addition to growth suppression. Thickening of the glomerular basement membrane, liver necrosis, and ceroid deposition in the liver, spleen, and kidney were the specific histological changes observed.

The limited information available regarding effects of gossypol in fish and the great variation in outcome of the performed studies suggest that the tolerance level for gossypol may vary between developmental stages and species. Duration of feeding may also play a role. It seems clear, however, that channel catfish and rainbow trout can tolerate much higher levels of gossypol in the diets than monogastric land animals. More research is needed, in particular regarding other cultivated species, to establish upper limits that may secure good growth and health in fish. Estimates of upper limits are needed for increased use of cotton seed meal in aquaculture.

12.2.1.10 Cyclopropene fatty acids

Two fatty acids, sterculic and malvalic acid (Fig. 12.12), which contain a highly reactive cyclopropene ring, are found in many plants. Oils from plant families of the order *Malvales,* to which cotton belongs, contain especially high levels. The cyclopropene fatty acids are present both in the oil and the meal. Also members of the family *Fabales,* e.g., cuillaja and *Sapindale,* e.g., mango and cashew, contain high levels of cyclopropene (reviewed by Bao et al., 2002). Their biological mechanism of action seems to be an irreversible, covalent binding between the ring structure of the fatty acids and a thiol moiety on important fatty acid desaturases. The result is inhibition of the desaturase activity, which was first documented in chickens (Johnson et al., 1969; Raju and Reiser 1967). The cyclopropene fatty acids are known as antiparasites and are also investigated as medicine for various metabolic diseases including cancer (Cao H et al., 2018; Ortinau et al., 2012; Pan Hao et al., 2016; Peláez et al., 2020).

Cottonseed meal, as discussed earlier, is a feed ingredient for some cultivated fish species and of interest for many others. It is therefore necessary to define what implications the content of cyclopropene fatty acids might have in various fish species. Most information available regarding implications in fish stems from experiments with rainbow trout. These document that fish can be affected by cyclopropenes that modify their desaturases, as shown

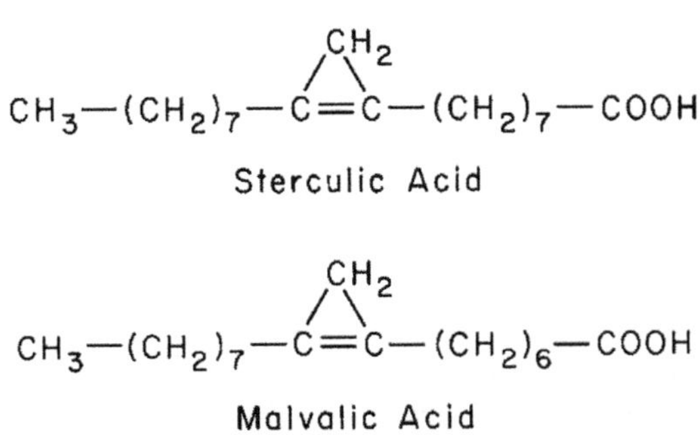

FIGURE 12.12 Schematic presentation of sterculic and malvalic acids, two cyclopropene fatty acids occurring in plants, e.g., in cotton seeds.

for the chicken, with effects on lipid metabolism (Hendricks, J. 2002; Roehm et al., 1970; Selivonchick et al., 1981; Struthers et al., 1975a; Struthers et al., 1975b). The studies also showed alterations in the mixed function oxidase system and other liver enzyme systems (Eisele et al., 1978; Selivonchick et al., 1981) with consequences for the structure and function of internal organs. Altered histological appearance was observed in the liver along with abnormal accumulation of glycogen, necrosis, and fibrosis of the hepatocytes. In the studies mentioned above, inclusion levels of cyclopropene fatty acids in the diets were in the range 0–500 mg/kg. No clear answer to which level would be acceptable or upper limit can be drawn from these experiments. Interaction between the cyclopropene fatty acids and aflatoxin, another liver carcinogen, has been studied showing synergistic effects. The effects depended, however, on the aflatoxin species used. The aflatoxin B1 was much more potent than the Q1 (Hendricks et al., 1980; Sinnhuber et al., 1974). Also in a study with sockeye salmon, carcinogenic effects of cyclopropene have been observed, but only when supplied together with aflatoxin B1. In channel catfish, sterculic acid has been observed to cause growth retardation (Hendricks, 2002).

Although the knowledge basis for conclusions regarding the antinutritional effects of the cyclopropenes is weak for most cultivated fish species, they should be considered as having the potential for important negative effects in all fish species until further observations are made. Care should therefore be taken when using cottonseed meal and oil. As the very young fish are considered more vulnerable than older, caution should be exercised in particular if cotton seed meal is included in broodstock diets and in start feeds (Hendricks, 2002).

12.2.1.11 Other plant antinutrients

In the present situation with increasing demand for cultivated fish worldwide and increasing focus on sustainability of feed ingredients for all production animals, ingredients which until now have been used at low levels will be used at higher levels, and new ingredients will be developed and used. Some of these contain known compounds, which give antinutritional effects when fed at higher levels, and others may contain unknown compounds, which turn out to be antinutrients. Among the former are arginase inhibitors, quinolizidine alkaloids, and cyanide-releasing compounds. A comprehensive review of these is presented in NRC's Nutrient Requirement of Fish and Shrimp (NRC 2011).

Arginase inhibitors are found in sunflower, soybeans, and potatoes as well as in some plants of little relevance for today's fish cultivation. They belong to a group of related polyphenol esters called chlorogenic acids. Some of them inhibit not only arginase but also other enzymes, such as glycerol-3-phosphate dehydrogenase of the hepatic and renal glucose-6-phosphatase systems (Arion et al., 1998). There is a lack of studies of the effects of the pure compounds in fish. However, studies with sunflower meal indicate that arginase inhibitors may be the cause of the negative effects on protein digestibility observed at sunflower levels above 20% in diets for Atlantic salmon. This is a result in line with the in vitro effects on bioavailability of amino acids (Abal et al., 2017; Aslaksen et al., 2007; Gill et al., 2006; Pierpoint 1970). Recently, they have been investigated for their potential therapeutic use in human medicine (Naveed et al., 2018).

Quinolizidine alkaloids (Fig. 12.13) are found in a wide variety of plants and are the predominant antinutrient in lupins (genus *Lupinus*). They are characterized as inhibitors of antimuscarinic acetylcholine receptors, which means that they reduce the activation of the

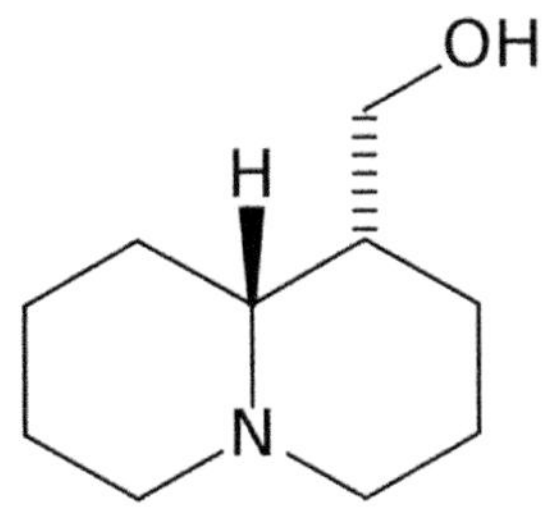

FIGURE 12.13 Schematic presentation of the structure of lupinine which, with its derivatives, represents a group quinolizidine alkaloid found in lupins.

parasympathetic nervous system. Some also inhibit the action of acetylcholine (Wink et al. 1995, 1998). In mammals, quinolizidine alkaloids cause loss of motor coordination and muscular control with trembling, shaking, excitation, and convulsion (Pothier et al., 1998). Information on effects of quinolizidine alkaloids in fish is limited. Glencross et al. conducted a series of experiments with Atlantic salmon and rainbow trout fed diets with purified alkaloids (Glencross et al. 2004a, 2004b, 2006). However, no clear effects were seen except for reduced feed intake, which can be related to the bitterness of the lupin meal. Through breeding, low alkaloid varieties of lupin have been developed, eliminating the effects on feed intake.

Cyanogenic glucosides are compounds found in a wide range of plants, which are highly toxic to animals. Among the many plant species containing compounds, which can release HCN, are important feed and food plants such as cassava, bamboo, linseed, sorghum, peas, maize, barley, wheat, peanut, almonds, and rapeseed (Jones, 1998; Poulton, 1990; Lechtenber and Nahrstedt, 1999). Cassava and bamboo contain the highest levels and need detoxification before they can be eaten or fed to animals. The compound in cassava is named linamarin, and in bamboo, taxiphyllin. The HCN is released from the glycoside when the plant cells open, e.g., by chewing or grinding. Hydrogen cyanide is readily absorbed from the intestinal tract, lungs, and skin. Among other effects, most importantly, it inactivates cytochrome oxidase in the mitochondria by binding to the iron contained in the enzyme. This causes a decrease in the utilization of oxygen in the tissues and a shift from aerobic to anaerobic metabolism. The effects if cyanide will be most immediate on the respiratory system and the heart, and may be fatal within a short time (Vetter, 2017). In experimental animals, chronic sublethal dietary cyanide has caused reproductive effects including lower birth rates and an increased number of neonatal deaths, impaired thyroid function, and behavioral effects including increasing ambivalence and slower response time. The European recommendations for maximum content of cyanogenic glycosides in feeds for animals vary between 10 and 50 mg/kg, depending on the animal species (European Parliament, 2002). No scientific papers on effects of cyanogenic glucosides in fish have been found. The aforementioned information is a reminder that previously unused resources, which possibly can be used in the future, may contain highly toxic compounds.

There are a number of other toxins in plants, which can have significant impacts on animals, including fish. Examples are favatoxin, oxalates, toxic amino acids, phallotoxins, indoles, and potato glycoalkaloids, but the likelihood that plants containing these toxins will be used in fish diets is low. Furthermore, there is little if any experimental evidence for the effects of these compounds in fish.

12.2.2 Antinutrients and toxins of animal origin

Some animal products are well-known ingredients in fish feed. With the increased focus on reduction of waste and increased recycling of resources, more of such product will supposedly find their way into fish feed. The animal products that are used worldwide comprise fishmeal (herring, anchovy, menhaden, and whitefish), whey, blood meal, poultry byproduct meals, feather meal, meat meal, wet fish (hake, herring, bottom fish fillet scrap), shrimp or crab meal, krill, and fish oil. Most of these products do not represent challenges to the fish feed industry, if stored and processed properly, and contamination with microbes is avoided. However, two toxins found naturally in the liver, ovaries, and roe of certain few fish species could potentially enter fish diets if these fish species are inadvertently included with fish being prepared for fishmeal or wet fish. These are tetrodotoxin and dinogunellin. Also, dinoflagellate toxins may represent a challenge if fish, which have been exposed to algae toxins under algae blooms, are used in the production of fish feed. Moreover, the so-called thiaminase may induce thiamin deficiency if some species of fish and their by-products are used in the making of fish feed without proper storage and cooking.

12.2.2.1 Tetrodotoxin

Tetrodotoxin (Fig. 12.14) is a toxin that is lethal at very low doses to all vertebrates, which do not produce the toxin itself (Abal et al., 2017). It selectively binds and blocks voltage-gated sodium channels (Biessy et al., 2018). Tetrodotoxin is found in many fish species belonging to the phylum Cordata, such as puffer fish, as well as in bivalves and some other lower organisms (Chau et al., 2011; Kalaitzis et al., 2010). This complex cyclic compound is produced and stored in the fish. It is conceivable that these fish could enter the fishery and end up in fishmeal or wet fish, but the risk in fish diets is low. In humans and laboratory mammals, it causes numbness of the lips, tongue, and fingers, muscular paralysis, and eventual respiratory paralysis and death (Abal et al., 2017; Nieto et al., 2012). No information on effects and toxicity of tetrodotoxin in fish has been found.

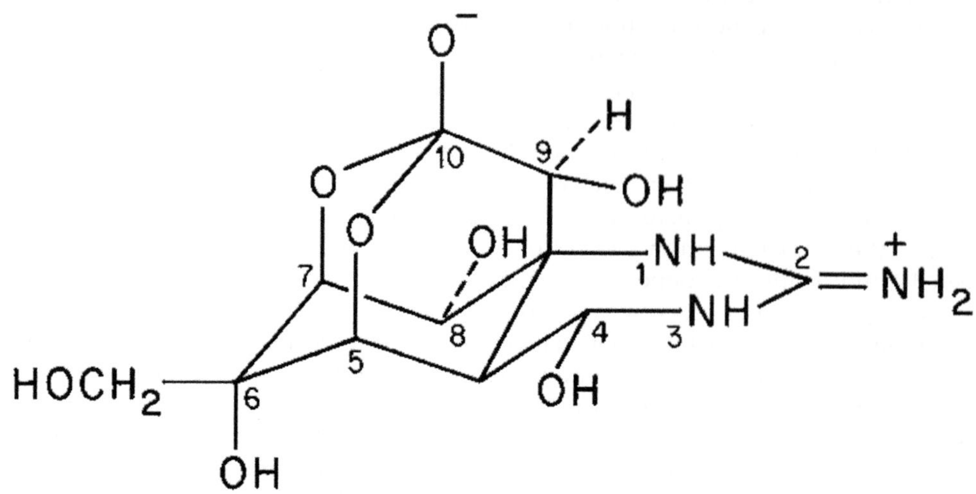

FIGURE 12.14 Schematic presentation of the structure of tetrodotoxin.

12.2.2.2 *Dinogunellin*

Dinogunellin is a toxic lipoprotein that is produced in the roe of several fish species including the northern blenny (*Stichaeus grigorgjewi*) from northern Japan, and the cabezon or marbled sculpin (*Scorpaenichthys marmoratus*), a fish common to the Pacific coast of North America. This toxin is not as potent as tetrodotoxin, but it causes diarrhea, nausea, vomiting, epigastric distress, and liver and spleen necrosis in humans (Matsunaga et al., 2009). Some of these fish and their by-products might end up in the fillet scrap of groundfish used as wet fish in moist pellets. However, no information has been found regarding effects in fish of dinoguellin.

12.2.2.3 *Dinoflagellate toxins*

Several bloom-forming algae contain toxins and can create dangerous conditions for fish as well as humans and other terrestrial animals (Burkholder 2009; Hambright et al., 2014; Wang 2008). They are phospholipids, some very complex, which act in the nervous system, via blockage of voltage-gated sodium, calcium, or potassium channels and other pathways important for health and survival (Duran-Riveroll et al., 2019; Lee et al., 2016; Raposo et al., 2020). Mollusks filter these organisms from the water, digest them, and accumulate the toxin in various tissues (Blanco, 2018). Mollusks are not affected by the toxins, but humans and other mammals are very sensitive. Mortalities in humans, livestock, wild animals, and pets have resulted from eating marine animals, which have eaten poisonous algae. Fish are also sensitive to these toxins, as evidenced by massive fish kills in algae bloom areas (Mincarelli et al., 2018). At lower concentrations, changes in mobility are a common observation (*Diplodus cervinus*) (Corriere et al., 2020). In contrast to mollusks, fish do not accumulate these toxins in their flesh. However, in the intestine and liver, sufficiently high levels may accumulate to cause disease if the fish are eaten whole (Starr et al., 2017). The potential for these dinoflagellate toxins to appear in formulated fish diets is increasing. One example is the use of by-products from the production of shellfish, which has become current. Only one report has been found on effects of dinoflagellate toxin in fish. In this study with rainbow trout, no signs of toxicity were observed. The toxin did not accumulate in edible tissues but remained in the gastrointestinal tract (Hardy et al., 1995).

12.2.2.4 *Thiaminase*

Several fish species such as herring, sardine, mackerel, and carp, capelin, channel catfish, some tuna species, white bass, as well as mussels, clams, and shrimp, contain components with thiaminase activity. A more extensive list of fish showing such enzyme activity is given in NRC (1983). The enzyme is mainly located in the intestine of these fish. It is active in live fish and in raw fish products. It is a protein that is inactivated by heat. Thiaminase cleaves the rather unstable bridge between the two ring structures of the thiamin molecule (Fig. 12.15). Thiamin deficiency, due to thiaminase in the feed, is a major problem in wild fish, as well as in birds and animals living on these fish, and the problem seems to increase in frequency worldwide (Fisher et al., 1998; Harder et al., 2018; Riley and Evans, 2008; Riley et al., 2008; Tillitt et al., 2005). The symptoms of thiamin deficiency may vary between fish species. An overview of results of experiments conducted to find thiamine requirement in various fish species is given in NRC (2011b). The following symptoms have been observed in these

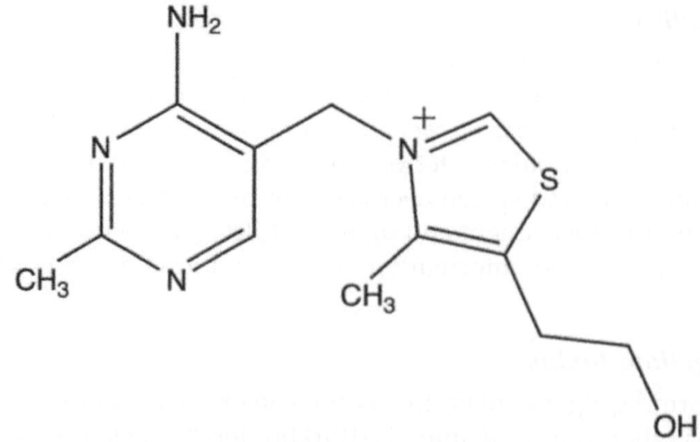

FIGURE 12.15 Schematic presentation of thiamin. Thiaminase breaks the bridge between the two ring structures and thereby inactivates the thiamin.

studies: reduced growth, increased mortality, hyperirritability, neurological disorders, subcutaneous hemorrhages, and congested fins. If uncooked fish, whole or by-products, of species containing thiaminase are used, the feed must be fed to the fish the same day. Use of raw fish in semimoist diets caused many cases of thiamine deficiency in the early days of Atlantic salmon production in Norway. Feed was often made to last for several days. After a day in cold storage, all thiamin in the feed mixture was inactivated, and the fish feed became thiamin deficient. The most prominent symptom of thiamin deficiency in the fish was aberrant swimming.

12.3 Adventitious toxins

Adventitious toxins are elements and compounds that are not normal constituents of feed ingredients and can cause harm to animals, including fish, when they contaminate feeds. Adventitious toxins can be natural, biological compounds, synthetic chemicals, or compounds formed through chemical processes during abusive storage conditions. The effects of adventitious toxins range from mild to lethal depending on the compound, dose, duration of exposure, life stage of the fish, and presence of other confounding biological or environmental factors that may amplify or mitigate toxicity. Preventing exposure is the most effective means of protecting fish from the effects of adventitious toxins, although some adventitious toxins are ubiquitous in the environment, making it impossible to completely eliminate them from feeds. Regulatory agencies have established maximum allowable levels in feeds and/or ingredients, and routine monitoring of the most common adventitious toxin in susceptible ingredients is advised. Thorough reviews of risk assessments of adventitious toxins in fish feeds have recently been published (Glencross et al., 2020; Pietsch, 2020).

12.3.1 Oxidized lipids

It is commonly accepted that oxidized fish or plant oils should not be used in feeds. However, oxidized lipids are not toxic, per se, providing they are fully oxidized. Rather, fatty acids undergoing the process of autoxidation are the products of concern in feeds. Lipid autoxidation is a multistep chemical process that converts double bonds in unsaturated fatty acids to single bonds through the addition of hydrogen, a process called hydrogenation. Fish oils typically contain high proportions of polyunsaturated fatty acids having numerous double bonds, making them highly susceptible to autoxidation. However, any oil containing unsaturated fatty acids can undergo autoxidation when exposed to light or metal catalysts, particularly iron (Fe), in the presence of oxygen. Oxygen by itself cannot start autoxidation.

Intermediate products of autoxidation reactions are problematic in fish nutrition for several reasons. The initial step in autoxidation, called the initiation phase, involves the conversion of Fe^{2+} to Fe^{3+}, the active form of iron, forming superoxides in the process (Fig. 12.16). Superoxides react with hydrogen to form hydrogen peroxide. Fe^{2+} then reacts with hydrogen peroxide to form Fe^{3+} plus a hydroxy (free) radical (^{-}OH). Natural and supplemental antioxidants in oils interrupt further chemical reactions and keep the autoxidation process in its initial phase until remaining antioxidant levels are insufficient to keep up with free radical production. This is the end of the initiation stage and the beginning of the propagation stage where the process of free radical production and oxidation becomes self-sustaining, and toxic intermediate products, such as peroxides, aldehydes, peroxides, ketones, furans, numerous alcohols, and epoxides, are formed. When feeds containing oxidizing lipids are consumed, free radical levels in tissues increase, activating cellular detoxifying mechanisms that protect tissues. These rely on vitamins C and E. Tissue levels of vitamins C and E become depleted, causing a range of clinical and pathological problems including liver degeneration, anemia, and spleen abnormalities (Hung et al., 1981; Moccia et al., 1984; Rehulka, 1990; Smith, 1979). Because autoxidation in oils or feeds consumes natural antioxidants, levels of vitamins C and various forms of tocopherol in feeds are reduced, lowering intake by fish. Thus, effects of oxidizing lipids of fish health are due to depletion of tissue levels of antioxidants, intake of toxic products formed during autoxidation, and alteration of normal cellular processes associated with membrane integrity.

The most serious clinical problem associated with feeding oxidizing lipids is lipoid liver disease. Fish suffering from lipoid liver disease have extreme anemia (manifested by pallor of the gills and erythrocyte fragility), a bronzed, rounded heart, and a swollen liver with rounded edges (Smith, 1979). Vitamin E deficiency in fish is characterized by erythrocyte fragility leading to anemia. Histologically, the main feature associated with lipoid liver disease is extreme infiltration of hepatocytes by lipid, which causes loss of cytoplasmic staining and distortion of hepatic muralia. There is degeneration of splenic and renal hemopoietic tissue with high levels of pale-staining pigment in melanomacrophage centers. There is also often auxiliary hemopoiesis in the subepicardial tissues and the periportal areas. Depending on the length of time the condition has been extant and the degree of oxidation and type of fat in the diet, there is a varying degree of infiltration of the liver by macrophages containing ceroid, a pigmented breakdown product of phospholipid metabolism (Hardy, 2001). Slightly affected fish are usually capable of complete recovery, but once there is severe anemia and

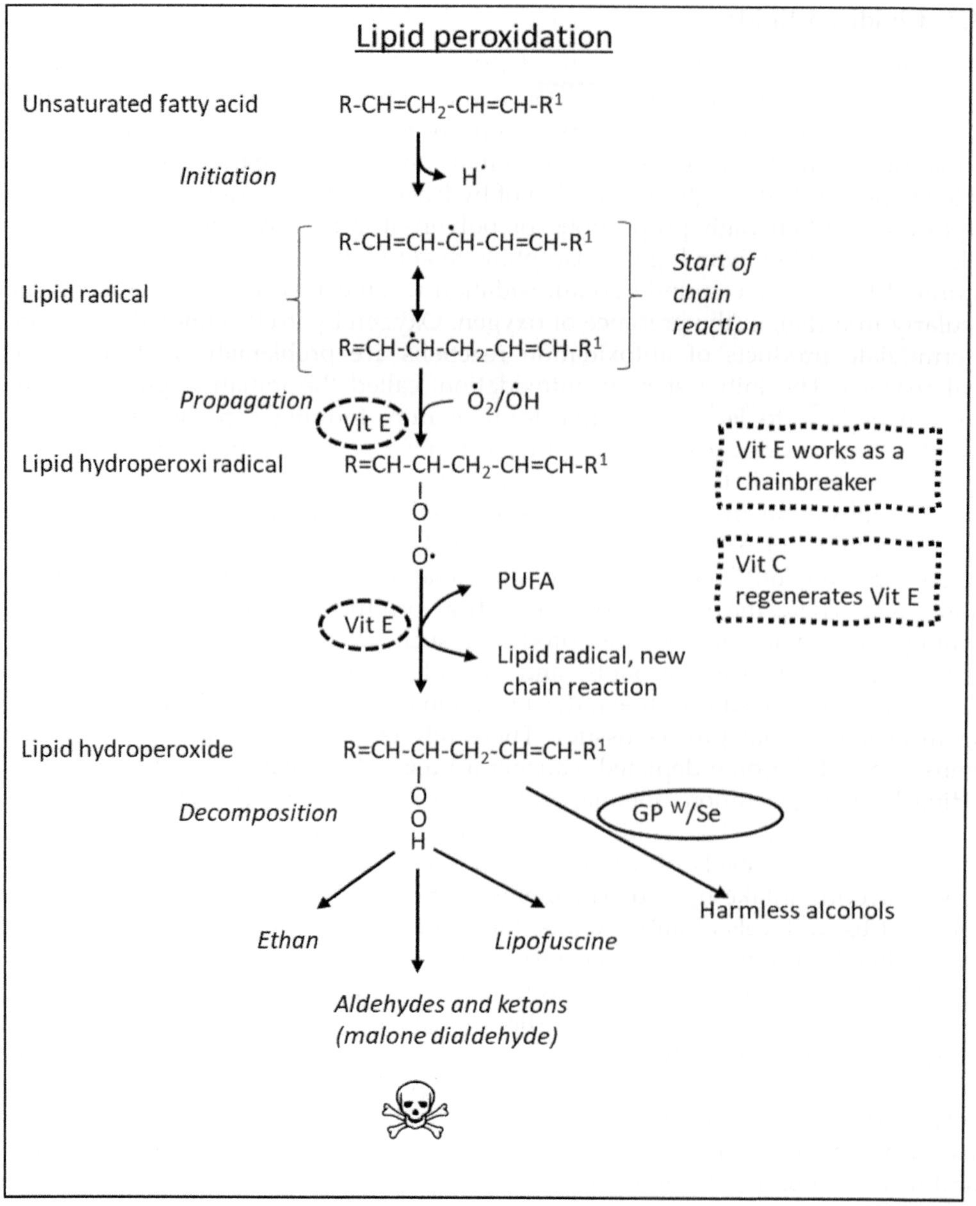

FIGURE 12.16 Indication of steps in the autooxidation of polyunsaturated fatty acids such as EPA and DHA.

hepatic ceroidosis has developed, fish are incapable of recovering to normal growth and feed efficiency. Other conditions reported to be associated with the feeding of oxidizing lipids include exophthalmia, steatitis, darkening, splenic hemosiderosis, and skeletal myopathy (Moccia et al., 1984; Soliman et al., 1983; Murai and Andrews, 1974; Park, 1978).

Because the primary health effects of feeding oxidizing fish oil result from depletion of tissue vitamins C and E, it is logical to conclude that elevating dietary intake of these vitamins might alleviate the condition. In fact, in catfish and trout, many of the pathological effects associated with feeding oxidizing fish oil can be prevented by increasing the dietary intake of tocopherol (Moccia et al., 1984; Murai and Andrews, 1974). However, acute toxicity is not always prevented by increasing dietary tocopherol intake, especially in small fish that have not yet accumulated substantial tissue stores of ascorbic acid and tocopherol. Growth rates of Atlantic and coho salmon fry are reported to be inversely proportional to the level of oxidation in starter feeds, with reduced growth being observed at levels of oxidation that also reduce growth in larger fish (Hung et al., 1981; Ketola et al., 1989). There is also some indication that there may be species differences in sensitivity to oxidizing feeds (Ketola et al., 1989). The link between tocopherol deficiency signs and those associated with consumption of oxidizing lipids is weak. If oxidizing lipids caused pathological conditions in fish solely by reducing tissue tocopherol levels, one might expect that other common signs of tocopherol deficiency, e.g., muscular dystrophy, exudative diathesis, and depigmentation, would be prevalent in fish consuming feeds containing oxidizing lipids. However, these conditions are not observed in fish fed oxidizing lipids.

12.3.1.1 Biogenic amines

Biogenic amines are compounds formed as a result of bacterial degradation of free amino acids in raw materials used to produce fishmeal and land animal protein meals. As such, they are correlated with raw material freshness, and detection of elevated levels in feed ingredients is an indicator of reduced product quality. Cadaverine, putrescine, histamine, tyramine, spermidine, spermine, and phenethylamine are biogenic amines that are produced by decarboxylation of lysine, arginine, histidine, tyrosine, methionine, and phenylalanine, respectively (Sander et al., 1996).

Young chickens are acutely sensitive to histamine in feeds; histamine causes excessive gastric acid secretion with associated pathologies including hemorrhaging in the stomach and gizzard. Effects of biogenic amines on fish were unclear until researchers showed that biogenic amines, per se, were not particularly toxic, although fishmeal containing high levels of biogenic amines could be damaged in other ways that affected fish health and performance. Feeding fishmeal that was toxic to chickens had no effects on fish growth, feed efficiency, or mortality of rainbow trout (Fairgrieve et al., 1994).

Supplementing a semipurified diet with levels of histamine, putrescine, and cadaverine equivalent to poor quality fishmeal did not reduce feed intake, growth, or feed efficiency, but the histidine supplemented diet caused transient stomach distention that rapidly resolved when feed was changed. Subsequent research with Atlantic salmon, Atlantic halibut, and shrimp showed that feeding diets containing fishmeal with high biogenic amine levels reduced feed intake and growth (Aksnes and Mundheim, 1997; Anderson et al., 1997) (Rique-Marie et al., 1998). Feeding Atlantic salmon with high-quality fishmeal supplemented with biogenic amines or poor-quality fishmeal containing high levels of biogenic amines confirmed that biogenic amines do not affect fish performance, whereas fish fed the diet with poor quality fishmeal reduced performance (Opstvedt et al., 2000). This suggests that factors other than biogenic amines in poor-quality fishmeal quality reduce nutritional quality.

12.3.1.2 Mycotoxins

Mycotoxins are compounds produced by fungi that are toxic to humans and animals, including fish (Matejova et al., 2017). Mycotoxins are produced by mold, yeast, and mushrooms, but the term mycotoxin usually refers to metabolites of molds that contaminate crops, especially grains and oilseeds. Mold spores are ubiquitous in the environment, but mold spores only flourish and produce mold colonies when their environment is favorable. In addition to having a source of organic matter to provide nutrients and access to oxygen, molds require moisture and favorable ambient temperatures. It follows that mold growth in foods and feeds can be prevented by controlling temperature, moisture content, and access to oxygen. Toxin production by mold colonies depends on a combination of factors and is thought to be at least partially associated with defense against competing organisms. There are many species of molds capable of producing mycotoxins, and some mold species can produce more than one mycotoxin. Mold can grow on crops in fields, on harvested crops during storage, on feed ingredients produced from grains and oilseeds, and on finished feeds if environmental conditions favor their growth. Therefore, vigilance is required at all stages of harvesting, storage, and shipping of grains and other susceptible ingredients used in fish feeds, as well as storage and handling of pelleted feeds. From a practical point of view, exposure to moisture is the factor most likely to cause ingredients or feeds to develop mold colonies that produce mycotoxins. Grains and oilseeds harvested during high moisture periods must be dried before being stored, ingredients must be kept dry during storage, and feeds must be dried to less than 12% moisture after pelleting to prevent growth of mold colonies. Mold growth in feeds containing over 12% moisture can be prevented by freezing or packaging feeds in oxygen purged and sealed bags.

Mycotoxin contamination of fish feeds is common throughout the world, although the prevalence of various mycotoxins varies with region. Assessment of ingredients used in European fish feeds indicated that corn products present the highest risk as sources of mycotoxins in feeds, followed by distiller's dried grains with solubles (DDGS), wheat bran, wheat flour, barley, and soybean products (Pietsch et al., 2020). Similar findings have been reported for ingredients in North America, South America, Africa, and Southeast Asia (Goncalves et al., 2018; Oliveira and Vasconcelos, 2020; Pinotti et al., 2016). The potential effects of climate change on prevalence of mycotoxins are a concern. Aquaculture feed formulations have changed over the past 20 years toward lowering fishmeal levels and increasing plant protein levels (Naylor et al., 2020). This may increase the risk of mycotoxin contamination of fish feeds.

Prior to the adoption of dry, pelleted feeds in aquaculture in the 1960s, documented cases of fish losses due to mycotoxin contamination of feeds were rare. This changed in the 1960s with the occurrence of an epizootic of liver cancer in hatchery rainbow trout (Hendricks, 2002). The outbreak was traced to use of feeds containing cottonseed meal contaminated with *Aspergillus flavus*, a mold species known to produce aflatoxins. Aflatoxins are the most common mycotoxin found in feeds (Pietsch et al., 2020). There are four aflatoxins that differ in structure and toxicity, namely B_1, B_2, G_1, and G_2 (Fig. 12.17); aflatoxin B_1 is the most toxic to fish (Hendricks, 2002). Other mycotoxins that can contaminate fish feeds are fumonisins, ochratoxin A, citrinin, trichothecenes, and zearalenone. Fumonisins, produced by *Fusarium* spp., mainly contaminate crops prior to harvest, while *Aspergillus*

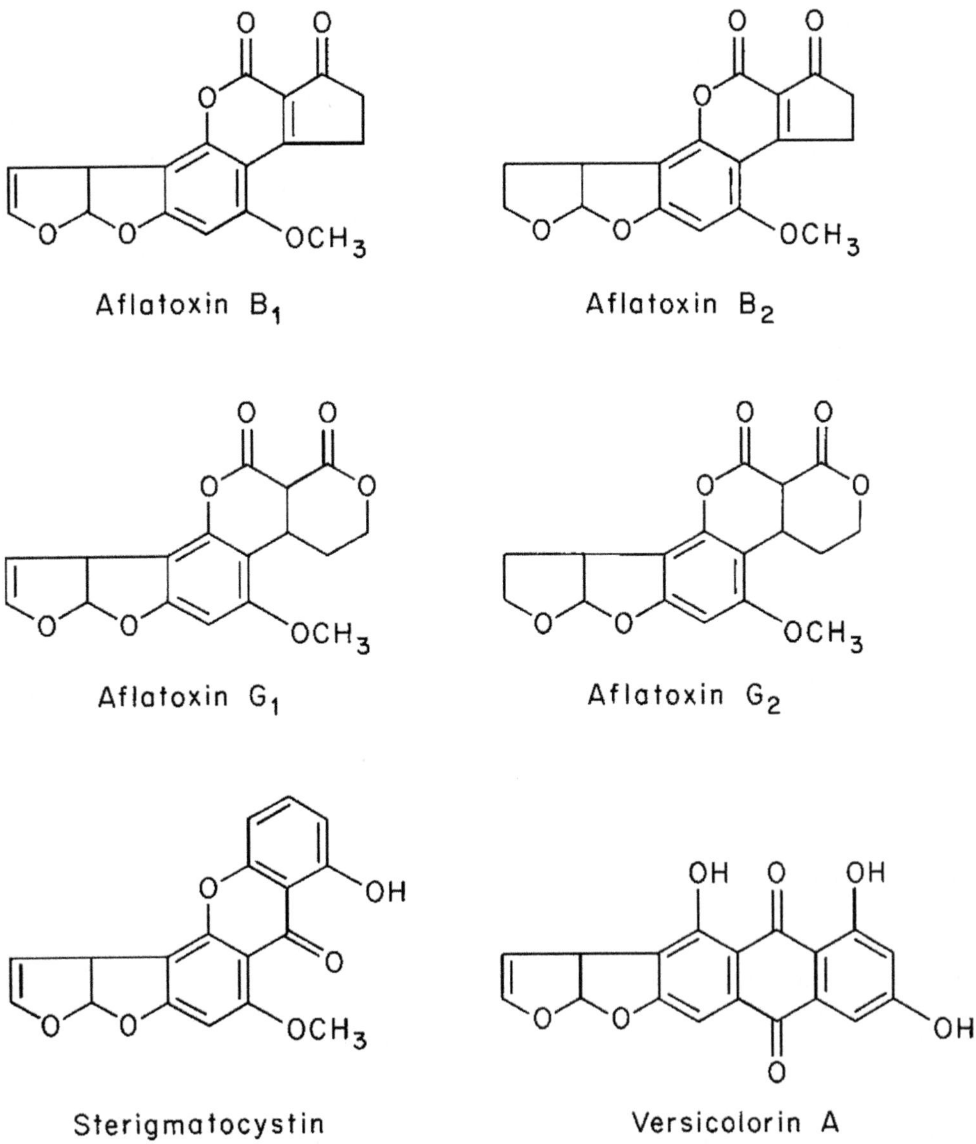

FIGURE 12.17 Schematic presentation of naturally occurring aflatoxins.

spp. and *Penicillium* spp. contaminate crops after harvest during storage and ingredients produced from these crops (Oliveira and Vasconcelos, 2020).

Mycotoxin levels in ingredients can be reduced by normal processing steps, including sorting, cleaning, or extrusion. Mycotoxins vary in thermal stability, with some, such as fumonisins and zearalenone, being somewhat destroyed by extrusion over 150°C, while others, notably, aflatoxins, are unaffected by temperatures associated with normal feed processing conditions (Bullerman and Bianchini, 2007). Therefore, preventing mycotoxin contamination of feeds in the first place is the primary strategy to avoid exposure of animals to these toxic compounds. Screening feed ingredients for the presence of mycotoxins and avoiding contaminated batches is a quality control step that fish feed producers should follow. Mycotoxin levels in feed ingredients can be reduced by water or solvent extraction and by inoculating ingredients with specific bacterial strains that produce enzymes capable of inactivating mycotoxins. The risk of mycotoxin exposure of fish fed contaminated feed can be reduced by including bentonite in the feed formulation; bentonite binds mycotoxins in the gut and prevents absorption. Although feeds are dried to below 12% moisture before bulk storage or bagging to inhibit mold growth during storage, feed manufacturers commonly add mold inhibitors to feeds as an additional precaution, especially in areas where temperatures and humidity are high.

Aflatoxin was the first mycotoxin to be identified as a health threat to fish. Farmed fish species differ in sensitivity to aflatoxins, and toxicity may vary with life history stage, with fry being more sensitive than juvenile or postjuvenile fish. In general, species reared in warm, freshwater habitats, such as catfish and tilapia, tolerate higher levels of aflatoxins in feeds than freshwater species native to cool water habitats, or farmed marine species. Exposure to low levels of aflatoxin over long periods causes several nonspecific signs of disease, such as reduced growth, anemia, and reduced survival. Among farmed fish, rainbow trout are the most sensitive species tested to date. Rainbow trout develop liver carcinomas when fed as little as 0.4 μg of aflatoxin B_1 per kg of feed over an extended period (Halver and Mitchell, 1969), whereas channel catfish do not appear to develop liver carcinomas when fed diets containing relatively high levels of aflatoxin B_1 (Manning, 2001). Tilapia developed precancerous lesions when fed 1.88 mg aflatoxin B_1 per kg but did not develop carcinomas over a 75-day experimental period. Acute toxicity in trout is induced by feeding diets containing 80 μg aflatoxin B_1 per kg of feed for short periods (Roberts, 2002). The acute LD_{50} of rainbow trout for aflatoxin B_1 is 810 μg/kg, and for aflatoxin G_1, the LD_{50} is about double at 1900 μg/kg (Hendricks, 2002). In contrast, LD_{50} value of coho salmon to aflatoxin B_1 was estimated at 5–10 mg/kg, similar to that of rats. In contrast, no acute LD_{50} has been reported for channel catfish (Manning, 2001). Differences in toxicity among fish species are thought to be associated with differences in activity of glucuronyltransferase and sulfotransferase detoxification systems. However, evaluating differences in toxicity among species is complicated by differences in dose and duration of exposure among (Deng et al., 2010).

Fumonisins (Fig. 12.18) are mycotoxins produced by several *Fusarium* species, most commonly *Fusarium verticillioides*, and are generally limited to contamination of corn in fields prior to harvest. The toxins are carried over into corn products used in feeds. Farmed omnivorous species, specifically carp, catfish, and tilapia, are the main fish at risk of exposure to fumonisins in feeds because their feeds are more likely to contain corn products than are carnivorous farmed species such as salmon, sea bass, and others. Fumonisin B_1 is the most toxic form of this mycotoxin, and signs of toxicity are nonspecific and include reduce weight gain, hematocrit, and survival in catfish, tilapia, and carp (Oliveira and Vasconcelos, 2020).

Other mycotoxins likely to contaminate fish feeds are ochratoxin (Fig. 12.19), trichothecenes (vomitoxin), and zearalenone. Ochratoxin is most commonly produced by *Penicillium* and *Aspergillus* species on cereal grains during storage. Because it is a stable compound resistant to heat treatment or other common processes, it is retained in ingredients and therefore

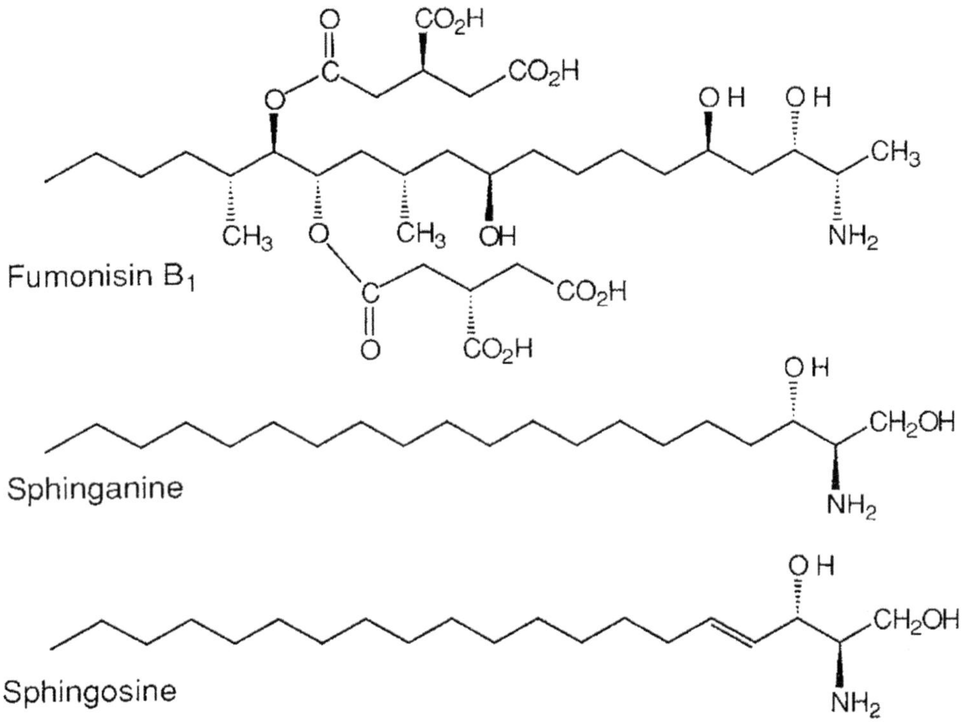

FIGURE 12.18 Schematic presentation of the structure of fumonicines.

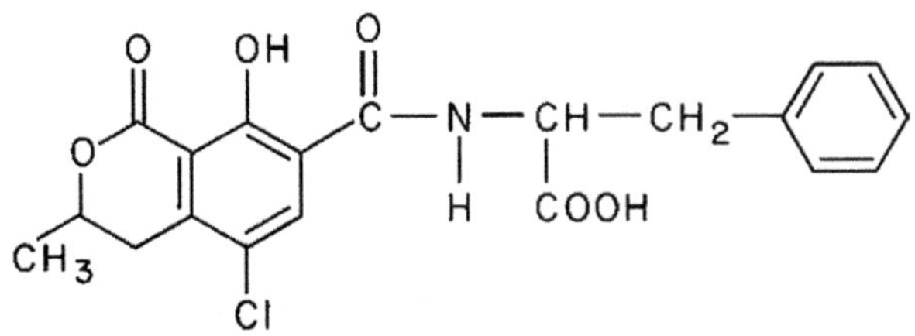

FIGURE 12.19 Schematic presentation of ochratoxin.

can contaminate fish feeds (Oliveira and Vasconcelos, 2020). Sea bass are highly susceptible to ochratoxin toxicity, as are Atlantic salmon, catfish, and tilapia. Toxicity in other farmed species has not been extensively studied. Growth reduction, loss of equilibrium, fin erosion, liver, kidney and spleen lesions, and increased mortality are associated with ochratoxin toxicity. In contrast, fish respond to feeds contaminated with trichothecenes, specifically deoxynivalenaol and T-2 toxin, by rejecting feed, leading to slow growth and associated increases in feed conversion ratios. In contrast, trichothecene consumption in humans from consumption of bread made from moldy wheat is associated with acute gastrointestinal distress. Finally, zearalenone, produced by *Fusarium* spp., is a potent estrogenic compound. Thus, its effects on fish are associated with reproduction problems, although at low doses, it increases growth in rainbow trout.

12.3.1.3 Pesticides

Synthetic pesticides are widely used to control insects, weeds, fungi, and other organisms that cause losses to agricultural production, most importantly crops. Most pesticides are nonspecific in action, meaning they kill target pests as well as nontarget organisms. Furthermore, many are persistent in the environment and bioaccumulate in animal tissues. They are widely distributed throughout the environment, including soils, plants, and animals. Pesticides used to control insect pests are organochlorines (such as DDT (Fig. 12.20), DDD, eldrin, dieldrin (Fig. 12.21)), lindane, toxaphene, and endosulfan), organophosphates (methyl parathion, malathion, trichlrofon), carbamates, pyrethroids, phenyl amides, and phenoxy alkanoates. Herbicides include diquat, paraquat, 2,4-D, 2,4,5-T, and many others (Jayaraj et al., 2016). Many of these compounds, notably DDT and toxaphene, have been banned in North American and European countries for decades, but they are still widely used in developing countries in Asia. As a result, the aquatic environment is widely contaminated with persistent pesticide compounds. Fish can be exposed to pesticides from drifting spray after application on nearby fields or from runoff of irrigation or rainwater from fields into waterways. Effects on fish can be acute if exposure is extreme, but for the most part, fish are subject to chronic exposure via the food chain. For farmed fish, the risk of exposure to pesticides and herbicides is from overspray or contamination of pond water sources. Exposure to pesticides from feeds via use of contaminated feed ingredients produced from crops, mainly grains and oilseeds, in feeds is possible, but levels in feeds are low. Nevertheless, pesticides that persist in the environment, such as DDT or toxaphene, accumulate in the body over time and therefore pose a risk to farmed fish and consumers. Most of the scientific literature on toxic effects of pesticides is associated with water-borne exposure and in the realm of aquatic toxicology. Information associated with effects of pesticide intake by fish is limited to acute dietary exposure to high pesticide levels and thus not informative regarding effects of exposure to low levels in feeds (Hendricks, 2002).

12.3.1.4 Persistent organic pollutants

Persistent organic pollutants (POPs) are synthetic chemicals widely used in a number of industrial and consumer products and include dioxin plus dioxin-like compounds including polychlorinated biphenyls (PCBs), polybrominated diphenyl esters (PBDEs), and literally

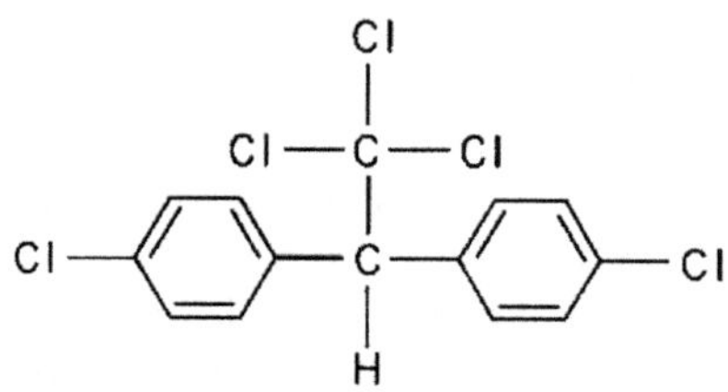

FIGURE 12.20 Schematic presentation of DDT (1,1-bis(4-chlorophenyl)-2,2,2-trichloroethane).

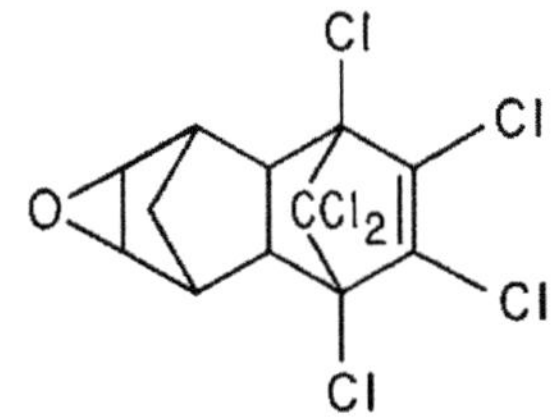

FIGURE 12.21 Schematic presentation of dieldrin.

hundreds of other compounds and congenators. These lipophilic compounds are environmental contaminants that resist degradation and are detected almost everywhere on the planet. They are not metabolized by vertebrates and accumulate in tissue lipid reserves. They express toxicity in many ways, but, in general, toxicity in fish involves developmental abnormalities in embryos as well as interference in reproduction.

The primary risk of contamination of farmed fish feeds is through use of contaminated fish oil and residual fish oil in fishmeals produced from fish caught in areas where industrial contamination is high, such as certain areas of the Baltic and North Seas. Fish oil can also be accidently contaminated through fish oil distribution chains by transporting fish oil in POP-contaminated shipping tanks. This is extremely rare, but incidents of this route of contamination have occurred in recent years.

POP levels in fish oil can be reduced by treatment of fish oil by distillation (Berntssen et al., 2010), by treatment with activated carbon filtering (Oterhals et al., 2007), or by using a combination of both methods. Because these treatments may also damage or remove essential fatty acids and fat-soluble vitamins, the conditions of processing must be carefully controlled (Oterhals and Berntssen, 2010). Salmon fillets contain high levels of oil, making them a species of concern regarding accumulation of POPs in edible products. POP levels in salmon fed diets containing decontaminated fish oil over a production cycle reduced POP content of fillets to levels below regulatory standards. Blending fish oils with plant-derived oils is another effective approach to reducing POP levels in farmed salmon (Sprague et al., 2010). In recent years, fish oil and fishmeal have, to a large extent, been replaced with rapeseed oil and plant ingredients. In parallel, a gradual reduction in POP level in fish diets, in particular diets for fatty fish, has lowered levels of POPs in fillets (VKM, 2014). Screening methods to detect POPs in fish oil are now routinely available (Pereira et al., 2017).

12.3.1.5 *Heavy metals*

Certain minerals are toxic to fish when present at elevated levels in feeds. Several minerals, such as copper, iron, manganese, selenium, and zinc, are essential dietary nutrients for optimal fish growth and health but are toxic when feed levels become excessive (NRC, 2011). They are common constituents of mineral premixes used to supplement fish feeds, especially low fishmeal feeds using plant protein concentrates. Contamination can occur from over supplementation, or release from feed manufacturing equipment. For all but selenium, the feed level that induces toxicity in fish is much higher than dietary required levels.

Possible adverse effects and some threshold limits for farmed fish have been recently reviewed (Baeverfjord et al., 2019).

Cadmium, lead, and mercury are nonessential minerals that are the chief concerns regarding metal contamination of fish feeds. In fish nutrition terminology, they are referred to as heavy metals. Toxicity of these minerals depends on chemical form, with the inorganic form of arsenic being more toxic than organic forms, while the opposite is true for lead and mercury. For farmed fish, the risk of contamination of feeds with heavy metals is relatively small. They occur naturally at low levels in the environment, but all can be found in much higher concentrations in areas where industrial pollution is high. Furthermore, all are concentrated via the food chain, resulting in potential health risks for people consuming large amounts of top-level predator fishes. Most fishmeal is produced from small, short-lived species of fish, i.e., herring, sardine, anchovy, or pilchard, which do not live long enough to accumulate high levels of mercury from the environment. Consequently, the risk of contamination of fish feeds with cadmium, lead, or mercury is relatively low (Adamse et al., 2017). However, fishmeal produced from long-lived fish from polluted environments can contain elevated levels of mercury. Fishmeal produced from dogfish (*Squalus acanthias*) from a high mercury environment was fed to coho salmon for 210 days (Spinelli and Mahnken, 1976). Mercury levels in fish tissues were elevated but remained below regulatory action levels. Fish grew normally and did not exhibit signs of toxicity. Monitoring efforts have detected heavy metal contamination of feeds in Bangladesh, suggesting that the possibility of feed contamination is a possibility (Islam et al., 2017; Kundu et al., 2017; Liu et al., 2017; Uchida et al., 2016).

12.3.1.6 Other residues in farmed fish

Medications supplemented to feeds to treat disease outbreaks or parasite infestation of farmed fish are highly regulated in Europe, North American, and other areas (Costello et al., 2001; Tacon and Metian, 2008). Antibiotics and antibiotic metabolites are the most common medicinal residues detected in farmed fish products. Monitoring of farmed fish and shrimp for drug residues periodically detects antibiotic residues, notably in products from Asia, including several medications that are banned in developed countries (Bakar et al., 2010; He et al., 2016; Turnipseed et al., 2018). Medications that are detected in farmed fish include antibiotic residues used to treat bacterial diseases, and chemicals used to treat fungi on eggs and skin of fish, such as malachite green. The effect of these residues on farmed fish health is likely negligible; the concern regarding residues is of potential exposure of consumers. Illegal use of antibiotics or mistakes made during production of medicated feed are the means by which antibiotic residues may occur in feeds. Antibiotic contamination of fish or shellfish can occur postharvest in an attempt to extend shelf life of aquaculture products. This practice is, of course, illegal in most jurisdictions.

Ethoxyquin is a synthetic antioxidant that was commonly used to prevent oxidative rancidity in fishmeal. It is now banned due to concerns about toxicity to all animals (EFSA et al., 2015) but may still be found in fishmeal produced from some areas. Toxicity to fish has been recorded when ethoxyquin is added to diets above 50 mg EQ/kg diet (Wang et al., 2014).

References

Abal, P., Louzao, M.C., Antelo, A., Alvarez, M., Cagide, E., Vilariño, N., Vieytes, M.R., Botana, L.M., 2017. Acute oral toxicity of tetrodotoxin in mice: determination of lethal dose 50 (LD50) and No observed adverse effect level (NOAEL). Toxins 9, 75.

Adamse, P., Van der fels-Klerx, H.J., de Jong, J., 2017. Cadmium, lead, mercury and arsenic in animal feeds and feed materials - trend analysis of monitoring results. Food Addit. Contam. 34 (8), 1298–1311.

Agboola, J.O., Øverland, M., Skrede, A., Hansen, J.Ø., 2020. Yeast as major protein-rich ingredient in aquafeeds: a review of the implications for aquaculture production. Rev. Aquacult. 13, 949–970. https://doi.org/10.1111/raq.12507.

Aksnes, A., Mundheim, H., 1997. The impact of raw material freshness and processing temperature for fish meal on growth, feed efficiency and chemical composition of Atlantic halibut (*Hippoglossus hippoglossus*). Aquaculture 149, 87–106.

Anderson, J.S., Higgs, D.A., Beames, R.M., Rowshandeli, M., 1997. Fish meal quality assessment for Atlantic salmon (*Salmo salar* L.) reared in sea water. Aquacult. Nutr. 3, 25–38.

Anderson, R.L., Rackis, J.J., Tallent, W.H., 1979. Biologically active substances in soy products. In: Wilcke, H.L., Hopkins, O.T., Waggle, D.H. (Eds.), Soy Protein and Human Nutrition. Academic Press, New York, pp. 209–234.

Arion, W.J., Canfield, W.K., Ramos, F.C., Su, M.L., Burger, H.J., Hemmerle, H., Schubert, G., Below, P., Herling, A.W., 1998. Chlorogenic acid analogue S 3483: a potent competitive inhibitor of the hepatic and renal glucose-6-phosphatase systems. Arch. Biochem. Biophys. 351, 279–285.

Aslaksen, M.A., Kraugerud, O.F., Penn, M., Svihus, B., Denstadli, V., Jorgensen, H.Y., Hillestad, M., Krogdahl, A., Storebakken, T., 2007. Screening of nutrient digestibilities and intestinal pathologies in Atlantic salmon, *Salmo salar*, fed diets with legumes, oilseeds, or cereals. Aquaculture 272 (1–4), 541–555.

Atanasov, A.G., Waltenberger, B., Pferschy-Wenzig, E.M., Linder, T., Wawrosch, C., Uhrin, P., Temml, V., Wang, L.M., Schwaiger, S., Heiss, E.H., et al., 2015. Discovery and resupply of pharmacologically active plant-derived natural products: a review. Biotechnol. Adv. 33 (8), 1582–1614.

Atta-ur-Rahman, 2003. Bioactive Natural Products (Part J) Studies. In: Natural Products Chemistry, vol. 29. Elsevier, Amsterdam, 902 pp.

Baeverfjord, G., Krogdahl, Å., 1996. Development and regression of soybean meal induced enteritis in Atlantic salmon, *Salmo salar* L, distal intestine: a comparison with the intestines of fasted fish. J. Fish. Dis. 19 (5), 375–387.

Baeverfjord, G., Prabhu, A.P., Fjelldal, P.G., Albrektsen, S., Hatlen, B., Denstadli, V., Ytteborg, E., Takle, H., Lock, E.-J., Berntssen, M.H.G., et al., 2019. Mineral nutrition and bone health in farmed salmonids - a review. Rev. Aquacult. 11 (3), 740–765.

Bakar, I.A., Ayub, M.K., Yatim, A.M., Sani, N.A., 2010. Pesticide and antibiotic residues in freshwater aquaculture fish: chemical risk assessment from farm to table. Asian J. Food Agro. Ind. 3, 328–334.

Bala, M., Tushir, S., Tyagi, S.K., Gupta, R.K., 2015. Antinutrients in oilseed Brassica: uses and potential applications. Anim. Nutr. Feed Technol. 15 (2), 295–310.

Bao, X.M., Katz, S., Pollard, M., Ohlrogge, J., 2002. Carbocyclic fatty acids in plants: biochemical and molecular genetic characterization of cyclopropane fatty acid synthesis of *Sterculia foetida*. Proc. Natl. Acad. Sci. U. S. A. 99 (10), 7172–7177.

Belitz, H.-D., Grosch, W., 1999a. Amino acids, peptides, Proteins. In: Belitz, H.-D., Grosch, W. (Eds.), Food Chemistry. Springer-Verlag, Berlin, pp. 8–91.

Belitz, H.-D., Grosch, W., 1999b. Legumes. In: Food Chemistry. Springer-Verlag, Berlin, pp. 693–715.

Bennetau-Pelissero, C., Breton, B.B., Bennetau, B., Corraze, G., Le Menn, F., Davail-Cuisset, B., Helou, C., Kaushik, S.J., 2001. Effect of genistein-enriched diets on the endocrine process of gametogenesis and on reproduction efficiency of the rainbow trout *Oncorhynchus mykiss*. Gen. Comp. Endocrinol. 121 (2), 173–187.

Berardi, L.C., Goldblatt, L.A., 1980. Gossypol. In: Liener, I.E. (Ed.), Toxic Constituents of Plant Foodstuffs. Academic Press, New York, pp. 183–237.

Berg-Lea, T., Brattås, L.-E., Krogdahl, A., 1989. Soybean proteinase inhibitors affect nutrient digestion in rainbow trout. In: Huisman, J., van der Pool, T.F.B., Liener, I. (Eds.), Recent Advances of Research in Antinutritional Factors in Legume Seeds. Pudoc, Wageningen, pp. 99–102.

Berntssen, M.H.G., Olsvik, P.A., Torstensen, B.E., Julshamm, K., Midtun, T., Goksoyr, A., Johansen, J., Sigholt, T., Joerum, N., Jakobsen, J.-V., et al., 2010. Reducing persistent organic pollutants while maintaining long chain omega-3 fatty acid in farmed Atlantic salmon using decontaminated fish oils for an entire production cycle. Chemosphere 81 (2), 242–252.

Bhat, P., Jacob, R., Pattabiraman, T., 1981. Enzyme inhibitors from plants: enterokinase inhibitors in tubers and seeds. J. Biosci. 3, 371–377.

Biessy, L., Smith, K.F., Boundy, M.J., Webb, S.C., Hawes, I., Wood, S.A., 2018. Distribution of tetrodotoxin in the New Zealand clam, *Paphies australis*, established using immunohistochemistry and liquid chromatography-tandem quadrupole mass spectrometry. Toxins 10 (7), 14.

Birk, Y., 1985. The Bowman-Birk inhibitor. Int. J. Pept. Protein Res. 25, 113–131.

Blanco, J., 2018. Accumulation of dinophysis toxins in bivalve molluscs. Toxins 10 (11).

Blažević, I., Montaut, S., Burčul, F., Olsen, C.E., Burow, M., Rollin, P., Agerbirk, N., 2020. Glucosinolate structural diversity, identification, chemical synthesis and metabolism in plants. Phytochemistry 169, 112100.

Bullerman, L.B., Bianchini, A., 2007. Stability of mycotoxins during food processing. Int. J. Food Microbiol. 119 (1–2), 140–146.

Bureau, D.P., Harris, A.M., Cho, C.Y., 1998. The effects of purified alcohol extracts from soy products on feed intake and growth of chinook salmon (*Oncorhynchus tshawytscha*) and rainbow trout (*Oncorhynchus mykiss*). Aquaculture 161 (1–4), 27–43.

Burel, C., Boujard, T., Kaushik, S.J., Boeuf, G., Mol, K.A., Van der Geyten, S., Darras, V.M., Kuhn, E.R., Pradet-Balade, B., Querat, B., et al., 2001. Effects of rapeseed meal-glucosinolates on thyroid metabolism and feed utilization in rainbow trout. Gen. Comp. Endocrinol. 124 (3), 343–358.

Burkholder, J.M., 2009. Harmful algal blooms. In: Likens, G.E. (Ed.), Encyclopedia of Inland Waters. Academic Press, pp. 264–285.

Buttle, L.G., Burrells, A.C., Good, J.E., Williams, P.D., Southgate, P.J., Burrells, C., 2001. The binding of soybean agglutinin (SBA) to the intestinal epithelium of Atlantic salmon, *Salmo salar* and Rainbow trout, *Oncorhynchus mykiss*, fed high levels of soybean meal. Vet. Immunol. Immunopathol. 80 (3–4), 237–244.

Cai, Y., Xie, Y., Liu, J., 2020. Glandless seed and glanded plant research in cotton. A review. Agron. Sustain. Dev. 30, 181–190.

Cantoral, R., FernandezQuintela, A., Martinez, J.A., Macarulla, M.T., 1995. Composition and nutritional value of legume seeds and protein concentrates. A comparative study. Arch. Latinoam. Nutr. 45 (3), 242–248.

Cao, H., Sethumadhavan, K., JM, B., 2018. Isolation of cottonseed extracts that affect human cancer cell growth. Sci. Rep. 11, 10458.

Chau, R., Kalaitzis, J.A., Neilan, B.A., 2011. On the origins and biosynthesis of tetrodotoxin. Aquat. Toxicol. 104 (1–2), 61–72.

Chikwati, E., 2007a. Effects of Soy Saponins and Phytosterols on Structure, Function and Dietary Lipid Digestion and Uptake by Atlantic Salmon Intestine. Norwegian School of Veterinary Science.

Chikwati, E.M., 2007b. Effects of Soyasaponins, Phytosterols, Chitosan and Orlistat on Digestive Function and Histomorphology of the Intestinal Tract of Atlantic salmon (*Salmo salar* L.). Monography. School of Veterinary Sciences, Department of Basic Sciences and Aquatic Medicine, Oslo, Norway.

Chikwati, E.M., Venold, F.F., Penn, M.H., Rohloff, J., Refstie, S., Guttvik, A., Hillestad, M., Krogdahl, A., 2012. Interaction of soya saponins with plant ingredients in diets for Atlantic salmon, *Salmo salar* L. Br. J. Nutr. 107 (11), 1570–1590.

Corriere, M., Baptista, M., Paula, J.R., Repolho, T., Rosa, R., Costa, P.R., Solino, L., 2020. Impaired fish swimming performance following dietary exposure to the marine phycotoxin okadaic acid. Toxicon 179, 53–59.

Costas, B., Couto, A., Azeredo, R., Machado, M., Krogdahl, A., Oliva-Teles, A., 2014. Gilthead seabream (*Sparus aurata*) immune responses are modulated after feeding with purified antinutrients. Fish Shellfish Immunol. 41 (1), 70–79.

Costello, M.J., Grant, A., Davies, I.M., Cecchini, S., Papoutsoglou, S., Quigley, D., Saroglia, M., 2001. The control of chemicals used in aquaculture in Europe. J. Appl. Ichthyol. 17, 173–180.

Couto, A., Kortner, T.M., Penn, M., Bakke, A.M., Krogdahl, A., Oliva-Teles, A., 2014a. Effects of dietary phytosterols and soy saponins on growth, feed utilization efficiency and intestinal integrity of gilthead sea bream (*Sparus aurata*) juveniles. Aquaculture 432, 295–303.

Couto, A., Kortner, T.M., Penn, M., Bakke, A.M., Krogdahl, A., Oliva-Teles, A., 2014b. Effects of dietary soy saponins and phytosterols on gilthead sea bream (*Sparus aurata*) during the on-growing period. Anim. Feed Sci. Technol. 198, 203–214.

Couto, A., Kortner, T.M., Penn, M., Bakke, A.M., Krogdahl, A., Oliva-Teles, A., 2015a. Dietary saponins and phytosterols do not affect growth, intestinal morphology and immune response of on-growing European sea bass (*Dicentrarchus labrax*). Aquacult. Nutr. 21 (6), 970–982.

Couto, A., Kortner, T.M., Penn, M., Ostby, G., Bakke, A.M., Krogdahl, A., Oliva-Teles, A., 2015b. Saponins and phytosterols in diets for European sea bass (*Dicentrarchus labrax*) juveniles: effects on growth, intestinal morphology and physiology. Aquacult. Nutr. 21 (2), 180–193.

D'Souza, N., Skonberg, D.I., Camire, M.E., Guthrie, K.E., Malison, J., Lima, L., 2005. Influence of dietary genistein levels on tissue genistein deposition and on the physical, chemical, and sensory quality of rainbow trout, *Oncorhynchus mykiss*. J. Agric. Food Chem. 53 (9), 3631–3636.

Dabrowski, K., Rinchard, K., Lee, K.-J., Blom, J.H., Ciereszko, A., Ottobre, J., 2000. Effects of diets containing gossypol on reproductive capacity of rainbow trout (*Oncorhynchus mykiss*). Biol. Reprod. 62, 227–234.

Deng, S.X., Tian, L.X., Liu, F.J., Jin, S.J., Liang, G.Y., Yang, H.J., Du, Z.Y., Liu, Y.J., 2010. Toxic effects and residue of aflatoxin B1in tilapia (*Oreochromis niloticus*×*O. aureus*) during long-term dietary exposure. Aquaculture 307, 233–240.

Dersjant-Li, Y., Awati, A., Schulze, H., Partridge, G., 2015. Phytase in non-ruminant animal nutrition: a critical review on phytase activities in the gastrointestinal tract and influencing factors. J. Sci. Food Agric. 95, 878–896.

DiMaggio, M., Kenter, L., Breton, T., Berlinsky, D., 2014. Effects of dietary genistein administration on growth, survival and sex determination in southern flounder, *Paralichthys lethostigma*. Aquacult. Res. 47.

Dixon, R.A., 2004. Phytoestrogens. Annu. Rev. Plant Biol. 55, 225–261.

Dorsa, W.J., Robinette, H.R., Robinson, E.H., Poe, W.E., 1982. Effects of dietary cottonseed meal and gossypol on growth of young channel catfish. Trans. Am. Fish. Soc. 111 (5), 651–655.

Duodua, K.G., Taylora, J.R.N., Beltonb, P.S., Hamakerc, B.R., 2003. Mini Review Factors affecting sorghum protein digestibility. J. Cereal. Sci. 38, 117–131.

Duran-Riveroll, L.M., Cembella, A.D., Okolodkov, Y.B., 2019. A review on the biodiversity and biogeography of toxigenic benthic marine dinoflagellates of the coasts of Latin America. Front. Marine Sci. 6, 25.

EFSA, 2008. Glucosinolates as undesirable substances in animal feed. In: S. P. o. C. i. t. F. (Ed.), Chain, pp. 1–76.

EFSA Panel, F., Panel, Animal, o. A. a. P. o. S. u. i, 2015. Scientific opinion on the safety and efficacy of ethoxyquin 96-ethoxy-1,2-dihydro-2,2,4-trimethylquinoline) for all animal species. EFSA J. 13 (11), 4272.

Eisele, T.A., Nixon, J.E., Pawlowski, N.E., Sinnhuber, R.O., 1978. Effects of dietary cyclopropene fatty acids on the mixed function oxidase system of the rainbow trout (*Salmo gairdneri*). J. Environ. Pathol. Toxicol. 773–778.

European Parliament, C. o. t. E. U, 2002. Directive 2002/32/EC of the European Parliament and of the Council of 7 May 2002 on Undesirable Substances in Animal Feed. Belgium, Denmark, Germany, Ireland, Greece, Spain, France, Italy, Luxembourg, Netherlands, Austria, Portugal, Finland, Sweden, United Kingdom.

Evans, J.J., Pasnik, D.J., Yildirim-Aksoy, M., Lim, C., Klesius, P.H., 2010. Histologic changes in channel catfish, *Ictalurus punctatus* Rafinesque, fed diets containing graded levels of gossypol-acetic acid. Aquacult. Nutr. 16 (4), 385–391.

Fairgrieve, W.T., Myers, M.S., Hardy, R.W., Dong, F.M., 1994. Gastric abnormalities in rainbow trout (*Oncorhynchus mykiss*) fed amine-supplemented diets or chicken gizzard-erosion-positive fish meal. Aquaculture 127, 219–232.

Fisher, J.P., Brown, S.B., Wooster, G.W., Bowser, P.R., 1998. Maternal blood, egg and larval thiamin levels correlate with larval survival in landlocked Atlantic salmon (*Salmo salar*). J. Nutr. 128 (12), 2456–2466.

Francis, G., Kerem, Z., Makkar, H.P.S., Becker, K., 2002. The biological action of saponins in animal systems: a review. Br. J. Nutr. 88 (6), 587–605.

Francis, G., Makkar, H.P.S., 2005. Quillaja saponins - a natural growth promoter for fish. Anim. Feed Sci. Technol. 121, 147–157.

Froystad, M.K., Lilleeng, E., Sundby, A., Krogdahl, A., 2006. Cloning and characterization of alpha-amylase from Atlantic salmon (*Salmo salar* L.). Compar. Biochem. Physiol. Mol. Integr. Physiol. 145 (4), 479–492.

Gan, L., Feng, L., Jiang, W.D., Wu, P., Liu, Y., Jiang, J., Kuang, S.Y., Tang, L., Zhou, X.Q., 2020. Erucic acid impairs intestinal immune function of on-growing grass carp (*Ctenopharyngodon idella*). Aquaculture 519.

Gatlin, D.M., Barrows, F.T., Brown, P., Dabrowski, K., Gaylord, T.G., Hardy, R.W., Herman, E., Hu, G.S., Krogdahl, A., Nelson, R., et al., 2007. Expanding the utilization of sustainable plant products in aquafeeds: a review. Aquacult. Res. 38 (6), 551–579.

Gee, J.M., Wortley, G.M., Johnson, I.T., Price, K.R., Rutten, A.A.J.J., Houben, G.F., Penninks, A.H., 1996. Effects of saponins and glycoalkaloids on the permeability and viability of mammalian intestinal cells and on the integrity of tissue preparations in vitro. Toxicol. Vitro 10 (2), 117–128.

Ghazarian, H., Idoni, B., Oppenheimer, S.B., 2011. A glycobiology review: carbohydrates, lectins and implications in cancer therapeutics. Acta Histochem. 113, 236–247.

Gill, N., Higgs, D.A., Skura, B.J., Rowshandeli, M., Dosanjh, B.S., Mann, J., Gannam, A.L., 2006. Nutritive value of partially dehulled and extruded sunflower meal for post-smolt Atlantic salmon (*Salmo salar* L.) in sea water. Aquacult. Res. 37 (13), 1348–1359.

Giovanni Amato, M., Biasato, I., Chiesa, S., Gasco, L., 2019. The potential role of insects as feed: a multi-perspective review. Animals 9.

Glencross, B., Carter, C.G., Duijster, N., Evans, D.R., Dods, K., McCafferty, P., Hawkins, W.E., Maas, R., Sipsas, S., 2004a. A comparison of the digestibility of a range of lupin and soybean protein products when fed to either Atlantic salmon (*Salmo salar*) or rainbow trout (*Oncorhynchus mykiss*). Aquaculture 237 (1–4), 333–346.

Glencross, B., Evans, D., Hawkins, W., Jones, B., 2004b. Evaluation of dietary inclusion of yellow lupin (*Lupinus luteus*) kernel meal on the growth, feed utilisation and tissue histology of rainbow trout (*Oncorhynchus mykiss*). Aquaculture 235 (1–4), 411–422.

Glencross, B., Evans, D., Rutherford, N., Hawkins, W., McCafferty, P., Dods, K., Jones, B., Harris, D., Morton, L., Sweetingharn, M., et al., 2006. The influence of the dietary inclusion of the alkaloid gramine, on rainbow trout (*Oncorhynchus mykiss*) growth, feed utilisation and gastrointestinal histology. Aquaculture 253 (1–4), 512–522.

Glencross, B.D., Baily, J., Berntssen, M.H.G., Hardy, R.W., MacKenzie, S., Tocher, D.R., 2020. Risk assessment of the use of alternative animal and plant raw material resources in aquaculture feeds. Rev. Aquacult. 12, 703–758.

Goncalves, R.A., Hofstetter, U.S., Jenkins, T., 2018. Mycotoxins in Southeast Asian aquaculture: plant-based meals and finished feeds. World Mycotoxin J. 11 (2), 265–275.

Granum, P.E., 1979. Studies on α-amylase inhibitors in foods. Food Chem. 4, 173–178.

Gu, M., Kortner, T.M., Penn, M., Hansen, A.K., Krogdahl, Å., 2014. Effects of dietary plant meal and soya-saponin supplementation on intestinal and hepatic lipid droplet accumulation and lipoprotein and sterol metabolism in Atlantic salmon (*Salmo salar* L.). Br. J. Nutr. 111 (03), 432–444.

Gu, M., Gu, J.N., Penn, M., Bakke, A.M., Lein, I., Krogdahl, A., 2015. Effects of diet supplementation of soya-saponins, isoflavones and phytosterols on Atlantic salmon (*Salmo salar*, L) fry fed from start-feeding. Aquacult. Nutr. 21 (5), 604–613.

Gu, M., Zhang, Z., Bai, N., Xu, X., Xu, B., 2018. Soya-saponins induce intestinal inflammation and barrier dysfunction in juvenile turbot (*Scophthalmus maximus*). Fish Shellfish Immunol. 77, 264–272.

Guclu-Ustundag, O., Mazza, G., 2007. Saponins: properties, applications and processing. Crit. Rev. Food Sci. Nutr. 47 (3), 231–258.

Gustafson, G., Ryan, C.A., 1974. Protease inhibitor in potato plants – isoinhibitor compartmentalization. Fed. Proc. 33 (5), 1446.

Food Standards Australia New Zealand, 2003. Erucic acid in food: a toxicological review and risk assessment. In: Food Standards Australia New Zealand. Technical Report Series, vol. 22.

Halver, J.E., Mitchell, I.A., 1969. Trout Hepatoma Research Conference Papers. Bureau Sport Fish. Wildlife Rep. 70, U.S. Dept. Interior, Washington, DC.

Hambright, K.D., Zamor, R.M., Easton, J.D., Allison, B., 2014. Algae. In: Wexler, P. (Ed.), Encyclopedia of Toxicology. Academic Press, pp. 130–141.

Hannoufa, A., Pillai, B.V.S., Chellamma, S., 2014. Genetic enhancement of *Brassica napus* seed quality. Transgen. Res. 23 (1), 39–52.

Harder, A.H., Ardren, W.R., Evans, A., Futia, M.H., Kraft, C.E., Marsden, J.E., Richter, C.A., Rinchard, J., Tillitt, D.R., Christie, M.R., 2018. Thiamine deficiency in fishes: causes, consequences, and potential solutions. Rev. Fish Biol. Fish. 28, 865–886.

Hardy, R.W., Scott, T., Hatfield, C.L., Barnet, H.J., Gauglitz, E.J., Wekell, J.C., Eklund, M.W., 1995. Domoic acid in rainbow trout (*Oncorhynchus mykiss*) feeds. Aquaculture 131, 253–260.

Hardy, R.W., 2001. Nutritional pathology of teleosts. In: Fish Pathology, third ed. W.B. Saunders, pp. 347–366.

He, X., Deng, M., Wang, Q., Yang, Y., Yang, Y., Nie, X., 2016. Residues and health risk assessment of quinolones and sulfonamides in cultured fish from Pearl River Delta, China. Aquaculture 458, 38–46.

Hellinger, R., Gruber, C.W., 2019. Peptide-based protease inhibitors from plants. Drug Discov. Today 24 (9), 1877–1889.

Hendricks, J., 2002. Adventious toxins. In: Halver, J.E., Hardy, R.M. (Eds.), Fish Nutrition. Academic Press, Amsterdam, pp. 603–649.

Hendricks, J.D., Sinnhuber, R.O., Loveland, P.M., Pawlowski, N.E., Je, N., 1980. Hepatocarcinogenicity of glandless cottonseeds and cottonseed oil to rainbow trout (*Salmo gairdneri*). Science 208, 309–311.

Hendricks, J.D., Bailey, G.S., 1989. Adventitious; toxins. In: Halver, J.E. (Ed.), Fish Nutrition. Academic Press Inc, New York, pp. 606–651.

Hendricks, J.D., 2002. Adventitious toxins. In: Fish Nutrition, third ed. Elsevier, Amsterdam, pp. 601–649.

Hendriks, H.G.C.J., van den Ingh, T.S.G.A., Krogdahl, Å., Olli, J.J., Koninkx, J.F.J.G., 1990. Binding of soybean agglutinin to small intestinal brush border membranes and brush border membrane enzyme activities in Atlantic salmon (*Salmo salar*). Aquaculture 91, 163–170.

Herman, R.L., 1970. Effect of gossypol on rainbow trout Salmo gairdneri Richardson. J. Fish. Biol. 2, 293–301.

Ho, M.F., Yin, X., Filho, F.F., Lajolo, F., Whitaker, J.R., 1994. Naturally occurring α-amylase inhibitors: structure/function relationships. In: Yada, R.Y., Jackman, R.L., Smith, J.L. (Eds.), Protein Structure-Function Relationships in Foods. Springer US, Boston, MA, pp. 89–119.

Hofer, R., Sturmbauer, C., 1986. Inhibition oaf trout and carp a-amylase by wheat. Aquaculture 48, 277–283.

Huang, Q., Liu, X., Zhao, G., Hu, T., Wang, Y., 2018. Potential and challenges of tannins as an alternative to in-feed antibiotics for farm animal production. Animal Nutr. 4, 137–150.

Hung, S.S.O., Cho, C.Y., Slinger, S.L., 1981. Effect of oxidized fish oil, DL-A-tocopheryl acetate and ethoxyquin supplementation on the vitamin E nutrition of rainbow trout (*Salmo gairdneri*) fed practical diets. J. Nutr. 111, 648–657.

Inudo, M., Ishibashi, H., Matsumura, N., Matsuoka, M., Mori, T., Taniyama, S., Kadokami, K., Koga, M., Shinohara, R., Hutchinson, T.H., et al., 2004. Effect of estrogenic activity, and phytoestrogen and organochlorine pesticide contents in an experimental fish diet on reproduction and hepatic vitellogenin production in medaka (*Oryzias latipes*). Comp. Med. 54 (6), 673–680.

Ireland, P.A., Dziedzic, S.Z., Kearsley, M.W., 1986. Saponin content of soya and some commercial soya products by means of high-performance liquid chromatography of the sapogenins. J. Sci. Food Agric.

Islam, G.M.R., Habib, M.R., Waid, J.L., Rahman, M.S., Kabir, J., Akter, S., Jolly, Y.N., 2017. Heavy metal contamination of freshwater prawn (*Macrobrachium rosenbergii*) and prawn feed in Bangladesh: a market-based study to highlight probably health risks. Chemosphere 170, 282–289.

Iwashita, Y., Yamamoto, T., Furuita, H., Sugita, T., Suzuki, N., 2008. Influence of certain soybean antinutritional factors supplemented to a casein-based semipurified diet on intestinal and liver morphology in fingerling rainbow trout *Oncorhynchus mykiss*. Fish. Sci. 74 (5), 1075–1082.

Jackson, L.S., Li, M.H., Robinson, E.H., 1996. Use of microbial phytase in channel catfish *Ictalurus punctatus* diets to improve utilization of phytate phosphorus. J. World Aquacult. Soc. 27, 309–313.

Jaffe, W.G., Moreneo, R., Vallis, V., 1973. Amylase inhibitors in legume seeds. Nutr. Rep. Int. 7, 169–174.

Jayaraj, R., Megha, P., Sreedev, P., 2016. Organochlorine pesticides, their toxic effects on living organisms and their fate in the environment. Toxicology 9 (3–4), 90–100.

Johnson, A.R., Fogerty, A.C., Pearson, J.A., Shenstone, F.S., Bersten, A.M., 1969. Fatty acid desaturase systems of hen liver and their inhibition by cyclopropene fatty acids. Lipids 4, 265–269.

Johny, A., Ivanova, L., Knutsdatter Ostbye, T.K., Faeste, C.K., 2020. Biotransformation of phytoestrogens from soy in enzymatically characterized liver microsomes and primary hepatocytes of Atlantic salmon. Ecotoxicol. Environ. Saf. 197, 110611.

Jones, D.A., 1998. Why are so many food plants cyanogenic? Phytochemistry 47, 155–162. In this issue.

Kako, K.J., Peckett, S.D., 1981. Effect of high fat/high erucic acid diet on phosphatidate synthesis and phosphatidate phosphatase in the subcellular fractions of rat heart and liver. Lipids 16 (1), 23–29.

Kalaitzis, J.A., Chau, R., Kohli, G.S., Murray, S.A., Neilan, B.A., 2010. Biosynthesis of toxic naturally-occurring seafood contaminants. Toxicon 56 (2), 244–258.

Kausch, U., Alberti, M., Haindl, S., Budczies, J., Hock, B., 2008. Biomarkers for exposure to estrogenic compounds: gene expression analysis in zebrafish (*Danio rerio*). Environ. Toxicol. 23 (1), 15–24.

Ketola, H.G., Smith, C.E., Kindschi, G.A., 1989. Influence of diet and oxidative rancidity on fry of Atlantic and coho salmon. Aquaculture 79, 417–423.

Kim, J.Y., Park, S.C., Kim, M.H., Lim, H.T., Park, Y., Hahm, K.S., 2005. Antimicrobial activity studies on a trypsin-chymotrypsin protease inhibitor obtained from potato. Biochem. Biophys. Res. Commun. 330 (3), 921–927.

Kim, J.Y., Park, S.C., Hwang, I., Cheong, H., Nah, J.W., Hahm, K.S., Park, Y., 2009. Protease inhibitors from plants with antimicrobial activity. Int. J. Mol. Sci. 10 (6), 2860–2872.

Kim, M.H., Park, S.C., Kim, J.Y., Lee, S.Y., Lim, H.T., Cheong, H., Hahm, K.S., Park, Y., 2006. Purification and characterization of a heat-stable serine protease inhibitor from the tubers of new potato variety "Golden Valley". Biochem. Biophys. Res. Commun. 346 (3), 681–686.

Knudsen, D., Ron, O., Baardsen, G., Smedsgaard, J., Koppe, W., Froklaer, H., 2006. Soyasaponins resist extrusion cooking and are not degraded during gut passage in Atlantic salmon (*Salmo salar* L.). J. Agric. Food Chem. 54 (17), 6428–6435.

Knudsen, D., Jutfelt, F., Sundh, H., Sundell, K., Koppe, W., Frokiaer, H., 2008. Dietary soya saponins increase gut permeability and play a key role in the onset of soyabean-induced enteritis in Atlantic salmon (*Salmo salar* L.). Br. J. Nutr. 100 (1), 120–129.

Knutsen, H.K., Alexander, J., Barregard, L., Bignami, M., Bruschweiler, B., Ceccatelli, S., Dinovi, M., Edler, L., Grasl-Kraupp, B., Hogstrand, C., et al., 2016. Erucic acid in feed and food. EFSA J. 14 (11).

Knutsen, H.K., Barregård, L., Bignami, M., Brüschweiler, B., Ceccatelli, S., Dinovi, M., Edler, L., Grasl-Kraupp, B., Hogstrand, C., Hoogenboom, L., et al., 2017. Presence of free gossypol in whole cottonseed. EFSA J. 15 (7), e04850.

Ko And, K., Malison, J.A., Reed, J.D., 1999. Effect of genistein on the growth and reproductive function of male and female yellow perch *Perca flavescens*. J. World Aquacult. Soc. 30 (1), 73–79.

Kong, W., Huang, C., Tang, Y., Zhang, D., Wu, Z., Chen, X., 2017. Effect of *Bacillus subtilis* on Aeromonas hydrophila-induced intestinal mucosal barrier function damage and inflammation in grass carp (*Ctenopharyngodon idella*). Sci. Rep. 6. Article number: 1588.

Konig, V., Vertesy, L., Schneider, T.R., 2003. Structure of the [alpha]-amylase inhibitor tendamistat at 0.93 A. Acta Crystallogr. D 59 (10), 1737–1743.

Kortner, T.M., Skugor, S., Penn, M.H., Mydland, L.T., Djordjevic, B., Hillestad, M., Krasnov, A., Krogdahl, A., 2012. Dietary soyasaponin supplementation to pea protein concentrate reveals nutrigenomic interactions underlying enteropathy in Atlantic salmon (*Salmo salar*). BMC Vet. Res. 8, 101.

Kortner, T.M., Bjorkhem, I., Krasnov, A., Timmerhaus, G., Krogdahl, A., 2014. Dietary cholesterol supplementation to a plant-based diet suppresses the complete pathway of cholesterol synthesis and induces bile acid production in Atlantic salmon (*Salmo salar* L.). Br. J. Nutr. 111, 2089–2103.

Krogdahl, A., Penn, M., Thorsen, J., Refstie, S., Bakke, A.M., 2010. Important antinutrients in plant feedstuffs for aquaculture: an update on recent findings regarding responses in salmonids. Aquacult. Res. 41 (3), 333–344.

Krogdahl, A., Gajardo, K., Kortner, T.M., Penn, M., Gu, M., Berge, G.M., Bakke, A.M., 2015. Soya saponins induce enteritis in Atlantic salmon (*Salmo salar* L.). J. Agric. Food Chem. 63 (15), 3887–3902.

Krogdahl, A., Kortner, T.M., Jaramillo-Torres, A., Gamil, A.A.A., Chikwati, E., Li, Y.X., Schmidt, M., Herman, E., Hymowitz, T., Teimouri, S., et al., 2020. Removal of three proteinaceous antinutrients from soybean does not mitigate soybean-induced enteritis in Atlantic salmon (*Salmo salar*, L). Aquaculture 514.

Krogdahl, Å., Holm, H., 1982. Activation and pattern of proteolytic enzymes in pancreatic tissue from rat, pig, cow, chicken, mink and fox. Compar. Biochem. Physiol. Compar. Physiol. 72 (3), 575–578.

Krogdahl, Å., Berg Lea, T., Olli, J.J., 1994. Soybean proteinase inhibitors affect intestinal trypsin activities and amino acid digestibilities in rainbow trout (*Oncorhynchus mykiss*). Compar. Biochem. Physiol. Compar. Physiol. 107A (1), 215–219.

Kundu, G.K., Alauddin, M., Akter, M.S., Khan, M.S., Islam, M.M., Mondal, G., Islam, D., Mohanta, L.C., Huque, A., 2017. Metal contamination of commercial fish feed and quality aspects of farmed tilapia (*Oreochromis niloticus*) in Bangladesh. Biores. Commun. 3, 345–353.

Kurzer, M.S., Xu, X., 1997. Dietary phytoestrogens. Annu. Rev. Nutr. 17, 353–381.

Lawrence, P.K., Koundal, K.R., 2002. Plant protease inhibitors in control of phytophagous insects. Electron. J. Biotechnol. 5 (1).

Lechtenber, M., Nahrstedt, A., 1999. Cyanogenic glycosides. Naturally Occurring Glycosides. John Wiley & Sons, Chichester, United Kingdom, pp. 147–191. In this issue.

Lee, T.C.-H., Fong, F.L.-Y., Ho, K.-C., Lee, F.W.-F., 2016. The mechanism of diarrhetic shellfish poisoning toxin production in *Prorocentrum* spp.: physiological and molecular perspectives. Toxins 8, 272.

Liland, N.S., Espe, M., Rosenlund, G., Waagbo, R., Hjelle, J.I., Lie, O., Fontanillas, R., Torstensen, B.E., 2013. High levels of dietary phytosterols affect lipid metabolism and increase liver and plasma TAG in Atlantic salmon (*Salmo salar* L.). Br. J. Nutr. 1–10.

Ling, W.H., Jones, P.J., 1995. Dietary phytosterols: a review of metabolism, benefits and side effects. Life Sci. 57 (3), 195–206.

Liu, H., Dong, X.H., Tan, B.P., Du, T., Zhang, S., Yang, Y.Z., Chi, S.Y., Yang, Q.H., Liu, H.Y., 2020. Effects of fish meal replacement by low-gossypol cottonseed meal on growth performance, digestive enzyme activity, intestine histology and inflammatory gene expression of silver sillago (*Sillago sihama* Forsskal) (1775). Aquacult. Nutr. 26 (5), 1724–1735.

Liu, X., Steele, J.C., Meng, X.-Z., 2017. Usage, residue, and human health risk of antibiotics in Chinese aquaculture: a review. Environ. Pollut. 223, 160–169.

Lopez Nadal, A., Peggs, D., Wiegertjes, G.F., Brugman, S., 2018. Exposure to antibiotics affects saponin immersion-induced immune stimulation and shift in microbial composition in zebrafish larvae. Front. Microbiol. 9, 2588.

Maclean, W.C., De Romana, G.L., Gastanaduy, A., Graham, G.G., 1983. The effect of decortication and extrusion on the digestibility of sorghum by preschool children. J. Nutr. 113, 2171–2177.

Manning, B.B., 2001. Mycotoxins in fish feeds. In: Webster, C.D. (Ed.), Nutrition and Fish Health. CRC Press, Boca Raton, pp. 267–287.

Matejova, I., Svobodova, Z., Vakula, J., Mares, J., Modra, H., 2017. Impact of mycotoxins on aquaculture fish species: a review. J. World Aquacult. Soc. 48 (2), 186–200.

Matsuda, K., Kaneko, M., Kusaka, K., Shishido, T., Tamaki, Y., 1998. Soy saponins as feeding stimulants to the oriental clouded larva, *Coliaserate poliographus*. Appl. Entomol. Zool. 33, 255–258.

Matsunaga, S., Takahashi, N., Fusetani, N., 2009. Dinogunellins A–D: putative ichthyootoxic phospholipids of northern blenny *Stichaeus grigoriewi* eggs. Pure Appl. Chem. 81 (6), 1001–1008.

Messina, M.J., 1999. Legumes and soybeans: overview of their nutritional profiles and health effects. Am. J. Clin. Nutr. 70 (3), 439S–450S.

Miller, M.R., Nichols, P.D., Carter, C.G., 2008. The digestibility and accumulation of dietary phytosterols in Atlantic salmon (*Salmo salar* L.) smolt fed diets with replacement plant oils. Lipids 43 (6), 549–557.

Milligan, S.R., Khan, O., Nash, M., 1998. Competitive binding of xenobiotic oestrogens to rat alpha-fetoprotein and to sex steroid binding proteins in human and rainbow trout (*Oncorhynchus mykiss*) plasma. Gen. Comp. Endocrinol. 112 (1), 89–95.

Mincarelli, L.F., Paula, J.R., Pousão-Ferreira, P., Rosa, R., Costa, P.R., 2018. Effects of acute waterborne exposure to harmful algal toxin domoic acid on foraging and swimming behaviours of fish early stages. Toxicon 156, 66–71.

Moccia, R.D., Hung, S.S.O., Slinger, S.J., Ferguson, H.W., 1984. Effect of oxidized fish oil, vitamin E and ethoxyquin on the histology and haematology of rainbow trout, *Salmo gairdneri* Richardson. J. Fish. Dis. 7, 269–282.

Morais, T., Inácio, A., Coutinho, T., Ministro, M., Cotas, J., Pereira, L., Bahcevandziev, K., 2020. J. Mar. Sci. Eng. 8.

Morales, G.A., Moyano, F.J., Marquez, L., 2011. In vitro assessment of the effects of phytate and phytase on nitrogen and phosphorus bioaccessibility within fish digestive tract. Anim. Feed Sci. Technol. 170, 209–221.

Murai, T., Andrews, J.W., 1974. Interactions of dietary alpha-tocopherol, oxidized menhaden oil and ethoxyquin on channel catfish (*Ictalurus punctatus*). J. Nutr. 104, 1416–1431.

Naumann, H.D., Tedeschi, L.O., Zeller, W.E., Huntley, N.F., 2017. The role of condensed tannins in ruminant animal production: advances, limitations and future directions. J. Animal Sci. 46 (12), 929–949.

Naveed, M., Hejazi, V., Abbas, M., Kamboh, A.A., Khan, G.J., Shumzaid, M., Ahmad, F., Babazadeh, D., FangFang, X., Modarresi-Ghazani, F., et al., 2018. Chlorogenic acid (CGA): a pharmacological review and call for further research. Biomed. Pharmacother. 97, 67–74.

Naylor, R., Hardy, R.W., Buschmann, A.H., Bush, S.R., Cao, L., Klinger, D.H., Little, D.C., Lubchenco, J., Shumway, S.E., Troell, M., 2020. A 20-year retrospective review of global aquaculture. Nature.

Ng, Y., Hanson, S., Malison, J.A., Wentworth, B., Barry, T.P., 2006. Genistein and other isoflavones found in soybeans inhibit estrogen metabolism in salmonid fish. Aquaculture 254, 658–665.

Nieto, F.R., Cobos, E.J., Tejada, M.A., Sanchez-Fernandez, C., Gonzalez-Cano, R., Cendan, C.M., 2012. Tetrodotoxin (TTX) as a therapeutic agent for pain. Mar. Drugs 10 (2), 281–305.

NRC, 1983. Antinutrients. In: NRC (Ed.), Nutrient Requirements of Warmwater Fishes and Shellfishes. National Academy Press, New York, pp. 59–70.

NRC, 2011a. Antinutritional factors and adventitious toxins. In: Nutrient Requirments of Fish and Shrimp. National Academic Press, Washington, DC, pp. 232–252.

NRC, 2011b. Vitamins. In: NRC (Ed.), NRC Nutrient Requirement of Fish and Shrimp. National Academic Press.

Oda, K., Matsuda, H., Murakami, T., Katayama, S., Ohgitani, T., Yoshikawa, M., 2000. Adjuvant and haemolytic activities of 47 saponins derived from medicinal and food plants. Biol. Chem. 381 (1), 67–74.

Odani, S., Ikenaka, T., 1972. Studies on soybean trypsin inhibitors. IV. Complete amino acid sequence and the anti-proteinase sites of Bowman-Birk soybean proteinase inhibitor. J. Biochem. 5, 839–848.

Oliveira, M., Vasconcelos, V., 2020. Occurrence of mycotoxins in fish feed and its effects: a review. Toxins 12.

Olli, J.J., Krogdahl, Å., Berg-Lea, T., 1989. Effects of Soybean Trypsin Inhibitor Activity on Nutrient Digesibility in Salmonids Fed Practical Diets Containing Various Soybean Meals. Laboratory of Fish Nutrition, Department of Aquatic Biosciences, Tokyo, Japan, pp. 263–271.

Olli, J.J., Hjelmeland, K., Krogdahl, Å., 1994. Soybean trypsin inhibitors in diets for Atlantic salmon (*Salmo salar*, L): effects on nutrient digestibilities and trypsin in pyloric caeca homogenate and intestinal content. Compar. Biochem. Physiol. Compar. Physiol. 109 (4), 923–928.

Opstvedt, J., Mundheim, H., Nygard, E., Aase, H., Pike, I.H., 2000. Reduced growth and feed consumption of Atlantic salmon (*Salmo salar* L.) fed fish meal made from stale fish is not due to increased content of biogenic amines. Aquaculture 188, 323–337.

Ortinau, L.C., Pickering, R.T., Nickelson, K.J., Stromsdorfer, K.L., Naik, C.Y., Haynes, R.A., Bauman, D.E., Rector, R.S., Fritsche, K.L., 2012. Sterculic oil, a natural SCD1 inhibitor, improves glucose tolerance in obese ob/ob mice. Endocrinology 947323.

Oterhals, A., Solvang, M., Nortvedt, R., Berntssen, M.H.G., 2007. Optimization of activated carbon-based decontamination of fish oil by response surface methodology. Eur. J. Lipid Sci. Technol. 109 (7), 691–705.

Oterhals, A., Berntssen, M.H.G., 2010. Effects of refining and removal of persistent organic pollutants by short-path distillation on nutritional quality and oxidative stability of fish oil. J. Agric. Food Chem. 58, 23.

Pan Hao, P., Alaraj, I.Q.M., Dulayymi, J.R., Tacon, A.G.J., Baird, M.S., Liu, J., Liu, Q., 2016. Sterculic acid and its analogues are potent inhibitors of toxoplasma gondii. Kor. J. Parasitol. 54, 139–145.

Park, S.I., 1978. Nutritional liver disease in cultured yellowtail, Seriola quinqueradiata, caused by feed deficiency. Bull. Kor. Fish. Soc. 11, 1–4.

Pastore, m. r., Negrato, E., Poltronieri, C., Barion, G., Maria, M., Francesca, T., Ballarin, C., Maccatrozzo, L., Radaelli, G., Daniela, B., 2018. Effects of dietary soy isoflavones on estrogenic activity, cortisol level, health and growth in rainbow trout, *Oncorhynchus mykiss*. Aquacult. Res. 49.

Peláez, R., Pariente, A., Pérez-Sala, Á., Larráyoz, I.M., 2020. Sterculic acid: the mechanisms of action beyond stearoyl-CoA desaturase inhibition and therapeutic opportunities in human diseases. Cells 140.

Pelissero, C., Bennetau, B., Babin, P., Le Menn, F., Dunogues, J., 1991. The estrogenic activity of certain phytoestrogens in the Siberian sturgeon Acipenser baeri. J. Steroid Biochem. Mol. Biol. 38 (3).

Penn, M.H., Bendiksen, E.A., Campbell, P., Krogdahl, Å., 2011. High level of dietary pea protein concentrate induces enteropathy in Atlantic salmon (*Salmo salar* L.). Aquaculture 310 (3–4), 267–273.

Pereira, E.R., Merib, J., Cadorim, H., Schneider, J., Carvalho, G.S., Duarte, F.A., Welz, B., Del Campo Menoya, J., Feldmann, J., 2017. Development of a fast screening method for the direct determination of chlorinated persistent organic pollutants in fish oil by high-resolution continuum source graphite furnace molecular absorption spectrometry. Food Contr. 78, 456–462.

Pierpoint, W.S., 1970. Formation and behaviour of o-quinones in some processes of agricultural importance. Rep. Rothams. Exp. Stn. 2, 199–218.

Pietsch, C., 2020. Risk assessment for mycotoxin contamination in fish feeds in Europe. Mycotoxin Res. 36, 41–62.

Pietsch, C., Muller, G., Mourabit, S., Carnal, S., Bandara, K., 2020. Occurrence of fungi and fungal toxins in fish feed during storage. Toxins 12, 171.

Pinotti, L., Ottoboni, M., Giromini, C., Dell'Orto, V., Cheli, F., 2016. Mycotoxin contamination in the EU feed supply: a focus on cereal byproducts. Toxins 8, 45.

Pothier, J., Cheav, L., Galand, N., Vie, I.C., 1998. A comparative study of the effects of sparteine, lupanine and lupin extract on the central nervous system of the mouse. J. Pharm. Pharmacol. 50, 949–954.

Poulton, J.E., 1990. Cyanogenesis in plants. Plant Physiol. 94, 401–405.

Pusztai, A., Grant, G., Duguid, T., Brown, D.S., Peumans, W.J., Vandamme, E.J.M., Bardocz, S., 1995. IInhibition of starch digestion by alpha-amylase inhibitor reduces the efficiency of utilization of dietary proteins and lipids and retards the growth of rats. J. Nutr. 125 (6), 1554–1562.

Raju, P.K., Reiser, R., 1967. Inhibition of fatty acyl desaturase by cyclopropene fatty acids. J. Biol. Chem. 242, 379–384.

Raposo, M.I.C., Gomes, M., Botelho, M.J., Rudnitskaya, A., 2020. Paralytic shellfish toxins (PST)-Transforming enzymes: a review. Toxins 12 (5).

Rathore, K.S., Pandeya, D., Campbell, L.M., Wedegaertner, T.C., Puckhaber, L., Stipanovic, R.D., Thenell, J.S., Hague, S., Hake, K., 2020. Ultra-low gossypol cottonseed: selective gene silencing opens up a vast resource of plant-based protein to improve human nutrition. Crit. Rev. Plant Sci. 39, 1–29.

Rehulka, J., 1990. Effect of hydrolytically changed and oxidized fat in dry pellets on the health of rainbow trout, *Oncorhynchus mykiss* (Richardson). Aqua. Fish. Mgmt. 21, 419–434.

Reseland, J.E., Holm, H., Jenssen, T., Jacobsen, M.B., Hanssen, L.E., 1995. Inhibitor-stimulated nonparallel pancreatic-secretion in man - hormonal and neural regulation. Scand. J. Gastroenterol. 30 (1), 72–80.

Richards, J., Zhao, J., Harrell, R., Atwell, C., Dibner, J., 2010. Trace mineral nutrition in poultry and swine. Asian-Australas. J. Anim. Sci. 23, 1527–1534.

Rietjens, I., Louisse, J., Beekmann, K., 2017. The potential health effects of dietary phytoestrogens. Br. J. Pharmacol. 174 (11), 1263–1280.

Riley, S.C., Evans, A.N., 2008. Phylogenetic and ecological characteristics associated with thiaminase activity in Laurentian Great Lakes fishes. Trans. Am. Fish. Soc. 137 (1), 147–157.

Riley, S.C., Munkittrick, K.R., Evans, A.N., Krueger, C.C., 2008. Understanding the ecology of disease in Great Lakes fish populations. Aquat. Ecosys. Health Manag. 11 (3), 321–334.

Rique-Marie, D.M., Abdo-De La Parra, I., Cruz-Suarez, L.E., Cuzon, G., Cousin, M., Pike, I.H., 1998. Raw material freshness, a quality criterion for fish meal fed to shrimp. Aquaculture 165, 95–109.

Roberts, R.J., 2002. Nutritional pathology. In: Fish Nutrition, third ed. Elsevier, Amsterdam, pp. 453–504.

Roehm, J.N., Lee, D.J., Sinnhuber, R.O., 1967. Accumulation and elimination of dietary gossypol in the organs of rainbow trou. J. Nutr. 92, 425–428.

Roehm, J.N., Lee, D.J., Wales, J.H., Polityka, S.D., Sinnhuber, R.O., 1970. The effect of dietary sterculic acid on the hepatic lipids of rainbow trout. Lipids 5 (1), 80–84.

Ryan, C.A., Clary, J.J., 1964. Some reactions of chicken chymotrypsin + Turkey trypsin with substrates + naturally occurring protease inhibitors. Arch. Biochem. Biophys. 108 (1), 169–&.

Ryan, C.A., 1990. Protease inhibitors in plants - genes for improving defenses against insects and pathogens. Annu. Rev. Phytopathol. 28, 425–449.

Sander, J.E., Cai, T., Dale, N., Bennett, L.W., 1996. Development of biogenic amines during fermentation of poultry carcasses. J. Appl. Poultry Res. 5 (2), 161–166.

Santigosa, E., de Rodriganez, M.A.S., Rodiles, A., Barroso, F.G., Alarcon, F.J., 2010. Effect of diets containing a purified soybean trypsin inhibitor on growth performance, digestive proteases and intestinal histology in juvenile sea bream (*Sparus aurata* L.). Aquacult. Res. 41 (9), e187–e198.

Schmidt, M.A., Hymowitz, T., Herman, E.M., 2015. Breeding and characterization of soybean triple null; a stack of recessive alleles of Kunitz trypsin inhibitor, soybean agglutinin, and P34 allergen nulls. Plant Breed. 134 (3), 310–315.

Selivonchick, D.P., Williams, J.L., Schaup, H.W., 1981. Alteration of liver microsomal proteins from rainbow trout (*Salmo gairdneri*) fed cyclopropenoid fatty acids. Lipids 16, 211–214.

Serrano Jr., A.E., 2013. Effects of Quillaja saponins on growth, feed efficiency, digestive enzyme activities and metabolism of common carp (*Cyprinus carpio* L). Aquacult. Nutr. 19 (4), 468–474.

Shen, J.F., Liu, H.Y., Tan, B.P., Dong, X.H., Yang, Q.H., Chi, S.Y., Zhang, S., 2020. Effects of replacement of fishmeal with cottonseed protein concentrate on the growth, intestinal microflora, haematological and antioxidant indices of juvenile golden pompano (*Trachinotus ovatus*). Aquacult. Nutr. 26 (4), 1119–1130.

Shi, J., Arunasalam, K., Yeung, D., Kakuda, Y., Mittal, G., Jiang, Y., 2004. Saponins from edible legumes: chemistry, processing, and health benefits. J. Med. Food 7 (1), 67–78.

Singh, G., Suresh, S., Bayineni, V.K., Kadeppafari, R.K., 2015. Lipase inhibitors from plants and their medical applications. Int. J. Pharm. Pharmaceut. Sci. 7, 5.

Sinnhuber, R.O., Lee, D.J., Wales, J.H., Landers, M.K., Keyl, A.C., 1974. Hepatic carcinogenesis of aflatoxin M1 in trout (*Salmo gairdneri*) and its enhancement by cyclopropene fatty acids. J. Natl. Cancer Inst. 53, 1285–1288.

Sissener, N., Liland, N., Holen, E., Stubhaug, I., Torstensen, B., Rosenlund, G., 2017. Phytosterols are not involved in the development of fatty liver in plant oil fed Atlantic salmon (*Salmo salar*) at high or low water temperature. Aquaculture 480.

Sissener, N., Rosenlund, G., Stubhaug, I., Liland, N., 2018. Tissue sterol composition in Atlantic salmon (*Salmo salar* L.) depends on the dietary cholesterol content and on the dietary phytosterol:cholesterol ratio, but not on the dietary phytosterol content. Br. J. Nutr. 119, 1–11.

Sissener, N.H., Ornsrud, R., Sanden, M., Froyland, L., Remo, S., Lundebye, A.K., 2018. Erucic acid (22:1n-9) in fish feed, farmed, and wild fish and seafood products. Nutrients 10 (10).

Slinger, S.J., 1977. Improving the nutritional properties of rapeseed. J. Am. Oil Chem. Soc. 54, 9A–99A.

Smichi, N., Fendri, A., Triki, S., Arondel, V., Rebai, A., Gargouri, Y., Miled, N., 2017. Biochemical characterization, cloning and molecular modeling of a digestive lipase from red seabream (*Pagrus major*): structural explanation of the interaction deficiency with colipase and lipidic interface. Eng. Life Sci. 17 (6), 664–677.

Smith, C.E., 1979. The prevention of liver lipoid degeneration (ceroidosis) and microcytic anaemia in rainbow trout *Salmo gairdneri* Richardson fed rancid diets: a preliminary report. J. Fish. Dis. 2, 429–437.

Soliman, A.K., Roberts, R.J., Jauncey, K., 1983. The pathological effects of feeding rancid lipid in diets for *Oreochromis niloticus* (Trewavas). In: Proc. Int. Symp. Tilapia in Aquaculture. Tel Aviv University Press, pp. 193–199.

Spankeren, M., Sibinga, N., Reinshol, Ø., Torstensen, B., Saele, Ø., Liland, N., 2019. A single meal containing phytosterols does not affect the uptake or tissue distribution of cholesterol in zebrafish (*Danio rerio*). Front. Marine Sci. 6, 92.

Spinelli, J., Mahnken, C., 1976. Effect of diets containing dogfish (*Squalus acanthias*) meal on the mercury content and growth of pen-reared coho salmon (*Oncorhynchus kisutch*). J. Fish. Res. Board Can. 33, 1771–1778.

Sprague, M., Bendiksen, E.A., Dick, J.R., Stracan, F., 2010. Effects of decontaminated fish oil or a fish and vegetable oil blend on persistent organic pollutant and fatty acid compositions in diet and flesh of Atlantic salmon (*Salmo salar*). Br. J. Nutr. 10, 1442–1451.

Starr, M., Lair, S., Michaud, S., Scarratt, M., Quilliam, M., Lefaivre, D., Robert, M., Wotherspoon, A., Michaud, R., Ménard, N., et al., 2017. Multispecies mass mortality of marine fauna linked to a toxic dinoflagellate bloom. PLoS One 12, e0176299.

Storebakken, T., Shearer, K.D., Roem, A.J., 1998. Availability of protein, phosphorus and other elements in fish meal, soy-protein concentrate and phytase-treated soy-protein-concentrate-based diets to Atlantic salmon, *Salmo salar*. Aquaculture 161 (1–4), 365–379.

Struthers, B.J., LEE, D.J., Sinnhuber, R.O., 1975a. Altered lipid-metabolism in livers of rainbow-trout fed cyclopropenoid fatty-acids. Exp. Mol. Pathol. 23 (2), 181–187.

Struthers, B.J., Wales, J.H., Lee, D.J., Sinnhuber, R.O., 1975b. Liver composition and histology of rainbow-trout fed cyclopropenoid fatty-acids. Exp. Mol. Pathol. 23 (2), 164–170.

Tacon, A.G.J., 1997. Fishmeal replacers: review of antinutrients within oilseeds and pulses. In: Tacon, A.G.J., Basurco, B. (Eds.), Limiting Factor for the Aquafeed Green Revolution? Feeding Tomorrow's Fish. CIHEAM, 22, Zaragoza, pp. 153–182.

Tacon, A.G.J., Metian, M., 2008. Aquaculture feed and food safety. The role of the Food and Agriculture Organization and the Codex Alimentarius. Ann. N. Y. Acad. Sci. 1140, 50–59.

Tahir, M., Shim, M.Y., Ward, N.E., Smith, C., Foster, E., Guney, A., Pesti, G.M., 2012. Phytate and other nutrient components of feed ingredients for poultry. Poultry Sci. 91, 928–935.

Tian, J., Wang, K., Wang, X., Wen, H., Zhou, H., Liu, C., Mai, K., He, G., 2018. Soybean saponin modulates nutrient sensing pathways and metabolism in zebrafish. Gen. Comp. Endocrinol. 257, 246–254.

Tillitt, D.E., Zajicek, J.L., Brown, S.B., Brown, L.R., Fitzsimons, J.D., Honeyfield, D.C., Holey, M.E., Wright, G.M., 2005. Thiamine and thiaminase status in forage fish of salmonines from Lake Michigan. J. Aquat. Anim. Health 17 (1), 13–25.

Tollefsen, K.E., Ovrevik, J., Stenersen, J., 2004. Binding of xenoestrogens to the sex steroid-binding protein in plasma from Arctic charr (*Salvelinus alpinus* L.). Comp. Biochem. Physiol. C Toxicol. Pharmacol. 139 (1–3), 127–133.

Tookey, H.L., van Etten, C.H., Daxenbichler, M.E., 1980. Glucosinolates. In: Liener, I.E. (Ed.), Toxic Constituents of Plant Foodstuffs. Academic Press, New York, pp. 103–142.

Torno, C., Staats, S., de Pascual-Teresa, S., Rimbach, G., Schulz, C., 2019. Effects of resveratrol and genistein on growth, nutrient utilization and fatty acid composition of rainbow trout. Animal 13 (5), 933–940.

Turnipseed, S.B., Storey, J.M., Wu, I.-L., Gieseker, C.M., Hastrouck, N.R., Crosby, T.X., Andersen, W.C., Lanier, S., Casey, C.R., Burger, R., et al., 2018. Application and evaluation of a high-resolution mass spectrometry screening method for veterinary drug residues in incurred fish and imported aquaculture samples. Anal. Bioanal. Chem. 410, 5529–5544.

Uchida, K., Konishi, Y., Harada, K., Okihashi, M., Yamaguchi, T., Do, M.H.N., Bui, L.T., Nguyen, T.D., Nguyen, P.D., Khong, D.T., et al., 2016. Monitoring of antibiotic residues in aquatic products in urban and rural areas of Vietnam. J. Agric. Food Chem. 64 (31), 6133–6138.

Van Damme, E.J.M., Peumans, W.J., Pusztai, A., Bardocz, S., 1997. Handbook of Plant Lectins: Properties and Biomedical Applications. John Wiley and Sons.

van den Ingh, T.S.G.A., Krogdahl, Å., Olli, J.J., Hendriks, H.G.C.J., Koninkx, J.G.J.F., 1991. Effects of soybean-containing diets on the proximal and distal intestine in Atlantic salmon (*Salmo salar*) - a morphological study. Aquaculture 94 (4), 297–305.

Verlhac-Trichet, V., Vielma, J., Dias, J., Rema, P., Santigosa, E., Wahli, T., Vogel, K., 2014. The efficacy of a novel microbial 6-phytase expressed in Aspergillus oryzae on the performance and phosphorus utilization of cold- and warm-water fish: rainbow trout, *Oncorhynchus mykiss*, and Nile Tilapia, *Oreochromis niloticus*. J. World Aquacult. Soc. 45 (4), 367–379.

Vetter, J., 2017. Cyanogenic glycosides in cassava and bamboo shots. A human health risk assessment. In: Gopalakrishnakone, P., Carlini, C., Ligabue-Braun, R. (Eds.), Plant Toxins. Toxinology. Springer, Dordrecht.

Vielma, J., Lall, S.P., Koskela, J., Schoner, F.J., Mattila, P., 1998. Effects of dietary phytase and cholecalciferol on phosphorus bioavailability in rainbow trout (*Oncorhynchus mykiss*). Aquaculture 163 (3–4), 309–323.

Vielma, J., Lall, S.P., Koskela, J., Mattila, P., 1999. Influence of low dietary cholecalciferol intake on phosphorus and trace element metabolism by rainbow trout (*Oncorhynchus mykiss*, Walbaum). Comp. Biochem. Physiol. Mol. Integr. Physiol. 122 (1), 117–125.

Vielma, J., Makinen, T., Ekholm, P., Koskela, J., 2000. Influence of dietary soy and phytase levels on performance and body composition of large rainbow trout (*Oncorhynchus mykiss*) and algal availability of phosphorus load. Aquaculture 183 (3–4), 349–362.

VKM, 2014. Benefit-risk Assessment of Fish and Fish Products in the Norwegian Diet - an Update. Norwegian Scientific Committee for Food Safety (VKM), Oslo, ISBN 978-82-8259-159-1, 293 pp.

Voelter, W., Wacker, R., Tsitsilonis, R., 2005. Mistletoe lectins, structure and function. Front. Nat. Prod. Chem. 1, 149–162.

von Danwitz, A., Schulz, C., 2020. Effects of dietary rapeseed glucosinolates, sinapic acid and phytic acid on feed intake, growth performance and fish health in turbot (*Psetta maxima* L.). Aquaculture 516, 7.

Wan-Loy, C., Siew-Moi, P., 2016. Marine algae as a potential source for anti-obesity agents. Mar. Drugs 14 (12).

Wang, D.-Z., 2008. Neurotoxins from marine dinoflagellates: a brief review. Mar. Drug. 6, 349–371.

Wang, J., Ai, Q., Mai, K., Xu, H., Zuo, R., Xu, W., Zhang, W., Zhang, C., 2014. Effects of dietary ethoxyquin on growth, feed utilization and residue in the muscle of juvenile Japanese seabass, Lateolabrax japonicus. Aquacult. Res. 45 (11), 2656–2664.

Wink, M., Meissner, C., Witte, L., 1995. Patterns of quinolizidine alkaloids in 56 species of the genus *Lupinus*. Phytochemistry 38 (1), 139–153.

Wink, M., Schmeller, T., Latz-Bruning, B., 1998. Modes of action of allelochemical alkaloids: interaction with neuroreceptors, DNA, and other molecular targets. J. Chem. Ecol. 24 (11), 1881–1937.

Yamaya, A., Endo, Y., Fujimoto, K., Kitamura, K., 2007. Effects of genetic variability and planting location on the phytosterol content and composition in soybean seeds. Food Chem. 102, 1071–1075.

Yan, W., Reigh, R.C., Xu, Z., 2002. Effects of fungal phytase on utilization of dietary protein and minerals, and dephosphorylation of phytic acid in the alimentary tract of channel catfish *Ictalurus punctarus* fed an all-plant-protein diet. J. World Aquacult. Soc. 33, 10–22.

Yao, J.T., Chen, P., Apraku, A., Zhang, G.G., Huang, Z.Y., Hua, X.M., 2019. Hydrolysable tannin supplementation alters digestibility and utilization of dietary protein, lipid, and carbohydrate in grass carp (*Ctenopharyngodon idellus*). Front. Nutr. 6, 12.

Yildirim-Aksoy, M., Lim, C., Wan, P., Klesius, P.H., 2004. Effect of natural free gossypol and gossypol-acetic acid on growth performance and resistance of channel catfish (*Ictalurus punctatus*) to *Edwardsiella ictaluri* challenge. Aquacult. Nutr. 10 (3), 153–165.

Zhang, Y., Chen, W., Xu, W., Ai, Q., Zhang, W., 2012. Effects of dietary soy isoflavones on feed intake, growth performance and digestibility in Juvenile Japanese Flounder (*Paralichthys olivaceus*). J. Ocean Univ. China 11.

Zhong, J.R., Feng, L., Jiang, W.D., Wu, P., Liu, Y., Jiang, J., Kuang, S.Y., Tang, L., Zhou, X.Q., 2019. Phytic acid disrupted intestinal immune status and suppressed growth performance in on-growing grass carp (*Ctenopharyngodon idella*). Fish Shellfish Immunol. 92, 536–551.

Zhong, J.R., Wu, P., Feng, L., Jiang, W.D., Liu, Y., Kuang, S.Y., Tang, L., Zhou, X.Q., 2020. Dietary phytic acid weakened the antimicrobial activity and aggravated the inflammatory status of head kidney, spleen and skin in on-growing grass carp (*Ctenopharyngodon idella*). Fish Shellfish Immunol. 103, 256–265.

Zhou, C., Lin, H., Ge, X., Niu, J., Wang, J., Wang, Y., Chen, L., Huang, Z., Yu, W., Tan, X., 2015. The effects of dietary soybean isoflavones on growth, innate immune responses, hepatic antioxidant abilities and disease resistance of juvenile golden pompano *Trachinotus ovatus*. Fish Shellfish Immunol. 43 (1), 158–166.

CHAPTER

13

Nutritional pathology

Ronald J. Roberts[1,2]

[1]Center for Sustainable Aquaculture, Hagerman Fish Culture Experiment Station, University of Idaho, Hagerman, Idaho, United States; [2]Institute of Aquaculture, University of Stirling, Scotland

13.1 Introduction

The tissues of the vertebrate body change significantly between early growth stages and sexual maturity changes, which occur independently of nutritional status. In poikilotherms (ectotherms), they are further modified as the animal progresses through the annual cycle of temperature change and feed availability. Drought or pollution may limit food availability and particular phenomena, such as the failure of oceanic currents may, on occasion, result in a collapse of planktonic growth and consequent starvation of their predators. Generally, however, in their natural environment, wild fish can acquire adequate levels of their nutritional requirements for survival. Absolute starvation per se does present distinctive clinical and pathological effects, but this is rare in wild fish.

When wild fish are deprived of food due to seasonal low availability or to hormonal effects, it is important to accept that the histopathological signs of inanition or starvation, while they resemble those found in unfed farmed fish are not pathological per se but rather completely natural changes. The inanition of the mature Atlantic salmon kelt in winter, after spawning—a fish bearing little resemblance to the magnificent fresh-run salmon—is merely an extreme example of the effects of hormonally induced starvation.

Dietary provision for farmed fish may take the form of a complete diet, incorporating the entire nutrient requirement, at appropriate levels for each component. Alternatively, in the case of pond fish, particularly in developing countries in the tropics, provision of a supplementary diet serves mainly to enhance the growth rate of fish, which are expected to obtain at least a proportion of their nutritional complement from the natural productivity of their ponds. These ponds will normally have been fertilized to increase their primary production (Hepher and Herman, 1988).

Although most deficiency diseases are associated with a complexity of marginal or absolute deficiencies, it is only by detailed experimental studies of single-deficiency conditions for

Fish Nutrition, Fourth Edition
https://doi.org/10.1016/B978-0-12-819587-1.00009-4

each species that a true understanding of the detailed biochemistry underlying nutritional pathologies of fish can be obtained. Such data are rarely available, apart from the pioneering experimental studies on single vitamin or mineral deficiency conditions, in salmonid species, described by Halver et al. (1957 *et seq.*) in which a wide range of specific, and often pathognomonic, signs were demonstrated.

In the clinical situation, nutritional deficiencies are rarely as specific as in these experimentally defined examples. Whether in cultured or wild fish, a more general syndrome is more common, often with inappetence, darkening skin color, and poor or deformed growth as the only obvious clinical features (Hardy, 2012).

Artificial diets intended as complete sources of nutrients, when produced by a major manufacturer, are usually of high quality. It is only rarely that a particular batch will be responsible for a problem. However, when home mixing of wet diets or trash fish feeding is practiced, the possibility of a deficiency or imbalance syndrome is much higher. Even the high-quality commercially compounded diet is susceptible to degeneration during storage. This is particularly likely under conditions of high temperature or humidity. Also, in the continuing search for cheaper formulations and alternative sources of the more expensive elements of the diet, unintentional pathological effects may be induced, often only affecting particular stages in the production cycle (Sullivan et al., 2007a, b).

Although adequate nutrition is essential for health and growth of fish, because they are poikilothermic (endothermic) and therefore their metabolic rate is closely related to water temperature, basal feeding requirements, as well as growth and reproduction needs, vary seasonally. This is particularly a feature with wild species, in temperate areas. This means that the observed pathological picture and the rate at which it develops will often vary depending on feed levels and temperature.

Occasionally, major environmental changes, such as the failure of *El Nino*, the Pacific Ocean current that brings nutrients to western American coastal waters, can lead to heavy mortalities of wild fish, or even failure of entire year classes from starvation. More commonly, there is merely a reduction in year-class numbers or average fish size, associated with changes in oceanographic conditions that lower food abundance. In some cases, uncontrolled and excessive harvesting of prey species such as sand eels can induce starvation pathologies in their predators. Anthropogenic effects, such as spraying of waterways or rice fields with insecticides, can also affect sensitive species by removing or greatly reducing their food supply, even if the agents are not themselves toxic to the fish.

Since many marine and freshwater species lose weight during a prolonged period of winter inanition, and in addition, fish often spawn in spring, tissue fat stores, and particularly the liver, which in many species will be extremely (normally) fatty in summer, may be greatly reduced. Fatty change in the liver is also an indicator for several dietary faults so great care needs to be taken in distinguishing between histopathological evidence of fatty change due to nutritional disease and manifestations of normal seasonal changes in the nutritional status. Basal metabolic rate and therefore the rate of catabolic tissue metabolism for energy in temperate fish is six times lower at 5°C than at 24°C. In any situation, therefore, where food availability is absolutely nil, starvation pathologies will manifest themselves as catabolism takes place.

Snieszko (1972) defined nutritional diseases as "the deficiency, excess, or improper balance of the components present in a fish's diet." Deficiency of necessary components is, however, only one possible cause of nutritional disease. Properly the definition should also include the presence of noxious or toxic components within the food, or endogenous antinutrient factors (DeSilva and Anderson 1995; Jauncey 1998).

The shift in feed formulation from overdependence on fishmeal to increasing use of plant proteins removes some of the risks from persistent marine toxins within the fish meals but brings with it the increasing possibility of inadvertent addition of adventitious terrestrial-derived toxins to feeds. These latter have become increasingly significant, as intensive aquaculture has developed. Thus, the importance of adventitious *Aspergillus* contamination, associated with vegetable protein sources has come to be recognized as a serious problem requiring constant monitoring (Halver, 1965; Olufemi and Roberts 1986).

Even diets that are only marginally compromised can increase the susceptibility of a fish population to infectious diseases. Since the infection signs are usually more obvious, they usually mask the underlying predisposing cause.

As knowledge of the dietary requirements of each species has grown, commercial feeds for farmed fish have improved. However, although deficiency conditions and toxic contamination are much less frequent than in the early days of fish feed production, they still occur. The type of nutritional deficiency or feed-related toxic condition has also changed. Before the nutritional requirements of most farmed fish were quantified and before stable forms of vitamins were developed, nutritional deficiencies often resulted from low levels of vitamins present in feeds at the time of feeding. Today, deficiencies and toxic conditions are more likely to be associated with oxidation of dietary lipids, mold growth associated with storage of feed exposed to moisture or high humidity (>12% moisture in feed), and deficiencies resulting from antagonistic interactions among feed ingredients in formulated feeds.

The fish species farmed are now also very different genetically from their original parental stcks, being highly selected for growth or modified by sex reversal or triploidy. They often grow very rapidly at certain stages in their life cycle, and a series of small ingredient changes, or failure to adjust the diet for the demands of such stages, of little obvious immediate significance for most stages, may induce minor pathologies that resonate seriously later in life (e.g., Roberts et al., 2001).

13.2 Principles of nutritional pathology

Starvation may be due to complete deprivation of food, to inadequate feeding levels, or to behavioral, physiological, or mechanical prevention of food intake. Complete deprivation may occur if fish are accidentally left in a facility, but it may also occur in blind fish with, for example, parasitic cataracts, which cannot feed (Karvonen et al., 2004). Inadequate feeding levels may be associated with poor husbandry, or, in the case of stocked sport fisheries, overstocking. Behavioral starvation is the term used to describe the condition often found when attempts are made to rear larval fish of species new to culture on artificial diets. Frequently the larvae do not recognize or accept the diet because of inadequate presentation, texture, or taste. This is a particular problem with attempts to wean certain larval fish such as Atlantic halibut, walleye, sea bass, sea bream, and turbot (Hardy, 2012). Nonfeeding larvae or fry may, by catabolizing structural nutrients, survive for a considerable time, but eventually they will appear dark and the head will usually be disproportionately large.

When larvae are starved, they eventually reach a stage at which the degree of body damage induced by the catabolic breakdown is such that, even when normal nutrition is resumed, the damaged tissues cannot recover. This stage, beyond which they cannot be restored, is known as the point of no return (PNR). Larval fish can, however, survive beyond this stage for a considerable time before they inevitably succumb (Ehrlich, 1974).

The process of physiological starvation is characterized, not just by loss of weight but also by change in gross morphology. Affected fish become thinner, and the head becomes relatively larger in proportion. Due to loss of control of dermal chromatophores, skin color becomes very dark, and eventually the fish loses appetite. At necropsy, there is usually pallor, associated with anemia, and no abdominal lipid is present. The liver and spleen are normally shrunken with sharp edges, and pyloric or intestinal caeca, when present, are pronounced.

Histologically the principal feature of starved fish, apart from lack of lipid tissue, is the distinctive granular degeneration of the skeletal muscle fibers, with central migration of sarcolemmal nuclei. They are predominantly pale staining and have their sarcoplasm condensed into strands and granular deposits in a clear stroma (Roberts, 2012). The digestive tract is condensed, with an apparent increased proportion of collagenous, fibroblastic, and nuclear elements. The hemopoietic tissue is dark and condensed, but a distinctive feature is the increased size of the melanomacrophage centers of spleen and kidney, with large deposits of lipofuscin within them, as well as an increase in the melanin content (Agius and Roberts, 1981). A particularly significant feature of the melanomacrophage centers of starved fish is the high level of hemosiderin sequestered there, indicating the premium placed on retaining iron complexes within the body when catabolizing structural elements, even in extremis (Agius and Roberts, 2003) (Fig. 13.1).

Stress is a major factor in fish diseases, and for nutritional diseases, its effects can be profound. The best definition of it for aquatic species is that of Brett (1958) who described it as "a stage produced by an environmental or other factor which extends the adaptive responses of an animal beyond the normal range or which disturbs the normal functioning to such an

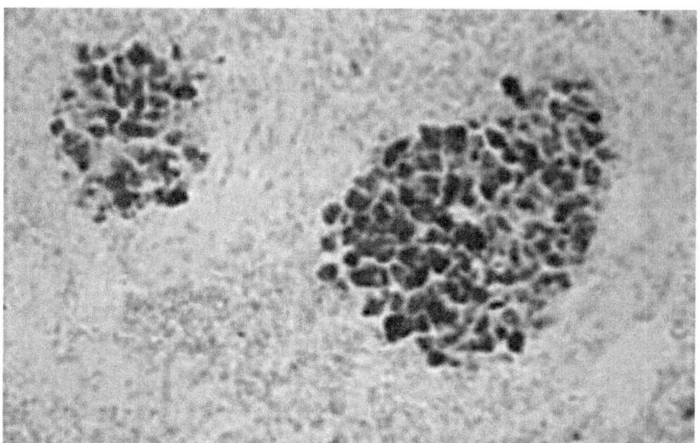

FIGURE 13.1 Melanomacrophage centers containing heavy deposits of hemosiderin in spleen of starved plaice. Stained by Perl's Prussian blue ×75.

extent that the chances of survival are significantly reduced." In farmed fish husbandry, such stressor factors, associated with density, environmental provision, or other husbandry features, are always present, and indeed, feed factors may be stressors per se (Pickering, 1978, 1998).

The primary function of the farmer is to attempt to maintain a system whereby the various components capable of acting as stressors are at the lowest possible level compatible with economic success. This has significant implications for growth and feed conversion as well as vulnerability to infection. Stress induces tissue catabolism and energy required for such degradation, and resynthesis is not available for growth (DeSilva and Anderson, 1995). A number of the stressors that can play a role in determining the levels of nonspecific stress in fish are related to food and feeding. Frankly inadequate deficient diets, as well as other factors such as irregular feeding, presentation of feed in an unsuitable form, and buildup of waste food detritus, can contribute to the sum total of pressures on the adaptive capacity of the animals.

The physiological and biochemical changes, which occur in the animal under the influence of environmental stressors and which serve to moderate the negative effects of the stresses, are termed the general adaptive syndrome (GAS). They are neither species nor stressor specific and are mediated by nervous and hormonal action. Stress is the end result, if the adaptive response is stimulated beyond physiological levels and adaptive exhaustion occurs. This is irrespective of whether the stimulus has an environmental or husbandry-related origin; the end result is that the fish are made increasingly vulnerable to infection. The infection may be by opportunist, obligate, or facultative pathogens in the fish's environment. Thus, many nutritional diseases in fish are compounded by bacterial or fungal infection. Indeed, it is often the assumption that the microbial agent is the primary pathogen. This will lead to failure to recognize and correct the underlying nutritional stressor (Moller, 1984).

An important aspect of nutritional pathology is the role of the metabolic cost referred to as the cost of specific dynamic action (SDA). This is the cost in energy terms of the metabolism associated with digestion, increased circulatory vascular activity in the gut and related organs, and the excretion of nitrogenous metabolites. Which processes contribute and their exact role is uncertain. However, it is well recognized that at high temperatures, when dissolved oxygen levels in water are reduced and even the general resting metabolic oxygen demand by a fish population is high, feeding to satiation can result in such a high oxygen demand to satisfy SDA that heavy and very rapid mortality can result.

The clinical features of such losses are typically those of oxygen starvation. Fish die about 2 h after feeding. They have open mouths, and considerable amounts of undigested food in the gut. The problems of feeding at high temperatures are greatly exacerbated if there is an underlying gill or blood pathology. Often, when fish have hyperplastic or telangiectatic gill damage, associated with pollution, parasitism, transport, or handling, they can survive until the dissolved oxygen levels fall below a threshold, but it is frequently a feeding episode, which provides the final metabolic oxygen demand, which cannot be met by the compromised respiratory system.

Other conditions that can increase the susceptibility to SDA-mediated mortalities include fatty liver syndrome, which inhibits hemopoiesis, and a number of chronic hemorrhagic conditions such as warfarin poisoning, vibriosis, chronic viral hemorrhagic septicemia (VHS), and infectious salmon anemia.

13.3 The deficiency and imbalance diseases

Deficiency diseases are of two types, deficiency or imbalance of the macronutrients in the diet—the protein, carbohydrate, lipid, and fiber—and deficiency of the micronutrients—the vitamins and minerals. In terms of macronutrients, it is usually in the lipid component that the most serious problems arise, whether in relation to deficiency, oxidation, or imbalance. Among the micronutrients, any of a wide range of components can exert an effect, especially in fast-growing, younger fish.

13.3.1 Major dietary components

13.3.1.1 Protein

A high level of protein is required by all fish species not only for provision of structural amino acids but also for gluconeogenesis. Feed manufacturers will generally economize as much as possible on protein level while still attempting to provide an adequate diet. The biological value of a protein is a function of its essential amino acid (EEA) content and EAA digestibility.

Protein is essential for maintenance, growth, reproduction and repletion of depleted tissues following migration and spawning. Normally, EAAs will be conserved, but if amino acid intake is restricted, then they may be metabolized to form nonessentials.

Deficiency diseases can still arise even in the presence of apparent *luxus*. This is because certain amino acids may be rendered biologically unavailable or inactive, even when still chemically measurable, during the course of processing of the diet. Lysine, for example, may form an addition compound with carbohydrate in the feed, rendering it unavailable to the fish (Hardy, 2012).

The EAAs, i.e., those that cannot be synthesized in cells from metabolic precursors by transamination, are the most critical factors in the nutritional value of a protein. It is phylogenetically significant that the same 10 amino acids that are essential for all fish studied are also essential for growing mammals and birds. These EAAs are arginine, histidine, isoleucine, leucine, lysine, methionine, phenylalanine, threonine, tryptophan, and valine.

Amino acid deficiencies: The most frequent indicator of an amino acid deficiency problem is impairment of growth, but several specific deficiency syndromes have been recorded under experimental conditions, which may well have counterparts in commercial farming. These include dorsal fin erosion, which has been associated with lysine deficiency; spinal deformities associated with tryptophan, leucine, lysine, arginine, or histidine deficiency; and a lenticular cataract associated with methionine and tryptophan deficiency (Walton et al., 1984; Ketola, 1983.). A very specific central lenticular cataract, due to histidine requirement above normal tissue metabolic requirements, occurs in certain strains of fast-growing Atlantic salmon (Sambraus et al., 2017).

It is the case, however, that most amino acid deficiencies are unlikely to occur under normal conditions, and if they do, the clinical signs are poor growth and darkening of the skin.

13.3.1.2 *Carbohydrate*

Salmonids have a much more limited capacity for metabolism of dietary carbohydrate than omnivorous birds and mammals. Most farmed fish species can efficiently utilize dietary carbohydrates for energy (Lovell, 1999); however, salmonids, tunas, and other top predators have a much more limited capacity for metabolism of dietary carbohydrate. Excessive dietary carbohydrate levels can result in hepatocyte degeneration and excessive glycogen deposition, particularly in ornamental cyprinids (Jauncey, 1982; Hess, 1935).

13.3.1.3 *Fats*

Problems associated with the lipid component of the diet appear to be among the most serious and prevalent of all nutritional problems. Fish tissues contain predominantly fatty acids of the ω-3 series and so fish diets must provide sufficient essential fatty acids of the ω-3 (linolenic) and ω-6 (linoleic) series as well as a general lipid contribution to calorie requirements. Deficiency syndromes result when sufficient levels of ω-3 or ω-6 or longer-chain members of the series are unavailable.

Fish appear capable of synthesizing fatty acids of the ω-7 and ω-9 series but, unfortunately, not the ω-6 (linoleic) or ω-3 (linolenic) series. Linolenic, or longer-chain, members of the ω-3 series appear to be particularly vital for normal growth and avoidance of pathology, the requirement being up to 1% of the diet.

The ability to elongate linolenic acid (18: ω-3) seems to vary among species, and this in turn affects the occurrence of lesions. Some can, at least to some degree, convert dietary oleate (18:1) or linoleate (18:2) to arachidonate (20:4). Similarly, they can convert linolinate (18:3) to eicosapentanoic and docosahexanoic acids.

Linolenic acid is particularly important for normal growth. In freshwater species such as rainbow trout, dietary oleate or linoleate can be converted to some extent to arachidonate (Castell et al., 1972). The studies of Cowey and Roberts (Cowey et al., 1976), however, show that, in turbot, a marine species, little conversion of linoleate to arachidonate is possible when they are fed diets high in such lipid, e.g., corn oil. In such circumstances, major pathological changes take place in the structure of the fat storage cells of the animal. The fat cells surrounding the lateral lymphatic sinus are particularly affected, with thickening and deformity of the lipid cell walls, increased vascularity, and deposition of a hyaline material between the cells (Fig. 13.2).

By far, the most significant problem associated with provision of the lipid requirements of farmed fish is the prevention of autooxidation of the high levels of unsaturated fatty acids, including of course the essential ω-3 and ω-6 fatty acids, by atmospheric oxygen. Such autooxidative rancidity reduces the availability of the fatty acids to the host. It also induces high levels of free radicals, peroxide aldehydes, and ketones. These are toxic to fish and also capable of reacting with other dietary components. Required levels of tissue and membrane antioxidants, such as α-tocopherol and ascorbic acid, are reduced when feeds containing oxidizing lipid are fed. This results in alterations to cell membrane permeability and fragility, particularly that of erythrocytes, leading to hemorrhagic anemia, with excessive levels of iron in the melanomacrophage centers.

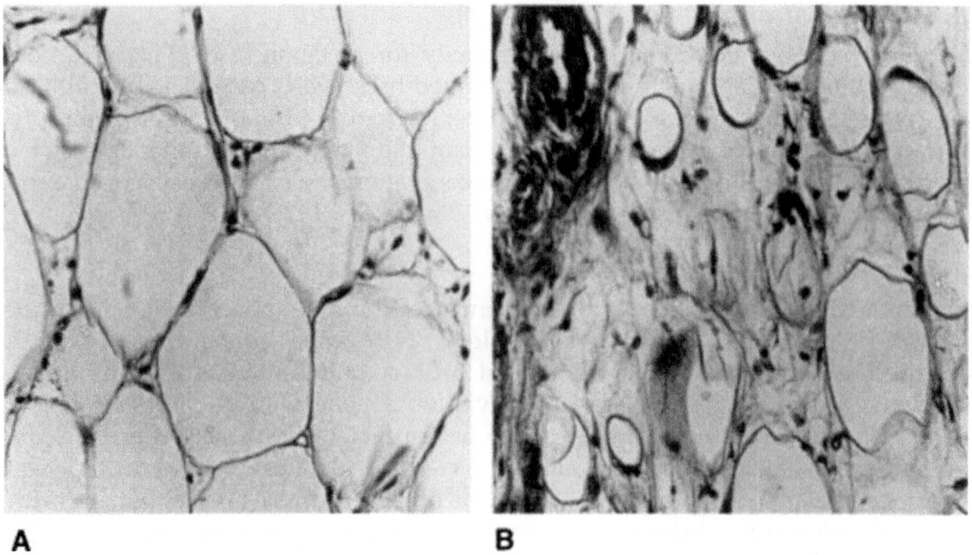

FIGURE 13.2 (A) Normal perilymphatic cells of a turbot. (B) Perilymphatic cells of a turbot on diet containing hydrogenated coconut oil. H & E ×300.

The most serious problem associated with rancidity is lipoid liver disease. This is usually associated with prolonged storage of feeds at high temperature, which accelerates the depletion of antioxidants. For this reason, farmed fish diets often include a significant luxus of α-tocopherol as an antioxidant. The protected form of α-tocopherol normally added as a vitamin E source to fish feeds (α-tocopherol acetate) does not act as an antioxidant within stored feeds, but when fed, it is biologically active and then serves as a tissue antioxidant.

Rancid lipids are toxic per se: They also react with protein to lower its biological value and have a deleterious effect on other vitamins, notably retinol and ascorbic acid.

Fish suffering from lipoid liver disease have extreme anemia characterized clinically by pallor of the gills. At necropsy, there is a bronzed, rounded heart, and a swollen liver with rounded edges (Fig. 13.3). Histologically, the main feature is the extreme infiltration of hepatocytes by lipid, causing loss of cytoplasmic staining and distortion of hepatic muralia (Fig. 13.3).

There is degeneration of splenic and renal hemopoietic tissue with erythrocyte fragility leading to high levels of pale-staining pigment in melanomacrophage centers. There is also often auxiliary hemopoiesis in the subepicardial tissues and the periportal areas (Roberts, 2012).

If the condition has been extant for any length of time, there is also, dependent on the degree of oxidation and type of fat in the diet, a varying degree of infiltration of the liver by macrophages containing ceroid, a pigmented breakdown product of phospholipid catabolism.

Other clinical conditions associated with the feeding of rancid lipids include exophthalmia, steatitis, darkening, splenic hemosiderosis, and skeletal myopathy (Soliman et al., 1983; Murai and Andrews, 1974; Park, 1978; Moccia et al., 1984).

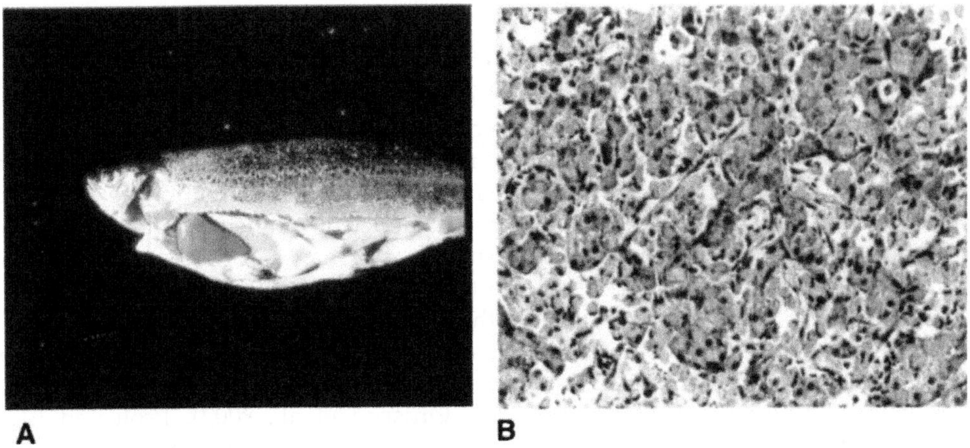

FIGURE 13.3 (A) Rainbow trout with lipoid liver disease. Liver is swollen and bronze and has rounded edges. (B) Liver cells are infiltrated by ceroid distorting liver structure and loaded with yellow pigment.

The condition of Atlantic salmon in Norway, known as 'Hitra disease' (Poppe et al., 1985), where lipidosis and also anemia and hemorrhage featured in what were often major mortalities, was considered to be associated with oxidized lipid or vitamin E deficiency. This is now, however, recognized as a good example of the difficulty of resolving complex pathologies where a microbial infection overlays a nutritional deficiency, and it is now generally accepted that the primary cause of the condition is vibriosis (Hjeltnes and Roberts, 1993), which is now eliminated by use of *Vibrio* vaccines, although feed technologies have also advanced significantly since that time.

13.3.1.4 Fiber

The role of fiber in fish nutrition is considered to be limited, and there are no pathologies specifically linked to fiber excess or deficiency although growth rate may be affected (Davies, 1985).

13.3.1.5 Vitamins

Although vitamins are only required in small amounts in the diet, requirements may increase during rapid growth and spawning, or in high-energy feeds. The vitamins are divided into two classes, fat-soluble and water-soluble, which also modifies the pathologies associated with their deficiency.

13.3.2 Fat-soluble vitamins

Fat-soluble vitamins are stored in the body, often in the liver. They are metabolized only slowly, and in contrast to the more readily metabolized water-soluble vitamins, this can lead to the possibility of cumulative hypervitaminoses if levels fed are excessive.

13.3.3 Retinol (vitamin A)

Retinol occurs in three active forms: retinol, retinal, and retinoic acid. Retinol is the principle form found in marine fish, while both retinol (A_1) and dehydroretinol (A_2) are found in freshwater fish. Plants contain precursors, the carotenoids, which are lipid-soluble pigments converted to retinol in the intestinal mucosa. Vitamin A is essential for normal vision, embryonic development, maintenance of mucous membranes, cellular membrane permeability, bone development, and corticosterone synthesis. Its deficiency in fish results in poor growth, keratomalacia leading to blindness, exophthalmia, hemorrhages at the base of the fins, and foreshortening of the gill arch and operculum (Kitamura et al., 1967; Aoe et al., 1969).

Hypervitaminosis A has been reported both experimentally and in farmed salmon fed a diet rich in whale liver. It results in epithelial squamous metaplasia, hepatomegaly, splenomegaly, osteopathy, and choroidal and corneal inflammation, but unless the condition is extreme, it is readily reversible (Poston, 1971; Ørnsrud et al., 2002). Protected forms of vitamin A, unaffected by oxidizing dietary lipids, are now used to supplement fish feeds, and deficiency conditions are not normally found.

13.3.4 Calciferol (vitamin D)

In mammals, these thermostable steroids are responsible, along with adequate mineral levels, for prevention of rickets. Deficiency has been demonstrated experimentally in salmonids and channel catfish (Leatherland et al., 1980; Lovell and Li, 1978).

Hypervitaminosis occurs experimentally only at very excessive levels, causing hypercalcaemia and increased hemotocrit (Poston, 1969). Hilton and Ferguson fed rainbow trout at 1,000,000 IU/kg without inducing any pathology.

13.3.5 Alpha-tocopherol (vitamin E)

This vitamin is one of the most critical in relation to commercial dietary formulation. Nutritional pathologies are well recognized in cases where it is deficient. The metabolism of the tocopherols is closely related to that of selenium, and generally manufacturers aim for a luxus in diets, because of its antioxidant properties, but this can change when a diet is already partly oxidized before feeding.

Deficiency will occur unless levels are adequate to serve both antioxidant and metabolic functions to the full. To ensure that adequate amounts of tocopherol reach the fish, vitamin premixes contain a protected form (α-tocopheryl acetate) that cannot oxidize during feed storage.

Tocopherol acts primarily as a sacrificial antioxidant by donating protons to free radicals, converting them to stable, innocuous compounds. In living tissues, tocopherol can be regenerated by a mechanism involving selenium. When fish are fed diets containing oxidizing lipid, however, the rate of tocopherol regeneration cannot keep up with tocopherol oxidation. This accelerates depletion of membrane and cellular tocopherol levels and the appearance of tocopherol deficiency signs. This is complicated, however, by the oxidized tocopherol regeneration mechanism catalyzed by selenium. This complex interplay between dietary lipid quality and quantity and the vitamin E and selenium levels in fish diets is also complicated by the moderating effect of environmental temperature on the pathological response, which is much more pronounced at lower temperatures. As in other animals, there is also evidence to suggest that vitamin E deficiency is also exercise related, adding a further variable to a very complex situation.

Clinical conditions associated with vitamin E deficiency have been widely described, and associated with skeletal, digestive and cardiac muscular dystrophy, anemia and ceroidosis (Fig. 10.5) (Pearse et al., 1974; Ferguson et al., 1986a,b), steatitis (Roberts et al., 1979) and with swim-bladder, digestive and cardiac muscle pathology, and anemia (Wang et al., 2016).

A very significant condition historically associated with severe myopathy and circulating vitamin E deficiency is the virus disease salmon alphavirus infection (also known as salmon pancreas disease; PD). Until the pioneering work of McLoughlin and her coworkers (McLoughlin et al., 1996; Jansen et al., 2017), this condition was widely considered to be a primary vitamin E deficiency syndrome, but it is now conclusively proven to be due to salmonid alpha virus X.

Affected fish are dark in color and inappetent and with shrunken abdominal musculature. Clinically, they cannot swallow food pellets due to myopathy of the esophageal musculature. Although an entire population may appear affected, and loss of growth may be very significant in economic terms, mortalities are generally low, and deaths are often associated with intercurrent secondary infections. Although dietary selenium and vitamin E levels appear to be normal, affected fish have low blood and liver tocopherol levels and high levels of creatinine kinase. The cause of the low tissue vitamin E levels in the presence of adequate levels in feed is difficult to explain, but a possible viral effect at the intestinal level preventing absorption is a possibility (Rodger et al., 1994).

Histologically, the condition is characterized by distinctive but variable polymyopathy. It resembles in many respects a similar virus disease, cardiomyopathy syndrome (CMS), caused by a totivirus (Rodger and Turnbull, 2000), but in PD, but not in CMS, there is also a degenerative pancreatic acinar lesion characterized by fibrosis of periacinar fat tissue and round cell infiltration (Ferguson et al. 1986a) associated with the virus activity.

13.3.6 Vitamin K

There are two naturally occurring forms of this fat-soluble vitamin. Both are methylnaphthoquinones. Phylloquinone is synthesized by green plants, and farnoquinone is synthesized by certain microorganisms. The primary function of vitamin K, as in all vertebrates, is as a component of the blood-clotting mechanism, but it is also bacteriostatic and has coenzyme properties. In fish, its deficiency results in prolonged clotting time and hemorrhages into muscles and viscera, coupled with anemia in chronic cases. The signs of chronic vitamin K deficiency are similar to those of infection with the rhabdovirus of VHS infection, from which it must be carefully differentiated.

Vitamin K has a well-defined set of competitive antagonists, the dicoumarols (warfarins). These are frequently used as rodenticides and can contaminate fish feed in store. The signs of food contamination with warfarin are similar to those of vitamin K deficiency, with particularly significant intraperitoneal hemorrhage in serious cases.

13.3.7 Water-soluble vitamins

Water-soluble vitamins (B-group and C-group vitamins) are usually readily available in fish diets, so that the clinical frequency of deficiencies is very low. Thus, although experimental investigations have yielded extensive information on specific clinical features of a wide range of water-soluble vitamin deficiencies; in practice, the general features of water-soluble vitamin deficiencies are poor growth and darkening of color. For some water-soluble vitamins, such as vitamin C, specific deficiency signs have been identified.

Since there are great differences in the rate of tissue turnover of the different water-soluble vitamins, the time required to induce a specific deficiency is also variable. They may develop rapidly, whenever a deficient diet is fed, or take months to develop. They also take two or three times longer to develop in older fish compared with fry or fingerlings. This is because younger stages have much lower tissue reserves and higher growth rates, especially at high water temperatures on high-energy diets (Hardy, 2012).

13.3.8 B-group vitamins

13.3.8.1 Thiamin (vitamin B1)

Thiamin is a coenzyme of many of the essential enzymes catalyzing carbohydrate metabolism. The coenzyme form, thiamin pyrophosphate, is involved in digestion, reproduction, and central and peripheral nerve function. Thiamin deficiency may be actual or induced by the presence within the diet of thiaminase, an enzyme capable of destroying thiamin (Seacock and Goodland, 1944). Thiaminases are common in the tissues of certain fish, notably the cyprinids and clupeids, and in certain legumes. These should not, therefore, be included as protein elements in fish diets unless temperature treated. Thiamin deficiency conditions have been reported more frequently in salmonids and eels than in species such as cyprinids, which have a higher proportion of carbohydrate in their diet, despite the importance of thiamin in carbohydrate metabolism (Murai and Andrews, 1978a).

Clinical signs include change of color and hemorrhage at the base of fins, but the most frequent aberration is hyperexcitability with paralysis or aberrant swimming. This is due to more rapid depletion of thiamin pyrophosphate from brain tissue than from muscle. Histologically, lesions are mainly found in the brain where hemorrhages and degeneration of specific nuclei of the periventricular areas are found. Histological correlation of such lesions was described by Blaxter et al., 1974, who reported hemorrhage and degeneration of specific nuclei of the periventricular areas of the brain of Atlantic herring experimentally fed a diet deficient in thiamin, a condition readily prevented by enhancement of dietary thiamin levels (Fig. 13.4).

Early mortality syndrome or EMS is a condition that affects salmonids in the Great Lakes of the United States and Canada. It resembles Cayuga syndrome of Atlantic salmon in the New York Finger Lakes and M74 syndrome that affects Atlantic salmon from the Baltic Sea. The three conditions share associated low thiamine concentrations in eggs, and they respond to therapeutic thiamine treatment. Clinical signs include loss of equilibrium, depression of respiration, and reduced heart rate. Similar clinical features can be induced with thiamine antagonists. Patterns of temporal variability of EMS among Great Lakes' salmonids over the past three decades have been similar. It is suggested that the deficiency coincides

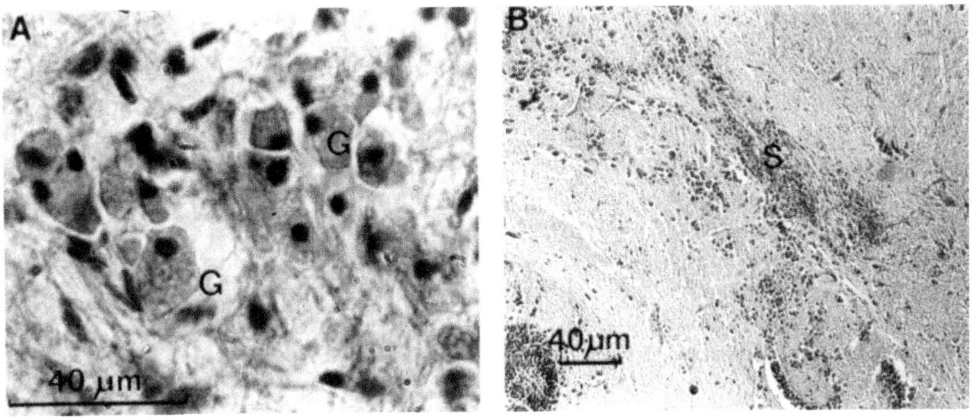

FIGURE 13.4 Brain lesions in thiamin deficient Atlantic herring. (A) Gemastocytes around area of neuronal necrosis. (B) Glial scar (S) in thalamic area in recovered herring. H&E.

with dietary availability variations, but histological confirmation of neuropathy and definitive correlation is still required (Honeyfield et al., 2005).

13.3.8.2 Riboflavin (vitamin B2)

Riboflavin is particularly active as a coenzyme for oxidase systems and thus is particularly important in the respiration of poorly vascularized tissues such as the cornea and lens of the eye. It is widely distributed in animal and plant tissue but is readily destroyed by ultraviolet light. In fish, it has also been associated with petechial hemorrhages in cyprinids (Ogino, 1967) and with congenital dwarfism in channel catfish (Murai and Andrews, 1978b). In the eye, the corneal lesion is similar to that of higher animals, namely, cloudiness, with growth into the substance of the cornea by limbal capillaries and hemorrhage. The cataract is usually bilateral and results in blindness.

13.3.8.3 Pyridoxine (vitamin B6)

Pyridoxine is a particularly important vitamin for fish, with their high protein requirement, since it is a major coenzyme in the deamination of amino acids by a wide range of deamination systems. Pyridoxine is widely distributed in yeasts, grains, and animal tissues, so it is unlikely to be deficient in most circumstances, but it is vulnerable to denaturation by ultraviolet light. The active form of pyridoxine in fish tissues, pyridoxal phosphate, is rapidly depleted from tissues during deficiency, resulting in deficiency signs in fry fed a completely deficient diet for as little as 3–4 weeks.

Renal, ovarian, and hepatic degenerative changes and hemopoietic hyperplasia, associated with clinical features of hyperexcitability and poor growth, have been described in farmed fish, and these conform with descriptions for the experimentally induced condition in salmonids (Halver, 1957), turbot (Adron et al., 1978), and gilthead bream (Kissil et al., 1981).

13.3.9 Pantothenic acid

As the name implies, this vitamin has a very wide distribution in foodstuffs but nevertheless deficiency disease does occur in fish. Its role is as a coenzyme in both lipid and carbohydrate metabolism. Apart from general inappetence, deficiency leads to "nutritional gill disease" a very characteristic lesion of the distal third of gill lamellae with development of primary lamellar hyperplasia and clubbing of secondary lamellae. The lamellar hyperplasia progresses from the periphery and coincides with anorexia, which may be a function of loss of respiratory capacity. Respiratory embarrassment, however, normally only supervenes when the oxygen demand increases. Thus, nutritional gill disease may only be apparent in cases of acute mortality in fish, which have been overfed at high temperatures.

The lamellar hyperplasia, accompanied by excessive mucus secretion, often results in secondary myxobacterial infection but should not be confused with primary flavobacterial gill disease. The latter is a proliferative response to flavobacterial infection, and it occurs in the absence of any pantothenic acid deficiency (Fig. 13.5).

Other clinical features associated with pantothenic acid deficiency include erosion of fins and barbels and superficial skin ulceration. Anemia has also been reported, but only toward the end of experimental studies, where it is probably a result of the concomitant anorexia (Karges and Woodward, 1984).

Replacement therapy generally allows recovery from pantothenic acid deficiency, with complete restoration of gill architecture, provided that the lesion is not too severe.

13.3.10 Inositol

Absence of inositol from the diet has been shown to reduce growth rate (Aoe and Masuda, 1967; Coates and Halver, 1958; Arai et al., 1972), but the only specific lesion reported was by

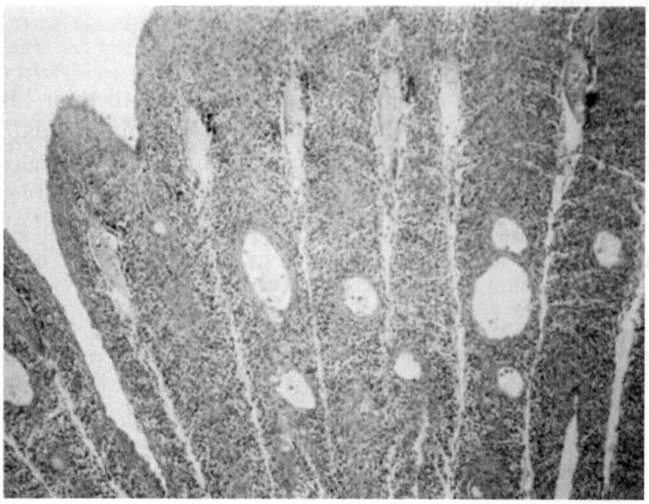

FIGURE 13.5 Gill from pantothenic acid deficient rainbow trout. H&E ×15.

McLaren et al., 1947, who showed that mild integumentary ulceration occurring in its absence was readily healed by supplementation.

13.3.11 Niacin

Widespread hemorrhage has been associated with niacin deficiency in cyprinids, catfish, and eels, and in salmonids, muscle spasm and edema of the gastric mucosa are reported experimentally (Halver, 1957; Poston and Di Lorenzo, 1973). The most common lesions associated clinically, however, are various dermatoses, in particular a predisposition to erosion of the dorsal epidermis following excessive solar ultraviolet irradiation, colloquially known as "backpeel" (De Long et al., 1958). Poston and Wolfe (1985) have shown that for the rainbow trout, a level of at least 18 ppm niacinamide is required to prevent this condition.

13.3.12 Biotin

Biotin is essential as a coenzyme for many aspects of lipid and carbohydrate metabolism, but it is rarely deficient. It is widely distributed in food ingredients and is also a product of bacterial metabolism in the gut. It is virtually impossible to induce a biotin deficiency in fish fed practical diets, but in experimental salmonid diets, its absence has been shown to be responsible for the typical B-group deficiency features of reduction in growth and food conversion, darkening, and anorexia (Poston and Page, 1982). Another more specific feature, recorded by Ogino et al. (1970) in common carp and by Castledine et al., 1978 in the rainbow trout, is increase of mucus cells of the epidermis, which secrete a much more tenacious cuticular glycoprotein.

13.3.13 Choline

The choline requirement of fish is high, and its deficiency results in the standard features of poor growth and feed conversion (Halver, 1957; Ketola, 1976). There is, however, some interspecific variation. Renal hemorrhage is consistently reported in deficient Chinook salmon but not in others, while sturgeon exhibited a thinning of the intestinal muscle wall and focal degeneration of pancreatic tissue (Hung, 1989).

13.3.14 Cyanocobalamin (vitamin B12)

As in higher animals, deficiency of cyanocobalamin is responsible for blood dyscrasia in fish, but very little is known about the mechanism. Halver (1953, 1957 demonstrated that for coho salmon the response to deficiency was anemia and poor growth accompanied by darkening in coloration and variable reduction in erythropoiesis, while John and Mahajan (1979) showed, in the *Labeo rohita*, that cyanocobalamin deficiency led to a significant leukocytopoiesis. This was related mainly to thrombocyte and, to a lesser degree, eosinophil numbers, with neutrophil and large lymphocyte numbers significantly depressed.

13.3.15 Folic acid

Folic acid deficiency in fish is related principally to an abnormal blood picture. Smith (1968) and Smith and Halver (1969) reported a megaloblastic, normochromic, macrocytic anemia in coho salmon, a finding confirmed by John and Mahajan (1979) in *Labeo rohita*. This differed from the latter authors' findings in the air-breathing snakehead *Channa punctata*, which had a hypochromic anemia under similar conditions (Mahajan and John, 1977). Those authors also showed that a combined deficiency of folic acid and cyanocobalamin produced a more severe anemia than each did separately.

13.3.16 Ascorbic acid (vitamin C)

The antiscorbutic vitamin plays an important role in fish diets. Two clinical disease syndromes, skeletal deformity and failure of wound healing, are associated with its deficiency. These both result from its fulcrum role as a cofactor for hydroxylation of proline to hydroxyproline, an essential component of collagen. This means it is involved in all aspects of connective tissue, bone matrix, and scar tissue elaboration.

Clinical conditions regularly associated with vitamin C deficiency include spinal lordosis and scoliosis, stress fracture, mandibular deformity, and opercular and lamellar deformation. Studies carried out by Kitamura (1965), Poston (1967), Halver (1969), and Soliman et al. (1985), working on rainbow trout, brook trout, coho salmon, and tilapias, respectively, all showed that after a period of exposure to a vitamin C-deficient diet, a deforming diathesis of cartilage, accompanied by osteoid replacement of bony tissue, took place, which rendered the tissue extremely vulnerable to fracture, typically at the proximal end of the spine. Generally the osteoid changes in the vertebrae were associated with a distinctive decrease in radioopacity, although occasional areas of considerable density were demonstrable.

The use of wound healing histology as a model for the collagen moderation effect of vitamin C deficiency was pioneered in fish by Halver et al., 1969, who created sutured surgical lesions in the dorsum or flank of salmonids and showed that healing did not occur in the absence of the vitamin. Jauncey et al., 1985, similarly, studied the effect at tropical temperatures, in small lesions in the Nile tilapia. Again the deficient lesion was characterized by loss of the capacity to initiate fibroplasia to elaborate cicatrix tissue. However, they showed that the immediate epidermal migration, characteristic of the teleost response to integumentary ulceration, and independent of fibroplasia, occurred without regard to ascorbate level (Fig. 13.6).

Although there is now an extensive literature on experimental vitamin C deficiency, it is only in culture of catfish, either channel catfish or the Asian catfish *Clarias lazera*, that clinical disease is reported to be of significance. Broken back syndrome in the channel catfish is, as its name suggests, a condition associated with spinal fracture as well as depigmentation and reduction in growth (Meyer, 1975). Crackhead, a condition characterized by bacterial infection of fractured or deformed cranial symphyses, often accompanied by spinal fracture or deformity, is not uncommonly seen in high-intensity pond culture of the Asian catfish. Bones of the head are less opaque than in normal fish, and the skull is characteristically resonant when tapped (Kamonporn et al., 1981). Catfish are, of course, advanced fish, so that their acellular bone can only heal with difficulty and requires a luxus of calcium.

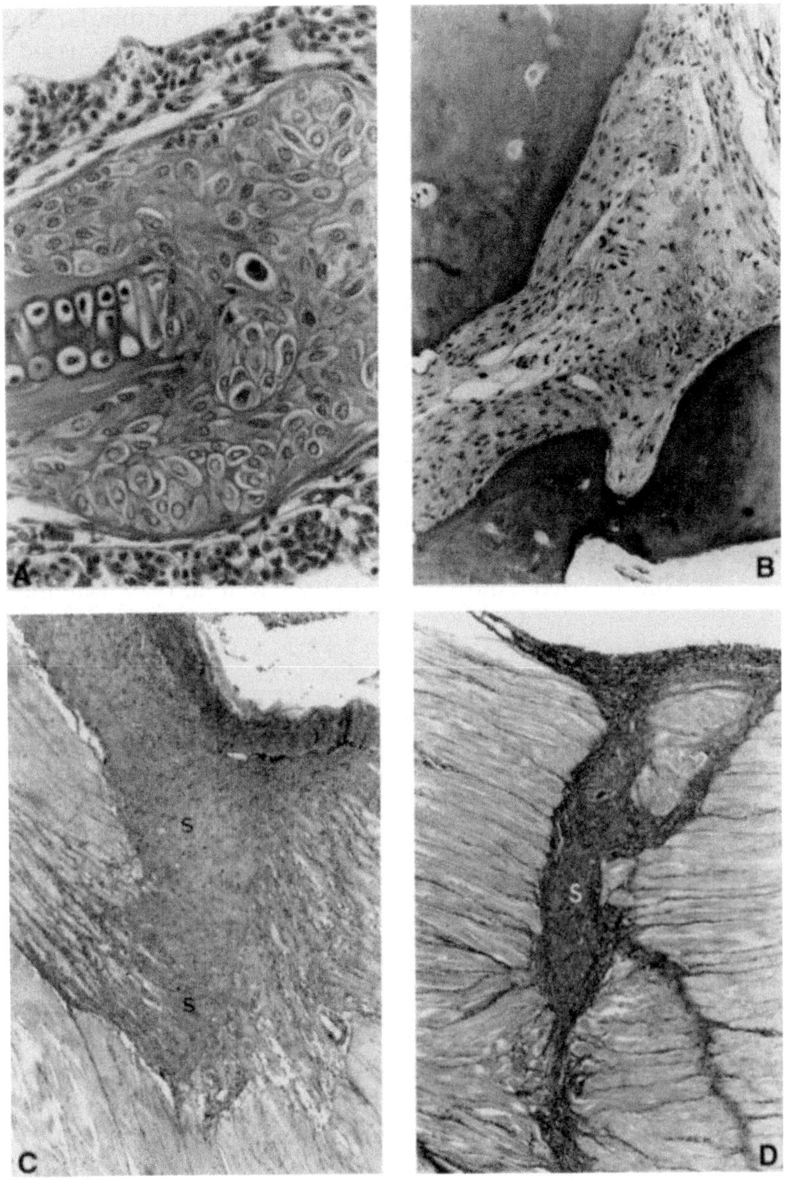

FIGURE 13.6 (A) Gill cartilage metaplasia (H&E ×150). (B) Fibrous osteoid replacement tissue (uncalcified) in vitamin C-deficient rainbow trout (H&E ×100). (C) Poorly healed scar tissue in vitamin C-deficient tilapia 16 days after induction (H&E ×60). (D) Contracted, collagenized, normal scar tissue in tilapia fed normal diet (H&E ×50).

Kamonpern et al., (1981) speculated that the reason for this facultative vitamin C deficiency was that normally catfish would ingest vitamin C with the viscera of their prey or the aquatic algae of the water, but in the farm situation with little in the way of prey species or primary production available, this is not possible.

(Vitamin C deficiency is also considered to play a role in the mandibular "lockjaw" known as "screamer disease" in fast-growing Atlantic and Pacific salmon as discussed under 'Section 13.3.21'.)

13.3.16.1 Mineral deficiencies and imbalances

Despite the fact that almost all necessary minerals are available in most practical fish diets and can also be absorbed from the water, mineral deficiencies do, on occasion, arise in farmed fish. They are almost invariably associated with a reduced bioavailability of the mineral rather than a frank deficiency. This may be associated with a dietary imbalance or with presentation of the element in an unsuitable form. It may also be associated with interaction with other dietary ingredients, such as vitamins and fibers (Tacon and De Silva, 1983). The absence of an acid-secreting stomach in cyprinid fish prevents the absorption of organically bound minerals, especially phosphorus, and within plant proteins, the presence of phytic acid may chelate trace elements such as iron, copper, and zinc (Satoh et al., 1989). Phytate is the storage form in which phosphorus is retained in seeds. Grains used in fish diets may therefore be rich in phytate-phosphorus, but this is unavailable to fish.

Fish meals used in commercial diets contain high levels of calcium or ash. These can significantly affect the availability of trace elements, which have to be heavily supplemented if deficiency conditions are to be avoided (Andrews et al., 1973; Ketola, 1979; Watanabe et al., 1980).

13.3.17 Copper

Copper deficiency is generally unlikely due to the ubiquity of the element in water. Its toxicity at high levels in water is well recognized, but another aspect of high copper exposure, whether in the diet or in the water, is the increased susceptibility it induces for bacterial infections such as vibriosis, where copper is often the limiting nutrient for bacterial growth and multiplication in vivo (Hilton and Hodson, 1983). However, there may be circumstances where supplementation is necessary as Satoh et al., 1983 reported that carp fed high-ash diets without copper supplementation developed cataracts.

13.3.18 Iodine

Fishmeal contains high levels of iodine, and since it often forms a significant element of commercial fish diets, iodine deficiency is very unlikely. In the past, salmonid diets were based primarily on terrestrial animal offals, and iodine-dependent goitres were not uncommon. There is confusion in the literature between such lesions (Marine, 1910) and thyroid adenocarcinomata (Gaylord and Marsh, 1914), but it is likely that this is due to the diffuse nature of the teleost thyroid, which when goitrous can resemble an invasive or metastatic adenocarcinoma (Davis, 1953). The recent trend toward replacement of fish meal with protein sources of plant origin in salmonid diets may, however, bring iodine-dependent goitres back.

13.3.19 Iron

Iron is more often a problem in relation to the toxicity of ferric ions at high levels in certain water supplies and in recirculation systems, rather than as a deficiency. Organic iron compounds, however, are conserved in conditions such as hemorrhagic anemia where the melanomacrophage system of the hemopoietic tissue plays a particularly avid role in its retention (Roberts, 2012). Hypochromic normocytic anemia has been demonstrated in iron-deficient brook trout (Kawatsu, 1972), yellowtail (Ikeda et al., 1973), red sea bream (Sakamoto and Yone, 1978), carp (Sakamoto and Yone, 1978), eel (Nose and Arai, 1979), and catfish (Gatlin and Wilson, 1986).

13.3.20 Manganese

Manganese is an essential element. When it is deficient, the usual features of poor growth and darker color are complemented by the development of cataracts. It generally occurs in diets with high levels of calcium or ash.

Satoh et al., 1983 and Yamamoto et al., 1983, working with rainbow trout and carp, showed that manganese supplementation in an available form was essential where fish meals were a major component of the diet. Ishac and Dollar (1968) showed that in tilapia the absence of manganese resulted in poor growth, reduced food consumption, and loss of equilibrium, but they did not report cataract development. Skeletal abnormalities associated with manganese deficiency have also been reported for rainbow trout (Ishac and Dollar, 1968), carp (Ogino and Yang, 1980), and tilapias.

13.3.21 Phosphorus

Phosphorus is one of the few minerals which fish cannot supply their metabolic needs from via the water. In the past, phosphorus deficiency was uncommon because fish feeds were largely fishmeal-based, which are high in phosphorus levels. There has long been concern, however, not only for the long-term availability of fish meals but also for phosphorus pollution from fish farm effluents. For these reasons, the trend has been for more sophisticated diets with proteins of plant origin and high-energy levels.

Fish that are deficient in phosphorus do exhibit the typical signs of anorexia and dark coloration, but there are also very specific clinical signs of phosphorus deficiency associated with its primary role in bone mineralization, such as deformities of the head, ribs, and vertebrae (Ogino and Takeda, 1976; Skonberg et al., 1997).

As salmonid culture has become more intensive, with fry, smolts, and growers all receiving high energy diets and with periods of very rapid growth associated particularly with the first feeding of larvae, smoltification, and sea transfer stages, it has become apparent that there is a particularly high requirement for available phosphate in modern feeds. If the balance of available phosphorus and calcium and the vitamin C levels are not readily available, certain minor changes can take place in certain critical vertebrae of first feeding larvae as they calcify. Often at these stages, they are only visible with X-ray examination. A critical factor in first-feeding diets is availability of adequate phosphorus levels in all of the very small pellets, which are provided at this stage. Often a larva will consume only a few pellets each day. Such consistent availability is not so important in older fish, as they have a larger food intake, but enhanced levels are still required. Usually the majority of affected fish will not be clinically obvious until they reach smoltification or on-growing, when deformity associated with these original lesions will lead to deformed fish and/or ready fracture of the spine, after even minimal handling (Sullivan et al., 2007a,b).

The importance of optimized phosphate levels in the freshwater diet in the triploid Atlantic salmon, which are being used increasingly in commercial culture, has been emphasized in the careful work of Fjelldal and his colleagues (Fjelldal et al., 2012, Fjelldal et al., 2015). They supported the contention of Sullivan et al. (2007) that it is the juvenile rearing period of fish with higher growth rates, which is the most sensitive period for P deficiency, with the clinical effects following in the fish during adult life. They also showed convincingly that further enhanced levels of dietary phosphate were necessary in triploid populations, to reduce deformities, improve bone mineralization, and enhance growth.

Phosphorus deficiency and combined phosphorus and vitamin C deficiencies are also prominent in a distinctive condition of fast-growing strains of Atlantic salmon, being reared on high-energy diets at high ambient temperatures, particularly in Chile. If enhanced levels are not supplied for the particularly high demands of recently seawater-transferred, very rapidly growing fish, there is failure to ossify or resorption of phosphorus from bones, which become soft and rubbery. Deformity develops in particular stress sites such as the articulation of the jaw, which may lock open. As this greatly reduces feeding ability, growth becomes limited, and the malformed bones can then manage to ossify, to produce characteristically deformed fish, with a wide open mouth. These are colloquially referred to as "screamers" from their uncanny resemblance to the famous painting entitled "The Scream" by Edvard Munch (Fig. 13.7). They also show degeneration, and often the collapse of one or more cervical or lumbar vertebrae, with resultant kyphosis or scoliosis. A particular feature of such cases is a twisted or curled shape to the dorsal vertebral spines, which are also frequently encrusted with mineralized plaques (Roberts et al., 2001).

13.3.22 Selenium

Growth depression has been reported as a result of selenium deficiency in rainbow trout and channel catfish, but pathological changes, such as exudative hemorrhage, do not occur unless vitamin E is also deficient (Hilton et al., 1980; Gatlin and Wilson 1984).

13.3.23 Zinc

Zinc deficiencies normally occur as a result of diets where zinc absorption is greatly inhibited by relatively high levels of ash, calcium, or phytate (Ketola, 1979; Hardy and Shearer, 1985; Richardson et al., 1985). General features of zinc deficiency have been described in most farmed species, but the usual deficiency signs of poor growth and skin and fin lesions are complemented by the important feature of bilateral lenticular cataract development. Other clinical signs of zinc deficiency in salmonids include erosion of fins, short-body dwarfism, and poor egg hatchability (Hardy, 2012).

FIGURE 13.7 Phosphorus/ascorbate deficiency in fast-growing Atlantic salmon producing 'lockjaw' and deformity of the head bones ("screamer fish").

Zinc deficiency is certainly not the only cause of cataract in fish, though it is probably the most frequent. They can be associated with riboflavin, methionine, and histidine deficiency (Phillips and Brockway, 1957; Poston et al., 1977; Taylor et al., 2014). They were also formerly associated with the incorporation of cataractogenic material in the feed from derived from certain mammalian offals in fish fed with older traditional feeds (vide infra).

13.4 Toxic components of the diet

13.4.1 Dietary mineral toxicity

Excessive levels of minerals in the diet or water can be responsible for a number of well-recognized disease conditions. These may be of natural origin due to selenium in areas with serpentine soils, associated with copper contamination of fermentation residues or zinc accumulation in feather meals. Unless the toxicant level is overwhelming, the clinical signs are those so often associated with nutritional disease, namely reduced growth and poor conversion, often with darkening of skin color. Lead toxicity, however, has been linked with specific features of blackening of the caudal area and scoliosis (Sippel et al., 1983). Cadmium toxicity is associated with a very specific syndrome of hypocalcemia, hyperexcitability, and osteoporosis (Roch and Maly, 1979). Iron toxicity causes pathological changes in liver, specifically vacuolated hepatocytes appearing to contain large amounts of fat and reduced levels of glycogen. Hepatocytes exhibit pleomorphic nuclei, with large intranuclear inclusions and peripheral clumping of chromatin (Desjardins et al., 1987).

There is great variation in the levels of minerals, which are toxic in the diet, depending principally on intake (and thus often temperature) but also on the presence of calcium in the diet. High levels of calcium, as well as helping to induce cataract by reducing bioavailability of minerals in normal diets, also affect the toxicity of excessive levels of other minerals. This is well documented in relation to lead and zinc (Takeda and Shimma, 1977; Varanasi and Gmur, 1978) and probably also applies in relation to other toxic minerals.

13.4.2 Selenium

Vitamin E and selenium are to an extent interrelated, and in fish, supplementation of vitamin E with selenium is linked to prevention of bland muscular dystrophy (Poston et al., 1976) and enhanced tocopherol activity in liver, though not in muscle. It is, however, highly toxic when supplied at slightly higher levels, especially when carbohydrate levels are also excessive. It has been associated by Hicks et al., 1984 with nephrocalcinosis and has been shown to react with copper to produce a significant increase in susceptibility to infection (Hilton and Hodson, 1983), but it is unlikely that such levels would be achieved in practical diets.

13.5 Calcium

Although there is a high requirement for calcium by fish, for elaboration of calcified tissues such as bones and scales as well as maintenance of electrolyte levels, calcium can be readily absorbed branchially in all but the softest waters. Thus, it is of importance in dietary terms principally in regard to imbalances with other dietary components. It can often lead to metastatic calcification and has a capacity to impair absorption of heavy metals such as iron, copper, and zinc.

Metastatic calcification can also occur in association with a number of other pathological processes, of which calcification of chronic granulomata is probably the most frequent, but the most common in fish under culture conditions are the two specific conditions visceral granuloma and nephrocalcinosis.

Visceral granuloma is normally described in the digestive tract of salmonids, where mineralized focal granulomata enlarge to form calcified nodules. These are very hard and can be palpated through the abdominal wall. The lesion begins as a small foreign body-type granuloma with giant cells, epithelioid cells, and early fibrosis (Fig. 13.8). Gradually, as it enlarges, it calcifies, and smaller granulomata develop in clusters around the primary. No specific cause for the condition has been defined, although it is possible that a mycotoxin, or possibly spicules of algae or dietary silica, may induce it. Dietary calcium certainly does not appear to affect it in any way, and the calcification appears to be of straightforward chronic inflammatory origin.

Pisiform granulomata with calcification have also been described elsewhere in the tissues of farmed fish. Paperna (personal communication, 1978) described such ectopic calcification in gilthead bream. The fish were fed on a wet trash fish diet of high mineral content, which developed nodular granulomata throughout the body, even in the eye and brain.

Nephrocalcinosis (precipitation of calcium complexes within the renal tubules), occurs in association with magnesium deficiency (Knox et al., 1981) and with selenium toxicity (Hilton and Hodson, 1983). It has also been shown to occur in association with nephrotoxic heavy

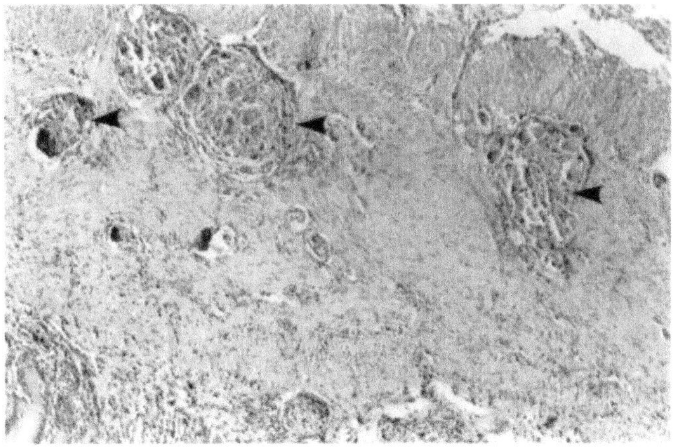

FIGURE 13.8 Visceral granulomata in the wall of the stomach of brook trout. H&E ×100.

metal complexes such as may be found in some wet-feed binders (Dick et al., 1976). It is also generally associated with excessive carbon dioxide levels in the water (Harrison and Richards, 1979; Smart et al., 1979).

Grossly, there is a white deposit, usually restricted to the ureters, on the surface of the kidney, which appear swollen and white (Fig. 13.9), but as the condition exacerbates, the entire kidney becomes swollen and gray with an irregular surface. By this stage, the ureters are more sinuous as well as thickened.

Histologically the first signs are the accumulation of irregular calcified inclusions within vacuolated proximal tubular epithelium cells. These slough into the lumen, where they

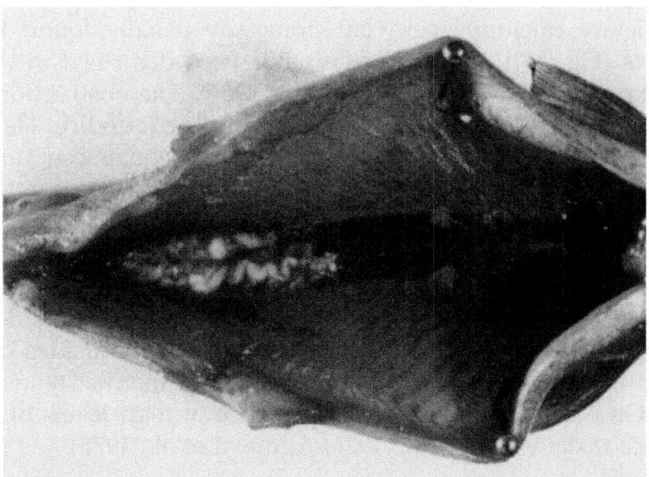

FIGURE 13.9 Gross lesions of nephrocalcinosis in the kidney of a rainbow trout.

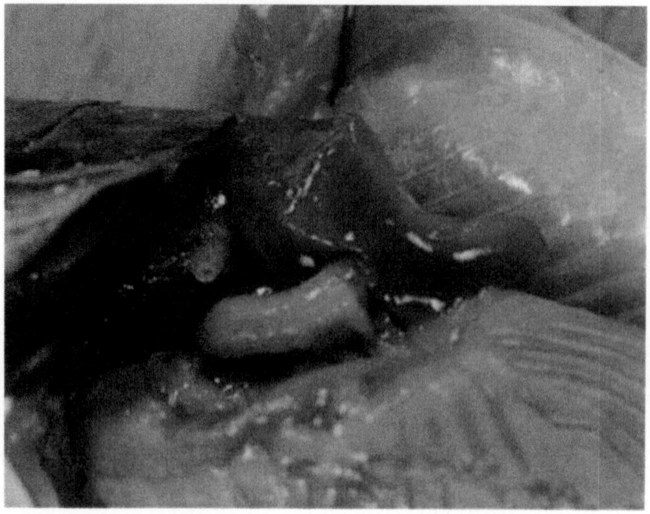

FIGURE 13.10 Cholelithiasis in an Atlantic salmon fed on a wet diet.

accumulate as a dense precipitate—the urolith, acting as a tubular cast around which the degenerating tubule condenses. Obstruction of urine flow leads to tubular distension and generalized tubular and glomerular degeneration. Characteristically the glomerulus becomes very small within a large, dilated, thin-walled, and darkly staining Bowman's capsule. In the final stages, most of the tubular tissue becomes replaced by calcifying fibrosis, and the hemopoietic tissue is also destroyed. Urolithiasis can be a particularly severe problem where pumped ground waters are used, and considerable engineering effort may be required to reduce dissolved carbon dioxide levels to acceptable levels.

Lithiasis is also associated with the biliary system, though again there is little direct connection with dietary calcium level. Gall stones are usually found in association with feeding of wet diets (Fig. 13.10). Cholelithiasis has been described in Atlantic and Pacific salmon and also experimentally in tilapias (Maier 1984). The obstruction leads to blockage of the bile duct and retrograde distension, with obstructive jaundice. Hepatic cirrhosis may ultimately supervene, but generally cholelithiasis is an occasional observation at postmortem.

13.6 Toxic organic products in the diet

Although naturally occurring toxicants, such as poisonous plant alkaloids, fungal metabolites, or anthropogenic compounds such as organochlorine pesticides or hormone disruptors, may be found in fish diets, the other components of the diet tend to reduce or dilute their toxic effects. On occasion, however, the presence of high levels of lipid, for example, can actually enhance toxicity (Lee et al., 1976; Agrawal et al., 1978).

13.6.1 Mycotoxins

Aflatoxins are a problem in formulated diets of all farmed animals. It was, however, the discovery of the toxicity of these metabolites of mutant strains of *Aspergillus flavus*, the blue-green mold that grows on oil seeds, for rainbow trout, which first highlighted their importance. Aflatoxin presence at levels as low as 0.01 ppb can induce neoplastic changes in rainbow trout over a relatively short period. This very high toxicity of aflatoxins for fish was first described by Ashley and his colleagues (Ashley et al., 1964; Halver, 1965), who showed that the aflatoxins were extremely carcinogenic and responsible for the widespread occurrence of hepatic carcinoma in farmed rainbow trout observed as the use of formulated feeds developed. When fed at higher levels (80ppb or more), the toxin also induces an acute toxic syndrome with massive focal hepatic necrosis, branchial edema, and a generalized hemorrhagic syndrome (Ashley, 1970).

The histology of the principal lesion is that of an invasive malignant trabecular hepatocarcinoma. It is relatively easy to distinguish, in sections of liver, even in the early stages, because of its darker staining reaction (Fig. 13.11). The lesion can also involve biliary and vascular tissue and develop a very vascular supporting stroma. As it grows, the lesion becomes friable, and death is generally from hemorrhage following infarctive necrosis of the center of the rapidly growing tumor, or traumatic rupture following handling. Occasionally, metastasis may occur to spleen, kidney, or gill, but generally the neoplasm is characterized by its locally invasive nature (Majeed et al., 1984).

Rainbow trout is the species for which aflatoxins have posed the most significant risk. Catfish and other salmonids seem less vulnerable. The condition is now rarely seen, because, as a result of the detailed work by the American workers (Ashley et al., 1964; Halver, 1965, Halver et al., 1969; Wales and Sinnhuber, 1966; Wales, 1979), there is an active understanding by feed compounders of the need for great care in the screening and storage of oil seeds.

In developing countries, particularly where storage of grains of necessity involves exposure to humid, tropical conditions, aflatoxicosis is still a serious problem. In tilapia culture,

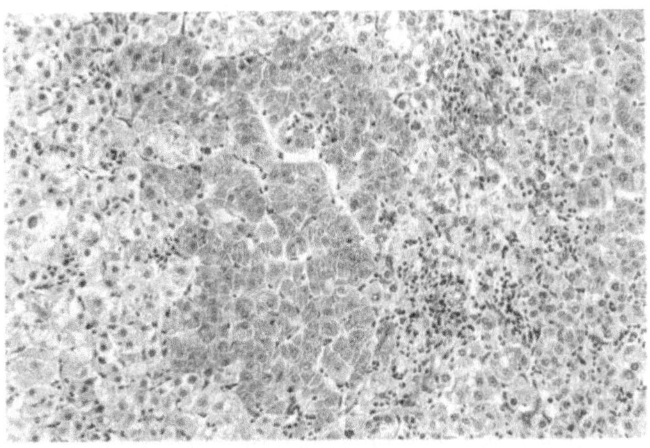

FIGURE 13.11 Early malignant mycotoxin-induced hepatoma in rainbow trout. The malignant tissue is darker and well delineated (H&E ×120).

the toxin is associated with a wide range of neoplasms, including renal tubular carcinoma, lymphoma, and hepatoma (Haller and Roberts, 1980).

Aflatoxin may be present in such minute quantities that the fish do not develop obvious pathological signs. Nevertheless, the growth rates, hemoglobin concentrations, and erythrocyte counts of such fish can be significantly lower than those of fish fed on diets totally free (Jantrarotai et al., 1990).

13.6.2 Toxic algae

The role of algal blooms in removing oxygen and causing fish kills is well known. Many of the freshwater algae are also highly toxic to higher animals (Codd, 1984). However, when known toxic and nontoxic freshwater algae were presented to rainbow trout and tilapias, the fish demonstrated a remarkable capacity for detoxication of orally presented algae, which were nevertheless highly toxic when presented parenterally (Phillips et al., 1985).

Toxic marine algae, the "red tide" organisms or dinoflagellates, are generally associated with acute necrotizing branchial lesions, but Roberts et al., 1983 have shown both clinically and experimentally that at least in the case of *Gyrodinium aureoleum*, one of the most common of the toxic marine algae, they are also ingested during an outbreak and induce severe enteritis and focal hepatic necrosis. Domoic acid, an algal toxin that is the causative agent of "amnesiac shellfish poisoning" (ASP) in humans, does not induce clinical signs of illness or other pathology when fed to rainbow trout, although it is quite toxic to birds and mammals (Hardy et al., 1995).

13.6.3 Plant toxins

13.6.3.1 Ragwort

Senecio alkaloids, the toxic components of the ragwort plant, *Senecio jacobaea*, have not been reported as causing clinical disease, but Hendricks et al., 1981 carried out a detailed study of the toxicity of these compounds to rainbow trout. They showed that even low doses in the diet led to necrotic megalocytosis of hepatocytes with biliary hyperplasia and cirrhosis.

13.6.3.2 Cottonseed

Cottonseed is a valuable oilseed for incorporation in fish diets, but as well as being a ready substrate for growth of *Aspergillus*, cottonseeds generally also contain two toxins of their own. The most important of these is gossypol, a yellow pigment which, when it is included in a diet, causes anorexia and accumulation of large lipid granules in the hepatocytes and renal tubules, which act as nuclei for granulomatosis (Herman, 1970).

Special strains of unpigmented cottonseed are available, but there is also a cyclopropenoid fatty acid in the kernel of the seed which is toxic per se, as well as acting as a powerful synergist of aflatoxin B_1 (Hendricks et al., 1980).

13.6.3.3 Ipil-ipil toxins

The ipil-ipil tree or *Leucaena* is a legume, which produces pods high in protein and valuable for inclusion in fish diets in the developing tropical countries, where other protein sources are difficult to obtain. Unfortunately, the protein contains a toxic amino acid, mimosine, which causes poor growth and cachexia. The level of mimosine varies with the strain of *Leucaena* used, and selective breeding for absence of mimosine would seem the best approach to allowing long-term use of this valuable plant (Jackson et al., 1982).

13.6.4 Pesticides

Terrestrial agriculture has been greatly enhanced by the significant developments that have taken place in control of insect pests by use of chemical pesticides. Some of these, such as the pyrethrins, are exceedingly toxic to fish even in low doses, by virtue of their inhibition of respiratory function. Others are of relatively lower toxicity and are ingested with food or, in the case of wild fish, accumulate in the prey species to affect top predators.

It is unusual for such chemicals to cause frank clinical disease or histopathological findings, the effects being generally more occult, *viz* reduction in breeding level or reduced growth. Occasionally, however, overdosing or dietary contamination can cause mortalities. Usually the features are those of an acute necrotizing toxaemia with massive liver and kidney necrosis. Surviving fish show fibrous replacement of necrotic areas and generally recover, at least clinically, within a relatively short period (Matthiessen and Roberts, 1982).

13.6.5 Photosensitizing chemicals

Photosensitizers are chemicals that have the property of releasing electrons when stimulated by ultraviolet light (UV). When incorporated into fish diets, they are ingested and circulate in the tissues. In the outer layers of the skin, they can be stimulated by UV light, and the release of energy causes localized tissue damage. The commonest photosensitizer for fish is phenothiazine, an anthelmintic drug occasionally used for treatment of *Octomitus* infection. The phenomenon was first recorded by Rucker (1957), who described a condition he called "sunburn" or "backpeel" in fish treated with phenothiazine. Bullock and Roberts (1979) defined the pathology of the condition for Atlantic salmon and described the distinctive "sunburn cells" of the epithelium of affected fish.

Photosensitization in higher animals is associated with a wide range of chemical molecules. With the increasing use of similar chemicals in agriculture, in water treatment and in fish therapeutics, the potential for the problem in fish seems considerable (Bullock, 1987).

13.6.6 Silkworm toxins (*Sekoke*)

Yokote (1974) has described a very distinctive syndrome in carp fed on diets with a high level of silkworm pupae incorporated in them. The condition closely resembles diabetes in the higher species. There is destruction of the endocrine pancreatic islets of Langerhans, concomitant lipid infiltration of parenchymatous organs, bilateral cataract, and degenerative alterations to the extrinsic eye muscles, the retina, and the choroid. Sekoke is now rare, as pelleted diets have gradually replaced the more traditional carp diets.

13.6.7 Nutritional cataracts

In the original diets for salmonids, devised in the 1870s, animal offals played a significant role. However, when spleens or horse liver was incorporated, eye conditions, principally characterized by the development of bizarre highly cellular bilateral cataracts, regularly occurred. These are very distinctive and easily differentiated from zinc or magnesium deficiency cataracts (Allison, 1950).

A particularly distinctive ophthalmia beginning as bilateral cataract but ultimately evolving to panophthalmia and evulsion of the orbit was described in rainbow trout fed a diet of minced leopard offals on the Zomba Plateau in Malawi (Lee et al., 1976).

In the 1990s, a new cataract condition became apparent in farmed Atlantic salmon, particularly in Scotland and Ireland. It was a posterior cortical bilateral cataract often leading to complete blindness (Wall, 1998). After considerable investigation, it was shown that the cause was deficiency of the amino acid histidine, which, when blood meals were removed from fish diets, was no longer available in *luxus* (Waagbø et al., 2010). The problem was more pronounced in triploid fish than in diploids (Taylor et al., 2015).

13.6.7.1 Drug toxicity

Before the advent of vaccines and better biosecurity and management techniques, antibiotics were incorporated into fish diets at significant levels, but it was only if they were included in very high levels, or if usage was prolonged, that pathological changes were observed. Sulfonamides, which were often used against gram-negative bacteria, produced toxic vacuolar degeneration of proximal renal tubular epithelium with renal casts and tubular necrosis and also focal hepatic necrosis and arterial sclerosis if used over a prolonged period (Wood et al., 1957).

References

Adron, J.W., Knox, D., Cowey, C.B., Ball, G.T., 1978. Studies on the nutrition of marine flatfish. The pyridoxine requirement of turbot (Scophthalmus maximus). Br. J. Nutr. 40, 261–268.

Agius, C., Roberts, R.J., 1981. Effects of starvation on the melano-macrophage centres of fish. J. Fish. Biol. 19, 161–169.

Agius, C., Roberts, R.J., 2003. Melano-macrophage centres and their role in fish pathology. J. Fish Dis. 26 (9), 499–509.

Agrawal, N.K., Juneja, C.J., Mahajan, C.L., 1978. Protective role of ascorbic acid in fishes exposed to organochlorine pollution. Toxicology 11, 369–375.

Allison, L.N., 1950. White blindness in hatchery brook trout. Prog. Fish-Cult. 12, 52.

Andrews, J.W., Murai, T., Campbell, C., 1973. Effects of dietary calcium and phosphorus on growth, food conversion, bone ash and hematocrit levels of catfish. J. Nutr. 103, 766–771.

Aoe, H., Masuda, I., 1967. Water-soluble vitamin requirements of carp. 2. Requirements for p-aminobenzoic acid and inositol. Bull. Jpn. Soc. Sci. Fish. 33, 674–680.

Aoe, H., Masuda, Mimura, T., Saito, T., Komo, A., Kitamura, S., 1969. Water-soluble vitamin requirements for carp. 6. Requirements for thiamin and effects of antithiamins. Bull. Jpn. Soc. Sci. Fish 35, 459–465.

Arai, S., Nose, T., Hashimoto, Y., 1972. Qualitative requirements of young eels, Anguilla japonica, for water-soluble vitamins and their deficiency symptoms. Bull. Freshwater Fish Res. Lab. Tokyo 22, 69–83.

Ashley, L.M., 1970. In: Snieszko, S.F. (Ed.), A Symposium on Diseases of Fishes and Shellfishes, pp. 366–379.

Ashley, L.M., Halver, J.E., Wogan, G.N., 1964. Hepatoma and aflatoxicosis in trout. Fed. Proc. Fed. Am. Soc. Exp. Biol. 23, 105.

Blaxter, J.H.S., Roberts, R.J., Balbontin, F., McQueen, A., 1974. B-group vitamin deficiency in cultured herring. Aquaculture 3, 387–394.

Brett, J.R., 1958. H. R. MacMillan Lectures on Fisheries. University of British Columbia Press, Vancouver.

Bullock, A.M., 1987. In: Muir, J.F., Roberts, R.J. (Eds.), Recent Advances in Aquaculture 3. Croom Helm, London, pp. 139–224.

Bullock, A.M., Roberts, R.J., 1979. Induction of UDN-like lesions in salmonid fish by exposure to ultraviolet light in the presence of phototoxic agents. J. Fish Dis. 2 (5), 439–441.

Castell, J.D., Sinnhuber, R.O., Wales, J.H., Lee, D.J., 1972. Essential fatty acids in the diet of rainbow trout (Salmo gairdneri): growth, feed conversion and some gross deficiency symptoms. J. Nutr. 102, 77–85.

Castledine, A.J., Cho, C.Y., Sunger, S.J., Hicks, B., Bayley, H.S., 1978. Influence of dietary biotin level on growth, metabolism and pathology of rainbow trout. J. Nutr. 108, 698–711.

Coates, J.A., Halver, J.E., 1958. Water-soluble vitamin requirements of silver salmon. U.S. Fish Wildl. Serv. Spec. Sci. Rep. 281.

Codd, G.A., 1984. Toxins of freshwater cyanobacteria. Microbiol. Sci. 1, 48–52.

Cowey, C.B., Adron, J.W., Owen, J.M., Roberts, R.J., 1976. The effect of different dietary oils on tissue fatty acids and tissue pathology in turbot Scophthalmus maximus. Comp. Biochem. Physiol. 53B (3), 399–403.

Davies, S.J., 1985. The role of dietary fibre in fish nutrition. Recent Adv. Aquacult. 2, 219–249.

Davis, H.S., 1953. Culture and Diseases of Game Fishes. University of California Press, Berkeley.

De Long, D.C., Halver, J.E., Yasutake, W.T., 1958. A possible cause of sunburn in fish. Prog. Fish-Cult. 20 (3), 111–113.

De Silva, S.S., Anderson, T.A., 1995. Fish Nutrition in Aquaculture. Chapman & Hall, London.

Desjardins, L.M., Hicks, B.D., Hilton, J.W., 1987. Iron catalyzed oxidation of trout diets and its effect on the growth and physiological response of rainbow trout. Fish Physiol. Biochem. 3, 173–182.

Dick, G.C., Machin, D., Roberts, R.J., Anderson, C.D., 1976. The hepato-renal syndrome of cultured turbot. Its effect on a 2-year trial of farm reared turbot (Scophthalmus maximus). Aquaculture 8, 241–249.

Ehrlich, K.F., 1974. Chemical changes during growth and starvation of larval *Pleuronectes platessa*. Mar. Biol. 24, 39–48.

Ferguson, H.W., Roberts, R.J., Richards, R.H., Collins, R.O., Rice, D.A., 1986a. J. Fish. Dis. 9, 95–98.

Ferguson, H.W., Rice, D.A., Lynas, J.K., 1986b. Vet. Rec. 119, 297–299.

Fjelldal, P.G., Hansen, T., Breck, O., Ørnsrud, R., Lock, E.J., Waagbø, R., Witten, P.E., 2012. Vertebral deformities in farmed Atlantic salmon (*Salmo salar* L.) - Aetiology and pathology. J. Appl. Ichthyol. 28, 433–440.

Fjelldal, P.G., Hansen, T.J., Lock, E.-J., Wargelius, A., Fraser, T.W.K., Sambraus, F., El-Mowafi, A., Albrektsen, A., Waagbø, R., Ørnsrud, R., 2015. Increased dietary phosphorous prevents vertebral deformities in triploid Atlantic salmon (*Salmo salar* L.). Aquacult. Nutr. 22, 72–90.

Gatlin III, D.M., Wilson, R.P., 1984. Dietary Selenium Requirement of Fingerling Channel Catfish. J. Nutr. 114 (3), 627–633.

Gatlin, D.M., Wilson, R.P., 1986. Characterization of iron deficiency and the dietary iron requirement of fingerling channel catfish. Aquaculture 52, 191–198.

Gaylord, H.R., Marsh, M.C., 1914. Carcinoma of the thyroid in the salmonid fishes. Bull. U.S. Bur. Fish 32, 363–524.

Haller, R.D., Roberts, R.J., 1980. Dual neoplasia in a specimen of *Sarotherodon spilurus spilurus* (Günther) (=*Tilapia spilurus*). J. Fish. Dis. 3, 63–66.

Halver, J.E., 1957b. Prog. Fish-Cult. 19, 112–118.

Halver, J.E., 1957. Nutrition of salmonid fishes. III. Water-soluble vitamin requirement of chinook salmon. J. Nutr. 62, 225–243.

Halver, J.E., 1965. Fed. Proc. 24, 169–185.

Halver, J.E., Ashley, L.M., Smith, R.R., 1969. Ascorbic acid requirements of coho salmon and rainbow trout. Trans. Am. Fish. Soc. 98, 762–771.

Halver, J.E., Coates, J.A., 1957. A vitamin test diet for long-term feeding studies. The Progressive Fish-Culturist 3, 112–118.

Hardy, R.W., 2012. In: Roberts, R.J. (Ed.), Fish Pathology, Fourth ed. John Wiley & Sons, Oxford, New York, pp. 402–424.

Hardy, R.W., Scott, T.M., Hatfield, C.L., Barnett, H.J., Gauglitz Jr., E.J., Wekell, J.C., Eklund, M.W., 1995. Domoic acid in rainbow trout (*Oncorhynchus mykiss*) feeds. Aquaculture 131, 253–260.

Hardy, R.W., Shearer, K.D., 1985. Effect of dietary calcium phosphate and zinc supplementation on whole body zinc concentration of rainbow trout (*Salmo gairdneri*). Can. J. Fish. Aquat. Sci. 42, 181–184.

Harrison, J.G., Richards, R.H., 1979. The pathology and histopathology of nephrocalcinosis in rainbow trout Salmo gairdneri Richarson in fresh water. J. Fish Dis. 2, 1–12.

Hendricks, J.D., Sinnhuber, R.O., Henderson, M.C., Buhler, D.R., 1981. Liver and kidney pathology in rainbow trout (Salmo gairdneri) exposed to dietary pyrrolizidine (Senecio) alkaloids. Exp. Mol. Pathol. 35, 170–183.

Hendricks, J.D., Sinnhuber, R.O., Loveland, P.M., Pawlowski, N.E., Nixon, J.E., 1980. Toxic compounds in cottonseed meal. Science 209, 308–311.

Hepher, B., 1988. Nutrition of Pond Fishes. Cambridge University Press, pp. 292–306.

Herman, R.L., 1970. Effects of gossypol on rainbow trot Salmo gairdneri Richardson. J. Fish. Biol. 2, 293–304.

Hess, W.N., 1935. Reduction of the islets of Langerhans in the pancreas of fish by means of diet, overeating and lack of exercise. J. Exp. Zool. 70, 187–195.

Hicks, B.D., Hilton, J.W., Ferguson, H.W., 1984. Influence of dietary selenium on the occurrence of nephrocalcinosis in the rainbow trout, *Salmo gairdneri* Richardson. J. Fish Dis. 7, 379–389.

Hilton, J.W., Hodson, P.V., 1983. Effect of increased dietary carbohydrate on selenium metabolism and toxicity in rainbow trout (Salmo gairdneri). J. Nutr. 113, 1241–1248.

Hilton, J.W., Hodson, P.V., Slinger, S.J., 1980. The requirement and toxicity of selenium in rainbow trout (Salmo gairdneri). J. Nutr. 110, 2527–2535.

Hjeltnes, B., Roberts, R.J., 1993. In: Inglis, V., Roberts, R.J., Bromage, N.R. (Eds.), Bacterial Diseases of Fish. Blackwell Scientific, Oxford, pp. 109–121.

Honeyfield, D.C., Brown, S.B., Fitzsimons, J.D., Tillitt, D.E., 2005. Early mortality syndrome in Great Lakes salmonines. J. Aquat. Anim. Health 17, 1–3.

Hung, S.S.O., 1989. Choline requirement of hatchery-produced juvenile white sturgeon (Acipenser transmontanus). Aquaculture 78 (2), 183–194.

Ikeda, Y., Ozaki, H., Uematsu, K., 1973. Effects of enriched diet with iron in culture of yellowtail. J. Tokyo Univ. Fish., 59: 91–99.

Ishak, M.M., Dollar, A.M., 1968. Studies on manganese uptake in Tilapia mossambica and Salmo gairdneri. 1. Growth and survival of Tilapia mossambica in response to manganese. Hydrobiologia 31, 572–584.

Jackson, A.J., Capper, B.S., Matty, A.J., 1982. Evaluation of some plant proteins in complete diets for the tilapia Sarotherodon mossambicus. Aquaculture 27, 97–109.

Jansen, M.D., Bang, J., McLoughlin, M.F., Rodger, H.D., Taksdal, T., Sindre, H., Graham, D.A., Lillehaug, A., 2017. The epidemiology of pancreas disease in salmonid aquaculture: a summary of the current state of knowledge. J. Fish Dis. 40, 141–155.

Jantrarotai, W., Lovell, R.T., Grizzle, J.M., 1990. Acute toxicity of aflatoxin B1 to channel catfish. J. Aquat. Anim. Health 2, 237–247.

Jauncey, K., 1982. Recent Adv. Aquacult 1, 215–263.

Jauncey, K., 1998. Tilapia Feeds and Feeding. " Pisces Press, Stirling, pp. 1–210.

Jauncey, K., Soliman, A., Roberts, R.J., 1985. Ascorbic acid requirements in relation to wound healing in the cultured tilapia Oreochromis niloticus (Trewavas). Aquacult. Fish Manag. 16, 139–149.

John, M.J., Mahajan, C.L., 1979. The physiological response of fishes to a deficiency of cyanocobalamin and folic acid. J. Fish Biol. 14, 127–133.

Kamonporn, L., Areerat, S., Booyaratpalin, S., Chinabut, S., McRae, I.M., Muir, J.F., Richards, R.H., Roberts, R.J., Sommerville, C., 1981. Diseases of Catfish Clarias Lazera. Natl. Inland Fish. Inst., Bangkok.

Karges, R.G., Woodward, B., 1984. Development of lamellar hyperplasia in pantothenic acid deficient rainbow trout Salmo gairdneri (Richardson). J. Fish Biol. 25, 57–62.

Kawatsu, H., 1972. Studies on the anemia of fish-V. Dietary iron deficient anemia in brook trout, Salvelinus fontinalis. Bull. Freshwater Fish Res. Lab. Tokyo 22, 59–67.

Ketola, H.G., 1976. Choline Metabolism and Nutritional Requirement of Lake Trout (Salvelinus namaycush). J. Anim. Sci. 43, 474–477.

Ketola, H.G., 1979. Influence of dietary zinc on cataracts in rainbow trout (Salmo gairdneri). J. Nutr. 109 (6), 965–969.

Ketola, H.G., 1983. Requirement for dietary lysine and arginine by fry of rainbow Ttout. J. Anim. Sci. 56, 101–107.

Kissil, G.W., Cowey, C.B., Adron, J.W., Richards, R.H., 1981. Pyridoxine requirements of the gilthead bream, Sparus aurata. Aquaculture 23, 243–255.

Kitamura, S., Suwa, T., Ohara, S., Nakagawa, K., 1967. Studies on vitamin requirements of rainbow trout. III. Requirement for vitamin A and deficiency symptoms. Bull. Japan. Soc. Sci. Fish 33, 1126–1131.

Kitamura, S., 1965. Studies on vitamin requirements of rainbow trout, Salmo gairdneri-I. On the ascorbic acid. Bull. Jpn. Soc. Sci. Fish 31, 818–826.

Knox, D., Cowey, C.B., Adron, J.W., 1981. Studies on the nutrition of salmonid fish. The magnesium requirements of rainbow trout (Salmo gairdneri). Br. J. Nutr. 45, 137–148.

Leatherland, J.F., Barnett, B.J., Cho, Y.C., Slinger, S.J., 1980. Effect of dietary cholecalciferol deficiency on plasma thyroid hormone concentrations in rainbow trout, *Salmo gairdneri* (Pisces, Salmonidae). Environ. Biol. Fish. 5, 167–173.

Lee, W.R., Roberts, R.J., Shepherd, C.J., 1976. Ocular pathology in rainbow trout in Malawi (Zomba disease). J. Comp. Pathol. 76, 221–233.

Lovell, R.T., 1999. Nutrition and Feeding of Fish, Second ed. Van Nostrand, Reinhold, N.Y.

Lovell, R.T., Li, Y.-P., 1978. Essentiality of Vitamin D in Diets of Channel Catfish (Ictalurus punctatus). Trans. Am. Fish. Soc. 107, 809–811.

Mahajan, C.L., John, M.J., 1977. In: Proc. All India Symp Comp. Physiol., p. 52.

Maier, K.J., 1984. Gallstone induction in tilapia, *Oreochromis mossambicus* (Peters). J. Fish Dis. 7, 521–524.

Majeed, S.K., Jolly, D.W., Gopinath, C., 1984. An outbreak of liver cell carcinoma in rainbow trout, Salmo gairdneri, in the U.K. J. Fish Dis 7, 165–168.

Marine, D., 1910. Observations And Experiments On The So-Called Thyroid Carcinoma Of Brook Trout (Salvelinus Fontinalis) And Its Relation To Ordinary Goitre. J. Exp. Med. 12, 311–337.

Matthiessen, P., Roberts, R.J., 1982. Histopathological changes in the liver and brain of fish exposed to endosulfan insecticide during tsetse fly control operations in Botswana. J. Fish Dis. 5, 153–159.

McLaren, B.A., Keller, E., O'Donnell, D.J., Elvehjem, C.A., 1947. The nutrition of rainbow trout. 1. Studies of vitamin requirements. Arch. Biochem. 15, 169–178.

McLoughlin, M.F., Nelson, R.T., Rowley, H.M., Con, D.J., Grant, A.N., 1996. Experimental pancreas disease in Atlantic salmon *Salmo salar* post-smolts induced by salmon pancreas disease virus (SPDV). Dis. Aquat. Org. 26, 117–124.

Meyer, F.P., 1975. In: Ribelin, W.E., Migaki (Eds.), Pathology of Fishes. University of Wisconsin Press, Madison, pp. 275–286.

Moccia, R.D., Hung, S.S.O., Sunger, S.J., Ferguson, H.W., 1984. Effect of oxidized fish oil, vitamin E and ethoxyquin on the histopathology and haematology of rainbow trout, *Salmo gairdneri* Richardson. J. Fish Dis. 7, 269–282.

Moller, H., 1984. Dynamics of fish diseases in the lower Elbe River. Helgolander. Meeresun. 37, 389–413.

Murai, T., Andrews, J.W., 1974. Interactions of dietary alpha-tocopherol, oxidized menhaden oil and ethoxyquin on channel catfish (Ictalurus punctatus). J. Nutr. 104, 1416–1431.

Murai, T., Andrews, J.W., 1978a. J. Nutr. 108, 176–180.

Murai, T., Andrews, J.W., 1978b. J. Nutr. 108, 1512–1517.

Nose, T., Arai, S., 1979. Recent advances on studies on mineral nutrition of fish in Japan. Advances in Aquaculture. Fishing News, Farnam, England, pp. 584–590.

Ogino, C., 1967. B-vitamin requirements of carp. 2. Requirements for riboflavin and pantothenic acid. Bull. Jpn. Soc. Sci. Fish 33, 351–354.

Ogino, C, Takeda, H., 1976. Mineral requirements in fish. e. Calcium and phosphorus requirements of carp. Bull. Jpn. Soc. Sci. Fish 42, 793–799.

Ogino, C., Watanabe, T., Kakino, J., Iwanaga, N., Mizund, M., 1970. Bull. Jpn. Soc. Sci. Fish. 36, 734–740.

Ogino, C., Yang, G.Y., 1980. Requirement of carp and rainbow trout for dietary manganese and copper. Bull. Jpn. Soc. Sci. Fish 46, 455–458.

Olufemi, B.E., Roberts, R.J., 1986. Induction of clinical aspergillomycosis by feeding contaminated diet to tilapia, *Oreochromis niloticus* (L.). J. Fish Dis. 9 (2), 123–128.

Ørnsrud, R., Graff, I.E., Høie, S., Totland, G.K., Hemre, G.I., 2002. Hypervitaminosis A in first-feeding fry of the Atlantic salmon (*Salmo salar* L.). Aquacult. Nutr. 8, 7–13.

Park, S.I., 1978. Nutritional liver disease in cultured yellowtail Seriola quinqueradiata caused by feed deficiencies. Bull. Korean Fish. Soc. 11, 1–4.

Pearse, L., McQueen, A., Roberts, R.J., 1974. Muscular dystrophy in cultued turbot (Scophthalmus maximus L.). Vet. Rec. 94, 435–437.

Phillips, A.M., Brockway, D.R., 1957. The nutrition of trout. IV. Vitamin requirements. Prog. Fish-Cult. 19, 119–123.

Phillips, A.M., Roberts, R.J., Stewart, J.A., Codd, G.A., 1985. The toxicity of the cyanobacterium Microcystis aeruginosa to rainbow trout Salmo gairdneri Richardson. J. Fish Dis. 8, 339–344.

Pickering, A.D., 1978. A note on the failure of vitamin A to influence the epidermis of the brown trout, *Salmo trutta* L. J. Fish Biol. 12, 441–447.

Pickering, A.D., 1998. In: Black, K.D., Pickering, A. (Eds.), Biology of Farmed Fish. Sheffield Academic Press, Sheffield, pp. 234–235.

Poppe, T.T., Hastein, T., Salte, R., 1985. In: Ellis, A.E. (Ed.), Fish and Shellfish Pathology. Academic Press, London, pp. 222–229.

Poston, H.A., 1969. Effects of massive doses of vitamin D3 on fingerling brook trout. Fish Res. Bull. N.Y. State Conserv. Dep. 29, 45–50.

Poston, H.A., 1971. Effect of feeding excess viamin A on the carbohydrate and lipid metabolism and growth of brook trout. Fish Res. Bull. N.Y. State Conserv. Dep. 34, 22–26.

Poston, H.A., Combs, G.F., Leibovitz, L., 1976. Vitamin E and Selenium Interrelations in the Diet of Atlantic Salmon (Salmo salar): Gross, Histological and Biochemical Deficiency Signs. J. Nutr. 106, 892–904.

Poston, H.A., Di Lorenzo, 1973. Tryptophan conversion to niacin in the brook trout (*Salvelinus fontinalis*). Proc. Soc. Exp. Biol. Med. 144, 110–112.

Poston, H.A., Page, J.W., 1982. Gross and histological signs of dietary deficiencies of biotin and pantothenic acid in lake trout, Salvelinus namaycush. Cornell Vet. 72, 242–261.

Poston, H.A., Riis, R.C., Rumsey, G.L., Ketola, H.G., 1977. The effect of supplemental dietary amino acids, minerals and vitamins on salmonids fed cataractogenic diets. Cornell Vet. 67, 472–509.

Poston, H.A., Wolfe, M.J., 1985. Niacin requirement for optimum growth, feed conversion and protection of rainbow trout, Salmo gairdneri Richardson, from ultraviolet-B irradiation. J. Fish Dis. 8, 451–460.

Poston, H.A., 1976. Effect of L-ascorbic acid on immature brook trout. Fish. Res. Bull. N.Y State Conservation Dept. 30, 46–52.

Richardson, N.L., Higgs, D.A., Beames, R.M., McBride, J.R., 1985. Influence of dietary calcium, phosphorus, zinc and sodium phytate level on cataract Incidence, growth and histopathology in Juvenile chinook salmon (Oncorhynchus tshawytscha). J. Nutr. 115, 553–567.

Roberts, R.J., 2012. In: Roberts, R.J. (Ed.), Fish Pathology, fourth ed. John Wiley & Sons, Oxford, pp. 62–143.

Roberts, R.J., Bullock, A.M., Turner, M., Jones, K., Tett, P., 1983. Mortalities of Salmo gairdneri exposed to cultures of Gyrodinium aureolum. J. Mar. Biol. Assoc. U.K. 63, 741–743.

Roberts, R.J., Hardy, R.W., Sugiura, S., 2001. Screamer disease in Atlantic salmon, *Salmo salar* L., in Chile. J. Fish Dis. 24, 543–549.

Roberts, R.J., Richards, R.H., Bullock, A.M., 1979. Pansteatitis in rainbow trout *Salmo gairdneri* Richardson: a clinical and histopathological study. J. Fish Dis. 2, 85–92.

Roch, M., Maly, E.J., 1979. Relationship of cadmium-induced hypocalcemia with mortality in rainbow trout (*Salmo gairdneri*) and the influence of temperature on toxicity. J. Fish. Res. Board Can. 36, 1297–1303.

Rodger, H.D., Turnbull, J.T., 2000. Cardiomyopathy syndrome in farmed Scottish salmon. Vet. Rec. 146 (17), 500–501.

Rodger, H.D., Turnbull, J.T., Richards, R.H., 1994. Myopathy and pancreas disease in salmon: a retrospective study in Scotland. Vet. Rec. 135, 234–235.

Rucker, R.R., 1957. Phenothiazine as a cause of 'back peel' in trout. Trans. Am. Fish. Soc. 87, 374–379.

Sakamoto, S., Yone, Y., 1978. Iron deficiency symptoms of carp. Bull. Jpn. Soc. Sci. Fish. 44, 1157–1160.

Sambraus, F., Fjelldal, P.G., Remø, S.C., Hevroy, E.M., Nilsen, T.O., Thorsen, A., Hansen, T.J., Waagbo, R., 2017. Water temperature and dietary histidine affect cataract formation in Atlantic salmon (*Salmo salar* L.) diploid and triploid yearling smolt. J. Fish Dis. 40, 1195–1212.

Satoh, S., Poe, W.E., Wilson, R.P., 1989. Effect of supplemental phytate and/or tricalcium phosphate on weight gain, feed efficiency and zinc content in vertebrae of channel catfish. Aquaculture 80, 155–161.

Satoh, S., Yamamoto, H., Takeuchi, T., Watanabe, T., 1983. Effect of growth and mineral composition of carp of deletion of trace elements or magnesium from fish mal diet. Bull. Jpn. Soc. Sci. Fish 49, 431–435.

Seacock, R.R., Goodland, R.L., 1944. Thiamine inactivation by the Chastek-Paralysis factor. Inhibition of thiamine inactivation. J. Am. Chem. Soc. 66, 507–510.

Sippel, A.J.A., Geraci, J.R., Hodson, P.V., 1983. Histopathological and physiological responses of rainbow trout (Salmo gairdneri richardson) to sublethal levels of lead. Water Res. 17, 1115–1118.

Skonberg, D.I., Yogev, L., Hardy, R.W., Dong, F.M., 1997. Metabolic response to dietary phosphorus intake in rainbow trout (*Oncorhynchus mykiss*). Aquaculture 157, 11–24.

Smart, G.R., Knox, D., Harrison, J.G., Ralph, J.A., Richards, R.H., Cowey, C.B., 1979. Nephrocalcinosis in rainbow trout *Salmo gairdneri* Richardson; the effect of exposure to elevated CO_2 concentrations. J. Fish Dis. 2, 279–289.

Smith, C.E., 1968. Hematological changes in coho salmon fed a folic acid deficient diet. J. Fish. Res. Board Can. 25, 151–156.

Smith, C.E., Halver, J.E., 1969. Folic Acid Anemia in Coho Salmon. J. Fish. Res. Board Can. 26, 111–114.

Snieszko, S.F., 1972. In: Halver, J.E. (Ed.), Fish Nutrition. Academic Press, New York.

Soliman, A.K., Roberts, R.J., Jauncey, K., 1983. In: Proc. Int. Symp. Tilapia Aquacult, pp. 193–199.

Soliman, A.K., Jauncey, K., Roberts, R.J., 1985. Aquacult. Fish. Manag. 16, 249–256.

Sullivan, M., Hammond, G., Roberts, R.J., Manchester, N.J., 2007a. J. Fish. Dis. 30, 745–747.

Sullivan, M., Reid, J.W.J., Ternent, H., Manchester, N.J., Roberts, R.J., Stone, D.A.J., Hardy, R.W., 2007b. J. Fish. Dis. 30, 759–767.

Tacon, A.G.J., DeSilva, S.S., 1983. Mineral composition of some commercial fish feeds available in Europe. Aquaculture 31, 11–20.

Takeda, H., Shimma, Y., 1977. Toxicity and availability of lead and zinc to fishes. Bull. Jpn. Soc. Sci. Fish. 27, 103–109.

Taylor, J.F., Waagbø, R., Diez-Padrisa, M., Campbell, P., Walton, J., Hunter, D., Matthew, C., Migaud, H., 2015. Adult triploid Atlantic salmon (*Salmo salar*) have higher dietary histidine requirements to prevent cataract development in seawater. Aquacult. Nutr. 21, 18–32.

Varanasi, U., Gmur, D.J., 1978. Influence of water-borne and dietary calcium on uptake and retention of lead by coho salmon (Oncorhynchus kisutch). Toxicol. Appl. Pharmacol. 46, 65–75.

Waagbø, R., Tröße, C., Koppe, K., Fontanillas, R., Breck, O., 2010. Dietary histidine supplementation prevents cataract development in adult Atlantic salmon, Salmo salar L., in seawater. Br. J. Nutr. 104, 1460–1470.

Wales, J.H., 1979. Induction of hepatoma in rainbow trout *Salmo gaivdneri* Richardson by the egg bath technique. J. Fish Dis. 2, 563–566.

Wales, J.H., Sinnhuber, R.O., 1966. An early hepatoma epizootic in rainbow trout, Salmo gairdneri. Calif. Fish Game 52, 85–91.

Wall, A.E., 1998. Cataracts in farmed Atlantic salmon (*Salmo salar*) in Ireland, Norway and Scotland from 1995 to 1997. Vet. Rec. 142, 626–631.

Walton, M.J., Cowey, C.B., Adron, J.W., 1984. Effects of biotin deficiency in rainbow trout (Salmo gairdneri) fed diets of different lipid and carbohydrate content. Aquaculture 37, 21–38.

Wang, K., Wang, E., Qin, Z., Zhou, Z., Geng, Y., Chen, D., 2016. Effects of dietary vitamin E deficiency on systematic pathological changes and oxidative stress in fish. Oncotarget 7, 83869–83879.

Watanabe, T., Takeuchi, J., Ogino, C., 1980. Effect on Rainbow Trout and Chum Salmon of Deletion of Trace Elements from Fish Meal Diet. Nippon Suisan Gakk. 46, 1521–1525.

Wood, E.M., Yasutake, W.T., Woodall, A.N., Halver, J.E., 1957. The nutrition of salmonoid fishes: I. Chemical and histological studies of wild and domestic fish. J. Nutr. 61, 465–478.

Yamamoto, H., Satoh, S., Takeuchi, T., Watanabe, T., 1983. Effects on rainbow trout [Salmo gairdnerii] of deletion of manganase or trace elements from fish meal diet. Bull. Jpn. Soc. Sci. Fish. 49, 287–293.

Yokote, M., 1974. In: Spec. Publ. Japan. Sea Fish Lab., pp. 67–74.

CHAPTER

14

Feeding and fish husbandry

D. Allen Davis[1] and Ronald W. Hardy[2]

[1]School of Fisheries, Aquaculture & Aquatic Sciences, Auburn University, Auburn, AL, United States; [2]Aquaculture Research Institute, Hagerman Fish Culture Experiment Station, University of Idaho, Hagerman, ID, United States

14.1 Introduction

Aquaculture presents unique challenges that terrestrial animal producers do not face, including feed management. First and foremost is the challenge of providing feed to fish and shrimp. Unlike terrestrial livestock and poultry where unlimited access to feed is common, fish and shrimp do not feed themselves. Feeding must be actively managed to achieve efficient growth, prevent feed waste, and maintain water quality. Feed management involves determining appropriate feeding levels and feeding frequency, plus employing feed delivery systems and practices that match the feeding behavior of each farmed species. Feeding level is usually based on feeding a percentage of pond or pen biomass, but to do so requires accurate estimates of the number of fish and their average weight in a large pond or pen that may contain up to a hundred thousand individuals, most of whom are not visible to the farmer. Feeding a large population of fish to apparent satiation is practically impossible and will reduce feed efficiency (FE), increase feed waste, and affect fish health from water pollution. Underfeeding can result in lost fish performance and increased size variation due to differences in feed access; large, aggressive individuals outcompete smaller, less aggressive individuals for feed pellets. In either case, farm economic performance is reduced. Given the fact that feed accounts for 50%–60% of the operating costs of raising fish, feeding practices and feed management are just as important as feed quality for aquaculture producers to be successful. The highest quality and most expensive feed cannot support production goals, such as rapid growth, unless the feed is provided properly to the fish and, conversely, proper feed management cannot overcome the use of low-quality feeds.

Feeds and feeding strategies used in aquaculture are influenced by many abiotic and biotic factors. Of the abiotic factors, water temperature is the most important. Aquaculture species are poikilothermic, so metabolic activity and feed intake increase with water temperature up to an upper thermal limit, at which point feed intake decreases. Thus, for a given species at a

Fish Nutrition, Fourth Edition
https://doi.org/10.1016/B978-0-12-819587-1.00015-X

given fish size, water temperature is the primary factor determining feeding level and feeding frequency. Other abiotic factors also influence feeding activity of farmed aquatic species and include photoperiod, husbandry practices, nutrient and energy content of feed, and other characteristics of feed, such as pellet density, handling stability, and water stability. Biotic factors that dictate feeding practices include species, feeding behavior and diel feeding rhythms, life history stage, rearing density, and in some species, social interactions. In nature, fish and shrimp have evolved to exploit a wide variety of habitats and natural food sources, leading to great differences in anatomy, digestive physiology, feeding behavior, and other factors associated with feeding. Given that there are several hundred fish and shrimp species being farmed for food production, it is an understatement to say that no single feeding protocol or method fits all species and production systems. Feeds must possess characteristics, such as type (live, fresh frozen, or compounded feeds), physical characteristics (size of feed particles, fineness of grind, density, or specific gravity), and attraction and palatability, which match the needs of various species and life stages. Feed management, i.e., ration size and feeding frequency, must also match the needs of various species and life stages to support rapid and efficient growth, health and quality of the animal, plus the economic viability of the aquaculture enterprise. Consequently, this chapter covers various aspects of feeding, nutrient delivery, and models to improve efficiency of nutrient delivery and utilization, cost-effectiveness, and environmental sustainability of production systems. Feed management practices for several life history stages and examples of feeding practices for several species of farmed fish and shrimp that represent different fish culture systems, e.g., marine or freshwater pens, ponds, and flow-through systems are provided as illustrations of the myriad of considerations associated with feed management in aquaculture production.

14.2 Feed utilization and fish growth

Growth is the deposition of new tissue, and growth rate is governed by genetic potential of the animal, nutrient intake, and environmental conditions. Growth is the net result of utilization of digested energy and nutrients after metabolism, maintenance, and voluntary activity needs have been met. Biomass (live weight) gain in animals is the sum of deposition of water, protein (amino acids), fat, and minerals as well as small amounts of other compounds. The quantity of nutrients deposited per unit of live weight gain is not constant but changes with biotic factors such as the weight of the animal, nutrient content of the feed, and physiological state (Shearer, 1994; Lupatsch and Kissil, 1998). Consequently, one unit of biomass gain (e.g., 1 g or 1 kg of biomass gain) results in different amounts of dry matter deposited as fish growth. Numerous other factors (nutrient composition of the feed, feeding level, and physiological and environmental factors) also affect the composition of fish. A change in biomass is, therefore, not of "absolute" or "constant" composition, as gain may represent varying quantities of a nutrient either as the animal grows or as conditions change.

The conversion of feed biomass to animal biomass gain over a fixed time period is termed the feed conversion ratio (FCR), defined as the amount (weight, as fed) of feed fed divided by wet body weight gain. The dry matter content of the feed is considered in FCR calculations. The reciprocal ratio, live body weight gain to feed intake, is termed FE and is expressed as a percentage. These ratios are useful to fish producers, as they provide insight into feed quality and effectiveness of feeding practices. Economic FCR is calculated by adding the cost of feed and the value of product into the calculation of FCR.

Many studies have been conducted to model the impact of feed inputs and various biotic (genetic, sex) and abiotic (environmental, husbandry) factors on feed efficiency. The results and models derived from these studies have been widely presented and are generally accepted (Fig. 14.1). Yet numerous contradictions exist among these studies. For example, some studies suggest that maximum FE is achieved at feeding levels below that required for maximum growth (G_{max}) (Brett and Groves, 1979). Conversely, Talbot (1993) suggested that maximum FE is attained at maximum feed intake (R_{max}) and G_{max}. The differences between these studies are largely related to the fact that parameters used in these analyses and models (mass of feed offered, biomass gain, G_{max}, R_{max}, growth rate, FE, and FCR) are all relative and/or semiquantitative parameters. Limitations of these parameters (growth rate, G_{max}, R_{max}) are discussed in various publications (National Research Council, 2011; Cho, 1992; Azevedo et al., 1998; Cho and Bureau, 1998; Bureau and Hua, 2008). Irrespective of theoretical models, as a rule, feeding just below maximum intake results in the best fish performance, as this typically yields the lowest FCR or highest FE, plus it results in sustained high feed intake by fish. Many species of fish are gluttonous feeders, meaning they will consume as much feed they can possibly consume over a short period. Feeding to maximum intake for several days is often followed by a period of reduced feed intake, resulting in wasted feed if feeding level is not reduced. More efficient feed utilization is achieved by feeding a set amount every day and making slight upward or downward adjustments based upon assessment of feeding activity.

The amount of feed required by a fish to achieve a certain amount of weight gain depends on nutrient density of the feed and feed intake, which in turn determines daily energy and nutrient intake. Consequently, a larger quantity of a low nutrient dense feed (or a feed with low digestibility) compared with a higher nutrient dense feed (or feed with a higher digestibility) will be required to achieve the same growth performance, assuming the feeds are nutritionally adequate and balanced. As a simple example, feeding 100% of a ration with a

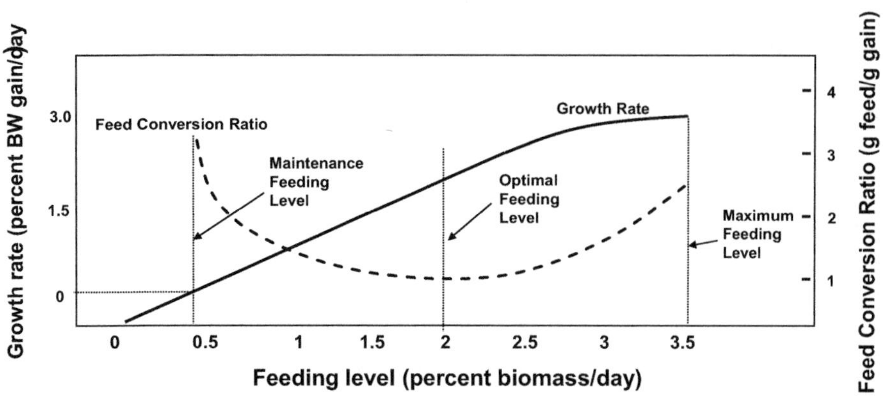

FIGURE 14.1 Schematic relationship between feed intake, fish growth rate, and feed conversion ratio (FCR).

30% protein diet to catfish will result in the same daily intake of protein (g) as feeding 75% of the ration using a diet with 40% protein (assuming the feeds are appropriately formulated). Clearly, as digestibility or nutrient density of a diet is reduced, more units of feed are required to sustain the same biomass gain, resulting in an increase in FCR or a decrease in FE. This is illustrated by research with Pacific white shrimp (Venero et al., 2007, 2008). The effect of nutrient density on feed intake and FCR is also evident from changes in salmon and trout diets. In the early 1980s, for example, commercial rainbow trout feeds contained relatively low protein (e.g., 35%–40%) and fat (e.g., 8%–10%) levels but were rich in starch (35%–40%). FCRs of 2–2.5 were commonly recorded for rainbow trout harvested at 0.5 kg live weight. Today, trout feeds have lower carbohydrate levels, higher digestibility, and higher energy levels (e.g., 34%–55% crude protein, 18%–32% fat, 10%–20% starch) resulting in FCRs of about 1.2 for rainbow trout harvested at 0.8–1.0 kg live weight.

FCR is best defined as quantity of feed consumed per unit of weight gain. However, this terminology is used with a range of alternative definitions. At the farm level, a true FCR is typically not calculated because accurate values for feed intake and biomass changes are not always known. Instead, an observed FCR or a pond conversion ratio (S, as described by Swingle (1968) but yet to be adopted) is often calculated based on the quantity of feed utilized on the farm and the quantity of product sold over a given time period (feed added per unit net fish produced). Such values represent a combination of factors including the true FCR of surviving animals, as well as contributions of natural food sources (resulting in a reduction in apparent FCR), feed that was not consumed and FCR of animals lost during the production cycle (mortality, theft, escape) resulting in an apparent increase in FCR. Consequently, on-farm observed FCR (feed in per unit of product out) is, in general, more a measure of the overall efficiency of a given aquaculture operation rather than solely a measure of the nutritional quality of the feed.

These measures can only be considered "semiquantitative" parameters because both "feed" and "fish" are variable in chemical composition. Hence, they are not good indicators of the efficiency of conversion of dietary nutrients into fish biomass. Caution should be used when comparing FCR values across studies, facilities, seasons, sampling intervals, etc. Describing and predicting FCR is achievable with different feeds and as a function of fish size or environment, but it is an imprecise or predictive measure. Probably the best use of FCR values is for economic analyses at the farm level, as it allows an estimation of the overall economic efficiency of the feed, feeding practices, and management practices as combined factors. Evaluation of specific nutrient retention values, e.g., apparent net protein retention, are better indices of the nutritional value of a feed.

14.2.1 Estimating biomass in aquaculture systems

In all production systems, accurate assessment of numbers of stocked animals and their average weight during rearing periods is critical to efficient feed management. In aquaculture, animals are reared in a wide variety of environments and housing infrastructures including tanks, raceways, ponds, and floating net pens. Due to water turbidity and distribution of the animals, a population of fish is rarely fully visible to caretakers, making it impossible to accurately estimate population size. Challenges associated with capture, handling, and counting of free-swimming animals limit the frequency and accuracy of directly estimating population numbers in production rearing systems. This complicates assessment of biomass, a value necessary to determine the most efficient ration, defined as amount fed per day.

Population numbers are more easily estimated in confined production systems (tanks, raceways, or cages) where the animals are visible, and mortalities can be removed and counted than in ponds or large pens. In confined systems, population numbers can be determined by physically counting animals when they are moved from one culture container to another or through nonintrusive measures such as herding them through a counting device. Such devices can use fluctuations in water conductivity, optical imagery, or hydroacoustics to determine fish numbers and have been recently reviewed by Zhou et al. (2018) and Li et al. (2020). Biomass estimates are also required to calculate feeding level. The most common way to estimate biomass is to weigh a subsample of the population to determine average fish weight; however, this can cause stress and damage to the fish and is labor intensive. Newer technologies including various forms of digital imagery (e.g., stereo photography) can be used to predict fish size (Ruff et al., 1995; Beddow et al., 1996; van Rooij and Videler, 1996; Petrell et al., 1997; Costa et al., 2006) without removing the fish from the culture environment.

In production systems when the animal numbers are not easy to measure (e.g., pond production systems, large marine pens), care during transportation and acclimation of animals during the stocking process is critical. If the animals are not accurately counted at stocking or if mortality occurs after stocking and is not quantified, population estimates will be inaccurate and feed management decisions are likely to overestimate feed needs, resulting in overfeeding. In production systems that are drain-harvested after one production season, such as shrimp or single batch fish systems, estimates of survival are typically based on subsampling (e.g., cast netting for shrimp), direct observation of mortality, and historical results. When dealing with pond systems that are not drain-harvested on a regular basis (e.g., multibatch catfish culture), inventory estimates are quite problematic. Inefficiencies in estimating population numbers, as well as the retention of fish larger than the targeted size (fish that escape top harvesting and grow beyond the minimal market size), lead to considerable inefficiencies of production systems. In general, the more frequently these systems are top-harvested, the more accurate the population estimates, the more uniform is the population (i.e., there are fewer oversized fish), and the shorter the period between complete harvest, the better the overall inventory control.

14.2.2 Feeding programs

The key to any feeding program is the proper delivery of the feed in adequate levels, and over a suitable time period, to meet the nutritional requirement of the species for a targeted growth rate. This may mean providing all the feed that an animal will consume (satiation) or providing a restricted ration (less than satiation). Satiation feeding in fish involves providing feed up to the point of fulfilling the desire to eat, evidenced by lack of a strong feeding response among visible fish. Hence, one is providing all the feed the animals will consume, albeit within a distinct number of meals or time period (e.g., all the feed that can be consumed within 15 min, two times a day). This feeding practice is commonly termed apparent satiation. Growth rates and survival are inherently challenging to forecast accurately, making it difficult to determine the exact amount of feed that should be fed. Feed inputs are designed to deliver a specific level of nutrients; hence, feed inputs are dependent on the quantity of feed offered in combination with the nutrient content of the feed. In other words, as the nutrient density or protein level of a feed decreases, more feed will be needed to obtain maximum growth. In addition to nutrient density of the diet, feeding rate can be affected by diet processing conditions (grinding, pelleting, and top coating), size and type of pellet (e.g., diameter/length and floating/sinking), number of feedings per day, as well as feed delivery mechanisms. Feed management will not only vary between types of cultures systems but will vary within a system making it more important to understand basic concepts that can be adjusted to fit specific aquaculture systems.

14.3 Pellet characteristics

To minimize economic inefficiencies and pollution loading of culture systems, feed must be presented in forms that promote rapid and complete consumption. The size and form of the feed depends on the cultured species, fish size, culture conditions, and feed delivery practices. Feed particle size should progressively increase with the size of the animal. An example of feed sizes for trout is provided in Table 14.1. Typically, feeds are selected that are 20%–30% of mouth diameter (gape) or, in the case of shrimp, a size easily handled by

TABLE 14.1 Guidelines for feed particle/pellet size for salmon and trout of various sizes. Feed size will vary with species and type of pellet (semimoist, extruded, pelleted).

Fish weight (g)	Particle/pellet size (mm)
<0.3	<0.5
0.3–0.5	0.5–0.8
0.5–0.9	0.8
0.9–2.0	1
2.0–4.0	1.5
4.0–8.0	2.5
8.0–20	3
20–50	3.5
50–125	4
125–500	4.5
500–1000	6
1000–2000	9

the shrimp (Obaldo and Tacon, 2001). Rapid feed consumption before pellets disintegrate is critical in shrimp farming. Therefore, it is often advantageous to use a small feed particle to minimize effects of uneaten feed on pond water quality (Peixoto et al., 2020). In the case of fish, another consideration when selecting feed particle or pellet size is that in any given culture situation, fish are not all the same size. If feed particle size is increased too soon, larger fish will easily consume the feed while the smaller fish will struggle, thereby increasing variability in fish size. Consequently, under some conditions (typically early juvenile stages), a mixture of feed sizes is used to transition fish to a larger particle size. Typically, larger aggressive fish in a population are likely to displace smaller individuals during feeding. Dominance and competition have also been reported for shrimp feeding behavior (Obaldo and Tacon, 2001). Therefore, enough pellets should be delivered to allow all fish and shrimp the opportunity to feed. When a limited number of daily feeding are offered, providing enough feed to reach all animal at each feeding is critical. However, when using automated feeding systems that provide many feedings per day, only a portion of the population actually consumes feed at each feeding. With careful feed management, the daily ration is provided across multiple feedings, and all fish have access to feed over the course of the day.

Feed density plays a critical role in how it is offered, as density determines if pellets sink or float. Sinking feeds have sufficient density to sink immediately when placed in water and are used in fish, shrimp, and early fry production. Sinking feeds are used when it is desirable for feed to pass through the water column and not remain on the surface. Sinking feeds allow fish feeding in the water column or shrimp feeding on the bottom to easily obtain the pellets. Floating feeds are designed to remain on the surface for a period of time. They include both flake feed and extruded feed. Extrusion is the forming process of choice for most production fish feeds, as it offers the possibility to control product density and to produce diets with high lipid content. Most floating and slow sinking feeds are produced using cooking-extrusion technology. For crustacean species such as shrimp as well as numerous fish species, water-stable sinking feeds are produced by compression pelleting. Irrespective of the process to produce the feed, it must be in a form that is suitable size, shape, and density for the species being grown.

Pellet durability refers to the amount of fines produced by a feed during manufacturing, storage, shipping, and on-farm handling (Storebakken, 1985). Most commercial feeds are moved by bulk handling systems, transported in large containers or bulk bags and fed using mechanical or pneumatic conveying systems. Fines in feed represent a direct economic loss, as they are not consumed by the fish but rather are a source of pollution as they add to nutrient loading of the system. They may also accelerate abrasion in bearings in automated feeding systems. Other factors relating to pellet characteristics include hardness, which is related to pelleting conditions and to content of binders in the formulation. Effects of binders in feeds vary with species. For rainbow trout, high levels of binders, high pellet durability, and water stability reduce feed intake, stomach evacuation rates, and digestibility (Hilton et al., 1981; Storebakken, 1985; Baeverfjord et al., 2006). Conversely, inadequate pellet stability of high lipid feeds has been reported to release oils into the stomach of trout or salmon, resulting in regurgitation as evidenced by an oil scum on water surfaces or sides of pens or nets (Baeverfjord et al., 2006).

14.4 Feed delivery systems

Feeding systems can be categorized as (1) nondemand systems in which the ration and time of delivery is predetermined or (2) demand or self-feeding systems for which the fish activate or provide feedback for feed delivery. Feed can be delivered by hand or by mechanical feeders using rotating belts or disks, vibration, or pneumatic forces. Feeding systems used in aquaculture have recently been reviewed (Lekang, 2020).

The choice of how to move feed from the mill to the farm is determined by local shipping resources. Bagged feed is the most common method but is also the most expensive. Whenever possible, bulk movement and storage of feed is recommended as more economical and efficient than moving and storing bagged feed. Similarly, feed can be quantified and dispersed using a range of systems depending on capital investments and local resources. The simplest involves weighing the feed and dispersing it by hand to the culture animals. The most complex are typically found in intensive operations where feed is quantified, conveyed, and dispersed using bulk-handling equipment.

14.5 Feeding in semi-intensive systems

Semi-intensive fish culture systems are used to produce fry, fingerling and fish for consumption with limited inputs of feed, aeration, and water exchange. Species produced in these systems typically feed low on the food chain (e.g., Pacific white shrimp, carp and tilapia) and consume natural food organisms in the culture systems. In terms of standing crop, production levels increase as system inputs are increased. Production is increased by adding inorganic fertilizers, organic materials such as cow dung, or adding food processing wastes such as rice bran to ponds. Production levels are further increased by supplementing semiintensive systems with farm-made or incomplete feed (Tacon and De Silva, 1997; El-Sayed, 2006). Further increases are made by employing improved feed management and providing complete feeds.

In the case of species that are capable of utilizing natural foods in culture systems (e.g., shrimp, tilapia, and carps), it is prudent to encourage the use of natural foods, as this reduces the investment for prepared feeds. Stable carbon isotope analysis is a useful tool to demonstrate the relative contribution of various food sources for species such as white shrimp (Anderson et al., 1987), golden shiner (Lochman and Phillips, 1996), as well as the common carp and tilapia (Schroeder, 1983). In these examples, natural productivity contributed from 40% to 80% of carbon deposition in the culture species. This clearly indicates the potential contribution of natural food sources even under semiintensive and in some cases intensive conditions. In well-fertilized production systems, natural foods are capable of supporting a given level of biomass often referred to as critical standing crop (El-Sayed, 2006). Until the critical standing crop is reached, providing prepared feeds does not increase production. Hence, eliminating or reducing feed inputs results in more efficient utilization of resources. For example, in the semiintensive culture of tilapia, fingerling fish are often stocked into well fertilized ponds in which natural foods are plentiful. Diana et al. (1996) demonstrated that at an initial stocking density of 3 fish/m^2, fish obtained enough natural food to grow rapidly until they were around 100 g after which prepared feeds had to be supplied to maximize production. Delaying feeding in well-fertilized ponds does not reduce fish yield but does reduce feed costs providing appropriate stocking densities are used (El-Sayed, 2006).

Feeding level is another important factor in semiintensive systems. Since natural foods are present, feeding to satiation is not likely to increase production but increases production costs. Diana et al. (1994) reported that reducing feed inputs in tilapia production ponds (3 fish/m^2) from satiation to 50% of satiation did not reduce production. This feeding level encouraged the fish to forage for natural foods and improved the observed conversion of feed inputs to harvest biomass. The consumption of natural food sources should be encouraged in all production systems for which it can be utilized. This simply means feeding below satiation. However, the quantity and quality of natural foods are quite variable within a site, between sites and across time. Hence, site-specific management practices must be developed.

14.6 Feeding intensive production systems

Intensive production systems are used almost exclusively in commercial aquaculture production, and economic considerations dictate how they are managed. Natural foods are lacking or limited in intensive systems, so providing nutritionally complete feeds is essential. Economic returns from investments in feeds and feed management are primary driving forces for management decisions. However, government regulations, waste production, and consumer demands for specific products (e.g., pigmented fillets) also influence the selection of feeds, their nutrient and ingredient compositions, and production practices.

Proper delivery of feed is one of the most complex elements of commercial aquaculture. Efficient feeding levels can be calculated if the population size, mean weight, and projected growth are known. Biomass gain (i.e., weight gain) depends on the growth potential of the animal and environmental factors, such as water temperature, photoperiod, salinity, and weather conditions. Consequently, biomass increases, and time needed to reach market weight cannot always be accurately predicted based on previous performance, published recommendations, or predictive models.

The same factors that hinder accurate assessments of inventory also complicate feeding. Feed is often delivered to a fish population that is only partly visible or, in many cases, delivered based on the fish's response to feed inputs (e.g., by satiation feeding using visual or acoustic measures or the use of mechanical demand feeders) without an accurate knowledge of the biomass. Feed delivered but not consumed is difficult to recover and quantify. Overfeeding results in economic loss and greater waste outputs. Feeding to near satiation or as low as 70% below satiation does not result in reduced nutrient retention efficiencies in rainbow trout (Azevedo et al., 1998). However, underfeeding limits the ability of the fish to express their growth potential, leading to economic loss (opportunity cost), and may lower profitability.

In pond culture, the use of floating feed allows for visual observation of fish feeding behavior and feed consumption. The response of the fish to offered feed is a critical management tool, as it gives an indication of water quality conditions (low dissolved oxygen or high ammonia levels will often result in reduced feed intake), the health or disease status of the fish (sick fish often reduce feed intake or refuse to feed) as well as fish appetite (feeding response slows as satiation is approached). Satiation feeding is the process of offering feed until fish no longer exhibit a strong feeding response. This is an easy way to provide the maximum ration that fish will consume on a given day and feed offering. This can be accomplished by visually observing feeding response or through the use of self-feeders (manual and automated feed delivery systems that rely on input from the fish). In the case of visual observation, feed is offered over a fixed time period (e.g., 15–30 min), during which time the fish are allowed to consume as much feed as they will eat. Feeding activity should be aggressive during the entire period in which feed is available to the fish, but as the satiation point is reached, feeding activity visible on the water surface will slow and ultimately the fish will cease feeding.

Slowly sinking feeds can also be fed to apparent satiation by direct observation in clear water systems or using a variety of feedback systems typically used in sea cages and other clear water applications. Such devices include video cameras, infrared photoelectric sensors, and sonar. In these systems, feeding of fish in the cages starts with automatic feeders delivering a certain quantity of feed at intervals. Feeding continues until feed pellets are observed (via computer vision or acoustic technology) passing through the schooling fish without being ingested, indicating fish are reaching satiation. In one system, feed reaching the bottom of the cage is collected and lifted to the surface by a pump for collection, recording, and possibly reuse. Such feedback allows dynamic feeding according to appetite. Similarly, when self-feeding mechanisms such as demand feeders (Fig. 14.2A–C) used in raceway or pond rearing systems are properly calibrated, fish trigger the release of feed until they have consumed as much as they would like to eat. This greatly reduces the amount of uneaten feed.

FIGURE 14.2A Demand feeder at Danish trout farm.

FIGURE 14.2B Demand feeder at Idaho trout farm.

FIGURE 14.2C Mechanical feeders in trout farm.

14.7 Observational feed inputs

The observation (direct or indirect) of feeding response is an invaluable tool that provides insight into fish wellness and allows daily adjustments of feed inputs based on actual demand. The use of floating feeds and feeding to satiation is a popular method of feeding that typically results in good growth and reasonably efficient feed utilization. This is a common method employed in production systems with inadequate inventory control such as multibatch catfish production systems or mixed sex pond culture of tilapia. One advantage is that the fish regulate their feed intake based on the environmental conditions, size, and to a certain extent, nutrient (or energy) content of feed. Knowledge of the population in terms of biomass, fish size, and number is not required. Feeding to satiation typically results in maximum growth rates, but it is difficult to execute and can result in elevated FCR values (Bureau et al., 2006). Furthermore, under commercial conditions, it is easy to overfeed if feed is applied too aggressively and outpaces feeding intensity, resulting in uneaten feed being wasted. Feeding to near satiation remains one of the most popular feeding strategies in fish culture.

In cases where inventory control is difficult (e.g., multibatch production systems) and one does not desire to feed to satiation, floating feeds can be utilized to estimate the amount of feed required to reach satiation. Feed inputs can be set at a level below satiation (e.g., 90%) for a period of time after which feed input levels are reevaluated with another satiation feeding followed by another adjustment of feed inputs. Similarly, in situations where feed inputs are predetermined using feed tables or computer programs that use a variety of techniques to predict feed inputs, the use of floating feeds allows the observation of feeding response and the estimation of feed not consumed (visual estimates of feed remaining on the surface). The use of a floating feed to visualize feeding is an important management tool in a number of feeding strategies and should be utilized whenever practical.

More recently, intelligent feed control systems have been used to monitor feeding behavior and adjust feeding rates accordingly (Zhou et al., 2018). These systems use cameras, ultrasound, and acoustic equipment to assess feeding activity in real time in marine salmon pens. Computer programs then alter feed provision to prevent overfeeding, i.e., uneaten pellets passing through pens. Such systems not only reduce feed waste but also reduce labor costs associated with in-person monitoring of feeding activity. Although these systems are expensive, they pay for themselves many times over by improving farm-based FCRs. Acoustic assessment of feeding behavior of shrimp is another example of intelligent feed control that has the potential to improve feed efficiency in pond culture where visual assessment of feeding activity is not possible (Ullman et al., 2019b; Napaumpaiporn et al., 2013; Smith and Tabrett, 2013; Darodes de Tailly et al., 2021).

14.8 Feed tables

Feed tables or computer-generated feeding rates are commonly utilized to estimate required daily rations, expressed on a percent biomass basis. These tables are based on animal size but should also account for water temperature and nutrient content of the diet. The advantage of "predetermining" the ration to be fed is that nutrient intake is controlled and is typically set below satiation. Predetermined feed inputs are guidelines, not absolute values, and therefore should be combined with other forms of observation to confirm that feed is being consumed and not being wasted. As an example, Table 14.2 presents site-specific feed inputs for various sizes of catfish, tilapia, and carps offered a 28% protein floating feed, reared under intensive culture conditions with a temperature of around 28°C. The daily ration is expressed as a percentage of the body weight (% body weight $\times$ total biomass $=$ daily ration) with the total ration divided into one or more feedings. Daily feed rate is reduced, as the animal size increases and the number of feedings per day is also reduced. As nutrient intake is directly proportional to growth rate, this is logical as the relative growth rate (% increase per day) is reduced as fish grow. Additional examples for gilthead sea bream and European seabass are presented in Tables 14.3 and 14.4, respectively.

TABLE 14.2 General daily feeding rates and frequency guide for the production of channel catfish, common carp, and Nile tilapia using a 32% protein floating feed at 28°C in single batch production systems (Schmittou et al., 1998).

	Channel catfish		Common carp		Nile tilapia	
Fish mean weight (g)	Feed rate (%)	Frequency (×/day)	Feed rate (%)	Frequency (×/day)	Feed rate (%)	Frequency (×/day)
25	4.0	2	4.5	3	4.5	3
50	3.5	2	4.0	3	3.7	3
75	3.2	2	3.6	3	3.4	3
100	3.0	2	3.3	3	6.2	3
150	2.8	1	3.1	2	3.0	2
200	2.5	1	3.0	2	2.8	2
250	2.2	1	2.6	2	2.5	2
300	2.0	1	2.4	2	2.3	2
400	1.6	1	2.1	2	2	2
500	1.4	1	1.7	2	1.7	2
600	1.2	1	1.4	2	1.4	2

Feed allowances should be adjusted for temperature shifts based on the following schedule. At ≤ 15°C feed at 1% rate 1 ×/day only 3 ×/week. At 16–19°C feed at 60% of calculated allowance 1 ×/day every day. At 20–24°C, feed at 80% of the calculated allowance 1 or 2/day every day. At 25–29°C, feed at 100% of the calculated allowance and maximum frequency every day. At 30–32°C, feed at 80% calculated allowance. At ≥ 33°C, feed only what fish are observed to consume.

TABLE 14.3 Example of feeding table for gilthead sea bream at different body weight (BW) and water temperatures.

BW (g)	Pellet size (mm)	13°C	15°C	17°C	19°C	21°C	23°C	25°C	27°C	29°C
0.3–0.5	0.5 mm					ad libitum				
0.5–1.0	0.8 mm					ad libitum				
1.0–3	1.1 mm	1.06	2.47	3.87	5.28	6.26	6.66	6.79	6.43	5.28
3–8	1.5 mm	0.83	1.94	3.05	4.16	4.92	5.24	5.34	5.06	4.16
8–15	1.9 mm	0.69	1.61	2.53	3.46	4.09	4.36	4.44	4.20	3.46
15–20	1.9 mm	0.60	1.40	2.20	3.00	3.56	3.79	3.86	3.65	3.00
20–40	3.0 mm	0.55	1.29	2.03	2.77	3.28	3.49	3.56	3.37	2.77
40–60	3.0 mm	0.47	1.09	1.72	2.34	2.77	2.95	3.01	2.85	2.34
60–100	4.5 mm	0.39	0.91	1.42	1.94	2.30	2.45	2.49	2.36	1.94
100–160	4.5 mm	0.32	0.75	1.17	1.60	1.89	2.01	2.05	1.94	1.60
160–300	4.5 mm	0.26	0.60	0.95	1.29	1.53	1.63	1.66	1.57	1.29
300–450	6.5 mm	0.19	0.44	0.7	0.95	1.12	1.2	1.22	1.15	0.95
450–700	6.5 mm	0.15	0.36	0.57	0.77	0.92	0.98	0.99	0.94	0.77
700–1200	9.0 mm	0.12	0.28	0.43	0.59	0.70	0.75	0.76	0.72	0.59

Source: http://www.fao.org/fileadmin/user_upload/affris/docs/Gilthead_Seabream/English/table_13.htm.

TABLE 14.4 Example of feeding table for European sea bass[a] at different body weight (BW) and water temperatures.

Body weight (g)	Pellet size (mm)	13°C	15°C	17°C	19°C	21°C	23°C	25°C	27°C	29°C
0.3–0.5	0.5					ad libitum				
0.5–1.0	0.8					ad libitum				
1.0–3	1.1	1.21	2.42	3.63	4.84	5.41	5.81	5.85	5.24	3.52
3–8	1.5	0.91	1.83	2.74	3.66	4.09	4.39	4.42	3.95	2.66
8–15	1.9	0.74	1.48	2.22	2.96	3.31	3.55	3.58	3.20	2.15
15–20	1.9	0.63	1.27	1.90	2.53	2.83	3.04	3.06	2.74	1.84
20–40	3	0.52	1.03	1.55	2.07	2.31	2.48	2.50	2.24	1.50
40–60	3	0.43	0.86	1.29	1.72	1.93	2.07	2.08	1.87	1.25
60–100	4.5	0.35	0.71	1.06	1.41	1.58	1.69	1.70	1.53	1.02
100–160	4.5	0.29	0.57	0.86	1.14	1.28	1.37	1.38	1.24	0.83
160–300	4.5	0.23	0.46	0.69	0.91	1.02	1.10	1.11	0.99	0.66
300–450	6.5	0.16	0.33	0.49	0.66	0.74	0.79	0.80	0.71	0.48
450–700	6.5	0.13	0.26	0.40	0.53	0.59	0.63	0.64	0.57	0.38
700–1200	9	0.10	0.20	0.29	0.39	0.44	0.47	0.47	0.42	0.28

[a]*Source: Different feed companies. http://www.fao.org/fileadmin/user_upload/affris/docs/European_Seabass/English/table_13.htm.*

14.9 Feeding frequency

Feeding frequency refers to the number of times per day that feed is provided. The need for multiple feedings is often driven by the physical capacity of the stomach to hold feed and nutrient demands for growth. From a practical standpoint, the number of feedings utilized is tempered by nutrient loading (e.g., more small feedings spread's out nutrient loading over time), water quality (e.g., prevents excessive oxygen consumption associated with feeding activity at high fish densities), feed delivery cost (labor or investments in technology), and the expectation of improved growth rates as well as uniformity of the population. Feeding frequency varies with size or life stage from near continuous inputs for newly hatched larvae and fry to three to four times per day for fingerlings and one to three times per day for fish during the grow-out stage.

In terms of grow-out of juveniles to marketable size, feeding frequency or time of fed application has limited effects on food-size fish for some species, for example, catfish reared in ponds, cages, and raceways (Jarboe and Grant, 1996; Li and Lovell, 1992; Webster et al., 1992; Wu et al., 2004). However, other species such as trout and salmon respond to an increased number of feedings, for example, two to three feedings per day (Thomassen and Fjæra, 1996; Ruohonen et al., 1998). Feeding multiple times to aggressive feeding species ensures sufficient access to feed by all fish rather than just the largest and most aggressive individuals. For these species at the grow-out stage, the amount of feed provided at each feeding should be sufficient to ensure that all fish have access to feed. This can be judged by observing feeding activity. More frequent daily feeding is more likely to improve the growth rates of species with small stomachs and continuous foraging behavior (e.g., common carp, tilapia, and shrimp) than for species that are aggressive feeders. For example, during the grow-out phase of shrimp production, increasing the number of feedings results in improved production (Robertson et al., 1993; Carvalho and Nunes, 2006; Reis et al., 2020a,b; Ullman et al., 2019a,b). In shrimp, increasing the number of feeding allows for more feed to be offered. This increases feed consumption and reduces nutrient leaching associated with prolonged exposure of pellets in water, thereby improving production. Given the economic advantage of increased number of feedings, the shrimp industry is transitioning from hand feeding several times a day to using automated feeders.

In outdoor pond production systems covering a large area or large marine pens, labor and fuel costs determine which feed delivery system is used. If feed is delivered by hand or a small vessel with a blower that moves from pen to pen, this will limit the number of feedings to one to two times per day. Salmon farms using pneumatic mechanical feeders (Fig. 14.3)

FIGURE 14.3 Pneumatic feeding system in salmon farm (three photos including control barge).

that feed heavily in early morning and late afternoon and blow small amounts of feed onto the water surface of pens periodically throughout the day, even to large fish. Feed should be distributed over a relatively large area to allow equal access of the fish to the feed. In cases where the fish are concentrated in a relatively small area such as a cage or tank, labor and fuel costs are often reduced, and there is an opportunity to utilize self-feeding and automated systems. In such cases, feed inputs can be regulated either through the use of self-feeding systems (demand feeders), automated feedback systems, or preprogrammed feeding regimes or by hand feeding using visual observations of the animals. Irrespective of the method and number of feedings, observation of feeding activity is a critical component of all feed management strategies, allowing fish feeder to adjust to daily fluctuations in feeding activity in ponds or marine pens.

14.10 Abiotic and biotic factors affecting feed management

There are several factors effecting feed intake. As you can see for Tables 14.2–14.3, feed table should be adjusted based on temperature. Table 14.5 provides an example of feed intake for a single batch of channel catfish grown from advanced fingerlings to marketable size in temperate outdoor ponds in the southern United States. The feed rate (% body weight or biomass) does not shift as much as one would expect based on the fish weight. This is due to the interaction of fish weight and temperature on feed intake and is typical for pond

TABLE 14.5 Example of feed consumption rates for channel catfish reared in outdoor ponds from advanced fingerlings to marketable size.

Date	AM water Temperature C	Fish size	Feeding rate (% body weight)
May 1	18.9	50	2.1
May 15	22.2	62	3.4
June 1	21.1	82	2.9
June 15	27.2	111	3.2
July 1	27.2	143	2.7
July 15	27.8	176	2.4
August 1	27.8	233	1.8
August 15	27.2	285	2.0
September 1	25.0	335	1.5
September 15	25.0	382	1.3
October 1	18.9	462	1.1

Adapted from Robinson, E.H., Li, M.H., Manning, B.E., 2001. A Practical Guide to Nutrition, Feeds, and Feeding of Catfish. Office of Agricultural Communications, Division of Agriculture, Forestry.

systems in temperate climates. This is an excellent example of why feed management must be adjusted for temperature.

Feeding level and feeding frequency decrease as fish increase in size. However, both are affected by water temperature, which is illustrated by studies with rainbow trout reared at different temperatures (Fig. 14.4). Trout during the juvenile and early postjuvenile stages reared at 15°C consume approximately 1% of their body weight at a single feeding. Feeding this amount will ensure that sufficient feed is provided so that all fish consume feed, not just the most aggressive individuals. Feeding frequency should be adjusted accordingly. For example, the recommended feeding level for 30 g trout is 3% of body weight per day; feeding 1% body weight at a single feeding means that the ration should be provided in three feedings. For 100 g fish, the recommended feeding level at 15°C is 2% per day; the daily ration should be divided into two feedings.

Photoperiod is another abiotic factor that affects feeding practices in temperate species, notably salmonids. Photoperiod is often overlooked as a determinate of feeding level because it increases and decreases in parallel with water temperature in temperate geographical areas. However, in outdoor aquaculture systems suppled with constant temperature water, the connection between declining photoperiod and water temperature is uncoupled, altering the normal seasonal decrease in feed intake exhibited by trout, for example. However, circulating levels of metabolic hormones respond to a declining photoperiod in freshwater fish from temperate zones, leading to accumulation of body lipid and a reduction in protein

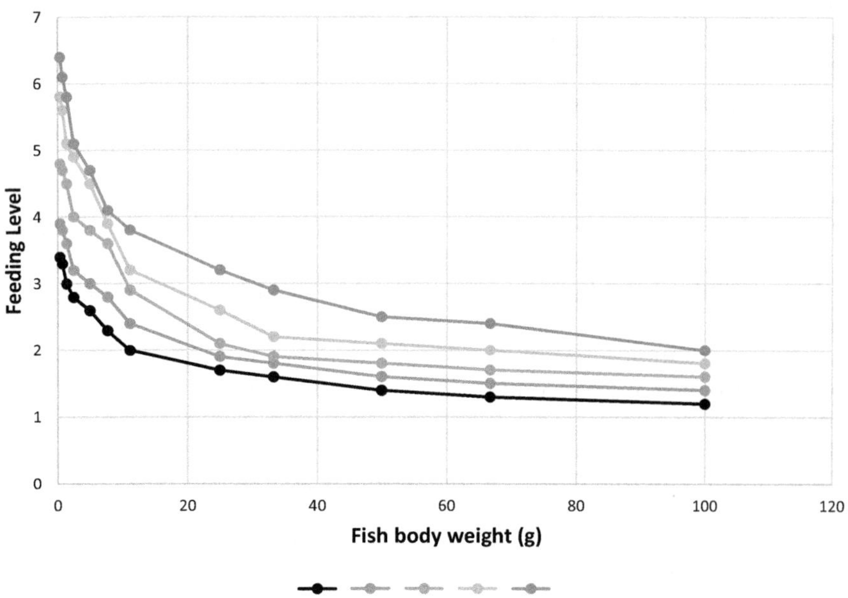

FIGURE 14.4 Relationship between fish size (g) and feeding level (% BW/day) for rainbow trout at different water temperatures.

synthesis (muscle growth) regardless of the protein content of the diet (Overturf and Hardy, 2001). Conversely, when photoperiod increases, voluntary feed consumption and protein accretion increase, as does protein retention. Feed composition and feeding levels should be adjusted accordingly. For practical reasons, many farmed marine species, such as sea bass or sea bream, are only fed once or twice a day, generally by hand or using blowers from boats. Feeding level is set at a fixed level based upon recommendations and experience and feeding activity by the fish. Feeding activity is affected by weather, ocean conditions (turbidity, algae blooms), and other factors and can vary from day to day. Thus, it is important that feeders are trained to observe fish feeding behavior and adjust feeding level accordingly.

Feeding activity in many fish species follows a daily pattern. In salmonids, feeding activity of fish recorded using self-feeders is greatest at daylight and twilight. For this reason, automatic feeders used in marine salmon farms are programmed to deliver most of the daily ration in early morning and evening.

14.11 Predicting growth, nutrient input, and feed utilization

Bioenergetic approaches have been effectively used to predict growth, optimum feed ration, FCR, and waste outputs of fish fed diets of varying nutritional composition and reared under different environments. Fish growing at different rates will also retain nutrients at different rates; thus, they will have different energy and feed requirements. Consequently, the daily energy requirement should be calculated based on a given level of performance (e.g., expected or achievable level of growth), feed composition (nutrient density), and life stage. This can be done using factorial approaches (Cho and Bureau, 1998; Lupatsch and Kissil, 1998) that divides the organism's energy and/or nutrient requirements into its different components or fractions.

This concept has been incorporated into the "Fish-PrFEQ" model (Cho and Bureau, 1998), which predicts energy requirement of fish based on expected level of performance, dietary DE content, and expected body energy deposition. Further development or various adaptations of the Fish-PrFEQ model have been proposed by Kaushik (1998), Bureau et al. (2002, 2003), Papatryphon et al. (2005), and Zhou et al. (2005).The main premise is that "animals will seek to eat a sufficient amount of an appropriately balanced diet to achieve their target or preferred performance unless limited by constraints or overridden by an externally managed intervention" (Oldham et al., 1997). The Fish-PrFEQ model estimates a DE requirement to achieve certain "desired" performance (determined by fish genetic potential and prevailing conditions). The calculated DE requirement is translated into a feed requirement by considering the DE content (kJ/g feed) of the feed being fed.

Bioenergetics models are very useful and practical, as they allow modeling of feed requirement and prediction of feed efficiency simply based on relatively straightforward estimations of the DE requirement of the fish, rather than through a detailed analysis of nutrient deposition and nutritional composition of the feed. Comparison of feeds can also be done simply based on DE content. The suitability of comparing feeds on the basis of their DE content has been demonstrated at numerous occasions, although a number of studies have shown that comparisons need to be made carefully, notably when dealing with feeds with high digestible carbohydrate content (Bureau et al., 1998) or comparing different fish species or fish at different life stages (Azevedo et al., 2004a,b).

The user of bioenergetics models should always remember that bioenergetics is a "methodology" or "system." Animals do not metabolize "energy" per se but metabolize "nutrients." The amount of energy effectively "retained" or "harnessed" during life processes (efficiency of energy utilization) is governed by which and/or how these nutrients are utilized. More importantly, most nutrients have very specific roles in metabolism, and many nutrients cannot be substituted by one another. Studies indicate that utilization of energy-yielding nutrients is highly dependent on the type of nutrients (not simply their "energy" value), the balance between these nutrients, and the genetic and physiological state of the fish (Azevedo et al., 2004a,b). Considering nutrient inputs solely on the basis of their "free energy" content is, therefore, largely irrational.

Arguably the most significant limitation of bioenergetics models is that they are based on "hierarchy of energy allocation." Growth is the surplus of energy after all other components of the energy budget have been covered or satisfied. Bioenergetics models consequently predict that when body energy gain (recovered energy [RE]) is nil ($RE = 0$), growth (live weight gain) and feed efficiency (gain/feed) should also be nil. Bioenergetics models consequently assume that the relationship between "energy deposition" and growth (biomass gain) is absolute, an assumption that has been shown to be inaccurate. RE reflects the "weighed average" of the deposition of energy-yielding nutrients (mostly protein and lipids) in the body of the animal. Volumes of evidence indicate that animals (pigs, chicken, fish) fed a ration allowing an $RE = 0$ can still deposit body protein (positive nitrogen balance and associated energy gain) and gain weight, while mobilizing body lipids. Live weight gain is consequently driven by protein deposition since there are three to six grams of water associated with each gram of tissue protein deposited (Cho and Kaushik, 1990; Bureau et al., 2002). Lipid gain results in no or insignificant live weight gain since, in fish, lipids are stored in tissues by substituting for water (Shearer, 1994).

Current bioenergetic models are very useful and have been utilized with a wide range of fish and crustacean species. However, they are not robust enough to describe adequately and predict efficiency of conversion of dietary inputs into biomass under a wide variety of conditions (Bureau and Hua, 2008). Hence, there is a need for more advanced models that describe metabolic uses of specific nutrients as opposed to simple aggregates of nutrient in the form of energy. To meet this need, there is a trend to move toward nutrient flow models. In these models, nutrient transformations are represented at the whole animal level based on biological principles. Such models will be more robust toward nutrient utilization and allow the evaluation of the relationships between nutrition, genetics, health, animal management, growth performance, and nutrient excretion to be established.

Irrespective of the mechanism of the model, biological models to predict current and future growth, survival, and feed inputs are excellent tools to improve the management efficiency and profitability of production systems. Quite often, outputs from biological models lend themselves to merging with financial (cash) and economic (cash and noncash resource) analyses to facilitate future production and or production strategies. For example, Cacho et al. (1990) utilized a combined model to evaluate the effects of dietary protein and ration size on cost structures for channel catfish production. If suitable models can be built that utilize past inputs and predict future outputs for not only the biological response of the animal (growth, size distribution, feed utilization, waste outputs, and survival), the culture system (e.g., level of supplemental aeration), and variable and fixed costs, then a more rational decision can be made with regard to nutrient content of the feed, feed inputs as well as other management and marketing strategies.

14.12 Feeding early life stages (larvae and fry)

Newly hatched fish and crustacean species include a wide range of physiological forms, which for simplicity will be generically described as larvae. The successful use of live and artificial feeds is related to the size of the larvae and the type or developmental stage of its digestive system. Larvae having immature digestive systems that are not fully developed when larvae begin to feed are typically the most difficult to feed and commonly require provision of live feeds. Those with developed digestive systems at the time of first feeding, such as salmonids, can be reared solely on manufactured feeds from first feeding.

14.12.1 Live foods

Live food organisms that are natural prey are the preferred live feed for larvae. However, they are not always easy to identify or culture, so choices have to be made concerning substitution by practical alternative organisms that larvae will consume and that will provide appropriate nutrient and energy intake. Phytoplankton make up the base of the aquatic food web and are essential for larval rearing of many species. It is the primary food for bivalve mollusks, early larval stages of many crustacean species, early growth stages of some fish species, and the culture of numerous zooplankton species used as food for larval fish and shrimp (Couteau, 1996). Algae species used to culture zooplankton differ in size, nutrient content, growth rate, and ease of culture. Hence, choice of what species to use is often based on a compromise between nutrition and practicality. The rotifer (*Brachionus plicatilis*) and brine shrimp (*Artemia* spp.) are the primary types of zooplankton produced in mass quantities as live food for larvae. Phytoplankton and zooplankton from different geographic origins and cultured under different conditions will differ in nutritional quality, primarily n-3 and n-6 polyunsaturated fatty acid (PUFA) content (Napiorkowska-Krzebietke, 2017; Watanabe et al., 1983; Rainuzzo et al., 1997; Volkman et al., 1989). Fish larvae are often sensitive to a deficiency of PUFAs (Watanabe et al., 1983, Rainuzzo et al., 1997; Sargent et al., 1999; Kandathil Radhakrishnan et al., 2020); thus, the quantity and composition of fatty acids in larval foods are critical. Zooplankton is enriched with PUFAs by (1) feeding the zooplankton for a period of 24 h on marine algae such as *Chlorella* spp. or yeast containing a high concentration of PUFAs, or (2) by exposing zooplankton to a suspension of lipid rich in PUFA, such as fish oil, and an emulsifying compound for 3–12 h before being offered to the larvae (Sorgeloos et al., 2001; Sorgeloos, 1980; Kandathil Radhakrishnan et al., 2020).

Microalgae are commonly utilized as a food source and/or enrichment medium in the production of zooplankton. The nutrient content of zooplankton will quickly shift if food is not available to them, so algae can serve as a nutrient source to both the larvae and zooplankton provided to larvae as live food organisms (Reitan et al., 1997). The use of algae in intensive culture systems has a number of advantages: They most likely modify the bacterial population of the water, enhance water quality, stabilize the nutrient content of live foods, modify light levels, stimulate the digestive process in the cultured organism, and contribute to the establishment of early gut microbiota (Egerton et al., 2018; Reitan et al., 1997; Liao et al., 2001; Palmer et al., 2007; Vadstein et al., 2018).

Larvae are in log phase growth while at a stage with minimal nutritional reserves. Consequently, they need highly nutritious and constantly available food sources. Live prey items must be of optimal size, quantity, and quality for the larvae species being grown. The management of live food production systems as well as live food concentration or density in the culture system is critical to success. Larvae can obtain live food in either extensive low-density systems such as ponds or large tanks or intensive culture systems. From a nutritional perspective, intensive culture systems where live food is added to the system provide more control of the quality and quantity of live foods than do extensive systems where live food is produced naturally. Common prey densities (e.g., 10–20 rotifers per ml) seem to be effective across a number of larval fish species. The most appropriate prey density will vary somewhat with the species cultured, size of prey, type of culture system, and management considerations (Lee and Ostrowski, 2001). The most common way to evaluate densities is to monitor both intake by the larvae and density of prey in the culture system to avoid under- and overfeeding (Palmer et al., 2007). Underfeeding slows larval development, whereas overfeeding can also result in reduced performance by reducing prey capture success, increasing retention time of prey in the culture system, and lowering water quality. Quite often, food for zooplankton prey is not added to intensive larval rearing systems; hence, the nutritional quality of the zooplankton decreases rapidly over time. In the case of *Artemia*, they may outgrow and hence compete with the culture species for phytoplankton (algae) (Lee and Ostrowski, 2001).

14.12.1.1 Microparticulate diets for larvae and fry

Fish that produce offspring with well-developed digestive systems can be reared from first feeding on prepared feeds. However, relatively few larval species can be reared from hatch exclusively on prepared feeds and require provision of live prey organisms. After a brief period of feeding live organisms, the larvae of many species can be reared using cofeeding practices in which a combination of live and prepared feeds is provided. It is common knowledge that larval fish have higher metabolic rates than fry and thus benefit from a higher concentration of nutrients and energy in their diet (Dabrowski, 1986). Such diets must meet the nutritional requirements of the species, be of a size appropriate for ingestion, and possess physical properties, such as buoyancy, texture, and color, which are appropriate for larvae. Larvae feeds differ from those of the juvenile stages primarily in the need for phospholipids, essential fatty acids, and free amino acids or short peptides (Jones et al., 1997; Jones, 1998; Cahu and Zambonino Infante, 2001; Holt, 2011).

As larvae grow, so does the gape of the mouth. It is important that feed particle size increases in proportion to fish size and does not exceed 20% of the mouth opening gape (Dabrowski and Bardega, 1984). Frequent feeding is important when rearing larvae as well as fry or postlarval stages; feed should be offered frequently by hand or automatic feeders to increase the number of feeding (>10 per day), or fed on a continuous basis (Charlon and Bergot, 1984). At early larvae stages, the larvae-feed encounter rate is more important than feeding to a specific intake. Feed is often offered in excess to ensure that feed intake is high. However, when feed is provided in excess, the effect of prepared feeds on water quality must be considered. The small particle size of microdiets coupled with a high level of feeding increases leaching of nutrients into the water leading to deterioration of water quality and increased bacterial loading of the culture system (Jones et al., 1993). Consequently, the design of larval culture systems often includes water flow systems to resuspend particles to increase encounter rate, relatively higher water exchange rates, and mechanisms to remove solids and often time to control bacterial loads of the system.

Formulations for larvae and fry feeds generally include high-quality marine ingredients. These feeds are fed to fry of species that are large enough at first feeding to ingest relatively large feed particles. Microparticulate diets for marine fish and shrimp larvae have utilized micropulverized meals such as those from fish, crab, squid, mussel, yeast, chicken egg yolk, short-necked clam, and krill (Kanazawa, 1986, 1989, 2003; Cahu and Zambonino Infante, 2001) as well as semipurified ingredients such as casein and gelatin (Paibulkichakul et al., 1998).

14.13 Summary

The selection of the proper feed in combination with application of appropriate feed management strategies are both critical aspects of efficient aquaculture production. Matching feeding practices that account for nutrient density and physical characteristics of feed to provide the optimum daily supply of nutrients to optimize growth, and health of the culture species is one of the most difficult aspects of aquaculture production to manage. Yet it is a task that is often misunderstood and/or overlooked in terms of importance, especially at the farm level. Small errors that lead to delivering more feed than can be consumed by employing an excessive feeding level or the wrong feeding frequency cause direct economic loss and increased nutrient loading of culture systems. Feeding fish that exceed minimal market sizes, due to harvest inefficiencies especially in top-harvesting pond fish, often leads to poor conversion of feed to final product due to both inefficiencies of growing larger fish and increased likelihood that part of the population will be lost due to disease or some other problem. Accurate and up-to-date records of feed inputs, population numbers, and growth rates are critical control points in feed management. Feed management must also be tied to nutrient loading of the culture system, oxygen consumption after feeding, and levels of waste products released by the animals.

Precision feed formulation must be coupled with precision feeding practices for aquaculture production to become a more efficient and environmentally benign form of food production. The aquatic environment is a challenging setting to raise animals. Advances in detecting and assessing fish feeding behavior using new monitoring systems and machine learning may be one element that improves feeding efficiency. Another likely future development will be land-based, confinement aquaculture where biosecurity and control of abiotic factors will remove a number of uncontrolled variables that affect feeding practices and feed efficiency.

In addition to current technologies, as aquaculture moves toward more automated systems and follows the path of precision agriculture, fish farming systems will evolve and advance. As reviewed by Wei et al. (2020), considerable strides are being made with control technologies. The advent of electronic data collection also creates the ability to mine large data sets and use this information to monitor and improve fish farming technologies (Måløy et al., 2019).

References

Anderson, R.K., Parker, P.L., Lawrence, A., 1987. A $^{13}C/^{12}C$ tracer study of the utilization of presented feed by a commercially important shrimp *Penaeus vannamei* in a pond growout system. J. World Aquacult. Soc. 18, 148–155.

Azevedo, P.A., Cho, C.Y., Leeson, S., Bureau, D.P., 1998. Effects of feeding level and water temperature on growth, nutrient and energy utilization and waste outputs of rainbow trout (*Oncorhynchus mykiss*). Aquat. Living Resour. 11, 227–238.

Azevedo, P.A., Leeson, S., Cho, C.Y., Bureau, D.P., 2004a. Growth and feed utilization of large size rainbow trout (*Oncorhynchus mykiss*) and Atlantic salmon (*Salmo salar*) reared in freshwater: diet and species effects, and responses over time. Aquacult. Nutr. 10, 401–411.

Azevedo, P.A., Leeson, S., Cho, C.Y., Bureau, D.P., 2004b. Growth, nitrogen and energy utilization of juveniles from four salmonid species: diet, species and size effects. Aquaculture 234, 393–414.

Baeverfjord, G., Refstie, S., Krogedal, P., Åsgård, T., 2006. Low feed pellet water stability and fluctuating water salinity cause separation and accumulation of dietary oil in the stomach of rainbow trout (*Oncorhynchus mykiss*). Aquaculture 261, 1335–1345.

Beddow, T.A., Ross, L.G., Marchant, J.A., 1996. Predicting salmon biomass remotely using a digital stereo-imaging technique. Aquaculture 146, 189–203.

Brett, J.R., Groves, T.D.D., 1979. 6 - physiological energetics. In: Hoar, W.S., Randall, D.J., Brett, J.R. (Eds.), Fish Physiology. Academic Press, pp. 279–352.

Bureau, D.P., Hua, K., 2008. Models of nutrient utilization by fish and potential applications for fish culture operations. In: France, J., Kebreab, E. (Eds.), Mathematical Modelling in Animal Nutrition. CAB International, Wallingford, UK, pp. 442–461.

Bureau, D.P., Kirkland, J.B., Cho, C.Y., 1998. The partitioning of energy from digestible carbohydrate by rainbow trout (*Oncorhynchus mykiss*). In: McCracken, K.J., Unsworth, E.F., Wylie, A.R.G. (Eds.), Energy Metabolism of Farm Animals. CAB International Press, Wallingford, Oxon, UK, pp. 163–166.

Bureau, D.P., Kaushik, S.J., Cho, C.Y., 2002. Bioenergetics. In: Halver, J.E., Hardy, R.W. (Eds.), Fish Nutrition. Academic Press., San Diego, CA, pp. 2–62.

Bureau, D.P., Gunther, S.J., Cho, C.Y., 2003. Chemical composition and preliminary theoretical estimates of waste outputs of rainbow trout reared in commercial cage culture operations in Ontario. N. Am. J. Aquacult. 65, 33–38.

Bureau, D.P., Hua, K., Cho, C.Y., 2006. Effect of feeding level on growth and nutrient deposition in rainbow trout (*Oncorhynchus mykiss* Walbaum) growing from 150 to 600 g. Aquacult. Res. 37, 1090–1098.

Cacho, O.J., Hatch, U., Kinnucan, H., 1990. Bioeconomic analysis of fish growth: effects of dietary protein and ration size. Aquaculture 88, 223–238.

Cahu, C.L., Zambonino Infante, J.L., 2001. Substitution of live food by formulated diets in marine fish larvae. Aquaculture 200, 161–180.

Carvalho, E.A., Nunes, A.J.P., 2006. Effects of feeding frequency on feed leaching loss and grow-out patterns of the white shrimp *Litopenaeus vannamei* fed under a diurnal feeding regime in pond enclosures. Aquaculture 252, 494–502.

Charlon, N., Bergot, P., 1984. Rearing system for feeding fish larvae on dry diets. Trial with carp (*Cyprinus carpio* L.) larvae. Aquaculture 41, 1–9.

Cho, C.Y., 1992. Feeding systems for rainbow trout and other salmonids with reference to current estimates of energy and protein requirements. Aquaculture 100, 107–123.

Cho, C.Y., Bureau, D.P., 1998. Development of bioenergetic models and the Fish-PrFEQ software to estimate production, feeding ration and waste output in aquaculture. Aquat. Living Resour. 11, 199–210.

Cho, C.Y., Kaushik, S.J., 1990. Nutritional energetics in fish: energy and protein utilization in rainbow trout (*Salmo gairdneri*). World Rev. Nutr. Diet. 61, 132–172.

Costa, C., Loy, A., Cataudella, S., Davis, D., Scardi, M., 2006. Extracting fish size using dual underwater cameras. Aquacult. Eng. 35, 218–227.

Dabrowski, K., Bardega, R., 1984. Mouth size and predicted food size preferences of larvae of three cyprinid fish species. Aquaculture 40, 41–46.

Darodes de Tailly, J.B., Keitel, J., Owen, M.A.G., Alcaraz-Calero, J.M., Alexander, M.E., Sloman, K.A., 2021. Monitoring Methods of Feeding Behaviour to Answer Key Questions in Penaeid Shrimp Feeding. Reviews in Aquaculture.

Diana, J.S., Lin, C.K., Jaiyen, K., 1994. Supplemental feeding of tilapia in fertilized ponds. J. World Aquacult. Soc. 25, 497–506.

Diana, J.S., Lin, C.K., Yi, Y., 1996. Timing of supplemental feeding for Tilapia production. J. World Aquacult. Soc. 27, 410–419.

Egerton, S., Culloty, S., Whooley, J., Stanton, C., Ross, R.P., 2018. The gut microbiota of marine fish. Front. Microbiol. 9.

El-Sayed, A.F.M., 2006. Tilapia Culture. CABI Publisher, Wallingford, UK; Cambridge, MA.

Hilton, J., Cho, C., Slinger, S., 1981. Effect of extrusion processing and steam pelleting diets on pellet durability, pellet water absorption, and the physiological response of rainbow trout (*Salmo gairdneri*R.). Aquaculture 25, 185–194.

Holt, J., 2011. Larval Fish Nutrition. Wiley-Blackwell, Chichester, West Sussex, UK; Ames, Iowa.

Jarboe, H.H., Grant, W.J., 1996. Effects of feeding time and frequency on growth of channel catfish *Ictalurus punctatus* in closed recirculating raceway systems. J. World Aquacult. Soc. 27, 235–239.

Jones, D.A., 1998. Crustacean larval microparticulate diets. Rev. Fish. Sci. 6, 41–54.

Jones, D.A., Kamarudin, M.S., Vay, L.L., 1993. The potential for replacement of live feeds in larval culture. J. World Aquacult. Soc. 24, 199–210.

Jones, D.A., Yule, A.B., Holland, D.L., 1997. Larval nutrition. In: Akiyama, D.M., Conklin, D.B., D'Abramo, L.R. (Eds.), Crustacean Nutrition. World Aquaculture Society, Baton Rouge, LA, pp. 353–389.

Kanazawa, A., 1986. New developments in fish nutrition. In: Maclean, J.L., Dizon, L.B., Hsillos, L.V. (Eds.), The First Asian Fisheries Forum. Asian Fisheries Society, Manila, Philipines, pp. 9–14.

Kanazawa, A., 1989. Microparticulate Feeds for Penaeid Larvae.

Kanazawa, A., 2003. Nutrition of marine fish larvae. J. Appl. Aquacult. 13, 103–143.

Kandathil Radhakrishnan, D., AkbarAli, I., Schmidt, B.V., John, E.M., Sivanpillai, S., Thazhakot Vasunambesan, S., 2020. Improvement of nutritional quality of live feed for aquaculture: an overview. Aquacult. Res. 51, 1–17.

Kaushik, S.J., 1998. Nutritional bioenergetics and estimation of waste production in non-salmonids. Aquat. Living Resour. 11, 211–217.

Lee, C.S., Ostrowski, A.C., 2001. Current status of marine finfish larviculture in the United States. Aquaculture 200, 89–109.

Lekang, O.I., 2020. Aquaculture Engineering, third ed. Wiley-Blackwell.

Li, M., Lovell, R.T., 1992. Comparison of satiate feeding and restricted feeding of channel catfish with various concentrations of dietary protein in production ponds. Aquaculture 103, 165–175.

Li, D., Hao, Y., Duan, Y., 2020. Nonintrusive methods for biomass estimation in aquaculture with emphasis on fish: a review. Rev. Aquacult. 12, 1390–1411.

Liao, I.C., Su, H.M., Chang, E.Y., 2001. Techniques in finfish larviculture in Taiwan. Aquaculture 200, 1–31.

Lochman, R., Phillips, H., 1996. Stable isotopic evaluation of the relative assimilation of natural and artificial foods by golden shiners *Notemigonus crysoleucas* in ponds. J. World Aquacult. Soc. 27, 168–177.

Lupatsch, I., Kissil, G.W., 1998. Predicting aquaculture waste from gilthead seabream (*Sparus aurata*) culture using a nutritional approach. Aquat. Living Resour. 11, 265–268.

Måløy, H., Aamodt, A., Misimi, E., 2019. A spatio-temporal recurrent network for salmon feeding action recognition from underwater videos in aquaculture. Comput. Electron. Agric. 167, 105087.

Napaumpaiporn, T., Chuchird, N., Taparhudee, W., 2013. Study on the efficiency of three different feeding techniques in the culture of Pacific white shrimp (*Litopenaeus vannamei*). Kasetsart Univ. Fish. Res. Bull. 37, 8–16.

Napiorkowska-Krzebietke, A., 2017. Phytoplankton as a basic nutritional source in diets of fish. J. Elementol. 22, 831–841.

National Research Council, 2011. Nutrient Requirements of Fish and Shrimp. The National Academic Press, Washington DC.

Obaldo, L.G., Tacon, A.G.J., 2001. Manufacturing different diet sizes and its effect on pellet water stability and growth of three size classes of Pacific white shrimp, *Litopenaeus vannamei*. J. Appl. Aquacult. 11, 57–66.

Overturf, K., Hardy, R.W., 2001. Myosin expression in trout muscle: a new method for monitoring specific growth rates for rainbow trout *Oncorhynchus mykiss* Walbaum on varied planes of nutrition. Aquacult. Res. 32 (4), 315–322.

Paibulkichakul, C., Piyatiratitivorakul, S., Kittakoop, P., Viyakarn, V., Fast, A.W., Menasveta, P., 1998. Optimal dietary levels of lecithin and cholesterol for black tiger prawn *Penaeus monodon* larvae and postlarvae. Aquaculture 167, 273–281.

Palmer, P.J., Burke, M.J., Palmer, C.J., Burke, J.B., 2007. Developments in controlled green-water larval culture technologies for estuarine fishes in Queensland, Australia and elsewhere. Aquaculture 272, 1–21.

Papatryphon, E., Petit, J., Van Der Werf, H.M., Sadasivam, K.J., Claver, K., 2005. Nutrient-balance modeling as a tool for environmental management in aquaculture: the case of trout farming in France. Environ. Manag. 35, 161–174.

Peixoto, S., Soares, R., Allen Davis, D., 2020. An acoustic based approach to evaluate the effect of different diet lengths on feeding behavior of *Litopenaeus vannamei*. Aquacult. Eng. 91, 102114.

Petrell, R.J., Shi, X., Ward, R.K., Naiberg, A., Savage, C.R., 1997. Determining fish size and swimming speed in cages and tanks using simple video techniques. Aquacult. Eng. 16, 63–84.

Rainuzzo, J.R., Reitan, K.I., Olsen, Y., 1997. The significance of lipids at early stages of marine fish: a review. Aquaculture 155, 103–115.

Reis, J., Novriadi, R., Swanepoel, A., Guo, J.P., Rhodes, M., Davis, D.A., 2020a. Optimizing feed automation: improving timer-feeders and on demand systems in semi-intensive pond culture of shrimp *Litopenaeus vannamei*. Aquaculture 519.

Reis, J., Novriadi, R., Swanepoel, A., Jingping, G., Rhodes, M., Davis, D.A., 2020b. Optimizing feed automation: improving timer-feeders and on demand systems in semi-intensive pond culture of shrimp *Litopenaeus vannamei*. Aquaculture 519, 734759.

Reitan, K.I., Rainuzzo, J.R., Oie, G., Olsen, Y., 1997. A review of the nutritional effects of algae in marine fish larvae. Aquaculture 155, 207–221.

Robertson, L., Lawrence, A.L., Castille, F.L., 1993. Effect of feeding frequency and feeding time on growth of *Penaeus vannamei* (Boone). Aquacult. Fish. Manag. 24, 1–6.

Robinson, E.H., Li, M.H., Manning, B.E., 2001. A Practical Guide to Nutrition, Feeds, and Feeding of Catfish,. Office of Agricultural Communications, Division of Agriculture, Forestry.

Ruff, B.P., Marchant, J.A., Frost, A.R., 1995. Fish sizing and monitoring using a stereo image analysis system applied to fish farming. Aquacult. Eng. 14, 155–173.

Ruohonen, K., Vielma, J., Grove, D.J., 1998. Effects of feeding frequency on growth and food utilisation of rainbow trout (*Oncorhynchus mykiss*) fed low-fat herring or dry pellets. Aquaculture 165, 111–121.

Sargent, J., McEvoy, L., Estevez, A., Bell, G., Bell, M., Henderson, J., Tocher, D., 1999. Lipid nutrition of marine fish during early development: current status and future directions. Aquaculture 179, 217–229.

Schmittou, H.R., Jian, Z., Cremer, M.C., Association, A.S., 1998. Principles and Practices of 80:20 Pond Fish Farming. American Soybean Association.

Schroeder, G.L., 1983. Sources of fish and prawn growth in polyculture ponds as indicated by δC analysis. Aquaculture 35, 29–42.

Shearer, K.D., 1994. Factors affecting the proximate composition of cultured fishes with emphasis on salmonids. Aquaculture 119, 63–88.

Smith, D.V., Tabrett, S., 2013. The use of passive acoustics to measure feed consumption by *Penaeus monodon* (giant tiger prawn) in cultured systems. Aquacult. Eng. 57, 38–47.

Sorgeloos, P., 1980. The use of the brine shrimp Artemia in Aquaculture. In: Persoon, G.a.P.S. (Ed.), The Brine Shrimp Artemia, Ecology, Culture and Use in Aquaculture. Universa Press, Belgium, pp. 24–46.

Sorgeloos, P., Dhert, P., Candreva, P., 2001. Use of the brine shrimp, Artemia spp., in marine fish larviculture. Aquaculture 200, 147–159.

Storebakken, T., 1985. Binders in fish feeds.1. Effect of alginate and guar gum on growth, digestibility, feed-intake and passage through the gastrointestinal-tract of Rainbow-Trout. Aquaculture 47, 11–26.

Swingle, H.S., 1968. Estimation of Standing Crops and Rates of Feeding Fish in Ponds. FAO Fish Report, pp. 416–426.

Tacon, A.J., De Silva, S.S., 1997. Feed preparation and feed management strategies within semi-intensive fish farming systems in the tropics. Aquaculture 151, 379–404.

Talbot, C., 1993. Some aspects of the biology of feeding and growth in fish. Proc. Nutr. Soc. 52, 403–416.

Thomassen, J.M., Fjæra, S.O., 1996. Studies of feeding frequency for Atlantic salmon (*Salmo salar*). Aquacult. Eng. 15, 149–157.

Ullman, C., Rhodes, M., Hanson, T., Cline, D., Davis, D.A., 2019a. Effects of four different feeding techniques on the pond culture of Pacific white shrimp, *Litopenaeus vannamei*. J. World Aquacult. Soc. 50, 54–64.

Ullman, C., Rhodes, M.A., Allen Davis, D., 2019b. Feed management and the use of automatic feeders in the pond production of Pacific white shrimp *Litopenaeus vannamei*. Aquaculture 498, 44–49.

Vadstein, O., Attramadal, K.J.K., Bakke, I., Forberg, T., Olsen, Y., Verdegem, M., Giatsis, C., Skjermo, J., Aasen, I.M., Gatesoupe, F.-J., Dierckens, K., Sorgeloos, P., Bossier, P., 2018. Managing the microbial community of marine fish larvae: a holistic perspective for larviculture. Front. Microbiol. 9, 1820.

van Rooij, J.M., Videler, J.J., 1996. A simple field method for stereo-photographic length measurement of free-swimming fish: merits and constraints. J. Exp. Mar. Biol. Ecol. 195, 237–249.

Venero, J.A., Davis, D.A., Rouse, D.B., 2007. Variable feed allowance with constant protein input for the Pacific white shrimp *Litopenaeus vannamei* reared under semi-intensive conditions in tanks and ponds. Aquaculture 269, 490–503.

Venero, J.A., Davis, D.A., Rouse, D.B., 2008. Effect of the dietary ratio of digestible energy to crude protein on growth and feed conversion in juvenile Pacific white shrimp *Litopenaeus vannamei* under similar levels of daily protein consumption. N. Am. J. Aquacult. 70, 43–49.

Volkman, J.K., Jeffrey, S.W., Nichols, P.D., Rogers, G.I., Garland, C.D., 1989. Fatty-acid and lipid-composition of 10 species of microalgae used in mariculture. J. Exp. Mar. Biol. Ecol. 128, 219–240.

Watanabe, T., Kitajima, C., Fujita, S., 1983. Nutritional values of live organisms used in Japan for mass propagation of fish: a review. Aquaculture 34, 115–143.

Webster, C.D., Tidwell, J.H., Tidwell, J.H., Yancey, D.H., 1992. Effect of protein level and feeding frequency on growth and body composition of cage-reared channel catfish. Progress. Fish Cult. 54, 92–96.

Wei, Y., Wei, Q., An, D., 2020. Intelligent monitoring and control technologies of open sea cage culture: a review. Comput. Electron. Agric. 169, 105119.

Wu, G., Saoud, I.P., Miller, C., Davis, D.A., 2004. The effect of feeding regimen on mixed-size pond-grown channel catfish, *Ictalurus punctatus*. J. Appl. Aquacult. 15, 115–125.

Zhou, Z., Xie, S., Lei, W., Zhu, X., Yang, Y., 2005. A bioenergetic model to estimate feed requirement of gibel carp, *Carassius auratus* gibelio. Aquaculture 248, 287–297.

Zhou, C., Xu, D., Lin, K., Sun, C., Yang, X., 2018. Intelligent feeding control methods in aquaculture with an emphasis on fish: a review. Rev. Aquacult. 10, 975–993.

Further reading

Abdel Fattah, A.F., Ahmed, F.A., Said, E.N., Farag, M.R., 2021. Impact of feeding system on the behaviour and performance of Nile tilapia (*Oreochromis niloticus*). Aquaculture. https://doi.org/10.1016/j.aquaculture.2021.736514, 736514.

Index

Note: 'Page numbers followed by "*f*" indicate figures and "*t*" indicate tables.'

B

D

E

F

H

M

T

U

V